AF755589

KIRK-OTHMER

ENCYCLOPEDIA OF CHEMICAL TECHNOLOGY

Third Edition

VOLUME 11

Fluorine Compounds, Organic
to
Gold and Gold Compounds

KIRK-OTHMER

ENCYCLOPEDIA OF CHEMICAL TECHNOLOGY

THIRD EDITION

VOLUME 11

FLUORINE COMPOUNDS, ORGANIC
TO
GOLD AND GOLD COMPOUNDS

A WILEY-INTERSCIENCE PUBLICATION
John Wiley & Sons
NEW YORK • CHICHESTER • BRISBANE • TORONTO

Library of Congress Cataloging in Publication Data:

Main entry under title:
Encyclopedia of chemical technology.

At head of title: Kirk-Othmer.
"A Wiley-Interscience publication."
Includes bibliographies.
1. Chemistry, Technical—Dictionaries. I. Kirk, Raymond Eller, 1890–1957. II. Othmer, Donald Frederick, 1904— III. Grayson, Martin. IV. Eckroth, David. V. Title: Kirk-Othmer encyclopedia of chemical technology.

TP9.E685 1978 660'.03 77-15820
ISBN 0-471-02064-8

Printed in the United States of America

CONTENTS

EDITORIAL STAFF FOR VOLUME 11

CONTRIBUTORS TO VOLUME 11

Norman C. Abler, *Borg Textile Corporation, Delavan, Wisconsin,* Furs, synthetic

J. H. Adams, *Eagle-Picher Industries, Inc., Quapaw, Oklahoma,* Germanium and germanium compounds

Gary Astrologes, *Halocarbon Products Corporation, Hackensack, New Jersey,* Bromotrifluoroethylene under Fluorine compounds, organic

R. V. Berthold, *Union Carbide Corporation, South Charleston, West Virginia,* Ethylene glycol and propylene glycol under Glycols

David C. Boyd, *Corning Glass Works, Corning, New York,* Glass

Donald E. Brasure, *E. I. du Pont de Nemours & Co., Inc., Buffalo, New York,* Poly(vinyl fluoride) under Fluorine compounds, organic

E. S. Brown, *Union Carbide Corporation, South Charleston, West Virginia,* Ethylene glycol and propylene glycol under Glycols

Kees A. Bueters, *Combustion Engineering, Inc., Windsor, Connecticut,* Furnaces, fuel-fired

E. J. Butterfield, *Boyce Thompson Institute for Plant Research, Inc., Ithaca, New York,* Fungicides, agricultural

A. M. Chakrabarty, *University of Illinois at the Medical Center, Chicago, Illinois,* Genetic engineering

J. G. Cohn, *Englehard Minerals & Chemicals Corp., Edison, New Jersey,* Gold and gold compounds

B. B. Crocker, *Monsanto Company, St. Louis, Missouri,* Gas cleaning

John H. Davis, *Eastman Chemical Products, Inc., Kingsport, Tennessee,* Other glycols under Glycols

Pierre de la Bretèque, *Alusuisse France S. A., Marseille, France,* Gallium and gallium compounds

Julius E. Dohany, *Pennwalt Corporation, King of Prussia, Pennsylvania,* Poly(vinylidene fluoride) under Fluorine compounds, organic

Claudio L. Eberling, *BASF Aktiengesellschaft, Ludwigshafen, Federal Republic of Germany,* Formamide; Dimethylformamide both under Formic acid and derivatives

Takeshi Egami, *Institut für Physik, Max Planck Institut für Metallforschung, Stuttgart, Federal Republic of Germany,* Glassy metals

Robert H. Essenhigh, *The Ohio State University, Columbus, Ohio,* Furnaces, fuel-fired

D. F. Farkas, *University of Delaware, Newark, Delaware,* Food processing

Thomas Furia, *Intermark Corp., Palo Alto, California,* Food additives

S. V. Gangal, *E. I. du Pont de Nemours & Co., Inc., Parkersburg, West Virginia,* Polytetrafluoroethylene; Fluorinated ethylene–propylene copolymers both under Fluorine compounds, organic

H. R. Gerberich, *Celanese Chemical Company, Inc., Corpus Christi, Texas,* Formaldehyde

Charles O. Grigsby, *Los Alamos Scientific Laboratory, Los Alamos, New Mexico,* Geothermal energy

Gerald M. Halpern, *Exxon Research and Engineering Company, Linden, New Jersey,* Fusion energy

C. F. Hauser, *Union Carbide Corporation, South Charleston, West Virginia,* Ethylene glycol and propylene glycol under Glycols

John H. Hillard, *Mountain Fuel Supply Company, Salt Lake City, Utah,* Gas, natural

W. C. Hopkins, *Celanese Chemical Company, Inc., Corpus Christi, Texas,* Formaldehyde

Fred H. Hoskins, *Louisiana State University, Baton Rouge, Louisiana,* Food toxicants, naturally occurring

Jack Huebler, *Institute of Gas Technology, Chicago, Illinois,* Gaseous fuels under Fuels, synthetic

Henry I. Jacoby, *McNeil Laboratories, Fort Washington, Pennsylvania,* Gastrointestinal agents

John C. Janka, *Institute of Gas Technology, Chicago, Illinois,* Gaseous fuels under Fuels, synthetic

Richard L. Johnson, *E. I. du Pont de Nemours & Co., Inc., Parkersburg, West Virginia,* Tetrafluoroethylene copolymers with ethylene; Tetrafluoroethylene copolymers with perfluorovinyl ethers both under Fluorine compounds, organic

Charles D. Kalfadelis, *Exxon Research and Engineering Company, Linden, New Jersey,* Liquid fuels under Fuels, synthetic

Joyce C. Kern, *Glycerine Producers' Association, New York, New York,* Glycerol

Guy Kieckhefer, *Borg Textile Corporation, Dalavan, Wisconsin,* Furs, synthetic

Y. K. Kim, *Dow Corning Corporation, Midland, Michigan,* Polyfluorosilicones under Fluorine compounds, organic

Donald L. Klass, *Institute of Gas Technology, Chicago, Illinois,* Fuels from biomass

John C. Lane, *Ethyl Corporation, Ferndale, Michigan,* Gasoline and other motor fuels

John H. Litchfield, *Battelle Columbus Laboratories, Columbus, Ohio,* Foods, nonconventional

W. J. McKillip, *The Quaker Oats Company, Chicago, Illinois,* Furan derivatives

David Meidar, *University of Southern California, Los Angeles, California,* Friedel-Crafts reactions

Richard M. Miller, *Wellman Thermal Systems Corp., Shelbyville, Indiana,* Resistance furnaces under Furnaces, electric

Thomas F. Mitchell, *Darling and Company, Chicago, Illinois,* Glue

James C. Moyer, *New York State Agricultural Experiment Station, Geneva, New York,* Fruit juices

Kenneth R. Neel, *American Gilsonite Co., Salt Lake City, Utah,* Gilsonite

Dan Oakland, *Whiting Corporation, Harvey, Illinois,* Arc furnaces under Furnaces, electric

George A. Olah, *University of Southern California, Los Angeles, California,* Friedel-Crafts reactions

Theodore P. Perros, *George Washington University, Washington, D. C.,* Forensic chemistry

Harry Perry, *Resources for the Future, Washington, D. C.,* Fuels, survey

B. C. Ream, *Union Carbide Corporation, South Charleston, West Virginia,* Ethylene glycol and propylene glycol under Glycols

Lester E. Robb, *Pennwalt Corporation, King of Prussia, Pennsylvania,* Poly(vinylidene fluoride) under Fluorine compounds, organic

Sydney Ross, *Rensselaer Polytechnic Institute, Troy, New York,* Foams

L. R. Rothrock, *Union Carbide Corporation, San Diego, California,* Gems, synthetic

Henry Shaw, *Exxon Research and Engineering Company, Linden, New Jersey,* Liquid fuels under Fuels, synthetic

E. Sherman, *The Quaker Oats Company, Chicago, Illinois,* Furan derivatives

R. E. Skochdopole, *The Dow Chemical Company, Midland, Michigan,* Foamed plastics

A. L. Stautzenberger, *Celanese Chemical Co., Inc., Corpus Christi, Texas,* Formaldehyde

Eric W. Stern, *Englehard Minerals & Chemicals Corp., Edison, New Jersey,* Gold and gold compounds

Daniel R. Stewart, *Owens-Illinois, Toledo, Ohio,* Glass-ceramics

K. W. Suh, *The Dow Chemical Company, Granville, Ohio,* Foamed plastics

Mario Tama, *Ajax Magnethermic Corporation, Warren, Ohio,* Introduction; Induction both under Furnaces, electric

Jefferson Tester, *Los Alamos Scientific Laboratory, Los Alamos, New Mexico,* Geothermal energy

David A. Thompson, *Corning Glass Works, Corning, New York,* Glass

David A. Tillman, *Consultant, Bellevue, Washington,* Fuels from waste

D. C. Torgeson, *Boyce Thompson Institute for Plant Research, Inc., Ithaca, New York,* Fungicides, agricultural

Felix Viro, *Kind & Knox, Division of Knox Gelatine, Inc., Sioux City, Iowa,* Gelatin

Paul T. von Bramer, *Eastman Chemical Products, Inc., Kingsport, Tennessee,* Other glycols under Glycols

Frank S. Wagner, Jr., *Celanese Chemical Company, Inc., Corpus Christi, Texas,* Formic acid under Formic acid and derivatives; 1,3 Butylene glycol under Glycols

A. C. West, *3M Company, St. Paul, Minnesota,* Polychlorotrifluoroethylene under Fluorine compounds, organic

NOTE ON CHEMICAL ABSTRACTS SERVICE REGISTRY NUMBERS AND NOMENCLATURE

Chemical Abstracts Service (CAS) Registry Numbers are unique numerical identifiers assigned to substances recorded in the CAS Registry System. They appear in brackets in the *Chemical Abstracts* (CA) substance and formula indexes following the names of compounds. A single compound may have many synonyms in the chemical literature. A simple compound like phenethylamine can be named β-phenylethylamine or, as in *Chemical Abstracts,* benzeneethanamine. The usefulness of the Encyclopedia depends on accessibility through the most common correct name of a substance. Because of this diversity in nomenclature careful attention has been given the problem in order to assist the reader as much as possible, especially in locating the systematic CA index name by means of the Registry Number. For this purpose, the reader may refer to the CAS Registry Handbook-Number Section which lists in numerical order the Registry Number with the *Chemical Abstracts* index name and the molecular formula; eg, 458-8-8,Piperidine, 2-propyl-, (*S*)-, $C_8H_{17}N$; in the Encyclopedia this compound would be found under its common name, coniine [*458-88-8*]. The Registry Number is a valuable link for the reader in retrieving additional published information on substances and also as a point of access for such on-line data bases as Chemline, Medline and Toxline.

In all cases, the CAS Registry Numbers have been given for title compounds in articles and for all compounds in the index. All specific substances indexed in *Chemical Abstracts* since 1965 are included in the CAS Registry System as are a large number of substances derived from a variety of reference works. The CAS Registry System identifies a substance on the basis of an unambiguous computer-language description of its molecular structure including stereochemical detail. The Registry Number is a machine-checkable number (like a Social Security number) assigned in sequential order to each substance as it enters the registry system. The value of the number lies in the fact that it is a concise and unique means of substance identification, which is independent of, and therefore bridges, many systems of chemical nomenclature. For polymers, one Registry Number is used for the entire family; eg, polyoxyethylene (20) sorbitan monolaurate has the same number as all of its polyoxyethylene homologues.

Registry numbers for each substance will be provided in the third edition cumulative index and appear as well in the annual indexes (eg, Alkaloids shows the Registry Number of all alkaloids (title compounds) in a table in the article as well, but the intermediates have their Registry Numbers shown only in the index). Articles such as Analytical methods, Batteries and electric cells, Chemurgy, Distillation, Economic evaluation, and Fluid mechanics have no Registry Numbers in the text.

Cross-references are inserted in the index for many common names and for some systematic names. Trademark names appear in the index. Names that are incorrect, misleading or ambiguous are avoided. Formulas are given very frequently in the text to help in identifying compounds. The spelling and form used, even for industrial names, follow American chemical usage, but not always the usage of *Chemical Abstracts* (eg, *coniine* is used instead of (*S*)-*2-propylpiperidine, aniline* instead of *benzenamine,* and *acrylic acid* instead of *2-propenoic acid*).

There are variations in representation of rings in different disciplines. The dye industry does not designate aromaticity or double bonds in rings. All double bonds and aromaticity are shown in the *Encyclopedia* as a matter of course. For example, tetralin has an aromatic ring and a saturated ring and its structure appears in the

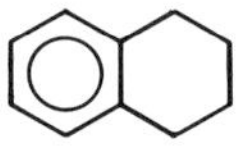

Encyclopedia with its common name, Registry Number enclosed in brackets, and parenthetical CA index name, ie, tetralin [*119-64-2*] (1,2,3,4-tetrahydronaphthalene). With names and structural formulas, and especially with CAS Registry Numbers the aim is to help the reader have a concise means of substance identification.

CONVERSION FACTORS, ABBREVIATIONS, AND UNIT SYMBOLS

SI Units (Adopted 1960)

A new system of measurement, the International System of Units (abbreviated SI), is being implemented throughout the world. This system is a modernized version of the MKSA (meter, kilogram, second, ampere) system, and its details are published and controlled by an international treaty organization (The International Bureau of Weights and Measures) (1).

SI units are divided into three classes:

BASE UNITS

length	meter† (m)
mass‡	kilogram (kg)
time	second (s)
electric current	ampere (A)
thermodynamic temperature§	kelvin (K)
amount of substance	mole (mol)
luminous intensity	candela (cd)

† The spellings "metre" and "litre" are preferred by ASTM; however "-er" are used in the Encyclopedia.

‡ "Weight" is the commonly used term for "mass".

§ Wide use is made of "Celsius temperature" (t) defined by

$$t = T - T_0$$

where T is the thermodynamic temperature, expressed in kelvins, and T_0 = 273.15 K by definition. A temperature interval may be expressed in degrees Celsius as well as in kelvins.

SUPPLEMENTARY UNITS

plane angle	radian (rad)
solid angle	steradian (sr)

DERIVED UNITS AND OTHER ACCEPTABLE UNITS

These units are formed by combining base units, supplementary units, and other derived units (2–4). Those derived units having special names and symbols are marked with an asterisk in the list below:

Quantity	*Unit*	*Symbol*	*Acceptable equivalent*
* absorbed dose	gray	Gy	J/kg
acceleration	meter per second squared	m/s^2	
* activity (of ionizing radiation source)	becquerel	Bq	1/s
area	square kilometer	km^2	
	square hectometer	hm^2	ha (hectare)
	square meter	m^2	
* capacitance	farad	F	C/V
concentration (of amount of substance)	mole per cubic meter	mol/m^3	
* conductance	siemens	S	A/V
current density	ampere per square meter	A/m^2	
density, mass density	kilogram per cubic meter	kg/m^3	g/L; mg/cm^3
dipole moment (quantity)	coulomb meter	C·m	
* electric charge, quantity of electricity	coulomb	C	A·s
electric charge density	coulomb per cubic meter	C/m^3	
electric field strength	volt per meter	V/m	
electric flux density	coulomb per square meter	C/m^2	
* electric potential, potential difference, electromotive force	volt	V	W/A
* electric resistance	ohm	Ω	V/A
* energy, work, quantity of heat	megajoule	MJ	
	kilojoule	kJ	
	joule	J	N·m
	electron volt†	eV†	
	kilowatt-hour†	kW·h†	

† This non-SI unit is recognized by the CIPM as having to be retained because of practical importance or use in specialized fields (1).

Quantity	*Unit*	*Symbol*	*Acceptable equivalent*
energy density	joule per cubic meter	J/m^3	
* force	kilonewton	kN	
	newton	N	$kg{\cdot}m/s^2$
* frequency	megahertz	MHz	
	hertz	Hz	1/s
heat capacity, entropy	joule per kelvin	J/K	
heat capacity (specific), specific entropy	joule per kilogram kelvin	J/(kg·K)	
heat transfer coefficient	watt per square meter kelvin	$W/(m^2{\cdot}K)$	
* illuminance	lux	lx	lm/m^2
* inductance	henry	H	Wb/A
linear density	kilogram per meter	kg/m	
luminance	candela per square meter	cd/m^2	
* luminous flux	lumen	lm	cd·sr
magnetic field strength	ampere per meter	A/m	
* magnetic flux	weber	Wb	V·s
* magnetic flux density	tesla	T	Wb/m^2
molar energy	joule per mole	J/mol	
molar entropy, molar heat capacity	joule per mole kelvin	J/(mol·K)	
moment of force, torque	newton meter	N·m	
momentum	kilogram meter per second	kg·m/s	
permeability	henry per meter	H/m	
permittivity	farad per meter	F/m	
* power, heat flow rate, radiant flux	kilowatt	kW	
	watt	W	J/s
power density, heat flux density, irradiance	watt per square meter	W/m^2	
* pressure, stress	megapascal	MPa	
	kilopascal	kPa	
	pascal	Pa	N/m^2
sound level	decibel	dB	
specific energy	joule per kilogram	J/kg	
specific volume	cubic meter per kilogram	m^3/kg	
surface tension	newton per meter	N/m	
thermal conductivity	watt per meter kelvin	W/(m·K)	
velocity	meter per second	m/s	
	kilometer per hour	km/h	
viscosity, dynamic	pascal second	Pa·s	
	millipascal second	mPa·s	
viscosity, kinematic	square meter per second	m^2/s	

Quantity	*Unit*	*Symbol*	*Acceptable equivalent*
	square millimeter per second	mm^2/s	
volume	cubic meter	m^3	
	cubic decimeter	dm^3	L(liter) (5)
	cubic centimeter	cm^3	mL
wave number	1 per meter	m^{-1}	
	1 per centimeter	cm^{-1}	

In addition, there are 16 prefixes used to indicate order of magnitude, as follows:

Multiplication factor	*Prefix*	*Symbol*	*Note*
10^{18}	exa	E	
10^{15}	peta	P	
10^{12}	tera	T	
10^9	giga	G	
10^6	mega	M	
10^3	kilo	k	
10^2	hecto	h[a]	[a] Although hecto, deka, deci, and centi are SI prefixes, their use should be avoided except for SI unit-multiples for area and volume and nontechnical use of centimeter, as for body and clothing measurement.
10	deka	da[a]	
10^{-1}	deci	d[a]	
10^{-2}	centi	c[a]	
10^{-3}	milli	m	
10^{-6}	micro	μ	
10^{-9}	nano	n	
10^{-12}	pico	p	
10^{-15}	femto	f	
10^{-18}	atto	a	

For a complete description of SI and its use the reader is referred to ASTM E 380 (4) and the article Units and Conversion Factors which will appear in a later volume of the *Encyclopedia*.

A representative list of conversion factors from non-SI to SI units is presented herewith. Factors are given to four significant figures. Exact relationships are followed by a dagger. A more complete list is given in ASTM E 380-76(4) and ANSI Z210.1-1976 (6).

Conversion Factors to SI Units

To convert from	*To*	*Multiply by*
acre	square meter (m^2)	4.047×10^3
angstrom	meter (m)	1.0×10^{-10} †
are	square meter (m^2)	1.0×10^2 †
astronomical unit	meter (m)	1.496×10^{11}
atmosphere	pascal (Pa)	1.013×10^5
bar	pascal (Pa)	1.0×10^5 †
barn	square meter (m^2)	1.0×10^{-28} †

† Exact.

To convert from	*To*	*Multiply by*
barrel (42 U.S. liquid gallons)	cubic meter (m^3)	0.1590
Bohr magneton, μ_β	J/T	9.274×10^{-24}
Btu (International Table)	joule (J)	1.055×10^3
Btu (mean)	joule (J)	1.056×10^3
Btu (thermochemical)	joule (J)	1.054×10^3
bushel	cubic meter (m^3)	3.524×10^{-2}
calorie (International Table)	joule (J)	4.187
calorie (mean)	joule (J)	4.190
calorie (thermochemical)	joule (J)	4.184†
centipoise	pascal second (Pa·s)	1.0×10^{-3} †
centistoke	square millimeter per second (mm^2/s)	1.0†
cfm (cubic foot per minute)	cubic meter per second (m^3/s)	4.72×10^{-4}
cubic inch	cubic meter (m^3)	1.639×10^{-5}
cubic foot	cubic meter (m^3)	2.832×10^{-2}
cubic yard	cubic meter (m^3)	0.7646
curie	becquerel (Bq)	3.70×10^{10}†
debye	coulomb·meter (C·m)	3.336×10^{-30}
degree (angle)	radian (rad)	1.745×10^{-2}
denier (international)	kilogram per meter (kg/m)	1.111×10^{-7}
	tex‡	0.1111
dram (apothecaries')	kilogram (kg)	3.888×10^{-3}
dram (avoirdupois)	kilogram (kg)	1.772×10^{-3}
dram (U.S. fluid)	cubic meter (m^3)	3.697×10^{-6}
dyne	newton (N)	1.0×10^{-5} †
dyne/cm	newton per meter (N/m)	1.0×10^{-3} †
electron volt	joule (J)	1.602×10^{-19}
erg	joule (J)	1.0×10^{-7} †
fathom	meter (m)	1.829
fluid ounce (U.S.)	cubic meter (m^3)	2.957×10^{-5}
foot	meter (m)	0.3048 †
footcandle	lux (lx)	10.76
furlong	meter (m)	2.012×10^{-2}
gal	meter per second squared (m/s^2)	1.0×10^{-2} †
gallon (U.S. dry)	cubic meter (m^3)	4.405×10^{-3}
gallon (U.S. liquid)	cubic meter (m^3)	3.785×10^{-3}
gallon per minute (gpm)	cubic meter per second (m^3/s)	6.308×10^{-5}
	cubic meter per hour (m^3/h)	0.2271
gauss	tesla (T)	1.0×10^{-4}
gilbert	ampere (A)	0.7958
gill (U.S.)	cubic meter (m^3)	1.183×10^{-4}
grad	radian	1.571×10^{-2}
grain	kilogram (kg)	6.480×10^{-5}
gram force per denier	newton per tex (N/tex)	8.826×10^{-2}
hectare	square meter (m^2)	1.0×10^4 †
horsepower (550 ft·lbf/s)	watt (W)	7.457×10^2

† Exact.

‡ See footnote on p. xiv.

To convert from	*To*	*Multiply by*
horsepower (boiler)	watt (W)	9.810×10^3
horsepower (electric)	watt (W)	$7.46 \times 10^{2\dagger}$
hundredweight (long)	kilogram (kg)	50.80
hundredweight (short)	kilogram (kg)	45.36
inch	meter (m)	$2.54 \times 10^{-2\dagger}$
inch of mercury (32°F)	pascal (Pa)	3.386×10^3
inch of water (39.2°F)	pascal (Pa)	2.491×10^2
kilogram force	newton (N)	9.807
kilowatt hour	megajoule (MJ)	$3.6^\dagger$
kip	newton (N)	4.48×10^3
knot (international)	meter per second (m/s)	0.5144
lambert	candela per square meter (cd/m^2)	3.183×10^3
league (British nautical)	meter (m)	5.559×10^3
league (statute)	meter (m)	4.828×10^3
light year	meter (m)	9.461×10^{15}
liter (for fluids only)	cubic meter (m^3)	$1.0 \times 10^{-3\dagger}$
maxwell	weber (Wb)	$1.0 \times 10^{-8\dagger}$
micron	meter (m)	$1.0 \times 10^{-6\dagger}$
mil	meter (m)	$2.54 \times 10^{-5\dagger}$
mile (U.S. nautical)	meter (m)	$1.852 \times 10^{3\dagger}$
mile (statute)	meter (m)	1.609×10^3
mile per hour	meter per second (m/s)	0.4470
millibar	pascal (Pa)	1.0×10^2
millimeter of mercury (0°C)	pascal (Pa)	$1.333 \times 10^{2\dagger}$
minute (angular)	radian	2.909×10^{-4}
myriagram	kilogram (kg)	10
myriameter	kilometer (km)	10
oersted	ampere per meter (A/m)	79.58
ounce (avoirdupois)	kilogram (kg)	2.835×10^{-2}
ounce (troy)	kilogram (kg)	3.110×10^{-2}
ounce (U.S. fluid)	cubic meter (m^3)	2.957×10^{-5}
ounce-force	newton (N)	0.2780
peck (U.S.)	cubic meter (m^3)	8.810×10^{-3}
pennyweight	kilogram (kg)	1.555×10^{-3}
pint (U.S. dry)	cubic meter (m^3)	5.506×10^{-4}
pint (U.S. liquid)	cubic meter (m^3)	4.732×10^{-4}
poise (absolute viscosity)	pascal second (Pa·s)	$0.10^\dagger$
pound (avoirdupois)	kilogram (kg)	0.4536
pound (troy)	kilogram (kg)	0.3732
poundal	newton (N)	0.1383
pound-force	newton (N)	4.448
pound per square inch (psi)	pascal (Pa)	6.895×10^3
quart (U.S. dry)	cubic meter (m^3)	1.101×10^{-3}
quart (U.S. liquid)	cubic meter (m^3)	9.464×10^{-4}
quintal	kilogram (kg)	$1.0 \times 10^{2\dagger}$

†Exact.

To convert from	*To*	*Multiply by*
rad	gray (Gy)	1.0×10^{-2}†
rod	meter (m)	5.029
roentgen	coulomb per kilogram (C/kg)	2.58×10^{-4}
second (angle)	radian (rad)	4.848×10^{-6}
section	square meter (m^2)	2.590×10^{6}
slug	kilogram (kg)	14.59
spherical candle power	lumen (lm)	12.57
square inch	square meter (m^2)	6.452×10^{-4}
square foot	square meter (m^2)	9.290×10^{-2}
square mile	square meter (m^2)	2.590×10^{6}
square yard	square meter (m^2)	0.8361
stere	cubic meter (m^3)	1.0†
stokes (kinematic viscosity)	square meter per second (m^2/s)	1.0×10^{-4}†
tex	kilogram per meter (kg/m)	1.0×10^{-6}†
ton (long, 2240 pounds)	kilogram (kg)	1.016×10^{3}
ton (metric)	kilogram (kg)	1.0×10^{3}†
ton (short, 2000 pounds)	kilogram (kg)	9.072×10^{2}
torr	pascal (Pa)	1.333×10^{2}
unit pole	weber (Wb)	1.257×10^{-7}
yard	meter (m)	0.9144†

† Exact.

Abbreviations and Unit Symbols

Following is a list of commonly used abbreviations and unit symbols appropriate for use in the *Encyclopedia.* In general they agree with those listed in *American National Standard Abbreviations for Use on Drawings and in Text* (*ANSI Y1.1*) (6) and *American National Standard Letter Symbols for Units in Science and Technology* (*ANSI Y10*) (6). Also included is a list of acronyms for a number of private and government organizations as well as common industrial solvents, polymers, and other chemicals.

Rules for Writing Unit Symbols (4):

1. Unit symbols should be printed in upright letters (roman) regardless of the type style used in the surrounding text.
2. Unit symbols are unaltered in the plural.
3. Unit symbols are not followed by a period except when used as the end of a sentence.
4. Letter unit symbols are generally written in lower-case (eg, cd for candela) unless the unit name has been derived from a proper name, in which case the first letter of the symbol is capitalized (W,Pa). Prefix and unit symbols retain their prescribed form regardless of the surrounding typography.
5. In the complete expression for a quantity, a space should be left between the numerical value and the unit symbol. For example, write 2.37 lm, *not* 2.37lm, and 35 mm, *not* 35mm. When the quantity is used in an adjectival sense, a hyphen is often used, for example, 35-mm film. *Exception:* No space is left between the numerical value and the symbols for degree, minute, and second of plane angle, and degree Celsius.
6. No space is used between the prefix and unit symbols (eg, kg).

7. Symbols, not abbreviations, should be used for units. For example, use "A," not "amp," for ampere.

8. When multiplying unit symbols, use a raised dot:

$$N{\cdot}m \text{ for newton meter}$$

In the case of W·h, the dot may be omitted, thus:

$$Wh$$

An exception to this practice is made for computer printouts, automatic typewriter work, etc, where the raised dot is not possible, and a dot on the line may be used.

9. When dividing unit symbols use one of the following forms:

$$m/s \textit{ or } m{\cdot}s^{-1} \text{ or } \frac{m}{s}$$

In no case should more than one slash be used in the same expression unless parentheses are inserted to avoid ambiguity. For example, write:

$$J/(mol{\cdot}K) \textit{ or } J{\cdot}mol^{-1}\cdot K^{-1} \textit{ or } (J/mol)/K$$

but *not*

$$J/mol/K$$

10. Do not mix symbols and unit names in the same expression. Write:

$$\text{joules per kilogram } \textit{or } J/kg \textit{ or } J{\cdot}kg^{-1}$$

but *not*

$$\text{joules/kilogram } \textit{nor} \text{ joules/kg } \textit{nor} \text{ joules}{\cdot}kg^{-1}$$

ABBREVIATIONS AND UNITS

A	ampere
A	anion (eg, H*A*)
a	atto (prefix for 10^{-18})
AATCC	American Association of Textile Chemists and Colorists
ABS	acrylonitrile–butadiene–styrene
abs	absolute
ac	alternating current, *n.*
a-c	alternating current, *adj.*
ac-	alicyclic
ACGIH	American Conference of Governmental Industrial Hygienists
ACS	American Chemical Society
AGA	American Gas Association
Ah	ampere hour
AIChE	American Institute of Chemical Engineers
AIP	American Institute of Physics
alc	alcohol(ic)
Alk	alkyl
alk	alkaline (not alkali)
amt	amount
amu	atomic mass unit
ANSI	American National Standards Institute
AOCS	American Oil Chemist's Society
AP	atomic orbital
APHA	American Public Health Association
API	American Petroleum Institute
aq	aqueous
Ar	aryl
ar-	aromatic
as-	asymmetric(al)

ASHRAE	American Society of Heating, Refrigerating, and Air Conditioning Engineers
ASM	American Society for Metals
ASME	American Society of Mechanical Engineers
ASTM	American Society for Testing and Materials
at no.	atomic number
at wt	atomic weight
av(g)	average
bbl	barrel
bcc	body-centered cubic
Bé	Baumé
bid	twice daily
Boc	*t*-butyloxycarbonyl
BOD	biochemical (biological) oxygen demand
bp	boiling point
Bq	becquerel
C	coulomb
°C	degree Celsius
C-	denoting attachment to carbon
c	centi (prefix for 10^{-2})
ca	circa (approximately)
cd	candela; current density; circular dichroism
CFR	Code of Federal Regulations
cgs	centimeter-gram-second
CI	Color Index
cis-	isomer in which substituted groups are on same side of double bond between C atoms
cl	carload
cm	centimeter
CMA	Chemical Manufacturers' Association
cmil	circular mil
cmpd	compound
COA	coenzyme A
COD	chemical oxygen demand
coml	commercial(ly)
cp	chemically pure
cph	close-packed hexagonal
CPSC	Consumer Product Safety Commission
D-	denoting configurational relationship
d	differential operator
d-	dextro-, dextrorotatory
da	deka (prefix for 10^1)
dB	decibel
dc	direct current, *n.*
d-c	direct current, *adj.*
dec	decompose
detd	determined
detn	determination
dia	diameter
dil	dilute
dl-; DL-	racemic
DMF	dimethylformamide
DMG	dimethyl glyoxime
DOE	Department of Energy
DOT	Department of Transportation
dp	dew point; degree of polymerization
DPH	diamond pyramid hardness
dstl(d)	distill(ed)
dta	differential thermal analysis
(*E*)-	entgegen; opposed
ϵ	dielectric constant (unitless number)
e	electron
ECU	electrochemical unit
ed.	edited, edition, editor
ED	effective dose
EDTA	ethylenediaminetetraacetic acid
emf	electromotive force
emu	electromagnetic unit
eng	engineering
EPA	Environmental Protection Agency
epr	electron paramagnetic resonance
eq.	equation
esp	especially
esr	electron-spin resonance
est(d)	estimate(d)
estn	estimation
esu	electrostatic unit
exp	experiment, experimental
ext(d)	extract(ed)
F	farad (capacitance)
F	faraday (96,487C)
f	femto (prefix for 10^{-15})
FAO	Food and Agriculture Organization (United Nations)

fcc face-centered cubic
FDA Food and Drug Administration
FEA Federal Energy Administration
fob free on board
fp freezing point
FPC Federal Power Commission
frz freezing
G giga (prefix for 10^9)
G gravitational constant
g gram
(g) gas, only as in $H_2O(g)$
g gravitational acceleration
gem- geminal
glc gas-liquid chromatography
g-mol wt; gmw gram-molecular weight
grd ground
Gy gray
H henry
h hour; hecto (prefix for 10^2)
ha hectare
HB Brinell hardness number
hcp hexagonal close-packed
Hb hemoglobin
HK Knoop hardness number
HRC Rockwell hardness (C scale)
HV Vickers hardness number
hyd hydrated, hydrous
hyg hygroscopic
Hz hertz
^{i}(eg, Pr^i) iso (eg, isopropyl)
i- inactive (eg, *i*-methionine)
IACS International Annealed Copper Standard
ibp initial boiling point
ICC Interstate Commerce Commission
ICT International Critical Table
ID inside diameter; infective dose
IPS iron pipe size
IPT Institute of Petroleum Technologists
IPTS International Practical Temperature Scale (NBS)
ir infrared
IRLG Interagency Regulatory Liaison Group
ISO International Organization for Standardization
Iu International Unit
IUPAC International Union of Pure and Applied Chemistry
IV iodine value
J joule
K kelvin
k kilo (prefix for 10^3)
kg kilogram
L denoting configurational relationship
L liter (for fluids only)(5)
l- *levo-*, levorotatory
(l) liquid, only as in $NH_3(l)$
LC_{50} conc lethal to 50% of the animals tested
LCAO linear combination of atomic orbitals
LCD liquid crystal display
lcl less than carload lots
LD_{50} dose lethal to 50% of the animals tested
LED light emitting diode
liq liquid
lm lumen
ln logarithm (natural)
LNG liquefied natural gas
log logarithm (common)
LPG liquefied petroleum gas
ltl less than truckload lots
lx lux
M mega (prefix for 10^6); metal (as in MA)
M molar
m meter; milli (prefix for 10^{-3})
m molal
m- meta
max maximum
MEK methyl ethyl ketone
meq milliequivalent
mfd manufactured
mfg manufacturing
mfr manufacturer
MIBC methyl isobutyl carbinol
MIBK methyl isobutyl ketone
MIC minimum inhibiting concentration
min minute; minimum
mL milliliter

MLD	minimum lethal dose
MO	molecular orbital
mo	month
mol	mole
mol wt	molecular weight
mom	momentum
mp	melting point
MR	molar refraction
ms	mass spectrum
mxt	mixture
μ	micro (prefix for 10^{-6})
N	newton (force)
N	normal (concentration)
N-	denoting attachment to nitrogen
n (as n_D^{20}	index of refraction (for 20°C and sodium light)
^{n} (as Bu^n), *n*-	normal (straight-chain structure)
n	neutron
n	nano (prefix for 10^{-9}
na	not available
NAS	National Academy of Sciences
NASA	National Aeronautics and Space Administration
nat	natural
NBS	National Bureau of Standards
neg	negative
NF	*National Formulary*
NIH	National Institutes of Health
NIOSH	National Institute of Occupational Safety and Health
nmr	nuclear magnetic resonance
NND	New and Nonofficial Drugs (AMA)
no.	number
NOI-(BN)	not otherwise indexed (by name)
NOS	not otherwise specified
nqr	nuclear quadrople resonance
NRC	Nuclear Regulatory Commission; National Research Council
NRI	New Ring Index
NSF	National Science Foundation
NTA	nitrilotriacetic acid
NTP	normal temperature and pressure (25°C and 101.3 kPa or 1 atm)
NTSB	National Transportation Safety Board
O-	denoting attachment to oxygen
o-	ortho
OD	outside diameter
OPEC	Organization of Petroleum Exporting Countries
OSHA	Occupational Safety and Health Administration
owf	on weight of fiber
Ω	ohm
P	peta (prefix for 10^{15})
p	pico (prefix for 10^{-12})
p-	para
p	proton
p.	page
Pa	pascal (pressure)
pd	potential difference
pH	negative logarithm of the effective hydrogen ion concentration
pmr	proton magnetic resonance
POP	polyoxypropylene
pos	positive
pp.	pages
ppb	parts per billion
ppm	parts per million
PPO	poly(phenyl oxide)
ppt(d)	precipitate(d)
pptn	precipitation
Pr (no.)	foreign prototype (number)
pt	point; part
PVC	poly(vinyl chloride)
pwd	powder
qv	quod vide (which see)
R	univalent hydrocarbon radical
(*R*)-	rectus (clockwise configuration)
rad	radian; radius
rds	rate determining step
ref.	reference
rf	radio frequency, *n.*
r-f	radio frequency, *adj.*

rh	relative humidity
RI	Ring Index
RT	room temperature
^{s} (eg, Bu^s); *sec*-	secondary (eg, secondary butyl)
S	siemens
(*S*)-	sinister (counterclockwise configuration)
S-	denoting attachment to sulfur
s-	symmetric(al)
s	second
(s)	solid, only as in $H_2O(s)$
SAE	Society of Automotive Engineers
SAN	styrene-acrylonitrile
sat(d)	saturate(d)
satn	saturation
SCF	self-consistent field; standard cubic feet
Sch	Schultz number
SFs	Saybolt Furol seconds
SI	Le Système International d'Unités (International System of Units)
sl sol	slightly soluble
sol	soluble
soln	solution
soly	solubility
sp	specific; species
sp gr	specific gravity
sr	steradian
std	standard
STP	standard temperature and pressure (0°C and 101.3 kPa)
SUs	Saybolt Universal seconds
syn	synthetic
^{t} (eg, Bu^t), *t*-, *tert*-	tertiary (eg, tertiary butyl)
T	tera (prefix for 10^{12}); tesla (magnetic flux density)
t	metric ton (tonne); temperature
TAPPI	Technical Association of the Pulp and Paper Industry
tex	tex (linear density)
tga	thermogravimetric analysis
THF	tetrahydrofuran
tlc	thin layer chromatography
TLV	threshold limit value
trans-	isomer in which substituted groups are on opposite sides of double bond between C atoms
TSCA	Toxic Substance Control Act
TWA	time-weighted average
Twad	Twaddell
UL	Underwriters' Laboratory
USDA	United States Department of Agriculture
USP	*United States Pharmacopeia*
uv	ultraviolet
V	volt (emf)
var	variable
vic-	vicinal
vol	volume (not volatile)
vs	versus
v sol	very soluble
W	watt
Wb	Weber
Wh	watt hour
WHO	World Health Organization (United Nations)
wk	week
yr	year
(*Z*)-	zusammen; together

Non-SI (Unacceptable and Obsolete) Units		*Use*
Å	angstrom	nm
at	atmosphere, technical	Pa
atm	atmosphere, standard	Pa
b	barn	cm^2

Non-SI (Unacceptable and Obsolete) Units		*Use*
bar†	bar	Pa
bbl	barrel	m^3
bhp	brake horsepower	W
Btu	British thermal unit	J
bu	bushel	m^3; L
cal	calorie	J
cfm	cubic foot per minute	m^3/s
Ci	curie	Bq
cSt	centistokes	mm^2/s
c/s	cycle per second	Hz
cu	cubic	exponential form
D	debye	C·m
den	denier	tex
dr	dram	kg
dyn	dyne	N
dyn/cm	dyne per centimeter	mN/m
erg	erg	J
eu	entropy unit	J/K
°F	degree Fahrenheit	°C; K
fc	footcandle	lx
fl	footlambert	lx
fl oz	fluid ounce	m^3; L
ft	foot	m
ft·lbf	foot pound-force	J
gf den	gram-force per denier	N/tex
G	gauss	T
Gal	gal	m/s^2
gal	gallon	m^3; L
Gb	gilbert	A
gpm	gallon per minute	(m^3/s); (m^3/h)
gr	grain	kg
hp	horsepower	W
ihp	indicated horsepower	W
in.	inch	m
in. Hg	inch of mercury	Pa
in. H_2O	inch of water	Pa
in.-lbf	inch pound-force	J
kcal	kilogram-calorie	J
kgf	kilogram-force	N
kilo	for kilogram	kg
L	lambert	lx
lb	pound	kg
lbf	pound-force	N
mho	mho	S
mi	mile	m
MM	million	M
mm Hg	millimeter of mercury	Pa
mμ	millimicron	nm
mph	miles per hour	km/h
μ	micron	μm
Oe	oersted	A/m
oz	ounce	kg
ozf	ounce-force	N
η	poise	Pa·s

† Do not use bar (10^5Pa) or millibar (10^2Pa) because they are not SI units, and are accepted internationally only for a limited time in special fields because of existing usage.

Non-SI (Unacceptable and Obsolete) Units		*Use*
P	poise	Pa·s
ph	phot	lx
psi	pounds-force per square inch	Pa
psia	pounds-force per square inch absolute	Pa
psig	pounds-force per square inch gauge	Pa
qt	quart	m^3; L
°R	degree Rankine	K
rd	rad	Gy
sb	stilb	lx
SCF	standard cubic foot	m^3
sq	square	exponential form
thm	therm	J
yd	yard	m

BIBLIOGRAPHY

1. The International Bureau of Weights and Measures, BIPM, (Parc de Saint-Cloud, France) is described on page 21 of Ref. 4. This bureau operates under the exclusive supervision of the International Committee of Weights and Measures (CIPM).
2. *Metric Editorial Guide* (*ANMC-75-1*), American National Metric Council, 1625 Massachusetts Ave. N.W.; Washington, D.C. 20036, 1975.
3. *SI Units and Recommendations for the Use of Their Multiples and of Certain Other Units* (*ISO 1000-1973*), American National Standards Institute, 1430 Broadway, New York, N. Y. 10018, 1973.
4. Based on *ASTM E 380-79* (*Standard for Metric Practice*), American Society for Testing and Materials, 1916 Race Street, Philadelphia, Pa. 19103, 1979.
5. *Fed. Regist.*, Dec. 10, 1976 (41 FR 36414).
6. For ANSI address, see Ref. 3.

R. P. LUKENS
American Society for Testing and Materials

F *continued*

FLUORINE COMPOUNDS, ORGANIC

POLYTETRAFLUOROETHYLENE

Polytetrafluoroethylene [*9002-84-0*], (PTFE) a perfluorinated straight-chain high polymer having the chemical formula $\leftarrow CF_2CF_2 \rightarrow_n$, is made by polymerizing the tetrafluoroethylene [*116-14-3*] (TFE) monomer. The white-to-translucent solid polymer has an extremely high molecular weight—in the 10^6–10^7 range—and, consequently, has a viscosity of about 10 GPa·s (10^{11} P) at 380°C. It is a highly crystalline polymer and goes through a melting-type transition. Its high thermal stability results from the strong carbon—fluorine bond and characterizes PTFE as a very useful high temperature polymer. Its heat resistance, chemical inertness, and electrical insulation properties, and its low coefficient of friction in a very wide temperature range make PTFE the most outstanding plastic in the industry.

The discovery of PTFE by Plunkett (1) in 1938 opened the commercial field of perfluoropolymers. Initial production of PTFE was directed toward the World War II effort and hence commercial production was delayed by DuPont until 1947. Commercially available PTFE is manufactured by two entirely different polymerization techniques that result in two completely different types of chemically identical polymers. Suspension polymerization produces a granular resin PTFE and emulsion polymerization produces the coagulated dispersion that is often referred to as a fine powder, and PTFE dispersion.

Because of its chemical inertness and high molecular weight, PTFE is a nonflowing material that does not undergo conventional polymer-fabrication techniques. Therefore, an extensive processing technology developed.

The suspension-polymerized PTFE polymer (referred to as granular PTFE) generally is fabricated using modified powder metallurgy techniques. Emulsion-polymerized PTFE behaves entirely differently from granular PTFE and coagulated dispersions of PTFE are processed by a cold extrusion process (like lead). Stabilized PTFE dispersions, made by emulsion polymerization, usually are processed according to latex processing techniques (see Latex technology).

Manufacturers of PTFE include Allied Chemical (Halon), Daikin Kogyo (Polyflon), DuPont (Teflon), Hoechst (Hostaflon), ICI (Fluon), Montecatini (Algoflon), Ugine Kuhlmann (Soreflon), and the USSR (Fluoroplast). The People's Republic of China recently introduced some PTFE products.

This article briefly reviews the preparation and properties of the monomer and the types of polymers and their preparations, and summarizes the outstanding properties of various forms of PTFE. Applications of PTFE and fabrication techniques are discussed also.

Monomer

Preparation. The tetrafluoroethylene manufacturing process involves the following steps:

$$CaF_2 + H_2SO_4 \rightarrow CaSO_4 + 2\ HF$$

$$CH_4 + 3\ Cl_2 \rightarrow CHCl_3 + 3\ HCl$$

$$CHCl_3 + 2\ HF \rightarrow CHClF_2 + 2\ HCl$$

$$2\ CHClF_2 \rightleftarrows CF_2{=}CF_2 + 2\ HCl$$

Hydrogen fluoride is manufactured by the first step and chloroform by the second step; the latter as a part of the chloromethanes production process. Chloroform is partially fluorinated with hydrogen fluoride to chlorodifluoromethane using antimony trifluoride as a catalyst. Chlorodifluoromethane then is converted to tetrafluoroethylene by pyrolysis (2–9). The last step (pyrolysis) is often conducted at a PTFE manufacturing site because of the difficulty of handling TFE.

Pyrolysis of chlorodifluoromethane is a noncatalytic gas-phase reaction carried out in a flow reactor at atmospheric or subatmospheric pressure. The pyrolysis reaction can be run with yields as high as 95% at 590–900°C. The economics of monomer cost is very sensitive to the yields of this process. It is obvious that a significant amount of hydrogen chloride waste product is generated during the formation of the carbon—fluorine bonds of tetrafluoroethylene.

A large number of by-products are formed in this process other than TFE and HCl. Most of the high boiling and low boiling by-products are in trace amounts; some of the significant ones are: hexafluoropropylene, perfluorocyclobutane, 1-chloro-1,1,2,2-tetrafluoroethane, and 2-chloro-1,1,1,2,3,3-hexafluoropropane. A small but very toxic by-product formed in chlorodifluoromethane pyrolysis is perfluoroisobutylene, $CF_2{=}C(CF_3)_2$ (2).

Studies of monochlorodifluoromethane pyrolysis showed that the use of diluents, eg, steam, to achieve subatmospheric partial pressures is effective. Because of the corrosive nature of the acids (HF and HCl) formed during pyrolysis, the use of a platinum-lined tubular reactor, atmospheric pressure reactions and a reactor made of nickel alloys for those reactions run in the presence of a diluent (such as steam) are suggested. Since the pyrolysate contains multitudes of by-products that adversely affect TFE polymerization, it is imperative that TFE be very pure. This refinement process is extremely complex and contributes to the high cost of the TFE monomer. After TFE purification, inhibitors are added to the monomer to avoid polymerization during storage. A variety of terpenes are effective inhibitors, eg, *d*-limonene and terpene B (10).

TFE was synthesized first by Ruff and Bretschneider in 1933 by decomposing tetrafluoromethane, CF_4, in an electric arc furnace (11). Since then a number of synthesis routes have been used (12–18). Depolymerization of PTFE by heating at about 600°C probably is the preferred means of obtaining small amounts of TFE on a laboratory scale, since it gives a 97% pure monomer (19–20). Care should be taken while handling depolymerization products since highly toxic perfluoroisobutylene is one of the by-products.

Properties. TFE is a colorless, tasteless, odorless, and relatively nontoxic gas at RT. Its physical properties are listed in Table 1 (21). TFE is stored as a liquid (its vapor pressure at −20°C is 1 MPa or 9.9 atm); however, it is generally polymerized above its critical temperature of 33.3°C and below its critical pressure of 3.94 MPa (38.9 atm). The polymerization reaction is highly exothermic.

TFE undergoes addition reactions typical of an olefin. It burns in air to form carbon tetrafluoride, carbonyl fluoride, and carbon dioxide (25). Under controlled conditions, oxygen can be added to TFE to produce epoxide (26),

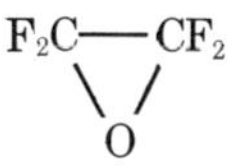

or explosive polymeric peroxide (25). TFE gives trifluorovinyl ethers ($RO{-}CF{=}CF_2$) from reaction with sodium salts of alcohols (27). An ozone–TFE reaction was accompanied by chemiluminescence (28). TFE dimerizes to form perfluorocyclobutane, $CF_2CF_2CF_2CF_2$, at 600°C; on further heating, it forms hexafluoropropylene, $CF_2{=}CFCF_3$; and with increasing contact time, it forms perfluoroisobutylene, $CF_2{=}C(CF_3)_2$ (29). Both gas–liquid and gas–solid chromatography are extremely valuable for determining the purity of the monomer. The ir spectrum of TFE is complex and, consequently, is of no use for detecting purposes.

Uses and Safety. Besides polymerizing TFE to various types of high PTFE homopolymer, TFE is utilized as a comonomer in a large number of other polymers. It is copolymerized with hexafluoropropylene (30), ethylene (31), perfluorinated ether (32), isobutylene (33), propylene (34), and in some cases it is used as a termonomer (35). TFE is used to prepare low molecular weight polyfluorocarbons (36) and carbonyl fluoride oils (37), as well as to obtain *in situ* PTFE on metal surfaces (38). TFE is used to prepare hexafluoropropylene (39–40), perfluorinated ethers, and other oligomers.

Table 1. Physical Properties of Tetrafluoroethylene

Property	Value	Ref.
molecular weight	100.02	
boiling point at 0.1 MPa (1 atm), °C	−76.3	22
freezing point, °C	−142.5	22
liquid density at t°C, g/mL		
$-100 < t < -40$	$= 1.202–4.14 \times 10^{-3}\, t$	22
$-40 < t < 8$	$= 1.1507–6.935 \times 10^{-3}\, t$	22
	$= 1.1507–3.76 \times 10^{-5}\, t^2$	
$8 < t < 30$	$= 1.1325–2.904 \times 10^{-3}\, t$	22
	$= 1.1325–2.556 \times 10^{-4}\, t^2$	
vapor pressure at T K, kPa[a]		
$196.85 < T < 273.15$	$\log_{10} P_{kPa} = 6.4593–875.14/T$	22
$273.15 < T < 306.45$	$\log_{10} P_{kPa} = 6.4289–866.84/T$	22
critical temperature, °C	33.3	22
critical pressure, MPa[b]	392	22
critical density, g/mL	0.58	22
dielectric constant at 28°C		
at 0.101 MPa (1 atm)	1.0017	22
at 0.858 MPa (8.5 atm)	1.015	22
thermal conductivity at 30°C, mW/(m·K)	15.5	22
heat of formation for ideal gas at 25°C, ΔH, kJ/mol[c]	−635.5	23
heat of polymerization at 25°C to solid polymer ΔH, kJ/mol[c] of TFE	−172.0	24
flammability limits in air at 101.3 kPa (1 atm), vol %	14–43	

[a] To convert kPa to psi, multiply by 0.145.
[b] To convert MPa to atm, divide by 0.101.
[c] To convert J to cal, divide by 4.184.

TFE can disproportionate violently in the absence of air to give carbon and carbon tetrafluoride and generates the same amount of energy as black-powder explosions. Since this type of decomposition is initiated thermally, hot spots in TFE handling equipment must be avoided. It burns when mixed with air, the flammability limits are 11–60% TFE, and forms explosive mixtures with air and oxygen. It can be stored in steel cylinders under controlled conditions when it is inhibited with a suitable stabilizer. The oxygen content of the vapor phase should not exceed 10 ppm.

Although TFE is relatively nontoxic, other fluorocarbon compounds that may be extremely toxic may contaminate it at high temperatures.

Manufacture of PTFE

Engineering problems involved in the production of monomer seem relatively simple when compared with those associated with polymerization and processing of various types of PTFE resins. TFE must be polymerized to extremely high molecular weight so that the desired properties are achieved. The low molecular weight polymer does not have the requisite strength needed in end-use applications.

PTFE is manufactured and sold in three forms each of which requires different fabrication techniques. The first form, referred to as granular, results from different polymerization processes than the other two forms, namely, fine powders and aqueous dispersions. Granular resins are manufactured in a wide variety of grades which describe different balances between powder flows and ultimate end-use properties (see Fig. 1). Fine powders that are made by coagulating aqueous dispersions also are available in various grades. Differences in fine powder grades correspond to their usefulness in specific end-use applications and to the ease of fabrication. Aqueous dispersions are sold in latex form and are available in different grades. A variety of formulation techniques are used to tailor these dispersions for specific applications.

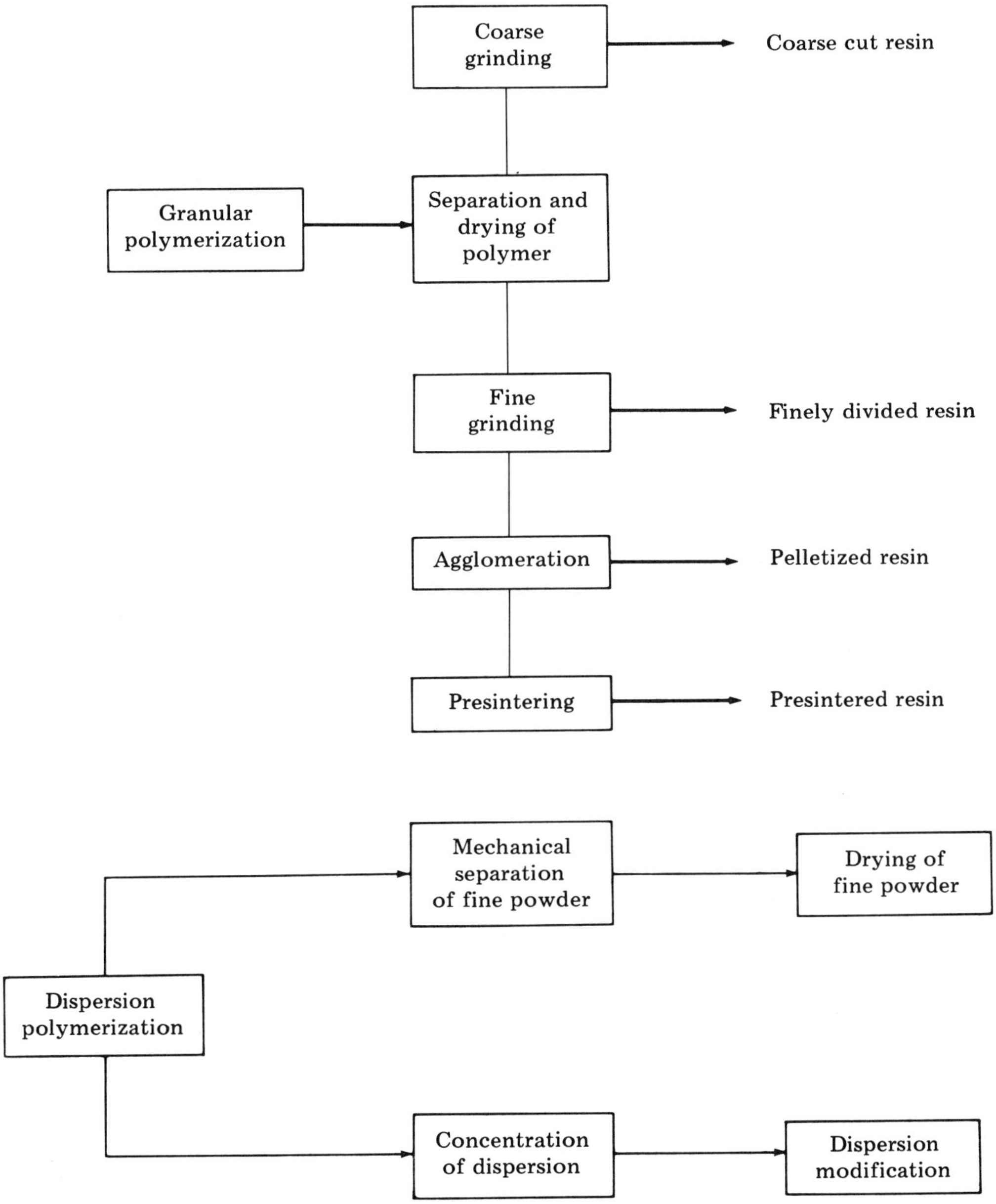

Figure 1. Granular, fine powder, and dispersion PTFE products.

Polymerization. Two distinctly different procedures result in the homopolymerization of TFE in aqueous medium. In one procedure, little or no dispersing agent is used and vigorous agitation is maintained to produce a precipitated resin, commonly referred to as granular resin. In the other procedure, sufficient dispersing agent is employed and agitation is very mild to produce small colloidal particles dispersed in the aqueous reaction medium. In the second procedure, called aqueous-dispersion polymerization, precipitation of the resin particles is avoided. The two procedures produce distinctly different products, even though both are chemically high molecular weight PTFE polymers. The granular product can be molded in various forms, whereas the resin produced by the aqueous dispersion method cannot be molded but is fabricated by dispersion coating or by conversion to powder for paste extrusion with a lubricant medium. In contrast, the granular resin cannot be paste-extruded or dispersion-coated.

Granular Resins. Granular resins are made by polymerizing TFE alone or in the presence of trace amounts of comonomers (41–42) with an initiator and with or without a small amount of dispersing agent. An alkaline buffer is occasionally used (43). In the early stages of polymerization, an unstable dispersion is formed but lack of dispersing agent, and vigorous agitation, cause the polymer to partially coagulate; the remainder of the polymerization process is fairly complex. The polymerized product is stringy, irregular and variable in shape. The dried granular polymer is ground to different average particle sizes depending on the product requirements, eg, the powder flow and developed properties. Coarser fabrication of polymer particles leaves a higher void content in the sintered article. A better balance between handleability and moldability (ability to mold and sinter into a void-free article) is achieved by agglomerating the finely divided resin to ca 400–800 μm (44). For ram extrusion of granular resin into long tubes and rods, a partially presintered resin is preferred.

Fine Powder Resins. Fine powder resins are made by polymerizing TFE in an aqueous medium with an initiator and emulsifying agents (45). The polymerization mechanism is not a typical emulsion type; however, some of the principles of emulsion polymerization apply to this process. The process, as well as the ingredients, has a significant effect on the product. It is extremely important that the dispersion remains sufficiently stable throughout polymerization so that it does not coagulate prematurely but unstable enough so that it can be subsequently coagulated into a fine powder. Stirring very gently during polymerization ensures the stability of the dispersion.

The thin dispersion rapidly thickens into a gelled matrix and coagulates into a nonwater-wet agglomeration which floats on the aqueous solvent as the mechanical agitation is continued. The agglomeration is dried very gently and care is taken to ensure absence of shearing.

Aqueous Dispersions. The dispersion is made by the same polymerization process as that used to produce fine powders. Raw dispersions are polymerized to different average particle sizes. The most commonly used dispersion, which has an average dispersion particle size of about 0.2 μm, probably is the optimum particle size for most applications. The raw dispersion is stabilized with a nonionic or anionic surfactant. This stabilized dispersion is concentrated to 60–65 wt % solids by electrodecantation, evaporation, or thermal concentration (46). This concentrated dispersion can be modified further with other chemical additives. The fabrication characteristics of these dispersions depend on polymerization conditions and chemical additives for modification of the dispersion behavior.

Filled Resins. A range of products is produced by PTFE manufacturers and processors, eg, fillers which have been compounded into all three types of PTFE polymers. The fillers used for compounding include glass fibers, graphite, asbestos, or powdered metals. Compounding is achieved by intimate mixing of the filler and the resin. Some compounding is obtained via co-coagulation of the polymer with a filler to produce a filled fine powder.

Properties

Basic Polymer Properties. The properties discussed in this section are those of polymers prepared by suspension polymerization, and those prepared by emulsion polymerization, since the latter relate to the basic structure of PTFE. The two valence forces binding the PTFE structure are the carbon—carbon bonds, which form the backbone of the polymer chain, and the carbon—fluorine bonds. Both of these chemical bonds are extremely strong and are the key contributors in imparting the outstanding combination of properties of PTFE. The fluorine atoms form a protective sheath over the chain of carbon atoms. If the atoms attached to the carbon-chain backbone were either smaller or larger than fluorine, the sheath would not form a regular, uniform cover. This sheath shields the carbon chain from attack by various chemicals and gives PTFE its chemical inertness and stability. It also lowers the surface energy giving PTFE a low coefficient of friction and nonstick properties (21).

PTFE does not dissolve in any common solvent; therefore, it is impossible to measure its molecular weight by the usual methods of molecular weight determination. A number-average molecular weight has been estimated using determinations of the concentration of end groups derived from the initiator. Earlier estimates, based on an iron–bisulfite system containing radioactive sulfur (^{35}S), ranged from 142,000–534,000 on low molecular weight polymer and application of the same technique to polymers of industrial interest gave molecular weights of 389,000–8,900,000 (47–48). In the absence of a normal molecular weight determination method, estimated relative molecular weight is used for all practical purposes and is obtained by measuring the specific gravity following a standardized fabricating and sintering procedure (49). Since the rate of crystallization decreases with increasing molecular weight, samples prepared from the high molecular weight polymer and cooled from the melt at a constant, slow rate have lower standard specific gravities than those prepared from low molecular weight polymer cooled at the same rate (50). The correlation between number-average molecular weight $\overline{M}_n$, based on end-group estimations, and standard specific gravity (SSG) is given by:

$$\text{SSG} = 2.612 - 0.058 \log_{10} \overline{M}_n$$

The SSG procedure assumes void-free samples (or constant void content). The presence of voids in samples results in low values for the measured specific gravity. The inaccuracies caused by voids can be corrected by applying ir techniques (51).

At ca 342°C, virgin PTFE changes from white crystalline material to almost transparent amorphous gel. Differential thermal analysis indicates that the first melting of virgin polymer is irreversible and that the subsequent remeltings occur at 327°C, which generally is reported as the melting point of PTFE (21). It should be noted that most of the studies reported in the literature are based on the previously sintered (ie, melted and recrystallized) polymer and very little work is reported on the virgin polymer. The melting of the polymer is accompanied by a volume increase of ca 30%. Since the viscosity of the polymer at 380°C is 10 GPa (10^{11} P), the melt is form-stable. The melting point of PTFE increases with increasing applied pressure at the rate of 1.52°C/MPa (0.154°C/atm) (52).

Virgin PTFE has a crystallinity in the range of 92–98%, which is a sign of unbranched chain structure.

The fluorine atoms are too large to allow a planar zig-zag structure which confers rigidity on the polymer (53). PTFE molecule has a regular folded structure which produces a laminar crystal. The crystallinity of the polymer has been estimated by x-ray, ir, and density methods by measurement of disorder in the crystal. It has been shown that a rapidly cooled polymer possesses the same molecular conformation and crystalline structure as does a slowly cooled one, even though the rapidly cooled sample has lower crystallinity (54).

Transitions. Extensive work has been reported on transitions (55–60). Effects of structural changes on properties, eg, specific heat, specific volume, and/or dynamic mechanical and electrical properties, are observed at various temperatures. A number of transitions observed by various investigators, their interpretation, and the modes of identification are listed in Table 2.

Besides the transition at the melting point, the transition at 19°C is of great consequence because it occurs around ambient temperature and significantly affects the product behavior. Above 19°C, the triclinic pattern changes to a hexagonal unit cell. Around 19°C, a slight untwisting of the molecule from a 180° twist per 13 CF_2 groups to a 180° twist per 15 CF_2 groups occurs. At the first-order transition at 30°C, the hexagonal unit cell disappears and the rodlike hexagonal packing of the chains in the lateral direction is retained (55). Below 19°C there is almost perfect three-dimensional order, between 19–30°C the chain segments are disordered and above 30°C, the preferred crystallographic direction is lost and the molecular segments oscillate above their long axes with a random angular orientation in the lattice (56–57).

Table 2. Transitions in Polytetrafluoroethylene

Type of transition	Temperature, °C	Region affected	Technique used	Ref.
1st order	19	crystalline, angular displacement causing disorder	thermal methods, x-ray, nmr	56
	30	crystalline, crystal disordering	thermal methods, x-ray, nmr	56
	90 (80–110)	crystalline	stress relaxation, Young's modulus, dynamic methods	59
2nd order	−90 (−110 to −73)	amorphous, onset of rotational motion around C—C bond	thermal methods, dynamic methods	60
	−30 (−40 to −15)	amorphous	stress relaxation, thermal expansion, dynamic methods	59
	130 (120–140)	amorphous	stress relaxation, Young's modulus, dynamic methods	59

PTFE transitions occur at specific combinations of temperature and at the frequency of mechanical or electrical vibrations. Transitions, sometimes called dielectric relaxations, can cause wide fluctuations in the value of the dissipation factor (see Electrical Properties).

Mechanical Properties. Mechanical properties of PTFE are highly influenced by many processing variables, eg, preforming pressure, sintering temperature and time, cooling rate, void content, and crystallinity level. Properties, such as low coefficient of friction, flexibility at low temperatures, and stability at high temperatures, are relatively independent of fabrication conditions. Molding and sintering conditions affect flex life, permeability, stiffness, resiliency, and impact strength. The physical properties of PTFE have been reviewed and compiled in several publications (58,61–62). Some of the more frequently-used average values for molded and sintered samples have been listed in Table 3.

A marked change in volume of 1.0–1.8% is observed for PTFE in the transition zone from 18–25°C. A part that has been machined on either side of this zone changes dimensions when permitted to go through the transition zone; hence, the final operating temperature of a precision part must be accurately determined. Parts fabricated of PTFE resins have high strength, toughness, and self-lubrication at low temperatures, are useful from 5 K and are highly flexible from 194 K. Articles made from PTFE polymers tend to return to their original dimensions after a deformation. At sintering temperature, the articles rapidly recover their original shapes. For most applications, decomposition rates are slightly below the maximum service temperature of 260°C

Table 3. Typical Mechanical Properties of PTFE Resins[a]

Property	Granular resin	Fine powder	ASTM method
tensile strength at 23°C, MPa[b]	7–28	17.5–24.5	D 638-61T
elongation at 23°C, %	100–200	300–600	D 638-61T
flexural strength at 23°C, MPa[b]	does not break		D 790-61
flexural modulus at 23°C, MPa[b]	350–630	280–630	D 747-61T
impact strength, J/m[c], at 21°C	106.7		D 256-56
24°C	160		
77°C	>320		
hardness durometer, D	50–65	50–65	D 1706-59T
compressive stress, MPa[b]			
at 1% deformation at 23°C	4.2		D 695-52T
at 1% offset at 23°C	7.0		D 695-52T
coefficient of linear thermal expansion per °C, 23–60°C	12×10^{-5}		D 696-44
thermal conductivity, 4.6 mm, W/(m·K)	0.24		Cenco-Fitch
deformation under load, %			D 621-59
26°C, 6.86 MPa[b], 24 h		2.4	
26°C, 13.72 MPa[b], 24 h	15		
water absorption	<0.01	<0.01	D 570-54T
flammability	nonflammable		D 635-56T
static coefficient of friction against polished steel	0.05–0.08		

[a] Ref. 62.
[b] To convert MPa to psi, multiply by 145.
[c] To convert J/m to lbf·ft/in., divide by 53.38.

for PTFE and no special precautions are necessary. PTFE resins have excellent impact strength over a wide range of temperatures. PTFE articles have a smooth, slippery surface. Static friction of these resins decreases with an increase in load. Static coefficient of friction is lower than the dynamic coefficient and avoids stick-slip problems.

Liquids having surface tensions below 18 mN/m (= dyn/cm) are spread completely on the PTFE surface; hence, solutions of various perfluorocarbon acids in water wet the polymer (63). Surface treatments involving alkali metals are used to promote the adhesion between PTFE and other substances (64); however, this treatment increases the coefficient of friction (65).

Filled Resin Properties. Besides retaining the desirable properties of uncompounded resins, filled compositions meet the requirements of an increased variety of mechanical, electrical, and chemical applications. Table 4 lists a number of physical properties of a variety of filled granular compounds (66).

Chemical Properties. ***Thermal Degradation.*** Vacuum thermal degradation of PTFE results in monomer formation. The rate of degradation is a first-order reaction (67). Mass spectroscopic analysis of PTFE thermal degradation shows degradation begins at ca 440°C, peaks at 540°C, and continues until 590°C (68). Table 5 is a list of rates of thermal degradation at different temperatures.

Effect of Radiations. Radiation degrades PTFE. In the absence of oxygen, stable secondary radicals are produced. An increase in stiffness in the vacuum-irradiated material indicates cross-linking (69). Degradation of PTFE results from random scission of the chain by the radiation, and the relative stability of the radicals *in vacuum* protect the materials from rapid deterioration. In air or oxygen, reactions take place and accelerated scission and rapid degradation occurs.

Table 4. Filled PTFE Compounds and Their Properties[a]

Property	Unfilled PTFE	15 wt % glass fiber	25 wt % glass fiber	15 wt % graphite	60 wt % bronze
specific gravity	2.18	2.21	2.24	2.16	3.74
tensile strength, MPa[b]	28	25	17.5	21	14
elongation, %	350	300	250	250	150
stress at 10% elongation, MPa[b]	11	8.5	8.5	11	14
thermal conductivity, mW/(m·K)	0.244	0.37	0.45	0.45	0.46
creep modulus, kN/m[c]	2	2.21	2.1	3.4	6.2
hardness, Shore durometer, D	51	54	57	61	70
Izod impact, J/m[d]	152	146	119		
PV[e] for 0.13 mm radial wear in 1000 h (unlubricated), (kPa·m)/s[f]	0.70	106	177	52	281
wear factor, 1/PPa[g]	5013	28.0	26.2	102	12
coeff. of friction					
static, 3.4 MPa[b] load	0.08	0.13	0.13	0.10	0.10
dynamic, PV[e] = 172 (kPa·m)/s[f]		0.15–0.24	0.17	0.15	0.15
V = 900 m/s	0.01		−0.24	−0.18	−0.22

[a] Ref. 66.
[b] To convert MPa to psi, multiply by 145.
[c] To convert kN/m to lbf/in., divide by 0.175.
[d] To convert J/m to lbf·ft/in., divide by 53.38
[e] PV = pressure × velocity.
[e] To convert kPa to psi, multiply by 0.145.
[g] To convert 1/PPa to cm·s/(bar·m·min), multiply by 6×10^{23}.

Table 5. Thermal Degradation of Polytetrafluoroethylene[a]

Temperature, °C	Rate of decomposition, %/h		
		Granular	
	Fine powder	Initial	Steady state
232	0.0001–0.0002	0.00001–0.00005	1×10^{-11}
260	0.0006	0.0001–0.0002	100×10^{-11}
316	0.005	0.0005	0.000002
371	0.03	0.004	0.0009

The current flowing through a PTFE sheet that was exposed to radiation (with dose rates at 10^{-5} to 10 Gy/s (10^{-3}–10^{3} rad/s) and energies of 15, 30, and 38 keV) in the time interval of 10–10^4 s after application of d-c voltage shows that these differences have no significant affect on the transient conductivity (70). The steady-state conductivity increases with dose rate and the effect of radiation on the dissipation factor of PTFE is less pronounced (71).

Polymer crystallinity has been studied by x-ray irradiation (72). An initial increase in crystallinity of PTFE caused by chain scission in the amorphous phase was followed (above 3 kGy or 3×10^5 rad) by a gradual decrease associated with a disordering of the crystallinities. The amorphous component showed a maximum of radiation-induced broadening in the nmr at 7 kGy (7×10^5 rad).

In vacuum, PTFE has a damage threshold in the range of $(2–7) \times 10^3$ Gy ($(2–7) \times 10^5$ rad). It retains 50% of initial tensile strength after a dose of 10^5 Gy (10 megarad), 40% of initial tensile strength after 8×10^6 Gy (8×10^8 rad), and 100% or more of ultimate elongation for doses up to $(2–5) \times 10^6$ Gy ($(2–5) \times 10^8$ rad). In air, PTFE has a damage threshold in the range $(2–7) \times 10^2$ Gy ($(2–7) \times 10^4$ rad), retains 50% of initial tensile strength after a dose of 10^4 Gy (1 megarad), retains 40% of initial tensile strength after a dose of 10^5 Gy (10^7 rad), and retains ultimate elongation of 100% or more for doses up to $(2–5) \times 10^3$ Gy ($(2–5) \times 10^5$ rad). During irradiation, the resistivity decreases, and the dielectric constant and the dissipation factor increase. After irradiation, these properties tend to return to their preexposure values. Dielectric properties at high frequency are less sensitive to radiation than are properties at low frequency. Radiation has very little effect on dielectric strength (73).

Absorption and Permeation. PTFE is chemically inert to essentially all industrial chemicals and solvents even at elevated temperatures and pressures. This chemical compatibility stems from the strong interatomic bonds, the almost perfect shielding of the carbon backbone by fluorine atoms, and the high molecular weight of the polymer. Under some strenuous conditions PTFE is not compatible with certain materials. It reacts with molten alkali metals, fluorine, strong fluorinating agents, and with sodium hydroxide at temperatures above 300°C. Shapes of small cross-section burn vertically upwards after ignition in 100% oxygen. Because gases may be evolved in the action, the weight loss during sintering of a blend of PTFE and white asbestos is many times greater than loss from pure PTFE. Finely-divided aluminum and magnesium that is thoroughly mixed with finely-divided PTFE react vigorously after ignition or at high temperatures (74). A long list of typical chemicals that are compatible with PTFE is given in ref. 74.

Absorption of a liquid is usually a matter of the liquid dissolving in the polymer; however, in the case of PTFE, there is essentially no chemical interaction between the polymer and other substances. Submicroscopic voids between the polymer molecules provide space for the material absorbed; this is marked by a slight weight increase and sometimes by discoloration. PTFE polymers absorb no common acids or bases at temperatures as high as 200°C (see Table 6). In general, aqueous solutions scarcely are absorbed by the polymer at atmospheric pressure. Even the absorption of solvents is slight, partially resulting from the low wettability of PTFE. Since absorption of chemicals or solvents has no substantial effect on the chemical bond within the fluorocarbon molecule, the absorption should not be confused with degradation;

Table 6. Absorption of Common Acids, Bases and Solvents in PTFE[a]

Liquid	Exposure temperature[b], °C	Exposure time	Increase[b], wt %
hydrochloric acid, 10%	25	12 mo	0
	50	12 mo	0
	70	12 mo	0
20%	100	8 h	0
	200	8 h	0
nitric acid, 10%	25	12 mo	0
	70	12 mo	0.1
sulfuric acid, 30%	25	12 mo	0
	70	12 mo	0
	100	8 h	0
	200	8 h	0.1
sodium hydroxide, 10%	25	12 mo	0
	70	12 mo	0.1
50%	100	8 h	0
ammonium hydroxide, 10%	25	12 mo	0
	70	12 mo	0.1
acetone	25	12 mo	0.30
	50	12 mo	0.4
	70	2 wk	0
benzene	78	96 h	0.5
	100	8 h	0.6
	200	8 h	1.0
carbon tetrachloride	25	12 mo	0.6
	50	12 mo	1.6
	70	2 wk	1.9
	100	8 h	2.5
	200	8 h	3.7
ethyl alcohol, 95%	25	12 mo	0
	50	12 mo	0
	70	2 wk	0
	100	8 h	0.1
	200	8 h	0.3
ethyl acetate	25	12 mo	0.5
	50	12 mo	0.7
	70	2 wk	0.7
toluene	25	12 mo	0.3
	50	12 mo	0.6
	70	2 wk	0.3

[a] Values are averages only and not for specification purposes. Exposures over the boiling point of the solvent are at its vapor pressure (74).

[b] Weight changes less than about 2% are probably not experimentally significant.

it is a reversible physical process. The polymer does not suffer loss of mechanical or bulk electrical properties unless subjected to severely fluctuating conditions (74).

PTFE finds extensive use in the chemical industry as an excellent barrier resin because it exhibits low permeability. Since similarity in the chemical composition of a barrier and a permeant results in a high rate of permeation, PTFE is a poor barrier for fluorocarbon oils. Most liquids and gases (other than fluorocarbons) have generally low permeability across highly crystalline PTFE since high crystallinity reduces permeability.

At 25°C the permeability of PTFE to water is 5.14 g·mm/(100 m^2·d); to ethanol, 7.9 g·mm/(100 m^2·d); to red fuming nitric acid, 177.75 g·mm/(100 m^2·d); to 98% sulfuric acid, 7.9×10^6 g·mm/(100 m^2·d); and to 20% hydrochloric acid, 0.4 g·mm/(100 m^2·d). The permeabilities of PTFE to carbon dioxide and nitrogen at 20–30°C are 27.65 and 3.95 g·mm/(100 m^2·d), respectively (75).

By diffusion, gases and vapors permeate PTFE at a significantly lower rate than most other polymers. The higher the crystallinity and the less space between polymer molecules, the slower the permeation. Voids of somewhat greater than molecular size cause an increase in permeability. Molding the resin to obtain low porosity and high density can control the permeability of finished parts. The optimum specific gravity for low permeability and good flexural properties is 2.16–2.195. Permeability increases with temperature, not only as a result of the increase in activity of the solvent molecules but because of the increase in vapor pressure of the liquids. Swelling of PTFE resins and film is very low. Table 7 lists permeability data of PTFE films to various vapors at their vapor pressures (see also Barrier polymers).

Electrical Properties. Mechanical strength and chemical and thermal stability in combination with excellent electrical properties characterize PTFE as a very versatile electrical insulator. It does not absorb water; therefore, volume resistivity remains unchanged even after prolonged soaking in water. Surface-resistivity values were taken at 100% rh (see Table 8). The dielectric constant remains constant at 2.1 for a temperature range of −40 to 250°C and a frequency range of 5 Hz–10 GHz. After heat-aging at 250°C, skived tape made from granular resin has a dielectric constant of 755,200 V/m at the end of 1 mo. The value drops to 73,744 V/mm after 6 mo. Comparable values for spaghetti tubing made from fine powder are 43,263 and 37,757 V/mm, respectively.

Table 7. Permeability of PTFE Resins to Vapors, g/(m^2·d)[a]

Solvent vapors	PTFE 23°C	PTFE 30°C
acetophenone	8.61	
benzene	5.54	12.31
carbon tetrachloride	0.92	
ethyl alcohol	1.99	
HCl, 20%	<0.15	
piperidine	1.08	
Skydrol hydraulic fluid	0.92	
NaOH, 50%	8×10^{-4}	
H_2SO_4, 98%	3×10^{-4}	
water		5.38

[a] Ref. 74; test method ASTME 96-35-T (at vapor pressure: for 25.4 μm film thickness). Values are averages only and not for specification purposes.

Table 8. Electrical Properties of Polytetrafluoroethylene[a]

Property	Granular	Fine powder	ASTM method
dielectric strength, short time, 2 mm, V/mm	23,600	23,600	D 149-55T
surface arc-resistance, s	>300	>300	D 495-55T
volume resistivity, Ω·cm	$>10^{18}$	$>10^{18}$	D 257-57T
surface resistivity, Ω/sq	$>10^{16}$		D 257-57T
dielectric constant, at 60 to 2×10^9 Hz	2.1	2.1	D 150-59T
dissipation factor, at 60 to 2×10^9 Hz	0.0003		D 150-59T

[a] Ref. 62.

Parts fabricated according to standard practice should have dielectric constants in the range of 2.05 ± 0.5 when tested at RT. The dielectric constant varies with density and with factors that affect density. Machined components of PTFE can be fabricated to a predetermined dielectric constant value by controlling the rod density during processing. The density is controlled by the preforming pressure on the resin and cooling after sintering. The dielectric constant and the density have a linear relationship. Predictable variations in the dielectric constant resulting from density changes that accompany thermal expansion occur with temperature. A timed exposure of PTFE showed no change in dielectric constant over 2–3 yr of measurements.

The dissipation factor (the ratio of the energy dissipated to the energy stored per cycle) of PTFE is affected by the frequency, the temperature, the crystallinity, and the void content of the fabricated structure. At particular combinations of temperature and frequency, the crystalline and amorphous regions become resonant. Because of the molecular vibrations, applied electrical energy is lost by internal friction within the polymer which results in an increase in the dissipation factor. The dissipation factor for these resins corresponds to well-defined transitions but the magnitude of the variation is minor as compared to other polymers. The low temperature transition at −97°C causes the only meaningful dissipation factor peak. The dissipation factor has a maximum of 10^8–10^9 Hz at RT. The dissipation factor peak at 10^8–10^9 Hz is eliminated if the crystallinity is high (93%).

As crystallinity increases, both the internal molecular friction and the dissipation factor decrease. Voids dilute the dissipation factor in proportion to the percentage of microvoids present. Certain extruded shapes utilize air to reduce the effective dielectric constant and dissipation factor of a coaxial cable. The dielectric strength of these resins is high and is unaffected by thermal aging at 200°C. Frequency has a marked effect on the dielectric strength since corona discharge becomes more continuous as frequency increases. If the voltage stress is not high enough to cause corona ignition, then a very long dielectric life is anticipated for PTFE at any frequency. Corona discharges on the surface or in a void initiate dielectric breakdown (76). Surface arc-resistance of these resins is high and is not affected by heat-aging. The resins do not track or form a carbonized conducting path when subjected to a surface arc in air. PTFE resins are capable of continuous service up to 260°C and can withstand much higher temperatures for limited periods of time. They do not melt or flow, and they retain some strength even in the gel state which begins at 327°C (see Insulation, electric).

Fabrication Techniques

Granular Resins. These resins are sold in different forms. Generally, an optimum balance between resin handleability and product properties is sought. A resin that can flow freely is used in small moldings and automatic moldings. A finely divided resin that is more difficult to handle provides a good leveling in large moldings and develops superior properties in sintered articles; it is used for large billet and sheet-molding operations. Reduced crystallinity and superior handleability make a presintered resin an ideal product for ram extrusion applications.

Virgin PTFE melts at about 342°C; however, its melt viscosity, even at 380°C, is 10 GPa·s (10^{11} P). This makes it impossible to process the polymer by normal thermoplastic processing techniques; hence, other fabrication techniques have been developed. The basic steps involved in fabricating these resins include: preforming the dry powder into a compressed form so it can be handled; heating the preform above the polymer melting point so that the polymer particles can coalesce into a strong homogeneous structure; and cooling the polymer at a controlled rate to achieve the desired degree of crystallinity development.

Molding. Many manufacturers of PTFE give detailed descriptions of molding equipment and molding procedures (77–83). The most frequently used molds are round piston molds for the production of solid or hollow cylinders. Since preforming is done generally at moderate temperatures (below 100°C), carbon steel is a suitable material of construction. The compression ratio (the bulk volume of the powder to the specific volume of the unsintered molding) for granular resins is 3:1 to 6:1. For large-area, thin-walled moldings of unfilled polymer, a short-stroke press with a working capacity of 19.6–34.3 MPa (194–339 atm) mold pressure is sufficient, and for tall moldings of filled compounds with a small cross-sectional area, a long-stroke press with a relatively low thrust is required. The powder should be evenly distributed and leveled in the mold (77).

To ensure adequate compression uniformly throughout the preform, maximum pressure should be maintained for a sufficient length of time, and then be released slowly (see Plastics technology).

Automatic Molding. Automatic molding permits high-speed mass production. It is also preferable to machining finished material. Automatic presses can be operated mechanically, pneumatically, or hydraulically. The mold is filled by means of a special metering system from a storage hopper containing a free-flowing-grade resin. Loading shoes that shuttle backwards and forwards over the single- or multicavity mold also are used. Since automatic molding requires short cycles, the powder usually is compressed at high speed with a high preform pressure. Small parts, such as rings, bushings, washers, gaskets, and ball-valve seats can be molded by this technique.

Isostatic Molding. Isostatic molding allows uniform compression from all sides. A flexible mold is filled with a free-flowing granular powder and then is evacuated, tightly sealed, and placed in an autoclave containing a liquid that can be raised to the pressure required for preforming. The moldings require subsequent finishing because close tolerance cannot be achieved.

Sintering. Electrical ovens with air circulation and service temperatures up to 400°C are satisfactory for sintering. Free sintering is the cheapest and the most widely used process. A preformed mold is placed in an oven having a temperature variance of ±2°C. In pressure sintering, the preform is not removed from the mold; instead the mold containing the preform is heated in an oven until the sintering temperature is reached. After reaching the sintering temperature, the mold again is placed under pressure (lower than the preform pressure) during sintering and cooling. Pressure-sintered products have internal stresses that can be relieved by subsequent annealing. The pressure-cooling procedure involves application at pressure on the molded article after it has reached sintering temperature and maintenance at that pressure throughout the cooling period. This type of molding has a lower void content than the free-sintered mold.

To improve homogeneous structure, the preformed article is heated to 370–390°C. The heating-up and sintering times depend on the dimensions of the molding; however, the cooling rate, which affects the crystallinity and hence the product properties, should be slow.

Free-sintered articles do not have the same dimensions as the mold cavity because they shrink at right angles to the direction of the preform pressure and grow in the direction of the applied pressure.

Processing after Sintering. The least expensive route for sintered PTFE production requires that a large billet be skived on a lathe after the billet has been sintered and cooled. High precision articles are machined from ram-extruded rods.

Another method, called coining, prepares articles that are too complicated to obtain by machining. A sintered molding is heated to its melting point, transferred to a mold, and quickly deformed at low pressure and held under pressure until it has cooled sufficiently to retain the improved shape. This method has a major defect: the coined molding, if reheated to a high temperature, returns to its original shape, and hence there is a limit upon the maximum service temperature appropriate to coined molding.

Ram Extrusion. Compression molding is not a suitable method for the manufacture of continuous long moldings such as pipes or rods. In ram extrusion a small charge of PTFE powder is successively preformed by a reciprocating ram and then sintered and the subsequent charges fused into the previously sintered charge, and this process continues to form homogeneous long rods (77,84–86). The die tube, which is made of a corrosion-resistant material, is heated by resistance heating. Since sintering occurs in this tube, good temperature control is essential. It is important that the melted and compacted powder does not pass any constrictions in its path. Thermal expansion and friction produce great resistance to movement and, as a result, a considerable force is required to push the polymer through the tube. A high quality surface finish on the inside of the tube helps to reduce the pressure. If adequate bond strength between successive charges is not developed, the extrudate may break at the interface (poker chipping). Free-flowing powders and presintered resins are preferred for ram extrusion. Ram-extruded rods are used for automatic screw machining, and tubing is used as pipe liners or stock from which seals, gaskets, and bellows are machined.

Fine Powder Resins. Fine powder resins are extremely sensitive to shear. As a result, the manufacturer and the processor must take great care to handle the powder gently. These resins can be smeared between the fingers or sheared if scooped roughly out of the drum. The sheared polymer cannot be processed. Since fine powder is suitable for the manufacture of tubing and wire insulation and since compression molding is not suitable for the preparation of thin-walled continuous lengths, a paste-extrusion process is used. This method may be applied to the fabrication of tubes with diameters ranging from fractions of a millimeter to about a meter, walls having thicknesses of 100–400 μm, thin rods having up to 50 mm diameters, and cable sheeting. Calendering unsintered extruded solid rods produces thread-sealant tape and gaskets.

The paste-extrusion process includes mixing of ca 16–25 wt % of the lubricant (usually a petroleum fraction), and rolling the mixture to obtain uniform distribution of the lubricant. This wetted powder then is shaped into a preform at low pressure (2.0–7.8 MPa or 19–77 atm). This preform is pushed through a die mounted in the extruder at ambient temperature. The high level of shear stress exerted on the powder during the extrusion process longitudinally fibrillates the polymer which, consequently, acquires strength in that direction. The lubricant is evaporated from the extrudate and the extrudate is subsequently sintered at ca 380°C.

The exact amount of lubricant required for extrusion depends on the design of the extruder, the reduction ratio (ratio of the cross-sectional preform area to the cross-sectional area in the die) and the quality of the lubricant. Too low a lubricant level results in a high extrusion pressure, whereas too high a lubricant level causes a poor coalescence of the polymer and generates defects in the extrudate.

Fine powder resins can be pigmented with pigments that can withstand the sintering temperature. The pigment should be thoroughly mixed with the powder by rolling the mixture prior to lubricating the fine powder. Detailed suggestions regarding the design parameters of the paste extruder are given in refs. 77, 87–91.

Drying and sintering of the extrudate are performed consecutively by passing the extrudate through a multistage oven that is located immediately after the extruder. Pipes and rods may be extruded first and dried and sintered in a separate oven. The sintering oven must be capable of reaching 380°C. The throughput rate in this extrusion depends on the length of the sintering oven. Residence time in the oven varies from a few seconds for thin-walled insulations on a wire to a few minutes for large diameter tubing. Temperatures up to ca 480°C for a short residence time are also used. The extrusion pressure depends on the reduction ratio, the extrusion rate, the lubricant content, and the characteristics of the extruder.

For unsintered tape production by paste extrusion, the fine powder is lubricated and preformed according to the procedure described above. The preform then is extruded in the form of rods, which are calendered on hot rolls to the desired width and thickness (92–93).

Different resins have been developed for use in different reduction-ratio application ranges (45,94–98). The powders that are suitable for high reduction-ratio applications, eg, wire coatings, are not necessarily suitable for the medium reduction-ratio-range applications, eg, tubings, or the low reduction-ratio applications, eg, thread-sealant tapes or pipe liners.

New applications and new processing techniques are developed continuously and exploit the unique combination of properties offered by PTFE in fine powder form (99–101).

Dispersion Resins. PTFE dispersions generally contain 30–60 wt % polymer particles and some surfactant in the aqueous medium. The type of surfactant and the particle characteristics depend on the dispersion application. These dispersions are applied to various substrates by spraying, flow-coating, dipping, co-coagulating, or electrodepositing.

Aqueous dispersion is sprayed on metal substrates to provide a desired chemical resistance and nonstick and low friction properties. The coated surface is dried and sintered.

Impregnation of fibrous or porous materials with these dispersions combines the properties of the materials used with those of PTFE. Some materials require only a single dipping, eg, asbestos. Generally, the material is dried after dipping. For high pressure sealing applications, sintering at 380–400°C increases strength and dimensional stability.

In order to cast films, the dispersion is poured on a smooth surface, the formed film is dried and sintered and subsequently is peeled from the supporting surface.

Aqueous dispersions are used for the spinning of PTFE fibers. The dispersion is mixed with a matrix-forming medium (102) and then is forced through a spinneret into a coagulating bath. The resultant fibers are heated to remove the matrix material, sintered, and finally drawn under molten condition to develop their full strength.

Effects of Fabrication on Physical Properties. The basic factors that affect physical properties are molecular weight, which can be decreased by degradation but cannot be increased during processing; void content; and crystallinity. These factors can be controlled during molding, through choice of resin and fabricating conditions. Void distribution (or size and orientation) also affects properties; however it is not easily measurable.

Preforming primarily affects void content, sintering controls molecular weight, and cooling determines crystallinity. Voids caused by insufficient consolidation of particles during preforming may appear in the finished articles. Densities below 2.10 g/cm^3 indicate a high void content for most applications. Electrical and chemical operations require a minimum density of 2.12–2.14 g/cm^3. Particle size, shape and porosity also are important in determining void content. Although void content is determined largely by particle characteristics and preforming conditions, sintering conditions also can have an effect. Sintering at too high or too low a temperature can increase void content. The initial molecular weight of the polymer primarily determines its molecular weight after fabrication; however, an excessively high sintering temperature can decrease the molecular weight. The final crystallinity of a molding depends upon the initial molecular weight of the polymer, upon the rate of cooling of the molding, and—to a lesser extent—upon sintering conditions. The level of crystallinity of moldings can be changed substantially by varying either the cooling or the annealing conditions.

Flexural modulus increases by a factor of five as crystallinity increases from 50–90% with a void content of 0–2%; however, recovery decreases with increasing crystallinity. Therefore, the application requirements can determine the balance between stiffness and recovery. Rapid cooling can reduce crystallinity, and slow cooling can increase the crystallinity of the article. Table 9 summarizes the effects of maximum changes in properties caused by increasing the molecular weight, crystallinity, and void content of the polymer (62).

Economics

The high cost of monomer preparation and purification, and of polymerization and post-treatments are the main contributors to PTFEs price. Depending on the type of resin marketed, the polymer was sold for $7.39–11.44/kg in 1978 (in the U.S.). Since the PTFE fabrication techniques are different from typical thermoplastics and generally involve batch operations, the cost of converting the polymer to an end-use article is also high. Hence, the final product is relatively expensive.

Table 9. Maximum Change in Properties Caused by Increasing Each Basic Factor[a]

Physical property	Molecular weight	Crystallinity	Void content
flex fatigue life	+100-fold	−100-fold	−1000-fold
compressive stress at 1% deformation	0	+50%	0
compressibility	0	−50%	
recovery	0	−70%	
permeability to carbon dioxide	0	−30-fold	+1000-fold
flexural modulus	0	+5-fold	−30%
hardness: Durometer	0	+20%	
Rockwell	0	−20%	−30%
Seleroscope	0	−70%	−10%
tensile impact strength	0	−15-fold	−80%
dielectric strength	0	0	−70%
proportional limit	0	+80%	−20%
yield stress	0	+15%	−20%
yield strain	0	−15-fold	0
tensile strength	+25%	−50%	−50%
ultimate strength	+50%	−70%	−50%
ultimate elongation	−20%	+100%[b]	−80%

[a] Ref. 62.
[b] Reaches maximum at 85% crystallinity.

Testing and Standards

A description of PTFE resins and their classification is available (49). A comprehensive listing of industrial and military specifications covering mechanical, electrical and chemical applications of PTFE can be found in ref. 103.

Safety and Health

Exposure to PTFE can arise from the ingestion, skin contact, or inhalation of the unheated material, which is physiologically inert, and from the inhalation of the heated polymer. The polymer is nonirritating to the skin, and test animals that have ingested the sintered polymer have not shown adverse reactions. Dust that is generated by grinding the unheated resin also is inert in test animals. Since this polymer is unreactive with most chemicals, there is little possiblity that toxic products could be formed by chemical reaction. Therefore, only the heated polymer is a source of a possible health hazard (104).

A number of publications have included discussions regarding safe handling practices during the fabrication of PTFE (21). Because PTFE resins decompose slowly, they may be heated to a high temperature. The toxicity of the pyrolysis products, which may be released as a small quantity of fumes, warrants care where exposure of personnel is likely to occur (104). In general, PTFE resins do not decompose unless heated above 200°C. Only above 230°C, do decomposition rates become sufficient to provide weight losses. Small amounts of the toxic perfluoroisobutylene have been isolated at 400°C and above. Free fluorine has never been found among the decomposition products. Above 690°C, the decomposition products burn but they do not support combustion if the heat is removed. Combustion products primarily consist of carbon dioxide, carbon tetrafluoride, and small quantities of hydrogen fluoride. PTFE resins are nonflammable and do not propagate flame. In very hot fires, which cause appreciable decomposition of PTFE, heat is absorbed from the surroundings as heat of decomposition (104).

Prolonged exposure to thermal decomposition products of PTFE causes a temporary influenza-like condition that is called polymer fume fever. The condition, which may be contracted by smoking tobacco that has been contaminated with the polymer, occurs several hours after exposure and passes within 36–48 h even in the absence of treatment. The temporary effects are not cumulative.

Thousands of metric tons of PTFE resins have been manufactured, processed at temperatures in excess of 370°C, and placed in various applications, including many that are above the rated use temperatures. There are no reported cases of serious injury, prolonged illness or death resulting from the handling of these resins.

Table 10. Polytetrafluoroethylene Resins and Their Uses

Type	Resin grade	Method of processing	Description	Main uses
granular	agglomerates	molding, preforming-sintering, ram extrusion	free-flowing powder	gaskets, packing, seals, electronic components, bearings, sheet, rod, heavy-wall tubing; tape and molded shapes for chemical, electrical, mechanical, and nonadhesive (adherent) applications
	coarse	molding, preforming-sintering	granulated powder	tape, molded shapes for chemical mechanical, electrical, and nonadhesive applications
	finely divided	molding, preforming-sintering	powder for use where highest quality, void-free moldings are required	molded sheets, tape, wire wrapping, tubing, gasketing
	presintered	ram extrusion	granular, free flowing powder	rods and tubes
fine powder	high reduction-ratio resin	paste extrusion	agglomerated powder	wire coating, thin-walled tubing
	medium reduction-ratio resin	paste extrusion	agglomerated powder	tubing, pipe, overbraided hose, spaghetti tubing
	low reduction-ratio resin	paste extrusion	agglomerated powder	thread-sealant tape, pipe-liners, tubing, porous structures
dispersion	general purpose	dip coating	aqueous dispersion	impregnation, coating, packing
	coating	dip coating	aqueous dispersion	film, coating
	stabilized	cocoagulation	aqueous dispersion	bearings

Applications

Consumption of PTFE continuously grows as new applications are developed. The major applications of these resins are electrical, mechanical, and chemical. Electrical applications consume half of the PTFE produced, and mechanical and chemical applications share equally the other half (21). Various grades of PTFE and their applications are shown in Table 10.

Electrical Applications. The largest application of PTFE is for hookup and hookup-type wire used in electronic equipment in the military and aerospace industries. Coaxial cables, the second largest application, use tapes made from fine powder resins and some from granular resin. Interconnecting wire applications include airframes. Other electrical applications include computer wire, electrical tape, electrical components, and "spaghetti" tubing (see Insulation, electric).

Mechanical Applications. Seals and piston rings, basic shapes and anti-stick uses constitute two-thirds of the resin consumed in mechanical applications (see Abherents). Bearings, mechanical tapes, and coated glass fabrics also consume a large amount of PTFE resins (see Bearing materials). Seals and piston rings, bearings, and basic shapes are manufactured from granular resins, whereas the dispersion is used for glass fabric coating and a number of anti-stick applications. Most pressure-sensitive mechanical tapes are made from the skived tape made from granular resins.

Chemical Applications. The chemical processing industry extensively uses PTFE and large amounts of granular resin. Fine powder resins also are used. Soft packing applications are manufactured from dispersions, and hard packings are molded, or machined from stocks and shapes made from granular resin (see Packing materials).

Overbraided hose liners are made from fine powder resins by paste extrusion, and thread-sealant tapes are produced from fine powder by calendering. Fabricated gaskets are made from granular resins and pipe liners are produced from fine powder resins. PTFE also is converted into fiber and filament forms.

Highly porous fabric structures, eg, Gore-Tex, that can be used as membranes for a variety of applications have been developed by exploiting the unique fibrillation capability of dispersion-polymerized PTFE (99).

Recently PTFE dispersions have been applied to heavy-duty glass fabric as a coating which protects the fabric strength and adds weather protection and resistance against chemical and microbial attack.

BIBLIOGRAPHY

"Tetrafluoroethylene Resins" in *ECT* 1st ed., Vol. 11, pp. 687–691, by B. E. Ely, E. I. du Pont de Nemours & Co., Inc.; "Polytetrafluoroethylene" under "Fluorine Compounds, Organic" in *ECT* 2nd ed., Vol. 9, pp. 805–831, by S. Sherratt, Imperial Chemical Industries, Ltd., Plastics Division.

1. U.S. Pat. 2,230,654 (Feb. 4, 1941), R. J. Plunkett (to Kinetic Chemicals, Inc.).
2. J. D. Park and co-workers, *Ind. Eng. Chem.* **39**, 354 (1947).
3. J. M. Hamilton in M. Stacey, J. C. Tatlow, and A. G. Sharpe, eds., *Advances in Fluorine Chemistry*, Vol. 3, Butterworth & Co., London, Eng., 1963, p. 117.
4. J. W. Edwards and P. A. Small, *Nature (London)* **202**, 1329 (1964); J. W. Edwards and P. A. Small, *Ind. Eng. Chem. Fundam.* **4**, 396 (1965).
5. F. Gozzo and C. R. Patrick, *Nature (London)* **202**, 80 (1964).
6. Jpn. Pat. 60 15,353 (Oct. 14, 1960), M. Hisazumi and H. Shingu.

7. U.S. Pat. 2,994,723 (Aug. 1, 1961), O. Scherer and co-workers (to Farbewerke Hoechst).
8. Br. Pat. 960,309 (June 10, 1964), J. W. Edwards, S. Sherratt, and P. A. Small (to ICI).
9. U.S. Pat. 3,459,818 (Aug. 5, 1969), H. Ukahashi and M. Hisasne (to Asahi Glass Co.).
10. U.S. Pat. 2,407,405 (Sept. 10, 1946), M. A. Dietrich and R. M. Joyce (to E. I. du Pont de Nemours & Co., Inc.).
11. O. Ruff and O. Bretschneider, *Z. Anorg. Allgem. Chem.* **210,** 173 (1933).
12. E. G. Locke, W. R. Brode, and A. L. Henne, *J. Am. Chem. Soc.* **56,** 1726 (1934).
13. O. Ruff and W. Willenberg, *Chem. Ber.* **73,** 724 (1940).
14. L. T. Hals, T. S. Reid, and G. H. Smith, *J. Am. Chem. Soc.* **73,** 4054 (1951); U.S. Pat. 2,668,864 (Feb. 9, 1954), (to Minnesota Mining and Manufacturing Co.).
15. U.S. Pat. 3,009,966 (Nov. 21, 1961), M. Hauptschein and A. H. Fainberg (to Pennsalt Chemicals Corp.).
16. U.S. Pat. 3,471,546 (Oct. 7, 1969), G. Bjornson (to Phillips Petroleum Co.).
17. U.S. Pat. 3,662,009 (May 9, 1972), W. M. Hutchinson (to Phillips Petroleum Co.).
18. U.S. Pat. 3,799,996 (Mar. 26, 1974), H. S. Bloch (to Universal Oil Products).
19. E. E. Lewis and M. A. Naylor, *J. Am. Chem. Soc.* **69,** 1968 (1947).
20. U.S. Pat. 3,832,411 (Aug. 27, 1974), B. C. Arkles and R. N. Bonnett (to Liquid Nitrogen Processing Co.).
21. D. I. McCane, "Tetrafluoroethylene Polymers" in N. M. Bikales, ed., *Encyclopedia of Polymer Science and Technology,* Vol. 13, John Wiley & Sons, Inc., New York, 1970, pp. 623–670.
22. M. M. Renfrew and E. E. Lewis, *Ind. Eng. Chem.* **38,** 870 (1946).
23. H. C. Duus, *Ind. Eng. Chem.* **47,** 1445 (1955).
24. W. M. D. Bryant, *J. Polym. Sci.* **56,** 277 (1962).
25. A. Pajaczkowski and J. W. Spoors, *Chem. Ind. (London),* (16), 659 (1964).
26. Br. Pat. 931,587 (July 17, 1963), H. H. Gibbs and J. L. Warnell (to E. I. du Pont de Nemours & Co., Inc.).
27. U.S. Pat. 3,159,609 (Dec. 1, 1964), J. F. Harris, Jr., and D. I. McCane (to E. I. du Pont de Nemours & Co., Inc.).
28. F. S. Toby and S. Toby, *J. Phys. Chem.* **80,** 2313 (1976).
29. B. Atkinson and V. A. Atkinson, *J. Chem. Soc.,* 2086 (1957).
30. U.S. Pat. 2,946,763 (July 26, 1960), M. I. Bro and B. W. Sandt (to E. I. du Pont de Nemours & Co., Inc.).
31. U.S. Pat. 3,847,881 (Nov. 12, 1974), M. Mueller and S. Chandrasekaran (to Allied Chemicals Co.).
32. U.S. Pat. 3,528,954 (Sept. 15, 1970), D. P. Carlson (to E. I. du Pont de Nemours & Co., Inc.).
33. U.S. Pat. 3,475,391 (Oct. 28, 1969), J. N. Coker (to E. I. du Pont de Nemours & Co., Inc.).
34. U.S. Pat. 3,846,267 (Nov. 5, 1974), Y. Tabata and G. Kojima (to Japan Atomic Energy Research Institute).
35. U.S. Pat. 3,467,636 (Sept. 16, 1969), A. Nersasian (to E. I. du Pont de Nemours & Co., Inc.).
36. U.S. Pat. 3,403,191 (Sept. 24, 1968), D. P. Graham (to E. I. du Pont de Nemours & Co., Inc.).
37. U.S. Pat. 3,404,180 (Oct. 1, 1968), K. L. Cordes (to E. I. du Pont de Nemours & Co., Inc.).
38. U.S. Pat. 3,567,521 (Mar. 2, 1971), M. S. Toy and N. A. Tiner (to McDonnell Douglas).
39. U.S. Pat. 3,446,858 (May 27, 1969), H. Shingu and co-workers (to Daikin Kogyo Co.).
40. U.S. Pat. 3,873,630 (Mar. 25, 1975), N. E. West (to E. I. du Pont de Nemours & Co., Inc.).
41. U.S. Pat. 3,855,191 (Dec. 17, 1974), T. R. Doughty, C. A. Sperati, and H. Un (to E. I. du Pont de Nemours & Co., Inc.).
42. U.S. Pat. 3,655,611 (Apr. 11, 1972), M. B. Mueller, P. P. Salatiello, and H. S. Kaufman (to Allied Chemicals Co.).
43. U.S. Pat. 3,419,522 (Dec. 31, 1968), P. N. Plimmer (to E. I. du Pont de Nemours & Co., Inc.).
44. U.S. Pat. 3,766,133 (Oct. 16, 1973), R. Roberts and R. F. Anderson (to E. I. du Pont de Nemours & Co., Inc.).
45. U.S. Pat. 2,612,484 (Sept. 30, 1952), S. G. Bankoff (to E. I. du Pont de Nemours & Co., Inc.).
46. U.S. Pat. 2,478,229 (Aug. 9, 1949), K. L. Berry (to E. I. du Pont de Nemours & Co., Inc.).
47. K. L. Berry and J. H. Peterson, *J. Am. Chem. Soc.* **73,** 5195 (1951).
48. R. C. Doban, and co-workers, paper presented at the *130th Meeting of the American Chemical Society, Atlantic City, N.J., Sept. 1956.*
49. *TFE Fluorocarbon Resin Molding and Extrusion Materials, ASTM D 1457-69,* American Society for Testing and Materials, Philadelphia, Pa., 1969.
50. C. A. Sperati and H. W. Starkweather, *Fortschr. Hochpolym. Forsch.* **2,** 465 (1961).

51. R. E. Moynihan, *J. Am. Chem. Soc.* **81,** 1045 (1959).
52. P. L. McGeer and H. C. Duus, *J. Chem. Phys.* **20,** 1813 (1952).
53. C. W. Bunn, *J. Polym. Sci.* **16,** 332 (1955).
54. E. S. Clark and H. W. Starkweather, *J. Appl. Polym. Sci.* **6,** S41 (1962).
55. R. H. H. Pierce and co-workers in ref. 48.
56. E. S. Clark and L. T. Muus, paper presented at the *133rd Meeting of the American Chemical Society, New York, Sept. 1957.*
57. E. S. Clark, paper presented at *Symposium on Helices in Macromolecular Systems, Polytechnic Institute of Brooklyn, Brooklyn, N.Y., May 16, 1959.*
58. C. A. Sperati in J. Brandrup and E. H. Immergut, eds., *Polymer Handbook,* 2nd ed., John Wiley & Sons, Inc., New York, 1975, pp. V-29–36.
59. Y. Araki, *J. Appl. Polym. Sci.* **9,** 3585 (1965).
60. N. G. McCrum, *J. Polym. Sci.* **34,** 355 (1959).
61. J. T. Milek, *A Survey Materials Report on PTFE Plastics, AD 607798,* U.S. Dept. of Commerce, Washington, D.C., Sept. 1964.
62. *Mechanical Design Data, Teflon Fluorocarbon Resins,* bulletin, E. I. du Pont de Nemours & Co., Inc., Wilmington, Del.
63. M. K. Bernett and W. A. Zisman, *J. Phys. Chem.* **63,** 1911 (1959).
64. U.S. Pat. 2,871,144 (Jan. 27, 1959), R. C. Doban (to E. I. du Pont de Nemours & Co., Inc.).
65. A. J. G. Allan and R. Roberts, *J. Polym. Sci.* **39,** 1 (1959).
66. *J. Teflon (DuPont)* **9**(2), (1972).
67. J. C. Siegle and co-workers, *J. Polym. Sci.* **A2,** 391 (1964).
68. G. P. Shulman, *Polym. Lett.* **3,** 911 (1965).
69. L. A. Wall and R. E. Florin, *J. Appl. Polym. Sci.* **2,** 251 (1959).
70. V. Adamec, *Dielectrics* **1,** 159 (Aug. 1963).
71. V. Adamec, *Nature (London)* **200,** 1196 (1963).
72. W. M. Peffley, V. R. Honnold, and D. Binder, *J. Polym. Sci.* **4,** 977 (1966).
73. *J. Teflon* **10,** (Jan.–Feb. 1969).
74. *The Journal of Teflon (DuPont)* **11,** (Jan.–Feb. 1970).
75. A. Lebovids, *Mod. Plast.* **43**(7), 139 (1966).
76. J. C. Reed, E. J. McMahon, and J. R. Perkins, *Insul. Mag.* **10**(5), 35 (1964).
77. *Hostaflon,* TF product information booklet, American Hoechst Corp., Somerville, N.J., 1970.
78. *Soreflon,* products information booklet, Ugine Kuhlmann, France.
79. *Teflon TFE–Fluorocarbon Resins Molding Techniques,* bulletin, 2nd ed., E. I. du Pont de Nemours & Co., Inc., Wilmington, Del., May 1966.
80. *The Moulding of Granular Polymers,* technical service note, ICI, Wilmington, Del., 1966.
81. J. A. Ross, *High Speed Molding of Teflon Tetrafluoroethylene Resins,* technical release, E. I. du Pont de Nemours & Co., Inc., Wilmington, Del., 1963.
82. *Automatic Molding with 'Halon' TFE,* Allied Chemical Corp., Morristown, N.J., 1967.
83. *Isostatic Molding of Teflon TFE–Fluorocarbon Resins,* preliminary information bulletin, E. I. du Pont de Nemours & Co., Inc., Wilmington, Del., 1969.
84. R. J. Dahlen, *The Ram Extrusion of Teflon,* technical release, E. I. du Pont de Nemours & Co., Inc., Wilmington, Del., 1969.
85. *The Granular Extrusion of 'Halon' TFE,* Allied Chemical Corp., Morristown, N.J., 1968.
86. *The Extrusion of Granular Polymers,* technical service note, 2nd ed., ICI, Wilmington, Del., 1966.
87. J. F. Lontz and co-workers, *Ind. Eng. Chem.* **44,** 1805 (1952).
88. *TFE–Fluorocarbon Resins Paste Extrusion of Wire Insulations,* bulletin, E. I. du Pont de Nemours & Co., Inc., Wilmington, Del., 1961.
89. *The Extrusion Coating of Wire,* technical service note, 2nd ed., Wilmington, Del., 1969.
90. *Teflon TFE–Fluorocarbon Resins Extrusion of Thin-Walled Tubing,* bulletin, E. I. du Pont de Nemours & Co., Inc., Wilmington, Del., 1957.
91. *The Extrusion of Thin Sections,* technical service note, 3rd ed., ICI, Wilmington, Del., 1969.
92. R. C. Ribbans, *Unsintered Tape Manufactured Calendering Round Rods,* technical release, E. I. du Pont de Nemours & Co., Inc., Wilmington, Del., 1966.
93. *The Manufacture of Unsintered Tape,* technical service note, 2nd ed., ICI, Wilmington, Del., 1966.
94. U.S. Pat. 3,142,665 (July 28, 1964), A. J. Cardinal, W. L. Edens, and J. W. Van Dyk (to E. I. du Pont de Nemours & Co., Inc.).

95. U.S. Pat. 4,036,802 (July 19, 1977), R. V. Poirier (to E. I. du Pont de Nemours & Co., Inc.).
96. U.S. Pat. 3,819,594 (June 25, 1974), D. A. Holmes and E. W. Fasig (to E. I. du Pont de Nemours & Co., Inc.).
97. U.S. Pat. 3,654,210 (Apr. 4, 1972), J. Kuhls, T. Martini, and A. Steininger (to Hoechst A.G.).
98. U.S. Pat. 4,038,231 (July 26, 1977), J. M. Douner, W. G. Rodway, and L. S. J. Shipp (to I.C.I.).
99. U.S. Pat. 3,962,153 (June 8, 1976), R. W. Gore (to W. L. Gore and Associates).
100. U.S. Pat. 3,993,584 (Nov. 23, 1976), J. E. Owen and J. W. Vogt (to Kewanee Oil Co.).
101. U.S. Pat. 3,704,171 (Nov. 28, 1972), H. P. Landi (to American Cyanamid Co.).
102. U.S. Pat. 3,051,545 (Aug. 28, 1962), W. Steuber (to E. I. du Pont de Nemours & Co., Inc.).
103. *J. Teflon (DuPont)* **8,** (Nov. 1967).
104. *Teflon Occupational Health Bull.* **17**(2), (1962) (published by Information Service Division, Dept. of National Health and Welfare, Ottawa, Can.).

S. V. GANGAL
E. I. du Pont de Nemours & Co., Inc.

FLUORINATED ETHYLENE–PROPYLENE COPOLYMERS

Low-melting perfluorinated copolymers of tetrafluoroethylene [*116-14-3*] (TFE) and hexafluoropropylene [*116-15-4*] (HFP) are produced as alternatives to polytetrafluoroethylene [*9002-84-0*] (PTFE) (qv), which has a high melt viscosity (1). The copolymers retain most of the desired properties of PTFE and, unlike the latter compound, can be processed by conventional melt-processing techniques. They also have excellent thermal stability and chemical inertness (2), and their dielectric constants and dissipation factors are low, and remain unchanged when subjected to wide ranges of temperature and frequency. The copolymers retain mechanical properties up to 200°C, even in continuous service, better than most plastics. They react with fluorine, fluorinating agents, and molten alkali metals at high temperatures. Fluorinated ethylene–propylene copolymers [*25067-11-2*] are commercially available under the DuPont trademark Teflon FEP fluorocarbon resin. A product similar to Teflon FEP is manufactured by Daikin Kogyo of Japan and sold under the trademark Neoflon.

Monomers

Some of the properties of HFP are listed in Table 1. It does not homopolymerize easily and, hence, can be stored as a liquid. HFP undergoes many addition reactions typical of an olefin. Some reactions that HFP undergoes include: preparation of linear dimers and trimers, and cyclic dimers (6–9); decomposition at 600°C with subsequent formation of octafluoro-2-butene and octafluoroisobutylene (10); oxidation with formation of the epoxide

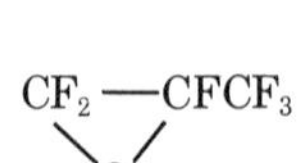

Table 1. Properties of Hexafluoropropylene

Property	Value
molecular weight	150.021
boiling point at 101 kPa (1 atm), °C[a]	−29.4
freezing point, °C	−156.2
critical temperature, °C	85
critical pressure, kPa[b]	3254
critical density, g/cm³	0.60
vapor pressure at K, kPa[b] $243.75 < T < 358.15$	$\log P_{kPa} = 6.6938 - 1139.156/T$
liquid density, g/cm³	
60°C	1.105
20°C	1.332
0°C	1.419
−20°C	1.498
heat of formation for ideal gas at 25°C, H, kJ/mol[c,d]	−1078.6
flammability limits in air at 101 kPa	nonflammable for all mixtures of air and hexafluoropropylene
heat of combustion, kJ/mol[c,d]	879
toxicity, LC_{50} (rat), 4 h, ppm[e]	3000

[a] Ref. 3.
[b] To convert kPa to mm Hg, multiply by 7.5. To convert kPa to psi, multiply by 0.145.
[c] To convert kJ to kcal, divide by 4.184.
[d] Ref. 4.
[e] Ref. 5.

(11), an intermediate for a number of perfluoroalkyl perfluorovinyl ethers (12–13); and homopolymerization to low molecular weight liquids (14–15) and high molecular weight solids (16–17). HFP also reacts with hydrogen (18), alcohols (19), ammonia (20), and the halogens and their acids, except I_2 and HI (18,21–23), and is used as a comonomer to produce elastomers (see Elastomers, synthetic) and other copolymers (24–28).

Preparation. Hexafluoropropylene (HFP) was initially prepared by pyrolysis of PTFE (29–30) and by fluorination followed by dehalogenation of 1,2,3-trichloropropane (3). A number of other routes to HFP are described in the patent literature (31–33). HFP can be prepared in high yield by thermally cracking TFE at reduced pressure at 700–800°C (34–35). The pyrolysis of PTFE at 860°C under vacuum produced HFP in a 58% yield (36). Fluorination of 3-chloropentafluoro-1-propene at 200°C over activated carbon catalyst yields HFP (37). Decomposition of fluoroform at 800–1000°C in a platinum-lined nickel tube is another route to HFP production (38). The thermal decomposition of sodium heptafluorobutyrate, $CF_3CF_2CF_2CO_2Na$ (39), copyrolyses of fluoroform and chlorotrifluoroethylene (40), and chlorodifluoromethane and 1-chloro-1,2,2,2-tetrafluoroethane (41) give good yields of HFP.

Copolymers

HFP–TFE were copolymerized with trichloroacetyl peroxide as the catalyst, at a low temperature (42). A variety of newer catalytic methods, including irradiation, have been used to achieve HFP–TFE copolymerization at different temperatures

(43–44). However, aqueous and nonaqueous dispersion polymerizations appear to be the most convenient routes to commercial production of HFP–TFE copolymers (1,45–46). The polymerization conditions used to produce these polymers are very similar to those used in TFE homopolymer dispersion polymerization. The copolymer of HFP–TFE is a true copolymer, and HFP units add to the growing chains at random intervals. The optimization composition of the copolymer requires that the retention of mechanical properties is in the usable range and the melt viscosity of the product is low enough for easy melt processing.

Fluorinated Ethylene–Propylene Copolymers

FEP copolymers are available in four forms: the low melt viscosity, extrusion grade, high melt viscosity resins, and the FEP copolymer dispersion.

The low melt viscosity (MV) resin has the outstanding properties of a fluorocarbon polymer and can be injection molded using conventional thermoplastic molding techniques. It is more suitable for injection molding than other FEP resins because of its lower melt viscosity (see Plastics technology).

The extrusion-grade resin is suitable for extrusion applications, eg, tubing, wire coating, and cable jacketing. It is less suitable for injection molding than the low MV resin because of its high melt viscosity.

The high MV resin is used as liners for process equipment. The melt viscosity of this resin is significantly higher than that of low MV and extrusion-grade resins and, therefore, is unsuitable for conventional injection molding. This product has a superior stress-crack resistance and higher mechanical properties than other two products (47). Typical properties of these products are listed in Table 2.

FEP copolymer dispersion is available as a 55 wt % aqueous dispersion containing 6% nonionic surfactant (on a solids basis) and a small amount of anionic dispersing agent. The average particle size of this dispersion is ca 0.2 μm.

Table 2. Typical Properties of FEP Copolymer Resins [a]

Property	Low MV resin	Extrusion grade	High MV resin	ASTM Method
specific gravity	2.12–2.17	2.12–2.17	2.12–2.17	D 792
melting point, °C	253–272	262–282	262–282	
tensile strength at 23°C, MPa [b]	20.68	20.68	20.68	D 1708
elongation at 23°C, %	300	300	300	D 1708
flexural modulus 23°C, MPa [b]	655	655	754	D 790
hardness, durometer	D55	D55	D55	D 2240
dielectric constant at 60 Hz to 1 GHz	2.1	2.1	2.1	D 150
dielectric strength for a short time at 0.254 mm, kV/mm	82.7	82.7	82.7	D 149
dissipation factor at 60 Hz to 1 GHz	10^{-4}–10^{-3}	10^{-4}–10^{-3}	10^{-4}–10^{-3}	D 150
volume resistivity, Ω·cm	$>10^{18}$	$>10^{18}$	$>10^{18}$	D 257
water absorption, %	<0.01	<0.01	<0.01	D 570
weather resistance	excellent	excellent	excellent	
chemical resistance	excellent	excellent	excellent	
flammability, cm/min	nonflammable	nonflammable	nonflammable	D 635

[a] Ref. 47.

[b] To convert MPa to psi, multiply by 145.

Properties. ***Structure.*** FEP polymer has a significantly lower crystallinity than that of PTFE (70% vs 98%) and has a structure similar to that of PTFE except for a random replacement of a fluorine atom by a perfluoromethyl group (CF_3). The crystallinity of FEP after processing depends on the rate of cooling of the molten polymer. The presence of HFP in the polymer chain tends to distort the highly crystallized structure of the polytetrafluoroethylene chain and results in a higher amorphous fraction.

Effect of Radiation. The primary microscopic effect of radiation is the formation of small molecules from the large polymer molecule. Molecular weight reduction can be minimized by excluding oxygen. However, if FEP is irradiated at elevated temperatures in the absence of oxygen, cross-linking occurs to offset molecular breakdown (48,50).

The degree to which radiation exposure affects FEP resins is determined by the energy absorbed regardless of the type of radiation. Changes in mechanical properties depend on total radiation dosage, but are independent of dose rate. The radiation tolerance of FEP in air is greater than that of PTFE by a factor of 10:1 (see Polytetrafluoroethylene). If both compounds are irradiated in the absence of oxygen, FEP tolerates a dose of up to 10 times more than PTFE.

Transitions and Relaxations. FEP has only one first-order transition, its melting point. Increasing the pressure on the polymer results in elevation of the melting point. At low pressures, the rate of increase in the melting point is ca 1.74°C/MPa (0.012°C/psi) and at high pressures the rate of increase in the melting decreases to ca 0.725°C/MPa (0.005°C/psi). Melting FEP produces an 8% volume increase of the compound (2). If HFP comonomer is present in the crystal, crystal distortion results along with an increase in intramolecular distance which, in turn, causes a decrease in the melting point of the copolymer (49).

The relaxation temperature for copolymers of TFE–HFP appears to increase with increasing HFP content. The relaxation involves 5–13 of the chain carbon atoms. Besides α and γ relaxations, one other dielectric relaxation was observed below −150°C, which did not vary in temperature nor in magnitude with comonomer content or copolymer density (50). The α relaxation (also called Glass I) is a high temperature transition (157°C) and γ relaxation (Glass II) (the internal friction maxima) occurs between −5° and 29°C.

Mechanical Properties. Extensive lists of the physical properties of FEP copolymers are available in refs. 51–56. Selected mechanical properties of FEP resin are summarized in Table 3 (53).

Most of the important application properties of FEP are similar to those of PTFE; the main difference is the somewhat lower continuous service temperature (204°C) of FEP. The flexibility at low temperatures and the low coefficients of friction and stability at high temperatures are the major mechanical properties of FEP that are relatively independent of fabrication conditions. FEP resins do not exhibit a marked change in volume at room temperature, unlike PTFE, because they do not have a first-order transition at 19°C. FEP resins are useful at temperatures at and above −267°C and are highly flexible at and above −79°C. The rates of decomposition of FEP at 232, 260, 316, and 370°C are 0.0004, 0.001, 0.02 and 0.3% per hour, respectively (57).

The static friction of FEP decreases with an increase in load, and the static coefficient of friction is lower than the dynamic coefficient.

Table 3. Mechanical Properties of FEP[a]

Property	Value	ASTM Test method
specific gravity	2.14–2.17	D 792-50
thermal conductivity, W/(m·K)		
−129 to 182°C	2.4	Cenco Fitch
−253°C	1.4	Cenco Fitch
water absorption, 24 h		
% wt increase, 3.175 mm thick	<0.1	D 570-547
dimensional change m/m, 23°C	none	D 570-547
coefficient of thermal expan., 10^{-5} cm/(cm·°C)		
>23°C	9.3	D 696-44
<23°C	5.7	D 696-44
specific heat, kJ/(kg·K)[b]		
20°C	1.09	
100°C	1.17	
260°C	1.30	
heat distortion, °C		
455 kPa (4.5 atm)	70	D 648-56
1820 kPa (18 atm)	51	D 648-56
tensile yield strength, av MPa[c]		
−251°C	165	D 638-527
−160°C	131	
−129°C	96	
−73°C	62	
−56°C	27	
0°C	14	
23°C	12	
70°C	7	
121°C	3.5	
tensile modulus, Pa[d]		
−251°C	57	
−160°C	50	
−129°C	36	
−73°C	24	
23°C	4	
100°C	1	
tensile elongation, %		
−251°C	4	D 638-527
−160°C	10	
−129°C	110	
−73°C	200	
23°C	350	
flexural modulus, Pa[d]		
−251°C	53	D 747-50
−160°C	47	
−101°C	32	
23°C	6.6	
55°C	3.4	
compressive strength, MPa[c]		
−251°C	251	D 695
−160°C	207	
23°C	15	
55°C	11	
100°C	3.4	
Izod impact strength, J/m, notched[e]		
23°C	no break	D 256-56

Table 3 *(continued)*

Property	Value	ASTM Test method
hardness, Durometer		
23°C	D 59	D 2240-T
Taber abrasion, g/MHz, 100 g load		
CS-17 wheel	7.5	

[a] These properties were measured on Teflon FEP T-100 (53).
[b] To convert kJ to kcal, divide by 4.184.
[c] To convert MPa to psi, multiply by 145.
[d] To convert Pa to mm Hg, multiply by 7.5×10^{-3}.
[e] To convert J/m to ft·lbf/in., divide by 53.4 (see ASTM D 256).

A limited number of fillers are used to increase wear resistance, and stiffness of FEP. Limitations on the variety of fillers used stems from processing difficulties. About 30% bronze-filled FEP, 20% graphite-filled FEP, and 10% glass-fiber-filled FEP have been made, and they offer high PV (pressure-velocity) values and are good bearing candidates.

	PV limit at 0.5 m/s
30% bronze-filled	420 kPa·m/s (200 psi·ft/s)
20% graphite-filled	350 kPa·m/s (166 psi·ft/s)
10% glass-fiber-filled	350 kPa·m/s (166 psi·ft/s)

Articles fabricated from FEP resins can be made bondable by modification of their surfaces. Surface treatment of FEP with a solution of sodium in liquid ammonia, or naphthalenylsodium in tetrahydrofuran (57) facilitates subsequent surface wetting. Exposing the FEP surface to corona discharge (58) or to amines at elevated temperatures in an oxidizing atmosphere (59) also makes the resins bondable.

FEP resins have excellent vibration dampening properties at sonic and ultrasonic frequencies; however, the thickness of the resin must be sufficient to absorb the energy produced and is usually determined experimentally.

Weathering. Parts fabricated from FEP are virtually unaffected by weather and their excellent resistance to extreme heat, cold, and uv irradiation qualifies their applications in radar and other electronic components (see Microwave technology). For example, after 15 years of solar exposure in Florida, the tensile strength (60) and the light transmission (96%) of a 25-μm thick film was unchanged and the film remained crystal clear. The film exhibited slightly increased elongation for the first 5–7 yr of outdoor exposure, probably as a result of stress relaxation. Beyond 10 yr, a very small decrease was observed (see also Film and sheeting materials).

Thermal Stability. FEP is a thermally stable polymer and can be processed at temperatures of ca 370°C. However, the thermal degradation is a function of both temperature and time and the stability, therefore, is limited. There is a significant increase in melt flow rate (thermal degradation) for short periods at temperatures above 380°C, and degradation occurs at lower temperatures with longer hold times. Degradation is not significant if the change in melt flow rate during molding is ≤10%.

Electrical Properties. The outstanding electrical properties of FEP distinguish this material as a valuable and versatile electrical insulator. PTFE and FEP have practically identical properties as electrical insulators when used within the recommended service temperature range. Volume resistivity remains unchanged even after prolonged soaking in water.

The dielectric constant of FEP is constant at lower frequencies (see Table 4). However, at ≥100 MHz the constant drops slightly with increasing frequency. As a true thermoplastic, FEP has a void content of zero and most of the fabricated material has a density range of 2.14–2.17 g/cm^3. The National Bureau of Standards has selected Teflon FEP resins for dielectric reference specimens because of the stability of FEP in dielectric constant measurements. The dissipation factor of FEP has several peaks as a function of temperature and frequency. The magnitude of the dissipation factor peak is greater for FEP than for PTFE because the former has a less symmetrical molecular structure. The dissipation factor of FEP is insignificantly affected by the irradiation–annealing procedure (61) and is unaffected by humidity. The dielectric strength of FEP resin is high and is unaffected by thermal aging at 200°C. At high frequencies, a lower value of dielectric life should be expected when corona is present. If the voltage stress is not high enough to cause corona ignition, then an infinitely long dielectric life is expected for FEP resin at any frequency. Corona discharges on the surface or in a void initiate dielectric breakdown (62). FEP resins are recommended for continuous service at temperatures up to 205°C, and although they begin to melt at 270°C, they retain some structural integrity at temperatures as high as 250°C (61).

Chemical Properties. FEP resin is chemically inert to essentially all industrial chemicals and solvents, even at elevated temperatures and pressures. However, it reacts with fluorine, molten alkali metal, and molten sodium hydroxide. FEP absorbs practically no common acids or bases at temperatures as high as 200°C and at exposures of as much as one year. Even the absorption of solvents, eg, acetone, benzene, ethyl alcohol, toluene, or ethyl acetate, is less than 1% at elevated temperatures and long exposure times. It should be noted that absorption of chemicals or solvents has no substantial effect on the chemical integrity of the FEP molecule and is a reversible physical process.

Gases and vapors permeate FEP resin at a rate that is considerably lower than that of most other plastics. Because FEP resins are melt-processed, permeation occurs only by molecular diffusion (see Barrier polymers). Variation in crystallinity and density is relatively limited except in unusual melt processing conditions.

FEP polymer is used extensively in the chemical industry because of its low

Table 4. Electrical Properties of FEP Copolymer[a]

Property	Value	ASTM Method
dielectric strength short time, 0.2 cm thickness, V/m	23.6	D 149-55T
surface arc-resistance, s	>300	D 495-55T
volume resistivity, Ω·cm	$>10^{18}$	D 257-57T
dielectric constant, 60 Hz to 2 GHz	2.1	D 150-59T
dissipation factor, 60 Hz to 2 GHz	$(2–12) \times 10^{-4}$	D 150-59T
surface resistivity, Ω/sq	10^{16}	D 257-57T

[a] Ref. 56.

permeability. The permeation characteristics of FEP are very similar to those of polytetrafluoroethylene polymers. The permeability constant of FEP to water and ethanol at 25°C is 7.9 mg·cm/(m^2·d); to red fuming nitric acid, 567 mg·cm/(m^2·d); to 98% sulfuric acid, 4×10^{-4} mg·cm/(m^2·d); and to 20% hydrochloric acid, 0.4 mg·cm/(m^2·d). The permeability constants of FEP to carbon dioxide, oxygen, nitrogen, and methane are 23.6, 55.1, 7.9, and 5.1 mg·cm/(m^2·d), respectively. Permeability of FEP resin to various vapors at their vapor pressure in g/(m^2·d) converted to 25.4 μm film thickness are summarized in Table 5 (53). The inverse relationship between permeability and film thickness applies to FEP.

Fabrication. Standard thermoplastic processing techniques can be used to fabricate useful products from FEP. Factors of primary importance in FEP fabrication include avoiding thermal degradation of the polymer, and maintaining a homogeneous structure and good surface quality (2).

Injection Molding. Compared to most thermoplastic products, the low MV resin has a significantly higher melt viscosity and, therefore, requires higher processing temperatures, slower injection rates, and special mold designs, eg, as corrosion-resistant construction material. When the flow velocity of a resin in melt processing exceeds a critical value, melt fracture occurs. The critical flow rate for FEP is much lower than for other thermoplastics.

The recommendations for materials of construction and the screw design, valves, smear heads, nozzle, operating conditions, and mold design are described in ref. 47.

Dry blending is a simple way to add pigments (thermally stable at processing temperature) to FEP prior to molding. Pigments at loadings from 0.1–1% have no appreciable effect on the dielectric strength, dielectric constant, or mechanical properties of FEP. The dissipation factor of pigmented resin varies with the species and with the concentration of the pigment used (63).

Extrusion. Conventional melt-extrusion equipment is used in processing FEP resins. Commercially available pigments can be mixed with the resin before extrusion. FEP can be extruded into wire coating, tubing, rods, molding, beading channels, etc.

Table 5. Permeability of FEP Fluorocarbon Resins to Vapors, g/(m^2·d)[a]

	23°C	35°C	50°C
acetic acid		6.5	
acetone	2.0		51.0
acetophenone	7.3		
benzene	2.3		
dibutyl ether	1.2		10.1
tetrachloride	1.7	4.8	
decane	11.2		16.0
dipentene	2.6		5.4
ethyl acetate	0.9	11.9	45.0
ethanol	1.7	9.8	
hydraulic fluid	0.8		
NaOH, 50%	6.2×10^{-4}		
H_2SO_4, 98%	1.2×10^{-4}		
toluene	5.7		
water	1.4	7.0	13.8

[a] Test method: ASTM E 96-35T (at vapor pressure: for 25.4 μm film thickness). Values are averages only and not for specification purposes.

Coating thicknesses of 0.076–2.54 mm have been extruded over such materials as silicone rubber, poly(vinyl chloride), glass braid, metal-shielded cables, twisted conductors, and parallel-multiconductor cables (see Insulation, electric).

When extruding FEP as a primary insulation or cable jacket, high production rates are achieved by extruding a tube of resin having a larger internal diameter than the base wire construction and a heavier wall thickness than the final insulation thickness, and then by drawing the larger tube down to desired size. An operating temperature of 315–400°C is preferred and depends on hold-up time. The surface roughness caused by melt fracture determines the upper limit of production rates under any specific extrusion condition (66). Because molten FEP corrodes metals, metals that offer corrosion resistance should be used for all parts of the extrusion equipment that come in contact with the molten copolymer. Detailed recommendations are given in ref. 64.

FEP tubing is made in wide range of sizes and is used as slip-on electrical insulation, as instrument tubing, and in hose applications. Small tubing, called spaghetti tubing, of FEP can be produced by a free-extrusion technique, whereas hose-size tubing is produced by conventional forming box techniques. FEP also is extruded into films.

Dispersion Processing. An aqueous dispersion of FEP is commercially available. It contains 55 wt % of hydrophobic, negatively charged FEP particles and ca 6 wt % (based on FEP) of a mixture of nonionic and anionic surface-active agents. The average particle size of dispersed FEP is ca 0.2 μm. The dispersion is processed by essentially the same technique as that used to process PTFE dispersion. The processing involves: coating the fabric with FEP dispersion, evaporation of water from the coating, removal of the wetting agent, and fusion of the FEP layer with the fabric.

This dispersion is used as a coating for glass fabric, chemical barriers, and wire-insulating tapes; as adhesive coatings for bonding seals and bearings of PTFE to metallic and nonmetallic components; and as antifriction or antistick coatings for metals. Successful fusion of FEP to provide a continuous film depends on a time–temperature relationship; 1 min at 400°C or 40 min at 290°C (65) is sufficient to achieve good fusion.

Other Techniques. FEP resin is bonded to metal surfaces by application of heat and pressure; it can be heat-sealed or hot-gas welded. FEP is also made into shrinkable tubing used for coating and is suited to extrusion laminating. Stress relieving, or annealing, can be done by heating FEP at 260°C and then by allowing it to cool slowly. FEP film is used as a bonding agent for PTFE resins.

Economics and Applications. FEP manufacturing costs include those involved in manufacturing PTFE and the hexafluoropropylene comonomer. As a result, FEP is more costly to manufacture than PTFE. 1978 prices for molding and extrusion grade FEP were 14.52–16.39 $/kg, depending on the type and quantity of resin; and, for the dispersion, $18.81/kg.

The principal electrical applications include hook-up wire, interconnecting wire, coaxial cable, computer wire, thermocouple wire, and molded electrical parts. Principal chemical applications are lined pipes and fittings, overbraided hose, heat exchangers, and laboratory ware. The major mechanical uses are antistick applications, such as conveyor belts and roll covers (see Abherents). A recent mechanical application of FEP film for solar collector windows indicates that many manufacturers are taking advantage of the light weight, excellent weatherability, and high solar transmission of FEP film to make solar collectors efficient, easy and relatively inexpensive to install (see Solar energy) (66).

Testing and Standards. FEP resin requirements for extrusion and molding grades are cited in ASTM specifications, D 2116-66 (67) and in Federal specification LP-389A, of May, 1964. For fabricated shapes, FEP film and sheet are covered by Aeronautical Material Specifications (AMS) 3647 and LP-523. Besides the specifications covered by the Fluorocarbons Division of the Society of the Plastics Industry, Inc., other specifications are listed in ref. 68.

Safety. The safety precautions required in handling TFE–HFP copolymers are the same as those used to handle PTFE. Large quantities have been processed safely by many different fabricators in a variety of operations. If proper ventilation is used, the polymer can be processed and used at elevated temperatures without hazard. The fumes from heated FEP or its thermal decomposition products are toxic in heavy concentrations just as are the fumes or decomposition products of many other polymers. Ventilation should be provided in areas where the resin is at processing temperature (270–400°C). Unheated FEP resin is essentially inert. If adequate ventilation is not provided and workers inhale the fumes of heated FEP resin in sufficient quantity, influenza-like symptoms may follow. These symptoms occur several hours after exposure and disappear within 36–48 h, even in the absence of treatment. Observation indicates that these effects are not cumulative (47). Such attacks usually follow exposure to vapors evolved from the polymer without adequate ventilation or from smoking tobacco or cigarettes contaminated with the polymer.

BIBLIOGRAPHY

1. U.S. Pat. 2,946,763 (July 26, 1960), M. I. Bro and B. W. Sandt (to E. I. du Pont de Nemours & Co., Inc.).
2. D. I. McCane, "Copolymers With Hexafluoropropylene" in N. M. Bikales, ed., *Encyclopedia of Polymer Science and Technology,* Vol. 13, John Wiley & Sons, Inc., New York, 1970, pp. 654–670.
3. A. L. Henne and T. P. Waalkes, *J. Am. Chem. Soc.* **68,** 496 (1946).
4. H. C. Duus, *Ind. Eng. Chem.* **47,** 1445 (1955).
5. J. W. Clayton, *Occup. Med.* **4,** 262 (1962).
6. U.S. Pat. 2,918,501 (Dec. 22, 1959), W. J. Brehm and co-workers (to E. I. du Pont de Nemours & Co., Inc.).
7. U.S. Pat. 2,983,764 (May 9, 1961), D. F. Knaack (to E. I. du Pont de Nemours & Co., Inc.).
8. M. Hauptschein, A. H. Fainberg, and M. Braid, *J. Am. Chem. Soc.* **80,** 842 (1958).
9. U.S. Pat. 3,316,312 (April 25, 1967), D. I. McCane and I. M. Robinson (to E. I. du Pont de Nemours & Co., Inc.).
10. R. A. Matula, *J. Phys. Chem.* **72,** 3054 (1968).
11. U.S. Pat. 3,358,003 (Dec. 12, 1967), H. S. Eleuterio and R. W. Meschke (to E. I. du Pont de Nemours & Co., Inc.).
12. U.S. Pat. 3,180,895 (April 27, 1965), J. F. Harris, Jr., and D. I. McCane (to E. I. du Pont de Nemours & Co., Inc.).
13. U.S. Pat. 3,291,843 (Dec. 13, 1966) C. G. Fritz and S. Selman (to E. I. du Pont de Nemours & Co., Inc.).
14. E. V. Volkova and A. I. Skobina, *Vysokomol. Soedin* **6,** 964 (1964).
15. Fr. Pat. 1,524,571 (May 10, 1968), S. W. Osborn and E. Broderich (to Thiokol Chemical Corp.).
16. D. Sianesi and G. Caporiccio, *Makromol. Chem.* **60,** 213 (1963).
17. U.S. Pat. 2,958,685 (Nov. 1, 1960), H. S. Eleuterio (to E. I. du Pont de Nemours & Co., Inc.).
18. I. L. Knunyants, E. I. Mysov, and M. P. Krasuskaya, *Izv. Akad. Nauk. SSSR Otdel. Khim. Nauk,* 906 (1958).
19. I. L. Knunyants, A. I. Shchekotikhin, and A. V. Fakin, *Izv. Akad. Nauk. SSSR Otdel. Khim. Nauk,* 282 (1953).

20. I. L. Knunyants, L. S. German, and B. L. Dyatkin, *Izv. Akad. Nauk. SSSR Otdel. Khim. Nauk,* 1353 (1956).
21. R. N. Haszeldine and B. R. Steele, *J. Chem. Soc.,* 1592 (1953).
22. W. T. Miller, Jr., E. Bergman, and A. H. Fainberg, *J. Am. Chem. Soc.* **79,** 4159 (1957).
23. I. L. Knunyants, V. V. Shokina, and N. D. Kuleshova, *Izv. Akad. Nauk. SSSR Otdel. Khim. Nauk,* 1693 (1960).
24. U.S. Pat. 3,467,636 (Sept. 16, 1969), A. Nersasian (to E. I. du Pont de Nemours & Co., Inc.).
25. U.S. Pat. 3,536,683 (Oct. 27, 1970), F. V. Bailor and J. R. Cooper (to E. I. du Pont de Nemours & Co., Inc.).
26. U.S. Pat. 3,790,540 (Feb. 5, 1974), J. E. Dohany and A. C. Whiton (to Pennwalt Corp.).
27. U.S. Pat. 3,817,951 (June 18, 1974), D. N. Robinson (to Pennwalt Corp.).
28. U.S. Pat. 3,868,337 (Feb. 25, 1975), P. Gros (to Society Superflexit).
29. U.S. Pat. 2,394,581 (Feb. 12, 1946), A. F. Benning, F. B. Dowing, and J. D. Park (to Kinetic Chemicals Inc.).
30. E. G. Young and W. S. Murray, *J. Am. Chem. Soc.* **70,** 2814 (1948).
31. U.S. Pat. 3,446,858 (May 27, 1969), H. Shinzu and co-workers (to Daikin Koygo Co.).
32. U.S. Pat. 3,459,818 (Aug. 15, 1969), H. Ukahashi and M. Hisasue (to Asahi Glass Co.).
33. U.S. Pat. 3,873,630 (March 25, 1975), N. E. West (to E. I. du Pont de Nemours & Co., Inc.).
34. U.S. Pat. 3,758,138 (Aug. 7, 1956), D. A. Nelson (to E. I. du Pont de Nemours & Co., Inc.).
35. B. Atkinson and A. B. Trenwith, *J. Chem. Soc.,* Pt. II, 2082 (1953).
36. U.S. Pat. 2,759,983 (Aug. 21, 1956), J. S. Waddell (to E. I. du Pont de Nemours & Co., Inc.).
37. U.S. Pat. 3,047,640 (July 31, 1962), R. F. Sweeny and C. Woolf (to Allied Chemical Corp.).
38. U.S. Pat. 3,009,966 (Nov. 21, 1961), M. Hauptschein and A. Fainberg (to Pennsalt Chemicals Corp.).
39. L. T. Hals, T. S. Reid, and G. H. Smith, *J. Am. Chem. Soc.* **73,** 4054 (1951); U.S. Pat. 2,668,864 (Feb. 9, 1954), (to Minnesota Mining and Manufacturing Co.).
40. G. Pass, *J. Chem. Soc.,* Pt. I, 824 (1965); Fr. Pat. 1,399,414 (May 14, 1965), (to Imperial Chemical Industries, Ltd.).
41. Ger. Offen. 1,236,497 (March 16, 1967), W. Oese, H. Dude, and F. Reinke (to VEB Fluorwerke Dohma).
42. U.S. Pat. 2,598,283 (May 27, 1952), W. T. Miller (to USAEC).
43. R. A. Naberezhnykh and co-workers, *Dokl. Akad. Nauk SSSR* **214,** 149 (1974).
44. A. S. Kabankin, S. A. Balabanova, and A. M. Markevich, *Vysokomol. Soedin. Ser. A* **12,** 267 (1970).
45. Brit. Pat. 781,532 (Aug. 21, 1957), C. G. Krespan (to E. I. du Pont de Nemours & Co., Inc.).
46. U.S. Pat. 3,132,124 (May 5, 1964), M. J. Couture, D. L. Schindler, and R. B. Weiser (to E. I. du Pont de Nemours & Co., Inc.).
47. *Teflon-FEP Fluorocarbon Resin, Techniques for Injection Molding,* Information Bulletin, 95d, E. I. du Pont de Nemours & Co., Inc., Wilmington, Del.
48. R. Y. M. Huang and P. J. F. Kanitz, *Polym. Prepr.* **10**(2), Sept. 2, 1969.
49. R. K. Eby, *J. Appl. Phys.* **34,** 2442 (1963).
50. R. K. Eby and F. C. Wilson, *J. Appl. Phys.* **33,** 2951 (1962).
51. R. J. Diamond, *Plastics* **27,** 109 (Sept. 1962).
52. J. Frados, ed., *Modern Plastics Encyclopedia,* Vol. 46, No. 10A, 1969, p. 974.
53. *J. Teflon* **11**(1), 8 (1970).
54. J. A. Brydson, *Plastics Materials,* Iliffe Books, Ltd., London, 1966, Chapt. 10, pp. 203–218.
55. *Teflon 100 FEP-Fluorocarbon Resin-Melt Processible Resin,* Information Bulletin, X-90a, E. I. du Pont de Nemours & Co. Inc., Wilmington, Del., 1960.
56. *Teflon Fluorocarbon Resins, Mechanical Design Data,* 2nd ed., E. I. du Pont de Nemours & Co., Inc., Wilmington, Del., 1965.
57. A. A. Benderly, *J. Appl. Polym. Sci.* **6,** 221 (1962).
58. Brit. Pat. 890,466 (Feb. 28, 1962), D. L. Ryan (to E. I. du Pont de Nemours & Co., Inc.).
59. U.S. Pat. 3,063,882 (Nov. 13, 1962), J. R. Chesire (to E. I. du Pont de Nemours & Co., Inc.).
60. *Teflon Solar Film for Solar Collectors,* E. I. du Pont de Nemours & Co., Inc., Wilmington, Del.
61. *Electrical/Electronic Design Data for Teflon,* E. I. du Pont de Nemours & Co., Inc., Wilmington, Del.
62. J. C. Reed, E. J. McMahon, and J. R. Perkins, *Insulation,* 35 (May 1964).
63. L. H. Gillespe, D. O. Saxton, and F. M. Chapman, *New Design Data for Teflon,* E. I. du Pont de Nemours & Co., Inc., Wilmington, Del., Jan. 1960.

64. *Teflon FEP-Fluorocarbon Resin—Techniques for Processing by Melt Extrusion,* 2nd ed., Information Bulletin, X-82, E. I. du Pont de Nemours & Co., Inc., Wilmington, Del., 1960.
65. *Properties and Processing Techniques for Teflon 120FEP—Fluorocarbon Resin Dispersion,* Preliminary Information Bulletin, No. 20, E. I. du Pont de Nemours & Co., Inc., Wilmington, Del., 1961.
66. *J. Teflon* **18**(1), 8 (1977).
67. *FEP-Fluorocarbon Molding and Extrusion Materials, ASTM D 2116-66,* American Society for Testing and Materials, Philadelphia, 1969.
68. *J. Teflon* **11**(3), 6 (1970).

S. V. Gangal
E. I. du Pont de Nemours & Co., Inc.

TETRAFLUOROETHYLENE COPOLYMERS WITH ETHYLENE

Copolymers of ethylene and tetrafluoroethylene (ETFE) [*25038-71-5*] have been known since the 1940s (1–5). It was soon recognized that the two monomers could be readily combined into a nearly 1:1 alternating structure of thermoplastic nature. Such polymers exhibited a unique combination of mechanical, chemical, and electrical properties along with nonflammability in air and excellent weatherability. However, the copolymer suffered poor thermal-stress crack resistance. Although thin-wall insulations, coatings, or films could be prepared, in general, thick cross-section moldings tended to crack upon standing at RT. The copolymers exhibit increasing elongation below 150°C; between 150 and 200°C elongation decreases drastically.

In 1970 DuPont introduced Tefzel fluoropolymer, a proprietary modified copolymer that not only exhibits properties nearly identical to those of the copolymer but also retains high ultimate elongation to 200°C. Tefzel is a high temperature (150°C rating) thermoplastic that is readily processed by conventional methods. It has gained acceptance in wire and cable applications as well as in chemical service and molded parts. In addition to being tough, it exhibits excellent electrical performance and good thermal properties and tolerates relatively large doses of radiation.

The 1979 price of Tefzel was $20/kg. Other ethylene–tetrafluoroethylene resins available are Aflon COP (Asahi) and Hostaflon ET (Hoechst).

Ethylene–tetrafluoroethylene copolymer is isomeric with poly(vinylidene fluoride). The tetrafluoroethylene segments of the polymer molecules account for 75 wt % of a 1:1 copolymer:

$\leftarrow CH_2CH_2CF_2CF_2 \rightarrow_n$	$\leftarrow CH_2CF_2 \rightarrow_n$
poly(ethylene-*co*-tetrafluoroethylene)	poly(vinylidene fluoride)

The properties of ETFE copolymers vary considerably with composition; those containing 40–90 wt % tetrafluoroethylene soften at 200–300°C, depending on their composition (6). These copolymers exhibit good dimensional stability below their softening points and have tensile strengths of 41–70 MPa (6,000–10,200 psi), ie, between two and three times the values of either polytetrafluoroethylene or polyethylene. Resistance to oxidation increases with increasing tetrafluoroethylene content. Since it is possible to produce a very wide range of compositions, it has been necessary to select an optimum monomer ratio to obtain the best compromise with regard to the final properties. For an article of commerce, a copolymer with a mole ratio of almost 1:1 of tetrafluoroethylene to ethylene gives the best balanced composition with respect to chemical resistance and physical characteristics. This polymer system is referred to here as ethylene–tetrafluoroethylene (ETFE) copolymer.

Monomer

Tetrafluoroethylene, perfluoroethylene, $CF_2{=}CF_2$, is prepared on a commercial scale by the pyrolysis of chlorodifluoromethane ($CHClF_2$). Material that is of purity suitable for granular or dispersion polymerizations is acceptable for copolymerization with ethylene. Polymerization-grade ethylene is suitable for copolymerization with tetrafluoroethylene.

Modifiers for ETFE polymers must generally undergo free-radical polymerization and must have such a structure that they do not lead to undesirable chain transfer or termination.

Tetrafluoroethylene may undergo spontaneous polymerization under certain conditions and is generally inhibited when stored. The hazards and safe handling of tetrafluoroethylene are described in ref. 7. However, it is generally not known that gaseous mixtures of tetrafluoroethylene and ethylene present a greater explosive potential than tetrafluoroethylene alone at the same temperature and pressure.

Copolymerization. Ethylene and tetrafluoroethylene are copolymerized by suspension or emulsion techniques, generally under such conditions that tetrafluoroethylene may homopolymerize but not ethylene. Bulk polymerization is commercially not feasible, not only because of heat transfer limitations but also because of the explosion hazard of the comonomer mixture. Polymerizations typically take place below 100°C and 5 MPa (50 atm). A variety of initiators has been used, including peroxides (1–2), redox systems (8), free-radical sources (9), and ionizing radiation (10) (see Initiators).

Polymerization systems in which water alone is used as the reaction medium give rise to copolymers that are not wetted by the reaction medium. The products tend to become consolidated into large agglomerates that may not only plug valves, etc, but also form tightly adhering masses on stirrer blades, thermocouples, or reactor walls. Some of these objectionable features can be overcome by conducting the polymerization in organic liquids or mixtures of these with water.

Reactivity ratios of ethylene and tetrafluoroethylene are given below (6):

Temperature, °C	$r_{(C_2F_4)}$	$r_{(C_2H_4)}$
−35	0.013 ± 0.008	0.10 ± 0.02
65	0.045 ± 0.010	0.14 ± 0.03

These values indicate strong alternation tendencies that decrease with increasing temperature. Computations have shown that 1:1 ETFE copolymers obtained at −30 and 65°C should have about 97 and 93%, respectively, of alternating sequences.

Ethylene–tetrafluoroethylene resins can be modified by addition of a third (or fourth) ethylenically unsaturated monomer during polymerization (11). Obvious considerations for such modifiers are relative reactivities, telogenic activity, and of course, effect on the properties of the product.

Polymer

Structure. The unit cell of an alternating copolymer of ethylene and tetrafluoroethylene has been determined by x-ray diffraction (12). The unit cell is probably either orthorhombic or monoclinic with the following parameters: a = 0.96 nm, b = 0.925 nm, c = 0.50 nm; (γ = 96°C). Internal motions of the alternating copolymer have been studied (13). At 1 Hz the α, β, and γ relaxations of a quenched sample were observed at 110, −25, and −120°C, respectively.

The tendency of the two monomers to alternate in an equimolar copolymer is generally 90% or more (14). Although the copolymer is isomeric with poly(vinylidene fluoride), it melts almost 100°C higher indicating marked differences in internal motions (15–16).

The crystalline nature of ETFE copolymers is evident by x-ray diffraction. The two main peaks at 2θ = 19.63° and 2θ = 21.00° correspond to Bragg distances of 0.452 nm and 0.423 nm. Compression-molded samples are generally 50–60% crystalline but this is greatly affected by quench rate.

Properties. Ethylene–tetrafluoroethylene resins are less dense, tougher, and stiffer and exhibit a higher tensile strength and creep resistance than polytetrafluoroethylene and fluorinated ethylene–propylene copolymer. They are, however, similarly ductile. Ethylene–tetrafluoroethylene copolymer displays the relatively nonlinear stress–strain relationships characteristic of nearly all ductile materials.

Thermal stability, chemical inertness, strength, toughness, nonflammability, clarity, excellent electrical properties, weatherability, and ease of processing are useful properties of these polymers. Table 1 summarizes the physical properties of Tefzel 200. Since some of these depend upon crystallinity (flex life, elongation, etc), these values vary according to fabrication conditions, particularly the quench-rate.

Stiffness and wear resistance can be markedly enhanced by the incorporation of fillers (17), such as glass fiber. Generally, polymer degradation because of chemical incompatibility with the filler does not occur.

In the range of 45–55 mol % tetrafluoroethylene, the copolymers melt between 260 and 270°C. Somewhat higher or lower tetrafluoroethylene ratios result in lower melting points. Moreover, the physical properties such as hardness, abrasion resistance, and cut-through (as a wire insulation) are nearly all at maximum at a tetrafluoroethylene content at 45–55 mol %.

Ethylene–tetrafluoroethylene copolymers have much in common with completely fluorinated resins as far as chemical inertness is concerned. None of a large number of representative organic and inorganic chemicals has shown any noticeable effect on the resin when tested at 50°C for one month. However, ETFE copolymers are sensitive to strong oxidizing agents (70% HNO_3, >100°C) and strong organic bases (n-butylamine, reflux) and react with molten alkali metals and elemental fluorine. There is no known solvent that dissolves the copolymers below 200°C. Water absorption is less than 0.1% at 100°C.

Electrical Properties. Ethylene–tetrafluoroethylene copolymers exhibit high resistivity and low losses. Some absorption and increase in the dissipation factor is apparent in the 1 GHz region. The dielectric constant is 2.6 at 100 Hz, 1 kHz, and 1 MHz.

Table 1. Properties[a] of Tefzel[b] 200

Property	ASTM method	Value
ultimate tensile strength, MPa[c]	D 638	44.8
ultimate elongation, %		200[d]
compressive strength, MPa[c]	D 695	48.4
shear strength, MPa[c]	D 732	41.3
heat deflection temperature, °C		
at 0.45 MPa (65.4 psi)[c]		104
at 1.81 MPa (263.1 psi)[c]		74
max continuous use temp, no load, °C		150
low temp embrittlement, °C	D 746	below 100
melting range, °C		255–270
specific heat, J/(mol·K)[e]		0.46–0.47
thermal conductivity, mW/(m·K)		56.9
tensile modulus, MPa[c]	D 638	827.4
flexural modulus, MPa[c]	D 790	1.379
impact strength notched Izod, J/m[f]		
at −54°C	D 256	>1050
at 23°C		no break
deformation under load, 140 kg/cm² (1991 psi), at 50°C, %	D 621	4.11
coefficient of linear expansion, cm/(cm·°C)		
at 20–30°C		9×10^{-5}
at 50–90°C		9.3×10^{-5}
at 140–180°C		14×10^{-5}
specific gravity, g/cm³	D 792	1.70

[a] At 25°C, 50% rh, unless otherwise stated.
[b] Trademark E. I. du Pont de Nemours & Co., Inc.
[c] To convert MPa to psi, multiply by 145.
[d] Elongations between 100 and 200% are achieved with different methods of sample fabrication.
[e] To convert J to cal, divide by 4.184.
[f] To convert J/m to ft·lbf/in., divide by 53.38 (see ASTM D 256).

The addition of glass reinforcement fibers modifies the electrical performance. In a 25 wt % glass-fiber ethylene–tetrafluoroethylene copolymer composition, the dielectric constant rises to 3.4 and the dissipation factor increases tenfold.

The dielectric strength is excellent and comparable to that of polyethylene or polytetrafluoroethylene. It is not affected by thermal aging until a physical break occurs in the material. A practical stress range for continuous operation is 1–2 kV/mm for uses where ionization could occur. Higher values can be used in the absence of ionization.

Tracking resistance is about 70 s by ASTM D 495. This is very good and comparable to materials considered to be nontracking but, like other nontracking materials (polytetrafluoroethylene and acrylics), it can be made to track under special conditions.

Thermal Stability. Ethylene–tetrafluoroethylene copolymers exhibit excellent high and low temperature properties and are suitable for sustained usage at 150°C. The dielectric strength and cut-through resistance are unaffected and weight loss is minimal after 10,000 h at 180°C (18). At processing temperatures of 300–320°C, minute amounts of hydrogen fluoride are released with little or no decrease in melt viscosity. Severe extrusion working or numerous reextrusions are reflected in deterioration of melt viscosity. The polymers do react slowly with air above their melting points to undergo oxidative cross-linking, melt-viscosity increases, and color deterioration. Above 400°C ETFE copolymer undergoes degradation in the form of random chain scission rather than unzippering. When subjected to a flame, it does not leave a char residue as is typical of poly(vinylidene fluoride). This nonchar behavior is typical of a polymer that degrades to low molecular weight fragments rather than splitting into small molecules, leaving a residual carbonaceous skeleton.

Ethylene–tetrafluoroethylene copolymer has excellent low temperature toughness. It does not break in the notched Izod impact test at RT or at −54°C. This corresponds to a very high impact strength of over 1068 J/m (20 ft·lbf/in.). The brittleness temperature is below −100°C. Thermal stability increases with increasing tetrafluoroethylene content.

Flammability. Ethylene–tetrafluoroethylene copolymer can be rated as either nonflammable or self-extinguishing in various flammability tests (see Flame retardants). For example, it passes such tests as UL STD 83, MIL-W-81004, MIL-W-22759, MIL-W-5086, and MIL-W-22759/13. Furthermore, it passes both vertical and horizontal flammability tests on insulated wire, both as a single wire and in bundles. By these tests, it is considered self-extinguishing; by ASTM D 635, it is rated nonburning. Afterburn, following removal of the ignition source, varies somewhat with flame temperature but remains within the limits of each specification test for a self-extinguishing material.

The ability to pass the most severe vertical flame test in air has been confirmed by the Limiting Oxygen Index Test (ASTM D 2863). It requires an oxygen concentration of at least 30% to sustain combustion. This is higher than the 21% oxygen concentration in normal air by a comfortable margin. The heat of combustion of ETFE copolymer is only 13.7 MJ/kg (5.9×10^3 Btu/lb), one half to one third that of other commonly used nonfluorinated insulating materials (see Insulation, electric).

Weight Loss. Under vacuum conditions, ETFE polymers generate very little volatile material at elevated temperatures. The loss rate is about one tenth of the generally acceptable maximum rates for spacecraft uses. Exposure of specimens 760 μm thick for 24 h in a vacuum of 13.3 μPa (10^{-7} mm Hg) at 149°C resulted in a weight loss of 0.07 wt %. In general, a high vacuum, maximum weight loss of 1.0 wt % is acceptable.

Because of the excellent thermal stability, the annual weight loss at elevated temperatures is very low. At 135 and 180°C the polymer loses 0.0006 g/g and 0.006 g/g per year, respectively, after an initial weight loss of 0.0013 g/g because of absorbed gases.

Hydrolytic Stability and Water Absorption. Using room temperature tensile strength and elongation as control properties, ETFE copolymer is unaffected after 3000 h exposure to boiling water. Water absorption is less than 0.03 wt % after 2 h in boiling water. This contributes to the outstanding dimensional stability of molded components as well as to the consistency of mechanical and electrical properties regardless of humidity.

Radiation Behavior. An ETFE copolymer can be cross-linked by ionizing radiation (19), despite the high proportion of tetrafluoroethylene units. Such cross-linking destroys its plasticity and provides enhancement of high temperature properties and overload (nondrip) performance. Irradiation also improves solder-iron resistance to the point that the resin can withstand a 400°C solder iron for over 10 min without noticeable effect.

Orientation. Tests on oriented ETFE films show a high degree of recoverable strain. The elastic recovery is high enough to yield shrink ratios of 2:1 when it is heated below its melting point, eg, 120°C. This ratio can be increased to as much as 5:1 by irradiation cross-linking before expansion when heated above its melting point. The balance of properties offered by ETFE copolymer, combined with its elastic memory, make these new fluoropolymers candidates for heat-shrinkable products, especially for encapsulating, protecting, and identifying irregular shapes.

Processing. Ethylene–tetrafluoroethylene copolymers are high temperature thermoplastics that can be readily processed by conventional methods including extrusion and injection molding. Different melt-viscosity grades are available and can be chosen to optimize fabrication in particular applications. Thin-wall sections (0.25 mm) can be achieved without difficulty by direct molding or drawn down in wire and tubing extrusion. Injection molded parts weighing up to 4.5 kg have been made with cycle times of 10 s to several minutes. Transfer molding has been used to line chemical service equipment such as pipes, valves, and pipe tees. Complex, difficult to mold, shapes can be made by rotomolding copolymer powders.

Although the resin melts at 250–270°C, extrusion temperatures are somewhat higher, generally 300–325°C. Good thermal stability allows considerable variation in extrusion and molding temperatures. A range of melt viscosities is available. Like other thermoplastics, these resins exhibit melt fracture above certain critical flow rates. In most cases, however, because of the considerable extrusion temperature range, melt fracture is not a problem. For high-rate extrusion, particularly in insulation of wire or production of tubing, streamlined die designs are used.

Molten ETFE resins are corrosive to mild and stainless steels, and Monel or Hastelloy C should be used for all parts of extrusion equipment that contact the molten resin. Successful extrusion performance has been demonstrated in both single- and twin-screw extruders from 12.5 to 83 mm dia. Heaters and controllers capable of operating to 350°C are required.

Unlike completely fluorinated melt-fabricable resins, ETFE copolymers react slowly with oxygen in the melt. Although normal start-up operations and process interruptions do not have detrimental effects, prolonged soaking should be avoided. Upon exposure to air at or above 300°C, ETFE resins undergo slow oxidative cross-linking and color deterioration and the melt viscosity increases under these conditions.

Health and Safety Factors. Initial tests in exposing rats to the pyrolysis decomposition products of Tefzel 200 at 400°C suggest a close comparison with Teflon FEP. Accordingly, all safety precautions recommended for processing of Teflon FEP should be followed (20). Unlike poly(vinylidene fluoride), the copolymer does not undergo an autocatalytic decomposition with elimination of hydrogen fluoride.

As with all organic polymers exposed to high temperatures, adequate ventilation is required. The heated fluoropolymer should be kept enclosed or exhaust ventilation should be used to prevent the inhalation of irritating or toxic fumes and gases. Incineration of the resin produces hydrogen fluoride.

Uses. Modified ETFE copolymers have gained rapid acceptance as a high performance insulation for wire and cable (21–24) (see Insulation, electric). Long melt extrusions are obtained with excellent uniformity and yield. The resin can be pigmented readily for color coding. It is rugged, with excellent cut-through and abrasion resistance, and is well suited to use in economical lightweight constructions such as airframe wire and computer hook-up wire. It is also gaining acceptance for use under stringent conditions such as wiring in locomotives and transit cars and in high-rise building and skyscraper fire alarm systems. The insulation can be applied over bare as well as plated conductors (single or multistrand) or used as a protective jacket over multiple bundles or coaxial cable.

BIBLIOGRAPHY

1. U.S. Pat. 2,468,664 (Apr. 26, 1949), W. E. Hanford and J. R. Roland (to E. I. du Pont de Nemours & Co., Inc.).
2. U.S. Pat. 2,479,367 (Aug. 16, 1949), R. M. Joyce, Jr. (to E. I. du Pont de Nemours & Co., Inc.).
3. Br. Pat. 1,166,020 (Oct. 1, 1969), M. Modena and co-workers (to Montecatini Edison, S.p.A.).
4. Jpn. Kokai 64 22,586 (Sept. 12, 1964), K. Hirose and co-workers (to Nitto Chemical Industry, Co., Ltd.).
5. Belg. Pat. 725,356 (Feb. 14, 1969), Z. Kenkyusho (to Asahi Glass Co., Ltd.).
6. M. Modena, C. Garbuglio, and M. Ragazzini, *Polym. Lett.* **10,** 153 (1972).
7. U.S. Pat. 3,009,966 (Nov. 21, 1961), M. Hauptschein and A. Faisberg (to Pennwalt Corp.).
8. Br. Pat. 1,353,535 (May 22, 1974), R. Hartwimmer (to Farbwerke Hoechst, A.G.).
9. U.S. Pat. 3,401,155 (Sept. 10, 1968), G. Borsini and co-workers (to Montecatini Edison S.p.A.).
10. *J. Polym. Sci. A* **2**(4), 1977 (1964).
11. U.S. Pat. 3,624,250 (Nov. 30, 1971), D. P. Carlson (to E. I. du Pont de Nemours & Co., Inc.); Jpn. Kokai 73 25,415 (July 28, 1973), (to Asahi Glass Co.).
12. F. C. Wilson and H. W. Starkweather, *J. Polym. Sci. Polym. Phys. Ed.* **11,** 919 (1973).
13. H. W. Starkweather, *J. Polym. Sci. Polym. Phys. Ed.* **11,** 587 (1973).
14. M. Modena, C. Garbuglio, and M. Ragazzini, *J. Polym. Sci. B* **10,** 153 (1972).
15. F. S. Ingraham and D. F. Wooley, Jr., *Ind. Eng. Chem.* **56**(9), 53 (1964).
16. S. Yano, *J. Polym. Sci. A-2* **8,** 1057 (1970).
17. Can. Pat. 900,075 (May 9, 1972), D. P. Carlson, J. A. Effenberger, and M. B. Polk (to E. I. du Pont de Nemours & Co., Inc.).
18. *Tefzel Fluoropolymer, Design Handbook,* E. I. du Pont de Nemours & Co., Inc., Wilmington, Del., 1973.
19. U.S. Pat. 3,738,923 (June 12, 1973), D. P. Carlson and N. E. West (E. I. du Pont de Nemours & Co., Inc.).
20. *Handling and Use of Teflon Fluorocarbon Resins at High Temperatures,* E. I. du Pont de Nemours & Co., Inc., Wilmington, Del., 1961.
21. *J. Teflon* **13**(1), (1972).
22. *J. Teflon* **13**(4), (1972).
23. E. A. Stecca, E. W. Fasig, Jr., and J. C. Chevrier, paper presented at *19th International Wire and Cable Symposium, Atlantic City, N.J., Dec. 3, 1970.*
24. J. C. Reed and J. R. Perkins, paper presented at *21st International Wire and Cable Symposium, Atlantic City, N.J., Dec. 6, 1972.*

RICHARD L. JOHNSON
E. I. du Pont de Nemours & Co., Inc.

TETRAFLUOROETHYLENE COPOLYMERS WITH PERFLUOROVINYL ETHERS

Teflon PFA [*26425-79-6*] fluorocarbon resins are thermoplastic polymers that combine the ease of melt fabrication with properties rivaling those of polytetrafluoroethylene (TFE). Introduced by DuPont in 1972, these polymers combine the carbon–fluorine backbone (—CF_2—) of polytetrafluoroethylene resins with a perfluoroalkoxy (PFA) side chain:

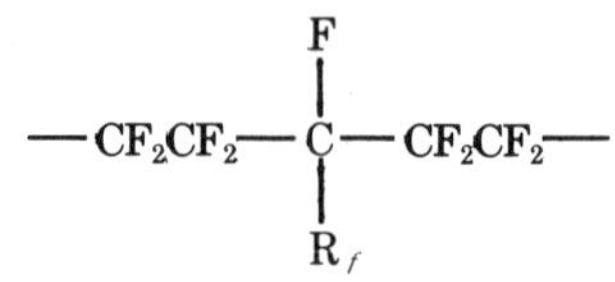

R_f = CF_3, C_2F_5, C_3F_7, etc
portion of a PFA molecule

The pendant perfluoroalkoxy groups exert their greatest effect on the crystallinity of the polymer. Unlike polytetrafluoroethylene homopolymers which have virtually no melt flow at or above their melting point, Teflon PFA resins are readily melt processible by conventional extrusion or injection and transfer molding.

Polytetrafluoroethylene, a polymer with excellent properties, was developed first. It was followed in 1960 by Teflon FEP fluorocarbon resin, a copolymer of tetrafluoroethylene and hexafluoropropylene. It was designed to give properties close to those of Teflon TFE, with certain modifications to achieve melt processibility. A major difference is in maximum continuous end-use temperature which is 205°C for Teflon FEP, compared with 250°C for Teflon TFE. In addition, electrical losses are slightly higher than those with polytetrafluoroethylene.

Teflon PFA resin combines easier processing with excellent properties and high temperature product performance. Its melting point is very close to that of polytetrafluoroethylene and well above that of Teflon FEP. Furthermore, its mechanical and chemical properties and its continuous operating temperature are very close to those of polytetrafluoroethylene and it can be melt-processed without difficulty. In short, Teflon PFA approximates rather closely the overall properties of polytetrafluoroethylene while offering the flexibility of thermoplastic processing.

Teflon PFA is offered by DuPont in two melt-viscosity grades: Teflon 340 (melt flow rate 7–18 g/10 min) and Teflon 350 (melt flow rate 1–3 g/10 min). Melt flow is determined according to ASTM 3307. Extruded strand-cut cubes, powder, and film forms are available. The polymer contains no plasticizers or stabilizers. As of January, 1980, Teflon 340 was priced at $34/kg.

Monomer

Preparation. The perfluorovinyl ethers, $R_fOCF{=}CF_2$, may be prepared by reaction sequences based on hexafluoropropylene epoxide (1–2). The epoxide (HFPO) and perfluorinated acyl fluorides produce a perfluoroacyl fluoride with an ether linkage (3–4):

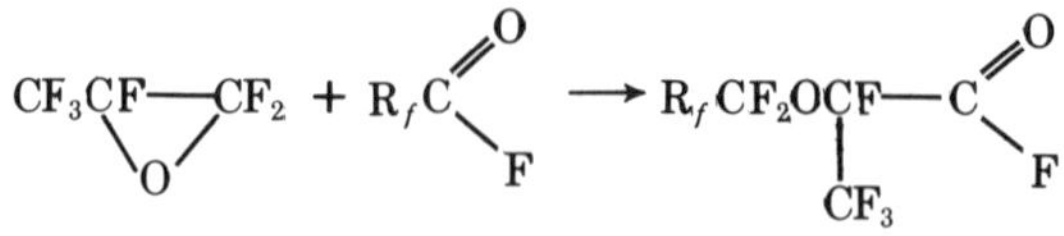

HFPO perfluoroacyl fluoride alkoxyacyl fluoride

R_f represents $F(CF_2)_n$, where $n = 0, 1, 2$, or 3, etc.

Alternatively, alkoxyacyl fluorides can be prepared by electrochemical fluorination of the appropriate alkoxycarboxylic acid (5).

The alkoxyacyl fluorides may be converted to vinyl ethers by base treatment at 250–300°C as follows (6):

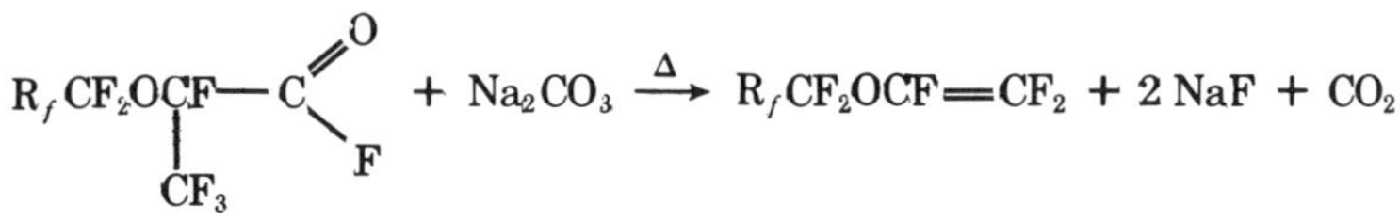

Another route to the vinyl ethers involves the reaction of 1,2-dichloro-1,2-difluoroethylene with trifluoromethyl hypofluorite, CF_3OF, to give $CF_3OCFClCF_2Cl$, which on treatment with zinc dust gives perfluoro(methyl vinyl ether) [*60917-27-3*] and zinc chloride (7).

Polymerization. Tetrafluoroethylene and perfluoroalkyl vinyl ethers copolymerize in either aqueous (8) or nonaqueous media (9). Soluble initiators are preferred, such as inorganic persulfates for aqueous media or fluorinated acyl peroxides for nonaqueous media (10).

For aqueous polymerizations, temperatures of 70–95°C and pressures of 1.48–2.51 MPa (200–350 psig) are employed, and for nonaqueous systems, temperatures of 45–65°C and pressures of 0.45–3.55 MPa (50–500 psig) are used (11).

Teflon PFA resins are not soluble in any solvent, at least below the melting point of the polymer. Polymerization, accordingly, results in a suspension of polymer particles in a liquid phase. Removal of the polymerization medium gives a finely divided powder. Typically, this powder is melt-extruded and converted to cubes or pellets suitable for injection-molding operations. The powder can be used for powder coating or rotational molding.

Polymer

Properties. Table 1 summarizes the physical properties of Teflon PFA resin. The mechanical characteristics of polytetrafluoroethylene, Teflon FEP, and Teflon PFA are very similar at RT. However, with increasing temperature, differences between Teflon FEP and Teflon PFA become readily apparent. Teflon FEP deteriorates above 200°C and it melts at 270–280°C. Teflon PFA melts above 300°C. For practical purposes, Teflon PFA resin can be used over the same temperature range as polytetrafluoroethylene. The polymer remains flexible and tough at −200°C and is strong and stiff at 260°C.

Teflon PFA fluorocarbon resin is inert to a variety of organic and inorganic chemicals commonly encountered in chemical service (12). However, like other perfluorinated polymers, it reacts with alkali metals and elemental fluorine.

Thermal Stability. Teflon PFA is an exceptionally stable polymer and can be processed up to 425°C. However, there are limits to this stability since thermal degradation of any polymer is a function of both temperature and time. The melt flow rate (an index of degradation) increases significantly for short times above 425°C and at lower temperatures with longer holdup times. Degradation of physical properties is not significant if the change in melt flow rate during molding is 20% or less. Degradation may also be indicated by the formation of small bubbles or discoloration;

Table 1. Physical Properties of Teflon PFA Fluorocarbon Resins

Property	ASTM method	Teflon 340	Teflon 350
nominal melting point, °C		302–306	302–306
specific gravity, g/cm^3		2.13–2.16	2.13–2.16
melt flow rate, g/10 min	D 3307	10.6	1.8
continuous use temperature, °C		260	260
tensile strength, MPa[a]			
23°C	D 1708	27.6	31.0
250°		12.4	13.8
tensile yield, MPa[a]			
23°C	D 1708	13.8	15.2
250°C		3.5	4.1
ultimate elongation, %			
23°C	D 1708	300	300
250°C		480	500
flexural modulus, MPa[a]			
23°C	D 790	655	690
250°C		55	69
creep resistance tensile modulus, MPa[a]			
23°C	D 695	270	270
250°C		41	41
hardness Durometer, D	D 2240	60	60
MIT[b] folding endurance at 180–200 μm, cycles		50,000	500,000
water absorption, %	D 570	0.03	0.03
coefficient of linear thermal expansion, cm/(cm·°C)			
20–100°C		12×10^{-5}	12×10^{-5}
100–150°C		17×10^{-5}	17×10^{-5}
150–210°C		20×10^{-5}	20×10^{-5}

[a] To convert MPa to psi, multiply by 145.
[b] MIT = Mass. Inst. of Tech., Flex Test. Apparatus by Tinius Olsen Mfg. Co., Willow Grove, Pa.

however, the use of high processing temperatures may cause slight discoloration without adversely affecting properties.

When Teflon PFA is aged at temperatures below but near its melting point, about 285°C, it actually becomes stronger. For example, samples aged in a circulating-air oven for 7500 h at 285°C have shown a pronounced decrease in melt flow number as defined by ASTM D 2116.

Hours	*g/min*
start of test	9.5
500	3.5
2000	1.5
7500	0.9

A decline in melt flow number is an indication of increasing average molecular weight, which is also manifested in a 25% increase in tensile strength and enhanced ultimate elongation. Another measure of toughness is MIT flex life, which improves severalfold with heat aging at 285°C.

Exposed to a fire, the polymer contributes little in fuel value and is self-extinguishing when the flame is removed. The fuel value is approximately 5.4 MJ/kg (2324 Btu/lb). It passes the UL 83 vertical flame test and is classified as 94VE-O according to UL 94. The limiting oxygen index (LOI) by ASTM D 2863 is greater than 95%.

Electrical Properties. Table 2 gives the electrical properties of Teflon PFA. The dielectric constant is slightly higher than that of Teflon TFE, and the dielectric losses are somewhat higher. The high temperature properties enhance the attractive electrical characteristics of the resin (15).

Optical Properties. In thin sections, such as extruded film, the resin is colorless and virtually transparent. In thicker sections, the resin becomes translucent. The resin is highly transparent to ir radiation. Uv absorption is low in thin sections; therefore, Teflon PFA is essentially unaffected by uv radiation and outdoor weathering. Weather-Ometer tests indicate virtually unlimited outdoor life.

Radiation Behavior. Fluorocarbon resins are generally not very resistant to radiation in air although Teflon FEP can be cross-linked to some degree under carefully controlled conditions. All fluorocarbons exhibit better radiation resistance *in vacuum* (as in space environment) than in air and Teflon PFA exhibits better radiation resistance in air or vacuum than polytetrafluoroethylene.

In general, Teflon PFA is little affected by low radiation dosages up to 30 kGy (3×10^6 rad); rapidly approaches the performance of polytetrafluoroethylene between 30 and 100 kGy ($3–10 \times 10^6$ rad); and embrittles above 100 kGy (10×10^6 rad); all the perfluorinated resins are severely degraded at 500 kGy (50×10^6 rad).

Wear and Friction Performance. The wear performance of unfilled Teflon PFA has been determined in mechanical applications such as bearings, seals, and films (13,16). Tests have been run on molded thrust bearings at 690 kPa (100 psi) against AISI 1018, Rc 20, and 16AA steels, at ambient conditions in air without lubrication. Results indicate a limiting performance value (PV) of 5000 but wear rate rather than PV is likely to be the critical parameter. At PV = 1000, for instance, Teflon PFA wears 47.6 mm/1000 h. Wear factors decrease over the PV range 1000–5000 from 1840×10^{-10} to 700×10^{-10}. The friction coefficient of Teflon PFA is 0.236 compared to a control of Teflon FEP at 0.333.

Cryogenic Properties. Tests made at liquid nitrogen temperature indicate that Teflon PFA performs well in cryogenic applications (see Cryogenics). The results of ASTM tests are shown in Table 3.

Pigmentation. Teflon PFA may be pigmented to a wide variety of colors, but pigments with good heat stability must be selected. Colorant concentrates containing approximately 2% pigment are available for use in injection molding. Teflon 340 can

Table 2. Typical Electrical Properties of Teflon 340

Property	ASTM method	Teflon 340	Ref.
dielectric strength, V/0.025 mm, short term (0.25 mm film)	D 149	2000	
surface resistivity, Ω/sq		>10	13
volume resistivity, Ω·cm	D 257	10	14
dielectric constant at 23°C	D 150		
at 1 kHz to 1 GHz		2.06	
at 240 GHz		2.049	
dissipation factor at 23°C	D 150		
at 1 kHz		0.000027	
at 1 MHz		0.000080	
at 1 GHz		0.00115	
at 240 GHz		0.00124	

Table 3. Cryogenic Properties of Teflon PFA

Property	ASTM method	Value	
		At 23°C	At −196°C
yield strength, MPa[a]	D 1708	14.5	
ultimate tensile strength, MPa[a]	D 1708	17.9	129
ultimate elongation, %	D 1708[b]	260	8
flexural modulus, MPa[a]	D 790-71[c]	558	5790
impact strength notched Izod, J/m[e]	D 256-72a[d]	no break	64
compressive strength, MPa[a]	D 695	na[f]	414
compressive strain, %	D 695	na[f]	35
modulus of elasticity, MPa[a]	D 695	na[f]	4690

[a] To convert MPa to psi, multiply by 145.
[b] Crosshead speed 1.3 mm/min used at both temperatures for more direct comparison.
[c] Method 1, procedure B.
[d] Method A, head weight is 4.5 kg at 23°C and 0.9 kg at 160°C.
[e] To convert J/m to lbf. ft/in. divide by 53.8.
[f] na = not available.

be dry blended with thermally stable inorganic pigments (see Pigments). At loadings of 0.1–1.0 wt %, pigments have no appreciable effect on the dielectric strength, dielectric constant, or mechanical properties of the resin. The dissipation factor of pigmented resin varies with the species and concentration of the pigment used.

Processing. Teflon PFA fluorocarbon resin can be fabricated by conventional melt-processing techniques (14,17) (see Plastics technology). Equipment should be constructed of corrosion-resistant materials and be capable of operating between 315 and 425°C.

Although Teflon PFA melts at about 307°C, extrusion temperatures are generally 35–50°C higher. The minimum melt temperature of the resin during processing should be 330°C. For short residence times, melt temperatures as high as 400°C may be used. The melt viscosity of Teflon 340 at its processing temperatures is many times higher than that of most other thermoplastics at their processing temperatures. As with other thermoplastics, Teflon 340 exhibits melt fracture above certain critical flow rates that are dictated by extrusion temperatures and die geometries. Teflon 340, a resin with a substantially lower melt viscosity compared to Teflon 350, can be processed more rapidly at lower temperatures and with much less tendency to melt fracture.

Extrusion and Injection Molding. Teflon PFA can be processed in the same type of equipment used for Teflon FEP. In general, it can be melt-extruded into rods, film, or tubing, drawn down onto wire as an insulation via tubing dies, or injection molded into complex shapes in a fashion very similar to Teflon FEP except that processing temperatures are 25–50°C higher (14,17).

Equipment. All parts of extrusion equipment coming into contact with the polymer melt must be made of special corrosion-resistant alloys such as Hastelloy C, Monel 400, and Xaloy 306. Resin degradation accelerates corrosion, and problems are likely to occur if processing temperatures are too high, residence times too long, or dead spots exist in the equipment.

Extruder. The same types of extruders used with Teflon FEP are also suitable for Teflon PFA. A 5-cm extruder with a barrel length-to-diameter ratio of about 20:1 provides good versatility for making a wide range of extruded products. Heaters and controllers capable of accurate operation in the range of 330–425°C are required. Extruder barrels should have 3 or 4 independently controlled heating zones, each equipped with its own thermocouple and temperature indicating controller. Any type of drive system with good regulation of speed of 3–50 rpm may be used.

Screw. The same type of screw is used in extruding Teflon FEP as with Teflon PFA. It consists of a feed section, a rapid transition section, and a metering section.

Breaker Plate. The breaker plate, used to convert the rotary motion of the melt into smooth, straight-line flow, should have as many holes as possible. Both ends of each hole should be countersunk for streamlined flow.

Melt Thermocouple. The temperature of the melt downstream from the breaker plate is of crucial importance. It may exceed the front barrel temperature because of the mechanical work put into the resin by the screw, and it varies with screw speed and flow rate. Since the melt temperature is not necessarily correctly indicated by barrel temperature indicators, it should be measured by a thermocouple located downstream from the breaker plate inserted into the melt.

Molds—Injection Molding. Since the mold is usually maintained at temperatures below the melting point of the resin, the corrosion rate on the mold surfaces is far less than that in the molding machine. Teflon 340 has no effect on molds with a high-quality hard chrome or nickel plating. Molds made of conventional tool steel or hardened stainless steel are adequate for short pilot-plant runs but should not be used in production.

Large diameter, full-rounded runners of the shortest possible length should be used. Runners must be smooth, free of restrictions or sharp corners, and blend smoothly with gates that are as large as practical. The land length of the gate should be very short.

In the mold cavities themselves generous filleting and streamlining of all angles and intersections is desirable.

Crosshead and Dies. Crossheads used for wire coating or tubing extrusions must be corrosion resistant and streamlined. Care should be taken to avoid holdup spots, particularly where the direction of melt flow is changed 90°C.

Long-land, tubing-type dies minimize surging and permit the highest extrusion rates.

The temperatures of the crosshead and any adapters should be maintained by an independently controlled heater backed with appropriate thermocouples and temperature controllers. An independently controlled heater capable of reaching 425°C is required for the die.

Exhaust. A hooded exhaust with sufficient capacity to vent any decomposition products should be placed over the extruder die and feed hopper whenever the extruder is heated.

Safety Practices; Shipping. Information with regard to the safe handling of polytetrafluoroethylene and Teflon FEP fluorocarbon resins (18) is available but not yet with regard to Teflon PFA. The usual safety practices applied to fluorocarbon resins are adequate for Teflon PFA. Adequate ventilation is essential in processing areas and contamination of smoking tobacco or cigarettes should be avoided.

Scrap should be disposed by landfill. Incineration would produce hydrogen fluoride and other toxic products.

Teflon PFA fluorocarbon resin is shipped by rail classified as Plastics, Synthetic,

OTL, NOIBN, by truck as Plastic Materials, Granules, and by express as Plastics, Synthetic.

Uses. Teflon 340 is the general-purpose grade of Teflon PFA designed for extended service in environments that demand high levels of chemical, thermal, and mechanical performance because it combines melt processibility, excellent flex life, high temperature strength, and chemical inertness. It is used in a variety of molding and extrusion applications such as extruded tubing and shapes, primary insulation, and jacketing for wire and cable, injection- and blow-molded components, and compression-molded articles (see Insulation, electric).

Teflon PFA is used in severe environments in the chemical processing industry in lined valves, pipes, and tanks and heat-exchange applications. In addition, Teflon PFA can be molded into electrical insulating parts such as connector inserts or insulating bushings and standoff insulators (see Electrical connectors) (19). Other uses include electrical spaghetti tubing and film for flexible circuits, flat cable, and melt bonding. With melt fabrication, it is possible to coat continuous lengths of wire in a wide variety of wall thicknesses (20). Current insulation uses range from hook-up wire to heater cables and heavy-wall conduit. Large diameter coaxial cables have been extruded and multiconductor cables jacketed. Although higher in melt viscosity than the general-purpose grade, Teflon 350 can also be used as an electrical insulation in a variety of wire and cable uses.

BIBLIOGRAPHY

1. U.S. Pat. 3,358,003 (Dec. 12, 1967), H. S. Eleuterio and R. W. Meschke (to E. I. du Pont de Nemours & Co., Inc.).
2. U.S. Pat. 3,180,895 (Apr. 27, 1965), J. F. Harris and D. I. McCane (to E. I. du Pont de Nemours & Co., Inc.).
3. U.S. Pat. 3,250,808 (Oct. 10, 1966), E. P. Moore, A. S. Milian, Jr., and H. S. Eleuterio (to E. I. du Pont de Nemours & Co., Inc.).
4. U.S. Pat. 4,118,421 (Oct. 3, 1978), T. Martini (to Hoechst Aktiengesellschaft).
5. U.S. Pat. 2,713,593 (July, 1955), T. J. Brice and W. H. Pearlson (to Minnesota Mining and Manufacturing Co.).
6. U.S. Pat. 3,291,843 (Dec. 13, 1966), C. G. Fritz and S. Selman (to E. I. du Pont de Nemours & Co., Inc.).
7. W. S. Durrell and co-workers, *J. Polym. Sci. A* **3,** 4065 (1965).
8. U.S. Pat. 3,132,123 (May 5, 1964), J. F. Harris and D. I. McCane (to E. I. du Pont de Nemours & Co., Inc.).
9. U.S. Pat. 3,536,733 (Oct. 27, 1970), D. P. Carlson (to E. I. du Pont de Nemours & Co., Inc.).
10. U.S. Pat. 2,792,423 (May 14, 1957), D. M. Young and W. N. Stoops (to Union Carbide & Carbon Corp.).
11. U.S. Pat. 3,635,926 (Jan. 18, 1972), W. F. Gresham and A. F. Vogelpohl (to E. I. du Pont de Nemours & Co., Inc.).
12. *Teflon PFA Fluorocarbon Resin—Chemical Resistance, PIB #2,* bulletin, E. I. du Pont de Nemours & Co., Inc., Wilmington, Del., 1972.
13. R. C. Bowers and W. A. Bisman, *Ind. Chem. Prod. Res. Dev.* **13**(2), 115 (1974).
14. *Teflon PFA Fluorocarbon Resin—Melt Processing of Teflon PFA TE-9705, PIB # 1,* bulletin, E. I. du Pont de Nemours & Co., Inc., Wilmington, Del., 1973.
15. E. W. Fasig, D. I. McCane, and J. R. Perkins, paper presented at the *22nd International Wire and Cable Symposium, Atlantic City, N.J., Dec. 1973.*
16. *Teflon PFA Fluorocarbon Resin—Wear and Frictional Data, APD #2,* bulletin, E. I. du Pont de Nemours & Co., Inc., Wilmington, Del., 1973.
17. *Teflon PFA Fluorocarbon Resin—Injection Molding of Teflon PFA TE-9704, PIB #4,* bulletin, E. I. du Pont de Nemours & Co., Inc., Wilmington, Del., 1973.

18. *Handling and Use of Teflon Fluorocarbon Resins at High Temperatures,* bulletin, E. I. du Pont de Nemours & Co., Inc., Wilmington, Del., 1961.
19. *Teflon PFA Fluorocarbon Resin—TE-9704 Extrusion Coating of Wire, PIB #3,* bulletin, E. I. du Pont de Nemours & Co., Inc., Wilmington, Del., 1972.
20. *Teflon PFA Fluorocarbon Resin—Typical Characteristics for 20 AWG Wire, APD #1,* bulletin, E. I. du Pont de Nemours & Co., Inc., Wilmington, Del., 1973.

RICHARD L. JOHNSON
E. I. du Pont de Nemours & Co., Inc.

POLYCHLOROTRIFLUOROETHYLENE

Since disclosure in 1937 (1), polychlorotrifluoroethylene [*9002-83-9*] (PCTFE) has been developed to serve many demanding industrial and military needs. The preparation and characterization of this high molecular weight thermoplastic was studied intensively during the Manhattan Project (2). The unique combination of chemical inertness, radiation resistance, low vapor permeability, and thermal stability of this polymer filled an immediate need for a construction material in the gaseous diffusion process for the separation of uranium isotopes (see Diffusion separation methods).

Properties

PCTFE properties are governed by the molecular weight and percent crystallinity. Since PCTFE has limited solubility in selected solvents, correlations of the number average molecular weight with the zero-strength time (ZST) (3–4) have been made. The high molecular weight thermoplastic has a melt temperature (T_m) of 211–216°C and glass transition temperature (T_g) of 71–99°C, and is thermally stable up to 250°C. Because of the loss of strength above the T_m, the useful range of operation is from −240 to 205°C. An increase of upper temperature service can be realized through use of commercially available filled versions of the plastic.

The specific gravity of PCTFE for the amorphous and crystalline polymers has been calculated to be 2.075 and 2.185, respectively (5–9). In practice, the final PCTFE parts have some degree of crystallinity ranging from about 45% (specific gravity of 2.10) for quick-quenched to 65% (specific gravity of 2.13) for slow-cooled parts. The higher crystalline forms are less transparent, with higher tensile modulus, lower elongation, and have more resistance to the penetration of liquids and vapors. The less crystalline form is optically clear, tough, and ductile.

The combination of mechanical properties that makes PCTFE a unique engineering thermoplastic is shown in Table 1. In addition, PCTFE has a low coefficient of thermal expansion. A general outline of cryogenic mechanical properties is shown in Table 2.

The high fluorine content contributes to resistance to attack by essentially all

Table 1. Mechanical Properties of Polychlorotrifluoroethylene

Property	Value
tensile strength, MPa[a]	32–39
compressive strength, MPa[a]	380
modulus of elasticity, MPa[a]	1423
hardness, Shore D	76
deformation under load, 0.3% at 24 h, MPa[a]	6.86
heat deflection temperature, °C at 1.78 MPa[a]	126

[a] To convert MPa to psi, multiply by 145.

Table 2. Typical Cryogenic Mechanical Properties of Polychlorotrifluoroethylene (ASTM D 1430 Type I, Grade I, Class C)

Property	PCTFE % crystallinity	Temp, °C	Value[a]
tensile: ultimate strength, MPa[b]	40	25	38.6
		−129	150
		−252	200
elongation, %	40	25	140
		−129	9
		−252	5
modulus of elasticity, MPa[b]	40	25	1521
		−129	5494
		−252	8682
impact strength notched Izod, J/m[c]	60	25	13.7
		−196	12.8
		−252	13.7

[a] Ref. 10.
[b] To convert MPa to psi, multiply by 145.
[c] To convert J/m to ft·lbf/in., divide by 53.38 (see ASTM D 256).

chemicals and oxidizing agents; however, PCTFE does swell slightly in halogenated compounds, ethers, esters, and aromatic solvents. PCTFE has the lowest water–vapor transmission rate of any plastic (11), is highly impermeable to gaseous vapors, and does not carbonize or support combustion (see also Barrier polymers).

This plastic is compatible with liquid oxygen, remains flexible at cryogenic temperatures (12–15), and retains its properties when exposed to either uv or gamma radiation. Unlike other fluoroplastics, PCTFE resists cold flow as reflected by its high compressive strength which qualifies this plastic as a construction material in many load-bearing structural parts.

Manufacture and Processing

The preparation of the high molecular weight thermoplastic from chlorotrifluoroethylene [*79-38-9*] has been carried out in bulk (16–20), solution (21–23), suspension (24–29), and emulsion (30–34) systems using free-radical initiators, uv, and gamma radiation. The polymer obtained from emulsion represents a more thermally stable system than bulk or suspension-derived polymers. Polymerizations can be carried out in glass or stainless steel agitated reactors. Pressures (0.34–1.03 MPa or 50–150 psi) and temperatures (21–52°C) normally employed present no unusual equipment design problems.

Following the polymerization step, the polymer is then isolated from the latex or suspension. The suspension polymer, already in powder form, is washed to remove initiator residues and then dried. The emulsion polymer is coagulated from the latex by freezing or by the addition of salts, acids, and solvents (see Latex technology). The isolated powder is then washed and dried. The dried powder from either process additionally can be chemically treated to remove trace impurities that can result in chain degradation during further processing; treatment with carboxylic acids (35), ozone in air (36), or chlorine (37) improves thermal stability, color, and light transmission of the final polymer. The stabilized polymer is then sold to molders as powder or melt-extruded pellets in filled or unfilled form.

The lower molecular weight oils, waxes, and greases of PCTFE can be prepared directly by telomerization of the monomer or by pyrolysis of the higher molecular weight thermoplastic (38–47). PCTFE plastics can be processed by the standard thermoplastic fabrication techniques eg, extrusion, injection, compression, and transfer molding (see Plastics technology). The control of processing temperatures is important to avoid prolonged overheating (above 260°C) which can result in degradation of the polymer. The plastic can be readily machined from molded billets or rod stock on standard equipment to fabricate more exacting part geometries.

Economic Aspects

At present, there are several commercial suppliers of PCTFE and vinylidene fluoride-modified copolymers [*9010-75-7*] as shown in Table 3. PCTFE plastics are sold in the range of $37–66/kg, depending on the molecular weight, grade, product form, and supplier. As a result, PCTFE thermoplastics are used in high-technology, specialty engineering areas where the unique combination of properties and longevity of service demands a high performance thermoplastic.

Table 3. PCTFE Manufacturers and Products

Trademark	Manufacturer	Product forms
Homopolymers		
Daiflon	Daikin Koygo Yodogawa, Osaka, Japan	molding powder, pellets, dispersion oils, and greases
Kel-F 81 Brand Plastic	3M Company, St. Paul, Minn.	molding powders, pellets, and filled pellets
Halocarbon oil	Halocarbon Products Corporation, Hackensack, N.J.	oils, waxes, and greases
Voltalef	Ugine Kuhlmann, Pierre-Benite, France	molding powders and pellets
Copolymers		
Aclon, Aclar	Allied Chemical, Morristown, N.J.	molding powders, pellets, and film
Kel-F 82; Kel-F 800	3M Company, St. Paul, Minn.	molding powders and latex

Specifications

The specifications met by PCTFE plastics are Military Specifications, MIL-P-46036 and MIL-P-21470; Federal Specification LP-385B; and ASTM 1430 Grades II and III. Standards for fabricated forms are available for compression molded heavy sections (AMS-3645), molded sheets (AMS-3646 and -3649), thin-walled tubing (AMS-3648), and rod, sheet, and molded shapes (AMS-3650). PCTFE plastics have been approved for use in contact with food by the FDA (48).

Test Methods

The test methods employed are: determination of molecular weight as measured by ZST (ASTM D 1430); specific gravity (ASTM D 792); tensile strength, elongation and modulus (ASTM D 638); compressive strength and modulus (ASTM D 695); modulus of rigidity (ASTM D 747); deformation under load (ASTM D 621); heat deflection (ASTM D 648); impact strength (ASTM D 256); flammability (ASTM D 2863); hardness (ASTM D 2240 and D 785); and coefficient of linear expansion (ASTM D 696).

Health and Safety Factors

In general, the PCTFE resins have been found to be low in toxicity and irritation potential under normal handling conditions. Specific toxicological information and safe handling procedures are provided by the manufacturer of specific PCTFE products upon request.

Uses

The major uses for PCTFE plastics today have been in the areas of electrical/electronics, cryogenic, chemical, and medical instrumentation industries. Applications include chemically-resistant electrical insulation and components; cryogenic seals, gaskets, valve seats (49–50) and liners; instrument parts for medical and chemical (51) equipment; medical packaging; fiber-optic applications (see Fiber optics); and seals for the petrochemical/oil industry.

The lower molecular weight PCTFE oils, waxes, and greases are used as inert sealants and lubricants for equipment handling oxygen and other oxidative or corrosive media. Other uses include gyroscope flotation fluids and plasticizers for thermoplastics. The greases have specialty lubrication uses in severe environments such as in corrosive or oxidative media.

BIBLIOGRAPHY

1. Brit. Pat. 465,520 (May 3, 1937), (to I. G. Farbenindustrie).
2. U.S. Pat. 2,564,024 (Aug. 14, 1951), W. T. Miller (to USAEC).
3. H. S. Kaufman, C. O. Kroncke, and C. K. Giannotta, *Mod. Plast.* **32,** 146 (1954).
4. E. K. Walsh and H. S. Kaufman, *J. Polym. Sci.* **26,** 1 (1957).
5. H. Matsuo, *J. Polym. Sci.* **21,** 331 (1956).
6. H. Matsuo, *J. Polym. Sci.* **25,** 234 (1957).
7. H. Matsuo, *Bull. Chem. Soc. Jpn.* **30,** 593 (1957).

8. J. D. Hoffman *J. Am. Chem. Soc.* **74,** 1696 (1952).
9. J. D. Hoffman and J. J. Weeks, *J. Res. Nat. Bur. Stand.* **60,** 465 (1958).
10. R. E. Mowers "Cryogenic Properties of Poly(Chlorotrifluoroethylene)," *Technical Document Report No. RTD-TDR-63-11,* Air Force Contract No. AF04(611)-6354, 1962.
11. J. D. Hoffman and J. J. Weeks, *J. Chem. Phys.* **37,** 1723 (1962).
12. A. W. Myers and co-workers, *Mod. Plast.* **37,** 139 (1960).
13. N. Brown, B. D. Metzger, and Y. Imai, *J. Polym. Sci.* **16,** 1085 (1978).
14. Y. Imai and N. Brown, *Polymer* **18,** 298 (1977).
15. N. Brown and S. Fischer, *J. Poly. Sci. Polym. Phys.* **13,** 1315 (1975).
16. U.S. Pat. 2,586,550 (Feb. 19, 1952), W. T. Miller, A. L. Dittman, and S. K. Reed (to USAEC).
17. U.S. Pat. 2,792,377 (May 14, 1957), W. T. Miller.
18. U.S. Pat. 2,636,908 (Apr. 28, 1953), A. L. Dittman and J. M. Wrightson (to M. W. Kellogg Co.).
19. Brit. Pat. 729,010 (Apr. 27, 1955), (to Farbenfabriken Bayer A.F.).
20. Fr. Pat. 1,419,741 (Dec. 3, 1965), (to Kureha Chemical Company).
21. U.S. Pat. 2,700,662 (Jan. 25, 1955), D. M. Young and B. Thompson (to Union Carbide-Carbon Co.).
22. U.S. Pat. 2,820,027 (Jan. 14, 1958), W. E. Hanford (to 3M).
23. M. Lazar, *J. Polym. Sci.* **29,** 573 (1958).
24. U.S. Pat. 2,613,202 (Oct. 7, 1952), G. F. Roedel (to General Electric Company).
25. U.S. Pat. 2,600,202 (June 10, 1952), D. W. Caird (to General Electric Company).
26. Fr. Pat. 1,155,143 (Apr. 23, 1958), (to Society d'Ugine).
27. U.S. Pat. 2,842,528 (July 8, 1958), R. L. Herbst and B. F. Landrum (to 3M).
28. U.S. Pat. 2,689,241 (Sept. 14, 1954), A. L. Dittman, H. J. Passino, and J. M. Wrightson (to M. W. Kellogg Co.).
29. J. M. Hamilton, *Ind. Eng. Chem.* **45,** 1347 (1953).
30. U.S. Pat. 2,569,524 (Oct. 21, 1951), J. M. Hamilton (to E. I. du Pont de Nemours & Co., Inc.).
31. U.S. Pat. 2,774,751 (Dec. 18, 1956), H. J. Passino and co-workers (to M. W. Kellogg Co.).
32. Brit. Pat. 840,735 (July 6, 1960), (to 3M).
33. U.S. Pat. 2,559,749 (July 10, 1951), A. F. Benning (to E. I. du Pont de Nemours & Co. Inc.).
34. U.S. Pat. 2,559,752 (July 10, 1951), K. L. Berry (to E. I. du Pont de Nemours & Co., Inc.).
35. U.S. Pat. 2,751,376 (June 19, 1956), R. M. Mantell and W. S. Barnhart (to 3M).
36. U.S. Pat. 2,902,477 (Sept. 1, 1959), E. Fischer, K. Weissermel, and G. Bier (to 3M).
37. U.S. Pat. 3,045,000 (July 17, 1962), R. R. Davis.
38. U.S. Pat. 2,770,659 (Nov. 13, 1956), W. S. Barnhart (to M. W. Kellogg Co.).
39. U.S. Pat. 2,786,827 (Mar. 26, 1957), W. S. Barnhart (to M. W. Kellogg Co.).
40. U.S. Pat. 2,664,449 (Dec. 29, 1953), W. T. Miller (to USAEC).
41. U.S. Pat. 2,636,907 (Apr. 28, 1953), W. T. Miller (to M. W. Kellogg Co.).
42. U.S. Pat. 2,636,908 (Apr. 28, 1953), A. L. Dittman and J. M. Wrightson (to M. W. Kellogg Co.).
43. U.S. Pat. 2,706,715 (Apr. 19, 1955), R. C. Conner (to M. W. Kellogg Co.).
44. U.S. Pat. 2,716,141 (Aug. 23, 1955), W. T. Miller.
45. U.S. Pat. 2,854,490 (Sept. 30, 1958), E. Fischer and H. Frey (to Farbwerke Hoechst).
46. U.S. Pat. 2,992,988 (July 18, 1961), C. D. Dipner (to 3M).
47. U.S. Pat, 3,076,765 (Feb. 5, 1963), F. W. West, R. J. Seffl, and L. J. Reilly (to 3M).
48. *Code of Federal Regulations,* **Title 21,** *Paragraph 177.1380,* revised April 1, 1979, p. 604.
49. W. Broadway, "A Pressure Sensitivity and Temperature Response Butterfly Valve for Cryogenic Service," *Petroleum Division of ASME, Energy Technology Conference and Exhibition, Houston, Nov. 5–9, 1978.*
50. W. Broadway, "Development of Cryogenic Butterfly Valve Seat," *AIChE 71st Annual Meeting, Cryogenic Equipment Session, Miami, Nov. 12–16, 1978.*
51. J. E. Harrar and R. J. Sherry, *Anal. Chem.* **47,** 601 (1975).

General References

R. P. Bringer, "Influence of Unusual Environmental Conditions on Fluorocarbon Plastics," *SAMPE (Society of Aerospace Material and Process Engineers) Symposium, St. Louis, May 7–9, 1962.*

R. P. Bringer and C. C. Solvia, *Chem. Eng. Prog.* **56**(10), 37 (1960).

R. E. Schamm, A. F. Clark, and R. P. Reed, *A Compilation and Evaluation of Mechanical, Thermal and Electrical Properties of Selected Polymers,* NBS Report, AEC SAN-70-113, SANL 807 Task 7, SANL Task 6, Sept. 1973, pp. 335–443.

C. A. Harper, ed., *Handbook of Plastics and Elastomers,* McGraw-Hill, New York, 1975.
W. T. Miller, "General Discussion of Chlorotrifluoroethylene Polymers" in C. Slesser and S. R. Schramm, eds., *Preparation, Properties and Technology of Fluorine and Organic Fluoro Compounds,* McGraw-Hill, New York, 1951.
L. A. Wall, ed., *Fluoropolymers,* Vol. XXV, Wiley-Interscience, New York, 1973, Chapts. 15 and 16.

A. C. WEST
3M Company

BROMOTRIFLUOROETHYLENE

Physical Properties

The monomer, bromotrifluoroethylene [*598-73-2*], $CF_2{=}CFBr$, is a colorless gas; bp −3.0°C at 101 kPa (754 mm Hg); 58°C at 790 kPa (100 psig); and d_4^{25} 1.86 g/cm^3. Since it is spontaneously flammable in air, its odor is that of its oxidation products, mixed carbonyl halides. The olefin can be distilled, but it polymerizes on standing at ambient temperature unless an inhibitor such as 0.1% tributylamine is added. If desired, the inhibitor can be readily removed by passing the gas through silica gel. Higher temperatures or ultraviolet light increase the polymerization rate. The nmr (1–2), ir (3), uv (4), and photoelectron spectra (5) of the monomer have been reported, and some thermochemical data have been calculated for it (3). Its electric dipole moment has been determined to be 2.54×10^{-30} C·m (0.76 D) (6).

Chemical Properties

Many reactions of bromotrifluoroethylene have been studied. Under basic conditions it adds alcohols such as methanol (7–8) or ethanol (8–9), forming ethers with the general formula, $ROCF_2CFBrH$. Similarly, diethylamine adds to it giving $(C_2H_5)_2NCF_2CFBrH$ (10). This addition is faster than diethylamine additions to tetrafluoroethylene or chlorotrifluoroethylene. Vapor-phase photochemical bromination of bromotrifluoroethylene gives the expected adduct, $CF_2BrCFBr_2$ (11). On the other hand, photochemical chlorination results in only 60% of the expected adduct and 40% scrambled bromo and chloro products. Hydrogenation of the double bond can be accomplished either catalytically (12) or with sodium borohydride (13). The former method also gives some trifluoroethylene and the latter method gives only a 37% yield of the pure product, $CF_2HCFBrH_2$. Other reagents that add across the double bond include S_2Cl_2, to give mainly a disulfide (14), and aqueous sodium nitrite (15).

Another class of reactions that bromotrifluoroethylene undergoes is cycloaddition with acetylenes (16) or olefins (17). When heated, the pure monomer, in addition to polymerizing, also dimerizes to *cis*- and *trans*-1,2-dibromohexafluorocyclobutane, which can be debrominated with zinc to perfluorocyclobutene (18). The monomer is useful as a source of the trifluorovinyl group in organic synthesis since treatment of it with magnesium metal (19) or methyllithium (20) yields a trifluorovinyl Grignard or lithium reagent which readily adds to ketones (20) or epoxides (19).

Manufacture

Although bromotrifluoroethylene was first prepared by Swarts (9) by the dehydrobromination of 1,2-dibromo-1,1,2-trifluoroethane, it is more conveniently prepared from chlorotrifluoroethylene by the following high-yield steps (21):

$$CF_2{=}CFCl + HBr \rightarrow CF_2BrCFClH$$

$$CF_2BrCFClH + Zn \rightarrow CF_2{=}CHF + ZnBrCl$$

$$CF_2{=}CHF + Br_2 \rightarrow CF_2BrCHFBr$$

$$CF_2BrCHFBr + KOH \rightarrow CF_2{=}CFBr + KBr + H_2O$$

Bromotrifluoroethylene is manufactured and sold in commercial quantities with a purity of 99.9% by the Halocarbon Products Corporation for about \$100/kg (1979).

Polymers

The olefin can be polymerized in trichlorofluoromethane solution at −5°C for 7 d with a halogenated acetyl peroxide such as 0.037% trichloroacetyl peroxide as the initiator (22). Alternatively, it may be polymerized in an aqueous suspension with 2 parts by weight distilled water, 0.01 part ammonium persulfate, 0.004 part sodium bisulfite, and 0.001 part hydrated ferrous sulfate present for each part of the monomer. Mixing for 24.5 h at 20°C gives a 52% conversion of the monomer to the homopolymer (22). Prepared either way, the homopolymer [*55157-25-0*] is a white powder which is soluble in acetone and useful as a hard, chemically-resistant coating for metal or fabric surfaces. The addition of small amounts of chain-transfer agents such as chloroform, carbon tetrachloride, bromotrichloromethane, and in particular, 0.01 to 0.3% 1-dodecanethiol to the polymerization mixture gives a lower molecular weight homopolymer that is softer and more soluble. Copolymers of bromotrifluoroethylene with many other compounds such as chlorotrifluoroethylene (23), tetrafluoroethylene (24), or trifluoronitrosomethane (25) have been reported.

Telomers. Telomers of bromotrifluoroethylene have been prepared using chain-transfer agents such as CF_3SSCF_3 (26), C_2F_5I (27), CBr_4 (28), or CBr_3F (28). For example, when the olefin is slowly added to tribromofluoromethane under light from sunlamps, a liquid is obtained which after saturation and distillation has a viscosity of 510 mm^2/s (= cSt) and a density of 2.65 g/cm^3 at 58.3°C. These and other bromotrifluoroethylene telomers are useful as flotation agents and damping fluids for gyroscopes and accelerometers in inertial guidance systems. These telomers are noncrystalline up to a higher degree of polymerization than those of chlorotrifluoroethylene, allowing the preparation of liquids of greater viscosity. The higher densities of the bromine-containing oils permit further miniaturization of the instruments floating in them. For these reasons, these oils complement their less expensive chloro analogs in this application. Commercially available bromotrifluoroethylene telomers have densities of 2.14–2.65 g/cm^3 and viscosities of 2–4000 mm^2/s (= cSt). These fluids are expensive but are made in small volume for the aerospace industry. They are nontoxic and noncorrosive to many common metals, alloys, and other materials.

BIBLIOGRAPHY

"Polybromotrifluoroethylene" in *ECT* 2nd ed., under "Fluorine Compounds, Organic," Vol. 9, pp. 833–835, by L. L. Ferstandig, Halocarbon Products Corporation.

1. D. D. Elleman and S. L. Manatt, *J. Chem. Phys.* **36,** 1945 (1962).
2. J. Reuben, Y. Shvo, and A. Demiel, *J. Am. Chem. Soc.* **87,** 3995 (1965).
3. D. E. Mann, N. Acquista, and E. K. Plyler, *J. Chem. Phys.* **22,** 1199 (1954).
4. J. Schander and B. R. Russell, *J. Mol. Spectrosc.* **65,** 379 (1977).
5. K. Wittel and H. Bock, *Chem. Ber.* **107,** 317 (1974).
6. E. J. Gauss and T. S. Gilman, *J. Phys. Chem.* **73,** 3969 (1969).
7. U.S. Pat. 3,666,864 (May 30, 1972), R. C. Terrel (to Airco, Inc.).
8. A. Demiel, *J. Org. Chem.* **25,** 993 (1960).
9. F. Swarts, *Chem. Zentr.* **II,** 281 (1899).
10. R. N. Sterlin and co-workers, *Izv. Akad. Nauk SSSR Otd. Khim. Nauk,* 2151 (1959); *Chem. Abstr.* **54,** 10848i (1960).
11. J. R. Lacher, R. D. Burkhart, and J. D. Park, *Univ. Colo. Stud. Ser. Chem. Pharm.* **4,** 8, 22 (1962); *Chem. Abstr.* **58,** 5185c (1963).
12. J. R. Lacher and co-workers, *J. Phys. Chem.* **61,** 1125 (1957).
13. A. L. Anderson, R. T. Bogan, and D. J. Burton, *J. Fluorine Chem.* **1,** 121 (1971).
14. U.S. Pat. 2,451,411 (Oct. 12, 1948), M. S. Raasch (to E. I. du Pont de Nemours & Co., Inc.).
15. A. M. Krzhizhevskii, Y. A. Cheburkov, and I. L. Knunyants, *Izv. Akad. Nauk SSSR Ser. Khim.,* 2144 (1974); *Chem. Abstr.* **82,** 86040r (1975).
16. J. C. Blazejewski, D. Cantacuzene, and C. Wakselman, *Tetrahedron Lett.,* 2055 (1974).
17. U.S. Pat. 3,954,893 (May 4, 1976), G. J. O'Neill, R. S. Holdsworth, and C. W. Simons (to W. R. Grace and Co.).
18. W. R. Cullen and P. Singh, *Can. J. Chem.* **41,** 2397 (1963).
19. R. O'B. Watts and co-workers, *J. Fluorine Chem.* **3,** 7 (1973).
20. P. Tarrant, P. Johncock, and J. Savory, *J. Org. Chem.* **28,** 839 (1963).
21. J. D. Park, W. R. Lycan, and J. R. Lacher, *J. Am. Chem. Soc.* **73,** 711 (1951).
22. U.S. Pat. 2,793,202 (May 21, 1957), J. M. Hoyt (to M. W. Kellogg Co.).
23. Brit. Pat. 593,605 (Oct. 21, 1947), (to E. I. du Pont de Nemours & Co., Inc.).
24. S. Afr. Pat. 69 05,518 (Mar. 3, 1970), J. Kuhls, H. Hahn, and A. Steininger (to Farbwerke Hoechst A.-G.).
25. P. Tarrant, E. C. Stump, Jr., and C. D. Padgett, *Polym. Prepr. Am. Chem. Soc. Div. Polym. Chem.* **12,** 391 (1971).
26. R. E. A. Bear and E. E. Gilbert, *J. Fluorine Chem.* **4,** 107 (1974).
27. Ger. Pat. 2,235,885 (Feb. 7, 1974), J. Kuhls, H. Fitz, and P. Haasemann (to Farbwerke Hoechst A.-G.).
28. U.S. Pat. 3,668,262 (June 6, 1972), A. L. Dittman (to Halocarbon Products Corp.).

GARY ASTROLOGES
Halocarbon Products Corporation

POLY(VINYL FLUORIDE)

Monomer

Properties. Vinyl fluoride [*75-02-5*], VF, is a colorless gas with an ethereal odor. At atmospheric pressure it is insoluble in water and slightly soluble in acetone and ethanol. Other properties are listed in Table 1 (1).

Preparation. Vinyl fluoride was first prepared by reaction of zinc with 1,1-difluoro-2-bromoethane (2). Modern preparation involves the reaction of acetylene (qv) and hydrogen fluoride, either in a one-step or a two-step process. Direct addition of one mole of HF to one mole of acetylene in the presence of a mercury- or aluminum-based catalyst gives high yields of vinyl fluoride (3–5):

$$CH{\equiv}CH + HF \xrightarrow{Hg} CH_2{=}CHF$$

Ethylene was converted directly to vinyl fluoride by reaction with HF in the presence of a catalyst system consisting of carbon impregnated with Pd and $CuCl_2$. The HF and $CH_2{=}CH_2$ (2:1 mole ratio) reaction mixture containing 35% oxygen was passed over the catalyst at 240°C to give vinyl fluoride at 20% conversion of $CH_2{=}CH_2$ and 92% selectivity (6).

In the two-step process, 1,1-difluoroethane is formed by adding hydrogen fluoride to acetylene; dehydrofluorination then is effected by pyrolysis with an aluminum salt catalyst (7). Vinyl fluoride is purified by passing it through soda-lime towers to remove traces of hydrogen fluoride and by scrubbing it with ammoniacal CuCl to remove acetylene. Oxygen is removed by distillation.

Polymerization. The unreactivity of vinyl fluoride, relative to other monomers, is attributed to the electron-withdrawing effect of the fluorine atom (8). Vinyl fluoride was first polymerized in 1931 by heating at 67°C under 600 MPa (87,000 psi) for 16 h (9). High pressure (30–100 MPa or 4350–14,500 psi), moderate temperature (60–140°C) polymerizations are carried out with free-radical catalysts (10); catalysts, eg,

Table 1. Physical Properties of Vinyl Fluoride

Property	Value
molecular weight	46.04
boiling point, °C	−72.2
freezing point, °C	−160.5
critical temperature, °C	54.7
critical pressure, MPa[a]	5.1
critical density, g/cm^3	0.320
liquid density	
at 25°C, g/cm^3	0.620
vapor pressure at 21°C, MPa[a]	2.5
solubility in water	
at 80°C, g/100 g H_2O	
at 3.4 MPa[a]	0.94
at 6.9 MPa[a]	1.54

[a] To convert MPa to atm, divide by 0.101.

benzoyl peroxide, diisopropyl peroxydicarbonate, 2,2′-azobisisobutyramidine hydrochloride, or ammonium persulfate, have been used either in batch or in continuous systems (11–12).

The effect of temperature, pressure, impurities, media, telogens, and types of initiator on the properties of the polymer has been studied. The molecular weight of the benzoyl peroxide-initiated polymer ranged from 60,000 to 180,000 (11).

One successful polymerization of VF was carried out at low temperatures and pressures with the catalyst, triisobutyl boron, which was activated by molecular oxygen. Copolymers of VF also were prepared with vinylidene fluoride (VF_2) and tetrafluoroethylene (TFE). The reactivity ratios, r_1 (VF) = 5.5 ± 0.5, r_2 (VF_2) = 0.17 ± 0.03 and r_1r_2 = 0.93 ± 0.15, implied that the copolymerization proceeded with random arrangement of monomer units. With TFE, r_1 (VF) = 0.27 ± 0.03, r_2 (TFE) = 0.05 ± 0.02, and r_1r_2 = 0.011 ± 0.005, a polymer of alternating units formed (13).

The copolymerization studies were extended to a wide variety of other monomers using both Ziegler-Natta and oxygen-activated triisobutylborane catalysts (14); with both catalyst systems, polymerization proceeded by a free-radical mechanism (see Ziegler-Natta catalysts). Q (mean reactivity of VF in copolymerization) and e (polar nature of VF) values were calculated according to the Alfrey-Price scheme: $Q = 0.010 \pm 0.005$ and $e = -0.8 \pm 0.2$ (15). The very low value of Q indicates that the fluorine atom provides little resonance stabilization to the radical on the adjacent carbon; consequently, the VF radical is highly unstable and, therefore, reactive. The lack of resonance stabilization explains the monomer's tendency to undergo chain transfer reactions with solvents and the need for reactive radicals (eg, the alkyl radical) for low temperature initiation. The hypothesis of resonance instability is also consistent with VF's tendency to undergo head-to-head additions during chain propagation. The high e values indicate that the VF monomer is electron-rich immediately prior to attack.

A modified catalyst (prepared from vanadium oxytrichloride, triisobutylaluminum, and tetrahydrofuran) was found to be the most satisfactory Ziegler-Natta system; however, the reactivity was much reduced from that of vinyl chloride with this same catalyst (8).

The first reported photopolymerization resulted in a 36% yield after two days at 27°C under 254 nm irradiation (16). In another experiment, PVF was produced in over 90% yield by irradiating VF in the presence of benzoyl peroxide (17). More recently, copolymers of vinyl fluoride and vinyl acetate were produced photochemically (18). VF that was exposed to a radio-frequency glow discharge operating at 3.14 MHz formed a cross-linked polymer containing considerable amounts of oxygen and unsaturation (19).

Polymer

Poly(vinyl fluoride) [*24981-14-4*] (PVF) is tough, transparent, high melting, inert, and highly resistant to weathering. This combination of properties accounts for most of its uses.

Properties. PVF is a semicrystalline polymer with a planar zig-zag chain configuration (20). X-ray studies indicate a hexagonal unit cell with dimensions $a = b = 0.493$ nm and $c = 0.253$ nm (21). However, wide-line nmr and x-ray diffraction studies show that there are two monomer units in the unit cell with cell dimensions $a = 0.857$ nm,

b = 0.495 nm, and c = 0.252 nm (22) and, based on similarity to the phase I crystal form of poly(vinylidene fluoride), that the unit cell is orthorhombic (23) (Cm2m; see ref. 24). An orthorhombic cell hypothesis also is supported by x-ray diffraction pole-figure measurements (25).

The degree of crystallinity depends on the method of polymerization and the thermal history with reported values of PVF ranging from 20–60% (26) (see Fluorine compounds, organic—poly(vinylidene fluoride)). Despite this high crystallinity, the polymer exhibits considerable disorder. Configurational disorder occurs because about one of every six monomer units is reversed, giving head-to-head structures; the ratio increases as polymerization temperature increases (27). Isomeric disorder was revealed by x-ray and Raman studies in which fluorine atoms were placed randomly along the chain, resulting in an atactic polymer (22,28). Presence of syndiotactic segments of four or five monomer units is also indicated by ir analysis (29–30). The continued high crystallinity despite the imposed disorder can be accounted for by the similarity in size of hydrogen and fluorine atoms.

Polymerization temperature also affects melting point. When the temperature drops from 85 to 0°C, the melting point increases from 197–205°C to 222–235°C, and the crystallinity increases correspondingly, from 37 to 50% (14).

PVF has low solubility in all solvents studied at temperatures below 100°C (30). Polymer with greater solubility was prepared by adding 0.1% 2-propanol as a polymerization modifier; the polymer was dissolved in N,N-dimethylformamide, which was made 0.1 N in LiBr, for characterization. For a series of such polymers, $\overline{M}_n$ (average molecular weight by osmometry) ranged from 76,000–234,000; M_s (molecular weight by sedimentation velocity), 143,000 to 654,000; and, assuming $\overline{M}_w \cong M_s/0.9$, $\overline{M}_w/\overline{M}_n$ = 2.47–5.59. Chain branching can account for the deviation from a normal distribution where $\overline{M}_w/\overline{M}_n = 2.0$. The Mark-Houwink relation, $(\eta) = 6.52 \times 10^{-5} M_s{}^{0.80}$, was calculated, with the exponent, 0.80, being typical of extended, polar polymers in good solvents (31).

Semicrystalline polymers such as poly(vinyl fluoride), poly(vinylidene fluoride), polyethylene, and poly(ethylene oxide) exhibit multiple glass temperatures. These multiple transitions result in confusion and controversy in the literature. In an extensive study a transition in PVF, $T_g(U)$ (upper) occurring at 40–45°C, was ascribed to relaxation in the amorphous regions under restraint by crystallites; and a lower transition, $T_g(L)$ (lower), at −20°C, to relaxation in amorphous regions essentially free from restraint (32). For $T_g(U)$, ΔH = 120–240 kJ/mol (28.7–57.4 kcal/mol); for $T_g(L)$, ΔH = 160 kJ/mol (38.2 kcal/mol). Two other transitions occurring at 150°C and −80°C are ascribed, respectively, to premelting intracrystalline relaxation and short chain relaxation in the amorphous region. Transitions have been studied using argon diffusion (33) and hydrogen and deuterium diffusion (34), dilatometry (35), torsion pendulum (26,36), dynamic mechanical (32), and dielectric methods (37).

Pyro- and piezoelectric responses exhibited by PVF were reported (38). The high dipole moment and the fact that the dipoles may be aligned normal to the surface by poling (ie, applying an electric field across the film thickness while heating the film to 100°C and subsequently cooling it under an applied field) results in a high pyroelectric coefficient. Values of 1×10^{-9} C/(cm²·K) for cast film and 1.8×10^{-9} C/(cm²·K) for laser-evaporated film have been attained (39). A sensitive PVF pyroelectric detector with rapid response time has been reported (40).

PVF is transparent to radiation in the uv, visible, and near ir regions of the solar spectrum, transmitting 90% of the radiation at 350–2500 nm. It absorbs radiation of 7000–12,000 nm (41). When exposed to electron beam radiation of 10 MGy (10^9 rad), PVF becomes embrittled, but it resists breakdown at lower dosages. For instance, it retains its strength at 0.32 MGy (32×10^6 rad), whereas polytetrafluoroethylene is degraded at 0.02 MGy (2×10^6 rad) (42).

Physical, chemical, and electrical properties of poly(vinyl fluoride) film are listed in Tables 2, 3, and 4 (41,43–44) (see also Barrier polymers).

Manufacture. PVF is converted into thin film by a plasticized melt extrusion (45). Although there are no solvents for commercial PVF at temperatures below 100°C, there are latent solvents that solvate the polymer at elevated temperatures and allow the polymer particles to coalesce. Pigments, stabilizers, and plasticizers can be incorporated in the film by mixing them with the organosol of the polymer and the latent solvent prior to extrusion. After extrusion the solvent is evaporated and recovered. Extruded film may be biaxially oriented to varying degrees, thereby producing different levels of tensile, tear, and impact strength (46). Adherability of the film may be altered by exposing it to boron trifluoride gas, flame, electric discharge, activated gas plasma, alkali metal in liquid ammonia solution, or to diazo compounds followed

Table 2. Physical and Thermal Properties of Poly(vinyl Fluoride) Film

Property	Value	Test method
bursting strength, kPa[a]	131–482	Mullen, ASTM D 774
coefficient of friction (film/metal)	0.15–0.30	Instron
density, g/cm^3	1.38–1.57	weighed samples
impact strength, kJ/m[b]	10–22	DuPont pneumatic tester
refractive index, n_D	1.46	ASTM D 542; Abbe refractometer (30°C)
tear strength, kJ/m[b]		
propagated	4.6–39	Elmendorf
initial	174–239	ASTM 1004
tensile modulus, MPa[c]	1700–2600	ASTM D 882, Method A (100% min, 23°C)
ultimate tensile strength, MPa[c]	48–120	ASTM D 882, Method A (100% min, 23°C)
ultimate elongation, %	115–250	ASTM D 882, Method A (100% min, 23°C)
ultimate yield strength, MPa[c]	33–41	ASTM D 882, Method A (100% min, 23°C)
linear coefficient of expansion, cm/(cm·°C)	0.00005	air oven, 30 min
useful temperature range, °C		
continuous use	−72–107	
short cycle (1–2 h)	175	
zero strength, °C	260–300	hot bar
thermal conductivity, W/(m·K) (1°C/cm)		
−30°C	0.14	
60°C	0.17	
self-ignition temperature, °C	390	ASTM D 1929
solar energy transmittance, % (over range 350–2500 nm)	90	ASTM E 427-71, selected ordinates

[a] To convert kPa to psi, multiply by 0.145.

[b] To convert kJ/m to ft·lbf/in., divide by 0.0534 (see ASTM D 256).

[c] To convert MPa to psi, multiply by 145.

Table 3. Permeability of Poly(vinyl Fluoride) Film

Property	Value	Test method
moisture absorption, %	<0.5	water immersion (23°C)
moisture vapor transmission, $g/(m^2 \cdot d)$ at 7.0 kPa[a], 39.5°C[b]	40–50	ASTM E 96-58T
gas permeability, $cm^3/(m^2 \cdot d \cdot kPa)$[a] at 98 kPa[a], 23.5°C[b]		
carbon dioxide	1.7	ASTM D 1434
helium	22.9	ASTM D 1434
hydrogen	8.9	ASTM D 1434
nitrogen	0.04	ASTM D 1434
oxygen	0.5	ASTM D 1434
vapor permeability, $g/(m^2 \cdot d)$ (at partial pressure of vapor at given temperature)[b]		
acetic acid at 23.5°C	10	ASTM E 96, modified
acetone at 23.5°C	2400	ASTM E 96, modified
benzene at 23.5°C	22	ASTM E 96, modified
carbon tetrachloride at 23.5°C	12	ASTM E 96, modified
ethyl acetate at 23.5°C	240	ASTM E 96, modified
ethyl alcohol at 23.5°C	8	ASTM E 96, modified
hexane at 23.5°C	13	ASTM E 96, modified
water at 39.5°C	43	DuPont permeability method

[a] To convert kPa to mm Hg, multiply by 7.5.
[b] Measurements made on films of nominal 25 μm thickness.

Table 4. Electrical Properties of Poly(vinyl Fluoride) Film

Property	Value	Test method
corona endurance, h at 60 Hz, 40 V/μm	2.5–6.0	ASTM D 2275
dielectric constant at 1 kHz at 23°C	8.5–9.9	ASTM D 150
dielectric strength, kV/μm	0.13	ASTM D 150
dissipation factor, %		
1 MHz at 23°C	1.4–1.6	ASTM D 150
1 MHz at 70°C	1.7–2.7	ASTM D 150
10 kHz at 23°C	3.4–4.2	ASTM D 150
10 kHz at 70°C	1.6–2.1	ASTM D 150
volume resistivity for transparent film, GΩ·m		
23°C	400	ASTM D 257
100°C	0.2	ASTM D 257

by uv exposure (47–52). Another method for making an adherable sheet is to coat the sheet, which still contains a small amount (0.5–0.6%) of latent solvent, with a polyurethane or a chlorinated PVF adhesive, and then to remove the remainder of the solvent (53–54).

PVF homopolymer and copolymer resins can also be applied directly to substrates with solvent-based systems, water dispersions, or electrostatic powder coatings. Water dispersions have been prepared by dispersing the polymer in water which contains polyalkylene glycols (55). Because of the high temperature required for curing PVF coatings, thermal stabilizers (see Heat stabilizers), eg, basic metal oxides (56), poly(vinylpyridine) compounds (57), and pentaerythritol (58), are required.

Economic Aspects. Poly(vinyl fluoride) is available from E. I. du Pont de Nemours & Co., Inc., both as transparent and pigmented films under the trademark Tedlar PVF film. Films are available in several tensile modifications and thicknesses, with either adherable or nonadherable surfaces. Current film prices are \$13–24/kg (\$6.6–11/kg in 1969), and prices for specially tailored films are significantly higher.

Health and Safety. Exposure of 200,000 ppm (20%) of the vinyl fluoride monomer for 2 h had no effect on rats (59). After 12.5 h at 800,000 ppm (80%), the rats were intoxicated but recovered fully in fresh air. The monomer is flammable in air between the limits of 2.6–22.0% by volume. Minimum ignition temperature for mixtures of air and VF is 400°C. Vinyl fluoride is inhibited against spontaneous polymerization by adding 0.2% of a mixture of terpenes; the DOT has classified the inhibited monomer as a flammable gas.

The self-ignition temperature of PVF film is 390°C. The limiting oxygen index (LOI), which is 22.6% for PVF (60), is raised to 30% for antimony-modified film (61) (see Flame retardants). The polymer's thermal degradation products are hydrogen fluoride and a mixture of aromatic and aliphatic hydrocarbons (62). The relative toxicity hazards of thermally degraded poly(vinyl fluoride) and other polymers have been reported (63).

Uses. Pigmented films, available in white and in a variety of colors, are used as surfacing for aluminum and galvanized steel for exterior residential and industrial building siding. The film first is laminated to flat continuous sheets of metal with specially developed adhesives and subsequently is formed to desired shape. PVF provides resistance to weather, chemicals, staining, and abrasion.

PVF is also laminated to cellulosics, eg, hardboard and paper; plastics, eg, vinyl, polystyrene, polyurethane; rubbers; and to resin-impregnated felt (qv). Applications of these laminations are board-and-batten building construction, stain-resistant wall coverings and aircraft cabin interiors, exterior automobile trim, pipe covering, and flexible duct liners (see Laminated and reinforced plastics; Plastic building products). Fiberglass-reinforced polyester panels that are surfaced with PVF resist erosion and glass-fiber blooming (lifting).

Nonadhering film is used as release sheets in plastic processing, particularly in high temperature pressing of epoxy resins (qv) for circuit boards and aerospace parts (see Abherents).

The combination of PVF's light weight and excellent weatherability and its light transmittance and resistance to breakage has led to extensive evaluation of PVF film as a cover for flat-plate solar energy (qv) collectors. The suitability of PVF's solar transmittance capability for efficient collectors was described (64); also the role that infrared transmittance has in panel performance has been analyzed (65) (see Film and sheeting materials).

PVF resin coatings on the interior of rigid metal pails and drums provide excellent resistance to corrosive chemicals (66); improved corrosion-resistant coatings with multiple layers of PVF have been described (56) (see Abherents; Coatings, resistant).

BIBLIOGRAPHY

"Poly(Vinyl Fluoride)" in *ECT* 2nd ed., under "Fluorine Compounds, Organic," Vol. 9, pp. 835–840, by L. E. Wolinski, E. I. du Pont de Nemours & Co., Inc.

1. *Technical Information Bulletin TD-39,* E. I. du Pont de Nemours & Co., Inc., Wilmington, Del., 1961.
2. F. Swarts, *Bull. Acad. Roy. Med. Belg.* **7,** 383 (1901); F. Swarts, *J. Chem. Soc. Abstr.* **82,** 129 (1902).
3. U.S. Pat. 2,419,010 (April 15, 1947), D. D. Coffman and T. A. Ford (to E. I. du Pont de Nemours & Co., Inc.).
4. U.S. Pat. 3,178,483 (April 13, 1965), F. J. Christoph, Jr., and G. Teufer (to E. I. du Pont de Nemours & Co., Inc.).
5. U.S. Pat. 3,607,955 (Sept. 21, 1971), L. E. Gardner (to Phillips Petroleum Company).
6. Jpn. Kokai 77 122,310 (Oct. 14, 1977), T. Kuroda and T. Yamamoto (to Onoda Cement Co., Ltd.).
7. U.S. Pat. 2,674,632 (April 1, 1954), B. F. Skiles (to E. I. du Pont de Nemours & Co., Inc.).
8. R. N. Haszeldine, T. G. Hyde, and P. J. T. Tait, *Polymer* **14,** 221 (1973).
9. H. W. Starkweather, *J. Am. Chem. Soc.* **56,** 1870 (1934).
10. U.S. Pat. 3,464,963 (Sept. 2, 1969), J. L. Hecht and C. T. Hughes (to E. I. du Pont de Nemours & Co., Inc.).
11. G. H. Kalb and co-workers, *J. Appl. Polym. Sci.* **4,** 55 (1960).
12. Jpn. Kokai 73 32,579 (Oct. 6, 1973), T. Nishida and Y. Iikubo (to Onoda Cement Co., Ltd.).
13. G. Natta and co-workers, *J. Polym. Sci. A* **3,** 4263 (1965).
14. D. Sianesi and G. Caporiccio, *J. Polym. Sci. A-1* **6,** 335 (1968).
15. T. Alfrey and C. C. Price, *J. Polym. Sci.* **2,** 101 (1947).
16. A. E. Newkirk, *J. Am. Chem. Soc.* **68,** 2467 (1946).
17. D. Raucher and M. Levy, *J. Polym. Sci. Polym. Chem. Ed.* **13,** 1339 (1975).
18. A. Korin, M. Levy, and D. Vofsi, *Polym. Prepr. Am. Chem. Soc. Div. Polym. Chem.* **20**(1), 672 (April 1979).
19. A. R. Westwood, *Eur. Polym. J.* **7,** 377 (1971).
20. G. Natta, *Makromol. Chem.* **35,** 94 (1960).
21. R. C. Golike, *J. Polym. Sci.* **42,** 583 (1960).
22. G. Natta, I. W. Bassi, and G. Allegra, *Atti. Accad. Naz. Lincei Cl. Sci. Fis. Mat. Nat. Rend.* **31,** 350 (1961).
23. J. B. Lando, H. G. Olf, and A. Peterlin, *J. Polym. Sci. A-1* **4,** 941 (1966).
24. *International Tables for X-ray Crystallography,* Vols. I, II, and III, Kynock Press, Birmingham, England, 1962.
25. R. Large, W. F. Maddams, and J. E. Preedy, *J. Appl. Polym. Sci.* **22,** 3031 (1978).
26. M. Goerlitz and co-workers, *Angew. Makromol. Chem.* **29/30**(371), 137 (1973).
27. C. W. Wilson III and E. R. Santee, Jr., *J. Polym. Sci.* **C8,** 97 (1965).
28. J. L. Koening and F. J. Boerio, *Makromol. Chemie* **125,** 302 (1969).
29. G. Zerbi and G. Cortelli, *Spectrochim. Acta* **26,** 733 (1970).
30. J. L. Koenig and J. J. Manion, *J. Polym. Sci. A-2* **4,** 401 (1966).
31. M. L. Wallach and M. A. Kabayama, *J. Polym. Sci. A-1* **4,** 2667 (1966).
32. R. F. Boyer, *J. Polym. Sci. Polym. Symp.* **50,** 189 (1975).
33. M. Fujii, V. Stannett, and H. B. Hopfenberg, *J. Macromol. Sci. Phys.* **B15,** 421 (1978).
34. K. D. Ziegel, H. K. Frensdorff, and D. E. Blair, *J. Polym. Sci. A-2* **7,** 809 (1969).
35. J. B. Enns and R. Simha, *J. Macromol. Sci. Phys.* **B13,** 11 (1977).
36. E. Sacher, *J. Polym. Sci. A-2* **6,** 1813 (1968).
37. S. Osaki, J. Uemura, and Y. Ishida, *J. Polym. Sci. A-2* **9,** 585 (1971).
38. H. Kawai, *Jpn. J. Appl. Phys.* **8,** 975 (1969).
39. A. W. Stephens and co-workers, *Thin Solid Films* **24,** 361 (1974).
40. J. Geist, H. J. Dewey, and M. A. Lind, *Appl. Phys. Lett.* **28,** 171 (1976).
41. *Technical Information Bulletin TD-31,* E. I. du Pont de Nemours & Co., Inc., Wilmington, Del., 1979.
42. R. Timmerman and W. Greyson, *J. Appl. Polym. Sci.* **6,** 456 (1962).
43. F. S. Cohen and P. Kraft in N. M. Bikales, ed., *Encyclopedia of Polymer Science and Technology,* Vol. 14, Wiley-Interscience, New York, 1971, pp. 522–540.
44. *Technical Information Bulletin TD-1A,* E. I. du Pont de Nemours & Co., Inc., Wilmington, Del., 1974.
45. U.S. Pat. 2,953,818 (Sept. 27, 1960), L. R. Bartron (to E. I. du Pont de Nemours & Co., Inc.).
46. U.S. Pat. 3,139,470 (June 30, 1964), R. S. Prengle and R. L. Richards, Jr. (to E. I. du Pont de Nemours & Co., Inc.).
47. U.S. Pat. 3,122,445 (Feb. 25, 1964), R. O. Osborn (to E. I. du Pont de Nemours & Co., Inc.).

48. U.S. Pat. 3,153,683 (Oct. 20, 1964), W. L. Bryan and D. E. Swarts (to E. I. du Pont de Nemours & Co., Inc.).
49. U.S. Pat. 3,274,088 (Sept. 20, 1966), L. E. Wolinski (to E. I. du Pont de Nemours & Co., Inc.).
50. C. A. L. Westerdahl and co-workers, *Picatinny Arsenal Technical Report No. 4279,* Dover, N.J., 1972.
51. Ger. Offen. 1,017,308 (Oct. 10, 1957), M. Reiher, D. Schleede, and K. Weissermel (to Farbwerke Hoechst).
52. U.S. Pat. 3,376,278 (April 2, 1968), A. W. Morgan and J. S. Swenson (to 3M Company).
53. U.S. Pat. 3,723,171 (March 27, 1973), O. Fuchs (to Dynamit Nobel A.G.).
54. U.S. Pat. 3,880,690 (April 29, 1975), O. Fuchs (to Dynamit Nobel A.G.).
55. Jpn. Kokai 77 50,323 (April 22, 1977), K. Nakano (to Onoda Cement Co., Ltd.).
56. U.S. Pat. 3,655,426 (April 11, 1972), O. Fuchs (to Dynamit Nobel A.G.).
57. U.S. Pat. 3,627,854 (Dec. 14, 1971) L. E. Scoggins (to Phillips Petroleum Co.).
58. Brit. Pat. 1,366,145 (Sept. 11, 1974), G. Heuser, G. Weisgerber, and E. Behr (to Dynamit Nobel A.G.).
59. D. Lester and L. A. Greenburg, *Arch. Ind. Hyg. Occ. Med.* **2,** 335 (1950).
60. C. P. Fennimore and F. J. Martin, *Combust. Flame* **10,** 135 (1966).
61. U.S. Pat. 3,963,672 (June 15, 1976), D. E. Brasure (to E. I. du Pont de Nemours & Co., Inc.).
62. I. N. Einhorn and co-workers, *FRC/UU-41, ETEC 75-022, NASA Contract No. NAS2-8244,* Final Report, Dec. 14, 1974.
63. C. H. Hilado and co-workers, *J. Combust. Toxicol.* **3,** 270 (1976).
64. A. Whillier, *Sol. Energy* **7,** 148 (1963).
65. S. Catalonotti and co-workers, *Sol. Energy* **17,** 83 (1975).
66. R. W. Miller and T. Podnar, Jr., *Package Eng.* **12,** 94 (1967).

DONALD E. BRASURE
E. I. du Pont de Nemours & Co., Inc.

POLY(VINYLIDENE FLUORIDE)

Poly(vinylidene fluoride) [*24937-79-9*] is the polymer of 1,1-difluoroethylene—vinylidene fluoride. Poly(vinylidene fluoride), PVF_2 or (preferably) PVDF, is a semicrystalline polymer containing 59.4% fluorine. The symmetrical arrangement of the hydrogen and fluorine atoms in the chain contributes to the unique polarity that influences the polymer's dielectric properties and solubility. The polymer is readily melt-fabricated by standard methods of molding and extrusion or solvated by polar solvents to form coatings. Poly(vinylidene fluoride) displays the stability characteristic of fluorocarbon polymers when exposed to harsh thermal, chemical, and uv environments. PVDF has many applications wherever high performance polymers are required, eg, construction or coating materials.

Monomer (VDF)

Vinylidene fluoride [*75-38-7*] or 1,1-difluoroethylene, VF_2 or (preferably) VDF, is a colorless and nearly odorless gas. Acute exposure to the monomer has not produced liver damage in rats (1). Some of the properties of the monomer are listed in Table 1; comprehensive thermodynamic properties have been reported (2,4).

Table 1. Properties of Vinylidene Fluoride

Property	Value	Ref.
molecular weight	64.038	
boiling point, °C	−85.7	2
freezing point, °C	−144	
critical temperature, °C	30.1	2
critical pressure, MPa[a]	4.433	2
critical density, g/cm^3	0.417	2
explosive limits, vol % in air	5.8–20.3	3
heat of formation, ΔH_f° (25°C), kJ/mol[b]	−345.2	4
heat of polymerization, ΔH_f°poly (25°C), kJ/mol[b]	−128.8	4
toxicity (rat, inhalation) LC_{50}, ppm	128,000	5

[a] To convert MPa to psi, multiply by 145.
[b] To convert kJ to kcal, divide by 4.184.

Preparation. Vinylidene fluoride monomer is prepared by dehydrohalogenation of 1-chloro-1,1-difluoroethane (6–20), 1-bromo-1,1-difluoroethane (21), or 1,1,1-trifluoroethane (22–25), or by dehalogenation of 1,2-dichloro-1,1-difluoroethane (26).

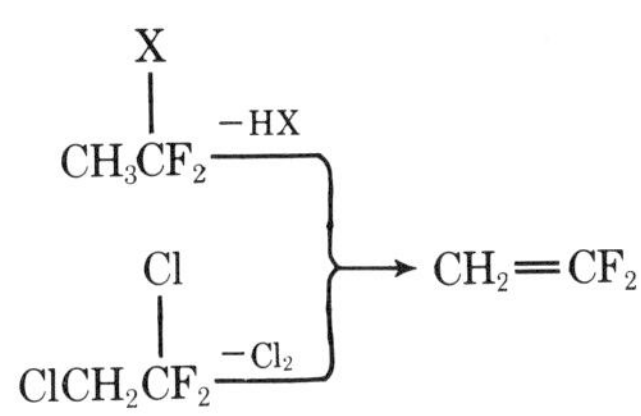

The preferred monomer precursor in commercial operations, 1-chloro-1,1-difluoroethane (also known as refrigerant 142b) is synthesized from 1,1-difluoroethane (10,14,17–19), vinylidene chloride (15), and 1,1,1-trichloroethane:

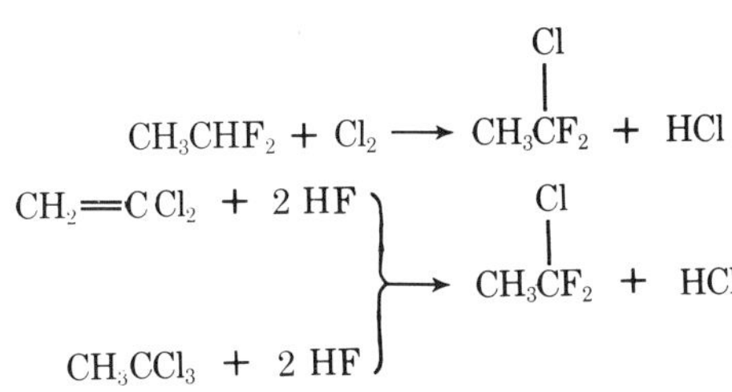

The monomer also can be prepared by pyrolytic dehydrofluorination of trifluoromethane, trifluorobromomethane, or trifluorochloromethane in the presence of a catalyst and either methane or ethylene (27–29). With methanol as the carbene source, vinylidene fluoride can be prepared continuously from difluorochloromethane (30). Co-pyrolysis of chlorodifluoromethane or dichlorodifluoromethane and either methane or methyl chloride yields vinylidene fluoride (31–33) as does electroreduction of 1,1-difluoro-1,2-dibromoethane (34). Vinylidene fluoride also can be obtained from 1,1,2,2-tetrafluorocyclobutane (35) or 1,1,2-trifluoro-2(trifluoromethyl)cyclobutane (36).

Vinylidene fluoride is usually stored and shipped without a polymerization inhibitor; however, if desired, terpenes can be added to inhibit polymerization.

Polymerization. The first successful radical-initiated, aqueous polymerization of vinylidene fluoride was reported in 1948 (37). Since then numerous polymerization processes have been disclosed using radical initiators at various temperatures and pressures in aqueous and nonaqueous mediums. Commercially significant are polymerizations that involve radical-initiated systems in emulsion or suspension.

Polymerization of vinylidene fluoride by emulsion requires the use of fluorinated surfactants to minimize chain transfer during the polymerization. The polymerization can be initiated both with water-soluble radical generators, eg, persulfate salts (38–39), hydrogen peroxide (40), alkyl hydroxy peroxide (41–43), and azo compounds (44), and with monomer-soluble initiators, eg, acyl peroxides (45), alkyl percarbonates (46–47), and alkyl peroxides (48–49) (see Initiators). Polymerization temperatures vary from 0° to 150°C, depending upon the initiator used, and pressures vary from 2.0 to 3.4 MPa (20–34 atm), depending upon the desired polymerization rate or molecular weight. Chain-transfer agents, eg, ketones and halogenated solvents, are often employed to regulate molecular chain length.

Suspension polymerization of vinylidene fluoride is carried out in an aqueous recipe using monomer-soluble, free-radical polymerization initiators similar to those used for the emulsion polymerization. Usually a water-soluble dispersing agent is added, which is usually a cellulose derivative, eg, methyl cellulose, and some solvents that may act either as reaction activators or chain-transfer agents (50–54). Solution polymerization of vinylidene fluoride in various solvents and initiation by radical initiators also has been reported (55).

Alkyl boron compounds activated by oxygen or peroxides initiate vinylidene fluoride polymerization in water either in the presence or absence of surface-active agents or in organic solvents from −50° to +50°C (56–57). Alkyl hyponitrites also initiate vinylidene fluoride polymerization (58).

Several investigators reported gamma radiation (^{60}Co) initiated polymerization of vinylidene fluoride in the gaseous and liquid phases (59–69). In addition the monomer can be polymerized *in situ* using plasma discharge (70). Coordination catalysts of the Ziegler-Natta type can produce poly(vinylidene fluoride), but yields are low (71–73).

Polymer (PVDF)

PVDF was introduced in 1961 and has been commercially available since 1965. Producers of PVDF and their trademarks are listed in Table 2.

Physical Properties. Poly(vinylidene fluoride) is a semicrystalline polymer whose melting point depends on the method of polymerization and/or its crystallization history (64,74–77). It has high levels of mechanical and impact strength, abrasion resistance, thermal stability, resistance to most chemicals and solvents, resistance to uv and nuclear radiation, and an unusually high dielectric constant. In addition, PVDF displays outstanding resistance to deformation under load (75) as well as exceptional ability to withstand fatigue-crack propagation (78–79). It possesses useful mechanical properties from −60° to 150°C. Physical property data of PVDF are listed in Table 3.

Table 2. Poly(vinylidene Fluoride) Producers

Manufacturer	Trade name
Dynamit Nobel A.G. (Troisdorf, FRG)	Dyflor
Kureha Chemical Co., Ltd. (Tokyo, Japan)	KF
Pennwalt Corporation (Philadelphia, Pa.)	Kynar
Produits Chimiques Ugine Kuhlmann (Paris, Fr.)	Foraflon
Solvay & Cie (Brussels, Belg.)	Solef
Suddeutsche Kalkstickstoff-Werke, A.G. (Trostberg, FRG)	Vidar

Table 3. Typical Properties of Poly(vinylidene Fluoride)

Property	Value	ASTM method
specific gravity	1.75–1.80	D 792
refractive index, n_D^{25}	1.42	D 542
melting point (T_m), °C	154–184	D 3418
glass transition (T_g), °C	−40	dilatometry
water absorption, %	0.04–0.06	D 570
tensile strength at yield, MPa[a]	36–59	D 638
tensile strength at break, MPa[a]	36–56	D 638
elongation at break, %	25–500	D 638
modulus of elasticity, MPa[a]		
in tension	1340–2000	D 638
in flexure	1200–2200	D 790
in compression	850–1275	D 695
in torsion	550–800	D 1043
Izod impact (notched), kJ/m[b]	150–530	D 256
Izod impact (unnotched), kJ/m[b]	1700–3100	D 256
hardness, Shore D	70–82	D2240
thermal conductivity, W/(m·K)	0.1–0.13	C 177
specific heat, J/(g·K)[c]	1.26–1.42	DSC
coefficient of linear expansion, K^{-1}	$7.9–15.7 \times 10^{-5}$	D 696
limiting oxygen index, %	43.6	D 2863
deflection temperature (1.82 MPa[a]), °C	112–150	D 648
low temperature embrittlement, °C	−62 to −64	D 746
Tabor abrasion, mg/kHz	7.0–10.0	Wheel CS-17, 1 kg
sand abrasion, m^3/mm	4.0	D 968

[a] To convert MPa to psi, multiply by 145.
[b] To convert kJ/m to ft·lbf/in., divide by 0.0534.
[c] To convert J to cal, divide by 4.184.

The properties of the polymer depend on molecular weight, molecular weight distribution, extent of irregularities along the polymer chain, and crystalline form; the first three factors are controlled by the polymerization process. Although vinylidene fluoride predominantly adds to the growing chain in head-to-tail sequence, 4–6% of monomer units add head-to-head and tail-to-tail, depending on the polymerization process (74,80).

Polymorphism of PVDF has been the subject of many investigations. The polymer crystallizes in three crystal forms: form I (β), form II (α), and form III (γ). The α form has a monoclinic crystal with a TGTG molecular conformation. The β form is an orthorhombic crystal and has a TT or planar zig-zag conformation. The γ form has a monoclinic crystal and planar zig-zag TT conformation. The ir and Raman spectra in the region of 4000–30 cm^{-1} of the α, β, and γ forms have been measured (81–84). The degree of crystallinity of PVDF ranges from 35–65% and depends on thermal history (85–89). The heat of fusion (ΔH_f) of the α form is 5.966 kJ/mol (1.426 kcal/mol) (87) and can be used to evaluate the degree of crystallinity by the differential scanning calorimetry (dsc) method.

Solution properties of PVDF were investigated and the Mark-Houwink relationship at 25°C in dimethylacetamide was derived (90):

$$\text{dL/g } (10^4 \text{ P}) = 1.93\, M^{0.677}$$

Three distinct relaxation processes of PVDF were identified as α, β, and γ and have been shown to be frequency dependent (91).

Polymorphism of PVDF and the two distinct dipole groups, CF_2 and CH_2, contribute to the polymer's exceptional dielectric properties. It has a very high dielectric constant which is a function of frequency, temperature, thermal history, and orientation (92–94). Electrical properties are listed in Table 4. Associated with the dielectric properties and the polymorphism is the extraordinarily large piezoelectric activity, which was discovered in 1969 (97–98). Many scientific and applied studies have been published on the piezoelectric and pyroelectric properties of poly(vinylidene fluoride), which are reviewed in recent articles (96).

Poly(vinylidene fluoride) cross-links when subjected to electron beam radiation without detrimental effects on physical properties (99). A similar discovery was made using high energy ionizing radiation from a ^{60}Co source (100). In general, PVDF responds readily to radiation (101–104). Addition of radiation sensitizers enhances

Table 4. Electrical Properties of Poly(vinylidene Fluoride)

Property	Value	ASTM method	Ref.
volume resistivity, Ω·cm	2×10^{14}	D 257	95
surface arc resistance, s	50–60	D 495	95
dielectric strength, MV/m			
short time 500 V/s		D 149	95
3.175 mm thickness	10.2		
0.203 mm thickness	50.4		
step by step, kV steps	37.4		
dielectric constant		D 150	93
60 Hz	8.40		
kHz	7.46		
MHz	6.10		
GHz	2.98		
dissipation factor		D 150	93
60 Hz	0.049		
kHz	0.019		
MHz	0.156		
GHz	0.110		
piezoelectric strain constant, pm/V[a]	20–50		96
piezoelectric stress constant (pressure), V/(m·Pa)[b]	0.23		96
pyroelectric coefficient, μC/(m^2·K)	40		96

[a] m/V = C/N.
[b] To convert Pa to mm Hg, divide by 133.3.

cross-linking efficiency and appreciably increases the high temperature modulus of PVDF without substantial degradation (105–107).

Several studies have shown that PVDF forms solid solutions or alloys with acrylic polymers (108–113). Recently, solution-cast blends of PVDF and poly(vinyl acetate) showed miscibility of the two polymers (114).

Fabrication and Processing. Poly(vinylidene fluoride) is available in a wide range of melt viscosities as required by the fabricator. The polymer is readily molded in conventional compression, transfer, and injection equipment (115–116). Because PVDF is a highly crystalline polymer, it shows a relatively high mold shrinkage of ca 3%. For optimum dimensional stability, post annealing of molded parts at 140–150°C is recommended (117).

Smooth PVDF profiles of all types—rod, sheet, film, pipe, tubing, monofilament, and wire insulation—can be melt extruded at satisfactory rates. In both molding and extrusion operations, care must be exercised to eliminate hang-up zones in the tooling where molten (260°C) resin can collect and thermally decompose with time (95). Water quenching is practiced for wire insulation, tubing, and pipe, whereas sheet and film that is extruded from slit dies are crystallized on polished steel rolls operating at 65–150°C. Woven glass fabrics can be continuously laminated to PVDF sheets by having the fabric impinge upon the melt just before it makes contact with the quench rolls (118). Extruded monofilament is postoriented at elevated temperatures using draw ratios of 3:1 to 5:1 and draw temperatures of 140–160°C to produce high strength filament having tenacities of 350–440 mN/tex (4–5 gf/den). In a somewhat similar fashion, extruded sheet can be uniaxially or biaxially oriented to various thicknesses in order to vastly increase mechanical strength, specular light transmission, and piezoelectric activity (119–120).

Parts fabricated from PVDF can be effectively cross-linked by treatment with high energy radiation to give the polymer unique properties; this method is applied to produce high temperature wire insulation and heat shrinkable tubing (121–122). PVDF also has been fabricated into porous structures by sintering very fine granules under controlled conditions (123).

Emulsion polymerization of PVDF produces a latex containing small spheres which have an average diameter of 0.2–0.5 μm. The small size in conjunction with the polymer's susceptibility to solvation by many polar solvents are important to the preparation and application of organosol coatings. Such organic dispersions are compounded with stable pigments and acrylic resin modifiers to form a wide variety of decorative and protective metal finishes (124–126). When applied by spray or roller to primed steel or aluminum surfaces and oven-fused at 230–260°C, the highly solvated particles coalesce to form continuous films that adhere tightly to the metal substrates. Applied as very thin films through high-speed continuous coil lines or by electrostatic spraying, these decorative finishes provide color durability and film integrity to wall panels and other metal building products. Other PVDF organic dispersion systems are compounded for spray applications of relatively thick coatings to protect metals that are exposed to highly corrosive environments. Extensions of dispersion technology include cast film by which a pigmented dispersion is continuously cast and fused onto release paper. Films treated with pressure-sensitive adhesives are used for exterior decorative laminates (127). PVDF dispersions are also used as saturants for glass and carbon cloth. The webs can be used individually or laminated together to form dimensionally stable, corrosion-resistant linings for steel or reinforced polyester vessels, tanks, and pipes (128–129). Low melt viscosity grades of PVDF are cryogenically ground to fine powders for applications as powder coatings (see Powder coating) (130–131).

Health and Safety. Based upon toxicity studies (including acute oral, subacute contact, and inhalation tests with rats), PVDF is considered nontoxic (132). A number of biological studies (including implantation, tissue culture, and systemic toxicity tests in various animals) show no adverse effects in terms of blood pressure, respiration, and hemolysis (133). Less than 1 ppm of monomer has been detected in PVDF powder (134). PVDF may be used in articles intended for repeated use in contact with food (135).

Thermal decomposition studies show that the polymer degrades rapidly in air at 375°C; the emission product is essentially anhydrous hydrogen fluoride (136). Certain plastic fillers, eg, silica, titanium dioxide, and antimony oxide, are known to catalyze the thermal decomposition of PVDF. Because HF is relatively toxic (TLV of 3 ppm), all precautions must be taken to prevent inhalation and physical contact in the event inadvertent decomposition takes place (137).

Uses. PVDF is the basic raw material used in many diverse industrial products that require high mechanical strength and resistance to severe environmental stresses.

Fluid handling systems in the chemical process and related fields have been a growth market for PVDF products, eg, solid and lined pipes, valves, pumps, tower packing, woven filter cloth, and tank linings (138–139). Extruded monofilament woven into coarse fabric is used widely for drum filtration during bleaching of wood pulp with chemicals, eg, sodium hypochlorite and caustic soda. Like other fluorocarbon polymers, PVDF is used as a binder for asbestos fiber used as a diaphragm in the electrolysis of brine to produce chlorine and caustic soda (140–141).

In the electronics field, PVDF insulated wire is used to connect computer panel boards and telecommunication systems (142). Other important wire constructions include irradiated (cross-linked) PVDF jackets for industrial control wiring and temperature self-limiting tracing tapes used for controlling the temperature of process equipment (143). Extruded and irradiated heat-shrinkable tubing is used to make termination devices for aircraft and electronic equipment (see Insulation, electric).

Pigmented thin film that is cast from PVDF dispersions has been specified for body trim by a major automobile manufacturer (144). Decorative thin film laminates are being evaluated for aircraft interiors because of the laminates' high flame resistance and low smoke generation (145). When PVDF film is stretched under controlled conditions, the molecular structure changes from the less polar α form to the highly polar β form. This polarity can be made permanent and is the basis for films having extremely high piezo- and pyroelectric activities. When subjected to a high dielectric field (poled) and metallized, the film becomes a very sensitive transducer, an element used to convert energy; current applications include pyroelectric detectors and audio devices, eg, stereo speakers and headphones. Many of the applications that are under study include pressure sensors for printing presses, hydrophones for long-range tracking in ocean depths, and many medical devices, eg, detectors for heart output and for ultrasonic imaging; the latter involve monitoring of mechanical waves within the body (96) (see Ultrasonics).

PVDF is used in large quantities as a base for long-lasting decorative finishes for architectural metals, eg, aluminum and galvanized steel. Accelerated weathering and field exposure tests along with case histories of buildings erected over the last twelve years show these finishes to be extremely durable. They are applied in a wide range of colors on exterior wall panels, roofing shingles, and aluminum extrusions to be used on industrial, commercial, monumental, and—to lesser extent—residential buildings, eg, power plants, schools, airport hangars, department stores, office buildings, and sports stadiums (146–147).

BIBLIOGRAPHY

"Poly(vinylidene Fluoride) under "Fluorine Compounds, Organic" in *ECT* 2nd ed., Vol. 9, pp. 840–847, by W. S. Barnhart and N. T. Hall, Pennsalt Chemicals Corporation.

1. D. Lester and L. A. Greenberg, *Arch. Ind. Hyg. Occup. Med.* **2,** 235 (1950).
2. W. H. Mears and co-workers, *Ind. Eng. Chem.* **47,** 1449 (1955).
3. A. N. Baratov and V. M. Kucher, *Zh. Prikl. Khim.* **38,** 1068 (1965).
4. D. R. Stull, E. F. Westrum, and G. C. Sinke, *The Chemical Thermodynamics of Organic Compounds,* John Wiley & Sons, Inc., New York, 1969, p. 502.
5. C. P. Carpenter, H. F. Smyth, and U. C. Pozzani, *J. Ind. Hyg. Toxicol.* **31,** 343 (1949).
6. U.S. Pat. 2,551,573 (May 8, 1951), F. B. Downing, A. F. Benning, and R. C. McHarness (to E. I. du Pont de Nemours & Co., Inc.).
7. U.S. Pat. 2,628,989 (Feb. 17, 1953), C. B. Miller (to Allied Chemical Corp.).
8. U.S. Pat. 2,774,799 (Dec. 18, 1956), R. M. Mantell and W. S. Barnhart (to M. W. Kellogg Co.).
9. U.S. Pat. 3,246,041 (Apr. 12, 1966), M. E. Miville and J. J. Earley (to Pennwalt Corporation).
10. U.S.S.R. Pat. 216,699 (Apr. 26, 1968), B. P. Zverev and co-workers.
11. Jpn. Pat. 68 11,202 (May 11, 1968), (to Kureha Chemical Co., Ltd.).
12. U.S. Pat. 3,600,450 (Aug. 17, 1971), F. Kaess, K. Lienhard, and H. Michaud (to Suddeutsche Kalkstickstoffwerke, A.G.).
13. Ger. Offen. 2,044,370 (Mar. 23, 1972), J. N. Meussdoerffer and H. Niederpruem (to Farbenfabriken Bayer, A.G.).
14. Belg. Pat. 788,277 (Dec. 18, 1972); Ger. Offen. 2,145,975 (Mar. 22, 1973); Br. Pat. 1,387,737 (Mar. 19, 1975), (to Kali Chemie, A.G.).
15. U.S. Pat. 3,723,549 (Mar. 27, 1973), F. Kaess, K. Lienhard, and H. Michaud (to Suddeutsche Kalkstickstoff-Werke, A.G.).
16. Ger. Offen. 1,959,343 (Aug. 23, 1973), (to Vulcan Materials).
17. U.S.S.R. Pat. 380,631 (Aug. 28, 1973), V. A. Korinko and co-workers.
18. Belg. Pat. 844,822 (Dec. 1, 1976), (to Solvay & Cie).
19. Ger. Offens. 2,625,470 (Dec. 15, 1977), 2,659,712 (July 6, 1978), N. Schultz, P. Martens, and H-J. Vahlensieck (to Dynamit Nobel A.G.).
20. U.S. Pat. 4,053,529 (Oct. 11, 1977), G. Martens (to Solvay & Cie).
21. Fr. Pat. 1,337,360 (Sept. 13, 1963), (to Produits Chimiques Pechiney Saint-Gobain).
22. Fr. Pat. 1,330,146 (June 21, 1963), A. E. Pavlath and F. H. Walker (to Stauffer Chemical Co.).
23. U.S. Pat. 3,189,356 (June 8, 1965), M. Hauptschein and A. H. Fainberg (to Pennwalt Corporation).
24. Jpn. Pat. 68 29,126 (Dec. 13, 1968), H. Ukihashi and M. Ichimura (to Asahi Glass Co., Ltd.).
25. E. Tschuikow-Roux and W. J. Quiring, *J. Phys. Chem.* **75,** 295 (1971).
26. U.S. Pat. 2,734,000 (Feb. 7, 1956), J. D. Calfee and C. B. Miller (to Allied Chemical Corporation).
27. U.S. Pat. 3,047,637 (July 31, 1962), F. Olstowski (to The Dow Chemical Co.).
28. U.S. Pat. 3,089,910 (May 14, 1963), F. Olstowski and J. D. Watson (to The Dow Chemical Co.).
29. Jpn. Pat. 65 22,453 (Oct. 5, 1965), S. Okazaki and N. Sakauchi (to Kureha Chemical Co., Ltd.).
30. Jpn. Pat. 68 10,602 (May 4, 1968), Y. Kometani and M. Takemoto (to Daikin Kogyo Co., Ltd.).
31. U.S. Pat. 3,073,870 (Jan. 15, 1963), D. M. Marquis (to E. I. du Pont de Nemours & Co., Inc.).
32. East Ger. Pat. 42,730 (Jan. 5, 1966), H. Madai.
33. U.S. Pat. 3,428,695 (Feb. 18, 1969), J. R. Soulen and W. F. Schwartz (to Pennwalt Corporation).
34. K. M. Smirnov and co-workers, *Zh. Prikl. Khim. (Lenningrad),* **51,** 703 (1978).
35. U.S. Pat. 3,996,301 (Dec. 7, 1976), G. B. Fozzard (to Phillips Petroleum Co.).
36. U.S. Pat. 4,086,407 (Apr. 25, 1978), G. B. Fozzard (to Phillips Petroleum Co.).

37. U.S. Pat. 2,435,537 (Feb. 3, 1948), T. A. Ford and W. E. Hanford (to E. I. du Pont de Nemours & Co., Inc.).
38. U.S. Pat. 4,025,709 (May 24, 1977), J. Blaise and E. Grimaud (to Produits Chimiques Ugine Kuhlmann).
39. U.S. Pat. 3,714,137 (Jan. 30, 1973), K. Lienhard and D. Ulmschneider (to Suddeutsche Kalkstickstoff-Werke, A.G.).
40. Jpn. Pat. 73 32,578 (Oct. 6, 1973), K. Shinoda and co-workers (to Kureha Chemical Co., Ltd.).
41. U.S. Pat. 3,640,985 (Aug. 2, 1972), H. C. Stevens (to PPG Industries, Inc.).
42. U.S. Pat. 3,642,755 (Feb. 15, 1972), J. A. Baxter and C. O. Eddy (to PPG Industries, Inc.).
43. U.S. Pat. 3,708,463 (Jan. 2, 1973), J. P. Stallings (to Diamond Shamrock Corporation).
44. U.S. Pat. 3,937,690 (Feb. 10, 1976), G. Weisgerber and W. Trautvetter (to Dynamit Nobel, A.G.).
45. U.S. Pat. 3,598,797 (Aug. 10, 1971), Y. Kometani and M. Okuda (to Daikin Kogyo Co., Ltd.).
46. U.S. Pat. 3,475,396 (Oct. 28, 1969), G. H. McCain (to Diamond Shamrock Corporation).
47. U.S. Pat. 3,857,877 (Dec. 31, 1974), J. E. Dohany (to Pennwalt Corporation).
48. U.S. Pat. 3,193,539 (July 6, 1965), M. Hauptschein (to Pennwalt Corporation).
49. U.S. Pat. 4,076,929 (Feb. 28, 1978), J. E. Dohany (to Pennwalt Corporation).
50. Fr. Pat. 1,419,741 (Oct. 25, 1965); Jpn. Pat. 43 4867 (Feb. 22, 1968), S. Okazaki and N. Sakauchi (to Kureha Chemical Co., Ltd.).
51. U.S. Pat. 3,553,785 (Jan. 5, 1971), Y. Amagi and N. Bannai (to Kureha Chemical Co., Ltd.).
52. U.S. Pat. 3,780,007 (Dec. 18, 1973), J. P. Stallings (to Diamond Shamrock Corporation).
53. U.S. Pat. 3,781,265 (Dec. 25, 1973), J. E. Dohany (to Pennwalt Corporation).
54. U.S. Pat. 3,925,339 (Dec. 9, 1975), H. Ishii, N. Bannai, and S. Nishita (to Kureha Chemical Co., Ltd.).
55. Ger. Offen. 1,806,426 (May 14, 1969), D. P. Carlson (to E. I. du Pont de Nemours & Co., Inc.).
56. Br. Pat. 1,004,172 (Sept. 8, 1965); Belg. Pat. 620,986 (Feb. 4, 1963), (to Deutsche Solvay-Werke, G.m.b.H.).
57. D. Sianesi, *Corsi Semin. Chim.* (8), 133 (1968).
58. Br. Pat. 1,149,451 (Apr. 23, 1969), R. G. Foster (to Imperial Chemical Industries, Ltd.).
59. E. V. Volkova, P. V. Zimakov, and A. V. Fokin, *Dokl. Akad. SSSR* **167,** 1057 (1966).
60. E. V. Volkova, P. V. Zimakov, and A. V. Fokin, *At. Energ.* **26,** 240 (1969).
61. S. S. Dubov and co-workers, *Zh. Fiz. Khim.* **43,** 1574 (1969).
62. L. A. Bulygina and E. V. Volkova, *Radiats. Khim. Polim. Mater. Simp. Moscow,* 122 (1964).
63. Br. Pat. 1,188,889 (Apr. 22, 1970), (to Asahi Glass Co., Ltd.).
64. W. W. Doll and J. B. Lando, *J. Appl. Polym. Sci.* **14,** 1767 (1970).
65. U.S. Pat. 3,616,371 (Oct. 26, 1971), H. Ukihashi and M. Ichimura (to Asahi Glass Co., Ltd.).
66. E. V. Volkova and co-workers, *Radiats. Khim.* **2,** 282 (1972).
67. *Ibid.,* 287 (1972).
68. Jpn. Pat. 73 08,755 (Mar. 17, 1973), K. Nakamura, M. Ichimura, and Y. Fukushima (to Asahi Glass Co., Ltd.).
69. Jpn. Pat. 73 37,754 (Nov. 13, 1973), K. Nakamura, M. Ichimura, and Y. Fukushima (to Asahi Glass Co., Ltd.).
70. U.S. Pat. 4,054,680 (Oct. 18, 1977), A. H. Sharbaugh and D. G. Shaw (to General Electric Co.).
71. U.S. Pat. 3,287,339 (Nov. 22, 1966), D. Sianesi and G. Caporiccio (to Montedison Co.).
72. U.S. Pat. 3,380,977 (Apr. 30, 1968), G. F. Helfrich and E. J. Rothermel (to The Dow Chemical Co.).
73. Fr. Pat. 1,566,523 (May 9, 1969), (to The Dow Chemical Co.).
74. I. W. Kuleshov, A. A. Remisova, and G. M. Gartenev, *Plaste Kautsch.* **24,** 635 (1977).
75. M. Goerlitz and co-workers, *Angew. Makromol. Chem.* **29–30,** 137 (1973).
76. K. Nakagawa and Y. Ishida, *J. Polym. Sci. Phys. Ed.* **11,** 1503 (1973).
77. E. L. Gal'perin and co-workers, *Vysokomol. Soedin. Ser. A* **12,** 1654 (1970).
78. R. W. Herzberg, J. A. Manson, and W. C. Wu, *ASTM, STP* **536,** 391 (1973).
79. R. W. Herzberg and J. A. Manson, *Plast. World,* 50 (May, 1977).
80. C. W. Wilson and E. R. Santee, Jr., *J. Polym. Sci. C* **8,** 97 (1965).
81. M. Kobayashi, K. Tashiro, and H. Tadokoro, *Macromolecules* **8**(2), 158 (1975).
82. F. J. Boerio and J. L. Koenig, *J. Polym. Sci. A-2* **7,** 1489 (1969); **9,** 1517 (1971).
83. S. Enomoto, Y. Kawai, and M. Sugita, *J. Polym. Sci. A-2* **6,** 861 (1968).
84. G. Cortili and G. Zerbi, *Spectrochim. Acta.* **23A,** 285 (1967).
85. K. Nakagawa and Y. Ishida, *J. Polym. Sci. Polym. Phys. Ed.* **11,** 2153 (1973).
86. K. Matsushige and T. Takemura, *J. Polym. Sci. Polym. Phys. Ed.* **16,** 921 (1978).

87. G. J. Welch and R. L. Miller, *J. Polym. Sci. Polym. Phys. Ed.* **14,** 1683 (1976).
88. G. Gianotti, A. Capizzi, and V. Zamboni, *Chim. Ind. (Milan)* **55,** 501 (1973).
89. S. Osaki and Y. Ishida, *J. Polym. Sci. Polym. Phys. Ed.* **13,** 1071 (1975).
90. G. J. Welch, *Polymer* **15,** 429 (1974).
91. J. B. Enns and R. Simha, *J. Macromol. Sci. Phys. B* **13,** 11 (1977).
92. H. Kakutani, *J. Polym. Sci. Pt. A-2,* **8,** 1177 (1970).
93. T. Wentink, Jr., *J. Appl. Phys.* **32,** 1063 (1961).
94. S. Yano, *J. Polym. Sci. Pt. A-2* **8,** 1057 (1970).
95. *Bulletin No. PL-148,* Pennwalt Corporation, Philadelphia, Pa., 1978.
96. P. E. Bloomfield and co-workers, *U.S. Naval Res. Rev.* **31**(5), 1 (1978); A. L. Robinson, *Science* **200,** 1371 (June 23, 1978).
97. H. Kawai, *J. Appl. Phys. (Japan)* **8,** 975 (1969).
98. H. Kawai, *Oyo Butsuri* **38,** 1133 (1969).
99. R. Timmerman and W. Greyson, *J. Appl. Polym. Sci.* **6** (22), 456 (1962).
100. G. B. Sands and G. F. Pezdirts, *NASA Report N66-13496,* U.S. Government Printing Office, Washington, D.C.; *Am. Chem. Soc. Polym. Prepr.* **6,** 987 (1965).
101. E. L. Gal'perin and B. P. Kosmynin, *J. Polym. Sci. USSR* **15,** 2895 (1974).
102. J. E. Dohany, K. N. Davis, and H. Stefanou, in *Proceedings of 24th Int. Wire & Cable Symp.,* U.S. Army Elec. Com., 1975, p. 340.
103. K. Makuuchi and co-workers, *Nippon Kagaku Kaishi* (11), 1990 (1975).
104. K. Makuuchi, M. Asano, and T. Abe, *Nippon Kagaku Kaishi* (4), 686 (1976).
105. U.S. Pat. 3,580,829 (May 25, 1971), V. L. Lanza (to Raychem Corporation).
106. K. Makuuchi, F. Yoshii, and T. Abe, *Nippon Kagaku Kaishi* (10), 1828 (1975).
107. K. Makuuchi, M. Asano, and K. Araki, *Kobunshi Ronbunshu,* **34** 515 (1977).
108. U.S. Pat. 3,253,060 (May 24, 1966), F. F. Koblitz and co-workers, (to Pennwalt Corporation).
109. J. C. Noland and co-workers, *Adv. in Chem. Ser.* **99,** 15 (1971).
110. T. Nishi and T. T. Wang, *Macromolecules* **8,** 909 (1975).
111. D. C. Douglass and V. J. McBrierty, *Macromolecules* **11,** 766 (1978).
112. D. C. Wahrmund and co-workers, *Polym. Eng. Sci.* **18,** 677 (1978).
113. T. K. Kwei and co-workers, *Macromolecules* **10,** 157 (1977).
114. R. E. Bernstein, D. R. Paul, and J. W. Barlow, *Polym. Eng. Sci.* **18,** 683 (1978).
115. A. A. Dukert, paper presented at *SPE, 18th ANTEC, Pittsburgh, Pa., Jan. 1962.*
116. J. P. Stallings and S. G. Howell, *Polym. Eng. Sci.* **11,** 507 (1971).
117. *Bulletin No. Br 1126,* Part III, Soltex Polymer Corp., Houston, Tex., 1976.
118. U.S. Pat. 3,922,186 (Nov. 25, 1975), M. Segawa, Y. Kawakami, and M. Itoh (to Kureha Chemical Co., Ltd.).
119. U.S. Pat. 3,197,538 (July 27, 1965), N. Capron, R. A. Ferren, and W. S. Barnhart (to Pennwalt Corporation).
120. U.S. Pat. 4,011,874 (Mar. 15, 1977), M. Segawa and co-workers (to Kureha Chemical Co., Ltd.).
121. U.S. Pat. 3,142,629 (July 28, 1964), Re. 25,904 (Nov. 16, 1965), R. Timmerman (to Radiation Dynamics Corporation).
122. U.S. Pat. 3,269,862 (Aug. 30, 1966), V. L. Lanza and E. C. Stivers (to Raychem Corporation).
123. U.S. Pat. 3,896,196 (July 22, 1975), C. A. Dickey and J. E. McDaniel (to Glassrock Products, Inc.).
124. U.S. Pat. 3,324,069 (June 6, 1967), F. F. Koblitz and R. G. Petrella (to Pennwalt Corporation).
125. U.S. Pat. 3,340,222 (Sept. 5, 1967), J. C. Fang (to E. I. du Pont de Nemours & Co., Inc.).
126. U.S. Pat. 3,449,466 (June 10, 1969), G. Gerek and L. K. Buell (to PPG Industires, Inc.).
127. *Fluorex Poly(vinylidene fluoride),* film data sheet, Rexham Corporation, Matthews, N.C., 1977.
128. *Chemtherm Poly(vinylidene fluoride) Laminates,* bulletin, Dodge Fluorglas Div., Oak Materials Group, Inc., Hoosick Falls, N.Y., 1976.
129. A. A. Dukert and A. Christofas, paper presented at *SPI, Reinforced Plastics Conf., Washington, D.C., Feb. 1974.*
130. *Powder Finish. World,* (2nd Quarter Issue), (1975).
131. *Bulletin No. CP7-1277, Liquinite Coating Powders,* L. N. P. Corporation, Malvern, Pa.
132. A. R. Latvin, "Poly(vinylidene fluoride) Acute and Subacute Toxicity in Rats," internal report to Pennwalt Corporation, King of Prussia, Pa., 1959.
133. W. L. Guess and J. Autian. *J. Oral Ther. Pharmacol.* **3**(2), 116 (1966).
134. *Kynar—Monomer Content,* internal report, Pennwalt Corporation, Philadelphia, Pa., 1975.

135. *Code of Federal Regulations, FDA Title 21,* Part 177, U.S. Government Printing Office, Washington, D.C., 1977, p. 2510.
136. S. L. Madorsky, *Thermal Degradation of Organic Polymers*, Interscience-Wiley, New York, 1964, p. 315.
137. *Bulletin No. TD-1, Kynar Poly(vinylidene fluoride)*, product safety bulletin, Pennwalt Corporation, King of Prussia, Pa., 1976.
138. *ASTM F 491, Specification for Poly(vinylidene fluoride)-Lined Metal Pipe and Fittings,* ASTM, Philadelphia, Pa., 1977.
139. L. E. Robb, paper presented at *NACE Tech. Conf., San Francisco, Calif. Mar. 1977.*
140. U.S. Pat. 4,065,534 (Dec. 27, 1977), T. A. Rechlicz and B. A. Maloney (to PPG industries, Inc.).
141. U.S. Pat. 4,093,533 (June 6, 1978), R. N. Beaver and C. W. Becker (to The Dow Chemical Co.).
142. A. A. Dukert and J. H. Houser, *Comput. Des.* (Jan. 1970).
143. *Plast. Des. Forum,* (Nov.–Dec. 1976).
144. *Bulletin No. PL-149-1177, Spotlight Kynar* **5**(2), Pennwalt Corporation, Philadelphia, Pa.
145. *Investigation of Surface-Burning Characteristics of Building Materials, Project No. 3-3503-34,* Southwest Research Institute, Fire Research Section, Report to Pennwalt Corporation, Oct. 1972.
146. *Buildings* **72**(10), 76 (1978).
147. R. Yee, *Prog. Archit.,* **12:74,** 100 (1974).

JULIUS E. DOHANY
LESTER E. ROBB
Pennwalt Corporation

POLY(FLUOROSILICONES)

The role of fluorine and the carbon–fluorine bond in achieving a high degree of solvent resistance and stability is well known in organic polymer systems. About 20 years ago the two properties were combined in silicone polymers by incorporation of fluorine into polyalkylsiloxane systems.

Structural Factors

Incorporation of the carbon–fluorine bond rather than the silicone–fluorine bond is a principal factor influencing the choice of structure for polyalkylsiloxane systems. The latter, although of high thermal stability, is subject to hydrolysis and, therefore, is less useful than the former. The second structural consideration is the location of fluorine relative to silicone in the alkyl substituent. Because of the great electronegativity and strong inductive effect of fluorine, the position alpha to silicon, eg, $CF_3SiRR'R''$, suffers a major disadvantage: hydrolytic cleavage of the silicon–carbon bond. In addition, thermal rearrangement is possible with the formation of silicon–fluorine bonds:

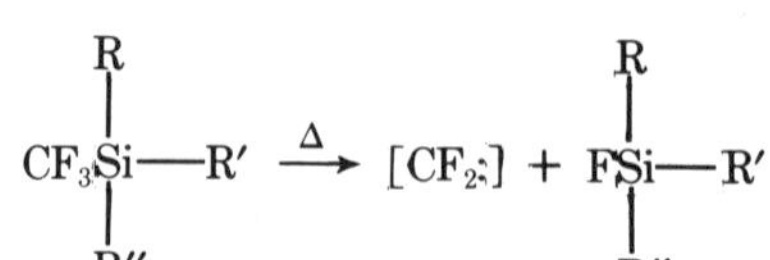

This rearrangement appears to be similar to the alpha elimination mechanism proposed for one mode of carbene formation (1).

The position beta to silicon, $CF_3CH_2SiRR'R''$, also suffers the same disadvantages, hydrolytic instability and thermal rearrangement. In the latter, the side chain is eliminated as an olefin:

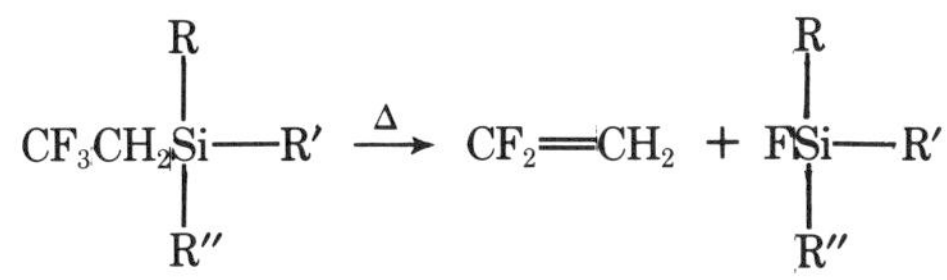

Therefore, the gamma position, $CF_3CH_2CH_2SiRR'R''$, is the obvious choice for maximum stability and ease of preparation. Positions beyond gamma are also suitable but their usefulness is limited by the oxidation of the CH bonds in the alkyl group. The polymer chosen for industrial production was $[CF_3CH_2CH_2Si(CH_3)O]_n$ [*25791-89-3*], since it demonstrates excellent solvent resistance and stability.

Properties

Fluid and Chemical Resistance. Fluorosilicone elastomers are especially suited for applications involving exposure to fuels, oils, solvents, hydraulic fluids, and various other chemicals. Fluid resistance is excellent to almost all solvents with a few exceptions, eg, some esters and ketones. Even at elevated temperatures, prolonged immersion causes only slight swelling. Table 1 describes the hardness change and swelling of fluorosilicones that result from immersion in a variety of fluids at various temperatures. These values are characteristic of the entire fluorosilicone family of elastomers and indicate that the fuel, oil, and solvent resistance of fluorosilicones is among the highest of all elastomers. Even dilute caustic solutions, nitric acid, hydrochloric acid, and sulfuric acid have little effect on fluorosilicone rubber.

Heat Resistance. Fluorosilicone elastomers give long-lasting dependability in both static and dynamic applications over a wide temperature range (2). Most fluorosilicones do not become brittle when subjected to repeated thermal cycling. The high temperature range of fluorosilicone elastomers can be increased significantly with 0.5–3 parts of iron oxide. Table 2 lists properties at RT both before and after exposure to elevated temperatures. The stability of the fluorosilicone elastomer is evident in the small changes in hardness resulting from aging at elevated temperatures.

Fluorosilicone elastomers possess outstanding ability to retain a high percentage of their RT properties at extreme operating temperatures, indicating their suitability for dynamic applications at high temperatures. Typical properties of these elastomers at RT and at several higher temperatures are given in Table 3.

Low Temperature Properties. Low temperature properties of fluorosilicone elastomers are outstanding. The TR-10 of these elastomers is typically −59°C, ie, the temperature at which the 100% elongated and frozen samples retract to 10% of their original extension. No other commercially available fluorocarbon elastomers have this low TR-10 value. The brittle point is −66°C, which suggests static applications with fluorosilicones are possible at temperatures well below the TR-10.

Table 1. Hardness and Volume Changes of Typical Fluorosilicone Elastomers in Liquid Media

Liquid	Immersion conditions	Hardness change Shore A Durol meter points	Percent swell
oil and fuels			
ASTM oil #1	70 h, 149°C	nil	nil
ASTM reference fuel B	70 h, 25°C	−10	+22
ASTM oil #3	70 h, 149°C	nil	+4
ASTM service fuel 101	48 h, 149°C	−7	+9
jet fuel JP-4	70 h, 25°C	−6	+10
Hypoid EP lubricant	336 h, 149°C	−20	nil
gasoline, low lead	24 h, 25°C	−12	+21
Hydraul 60 fluid	70 h, 121°C	−15	+5
Skydrol 500 A fluid	168 h, −54°C	nil	+5
Skydrol 500 A fluid	168 h, 100°C	−20	+25
aircraft turbine oil	140 h, 150°C	−8	+7
silicone fluid, 100 CS	168 h, 150°C	nil	nil
crude oil	72 h, 150°C	−3	+2
solvents and chemicals			
benzene	72 h, 70°C	−10	+20
butanol	168 h, 24°C	nil	+10
xylene	168 h, 24°C	−10	+20
nitric acid, 70% (conc)	168 h, 24°C	nil	+5
sulfuric acid, 50%	72 h, 65°C	+5	nil
hydrochloric acid, 10%	168 h, 24°C	−5	nil
sodium hydroxide, 50%	168 h, 24°C	−5	nil
methanol	336 h, 24°C	−10	+4

Table 2. Physical Properties of Fluorosilicone Elastomers Heat-Aged at 204°C

Test properties and conditions	Silastic LS-2311[a]	Silastic LS-2332[b]	Silastic LS-2380
original properties			
Durometer (Shore A-2)	75	52	70
tensile strength, MPa[c]	6.2	8.6	6.9
elongation, %	100	500	150
100% modulus, MPa[c]	6.1	1.0	4.8
tear strength kN/m[d], die B	8.8	46.4	
aged 70 h, 204°C, change			
Durometer, points	+4	+2	nil
tensile strength, %	nil	−9	
elongation, %	−20	−20	+6
wt loss, %	+2		
tear, %		−30	

[a] Molded 10 min, 171°C; postcured 8 h, 204°C (with benzoyl peroxide).
[b] Molded 5 min, 127°C; postcured 8 h, 204°C (with Varox).
[c] To convert MPa to psi, multiply by 145.
[d] To convert kN/m to ppi, divide by 0.175.

Electrical Properties. Like silicone polymers, fluorosilicone elastomers have inherently good electrical insulating properties. Their dielectric properties are excellent and remain relatively unchanged when the elastomer is exposed to severe environments. The arc resistance of fluorosilicones also is excellent (see Insulation, electric).

Table 3. Properties of Fluorosilicone Elastomers at Various Temperatures

Fluorosilicone	25°C	100°C	177°C	204°C
general purpose				
tensile strength, MPa[a]	8.0	6.2	4.4	3.5
elongation, %	480	480	370	300
modulus at 100% elongation, MPa[a]	1.2	0.96	1.0	1.0
high tear strength				
tensile strength, MPa[a]	8.0	5.5		3.2
elongation, %	420	450		325
tear strength, kN/m[b]	47	32		8.8
high modulus				
tensile strength, MPa[a]	7.6	5.5	4.1	3.2
elongation, %	350	260	200	160
modulus at 100% elongation, MPa[a]	2.1	1.9	2.0	1.9

[a] To convert MPa to psi, multiply by 145.
[b] To convert kN/m to ppi, divide by 0.175.

Bondability. The bondability of fluorosilicone elastomers to metals and to other polymeric composites can be significantly improved with Silastic A-4040 primer. A thin coat of this unique primer, applied by any conventional method that gives a smooth coverage, has adhesion values in the range of 1.8–12.2 kN/m (10–70 ppi), depending on the elastomer.

Other Properties. In general, other properties of fluorosilicone elastomers are similar to general purpose dimethylsilicone elastomers. Fluorosilicones can undergo repeated flexing without tearing. Chemical stress relaxation studies of Silastic LS-420 vulcanizates show that, at RT, fluorosilicone rubber shows negligible chemical degradation for very long times, ie, several decades (3).

Manufacture

MONOMER

The dichloro compound, $CF_3CH_2CH_2Si(CH_3)Cl_2$ [*675-62-7*], is the principal starting material for the fluorosilicone polymer system and is prepared as follows (4):

$$CF_3CH{=}CH_2 \xrightarrow{CH_3SiHCl_2} CF_3CH_2CH_2Si(CH_3)Cl_2$$

The desired chlorosilane is purified by distillation and then is converted to cyclotrisiloxanes (cis and trans) (5):

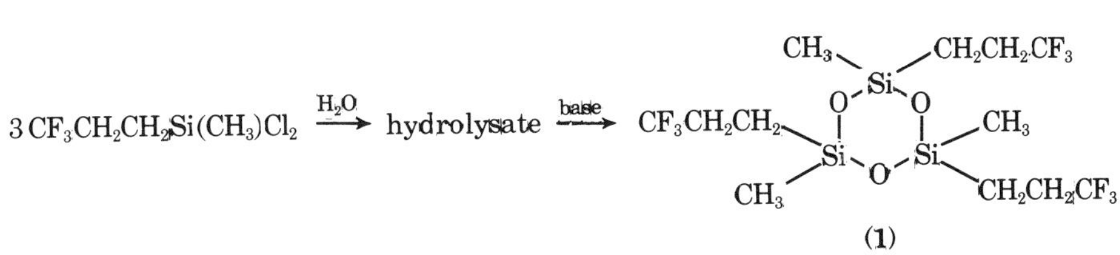

(**1**)
[*2374-14-3*]

The cyclic siloxane is purified by distillation and serves as the monomer for polymer synthesis.

POLYMERS

Preparation of fluorosilicone polymers from the cyclic trimer is best accomplished by base catalysis at elevated temperatures in the absence of solvent (6). The system involves the following equilibria (7):

$$\text{cyclic trimer} \rightleftarrows \text{polymer} \rightleftarrows \underset{[429\text{-}67\text{-}4]}{\text{cyclic tetramer}}$$

Generally, polymerization is stopped before equilibrium is reached. Careful control of the reaction time and temperature is essential for good yields of high purity polymer. Currently (1979) the price of \$55–220/kg depends on quantity and type of product.

Commercial production yields polymers having a degree of polymerization (DP) of ca 5000 which are suitable for most rubber applications. The proper choice of conditions and the use of end-blocking agents can effect lower molecular weight polymers that are useful in sealant and fluid applications (see also Elastomers, synthetic-fluorinated).

The silicone–oxygen backbone provides excellent thermal stability and good physical and electrical properties over a very wide temperature range (8). Products based on fluorosilicone polymers are produced commercially as rubbers, sealants, and fluids. Fluorosilicone elastomers are available as gums, compounding bases, and uncatalyzed fully compounded stocks and are characterized by a hardness of 40–80 Shore A durometer and other chemical and physical properties. Fluorosilicone elastomers are available in heat-vulcanized (rubbers) or room temperature vulcanized versions (RTV sealants).

Processing. Fluorosilicone elastomers can be formulated to provide specific properties, eg, high or low hardness, high tear strength, or high modulus. Materials that are designed to resist tearing, exhibit tear strength of up to 44 kN/m (250 ppi). Moduli may range from 1.0–6.0 MPa (150–900 psi) at 100% elongation. Compression sets as low as 10% (22 h/177°C) can be achieved.

Compounding. In the production of various rubber products, fluorosilicone gums are compounded generally with fumed and/or precipitated silica fillers, hydroxy-containing low viscosity silicone oils, and readily available peroxides.

These rubber products are based on very high molecular weight ($\overline{M}_w > 700{,}000$) polymers; but because of the freedom of rotation of the silicon–oxygen bond, they are soft and easily processed in conventional internal mixers, water-cooled mills, and calenders. Gum plasticities are generally 2.5–3.8 mm and fully compounded bases are in the 3.0–5.6 mm range. Processing characteristics can be modified with low viscosity hydroxy-containing silicone oils, silicone gums, 1–10 parts of dimethylsilicone softeners, and 1–50 parts of fumed or precipitated silica fillers. Most fluorosilicone composites can be colored with standard pigments (qv).

Vulcanization. Fluorosilicone elastomers are peroxide-vulcanized by a free-radical mechanism using unsaturated vinyl side groups that have been incorporated into the basic polymer structure. Peroxide-initiated cross-linking results in a carbon–carbon bond created by reaction of free radicals. The free radicals are generated in the polymer via the peroxide radicals, either by abstracting hydrogen from methyl groups or by adding to a vinyl group.

During vulcanization, the only significant volatile species formed are by-products of the peroxide. No fluorine-containing by-products are formed. Typical cure cycles are 5–10 min at 115–170°C, depending on the choice of peroxide. With most fluorosilicones (as well as with other fluoroelastomers), a postcure of 4–24 h at 150–200°C is recommended to maximize long-term thermal aging properties. This postcure completes reactions of the unsaturated side groups and results in increased tensile strength and higher cross-link density.

Fabrication. Fluorosilicone elastomers can be molded, extruded, or calendered by any of the conventional methods employed in the industry. Compression molding is the most widely used method and is ideal for a great many fabrications at 115–170°C and 5.5–10.3 MPa (800–1500 psi). Injection molding becomes increasingly important for high production operations and generally requires higher temperatures and pressures than compression molding. Transfer molding is particularly useful for molding complex parts in multicavity presses. Where dimensional accuracy of molded parts is important, shrinkage of the parts must be considered in the design of the mold. Linear shrinkage of most fluorosilicones is 2.5–3.5%.

Extrusion techniques are used to make tubes, rods, gaskets, preforms, etc. Standard rubber equipment may be used to extrude fluorosilicone elastomers. The green strength of fluorosilicone elastomers is less than that of typical fluorocarbon elastomers, and this should be considered when designing the feed system.

Calendering is used to produce long, thin sheets of fluorosilicone elastomers and to coat fluorosilicones on reinforcing substrates, eg, fiber glass, nylon, Dacron, etc, or on other substrates, eg, certain polymers and metals, to provide the protection of fluorosilicones at minimum cost. When very thin films are desired or when the vulcanizing temperature of typical fluorosilicones is too high for the supporting substrate, RTV dispersions are commercially available. Dispersions of fluorosilicone gums, bases, and fully compounded stocks—with or without catalysts—can be used to coat many fibrous metal or polymeric substrates. It is necessary to maintain dispersion neutrality to maximize the physical properties of the coating. Usually Silastic A-4040 is recommended to prepare dense polymeric or metal substrate surfaces for bonding (see Coatings; Film).

Uses

Sealants. Formation of a low molecular weight polymer with appropriate fillers, chain extenders, cross-linking agents, and catalysts gives a low viscosity stock that has excellent extrusion characteristics and cures at RT (ie, an RTV system). Vulcanization results from the interaction of reactive groups in the polymer chain with a cross-linking agent. One type of curing mechanism requires that moisture in the air reacts with an acetoxy group to produce the cross-link. This reaction liberates acetic acid and such sealants are used only if by-products do not interfere with the application. A second method of curing depends on the reaction of a vinyl group with a silicon hydride and liberates no volatile products. Use of this type of sealant lends itself to thick-section applications.

The physical properties of fluorosilicone sealants are quite comparable to those of the conventional dimethylsiloxane sealants. The superior solvent resistance of fluorosilicone elastomers also is evident in the sealants. Perhaps the most outstanding characteristic of the sealants is their resistance to degradation at elevated temperatures in fuels, as shown in Table 4. At even higher temperatures, both the tensile strength and elongation retention are excellent (Figs. 1–2).

Because of these unique properties, the trifluoropropylmethylsiloxane sealants are used extensively in both subsonic and supersonic aircraft construction. In these applications, the sealant must withstand temperature extremes from −57° to 232°C and continuous contact with aircraft fuel. The primary applications for the fluorosilicone sealants on aircraft are as filleting sealants for the integral fuel tanks built into the wings and as noncurable channel sealants. They are used also in ground-support fuel-handling systems and in areas of the aircraft where solvents are used for cleaning and wash-down procedures.

Fluorosilicone sealants also are useful in bonding, sealing, caulking, encapsulating and potting applications (see Sealants; Embedding).

Table 4. Fuel Resistance of Fluorosilicone Sealants (in N_2)

		JP-4, 121°C		JP-4, 232°C	
	Original	4 wk	8 wk	4 wk	8 wk
Durometer, Shore A	42	32	33	38	33
tensile strength, MPa[a]	3.80	3.42	3.66	4.14	3.73
elongation, %	500	420	310	265	270

[a] To convert MPa to psi, multiply by 145.

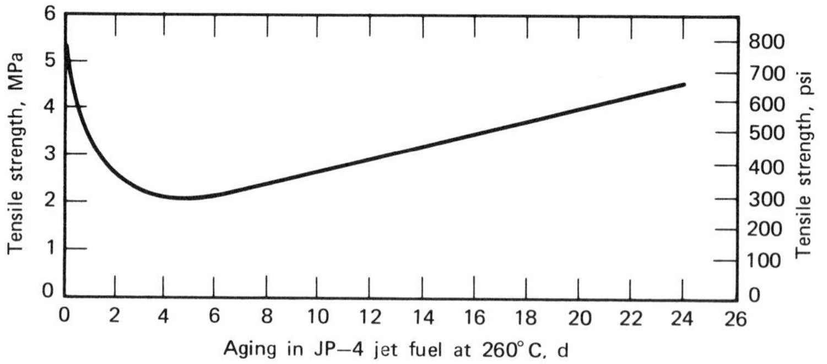

Figure 1. Tensile strength of fluorosilicone sealant exposed to jet fuel at 260°C in pressure bomb.

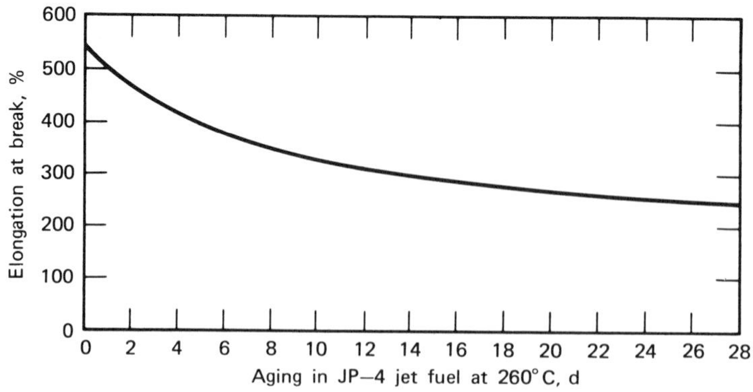

Figure 2. Elongation of fluorosilicone sealant exposed to jet fuel at 260°C in pressure bomb.

Table 5. Main Applications of Fluorosilicone Rubbers, Sealants, and Fluids (Oils)

automotive
exhaust-gas recirculating diaphragms
transmodulator diaphragms
umbrella valve
duckbill valves
aerospace
O-rings for fuel systems
jet-aircraft electrical connectors
jet-fuel seals
aircraft integral fuel-tank sealants (filleting, faying and channel)
other applications
potting sealants
dampers
fusing blankets
sealing compounds
lubricating grease and oil

Fluids (Oils). By careful control of the polymerization conditions and the use of suitable end-blocking agents, eg, hexamethyldisiloxane, various degrees of polymerization can be obtained to yield any desired viscosity of fluorosilicone oil.

In industry, fluorosilicone fluids are used mainly for defoaming of solvent-based wash solutions or processes (see Defoamers), and lubrication (qv).

The chemical stability and heat stability of fluorosilicone oils, coupled with their good lubricity, accounts for their use as lubricating oils in chemical compressors, and in vacuum pumps exposed to chemical fumes. Fluorosilicone fluids also are formulated with various thickeners to make greaselike sealing compounds and lubricating greases.

Other Uses. Some of the other applications of the rubbers, sealants, and fluids are shown in Table 5.

BIBLIOGRAPHY

1. R. N. Haszeldine and J. C. Young, *Proc. Chem. Soc.*, 394 (1959).
2. M. A. Cabey and G. E. Vogel, *Lubrication Eng.* **34,** 546 (1978).
3. S. H. Kalfayan, A. A. Mazzeo, and R. H. Silver, *Jet Propulsion Lab. Q. Rev.*, **1**(3), 38 (1971).
4. O. R. Pierce and Y. K. Kim, *J. Elastoplast.* **3,** 82 (1971).
5. L. L. Burshtein and co-workers, *Zh. Obshch. Khim.* **33,** 2789 (1963).
6. O. R. Pierce and co-workers, *Ind. and Chem.* **52,** 783 (1960).
7. E. D. Brown and J. B. Carmichael, *Polym. Lett.* **3,** 473 (1965).
8. *Fluorosilicone Bulletin Nos. 09-237,* 1968; *09-278,* 1968–1969; *02-028,* 1969; *63-005,* 1970; *17-052A,* 1971; *17-158;* 1971; *17-167,* 1972; *17-182,* 1973; *17-187,* 1973; etc, Dow Corning Corporation, Midland, Mich.

Y. K. KIM
Dow Corning Corporation

FLUORINE COMPOUNDS, ORGANIC, SURFACE CHEMISTRY. See Surfactants and detersive systems.

FLUORINE COMPOUNDS, SURFACE CHEMISTRY OF FLUOROCHEMICALS. See Surfactants and detersive systems.

FLUORITE, FLUOROSPAR, CaF_2. See Fluorine compounds, inorganic–Calcium fluoride.

FLUOROALUMINATES, FLUOROBERYLLATES, FLUOROPHOSPHATES, FLUOROSILICATES, AND SIMILAR ENTRIES. See Fluorine compounds, inorganic.

FLUOROCHEMICALS. See Fluorine compounds, organic.

FLUOROMETRY. See Analytical methods; Color; Dyes, application and evaluation.

FOAMED PLASTICS

Foamed plastics, otherwise known as cellular plastics or plastic foams, have been important to human life since primitive man began to use wood, a cellular form of the polymer cellulose. Cellulose (qv) is the most abundant of all naturally occurring organic compounds, comprising approximately one third of all vegetable matter in the world (1). Its name is derived from the Latin *cellula,* meaning very small cell or room, and most of the polymer does indeed exist in cellular form as in wood, straws, seed husks, etc. The high strength-to-weight ratio of wood, good insulating properties of cork and balsa, and cushioning properties of cork and straw have contributed both to the incentive to develop and to the background knowledge necessary for development of the broad range of cellular synthetic polymers in use today.

The first cellular synthetic plastic was an unwanted cellular phenol–formaldehyde resin produced by early workers in this field. The elimination of cell formation in these resins, as given by Baekeland in his 1909 heat and pressure patent (2), is generally considered the birth of the plastics industry. The first commercial cellular polymer was sponge rubber, introduced between 1910 and 1920 (3).

Many cellular plastics that have never reached significant commercial use have been introduced or their manufacture described in literature. Examples of such polymers are chlorinated or chlorosulfonated polyethylene (4), a copolymer of vinylidene fluoride and hexafluoropropylene [*9011-17-0*] (4), polyamides (4), polytetrafluoroethylene [*9002-84-0*] (5), styrene–acrylonitrile copolymers [*9003-54-7*] (6–7) and ethylene–propylene copolymers [*9010-79-1*] (8).

Cellular polymers have been commercially accepted in a wide variety of applications since the 1940s (9–18). The total usage of foamed plastics in the United States has risen from 245×10^3 metric tons in 1965 to over 1.4×10^6 metric tons in 1976, and has been projected to rise to about 6.6×10^6 t in 1990 (19). The markets and applications that make up this total usage are covered in detail in the section on applications.

Nomenclature

A cellular plastic has been defined as a plastic the apparent density of which is decreased substantially by the presence of numerous cells disposed throughout its mass (20). In this article the terms *cellular polymer, foamed plastic, expanded plastic,* and *plastic foam* are used interchangeably to denote all two-phase gas–solid systems in which the solid is continuous and composed of a synthetic polymer or rubber.

The gas phase in a cellular polymer is usually distributed in voids or pockets called cells. If these cells are interconnected in such a manner that gas can pass from one to another, the material is termed *open-celled.* If the cells are discrete and the gas phase of each is independent of that of the other cells, the material is termed *closed-celled.*

The nomenclature of cellular polymers is not standardized; classifications have been made according to the properties of the base polymer (21), the methods of manufacture, the cellular structure, or some combination of these. The most comprehensive classification of cellular plastics, proposed by Cooper in 1958 (22), has not been adopted and is not consistent with some of the currently common names for the more important commercial products.

A particular ASTM test procedure has suggested (23) that foamed plastics be classified as either rigid or flexible, a *flexible foam* being one that does not rupture when a 20 × 2.5 × 2.5 cm piece is wrapped around a 2.5-cm mandrel at a uniform rate of 1 lap/5 s at 15–25°C. *Rigid foams* are those that do rupture under this test. This classification is used in this article.

In the case of cellular rubber, the ASTM uses several classifications based on the method of manufacture (24–25). These terms are used here. *Cellular rubber* is a general term covering all cellular materials that have an elastomer as the polymer phase. *Sponge rubber* and *expanded rubber* are cellular rubbers produced by expanding bulk rubber stocks and are open-celled and closed-celled, respectively. *Latex foam rubber,* also a cellular rubber, is produced by frothing a rubber latex or liquid rubber, gelling the frothed latex, and then vulcanizing it in the expanded state.

The term *structural foams* has not been exactly defined but are used here to refer to those rigid foams produced at greater than about 320 kg/m^3 density.

Theory of the Expansion Process

Foamed plastics can be prepared by a variety of methods. The most important process, by far, consists of expanding a fluid polymer phase to a low density cellular state and then preserving this state. This is the foaming, or expanding, process. Other methods of producing the cellular state include leaching out solid or liquid materials that have been dispersed in a polymer, sintering small particles, and dispersing small cellular particles in a polymer. The latter processes, however, are relatively straightforward processing techniques but are of minor importance.

The expansion process consists of three steps: creating small discontinuities or cells in a fluid or plastic phase; causing these cells to grow to a desired volume; and stabilizing this cellular structure by physical or chemical means.

Initiation and Growth of Cells. The initiation, or nucleation, of cells is the formation of cells of such size that they are capable of growth under the given conditions of foam expansion. The growth of a hole or cell in a fluid medium at equilibrium is controlled by the pressure difference ΔP between the inside and the outside of the cell, the surface tension of the fluid phase γ, and the radius r of the hole, according to equation 1:

$$\Delta P = 2\,\gamma/r \tag{1}$$

The pressure outside the cell is the pressure imposed on the fluid surface by its surroundings. The pressure inside the cell is the pressure generated by the blowing agent dispersed or dissolved in the fluid. If blowing pressures are low, the radii of initiating holes must be large. The hole that acts as an initiating site can be filled with either a gas or a solid which breaks the fluid surface and thus enables blowing agent to surround it (26–29).

During the time of cell growth in a foam, a number of properties of the system change greatly. Cell growth can, therefore, be treated only qualitatively. The following considerations are of primary importance: (*1*) the fluid viscosity is changing considerably, this tends to influence both the cell growth rate and the flow of polymer from cell walls to intersections leading to collapse; (*2*) the pressure of the blowing agent decreases, falling off less rapidly than an inverse volume relationship because new blowing agent diffuses into the cells as the pressure falls off according to equation 1; (*3*) the rate of growth of the cell depends on the viscoelastic nature of the polymer phase, the blowing-agent pressure, the external pressure on the foam, and the permeation rate of blowing agent through the polymer phase; and (*4*) as a consequence of equation 1, the pressure in the cell of small radius r_2 is greater than that in the cell of larger radius r_1. There will thus be a tendency to equalize these pressures either by breaking the wall separating the cells or by diffusion of the blowing agent from the small to the large cells. The pressure difference ΔP between cells of radii r_1 and r_2 is shown in equation 2:

$$\Delta P = 2\,\gamma \left(\frac{1}{r_2} - \frac{1}{r_1}\right) \tag{2}$$

Stabilization of the Cellular State. The increase in surface area corresponding to the formation of many cells in the plastic phase is accompanied by an increase in the free energy of the system; hence the foamed state is inherently unstable. Methods of stabilizing this foamed state can be generally classified as chemical, ie, the polymerization of a fluid resin into a three-dimensional thermoset polymer, or physical, ie, the cooling of an expanded thermoplastic polymer to a temperature below its second-order transition temperature or its crystalline melting point to prevent polymer flow.

The important considerations in the chemical method of dimensional stabilization are the following:

Chemistry of the System. This determines both the rate at which the polymer phase is formed and the rate at which it changes from a viscous fluid to a dimensionally stable thermoset phase. It also governs the rate at which the blowing agent is activated, whether it be owing to temperature rise or to insolubilization in the liquid phase.

Type and Amount of Blowing Agent. These factors govern the amount of gas generated, the rate of generation, and the pressure that can be developed to expand the polymer phase. They also determine the amount of gas lost from the system relative to the amount retained in the cells.

Additives to the Foaming System. Additives (cell-control agents) can greatly influence nucleation of foam cells, either through their effect on the surface tension of the system, or by acting as nucleating sites from which cells can grow. They can influence the mechanical stability of the foam structure considerably by changing the physical properties of the plastic phase and by creating discontinuities in the plastic phase that allow blowing agent to diffuse from the cells to the surroundings.

Environmental factors such as temperature, geometry of foam expansion, and pressure also influence the behavior of thermoset foaming systems.

In physically stabilized foaming systems the factors are essentially the same as for chemically stabilized systems but for somewhat different reasons:

Chemical Composition of the Polymer Phase. The type of polymer determines the temperature at which foam must be produced, the type of blowing agent required, and the type of cooling of the foam necessary for dimensional stabilization.

Blowing-Agent Composition and Concentration. These control the rate at which gas is released, the amount of gas released, the pressure generated by the gas, escape or retention of gas from the foam cells for a given polymer, and heat absorption or release owing to blowing-agent activation.

Additives. Additives have the same effect on thermoplastic foaming processes as upon thermoset foaming processes.

Environmental Conditions. Environmental conditions are important in this case because of the necessity of removing heat from the foamed structure in order to stabilize it. The dimensions and size of the foamed structure are important for the same reason.

Manufacturing Processes for Cellular Polymers

Cellular plastics and polymers have been prepared by a wide variety of processes involving many methods of cell initiation, cell growth, and cell stabilization. The most convenient method of classifying these methods appears to be based on the cell growth and stabilization processes. According to equation 1, the growth of the cell depends on the pressure difference between the inside of the cell and the surrounding medium. Such pressure differences may be generated by lowering the external pressure (decompression) or by increasing the internal pressure in the cells (pressure generation). Other methods of generating the cellular structure are by dispersing gas (or solid) in the fluid state and stabilizing this cellular state, or by sintering polymer particles in a structure that contains a gas phase.

Foamable compositions in which the pressure within the cells is increased relative to that of the surroundings have generally been called expandable formulations. Both chemical and physical processes are used to stabilize plastic foams from expandable formulations. There is no single name for the group of cellular plastics produced by the decompression processes. The various operations used to make cellular plastics by this principle are extrusion, injection molding, and compression molding. Either physical or chemical methods may be used to stabilize products of the decompression process.

EXPANDABLE FORMULATIONS

Physical Stabilization Process. Cellular polystyrene [*9003-53-6*], poly(vinyl chloride) [*9002-86-2*], copolymers of styrene and acrylonitrile, and polyethylene [*9002-88-4*] can be manufactured by this process.

Polystyrene. Polystyrene is the outstanding example of a material that can be fabricated into cellular form by this method. In general, there are two types of expandable polystyrene processes, depending on the applications—expandable polystyrene for molded articles, and expandable polystyrene for loose-fill packing materials (see Styrene plastics).

Expandable polystyrene for loose-fill packing materials is available in various sizes and shapes varying from round disks to S-shaped strands. These particles can These particles are prepared either by heating polymer particles in the presence of a blowing agent and allowing the blowing agent to penetrate the particle (30) or by polymerizing the styrene monomer in the presence of blowing agent (31) so that the blowing agent is entrapped in the polymerized bead. Typical blowing agents used in such processes are the various isomeric pentanes and hexanes, halocarbons, and mixtures of these materials (32). The same agents are used for styrene–acrylonitrile foams.

The fabrication of these expandable particles into a finished cellular-plastic article is generally carried out in two steps (33–36). In the first step the particles are expanded by means of steam, hot water, or hot air into low density replicas of the original material, called prefoamed or preexpanded beads. After proper aging enough of these prefoamed beads are placed in a mold to just fill it; the filled mold is then exposed to steam. This second expansion of the beads causes them to flow into the spaces between beads and fuse together, forming an integral molded piece. Stabilization of the cellular structure is accomplished by cooling the molded article while it is still in the mold. The density of the cellular article can be adjusted by varying the density of the prefoamed particles.

Expandable polystyrene for loose-fill packing materials is available in various sizes and shapes varying from round disks to S-shaped strands. These particles can be prepared either by deforming the polystyrene under heat and impregnating the resin with a blowing agent in an aqueous suspension (37) or by the extrusion method with various die orifice shapes (38).

The expansion of these particles into a product is usually carried out in two or three expansions by means of steam with at least one day of aging after each expansion (39). Stabilization is accomplished by cooling the polymer phase below its glass transition temperature during the expansion process.

Poly(vinyl Chloride). Cellular poly(vinyl chloride) can be produced from several expandable formulations as well as by decompression techniques. Rigid or flexible products can be made depending upon the amount and type of plasticizer used (40) (see Vinyl polymers).

Polyethylene. Polyethylene has a sharp melting point and the viscosity of polyethylene decreases very rapidly over a narrow temperature range above the melting point. This makes it very difficult to produce a low density polyethylene foam with nitrogen or chemical blowing agents because the foam collapses before it can be stabilized. This problem can be eliminated by cross-linking the resin before it is foamed, which slows the viscosity decrease above the melting point as shown in Figure 1 and

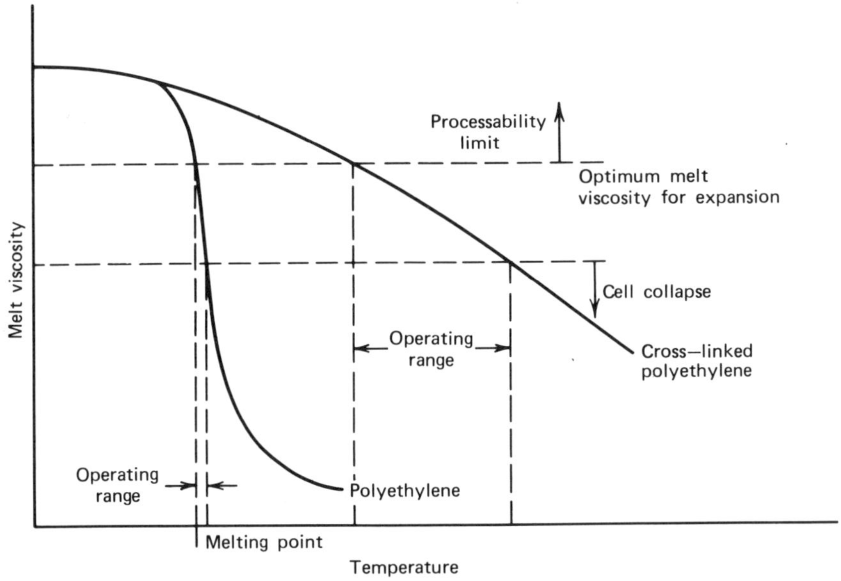

Figure 1. Effect of cross-linking on melt behavior of polyethylene (41).

allows the foam to be cooled without collapse of cell structure (see Olefin polymers).

Cross-linking of polyethylene can be accomplished either chemically or by high energy radiation. Radiation cross-linking is usually accomplished by x rays (42) or electrons (43–44). Chemical cross-linking of polyethylene is accomplished with dicumyl peroxide (45), di-*tert*-butyl peroxide (46), or other peroxides. Radiation cross-linking (41) is preferred for thin foams, and chemical cross-linking for the thicker foams.

Expandable polyethylene foam sheet can be made by a four-step process: (*1*) mixing of polyethylene, chemical blowing agent and cross-linking agent (in the case of chemical cross-linking) at low or medium temperature (examples of decomposable blowing agents used for expandable polyethylene are azodicarbonamide, 4,4′-oxybis(benzenesulfonyl hydrazide), and dinitrosopentamethylenetetramine (32)); (*2*) shaping at low or medium temperature; (*3*) chemical cross-linking at medium temperature or radiation cross-linking; and (*4*) heating and expanding at high temperature. Expansion of the cross-linked, expandable polyethylene sheet can be accomplished either by floating the sheet on the surface of a molten salt bath at 200–250°C and heating from above with ir heaters or circulating hot air or by expanding in the mold with a high pressure steam (see also Film and sheeting materials).

Chemical Stabilization Processes. This method has proved the most versatile in producing cellular plastics in that it has been used successfully for more materials than the physical stabilization process. It has generally been more adaptable for condensation polymers than for vinyl polymers because of the fast yet controllable curing reactions and the absence of atmospheric inhibition.

Polyurethane Foams. The most important commercial example of the chemical stabilization process is the production of polyurethane [*27416-86-0*] foams. More literature has been published on the chemical and physical aspects of the polyurethane foam process than on all other plastic foam processes combined in spite of the relatively short commercial lifetime of polyurethane. An excellent summary of the chemistry and technology of these polymers has been published (12,47) (see Urethane polymers).

The chemical ingredients of a polyurethane foam system are a polyfunctional isocyanate (**1**) and a hydroxyl-containing polymer (**2**), along with the catalysts necessary to control the rate and type of reaction and other additives to control the surface chemistry of the process. A number of competing reactions can occur when (**1**) and (**2**) are brought together but the main product, shown in equation 3, is a polyurethane (**3**):

$$\underset{(\mathbf{1})}{\mathrm{OCN{-}R{-}NCO}} + \underset{(\mathbf{2})}{\mathrm{HO{-}R'{-}OH}} \longrightarrow \underset{(\mathbf{3})}{\sim\!\left[\overset{\mathrm{O}}{\overset{\|}{\mathrm{C}}}\mathrm{NH{-}R{-}NH}\overset{\mathrm{O}}{\overset{\|}{\mathrm{C}}}\mathrm{{-}O{-}R'{-}O}\right]_n\!\sim} \qquad (3)$$

Originally carbon dioxide was generated *in situ* (by the reaction of isocyanate with water) as a blowing agent for both rigid and flexible polyurethane foams. It is still common practice to rely partly on this method of gas generation for flexible materials. Rigid cellular polyurethanes are now produced using volatile liquids, usually halocarbons, as expanding agents, producing gas as the foaming mixture heats up. These materials remain in the foam, and their presence as gas in the cells lowers the thermal conductivity of the foam considerably.

The general method of producing a cellular polyurethane is to mix the above ingredients and adjust the conditions and reactants such that the exotherm of the reaction causes expansion of the blowing agent and resultant foaming of the resins. The change of physical properties of the mixture is timed to coincide with the expansion of the foaming mixture so that dimensional stabilization by cross-linking of the polymer occurs at the time corresponding to minimum density.

The physical properties of the final cellular material can be varied broadly by controlling the degree of cross-linking in the final polymer as well as the structure of R and R′ in equation 3. The average molecular weight between cross-links is generally 400–700 for rigid polyurethane foams, 700–2,500 for semirigid foams, and 2,500–20,000 for flexible polyurethane foams (12). The variation of cross-link density is normally obtained by varying the structure of the hydroxyl-containing component (**2**), since the structure of the diisocyanate is limited to some 6 or 8 commercially available compounds (12,48). Because the variation between cross-links is controlled by the structure of the polyhydroxy resin component, it is common to use the equivalent weight per hydroxyl unit in the resin as a criterion for rigid, semirigid, and flexible foams. In these terms, the equivalent weights of resins used for rigid foams vary from 100 to 150, for semirigid between 200 and 1,000, and for flexible from 1,000 to 10,000. The equivalent weight in this case is defined as the ratio of molecular weight to hydroxyl functionality of the polyhydroxy resin.

Two general types of processes have been developed for producing cellular polyurethanes on a commercial scale (47). These two processes are commonly called the one-shot process and the prepolymer processes. In the one-shot process all the necessary ingredients for producing the foam are mixed and then discharged from the mixer onto a suitable surface. The reactions begin immediately and proceed at such a rate that expansion starts in about 10 s, the entire expansion is completed in 1–2 min. Curing may take several days.

In the prepolymer process the polyhydroxy component reacts with enough polyisocyanate to form a prepolymer with isocyanate end groups plus excess isocyanate:

$$\text{OCN—R—NCO} + \text{HO—R}'\text{—OH} \longrightarrow \underbrace{\text{OCN—R—NH—}\overset{\overset{\displaystyle O}{\|}}{\text{C}}\text{O—R}'\text{—O}\overset{\overset{\displaystyle O}{\|}}{\text{C}}\text{—}\overset{\displaystyle H}{\text{N}}\text{—R—NCO}}_{\text{prepolymer}} \qquad (4)$$

R is typically a polyether structure but can also be a polyester. The prepolymer mixture then reacts with water to release carbon dioxide for expansion and to link the chains into a cross-linked matrix. This method is sometimes used for flexible foams.

In a semiprepolymer process, which has become more extensively used, a prepolymer containing excess isocyanate is mixed with more polyhydroxy resin and a separate blowing agent such as a halocarbon. In this case the prepolymer may contain only a few percent of the total polyhydroxy resin. This method is particularly useful for producing rigid foams.

After mixing, the resin can be dispensed by several different methods (47). Typical methods are (*1*) as an expandable liquid, (*2*) as a spray of small droplets of mixed resins which adhere to surfaces and foam on these surfaces, and (*3*) as a froth into which some gas has been mixed prior to exiting from the mixing head, which causes the liquid mixture to froth as its pressure is decreased to atmospheric pressure. A modification of method (*3*) allows the froth to be sprayed in chunks upon a surface with subsequent additional expansion during curing.

Polyisocyanurates. The isocyanurate ring formed by the trimerization of isocyanate has been known to possess high thermal and flammability resistance and low smoke generation (49–52). Because of these characteristics, aluminum foil-faced sheets of isocyanurate foam are now widely used as an insulation material. Cross-linking via the high functionality of isocyanurates causes an inherent friability of the foam. Chemical modifications of the foam with urethane groups is usually made to circumvent the friability of pure isocyanurate foam. When introducing the urethane group into the polymer backbone, the proper catalyst must be selected to favor or promote formation of desired compounds from the excess isocyanate during polymerization (53–54). Sodium or potassium coordination compounds with organic borate esters have been suggested as suitable catalysts for isocyanurate-modified polyurethane foams (55). Another isocyanurate catalyst consists of a mixture of trialkylaminohexahydrotriazine and a 1,2-epoxide (56). The manufacturing process for isocyanurate foam is similar to that for polyurethane foam (see Cyanuric and isocyanuric acids).

Polyphenols. Another increasingly important example of the chemical stabilization process is the production of phenolic foams (57–60). Phenolic foams are produced by cross-linking polyphenols (resoles and novolacs). The principal features of phenolic foams are low flammability, solvent resistance, and excellent dimensional stability over a wide temperature range (57). Because of these unique properties, phenolic foams are good thermal insulating materials (see Phenolic resins).

Most phenolic foams are produced from resoles and acid catalyst. Suitable water-soluble acid catalysts are mineral acids (such as hydrochloric acid or sulfuric acid) and aromatic sulfonic acids (61). Phenolic foams can be produced from novolacs with more difficulty than from resoles (57). Novolacs are thermoplastic and require a source of methylene group to permit cure. This is usually supplied by hexamethylenetetramine (62).

A typical phenolic foam system consists of liquid phenolic resin, blowing agent, catalyst, surface-active agent, and modifiers. Various formulations and composite systems (63–65) can be used to improve one or more properties of the foam in specific applications such as insulation properties (61,66–69), flammability (70–73), and open cell (74–76) (quality).

Several manufacturing processes can be used to produce phenolic foams (57,77): continuous production of free-rising foam for slabs and slab stock similar to that for polyurethane foam (59,78); foam-in-place batch process (59,79); sandwich paneling (61,80–81); and spraying (68,82).

Other Materials. Foams from epoxy resins (qv) (57–58,83–84), and silicone resins (see Silicon compounds) (29,58,85–86) can also be formed by chemical stabilization processes.

DECOMPRESSION EXPANSION PROCESSES

Physical Stabilization Process. Cellular polystyrene, cellulose acetate, polyolefins, and poly(vinyl chloride) can be manufactured by this process.

Polystyrene. The extrusion process for producing cellular polystyrene is probably the oldest process utilizing physical stabilization in a decompression expansion process (87). A solution of blowing agent in molten polymer is formed in an extruder under pressure. This solution is forced out through an orifice onto a moving belt at ambient temperature and pressure. The blowing agent then vaporizes and causes the polymer to expand. The polymer simultaneously expands and cools under such conditions that it has developed enough strength to maintain dimensional stability at the time corresponding to optimum expansion. In this case the stabilization is owing to cooling of the polymer phase to a temperature below its glass transition temperature. Cooling comes from the vaporization of the blowing agent, gas expansion, and heat loss to the environment. Polystyrene foams produced by the decompression process are commercially offered in the density range of 27–53 kg/m^3 (1.7–3.3 lb/ft^3) as well as at higher densities (88).

Several procedures can be used to produce polystyrene foam sheet for various applications. The extrusion of expandable polystyrene beads or pellets containing pentane blowing agent was originally used to produce low density foam sheet (89–90). The current method is to extrude polystyrene foam in a single-screw tandem line or twin-screw extruder and produce foam sheet by addition of pentane or fluorocarbon blowing agents into the extruder (91–92). Two general methods are used for the production of low density polystyrene foam sheet. For sheet thicknesses of less than 500 μm (20 mil), the blown-bubble method is normally used. This method involves blowing a tube from a round or annular die, collapsing the bubble, and then slitting the edges to obtain two flat sheets. For greater sheet thicknesses the sheet is pulled over a sizing mandrel and slit to obtain a flat sheet. Cooling of the expanded material by the external air is necessary to stabilize the foam sheet with a good skin quality.

Cellular polystyrene can also be produced by an injection-molding process. Polystyrene granules containing dissolved liquid or gaseous blowing agents are used as feed in a conventional injection-molding process (93). With close control of time and temperature in the mold and use of vented molds, high density cellular polystyrene moldings can be obtained (see also Styrene; Styrene polymers).

Cellulose Acetate. The extrusion process has also been used to produce cellular cellulose acetate [*9004-35-7*] (94) in the density range of 96–112 kg/m^3 (6–7 lb/ft^3). A hot mixture of polymer, blowing agent, and nucleating agent is forced through an orifice into the atmosphere. It expands, cools, and is carried away on a moving belt (see Cellulose acetate).

Polyolefins. Cellular polyethylene and polypropylene are prepared by both extrusion and molding processes. High-density polyolefin foams in the density range of 320–800 kg/m^3 are prepared by mixing a decomposable blowing agent with the polymer and feeding the mixture under pressure through an extruder at a temperature such that the blowing agent is partially decomposed before it emerges from an orifice into a lower pressure zone. Simultaneous expansion and cooling take place, resulting in a stable cellular structure owing to rapid crystallization of the polymer, which increases the modulus of the polymer enough to prevent collapse of cell structure (26,36,95). This process is widely used in wire coating and structural foam products. These products can also be produced by direct injection of inert gases into the extruder (96–97).

Low density polyethylene foam products (thin sheets, planks, rounds, tubes) in the range of 32–160 kg/m^3 (2–10 lb/ft^3) have been prepared by an extrusion technique using various gaseous fluorocarbon blowing agents (98–99). The techniques are similar to those described earlier for producing extruded polystyrene foam planks and foam sheets (see also Olefin polymers).

Structural Foams. Structural foams having an integral skin, cellular core, and a high strength-to-weight ratio are formed by means of injection molding, extrusion, or casting, depending on product requirements (100–101). Two processes are most widely used among the existing injection molding processes: the Union Carbide low pressure process (102) and the USM high pressure process (103) (see also Plastic building products; Plastics processing).

In the low pressure process, a short shot of a resin containing a blowing agent is forced into the mold where the expandable material is allowed to expand to fill the mold under pressures of 690–4140 kPa (100–600 psi). This process produces structural foam products with a characteristic surface swirl pattern produced by the collapse of cells on the surface of molded articles.

In the high pressure process, a resin melt containing a chemical blowing agent is injected into an expandable mold under high pressure. Foaming begins as the mold cavity expands. This process produces structural foam products with very smooth surfaces since the skin is formed before expansion takes place.

Extruded structural foams are produced with conventional extruders and a specially designed die. The die has an inner, fixed torpedo located at the center of its opening, which provides a hollow extrudate. The outer layer of the extrudate cools and solidifies to form solid skin; the remaining extrudate expands toward the interior of the profile. One of the most widely used commercial extrusion processes is the Celuka process developed by Ugine-Kuhlman (104).

Large structural foam products are produced by casting expandable plastic pellets containing a chemical blowing agent in aluminum molds on a chain conveyor. After closing and clamping the mold, it is conveyed through a heating zone where the pellets soften, expand and fuse together to form the cellular products. The mold is then passed through a cooling zone. This process produces structural foam products with uniform, closed-cell structure but no solid skin.

Polyurethane structural foam produced by reaction injection molding (RIM) is a rapidly growing product that provides industry with the design flexibility required for a wide range of applications. This process is more efficient than conventional methods in producing large area, thin wall, and load-bearing structural foam parts. In the RIM process, polyol and isocyanate liquid components are metered into a temperature controlled mold that is filled 20–60%, depending on the density of structural foam parts (105). When the reaction mixture then expands to fill the mold cavity, it forms a component part with an integral, solid skin and a microcellular core. The quality of the structural part depends on precise metering, mixing, and injection of the reaction chemicals into the mold.

Poly(vinyl Chloride). Cellular poly(vinyl chloride) is prepared by many methods (106), some of which utilize decompression processes. In all reported processes the stabilization process used for thermoplastics is to cool the cellular state to a temperature below its second-order transition temperature before the resin can flow and cause collapse of the foam.

A new type of physical stabilization process, unique for poly(vinyl chloride) resins, is the fusion of a dispersion of plastisol resin in a plasticizer. The viscosity of a resin–plasticizer dispersion shows a sharp increase at the fusion temperature. In such a system expansion can take place at a temperature corresponding to the low viscosity; the temperature can then be raised to increase viscosity and stabilize the expanded state.

Extrusion processes have been used to produce high and low density flexible cellular poly(vinyl chloride). A decomposable blowing agent is usually blended with the compound prior to extrusion. The compounded resin is then fed to an extruder where it is melted under pressure and forced out of an orifice into the atmosphere. After extrusion into the desired shape, the cellular material is cooled to stabilize it and is removed by a belt.

Another type of extrusion process involves the pressurization of a fluid plastisol at low temperatures with an inert gas. This mixture is subsequently extruded onto a belt or into molds, where it expands (107–108). The expanded dispersion is then heated to fuse it into a dimensionally stable form.

Injection molding of high density cellular poly(vinyl chloride) can be accomplished in a manner similar to extrusion except that the extrudate is fed for cooling into a mold rather than being maintained at the uniform extrusion cross section.

Chemical Stabilization Processes. Cellular rubber and ebonite are produced by chemical stabilization processes.

Cellular Rubber. The term cellular rubber refers to an expanded elastomer produced by expansion of a rubber stock, in contrast to latex foam rubber, which is produced from a latex. The following general procedure applies to production of cellular rubbers from a variety of types of rubber (109). A decomposable blowing agent, along with vulcanizing systems and other additives, is compounded with the uncured elastomer at a temperature below the decomposition temperature of the blowing agent. When the uncured elastomer is heated in a forming mold, it undergoes a viscosity change, as shown in Figure 2. The blowing agent and vulcanizing systems are chosen to yield open-celled or closed-celled cellular rubber. Although inert gases such as nitrogen have been pressurized into rubber and the rubber then expanded upon release of pressure, the current cellular rubbers are made almost entirely with decomposable blowing agents. The various types of such decomposable blowing agents used are ex-

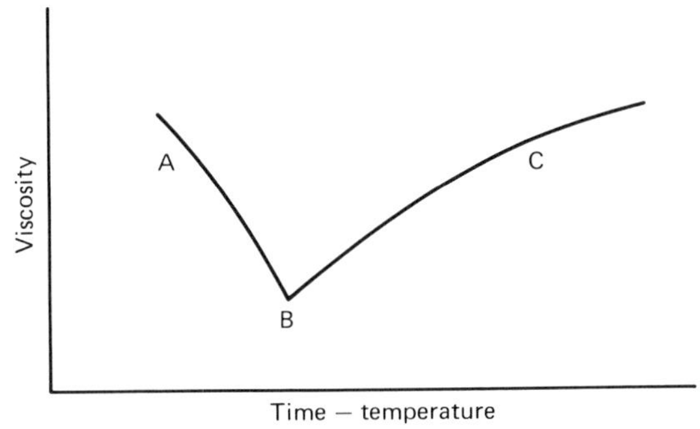

Figure 2. Viscosity of cellular rubber stock during production cycle (109).

emplified by sodium bicarbonate, 2,2′-azobisisobutyronitrile, azodicarbonamide, 4,4′-oxybis(benzenesulfonyl hydrazide), and dinitrosopentamethylenetetramine. Each of these blowing agents is typical of a class of compounds useful for expanding cellular rubber; the compound named is the most important commercial compound in its particular class (see Rubber; Elastomers, synthetic).

Open-Celled Cellular Rubber. To produce this type of rubber the blowing agent is decomposed just prior to point A in Figure 2, so that the gas is released at the point of minimum viscosity. As the polymer expands the cell walls become thin and rupture; however, the connecting struts have developed enough strength to support the foam. This process is ordinarily carried out in one step inside a mold under pressure.

Closed-Celled Cellular Rubber. The timing for blowing-agent decomposition is more critical in this case—it must occur soon enough after point A to cause expansion of the elastomer but far enough past point A to allow the cell walls to become strong enough not to rupture under the blowing stress. The expansion of closed-celled rubber is often carried out in two main steps: a partial cure is carried out in a mold that is a reduced-scale replica of the final mold; removed from this mold, it expands partly toward its final form. It is then placed in a mold of exactly the desired size and heated to complete the expansion and cure.

Most elastomers can be made into either open-celled or closed-celled materials. Natural rubber, SBR, nitrile rubber, polychloroprene, chlorosulfonated polyethylene, ethylene–propylene terpolymers, butyl rubbers, and polyacrylates have been successfully used (4,109–110).

A continuous extrusion process, as well as molding techniques, can be used as the thermoforming method. A more rapid rate of cure is then necessary to ensure the cure of the rubber before the cellular structure collapses. The stock is ordinarily extruded at a temperature high enough to produce some curing and expansion and then oven-heated to complete the expansion and cure.

A unique process for chemical stabilization of a cellular elastomer upon extrusion has been shown for ethylene–propylene rubber: the expanded rubber obtained by extrusion is exposed to high-energy radiation to cross-link or vulcanize the rubber and give dimensional stability (8).

Cellular Ebonite. Cellular ebonite [*12765-25-2*] is the oldest rigid cellular plastic. It was produced in the early 1920s by a process similar to the processes described for making cellular rubber. The formulation of rubber and vulcanizing agent is changed to produce an ebonite rather than rubber matrix (111).

DISPERSION PROCESSES

In several techniques for producing cellular polymers, the gas cells are produced by dispersion of a gas or solid in the polymer phase followed, when necessary, by stabilization of the dispersion and subsequent treatment of the stabilized dispersion. In frothing techniques a quantity of gas is mechanically dispersed in the fluid polymer phase and stabilized. In another method, solid particles are dispersed in a fluid polymer phase, the dispersion stabilized, and then the solid phase dissolved or leached, leaving the cellular polymer. Still another method relies on dispersing an already cellular solid phase in a fluid polymer and stabilizing this dispersion. This results directly in cellular polymers called syntactic foams.

Frothing. The frothing process for producing cellular polymers is an exact duplicate of the process used for making meringue topping for pies. A gas is dispersed in a fluid that has surface properties proper to produce a foam of transient stability. The foam is then permanently stabilized by chemical reaction. The fluid may be a homogeneous material, a solution, or a heterogeneous material.

Latex Foam Rubber. Latex foam rubber was the first cellular polymer to be produced by frothing. The basic steps in this process are as follows: (*1*) a gas is dispersed in a suitable latex; (*2*) the rubber latex particles are caused to coalesce and form a continuous rubber phase in the water phase; (*3*) the aqueous soap film breaks owing to deactivation of the surfactant in the water, breaking of the latex film and causing retraction into the connecting struts of the bubbles; and (*4*) the expanded matrix is cured and dried to stabilize it.

The earliest frothing process developed was the Dunlop process, which made use of chemical gelling agents (eg, sodium fluorosilicate) to coagulate the rubber particles and deactivate the soaps. The Talalay process, developed later, employs freeze–coagulation of the rubber followed by deactivation of the soaps with carbon dioxide. The basic processes and a multitude of improvements on each are discussed extensively in ref. 3. A discussion more oriented to current use of these processes is given in ref. 112.

Latex rubber foams are generally prepared in slab or molded forms in the density range 64–128 kg/m^3 (4–8 lb/ft^3). Synthetic SBR latexes have now replaced natural rubber latexes as the largest-volume raw material for latex foam rubber. Other elastomers used in significant quantities are polychloroprene [*9010-98-4*], nitrile rubbers, and synthetic *cis*-polyisoprene [*9003-31-0*] (112).

A recent method (113) of producing cellular polymers from a variety of latexes uses primarily latexes of carboxylated styrene–butadiene copolymers, although other elastomers such as acrylic elastomers, nitrile rubber, and vinyl polymers can be employed.

Urea–Formaldehyde Resins. Cellular urea–formaldehyde resins can be prepared in the following manner: An aqueous solution containing surfactant and catalyst is made into a low density, fine-celled foam by dispersing air into it mechanically. A second aqueous solution consisting of partially cured urea–formaldehyde resin is then mixed into the foam by mechanical agitation. The catalyst in the initial foam causes the dispersed resin to cure in the cellular state. The resultant hardened foam is dried at elevated temperatures. Densities as low as 8 kg/m^3 can be obtained by this method (114) (see Amino resins).

Syntactic Cellular Polymers. Syntactic cellular polymer is produced by dispersing rigid, foamed, microscopic particles in a fluid polymer and then stabilizing the system. The particles are generally spheres or microballoons of phenolic resin, urea–formaldehyde resin, glass, or silica, ranging 30–120 μm dia. Commercial microballoons have densities of approximately 144 kg/m^3 (9 lb/ft^3). The fluid polymers used are the usual coating resins, eg, epoxy resin, polyesters, and urea–formaldehyde resin.

The resin, catalyst, and microballoons are mixed to form a mortar which is then cast into the desirable shape and cured. Very specialized electrical and mechanical properties may be obtained by this method but at higher cost. This method of producing cellular polymers is quite applicable to small quantity, specialized applications because it requires very little special equipment.

In a variation on the usual methods for producing syntactic foams (115–116), expandable polystyrene or styrene–acrylonitrile copolymer particles (in either the unexpanded or prefoamed state) are mixed with a resin (or a resin containing a blowing agent) which has a large exotherm during curing. The mixture is then placed in a mold and the exotherm from the resin cure causes the expandable particles to foam and squeeze the resin or foamed matrix to the surface of the molding. A typical example is Dow Chemical Company's Voraspan, expandable polystyrene in flexible polyurethane foam matrix (117). These foams are finding acceptance in cushioning applications such as automobile crash-pad and molded furniture applications.

OTHER PROCESSES

Some plastics cannot be obtained in a low viscosity melt or solution that can be processed into a cellular state. Two other methods have been used to achieve the needed dispersion of gas in solid: sintering of solid plastic particles and leaching of soluble inclusions from the solid plastic phase.

Sintering. Sintering has been used to produce a porous polytetrafluoroethylene (15) (see Fluorine compounds, organic). Cellulose sponges are the most familiar cellular polymers produced by the leaching process (118). Sodium sulfate crystals are dispersed in the viscose syrup and subsequently leached out. Polyethylene (119) are poly(vinyl chloride) can also be produced in cellular form by the leaching process. The artificial leatherlike materials used for shoe uppers are rendered porous by extraction of salts (120) or by designing the polymers in such a way that they precipitate as a gel with many holes (121) (see Leatherlike materials).

Microporous polymer systems consisting of essentially spherical, interconnected voids, with a narrow range of pore and cell-size distribution have been produced from a variety of thermoplastic resins by the phase-separation technique (122). If a polyolefin or polystyrene is insoluble in a solvent at low temperature but soluble at high temperatures, the solvent can be used to prepare a microporous polymer. When the solutions, containing 10–70% polymer, are cooled to ambient temperatures, the polymer separates as a second phase. The remaining nonsolvent can then be extracted from the solid material with common organic solvents. These microporous polymers may be useful in microfiltrations or as controlled-release carriers for a variety of chemicals (see Microencapsulation; Pharmaceuticals, controlled release; Ultrafiltration).

A summary of the methods for commercially producing cellular polymers is presented in Table 1. This table includes only those methods thought to be commercially significant and certainly is not inclusive of all methods known to produce cellular products from polymers.

Table 1. Methods for Production of Cellular Polymers

Type of polymer	Extrusion	Expandable formulation	Spray	Froth foam	Compression mold	Injection mold	Sintering	Leaching
cellulose acetate	X							
epoxy resin		X	X	X				X
phenolic resin		X						
polyethylene	X	X			X	X	X	X
polystyrene	X	X				X	X	
silicones		X						
urea–formaldehyde resin				X				
urethane polymers		X	X	X		X		
latex foam rubber				X				
natural rubber	X	X			X			
synthetic elastomers	X	X			X			
poly(vinyl chloride)	X	X		X	X	X		X
ebonite					X			
polytetrafluoro-ethylene							X	

Properties of Cellular Polymers

Test Methods. Several countries have developed their own standard test methods for cellular plastics, and the International Organization for Standards (ISO) Technical Committee on Plastics T-61 has been making considerable progress on developing international standards (123). Information concerning the test methods for any particular country or the ISO procedures can be obtained in the United States from the American Standards Association (124). The most complete set of test procedures for cellular plastics is that developed by the ASTM. These procedures, the most used of any in the world, are published in new editions each year (125). There have been several reviews of those ASTM methods and others pertinent to cellular plastics (29,57,126–128).

Properties of Commercial Foamed Plastics. The properties of commercial rigid foamed plastics are presented in Table 2. The data shown are meant to demonstrate the broad ranges of properties of commercial products rather than a very accurate set of properties on a specific few materials. Specific producers of foamed plastics should be consulted for properties on a particular product (132,134, 139).

The properties of commercial flexible foamed plastics are presented in Table 3. The definition of a flexible foamed plastic is that recommended by the ASTM Subcommittee D 11-22. Again, these values are general in nature, and specific properties of specific foams should be obtained from the manufacturer of that product (132,139).

The properties that are achieved in commercial structural foams (density >0.3 g/cm^3) are shown in Table 4. Because these values depend upon several structural variables, they can be used only as general guidelines of mechanical properties from these products. Specific properties must be determined on the particular part to be produced because of the number of process and structural variables that influence these properties. A good guide recently has been published (142).

Structural Variables. The properties of a foamed plastic can be related to several variables of composition and geometry often referred to as structural variables. The particular set presented here was chosen (21) because it best meets the criteria of simplicity, completeness, and maximum compatibility with those structural variables reported in literature.

Polymer Composition. The properties of foamed plastics are influenced both by the foam structure and, to a greater extent, by the properties of the parent polymer. The polymer phase description must include the additives present in that phase as well.

Polymer State. The condition or state of the polymer phase (orientation, crystallinity, previous thermal history), as well as its chemical composition, determines the properties of that phase. The polymer state and cell geometry are intimately related because they are determined by common forces exerted during the expansion and stabilization of the foam.

Density. Density is the most important variable in determining mechanical properties of a foamed plastic of given composition. Its effect has been recognized since foamed plastics were first made and has been extensively studied.

Cell Structure. A complete knowledge of the cell structure of a cellular polymer requires a definition of its cell sizes, cell shapes, and location of each cell in the foam. Because such a specific definition is impractical, one must be content with measurable quantities that approach this absolute definition.

Cell Size. Cell size has been characterized by measurements of the cell diameter in one or more of the three mutually perpendicular directions (146) and as a measurement of average cell volume (147–148). Mechanical, optical, and thermal properties of a foam are all dependent upon the cell size.

Cell Geometry. The shape of the cells is governed predominantly by the final foam density and the external forces exerted on the cellular structure prior to its stabilization in the expanded state. In a foam prepared without such external forces, the cells tend to be spherical or ellipsoidal at gas volumes less than 70–80% of the total volume, and they tend toward the shape of packed regular dodecahedra at greater gas volumes. These shapes have been shown to be consistent with surface chemistry arguments (147,149–150). Photographs of actual foam cells (Fig. 3) show the general shapes to be consistent with the above argument but there is a broad range of variations in shape.

In the presence of external forces, plastic foams in which the cells are elongated or flattened in a particular direction may be formed. This cell orientation can have a marked influence on many properties. The results of a number of studies have been recently reviewed (57,151) (see also Foams).

Fraction Open Cells. An important character of the cell structure is the extent to which the gas phase of one cell is in communication with other cells. This is most commonly expressed as fraction open cells. When a large portion of cells are interconnected by gas phase, the foam has a large fraction of open cells, or is an open-celled foam. Conversely, a large proportion of noninterconnecting cells results in a small fraction of open cells, or a closed-cell foam.

The nature of the opening between cells determines how readily different gases and liquids can pass from one cell to another. Because of variation in flow of different liquids or gases through the cell-wall openings, a single measurement of fraction open cells does not fully characterize this structural variable, especially in a dynamic situation.

Table 2. Physical Properties of Commercial Rigid Foamed Plastics

Property	ASTM test	Cellulose acetate	Epoxy		Phenolic		Extruded plank	
density, kg/m³ [a]		96–128	32–48	80–128	32–64	112–160	35	53
mechanical properties								
compressive strength, kPa [b] at 10%	D 1621	862	138–172	414–620	138–620		310	862
tensile strength, kPa [b]	D 1623	1,172		345–1,240	138–379		517	
flexural strength, kPa [b]	D 790	1,014		1,380–5,516	172–448		1,138	
shear strength, kPa [b]	C 273	965			103–207		241	
compression modulus, MPa [c]	D 1621	38–90	3.9	14.5–44.8			10.3	
flexural modulus, MPa [c]	C 790	38		17.2–41.4			41	
shear modulus, MPa [c]	C 273				2.8–4.8		10.3	
thermal properties								
thermal conductivity, W/(m·K) [c]	C 177	0.045–0.046	0.016–0.022	0.035–0.040	0.029–0.032	0.035–0.040	0.030	
coefficient of linear expansion, 10^{-5}/°C	D 696		1.5	4.1	0.9	4.5	6.3	6.3
max service temperature, °C		177	205–260	205–260	132	205	74	
specific heat, kJ/(kg·K) [d]	C 351						1.1	
electrical properties								
dielectric constant	D 1673	1.12			1.19–1.20	1.19–1.20	<1.05	<1.05
dissipation factor		20			0.028–0.031	0.028–0.031	<0.0004	<0.0004
moisture resistance								
water absorption, vol %	C 272	4.5			13–51	10–15	0.02	0.05
moisture vapor transmission, g/(m·s·GPa) [e]	C 355		58				35	
references		21	129	129	130–131	130–131	21, 133	21

[a] To convert kg/m³ to lb/ft³, multiply by 0.0624.
[b] To convert kPa to psi, divide by 6.895.
[c] To convert MPa to psi, multiply by 145.
[d] To convert kJ/(kg·K) to Btu/(lb·°F), divide by 4.184.
[e] To convert GPa to psi, multiply by 145,000.

Table 3. Physical Properties of Commercial Flexible Foamed Plastics

Property	ASTM test	Expanded acrylonitrile–butadiene rubber	Expanded butyl rubber		Expanded natural rubber		Expanded neoprene		Expanded SBR	Latex	Foam	Rubber
density, kg/m³ [a]		160–400	128–144	224–304	56	320	112	192	72	80	130	160
cell structure		closed	closed	closed	closed	closed	closed	closed	closed	open	open	open
compressive strength												
10% deflection, kPa [b]	D 1621											
25% deflection, kPa [b]	D 1621		11–13.8	25–38				52				
tensile strength, kPa [b]	D 1564	275				206		758	551	103		
tensile elongation, %	D 1564							500		310		
rebound resilience, %	D 1054		39–36	30–16						73		
tear strength, (N/m) [c] × 10^2	D 624											
max service temperature, °C		100			70	70	105	70	70			
thermal conductivity, W/(m·K)	C 177	0.036–0.043			0.036	0.043	0.040	0.065	0.030		0.050	
electrical properties refs.											3	3
cushioning properties ref.												
mechanical properties refs.		128	128		128		128		128	128	128	128

[a] To convert kg/m³ to lb/ft³, multiply by 0.0624.
[b] To convert kPa to psi, divide by 6.895.
[c] To convert N/m to lbf/in., divide by 1.75.

Table 2 (*continued*)

Polystyrene							Polyurethane						
Expanded plank			Extruded sheet		Poly(vinyl chloride)		Polyether			Isocyanurate			Urea formal-dehyde
										Bun	Lami-nate		
16	32	80	96	160	32	64	32–48	64–128	144–192	304–400	32	32	13–19
90–124	207–276	586–896	290	469	345	1,035	138–344	482–1,896	2,000–3,800	8,270–13,800	117–206	97–282	34
145–193	310–379	1,020–1,186	2,070–3,450	4,137–6,900	551	1,207	138–482	620–2,000	1,585–3,100	4,800–8,960	248–290		
193–241	379–517				586	1,620	413–689	1,380–2,400		4,825–13,800			
	241				241	793	138–207	413–896		52,400	117		
	3.4–14				13.1	35	2.0–4.1	10.3–31					
	9.0–26				10.3	36	5.5–6.2	5.5–10.3					
	7.6–11.0				6.2	21	1.2–1.4	3.4–10.3			1.7		
0.037	0.035	0.035	0.035	0.035	0.023		0.016–0.025	0.022–0.030	0.027–0.036	0.049–0.060	0.019	0.020	0.026–0.030
5.4–7.2	5.4–7.2	5.4–7.2					5.4–7.2	7.2	7.2	7.2			
74–80	74–80	74–80	77–80	80			93–121	121–149	121–149	121–149	149	149	
							ca 0.9	ca 0.9	ca 0.9	ca 0.9			
1.02	1.02	1.02	1.27	1.28			1.05	1.1	1.2	1.4			
0.0007	0.0007	0.0007	0.00011	0.00014			13	18	32				
1–4	1–4	1–4											
<120	35–120	23–35	86	58	15		35–230	50–120	12		230		1,610–2,000
	131–132	134			135	135	136	136	136	136	137	137	138

Table 3 (*continued*)

Polyethylene extruded plank				Polyethylene sheet		Polypropylene			Polyurethane				Poly(vinyl chloride)			Silicone		
				Extru-ded	Cross-linked	Un-modi-fied	Modi-fied	Sheet	Super-soft		Poly-ether slab					Liquid	Sheet	Sheet
35	64	96	144	43	26–38	64–96	64–96	10	24	18	29	38	56	112	96	272	160	400–544
closed	closed	closed	closed	closed	closed	closed	closed		open	open	open	open	closed	closed	open	open	open	closed
48	75	124	360			550	206	4.8										
138	241	413	690	41		830	344		2.2	1.9	5.0	7	10.3	24	3.4	36 at 20%		
60	60	60	60	276	276–480	1100	1380	138–275	82	62	140	115			220	227	310	550–700
				50		25	75		190	220	258	163						
									31		49	55						
10.5	17.5	26	51	26							6	3						
82	82	82	82	82	79–93	135	135	121			125					350	260	232
0.053	0.058	0.058	0.058	0.040–0.049	0.036–0.040	0.039	0.039	0.039			0.040		0.035	0.040		0.078	0.086	
140	140	140	140	21, 140				131	141	43	43,	43	143	143		144	131	131
				140							142							
140	140	140	140	140	131	145	145	131	128	128	128	128	128	128	128	144	131	131

Table 4. Typical Physical Properties of Commercial Structural Foams

Property	ASTM test	ABS		Noryl	Nylon-6,6 glass-reinforced	Poly-carbonate	Thermo-plastic poly-ester	Poly-ethylene, high density	Poly-propylene		High impact polystyrene		Polyurethane			Poly-(vinyl chloride)
glass reinforced		no	yes	no	yes	no	30%	no	no	20%	no	20%	no	no	no	no
density, g/cm³		0.80	0.85	0.80	0.97	0.80	1.10	0.60	0.60	0.73	0.70	0.84	0.40	0.50	0.60	0.50
tensile strength, kPa[a]	D 1623	18,600	48,000	22,700	101,000	37,900	76,000	8,900	13,800	20,700	12,400	34,500	11,000	17,200	23,400	6,900
compression strength, kPa[a] at 10% compression	D 1621	6,900		34,500		51,700	76,000	8,900					5,500	12,400	19,300	
flexural strength, kPa[a]	D 790	25,500	82,700	41,400	172,000	68,900	137,900	18,800	22,000	41,400	31,000	58,600	22,000	31,700	41,400	
flexural modulus, GPa[b]	D 790	0.86	5.2	1.7	5.2	2.1	6.6	0.83	0.83	2.8	1.4	5.2	0.7	0.9	1.1	
max use temperature, °C		82		96	203	132	193	110	115							
ref.													141	141	141	

[a] To convert kPa to psi, divide by 6.895.
[b] To convert GPa to psi, multiply by 145,000.

Gas Composition. In closed-celled foams, the gas phase in the cells can contain some of the blowing agent (called captive blowing agent), gas components of air which have diffused in, or other gases generated during the foaming process. Such properties as thermal and electrical conductivity can be profoundly influenced by the cell gas composition. In open-celled foams the presence of air exerts only a minor influence on the static properties but does affect the dynamic properties such as cushioning.

Mechanical Properties. The mechanical properties of rigid foams vary considerably from those of flexible foams. The tests used to characterize these two classes of foams are, therefore, quite different, and the properties of interest from an application standpoint are also quite different. In this discussion the ASTM definition of rigid and flexible foams given earlier is used. A separate class of high density, rigid cellular polymers has grown rapidly in the last 5 to 7 years to become significant commercially. These are the structural foams with a density >0.3 g/cm^3. This class of polymer is treated as a separate class of rigid foams.

Rigid Cellular Polymers. Compressive strength and modulus are readily determined and are important in many applications. They have been widely used, therefore, as general criteria to characterize the mechanical properties of rigid plastic foams.

Rigid cellular polymers generally do not exhibit a definite yield point when compressed but instead show an increased deviation from Hooke's law as the compressive load is increased. Figure 4 demonstrates this behavior for several cellular polymers. For precision the compressive strength is usually reported at some definite deflection (commonly 5 or 10%). The compressive modulus is reported as extrapolated to 0% deflection unless otherwise stated. Structural variables that affect the compressive strength and modulus of a rigid plastic foam are, in order of decreasing importance: plastic-phase composition, density, cell structure, and plastic state. The effect of gas composition is minor, with a slight effect of gas pressure in some cases.

Figures 5 and 6 illustrate the large effect density and polymer composition have on the compressive strength and modulus. The dependence of compressive properties on cell size has been discussed (21). The cell shape or geometry has also been shown important in determining the compressive properties (21,57,151,158–159). In fact, the foam cell structure is controlled in some cases to optimize certain physical properties of rigid cellular polymers.

Strengths and moduli of most polymers increase as the temperature decreases (160). This behavior of the polymer phase carries over into the properties of polymer foams, and similar dependence of the compressive modulus of polyurethane foams on temperature has been shown (154).

Tensile strength and modulus of rigid foams have been shown to vary with density in much the same manner as the compressive strength and modulus. General reviews of the tensile properties of rigid foams are available (21,57,128,151,161).

Those structural variables most important to the tensile properties are polymer composition, density, and cell shape. Variation with use temperature has also been characterized (162).

Flexural strength and modulus of rigid foams both increase with increasing density in the same manner as the compressive and tensile properties. More specific data on particular foams are available from manufacturers literature and in references 21, 57–58, 128, and 161.

Shear strength and modulus of rigid foams depend on the polymer composition and state, density, and cell shape. The shear properties increase with increasing density

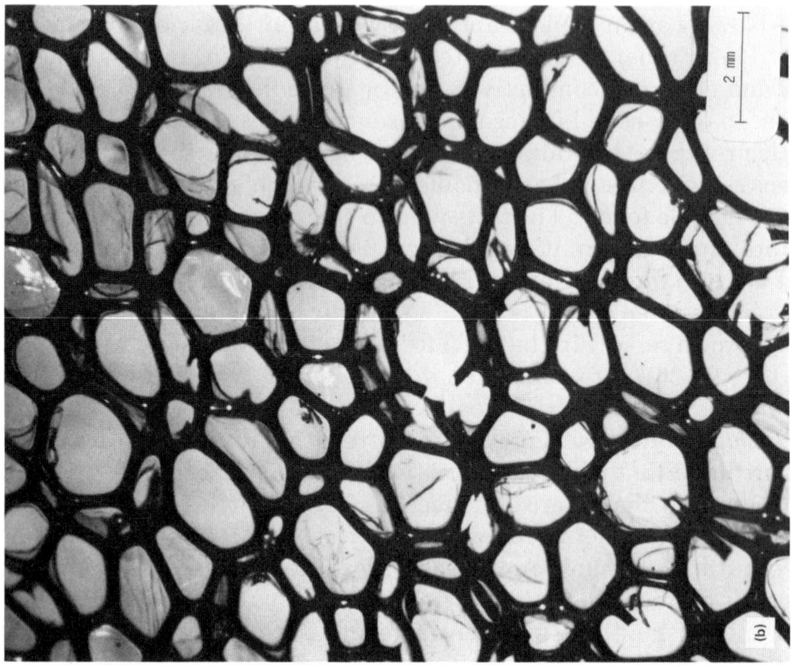
2 mm
(b)

1 mm
(a)

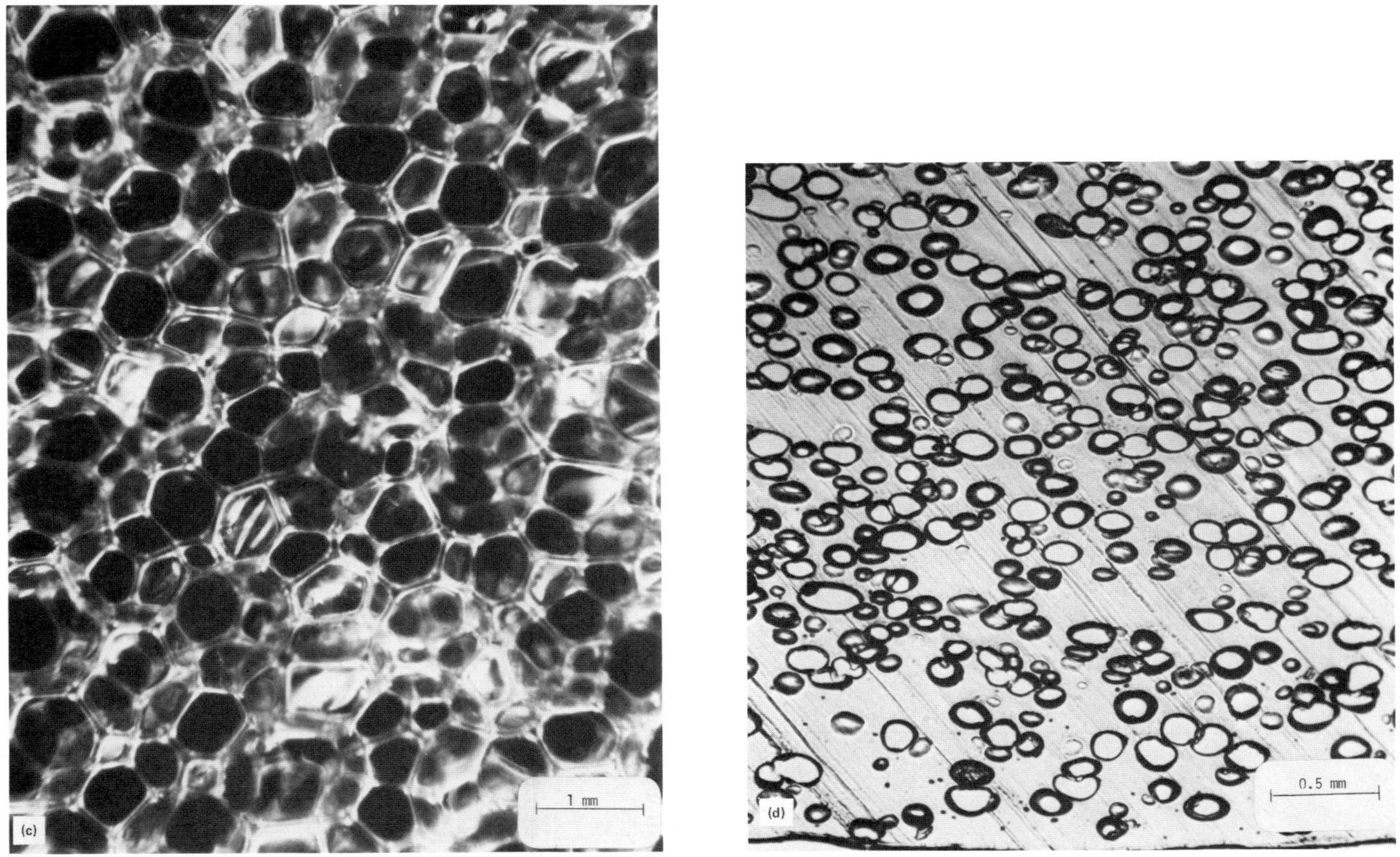

Figure 3. Photomicrographs of foam cell structure: (**a**) extruded polystyrene foam, reflected light, 26×; (**b**) polyurethane foam, reflected light, 12×; (**c**) polyurethane foam, transmitted light, 26×; (**d**) high density plastic foam, transmitted light, 50× (21). Courtesy of Van Nostrand Reinhold Publishing Corp.

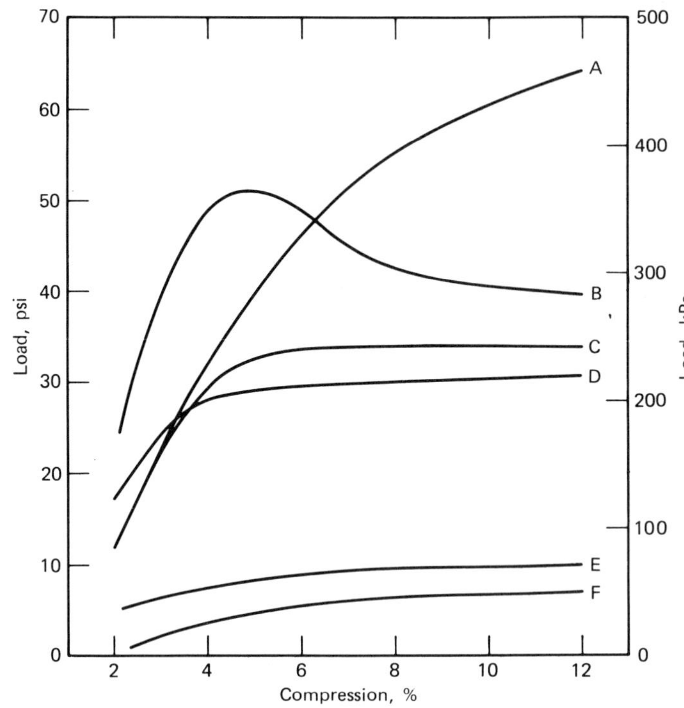

Figure 4. Compressive load vs deflection for rigid cellular polymers. A, polystyrene, 64 kg/m^3 (152); B, poly(vinyl chloride), 37 kg/m^3 (152); C, polyether–polyurethane, 30.5 kg/m^3 (152); D, polystyrene, 32 kg/m^3 (152); E, urea–formaldehyde, 29 kg/m^3 (152); F, polyethylene, 32 kg/m^3 (153). To convert kg/m^3 to lb/ft^3, multiply by 0.0624.

and with decreasing temperature (162). Specific values for a commercial rigid foam should be obtained from the manufacturer. Typical values are in Table 2.

Creep. The creep characteristic of plastic foams must be considered when they are used in structural applications. Creep is the change in dimensions of a material when it is maintained under a constant stress. The data of Brown (163) on the deformation of polystyrene foam under various static loads are shown in Figure 7. He states that there are two types of creep in this material: short-term and long-term. Short-term creep exists in foams at all stress levels, however, a threshold stress level exists below which there is no detectable long-term creep. Brown reports the minimum load required to cause long-term creep in molded polystyrene foam as shown in Table 5.

The successful application of time–temperature superposition (164) for polystyrene foam is particularly significant in that it allows prediction of long-term behavior from short-term measurements. This is of interest in building and construction applications where creep is a major consideration.

Structural Foams. Structural foams are usually produced as fabricated articles in injection molding or extrusion processes. The optimum product and process match differs for each fabricated article, so there are no standard commercial products for one to characterize. Rather there are a number of foams with varying properties. The properties of typical structural foams of different compositions are reported in Table 4.

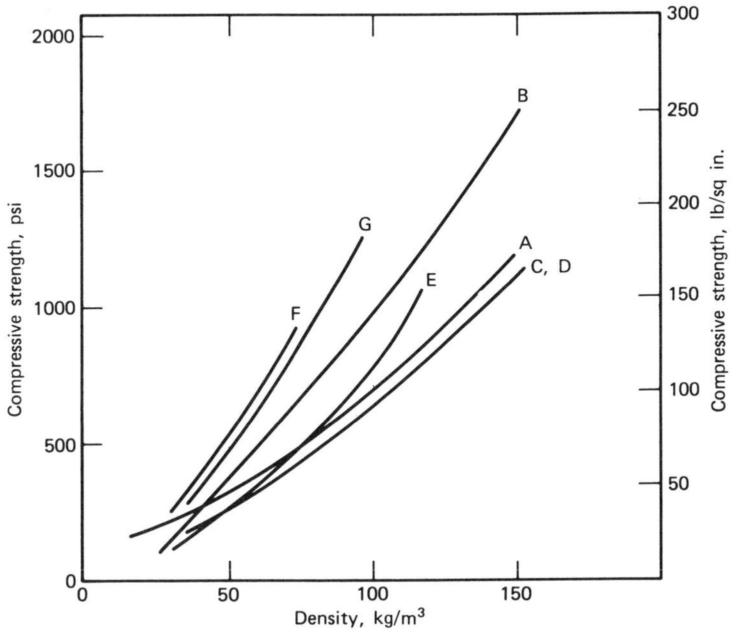

Figure 5. Effect of density on compressive strength of rigid cellular polymers. A, expanded polystyrene at 10% deflection (83); B, polyether polyurethane at yield and measured parallel to foam rise (154); C, phenol–formaldehyde foam at yield (56); D, expanded ebonite at 10% deflection (155); E, epoxy foam at yield (129); F, extruded polystyrene at 10% deflection (156); G, poly(vinyl chloride) (157). To convert kg/m^3 to lb/ft^3, multiply by 0.0624.

Table 5. Minimum Load to Cause Long-Term Creep in Molded Polystyrene[a]

Foam density, kg/m^3 (lb/ft^3)	Load, kPa (psi)
16 (1)	50 (7.3)
96 (6)	165 (24)
160 (10)	455 (66)

[a] Ref. 163.

The most important structural variables are again polymer composition, density, and cell size and shape. Since structural foams are generally not uniform in cell structure, they exhibit considerable variation in properties with particle geometry (142). Throne has extensively reviewed the mechanical properties of structural foams and their variation with polymer composition and density (142). Table 6 summarizes his conclusions as to the variation of structural foam mechanical properties with density as a function of polymer properties. The reader is cautioned to exercise care in the use of this information since in some cases the data are very limited. The design engineer should remember that the data are extracted from stress–strain curves and, owing to possible anisotropy of the foam, must be considered apparent values. These relations can provide valuable guidance toward arriving at an optimum structural foam, however.

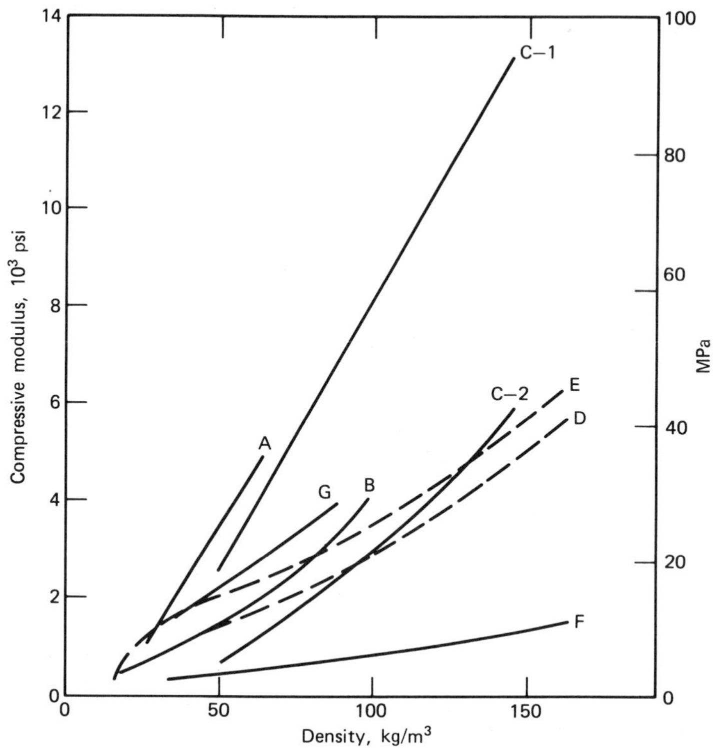

Figure 6. Effect of density on compressive modulus of rigid cellular polymers. A, extruded polystyrene (156); B, expanded polystyrene (155); C-1, polyether polyurethane (154); C-2, polyether polyurethane (154); D, phenol–formaldehyde (155); E, ebonite (155); F, urea–formaldehyde (155); G, poly(vinyl chloride) (157). To convert kg/m^3 to lb/ft^3, multiply by 0.0624.

Flexible Cellular Polymers. The application of flexible foams has been predominantly in comfort cushioning, packaging, and wearing apparel (165). This has resulted in emphasis on a different set of mechanical properties than for rigid foams. The compressive nature of flexible foams (both static and dynamic) is their most significant mechanical property for most uses. Other important properties are tensile strength and elongation, tear strength, and compression set. These properties can be related to the same set of structural variables as those for rigid foams.

Compressive Behavior. The most informative data in characterizing the compressive behavior of a flexible foam are derived from the entire load-deflection curve of 0–75% deflection and its return to 0% deflection at the speed experienced in the anticipated application. Various methods have been reported (3,165–169) for relating the properties of flexible foams to desired behavior in comfort cushioning. Other methods to characterize package cushioning have been reported (see Applications, Comfort Cushioning). The most important variables affecting compressive behavior are polymer composition, density, and cell structure and size.

That the polymer composition is the most important structural variable can be seen from Figure 8. Although the polystyrene and polyethylene foams are approximately the same density and the open-cell latex foam significantly more dense, all three show markedly different compressive strengths.

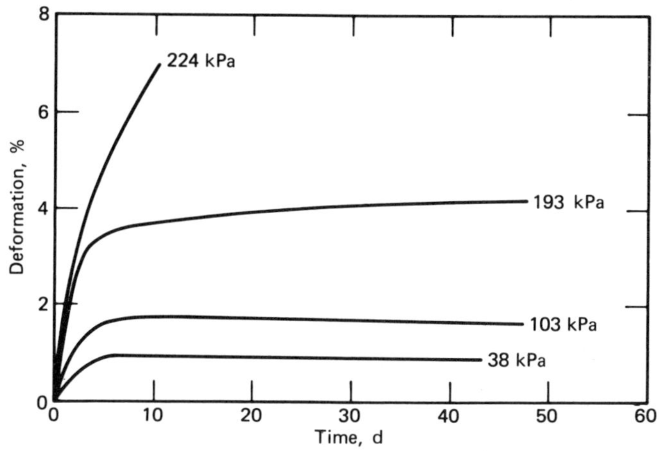

Figure 7. Deformation under various static loads for 48 kg/m^3 (3 lb/ft^3) molded polystyrene foams (163). To convert kPa to psi, multiply by 0.145.

Table 6. Design Criteria for Low Pressure Structural Foam[a]

Mechanical property	Equation[b]	Comments
tensile strength	$T_{foam} = T_o(\rho_{foam}/\rho_o)^2$	coefficient can increase with temperature
compressive modulus	$C_{foam} = C_o(\rho_{foam}/\rho_o)^2$	most data on short columns
shear modulus	$Sh_{foam} = Sh_o(\rho_{foam}/\rho_o)^2$	limited data
	$S_{foam} = S_o(\rho_{foam}/\rho_o)^{3/2}$	empirical, uncontrolled skin thickness, low foam density
flexural modulus	$S_{foam} = S_o(\rho_{foam}/\rho_o)$	law of mixtures, thick skins, high foam density
impact	$I_{foam} = I_o(\rho_{foam}/\rho_o)^4(t/t_o)^2$	tentative, few data from many test procedures
fatigue	$F_{foam} = F_o(\rho_{foam}/\rho_o)^2$	very tentative, rule-of-thumb
creep	$Cr_{foam} = Cr_o(\rho_{foam}/\rho_o)^2$	very tentative, rule-of-thumb

[a] Ref. 142.

[b] Where ρ_{foam} and ρ_o are the density of the foam and polymer, respectively, and t, t_o are the thickness of the foamed and unfoamed part.

The compressive behavior of latex rubber foams of various densities (3,170) is illustrated in Figure 9. Similar relationships undoubtedly hold for vinyl and flexible polyurethane foams as well. In the case of flexible polyurethane foams, there are many formulation variables that in addition to density can be changed to affect the compressive behavior of the foam (29,47,151). Figure 10 shows the compressive behavior of a good cushion latex foam rubber and of two polyurethane foams that have been formulated to match the latex foam performance (47).

Another variable that can exert considerable influence on the compressive behavior of flexible foams is the geometry of the sample. It has been demonstrated that various geometric coring patterns in polyurethanes (168,172) and in latex foam rubber (173) exert major influences on their compressive behavior.

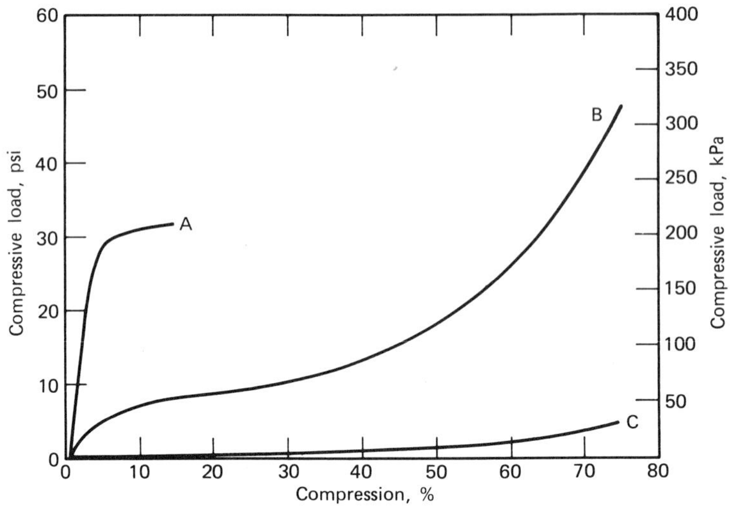

Figure 8. Load vs compression for plastic foams (153). A, polystyrene, 32 kg/m^3 (2 lb/ft^3); B, polyethylene, 32 kg/m^3; C, latex rubber foam. To convert kg/m^3 to lb/ft^3, multiply by 0.0624.

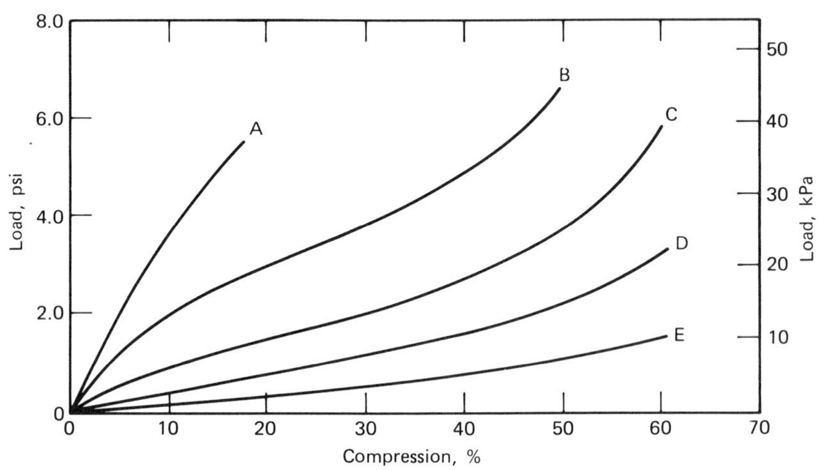

Figure 9. Effect of load on compression for latex foams of different densities (3,170): A, 304 kg/m^3; B, 208 kg/m^3; C, 179 kg/m^3; D, 139 kg/m^3; E, 99 kg/m^3. To convert kg/m^3 to lb/ft^3, multiply by 0.0624.

A good discussion of the effect of cell size and shape on the properties of flexible foams is contained in references (151,161). The effect of open-cell content is demonstrated in polyethylene foam (170).

Tensile Strength and Elongation. The tensile strengths of latex rubber foams has been shown to depend on the density of the foam (153,174) and on the tensile strength of the parent rubber (174–175). At low densities the tensile modulus approximates a linear relation with density but increases with a higher power of density at higher densities. Similar relations hold for polyurethane and other flexible foams (161,176–177).

The tensile elongation of solid latex rubber has been shown to correlate well with the elongation of foam from that latex (175). The elongation of flexible polyurethane has been related to cell structure (177–178).

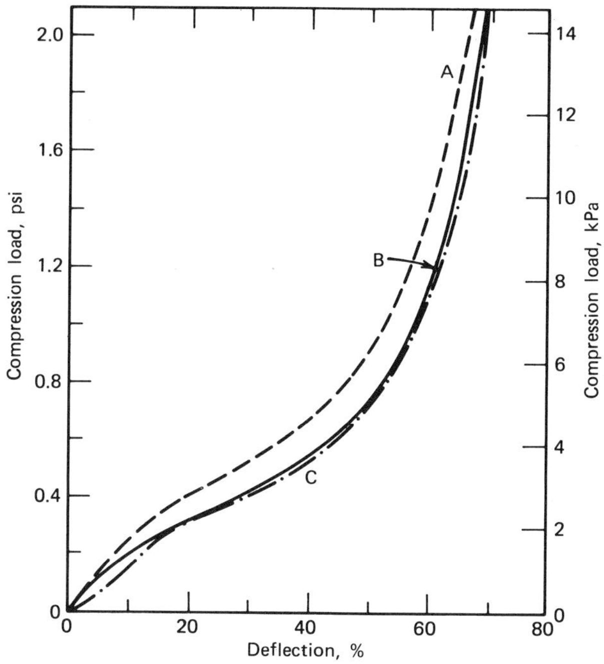

Figure 10. Compression–deflection characteristics of polyurethane and latex foams (171). A, polypropylene glycol urethane, 56 kg/m^3 (3.5 lb/ft^3); B, latex rubber, 101 kg/m^3 (6.3 lb/ft^3); C, polypropylene glycol urethane, 51 kg/m^3 (3.4 lb/ft^3).

Tear Strength. Gent and Thomas (174) have developed a relation for the tearing stress of flexible foams that predicts linear increase in the tearing energy with density; they also predict increased tearing energy with cell size. Both relationships are verified to a limited extent by experimental data.

Flex Fatigue. Considerable information on the measurement and cause of flex fatigue in flexible foams has been published (179–181). Changing compressive strength and volume upon repeated flexing over long periods of time is a major deterrent to the use of polyurethane foam in many cushioning applications. For polyurethane foams these changes have been correlated mainly with changes in chemical structure.

Compression Set. The compression set is an important property in cushioning applications. It has been studied for polyurethane foams (182–183), and is also discussed in recent reviews (29,58,161). Compression set has been described as flex fatigue and creep as well.

The thermal, electrical, acoustical, and chemical properties of all cellular polymers are of such a similar nature that the discussions of these properties are not separated into rigid and flexible groups.

Thermal Properties. ***Thermal Conductivity.*** More information is available relating thermal conductivity to structural variables of cellular polymers than for any other property. Several papers have discussed the relation of the thermal conductivity of heterogeneous materials in general (184–185) and of plastic foams in particular (129,146,154,186–188) with the characteristic structural variables of the systems. These references may be consulted for information more specific than the general treatment presented here. The manner in which each heat-transfer mechanism is applicable to cellular polymers is discussed, and the effect of each structural variable on these mechanisms is analyzed.

The following separation of the total heat transfer into its component parts, even if not completely rigorous, proves valuable to understanding the total thermal conductivity k of foams:

$$k = k_s + k_g + k_r + k_c \tag{5}$$

where k_s, k_g, k_r, and k_c are the components of thermal conductivity attributable to solid conduction, gaseous conduction, radiation, and convection, respectively.

As a good first approximation (184), the heat conduction of low density foams through the solid and gas phases can be expressed as the product of the thermal conductivity of each phase times its volume fraction. Most rigid polymers have thermal conductivities of 0.07–0.28 W/(m·K) and the corresponding conduction through the solid phase of a 32 kg/m^3 (2 lb/ft^3) foam (3 vol %) ranges 0.003–0.009 W/(m·K). In most cellular polymers this value is determined primarily by the density of the foam and the polymer-phase composition. Smaller variations can result from changes in cell structure.

Although conductivity through gases is much lower than that through solids, the amount of heat transferred through the gas phase in a foam is generally the largest contribution to the total heat transfer. This is because the gas phase is such a major part of the total value (ca 10 vol % in a 32 kg/m^3 foam). Table 7 lists values of the thermal conductivity for several gases that occur in the cells of cellular polymers.

As seen in Table 7, the thermal conductivities of the halocarbon gases are considerably less than those of oxygen and nitrogen. It has, therefore, proved advantageous to prepare cellular polymers using such gases that measurably lower the k of the polymer foam. Upon exposure to air the gas of low thermal conductivity in the cells can become mixed with air, and the k of the mixture of gases can be estimated by means of equation 6:

$$k_m = N_1k_1 + N_2k_2 \tag{6}$$

where k_m is the k of the gaseous mixture; N_1 and N_2 are the mole fractions of gases 1 and 2; and k_1 and k_2 are the thermal conductivity of pure gases 1 and 2.

Table 7. Thermal Conductivity at 20°C of Gases Used in Cellular Polymers

Compound	Thermal conductivity, W/(m·K)	Ref.
trichlorofluoromethane	0.0084	189
dichlorodifluoromethane	0.0098	189
trichlorotrifluoroethane	0.0072	189
dichlorotetrafluoroethane	0.0104	189
dichlorofluoromethane	0.0112	189
dichloromethane	0.0063	190
methyl chloride	0.0105	190
2-methylpropane	0.0161	190
carbon dioxide	0.0168	190
air	0.0259	190

Changes in total k calculated by equations 5 and 6 with change in gas composition agree well with experimental measurements (147,188,191–192).

There is ordinarily no measurable convection in cells of diameter less than about 4 mm (146). Theoretical arguments have been in general agreement with this work (154,188). Since most available cellular polymers have cell diameters smaller than 4 mm, convection heat transfer can be ignored with good justification. Studies of radiant heat transfer through cellular polymers have been made (146,154,188,193–194).

The variation in total thermal conductivity with density has the same general nature for all cellular polymers, as illustrated in Figure 11. The increase in k at low densities is owing to an increased radiant heat transfer; the rise at high densities to an increasing contribution of k_s.

The thermal conductivity of most materials decreases with temperature. When the foam structure and gas composition are not influenced by temperature, the k of the cellular material decreases with decreasing temperature. When the composition of the gas phase may change (ie, condensation of a vapor), then the relationship of k to temperature is much more complex (146,188,196).

It should be noted that the thermal conductivity of a cellular polymer can change upon aging under ambient conditions if the gas composition is influenced by such aging. Such a case is evidenced when oxygen or nitrogen diffuses into polyurethane foams that initially have only a fluorocarbon blowing agent in the cells (29,127,146,187–188,196–199).

Thermal conductivity of foamed plastics recently has been shown to vary with thickness (194). This has been attributed to the boundary effects of the radiant contribution to heat-transfer.

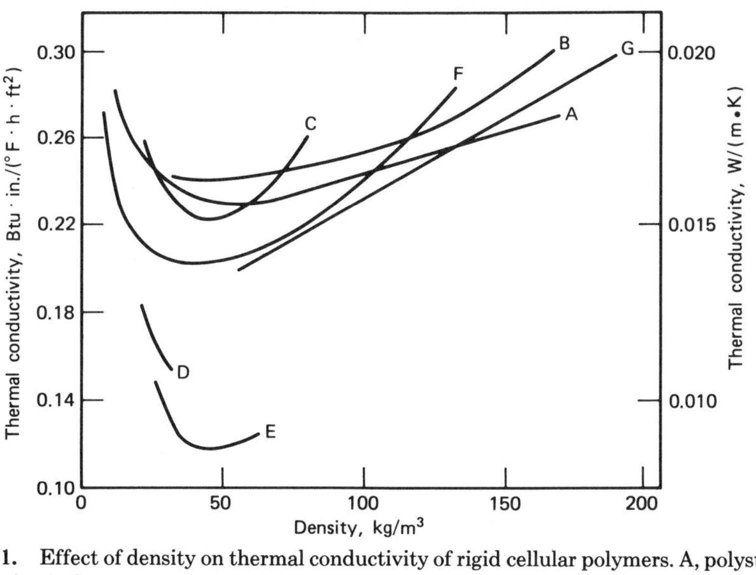

Figure 11. Effect of density on thermal conductivity of rigid cellular polymers. A, polystyrene (146); B, polystyrene (186); C, polyurethane–air (186); D, polyurethane–F-11 (CCl_3F) (195); E, polyurethane (186); F, phenol–formaldehyde (186); G, ebonite (186). To convert kg/m³ to lb/ft³, multiply by 0.0624.

Specific Heat. The specific heat of a cellular polymer is simply the sum of the specific heat of each of its components. The contribution of the gas is small and can be neglected in many cases.

Coefficient of Linear Thermal Expansion. The coefficients of linear thermal expansion of polymers are higher than those for most rigid materials at ambient temperatures because of the supercooled-liquid nature of the polymeric state. This large coefficient is carried over directly to the cellular state. A variation of this property with density and temperature has been reported for polystyrene foams (200) and for foams in general (21).

When cellular polymers are used as components of large structures, the coefficient of thermal expansion must be considered carefully because of its magnitude compared with those of most nonpolymeric structural materials. Many designs have been successful (201) when this fact has been considered.

Maximum Service Temperature. The maximum service temperature of cellular polymers cannot be defined precisely since the cellular materials, like their parent polymers (202), gradually decrease in modulus as the temperature rises rather than undergoing a sharp change in properties. The upper temperature limit of use for most cellular polymers is governed predominantly by the plastic phase. The act of fabricating the polymer into a cellular state normally builds some stress into the polymer phase; this may tend to relax at a temperature below the heat-distortion temperature of the unfoamed polymer. Of course, additives in the polymer phase or a plasticizing effect of the blowing agent on the polymer effect the behavior of the cellular material in the same way as the unfoamed polymer. Typical maximum service temperatures are given in Tables 2, 3, and 4.

Flammability. There has been an increased awareness on the part of industry, the government, and the consumer of the behavior of plastic foams in fire. The results of small-scale laboratory tests have been recognized as not predictive of the true behavior of plastic foams in other fire situations (203). Work now underway in industry and government is aimed at developing tests to evaluate the performance of plastic foams in actual fires. All plastic foams are combustible, some burning more readily than others when exposed to fire. Some additives (128,131), when added in small quantities to the polymer, markedly improve the behavior of the foam in the presence of small fire sources. All plastic foams should be used properly following the manufacturers recommendations and any applicable regulations (see also Flame retardants).

Moisture Resistance. Plastic foams have proven advantageous over other thermal insulations in several applications where they are exposed to moisture pickup. This is particularly true where they are subjected to a combination of thermal and moisture gradients in use. In some cases the foams are exposed to freeze–thaw cycles as well. The behavior of plastic foams has been studied under laboratory conditions simulating these use conditions as well as under the actual use conditions.

Hedlin (204) studied the moisture gain of foamed plastic roof insulations under controlled thermal gradients. His apparent permeability values are greater than those predicted by regular wet-cup permeability measurements. The moisture gains found in polyurethane are greater than those of bead polystyrene and much greater than those of extruded polystyrene.

Dechow and Epstein (205) reported moisture pickup and freeze-thaw resistance of various insulations and the effect of moisture on the thermal performance of these insulations. They showed that in protected membrane roofing applications the order of preference for minimizing moisture pickup is extruded polystyrene ≫ polyurethane > molded polystyrene. Water-pickup values for insulation in use for five years were extruded polystyrene 0.2 vol %, polyurethane without skins 5 vol %, molded polystyrene 8–30 vol %. These correspond to increases in K of 5–265%. For below-grade applications extruded polystyrene was better than molded polystyrene or polyurethane without skins in terms of moisture absorption resistance and retention of thermal resistance. Dechow and Epstein as well as a number of other investigators have related the increased water content with increased thermal conductivity of the insulations (206–210).

Electrical Properties. Cellular polymers have two important electrical applications. One takes advantage of the combination of inherent toughness and moisture resistance of polymers along with the decreased dielectric constant and dissipation factor of the foamed state to use cellular polymers as electrical-wire insulation (95). The other combines the low dissipation factor and the rigidity of plastic foams in the construction of radar domes. Polyurethane foams have been used as high voltage electrical insulation (211). See reference 21 for a compilation of the electrical properties of specific cellular polymers (see Insulation, electric).

Environmental Aging. Environmental aging of cellular polymers is of major importance to most applications of plastic foams. The response of cellular materials to the action of light and oxygen is governed almost entirely by the composition and state of the polymer phase (21). Expansion of a polymer into a cellular state increases the surface area; reactions of the foam with vapors and liquids are correspondingly faster than those of solid polymer.

Foams prepared from phenol–formaldehyde and urea–formaldehyde resins are the only commercial foams that are affected by water to any great extent (21). Polyurethane foams exhibit a deterioration of properties when subjected to a combination of light, moisture, and heat aging (47,197). All cellular polymers are subject to a deterioration of properties under the combined effects of light or heat and oxygen. A great deal of work has been done to develop additives that successfully eliminate this degradation (212). The best source of information on the use of specific additives is the individual manufacturer of the foam in question (see Plastics, environmentally degradable).

Miscellaneous Properties. The acoustical properties of polymers are altered considerably by their fabrication into a cellular structure. Sound transmission is altered only slightly because it depends predominantly upon the density of the barrier (in this case, the polymer phase). Cellular polymers by themselves are, therefore, very poor materials for reducing sound transmission. They are, however, quite effective in absorbing sound waves of certain frequencies (155). Materials with open cells on the surface are particularly effective in this respect. The combination of other advantageous physical properties with fair acoustical properties has led to the use of several different types of plastic foams in sound-absorbing constructions (213–214). The sound absorption of a number of cellular polymers has been reported (21,155,213,215) (see Insulation, acoustic).

The permeability of cellular polymers to gases and vapors depends upon the fraction of open cells as well as the polymer-phase composition and state. The presence of open cells in a foam allows gases and vapors to permeate the cell structure by effusion and convection flow, yielding very large permeation rates. In closed-celled foams the permeation of gases or vapors is governed by composition of the polymer phase, gas composition, density and cellular structure of the foam (191,197,213,216). Reference 217 contains a comprehensive treatment of the role of polymer and gas composition in permeation (see also Barrier polymers).

The penetration of visible light through foamed polystyrene has been shown to follow approximately the Beer-Lambert law of light absorption (21). This behavior presumably is characteristic of other cellular polymers as well.

Cellular polymers have been shown (155) to be of little or no food value for rodents but they do not act as a barrier to rodents; ie, rodents will eat through cellular polymers but they will not eat large quantities of the foam as a foodstuff. The resistance to rot, mildew, and fungus of cellular polymers can be related to the amount of moisture that can be taken up by the foam (155). Therefore, open-celled foams are much more likely to support growth than are closed-celled foams. Very high humidity and high temperature are necessary for the growth of microbes on any plastic foam.

Applications

Concern over energy conservation and consumer safety has provided a rapid growth in applications for insulation and cushioning in transport. A healthy and affluent economy is also expected to increase the consumer demand for comfort cushioning in furniture, bedding, and flooring, as well as for packaging such as drinking cups, and packaging inserts for consumer products (molded and loose-fill materials). The cost of finished wood articles has forced the industry to develop structural foams which are finding major applications as replacements for wood, metal, or solid plastics (see Engineering plastics).

The large growth in use of foamed plastics is based on economic, competitive, and technological factors. Because of the rapidly expanding and shifting market area served by the cellular polymers, applications of cellular polymers are discussed in general, and a minimum discussion is accorded to specific applications.

Comfort cushioning is the largest, single application of cellular polymers. As might be expected, the flexible foams are the major contributors to this field. Historically, cushioning in particular and flexible foams in general have constituted the greatest volume of cellular polymers. However, the growth rate of structural, packaging, and insulation applications is expected to bring their volume up to that of flexible foams within the next few years.

Good growth is expected for structural foams, which are projected to comprise more than one third of the total plastic foam consumption by 1990, up from 10% in 1976. On the other hand, cushioning, which now accounts for more than 40% of the total plastic foam consumption, will be expected to drop to less than 20% by 1990. Table 8 shows United States consumption of foamed plastics by resin, market, and function (19).

Comfort Cushioning. The properties of greatest significance in the cushioning applications of cellular polymers are compression–deflection behavior, resilience, compression set, tensile strength and elongation, and mechanical and environmental aging. Of these, compression–deflection behavior is undoubtedly the most important (see Mechanical Properties). The broad range of compressive behavior of various types of flexible foam is one of the strong points of cellular polymers, since the needs of almost any cushioning application can be met by changing either the chemical nature

Table 8. Market for Cellular Polymers, Thousands of Metric Tons[a]

Item	1963–1965	1976	1990[b]	Annual growth, % Historic	Annual growth, % Projected
Resin					
poly(vinyl chloride)	18	246	1871	24.3	15.6
polystyrene	68	370	1393	15.2	9.9
flexible polyurethane	120	542	974	13.4	4.3
rigid polyurethane	27	214	848	18.9	10.3
polyolefins		46	1077		25.3
ABS and other	11	46	479	12.3	18.2
Total	*244*	*1464*	*6642*	*16.1*	*11.4*
Market					
construction					
pipe		16	907		33.5
insulation	39	172	667	13.1	10.1
flooring	6	139	362	29.3	7.1
wood substitutes		56	612		18.6
wire		15	127		16.8
furniture					
structural–ornamental	2	48	306	31.4	14.1
cushioning and upholstery	80	302	547	11.7	4.3
packaging					
inserts	10	86	243	19.7	7.7
pallets		11	302		26.4
cups, trays, other	7	139	721	28.6	12.5
other					
transportation	37	234	470	16.6	5.1
appliances	7	37	293	14.6	15.9
consumer	5	40	306	19.8	15.8
textile	15	44	90	9.7	5.2
other	36	125	689	10.9	13.0
Total	*244*	*1464*	*6642*	*16.1*	*11.4*
Function					
cushioning	115	633	1269	15.3	5.1
insulation	52	230	920	13.2	10.4
extenders	2	153	2304	44.7	21.4
product protection	21	253	1291	22.9	12.4
other	54	195	858	11.2	11.2
Total	*244*	*1464*	*6642*	*16.1*	*11.4*

[a] Ref. 19.
[b] Projected.

or the physical structure of the foam. For example, flexible urethanes, vinyls, latex foam rubber, and olefins are used to make foamed plastic cushioning for automobile padding and seats, furniture, flooring, mattresses, and pillows. These materials compete with felt, fibers, innerspring, and other filling materials.

Flexible polyurethane foams are open-cell materials that allow free movement of air inside the foam when it is flexed. They are commercially available in densities of 13–80 kg/m^3 (0.8–5 lb/ft^3). Flexible foam for cushioning requires careful control of the load bearing characteristics, which depend primarily upon density. The indentation load deflection (ILD) test, ASTM D 1564-59T is often used as a criterion.

Fillers (qv) are often added to flexible urethane foams for improved cushioning and density control. The semiflexible polyurethane foams used extensively in automobile seats and interior safety applications have densities of up to 240 kg/m^3 (15 lb/ft^3) as compared to densities as low as 24 kg/m^3 (1.5 lb/ft^3) for the flexible foams used in household mattresses. The other flexible polyurethane foam markets include shoe liners, heating pads, furniture (117), and sponges.

Latex foam rubber was initially accepted as a desirable comfort-cushioning material because of its softness to the touch and its resilience (equal to that of a steel spring alone but with better damping qualities than the spring). It also has good endurance under repeated flexings and for extended periods of time. These advantages in physical properties and the ability to mold easily into a wide variety of shapes and sizes make latex foam rubber successful in the comfort cushioning field in spite of a higher price than competing noncellular cushioning materials.

Flexible cellular poly(vinyl chloride) was developed as a comfort cushioning material with compression–deflection behavior similar to latex rubber foam, and with the added feature of flame retardancy (40). It has a larger compression set than either latex rubber or polyurethane foams. The fact that the plasticizer in flexible vinyl foams can migrate to the surface restricts flexible vinyl foams in some applications. Furniture and motor vehicle upholstery is currently the largest market for flexible vinyl foams. Because of their better esthetics (leatherlike plastics), comfort, and favorable pricing, they are expected to show good growth in upholstery, carpet backing, resilient floor coverings, outerwear, footwear, luggage, and handbags.

Cellular rubber has been used extensively as shoe soles, where its combination of cushioning ability and wear resistance, coupled with desirable economics, has led to very wide acceptance. In this case the cushioning properties are of minor importance compared with the abrasion resistance and cost. Other major cushioning applications for cellular rubbers and latex foam rubbers are as carpet underlay and as cushion padding in athletic equipment (see Leatherlike materials).

Thermal Insulation. Thermal insulation is the second largest application of cellular polymers, and the largest application for the rigid materials. The properties of greatest importance in determining the applicability of rigid foams as thermal insulants are thermal conductivity, ease of application, cost, moisture absorption and transmission permeance, and mechanical properties (see Insulation, thermal).

The low thermal conductivity of low density cellular polymers has obviously been the main factor in their use as insulating materials. The thermal conductivities of other commercial insulating materials are compared with those of rigid cellular plastics in Table 9. The data show that those plastic foams containing a captive blowing agent have considerably lower thermal conductivities than other insulating materials, whereas other rigid cellular plastics are roughly comparable with the latter. Cellular plastics have been broadly accepted in various applications, not only because of their low thermal conductivities but because of other important considerations. For certain major applications these properties are outlined below.

Domestic Refrigeration. The very low thermal conductivity of polyurethanes plus the ease of application and structural properties of foamed-in-place materials gives refrigeration engineers considerable freedom of styling. This has resulted in an increasingly broad use of rigid polyurethane foams in home freezers and refrigerators which has displaced conventional rock wool and glass wool. Today, polyurethanes account for more than half the total insulation market for household refrigerators and

Table 9. Thermal Conductivity of Commercial Insulating Materials[a]

Material	Density, kg/m^3 [b]	Thermal conductivity at 20°C, W/(m·K)
sawdust	16	0.065
glass foam	144	0.058
vegetable fiberboard	208	0.045
wood fiberboard	368	0.058
corkboard	112	0.039
rock wool	32	0.048
perlite	104	0.053
vermiculite	80	0.064
cellulose	44	0.039
urea–formaldehyde	13	0.035
fiber glass	10	0.043
polystyrene foam (molded)	16	0.041
polystyrene foam (extruded)	32	0.029
polyurethane (or polyisocyanurate) foam	24	0.023

[a] Ref. 218.
[b] To convert kg/m^3 to lb/ft^3, multiply by 0.0624.

more than 85% of the total for household freezers (19). The remaining market is shared by other competitive materials such as polystyrene foam, glass fiber, and cork. Low moisture sensitivity and permanence are also very important (see Refrigeration).

Commercial Refrigeration. In this case again low thermal conductivity is important, as are styling and cost. Application methods and mechanical properties are of secondary importance because of design latitude available in this area. For example, large institutional chests, commercial refrigerators, freezers, and cold storage areas, including cryogenic equipment and large tanks for industrial gases, are insulated with polystyrene or polyurethane foams (see Cryogenics). Polystyrene foam is still popular where cost and moisture resistance are important; polyurethane is used where spray application is required. Polystyrene foam is also widely used in load-bearing sandwich panels in low temperature space applications.

Refrigeration in Transportation. Although styling is unimportant, the volume of insulation and a low thermal conductivity are of primary concern. Volume is not large, so application methods are not of prime importance. Low moisture sensitivity and permanence are again necessary. The mechanical properties of the insulant are quite important owing to the continued abuse the vehicle undergoes. Cost is of less concern here than in other applications (see Transportation).

Residential Construction. Owing to rising energy costs in recent years, the cost and low thermal conductivity are of prime importance in the wall and ceiling insulation of residential buildings. The cost of insulation to the customer per unit area of insulation and cost per R-value (thickness of insulation/K-factor) of insulation are measures of the cost–performance ratio of insulation materials. In the past, cellular plastics have not seriously penetrated this application, although their use has increased more than 200% over the past decade. For example, the combination of insulation efficiency, desirable structural properties, ease of application, ability to reduce air infiltration, and moisture resistance has led to use of extruded polystyrene foam in residential construction as sheathing, as perimeter and floor insulation under concrete and as a combined plaster base and insulation for walls.

The residential construction application may be divided into three major areas: insulation of new houses, retrofit insulation of old houses with inadequate insulation, and insulation of mobile homes. In residential sheathing insulation, fiberboard is still the most widely used product, although the use of extruded and molded polystyrene foam and of foil-faced isocyanurate foam is increasing depending on the cost, the amount of insulation required, and compatibility of insulation with other construction systems. In cavity-wall insulation, glass wool, polyurethane, urea-formaldehyde, and fiberglass are widely used, although fiberglass batt is the most economical insulation for stud-wall construction. In mobile and modular homes, cellular foams are used widely because of their light weight and more efficient insulation value.

Commercial Construction. The same attributes desirable on residential construction applications hold for commercial construction as well but insulation quality, permanence, moisture insensitivity, and resistance to freeze–thaw cycling in the presence of water are stressed more in commercial construction. For this reason cellular plastics have greater application here. Both polystyrene and polyurethane foams are highly desirable roof insulations in commercial as in residential construction.

Cellular polymers are also widely used for pipe and vessel insulation. Spray and pour-in-place techniques of application are particularly suitable, and polyurethane and epoxy foams are widely used. Ease of application, fire properties, and low thermal conductivity have been responsible for the acceptance of cellular rubber and cellular poly(vinyl chloride) as insulation for smaller pipes.

The insulating value and mechanical properties of rigid plastic foams have led to the development of several novel methods of building construction. The use of polyurethane foam panels as unit structural components presents interesting possibilities (219). Another development in building construction is the use of expanded polystyrene as a concrete base in thin-shell construction (220).

Packaging. Because of the extremely broad demands on the mechanical properties of packaging materials, the entire range of cellular polymers from rigid to flexible is used for packaging. The most important considerations in packaging are mechanical properties, cost, ease of application or fabrication, moisture susceptibility, thermal conductivity, and esthetic appeal.

The proper mechanical properties, particularly compressive properties, are the primary requirements for a cushioning foam (Fig. 12). The information necessary for proper design of the protective package should include: (*1*) the shock to which the item being packaged might be subjected, (*2*) an estimate of the type and degree of disturbances to which the entire package might be subjected, and (*3*) a definition of the exact capabilities of the cushioning material. Considerable progress has been made in some areas of this general problem. The reader is referred to the following sources for more specific information: package design (223); general vibration and shock isolation (224); protective package design (225); selection of cushioning material (221,226); and characterization of cellular polymers for cushioning applications (222,224–225).

Creep of a cushion packaging material when subjected to static stresses for long periods of storage or shipment is also an important consideration. Brown (163) showed that polystyrene foam does have considerable creep at high static loadings but that creep is insignificant under loadings in the static stress region of optimum package design (21) (see Fig. 7). The ability of polystyrene foam to withstand repeated impacts has also been studied (163,227).

The low density of most cellular plastics is important because of shipping costs

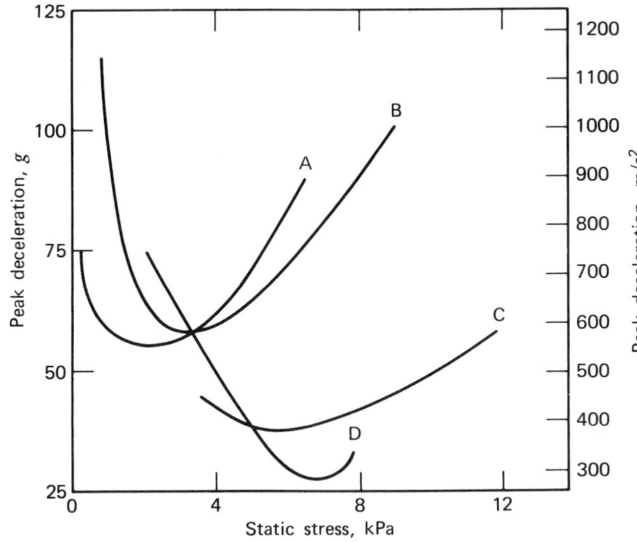

Figure 12. Dyanamic cushioning curves for cellular polymers at 76.2 cm drop height, 5.08 cm thickness. A, Polyurethane (221); B, polyethylene (222); C, expanded polystyrene (222); D, extruded polystyrene (222). To convert kPa to psi, multiply by 0.145.

for the cushioning in a package. Pelaspan-Pac (Dow), a low density, S-shaped, loose-fill, foamed polystyrene is a highly effective packing material (228). Other materials with a variety of shapes and sizes compete in this market.

The inherent moisture resistance of cellular plastics is of added benefit where packages may be subjected to high humidity or water. Many military applications require low moisture susceptibility. Foamed polystyrene is used as packaging inserts and as containers such as food trays, egg cartons, and drinking cups which require moisture resistance, rigidity, and shock resistance. Foamed polyurethane is also used as specialty packaging materials for expensive and delicate equipment.

The clean, durable, nondust-forming character of polyethylene foam has led to its increasing acceptance in packaging missile parts (229). Polyethylene foam sheet is also gradually displacing polystyrene foam sheet for packaging glass bottles and containers because of its greater resiliency and tear resistance (225) (see Film and sheeting materials).

Structural Components. In most applications structural foam parts are used as direct replacements for wood, metals or solid plastics. Structural foam parts find wide acceptance in appliances, automobiles, furniture, materials-handling equipment, and in construction. Use in the building and construction industry account for more than one half of the total volume of structural foam applications. Currently, high impact polystyrene is the most widely used structural foam, followed by polypropylene, high density polyethylene, and poly(vinyl chloride) (see Engineering plastics). The construction industry is also the area of application that exhibits the greatest growth potential for cellular plastics. Several of the more advanced applications which depend to a considerable extent on the excellent thermal properties of cellular plastics have been mentioned in the section on thermal insulation. These systems, such as the foam-cored panel (219) and thin-shell design utilizing cellular plastics (220), all make use of the structural properties of these materials. The design of entire structural systems that advantageously use such properties as the high bending stiffness-to-weight ratio of sandwich construction have been discussed recently (21,230). One obvious disadvantage of their use in such applications has been a lack of engineering design data on cellular plastics. As such data become available, the design engineer will be able to use cellular plastics as structural components, producing, no doubt, a large increase in the use of cellular plastics in construction applications.

For some structural applications an integral expandable sheet is prepared from high impact acrylonitrile–butadiene–styrene terpolymers [*9003-56-9*]. The sheet consists of solid ABS skins and an expandable core. Upon heating, the core expands and the entire composite can be formed into shapes. The resulting composite has the advantages both of chemical resistance of the plastic skins and the thermal insulation and buoyancy associated with the core, giving it the advantages of a sandwich construction as well as the immense latitude in design and use of a thermoformable material. Such items as automobile body parts, house trailers, and boats have been satisfactorily produced from this material (231). The sandwich-type structure of polyurethanes with a smooth integral skin produced by the reaction injection molding process provides a high degree of stiffness as well as excellent thermal and acoustical properties necessary for its use in housing and load-bearing structural components for the automotive, business machine, electrical, furniture, and materials-handling industry. A significant growth of this material is expected in many applications.

Buoyancy. The low density, closed-celled nature of many cellular polymers coupled with their moisture resistance and low cost resulted in their immediate acceptance for buoyancy in boats and floating structures such as docks and buoys. Since each cell in the foam is a separate flotation member, these materials cannot be destroyed by a single puncture.

The combination of structural strength and flotation has stimulated the design of pleasure boats using a foamed-in-place polyurethane between thin skins of high tensile strength (232). Other cellular polymers that have been used in considerable quantities for buoyancy applications are those produced from polyethylene, poly(vinyl chloride), and certain types of rubber. The susceptibility of polystyrene foams to attack by certain petroleum products that are likely to come in contact with boats led to the development of foams from copolymers of styrene and acrylonitrile which are resistant to these materials (6–7).

Electrical Insulation. The substitution of a gas for part of the solid polymer usually results in rather large changes in the electrical properties of the resulting material. The dielectric constant, dissipation factor, and dielectric strength are all generally lowered in amounts roughly proportional to the amount of gas in the foam (see Insulation, electric).

For low frequency electrical insulation applications, the dielectric constant of the insulation is ideally as low as possible. The lower the density of the cellular polymer, the lower the dielectric constant and the better the electrical insulation. The dielectric strength is also reduced by reducing the density; the insulation is, therefore, susceptible to breakdown from voltage surges from such sources as lightning and short circuits. The physical properties are also reduced proportionally to density, so that optimum density is determined by a compromise in properties. For many applications this compromise has been at an expansion of two or three volumes, mainly because the minimum physical properties required for fabrication and use are obtained at that point. Polyolefin foams have been most used as low frequency electrical insulation; poly(vinyl chloride) and polystyrene foams are used also. Ability to produce a completely homogeneous, closed-cell foam at lower densities in high speed wire-coating apparatus has been a problem at lower densities.

In high frequency applications, the dissipation factor is of greater importance. Coaxial cables using cellular polyolefins have been quite successfully used for frequencies in the megahertz range and above. Cellular plastics have also been used as structural materials in constructing very large radar-receiving domes (233). The very low dissipation factor of these materials makes them quite transparent to radar waves (see also Insulation, electric; Microwave technology).

Space Filling and Seals. Cellular polymers have become common for gasketing, sealing, and space filling. Cellular rubber, poly(vinyl chloride), silicone (144), and polyethylene are used extensively for gasketing and sealing of closures in the automotive and construction trade (109). Most cellular materials must be predominantly closed-celled in order to provide the necessary barrier properties. The combination of chemical inertness, excellent conformation to irregular surfaces, and ability to be compressed to greater than 50% with relatively small pressures and still function satisfactorily contribute to the acceptance of cellular polymers in these applications.

In the construction industry, cellular polymers are used as spacers and sealant strips in windows, doors, and closures of other types, as well as for backup strips for other sealants (qv).

Miscellaneous Applications. Cellular plastics have always been used for display and novelty pieces. Polystyrene foam combines ease of fabrication with lightweight, attractive appearance, and low cost to make it a favorite in these uses. Phenolic foam, on the other hand, enjoys its major use in floral displays. Its ability to hold large amounts of water for extended periods is used to preserve cut flowers. Cellular poly(vinyl chloride) is used in toys and athletic goods, where its toughness and ease of fabrication into intricate shapes have been valuable.

Cellular urea–formaldehyde and phenolic resin foams have been used to some extent in interior sound-absorbing panels and, in Europe, expanded polystyrene has been used in the design of sound-absorbing floors (234). In general, cost, flammability, and cleaning difficulties have prevented major penetration of the acoustical tile market. The low percent of reflection of sound waves from plastic foam surfaces has led to their use in anechoic chambers (214) (see Noise pollution).

Health and Safety

Plastic foams are organic in nature and, therefore, are combustible when exposed to open flames. They do vary in their response to small sources of ignition because of composition and/or additives (235). The products of combustion have been studied recently for a number of plastic foams (236). All plastic foams should be handled, transported, and used according to manufacturers' recommendations as well as applicable local and national codes and regulations.

The presence of additives or unreacted monomers in certain plastic foams can limit their use where food or human contact is anticipated. The manufacturers' recommendations or existing regulations again should be followed for such applications.

BIBLIOGRAPHY

"Foamed Plastics" in *ECT* 2nd ed., Vol. 9, pp. 847–884, by R. E. Skochdopole, The Dow Chemical Company.

1. E. Ott, H. M. Spulin, and M. W. Grafflin, eds., *Cellulose and Cellulose Derivatives,* 2nd ed., Part I, Interscience Publishers, Inc., New York, 1954, p. 9.
2. U.S. Pat. 942,699 (Dec. 7, 1909), L. H. Baekeland.
3. E. W. Madge, *Latex Foam Rubber,* John Wiley & Sons, Inc., New York, 1962.
4. A. Cooper, *Plast. Inst. Trans. J.* **29,** 39 (1961).
5. U.S. Pat. 3,058,166 (Oct. 16, 1962), R. T. Fields (to E. I. du Pont de Nemours & Co., Inc.).
6. A. R. Ingram, *J. Cell. Plast.* **1**(1), 69 (1965).
7. *Tyril Foam 80, Technical Data Sheet No. 2-1,* The Dow Chemical Co., Midland, Mich., Jan. 1964.
8. U.S. Pat. 3,062,729 (Nov. 6, 1962), R. E. Skochdopole, L. C. Rubens, and G. D. Jones (to The Dow Chemical Co.).
9. H. Junger, *Kunststoff-Rundschau* **9,** 437 (1962).
10. W. C. Goggin and O. R. McIntire, *Br. Plast.* **19**(223), 528 (1947).
11. A. F. Randolph, ed., *Plastics Engineering Handbook,* 3rd ed., Reinhold Publishing Corp., New York, 1960, pp. 137–138.
12. J. H. Saunders and K. C. Frisch, *Polyurethanes, Chemistry and Technology,* Vol. I, Interscience Publishers, a division of John Wiley & Sons, Inc., New York, 1963.
13. A. E. Lever, *Plastics (London)* **18**(193), 274 (1953).
14. *Plast. World,* 1 (May 1953).
15. *Ethafoam,* bulletin, The Dow Chemical Co., Midland, Mich., Apr. 1963.
16. U.S. Pat. 3,058,161 (Oct. 16, 1962), C. E. Beyer and R. B. Dahl (to The Dow Chemical Co.).
17. L. Nicholas and G. T. Gmitter, *J. Cell. Plast.* **1**(1), 85 (1965).
18. T. Wirtz, "Integral Skin Foam, A Progress in Urethane Molding," *paper presented at the 2nd SPI International Cellular Plastic Conference, New York, Nov. 8, 1968.*
19. G. P. Kratzschmer, *Plastic Trends, T44 Plastic Foams,* Predicasts Inc., Cleveland, Ohio, 1977.
20. *ASTM D 883-62T,* American Society for Testing and Materials, Philadelphia, Pa., 1962.
21. J. D. Griffin and R. E. Skochdopole, "Plastic Foams" in E. Baer, ed., *Engineering Design for Plastics,* Reinhold Publishing Corp., New York, 1964.
22. A. Cooper, *Plast. Inst. (London) Trans.* **26,** 299 (1958).
23. *ASTM D 1566-78,* Vol. 37, ASTM, Philadelphia, Pa., 1978.
24. *ASTM D 1056-62T,* ASTM, Philadelphia, Pa., 1962.
25. *ASTM D 1055-62T,* ASTM, Philadelphia, Pa., 1962.
26. R. H. Hansen, *SPE J.* **18,** 77 (1962).
27. A. R. Ingram and H. A. Wright, *Mod. Plast.* **41**(3), 152 (1963).
28. K. Hinselmann and J. Stabenow, *Ver. Dtsch. Ing.* **18,** 165 (1972).
29. K. C. Frisch and J. H. Saunders, *Plastic Foams,* Vol. 1, Part 1, Marcel Dekker, Inc., New York, 1972.
30. U.S. Pat. 2,681,321 (June 15, 1954), F. Stastny and R. Gaeth (to BASF).
31. U.S. Pat. 2,983,692 (May 9, 1961), G. F. D'Alelio (to Koppers Co.).
32. H. R. Lasman, *Mod. Plast.* **42**(1A), 314 (1964).
33. S. J. Skinner, S. Baxter, and P. J. Grey, *Trans. J. Plast. Inst.* **32,** 180 (1964).
34. *Ibid.,* p. 212.
35. S. J. Skinner and S. D. Eagleton, *Trans. J. Plast. Inst.* **32,** 321 (1964).
36. K. Goodier, *Br. Plast.* **35,** 349 (1962).
37. U.S. Pat. 3,697,454 (Oct. 10, 1972), R. L. Trimble (to Sinclair Koppers).
38. U.S. Pat. 3,066,382 (Dec. 4, 1962), M. L. Zweigle and W. E. Humbert (to The Dow Chemical Co.).
39. *Pelaspan-Pac,* bulletin, The Dow Chemical Co., Midland, Mich., 1974.
40. R. J. Meyer, *SPE J.* **18,** 678 (1962).
41. A. Osakada and M. Koyama, *Jpn. Chem. Q.* **5,** 55 (1969).
42. Jpn. Kokai 72 43,059 (June 22, 1972), J. Morita and co-workers (to Nitto Electric Industrial Co. Ltd.).
43. Jpn. Pat. 73 04,868 (Feb. 12, 1973), S. Nakada and co-workers (to Sekisui Chemical Co. Ltd.).
44. Br. Pat. 1,333,392 (Oct. 10, 1973), S. Minami and co-workers (to Toray Industries, Inc.).
45. U.S. Pat. 3,812,225 (May 21, 1974), K. Hosoda and co-workers, (to Furukawa Electric).
46. Fr. Pat. 1,446,187 (July 15, 1966), J. Zizlsperger and co-workers (to BASF).

47. J. H. Saunders and K. C. Frisch, *Polyurethanes, Chemistry and Technology,* Vol. II, Interscience Publishers, New York, 1964.
48. *Bayer Pocket Book for the Plastics Industry,* 3rd ed., Farbenfabriken Bayer, A.G., Leverkusen, FRG, Oct. 1963.
49. K. C. Frisch, K. J. Patel, and R. D. Marsh, *J. Cell. Plast.* **6**(5), 203 (1970).
50. R. Merten and co-workers, *J. Cell. Plast.* **4**(7), 262 (1968).
51. H. J. Papa, *Ind. Eng. Chem. Rod. Res. Dev.* **9,** 478 (1970).
52. R. H. Fish, *Proc. NASA Conf. Materials for Improved Fire Safety* **11,** 1 (1970).
53. D. E. Hipchen, "Modified Isocyanurates," paper presented at *Foams of the Future Seminar,* sponsored by Technomic Publishing Company, Inc., Pittsburgh, Pa., Mar. 17, 1976.
54. U.S. Pat. 3,799,896 (Mar. 26, 1974), E. K. Moss (to Celotex Corporation).
55. Ger. Pat. 2,034,166 (Jan. 21, 1971), G. M. Rambosek and co-workers (to 3M Co.).
56. Fr. Pat. 1,441,565 (June 10, 1966), L. Nicolas (to General Tire and Rubber Co.).
57. K. C. Frisch and J. H. Saunders, *Plastic Foams,* Vol. 1, Part 2, Marcel Dekker, Inc., New York, 1973.
58. C. J. Benning, *Plastic Foams,* Vol. 1, Wiley-Interscience, New York, 1969.
59. H. Weissenfeld, *Kunststoffe* **51,** 698 (1961).
60. Ref. 11, p. 149.
61. U.S. Pat. 3,821,337 (June 28, 1974), E. J. Bunclark and co-workers (to Esso Research Engineering).
62. *Japan Plastics Industry Annual,* 17th ed., Plastic Age, Tokyo, Japan, 1974.
63. U.S. Pat. 3,726,708 (Apr. 10, 1973), F. Weissenfels and co-workers (to Dynamit Nobel).
64. Fr. Pat. 2,185,488 (Feb. 8, 1974), J. Dugelay.
65. U.S. Pat. 3,830,894 (Aug. 20, 1974), H. Juenger and co-workers (to Dynamit Nobel).
66. U.S. Pat. 3,807,661 (Mar. 11, 1975), P. J. Crook and S. P. Riley (to Pilkington Brothers, Ltd.).
67. U.S. Pat. 3,907,723 (Sept. 23, 1975), M. Pretot (to Certain-teed Products Corp.).
68. Ger. Offen. 2,204,945 (Sept. 21, 1972), J. Tardy and co-workers.
69. Fr. Pat. 2,190,613 (Mar. 8, 1974), J. Tardy.
70. U.S. Pat. 3,766,100 (Oct. 16, 1973), H. A. Meyer-Stoll and co-workers (to Deutsche Texaco).
71. U.S. Pat. 3,741,920 (June 26, 1973), F. Weissenfels and co-workers (to Dynamit Nobel).
72. U.S. Pat. 3,694,387 (Sept. 26, 1972), H. Junger and co-workers (to Dynamit Nobel).
73. USSR Pat. 358,342 (Nov. 3, 1972), I. F. Ustinova and co-workers (to Kucherenko, V. A., Central Scientific Research Institute of Building Structures).
74. U.S. Pat. 3,101,242 (Aug. 20, 1963), J. M. Jackson, Jr. (to V. L. Smithers Manufacturing).
75. Jpn. Pat. 72 19,624 (June 5, 1972), M. Asaoka and co-workers (to Hitachi Chemical).
76. Br. Pat. 1,268,440 (Mar. 29, 1972), D. J. Rush and co-workers (to Midwest Research Institute).
77. K. Murai, *Plast. Age* **18**(6), 93 (1972).
78. V. D. Valgin and co-workers, *Plast. Massy* **1,** 28 (1974).
79. V. C. Valgin, *Europlast. Mon.* **46**(7), 57 (July 1973).
80. U.S. Pat. 3,400,183 (Sept. 3, 1968), P. I. Vidal (to Rocma Anstalt).
81. U.S. Pat. 3,214,793 (Nov. 2, 1965), P. I. Vidal (to Rocma Anstalt).
82. U.S. Pat. 3,122,326 (Feb. 25, 1964), D. P. Cook (to Union Carbide).
83. Ref. 11, p. 140.
84. *Epon Foam Spray 175, A Low Temperature Insulation,* bulletin, Shell Chemical Co., Houston, Tex., Nov. 1963.
85. Ref. 11, p. 164.
86. H. Vincent and K. R. Hoffman, *Mod. Plast.* **40**(1A), 418 (1962).
87. U.S. Pat. 2,515,250 (July 18, 1950), O. R. McIntire (to The Dow Chemical Co.).
88. V. L. Gliniecki, *Seventh Ann. Tech. Conf., Proc., SPI, New York, Apr. 24–25, 1963.*
89. *Mod. Plast.* **51**(1), 36, 40 (Jan. 1974).
90. T. P. Martens and co-workers, *Plast. Technol.* **12**(9), 46 (1966).
91. D. A. Knauss and F. H. Collins, *Plast. Eng.* **30**(2), 34 (1974).
92. F. H. Collins and co-workers, *Soc. Plast. Eng. Tech. Pap.* **19,** 643 (1973).
93. L. W. Meyer, *SPE J.* **18,** 1341 (1962).
94. Ref. 11, p. 139.
95. W. T. Higgins, *Mod. Plast.* **31**(7), 99 (1954).
96. U.S. Pat. 3,251,911 (May 17, 1966), R. H. Hansen (to Bell Telephone Lab.).
97. U.S. Pat. 3,268,636 (Aug. 23, 1966), R. G. Angell Jr. (to Union Carbide Corp.).
98. U.S. Pat. 3,065,190 (Nov. 20, 1962), D. S. Chisholm and co-workers (to The Dow Chemical Co.).

99. U.S. Pat. 3,067,147 (Dec. 4, 1962), L. C. Rubens and co-workers (to The Dow Chemical Co.).
100. J. L. Throne, *J. Cell. Plast.* **12**(5), 264 (1976).
101. R. W. Freund and co-workers, *Plast. Technol.*, 35 (Nov. 1973).
102. U.S. Pat. 3,436,446 (Apr. 1, 1969), R. G. Angell, Jr. (to Union Carbide Corp.).
103. *USM Foam Process, Technical Bulletin No. 653-A,* Farrel Company Division, Ansonia, Conn.
104. Fr. Pat. 1,498,620 (Sept. 11, 1967), P. Boutillier (to Ugine-Kuhlman).
105. J. J. Kolb in B. C. Wendle, ed., *Engineering Guide to Structural Foam,* Technomic Publishing Co., Inc., Westport, Conn., 1976, p. 161.
106. Ref. 11, p. 189.
107. U.S. Pat. 2,666,036 (Jan. 12, 1954), E. H. Schwencke (to Elastomer Chemical Co.).
108. U.S. Pat. 2,763,475 (Sept. 18, 1956), I. Dennis.
109. R. C. Bascom, *Rubber Age* **95,** 576 (1964).
110. L. Spenadel, *Rubber World* **150**(5), 69 (1964).
111. A. Cooper, *Plast. Inst. Trans.*, 51 (Apr. 1948).
112. T. H. Rogers, "Plastic Foams," paper presented at *Regional Tech. Conf., Palisades Section, Society of Plastics Engineers, New York, Nov. 1964.*
113. *Dow Latex Foam Process,* bulletin, The Dow Chemical Co., Midland, Mich., 1964.
114. *Iporka,* bulletin, Badische Anilin- und Soda-Fabrik A.G., Ludwigshafen am Rhein, FRG, July 1953.
115. *Dow Low Temperature Systems, 179-2086-77,* bulletin, The Dow Chemical Co., Midland, Mich., 1977.
116. U.S. Pat. 2,959,508 (Nov. 8, 1960), D. L. Graham and co-workers (to The Dow Chemical Co.).
117. J. B. Brooks and L. G. Rey, *J. Cell. Plast.* **9**(5), 232 (1973).
118. *Chem. Week* **19**(17), 43 (1962).
119. *Chem. Eng. News* **37**(36), 42 (1959).
120. U.S. Pat. 2,772,995 (Dec. 4, 1956), J. D. C. Wilson II (to E. I. du Pont de Nemours & Co., Inc.).
121. Can. Pat. 762,421 (July 4, 1967), M. E. Baguley (to Courtaulds Ltd.).
122. W. Worthy, *Chem. Eng. News,* 23 (Dec. 11, 1978).
123. *Mod. Plast.* **42**(4), 161 (1964).
124. W. E. Brown and F. C. Frost, in J. E. Schmitz, ed., *Testing of Polymers,* Wiley-Interscience, New York, 1965, p. 15.
125. *1978 Annual Book of ASTM Standards,* ASTM, Philadelphia, Pa., 1978.
126. W. H. Touhey, *J. Cell. Plast.* **4**(10), 395 (1968).
127. C. J. Hilado, *J. Cell. Plast.* **3**(11), 502 (1967).
128. R. E. Skochdopole, "Cellular Materials" in N. M. Bikales, ed., *Encyclopedia of Polymer Science and Technology,* Vol. 3, John Wiley & Sons, Inc., New York, 1965, p. 125.
129. R. P. Toohy, *Chem. Eng. Prog.* **57**(10), 60 (1961).
130. R. J. Bender, ed., *Handbook of Foamed Plastics,* Lake Publishing Co., Libertyville, Ill., 1965.
131. "Foam Plastics Chart," *Modern Plastics Encyclopedia* **54**(10A), 485 (1977–1978).
132. *Ibid.,* p. 776.
133. *Styrofoam SI Brand Plastic Foam,* bulletin, Dow Chemical U.S.A., Midland, Mich., 1978.
134. *Sweets Catalog File 7, Thermal and Moisture Protection,* Sweets Div., McGraw-Hill Information System Co., New York, 1978.
135. Y. Landler, *J. Cell. Plast.* **3**(9), 400 (1967).
136. R. K. Traeger, *J. Cell. Plast.* **3**(9), 405 (1967).
137. H. E. Reymove and co-workers, *J. Cell. Plast.* **11**(6), 328 (1975).
138. C. A. Schutz, *J. Cell. Plast.* **4**(2), 37, (1968).
139. *U.S. Foamed Plastics Market and Directory,* Technomic Publishing Co., Stamford, Conn., 1973.
140. *Ethafoam Brand Plastic Foam,* bulletin, Dow Chemical U.S.A., Functional Products and Systems Dept., Midland, Mich., 1976.
141. J. L. Eakin, *Plast. Eng.* **34**(7), 56 (1978).
142. J. L. Throne, B. C. Wendle, ed., *Engineering Guide to Structural Foams,* Technomic Publishing Co., Westport, Conn. 1976.
143. E. C. Van Buskirk and W. Pooley, paper presented at *American Chemical Society Meeting, Div. Paints, Plastics, Printing Ink, Chicago, Ill., Sept. 1958.*
144. C. E. Lee and co-workers, *J. Cell. Plast.* **13**(1), 62 (1977).
145. H. H. Lubitz, *J. Cell. Plast.* **5**(4), 221 (1969).
146. R. E. Skochdopole, *Chem. Eng. Prog.* **57**(10), 55 (1961).

147. R. H. Harding, *Mod. Plast.* **37**(10), 156 (1960).
148. D. M. Rice and L. J. Nunez, *SPE J.* **18,** 321 (1962).
149. *Br. Plast.* **35,** 18 (1962).
150. A. J. deVries, *Meded. Rubber Sticht. Delft* **326,** 11 (1957); *Rec. Trav. Chim.* **77,** 81, 209, 283, 383, 441 (1958).
151. C. J. Benning, *Plastic Foams,* Vol. I, Wiley-Interscience, New York, 1969.
152. T. L. Phillips and D. A. Lannon, *Br. Plast.* **34,** 236 (1961).
153. R. E. Skochdopole and L. C. Rubens, *J. Cell. Plast.* **1**(1), 91 (1965).
154. D. J. Doherty, R. Hurd, and G. R. Lester, *Chem. Ind. (London),* 1340 (1962).
155. A. Cooper, *Plast. Inst. Trans.* **26,** 299 (1958).
156. Ref. 128, pp. 103–104.
157. J. F. Hawden, *Rubber Plast. Age* **44,** 921 (1963).
158. R. H. Harding, *J. Cell. Plast.* **1**(3), 385 (1965).
159. R. H. Harding, *Resinography of Cellula Materials, ASTM Tech. Publ. 414,* ASTM, Philadelphia, Pa., 1967.
160. R. J. Corruccini, *Chem. Eng. Prog.* **53,** 397 (1957).
161. E. A. Meinecke and R. E. Clark, *Mechanical Properties of Polymeric Foams,* Technomic Publishing Co., Stamford, Conn., 1972.
162. R. M. McClintock, *Adv. Cryog. Eng.* **4,** 132 (1960).
163. W. B. Brown, *Plast. Prog.* **1959,** 149 (1960).
164. G. M. Hart, C. F. Balazs, and R. B. Clipper, *J. Cell. Plast.* **9**(3), 139 (1973).
165. G. H. Smith, *Rubber Plast. Age* **44**(2), 148 (1963).
166. G. J. Bibby, *Rubber Plast. Age* **45**(1), 52 (1964).
167. *Plast. Technol.* **8**(4), 26 (1962).
168. J. H. Saunders and co-workers, *J. Chem. Eng. Data* **3,** 153 (1958).
169. R. P. Marchant, *J. Cell. Plast.* **8**(2), 85 (1972).
170. T. H. Rogers and K. C. Hecker, *Rubber World* **139,** 387 (1958).
171. J. H. Saunders and co-workers, *J. Chem. Eng. Data* **3,** 153 (1958).
172. J. M. Buist and A. Lowe, *Trans. Plast. Inst.* **27,** 13 (1959).
173. J. Talaly, *Ind. Eng. Chem.* **46,** 1530 (1954).
174. A. N. Gent and A. G. Thomas, paper presented at *Proc. 7th Ann. Tech. Conf., Cellular Plastics Div., Soc. Plastics Ind., New York, Apr. 1963.*
175. L. Talalay and A. Talalay, *Ind. Eng. Chem.* **44,** 791 (1952).
176. M. A. Mendelsohn and co-workers, *J. Appl. Polym. Sci.* **10,** 443 (1966).
177. E. A. Blair, *Resinography of Cellular Materials, ASTM Tech. Publ. 414,* ASTM, Philadelphia, Pa., 1967, p. 84.
178. J. H. Saunders, *Rubber Chem. Technol.* **33,** 1293 (1960).
179. B. Beals, F. J. Dwyer, and M. A. Kaplan, *J. Cell. Plast.* **1**(1), 32 (1965).
180. R. P. Kane, *J. Cell. Plast.* **1**(1), 217 (1965).
181. *Plast. Technol.* **8**(4), 26 (1964).
182. S. M. Terry, *J. Cell. Plast.* **7**(5), 229 (1971).
183. *Ibid.,* **12**(3), 156 (1976).
184. R. L. Gorring and S. W. Churchill, *Chem. Eng. Prog.* **57**(7), 53 (1961).
185. M. E. Stephenson, Jr. and M. Mark, *ASHRAE J.* **3**(2), 75 (1961).
186. F. O. Guenther, *SPE Trans.* **2,** 243 (1962).
187. R. E. Knox, *ASHRAE J.* **4**(10), 3 (1962).
188. R. H. Harding, *Ind. Eng. Chem. Proc. Des. Dev.* **3,** 117 (1964).
189. *Freon Technical Bulletin #B-2,* E. I. du Pont de Nemours & Co., Inc., Wilmington, Del., 1975.
190. J. H. Perry, *Chemical Engineers Handbook,* 4th ed., McGraw-Hill Book Co., New York, 1963.
191. F. J. Norton, *J. Cell. Plast.* **3**(1), 23 (1967).
192. R. M. Lander, *Refrig. Eng.* **65**(4), 57 (1957).
193. B. K. Larkin and S. W. Churchill, *AIChE J.* **5,** 467 (1959).
194. B. Y. Lao and R. E. Skochdopole, paper presented at *4th SPI International Cellular Plastics Conference, Montreal, Can., Nov. 1976.*
195. L. E. LeBras, *SPE J.* **16,** 420 (1960).
196. G. A. Patten and R. E. Skochdopole, *Mod. Plast.* **39**(11), 149 (1962).
197. C. J. Hilado, *J. Cell. Plast.* **3**(4), 161 (1967).
198. R. R. Dixon, L. E. Edleman, and D. K. McLain, *J. Cell. Plast.* **6**(1), 44 (1970).

199. G. W. Ball, R. Hurd, and M. G. Walker, *J. Cell. Plast.* **6**(2), 66 (1970).
200. L. Vahl in ref. 115, p. 267.
201. C. H. Wheeler, "Foamed Plastics," *Proceedings of a Conference, Apr. 22–23, 1963, U.S. Army Natick Labs and Committee on Foamed Plastics, U.S. Dept. Comm. Office Tech. Serv. PB Rept. 181576,* p. 164.
202. G. A. Patten, *Mater. Des. Eng.* **55**(5), 117 (1962).
203. *FTC Consent Agreement, File #7323040,* The Dow Chemical Co., Midland, Mich., 1974.
204. C. P. Hedlin, *J. Cell. Plast.* **13**(5), 313 (1977).
205. F. J. Dechow and K. A. Epstein, *ASTM STP 660, Thermal Transmission Measurements of Insulation,* ASTM, Philadelphia, Pa., 1978, p. 234.
206. M. M. Levy, *J. Cell. Plast.* **2**(1), 37 (1966).
207. I. Paljak, *Mater. Constr. (Paris)* **6,** 31 (1973).
208. H. Mittasch, *Plaste Kautsch.* **16**(4), 268 (1969).
209. J. Achtziger, *Kunststoffe* **23,** 3 (1971).
210. C. W. Kaplar, *CRREL Internal Report No. 207,* U.S. Army CRREL, Hanover, N.H., 1969.
211. P. J. Palmer, *J. Cell. Plast.* **9**(4), 182 (1973).
212. N. Z. Searle and R. C. Hirt, *SPE Trans.* **2,** 32 (1962).
213. A. Cooper, *Plastics* **29**(321), 62 (1964).
214. *Mod. Plast.* **39**(8), 93 (1962).
215. G. L. Ball, III, M. Schwartz, and J. S. Long, *Off. Dig. Fed. Soc. Paint Technol.* **32,** 817 (1960).
216. E. F. Cuddihy and J. Moacanin, *J. Cell. Plast.* **3**(2), 73 (1967).
217. C. E. Rogers, "Permeability and Chemical Resistance" in E. Baer, ed., *Engineering Design for Plastics,* Reinhold Publishing Corp., New York, 1964.
218. *ASHRAE Handbook & Product Directory, 1977 Fundamentals,* ASHRAE, Inc., New York, 1977.
219. S. C. A. Paraskevopoulos, *J. Cell. Plast.* **1**(1), 132 (1965).
220. *Forming Thin Shells,* bulletin, The Dow Chemical Co., Midland, Mich., 1962.
221. R. K. Stern, *Mod. Packag.* **33**(4), 138 (1959).
222. R. G. Hanlon and W. E. Humbert, *Mod. Packag.* **35**(10), 158 (1962).
223. R. D. Mindlin, *Bell System Tech. J.* **24,** 353 (1945).
224. R. G. Hanlon and W. E. Humbert, *Package Eng.* **7**(4), 79 (1962).
225. K. Brown, *Package Design Engineering,* John Wiley & Sons, Inc., New York, 1959.
226. A. R. Gardner, *Prod. Eng.* **34**(25), 114 (1963).
227. C. Kienzle, "Plastic Foams," *paper presented at Regional Technical Conference, Buffalo, N.Y., Society of Plastics Engineers, Inc., Oct. 5, 1961,* p. 93.
228. P. W. Hemker, *J. Cell. Plast.* **5**(5), 295 (1969).
229. *Plast. World,* 15 (Mar. 1964).
230. *Styrofoam Sandwich Construction,* bulletin, The Dow Chemical Co., Midland, Mich., 1960.
231. *Plast. World,* 38 (Aug. 1963).
232. *Mod. Plast.* **42**(1A), 294 (1964).
233. E. B. Murphy and W. A. O'Neil, *SPE J.* **18,** 191 (1962).
234. F. Stastny, *Baugewerbe* **19,** 648 (Apr. 1957).
235. C. J. Hilado and R. W. Murphy, *ASTM Spec. Tech. Publ. (STP 685 Des. Build. Fire Safe), 16-105* ASTM, Philadelphia, Pa., 1979.
236. C. J. Hilado, H. J. Cumming, and C. J. Casey, *J. Cell. Plast.* **15**(4), 205 (1979).

K. W. Suh
R. E. Skochdopole
The Dow Chemical Company

FOAMS

Foams are coarse dispersions of gas in a relatively small amount of liquid. Bubbles vary in size from ca 50 μm (microbubbles) to several mm. Foam densities range from ca 20 g/L to ca 500 g/L. Foams are related to concentrated emulsions, sometimes called biliquid foams, in which the dispersed phase is another liquid rather than a gas (see Emulsions). In solid foams, the continuous phase is a solid (see also Foamed plastics). Certain characteristics of foams are desired for particular applications. For example, shampoos, shaving creams, and bubble-bath compositions form slow-draining and long-lived foams (see Aerosols; Cosmetics). For fire extinguishing the foam should resist destruction by contact with the fuel and exposure to high temperatures. The low density ensures that it floats on burning oil or gasoline (see Plant safety). In other processes, eg, machine laundering, too much foam should be avoided (see Drycleaning and laundering). Not only formation but the control, inhibition, or destruction of foam are often important in industrial processes (see Defoamers). Foam properties depend primarily on the chemical composition and properties of the adsorbed films and are influenced by numerous factors such as the extent of adsorption from solution at the liquid–gas surface, the rheology of the adsorbed layer, gaseous diffusion out of and into bubbles, size distribution of the bubbles, surface tension of the liquid, and external pressure and temperature (see also Sprays; Surfactants).

Much of the scientific work on foams deals with the behavior of liquid films. In examining the interference colors in soap films, in 1672 Hooke observed holes in the films, and Newton described black films of different shades. In 1869 Plateau's studies of flow behavior in films led to the concept of a surface viscosity (1). A number of books and reviews on liquid films and foams provide considerable material on both technological and theoretical aspects; these are listed under General References.

Mechanism

Foam cannot be formed from pure liquids. This fact is confirmed by a thermodynamic argument. A foam has sufficient surface area to make its surface energy a significant contribution to the total energy. The Helmholtz function for foam should, therefore, include surface area as an extensive variable (2):

$$dF = -pdV - SdT + \sigma dA + \mu_1 dn_1 + \mu_2 dn_2 \tag{1}$$

where σ is the surface tension of the liquid, A its surface area, and μ and n refer to the chemical potentials and concentrations, respectively, of its various components (here restricted to the minimum of two). Integrating equation 1 gives:

$$\Delta F = \sigma \Delta A - \int pdV \tag{2}$$

where ΔF is the change of the Helmholtz free energy at constant T, n_1, and n_2. Equation 2 applied to a foam shows that a decrease of the free energy results both from a loss of area and from an expansion of the gas, both of which come about by coalescence of bubbles; hence, a foam composed of gas in a pure liquid is thermodynamically unstable. The foams that can remain stable for an appreciable time are not created from pure liquids. The third component is usually a solute in the liquid, although it may also be a finely divided solid, a liquid–crystal phase, or an insoluble monolayer.

If the additional component is a foam stabilizer, it must be surface active, ie, it must spontaneously segregate at the liquid surface. If it is a solute, it must be positively adsorbed; if it is a finely divided solid, it must remain at the surface by virtue of a finite angle of contact with the liquid; and if it is a liquid crystal, it must have a positive spreading coefficient with respect to the liquid surface (an insoluble monolayer by virtue of its very existence necessarily meets that condition) (see Liquid crystals). These requirements all have in common spontaneous segregation of the third component at the liquid surface, so that work has to be done on the system in order to transfer the component out of the surface. This work, the work of desorption, is added as an extra term in equation 2, where ΔF represents the Helmholtz free energy of coalescence of bubbles, whereby ΔA cm^2 of liquid surface is destroyed and the adsorbate initially on that area is removed:

$$\Delta F = \sigma \Delta A - \int p dV + W(\text{desorption}) \qquad (3)$$

The first two terms on the right-hand side of equation 3 are negative and still further terms pertaining to the capillary drainage of the liquid in the lamellae and the diffusion of gas out of the bubbles are also negative; only the W (desorption) is a positive term. The net total may, however, be positive; in practice, nearly all liquid foams are manifestly unstable. Extremely long-lived, quasisolid foams may, however, be thermodynamically stable.

Even when not able to confer thermodynamic stability, an adsorbed component acts to reduce the decay rate of foam. Liquid films are subject to external stresses imposed by gravity or capillary forces that pull the liquid downward, or outward into regions of convex curvature in the foam structure. The film exists only momentarily, unless these forces are opposed by another force that checks the effluence of the liquid. The opposing force is the higher surface tension that is created when the liquid begins to flow. A newly formed surface created on a pure liquid reaches equilibrium very rapidly, within microseconds. If, however, the liquid contains an adsorbable solute, then the newly formed surface comes to equilibrium relatively slowly, more time being required for the adsorption of the solute to occur. The high surface tension developed initially, before adsorption takes place, diminishes slowly to its equilibrium value, which is reached only after periods of time ranging from a few seconds to several hours. Wherever a higher surface tension appears in a liquid lamella, contraction of the surface immediately ensues; surrounding areas of lower surface tension are pulled in toward the high-tension spot until the difference in tension is effaced. The movement of the liquid from areas of lower to areas of higher surface tension is accompanied by the flow of thick layers of underlying fluid, amounting to several micrometers in depth, which are brought into motion by the viscous drag of the surface. Thus the hydrodynamic drainage or the capillary flow of the liquid in the foam lamellae, which gave rise to the local difference of surface tension in the first place, is opposed by a counterflow of liquid at the surface. The restoration of the liquid in the lamellae by this mechanism is called the Marangoni effect (3–4).

In spite of the Marangoni effect, gravitational and capillary forces on the liquid lamellae cause liquid finally to drain away. Unstable or evanescent foams, such as those on champagne, soda water, or soft drinks, are typical of this behavior. For longer-lived foams, such as beer, whipped cream, etc, an additional stabilizing mechanism comes into play (see Food processing). A surface-active solute acts simultaneously to reduce the surface tension of the solution and to adsorb at the surface. The two effects are linked thermodynamically and are expressed quantitatively by the Gibbs adsorption theorem (4). Both effects contribute to Marangoni flow but the adsorption itself may also bring to the surface adsorbed layers that thicken into plastic skins on both sides of the foam lamellae and may even extend from both surfaces to a depth that effectively gelates the whole volume of solution lying between them. The separation of a skin at the solution–air interface does not necessarily depend on adsorption alone but may arise from an actual phase separation in the classical sense of the term. In multicomponent systems, surface activity is close, in terms of composition, to the separation of a surface-active phase, ie, a phase that has a positive spreading coefficient with respect to its matrix. When such a phase has both the fluidity of a liquid and the coherence of a crystal, it separates as a stabilizing layer on the foam lamellae. Foams have some rigidity even when their lamellae are completely liquid. The rigidity is greatly increased when the lamellae are plastic or even when only their surface layers are plastic. A secondary effect is the retardation of drainage (or syneresis) of liquid out of the foam. The slow rate of drainage keeps the foam lamellae thick for extended periods of time. When the lamellae are completely rigid, neither gas diffusion nor rupture is possible and the foams are indefinitely stable. These foams are solid rigid foams (see Foamed plastics).

Bubble Geometry and Equation of State. The two laws of bubble geometry (Plateau's laws) that hold for all assemblies of bubbles and for the morphology of foams are based on the minimizing of surface area of liquid films, which is the direct result of the tension of liquid surfaces (5): (*1*) along an edge three, and only three, liquid lamellae meet; the three lamellae are equally inclined to one another all along the edge; hence, their mutual or dihedral angles of inclination equal 120°; (*2*) at a point four, and only four, of those edges meet, the four edges are equally inclined to one another in space; hence, the angle at which they meet is the tetrahedral angle (109° 28′ 16″). These two statements are not independent, either follows from the other. Plateau's laws describe all foam structures where nonspherical bubbles are in contact. If the structure is disturbed by the rupture of a liquid film, the bubbles rearrange in such a way to conform to Plateau's laws. Clusters of a few bubbles, known as composite bubbles, demonstrate the laws very clearly; they also show striking structural similarities to collocations of some biological cells (6).

Plateau's laws do not exhaust the description of foam behavior in mathematical terms. Foam is a system that stores potential energy in the compressed gas inside the bubbles and in the extended surface of the liquid that encloses the gas. The potential energy of the compressed gas is approximately two thirds of that stored in the extended area of liquid surface. Foams can change with time, losing their potential energy both by expansion of the contained gas and contraction of the liquid films. This is brought about spontaneously by the rupture of the lamellae between bubbles, and by the transfer of gas between bubbles by diffusion. Some liquid foams are resistant to lamellae rupture but none is immune to the second phenomenon. Whatever the mechanism and the shape of the bubbles or their size distribution, the foam at any stage of its existence can be described by an equation of state (6–8) that relates the external pressure P, the volume of gas in the foam V, the moles of gas in the foam n, the area of the liquid films A, the surface tension of the liquid σ, and the temperature T. The contained gas is taken to be an ideal gas:

$$PV + \frac{2}{3}\sigma A = nRT \qquad (4)$$

This equation (6–7) can be applied in several different ways (8). For any spontaneous isothermal process of foam degradation as long as no gas is lost to the atmosphere and the external pressure remains constant:

$$3\,P\Delta V + 2\,\sigma\Delta A = 0 \tag{5}$$

For foam degradation at constant volume, the equation gives:

$$3\,V\Delta P + 2\,\sigma\Delta A = 0 \tag{6}$$

The pressure changes in a closed vessel containing foam can be monitored readily, and by means of equation 7 the rate of change of area in the foam can be found:

$$\frac{dA}{dt} = -\frac{3\,V}{2\,\sigma}\frac{dP}{dt} \tag{7}$$

The measurement of the pressure change and the use of equation 7 to give change of surface area is more convenient and precise than direct observation of area.

Evanescent Foams

Evanescent foams are those stabilized by the Marangoni effect alone. The simplest are single bubbles stabilized by an insoluble monomolecular layer on the surface of water. Gibbs' theory of film elasticity postulates that on expansion of local areas of the lamella its surface concentration decreases and its surface tension increases, thus causing a Marangoni flow to offset the hydrodynamic drainage and restoring the thickness of the lamella. Gibbs elasticity is defined as the ratio of the increase in the film tension resulting from an infinitesimal increase in area and the relative increment of the area. For a lamella with only an adsorbed monolayer on one side, the elasticity E is given by:

$$E = \frac{d\sigma}{d\,ln\,A} \tag{8}$$

The greater the elasticity the more stable is the bubble. To obtain relatively stable bubbles under a lamella that has an insoluble monolayer adsorbed on it, the monolayer has to be compressed to its maximum value of $d\sigma/d\,ln\,A$. The greatest bubble stability would occur, therefore, where $d\sigma/d\,ln\,A$ is greatest, ie, where the monolayer is least compressible. A condensed monolayer is, therefore, more effective than a gaseous monolayer. Comparing one adsorbate with another, those that form the least compressible monolayers are the most effective as bubble stabilizers.

Although evanescent foams have a stability of only a few minutes or less, they are the source of severe foaming problems in distillation (qv) or fractionation towers, where rapid evolution of vapor may build up a large volume of dynamic foam (9). Gas–liquid contacting is the basis of many production processes. For example, in the Girdler sulfide (GS) process for producing heavy water, liquid water and gaseous hydrogen sulfide are brought into contact at high temperature and pressure over sieve trays (10–11) (see Deuterium and tritium). In all such processes, an adequate interfacial area is required for good mass transport but excessive foaming may lead to instability, especially in equipment such as sieve trays. Foaming is indeed so prevalent that design engineers routinely provide extra capacity in gas–liquid contact towers. Problems of this kind can be anticipated and avoided by studying the phase diagram of the system being treated. The surface activity of a solute is primarily not the result of its amphipathic molecular structure but of its weak interaction with its solvent. This interaction must of course still be sufficient to dissolve the solute but need not be greater than the least degree required to do so. A tendency toward phase separation is a general indicator of surface-active behavior. This statement is akin to Lundelius' rule (12) that the least soluble materials are the most readily adsorbed. Thus, eg, the foam stability of gelatin (qv) solutions is greatest at the isoelectric point where the gelatin is the least soluble. Similarly, the foaminess of polymer solutions is maximum in poorer solvents, declining again, however, when insolubility supervenes (13). Not surprisingly, therefore, the surface activity displayed by solutions or mixtures can be related to the phase diagram, especially at compositions close to the critical temperature or to the plait-point composition. (The critical temperature is the temperature at which a two-component system forms one phase. The plait point is the composition at which, in a three or higher component system, two phases form one phase.) The thesis that surface activity is the precursor of phase separation is confirmed by observations of increasing foaminess of the unsaturated solutions as they approach saturation, with maximum foaminess near the critical solution point (14).

Homogeneous Solutions. A diagram for the two-component system, 2,6-dimethyl-4-heptanol and ethylene glycol, is shown in Figure 1; interpolated lines of equal foam stability (isaphroic lines, from aphros the Greek word for foam) are superimposed. The isaphroic lines center about a point close to the critical point as a maximum and decrease in value the farther they are from it. Foam stabilities of three-component

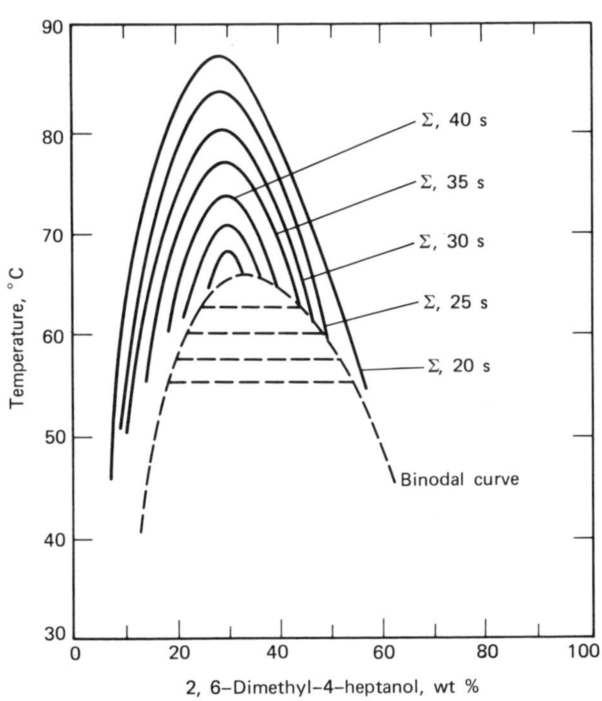

Figure 1. Phase diagram and superimposed thereon, isaphroic contours of the two-component system 2,6-dimethyl-4-heptanol and ethylene glycol [incorrectly described in the original publication (13)] showing maximum foaminess near the critical-solution point.

systems also show maximum foaming in the one-phase region near the plait point, with a gradual decrease of foaming in the one-phase region with increasing distance from the plait point. Measurements of foam stability, reported by superimposition of isaphroic contours on a phase diagram, markedly resemble contours of surface activity, named cosorption lines, obtained by plotting the Gibbs excess surface concentration Γ_2 superimposed on the same phase diagram. The Gibbs excess is calculated from data on surface tension vs activity at constant temperature. If C, C_S are the concentrations in the bulk of the solution and in the surface layer, respectively, the adsorbed excess $\Gamma_2 = C_s - C$ is given by Gibbs' equation, which for solutions of two components is

$$\Gamma_2 = -\ (1/RT)\ d\sigma/d\ ln\ a_2 \tag{9}$$

where Γ_2 is the Gibbs excess of solute 2 in mole/m^2, σ is the surface tension in N/m, and a_2 is the activity of the solute. Figure 2 shows a system of cosorption lines superimposed on a two-component phase diagram that has a liquid-miscibility gap. These

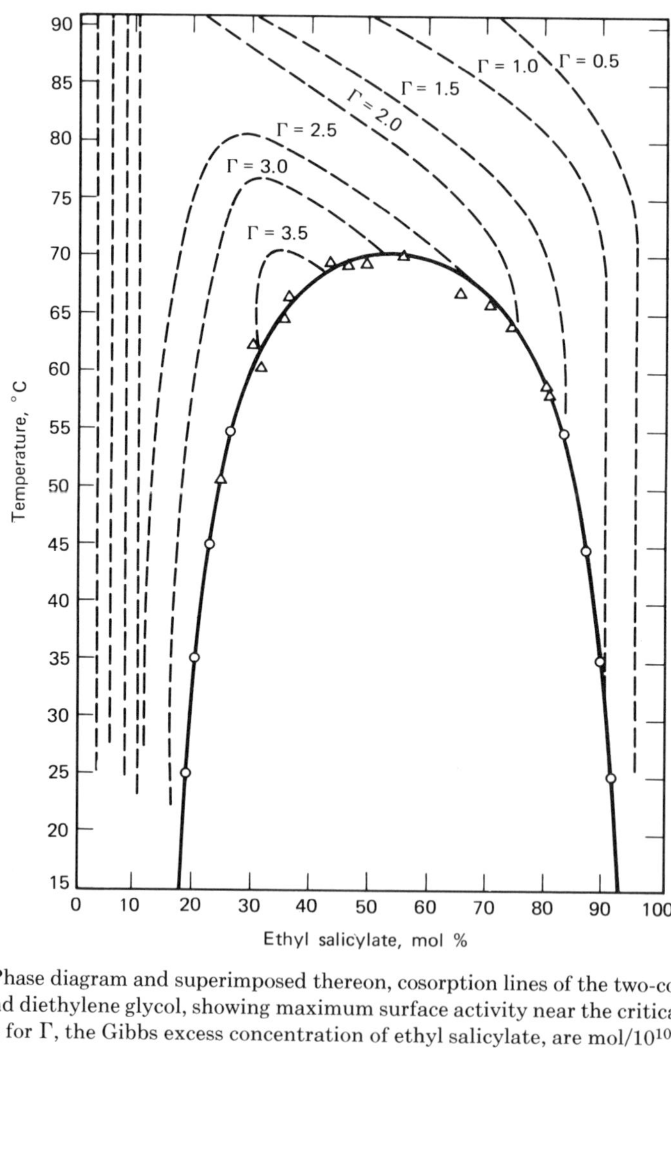

Figure 2. Phase diagram and superimposed thereon, cosorption lines of the two-component system ethyl salicylate and diethylene glycol, showing maximum surface activity near the critical-solution point. The values shown for Γ, the Gibbs excess concentration of ethyl salicylate, are mol/10^{10} cm^2 (15).

cosorption contours were determined experimentally from surface-tension isotherms (15). For evanescent foams at least, a direct relation holds between surface-excess concentration of solute and foam stability; the more stable foams, affected by additional factors, are more complex and simple correlations between foam stability and other physical properties of the solution have not been observed.

The current literature (16) on foaming problems that occur during fractionation or distillation of liquids emphasizes the surface-tension relation of solutions containing volatile constituents, and recommends the following rule of thumb: if the loss of volatile constituents causes the surface tension of the remaining solution to increase, a foam problem can be anticipated; if the loss of volatile constituents causes the surface tension to decrease, then no foam problem need be expected. The normal physical properties of liquids are such that, in general, involatility is correlated with high surface tension. Therefore, with certain exceptions such as the polydimethylsiloxanes, the usual behavior of solutions leads to the surface tension of the remaining solution being higher after the loss of its more volatile constituents. According to the rule cited above, that should lead to foaming inside a fractionation tower but it does not always occur in practice. A more useful rule can be obtained from the phase diagram; foaming of a solution reaches its maximum at the temperature and concentration where a transition into two separate liquid phases is imminent. Maximum foam stability of two-component systems occurs at a temperature and composition near the critical point, as long as the system is maintained as a homogeneous one-phase solution. Conditions of states further removed from that point show decreasing foam stability, which may even become negligible, although the more volatile constituent *has* a lower surface tension than the less volatile constituent. Here is a case, therefore, where the above surface-tensions rule does not apply. Again, should the slightest degree of phase separation occur, one of the separated liquid phases may act as a defoamer for its foamy conjugate (see below). Polycomponent systems behave in a similar way, showing maximum foam stability near a consolute point and also the possibility of innate defoaming action once the slightest degree of phase separation occurs.

Heterogeneous Mixtures. The heterogeneous portions of polycomponent system phase diagrams denote two conjugate solutions in equilibrium at a given temperature. Foam inhibition occurs when the dispersed liquid phase can spread spontaneously on a foamable solution (8,14). The lateral flow of a droplet of the spreading liquid communicates a shearing stress to the boundary fluid, carrying substrate away to a depth of several micrometers by viscous drag as the liquid on top advances. A single drop of the spreading liquid, once it arrives at the surface of the foam lamella, performs in effect as a venturi pump, ejecting on every side all the liquid lying beneath it and causing the lamella to break by the mechanical perturbation produced. The process described can occur only when certain relations exist between the various surface and interfacial tensions. A liquid spreads spontaneously if the value of the spreading coefficient S is positive, where S is given by

$$S = \sigma_m - \sigma_a - \sigma_{int} \quad (10)$$

where σ_m is the surface tension of the foamy liquid or medium, σ_a is the surface tension of the foam-inhibiting fluid or agent, and σ_{int} is the interfacial tension between them. Thus one conjugate solution may act as a foam-inhibiting agent for its conjugate partner, but if so the action cannot be reciprocated. These behavior patterns are shown by the partially miscible liquids in the ternary system benzene–water–ethanol in which regions of composition of foamable and foam-inhibited mixtures are mapped on the miscibility gap of the phase diagram (Fig. 3). The compositions along the lines ab and Pb', marked FIC (foam-inhibiting conjugate), are solutions able to spread spontaneously when dispersed in a matrix of their foamy conjugate solution, and under those conditions, by the mechanism described above, act as foam inhibitors. The shaded areas in Figure 3 represent mixtures in which this action occurs and foam is innately inhibited. The line IP is the locus of inversion points of the dispersion; the white areas subtended by the arcs ab and Pb' contain compositions of inverted dispersion, which are not foam inhibited. Measurements of foam stability made on compositions represented by a tie line just below bb′ in Figure 3 are reported in Figure 4, which shows two compositions where foam stability changes abruptly to foam inhibition. One of these is at composition c′, at which point droplets of the water-rich phase first appear in the benzene-rich medium; and the other is at the point I, the point of inversion of the liquid–liquid dispersion. These results point to the water-rich conjugate as the foam inhibitor, which agrees with the condition that the foam inhibitor have positive spreading coefficient with respect to the medium (this was measured independently) (see Defoamers).

Slow-Draining Lamellae

Marangoni effects, depending on Gibbs elasticity, do not confer more than an evanescent stability to a foam. The instability is caused by the bursting of relatively thick liquid lamellae. The duration of the prerupture regime is controlled by hydrodynamics, ie, a form of Stokes law applied to the rate of rise of a bubble. Thus the significant factors are bubble size, density difference, and viscosity of the medium. In highly viscous media, the separation rate of the gas from the liquid is greatly reduced

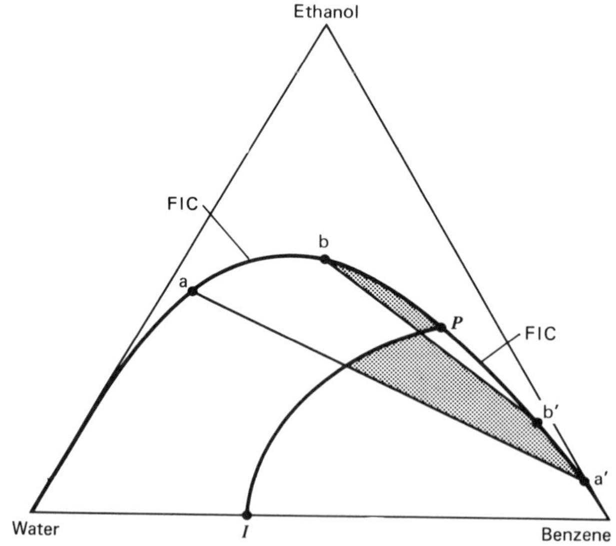

Figure 3. Phase diagram of benzene–water–ethanol at 20°C. The shaded regions of concentration represent mixtures where foaming is inhibited by a dispersed solution that can spread on its foamable conjugate (14). FIC = foam-inhibiting conjugate.

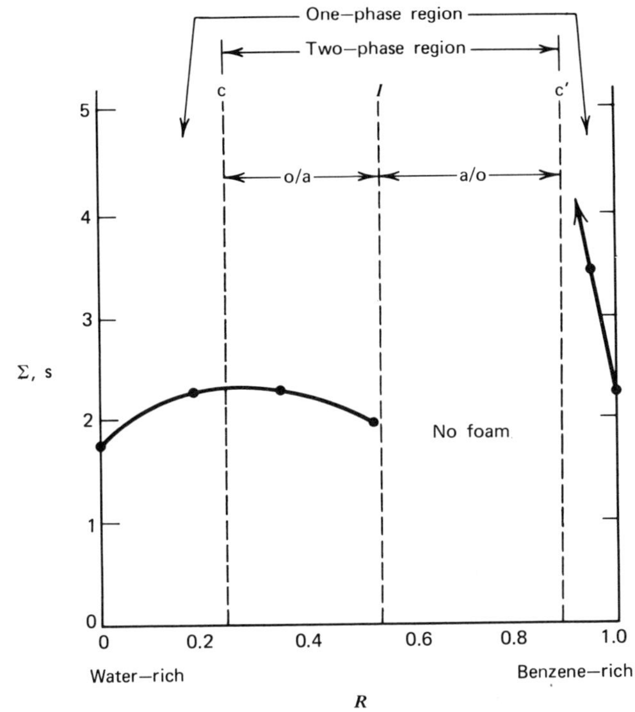

Figure 4. Foam stability, Σ, in seconds, of various compositions of benzene–water–ethanol (14). *R* is the relative position on an extended tie line through the points dd′ on the binodal curve. This tie line lies just below bb′ in Fig. 3. Foam stability drops to zero at c′, where the water-rich conjugate, acting as foam inhibitor, first separates out; and the foaminess is restored at *I*, the composition at which the dispersion inverts. o/a = oil phase dispersed in aqueous phase; a/o = aqueous phase dispersed in oil phase.

but the intrinsic stability of the liquid lamellae is not increased. In spite of their longer life, therefore, foams stabilized in this way are to be classified with the evanescent foams, as they share the same mechanism of formation and structure, differing only in the rate of decay.

Another mechanism, leading to a different foam morphology, comes into operation when the lamellae can drain to extremely thin films before rupture. This effect is a surface phenomenon of solutions of polymers such as proteins, or of long-chain solutes or mixtures of solutes in which adsorption at the surface produces coherent monolayers. The adsorbed surface layer of a solution may have mechanical or flow properties that differ from those of the bulk solution. Surface viscosity denotes the shear viscosity associated with the lateral shear of the surface of a liquid. The enhanced viscosity is the result of adsorbed solute. A mobile monolayer can move freely on the surface but when it is condensed, it becomes semirigid. The surface viscosity then shows a significant increase over the bulk viscosity of the liquid substrate. At higher shear rates the surface shows shear thinning, a form of non-Newtonian behavior. Surface viscosity is an empirical observation, which may be detected by a torsional pendulum seated at the surface of a liquid (17), or measured in absolute units by a more carefully controlled device (18). It can be ascribed to a semirigid compressed monolayer, as already mentioned, but independent evidence, based on the adsorption isotherm of the solute at the liquid surface, should be adduced to back up this interpretation (19); other effects can also give rise to an observed increase in viscosity at the surface. Among these is the Gibbs elasticity which causes Marangoni effects on dilatation. Even with compressed monolayers at the surface, a pertinent question is the extent to which the overlap of electrical double layers extending into the solution contributes to the effect. Many investigators have observed correlations between surface viscosity and the drainage rate of the liquid out of the lamella, but few have studied the effects of added neutral salts. Unpublished results (20) on lamellae stabilized with a mixture of sodium lauryl sulfate and lauryl alcohol, in a composition corresponding to a surface viscosity of 4 Pa·s (40 P) confirm that drainage is indeed very slow and the equilibrium thickness large (>100 nm) if no electrolyte is present. Upon addition of electrolyte (0.1 *M*) the draining velocity rises to the Reynolds velocity and the equilibrium thickness decreases. Therefore, the observed low drainage rates, as well as the observed surface viscosity, at least in some cases, are the result of a positive, electrostatic disjoining pressure caused by adsorbed ions. It is analogous to the electroviscous effect in latexes of low ionic strength. The presence of an electrical disjoining pressure can be readily confirmed by observing the effect on surface viscosity of added electrolyte, which by virtue of the contraction of the double layer, would eliminate any electroviscous effect. Finally, the presence of a macroscopic liquid–crystalline phase at the surface is another cause of greatly enhanced viscosity and non-Newtonian flow (see also Rheological measurements).

The drainage rates of liquid films and of foams depend on both bulk and surface viscosity (21). For some detergents the addition of several percent (on a solids basis) of water-insoluble long-chain fatty alcohols or acids produces a marked increase in the surface viscosity and a corresponding decrease in the drainage rate of foams deriving from the solutions. When a slow-draining film is formed, a sufficient increase in temperature causes a change to a fluid type of draining. The transition occurs over such a slight range of temperature, ie, within 0.2°C, that it resembles a phase transition, like the melting of a condensed monolayer. This behavior is faithfully reflected in the foam stability, ie, foams with slow-draining lamellae are stable below the transition temperature, and suddenly become evanescent foams when the temperature is raised above that point. The phenomenon is not restricted to aqueous solutions but has been observed, eg, in a heating oil that posed a troublesome foaming problem on cold winter days but not on warmer spring days. Measurements of surface viscosity showed a transition temperature of −3°C from semirigid to mobile surface films.

Sodium dodecyl sulfate (SDS) has been extensively investigated with respect to its surface and micellar properties, and is a standard water-soluble, surface-active solute in the same way that stearic acid has become a standard monolayer-forming material. The only known measurements of Gibbs surface elasticity have been made on aqueous SDS solutions (22). Theoretical calculations have also been made of the Gibbs elasticity of lamellae 1 μm thick containing SDS (23). The theory is in general agreement with the few observations available. Using a new, high sensitivity surface viscosimeter, the surface shear viscosity of SDS was measured as a function of increasing bulk concentration (19). The surface excess concentration Γ_2 was determined by a radioactive counting technique. Rise in surface viscosity depends on the increasingly dominant van der Waals' attraction between the paraffin chains. The Gibbs elasticity then declines from its maximum value all through the range of concentrations where the surface viscosity increases, showing that the contribution of elasticity to the viscosity is less significant than the growing coherence of the adsorbed monolayer. We should expect evanescent foam stability to correspond to the Gibbs elasticity in the dilute range of concentration, and a more stable foam to correspond to the growth of surface viscosity at a higher range of concentration. These data are not available for SDS solutions but the expected behavior is demonstrated by the foam stability of aqueous solutions of butyric acid (24), which show the separate influence of the two factors: a maximum in foam stability occurs at a concentration about half that of the critical micelle concentration (cmc); the foam stability then declines nearly to zero at the cmc as would be expected from the corresponding decline of the Gibbs elasticity to zero at that concentration; the foam stability reaches a second and higher maximum at a concentration above the cmc, also corresponding to a range of concentrations where the drainage rate from the foams of butyric acid solutions is observed to be unusually slow (25). Finally, in solutions containing an appreciable concentration of micelles, solubilization of the adsorbed film takes place; its cohesion is reduced by this attrition, and the foam stability declines again.

Multiphase Stabilization Factors

Finely divided solids that are poorly wetted by the liquid can stabilize foam. Soot shaken up in tap water gives a persistent foam; a pinch of clay added to a dilute solution of sodium oleate, too dilute to form a foam by itself, gives a stable foam on shaking; insoluble gypsum formed in the wet-process method of producing phosphoric acid stabilizes a foam from the evolved carbon dioxide and creates problems. The contact angle of the liquid on the surface of the particle is the determining property that enables particles to adhere to the liquid–air surface in seeming defiance of gravity. The work of adhesion is given by W (adhesion) $= \sigma_L(1 + \cos\theta)$. The contact angle may be either too high or too low to stabilize the foam. When too low, the particles are wetted by the liquid and sink; when too high, their adhesion to the lamellae is insufficient to stabilize it against the cohesion of the liquid component. Contact angles between 40 and 70° are optimum. In the process of froth flotation, whereby nonhydrophilic sulfides or selenides are removed from hydrophilic silicates, carbonates, or oxides, if the contact angle is too low, a suitable collector (eg, pine oil) raises it to a more effective value (see Flotation). The particles can also be too large to stabilize foam, and the surface-tension component cannot overcome gravity. The way in which poorly wetted solid particles are retained at a liquid–gas or liquid–liquid interface is analogous to the adsorption of solute molecules; in both cases, work is required to transfer the material out of the surface into the bulk solution. This work (of desorption) is the term that confers thermodynamic stability to the foam.

A good example of multiphase stabilization is aqueous soap solutions with a rigid surface layer as thick as 40 μm (26). In this case it is likely that a separate liquid–crystal phase spreads over the surface of an isotropic solution. Surface-denatured protein, such as stabilizes marshmallow whip, meringues, or whipped cream, is another example.

Liquid Crystals as Stabilizers. The stability of foams containing a liquid crystalline phase in equilibrium with an isotropic liquid phase has been investigated in a few cases (27). To obtain a stable foam, a combination of the isotropic solution and a liquid crystal is necessary. The solution (Fig. 5, L) or the liquid crystal (Fig. 5, LC) alone do not give stable foams; only compositions in the two-phase range of compositions, L

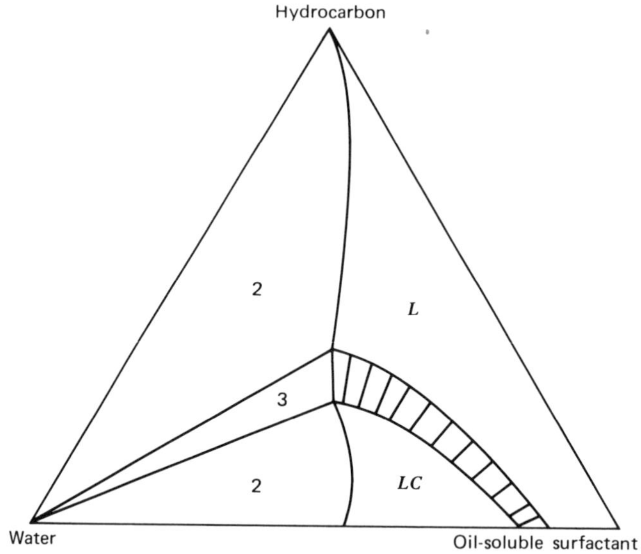

Figure 5. Generalized three-component phase diagram of an oil-soluble surfactant, hydrocarbon, and water. *LC* represents liquid-crystal phase; *L* represents isotropic liquid phase; the numbers 2 and 3 represent two- and three-phase compositions, respectively. The striped area represents two-phase compositions (*L* and *LC*) that give stable foams (27).

+ *LC* (Fig. 5, area with tie lines between *L* and *LC*) give foams stable for hours. At compositions outside this two-phase area, foam life is a matter of seconds. Foaming of organic liquids by means of organic solutes is not readily accomplished, as such solutes do not reduce the surface tension of liquids already of low surface tension. Yet, the foaming of polymer melts or monomer mixes is a first step in the production of cellular plastics. Recourse is usually possible to foam stabilizers such as polysiloxane copolymers or perfluorinated solutes. Figure 5 indicates how to accomplish the objective with low-priced ingredients. A ternary system composed of a water-soluble, surface-active solute water, and fatty alcohol has a phase diagram that also shows a liquid–crystalline phase. The isotropic aqueous solution already described has the foam stabilized by an adsorbed mixed monolayer of the two organic components; greater stability results when the total composition forms a liquid–crystal phase in equilibrium with the isotropic solution. The liquid crystal by itself does not foam (28). The viscous liquid crystal spreads over the surface of the lamellae, creating a layer of oriented molecules which gives the needed rigidity, analogous in this respect to the compressed monolayer that stabilizes the lamellae in systems consisting of only a single condensed phase. This phenomenon explains the high stability of shaving foams from aerosol extrusions (29–30). Still higher concentrations of the alcohol give an inverse micellar phase, which destabilizes the foam. Thus, the same compound, eg, a medium-chain-length alcohol, can act as a stabilizer or a destabilizer for a foam. Mixed into the aqueous solution before foaming, it confers stability; added as a spray to the foam, local high concentrations create the inverse micellar solution that causes the lamellae to break. This effect is inexplicable without the insight obtained from a knowledge of the phase relations of the three components (see Liquid crystals).

Desorption of Adsorbed Solute. The reduction of surface area, consequent to rupture of the lamellae, requires that adsorbed solute be returned to the bulk solution. As the adsorption is spontaneous, desorption requires work, which is an increase of the free energy. The more strongly adsorbed the solute, the larger is this contribution to the stability of the lamella. This purely thermodynamic generalization (see eq. 3) provides no explicit mechanism but supplements the idea (31) of film stabilization caused by positive or negative adsorption.

Drainage of Liquid from Lamellae. The thinning of a liquid lamella by drainage is a destabilizing factor. Thick lamellae are elastic but change to brittle when thinned (32). The negative radius of curvature at the Plateau border (1) gives rise to a lower pressure there. The pressure difference owing to the curvature of a liquid surface is:

$$\Delta P = \sigma(1/R_1 + 1/R_2) \tag{11}$$

in which R_1 and R_2 are the principal radii of curvature at any point. From the geometry of the border $R_1 \gg R_2$. The hydrostatic pressure is given by

$$\Delta P = \sigma/R \tag{12}$$

where R is the shorter radius of curvature, ie, in a plane normal to the lamella and to the border. The flow of liquid in response to the ΔP of equations 11 or 12 is called capillary flow. As in capillary flow, the gravitational field promotes drainage of the liquid from the foam lamellae.

Marangoni Flow. The draining of a liquid, including also the motion of a surface that carries an adsorbed layer, creates fresh surface of higher surface tension, a non-equilibrium temporary condition. The local difference of surface tension is removed by the motion of the surface from regions of lower to that of higher tension. This flow (Marangoni flow) opposes the movement of drainage and acts as a stabilizing factor to restore the thickness of the lamella.

Negative Pressure from the Electric Double Layer. Adsorbed ions establish electrostatic repulsion between the two surfaces of the lamella. The presence of excess counterions in the interstitial solution creates a hydrostatic pressure forcing the surfaces apart and thus maintains the lamella thickness. As the overlap of the diffuse double layers is responsible for this disjoining pressure, it is particularly sensitive to the ionic strength of the solution. The observation of notable decline in foam stability after addition of electrolyte to the solution demonstrates the influence of this factor.

Surface Viscosity and Slow Draining. The draining rate is reduced by the overlapping of electrical double layers, and by the presence of coherent, tightly packed, adsorbed monolayers of solute. An adsorbed monolayer must serve as an anchor for chains of oriented solute molecules extending across the lamella like iron filings held by a magnet (33). The structure of adsorbate postulated here, based on relayed forces emanating from a close-packed, oriented monolayer, is the precursor of the liquid–crystal phase, the separation of which, as shown on an adjacent portion of the ternary phase diagram (Fig. 5) is imminent. This is another example of the general rule that the structure and composition of the adsorbate before phase separation foreshadow the phase about to be nucleated. When the monolayer melts, the induced structure is disoriented and fades away. This is a lesser effect in foam lamellae, significant only in films thinner than 2.0 nm (34) but no less important in contributing to the structure of liquid crystals. The ordering of water between the monolayers has been discussed (35) recently.

Stability Measurement

The stability of short-lived or evanescent foams is measured by the dynamic foam meter (9) (see Fig. 6). Foam is made by passing gas through a porous plate into a solution. The foam height or volume is measured while bubbling continues. Nitrogen is passed through a suitable volume of liquid at measured rates, vol/T (cm^3/s) (T = time) and the steady-state volume of foam v_0 is measured. The height H of the liquid layer above the porous plate should be adequate to ensure a result independent of H, and a gas velocity not so slow that evaporation becomes appreciable. The ratio $v_0{:}\Sigma$ is almost independent of vol/T and may be used as a measure of foam stability; for instance, $\Sigma = 6$ s for 1% aqueous 1-butanol and ca 140 s for 0.40% solution of egg albumen at pH 4.8. The column is about 70 cm high and should have a diameter of at least 3 cm to avoid wall effects. Temperature is controlled by means of oil circulating from a thermostat bath. The apparatus shown in Figure 6 is designed for measurements at temperatures up to 70°C. The temperature of the solution is read directly by means of a thermocouple placed inside the sample space and can be kept to within 0.2°C.

The stability of slow-draining foams may be several orders of magnitude greater than that of evanescent foams. Stable foams are measured by means of a static foam meter (36) designed to take into account both the ability to produce foam and its stability once produced. The method is arbitrary and various conditions have to be carefully defined and controlled. A volume of 200 mL of a solution of prefoamer is contained in a pipette of specified dimensions. The end of the pipette is untapered to ensure turbulent exit flow of the solution; it has an orifice of 2.9 mm ID. The solution

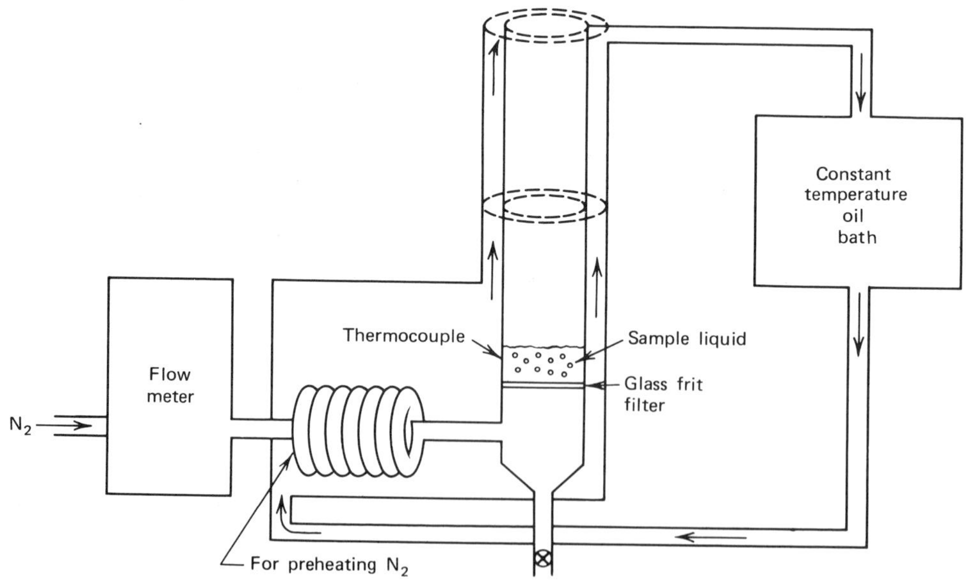

Figure 6. Dynamic foam meter equipped for temperature control of entering gas and of the entire foam column (8,13).

is allowed to fall a distance of 90 cm into 50 mL of the same solution contained in a cylindrical vessel maintained at a given temperature (ca 60°C) by means of a water jacket. The height of the foam produced in the cylindrical vessel is read immediately after all the solution has run out of the pipette (initial foam height) and then again after a lapse of time (generally, 5 min). Numerous data pertaining to solutions of soaps and detergents obtained by this method are assembled in ref. 37. In distilled water at room temperature, sodium alkyl sulfates and soaps with saturated straight-chain lipophiles of 12–14 carbon atoms have the best foaming capacities. At higher temperature, C_{16}-chains (at 60°C) and C_{18}-chains (at 100°C) are best. Nonionic surfactants generally produce less foam and much less stable foam than ionic surfactants in aqueous solutions.

For foams of high stability, observations that depend on waiting for rupture of the lamellae may be too protracted to be practicable; moreover, other changes occur in the foam from its inception, namely, gas diffusion, loss of interfacial area, change of bubble-size distribution, and drainage of interstitial liquid out of the lamellae. Of these, drainage is the least directly dependent on the foam stability and so is least informative, although most readily accessible to measurement. Nevertheless, the drainage rate is a standard measurement for the stability of beer foams (38). The rate of loss of interfacial area is a direct indicator of foam stability, and is a fundamentally significant parameter. Photographic records of foam at various stages of decay offer a direct method of observation (39) but the technique is too complicated. An alternative method is the absorption of light, which is dissipated by multiple internal reflections and refractions by the foam lamellae and consequently is related to the number of such lamellae in the light path (40). The rate of loss of interfacial area as foams decay can also be followed (41). Other parameters related to the interfacial area are given by the equation of state of foam in its various forms (see eqs. 4–7). Equation 7 is the best basis for the construction of an apparatus to measure the kinetics of foam decay by determining the rate of the build-up of gas pressure at constant volume and temperature in the head-space above the foam (42). Such an instrument is shown in Figure 7. A computer model of foam decay with time (43), based on calculations of changes in bubble-size distribution because of interbubble gas diffusion, gives a theoretical description but does not take into account any reduction of thickness of the foam lamellae as a result of drainage of liquid. This can be allowed for, however, and measurements of dP/dT can then be matched against the computed change of surface area with time to evaluate the original and subsequent distributions of bubble size and the diffusion constant of the gas through the foam lamellae.

Uses

Foam applications are based on the following properties (44).

Difference in composition between the bulk liquid and the collapsed foam. Froth flotation is the main application; foam fractionation is another.

Low Density. This property is required of fire-extinguishing foams, which are thus able to float on burning hydrocarbon fluids. It is, of course, applied wherever foam separates from mother liquid, as in froth flotation, foam fractionation, and gas–liquid extraction.

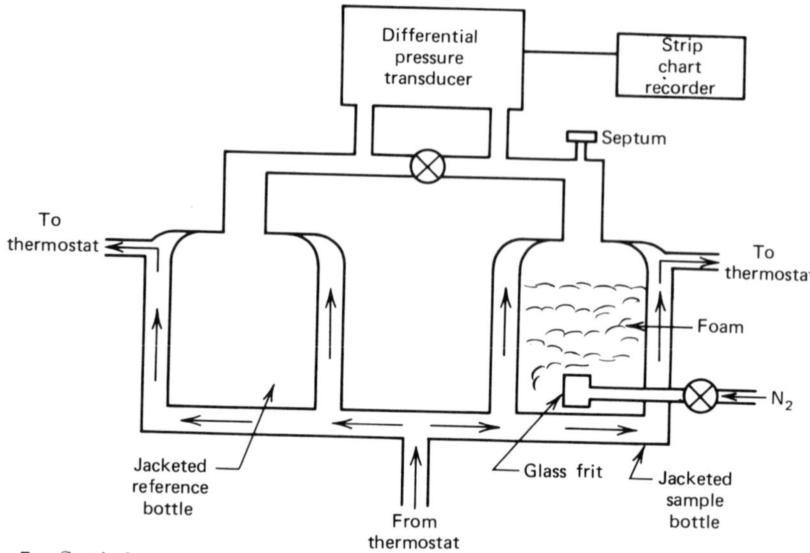

Figure 7. Static foam meter to determine the rate of pressure increase at constant volume during the decay of a stable foam (15).

Large Surface. Foam fractionation depends on the adsorption per square centimeter times the area of liquid surface created.

Mechanical Properties. The capillary pressure of foams leads to the Jamin effect, whereby gas under pressure can be contained in cavities filled with foam. At the same time foam is easy to pump and it is inexpensive to produce; consequently, stable preformed foam can replace drilling mud as a circulating medium in oil-well completion (see Petroleum operations below), and can remove sand and scale file in plugged well bores (45).

Immobilization of gas and liquid (sometimes also of solids) in foams. An interesting application in this group is the retention of air-borne particulate matter (eg, coal dust) on passage through a foam (46).

Thermal Properties. Convection effects are weak in foam, hence it is an effective heat insulator. This property is widely used in different kinds of solid foam (see Foamed plastics).

Froth Flotation. In flotation (qv), metallic minerals are separated from gangue particles by making the mineral particles hydrophobic and bringing them to the surface in a foam of air bubbles. A common agent for this purpose (the collector) is pine oil, an impure terpineol; others are cresols, amyl alcohol, and higher alcohols. The frothing agent may also be a detergent and, if present in excess, could remove the collector from the surface of the particles; excess collector acts also as defoamer. Collector and frother have to be adjusted with regulators, depressing agents, and activating agents. Molecular interactions between frothers and collectors take place at solid–liquid–air interfaces (47). The actual mechanism of adherence of an air bubble to a suitably collector-coated particle is established by the penetration of the monolayer at the solid–liquid interface by the monolayer at the air–liquid interface.

Fire-Extinguishing Foams. Foams are effective in extinguishing gasoline fires (48) by providing a semipermanent aqueous blanket (see also Plant safety). Fluoroaliphatic surface-active solutes and auxiliary foaming agents form a durable and resistant foam that spreads spontaneously and rapidly over the surface of a burning fuel and suppresses hydrocarbon vaporization (49).

Foam Fractionation. This has limited application, mostly on a laboratory scale. For multicomponent solutes, adsorption at gas–liquid interfaces frequently shows considerable selectivity. Examples of foam separation are given in refs. 50–51.

Petroleum Operations. The rheological properties (52) of aqueous foams are of great importance for a number of applications in petroleum production. These properties include flow rates in tubes or pipes and in porous media such as oil-bearing sandstones. In the drilling of oil wells, a fluid, generally a clay–water slurry, is circulated down the drill pipe, through the jets of the drill bit and back to the surface of the ground in the annular space between the drill pipe and the wall of the hole. Aqueous foam has some advantage in certain oilfields, such as those in the Central Valley of California, that contain viscous oil in low pressure reservoirs, where productivity is diminished by the clay–water slurries but enhanced by the foam. The foam raises to the surface not only sand pulverized by the bit but also pebbles and rocks up to the size of baseballs. Hundreds of wells are drilled with foam every year. Furthermore, in the secondary and tertiary recovery of oil (see Petroleum, enhanced oil recovery), fluids are often injected into one set of wells in order to drive or push the oil to a second set of wells. In general, water is used for this purpose but recently many other fluids have been tried. Here the flow (or in some cases the lack of flow) of fluids such as foam through porous media rather than through pipes is applied. Foam is applied in cases where it should selectively block the gas or vapor flow in highly conductive porous strata and redirect it into less conductive strata. Foams resistant to high temperatures have been developed for use with steam injected into reservoirs containing viscous crude oil (see Petroleum, drilling fluids).

Foams have rheological properties highly dependent on the shear or flow rate and also on the foam quality, ie, the volumetric concentration of gas in the foam. Most of the foams used in petroleum applications contain 70–90% gas. At very low shear stress they usually exhibit the yield value characteristic of a plastic. At high shear rates they behave like pseudoplastic fluids when flowing in tubes, with a pluglike flow with little shear near the center of the tube and some slippage along the wall. At low but finite rates of shear, foam can behave like a Newtonian fluid whose viscosity increases greatly, however, with foam quality. Flow behavior in porous media is very complex. Besides its movement as an essentially uniform fluid, continuous breakdown and regeneration of the foam in the pores which can act as very effective foam generators, may take place. The type of flow mechanism is chiefly determined by the nature and concentration of the chemical foaming agent used.

The following are definitions of technical terms.

Adsorption is the absorption of a component of the system at an interface.

Adsorbed films are layers of segregated solute a few molecular diameters in thickness occurring at a liquid–gas interface.

Lamella (pl. *lamellae*) is a thin sheet of liquid as found in bubbles or foam.

Spreading is the act of wetting of an interface by an insoluble component.

Wetting is the occupying of an interface by an insoluble component.

BIBLIOGRAPHY

"Foams" in *ECT* 1st ed., Vol. 6, pp. 772–778, by E. I. Valko, Polytechnic Institute of Brooklyn; "Foams" in *ECT* 2nd ed., Vol. 9, pp 884–901, by Leo Shedlovsky, Colgate-Palmolive Company.

1. J. A. F. Plateau, *Statique des Liquides,* Vol. 1, Paris, Fr., 1873, Chapt. 5.
2. H. C. van Ness, *Ind. Eng. Chem. Fundam.* **8,** 464 (1969).
3. C. Marangoni, *Nuovo Cimento* **2,** 5, 239 (1871); **3,** 50, 97, 192 (1878).
4. J. W. Gibbs, *Scientific Papers,* Vol. 1, Longmans, Green, and Co., London, Eng., 1906, pp. 300–314.
5. D. W. Thompson, *On Growth and Form,* 2nd ed., Cambridge University Press, Cambridge, Eng., 1942.
6. P. G. Tait, *Proc. R. Soc. Edinburgh* **6,** 292 (1867–1868); *Properties of Matter,* A. and C. Black, Edinburgh, Scotland, 1885, Chapt. 12.
7. B. V. Deryagin, *J. Phys. Chem. USSR* **2,** 6, 745 (1931); **7,** 3, 297 (1936).
8. S. Ross, *Ind. Eng. Chem.* **61**(10), 48 (1969).
9. J. J. Bikerman, *Trans. Faraday Soc.* **34,** 634 (1938).
10. E. W. Becker, *Heavy Water Production,* International Atomic Energy Agency, Vienna, Austria, 1962, pp. 33–43; L. R. Haywood and P. B. Lumb, *Chem. Can.* **27**(3), 19 (1975).
11. N. H. Sagert and M. J. Quinn, *J. Colloid Interface Sci.* **61,** 279 (1977).
12. E. F. Lundelius, *Kolloid Z.* **26,** 145 (1920).
13. S. Ross and G. Nishioka, *J. Phys. Chem.* **79,** 1561 (1975); S. Ross and G. Nishioka, *Colloid Polym. Sci.* **255,** 560 (1977); S. Ross and G. Nishioka in R. J. Akers, ed., *Foams,* Academic Press, London, Eng., 1976, pp. 17–31.
14. S. Ross and R. E. Patterson, *J. Phys. Chem.* **83,** 2226 (1979).
15. G. Nishioka, private communication, 1979.
16. F. J. Zuiderweg and A. Harmens, *Chem. Eng. Sci.* **9,** 89 (1958).
17. E. J. Burcik, *J. Colloid Sci.* **5,** 421 (1950); R. Bulas and C. A. Kumins, *J. Colloid Sci.* **13,** 429 (1958).
18. F. C. Goodrich, L. H. Allen, and A. Poskanzer, *J. Colloid Interface Sci.* **52,** 201 (1975); A. Poskanzer and F. C. Goodrich, *J. Colloid Interface Sci.* **52,** 213 (1975).
19. A. Poskanzer and F. C. Goodrich, *J. Phys. Chem.* **79,** 2122 (1975).
20. I. Ivanov, private communication, 1979.
21. G. D. Miles, L. Shedlovsky and J. Ross, *J. Phys. Chem.* **49,** 93 (1945).
22. K. J. Mysels, M. C. Cox, and J. D. Skewis, *J. Phys. Chem.* **65,** 1107 (1961).
23. M. J. van den Tempel, J. Lucassen, and E. H. Lucassen-Reynders, *J. Phys. Chem.* **69,** 1798 (1965).
24. A. Sheludko, *C. R. Acad. Bulg. Sci.* **9**(1), 11 (1956); J. J. Bikerman, *Foams,* Springer-Verlag, New York, 1973, p. 106.
25. A. Sheludko and D. Ekserova, *Izv. Khim. Inst. Bulg. Akad. Nauk.* **105,** (1959); J. J. Bikerman, *Foams,* Springer-Verlag, New York, 1973, p. 106.
26. R. E. Wilson and E. D. Ries, *Colloid Symp. Monogr.* **1,** 145 (1923).
27. S. Friberg, *Liquid Crystals and Foams* in *Advances in Liquid Crystals,* Vol. 3, Academic Press, Inc., New York, 1978, pp. 149–165.
28. S. Friberg and S. I. Ahmad, *J. Colloid Interface Sci.* **35,** 175 (1971).
29. S. Friberg, L. Rydhag, and G. Jederström, *J. Pharm. Sci.* **60,** 1883 (1971).
30. G. Jederström, L. Rydhag, and S. Friberg, *J. Pharm. Sci.* **62,** 1979 (1973).
31. C. W. Foulk and J. N. Miller, *Ind. Eng. Chem.* **23,** 1283 (1931); C. W. Foulk, *Ind. Eng. Chem.* **33,** 1086 (1941); C. W. Foulk and J. E. Barkley, *Ind. Eng. Chem.* **35,** 1013 (1943).
32. J. J. Bikerman, *J. Phys. Chem.* **56,** 164 (1952); T. B. Thomas and J. T. Davies, *J. Colloid Interface Sci.* **48,** 427 (1974).
33. W. B. Hardy, *J. Gen. Physiol.* **8,** 641 (1927); J. W. McBain and G. P. Davies, *J. Am. Chem. Soc.* **49,** 2230 (1927); J. W. McBain, *Colloid Science,* D. C. Heath and Co., Boston, Mass., 1950, p. 33.
34. D. M. de Neveu, R. P. Rand, and V. A. Parsegian, *Nature (London)* **259,** 601 (1976).
35. J. N. Israelachvili and B. W. Ninham, *Colloid Interface Sci.* **1,** 15 (1977).
36. J. Ross and G. D. Miles, *Oil Soap* **18,** 99 (1941); *ASTM Standard Method D 1173-53,* ASTM, Philadelphia, Pa., reapproved 1970.
37. M. J. Rosen, *Surfactants and Interfacial Phenomena,* Wiley-Interscience, New York, 1978, pp. 214–216.
38. A. D. Rudin, *J. Inst. Brew.* **63,** 506 (1957).
39. N. O. Clark and M. Blackman, *Trans. Faraday Soc.* **44,** 13 (1948); A. J. de Vries, *Rec. Trav. Chim.* **77,** 81, 209, 283, 383, 441 (1958); E. M. Savitskaya, *Kolloid Zh.* **13,** 4, 309 (1951).
40. N. O. Clark and M. Blackman, *Trans. Faraday Soc.* **44,** 7 (1948).
41. S. Ross and M. J. Cutillas, *J. Phys. Chem.* **59,** 863 (1955).
42. P. N. Kruglyakov and P. R. Taube, *J. Appl. Chem. USSR* **38,** 2258 (1965).
43. R. Lemlich, *Ind. Eng. Chem. Fundam.* **17,** 89 (1978).
44. J. J. Bikerman, *Foams,* Springer-Verlag, New York, 1973, p. 263.

45. S. O. Hutchison and G. W. Anderson, *Oil Gas J.*, 74 (May 15, 1972).
46. F. H. Price, *Trans. Inst. Min. Eng. (London)* **105,** 244 (1946); M. E. Pozin, I. P. Mukhlenov, and E. Ya. Tarat, *Foam Gas Purifiers, Heat Exchangers, and Absorbers* (in Russian), Goskhimizdat, Leningrad, USSR, 1959.
47. J. Leja and J. H. Schulman, *Trans. Am. Inst. Min. Eng.*, 221 (1954).
48. J. M. Perry, "Fire-Fighting Foams" in J. J. Bikerman, ed., *Foams, Theory and Industrial Applications,* Reinhold Publishing Corp., New York, 1953, pp. 189–242.
49. U.S. Pats. 3,562,156 (Feb. 9, 1971); 3,792,195 (Nov. 13, 1973), V. I. Francen (to 3M Co.).
50. E. Rubin and E. L. Gaden, Jr., "Foam Separation" in H. M. Schoen, ed., *New Chemical Engineering Separation Techniques,* Wiley-Interscience, New York, 1962, pp. 359–364.
51. J. J. Bikerman in ref. 44.
52. S. S. Marsden, Jr., and co-workers, *Proc. 7th World Pet. Congr.* **3,** 235 (1967); S. S. Marsden and S. A. Khan, *Trans. AIME* **237,** 17 (1966); S. H. Raza, *Soc. Pet. Eng. J.* **10**(4), 328 (1970).

General References

R. J. Akers, ed., *Foams: Proceedings of a Symposium organized by the Society of Chemical Industry, Colloid and Surface Chemistry Group, and held at Brunel University, Sept. 1975,* Academic Press, London, Eng., 1976.

S. Berkman and G. Egloff, *Emulsions and Foams,* Reinhold Publishing Corp., New York, 1941.

J. J. Bikerman, *Foams, Theory and Industrial Applications,* Reinhold Publishing Corp., New York, 1953.

J. J. Bikerman, *Foams,* Springer-Verlag, New York, 1973.

C. V. Boys, *Soap Bubbles and the Forces Which Mold Them,* Dover Publications, Inc., New York, 1959 (originally published by the Society for Promoting Christian Knowledge, London, Eng., 1890).

D. G. Dervichian, "Les Mousses: Formation, Stabilité, Destruction," *Bull. Soc. Chim. Fr.,* 15 (1956).

J. T. Davies, *Rep. Prog. Appl. Chem.* **45,** 588 (1960).

C. Isenberg, *The Science of Soap Films and Soap Bubbles,* Tieto Ltd., Clevedon, Eng., 1978.

J. A. Kitchener, "Foams and Free Liquid Films" in J. F. Danielli, K. G. A. Pankhurst, and A. C. Riddiford, eds., *Recent Progress in Surface Science,* Vol. 1, Academic Press, Inc., New York, 1964, pp. 51–90.

J. A. Kitchener and C. F. Cooper, "Current Concepts in the Theory of Foaming," *Q. Rev. London* **13,** 71 (1959).

A. S. C. Lawrence, *Soap Films,* G. Bell and Sons Ltd., London, Eng., 1929.

R. Lemlich, ed., *Adsorptive Bubble Separation Techniques,* Academic Press, Inc., New York, 1972.

E. Manegold, *Schaum,* Strassenbau, Chemie und Technik, Verlagsgesellschaft m.b.H., Heidelberg, FRG, 1953.

K. J. Mysels, K. Shinoda, and S. Frankel, *Soap Films, Studies of Their Thinning and a Bibliography,* Pergamon Press, Inc., New York, 1959.

E. Rubin and E. L. Gaden, Jr., "Foam Separation" in H. M. Schoen, ed., *New Chemical Engineering Separation Techniques,* Interscience Publishers, Inc., New York, 1962, pp. 359–364.

A. M. Schwartz and J. W. Perry, *Surface Active Agents,* Interscience Publishers, Inc., New York, 1949.

A. M. Schwartz, J. W. Perry, and J. Berch, *Surface Active Agents and Detergents,* Interscience Publishers, Inc., New York, 1958.

A. J. de Vries, *Foam Stability,* Rubber Stitching, Delft, The Netherlands, 1957.

SYDNEY ROSS
Rensselaer Polytechnic Institute

FOOD ADDITIVES

Food additives have been the topic of vast attention over the past decade by large numbers of concerned groups. Once the principal domain of scientific endeavors, food additives are now *de rigueur,* a lively conversational subject for the public at large. No longer an exclusive issue within the United States, concerns over the safety of food additives have been voiced on a global scale by the scientific ccommunity, the food and chemical industries, regulatory agencies, the legislative branches of governments, activist groups and, ultimately, the consuming public. In the United States, occasional congressional intervention threatens the authority of the regulatory agency (FDA) empowered to safeguard the public health (see Regulatory agencies).

Compounds that produce obvious toxic responses are not at issue. These are usually weeded out very early in screening programs and rarely reach the petitioning stage to receive approval for use in foods. The central question concerns the long-term safety of food additives in man. And long-term safety really means—will it produce cancer? The word cancer is charged with emotion and, unfortunately, although biomedical science knows a great deal about the subject, it is not currently very adept in making hard prophetic assessments for anything but the most blatant cases of carcinogenic substances. Ethical constraints preclude tests in humans, yet it is the central question of the safety of food additives in humans that needs resolution.

Epidemiological tests are similar to experiments on humans in that they are surveys of human self-experiment (1). Reference 2 describes several recent epidemiological tests of consumption of artificial sweeteners, and bladder cancer (see Sweeteners).

Animal models, which circumvent tests in man, depend on lifetime feeding studies where substances are introduced in diets at levels often exceeding by several orders of magnitude the actual intended use levels in humans. It is not unusual for additives intended for use in food at 100 ppm to be tested in animal diets at 5–10% (50,000–100,000 ppm) in the quest to magnify and hasten a potential carcinogenic response. But the models are flawed by numerous obstacles. For example, there are fundamental physiological, biochemical, and genetic differences among species. In the United States, the Delaney Clause (a portion of the 1958 *Food Additives Amendment* to the *U.S. Food, Drug, and Cosmetic Act of 1938* introduced by Congressman J. Delaney (N.Y.) in February 1957) bans use in foods of any chemical shown to produce cancer in man or animals, at any level. In the latest *cause célèbre* surrounding saccharin, the statistical interpretation of lifetime feeding studies are being severely questioned. For example, the incidence of cancer increases with increasing age; this confuses the interpretation of lifetime feeding studies since animals on control diets without the test substance show a spontaneous increase in the incidence of cancer with increasing age, and toxicologists are asked to assign significance between animal groups sharing some evidence of carcinogenicity. Under these conditions, saccharin is now viewed as a moderate carcinogen (2).

Food additives serve an invaluable role in maintaining the world's food production and safeguarding man's nutrition. A case in point is the issue currently surrounding the benefits derived by man from the use of nitrates and nitrites in meats for preventing botulism vs the cancer risks from the production of nitrosamines (see *N*-Nitrosamines). Nitrosamines are not completely foreign to man's environment nor directly attributable to the use of nitrates and nitrites in food. Nitrates occur naturally in spinach, lettuce, and celery. Nitrates and nitrites are a natural component of saliva, the stomach, and the intestines. Nitrosamines can be generated by the action of intestinal microflora on nitrates and nitrites from both saliva, and the food and water supply.

Clearly, the addition of chemicals to foods is necessary in the production, processing and marketing of many foods. When properly used, they are not harmful and contribute greatly to the abundance, variety, stability, flavor, and appearance of the food supply. The problem is to ensure that such chemicals are safe for continuous use, that they serve a useful technological purpose, and that they are not employed to mask with fraudulent intent a defective and inferior product.

Definition

Defining food additives is not a straightforward task since there are no less than two important factors that can affect description. In the technical sector, the term food additives has been used to describe broadly any chemical substance used to augment or modify the characteristics of a food product. To the food scientist, colorants, preservatives (including antimicrobial agents, antioxidants, and sequestrants), emulsifiers, thickeners, acidulants, enzymes, flavors, and nonnutritive sweeteners are traditionally included in the technical definition. Basic foodstuffs with nutritional value are generally excluded from the technical definition although some view vitamins, minerals, amino acids, and protein fractions as additives when added to foods to upgrade nutritional value, ie, fortification. For the most part, the food science community views food additives as useful chemicals of natural or synthetic origins employed in small amounts—a few ppm to perhaps as much as 0.5–1.0%—although there are exceptions (eg, acidulants such as citric acid can be used at a few percent in some products). Present wording suggests that food additives and food ingredients are synonymous. The latter generally is considered to be the broader term, covering both additive and nonadditive materials in a food product.

The legal definition of food additives is much more complex and subject to controversy, interpretation, and opinion. During recent years many countries and organizations have written definitions for food additives to suit their specific purposes. In 1959 the Food Protection Committee of the National Research Council (NRC) defined a food additive as a substance or mixture of substances, other than a basic foodstuff, that is present in food as a result of any aspect of production, processing, storge, and packaging. The term does not include chance contaminants. An additive may either be nutritive or nonnutritive; it may be physiologically active or inert; it may be present intentionally, to achieve some modification in the foods, or incidentally and serving no useful purpose in the final product (3). Canada, on the other hand, has defined a food additive as any substance, including any source of radiation, the use of which results, or may reasonably be expected to result, in it or its by-products becoming a part of, or affecting the characteristics of a food (4). The Canadian definition then proceeds to make certain exceptions, such as foods themselves, and vitamins and pesticides which are adequately controlled by other sections of that country's food legislation. The UK , FRG, Italy, France, the Scandinavian nations, and Japan have modified the definition to suit their purposes. Many have published approved (white lists) and nonapproved lists to better define what additive is or is not permitted. Interestingly, almost all nations feel it is totally rational to approve the use of a substance in certain foods but not others. In the United States, an approved food additive may be precluded from use in certain foods which are characterized by a standard of identity unless specifically included in the standard. If the substance is not specified in the standard of identity (usually as an optional ingredient), the standard can be amended via petitioning action. In other instances, additives are precluded from use in certain food categories simply because of the time required for petitioning, the petitioner did not seek specific approval. The regulations covering the use of grape skin extract (*enocianina*) as a colorant is a case in point. Grape skin extract is approved only for use in coloring beverages and is not permitted in other food products simply because extended usage was not sought.

In most nations, direct food-additive legislation limits the use of a substance to maximum doses in specified foods and are subject to premarketing approvals. In the United States, the FDA has primary jurisdiction over food additives; other regulatory agencies exercise authority over certain products. For example, the USDA has jurisdiction over additives and ingredients in meat and poultry, and the Bureau of Alcohol, Tobacco, and Firearms of the U.S. Department of the Treasury controls ingredients in wine (qv), beer (qv), and spirits (see Beverage spirits, distilled).

Typically, a new substance gains approval for use in foods through the successful submission of a Food Additive Petition (FAP) in which the following information is documented: safety, including negative carcinogenic studies; declaration of intended use(s); efficacy data at specific levels in specified food systems; manufacturing details including intermediates and final product specifications backed by analytical methods; methods for analyzing the substance in foods; and an environmental impact statement.

The time and financial requirements needed to submit an FAP are enormous. Since 1968 only a few new food additives have undergone the regulatory approval route in the United States. These include: *tert*-butylhydroquinone (TBHQ) [*1948-33-0*] (an antioxidant (qv)), xanthan gum [*11138-66-2*] (a hydrocolloid (see Gums)), and FD&C Red No. 40 [*25956-17-6*] (see Colorants for foods, drugs, and cosmetics), as well as aspartame [*53906-69-7*] (a nonnutritive sweetener) which, although temporarily approved by FDA in 1974, is not approved at present. FDA stayed the order on aspartame on December 5, 1975, pending resolution of safety questions (see Sweeteners).

In the United States, many food additives are old (traditional) substances for which there is a long history of use and, as such, they have been deemed GRAS (generally recognized as safe). GRAS substances are not usually supported by a food additive petition (FAP); however, they are currently the subject of a systematic review by the Select Committee on GRAS Substances (SCOGS) of the Federation of Associated Societies for Experimental Biology (FASEB) under contract from FDA (5). SCOGS collects and reviews data from the literature and makes recommendation to FDA which has ultimate responsiblity of affirming or rejecting the GRAS status of the substance. To date, many GRAS substances have received affirmation of GRAS status as a result of comprehensive reviews; individual reviews are published as monographs.

It is important to note that, in the United States, colorants are handled in a special way. A new colorant intended for use in foods is the subject of a color additive petition (CAP). This follows the basic format of an FAP but does not require specific proof of efficacy, other than imparting color, and is not stringently limited to specified use levels. Moreover, colorants are classified as either certified or uncertified. The certification of colorants is the only instance in which the FDA constantly participates with industry in accepting or rejecting every lot of color manufactured. Food colorants undergoing certification proceedings are termed FD&C dyes and can be used by the food, drug and cosmetic industries; D&C dyes are also certified but limited to use by the drug and cosmetic industries only. The manufacturer pays the FDA a certification fee of ca $0.38/kg of dye certified; this covers analytical and administrative costs incurred by the agency.

Finally, substances that are not intentionally added to foods but nonetheless find their way into foods from packaging materials, agricultural chemicals residues, etc, are also regulated by FDA and the agencies of other nations. In the United States, these are termed indirect food additives and are subject of yet other petitioning procedures. For the most part, the issue in this instance centers on proof of safety.

Safety. Safety aspects of food additives are described in the *Food Chemicals Codex* (6). Other publications that have some descriptions of food additives and their safety are given in reference 7.

Recent studies have disproved the harmful effects (the Feingold hypothesis) of food colorants and salicylates on hyperactive children (8).

International Activities. In some European nations, the concept of nature-identical substances is taking on special meaning. In the Federal Republic of Germany, eg, synthetic flavor chemicals, which have been shown to stem from naturally occurring substances traditionally consumed by man, are viewed less stringently than totally new compounds.

The term food additives often evokes a negative reaction in the public, although various treatments of food necessitate the use of chemicals to ensure adequacy and safety. Although the safety and suitability of additives are evaluated by most governments and by the Joint FAO/WHO Expert Committee on Food Additives in most parts of the world, consumers have been alarmed enough to demand that the food industry respond by omitting additives (9).

A recent report contains the work of an international group of experts in their evaluation of certain food additives (10).

Categories

Depending on the nation, food ingredients and additives are described in a variety of categories. An industrial publication describes the following 52 food additive categories (11): acidulants, amino acids, anticaking agents (drying agents), antimicrobial agents (fumigants), antioxidants, baking aids (yeast foods), binders (plasticizers, fillers), bleaching agents (oxidizing agents), buffering agents, bulking agents, carmelization aids, carriers (disintegrating agents, dispersing agents), clarifying agents, clouding agents (density adjusters for essential oils), colorants (pigments, lakes), color fixatives (color retention aids), cooking media, defoamers, dough conditioners, emulsifiers (foaming agents, whipping agents), enzymes, filter aids, firming agents, flavor enhancers (flavor intensifiers), flavors, foam stabilizers, food starch modifiers (starch modifiers), gums, humectants (moisturizing agents), hydrocolloids, leavening agents, lubricants (antistick agents, release agents), masticatory substances (chewing gum base), maturing agents, microorganisms (molds, yeasts), neutralizers (acids, bases, alkalies), nonnutritive sweeteners, nutrients and dietary supplements, preservatives, propellants, protein sources, refrigerants, salt substitutes, sequestrants (chelating agents), solvents (solubilizers, vehicles, diluents, extractants), stabilizers (suspending agents), surface-active agents (wetting agents), surface-finishing agents (glazes, coating agents, film forms, microencapsulating agents), sweeteners (carbohydrates), texturizers (texture modifiers), thickeners (gelling agents), and vitamins. The FDA lists 32 functional effects of food additives (12). Other publications show the current status of food additives and colorants (13–14). The number of ingredients and additives within each food additive category is decidedly skewed with synthetic flavor chemicals comprising the largest category followed closely by the natural flavors (see Flavors and spices). Table 1 shows a representative distribution of ingredients and additives within some major categories. Table 2 is a list of a few chemical additives and their estimated levels of use (15).

Technical Descriptions

Comprehensive technical and regulatory descriptions of food additives are references 16–18. The following sections are brief descriptions of the various food additives.

Anticaking Agents. As the name implies, these substances are added to finely powdered or crystalline food products to prevent their caking and, hence, to maintain their free-flowing properties. Such chemicals as calcium aluminum silicate [*1327-39-5*], calcium phosphate tribasic [*7758-87-4*], calcium silicate [*1344-95-2*], calcium stearate [*1592-23-0*], magnesium carbonate [*546-93-0*], magnesium silicate [*10034-94-3*], magnesium stearate [*557-04-0*], and sodium aluminum silicate [*1344-00-9*] at a level

Table 1. Representative Distribution of Food Additives Within Major Categories[a]

Category	Number of Additives
acidulants	9
antimicrobials	20
antioxidants	31
bulking agents	16
colorants	35
emulsifiers	84
enzymes	13
flavors	
synthetic	>1050
natural	320
enhancers	9
gums	11
hydrocolloids	7
nonnutritive sweeteners	6
sequestrants	34
stabilizers	61
sweeteners	6
thickeners	41
vitamins	13

[a] Many additives are classified in multiple categories.

Table 2. Selected Chemical Additives and Estimated Level of Use (1975) in the U.S. Food Supply [a]

Type of additive	Use per year (est), 1000 metric tons	Examples
antioxidants	7–9	butylated hydroxyanisole, erythorbic acid, nitrogen gas
colors	40	caramel, certified colors, beet powder
emulsifiers and surface-active agents	105	mono- and diglycerides, propylene glycol sorbitol, glycerol monostearate
enzymes	7–10	papain, rennet, glucose–isomerase, pectinmethylesterase
flavor enhancers	25	monosodium glutamate
flavors and spices	115	mustard, pepper, ethyl acetate, carbon dioxide, vanillin
leavening agents and baking aids	55–60	benzoyl peroxide, sodium acid pyrophosphate, various carbonates
nonnutritive sweeteners	1–2	sodium saccharin
nutrient supplements	10–15	ascorbic acid, niacin, ferrous sulfate, thiamin hydrochloride, zinc sulfate
pH control agents	215	acetic acid, calcium carbonate, calcium phosphates, citric acid, phosphoric acid, sodium bicarbonate, malic acid
preservatives	20–25	sodium benzoate, sulfur dioxide, sorbic acid, propionic acid
stabilizers, thickeners	200	modified food starch, various gums, carboxymethyl cellulose, cellulose
anticaking agents	4–5	calcium silicate, magnesium stearate
stimulants	1	caffeine
protein hydrolysate	20	
sequestrants	0.5	citric acid, disodium ethylenediaminetetraacetate

[a] Ref. 15.

of 1.0% have been suggested as effective anticaking agents in free-running salt and in dry powdered mixes. Colloidal silicon dioxide [*7631-86-9*] is also used at a level of 1.0% as an anticaking agent in salt. The decahydrate of sodium ferrocyanide [*14434-22-1*] at the extremely low level of 5 ppm is reported to be an effective anticaking agent for salt; when used at levels of 13 ppm, this chemical causes the formation of star-shaped crystals of sodium chloride rather than the usual cubical ones. The star-shaped crystals known in the trade as dendritic salt possess nonsegregating properties, hence, they are less likely to cake (19).

Antifoaming Agents. A troublesome frothing or foaming occurs during certain processing steps in the manufacture of many food items. Fortunately, foaming may be largely suppressed or completely eliminated by the use of small quantities, generally about 10 ppm, of dimethylpolysiloxane as an antifoaming agent. This compound has been suggested for use in the preparation of certain meat products, bakery products, confections, dairy products, fats and oils, fruit juice products, jams and jellies, molasses, soups, starches, syrups, wines, and pickles. The mono- and diglycerides have also been suggested as antifoaming agents (see Defoamers).

Antioxidants. (See Preservatives.)

Bleaching, Maturing, and Dough-Conditioning Agents. Because some chemicals serve as both bleaching and maturing agents, and others are referred to as dough-conditioning agents or bread improvers, it is perhaps desirable to consider all of them under one heading.

Bleaching agents (qv) are used in the production of certain cheeses, processed fruits, and meat products (qv) and to neutralize color that may be present naturally in fats and oils. However, in view of the fact that they are of special importance to the flour-milling and baking industry, these chemicals are discussed here only as they are used in these foods. Freshly milled wheat flour has a yellowish color, caused by the presence of small quantities of carotenoid and other pigments. This flour also lacks the quality necessary to make an elastic, stable dough (20). When such flour is stored and allowed to age for several months, it gradually becomes whiter because of oxidation, and matures, making it acceptable for baking. For many years, this natural bleaching and aging was the only means millers and bakers had of producing the desired material. These natural processes were slow and costly and did not always yield a consistently satisfactory product. Then it was discovered that certain oxidizing agents incorporated into the flour in small amounts brought about rapid improvements in its color and breadmaking properties. The chemistry of flour bleaching is reviewed (21), and references 22–24 discuss the reaction mechanisms in dough of some bleaching and maturing agents.

Table 3 lists some of the chemicals used as bleaching and maturing agents, together with the treatment level. The following have also been suggested as bleaching, maturing, or dough-conditioning agents: ammonium peroxysulfate [*7727-54-0*], *Aspergillus flavus-oryzae* enzyme [*9001-19-8*], calcium peroxide [*1305-79-9*], calcium stearoyl-2-lactyllate [*5793-94-2*], chlorine [*7782-50-5*], L-cysteine [*52-90-4*], nitrosyl chloride [*2696-92-6*], oxides of nitrogen (N_2O [*10024-97-2*] and NO [*10102-43-9*]), and potassium peroxysulfate [*7727-21-1*] (see Bakery processes and leavening agents).

Colorants. Colorants, both synthetic and derived from natural sources, are used in many foods to enhance their appearance. Although current estimates indicate that 80% of the food colorants are synthetic, those of natural origin are gaining popularity principally owing to adverse industry and consumer reactions to the safety issues

Table 3. Examples of Bleaching and Maturing Agents Used in Flour

Chemical	CAS Registry No.	Function	Treatment level, ppm: Range	Treatment level, ppm: Normal
acetone peroxides		bleaching and maturing	5–45	20
azodicarbonamide	[*123-77-3*]	maturing	5–45	10
benzoyl peroxide	[*94-36-0*]	bleaching	25–200	50
potassium bromate	[*7758-01-2*]	maturing	5–45[a]	15[a]
bromate–iodate (4:1) for continuous dough-making process		maturing	25–75	60
chlorine dioxide	[*10049-04-4*]	bleaching and maturing	10–40	20
calcium or potassium iodate	[*7789-80-2, 7758-05-6*]	maturing	5–20[a]	10[a]

[a] Expressed as bromate and iodate.

surrounding the synthetic colorants. Because of their high tinctorial strength, synthetic colorants of the azo and triphenylmethane class are used at levels of a few ppm to ca 300–500 ppm whereas naturally derived colorants require much higher levels, often a few percent as in the case of caramel color. A recent publication reviews the recent technology and regulatory aspects of food colorants as presented by various experts in an ACS symposium (25). Food colorants currently permitted in the United States are shown in the article on Colorants for foods, drugs, and cosmetics.

The use of colorants in foods has been plagued with controversy for well over a decade principally because of questionable safety. In recent years, the synthetic (certified) colorants have been subjected to rigorous toxicological examination with the result that most nations now have a very limited number of approved colorants from which to choose. For example, FD&C Red No. 2 [*915-67-3*] (amaranth) and FD&C Violet No. 1 [*1694-09-3*] have been banned in the United States and many other nations. FD&C Yellow No. 5 [*1934-21-0*] (tartrazine) has been associated with hypersensitivity reactions (see Safety). A recent ruling by FDA requires that foods colored with FD&C Yellow No. 5 be declared as such on product labels (26).

The most recent addition to the permitted list of certified food colorants, FD&C Red 40, has been a suspected carcinogen *per se* (although this opinion has been recently changed) because *p*-cresidine, an intermediate used in its manufacture, is a known carcinogen. Nor have colorants derived from natural sources escaped scrutiny from regulatory agencies and consumer activists. In adhering to the letter of the law, the use of beet color is restricted to dehydrated beet powder and juice concentrate; spray drying the juice into a food-grade soluble carrier is not permitted under present United States regulations. Similarly, grape skin extract (*enocianina*) is restricted to use in beverages (see Fruit juices).

Although the food colorant market was valued at 43.2×10^6 dollars in 1976 and expected to increase to 70.4×10^6 dollars in 1985, regulations inhibit introduction of new candidates. A notable exception is the development of polymer-bound colorants which are reported to be nonabsorbed from the gastrointestinal tract upon ingestion (27–28). Currently undergoing long-term chronic feeding studies in animals, the polymeric food dyes are scheduled for submission to regulatory review in 1981. Another development of significance is the reported formulation of the carotenoid food colorants into a product form that will render them optically clear (transparent) in beverages (25). To date, this development is not commercially available.

Emulsifying, Gelling, Stabilizing, and Thickening Agents. This group of food additives contains a large number of compounds that influence the texture, homogeneity, or stability of the foods in which they are used. Because in many cases the distinction between an emulsifier, a gelling agent, a stabilizer, or a thickener is not too easily defined, it is desirable to consider them together.

The texture of ice cream and other frozen desserts is dependent in part upon the size of the ice crystals in the product (see Milk products). Smoothness of texture results from rapid freezing with the formation of numerous small crystals; coarseness of texture results when the crystals form slowly and are large. The growth of large crystals is minimized by the use of stabilizers that increase viscosity (29). The viscosity of chocolate milk is increased by the use of stabilizers to prevent sedimentation of the dense cocoa particles (see Chocolate and cocoa).

The use of certain emulsifiers in baked goods is reported to result in improved texture and eating quality, prolonged palatability and other keeping qualities, improved flavor, and improved appearance and uniformity (30).

Process cheese is a blend of selected cheeses comminuted and melted into a plastic mass that firms upon cooling. Unless the butterfat is emulsified, it may separate as an oil during the heating process and prevent the blending of the components. Emulsification is aided by the presence of certain soluble salts which are not true emulsifiers but cause emulsification indirectly through their effect on the proteins. These salts also impart smoothness of texture, improved slicing properties, and freedom from fat leakage in the finished cheese (29).

Gelling agents are used in jams, jellies, and marmalades to provide a product of satisfactory consistency. Soft drinks may contain thickeners to provide a product with body or consistency. Brewed beverages often require a stabilizer to regulate their foaming qualities. Emulsifiers improve the keeping qualities and homogeneity of candies and confections. Chocolate candy tends to bloom or change in surface color if exposed to temperature changes because the cocoa butter tends to separate from the chocolate; an emulsifier helps to prevent this condition by keeping the fat in a more stable emulsion within the chocolate (20). Some of the compounds commonly used as emulsifiers are discussed below (see Emulsions).

The polysaccharides that are used widely in foods may be classified according to origin (31): (*1*) plant and tree exudates, eg, acacia gum [*9000-01-5*] (gum arabic), karaya gum [*9000-36-6*], tragacanth gum [*9005-65-1*]; (*2*) marine plant extracts (seaweed), eg, agar–agar [*9002-18-0*], algin [*9005-38-3*] and its derivatives, carrageenin [*9000-07-1*] (Irish moss extractive), furcellaran [*9000-21-9*]; (*3*) seed extracts, eg, carob bean gum [*9000-40-2*] (locust bean gum), guar gum [*9000-30-0*], oat gum; (*4*) fruit and vegetable extract, eg, pectin [*9000-69-5*]; and (*5*) fermentation products, eg, xanthan gum [*11138-66-2*] (see Gums).

These gums and pectin are used in amounts of 0.1–1.0%, with some specialty foods containing up to 5% or higher. They are added to such foods as cream cheese, chocolate drinks, French dressing, pickles and relishes, salad dressing, ice cream, sherbet, certain meat products, jams and jellies, pie fillings, fountain syrups, confectionery, dessert gels, and pudding mixes.

Cellulose derivatives (qv), such as methyl cellulose [*9004-67-5*], carboxymethyl cellulose [*9000-11-7*], (hydroxypropyl)methyl cellulose [*9004-65-3*], and sodium carboxymethyl cellulose [*9004-32-4*], are being used increasingly. In such foods as dressings, chocolate drinks, process cheese, ice cream, and sherbet, the amount generally added is ca 1%.

Various mineral emulsifiers suggested for process cheese include calcium citrate [*813-94-5*], tricalcium phosphate [*7758-87-4*], potassium citrate [*866-84-2*], dipotassium hydrogen phosphate [*7758-11-4*], sodium hydrogen pyrophosphate [*7758-16-9*], sodium citrate [*68-04-2*], sodium gluconate [*527-07-1*], sodium hexametaphosphate [*10124-56-8*], mono-, di-, and trisodium phosphates [*7558-80-7, 8027-28-9, 7601-54-9*], sodium potassium tartrate [*304-59-6*], tetrasodium pyrophosphate [*7722-88-5*], and sodium tartrate [*868-18-8*].

The mono- and diglycerides of fatty acids, as well as acetylated monoglycerides, acetylated tartaric acid esters of mono- and diglycerides, and lactylated mono- and diglycerides, are used in foods as emulsifiers and stabilizers. The mono- and diglycerides themselves are more widely used than the derivatives, being added to such foods as bread, shortening, process cheese, ice cream, chocolate milk, and confections in amounts ranging up to 20%.

The so-called synthetic emulsifying agents include such products as polyoxyethylene sorbitan monooleate [*9005-65-6*], polyoxyethylene sorbitan monostearate [*9005-67-8*], polyoxyethylene stearate [*9004-99-3*], and sorbitan monostearate [*1338-41-6*], which are added in amounts of 1% or less to such foods as ice cream, cakes, confectionery icings, pie fillings, whipped vegetable oil toppings, chocolate drinks, and beverage bases (see Emulsions; Surfactants).

Gelatin (qv) is widely employed in the food industry because of its elastic consistency, because it has the property of holding air and water, or because it can inhibit the crystallization or graining of sugar. It is used in certain meat products, process cheese, frozen desserts, confectionery, etc. The amounts required for good manufacturing practice may be only 0.5% for certain foods, but for others as much as 10% may be required.

For proper gel formation of certain food products, a source of calcium ions is required. Calcium salts used for this purpose include calcium citrate, calcium gluconate [*299-28-5*], calcium glycerophosphate [*27214-00-2*], calcium hypophosphite [*7789-79-9*], di- and tricalcium phosphates [*7757-93-9, 7758-87-4*], and calcium sulfate [*7778-18-9*].

Other emulsifiers, stabilizers, thickening agents, or gelling agents are cholic acid [*81-25-4*], deoxycholic acid [*83-44-3*], and glycocholic acid [*475-31-0*] (or bile extract); as well as sodium lauryl sulfate [*151-21-3*] and triethyl citrate [*77-93-0*], used in the processing of egg whites (0.1–0.25%); lecithin [*8002-43-5*], used in bread, margarine, process cheese, ice cream, etc; and stearyl monoglyceridyl citrate, used in shortening.

The Joint FAO/WHO Expert Committee on Food Additives (32) wrote specifications for and evaluated the toxicological data of 39 emulsifiers and stabilizers. The committee also established acceptable daily intake for man for these compounds.

Food Enzymes. Enzymes (qv) play a significant role in a very wide range of food processing (qv) technology. Cheesemaking, breadmaking, brewing, glucose production and the confection industry, fruit juice and soft drink manufacture, meat tenderizing, egg white preparation, and the manufacture of cereal products are examples of food processes utilizing enzymes (see Enzymes, industrial).

Proteases. Proteolytic enzymes, such as ficin [*9001-33-6*], papain [*9001-73-4*], bromelain [*9001-00-7*], and fungal proteases are used for the tenderizing of meat. One such technique involves the injection of papain into cattle just prior to slaughter; this practice is reported to provide uniform distribution of the enzyme to all parts of the animal body. Pancreatin [*8049-47-6*], a mixture of two proteinases as well as other enzymes, is used in the production of gelatin and protein hydrolyzates, and rennin [*9001-98-3*] is used in the manufacture of certain cheeses.

Pepsin [*9001-75-6*], papain, and other enzymes are used in the brewing industry to stabilize and chillproof beer (qv). Proteolytic enzymes aid in the separation of residual meat from the bones of animal carcasses, they render cereal proteins soluble in the production of instant cereals, and they help to wet milk and similar solids for quicker reconstitution. Fungal proteases improve bread grain, texture, and compressibility of bread crumb. They also improve color and increase loaf volume (33).

Carbohydrases. Invertase [*9001-57-4*] is used in the confection industry to make syrupy centers for chocolate and for the production of invert sugar, which is increasing in usefulness in the food industry. Amylases break down gelatinized starches in the production of corn syrup and dextrose sugars. They also convert acid-modified starches to sweet syrups. Fungal amylases liquefy cocoa carbohydrates to dextrin and sugar. Heat-stable amylases increase speed and completeness in liquefying brewer's cereal adjuncts. Enzyme mixtures containing cellulases, hemicellulases, and pectinases are used to clarify wines, as well as grape, apple, and citrus juices, and to give higher yields (see Fruit juices; Wine). Hemicellulases degrade gums that cause coffee (qv) concentrates to gel during the manufacture of the instant product. Pectinases are also employed for the clarification of wines and fruit juices; they tend to lower the viscosity of pectin [*9000-69-5*] and fruit juices, thus facilitating filtration. In the production of jellies, jams, purees, and pastes, pectinases are used to remove pectins from fruit juices in order that evaporation can occur without the formation of troublesome gels (33).

Glucose Oxidases. Glucose oxidases are used to stabilize egg solids (by desugaring yolks, whites, and whole eggs (qv)); to remove oxygen from carbonated (see Carbonated beverages) and still beverages, and food dressings; and to prevent undesirable flavor changes (see Flavor characterization).

Catalases. Catalases destroy residual hydrogen peroxide employed in the cold sterilization of milk for the manufacture of cheddar, colby, granular, Swiss, and washed-curd cheese.

Lipases. Lipases have been used for the treatment of butterfat to produce flavored substances that in turn are used to flavor bakery products and confections.

The latest use of enzymes by the food industry is in the conversion of corn syrup to high fructose syrups (qv). Enzymes are insolubilized by binding onto various carriers. Following hydrolysis, the corn syrup is contacted with the bound enzyme which converts glucose into fructose. The process is unique in that the enzyme used in the process does not become part of the finished product. This and other developments are expected to result in a large growth of food-grade enzymes, 58×10^6 dollars currently (1979) to $>128 \times 10^6$ dollars by 1985 (see Enzymes, immobilized).

Firming Agents. Fruits and vegetables contain pectin components that are relatively insoluble and which form a firm gel around the fibrous tissues of the fruit and prevent its collapse. Addition of calcium salts causes the formation of calcium pectate gel, which supports the tissues and affords protection against softening during processing. The calcium salt is sometimes added to the canned vegetable in the form of a combination tablet containing both salt and calcium chloride [*10043-52-4*].

Canned vegetables, canned apples, frozen apples, and canned tomatoes are sometimes treated during processing with calcium chloride, calcium citrate, monocalcium dihydrogen phosphate [*7758-23-8*], or calcium sulfate [*7778-18-9*] to prevent them from becoming soft and disintegrating. One suggested level of use of these calcium salts is 0.026%, calculated as calcium in the final food product. In canned potatoes, calcium chloride and calcium citrate at a level of 0.5% (calculated as calcium) are reportedly used.

Aluminum sulfate [*10043-01-3*], ammonium aluminum sulfate [*15710-63-1*], potassium aluminum sulfate [*10043-67-1*], and sodium aluminum sulfate [*10102-71-3*] are reported to be of value as firming agents in pickles and relishes. A more recently introduced firming agent is aluminum sulfate for canned crabmeat, lobster, salmon, shrimp, and tuna. Calcium chloride is reported to act as a canning agent in cheddar and cottage cheese.

Flavoring Agents. Because of the importance of the flavor of a food product, food manufacturers make use of a variety of flavoring substances (see Flavors and spices). As a result, flavoring agents, both natural and synthetic, make up one of the largest groups of food additives. For a recent listing of a very large number of flavoring substances, see reference 34.

Flavor Enhancers. Although monosodium glutamate [*142-47-2*] (MSG) has been recognized and used in the food industry as a flavor-enhancing product for many years, two 5′-ribonucleotides, disodium inosinate [*4691-65-0*] and disodium guanylate [*5550-12-9*], were introduced by Japanese scientists in 1964 (35–36) (see Amino acids, monosodium glutamate). Like MSG, these compounds enhance the natural flavoring substances of a food without contributing a characteristic flavor of their own. With MSG, they are used principally in meat and vegetable processed foods at levels of from 0.01 to 0.03%. Maltol [*118-71-8*] and ethyl maltol [*4940-11-8*] are used to enhance the flavor of fruits and fruital food compositions.

Glazing and Polishing Agents. Glazes and polishes are used on coated confections to give luster to the otherwise dull coating. Chemicals that are used for this purpose include acetylated monoglycerides, beeswax, carnauba wax, gum arabic, magnesium silicate [*1343-88-0*], mineral oil, petrolatum, shellac, spermaceti wax, and zein, generally at levels of ca 0.4%, with the exception of mineral oil and petrolatum which are used at 0.15%, and zein at 1.0%.

Humectants. These are hygroscopic materials which prevent loss of moisture when incorporated into foodstuffs. The principal humectants include glycerol [*56-81-5*] (qv), mono- and diglycerides of fatty acids, propylene glycol [*57-55-6*] (see Glycols), and sorbitol [*50-70-4*] (see Alcohols, polyhydric). Pectins also act as humectants. Shredded coconut, marshmallows, and certain confections are examples of foods in which humectants are employed.

Nonnutritive Sweeteners. For nearly a decade this particular category of food additives has been charged with controversy and emotion owing to the implied safety risks associated with their use by a large segment of the population. Originally intended for those who had to restrict the intake of carbohydrates, the use of nonnutritive sweeteners by weight-conscious individuals has grown to unprecedented levels (see Sweeteners). In 1978 the use level of saccharin in the United States was estimated at ca 3700 metric tons, valued at 24.8×10^6 dollars. If there is no ban, saccharin use in the United States is expected to be >4500 t by 1983.

Following the ban of cyclamate in the United States in 1970, saccharin remains the only approved nonnutritive sweetener. However, numerous safety tests have since shown that saccharin produces cancer in animals (2). As a result, foods and beverages sweetened with saccharin must carry a warning label in order to be marketed in the United States. An 18-mo stay in the FDA ban order on saccharin by the Congress of the United States expired May 1979 but the stay is expected to be extended through mid-1982.

Nonnutritive sweetener research is progressing at a brisk pace although there are no clear chemical models to predict sweetness (see Flavor characterization).

Nutrients. Where a group of people or a population is deficient in a specific essential nutrient, its addition to the diet can be justified, eg, potassium iodide [*7758-05-6*] (20) (see Iodine and iodine compounds) and thiamine (vitamin B_1) [*59-43-8*] (37) (see Vitamins).

The enrichment of flour and bread products is now a widely accepted procedure (see Bakery processes). In the United States, the standard for enriched flour reads in part as follows (38): (*1*) it contains in each kilogram 6.4 mg (2.9 mg/lb) of thiamine, 4.0 mg (1.8 mg/lb) of riboflavin, 53 mg (24 mg/lb) of niacin, and not less than 28.7 mg and not more than 36.4 mg (13.0–16.5 mg/lb) of iron; (2) it may contain added calcium in such quantity that the total calcium content is 2.12 g/kb (960 mg/lb). Fortunately, the technological means whereby these materials may be uniformly added to foods are available to the food processor.

pH Adjusting Agents. Another large group of chemical additives which are widely used in foods might be considered under the broad heading of pH adjusting agents. Other terms that describe these chemicals include acidulants, acids, alkalies, buffers, and neutralizers. These chemicals are used in the baking industry as chemical leavening agents; in soft drinks to provide tartness; in certain dairy products to adjust the acidity; in cheese spreads for emulsification; in confectionery products as flavoring, to control the degree of inversion of sugars, and to control the texture in the processing of chocolates; and in jams, jellies, and marmalades to provide for proper gel formation.

Preservatives. These chemicals may be divided into two main groups for discussion purposes: antioxidants (qv) and antimicrobials (see Industrial antimicrobial agents; also Disinfectants and antiseptics). It should be clearly understood, however, that although preservatives serve a very useful purpose in foods, they cannot take the place of cleanliness in food processing. The use of preservatives will not improve the quality of inferior material, nor will preservatives, once spoilage has set in, enable a processor to make a wholesome product out of an unwholesome one. The Joint FAO/WHO Expert Committee on Food Additives has prepared specifications of identity and purity and has evaluated the toxicity of a number of antimicrobials and antioxidants used in food preservation (39–40).

Antimicrobial Agents. *Benzoic acid* [*65-85-0*] (qv) and its sodium salt are among the bacteriostatic or germicidal agents most widely used in foods. They are included on the permitted lists of at least thirty countries throughout the world for a great variety of foods. Calcium and potassium salts are on the permitted list of certain European countries. Actually, it is the benzoic acid which is the effective agent, since sodium benzoate must be used in an acid medium to be effective. The acidity of the medium is significant; eg, a decrease in pH from 7 to 3.5 may increase five- to tenfold the antiseptic and bactericidal action. Benzoic acid is more effective against yeasts than against molds. Benzoic acid or sodium benzoate is used in several countries in such foods as soft drinks, including fruit drinks, margarine, certain fish products, roe, fruit juices, jams, marmalade, pickles, and certain vegetable products. Benzoic acid is generally used at a level of about 1000 ppm but as little as 100 ppm appears to be effective in certain foods.

p-Hydroxybenzoic Acid Esters. Butyl, ethyl, methyl, and propyl *p*-hydroxybenzoates [*94-26-8, 120-47-8, 99-76-3, 94-13-3*] and the sodium derivatives of the methyl and propyl esters (p-$NaOC_6H_4COOR$ [*5026-62-0, 35285-69-9*]) have been suggested as antimicrobials in foods; the methyl ester is more effective against molds, whereas the propyl ester is more active against yeasts (41). These esters have been suggested for use as an alternative preserving agent for benzoic acid in the same foods and at approximately the same levels.

Propionates. Although the baking process destroys the spores of molds and most species of bacteria present in flour and other ingredients, bakery products are constantly exposed to spores present in the air and on baking equipment (see Bakery processes). Despite sanitary precautions, molds frequently develop during the summer when the temperature and humidity favor their growth. Flour contains the spores of the bacterium, *Bacillus mesentericus*, which are not killed by the baking temperature. Under summer storage conditions, these bacteria become active and produce a condition called *rope*, which renders bread inedible (29). The calcium and sodium salts of propionic acid [*4075-81-4, 137-40-6*] are widely used as mold and rope inhibitors in bread and bakery products at levels of ca 2000 ppm. The propionates have also found use in the prevention of mold in certain cheeses and on certain fruit and vegetable products.

Sulfur Dioxide and Sulfites. These compounds have been employed as food preservatives since the time of the early Egyptians and Romans. Sulfur dioxide [*7446-09-5*] is unique in being the most effective inhibitor of the deterioration of dried fruits and fruit juices. It is used widely in the fermentation industry to prevent spoilage by microorganisms and as a selective inhibitor of undesirable organisms. It is also used as an antioxidant and as an inhibitor of enzyme-catalyzed oxidative discoloration and of nonenzymic browning during the preparation, storage, and distribution of many food products. Reference 42 is an excellent review of the chemistry and technology of the preservation of fruit and vegetable products with sulfur dioxide and sulfites.

It is desirable that the initial sulfur dioxide levels in dried, cut fruits be 2000–2500 ppm to provide satisfactory storage life and to maintain a high quality during prolonged storage. During storage the sulfur dioxide content decreases, owing in part to the volatilization of the gas and in part to oxidation of the sulfite to the inert sulfate. When cut fruits are prepared for consumption by cooking, the level of SO_2 is still further reduced. Levels of 1000 ppm or less of sulfur dioxide are used on dried vegetables.

In the production of wine (qv), sulfur dioxide serves not only as an antimicrobial but also as an antioxidant and as a specific remedy for casse brune, a brown discoloration. Residual sulfur dioxide is present in certain wines up to a level of 350 ppm.

In addition to sulfur dioxide, the following sulfur compounds are reported to be used as antimicrobial agents: potassium bisulfite [*7773-03-7*], potassium metabisulfite [*16731-55-8*], sodium bisulfite [*7631-90-5*], sodium metabisulfite [*7681-57-4*], sodium sulfite [*7757-83-7*], and sodium dithionite [*7775-14-6*].

Sorbic acid [*110-44-1*] (qv) and its calcium, potassium, and sodium salts [*7492-55-9, 24634-61-5, 7757-81-5*] are selective growth inhibitors for certain molds, yeasts, and bacteria that cause spoilage in food products. Sorbic acid or its salts are reported to be effective antimicrobial agents in such foods as cheese products, pickles, certain fish products, carbonated beverages, margarine, and certain fruit and vegetable products, including wines. The levels of use are in the range of 100–1000 ppm.

Nitrates and Nitrites of Potassium and Sodium. These salts are often present in the brines used for the curing of certain meat products, and appear to act as a color fixative as well as to retard microbial growth (KNO_2 [*7758-09-0*], $NaNO_2$ [*7632-00-0*], KNO_3 [*7757-79-1*], and $NaNO_3$ [*7631-99-4*]). Nitric oxide formed from nitrite combines with the pigment, myoglobin [*62602-69-1*], of meat to form nitric oxide myoglobin [*62602-69-1*], which, when heated, is converted into a more stable compound that gives to cured meats the desired pink color (29). As a result of the cancer risks associated with the formation of nitrosamines, nitrate and nitrite levels are being reduced and alternative preservatives are being sought (see *N*-Nitrosamines).

Diethyl Pyrocarbonate. Diethyl pyrocarbonate [*1609-47-8*], $C_2H_5OC(O)$-$OC(O)OC_2H_5$, has been suggested as an antimicrobial agent (43). It is reported to be effective against the common spoilage organisms of various beverages at levels of 10–100 ppm. It is different from other preservatives, however, in that it reacts rapidly with water, thereby decomposing into ethyl alcohol and carbon dioxide. Many of the applications of this chemical have been in wines and, to a lesser extent, in beers, fruit juices, and carbonated beverages. However, there is evidence to show that diethyl pyrocarbonate does not hydrolyze completely to CO_2 and ethanol. Intermediate alkylating agents are also formed and this presents a serious safety risk. Consequently, it is not permitted for use in foods in the United States and many other nations.

Miscellaneous Antimicrobial Agents. Mention should be made of sodium diacetate [*126-96-5*] as a mold and rope inhibitor in baked goods; diphenyl [*92-52-4*] as a fungistatic which migrates from wrapping material to inhibit growth of mold causing decay of citrus fruits; hexamethylenetetramine [*100-97-0*] as a preservative for certain fish products; and *o*-phenylphenol [*90-43-7*] and sodium *o*-phenylphenoxide [*132-27-4*] in the postharvest treatment of fruits and vegetables to protect against microbial damage.

Antioxidants. A serious type of spoilage in foods containing oils or fats is the development of rancidity. Though rancidity may arise because of oxidative, hydrolytic, or ketonic degradation, it is generally owing to oxidation through the formation of peroxides at the double bonds of fat molecules with subsequent breakdown of these peroxides to form aldehydes, ketones, and acids of lower molecular weight. The substances used to retard oxidation in foods may be classified into two groups, antioxidants and synergists. For a review, see references 16–18.

Among the more common oil-soluble antioxidants (qv) are: butylated hydroxytoluene [*128-37-0*] (BHT), butylated hydroxyanisole [*25013-16-5*] (BHA), *tert*-butylhydroquinone (TBHQ), and propyl gallate [*121-79-9*] (PG); in Europe, other gallate esters are also permitted (eg, isopropyl [*1138-60-9*], octyl [*1034-01-1*], and dodecyl [*1166-52-5*]). TBHQ is the most recently approved antioxidant for use in foods in the United States, and nordihydroguaiaretic acid [*500-38-9*] (NDGA) was delisted some years ago. Other antioxidants used by the food industry include: ascorbic acid [*50-81-7*] (water soluble) and its oil-soluble esters (eg, ascorbyl palmitate [*137-66-6*]), the tocopherols, gum guaiac [*9000-29-7*], and lecithin.

For the most part, BHT, BHA, and TBHQ are employed alone, as blends with one another, and in conjunction with synergists such as citric acid or other metal chelating agents. Use levels are generally set at 100–200 ppm of total antioxidant based on the fat or oil content of foods. However, certain foods have specific exemptions permitting higher use levels.

The safety of BHT is under investigation and it is currently on interim status in the United States, pending further information (44). However, industrial use of BHT has been replaced largely by BHA and TBHQ. The gallates, although quite effective, are prone to discoloring food substrates; they also may impart off-flavor. BHT, BHA, and TBHQ have low temperature volatility (150–170°C) and are readily lost from thermal processes generating steam.

Synergists used in conjunction with antioxidants are of two types: metal-chelating agents and chain terminators. In the first instance, metals that act as pro-oxidant catalysts (eg, copper, iron, etc) are effectively neutralized using citric acid, phosphoric acid, EDTA, etc. The most popular metal chelator used with oil-soluble antioxidants is citric acid (see Sequestering Agents below). Since this is water soluble, it is incorporated into fats and oils via special antioxidant formulations (liquid concentrates) containing glycerol monooleate and propylene glycol as cosolvents. These solvents also afford the opportunity to prepare blends of various antioxidants. Less popular is the use of sulfur-containing chain terminators as synergists, eg, thiodipropionic acid [*111-17-1*] and dilauryl thiodipropionate [*123-28-4*].

Release Agents. Release agents are used to prevent confectionery and, to a lesser extent, baked goods from sticking to the container in which they are heated. They are also used to prevent pieces of confection from adhering to each other. Included are acetylated monoglycerides, calcium stearate [*1592-23-0*], magnesium carbonate [*546-93-0*], magnesium silicate [*1343-88-0*], magnesium stearate [*557-04-0*], mannitol [*69-65-8*], mineral oil, mono- and diglycerides, sorbitol [*50-70-4*], and stearic acid [*57-11-4*] (see Abherents).

Sequestering Agents. Ethylenediaminetetraacetic acid [*60-00-4*] (EDTA) is a sequestering, or metal-complexing, agent which has taken its place among the other sequestering (stabilizing) agents as a full-fledged tool of the food industry (45) (see Chelating agents). This compound, as the disodium or calcium disodium salt, promotes color retention in dried bananas, beans, chick peas, canned clams, pecan pie filling, frozen potatoes, and canned shrimp; it improves flavor retention in canned carbonated beverages, dressings, mayonnaise, margarine, and sauces; it retards struvite formation in canned crabmeat and shrimp; and protects against rancidity in dressings, mayonnaise, sauces, and sandwich spreads. It is used at levels of 33–800 ppm. Other chemicals which may be included in this category are citric acid [*77-92-9*], stearyl citrate [*10061-30-0*], and calcium monoisopropyl citrate [*25168-59-6*], used in margarine; sodium hexametaphosphate [*10124-56-8*], phosphoric acid [*7664-38-2*], and various calcium, potassium, and sodium phosphates.

Starch-Modifying Agents. The food processor who wishes to use starch as an ingredient of a food product is no longer restricted to the starch as it was obtained from the plant source. A large variety of modified starches are available, each altered in a way that makes it suitable for a specific food purpose.

In many food-manufacturing procedures, starches are required that retain their thickening qualities under a variety of conditions. If heat sterilization of a food product is employed, a starch with high temperature stability is required. In foods such as fruit-pie fillings and salad dressings, a thickener is needed that will not be adversely affected by the fruit acids or vinegar present. There are cases in which the thickened food product receives mechanical agitation sufficient to decrease the thickening properties of ordinary starches. Homogenized gravies and sauces, eg, require a thickener that will not break down under severe mechanical agitation; the so-called cross-linked starches provide a stable thickener required in foods of this type. The chemicals, eg, vinyl acetate [*108-05-4*], acetic anhydride [*108-24-7*], and succinic anhydride [*108-30-5*], used to modify the starches, or the modified starches themselves, may be considered to be food additives. However, these thickeners are either completely removed from the starches, or remain only as traces after processing.

Water-Correcting Agents. Water used in the brewing industry is often corrected to a uniform mineral salt content that corresponds to water known to give the most satisfactory final product (see Beer). A wide variety of salts are used for this purpose, including mono- and diammonium phosphates [*7722-76-1, 7783-28-0*], calcium chloride, calcium hydroxide [*1305-62-0*], calcium oxide [*1305-78-8*], calcium dihydrogen phosphate, calcium sulfate [*7778-18-9*], magnesium sulfate [*7487-88-9*], potassium aluminum sulfate [*10043-67-1*], potassium chloride [*7447-40-7*], potassium sulfate [*7778-80-5*], sodium bisulfate [*7681-38-1*], and mono-, di-, and trisodium phosphates. Some of the chemicals, in addition to standardizing the salt content, also control the acidity, thus providing uniform conditions for yeast fermentation.

Economic Aspects

Growth of food additives is estimated to be 11%/yr during the 1980s (46). Food additives (including all additives described above) production in 1978 was 481,000 metric tons ($\$1.3 \times 10^9$). Estimated production in 1990 is 769,000 t ($\$4.5 \times 10^9$).

BIBLIOGRAPHY

"Food Chemicals" in *ECT* 1st ed., Vol. 6, pp. 835–848, by M. B. Jacobs, Polytechnic Institute of Brooklyn; "Food Additives" in *ECT* 2nd ed., Vol. 10, pp. 1–22, by Douglas G. Chapman, Food and Drug Directorate, Department of National Health and Welfare, Canada, and Z. I. Kertesz, Food and Agriculture Organization of the United Nations.

1. S. E. Epstein, *The Politics of Cancer,* Sierra Club Books, San Francisco, Calif., 1978.
2. I. I. Kessler, *J. Urol.* **15,** 143 (Feb. 1976); B. Armstrong and co-workers, *Br. J. Prev. Soc. Med.* **30**(3), 151 (1976); G. R. Howe and co-workers, *Lancet ii,* 578 (1977); E. L. Wynder and R. Goldsmith, *Cancer* **40**(3), 1246 (1977); R. N. Hoover and co-workers, *Progress Report to the FDA from the NCI Concerning the National Bladder Cancer Study,* FDA, Washington, D.C., Dec. 21, 1979; E. L. Wynder and S. D. Stellman, *Science* **207,** 1214 (1980); *Science* **208,** 154 (1980); A. S. Morrison and J. E. Buring, *New Engl. J. Med.* **302,** 537 (1980).
3. *Publ. 750,* Food Protection Committee, Food and Nutrition Board, NAS/NRC, Washington, D.C., 1959, p. 1.
4. *Can. Gaz.* **98,** 998 (Sept. 23, 1964); *Food and Drugs Act of Canada, R.S.C. F-27,* amended by 1976–1977 *C. 28,* Minister of Supply and Services, Ottawa, Can., 1978.
5. R. G. H. Siu and co-workers, *Fed. Proc. Fed. Am. Soc. Exp. Biol.* **36,** 2519 (1977).
6. *Foods Chemicals Codex,* 3rd ed., NAS/NRC, Washington, D.C., 1980.
7. V. Packard, *Processed Foods and the Consumer,* University of Minnesota Press, Minneapolis Minn., 1976, pp. 52–59; B. L. Oser in N. I. Sax, ed., *Dangerous Properties of Industrial Materials,* 4th ed., Van Nostrand, New York, 1975, Section 10.
8. J. I. Williams and co-workers, *Pediatrics* **61,** 811 (1978); J. P. Harley and co-workers, *Pediatrics* **61,** 818 (1978).
9. *Food and Nutrition,* Vol. 3, FAO, New York, 1976.
10. *Evaluation of Certain Food Additives,* FAO/WHO, Geneva, Switz., 1976.
11. *Food Ingredients Directory,* Intechmark Corp., Palo Alto, Calif.
12. CFR, *Title 21,* FDA, Washington, D.C., 1979, Section 170.3, pp. 409–410.
13. *Food Chemical News, Food Chemical News Guide,* Washington, D.C.
14. S. A. Miller, *Handbook 300,* New York Academy of Sciences, New York, 1977, p. 397.
15. *Poundage of Food Chemicals Generally Recognized as Safe (GRAS),* Food and Nutrition Board, Div. of Biological Sciences, NRC, Washington, D.C., Nov. 1978; *PB-288-081,* U.S. Department of Commerce, National Technical Information Service, Springfield, Va., 1978.
16. T. E. Furia, *Handbook of Food Additives,* 1st ed., CRC Press, Inc., West Palm Beach, Fl., 1968.
17. T. E. Furia, *Handbook of Food Additives,* 2nd ed., Vol. I, CRC Press, Inc., West Palm Beach, Fl., 1972.
18. *Ibid.,* Vol. II, 1980.
19. *Food Eng.* **35**(1), 109 (1963).
20. *Food Additives: What They Are—How They are Used,* MCA, Washington, D.C., 1961.
21. W. F. Sollars, *Baker's Dig.* **35,** 48 (1961).
22. C. C. Tsen, *Cereal Chem.* **40,** 638 (1963).
23. *Ibid.,* **41,** 22 (1964).
24. C. C. Tsen and W. Bushuk, *Cereal Chem.* **40,** 399 (1963).
25. T. E. Furia, *Current Aspects of Food Colorants,* CRC Press, Inc., West Palm Beach, Fl., 1977.
26. *Fed. Reg.* **44**(124), 37212 (June 20, 1979).
27. T. E. Furia, *Food Technol.* **31**(5), 34 (1977); *Chem. Eng. News,* 20 (Oct. 22, 1979).
28. N. Bellanca, *Proceedings of the American Soft Drink Tech., 25th Annual Meeting,* 1978.
29. *The Use of Chemical Additives in Food Processing, Publ. 398,* Food Protection Committee, Food and Nutrition Board, NAS/NRC, Washington, D.C., 1956; *The Use of Chemicals in Food Production,*

Processing, Storage, and Distillation, Committee on Food Protection, NAS/NRC, Washington, D.C., 1973.
30. C. C. Pratt, *Food Technol.* **5,** 425 (1952).
31. M. B. Jacobs, *The Chemical Analysis of Foods and Food Products,* 3rd ed., D. Van Nostrand Co., Inc., New York, 1958.
32. Joint FAO/WHO Expert Committee on Food Additives, "Specifications for the Identity and Purity of Food Additives and Their Toxicological Evaluation: Emulsifiers, Stabilizers, Bleaching and Maturing Agents," *7th Report, FAO Nutrition Meetings Report Ser. No. 35, Rome, Italy 1964.*
33. *Food Eng.* **36**(5), 80 (1964).
34. R. L. Hall, B. L. Oser, and R. A. Ford, *Food Technol.* **14**(10), 488 (1960); **15**(12), 20 (1961); **19**(2), 151 (1965); **24**(5), 25 (1970); **26**(5), 35 (1972); **27**(1), 64 (1973); **27**(11), 56 (1973); **28**(9), 76 (1974); **29**(8), 70 (1975); **31**(1), 65 (1977); **32**(2), 60 (1978); **33**(7), 65 (1979) (these are lists of all the flavor compounds described by the Flavoring Extract Manufacturers' Association (FEMA)).
35. H. Shimazono, *Food Technol.* **18**(3), 36 (1964).
36. A. Kuminaka, M. Kibi, and K. Sakaguchi, *Food Technol.* **18** (3), 29 (1964).
37. R. R. Williams, *Toward the Conquest of Beriberi,* Harvard University Press, Cambridge, Mass., 1961.
38. Ref. 12, Section 137.165.
39. Joint FAO/WHO Expert Committee on Food Additives, "Evaluation of the Toxicity of a Number of Antimicrobials and Antioxidants," *6th Report, FAO Nutrition Meetings Report Ser. No. 31, Rome, Italy, 1962.*
40. *Specifications for Identity and Purity of Food Additives,* Vol. I, Food and Agriculture Organization, Rome, Italy, 1962.
41. M. B. Jacobs, *The Chemistry and Technology of Food and Food Products,* 2nd ed., Vol. III, Interscience Publishers, Inc., New York, 1951, p. 1942.
42. M. A. Joslyn and J. B. S. Braverman, *Adv. Food Res.* **5,** 97 (1954).
43. H. B. Hawley, "The Cold Sterilization of Beverages" in *Proceedings of the International Food Industries Congress, June 1964,* Food Manufacture, London, Eng., 1965, p. 97.
44. Ref. 12, Section 172.115.
45. T. E. Furia, *Food Technol.* **18**(12), 50 (1964).
46. *Chem. Week,* 55 (Feb. 20, 1980).

THOMAS FURIA
Intechmark Corp.

FOOD PROCESSING

The food industry is a specialized chemical processing industry. Individual chemical nutrients are dispersed among thousands of additional compounds arranged in highly individualistic forms characteristic of the foods consumed by various geographic, ethnic, or political population groups (see Amino acids; Mineral nutrients; Vitamins). These materials must be protected against damage resulting from exposure to light, heat, cold, microbes, oxygen, animal pests, bruising, and chemical contamination. Protection begins with raw materials often available only a few days each year at locations far from population centers. Product varies according to weather, local labor supplies, and political conditions. Approximate wholesale prices for selected commodities and their U.S. per capita consumptions are shown in Table 1.

The technologies used to manufacture the hundreds of individual foods for a varied diet are too numerous to describe in detail. They cover the principles involved in food process engineering, from harvesting to retail distribution and those leading to a stable, nutritious, and safe food supply. Detailed coverage of specific technologies is covered in the literature (see General References).

Food processing, regardless of the type of food, can be divided into three classes: separation, assembly, and preservation (1). These can occur at harvest, at the food processing plant, or even at the point of retail sales. Increasing emphasis is being placed on field processing as exemplified by the widespread use of mechanical harvesters fitted with cleaning, sorting, color measurement, and size grading systems. There are ca 150 named processing operations and they are grouped in Tables 2, 3, and 4.

Foods can be categorized as living tissue or raw foods, and nonliving tissue. Living tissue foods include fresh fruits, vegetables, meats, and grains. Processes for preserving living tissue are considered separately from those used for the preservation of nonliving tissue, eg, canned, frozen, and dried foods.

Table 1. Approximate U.S. Per Capita Consumption of Several Food Groups, Unit Price (Data from Agricultural Statistics, 1978, USDA)

Food group[a]	Consumption, kg/yr	Approximate wholesale cost, \$/kg (1977)
meat	76	1.55
poultry	25	0.55
fish	8	2.60
eggs	16	0.80
dairy	160	0.20
fats and oils	26	0.90
fruits	75	0.20–0.45
vegetables	95	0.10–0.25
potatoes	40	0.05–0.10
cereals	63	0.20–0.35
dry beans	8	0.55
sweeteners	60	0.35
coffee, tea, cocoa	5	1.30–4.50

[a] See Meat products; Aquaculture; Eggs; Milk products; Synthetic dairy products; Fats and fatty oils; Fruit juices; Vegetable oils; Dietary Fiber; Wheat and other cereal grains; Soybeans and other seed proteins; Sugar; Sweeteners; Coffee; Tea; and Chocolate and cocoa.

Table 2. Separation Unit Operations of the Food Industry Arranged by Mode of Separation and Phases Being Separated

Phases to be separated	Mode of separation: Physical	Chemical	Mechanical
gas–gas			
gas–liquid	condensing, blanching		deaeration
gas–solid	blanching		controlled atmosphere, deaeration, compression, densifying
liquid–liquid	distillation, membrane filtration	solvent extraction	centrifugation
liquid–solid	foaming, coagulation, drying	ion exchange, solvent extraction, flocculation	11 unit operations[a]
solid–solid	dialysis, freeze concentration sublimation		34 unit operations[b]

[a] Mechanical separations include: churning, centrifugation, clarification, draining, expelling, filtration, flotation, pressing, racking, rendering, and skimming.
[b] Mechanical separations include: abrading, boning, cutting, coring, crushing, dividing, defeathering, eviscerating, flaking, finishing, filleting, grinding, harvesting, husking, hulling, inspection, milling, peeling, picking, pitting, pulverizing, slicing, sieving, sorting, shredding, sizing, sifting, shelling, scarifying, sampling, shucking, stemming, vining, and winnowing.

Table 3. Assembly Unit Operations of the Food Industry Arranged by Mode of Assembly and Phases Being Assembled

Phases to be assembled	Mode of assembly: Physical	Chemical	Mechanical
gas–gas			
liquid–gas			aerating, foaming, carbonating, whipping
solid–gas	baking, puffing, agglomeration	humidifying, oxidizing, proofing	extruding, aerating
liquid–liquid		neutralization, acidification	emulsifying, homogenization, dispensing, mixing, pumping
liquid–solid	crystallization	soaking, malting, rehydration, acidification	dispersing, dissolving, immersion, mixing, pumping
solid–solid	braising, roasting	aging	7 unit operations[a]
gas–liquid–solid		sprouting	weighing, blending

[a] Mechanical assembly operations for solid–solid mixtures, include: coating, enrobing, filling, forming, molding, pelleting, and stuffing.

Food Composition and Standards

Chemical Composition. Virtually all foods are derived from living tissues, although individual nutrients and additives, including lipids, carbohydrates, amino acids, and vitamins, can be synthesized. A nutritionally adequate dietary regimen of pure nutrients would be neither economically nor esthetically useful except for medical applications such as intravenous feeding. Foods consist of hundreds of compounds since

Table 4. Preservation Unit Operations of the Food Industry Arranged by Mode of Preservation and Spoilage Vectors

Spoilage vector	Mode of preservation				
	Physical				
	Heat	Cold	Drying	Chemical	Mechanical
microbes	frying, boiling, pasteurizing, retorting	cooling,freezing	dehydration, desiccation, evaporation, lyophilization	acidification, brining, fermentation, irradiation, isomerization, pickling, smoking	cleaning, centrifugation, washing
enzymes	blanching, boiling, scalding	hydrocooling		acidification	
chemical	exhausting, deodorizing	chilling		hydrogenation, esterification, smoking	cleaning, degassing, washing
economic pests	pasteurization	freezing	desiccation	fumigating	cleaning, compressing, washing
mechanical damage		freezing			aspiration, cleaning, centrifugation, sorting, washing

they are derived from the life processes of living tissues. The components of food composition studies are the nutrients that are essential to sustain life processes. Information on specific compounds in foods can be obtained from the literature, as can tables listing the nutrient composition of commodities, refined food components (eg, sucrose and gelatin), and processed as well as standardized formulated foods, eg, bread and margarine (2). Tables of nutrient composition are available for most foods found worldwide and contain quantitative data on moisture and caloric value (kJ or kcal × 4.184), and on lipid, protein, carbohydrate, nonnutritive fiber, ash, mineral, and vitamin contents. Tables of food composition are useful only as a first approximation to the actual nutritive value and gross chemical composition of specific foods. Foods derived from living tissue are of extremely variable composition. The concentration of any nutrient is affected by horticultural, genetic, harvesting, handling, storage, and distribution factors. Furthermore, chemical analytical procedures may not correlate with the biological activity of available nutrients in the food. Tabular data should be used only as a first approximation of the composition and nutritive values of foods.

Functional and Conformational Data. Foods are selected not only on the basis of nutritional content but for their functional and conformational attributes, which include: toughness, tenderness, and fiber content: style of cut; color and surface appearance; odor and flavor; microbial content; defects and extraneous matter; adaptability to freezing or heat treatment; genetic and varietal factors; geographical or regional production area; date of production (vintage); portion of plant or animal used; method of preservation; key ingredients; emulsifying capacity; water binding capacity; and foaming capacity.

Food Standards and Food Grades. Food standards are legal regulations specifying the wholesomeness, labeling requirements, and for certain foods, the composition, nutritive value, packaging, starting materials, methods of production, storage, and distribution. All foods entering interstate commerce in the United States are covered by food and drug regulations which specify levels of wholesomeness and labeling declarations, and prohibit the use of certain chemicals and manufacturing practices. In addition, food and drug regulations specify the composition of a number of foods, thereby giving them a standard of identity. Foods manufactured to meet the standard of identity need not carry a list of ingredients. Manufactured foods meeting the performance characteristics of the standardized food but not the compositional requirements must be labeled as imitation. Food and drug regulations pertinent to food manufacturing practice, nutritive requirements, standards of identity, labeling, and other regulations are available in CFR (3). In the United States, regulations are enforced by periodic plant inspection and product sampling at all levels of production and distribution.

The FDA has, in principle, designated the regulation of the meat and poultry processing industry to the Food Safety and Quality Service (FSQS) of the USDA. The FSQS has complete authority for the review, approval, and regulation of all aspects of the meat and poultry industry. Detailed regulations exist for equipment, plant layout, construction, processing operations, packaging, labeling, product formulation, storage, quality assurance, and distribution practices. Regulations are enforced by resident inspectors and those covering meat and poultry processing inspection procedures are published (4).

Grades and Grading. The FSQS also provides grading services for fruits, vegetables, meats, poultry, dairy products, and grains. Services may be obtained on a contractual basis for inclusive grading or for grading of selected lots of product. Grades are not mandatory as are identity standards. Grades provide a rational basis for evaluating the worth of processed foods for trading, financial, or contractual purposes where the buyer or seller cannot sample and inspect the products themselves. All grades and grading procedures, fees, labeling and identification procedures, and sampling plans, including those for canned, frozen, and dried fruits and vegetables, are specified (5). Because changes in regulations take place frequently, persons contemplating the processing of foods should consult their regional or the Washington office of the FDA or the FSQS for the latest pertinent regulations.

Preservation of Living Tissue

The genetically controlled taste, color, nutritive value, cell structure, and conformation of fresh foods usually are best at harvest. Protection of these qualities is achieved by retarding or inhibiting the detrimental action of microbes and enzymes, and other chemical degradation, while maintaining the integrity of the membranes, enzyme systems, and gross structure of the food. Specifically, the following must be accomplished:

(*1*) The growth rate of microbes normally associated with the product at harvest or during handling and transport must be inhibited or reduced.

(*2*) Production of deteriorative enzymes or enzyme systems responsible for softening, color loss, or flavor changes must be inhibited or their rate of action must be reduced. Enzymes (qv) may be associated with the product or with incidental microbial contamination (see Fermentation; Microbial transformations).

(3) Loss of moisture and deterioration by chemical contamination must be prevented. Volatile organic compounds generated through respiratory action or present in storage often must be removed.

(4) Higher life forms, insects, mites, etc, must be inactivated or eliminated.

(5) Bruising and various forms of mechanical damage must be prevented.

Processing operations required for the distribution of living tissue depends on the following factors: inherent storage potential of the tissue—this can be improved through genetic engineering, eg, by improving skin strength, resistance to microbes or reducing respiration rate, etc; desired storage life; degree of handling and shipping required; and intended use of the tissue, eg, further processing, food service, or retail consumption (see Genetic engineering).

The intensity of processing operations is limited by the tissue itself in terms of temperature range, water activity, respiratory gas composition, mechanical stresses, and concentrations of chemicals. Table 5 shows typical storage conditions and storage lives for several commodities. Temperature has the greatest effect on shelf life since the increased activation energies (and hence the reaction rates) for microbial growth and enzyme activity are two to five times greater than those of most deteriorative chemical reactions. Except for certain fruits and vegetables, storage should be to 0°C without ice formation.

When shipping distances are great, expensive and highly perishable commodities are shipped routinely by air freight.

Preservation of Nonliving Tissue

Most processed foods, eg, canned, frozen, and dried products, are marketed as nonliving tissue or manufactured foods. Separation, assembly, and preservation operations determine the final quality of the product. Nonliving tissue products must be defined in terms of composition, absence of defects, nutritional value, fill of container, etc, rather than of genetic quality.

Typical processing temperatures, pressures, and pH values are shown in Table 6. Because most preservation operations allow little or no residual microbial or enzyme activity, deterioration during storage is chemical. The relatively low activation energies of deteriorative chemical reactions compared to enzyme reactions makes practical the storage of sterile foods at 25°C for months without appreciable loss of original nutrients and quality factors (6). Often foods are manufactured with a specified economic shelf life. Expiration dates are used to ensure adequate quality while minimizing inventory costs. The effect of various processing operations on the nutritive content of foods has been reported (7). Measurable changes in the vitamin content of foods usually indicate significant changes in flavor, color, and structure.

Heat Treatment

Populations of all life forms show characteristic death rates when exposed to elevated temperatures, pressures, and concentrations of certain chemicals. Enzymes show similar inactivation kinetics. The benefits of heat preservation, eg, the high inactivation rates of microbes and deteriorative enzymes, considerably outweigh the drawbacks, eg, heat-induced losses of desirable food nutrients, structures, colors, and flavors (8).

Table 5. Storage Conditions and Approximate Useful Storage Life for Selected Living-Tissue Foods

Food	Temperature, °C	Relative humidity (rh), %	Atmosphere composition, vol %	Useful storage, d
apples	0	90–95	air	100–250
	0		1–5% CO_2, 2–3% O_2	over 180
bananas	14	90–95	air	7–10
snap beans	6	95	air	7–10
beets (topped)	0	>95	air	90–150
broccoli	0	>95	air	10–14
	0	>95	5% CO_2, 3.5% O_2	40
cabbage	0	>95	air	30–180
carrots (topped)	0	>95	air	30–270
celery	0	95	air	30–60
cherries	0	>90	air	15
corn, sweet	0	95	air	4–8
cranberries	3	90–95	air	60–120
cucumbers	10	90–95	air	10–14
eggs	0	85–92	air	150–180
grapefruit	10–15	85–90	air	40–60
grapes	0	80–95	air	20–180
lettuce	0	>95	air	15–20
melons	7–10	90–95	air	5–15
meat	0	95	air	3–15
onions	0	60–75	air	200
oranges	0	85–90	air	50–80
oysters (shucked)	0	100 (wet ice)		7–10
papaya	10	85–90	air	5–20
peaches	0	90	air	15–30
pears	0	90–95	air	60–180
peas	0	90–95	air	5–15
peppers, green	10	92–95	air	15
potatoes	4	90	air	150–250
poultry	0	>95	air	15
pumpkin	10–12	70–75	air	90
radishes	0	>90	air	60–120
raspberries	0	>90	air	2
shrimp	0	ice	air	10–12
spinach	0	90+	air	10
strawberries	0	90+	air	5
sweet potato	12–16	85–90	air	90–200
tomatoes, green	12–16	90	air	15–20
turnips	0	90+	air	100–150

The heat treatment needed to inactivate microbes or enzymes can be calculated from the energy for activation and the known rate of inactivation at a given temperature, usually 121°C. These values depend on the food system and its pH, water activity, and chemical profile. Typical activation energies and rates at 121°C for heat-resistant microbes are 209–335 kJ/mol (50–80 kcal/mol) and 0.1–10 per minute (one $\log_{10}$ cycle reduction in 0.1–10 min). Commercial heat preservation operations assume a starting concentration of *Clostridium botulinum* spores (mixed varieties) of 10^{12}/g and an inactivation rate of 5–10/min at 121°C.

Preservation heating times at any temperature can be determined by integrating

Table 6. Typical Temperatures, Pressures, and pH Values in Food-Processing Operations

Temperature	Example of use	Limits of application
Heat-transfer media		
cryogenic freezing		
liquid nitrogen, −196°C	rapid freezing to minimize cell damage, moisture loss	cost, stress cracking during freezing
solid carbon dioxide, −78.5°C	rapid freezing	cost
air, plate, and aqueous base freezants, −40 to −5°C	commercial freezing; parasite and insect destruction, commercial freezing, storage; freeze drying	foods not suitable for freezing
refrigerated storage, 0–10°C	commercial refrigerated storage	microbial growth possible
room temperature storage, air, 20–40°C	canned and dry food storage	suitable only for preserved and packaged foods
water; air; atmospheric steam, 50–100°C	pasteurization, milk, eggs, blanching, sterilization of acid foods (pH 4.5); air drying (product temperature); "cooking"	excessive times will cause poor color, flavor, and structure
steam, 110–130°C	thermal sterilization of nonacid foods; destruction of antinutritive factors	rapid thermal degradation of nutrients, pigments, structure, flavors
oil, steam, infrared radiation, 180°C	frying, roasting, baking, generation of browning reaction products	short duration surface treatments
various infrared radiation sources, 180°C	surface heating for flash drying, peeling	charring, pyrolysis without precision control
Pressure[a]		
<0.5 kPa	freeze-drying	cost
0.5–4 kPa	hypobaric storage, deaeration, vacuum concentration, vacuum cooling	cost
101 kPa	most processing operations	
0.1–1 MPa	steam sterilization, over pressure for glass and pouch packs; carbonated and aerosol packaged foods	cost
>1 MPa	extruders, hydraulic pressing, homogenization, potential microbial and enzyme inactivation at 1 GPa (1000 atm) and higher	cost, specialized products
Hydrogen-ion concentration		
pH 1	acid hydrolysis	neutralize to pH range of 7
pH 2.5	lemons, limes, vinegar, organic acids	taste
pH 3–4.5	fruits, acidified foods	noncompatability of foods; protein denaturation

Table 6 (*continued*)

Temperature	Example of use	Limits of application
pH 7–3.5	normal pH of most foods	
pH 8–9	solubilization of certain proteins for extraction, alkali process cocoa	neutralize to normal pH for use
pH 12	limed corn	color, flavor, nutrient loss
33% sodium hydroxide	peeling of fruits and vegetables	surface treatment only, neutralize to normal pH

[a] To convert kPa to mm Hg, multiply by 7.5; to convert MPa to atm, divide by 0.101.

the lethal temperature–time effects at the slowest heating point in the package. Thus if the geometry of the container and the thermophysical properties of the food are known, it is possible to calculate the heating time necessary to ensure a safe level of microbes or residual enzyme activity in the food and the residual concentration of desired nutrients. In general, higher life forms, eg, insect eggs, mites, etc, are killed by even mild heat treatments. Computer programs are available for determining safe heat-preservation operations given information about the product: its initial temperature, container characteristics, fill, and heat-exchange system (pure steam, steam–air, water, etc).

Hydrogen ion is extremely toxic to most microbes. Foods having a pH below 4.5 can be sterilized by heating to 100°C with a very limited holding period (see also Chemical Preservation Methods).

Aseptic preservation involves performing separate heat preservation operations on the food and on the containers prior to assembly. Liquid products can be heated and cooled rapidly under optimum conditions of heat transfer in specialized heat exchangers. The sterile product is packaged under sterile filling conditions into sterile unit retail packages, drums, or bulk-storage tanks (see also Sterile techniques).

Freezing

The rate of loss of color, flavor, structure, and nutrients in foods is a function of temperature; thus lower storage temperatures prolong the useful life of foods. However, below 0°C, the free water in food forms ice crystals as a function of moisture content, solute composition, and storage temperature. Supercooling is a transient phenomenon.

Ice formation is both beneficial and detrimental. Benefits include strengthening of structures and removal of free moisture which reduces water activity. Benefits, however, are often far outweighed by the deleterious effects of ice crystal formation, the partial dehydration of the tissue surrounding the ice crystal, and the freeze concentration of potential reactants. Ice crystals disrupt cell structures mechanically, and the increased concentration of cell electrolytes can result in the chemical denaturation of proteins.

The technology of food-freezing emphasizes as short a passage time through the temperature zone of maximum ice crystal formation as possible (9–10). The formation of as small an ice crystal as possible minimizes the mechanical disruption of cells and possibly reduces the effects of solute concentration damage. Rapid freezing can only be accomplished by large temperature differences and high heat transfer coefficients.

Many processed foods and certain animal products tolerate freezing and thawing since their structures can accommodate ice crystallization, movement of water, and the related changes in solute concentrations. Starches can be modified to form gels that accept several freezing and thawing cycles without breakdown. By contrast, most fruits and vegetables lose significant structural quality on freezing because their rigid cell structures fail to accommodate to ice crystal formation. However, it is not possible to store foods at temperatures low enough to ensure complete conversion of all water to ice; as a result, commercial frozen food storage temperatures represent an economic balance between storage costs (time, energy, and capital investment) and projected shelf life.

The freezing process disrupts tissue structures and allows cell contents to become mixed so that undesirable enzyme reactions can take place at significant rates even at storage temperatures of −18°C. These reactions can generate off-flavors, reduce nutrient concentrations, and cause major changes in the structure and appearance of foods. The amount of free liquid or "drip" found after a freeze–thaw cycle is a good indication of the structural damage.

Heat treatment (blanching) prior to freezing eliminates enzyme-mediated changes in color, flavor, and structure. Most deteriorative enzymes are inactivated by exposure to a temperature of 100°C for 1–5 min. The enzymatic oxidative deterioration of frozen fruits can be inhibited with sulfur dioxide, sucrose, and combinations of citric acid, sodium chloride, and ascorbic acid (preceded by vacuum removal of oxygen if heat is not used). Newer methods for blanching are discussed in ref. 11.

Most frozen foods have a useful storage life of one year at −18°C; however, foods high in fat, eg, sausage products, may become rancid in two weeks. The time and temperature tolerances of frozen foods are described in ref. 12. Frozen storage can result in moisture loss from the food through a freeze-drying process since the heat-transfer surfaces used to maintain storage temperatures are at a lower temperature than the storage area. For this reason, frozen foods must be protected against drying by a moisture barrier. In addition, foods subject to oxidative deterioration must be protected from air.

Freezing preservation equipment can be classified by method of heat transfer. Usually air is the heat-exchange medium for freezing foods. Foods are loaded on a belt or vibrating conveyor and passed through air flowing upwards at up to 5 m/s at temperatures as low as −40°C. The air is recycled through coils and fans located next to the conveyor and returned through the conveyor. Since the air has a partial water vapor pressure lower than the food, freeze-drying can occur and some of the water in the product is removed and deposited as ice on the heat-exchange coils. As a freezing medium, air has other drawbacks. The low gas–solid heat-transfer coefficient and heat capacity of air require either very low temperatures or high velocities to obtain needed high heat fluxes. Low air temperatures increase refrigeration costs and high air velocities generate additional fan heat loads. For these and other reasons, freezing by conduction or by liquid heat-transfer methods are less costly in terms of capital investment and energy.

Liquid heat-transfer media that are used for direct-immersion freezing include food-grade dichlorodifluoromethane, nitrous oxide, and water solutions of various edible salts, sugars, alcohols, acids, and esters. Liquid heat-transfer agents offer a high heat-transfer coefficient and reduced pumping costs, eliminate product desiccation, and allow operation at high lowside equipment temperatures (see Cryogenics; Heat-exchange technology, heat-transfer media). Major drawbacks are possible changes in food flavor and costs of processing. Although dichlorodifluoromethane offers major operating advantages as compared to other heat-transfer media, cost and environmental concerns have reduced its potential usefulness as an ideal direct-immersion freezant. There is a need for a direct-immersion liquid freezant that is safe, low-cost, thermodynamically efficient, and compatible with foods with respect to flavor, color, and odor.

Conduction freezing between chilled plates is a very cost-effective method of heat removal provided the product can be assembled in a geometry compatible with the plate surfaces. Packages having a semi-infinite slab geometry are loaded between stacks of platens through which refrigerant is circulated. Good heat transfer is maintained by maintaining a pressure on the stack of platens.

Other freezing methods use direct immersion in liquid nitrogen, exposure to solidified carbon dioxide, and immersions of packaged products in liquid freezants, eg, sodium or calcium chloride brines, methanol, or propylene glycol solutions (see Refrigeration).

The quality of frozen food is related to storage temperature but, since constant storage temperatures are not always feasible, the shelf life is often determined by the highest temperature and total length of time that food is exposed to that temperature before use. Maintaining a −18°C storage environment from time of freezing until use continues to be a major technical problem facing the frozen-food industry.

Reduction of Water Activity (Dehydration)

Microbes require a specific minimum level of water activity a_w, defined as the relative humidity (measured in equilibrium with the food) for growth and reproduction at a given temperature and substrate composition. As shown in Figure 1, all forms of microbial life are inhibited below an a_w of 0.65 at storage temperatures in the range of 0–40°C.

Foods possess characteristic equilibrium relationships between water activity and moisture content at given temperatures. Preservation against microbial spoilage by dehydration requires a moisture content equal to a water activity below 0.65. This contrasts with food concentration where water is removed for reasons other than for effective reduction in water activity.

Living tissue, when dried to a water activity below 0.97, suffers irreversible disruption of metabolic processes; however, Figure 1 shows that individual enzyme systems can retain activities almost to dryness, ie, a_w <0.2. Deteriorative chemical reactions, enzyme-catalyzed or not, are generally a function of water activity and reactant concentration. Thus nonenzymatic browning (Maillard reaction), oxidation, and internal rearrangements, eg, staling and protein cross-linking, can increase in rate as water activity is reduced. Many reactions show a minimum rate in the range of a_w = 0.4. The role of water activity and product stability has been reported (13–14).

Prior to dehydration, foods are usually heat-treated to inactivate enzyme systems. Those foods susceptible to rapid nonenzymatic browning resulting from high concentrations of reducing sugars are treated with sulfur dioxide, and products subject

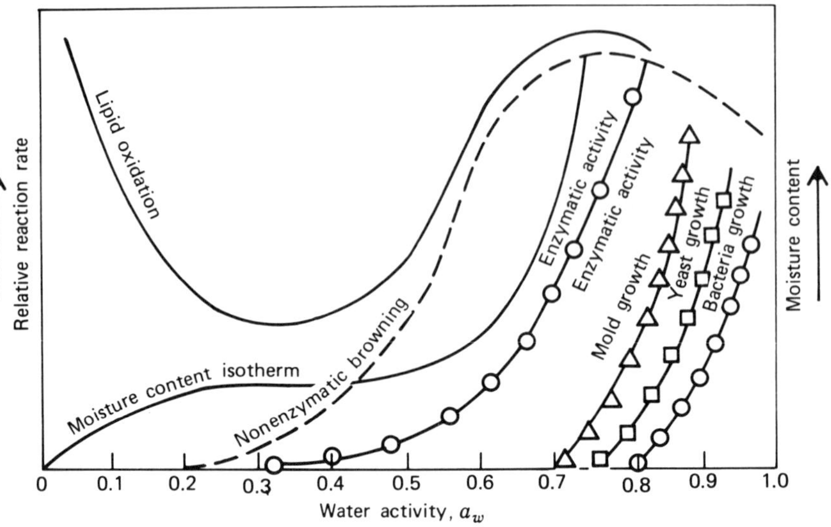

Figure 1. Relative rate of food deteriorative factors as a function of water activity a_w (13).

to oxidative rancidity can be treated with antioxidants and packaged to prevent exposure to oxygen. Low temperature storage (5°C) reduces chemical deterioration. Blends of dehydrated products must be assembled from ingredients having the same water activity (see Food additives).

High-quality dried foods can be obtained only if the drying system is designed to match the rate of release of moisture from the food at a predetermined maximum wet-bulb temperature. Typically, continuous-belt dryers are staged to provide three or more zones into which the product is repiled into progressively deeper beds. Each zone operates at a dry- and a wet-bulb temperature, a through-flow (upflow and downflow through a bed of materials) air velocity, and a bed depth which optimize product quality, energy use, and production rate. Ref. 15 gives a detailed description of food dehydration technology (see also Drying).

Liquids and pastes are commonly dried in spray, drum, or freeze dryers. Particulate foods can be dried in continuous conveyor systems, fluidized beds, freeze dryers, or batch tunnel systems. Some fruits, salted animal products, nuts (qv), berries, and many field crops are sun-dried.

Mechanical Preservation Methods

Mechanical food preservation methods are characterized by separation operations, which tend to affect food quality factors less than other methods, as exemplified by the production of sterile draft beer by filtration (see Ultrafiltration).

Packaging is the primary mechanical preservation method; barrier materials protect preserved foods from external spoilage factors during storage and distribution (see Barrier polymers; Film and sheeting materials). Other mechanical preservation methods include sieving, filtration, air classification, washing, sorting, ultrafiltration, centrifugation, extraction, stripping, flotation, etc. Size grading, defect sorting, and washing are used routinely to reduce or eliminate microbes, enzymes, chemicals, insects, and other matter. Virtually all food materials require repetition of these operations between harvest and preservation.

Washing equipment is designed to accommodate the type of food being washed. Leafy vegetable washers, for example, use a series of rotating paddles which transfer the material along the water surface in a tank so that sand can settle to the bottom of the tank. Fruit washers have reels or belts fitted with high pressure water sprays that mechanically loosen dirt and debris. Recently, a new type of washer was developed for tomatoes, tree fruit, and root vegetables (16). The unit has spinning soft rubber disks that are mounted in rows to give an intensive wiping action at the surface of the food as it rolls over the disk array. Disk washers reduce water consumption in tomato washing from 1500 L/t (360 gal/short ton) to as little as 20 L/t (5 gal/short ton). Other food washing equipment use combinations of sprays, brushes, and mechanical agitators. For example, froth flotation (qv) removes foreign particulate organic matter from cut corn and peas. Approved wetting agents, eg, sodium lauryl sulfate, are often incorporated in wash water to improve the efficiency of the washing operation (see Surfactants and detersive systems). The proper chlorination of wash and transport water also is extremely important in maintaining low levels of microbes. Breakpoint chlorination ensures adequate but not excessive residual free chlorine (since >5 ppm of free chlorine can spoil flavors of some foods). This method of chlorination is implemented by adding sufficient chlorine to oxidize organic materials in the water and leave a residual concentration to ensure antimicrobial activity. Water used in the final washing of food materials processed under FDA jurisdiction must be suitable for drinking (see Bleaching agents; Disinfectants).

Liquid foods free of suspended solids may be sterilized by filtering spoilage microbes. The filtered product, eg, wine (qv), beer (qv), and certain fruit juices (qv), is packaged under aseptic conditions in a sterilized container. Centrifugation can be used to reduce microbial contamination and to remove extraneous organic matter in liquid foods such as milk.

Ultrahigh pressure causes the death of microbes and inactivates enzymes at a rate proportional to concentration and temperature. Significant rates have been achieved experimentally at temperatures only slightly above 25°C; however, no commercial applications have been reported (see Sterile techniques).

Chemical Preservation Methods

Addition of chemicals usually occurs in conjunction with other preservation methods since they enhance the effectiveness of heat, refrigeration, drying, or packaging. Examples include the addition of antioxidants (qv) to fried foods to reduce the need for more expensive oxygen-impermeable flexible packaging, and the use of sulfur dioxide to retard the development of brown color in frozen or dried apples (see Food additives).

The extreme toxicity of the hydrogen ion towards food-spoilage microbes and its tolerance by humans has made it a preferred food preservative. Studies have shown that various organic acids exert a strong inhibitory effect on the growth of microbes beyond the activity of the hydrogen ion itself (see Carboxylic acids; Hydrogen-ion concentration). Lactic, propionic, acetic, sorbic, and benzoic acids, their sodium or potassium salts, and certain derivatives of benzoic acid find extensive use as yeast, mold, and bacterial inhibitors in bread, beverages, and sauces. Combinations of hydrogen ion, organic acids, mild heat treatment, and exclusion of oxygen have allowed the simple and safe preservation of foods not tolerant of conventional heat sterilization above 100°C, freezing, or dehydration.

Traditional food preservation methods were based largely on drying in conjunction with preservation with locally available chemical preservatives. These included salt, organic acids generated by a lactic or acetic acid fermentation (as in pickling), wood smoke, and ethanol derived from the fermentation of substances containing sugar. Many traditionally preserved products are available in the marketplace, although modern preservation methods predominate because they are convenient.

Chemical preservatives and chemical preservation methods used in conjunction with other preservation methods for meat and poultry are controlled by USDA regulations. The FDA has jurisdiction over the use of chemical preservatives in other foods. Additives are reviewed continually for health safety. The use of nitrite to inhibit *Clostridium botulinum* in cured meats is currently under review (1980) because nitrite can combine with amines present in food to form highly carcinogenic nitrosamines (qv).

The FDA has reviewed the use of ionizing radiation as a food preservation method since the 1960s. Ionizing radiation is considered to be a chemical preservation method and must be cleared for use not only on a product-by-product basis but also on a dosage basis. In the United States, a dose of up to 100 Gy (10,000 rad) may be used to inhibit potato sprouting, and a dose of as much as 500 Gy may be used to destroy insect life in grain products. There have been feeding studies to obtain clearance for beef sterilization using doses up to 6×10^4 Gy.

The status of specific chemicals can be checked in the *Food and Color Additives Directory* (17). A worldwide code is being prepared covering classes of foods and additives (18).

Toxicological Implications of Food Deterioration and Spoilage

A primary objective of food processing is to ensure a safe food supply. Certain harvested foods can contain naturally occurring poisonous substances, compounds with pharmacological effects, or compounds that can interfere with the utilization of nutrients present in other foods when they are consumed together. Table 7 provides information on some of these naturally occurring compounds (see also Food toxicants, naturally occurring).

Foods just harvested may contain or become contaminated with spore or vegetative forms of microbes, parasites, or the wastes of higher life forms. Careless post-harvest handling, storage, processing, and distribution operations provide additional opportunities for microbial growth or contamination. Microbial growth can result in the release of highly toxic substances which cannot be removed from the food in an economic manner. Since the quality characteristics of the food may not be altered during microbial growth and release of toxins, the consumer has no way of identifying contaminated foods and, consequentially, food-poisoning outbreaks occur. Fortunately, some toxins are heat sensitive and are inactivated during the cooking process. However, only good manufacturing, handling, and distribution practices ensures the prevention of microbial growth. These practices must be carried out from the time of harvest to the time of consumption. Table 8 provides information on some food-borne microbes and higher life forms that present significant health hazards (see Fermentation; Microbial transformations; Yeasts).

Table 7. Some Toxic Substances Occurring Naturally in Foods[a]

Compound	Effect	Food source (comments)
amines (eg, serotonin, histamine)	vasoactive	fermented foods, bananas, cheese, chocolate, and fish (arise from decarboxylation of amino acids)
myristicin	hallucinogen	nutmeg
Toxic compounds found in plants		
lathyrogens, (*L*)-α,γ-diaminobutyric acid and similar compounds	muscular weakness, leg paralysis	seeds of *Lathyrus sativas*
compounds associated with fava beans	hemolysis 5–24 h ("Favism" reduces glucose-6-phosphate dehydrogenase activity.)	broad beans, *Vicia faba*
gossypol	species dependent for domestic animals, no toxicity information for humans	cottonseed meal (heat obviates activity; maximum allowed free in meals, 450 mg/kg)
cyanogenetic glycosides	hydrogen cyanide released by acid or enzyme treatment, respiratory inhibitor	seeds or pit of peach, apple, pear, apricot
linamarin	hydrogen cyanide released by acid or enzyme treatment, respiratory inhibitor	cassava, manioc
protein-complexing protein, mol wt ca (15–30) $\times 10^3$	inhibits pancreatic proteolytic enzymes; trypsin inhibitor	soy, lima, pea beans, raw egg white
Toxic substances found in animals		
saxitoxin and similar dinoflagellate toxins	paralysis	shellfish containing various species of *Gonyaulax*
ciguatera	neurological complications within 6 h	fish or seafood feeding on marine organisms containing toxin

[a] For description of others, see Food toxicants, naturally occurring.

A third source of potential health hazards in food supplies are toxic compounds formed during the chemical degradation of processed and stored foods. Oxidative breakdown products of unsaturated fatty acids have been implicated as potential carcinogens, eg, malonaldehyde.

In addition to the above, foods must be protected from contamination by pesticides and other chemicals used for crop protection (see Fungicides; Insect control technology); from exposure to hazardous chemicals and extraneous materials, eg, asbestos (qv) in air and water (qv) supplies, and radioactive isotopes (see Radioisotopes). Many toxic compounds can be found as soil contaminants or as contaminants in lakes, oceans, or ground-water supplies and can be concentrated in the edible tissues of plants and animals. The most effective method to ensure the absence of these hazards is chemical analysis of representative samples of the raw material prior to purchase. Analyses can lead to prevention of contamination above prescribed levels by heavy metals (mercury, lead), DDT, polychlorinated biphenyls (PCBs), strontium-90, growth-promoting estrogens in meat, and other contaminants (see Trace and residue analysis).

Table 8. Microbes and Other Life Forms Constituting Food Safety Problems

Microbes	Infective action and food vector	Comments
Staphylococcus aureus	Protein enterotoxin, mol wt ca 30,000 (5 types). Diarrhea, vomiting 1–4 h. Toxin stable to 100°C for 1 h.	Hold foods below 5°C to prevent growth. Microbes can be inactivated above 60°C by one-hour heat treatment. No growth below pH 4.5 (acid food).
Clostridium botulinum (several types A, B, E)	Polypeptide nerve toxin in low acid canned foods. Inactivated by heating at 100°C for 10 min.	Spores require heat treatment equal to 121°C for greater than 2.5 min. No growth below pH 4.5 or below 5°C.
Salmonella sp. (various serotypes)	Gastroenteritis 12–48 h after ingestion of organisms, improperly cooked protein foods.	Inactivated by heating at 60°C for 20 min. 65°C for 2.5 min. No growth below 5°C.
Clostridium perfringens	Found in cooked animal products. Toxin. Intestinal cramps, diarrhea 8–15 h.	Hold food below 5°C. Cook adequately above 100°C.
Bacillus cereus	Cooked meat products. Intestinal cramps, diarrhea in 8–15 h.	Hold food below 5°C. Cook adequately.
Shigella sp.	Uncooked foods, dysentery.	Avoid contaminated foods and water; cooking inactivates.
Endamoeba histolytica	Uncooked vegetables, amoebic dysentery.	Avoid contaminated foods and water.
Molds, various species *Aspergillus* sp. *Fusarium* sp. *Penicillium* sp.	Mycotoxins, citrinin, aflatoxins, patulin growth of mold on moist grains, nuts before, during, or after harvest. Cause liver cancer, anemia, leucopenia LD_{50} (mg/kg) 0.5–5.	Mold growth prevented by water activity below 0.6 screen foods for toxins.
Claviceps sp.	Ergot alkaloids, mold growth on wheat, rye, ergotism.	Keep water activity below 0.6.
Vibrio parahaemolyticus	Consumption of organisms produces gastroenteritis in 12 h. Raw marine products.	Cooking destroys organisms.
Viruses		
infectious hepatitis; poliomyelitis	Food contaminated by food handlers	Each virus is destroyed by cooking, drying.
Parasites		
Trichinella spiralis	Trichinosis from cysts in pig muscle, gastroenteritis in 4–5 d, followed by fever, muscle pain	Cysts are destroyed by heating to 60°C, curing, or by freezing storage at −15°C for 20 d.
Taenia saginata (beef tapeworm)	Infected beef muscle, causes loss of weight, causes cysticercosis	Inactivated by cooking to 60°C, or freezing storage at −15°C for 20 d.
Trichuris trichiura, Ascaris lumbracoides	Eggs contaminate raw foods, water, gastrointestinal upset, diarrhea.	Eggs are inactivated by cooking or freezing.

Food-Processing Facilities

Factors important in the selection of sites for food processing operations include: availability of raw materials; abundant potable water supply; low-cost waste disposal facilities; adequate, low-cost energy supply; adequate seasonal and nonseasonal labor supply; ease of access to rail and truck transportation; proximity to consumers; and adequate storage areas. The relative importance of each of these factors depends on the length of the processing season, the ease of storage and handling of raw materials, the perishability of the finished products, and the complexity of the processing operation.

Highly perishable salad vegetables can be trimmed, inspected, bagged, and vacuum-cooled using mobile equipment in or at the edge of the growing area. Processing of frozen french fries requires raw-potato-storage capacities of up to ten months; thus plants tend to locate near major processing potato-growing areas, eg, the northwestern U.S. Continued developments in mechanical harvesting and mobile harvester separation operations for fruits and vegetables will cause further processing plant shifts toward concentrated growing areas where irrigation, climate, and level terrain make mobile processing feasible.

As processing technology improves, an increasing number of plants are relocating away from eastern U.S. population centers toward sources of raw materials. This shift is particularly noticeable in the meat industry and has occurred concurrently with the development of vacuum-packed, cut beef which provides more than a 50% reduction in storage and shipment space over conventional quarters or sides (see Meat products).

Certain food products are highly perishable, have marginal value, or have only regional acceptance. Dairy products, bread, soft drinks, beer, glass-packed items, and certain cured-meat products are usually manufactured from more stable raw materials or are locally produced starting materials and are distributed daily. With such distribution, production schedules are determined by local weather forecasts, by day of the week and holidays, and by seasonal factors so that returns of stale or overaged products can be minimized. Considerable progress has been made in the analysis of perishable-food distribution practices. A universal product code applicable to all processed foods for retail sale will further improve distribution efficiencies by allowing automatic retail checkout and continuous inventory review (19).

Equipment Standards

Standards for food-processing equipment are specified by the USDA in the dairy-, meat-, poultry-, and egg-processing industries. Equipment and processing methods used in the manufacturing of other foods are also covered (20). Where standards exist, eg, 3-A Sanitary Standards (21), they have taken the form of specifications for the design of food-processing equipment and for certain operational practices. Requirements state that surfaces must not harbor food materials or prevent the easy removal of these materials during the cleaning process; there should be no openings where food can accumulate and allow microbial growth, particularly openings not easily cleaned during routine equipment cleaning operations; equipment must be capable of visual inspection throughout the food contact area and must be easily disassembled (if necessary) for cleaning; all materials of construction which contact the food must be approved for use, or be inert to migration into the food and *vice versa*; interior equipment surfaces should be self-draining; and exterior surfaces should be easily cleaned and free of openings where material may accumulate.

The FDA does not approve specific pieces of food-processing equipment but provides approval of a process or equipment after reviewing pertinent information. Meat and poultry processing operations must have specific Food Safety and Quality

Service (FSQS) approved equipment and processes. The FSQS approves specific pieces of equipment meeting their design, operating, and sanitation requirements. Food processors and their equipment suppliers operating under the jurisdiction of FSQS regulations should contact their regional office or the headquarters of FSQS for further information.

Food Packaging

Glass-, woven-, natural fiber-, and ceramic containers have been used to store food since antiquity. However, the invention of tin-plated steel containers (ca 1800) provided the food industry with its first functional and disposable packaging system. Advances in metal container manufacturing continue to maintain the status of the cylindrical can as the primary package for heat-processed, shelf-stable foods and beverages. Notable developments include substitution of electrolytic tin-plating for hot-dipping, new organic linings to greatly reduce the need for tin, reduced steel thickness, and improved steel composition for better strength and corrosion resistance. Developments in coating technology have resulted in the availability of two-piece drawn aluminum and steel containers and of numerous easy-open systems. New glass containers have increased strength and have continued to reduce package weight and cost (see Glass). Can-manufacturing technology is described in ref. 22 (see also Coating processes).

Rigid glass or metal containers have a major processing advantage: they can be filled and sealed at speeds in excess of one thousand per minute. Thus the relatively higher container cost can be recovered to some degree by higher production rates. Their rigidity (ie, sturdiness) facilitates stacking and handling during distribution and storage.

Paper, plastics, and aluminum have provided the food industry with container material that can be tailored to specific barrier needs. Ease of forming, sealing, opening, and decorating; strength; and low weight have made them preferred materials for refrigerated, dry, and frozen products.

Resistance to thermal treatment and desired barrier and mechanical qualities are obtained by laminating or coextruding appropriate combinations of paper, aluminum foil, and plastic. Lamination and coextrusion seal transparent packages against moisture and gas permeation, and strengthen the packaging.

Pouches formed by heat-sealing paper–foil–plastic laminates on three sides, filling, and then heat-sealing the fourth side have found extensive use with moisture- and oxygen-sensitive dry foods as well as with high moisture thermally-processed foods. Pouches capable of withstanding saturated steam or steam–air mixtures at 121°C for several hours are finding greater use as substitutes for metal and glass containers. Pouches have a major advantage for heat sterilization in that their geometry is that of a slab rather than a cylinder. Their thinner cross section allows a shorter heat-processing time than cylindrical containers of the same capacity. Mechanical advances in forming, filling, sealing, and handling pouches will result in continued substitution of laminates for rigid metal and glass containers (see Film and sheeting materials).

Plastics and plastic-coated papers are widely used for beverages including carbonated drinks (see Carbonated beverages; Milk products). Several aseptic systems are marketed for bulk packing of liquids in multiliter plastic bags. Table 9 lists flexible packaging materials, and some of their desirable barrier and physical properties, and typical uses.

Table 9. Flexible Packaging Materials and Barrier Properties

Material	Approximate specific gravity, g/cm^3	Relative[a] tensile strength	Relative[b] water vapor transmission rate	Relative[c] oxygen transmission rate	Relative[c] carbon dioxide transmission rate	Usable temperature range, °C	Use in food packaging
acrylonitrile–butadiene–styrene (ABS) polymer	1.1	medium	low	low	low	to 90	bottles
cellulose acetate	1.3	high	high	high	high	RT	blister packs, overwraps
regenerated cellulose (cellophane)	1.5	high	high	low	low	−40 to RT	
ionomer	1.0	medium	low	high	high	−40 to RT	as coextruded film meat wraps, boil-in-bags
polyamide							
(nylon-6)	1.1	high	high	low	low	−40 to 130	wide range of products, frozen, retort pouch, bags
(nylon-12)	1.0	high	high	medium	high	−40 to 130	
polyester (poly(ethylene terephthalate))	1.4	high	medium	low	low	−40 to 130	boil-in-bags, retort pouches, bags
polyethylene							
low density	0.92	low	low	high	high	−40 to RT	bags, wraps, molded containers, laminates, copolymers
high density	0.95	low	low	high	high	−40 to RT	
polypropylene	0.9	medium	low	high	high	−40 to RT	bags, overwraps as coated polymers
polystyrene	1.0	medium	high	high	high	refrigerated RT	trays, foams, windows
poly(vinyl chloride)	1.3	medium	medium	high	high	refrigerated RT	bottles, formed packages, laminates, bags, shrink wraps
vinylidene chloride, vinyl chloride copolymer (5–50%, Saran)	1.6	high	low	low	low	−40 to RT	shrink films, coatings, laminates

[a] Tensile strength, MPa (psi): low, <35 (<5000); medium, 35–70 (5–10 × 10^3); high, >70 (>10^4).
[b] Moisture vapor transmission, $g \cdot m/(m^2 \cdot d)$ at 38°C (90% rh) (for 25 μm thickness): low, <20; medium, 20–100; high, >100, 101.3 kPa (1 atm).
[c] Gas transmission, $(m^3 \cdot m)/(m^2 \cdot d \cdot kPa)$ (per 25 μm thickness at 100 kPa pressure differential): low, <400; medium, 400–4000; high, >4000. To convert kPa to bar, divide by 10^2, 50% rh.

Semirigid containers are also gaining increased acceptance, particularly as replacements for the no. 10 can (2.72 kg or 96 oz). Initial developments have produced a lightweight drawn metal container measuring 24 × 30 × 6 cm with a double seamed top. The thickness of this container (6 cm) is such that during heat sterilization heat travels only 3 cm by conduction vs about 8 cm in the conventional no. 10 can. Shorter heat treatments allow the preservation of a greater variety of specialty products. The half steam-table size eliminates cleaning of pans, since the opened unit serves as a serving tray. Other semirigid systems have been developed to replace the steel can for single-portion service. These containers generally use a heat-sealed or glued paper–foil–plastic laminated closure.

Aerosol containers have played a specialized role in the food industry as dispensers for such foods as whipped toppings (using nitrous oxide as a propellant), cake frostings, and barbecue sauces. Special filling techniques and product formulations are required since the aerosol package cannot be heat-sterilized once it has been sealed (see Aerosols).

Packaging forms an integral part of most food processing. For this reason, studies of storage conditions over a range of temperatures, relative humidities, and handling conditions must be made to determine the suitability of a package. Packaging in contact with foods must be tested for migration of packaging components under actual conditions of use, and pickup of off-flavor must be evaluated on a product-by-product basis. Because of potential migration of packaging materials into foods, current food and drug regulations should be consulted prior to marketing foods packaged in nonstandard materials.

BIBLIOGRAPHY

"Food and Food Processing" in *ECT* 1st ed., Vol. 6, pp. 785–818, by Z. I. Kertesz, New York State Agricultural Experiment Station, Cornell University; "Food and Food Processing" in *ECT* 2nd ed., Vol. 10, pp. 23–61, by Z. I. Kertesz, Food and Agriculture Organization of the United Nations.

1. D. F. Farkas, *Chem. Technol.* **7,** 428 (1977).
2. *Composition of Foods,* Agricultural Handbook No. 8, USDA, Washington, D.C., 1976.
3. *Code of Federal Regulations* (CFR), Title 21, Subchapter B, parts 100–199.
4. *Code of Federal Regulations* (CFR), Title 9, Chapt. III, Subchapter A, parts 301–335, and Subchapter C, part 381.
5. *Code of Federal Regulations* (CFR), Title 7, parts 2800–2853.
6. T. P. Labuza, M. Shapero, and J. Kamman, *J. Food Proc. Pres.* **2**(2), 91 (1978).
7. R. S. Harris and E. Karmes, eds., *Nutritional Evaluation of Food Processing,* Avi Publishing Co., Westport, Conn., 1975.
8. D. B. Lund in R. S. Harris and E. Karmes, eds., *Nutritional Evaluation of Food Processing,* Avi Publishing Co., Westport, Conn., 1975, pp. 211–213.
9. O. R. Fennema, W. D. Powrie, and E. H. Marth, *Low-Temperature Preservation of Foods and Living Matter,* Marcel Dekker, Inc., New York, 1973, Chapt. 4.
10. D. K. Tressler, W. B. Van Arsdel, and M. J. Copley, eds., *The Freezing Preservation of Foods,* 4th ed., Vols. 1–4, Avi Publishing Co., Westport, Conn., 1979.
11. J. L. Bomben, "Waste Reduction and Energy Use" in J. H. Green and A. Kramer, eds., *Food Processing Waste Treatment,* Avi Publishing Co., Westport, Conn., 1979.
12. W. B. Van Arsdel, M. J. Copley, and R. L. Olson, eds., *Quality and Stability of Frozen Foods,* Wiley-Interscience, New York, 1969.
13. T. P. Labuza, *J. Food Proc. Pres.* **1**(2), 167 (1977).
14. R. Davies, G. G. Birch, and K. J. Parker, eds., *Intermediate Moisture Foods,* Applied Science Publishers, Ltd., London, 1976.

15. W. B. Van Arsdel, M. J. Copley, and A. I. Morgan, Jr., *Food Dehydration,* 2nd ed., Vols. 1 and 2, Avi Publishing Co., Westport, Conn., 1973.
16. J. M. Krochta, R. P. Graham, and W. W. Rose, *Food Technol.* **28**(12), 26 (1974).
17. *Food and Color Additives Directory,* Hazleton Laboratories, Inc., Vienna, Va., updated as new regulations appear.
18. *Recommended International Code of Practice for the Processing and Handling of Quick Frozen Foods,* issued by the Secretariate of the Joint FAO/WHO Food Standards Program, FAO, Rome, 1976.
19. R. C. Griffen and S. Sacharow, *Principles of Package Development,* Avi Publishing Co., Westport, Conn., 1972.
20. *Code of Federal Regulations* (CFR), Title 21, Subchapter B, part 110.
21. *3-A Sanitary Standards,* International Association of Milk, Food and Environmental Sanitarians, Inc., c/o *Journal of Milk and Food Technology,* Ames, Iowa.
22. S. Sacharow, *Handbook of Packaging Materials,* Avi Publishing Co., Westport, Conn., 1976.

General References

Agricultural Statistics, USDA, published yearly and available from the U.S. Government Printing Office; contains complete production and consumption data on all food commodities.

M. A. Amerine, H. W. Berg, and W. V. Cruess, *Technology of Wine Making,* 3rd ed., Avi Publishing Co., Westport, Conn., 1972.

Fundamentals, Systems, Equipment, Applications, ASHRAE Handbook Series, American Society of Heating, Refrigerating and Air-Conditioning Engineers, Inc., New York, 1977.

Canned Foods, Principles of Thermal Process Control and Container Closure Evaluation, National Food Processors Association, Western Research Laboratory, Berkeley, Calif., 1973.

The Commodity Storage Manual, Refrigeration Research Foundation, revised periodically.

Composition of Foods, Agricultural Handbook No. 8, USDA, Washington, D.C., 1976; Handbook 8 is being revised by sections; No. 8-1 covers dairy and egg products.

Food Chemicals Codex, 2nd ed. with three supplements (1974, 1975, 1978), Committee on Specifications, Food Chemicals Codex; Committee on Food Protection, National Research Council, National Academy of Sciences, Washington, D.C., 1972; 3rd ed., 1980.

W. J. Harper and C. W. Hall, *Dairy Technology and Engineering,* Avi Publishing Co., Westport, Conn., 1976.

J. H. Heid and M. A. Joslyn, eds., *Fundamentals of Food Processing Operations,* Vols. I, II, and III, Avi Publishing Co., Westport, Conn., 1967.

M. Karel, "Packaging Protection for Oxygen-Sensitive Products," *Food Technol.* **28**(8), 50 (1974).

W. E. Kramlich, A. M. Pearson, and F. W. Tauber, *Processed Meats,* Avi Publishing Co., Westport, Conn., 1973.

D. N. Lapedes, ed., *Encyclopedia of Food, Agriculture, and Nutrition,* McGraw-Hill Book Co., New York, 1977.

S. A. Matz, *Bakery Technology and Engineering,* 2nd ed., Avi Publishing Co., Westport, Conn., 1972.

G. J. Mountney, *Poultry Products Technology,* 2nd ed., Avi Publishing Co., Westport, Conn., 1976.

H. W. Ockerman, *Source Book for Food Scientists,* Avi Publishing Co., Westport, Conn., 1978.

A. L. Ryall and W. J. Lipton, eds., *Handling, Transportation and Storage of Fruits and Vegetables,* Vol. 1, Avi Publishing Co., Westport, Conn., 1972.

A. L. Ryall and W. T. Pentzer, eds., *Handling, Transportation and Storage of Fruits and Vegetables,* Vol. 2, Avi Publishing Co., Westport, Conn., 1974.

A. Woollen, ed., *Food Industries Manual,* 20th ed., Chemical Publishing Co., Inc., New York, 1970.

D. K. Tressler and M. A. Joslyn, *Fruit and Vegetable Juice Processing Technology,* 2nd ed., Avi Publishing Co., Westport, Conn., 1971.

J. G. Woodroof and G. F. Phillips, *Beverages: Carbonated and Noncarbonated,* Avi Publishing Co., Westport, Conn., 1974.

J. A. Troller and J. H. B. Christian, *Water Activity and Food,* Academic Press, New York, 1978.

H. Riemann, *Food-Borne Infections and Intoxications,* Academic Press, New York, 1969.

Toxicants Occurring Naturally in Foods, 2nd ed., Committee on Food Protection, Food and Nutrition Board, National Research Council, National Academy of Sciences, Washington, D.C., 1973.

D. F. FARKAS
University of Delaware

FOODS, DIET. See Sweeteners.

FOODS, NONCONVENTIONAL

Nonconventional foods differ from the usual materials of plant and animal origin that are used for human food or animal feed in that they can be produced from chemicals such as carbohydrates, hydrocarbons, or industrial organic chemicals by processes such as microbiological, enzymatic, or chemical synthesis, or from existing natural products containing carbohydrates, proteins, and fats and by physical, chemical, microbiological, or enzymatic modification.

Examples of nonconventional foods include: *single-cell proteins*—dried cells of microorganisms such as algae bacteria, actinomycetes, yeasts, molds, and higher fungi, or protein concentrates and isolates derived from them; *derived plant and animal products*—leaf meals and leaf protein concentrates, seed meals and seed meal proteins (see Nuts; Soybeans and other seed proteins), concentrates and isolates (eg, soy, cottonseed, peanut, etc), plant cells grown in tissue culture, fish, and meat protein concentrates and isolates (see Aquaculture; Meat products); *synthetic products*—carbohydrates (qv), fats (qv), proteins (qv), peptides (see Polypeptides), amino acids (qv), and vitamins (qv) prepared by chemical, microbiological, or enzymatic synthesis; and *manufactured or combination foods*—"engineered" or textured foods, and formulated foods (see Food processing).

Each of these general classes of nonconventional foods has been developed to meet specific applications. For example, single-cell protein (SCP) products were designed to provide a source of protein for use in animal feeds in those regions of the world where conventional sources of protein feedstuffs, such as soybean meal or fish meal, are periodically in short supply or available only at very high prices. Also, SCP products have applications in human foods as protein sources or as functional food ingredients for their flavoring, water- or fat-binding, stabilizing, and thickening characteristics.

Derived plant and animal products were developed to make better use or to upgrade the nutritional quality of already existing materials or products, whereas synthetic and manufactured products arose from new knowledge of human and animal nutrition that involved more precise definition of nutrient requirements for growth, reproduction, lactation, and body maintenance in both humans and domestic livestock (see also Pet and other livestock feeds). Recently, there has been an interest in developing food products to meet human needs under abnormal environments, such as military rations for arctic, tropical, or desert environments, and special products for manned space flights.

Numerous reviews have been published on various aspects of nonconventional foods in recent years (1–10). This discussion presents a summary of the historical development and current status of nonconventional food products and processes, with emphasis on proteins.

Single-Cell Protein (SCP)

Historical Development. Cells of microorganisms have constituted a portion of human food since ancient times. Yeast-leavened baked products contain the residual nutrients from the yeast cells destroyed during baking (see Bakery processes). Cultured dairy products, such as yogurt, buttermilk, and sour cream, contain up to 300×10^6 cells per gram of lactic acid bacteria (11) (see Milk products). Other examples of fermented foods that have been consumed since early times include fermented meats, fish and soybean products (see Fermentation).

Modern technology for producing microbial cells for human food or animal feed emerged in Germany during World War I. Bakers' yeast, *Saccharomyces cerevisiae*, was grown in aerated tanks using incremental feeding of molasses as the carbon and energy source, and ammonium salts as the nitrogen source (12).

In the period between World Wars I and II, processes were developed in Germany for producing fats from sulfite waste liquor from paper manufacture using *Endomyces vernalis* and for producing *Saccharomyces cerevisiae* for use as fodder yeast from wood hydrolysates (Scholler-Tornesch process). Also during this period, the Heiskenskjold process for propagating *S. cerevisiae* from sulfite waste liquor was introduced in Finland (11–12).

During World War II, major effort was undertaken in Germany to produce food and fodder yeast from waste products such as sulfite liquor. The yeast *Candida utilis*, known as Torula yeast, was used since it grows on pentoses, such as D-xylose and D-arabinose present in sulfite waste liquor, as well as on glucose. The Waldhof fermentor, introduced during this period, provided for both agitation and aeration. This fermentor was a significant advance in microbial cell production technology, since it enabled improved rates of oxygen transfer to growing cells, resulting in faster growth rates than had been achieved previously.

Current Technology. Two broad classes of microorganisms are of interest for single-cell protein (SCP) production: photosynthetic organisms including the algae and certain bacteria; and nonphotosynthetic organisms, including bacteria, actinomycetes, yeasts, molds, and higher fungi. In addition, two different uses of SCP are distinguished: food for humans and feed for animals.

Photosynthetic Organisms. Mass cultivation of algae in ponds or tanks under photosynthetic conditions, using incident sunlight as the energy sources and CO_2 as the carbon source, has been investigated in Japan, Mexico, Algeria, and India (13–16), and in California in the United States (17). Artificial illumination systems have been used for experimental mass cultivation of algae in Japan and Czechoslovakia (13) and in bioregenerative systems for converting CO_2 and human wastes into breathable oxygen and food as part of life-support systems for long duration, space exploration missions (18–19).

Considerable research has been conducted in Japan on growth of blue–green algae of the genus *Chlorella* heterotrophically in the dark, using organic carbon and energy sources such as glucose or acetate (13,20). The objective of these efforts has been to utilize strains having higher growth rates under heterotrophic conditions than under photosynthetic–autotrophic conditions.

Table 1 summarizes the more important algal species that have been investigated. The nitrogen requirements for algal growth can be met by either ammonium salts, urea, or nitrates. Bacterial action in sewage oxidation ponds may also liberate sufficient

Table 1. Photosynthetic Microorganisms in SCP Production

Algae	Scale[a], and growth conditions	Temperature, °C	pH	Cell production: Cell density (specific growth rate, h^{-1})	Cell production: Yield[b] (dry wt)	Refs.
Chlorella pyrenoidosa 71105	2.27-kL tank, sunlight or fluorescent lamps, CO_2 (9.7% in air)	39	4.5–5.8	0.28 vol % (0.0625)	1 kg/d	19
Chlorella regularis	autotrophic, 1-L jar fermentor, fluorescent lamp, CO_2 (5% in air)	36	6.5	6 mL/L		20
	heterotrophic, (dark) 20-L jar fermentor (13 L of medium), acetic acid (10%)	36	6.8	15 mL/L (0.28)	0.48 g/g acetate utilized	20
Scenedesmus acutus 276-3A	shallow tanks, 55 m^2, 30 cm deep, sunlight	ambient (30)	7–8		20 g/(m^2·d)	16
Scenedesmus quadricauda	10^6-L pond, CO_2, sunlight	25–35	8.5–9.5	1 g/L	12.6 g/(m^2·d)	17
S. quadricauda and *S. obliquus*	54-kL (900 m^2) pond, sunlight, CO_2 (0.5%)	34	6.5–7.0	1.5–2.0 g/L	10 g/(m^2·d)	13
Spirulina maxima	700-m^2 pond, sunlight, CO_2 (0.5%)	ambient			15 g/(m^2·d)	14–15

[a] To convert L to gal, divide by 3.785; to convert m^2 to ft^2, multiply by 10.76.
[b] To convert g/(m^2·d) to short tons per acre·year, multiply by 1.629.

ammonia for algal growth. Most natural wastes supply sufficient quantities of inorganic nutrients but additional phosphorus may be needed in some regions for their utility in SCP production.

Key factors influencing growth include temperatures, pH, availability of CO_2, nitrogen, phosphorus, and other inorganic nutrients, and availability of sunlight as influenced by latitude, cloud cover, and depth of the culture pond or tank. Slow, erratic growth results from wide fluctuations between day and night temperatures in outdoor ponds, and from season to season.

The availability of CO_2 and the pH are intimately related since the preferred pH range for growth of many species, such as *Chlorella,* is pH 6.5–7.0 and most of the CO_2 is bound as bicarbonate (HCO_3^-) in solution. Additional CO_2 beyond that present in air (0.003%) must be provided to attain optimum growth.

Organic sludges on the bottom of sewage oxidation ponds liberate CO_2 and ammonia slowly and uniformly by bacterial action, and provide a uniform supply of nutrients (17).

At Lake Texcoco, Mexico, bicarbonate is available in the alkaline waters from soda ash (sodium carbonate) deposits (see Alkali and chlorine products). This supply of carbon source is adequate for growing *Spirulina maxima* which tolerates alkaline pH values (14). However, in Algeria and France, combustion gases were used to grow this organism (15).

Sunlight availability is critical for algal growth. Outdoor algal cultivation is considered practical only in regions between latitudes 35° N and S, where cloud cover and variations in the length of day and night are minimized (13). Depth of the culture pond also affects the availability of sunlight and depths of 0.2–0.3 m are optimum for culturing *Scenedesmus* species (16–17).

Algal cultures must be agitated during growth to maintain the cells in suspension and to prevent sedimentation. Methods used for agitation include rotary-distributor arms (21), paddle wheels (16), recirculation alone (13,17), or in combination with injection of CO_2 (15).

Table 1 shows that algal densities in culture ponds are in the range of 1–2 g/L (dry wt basis). Consequently, large volumes of water must be handled in harvesting, dewatering and drying algal cells.

Centrifugation, flocculation with $Al_2(SO_4)_3$, $Ca(OH)_2$, or cationic polymers, treatment with ion-exchange resins and drum, sand bed, and sun drying have been investigated for separating, concentrating, and drying algal cells (22). All of these methods add significantly to the cost of the product.

Microstrainers and autoflocculation at pH 9.5 or above have been investigated on an experimental scale as means for separating *Chlorella* and *Scenedesmus* species. *Spirulina maxima* floats on the surface of the culture medium in large clumps when growth ceases, and the cells can be skimmed.

Yields of algae grown in outdoor pond cultures given in Table 1 are on the order of 12–15 g/(m^2·d) 20–24 short ton/(acre·yr). Higher yields can be obtained under artificial illumination but growth of algae under these conditions is not economically feasible.

Product quality is an important consideration in producing algae for food or feed use. Algal cells must be dried using time–temperature combinations sufficient to destroy pathogenic bacteria and viruses that may be present in ponds; particularly in those culture systems based on sewage (see Water, sewage). The possible presence of heat-stable algal toxins must also be considered.

Tables 2, 3, and 4 list proximate and amino acid analyses, and protein quality of selected algae as determined by animal feeding studies, respectively. More extensive compilations of compositional and nutritional data on algae are given in reference 29. It is important to note that the N × 6.25 factor used to calculate the crude protein contents of the algae does not reflect accurately the true protein content because algal cells may contain nonprotein nitrogen substances, eg, 4–6% nucleic acids (dry wt basis).

Algae tend to have lower contents of methionine than is desirable in human and animal nutrition and supplementation with this amino acid is necessary with many species. *Spirulina maxima* apparently has a higher protein quality than other species

Table 2. Proximate Analyses of Selected Algae Grown Photosynthetically

	Content, % (dry wt basis)				
Organism	Nitrogen	Crude Protein N × 6.25	Fat	Ash	Ref.
Chlorella pyrenoidosa 71105 (*Sorokiniana*)	9.6	60	8.1	8.9	23
Chlorella regularis	9.3	58	16	6.7	20
Chlorella ellipsoidea and *Scenedesmus obliquus*	9.7	61	19.5	13.4	24
Scenedesmus acutus 276-3A	8–8.8	50–55	12–19	6–8	16
Spirulina maxima	9.9	62	2–3		25

Table 3. Amino Acid Composition of Selected Algae Grown Photosynthetically

	Composition, g/16 g nitrogen			
Amino acid	*Chlorella pyrenoidosa* (*Sorokiniana*)[a]	*Chlorella regularis*[b]	*Spirulina maxima*[c]	F.A.O. reference pattern[d]
alanine	5.9	7.3	6.5	
arginine	5.6	5.8	6.6	
aspartic acid	5.9	8.8	8.6	
cystine		0.7	0.4	
glutamic acid	9.3	11.8	13.6	
glycine	4.8	5.4	4.6	
histidine	1.4	1.8	1.7	
isoleucine	3.4	4.2	5.8	4.2
leucine	4.0	8.1	7.8	4.8
lysine	7.9	7.7	4.8	4.2
methionine	1.8	1.3	1.5	2.2
phenylalanine	4.5	5.1	4.6	2.8
proline	4.0	4.3	3.8	
serine	2.2	3.0	4.3	
threonine	3.2	3.6	4.6	2.8
tryptophan	1.4	1.5	1.3	1.4
tyrosine	2.7	2.6	3.9	
valine	5.1	5.9	6.3	4.2

[a] Ref. 23.
[b] Ref. 20.
[c] Ref. 25.
[d] Ref. 26.

Table 4. Protein Quality and Digestibility of Selected Algae

Organism	Treatment	Animal	Protein in diet, %	Protein digestibility, %	PER[a]	Biological value	Ref.
Chlorella pyrenoidosa (*Sorokiniana*)	dried, dried and 20% L-methionine	rat	10	86	2.19		23
		rat	10	86	2.90		
Chlorella ellipsoidea and *Scenedesmus obliquus*	dried	rat	18		1.38		27
	dried	chick	18		1.55		
Scenedesmus acutus	drum dried and 20% L-methionine	rat	10	81	2.24	80.8	16
		rat	10		2.45		
Spirulina maxima	dried	rat	10	84	2.3–2.6	72	28

[a] Protein Efficiency Ratio. PER = weight gain (g) for a 10% protein level in the diet of rats as compared to the standard (casein = 2.5).

since its protein efficiency ratio (PER) is nearly equivalent to that of casein (2.5%) (see Table 4).

Dried algae are not accepted readily by livestock. Palatability of dried algae for swine was improved by pelletizing with steam-rolled barley (17).

There is considerable anecdotal information on the history of human consumption of *Spirulina maxima* as a source of protein in Mexico, and in the region of Lake Chad in Africa. However, relatively few controlled human feeding studies have been conducted using algae as a significant source of protein in the diet. Consumption of 100 g/d of a mixture of *C. ellipsoidea* and *S. obliquus* resulted in gastrointestinal distress attributed to toxins in the algae (27). Partial substitution of proteins in eggs and fish with *C. pyrenoidosa* did not reduce human nitrogen retention but digestibility was low (30). In general, many species of algae have cell walls that are resistant to digestive enzymes, have dark colors, and bitter flavors; all of which must be altered to make an acceptable food or feed product.

Photosynthetic bacteria have been studied only to a limited extent for their applicability to SCP production. *Rhodopseudomonas gelatinosa* was grown under artificial illumination in a bicarbonate–wheat-bran medium on a laboratory scale but productivities (dry weight of cells per unit volume per unit time) were too low to warrant further development (31).

Nonphotosynthetic Organisms. Nonphotosynthetic microorganisms of interest in SCP production include bacteria, actinomycetes, yeasts, molds, and higher fungi. Carbon and energy sources that have been considered for growing these organisms include carbohydrates such as simple sugars, starches, cellulose, and agricultural, forestry, pulp, and paper, or food processing wastes containing these carbohydrates, hydrocarbons, and chemicals derived from them including alcohols and organic acids. Table 5 presents data on the growth characteristics of selected nonphotosynthetic microorganisms that have been grown on a significant scale for SCP production.

Commercial-scale operations are conducted in batch-semicontinuous or continuous culture systems. Fermentation vessels include the conventional baffled aerated tank, with or without impeller agitation (43), and air-lift tower fermentors in which air is sparged into an annular space between the fermentor wall and an internal cylinder (52). Usually, a corrosion-resistant grade of stainless steel (316L) is used for fermentor construction, however, wood or concrete tanks have been used with agricultural or food wastes.

In batch systems, the concentration of the carbon and energy source for growth is 1–10%. In continuous culture systems, it is usually less than 1%. Suitable nitrogen sources include anhydrous ammonia or ammonium salts. Feed-grade phosphate is used as the source of phosphorus. Mineral-nutrient requirements vary among different organisms and are usually added to make up deficiencies in the water supply (see Mineral nutrients). Sulfates are used rather than chlorides to minimize corrosion. Carbon–nitrogen ratios should be 7:1–10:1 for yeasts to favor high protein contents and minimize fat synthesis in the cells that occurs at higher C:N ratios.

Temperature and pH conditions for optimum growth rates and productivities (dry weight of cells per unit volume per unit time) vary widely but are generally 25–40°C and pH 3.0–7.0, respectively. It is desirable to use strains of microorganisms that tolerate higher temperatures in this range since considerable quantities of heat are liberated during aerobic growth of microorganisms on either carbohydrates or hydrocarbons. Typical values are 15–34 kJ/g (3.6–8.1 kcal/g) of cells (dry wt) depending upon yield from a given substrate. In many geographical regions, cooling water is not available at a temperature below 20°C and refrigeration must be provided to control the temperature in the fermentor.

Production of food-grade SCP products requires operation under aseptic conditions in which the air, growth medium and equipment are sterilized. Feed-grade SCP can be produced under clean but nonsterile conditions provided that a pH of 3.0–4.5 and a large inoculum are used.

Transfer of oxygen and substrates to and across the cell surface is an important factor affecting growth rate, yield, and productivity in SCP processes. For yeasts, oxygen requirements range from 1 g/g dry wt of cells with carbohydrates to 2 g/g dry wt of cells for hydrocarbons (5).

A number of full-scale commercial plants for SCP production have been constructed but not all are presently in operation. Imperial Chemical Industries, Ltd. has constructed a 50,000–75,000-t/yr plant for producing the bacterium *Methylophilus methylotrophus* from methanol (qv). This process utilizes an air-lift pressure cycle fermentor, and a proprietary system for separating the cells from the growth medium by agglomeration (34–35).

Large-scale SCP production processes for growing yeasts of the genus *Candida* from hydrocarbon substrates have been developed by British Petroleum Co., Ltd. (37–39) and Kanegafuchi Chemical Industry, Ltd. of Japan (53). However, 100,000 t/yr capacity plants based on these processes that were constructed in Sardinia and Italy were abandoned because of regulatory agency questions regarding the residual hydrocarbon contents of the products (5). Other commercial-scale processes for yeast SCP production based on hydrocarbons or ethanol include several plants in the USSR using hydrocarbons (5) and a 4540-t/yr plant for producing *Candida utilis* (Torula yeast) using ethanol at Hutchinson, Minnesota (Amoco Foods Company) (5,41) (see Ethanol). *C. utilis* is also produced from sulfite liquor in a 4540-t/yr plant at Salem, Oregon (54). *Klusyeromyces fragilis* is used to convert lactose cheese whey to SCP in a 4540-t/yr plant at Juneau, Wisconsin. This process also can be operated under anaerobic conditions to produce ethanol (55).

Typical SCP processes that have been operated on an experimental scale are based on conversion of methanol by *Hansenula polymorpha* (56) and ethanol by *Candida ethanothermophilum* (57). Neither of these procedures has been scaled up to a commercial-sized operation. The Symba process, developed in Sweden, for the treatment of potato processing wastes uses a symbiotic combination of two yeasts, *Saccharomycopsis (Endomycopsis) fibuligera* and *Candida utilis.* The former organism produces α- and β-amylases that convert starch (qv) to a mixture of lower saccharides, mostly glucose and maltose. The latter organism utilizes these sugars for growth. This process has been operated at the pilot-plant production level of 40–100 kg dry yeast/d (5).

In Finland, sulfite waste liquor is converted by the Pekilo process to SCP using *Paecilomyces varioti*. The plant capacity is designed for a scale of operation of 10,000 t/yr (48–49).

Pilot-plant scale SCP processes utilizing fungi include the conversion of carob bean waste by *Aspergillus niger* (45), corn- and pea-processing wastes by *Giotrichium* sp (46), and coffee-processing wastes by *Trichoderma harzianum* (50–51).

The mycelium of the common mushroom, *Agaricus campestris,* has been grown in submerged cultures using a variety of carbohydrates as substrates. However, the product lacked the desired flavor and this process was not developed on a commercial

Table 5. Selected SCP Production Processes Based on Nonphotosynthetic Microorganisms

Organism, substrate	Scale[a]	Temperature, °C	pH	Cell production: Cell density g/L dry wt (specific growth rate, h^{-1})	Cell production: Yield, g/g substrate utilized on dry wt basis	Refs.
Bacterial processes						
Acinetobacter (Micrococcus) cerificans, *n*-alkanes	7.5-L batch fermentor	30	6.8	(1.33)	1.20	32
Cellulomonas sp and *Alcaligenes* sp, sugar cane bagasse	5.30-L fermentor, batch and continuous	34	6.6–6.8	16 batch, 10 continuous (0.20–0.29)	0.44–0.50	33
Methylophilus (Pseudomonas) methylotrophus, methanol	1000-t/yr pilot plant, continuous air-lift pressure cycle fermentor	35–40	6.0–7.0	0.30 (0.38–0.50)	0.50	34–35
Pseudomonas sp *Hyphomicrobium* sp *Acinetobacter* sp *Flavobacterium* sp (mixed culture), methane	large pilot plant	32	5.7	0.8	0.99	36

Yeast processes						
Candida lipolytica, *n*-alkanes	1.8-kL working volume fermentor, continuous	32	5.5	23.6	0.88	37
gas oil	12-m^3 fermentor, 2.4 kL of medium, continuous	30	4.0	25	0.18	38–39
Candida utilis, sulfite waste liquor or ethanol	Waldhof fermentor	32	5.0	(0.5)	0.39	40
	4540-t/yr plant	30	4.6	6–7 (0.3)		41
Kluyveromyces (Saccharomyces) fragilis, cheese whey	6.06-kL batch fermentor	32	5.5	14.6 (0.66)	0.55	42
Saccharomyces cerevisiae, molasses	75–225-m^3 batch fermentors, or 34.07-kL	30	4.5–5.0	40–45 (0.20–0.25)	0.50	43
	continuous fermentors	30	4.5–5.0	70 (0.14)		44
Mold and fungal processes						
Aspergillus niger, carob bean extract	3-kL fermentor (total capacity)	30–36	3.4	31.5 (0.25)	0.45	45
Geotrichium sp, corn and pea waste	37.85-kL aeration pool	ambient	3.7	0.75–1.0		46
Morchella hortensis, glucose	7.57-kL tank	2.5–3.0	6.5	24–30		47
Paecilomyces varioti (Pekilo), spent sulfite liquor	360-m^3 fermentor			17	0.55	48–49
Trichoderma harzianum, coffee waste	18.93-kL batch tanks, 28.39-kL continuous tanks	ambient	3.5	(0.10–0.19)	0.31–0.47	50–51

[a] To convert L to gal, divide by 3.785; to convert m^3 to ft^3, multiply by 35.31.

scale (58). Morel mushroom mycelium can also be produced in aerated fermentors. Mycelia of various species of the genus *Morchella,* especially, *Morchella crassipes,* and *M. hortensis* grow on glucose, maltose, or lactose. A process was developed for growing the mycelium of these organisms on a commercial scale in the United States for use as a food-flavoring ingredient rather than as a source of protein (47,59). This process is no longer practiced because of the relatively high production costs as compared with the costs of imported dried mushrooms.

Dry weight yields of bacteria and yeasts grown on hydrocarbons and methanol are ca 1.0 and 0.5 g/g substrate utilized, respectively. For yeasts, molds, and higher fungi grown on carbohydrate substrates, dry weight yields are 0.5–0.6 g/g substrate utilized.

Yeast cells are harvested readily by centrifugation. Molds and higher fungi grow in either pellet or filamentary forms. These organisms can be separated from the growth medium and dewatered by screens, filter processes, or basket centrifuges.

It is very costly to separate bacteria from the growth medium by centrifugation because of their small size (1–2 μm) and densities similar to that of water. Bacterial cells can be concentrated by agglomeration or electrocoagulation prior to centrifugation. The resulting wastewater and residual substrates are purified and recycled, particularly in processes based on hydrocarbons, methanol, or ethanol (5).

The product quality considerations for nonphotosynthetic microorganisms are similar to those mentioned previously for algae. Tables 6 and 7 present typical proximate analyses and amino acid analyses for selected bacteria, yeasts, molds, and higher fungi that have been produced on a large pilot-plant or commercial scale. Table 8 summarizes the results of animal feeding studies reported in the literature.

It is apparent that most of the bacteria, yeasts, molds, and higher fungi of interest for SCP production are deficient in methionine and must be supplemented with this amino acid to be suitable for animal feeding or human food applications. Also, lysine–arginine ratios should be adjusted in poultry rations in which yeast SCP is used (68). Human feeding studies have shown that only limited quantities of yeast such as *Candida utilis* can be added to food products without adverse effects on flavor (5).

Nucleic acid contents of SCP products range up to 16% in bacteria and 6–11% in yeasts. Nucleic acid contents must be reduced by processing so that intakes would be less than 2 g/d to prevent kidney stone formation or gout. Adverse skin and gastrointestinal reactions have also been encountered as a result of human consumption of some SCP products. Also, some fungi, such as *Fusarium graminearum,* may produce toxic substances. These species should not be used for SCP production.

The FDA regulations provide for the use of dried cells of the yeasts *S. cerevisiae, K. fragilis,* and *C. utilis* in foods. Folic acid contents must not exceed 0.04 mg/g (70). Functional concentrates and solids can be prepared from dried microbial cells by disrupting the cell walls using mechanical means or acid, alkaline, and enzyme hydrolysis; removing the cell walls and reducing nucleic acid contents by chemical or enzymatic methods. Also, microbial proteins can be spun into fibers (5). Bakers' yeast protein concentrate has been approved by the FDA for use as a functional food additive (70).

Table 6. Proximate Analyses of Selected Nonphotosynthetic Microorganisms Grown on Various Substrates

		Composition, g/100 g dry wt						
Organism	Substrate	Nitrogen	Protein	Fat	Crude fiber	Ash	Energy, kJ/g (kcal/g)	Ref.
Bacteria								
Acinetobacter (*Micrococcus*) *cerificans*	hexadecane	11	72					32
Cellulomonas sp	bagasse	14	87	8		7		33
Methylophilus (*Pseudomonas*) *methylotrophus*	methanol	13	83	7	<0.05	8.6	12.6 (3.0)	34
Yeast								
Candida lipolytica (Toprina)	*n*-alkanes	10	65	8.1		6		60
Candida lipolytica (Toprina)	gas oil	11	69	1.5		8		60
Candida utilis	ethanol	8.3	52	7	5	8		61
	sulfite liquor	9	55	5		8		62
Kluyveromyces (*Saccharomyces*) *fragilis*	cheese whey	9	54	1		9		62
Saccharomyces cerevisae	molasses	8.4	53	6.3		7.3		62
Molds and higher fungi								
Morchella hortensis	glucose	5.4	34	1.4				63
Paecilomyces varioti (Pekilo)	sulfite waste liquor	9.1–10.11	57–63					48

Table 7. Amino Acid Analyses of Selected Nonphotosynthetic Microorganisms Grown on Various Substrates

Organism	Substrate	Amino acid contents, g/16 g N (dry wt basis)[a]																		Ref.
		Ala	Arg	Asp	Cys	Glu	Gly	His	Ile	Leu	Lys	Met	Phe	Pro	Ser	Thr	Try	Tyr	Val	
Bacteria																				
Acinetobacter (Micrococcus) cerificans	*n*-alkanes	6.7	4.7	8.0		10.7	5.0	1.7	4.3	6.5	5.2	1.8	3.6	3.1	2.8	4.1	2.8	1.3	5.4	64
Cellulomonas-Alcaligenes	bagasse		6.5					7.8	5.4	7.4	7.6	2.0	4.7			5.5			7.1	33
Methylophilus (Pseudomonas) methylotrophus	methanol	6.8	4.5	8.5	0.6	9.6	4.9	1.8	4.3	6.8	5.9	2.4	3.4	3.0	3.4	4.6	0.9	3.1	5.2	34
Yeasts																				
Candida lipolytica (Toprina)	*n*-alkanes	7.4	4.8	10.2	1.1	11.3	4.8	2.0	4.5	7.0	7.0	1.8	4.4	4.4	4.8	4.9	1.4	3.5	5.4	65
Candida lipolytica	gas oil	5.8	5.0	10.0	0.9	12.1	4.5	2.1	5.3	7.8	7.8	1.6	4.8	3.7	5.1	5.4	1.3	4.0	5.8	65
Candida utilis	sulfite waste	5.8	5.4	9.2		15.6	3.6	1.2	3.8	7.6	4.8	1.1	8.6	6.0	5.0	5.4	2.4	6.2	3.8	66
Candida utilis	ethanol	5.5	5.4	8.8	0.4	14.6	4.5	2.1	4.5	7.1	6.6	1.4	4.1	3.4	4.7	5.5	1.2	3.3	5.7	61
Kluyveromyces (Saccharomyces) fragilis	cheese whey							2.1	4.0	6.1	6.9	1.9	2.8			5.8	1.4	2.4	5.4	55
Saccharomyces cerevisiae	molasses		5.0		1.6			4.0	5.5	7.9	8.2	2.5	4.5			4.8	1.2	5.0	5.5	43
Morchella hortensis	glucose	4.5	4.0	4.6	0.4	15.4	3.0	1.9	2.4	5.0	3.0	0.7	2.3	4.5	2.8	2.7	1.0	1.9	2.9	63
Paecilomyces varioti	spent sulfite liquor				1.1				4.3	6.9	6.4	1.5	3.7			4.6	1.2	3.4	5.1	48
FAO reference									4.2	4.8	4.2	2.2	2.8				1.4		4.2	26

[a] Amino acid abbreviations are as follows: Ala = alanine, Arg = arginine, Asp = aspartic acid, Cys = cystine, Glu = glutamic acid, Gly = glycine, His = histidine, Ile = isoleucine, Leu = leucine, Lys = lysine, Met = methionine, Phe = phenylanine, Pro = proline, Ser = serine, Thr = threonine, Try = tryptophan, Tyr = tyrosine, Val = valine.

Table 8. Protein Quality and Digestibility of Selected Nonphotosynthetic Microorganisms

Organism and substrate	Treatment	Animal	Percent in diet	Protein digestibility	PER[a]	Biological value	Feed conversion ratio, kg/kg wt gain	Ref.
Bacteria								
Acinetobacter (*Micrococcus*) *cerificans* (*n*-hexadecane)	dried	rat		83.4		67		67
Methylophilus methylotrophus (methanol)	dried	broiler chicken	9.8				2.30 (2.33)[b]	34
	dried	pig	6.7				3.13 (3.34)[b]	
Yeasts								
Candida lipolytica (*n*-alkanes)	dried	rat		96		61		68
	dried + 0.3% DL-methionine	rat		96		91		68
	dried	broiler chicken	10	88			2.58 (2.68)[b]	68
	dried	pig	7.5	92			3.04 (3.11)[b]	68
Candida utilis (sulfite waste liquor)	dried	rat		85–88	0.9–1.4	32–48		69
	dried + 0.5% DL-methionine	rat		90	2.0–2.3	88		69
Kluyveromyces fragilis (cheese whey)	dried	rat			2.26			55
Saccharomyces cerevisiae (molasses)	dried	rat		80–90	1.7	58–69		69
Fungi								
Aspergillus niger (carob bean extract)	dried	broiler chicken			2.50		2.02 (2.12)[b]	45

[a] Protein efficiency ratio. See Table 4.
[b] Control diet.

Derived Plant and Animal Products

Plant Protein Products—Leaf Protein Concentrates. Leaf protein concentrates (LPC) are prepared by crushing plant material, extracting the juices, using the juice *per se* or recovering protein from the juice by heating or chemical precipitation. Dehydrated alfalfa (lucerne) has a long history of use as a source of plant protein for animal feeds. The leaves of alfalfa and other crops are a source of proteins that can be extracted to give a concentrated product having increased protein and decreased fiber contents suitable for animal feeding. Many plants other than alfalfa have been considered as sources of LPC including clover, field beans, mustard, kale, fodder radish, banana leaves, and aquatic plants (7). Requirements for crops suitable for LPC production are as follows: rapid growth with high yields of protein (1600 kg/hm^2 or 1430 lbs/acre) during the growing season; absence of mucilagenous sap which makes it difficult to separate juice from the fiber, or acidic or high tannin saps which prevent extraction of protein into the juice because of precipitation in the pulp; and absence of toxic materials such as glycosides that could be carried into the final product. In addition, the entire plant should be utilizable for LPC, from an economic standpoint.

LPC Processes. Process development for LPC production dates from research in the United Kingdom and Hungary during 1920–1940 (7). Table 9 presents some of the processes that are currently being used or are under development.

Reference 7 is a review of the research on LPC, conducted at Rothamsted Experiment Station in England, that has been directed toward products suitable for food use. Drying conditions are important in determining the quality of the product. Air drying below 80°C minimizes damage to proteins but results in a hard, dark-colored product.

Solvent extraction removes chlorophyll and other color bodies to give a light-colored product but increases processing costs. Furthermore, solvent extraction removes β-carotene and reduces vitamin A activity (7) (see Terpenoids; Vitamins).

Leaf materials also contain lipoxidases and highly unsaturated lipids. The LPC process conditions should inactivate lipoxidases to obtain a stable product (7,74).

The USDA Western Regional Research Center has developed an improvement to the Pro-Xan process (Table 9), termed Pro-Xan II for preparing a bland, colorless LPC product from alfalfa suitable for human consumption (73). Aqueous sodium

Table 9. Selected Processes for Leaf Protein Concentrate Production

Process	Description	Refs.
Rothamsted (Pirie)	pulping in ribbed rollers, precipitating protein at 80°C, or at pH 4.0, drying in air below 80°C	7
Chayen	impulse rendering with 20-cm blades at 8000 rpm, coagulation at pH 4, centrifugation, drying	71
Wisconsin	cell rupture by forcing through die, separation, heat treatment, dewatering	72–73
Pro-Xan (USDA)	chopping, ammoniation to pH 8.5, roll or twin-screw pressing, sieve purification, coagulation with steam at 85°C, dewatering, drying	73–74
Vepex (Hungary)	mechanical disintegration, multistage pressing, add-back of liquor to press cake, coagulation at 80°C, centrifugation, evaporation drying	75
Polyprotein (Pisa, Italy)	screw pressing, acidification to pH 4.5, coagulation with polyelectrolyte filtration	76

metabisulfite ($Na_2S_2O_3$) is added to alfalfa prior to expressing the juice to lighten the color of the LPC and protect cystine and methionine from oxidation. The juice, after expressing, is heated by steam injection to 60°C, for 20 s and cooled to 45°C, followed by removing the proteins associated with the chloroplasts by centrifugation.

The suspended solids are removed from the liquid phase in the centrifuge by a plate-and-frame filter press, and the proteins in the filtrate are precipitated at 80°C for 2–4 min. The protein precipitate is removed by centrifugation, the protein is washed at pH 4–5, and spray dried to give a white LPC having potential food applications. The chloroplast protein fraction is adjusted to pH 8.5, and heated to 90–95°C to coagulate the protein. The protein is dewatered by centrifugation, granulated and dried in a rotary dryer to give the feed-grade Pro-Xan II LPC.

LPC Product Quality. Table 10 gives proximate analyses of several LPC products. Amino acid analyses of several LPC products, including those from alfalfa, wheat leaf, barley, lupin, and chinese cabbage, are given in reference 79, and amino acid analyses, for Pro-Xan LPC products are given in reference 73. In general, the composition of LPCs varies widely with the raw materials and processes used. Amino acid profiles are generally satisfactory except for methionine which is limiting.

Enzyme degradation of leaf protein may occur during crushing and separation from the fiber. The amino acids produced by this enzyme action are soluble in the juice and may be lost unless all the juice is recovered.

Typical protein efficiency ratio values for LPCs derived from alfalfa range from 1.87 without supplementation to 2.57 when 0.4% methionine was added as compared with casein adjusted to a PER of 2.50 (73). Data on the protein quality of a variety of LPC products obtained from rat-feeding studies are presented in reference 80.

Human-feeding studies on LPC have been conducted in Jamaica (81). Diets for malnourished infants contained half of the nitrogen as LPC, and nitrogen retention was equivalent to that obtained with milk (81).

In India, LPC added to low protein diets (12.6 g protein/d) to provide an intake of 45–48.8 g protein/d resulted in improved nitrogen retention and digestibility in 10–12-yr old children (81). Allergic reactions including facial edema were common (1 out of 8 children). It is apparent that there are definite limits to the use of LPC products in human diets and that raw materials used for LPC production should be evaluated for possible allergenic problems.

The present market for LPC-type products in the United States is for dried alfalfa meal for animal feed. This product is sold for both protein and carotenoid content. The USDA Pro-Xan product attempts to obtain improved xanthophyll contents for use in egg laying ratios in addition to protein contents. The limitations to commercial development of LPC products are high capital costs as compared with the low yields of protein, seasonal availability of raw materials, and the need in the United States for FDA approval of the product.

Table 10. Proximate Analysis of Leaf Protein Products

	Composition, % (dry wt)				
Leaf protein	Protein N × 6.25	Fat	Fiber	Ash	Reference
Pro-Xan II (alfalfa)	47	14	4	16	74
cauliflower (pressed leaves)	26	2.9	12.5	10.7	77
pea vine (spray-dried juice)	18.5	2.5	1.4	12.8	78

Plant Protein Products—Seed-Meal Concentrates and Isolates. Seed-meal protein concentrates and isolates, particularly soy protein products, can be used as extenders for meat, seafood, poultry, eggs, or cheese. In the United States, FDA regulations cite the requirements for using vegetable protein products as substitutes in whole or in part for conventional animal proteins (82) (see Soybeans and other seed proteins).

Products resembling chicken, ham, frankfurters, and bacon prepared from soy protein products are now available commercially. Also, soy protein isolates are used in place of milk proteins or sodium caseinate in products such as coffee creamers, whip toppings, yogurt, and infant formulas (see Synthetic dairy products). Vegetable protein products are also used in snacks and in baked foods.

Functional characteristics of vegetable protein products are important factors in determining their uses in food products. Hydration, water, fat, and flavor binding, gelation, emulsifying, foaming, and whipping characteristics vary among different soy protein products and complete substitution of animal proteins by these products is not always possible (83–85).

FDA regulations require that a vegetable protein product must be nutritionally equivalent to the protein food for which it is substituted. For breakfast or luncheon meat products, the food substitute must contain at least 13% protein, and for other meats, poultry, or seafood, at least 18% protein. Minimum vitamin and mineral contents are specified for each type of product.

Soy protein products may impart a beany flavor to foods when used at levels greater than 20%. Undesirable components are present in the beans prior to processing and also may be generated during processing. Off-flavors and odors also may arise from oxidation of lipid components and from degradation of phenolic compounds during thermal processing of food containing soy proteins. Recent research indicates that these undesirable flavor components can be diminished but not completely eliminated by extraction with alcohols such as methanol, ethanol, or isopropanol, or an azeotropic mixture of hexane and an alcohol (86).

Vegetable proteins other than soy protein also have potential applicability in food products.

Cottonseed, peanut, rapeseed, and sunflower proteins have been evaluated for functional characteristics (87), and for utility in protein components of baked products (88–89), meat products (90) and milk-type beverages (91) (see Synthetic dairy products).

Animal Proteins—Fish Protein Concentrates. Fish protein concentrates (FPC) are distinguished from fish meal and fish solubles in that the former are produced for human food use from whole edible species of fish using sanitary processing methods rather than for animal feed (92) (see Aquaculture). FPC products are of relatively recent origin, dating from World War II, as compared with fish meal and fish solubles. An extensive discussion on fish protein concentrates is given in reference 9.

FPC Processes. Table 11 summarizes some of the more important processes for producing FPC. These processes are based on solvent extraction (National Marine Fisheries Service, Viobin, and Wiking-Eiweiss), on fluidization with added fish oil (Carver-Greenfield), or on enzyme hydrolysis (Keyes-Meinke). The National Marine Fisheries Service of the U.S. Department of Commerce has conducted extensive development studies on modifications of the Guttman-Vandenheuvel isopropanol extraction process originally developed in Canada. The process has been scaled up to the pilot-plant stage and the product has been evaluated for nutritional value and for

Table 11. Selected Fish Protein Concentrate Processes

Process	Description	Ref.
National Marine Fisheries Service	deboning, 3-stage countercurrent extraction with isopropanol, desolventization by stripping with steam	92
Viobin	dehydration and defatting by heating at 60–90°C with ethylene dichloride, azeotropic distillation, desolventization, under vacuum, solvent recovery	93
Carver-Greenfield (heat-transfer method)	fluidization with added fish oil, 2-stage evaporation, centrifugation, solvent extraction, recycle of oil	94
Wiking-Eiweiss	extraction of filleted fish with acetic acid, washing, pressing, ethanol extraction, drying, grinding, alkaline hydrolysis at 80°C, neutralization with acetic acid, spray drying	92
Astra	cooking and dewatering, isopropanol extraction	92
Keyes-Meinke (enzyme hydrolysis)	(*1*) combined autolytic and added proteolytic enzyme hydrolysis of comminuted fish in presence of chelating agent; (*2*) pepsin hydrolysis of ground fish waste at pH 2.6, inactivation of enzyme at 80°C, neutralization, concentration, spray drying	95 96

food ingredient applications (92,94,97). Solvent recovery must be provided in the process, and bone must be removed mechanically to reduce fluoride levels in the product below 100 ppm.

The Viobin process has been operated on a commercial scale. It involves dehydration of ground fish using an azeotropic mixture of ethylene dichloride and water, which boils at 71°C, followed by solvent separation, product washing, and drying. The dark color and undesirable flavor, of the product, together with ethylene dichloride residues, can be removed by subsequent extraction with isopropanol.

The Carver-Greenfield process was originally developed for by-products of the meat-rendering industry, and is used for manufacturing fish meal (92,94). It has the advantage of not requiring solvents for the initial extraction.

Other solvent extraction processes include the Astra process developed in Sweden, the Fishing Industry Research Institute process in South Africa, and the UNICEF (Dabsch) process in Chile (92,94). The Wiking-Eiweiss process was developed in Germany during World War II for producing a functional egg-white substitute. However, this product has not been produced commercially since that time.

A plant for producing 9080 t/yr using the Keyes-Meinke process is being designed by de Laval Separator Co. for installation at Biloxi, Mississippi (98). In this process the fish are broken and then comminuted in a hammer mill, followed by proteolytic enzyme digestion, bone fraction separation, product concentration by evaporation, and drying. The product will be suitable for animal feeds but will not meet the requirements for human food applications.

FPC Product Quality. FDA regulations for FPC require a minimum protein content of 75% and maximum moisture and fat contents of 10 and 0.5 wt %, respectively. Hake and hakelike fish, herring of the genera *Clupea*, menhaden, and anchovy of the genus *Engraulis mordax*, are permitted. Residues of isopropyl alcohol or ethylene dichloride cannot exceed 250 and 5 ppm, respectively, and fluorides cannot exceed 100 ppm as F. The limit for total bacterial count is 10,000/g and *Escherichia coli*, *Salmonella*, and pathogenic organisms must be absent (99).

A typical analysis of hake FPC on a dry basis is 80.9 wt % protein, 0.18 wt % fat, and 13.5 wt % ash (93). Amino acid profiles are excellent and compare favorably with whole egg except for tryptophan and lysine. Typical amino acid values for FPC are listed in reference 98. Hake and Atlantic FPCs prepared by isopropanol extraction had PERs of 3.29 and 3.05, respectively, as compared with 3.00 for casein (98). Numerous human feeding studies have been conducted with FPC. The results indicate that high-quality, bland FPC products can be used as protein supplements but they are not suitable for use as a sole source of protein.

Fish protein concentrates vary widely in functional characteristics, ranging from those having high protein content and low water solubility to those having lower protein contents, but improved water solubility. Attempts have been made to improve the functional properties of fish protein by enzyme hydrolysis (100), or by modification of the myofibrillus protein by succinylation (101). The prospects for fish protein concentrate depend upon the availability of the desired species of fish, and the ability of processors to produce FPC having desirable functional and nutritional characteristics at prices competitive with those of conventional protein sources.

Synthetic Protein Products

Plastein Synthesis. Plasteins are mixtures of high molecular weight proteinaceous peptides. They are synthesized by enzyme-catalyzed growth of peptide chains from lower molecular weight peptides. The process by which plasteins are formed is called the plastein reaction and is the reverse of the proteolytic enzyme hydrolysis of peptide bonds of proteins (102). Japanese investigators conducted extensive studies on the utility of the plastein reaction and of plasteins in food technology. They investigated the enzymatic modification of proteins from such products as soy beans, codfish, algae (*Chlorella*), wheat, milk, and bakers' and hydrocarbon-grown yeasts followed by plastein synthesis (103–104). The plasteins prepared were bland and did not have objectionable tastes, odors, or colors.

Plasteins were formed from soy protein hydrolysates with a variety of microbial proteases (103). Preferred conditions for hydrolysis and synthesis were obtained with an enzyme to substrate ratio of 1:100, and a temperature of 37°C for 24–72 h. A substrate concentration of 30 wt % (80% hydrolyzed) gave an 80% net yield of plastein from the synthesis reaction. However, only a 1% protein solution was used in the hydrolysis step which would be too low for an economical process (see Microbial transformations).

Fish protein concentrate and soy protein concentrate were used to prepare a low phenylalanine, high tyrosine peptide for use with phenylketonuria patients (104). The process steps include pepsin hydrolysis at pH 1.5, pronase hydrolysis at pH 6.5 (to liberate aromatic amino acids), gel filtration (Sephadex G-15) (to remove aromatic amino acids), incubation with papain and ethyl esters of L-tyrosine and L-tryptophan (plastein synthesis), and ultrafiltration (qv). The plastein had a bland taste and odor, and did not contain free amino acids. The yields were 69.3 and 60.9% from FPC and soy protein concentrate, respectively.

The sulfur amino acid content of soy protein can be enhanced by preparing plasteins from soy protein hydrolysate and sources of methionine or cystine such as ovalbumin hydrolysate (Plastein AB), wool keratin hydrolysate (Plastein AC), or L-methionine ethyl ester (alkali saponified plastein) (105). Typical PER (see Table

4, footnote) values for a 1:2 mixture of plastein AC and soybean and a 1:3 mixture of alkali-saponified plastein and soybean protein were 2.86 and 3.38, respectively, as compared with 1.28 for the soy protein hydrolysate and 2.40 for casein.

At present plasteins are in the experimental stage of development. Further work is needed on the scale-up of processing conditions for plastein synthesis which would lead to a commercially useful product and on functional properties of plasteins in foods.

Synthetic Proteins. Reference 106 is a review of syntheses of proteinlike polypeptides from ammonia, water, and carbon dioxide, or from mixtures of amino acids which are now manufactured by chemical or microbiological synthesis. Polyamino acids can be produced in the laboratory by simultaneous polymerization of mixtures of amino acids at 180°C for 3–6 h under dry conditions. Protenoids, containing all of the common amino acids in peptide linkages, can be obtained if sufficient amounts of aspartic and glutamic acids are included in the reaction mixture. Apparently, these protenoids are digestible by mammalian proteinases and can serve as sources of nutrients for *Lactobacillus arabinosus* and *Proteus vulgaris*. The possibility exists that products could be nutritionally imbalanced, have mammalian toxicity and undesirable tastes, odors and stability. These problems must be investigated further before any assessment of the utility of polyamino acids or protenoids can be made.

Textured and Formulated Foods

The availability of protein concentrates and isolates prepared from soy bean protein has made it possible to produce a wide range of textured protein analogue foods in which the soy protein is substituted for animal protein. The technologies involved in producing oil seed proteins such as those from soy beans, cottonseed, peanuts, and sunflower are described in reference 107.

The technologies involved in preparing meat extenders and substitutes from soy beans are reviewed in reference 108. Soy protein concentrates contain 70% protein or greater (dry wt basis) and are prepared by desugaring solvent-extracted defatted soybean flakes or flours using immobilization of soluble proteins by acidification to pH 4.5, by adding 80–90% ethanol, or by denaturing the protein with moist heat prior to extraction. Soy protein isolates, containing not less than 90% protein, are prepared by extracting defatted soybean flakes at pH 7.0 to 8.0–8.5, followed by separating the fibrous residue, isoelectric precipitation of the protein at pH 4.5, concentration, neutralization, and drying (see also Enzymes, immobilized).

Soy bean protein can be formed into fibers by forming a spinning dope from a slurry at pH 10–11, aged at 40–50°C, which is forced through a spinnerette into an acid coagulating bath. The fibers are heated to reduce the diameter to about 75 μm. The fibers can be formed with binders, such as egg albumin and colored and flavored to give the desired product characteristics (109).

Textured soy protein products can be made by thermoplastic extrusion of defatted soy flakes through an orifice under selected temperature and pressure conditions (4). Textured soy protein isolates can be made by heating a slurry of the solute containing 35% solids in the presence of hydroxides, phosphates, and polyvalent linking agents to form elongated filaments. The elongated fibers are passed through an orifice and cooled. Textured soy protein isolates give structure and texture to meat products and to meat analogues.

Reference 4 is a review of the extensive patent literature in the vegetable protein processing field, including processes for manufacturing protein concentrates and isolates, extrusion processes such as spinning of protein fibers, as well as textured protein gels, expanded products, and consumer products.

The Protein Advisory Group, *ad hoc,* is the working group of the WHO United Nations system involving WHO, FAO, and UNICEF. It has developed guidelines for the evaluation of novel sources of protein, eg, single-cell protein, clinical testing of novel sources of protein, human testing of supplementary food mixtures, and nutritional and safety aspects of novel protein sources for animal feeding (110).

Economic Aspects

Table 12 presents typical price ranges (1979 U.S. dollars) for selected nonconventional and conventional protein products used in human foods or animal feeds. Certain yeast products such as *Candida utilis*, (Torula yeast), *Kluyveromyces fragilis* (Fragilis yeast), and *Saccharomyces cerevisiae* (primarily grown on dibittered brewers' yeast) have been used widely as food or feed ingredients. However, markets have not yet been established for some of the newer single-cell protein products such as the bacterium *Methylophilus methylotrophus*, grown on methanol, or molds such as *Paecilomyces varioti* grown on sulfite waste liquor.

Dehydrated alfalfa is used in animal feeds but currently (1980) there are no markets for a food-grade leaf protein concentrate from alfalfa. An estimated price for a feed-grade leaf protein concentrate based on the USDA Pro-Xan process is ca $0.33/kg.

Limited marketing of fish protein concentrates for food use during the 1970s indicated selling prices of ca ≥$1.00/kg. The price structure of fish protein concentrate will be influenced by fluctuations in the availability and costs of suitable species of fish to producers.

Table 12. Comparison of Selling Price Ranges for Selected Microbial, Plant, and Animal Protein Products

Product, substrate, and quality	Crude protein content, N × 6.25 (dry wt basis)	Price range, $/kg (1979) (dry wt product basis)
Single-cell proteins		
Candida utilis		
ethanol, food grade	52	1.32–1.35
Kluyveromyces fragilis		
cheese whey, food grade	54	1.32
Saccharomyces cerevisiae		
brewers, debittered, food grade	52	1.00–1.20
feed grade	52	0.39–0.50
Plant proteins		
alfalfa, dehydrated	17	0.12–0.13
soybean meal, defatted	49	0.20–0.22
soy protein concentrate	70–72	0.90–1.14
soy protein isolate	90–92	1.96–2.20
Animal proteins		
fish meal, Peruvian	65	0.41–0.45
meat and bone meal	50	0.24–0.25
dry skim milk	37	0.88–1.00

Soybean meal is an important protein component of animal feeds. Soy protein concentrate and isolates now enjoy increasing uses as functional ingredients in textured and engineered foods. Other plant protein products derived from glandless cottonseed, peanut, or other seed meals may become available.

In general, nonconventional protein foods must be competitive with conventional plant and animal protein sources on the bases of cost delivered to the consumer, nutritional value to humans or animals, functional value in foods, sensory quality, and social and cultural acceptability. Also, requirements of regulatory agencies in different countries for freedom from toxins or toxic residues in single-cell protein products, toxic glycosides in leaf protein products, heavy metals and toxins in fish protein concentrates, or inhibitory or toxic peptide components in synthetic peptides must be met before a new nonconventional food or feed protein product can be marketed.

BIBLIOGRAPHY

"Proteins from Petroleum" in *ECT* 2nd ed., Suppl. Vol., pp. 836–854, by E. R. Elzinga and A. I. Laskin, Esso Research and Engineering Co.

1. M. Milner, N. S. Scrimshaw, and D. I. C. Wang, eds., *Protein Resources and Technology: Status and Research Needs,* Avi Publishing Co., Westport, Conn., 1978.
2. P. Davis, ed., *Single-Cell Protein,* Academic Press, Inc., New York, 1974.
3. H. Gounelle de Pontanel, ed., *Proteins from Hydrocarbons,* Academic Press, Inc., New York, 1973.
4. L. P. Hanson, *Vegetable Protein Processing,* Noyes Data Corp., Park Ridge, N.J., 1974, 308 pp.
5. J. H. Litchfield in H. J. Peppler and D. Perlman, eds., *Microbial Technology,* 2nd ed., Vol. 2, Academic Press, Inc., New York, 1979, pp. 93–155.
6. R. I. Mateles and S. R. Tannenbaum, eds., *Single-Cell Protein,* MIT Press, Cambridge, Mass., 1968.
7. N. W. Pirie, ed., *Leaf Protein: Its Agronomy, Preparation, Quality, and Use, IBP Handbook No. 20,* Blackwell Scientific Publications, Oxford, Eng., 1971, pp. 53–62, 86–91, 155–163.
8. P. J. Rockwell, *Single-Cell Proteins from Cellulose and Hydrocarbons, 1976,* Noyes Data Corp., Park Ridge, N.J., 1976.
9. S. R. Tannenbaum, B. R. Stillings, and N. S. Scrimshaw, eds., *The Economics Marketing and Technology of Fish Protein Concentrate,* MIT Press, Cambridge, Mass., 1974.
10. S. R. Tannenbaum and D. I. C. Wang, eds., *Single-Cell Protein II,* MIT Press, Cambridge, Mass., 1975.
11. J. H. Litchfield, *Chem. Tech.* **8,** 218 (1978).
12. S. C. Prescott and C. G. Dunn, *Industrial Microbiology,* 3rd ed., McGraw-Hill Book Company, Inc., New York, 1959, Chapt. 3.
13. L. Enebo, *Chem. Eng. Prog. Symp. Ser.* **65**(93), 60 (1969).
14. H. Durand-Chastel and G. Clement in A. Chavez, H. Bourges, and S. Basta, eds., *Proceedings of the 9th International Congress on Nutrition,* Vol. 3, S. Karger, Basel, Switz., 1975, pp. 85–90.
15. G. Clement in ref. 10, pp. 467–474.
16. E. W. Becker and L. V. Venkataraman, *A Manual on the Cultivation and Processing of Algae as a Source of Single-Cell Protein,* Central Food Technological Research Institute, Mysore, India, 1978.
17. W. J. Oswald and C. G. Golueke in ref. 6, pp. 271–305.
18. P. A. Lachance in ref. 6, pp. 122–152.
19. R. P. Casey, J. A. Lubitz, R. J. Benoit, B. J. Weissman, and H. Chau, *Food Technol.* **17,** 1039 (1963).
20. H. Endo and M. Shirota, *Proc. 4th Int. Fermentation Symp., Kyoto, Japan, 1972,* p. 533.
21. H. Tamiya, *Annu. Rev. Plant. Physiol.* **8,** 309 (1957).
22. E. S. Lipinsky and J. H. Litchfield, *Crit. Rev. Food Technol.* **1,** 580 (1970).
23. J. A. Lubitz, *J. Food Sci.* **28,** 229 (1963).
24. R. C. Powell, E. M. Nevels, and M. E. McDowell, *J. Nutr.* **75,** 7 (1961).
25. G. Clement, C. Giddey, and R. Menzi, *J. Sci. Food Agric.* **18,** 497 (1967).

26. Food and Agricultural Organization of the United Nations, *FAO Nutr. Studies No. 16 (Rome)*, (1957).
27. G. A. Leveille, H. E. Sauberlich, and J. Shockley, *J. Nutr.* **76,** 423 (1962).
28. H. Bourges, A. Sotomayer, and E. Mendoza, *Abstr. 9th Int. Cong. Nutr. Mexico City,* 1972, p. 85.
29. C. I. Waslien, *Crit. Rev. Food Sci. Nutr.* **6,** 77 (1975).
30. S. K. Lee, H. M. Fox, C. Kies, and R. Dam, *J. Nutr.* **92,** 281 (1967).
31. R. H. Shipman, I. C. Kao, and L. T. Fan, *Biotechnol. Bioeng.* **17,** 1561 (1975).
32. U.S. Pat. 3,384,491 (May 21, 1968), K. R. Guenther and M. B. Perkins (to Exxon Inc.).
33. Y. W. Han, E. E. Dunlap, and C. D. Callihan, *Food Technol.* **25,** 130, 154 (1971).
34. J. S. Gow, J. D. Littlehailes, S. R. L. Smith, and R. B. Walter in ref. 10, pp. 370–384.
35. D. G. MacLennan, J. S. Gow, and D. A. Stringer, *Process Biochem.* **8**(6), 22 (1973).
36. T. G. Wilkinson, H. H. Topiwala, and G. Hamer, *Biotechnol. Bioeng.* **16,** 41 (1974).
37. U.S. Pat. 3,846,238 (Nov. 5, 1974), G. H. Evans and J. L. Shennan (to British Petroleum Co., Ltd.).
38. U.S. Pat. 3,560,341 (Feb. 2, 1971), A. Champagnat and J. Filosa (to British Petroleum Co., Ltd.).
39. B. M. Laine and J. du Chaffaut in ref. 10, pp. 424–437.
40. E. E. Harris, G. J. Hajny, and M. C. Johnson, *Ind. Eng. Chem.* **43,** 1593 (1951).
41. U.S. Pat. 3,865,691 (Feb. 11, 1975), J. A. Ridgeway, Jr., T. A. Lappin, B. M. Benjamin, J. B. Corns, and C. Akin (to Standard Oil Co., Indiana).
42. A. E. Wasserman, J. W. Hampson, and N. F. Alvare, *J. Water Pollut. Control Fed.* **33,** 1090 (1961).
43. G. Reed and H. J. Peppler, *Yeast Technology,* Avi Publishing Co., Westport, Conn., 1973, pp. 328–354.
44. A. J. C. Olson, *Soc. Chem. Ind. Monogr.* **12,** 81 (1961).
45. F. K. E. Imrie and A. J. Vlitos in ref. 10, pp. 223–243.
46. B. D. Church, E. E. Erickson, and C. M. Widmer, *Food Technol.* **27**(2), 36 (1973).
47. R. F. Robinson and R. S. Davidson, *Adv. Appl. Microbiol.* **1,** 261 (1959).
48. H. Romantschuk in ref. 10, pp. 344–355.
49. H. Romantschuk and M. Lehtomaki, *Process Biochem.* **13**(3), 16, 17, 23 (1978).
50. R. Espinosa, O. Maldonado, J. T. Menchu, and C. Rolz in A. E. Humphrey and E. L. Gaden, Jr., eds., *Single-Cell Protein from Renewable and Nonrenewable Resources, Biotechnol. Bioeng. Symp. No. 7,* John Wiley & Sons, Inc., New York, 1977, pp. 35–44.
51. F. Aguirre, O. Maldonado, C. Rolz, J. F. Menchu, R. Espinosa, and S. de Cabrera, *Chem. Tech.* **6,** 636 (1976).
52. R. T. Hatch in ref. 10, pp. 46–68.
53. M. Kanazawa in ref. 10, pp. 438–453.
54. R. Anderson, R. B. Weisbaum, and K. Robe, *Food Process* **35**(7), 58 (1974).
55. S. Bernstein, G. H. Tzeng, and D. Sisson in ref. 51, pp. 1–9.
56. C. L. Cooney, D. W. Levine, and B. Snedecor, *Food Technol.* **29**(2), 33 (1975).
57. U.S. Pat. 3,868,305 (Feb. 25, 1975), Y. Masuda, K. Kato, Y. Takayama, K. Kida, and M. Nakanishi (to Mitsubishi Petrochemical Co., Ltd.).
58. J. H. Litchfield, *Biotechnol. Bioeng.* **9,** 289 (1967).
59. J. H. Litchfield in ref. 6, pp. 309–329.
60. C. A. Shacklady, *World Rev. Nutr. Diet.* **14,** 154 (1972).
61. *Torutein Product Bulletin,* Amoco Foods Co., Chicago, Ill., 1974.
62. H. J. Peppler in ref. 6, pp. 229–242.
63. J. H. Litchfield, V. G. Vely, and R. C. Overbeck, *J. Food. Sci.* **28,** 741 (1963).
64. R. J. Ertola, R. F. Segovia, M. Cambaruto, J. C. Monsiglio, and C. Artuse, *Dev. Ind. Microbiol.* **12,** 72 (1971).
65. G. H. Evans in ref. 6, pp. 243–254.
66. H. J. Peppler, *J. Agric. Food. Chem.* **13,** 34 (1965).
67. R. J. Ertola, L. A. Mazza, A. P. Balatti, and J. Sanahuja, *Biotechnol. Bioeng.* **11,** 409 (1969).
68. C. A. Shacklady and E. Gatumel in ref. 3, pp. 27–52.
69. R. Bressani in ref. 6, pp. 90–121.
70. *CFR Title 21, Food and Drugs 172.325, 172.896,* U.S. Government Printing Office, Washington, D.C., 1979.
71. I. R. Chayen, R. H. Smith, G. R. Tristam, D. Thirkell, and T. Webb, *J. Sci. Food Agric.* **12,** 502 (1961).
72. R. G. Koegel, V. I. Fomin, and H. D. Bruhn, *Trans ASAE* **16,** 716 (1973).

73. G. O. Kohler, S. G. Wildman, N. A. Jorgenson, R. V. Enochian, and W. J. Bray in ref. 1, pp. 543–569.
74. G. O. Kohler and E. M. Bickoff in ref. 7, pp. 69–77.
75. J. Hollo and L. Koch in ref. 7, pp. 63–68.
76. G. Anelli, R. Fiorentini, L. Massignan, and C. Caloppini, *J. Food Sci.* **42,** 1401 (1977).
77. A. L. Livingston, R. E. Knowles, J. Page, D. D. Kuzmicky, and G. O. Kohler, *J. Agric. Food Chem.* **20,** 277 (1972).
78. G. H. Hartman, Jr., W. R. Akeson, and M. A. Stahmann, *J. Agric. Food Chem.* **15,** 74 (1967).
79. M. Byers in ref. 7, pp. 95–114.
80. A. A. Woodham in ref. 7, pp. 115–130.
81. N. Singh in ref. 7, pp. 131–134.
82. *Fed. Reg.* **43**(136), 102.76 (July 14, 1978).
83. J. M. Harper, *Crit. Rev. Food Sci. Nutr.* **11,** 155 (1979).
84. J. E. Kinsella, *Crit. Rev. Food Sci. Nutr.* **10,** 147 (1978).
85. J. E. Kinsella, *J. Am. Oil Chem. Soc.* **56,** 242 (1979).
86. J. J. Rackis, D. J. Sessa, and D. N. Honig, *J. Am. Oil Chem. Soc.* **56,** 262 (1979).
87. W. H. Martinez, *J. Am. Oil Chem. Soc.* **56,** 280 (1979).
88. J. J. Spadaro and H. K. Gardner, Jr., *J. Am. Oil Chem. Soc.* **56,** 422 (1979).
89. E. W. Lusas, *J. Am. Oil Chem. Soc.* **56,** 425 (1979).
90. R. Ohlson and K. Anjou, *J. Am. Oil Chem. Soc.* **56,** 431 (1979).
91. F. Sosulski, *J. Am. Oil Chem. Soc.* **56,** 438 (1979).
92. R. Finch, *Crit. Res. Food Technol.* **1,** 519 (1970).
93. E. Levin, *Food Technol.* **13,** 132 (1959).
94. T. L. Mead, *Crit. Rev. Food Technol.* **2,** 1 (1972).
95. U.S. Pat. 3,249,442 (May 3, 1966), C. W. Keyes and W. W. Meinke.
96. W. W. Meinke and K. F. Mattil, *J. Food Sci.* **38,** 864 (1973).
97. V. D. Sidwell, B. R. Stillings, and G. M. Knobl, Jr., *Food Technol.* **24,** 876 (1970).
98. *Food Eng.* **49**(11), 66 (1977).
99. *CFR Title 21, Food and Drugs 172.385,* U.S. Government Printing Office, Washington, D.C., 1979.
100. C. Cheftel, M. Phern, D. I. C. Wang, and S. R. Tannenbaum, *J. Agric. Food Chem.* **19,** 155 (1971).
101. R. Miller and H. S. Groninger, Jr., *J. Food Sci.* **41,** 268 (1976).
102. S. Eriksen and I. S. Fagerson, *J. Food Sci.* **41,** 490 (1976).
103. M. Fujimaki, H. Kato, S. Arai, and M. Yamashita, *J. Appl. Bacteriol.* **34,** 119 (1971).
104. M. Yamashita, S. Arai, and M. Fujimaki, *J. Food Sci.* **41,** 1029 (1976).
105. M. Yamashita, S. Ari, S. J. Tsai, and M. Fujimaki, *J. Agric. Food Chem.* **19,** 1151 (1971).
106. S. W. Fox, J. W. Frankenfeld, D. Romsos, B. M. Robinson, and S. A. Miller in ref. 1, pp. 569–583.
107. C. M. Cater, W. W. Cravens, F. E. Horan, C. J. Lewis, K. F. Mattil, and L. D. Williams in ref. 1, pp. 278–301.
108. A. M. Pearson, *Bioscience* **26,** 249 (1976).
109. U.S. Pat. 2,682,466 (June 29, 1954), R. A. Boyer.
110. Protein Advisory Group of the United Nations, *PAG Bulletin, Policy Statement No. 4, Single-Cell Protein, 1970; Guidelines No. 6, Preclinical testing of novel sources of protein,* 1970; *No. 7, Human testing of supplementary food mixtures,* 1970; *No. 8, Protein-rich mixtures for use as supplementary foods,* 1971; *No. 12, Production of single cell protein for human consumption,* 1972; *No. 15, Nutritional and safety aspects of novel protein sources for animal feeding,* 1974; FAO/WHO/UNICEF, United Nations, New York.

JOHN H. LITCHFIELD
Battelle, Columbus Laboratories

FOOD STANDARDS. See Food processing; Regulatory agencies.

FOOD TOXICANTS, NATURALLY OCCURRING

Food products are fundamentally mixtures of chemical compounds and elements; certain of these compounds may produce human toxicities. Because human diets are normally composed of large numbers of different foods, only minute quantities of any specific dangerous material are consumed; thus dilution exerts a significant protective effect against acute intoxication. Toxicants are substances which, upon ingestion, produce changes in homeostasis that are threatening to the normal function of the organism. There are substantial differences in the toxicity thresholds of individuals to specific agents. Factors affecting toxicity include body weight, sex, age, general state of health, and the presence of potentiating or inhibitory substances. Some naturally occurring food toxicants are shown in Table 1 (1–11) (see also ref. 1).

Toxic Proteins, Peptides, Amides, and Amino Acids

Nitrogenous compounds are present in every living cell, and are the most frequently implicated natural toxicants in foods. Their mode of action is often obscure but they may be loosely grouped according to either gross manifestations or to specific structural characteristics. Accordingly, vitamin-destroying enzymes, hemaglutenins, enzyme inhibitors, and many hepatotoxins (many of which are carcinogens) are of protein, peptide, or amino acid composition.

Enzyme inhibitors of a protein nature are of significant concern because of their widespread occurrence. The most common of these affect the pancreatic enzymes, trypsin [*9002-07-7*] and chymotrypsin [*9004-07-3*], and are found in legumes, eg,

Table 1. Some Common Naturally Occurring Food Toxicants

Compound, structure no.	CAS Registry No.	Toxin classification	Typical food sources
aflatoxin B_1 (**1**)	[*1162-65-8*]	mycotoxin	moldy grains, nuts, oilseeds
amygdalin (**2**)	[*29883-15-6*]	cyanogenic glycoside	apricot pits, peach pits
avidin (chicken), mol wt ca 60,000	[*1405-69-2*]	forms insoluble complex with biotin	raw egg white
caffeic acid (**3**)	[*331-39-5*]	destroys thiamine	bracken fern
caffeine (**4**)	[*58-08-2*]	alkaloid, stimulant	coffee, tea, cola nuts
capsaicin (**5**)	[*404-86-4*]	amide	*Capsicum* peppers
goitrin (**6**)	[*500-12-9*]	goitrogen	cabbage, kale, onions, cress, cauliflower, broccoli, turnips
myristicin (**7**)	[*607-91-0*]	alkaloid, psychoactive	nutmeg, mace, carrots
oxalic acid (**8**)	[*144-62-7*]	reacts with calcium to reduce availability	rhubarb
phytic acid (**9**)	[*83-86-3*]	reacts with calcium to reduce availability	oats
solanine (**10**)	[*20562-02-1*]	glycoalkaloid, antiacetylcholinesterase	potatoes, tomatoes, apples, eggplant
thiaminase, mol wt ca $(75–100) \times 10^3$	[*9030-35-7*]	enzyme, inactivates thiamine	raw fish
tyramine (**11**)	[*51-67-2*]	vasoactive amine	cheeses, bananas, plantains, tomatoes, pineapple

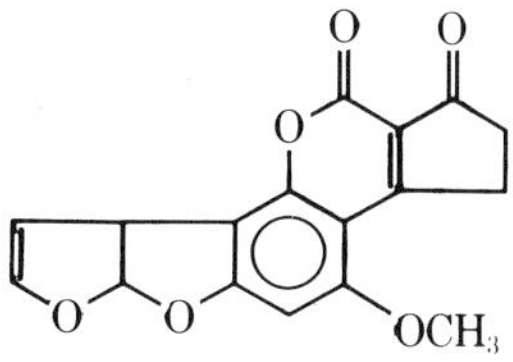

(1)

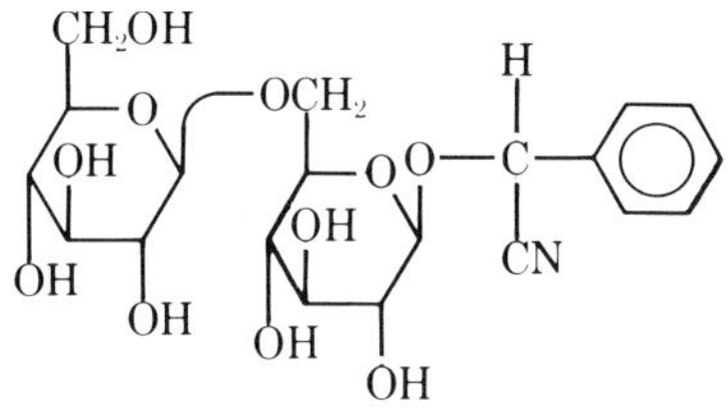

(2)

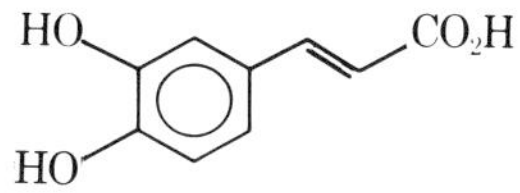

(3)

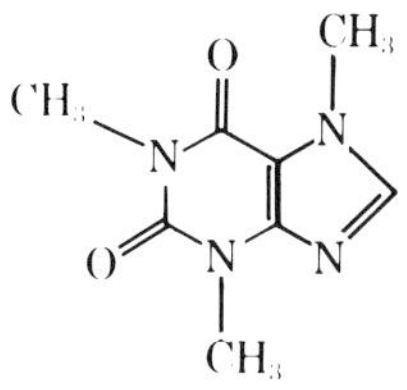

(4)

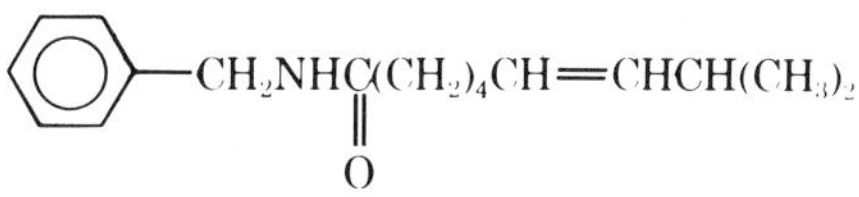

(5)

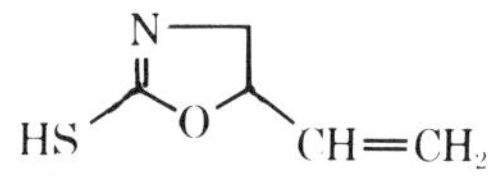

(6)

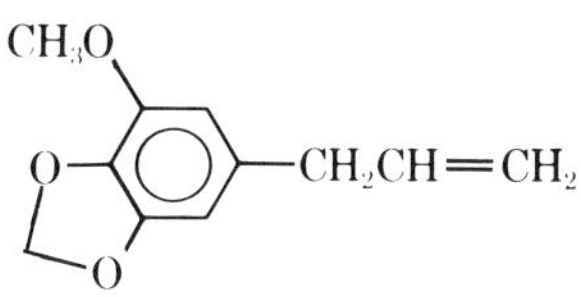

(7)

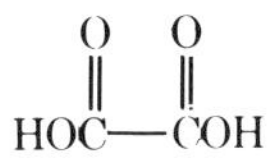

(8)

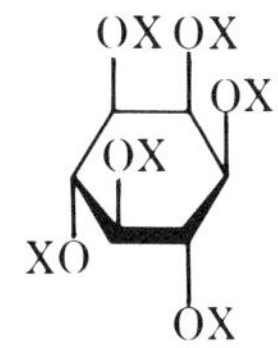

X = OP(O)OH$_2$

(9)

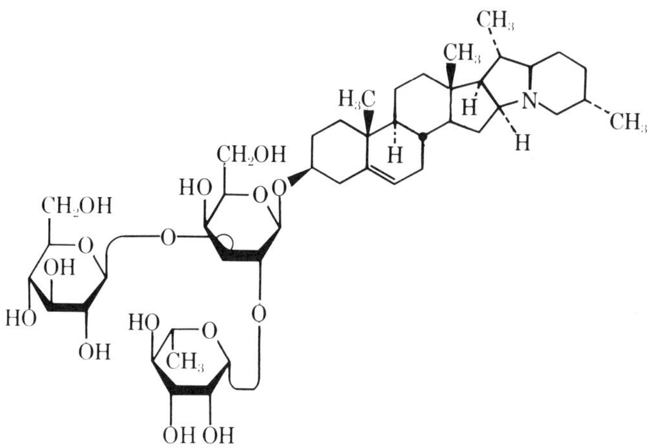

(10)

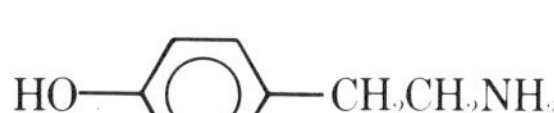

(11)

soybeans, peas, lima beans, kidney beans, and blackeyed peas, as well as in egg whites and potatoes (2–6). Although their mode of action is difficult to characterize—dozens of protein molecules of molecular weights of ca 15,000–30,000 are included in this group—they do share some functions: they form strong proteolysis-resistant complexes with their enzyme substrates, and consequently, block trypsin and chymotrypsin activity. They also enhance the production of amylase by the pancreas or, at least, stimulate its release by as much as 300%. Pancreatic hypertrophy, which has been hypothesized to produce increased secretion of cystine-rich enzymes, occurs and leads to a net loss of *S*-containing amino acids from the body. These inhibitors also seem to block the feedback inhibition of pancreatic enzyme production which is normally based on intestinal concentration. Thus, a cyclic and accelerating effect is produced whereby trypsin and chymotrypsin increase almost without limit, but their specific activities are blocked (7–8). Protein inhibitors often are active against a variety of enzymes. For example, many trypsin and chymotrypsin inhibitors are identical compounds, although each molecule may possess a separate and very distinct binding site for each enzyme (9).

Many protein inhibitors cause little nutritional difficulty because they are heat labile under ordinary processing and cooking procedures, and significant numbers are water soluble. Because many are found in highest concentrations in outer portions of plants, eg, wheat bran, normal peeling and milling operations give some protection.

Trypsin inhibitors occurring in soybean have been the most widely studied. The fact that both inhibitor and enzyme are proteinaceous makes this an example of one of the few cases of pure protein interaction. More than 30 separate trypsin inhibitors have been reported or suggested to be in legumes. The most extensively studied is the classic Kunitz inhibitor of soybeans. Since it is heat labile, it is extracted with 0.25 *N* H_2SO_4, adsorbed on bentonite clay, eluted with pyridine, and dialyzed at pH 4.65 at 5–10°C. Final crystallization occurs at pH 5.2 at 36°C (10). The alcohol-soluble, acetone-insoluble factor is perhaps the next most widely studied antitryptic compound (11). The use of DEAE–cellulose chromatography, zone electrophoresis, and ultracentrifugation has resulted in the separation and isolation of many new trypsin inhibitors (12).

Capsaicin (**5**), the amide responsible for the pungency of *Capsicum* peppers, is highly irritating at concentrations of 1:1,000,000 and highly toxic when taken in large doses (13). Reports of specific amino acid toxicities from normal eating patterns are rare, yet a few naturally occurring amino acids are sufficiently toxic in high concentrations to warrant discussion. Of the essential amino acids, methionine [*63-68-3*] seems the least tolerated, especially on restricted protein intakes, where toxic doses have been reported in the range of 10–46 g/d (14). Tyrosine [*60-18-4*] behaves similarly, and both result in hepatic and neurological lesions when fed to rats at toxic levels. Leucine [*61-90-5*] has been responsible for depressed NAD synthesis at levels of 14–20 g/d, whereas the toxicity of lysine [*56-87-1*], which causes abdominal cramps and diarrhea, occurs only with dose levels in excess of 64 g/d (14).

Of the nonessential amino acids, glycine [*56-40-6*] may be responsible for isolated toxicities at high dosages (>30 g/d) but it is nonetheless generally considered nontoxic. Recognition of the Chinese Restaurant Syndrome in 1968 stimulated increased interest in the toxicity of glutamic acid [*617-65-2*] (particularly as MSG). Numbness in the back and neck, weakness, palpitation, and other evidence of neurological block have been observed in susceptible individuals with intakes of 3–25 g/d of MSG. These effects seem specifically related to L-glutamic acid [*56-86-0*] intolerance, and interest remains high because many native proteins contain 30% or more of this amino acid (15). Although all amino acids except alanine [*56-41-7*] have been shown to be toxic, the probability of intoxication is very remote since humans seem able to tolerate all in excess of ten times the recommended intake (see Amino acids; Amino acids, monosodium glutamate).

All edible legumes contain proteinaceous substances that can agglutinate red blood cells from various animal species. Commonly referred to as hemagglutenins, these substances are also referred to as phytoagglutenins or lectins. Ranging in molecular weight from ca 30,000 to more than 125,000, these compounds are generally detoxified by heat, and most by fermentation. Since legume foods are normally processed by one of these methods, problems of toxicity are held to a minimum (see Food processing).

Acute toxicoses resulting from consumption of toxic mushrooms is infrequent, yet of increasing concern because of the practice of gathering fungi in the wild. The most serious of these toxicoses result from the Amanita family of mushrooms which contains several toxic peptides belonging to the amatoxin and phallotoxin groups. Other toxic fungi also contain an array of hepatotoxic peptides and hallucinogens, some having been utilized for their drug effects by central American indians for centuries. Most of these reactions are associated with progressive hepatic degeneration, and although only a few are fatal, their treatment is usually symptomatic.

Thiocyanates and Related *S*-Containing Compounds

Goitrogens are compounds that produce goiter by interferring with thyroxine synthesis in the thyroid gland. Food-borne goitrogens are often characterized by the presence of sulfur and most are thiocyanates or closely related compounds. Because of their widespread occurrence in *Cruciferae*, eg, cabbage, kale, onions, cress, broccoli, cauliflower, rutabaga, turnip and radish, goitrogens are among the most common and longest recognized substances of toxic nature in the human food supply (16). Thiocyanates actively compete with iodine for the same binding sites on the tyrosine molecule. Examples of goitrogenic compounds, other than thiocyanates, are derivatives of 2-3*H*-thiooxazolidinone (watercress and mustard), 5-phenyl-2-thiooxazolidinone (wintercress), and L-5-vinyl-2-thiooxazolidinone (**6**) (cabbage and kale). The precursor to the latter compound is structure (**12**) [*72918-22-0*] from which the active compound is formed by enzyme action. It has also been demonstrated that the quantity of the precursor found in plant materials is related to the available sulfur in the soil (17). The enzyme responsible for the conversion of the precursor into the active molecule is destroyed by the heat associated with normal cooking operations (18).

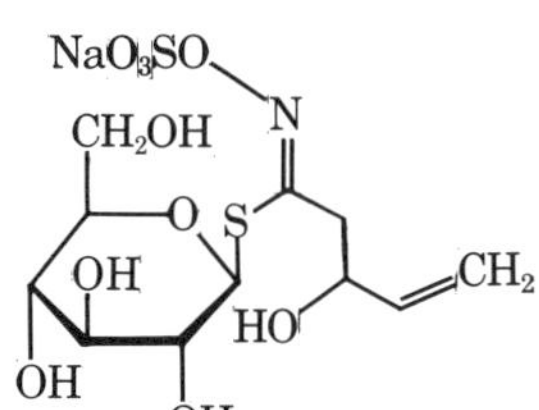

(**12**)

Oxalates, Phytates, and Other Chelates

Of nutrient chelates in the human diet, oxalates and phytates are the most common. Oxalic acid (**8**), found principally in spinach, rhubarb leaves, beet leaves, and mushrooms, is a primary chelator of calcium. The effects of oxalic acid in the diet may be twofold. First, it forms strong chelates with dietary Ca, thus rendering the complex unavailable for absorption and assimilation. Secondly, absorbed oxalic acid causes assimilated Ca to be precipitated as insoluble salts that accumulate in the renal glomeruli and contribute to the formation of renal calculi (19).

Phytic acid (**9**), although restricted to a more narrow range of food products (mainly grain products), complexes a broader spectrum of minerals than does oxalic acid. Decreased availability of P is probably the most widely recognized result of excessive intakes of phytic acid, yet Ca, Cu, Zn, Fe, and Mn are also complexed and rendered unavailable by this compound (20–21).

In addition to these purely nutritional effects, dietary chelates—especially oxalic acid—may produce more classic toxicity when ingested in excessive quantities. Complications, eg, corrosion and irritation in the mouth and intestinal tract, gastric hemorrhage, hematuria, and even convulsions, may occur in extreme cases. Although there is no doubt that intakes of 5 g or more of oxalic acid could be fatal to humans, the probability of such ingestion is remote: the consumption of some 4 kg of rhubarb leaves or roughly twice that quantity of spinach would be required. Therefore, concern regarding the possible hazardous effects of ingesting plant foods containing oxalate seems unwarranted. High intakes of both calcium and vitamin D help to offset the deleterious effects of oxalates (see Vitamins).

Vasoactive and Psychoactive Amines and Alkaloids

Most compounds producing hypertensive episodes are classified as amines and are found in greatest concentration in banana, plantain, tomato, avocado, pineapple, broad beans, and various cheeses. Amines that are vasoactive include dopamine [*51-61-6*], tyramine (**11**), histamine [*51-45-6*], tryptamine [*61-54-1*], noradrenaline [*51-41-2*], and dihydroxyphenylalanine [*59-92-7*] (DOPA).

Patients receiving monoamine oxidase inhibitors (MAOI) as antidepressant therapy are especially subject to the hypertensive effects of vasoactive amines (22). These dietary amines have also been implicated as causative agents in migraine (23). Other naturally occurring alkaloids have been recognized for centuries as possessing neurological stimulant and depressant properties (see Psychopharmacological agents).

Caffeine, a xanthine derivative, is perhaps the best known of the natural stimulants and is found in coffee (qv) beans, tea (qv) leaves, and cola nuts. Caffeine produces stimulatory effects by facilitating mental and muscular effort and diminishes drowsiness and fatigue. Although caffeine content in beverages is quite variable, it is generally accepted that in excess of 50 cups of coffee per day would be necessary to approach acutely toxic doses, and the fatal dose in man has been estimated at about 10 g (equivalent to ca 150–200 cups of coffee) in a 24-h period. Toxic symptoms involve insomnia, restlessness, mild delirium, muscular tremors, tachycardia, and increased respiration (24).

Depressant symptoms, which include burning abdominal pain, decreased excitability, convulsions, nausea, and coma, become the general syndrome for all oral alkaloid poisoning. Discorine [*3329-91-7*], an β,γ-unsaturated lactone found in yams, is an alkaloid (**13**) with the empirical formula $C_{13}H_{19}O_2N$. A similar alkaloid, $C_{13}H_{21}O_2N$, has been isolated from Nigerian yams, thus establishing the discorine-type alkaloids as representative of the *Dioscorea* genus of plants (25). Seeds of *Senecio* contain alkaloids belonging to the pyrrolizidine group and, in addition to producing characteristic signs of alkaloid toxicity, they are suspected of being hepatic carcinogens. Ergotism is the disease caused by intoxication by one of at least twelve alkaloid derivatives of lysergic acid produced by a parasitic fungus (ergot) on grains. Myristicin (**7**), found in both nutmeg and mace, is a psychoactive agent which may be fatal in infants who consume as little as two whole nutmegs. Its toxicity resembles alcohol intoxication.

Several glycoalkaloids present in food are of toxicological interest. Solanine (**10**), found in potatoes, tomatoes, apples, eggplant, and sugar beets, has been responsible

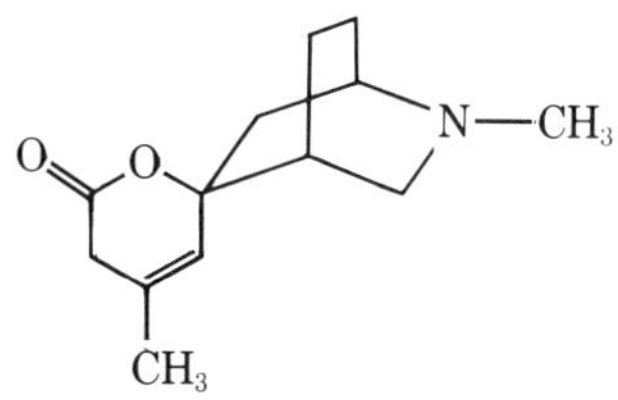

(**13**)

for several cases of moderate to severe poisoning. Solanine may act as a cholinesterase inhibitor (qv), and toxic doses are probably ca 200 mg. Market potatoes contain 2–13 mg solanine/100 g fresh weight. Since the greatest proportion of solanine occurs in the skin of potatoes, the practice of peeling diminishes the intake of this alkaloid. Also, it is poorly absorbed, rapidly excreted, and readily hydrolyzed to the less toxic soladinine [*80-78-4*] in the intestinal tract (26).

Antinutrients

Any substance that destroys, inactivates, or in other ways renders unavailable an essential dietary constituent can be termed an antinutrient. The most widely studied are antivitamins. Their presence in certain foods means that merely assuring an adequate intake of a vitamin is no guarantee that a deficiency state cannot exist physiologically. The enzyme thiaminase acts by either specific splitting of the thiamine molecule or nonspecific hydrolysis (27). Niacin inhibitors, acting through nicotinamide mononucleotide (NMN) depression in erythrocytes, have been studied in corn and millet; and a biotin antagonist, avidin, has long been recognized in raw egg white. Avidin apparently forms a stable complex with biotin, thus rendering the vitamin unavailable for metabolic reaction (28). Linatine [*10139-06-7*], found in flaxseed , is the only pyridoxine antagonist known and seems to function by the formation of a stable complex (29). Yeast and pea seedlings contain specific pantothenic acid antagonists, although their structure and mode of action remain unexplained (30). Riboflavin antagonism is also rather rare (found only in the Akee plum of Jamaica) but is of interest because it can be fatal (31).

Many plant substances possess antivitamin D activity but their mode of action and, in most cases, their identity remain unknown. Rachitogenic factors have been observed in yeast. Because of the metabolic interrelationships that exist between vitamin D, Ca, and P, it is sometimes difficult to differentiate between chelators of mineral elements and true antivitamins; such was the case with one early reported vitamin D antagonist in oats that was later identified as phytic acid (32).

The antagonisms that exist between unsaturated fatty acids, and carotene and vitamin E are complicated and largely undefined. Linoleic acid acts as an antivitamin to *dl-α*-tocopherol [*59-02-9, 1406-18-9, 10191-41-0*] (vitamin E) by reducing availability through direct intestinal destruction. Various lipoxidases destroy carotenes and vitamin A (33). Dicoumarol [*66-76-2*] (3,3′-methylenebis(4-hydroxycoumarin)) is a true antimetabolite of vitamin K but seems to occur only in clover and related materials that are used primarily as animal feeds (34).

At various times, antivitamin factors specific to vitamin B_{12}, folic acid, and choline have been reported. However, it is uncertain whether these are true antimetabolites or—like some reported vitamin D inhibitors—may result from metabolic interrelationships with other dietary constituents (see Vitamins).

Vitamin Toxicity

Reported cases of vitamin toxicity owing to overdose—once a phenomenon largely of the experimental laboratory—are becoming more widespread. Increased over-the-counter availability of supplemental vitamins and the often unwarranted advice of health food dealers (the majority of whom have no formal nutritional training) for indiscriminate supplementation have contributed to this observation. The common misconception that if a little is good, a lot is better, has compounded toxicological problems with the vitamins. Because fat-soluble vitamins tend to accumulate in the body with relatively inactive mechanism for excretion, they cause greater toxicological difficulties than do water-soluble vitamins.

Infants may be sensitive to doses of vitamin A [*11103-57-4*] in the range of 75,000–200,000 IU (22.5–60 mg), although the toxic dose in adults is probably 2–5 million IU (0.6–1.5 g). Intakes in this range from normal food supplies without oral supplements are simply beyond imagination (35). Vitamin D [*1406-16-2*] toxicity is much more difficult to substantiate clinically, since the body can synthesize active forms of the vitamin in the skin upon irradiation of 7-dehydrocholesterol. Toxic symptoms are relatively nonspecific, and dangerous doses seem to lie in the range of 1000–3000 IU/kg body wt (25–75 μg/kg body wt) (36). Cases of toxicity of both vitamins E and K [*12001-79-5*] have been reported but under ordinary circumstances they are considered relatively innocuous (37).

Of the water-soluble vitamins, intakes of nicotinic acid [*59-67-6*] on the order of ten to thirty times the recommended daily allowance (RDA) have been shown to cause flushing, headache, nausea, and moderate lowering of serum cholesterol with concurrent increases in serum glucose. Toxic levels of folic acid [*59-30-3*] are ca 20 mg/d in infants, and probably approach 400 mg/d in adults. The body seems able to tolerate very large intakes of ascorbic acid [*50-81-7*] (vitamin C) without ill effect but levels in excess of 9 g/d have been reported to cause increases in urinary oxalic acid excretion. Urinary and blood uric acid also rise as a result of high intakes of ascorbic acid, and these factors may increase the tendency for formation of kidney or bladder stones. All other water-soluble vitamins possess an even wider margin of safety, and present no practical problem (38).

Essential Minerals and Heavy Trace Elements

Toxic ingestion of essential minerals from naturally occurring foods is almost beyond comprehension. Ingestion of at least ten times normal levels would be required to approach toxic proportions of most essential minerals (see Mineral nutrients). Perhaps the only exceptions would occur in cases such as Mo [*7439-98-7*], Se [*7782-49-2*], and Cu [*7440-50-8*], where levels can reach toxic quantities when plant foods are grown on soils unusually high in these minerals. These are rare occurrences, however, and long experience with livestock as well as man has provided good information on traditional geographic areas of high mineral content.

Cases involving human toxicity from heavy trace elements, such as Pb [*7439-92-1*], Hg [*7439-97-6*], As [*7440-38-2*], and Cd [*7440-43-9*], are much more common but almost exclusively traced to accidental contamination rather than true natural occurrences. Attention was called to mercury poisoning with the report of Minimata disease in Japan over a decade ago. Consumption of seafoods taken from waters where industrial pollution had raised mercury levels was found responsible. Methylmercury acetate [*108-07-6*] seems to be the toxic compound of most interest to man, and is formed primarily from microbial action. Infrequent problems with lead and cadmiun have been traced to steels and equipment containing these elements coming in contact with foods during handling or processing. Naturally high levels of arsenic have been reported in some crustaceans and fishes, yet food problems to date have been restricted to meats taken from livestock treated with arsenic-containing antibiotics (39).

Heavy metals are of importance in human toxicity because the body possesses only inactive mechanisms for their excretion; thus chronic, low-level intakes can accumulate to toxic proportions. Treatment has likewise been relatively unsuccessful (except for symptomatic relief), inasmuch as no effective means has been discovered to increase their excretion.

Cyanogenic Glycosides

Complex glycosides, which upon hydrolysis yield hydrocyanic acid, are found commonly among plant materials (see Cyanohydrins). The toxicity of this class of compounds is directly related to their liberation of HCN upon digestive hydrolysis. They are found in the bitter almond, pits of stone fruits, sorghum, and lima beans.

Recent, unsubstantiated claims as to the efficacy of a component of apricot pits (sometimes referred to as Laetrile or "vitamin B_{17}") for cancer treatment has focused much attention on these compounds. Amygdalin [*29883-15-6*] (**2**), the cyanogenic glycoside of apricot pits, has never been associated with a medically verified cancer cure, and no trained nutritionist recognizes the compound as a vitamin. Indeed, many cases of poisoning, and even death have been ascribed to its use by the unscrupulous. Approximately 50 apricot pits contain a fatal dose of amygdalin (40).

The cyanogenic glycosides, phaseolunatin [*554-35-8*] and vicianin [*155-57-7*], have been isolated from lima beans and vetch, respectively. Several studies have reported that heating (cooking) acts to decrease the quantity of HCN liberated by these compounds upon enzymatic hydrolysis.

Nitrates, Nitrites, and Nitrosamines

The carcinogenicity of nitrosamines has created widespread concern over the safety of food products that are significant sources of nitrates and nitrites. Nitrosamines are readily formed by reaction of secondary amines with nitrites at acid pH, conditions which may occur in the gastrointestinal tract (see *N*-Nitrosamines).

Nitrates are found in fairly high concentrations in beets, spinach, kale, collards, eggplant, celery, and lettuce. Additionally, nitrates and nitrites are commonly used in the curing solutions of bacon, ham, and other cured meats. In cured meats, nitrates and nitrites control the growth of microorganisms, particularly *Clostridium botulinum*, and also serve as color preservatives.

Although the potentially carcinogenic nitrosamines may be present in foods, particularly cured meats, their occurrence has been reported only infrequently. The observation that ascorbic acid blocks the formation of carcinogenic *N*-nitroso compounds has led to suggested incorporation of the compound into pork-curing operations.

Nitrosamines vary widely in molecular structures, and some 100 or more may be capable of producing pathological effects. Most are specific hepatotoxins, producing hepatic parenchymal cell necrosis, and may also act in a synergistic capacity with other carcinogens, notably the polycyclic hydrocarbons (41).

Sodium Chloride

Sodium chloride [*7647-14-5*] is an essential dietary component. It is necessary for proper acid–base balance and for electrolyte transfer between the intra- and extracellular spaces. The adult human requirement for NaCl probably ranges between 5–10 g/d. The normal diet provides something in excess of 15 g/d NaCl, and adding salt during cooking or at the table increases this intake.

The fact that excessive intake of NaCl contributes to increased fluid retention has led to the conclusion that there may be a relationship between NaCl intake and hypertension. Few topics have caused such debate among nutritionists and physicians, with numerous reports in the literature both supporting and refuting such claims. However, few nutritionists can find any real purpose in consuming quantities of NaCl far in excess of requirements. There is no doubt that the relationship between hypertension and NaCl intake is of significance when compounded by obesity, or in susceptible individuals. This arises from the fact that the edema that results from excessive salt intake, especially in relatively inactive persons, is a major contributing factor in elevated blood pressure. The Senate Select Committee on Nutrition and Human Needs has recommended the reduction of sodium chloride intake in average American diets (42).

Mycotoxins

The condition produced by the consumption of moldy foods containing toxic material is referred to as mycotoxicosis. Molds and fungi fall into this category, and several derive their toxicity from the production of oxalic acid, although the majority of mycotoxins are much more complex (see also Fungicides).

Many parasitic fungi, such as *Fusarium*, *Phytophthora*, *Pythium*, *Helminthosporium*, *Periconia*, *Rhizoctonia*, *Verticullium*, and others have been shown to produce toxins. However, the toxins of *Aspergillus* and *Penicillium* perhaps have the greatest potency against humans. Polycyclic peptides are among the most dangerous toxic compounds produced by fungi. A number of mycotoxins contain the diketopiperazine nucleus, and all are composed of amino acid units. Islanditoxin [*10089-09-5*] **14**), produced by *Penicillium islandicum* on rice and other grains, is a Cl-containing bicyclic peptide which includes α-aminobutyric acid [*80-60-4*], β-phenyl-β-aminopropionic acid [*13921-90-9*], serine [*56-45-1*], and dichloroproline [*60548-67-6*]. Several anthraquinone derivatives, produced by members of the *Penicillium* group, are also of toxicological interest. Examples are skyrin [*602-06-2*], luteoskyrin [*21884-44-6*], and iridaskyrin [*568-42-3*]. Various mycotoxins also include alkaloids and xanthones, and coumarin and terpene derivatives. Some are thermolabile, others thermostable. The latter are of greatest importance in food products since routine processing and cooking do little to reduce their potential toxicity. Toxic metabolites are usually produced only after mycelia have become established in the substrate. For this reason, foods are usually free of toxins until such mycelial development occurs (43). The mode of action of mycotoxins varies according to their chemical nature. Some act by blocking cell wall synthesis, by disrupting membrane integrity, or by decoupling oxidative phosphorylation and inhibiting respiratory pathways. Others are chelates that inhibit synthesis of proteins or DNA.

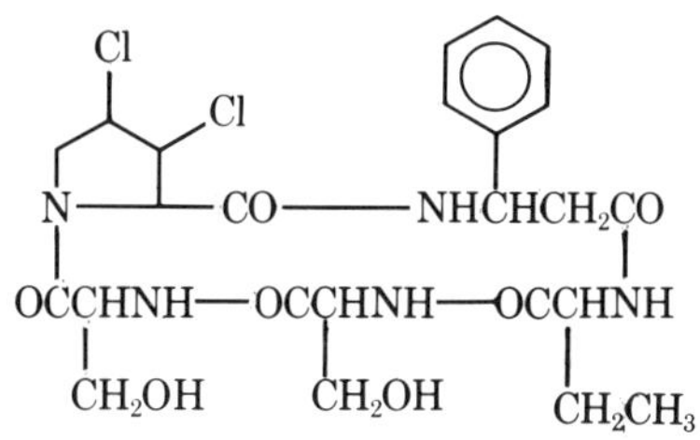

(14)

Aflatoxins, first noted as metabolites from *Aspergillus flavus,* are also produced by *A. parasiticus, A. niger, A. ruber, A. ostianus* Wehmer, *A. wentii, A. vesicolor, Penicillium puberculum, P. citrium, P. variable,* and *P. frequentans,* as well as species of *Rhizopus. Aspergillus flavus* is a common contaminant of a multitude of foods, including soybeans, groundnuts, cassava, peas, pears, cacao pods, Brazil nuts, pecans, millet, corn, and wheat. Because of this widespread occurrence on plant materials the rare isolation of aflatoxins in food products is not surprising (44).

The term aflatoxin is not compound-specific. Different toxic compounds have been designated as B_1 (1) and G_1 [*1165-39-5*], and B_2 [*7220-81-7*] and G_2 [*7241-98-7*] (dihydro derivatives of B_1 and G_1), as well as M_1 [*6795-23-9*], M_2 [*6885-57-0*], P_1 [*32215-02-4*], GM_1 [*23532-00-5*], and B_3 [*23315-33-5*]. Aflatoxin B_1 is of greatest occurrence in nature, followed by G_1, B_2, and G_2. Water content of medium, temperature, pH, and light are among those environmental factors that affect aflatoxin production.

Intoxication by aflatoxins is referred to as aflatoxicosis. Edema and necrosis of hepatic and renal tissues seem characteristic of aflatoxicosis, and hemorrhagic enteritis accompanied by nervous symptoms often appear in experimental animals. That they are specific hepatic carcinogens seems indisputable. It has been suggested that the most probable mode of action of aflatoxins involve an interaction with DNA and inhibition of the polymerases responsible for DNA and RNA synthesis (45).

Cereals, bakery products, and oilseed products are those foods that are of highest risk in regard to aflatoxin contamination. Specific links between liver carcinoma in man and aflatoxin ingestion have not been substantiated fully, and it is probable that only minimal dangers are to be encountered from commercially processed foods (see Food processing).

Many procedures have been studied for detoxification of aflatoxins, including heat, and treatment with ammonia, methylamine or sodium hydroxide coupled with extraction from an acetone–hexane–water solvent system. Since in detoxification it is important to free the toxin from cellular constitutents to which it is bound, a stabilization of proteins using a tanning compound such as acetaldehyde or glutaraldehyde may be an alternative solution to the problem (46).

Seafood Toxins

Virtually scores of fish and shellfish species have been reported to have toxic manifestations. However, since most toxicities have been shown to be microbiological in origin, they do not truly belong in a discussion of naturally occurring toxicants. A few that are natural components of seafoods, however, may be worthy of brief mention.

Several species of the moray eel (*Gymnothorax*) have caused toxic reactions, especially in Japan. The toxic principle appears to be proteinaceous, and is found predominately in the blood but it may occur also in the flesh as well. Its exact structure remains somewhat uncertain.

Pufferfish toxin, isolated from a dozen or more species, has been identified as having the empirical formula $C_{11}H_{17}N_3O_8$ but the exact structure is not well-established, nor is it certain that the same structure is universally responsible for poisoning, although this is assumed to be the case. The so-called paralytic shellfish poisoning reported in many areas of the world has now been recognized to have a microbiological etiology, and is thus more accurately a contamination rather than a natural toxicosis. The paralytic effects of the poisoning begin as a tingling sensation in the lips, tongue, and extremities, and gradually progress into nausea and convulsions. Japanese statistics indicate a mortality rate approaching 65% (47).

The liver of sharks, and other oily fishes sometimes accumulate toxic levels of vitamin A, and cases of acute poisoning have been reported both among Eskimos and the Japanese.

Other Toxins of Current Interest

A few additional classes of natural toxicants are worthy of mention because of their implications as mutagens and carcinogens. Some of these are found widely distributed in the plant kingdom as promutagens and procarcinogens. For further detailed information, see references 48 and 49.

Legislation and Regulatory Considerations

There exists little specific legislation dealing with natural toxicants in foods. The main exception is the Delaney clause to the *1958 Amendment of the Federal Food, Drug, and Cosmetic Act* (see Food additives). This stipulates that no substance that has been shown to be carcinogenic to either man or animals may be added to the food supply. Accordingly, those foods that contain carcinogens are subject to the Delaney clause, although specific interpretation of the law is far from clear (see Regulatory agencies). Maximum tolerances of heavy metals, such as Pb and Hg, have been established by FDA at 0.5 ppm in the food product. Also, maximum permissible levels are undergoing revision for nitrates and nitrites in curing solutions, and final limits can be expected to be well below current usage. For aflatoxins, there is presently a zero tolerance in effect (based on the Delaney clause), and screening is generally on a qualitative basis. With these exceptions, natural toxicants in food products are generally not treated by specific food legislation (50).

Naturally occurring toxicants have been reviewed in greater detail by several authors, see references 51–55.

BIBLIOGRAPHY

1. S. A. Miller, *Ann. N.Y. Acad. Sci.* **300,** 397 (1977).
2. Y. V. Wu and H. A. Scheraga, *Biochemistry* **1,** 698 (1962).
3. Y. Birk, A. Gertler, and S. Khalef, *Biochem. J.* **87,** 281 (1963).
4. J. J. Rakis and co-workers, *Arch. Biochem. Biophys.* **98,** 471 (1962).
5. M. Yamamoto and T. Ikenaka, *J Biochem.* **62,** 141 (1967).
6. V. Frattali and R. F. Steiner, *Biochemistry,* **7,** 521 (1968).
7. R. Huber and co-workers, *Naturwissenschaften* **57,** 389 (1970).
8. F. F. Buck and co-workers, *Arch. Biochem. Biophys.* **98,** 528 (1962).
9. F. H. Smith and A. J. Clawson, *J. Am. Oil Chem. Soc.* **47,** 443 (1970).
10. M. Kunitz, *J. Gen. Physiol.* **30,** 311 (1947).
11. D. E. Bowman, *Proc. Soc. Exp. Biol. Med.,* **63,** 547 (1946); Y. Birk, *Biochem. Biophys. Acta* **54,** 378 (1961).
12. V. Fratali, *J. Biol. Chem.* **244,** 274 (1969).
13. C. C. Toh, T. S. Lee, and A. K. Kiang, *Br. J. Pharmacol.* **10,** 175 (1955).
14. A. E. Harper in *Toxicants Occurring Naturally in Foods,* NAS/NRC, Washington, D.C., 1966, p. 221.
15. R. H. Kwok, *N. Engl. J. Med.* **278,** 796 (1968).
16. E. Josefson, *Phytochemistry* **6,** 1617 (1967); R. Gmelen, *Präparative Pharmazie* **5**(2), 17 (1969); **5**(3), 33 (1969).
17. E. P. Lichtenstein and co-workers, *J. Agric. Food Chem.* **10,** 30 (1962).
18. P. Friis and A. Kjaer, *Acta Chem. Scand.* **20,** 698 (1966).
19. A. Hodgkinson and P. M. Zarembski, *Calc. Tiss. Res.* **2,** 115 (1968).
20. R. E. Ferrel and co-workers, *Cereal Sci.* **14,** 110 (1969).
21. J. G. Reinhold and co-workers, *Lancet i,* 283 (1973).
22. N. Sapeika, *Trans. R. Soc. S. Afr.* **41,** 1 (1974).
23. B. Blackwell and co-workers, *Br. J. Psychol.* **113,** 349 (1967).
24. L. S. Goodwin and A. Gilman, *The Pharmacological Basis of Therapeutics,* 3rd ed., Macmillan, New York, 1965, 354.
25. C. W. L. Bevan and co-workers, *Nature (London)* **177,** 935 (1953).
26. K. Nishie and co-workers, *Toxicol. Appl. Pharmacol.* **19,** 81 (1971).
27. R. G. Green, *Minn. Wildl. Dis. Inv.* **3,** 83 (1937).
28. M. D. Melamad and N. M. Green, *Biochem. J.* **89,** 591 (1963).
29. N. D. Smashevskii, *Biol. Nauki (Moscow)* (9), 80 (1968); *Chem. Abstr.* **70,** 1157 (1969).
30. F. H. Kratzer, *Poult. Sci.* **26,** 90 (1947).
31. H. C. Fox and D. S. Miller, *Nature (London)* **186,** 561 (1960).
32. D. C. Harrison and E. Mellanby, *Biochem. J.* **33,** 1660 (1939).
33. H. Dam, *Pharmacol. Rev.* **9,** 1 (1957).
34. S. L. Balloun and E. L. Johnson, *Arch. Biochem. Biophys.* **42,** 355 (1953).
35. G. Bartolozzi, *Rev. Clin. Pediat.* **80,** 231 (1967).
36. D. Rosenberg and co-workers, *Pediatrics* **80,** 231 (1967).
37. L. Piedrabuena, *Nutr. Abstr. Rev.* **40,** 48 (1970).
38. M. P. Lamdew, *N. Engl. J. Med.* **284,** 336 (1970).
39. G. K. Murphy and co-workers, *Envir. Sci. Technol.* **5,** 436 (1971).
40. V. Herbert, *Am. J. Clin. Nutr.* **32,** 1121 (1979).
41. T. N. Blumer and co-eds, *Effects of Nitrates and Nitrites in Food as Related to Health,* Institute of Nutrition, University of North Carolina, 1973.
42. D. M. Hegsted, *Am. J. Clin. Nutr.* **31,** 1504 (1978).

43. R. B. Pringle and R. P. Scheefer, *Ann. Rev. Phytopath.* **II,** 133 (1964).
44. J. J. Langone and H. Van Vunakis, *J. Natl. Cancer Inst.* **56,** 591 (1976).
45. J. I. Clifford and K. R. Rees, *Nature (London)* **209,** 312 (1966).
46. M. S. Legator in L. A. Goldblatt, ed., *Aflatoxin,* Academic Press, Inc., New York, 1969, p. 107.
47. T. Kawabata, *Fish as Food,* Vol. 2, Academic Press, Inc., New York, 1962, p. 467.
48. N. R. Farnsworth and co-workers, *Cancer Treat. Rep.* **60,** 1171 (1976).
49. M. Nagoa and T. Sugimura, *Annu. Rev. Genet.* **12,** 117 (1978).
50. P. Handler in *How Safe is Safe? The Design of Policy on Drugs and Food Additives,* NAS, Washington, D.C., 1974.
51. F. H. Hoskins in C. L. Galli, R. Paolatti, and G. Vettorazzi, eds., *Chemical Toxicology of Foods,* Elsevier-North Holland, Amsterdam, The Netherlands, 1978, p. 233.
52. Food Protection Committee, *Toxicants Occurring Naturally from Foods,* 2nd ed., NAS/NRC, Washington, D.C., 1973.
53. C. Moreau, *Moulds, Toxins, and Food,* 2nd ed., John Wiley & Sons, Inc., New York, 1979.
54. G. N. Wogan, *J. Am. Diet. Assoc.* **49,** 95 (1966).
55. W. H. Lewis and M. P. F. Elvin-Lewis, *Medical Botany,* John Wiley & Sons, Inc., New York, 1977.

FRED H. HOSKINS
Louisiana State University

FORENSIC CHEMISTRY

Forensic chemistry is that field of knowledge which involves the application of the principles of chemistry and related sciences to the examination of physical evidence collected at the scene of a crime, and the presentation of the results of that examination in a court of law or public forum by an expert witness, a forensic chemist, who has either performed or supervised the tests (see Analytical methods).

The forensic laboratory is a facility of applied analytical chemistry with increasing emphasis on trace analysis employing the most modern techniques (see Trace and residue analysis). The reliability of scientific results often exceeds in value the contribution of eyewitness testimony, especially when trials occur many months after a crime has been committed and recollection of events by an eyewitness fades. Increasing reliance on scientific proof is a major contribution to the cause of justice.

The task of the forensic chemist is to analyze physical evidence found at the scene of a crime. A paint fragment, a strand of fabric, fire or bomb residues, blood stains, drug samples, and strands of hair are examples of the variety of evidence with which the forensic chemist is confronted. The objective is either to provide information that may link a suspect with the evidence collected at the scene of a crime or to answer questions about the crime. Was a fire started accidentally or was it arson? If the latter, what was the accelerant? Was a death natural or was it suicide, homicide, or an accident? Is it possible to obtain evidence that physically characterizes the criminal? If a weapon was used, can it be identified?

This examination is but one step in the process of solving a crime. Before the evidence is introduced in court, responsible officials must testify, outlining the custody from its collection and its transportation to the laboratory, its storage and the disposition of unused material. Any break in this chain of custody is grounds for denial of admission of evidence.

The court testimony of a forensic chemist may or may not be accepted. An expert's familiarity with the current state of the specialty may be questioned. Any failure to respond effectively creates an element of uncertainty which could influence the jury. Furthermore, the opinion of the expert may be questioned, suggesting that it is speculation rather than a conclusion dictated by the evidence. Occasionally, an attorney may construct a hypothetical situation and ask the expert for an opinion. The reply may seem to conflict with prior testimony, creating an opportunity for the attorney to weaken the expert's opinion. In addition, the use of inappropriate language may lead to reversal of a conviction. For example, in *State v. Holt* (1), an attorney asked an expert if two hair specimens were similar. The response was that the samples were "similar" and "likely to be" from the same source. The Ohio Supreme Court ruled that the word "likely" was weaker than "reasonably certain" and, hence, failed to reach the level of certainty demanded by law. The conviction was overturned.

New developments in chemistry or new methods of analysis are not accepted automatically by the courts. This is the essence of the landmark case of *Frye v. United States* (2) which established the guidelines for the acceptance of evidence based upon new techniques. The principle envisages a gray area between the experimental and demonstrable stages. To be accepted, the new development must be predicated upon concepts generally accepted and employed in the particular field by the scientific community. However, in a recent decision (3) a chemical test not generally known but founded on principles accepted by specialists was sufficient grounds for accepting the results of the chemical analysis.

The process by which a new method of analysis comes to be accepted in court is illustrated by the case of neutron activation analysis (naa) (see Analytical methods). In 1964 a report (4) was published dealing with the detection of gunshot residues by the use of naa. Following this publication, testimony citing naa was presented in a number of trials in federal courts. It was not until 1970, however, that an appellate court (5) declared that naa met the general acceptance principle of the Frye decision. Before this landmark case, trial and appellate court decisions were equivocal with respect to the general admissibility of naa. Since then, such analyses have been accepted in hundreds of cases as evidence.

The scope of physical evidence with which the forensic chemist deals is too broad and varied for complete identification in this discussion which is organized according to general classes of crime alphabetically rather than in order of incidence.

Arson

Arson, the malicious burning of a building or property, has been recognized as a crime for thousands of years. In recent years the incidence of arson in the United States has reached a level of national concern. In 1950 there were 5600 fires of suspicious origin amounting to a loss of 15 million dollars. However, 25 years later there were 144,100 fires of suspicious origin amounting to a loss of $633,900,000 (6). This trend is continuing.

Among the numerous motives for arson, insurance fraud is increasingly the most common cause (7). To some extent, the increasing incidence of fires identified as suspicious results from the greater knowledge and care with which the residues are examined. Several clues suggest arson. Fires that spread with exceptional rapidity for the type of structure in which they occur imply that an accelerant was used. A liquid accelerant such as gasoline, kerosene, or paint thinner may simply be poured, causing sites of contact to burn more rapidly than the adjacent areas. Surprising as it may seem, the charred residues contain detectable amounts of the liquid that can be identified by various analytical methods (8–10). Gas–liquid chromatography is by far the method preferred by forensic chemists. Another valuable method (11) is ir spectroscopy (see Analytical methods).

Crimes of Violence

For purposes of this discussion, crimes of violence include assault and robbery which can cause injury and even death. In the latter case, the crime becomes homicide, requiring the services of a forensic pathologist.

The crime-scene investigator is responsible for the integrity of the site, the custody of the evidence, and the search for suspects. It is a principle of crime investigation that every contact leaves a trace. Hence, every possible piece of evidence that could relate a suspect to the crime scene becomes important. Hair samples, fiber samples, body fluids, soils, tire prints, shoe prints, glass fragments, paint fragments, virtually any physical specimen can yield valuable information. Gunshot residues deposited on a hand that discharges a pistol consist primarily of traces of barium, antimony, and lead. These are carried by the gaseous decomposition products of the primer in the bullet. The amounts, on the order of micrograms, can be detected only by very sensitive analytical techniques such as neutron activation analysis and atomic absorption spectroscopy.

Neutron activation analysis was first considered in 1964 as a method of detection of gunshot residues (12). Though not sensitive to lead, barium and antimony can be determined in parts per million or less. Gunshot residues are readily removed from the hands by swabbing with 5% nitric acid (13–14). A more recent method is the application of a solution of cellulose acetate in acetone to the hands of the suspect. As the solvent evaporates, an easily removable film is deposited, containing the gunshot residues (15). The film is then exposed to a neutron flux in the core of a nuclear reactor. The barium and antimony become radioactive and emit gamma rays of different energies which can be measured by an appropriate detector system. The intensity of the radiation is proportional to the amount present.

Table 1 gives a representative set of values for gunshot residues from a suspect's and a victim's hands. The data indicate that John Doe fired the gun in his right hand

Table 1. Gunshot Residues

Location	Barium, μg	Antimony, μg
John Doe		
right back	2.59	1.76
right palm	10.40	8.65
left back	0.21	0.07
left palm	5.30	3.15
Jane Doe	0.00	0.00

with his left hand near to it. The absence of barium and antimony from Jane Doe's hands precludes suicide. Washing or normal use of the hands after a gun discharge can remove virtually all traces of the residues. For this reason, prompt collection of the sample is important.

Though naa is an excellent procedure for the detection of trace elements, the need for a nuclear reactor and expensive counting equipment limits its use.

Recently, flameless atomic absorption spectroscopy (faas) (see Analytical methods) has practically achieved the same sensitivity as naa (16). In addition to barium and antimony, faas detects lead. Instead of a flame atomizing the sample, an electrically heated tantalum plate is used with consistent results. The use of an electrically heated graphite plate gives low barium values, possibly because of the formation of barium carbide.

Photoluminescence (17) and the scanning electron microscope (18) have been used to detect gunshot residues but these instruments are expensive.

Fiber samples, frequently found at a crime scene, may be examined by microscope to distinguish natural from synthetic materials. There are two synthetic types: cellulose-based and noncellulose-based. They can be differentiated by their solubilities in different solvents. For example, viscose rayon and cellulose acetate are cellulose-based; the former dissolves in 60% sulfuric acid and the latter in glacial acetic acid. Table 2 shows examples of noncellulose-based fibers that can be identified by their solubilities in various solvents (see Fibers, chemical). Thin-layer chromatography (tlc) is used to identify the fiber dyes.

Strands of hair are often found at a crime scene. Microscopic examination can reveal a number of significant characteristics, and distinguish human from nonhuman hairs. If human, the race (Caucasian, Mongoloid, or Negroid), color, whether dyed or bleached, whether forcibly removed and damaged, and the body site (head, beard, or pubic) may be determined. Forensic chemists have been studying the use of naa in determining trace metal concentrations in hair (19).

Driving While Under the Influence of Intoxicating Beverages (DWI)

The frequency of DWI cases in the United States and other industrialized nations leads all other crimes. Furthermore, there has been a sharp rise in the number of drivers arrested who are under the simultaneous influence of alcohol and drugs.

Blood alcohol levels were established to distinguish between sobriety and intoxication. In 1962 the National Safety Council in conjunction with the American Medical Association defined the limit of 0.10% blood alcohol (10 mg/100 mL) as presumptive evidence of being under the influence of alcohol. Utah, Idaho, and the District

Table 2. Solubilities of Synthetic Fibers

	Acrylic			Polyester	
Solvent	Orlon	Acrilan	Nylon	Dacron	Kodel
formic acid	insol	insol	sol	insol	insol
ethylene carbonate	sol	sol	insol	insol	insol
hexafluoroisopropyl alcohol	slow	slow	sol	sol	sol

of Columbia are using a lower value of 0.08% whereas Maryland uses a limit of 0.15%.

Analysis of blood, urine, or breath determines the alcohol content, with blood analysis the most useful. It measures the alcohol in the blood stream which directly affects the central nervous system. When consumed, alcohol passes through the stomach and intestine where it diffuses unaltered into the blood stream and from there throughout the body. The rate at which this occurs depends, in part, upon the stomach contents. If food has first been consumed, the rate of entry of the alcohol into the blood stream is reduced. About 98% of the alcohol is metabolized in the liver with the remainder excreted unchanged with urine, perspiration, and respiration.

The constitutional issue of self-incrimination has been avoided in all 50 states through the application of the implied consent laws which require a DWI suspect to consent to give blood, urine, or breath samples for the privilege of receiving a driver's license.

When a suspect is apprehended, depending upon the requirements of the law in each state, he or she has a choice as to which test to take. With the rise in use of portable breath analyzers, this method has become common, with blood analysis a second choice.

The blood sample must be removed by a medically trained person in the presence of a law-enforcement officer. Analysis should be performed as quickly as possible to avoid two sources of error: (*1*) if the sample becomes contaminated and left unrefrigerated, bacterial growth may occur with formation of alcohol, yielding a higher result; (*2*) if the sample is not sealed, air absorbed by the blood can oxidize the alcohol slowly and lower the result.

Among the several methods for blood alcohol analysis, glc is the simplest and is now used almost universally. It is also used for urinalysis but errors may occur in the interpretation of results (20).

The commercial development of portable breath analyzers has made this method the commonest for DWI testing. The instrument is a photometer which measures the transmission of light through a reference solution of acidified potassium dichromate and a separate but identical solution through which a known volume of breath is passed. Alcohol present in the breath reacts with yellow dichromate(VI) solution to yield green chromium(III) ion. The instrument is calibrated to permit direct reading of blood alcohol. The potassium dichromate solutions are sold in vials with a removable tip to permit its insertion into the instrument. Each test requires a new solution.

This analysis is based on the finding that the alcohol content of 1.0 mL of venous blood is equal to that found in 2.1 mL of delivered deep-lung breath (21). The deep-lung (alveolar) breath is presumed to be in equilibrium with the alcohol-containing plasma circulating in the lung tissues. Alcohol evaporates from the blood of the lung tissues through the membranes into the lungs, in accordance with Henry's Law. Assuming constant temperature and equilibrium conditions, a steady state is reached between the alcohol in the lungs as the vapor at the lung-tissue interface and alcohol dissolved in the blood.

Breath analysis should not be performed immediately after drinking an alcoholic beverage or using an alcoholic mouth wash; a delay of 15 min is considered adequate.

Drug Offenses

Drug abuse has become a major concern in the United States and other industrial nations. The illicit trade in drugs is a multibillion dollar industry. Addicts often commit robbery to pay for their drugs. In the forensic science laboratories, the identification of drugs consumes more time than any other activity.

Today, marihuana is the commonest and most controversial drug of abuse. Its freedom from addiction has led many to urge its removal from the *Federal Index of Illicit Drugs*. In some states, discussions are being held to make the possession of marihuana in minimal quantity a misdemeanor rather than a felony. Uncertainty as to the long-range effects of the drug make the decision difficult.

Δ^9-Tetrahydrocannabinol (THC) is the chemical agent that produces the principal drug effect in marihuana; cannabidiol and cannabinol are inert constituents. The drug is sold in the dried form and consumed generally as a cigarette (joint) which contains up to about 5 mg of THC.

The forensic examination of marihuana requires several steps and begins with a low-power stereoscopic microscope. The leaves of marihuana have characteristic crystolithic hairs whose base contains a crystolith or stone which is calcium carbonate. If these crystoliths are treated with concentrated hydrochloric acid, carbon dioxide effervescence results.

Several color tests are available to screen samples for marihuana, the Duquenois–Levine test (22) is the most widely used. The sample is treated with diethyl ether which extracts THC. After evaporation of the solvent, a mixture of vanillin, acetaldehyde, and ethanol in hydrochloric acid is added. A purple vanillin–acetaldehyde–THC complex is formed which is extracted into chloroform. Certain other substances yield a purple color which is not extracted by chloroform. This test is highly specific for THC. These two tests are not conclusive for legal purposes and a third test, often glc, is routinely performed.

Table 3 lists the most commonly encountered drugs of abuse.

Confiscated drug samples are often adulterated. The components may be separated by the use of appropriate solvents. Generally, the steps taken for analysis proceed from microscopic examination to color tests followed by any one or a combination of the tests shown in Table 4 (see also Analytical methods). The last two methods listed, although effective and reliable, are too expensive for routine examinations.

After automobile or other accidents in which the possible presence of drugs or alcohol creates a legal issue, blood analyses are performed. Although tlc, glc, and uv spectrophotometry are the primary means of identifying drugs, gc-ms, nmr, and fluorometric analyses are likely to be used increasingly in the future.

Explosive Devices

The use of explosive devices in the United States, which reached the high level of 5.8 bombings per day in 1973 (28), makes the examination of bomb residues a common task for the forensic chemist (see Explosives). The scene of explosion is chaotic and scattered with debris. Bomb fragments are contaminated with material from the site. Evidence must be collected from a wide area.

The type of explosive used can frequently be identified but the chance of apprehending the criminal is much lower (ca 5%). The oldest and best known explosive

Table 3. Drugs of Abuse

Compound	CAS Registry No.	Source[a]	Other name or abbreviation
narcotics			
morphine	[57-27-2]	opium[b]	
diacetylmorphine	[561-27-3]	morphine	heroin
codeine	[76-57-3]	opium[b]	methylmorphine
hydromorphone	[466-99-9]	morphine	
oxycodone	[76-42-6]	codeine	Demerol
meperidine	[57-42-1]		
methadone hydrochloride	[1095-90-5]		
stimulants			
cocaine	[50-36-2]	*Erythroxylon coca* Lam.[b]	
amphetamine	[300-62-9]		Benzedrine
methamphetamine	[537-46-2]		Methedrine
dextroamphetamine	[51-64-9]		Dexedrine
phenmetrazine	[134-49-6]		Preludin
depressants			
diazepam	[439-14-5]		Valium
chlordiazepoxide	[58-25-3]		Librium
meprobamate	[57-53-4]		Equanil, Miltown
clorazepate	[20432-69-3]		Transzene
reserpine	[50-55-5]		Serpine
pentobarbital	[76-74-4]		Nembutal
secobarbital	[76-73-3]		Seconal
amobarbital sodium	[57-43-2]		Amytal
phenobarbital	[50-06-6]		Luminal
hallucinogens			
lysergic acid diethylamide	[50-37-3]		LSD
mescaline	[54-04-6]		peyote
psilocybin	[520-52-5]	*Psilocybe mexicana* Heim[b]	
psilocin	[520-53-6]	*Psilocybe mexicana* Heim[b]	
phencyclidine	[77-10-1]		angel dust
N,N-dimethyltryptamine	[61-50-7]		cohoba, DMT
N,N-diethyltryptamine	[61-51-8]		DET
2,5-dimethoxy-α,4-dimethylaniline	[15588-95-1]		DOM, STP

[a] By synthesis, unless otherwise stated.
[b] By extraction.

is black powder, consisting of potassium nitrate (74%), charcoal (15.6%), and sulfur (10.4%). It is manufactured in granular and pellet form. Its low sensitivity to shock and friction makes it safe to handle. Its color makes it easy to detect, and it can be identified by a simple and unique test (28): a particle on the tip of a dissecting needle when placed in a small flame burns instantly with a characteristic flash and puff of white smoke arising from potassium carbonate and sulfate.

Smokeless powder (nitrocellulose) can be extracted from ammunition for use in home-made explosive devices. It is manufactured by nitration of cellulose with a nitric–sulfuric acid mixture. On detonation it yields only gaseous products, mostly carbon monoxide, carbon dioxide, and water, with lesser amounts of hydrogen and nitrogen. It is identified by its appearance and instantaneous burning with a characteristic yellowish-white flame.

Table 4. Analytical Methods for Drugs of Abuse

Method	Abbreviation	Refs.
chromatography		
thin-layer	tlc	23
gas-liquid	glc	
high-pressure liquid	hplc	24–25
spectroscopy		
ultraviolet	uv	
visible	vis	
infrared	ir	22, 26
nuclear magnetic resonance	nmr	27
mass spectrometry	ms	

Military Explosives. Table 5 lists military explosives which are widely used by terrorists in Israel and other countries (29) but are seldom encountered in the United States. They are identified with the aid of tlc, glc (29), and ir (30).

Dynamite. This explosive is the most frequently encountered in criminal bombings in the United States. It is manufactured in a variety of compositions.

Straight Dynamites. These consist primarily of nitroglycerine (NG), sulfur, sodium nitrate, and filler (carbonaceous combustible material). The strength depends on the ratio of NG to sodium nitrate. Based on a strength of 100% NG, a mixture of 20.2% NG and 59.3% sodium nitrate is rated 20%, whereas 56.8% NG and 22.6% sodium nitrate is rated 60%.

The nitroglycerine contains ethylene glycol dinitrate (EGDN) which depresses the freezing point (31). Sodium nitrate lowers the cost and provides oxygen balance.

Straight gelatin dynamites contain the same ingredients plus soluble nitrocellulose. The most potent dynamite consists of 91.0% NG and 7.9% nitrocellulose.

Ammonia Dynamites. These dynamites contain ammonium nitrate in addition to the components of straight dynamite. The strength increases with increasing NG, increasing ammonium nitrate, and decreasing sodium nitrate.

Ammonia gelatin dynamites contain the same ingredients as gelatin dynamites plus ammonium nitrate. The strength of the dynamite increases with increasing content of NG and ammonium nitrate.

Table 5. Military Explosives

Name	CAS Registry No.	Abbreviation and other name
trinitrotoluene	[*118-96-7*]	TNT
pentaerythrityl tetranitrate	[*78-11-5*]	PETN, pentaerythritol tetranitrate
cyclotrimethylenenitramine	[*121-82-4*]	RDX or Cyclonite (U.S.)
		T4 (Italy)
		Hexogen (FRG)
plastic explosive[a]		C4

[a] 91% RDX, 9% plasticizer.

Detection. Traces of explosive material may be present on the hands or clothes of a suspect or in a vehicle that was used, and may be identified. The residues are dissolved with acetone, and the components separated by tlc (32).

A commercial device called the Vapor-Phase Analyzer (VPA, Hydronautics, Inc., Laurel, Md.) can be employed to identify explosion vapors and even residues on a suspect's hand. The device is based on gas chromatography (32) and is used by the Bureau of Alcohol, Tobacco, and Firearms. In another method, the bomb residue is first leached with hot acetone and then with hot water followed by a series of instrumental tests (29).

Murder

Deaths may be natural, accidental, suicidal, or homicidal. The coroner has the legal responsibility to issue the death certificate stating the cause. In doubtful cases a forensic pathologist may be required to make an autopsy to ascertain the cause of death. If there is evidence of poisoning, the forensic chemist or toxicologist analyzes blood and tissue samples for toxic substances by appropriate techniques discussed previously.

In general, suicides by shooting are easily confirmed. Gunshot residues are found on the hand of the victim. Further evidence is afforded by powder burns on the skin and the path of the bullet entry. The recovery of the bullet and its comparison with one fired from the gun found at the scene establishes if both came from the same weapon. The markings on a bullet are characteristic of the weapon from which it was fired.

Homicide by shooting is common. Even if a gun is not found at the scene of the crime, recovery of the bullet is important since it permits the identification of the type of weapon used. The distance between gun and victim, if it is short, may be estimated by means of controlled firings utilizing the suspect's gun and ammunition of the same make (33).

In homicide or assault, blood stains may be found at or beyond the scene of the crime, including the suspect's clothing and possessions. The blood may be that of the victim or of the suspect.

For proving that a stain is in fact blood, the Takayama reagent is very effective. It consists of a solution of sodium hydroxide, glucose, and pyridine, and it reacts with blood hemoglobin to yield salmon-colored crystals of pyridine hemochromogen.

The precipitin test (34) is the most widely used to distinguish between human and animal blood and is based on serological principles. The sample is tested with serum from a rabbit previously treated intravenously with human blood. If precipitation occurs, the sample blood is human.

Today the analysis of blood stains has gone far beyond the classification into the four blood groups O, A, B, and AB. Electrophoretic analysis yielding information as to enzymatic composition has become more and more valuable in establishing the origin of blood samples (34).

Offenses Against Property

A burglar is likely to leave traces of his presence and to take with him evidence of the site (33).

Paint fragments are first examined microscopically; then organic components are identified by pyrolysis gas chromatography, metals by optical emission spectroscopy. A variety of other techniques are also employed (see Analytical methods).

Glass fragments are frequently encountered in forced-entry and hit-and-run cases. In the latter, the clothing of the victim often contains paint and glass fragments, which may identify a suspect's vehicle. Glass is identified traditionally by refractive index and density. However, recent evidence has been presented that glasses with identical densities and refractive indexes may not have the same elemental composition. Energy-dispersive x-ray analysis, mass spectrometry (35), or atomic absorption spectroscopy (36) measure even trace amounts of elements.

Soil designates mineral substances found at a crime site, generally outdoors. It is identified by microscopic examination followed by other analytical techniques discussed here.

Questioned Documents

The basic issue relating to questioned documents is their integrity. Though the determination of forgery is the task of handwriting experts, the forensic chemist may investigate possible alteration of date or contents, eg, by the examination of the ink used (see Inks).

Ball-point pen inks are composed of vehicles and dyes. The vehicle consists of resins and a glycol base and includes small amounts of formaldehyde or phenol as preservatives. The dyes may be triphenylamines and phthalocyanines. A tlc method for analysis is available (37). Ball-point pen inks change in formulation slightly but detectably about every year. Manufacturers can furnish identification. Knowledge of the date of commercial introduction may give proof of back-dating of a document. Thus, if a document was dated 1975 but the ink had a composition consistent with one introduced in 1978, the document was clearly back-dated.

If the dyes are identical but the vehicle different, identification of the ink by tlc becomes difficult and hplc combined with uv and visible spectrophotometry is employed (38).

Sexual Offenses

In virtually all sexual offenses, the identification of seminal fluid is vital. Medical examination of the victim and microscopic examination of vaginal swabs to identify human spermatozoa are routine procedures. In addition, a sensitive color test for acid phosphatase, a major constituent of seminal fluid, is used. A reliable method uses a citric acid buffer solution, a solution of sodium α-naphthoxide and diazodiorthoanisidine reagent (Brentamine Fast Blue B) which produces an intense purple color with acid phosphatase (39). As in the case of blood, electrophoretic research on seminal fluid is being carried on with the goal of achieving genetic typing of semen.

BIBLIOGRAPHY

"Forensic Chemistry" in *ECT* 1st ed., Vol. 6, pp. 848–857, by R. L. Kirk, University of California.

1. *17 Ohio St. 2d 81, 246 N.E. 2d 365 (1969);* A. A. Moenssens, R. E. Moses, and F. E. Inbau, *Scientific Evidence in Criminal Cases,* The Foundation Press, Inc., Mineola, N.Y., 1973, p. 403.

2. *Frye v. United States, 293 Fed. 1013, 1014 (D.C. Cir. 1923).*
3. *Coppolino v. State, 223 So 2d 68 (Fl. App. 1968).*
4. R. Ruch, J. Buchanan, V. Guinn, S. Bellanca, and R. Pinker, *J. Forensic Sci.* **9,** 119 (1964).
5. *United States v. Stifel, 433 F. 2d 431, 441 (6th Cir. 1970).*
6. R. E. Carter, *Arson Investigation,* Glencoe Publishing Co., Inc., Encino, Calif., 1978, p. 4.
7. *Ibid.,* pp. 18–34.
8. C. R. Midkiff and W. D. Washington, *J. Assoc. Off. Anal. Chem.* **55,** 840 (1971).
9. P. M. Cain, *J. Forensic Sci. Soc.* **15,** 301 (1975).
10. I. H. L. Yip and E. G. Clair, *J. Can. Forensic Sci. Soc.* **9**(2), 75 (1976).
11. I. C. Stone, J. N. Lomonie, L. A. Fletcher, and W. T. Lowry, *J. Forensic Sci.* **23,** 78 (1978).
12. J. Schlesinger, J. Lukens, V. Guinn, R. Hackleman, and R. Korts, *Gulf-General Atomic Report GA-9829,* Gulf Publishing Co., San Diego, Calif., 1970.
13. C. Hoffman, *Identification News,* **18,** 7 (1968).
14. C. Hoffman, *J. Assoc. Off. Anal. Chem.* **56,** 1388 (1973).
15. K. Pillay, W. Jester, and H. Fox, III, *J. Forensic Sci.* **18,** 226 (1973).
16. J. Goleb and C. Midkiff, Jr., *Appl. Spectrosc.* **29**(1), 44 (1975).
17. P. F. Jones and R. S. Nesbitt, *J. Forensic Sci.* **20,** 231 (1975).
18. R. S. Nesbitt, J. E. Wessel, and P. F. Jones, *J. Forensic Sci.* **21,** 565 (1976).
19. R. Cornelis, *J. Radioanal. Chem.* **15**(1), 305 (1973).
20. H. Heise, *J. Forensic Sci.* **12,** 454 (1967).
21. M. R. Mason and K. M. Dubowski, *Clin. Chem.* **20**(2), 128 (1974).
22. T. J. Siek, *J. Forensic Sci.* **20,** 18 (1975).
23. S. A. Masoud, *J. Pharm. Sci.* **65,** 1585 (1976).
24. B. B. Wheals, *J. Chromatogr.* **122,** 85 (1976).
25. M. A. Peat and L. Kopjak, *J. Forensic Sci.* **24,** 46 (1979).
26. T. J. Siek, *J. Forensic Sci.* **19,** 193 (1974).
27. T. C. Kram, *J. Forensic Sci.* **23,** 456 (1978).
28. C. M. Hoffman and E. B. Byall, *J. Forensic Sci.* **19,** 54 (1974).
29. M. A. Kaplan and S. Zitrin, *J. Assoc. Off. Anal. Chem.* **60,** 619 (1977).
30. F. Prostera, M. Halek, A. Castelli, and W. Fredericks, *Anal. Chem.* **24,** 495 (1960).
31. F. C. Barnes and R. A. Helson, *J. Forensic Sci.* **19,** 448 (1974).
32. C. R. Midkiff, Jr., and W. D. Washington, *J. Assoc. Off. Anal. Chem.* **57,** 1092 (1974).
33. H. J. Walls, *Forensic Sciences,* 2nd ed., Praeger Publishers, Inc., New York, 1974, pp. 155–164.
34. E. T. Blake and G. E. Sensabaugh, *J. Forensic Sci.* **21,** 784 (1976).
35. J. V. Drayton, *Chem. Br.* **11**(12), 439 (1975).
36. J. C. Hughes, T. Catterick, and G. Southeard, *Forensic Sci.* **8,** 217 (1976).
37. R. L. Brunelle and M. J. Pro, *J. Assoc. Off. Anal. Chem.* **55,** 832 (1972).
38. L. F. Colwell, Jr., and B. L. Karger, *J. Assoc. Off. Anal. Chem.* **60,** 616 (1977).
39. A. F. Schiff, *J. Forensic Sci.* **23,** 833 (1978).

THEODORE P. PERROS
George Washington University

FORMALDEHYDE

Formaldehyde [*50-00-0*], $H_2C{=}O$, is the first of the series of aliphatic aldehydes. It was discovered by Butlerov in 1859 and has been manufactured since the beginning of this century. Annual worldwide production capacity now exceeds 12×10^6 metric tons (calculated as 37% solution). Because of its relatively low cost, high purity, and variety of chemical reactions, formaldehyde has become one of the world's most important industrial and research chemicals (1).

Physical Properties

At ordinary temperatures, pure formaldehyde is a colorless gas with a pungent, suffocating odor. Physical properties are summarized in Table 1; thermodynamic values for temperatures ranging from 0–6000 K are given in the *JANAF Interim Thermochemical Tables* (12–13). A number of other properties are listed in ref. 14.

Formaldehyde is produced and sold as water solutions containing variable amounts of methanol. These solutions are complex equilibrium mixtures of methylene glycol, ($CH_2(OH)_2$), poly(oxymethylene glycols), and hemiformals of these glycols.

Table 1. Properties of Monomeric Formaldehyde

Property	Value	Ref.
density, g/cm^3		
at −80°C	0.9151	2
at −20°C	0.8153	2
boiling point at 101.3 kPa^a, °C	−19	3
melting point, °C	−118	3
vapor pressure, Antoine constants, Pa^b		4
A	9.28176	
B	959.43	
C	243.392	
heat of vaporization, ΔH_v		
at 19°C, kJ/mol^c	23.3	3, 5
at −109 to −22°C, J/mol^c	27,384 + 14.56 T^d −0.1207 T^2	3
heat of formation, $\Delta H_f°$ at 25°C, kJ/mol^c	−115.9	6
std free energy, $\Delta G_f°$ at 25°C, kJ/mol^c	−109.9	6
heat capacity, $Cp°$, J/(mol·K)^c	35.4	6
entropy, $S°$, J/(mol·K)^c	218.8	6
heat of combustion, kJ/mol^c	561–571	6–9
critical constants		10, 14
temperature, °C	137.2–141.2	
pressure, MPa^e	6.784–6.637	
flammability in air		
lower/upper limits, mol %	7.0/73	11
ignition temperature, °C	430	11

[a] To convert kPa to mm Hg, multiply by 7.5.
[b] $\log_{10} P_{Pa} = A - B/(C + t)$; t = °C. To convert $\log_{10} P_{Pa}$ to $\log_{10\ mm\ Hg}$, subtract 2.1225 from A.
[c] To convert J to cal, divide by 4.184.
[d] T = K.
[e] To convert MPa to atm, divide by 0.101.

Ultraviolet spectroscopic studies (15–17) indicate that even in highly concentrated solutions the content of unhydrated HCHO is <0.04 wt %.

Density and refractive index are nearly linear functions of formaldehyde and methanol concentration. Based on available data (18–21), the density may be expressed by the following approximation:

$$\text{density}_{g/cm^3} = [1.119 + 0.003\,(F - 45) - 0.0027\,M][1.0 + 0.00055\,(55 - t)] \quad (1)$$

where F and M are the formaldehyde and methanol concentrations, wt %, respectively, and t is in °C.

The refractive index (22) may be expressed by a simple approximation for solutions containing 30–50 wt % HCHO and 0–15 wt % CH_3OH:

$$n_D^{18} = 1.3295 + 0.00125\,F + 0.000113\,M \quad (2)$$

Viscosities have been measured for representative commercial formaldehyde solutions (23). Over the ranges of 30–50 wt % HCHO, 0–12 wt % CH_3OH, and 25–40°C, viscosity may be approximated by

$$\text{viscosity} = (1.28 + 0.039\,F + 0.05\,M - 0.024\,t)\ \text{mPa·s}\ (= \text{cP}) \quad (3)$$

Significant discrepancies in formaldehyde partial pressures above aqueous solutions (24–25) can occur due to nonequilibrium conditions in the liquid phase. However, recently these problems have been overcome and more consistent results obtained (10,20,24,26–28).

In methanol–formaldehyde–water solutions, increasing the concentration of either methanol or formaldehyde reduces the volatility of the other. Vapor–liquid–equilibrium data (10,29) for several methanolic formaldehyde solutions are given in Table 2. The flash point decreases from 83 to 60°C with increasing formaldehyde and methanol concentrations (19–20).

Formaldehyde solutions exist as a mixture of oligomers, $HO(CH_2O)_nH$. Their distribution has been determined for 6–50 wt % HCHO solutions with low methanol using nmr and gas chromatographic techniques (30–31). Averages of the equilibrium constants for

$$HO(CH_2O)_{n-1}H + HOCH_2OH \overset{K_n}{\rightleftharpoons} HO(CH_2O)_nH + H_2O \quad (4)$$

are $K_2 = 7.1$, $K_3 = 4.7$, and $K_4 = 3.4$. The equilibrium constants appear to be nearly independent of temperature over the range of 30–65°C. Methanol stabilizes aqueous

Table 2. Vapor Compositions of Formaldehyde–Methanol Mixtures at 101.3 kPa (1 atm), Mol %

Wt % HCHO to wt % CH_3OH, liquid	HCHO	CH_3OH
30:7.5	17.0	12.8
30:15	16.3	23.0
37:7.5	20.3	18.7
37:15	11.9	21.9
50:7.5	26.7	25.8
50:15	11.4	21.7

formaldehyde solutions by decreasing the average value of n (30,32). Hence, methanolic solutions can be stored at relatively low temperatures without precipitation of polymer.

The following approximation was derived from data at 25–80°C (30,32–33) for solutions containing 7–55 wt % formaldehyde and 0–14 wt % methanol:

$$\text{monomer (mol \%)} = 100 - 12.3\sqrt{F} + (1.44 - 0.0164\,F)\,M \qquad (5)$$

where the monomer is the mole percent of formaldehyde present as formaldehyde hemiformal and methylene glycol, F is the wt % HCHO, and M is the wt % CH_3OH.

Commercial formaldehyde–alcohol solutions are clear and remain stable above 16–21°C. They are readily obtained by dissolving a high concentration formaldehyde in the desired alcohol.

Chemical Properties

Formaldehyde is noted for its reactivity and its versatility as a chemical intermediate. It is used in the form of anhydrous monomer, solutions, polymers, and derivatives.

Anhydrous, monomeric formaldehyde is not available commercially. The pure, dry gas is relatively stable at 80–100°C but slowly polymerizes at lower temperatures. Traces of polar impurities such as acids, alkalies, and water greatly accelerate the polymerization. When liquid formaldehyde is warmed to room temperature in a sealed ampul, it polymerizes rapidly with evolution of heat (63 kJ/mol or 15.05 kcal/mol).

Formaldehyde is surprisingly stable. Uncatalyzed decomposition is very slow below 300°C; extrapolation of kinetic data (34) to 400°C indicates that the rate of decomposition is ca 0.44%/min at 101 kPa (1 atm). The main products are CO and H_2. Metals such as platinum (35), copper (36), and chromia and alumina (37) also catalyze the formation of methanol, methyl formate, formic acid, carbon dioxide, and methane.

At ordinary temperatures, formaldehyde gas is readily soluble in water, alcohols, and other polar solvents. Its heat of solution in water and the lower aliphatic alcohols is approximately 63 kJ/mol (15 kcal/mol). The reaction of unhydrated formaldehyde with water is very fast; the first-order rate constant at 22°C is 9.8 s^{-1} (38).

Formaldehyde is readily reduced to methanol by hydrogen over many metal and metal oxide catalysts. It is oxidized to formic acid or carbon dioxide and water. The Cannizzaro reaction gives formic acid and methanol. Similarly, a vapor-phase Tischenko-type reaction is catalyzed by copper (36) and boric acid (39).

$$2\ HCHO \longrightarrow CH_3O\overset{\overset{\displaystyle O}{\|}}{C}H \qquad (6)$$

Formaldehyde condenses with itself in an aldol-type reaction to yield lower hydroxy aldehydes, hydroxy ketones, and other hydroxy compounds; the reaction is autocatalytic and is favored by alkaline conditions. Condensation with various compounds gives methylol ($—CH_2OH$) and methylene ($=CH_2$) derivatives. The former are usually produced under alkaline or neutral conditions, the latter under acidic conditions or in the vapor phase.

In the presence of alkalies, aldehydes and ketones containing α-hydrogen atoms undergo aldol condensation with formaldehyde to form mono- and polymethylol derivatives. Acetaldehyde gives pentaerythritol (PE) (see Alcohols, polyhydric).

$$CH_3CHO + 3\ HCHO \rightarrow C(CH_2OH)_3CHO \qquad (7)$$

$$C(CH_2OH)_3CHO + HCHO + NaOH \rightarrow \underset{\text{pentaerythritol}}{C(CH_2OH)_4} + NaO\overset{O}{\overset{\|}{C}}H \qquad (8)$$

In the vapor phase at 285°C, acetaldehyde gives acrolein (qv) (40):

$$HCHO + CH_3\overset{O}{\overset{\|}{C}}H \longrightarrow HOCH_2CH_2\overset{O}{\overset{\|}{C}}H \longrightarrow CH_2{=}CH\overset{O}{\overset{\|}{C}}H + H_2O \qquad (9)$$

Methylene derivatives are readily formed in the vapor phase with compounds having a hydrogen in the α position to an electron-withdrawing group. Acrylic (41–42) and methacrylic (43) acids (or esters) are produced at 300–425°C from acetic and propionic acids (or esters), respectively, using alkali and alkaline earth catalysts. The same catalysts are effective for producing acrylonitrile from acetonitrile (43). With ketones, addition of two formaldehyde units occurs (44); for example, 2-butanone is converted to 2-methyl-1,4-pentadiene-3-one.

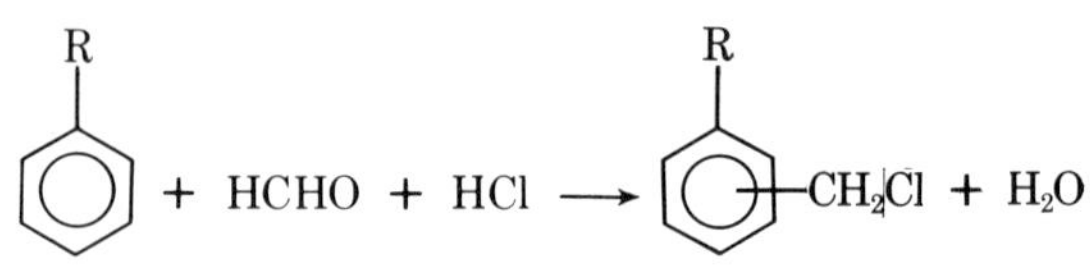

(10)

Chloromethylation occurs readily with phenolic esters, amides, and other compounds containing active hydrogen atoms in the presence of formaldehyde and hydrogen chloride.

The acid catalyzed addition of formaldehyde to olefins (Prins reaction) is carried out in an organic acid solvent.

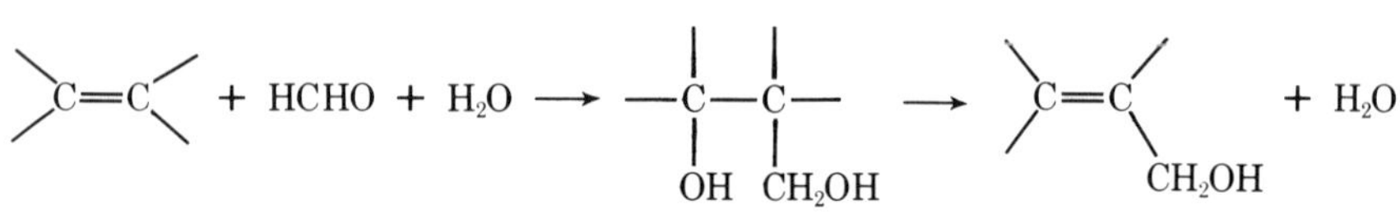

(11)

In aqueous acetic acid, styrene gives 1-phenyl-1,3-propanediol diacetate, and propylene gives 1,3-butanediol diacetate.

By a variation of this reaction, dienes are produced from olefins with addition of one carbon atom (45). Thus isoprene (qv) is obtained in good yields from isobutene:

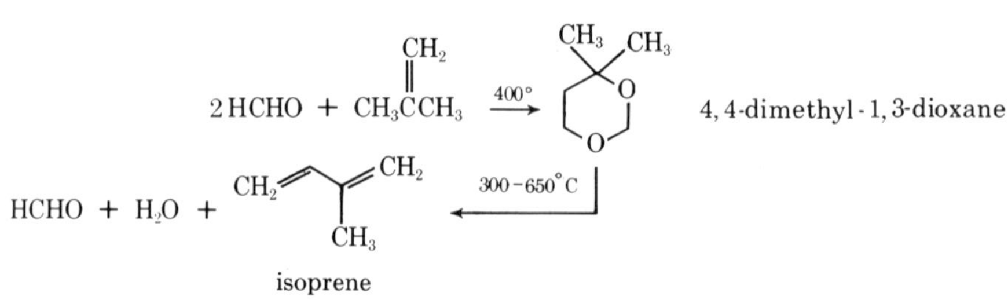

(12)

Liquid-phase condensation of formaldehyde with propylene, catalyzed by BF_3 or H_2SO_4, gives butadiene (qv) (46–47).

With acidic catalysts in the liquid phase, formaldehyde and alcohols give formals, eg:

$$HCHO + 2\,CH_3OH \rightarrow \underset{\text{dimethoxymethane}}{CH_3OCH_2OCH_3} + H_2O \qquad (13)$$

Under neutral or slightly alkaline conditions, only the unstable hemiformal (CH_3O-CH_2OH, methoxymethanol) is produced.

α-Chloromethyl ether is synthesized from aqueous formaldehyde, methanol, and hydrogen chloride (48). However, under anhydrous conditions, a carcinogenic by-product, bis(chloromethyl) ether is also formed (49).

Hydrogen sulfide precipitates cyclic trithiane (trithioformaldehyde) from formaldehyde solutions under strongly acidic conditions. Hydrogen cyanide reacts readily to form glycolonitrile (formaldehyde cyanohydrin), $HOCH_2CN$, in alkaline solution. Ketene reacts with anhydrous formaldehyde to give β-propiolactone.

Monosubstituted acetylenes add formaldehyde in the presence of copper, silver, and mercury acetylide catalysts to give acetylenic alcohols (50) (Reppe reaction). Acetylene itself adds two molecules (see Acetylene-derived chemicals).

$$2\,HCHO + HC{\equiv}CH \rightarrow \underset{\text{2-butyne-1,4-diol}}{HOCH_2C{\equiv}CCH_2OH} \qquad (14)$$

Primary and secondary amines readily give alkyl-aminomethanols; the latter condense upon heating or under alkaline conditions to give substituted methylene-amines (51).

$$3\,HCHO + 3\,CH_3NH_2 \rightarrow 3\,[CH_3NHCH_2OH] \rightarrow$$

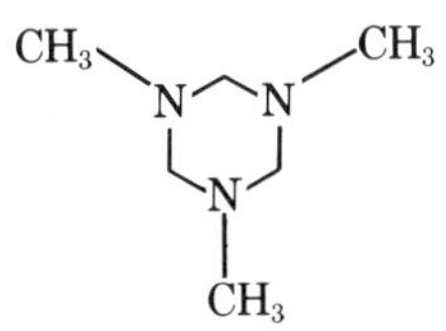

$+ 3\,H_2O$

hexahydro-1,3,5-trimethyl-1,3,5-triazine (15)

$$HCHO + (CH_3)_2NH \rightarrow [(CH_3)_2NCH_2OH] \xrightarrow{(CH_3)_2NH} \underset{\text{bis(dimethylamino)methane}}{(CH_3)_2NCH_2N(CH_3)_2} + H_2O \qquad (16)$$

With ammonia, the important industrial chemical, hexamethylenetetraamine, is produced. Tertiary amines do not react.

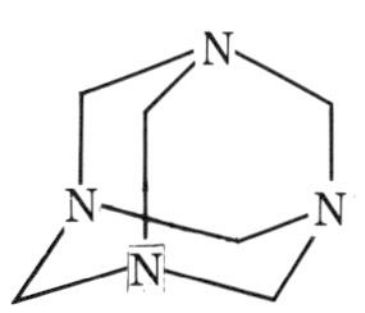

hexamethylene-
tetramine

The well-known Mannich reaction is illustrated by the following example:

$$CH_3COCH_3 + (CH_3)_2NH{\cdot}HCl + HCHO \rightarrow CH_3COCH_2CH_2N(CH_3)_2{\cdot}HCl + H_2O \qquad (17)$$

When formaldehyde is heated with amines or ammonium salts under acid conditions, alkylamines are formed:

$$R_2NH.HCl + 2HCHO \longrightarrow R_2\overset{\overset{CH_3}{|}}{N}.HCl + HCOOH \qquad (18)$$

With unsubstituted amides, relatively stable mono- and dimethylol derivatives are produced under neutral or alkaline conditions:

$$R\overset{\overset{O}{\|}}{C}NH_2 + HCHO \rightleftharpoons R\overset{\overset{O}{\|}}{C}NHCH_2OH \qquad (19)$$

$$R\overset{\overset{O}{\|}}{C}NHCH_2OH + HCHO \rightleftharpoons R\overset{\overset{O}{\|}}{C}N(CH_2OH)_2 \qquad (20)$$

Dimethylolurea, an item of commercial importance, is produced under controlled conditions,

$$2HCHO + NH_2\overset{\overset{O}{\|}}{C}NH_2 \xrightarrow[pH\,=\,7-9]{15-25^\circ C} HOCH_2NH\overset{\overset{O}{\|}}{C}NHCH_2OH \qquad (21)$$

Under acidic conditions, the methylolureas undergo condensation to urea–formaldehyde resins (see Amino resins). Melamine reacts similarly to urea.

A mixture of formaldehyde, methanol, acetaldehyde, and ammonia over a silica–alumina catalyst at 500°C gives pyridine and 3-picoline (52).

Since 1974, the cost of petrochemicals has increased rapidly. Thus there has been renewed interest in condensation products of formaldehyde with synthesis gas (CO, H_2), such as ethylene glycol (53):

$$HCHO + CO + H_2O \xrightarrow[>30\,MPa\ (300\,atm)]{H_2SO_4} \underset{\text{glycolic acid}}{HOCH_2\overset{\overset{O}{\|}}{C}OH} \longrightarrow (CH_2OH)_2 \qquad (22)$$

A variety of condensation catalysts, including cobalt carbonyl (54–55), rhodium oxide (55), hydrogen fluoride (56), and fluoroboric acid (56) are reported to be effective (see Glycols).

Manufacture

Historically, formaldehyde has been and continues to be manufactured from methanol. During the decades following World War II, however, as much as 20% of the formaldehyde produced in the United States was made by the vapor phase, noncatalytic oxidation of propane and butanes (57). This nonselective oxidation produces a broad spectrum of coproducts (58). Conversion of low molecular weight hydrocarbons in the oxidation step together with a complex separation system (59), results in an energy-intensive plant that, with increasing fuel costs, suffers economically when competing with newer, more selective manufacturing methods. Hence, the methanol process is preferred.

Methanol (qv) is manufactured from synthesis gas which, in turn, is usually produced from methane. Consequently, there has been a strong, continuing incentive to develop a useful single-step process that partially oxidizes methane to formaldehyde. A wide range of oxidation conditions and catalysts has been studied (60–62). Notwithstanding extensive effort, a successful commercial process is not known (63–64). Direct synthesis of formaldehyde from carbon monoxide and hydrogen is unfavorable at practical temperatures and pressures (65).

In 1966 a process to produce formaldehyde by the oxidation of dimethyl ether was commercialized in Japan, but operation was discontinued later (66–67). Today, all of the world's commercial formaldehyde is manufactured from methanol and air by an older process using a metal catalyst and a newer one using a metal oxide catalyst. Reactor feed to the former is on the methanol-rich side of a flammable mixture and virtually complete reaction of oxygen is obtained; conversely, feed to the metal oxide catalyst is lean in methanol and almost complete conversion of methanol is achieved.

Silver Catalyst Process. In early formaldehyde plants, methanol was oxidized over a copper catalyst, but in recent years this has been almost completely replaced with silver. Domestic licensors include Borden Chemical Company and Davy Powergas, Inc. (Imperial Chemical Industries Ltd., process) (68).

The silver-catalyzed reactions occur at essentially atmospheric pressure and 600 to 650°C (69) and can be represented by two simultaneous reactions:

$$CH_3OH + \tfrac{1}{2}\,O_2 \rightarrow HCHO + H_2O\ (\Delta H = -156\ \text{kJ or } -37.28\ \text{kcal}) \quad (23)$$

$$CH_3OH \rightarrow HCHO + H_2\ (\Delta H = +85\ \text{kJ or } 20.31\ \text{kcal}) \quad (24)$$

Between 50 and 60% of the formaldehyde is formed by the exothermic reaction (eq. 23) and the remainder by endothermic reaction (eq. 24) with the net result of a reaction exotherm. Carbon monoxide and dioxide, methyl formate, and formic acid are byproducts (69–70). In addition, there are also physical losses, liquid-phase reactions, and small quantities of methanol in the product, resulting in an overall plant yield of 86–90% (based on methanol).

Figure 1 is a flow diagram of a typical formaldehyde plant (69,71–74) employing silver catalyst. A feed mixture is generated by sparging air into a pool of heated methanol and combining the vapor with steam. The mixture passes through a superheater to a catalyst bed of silver crystals or layers of silver gauze. The product is then rapidly cooled in a steam generator and then in a water-cooled heat exchanger and fed to the bottom of an absorption tower. The bulk of the methanol, water, and formaldehyde is condensed in the bottom water-cooled section of the tower, and almost complete removal of the remaining methanol and formaldehyde from the tail gas occurs in the top of the absorber by countercurrent contact with clean process water. Absorber bottoms go to a distillation tower where methanol is recovered for recycle to the reactor. The base stream from the distillation, an aqueous solution of formaldehyde, is usually sent to an anion exchange unit which reduces the formic acid to specification level. The product contains 55% formaldehyde and less than 1.0% methanol.

A typical catalyst bed is very shallow (10 to 50 mm) (69,72). In some plants, the

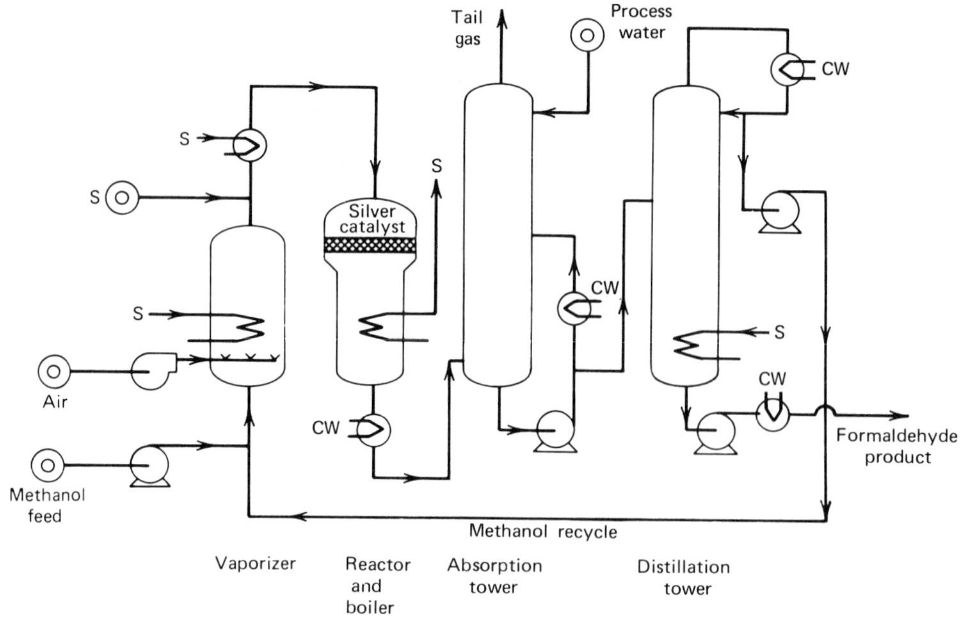

Figure 1. Flow scheme of a typical silver-catalyst process. S = steam; CW = cooling water.

catalyst is contained in numerous small parallel reactors; in others, catalyst-bed diameters up to 1.7 and 2.0 m (72,75) and capacities of up to 135,000 t/yr per reactor are reported (73). The silver catalyst has a useful life of three to eight months and can be recovered. It is easily poisoned by traces of transition group metals and by sulfur. An alkali wash to remove catalyst poisons from the air has been described (76).

The reaction occurs at essentially adiabatic conditions with a large temperature rise at the inlet surface of the catalyst. The predominant temperature control is thermal ballast in the form of excess methanol or steam, or both, which is in the feed (71). If a plant is to produce a product containing 50 to 55% formaldehyde and no more than 1.5% methanol, the amount of steam that can be added is limited, and both excess methanol and steam are needed as ballast. Recycled methanol required for a 50–55% product is 0.25–0.50 parts per part of fresh methanol (69,72).

With increasing energy costs, maximum methanol conversion is desirable, eliminating the need for the energy-intensive distillation for methanol recovery. If a dilute product containing 40 to 44% formaldehyde and 1.0 to 1.5% methanol is acceptable, then the ballast steam can be increased to a level where recycled methanol is eliminated with a significant savings in capital cost and energy (63,77). In another process, tail gas from the absorber is recycled to the reactor. This additional gas plus steam provides the necessary thermal ballast without the need for excess methanol (77–78). This process can produce 50% formaldehyde with as little as 1.2% methanol without a distillation tower (77). Methanol recovery can be obviated in two-stage oxidation systems, where for example, part of the methanol is converted with a silver catalyst, the product is cooled, excess air is added, and the remaining methanol is converted over a metal oxide catalyst such as that described below (79). In another two-stage process, both first and second stages use silver catalysts (80–82). Formaldehyde–methanol solutions can be made directly from methanol oxidation product by absorption in methanol (83).

The absorber tail gas contains about 20 mol % hydrogen (69) and has a net heating value of ca 1970 kJ/m^3 (53 Btu/ft^3) (77). With increased fuel costs and increased attention to the environment, tail gas is being burned for the two-fold purpose of generating steam and eliminating trace emissions of organic compounds.

Aqueous formaldehyde is corrosive to carbon steel, but formaldehyde in the vapor phase is not. All parts of the manufacturing equipment exposed to hot formaldehyde solutions must be a corrosion-resistant alloy such as type-316 stainless steel. Theoretically, the reactor and upstream equipment can be carbon steel, but in practice alloys are required in this part of the plant to protect the sensitive silver catalyst from metal contamination.

Metal Oxide–Catalyst Process. Oxidation of methanol to formaldehyde with vanadium pentoxide catalyst was first patented in 1921 (84), followed in 1933 by a patent for an iron oxide–molybdenum oxide catalyst (85), which is still the choice today. Catalysts are improved by modification with small amounts of other metal oxides (86), support on inert carriers (87), and methods of preparation (88–89) and activation (90). In 1952 the first commercial plant using an iron–molybdenum oxide catalyst was put into operation (91). By 1978, the metal oxide processes accounted for ca one-fifth of the formaldehyde made in the United States.

In contrast to the silver process, all of the formaldehyde is made by the exothermic reaction (23) at atmospheric pressure and at 300–400°C. By proper temperature control, a methanol conversion greater than 99% can be maintained. By-products are carbon monoxide and dimethyl ether, in addition to small amounts of carbon dioxide and formic acid. Overall plant yields are 88–91%.

Figure 2 is a flow scheme for a typical metal oxide–catalyst formaldehyde plant (92–95). Vaporized methanol is mixed with air and optionally recycled tail gas and passed through catalyst-filled tubes in a heat-exchanger reactor. Heat released by the exothermic reaction is removed by vaporization of a high-boiling heat-transfer fluid

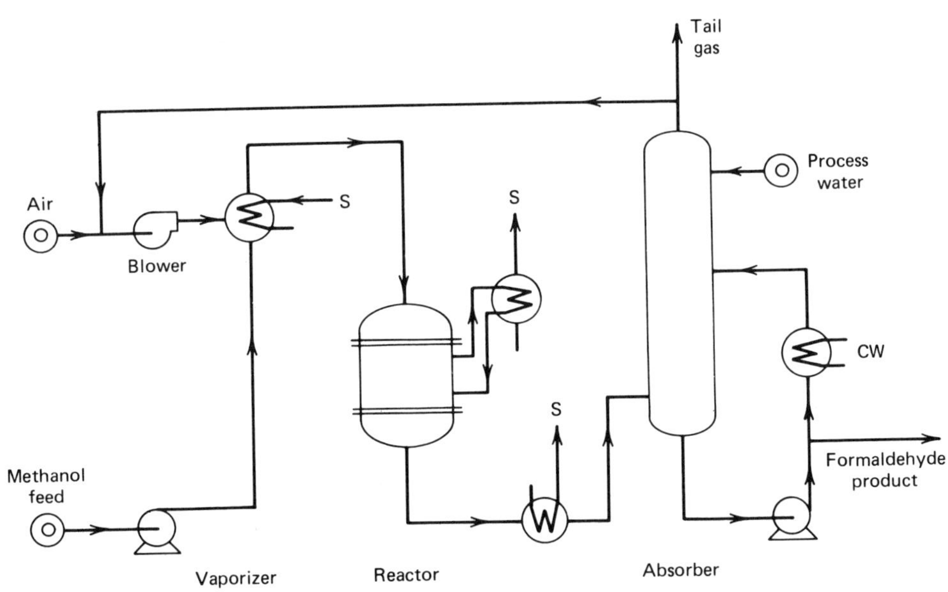

Figure 2. Flow scheme of a typical metal oxide catalyst process. S = steam; CW = cooling water.

on the outside of the tubes. A typical reactor has short tubes of 1.0 to 1.5 m and a large shell diameter of 2.5 m or more. Product leaving the bottom of the reactor is cooled, and passed to the base of an absorber. Formaldehyde concentration in the product is adjusted by controlling the amount of water added to the top of the absorber. A product with as much as 55% formaldehyde and less than 1% methanol can be made. Formic acid is removed by ion exchange.

The absence of a methanol recovery tower is an obvious advantage over the conventional silver process (Fig. 1). However, the equipment has to be large to accommodate the greater flow of gas. The air–methanol feed mixture must be on the methanol lean side of a flammable mixture, at least ca 13 mol per mol of methanol. However, if the oxygen in the total reactor feed is reduced to about 10 mol % by partially replacing air with recycled tail gas, the methanol in the feed can be increased without the danger of forming an explosive mixture (96), and for a given quantity of production, gas flow (air plus recycle) can be reduced by 17 to 37% (97). Even with gas recycle, the metal oxide process must handle a substantial volume of gas which is 3.0 to 3.5 times the gas flow of a conventional silver-catalyzed process.

A typical metal oxide catalyst has an effective life of 12 to 18 months (68–69,95). Compared to silver, it is much more tolerant to trace contamination. It requires less frequent change-outs, but a longer down time for each replacement. In contrast to a silver-catalyst plant, there is no economic justification to incinerate the metal-oxide plant tail gas for the purpose of generating steam. The tail gas is essentially nitrogen and oxygen with combustible components (dimethyl ether, carbon monoxide, formaldehyde, methanol) representing only a few percent of the total. In recent years there has been an increasing environmental incentive to reduce the emissions of these combustibles to the atmosphere. With the addition of auxiliary fuel, they can be oxidized by thermal incineration at temperatures of 700 to 900°C. As an alternative, they can be oxidized at 450 to 550°C in a catalytic incinerator. A catalytic system with proper heat exchange can be thermally self-sufficient with supplemental fuel required only for startup or abnormal operating conditions (97). If a metal oxide plant is to use either of these two tail gas incineration methods, there is a compelling justification to use tail gas recycle as described above. This recycle results in a several-fold reduction of net tail gas leaving the unit and thus greatly reduces the cost of incineration equipment (see Exhaust control, industrial).

The requirements for the material of construction are the same as for the silver catalyst process except the use of alloys to protect the catalyst is not as important.

Economic Aspects

Essentially all formaldehyde is produced as aqueous solutions containing 30–56 wt % HCHO and 0.5–15 wt % CH_3OH. All information on capacity, demand, and prices is reported on a 37 wt % formaldehyde basis.

Plant capacities in the United States are shown in Table 3. Capacity, production, and sales data between 1960 and 1978 are given in Table 4. Since the late 1960s, production has averaged 71% of capacity; however, it dropped to as low as 54% during the 1974–1975 recession. Sales have averaged about 35% of production, illustrating the large extent of captive use. Distribution of formaldehyde production in the United States according to uses is shown in Table 5. Consumption for the manufacture of the important products is shown in Table 6.

Table 3. Capacity (1978)[a] of U.S. Formaldehyde Producers[b]

Producer	10^3 t/yr
Allied Chemical Corp.	141
Borden, Inc.	714
Celanese Corp.	787
Chembond Corp.	132
E. I. du Pont de Nemours & Co., Inc.	667
GAF Corp.	45[c]
Georgia-Pacific Corp.	454
Gulf Oil Corp.	27
Hercules, Inc.	122
Hooker Chemicals & Plastics, Inc.	61
International Minerals & Chemicals Corp.	43
Monsanto Company	313
Pacific Resins & Chemicals, Inc.	43
Reichhold Chemicals, Inc.	372
Tenneco, Inc.	129
Wright Chemical Corp.	36
Total	*4086*

[a] 37% basis.
[b] Ref. 98.
[c] GAF Corp. plans to increase capacity to 90 × 10^3 t/yr (100).

Table 4. U.S. Formaldehyde Capacity, Production, and Sales[a]

Year	Capacity[b], 10^3 t/yr	Production[c], 10^3 t/yr	Sales[c], 10^3 t/yr	Price[b,d], \$/kg	Unit sales value[c], \$/kg
1960	1111	848	308	0.083	0.074
1965	1613	1409	540	0.038	0.056
1970		2008	626	0.071	0.053
1975	3803	2067	725	0.088	0.084
1977	4005	2742	na[f]	0.112	na[f]
1978	4086	2948[e]	na[f]	0.122	na[f]

[a] Based on 37 wt % formaldehyde (2 wt % methanol).
[b] Ref. 99.
[c] Ref. 100.
[d] Delivered bulk price for uninhibited 37% HCHO.
[e] Estimated from U.S. International Trade Commission data.
[f] Not available.

The growth rate of formaldehyde uses has declined over the last 20 years as shown below:

Period	*Annual growth rates, %*
1958–1968	11.7
1964–1974	7.4
1968–1977	3.5

Table 5. U.S. Distribution of Formaldehyde Production According to Uses, 1963–1977[a]

	Percent of consumption				
	1963	1969	1972	1975	1977
phenol–formaldehyde resins	22	22	26	22	25
urea–formaldehyde resins	21	25	26	25	25
acetal resins	4	6	7	7	9
acetylenics[b]	na[c]	2	4	7	6
melamine resins	6	7	6	3	5
pentaerythritol	9	7	7	7	5
hexamethylenetetramine	6	9	6	5	5
fertilizer[d]	3	3	4	4	na[c]
trimethylolpropane	na[c]	1.1	1.5	1.5	na[c]
ethylene glycol	12	4	0	0	0
miscellaneous	17	14	12	18	20

[a] Refs. 98–99, 101.
[b] Includes derivatives such as 1,4-butanediol and tetrahydrofuran.
[c] Not available.
[d] Urea–formaldehyde concentrate with a ratio of urea to formaldehyde about 1.3 to 1.

Table 6. Formaldehyde Consumption, 1978

Product	10^3 t
amino resins[a]	750
amino resins molding	59
phenolic resins	650
phenolic molding resins	66
fertilizers	180
textile finishes	60
acetal resins	180
1,4-butanediol	200
pentaerythritol	180
pyridines	40
methylenediphenyl isocyanate	65
trimethylolpropane	35
hexamine	150
(not available for paraformaldehyde)	

[a] Including urea and melamine resins.

Worldwide production capacity in 1977 was estimated to be over 12.6×10^6 t as 37 wt % formaldehyde. The United States, Western Europe, and Japan account for nearly 90% of the total capacity. Worldwide demand for formaldehyde in 1976 was estimated (101) to be about 7.5×10^6 t or 60% of capacity.

Formaldehyde prices for 1971–1978 are summarized in Table 7. Before the decision by the Organization of Petroleum Exporting Countries (OPEC) cartel to quadruple oil prices in 1973, the worldwide prices were fairly uniform. This action, followed by the recession of 1974–1975, caused a substantial variation of prices in U.S. dollars.

Table 7. Formaldehyde[a] Prices[b], $/100 kg

	1971	1974	1976	1978
United States	7.1	7.7	10.1	12.2
Belgium	7.1	8.9	10.3	12.9
FRG	8.2	15.7	14.7	18.9
Netherlands	7.7	11.5	11.7	14.7
Italy	7.5	9.7	10.5	9.5
UK	7.7	10.0	11.7	14.1
Japan	na[c]	10.8	14.7	20.1

[a] 37% basis.
[b] Currency conversions were based upon exchange rates for the date corresponding to the price quotation.
[c] Not available.

Specifications; Quality Control

Formaldehyde is sold in aqueous solutions with concentrations ranging from 30 to 56 wt % HCHO. Product specifications for typical grades (19–21,102–106) are summarized in Table 8.

Formaldehyde is sold as low methanol (uninhibited) and high methanol (inhibited) grades. Methanol is used as stabilizer.

Procedures for determining the quality of formaldehyde solutions are outlined by ASTM (107). The specifications in Table 8 are determined as follows: formaldehyde by the sodium sulfite method (108); methanol by specific gravity (109); acidity as formic acid by titration with sodium hydroxide (109); iron by colorimetry; and color (APHA) by comparison to platinum–cobalt color standards.

Specifications of commercial formaldehyde–alcohol solutions are given in Table 9.

Storage and Handling

Solutions of formaldehyde are unstable. Both formic acid (acidity) and paraformaldehyde concentrations increase with time and depend on temperature. Formic acid concentration increases at a rate of 1.5–3 ppm/d at 35°C and 10–20 ppm/d at 65°C

Table 8. Formaldehyde Specifications

	Methanol inhibited grades		Uninhibited, low methanol grades				ASTM Method
formaldehyde, wt %	37	44	37	44	50	56	D 2194
methanol, wt % (max)[a]	5–8	6–7	1.8	2.0	2.0	2.0	D 2380
acidity, wt % (max)[b]	0.03	0.03	0.03	0.04	0.05	0.04	D 2379
iron, ppm (max)	1	0.5	1	1	1	0.75	D 2087
color, APHA (max)	10	10	10	10	10	10	D 1209
turbidity, ppm (max)	10	na[c]	10	10	10	10	
ash, ppm (max)	40	na[c]	60	60	60	60	

[a] For inhibited grades, the range is the high and low specifications of producers surveyed.
[b] As wt % formic acid.
[c] Not available.

Table 9. Specifications of Commercial Formaldehyde–Alcohol Solutions

Assay	Methanol	*n*-Butanol	2-Methyl-1-propanol
alcohol, %	35	53	52
HCHO, %	55	40	40
H_2O, %	10	7	8
bp, °C	102	107	104

(19–20). Although low storage temperature minimizes acidity, it also effects polymerization. The latter is inhibited by addition of methanol or of stabilizers (102,110–111) such as hydroxypropyl methyl cellulose, methyl and ethyl celluloses, or isophthalobisguanamine. Stabilized solutions of formaldehyde contain up to 100 ppm of the additive. Inhibited formaldehyde typically contains 5–15 wt % methanol.

Most producers recommend a minimum storage temperature for both stabilized and unstabilized solutions. Figure 3 is a plot of data (19–20,102–103) for uninhibited (<2.0 wt % methanol) formaldehyde. Inhibition by methanol decreases the minimum storage temperature by ca 2.3°C per wt % CH_3OH for unstabilized solutions and ca 1.3°C per wt % CH_3OH for stabilized solutions.

Materials of construction preferred for storage vessels (19–21,111) are 304, 316, and 347-type stainless steels or lined carbon steels.

Health and Safety Factors

Table 10 shows health and safety data. The carcinogenic potential of formaldehyde has not been determined. Results from various studies differ (116–118). At

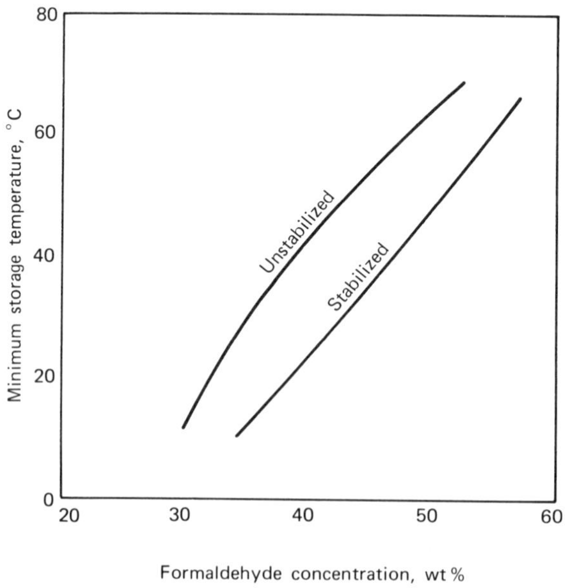

Figure 3. Storage temperature vs formaldehyde concentration (≤2.0 wt % CH_3OH).

Table 10. Health and Safety Factors for Formaldehyde Concentrations

	Value	Reference
exposure limits		
8-h time-weighted average, ppm	3	112
ceiling	5	
max peak above ceiling, ppm for 30 min	10[a]	
minimum levels for sensory detection		
airborne, odor detectable at ppm	1	113
in water, odor detectable at mg/L	20–50	114
profuse lachrymation, ppm	20	115
eye irritation[b], ppm	0.05–0.5	

[a] NIOSH (1977) recommends for revised Federal Standards 1 ppm (1.2 mg/m^3) for 30 min.
[b] Experienced by some persons.

present, it is on the *Suspected Carcinogens: A Subfile of the NIOSH Toxic Substances List* (119).

Mutagenic effects were noted in several organisms including *E. coli* and several viruses (120–121), but not on mammalian cells (122). Formaldehyde is not believed to be an active teratogen in animals (123).

Uses

As a basic chemical building unit, formaldehyde is an intermediate in a large variety of organic compounds ranging from amino and phenolic resins (qv) to slow-release fertilizers (101) (see Fertilizers; Amino resins).

Amino and Phenolic Resins. The largest use of formaldehyde is in the manufacture of amino and phenolic resins (qv), accounting for ca 55% of the total demand. Other important uses include wood-industry products, molding compounds, foundry resins, and adhesives (qv) for insulation.

Wood products account for about 36% of the total formaldehyde demand, with particle board (chips and sawdust with resin binder) first and plywood second. Most particle board products (ca 95%) are based on urea–formaldehyde resins, the rest on phenolics. Under certain conditions, urea–formaldehyde resins release formaldehydes which could create health or environmental problems. Almost 95% of the formaldehyde demand for wood products is captive (see Wood composites).

Amino and phenolic resins are both used as molding compounds. The bulk of the amino resins is melamine resin used in dinnerware. Phenolic molding resins are used in appliances, electric controls, telephones, and wiring services.

Phenolic resins are used as a binder for thermal and sound insulation, in both glass fiber and mineral wool. This is expected to be the fastest growing formaldehyde use to offset increasing energy costs (see Insulation, acoustic; Insulation, thermal; Foamed plastics).

Slow-Release Fertilizers. Demand for slow-release fertilizers is estimated at 900 $\times 10^3$ t/yr. Approximately 80% of the slow-release fertilizer market is based on urea–formaldehyde-containing products.

Chelating Agents. These represent a modest but important market for formaldehyde. The manufacture of ethylenediaminetetraacetic acid (EDTA) consumes about 75% of the formaldehyde used in the synthesis of chelating agents. The other 25%

is used to produce nitrilotriacetic acid (NTA), primarily for export (see Chelating agents).

Textile Finishes. Formaldehyde is used for the manufacture of permanent-press finishes of cellulosic fabrics. Formaldehyde-based textile finishes include melamine–formaldehyde, urea–formaldehyde, triazones, triazine, and glyoxal (see Textiles).

Acetal Resins. Manufacture of acetal (polyoxymethylene) resins is a captive formaldehyde use of DuPont and Celanese. The DuPont resin, Delrin, is based on anhydrous formaldehyde. The Celanese resin, Celcon, is based on trioxane. These resins are mainly used for automotive and building products (see Acetal resins).

Formaldehyde–Alcohol Solutions. Solutions of formaldehyde and various alcohols are sold for specific uses, particularly resin processes, where the overall amount of contained water and the additional water from resin condensation may not have to be removed from the product. They can partially replace paraformaldehyde in the production of urea–formaldehyde resins where they provide enough liquid for a workable reaction mixture without the addition of water; thus higher-solids resins can be obtained.

Other Applications. Formaldehyde is used for the manufacture of a great variety of chemicals, including elastomeric sealants (qv), herbicides (qv), fertilizer coatings, and pharmaceuticals (qv). Pyridines and isocyanates (qv) consume large amounts of formaldehyde (see Table 6). Methylenediphenyl isocyanate (MDI), used in the production of polyurethane foams, coatings, and elastomers, is expected to be one of the fastest growing consumers of formaldehyde. Pentaerythritol is chiefly applied in alkyd paint coatings, synthetic lubricants, and multifunctional acrylates. Trimethylolpropane is used as a specialty polyol in urethanes, alkyds, synthetic lubricants, multifunctional acrylates, flame retardants (qv), and textile chemicals (see Alkyd resins; Urethane polymers).

Derivatives

1,4-Butanediol. 1,4-Butanediol [*110-63-4*] is made from formaldehyde and acetylene. It is used in the manufacture of tetrahydrofuran (60%), as intermediate in polyvinylpyrrolidinone chemistry (20%), as chain extender for polyurethanes (10%), and for the engineering plastic poly(butylene terephthalate) (2%); 8% of production is exported (see Acetylene-derived chemicals; Engineering plastics).

Hexamethylenetetraamine. Pure hexamethylenetetraamine [*100-97-0*] (also called hexamine) is a colorless, odorless, crystalline solid of adamantane-like structure (124). It sublimes with decomposition >200°C but does not melt. Its solubility in water varies little with temperature; and at 25°C it is 46.5% in the saturated solution. It is a weak monobase; aqueous solutions are in the pH 8–8.5 range (125). Hexamethylenetetramine is readily prepared by treating aqueous formaldehyde with ammonia followed by evaporation and crystallization of the solid product. The reaction is fast and essentially quantitative (125).

Annual production is estimated at 36×10^3 metric tons. Hexamethylenetetramine is used for the manufacture of RDX (cyclonite) high explosive (see Explosives) and as a thermosetting catalyst for phenolic resins (qv). Hexamethylenetetraamine's largest use is in solid phenolic resins. It is also applied in molding compounds and in phenolic resins used in the foundry industry.

Paraformaldehyde. Paraformaldehyde [*9002-81-7*], or paraform, is a solid mixture of linear poly(oxymethylene glycols) of fairly short chain length, $HO(CH_2O)_nH$ (19,126) (the range of n is 8–100). The average degree of polymerization is only roughly given by the formaldehyde content. The properties of commercial paraformaldehyde are given in Table 11. Gaseous formaldehyde can be generated from paraformaldehyde by heating (127).

A current process for paraformaldehyde is based on continuous, staged vacuum evaporations, starting with 50% aqueous formaldehyde (128–129).

Catalytic production of paraformaldehyde from aqueous formaldehyde solution is a potentially attractive low-energy route; guanidine carbonate (130) and tertiary alkylamines (131) have been claimed as catalysts.

The rate at which paraformaldehyde dissolves (hydrolyzes) in water is at a minimum at pH 3–5; it increases rapidly at lower or higher pHs. Once dissolved in water, paraformaldehyde behaves like methanol-free formaldehyde of the same concentration.

Paraformaldehyde is used by resin manufacturers seeking low water content or more favorable control of reaction rates. It is often used in phenolic, urea, resorcinol, and melamine resins where high solids content is desired. Paraformaldehyde is also used as a fungicide and bactericide (see Disinfectants).

Trioxane and Tetraoxane. The cyclic symmetrical trimer of formaldehyde, trioxane [*110-88-3*], is prepared by acid-catalyzed liquid- or vapor-phase processes (132–136). It is a colorless crystalline solid that boils at 114.5°C and melts at 61–62°C (19,137). The heat of formation from monomeric formaldehyde is −176.9 kJ/mol (−42.28 kcal/mol) and from 60% aqueous formaldehyde −88.7 kJ/mol (−21.19 kcal/mol). It is produced by continuous distillation of 60% aqueous formaldehyde containing 2–5% sulfuric acid. Trioxane is extracted from the distillate with benzene or methylene chloride and recovered by distillation (138) or crystallization (139).

Trioxane is mainly used for the production of polyacetal plastic. It can be depolymerized under acid conditions (149), releasing formaldehyde at a controlled rate for the preparation of intermediates and phenolic resins. An effective way of generating monomeric formaldehyde is to depolymerize trioxane in the vapor phase over phosphoric acid on silicon carbide at 150–200°C (19).

A practical synthesis has been claimed for the cyclic tetramer of formaldehyde, 1,3,5,7-tetroxocane [*293-30-1*], bp, 175°C, mp, 112°C (140). It has found some use in textile treatment in Japan.

Table 11. Specifications of Commercial Paraformaldehyde[a]

	Flake	Powder
assay, as formaldehyde, wt %, min	91.0	95.0
av molecular wt	600	
acidity as formic, wt % max	0.03	0.03
water, wt %, max	9.0	5.0
% through mm (200 mesh) screen, min		90
flash point, Tag open cup, °C		93[b]
ignition temperature, °C		300[b]

[a] Ref. 19.
[b] Approximate.

BIBLIOGRAPHY

"Formaldehyde" in *ECT* 1st ed., Vol. 6, pp. 857–875, by J. F. Walker, E. I. du Pont de Nemours & Co., Inc.; "Formaldehyde" in *ECT* 2nd ed., Vol. 10, pp. 77–99, by J. Frederic Walker, Consultant.

1. J. F. Walker, *Formaldehyde,* 3rd ed., Reinhold Publishing Corp., New York, 1974.
2. A. Kekule, *Ber.* **25,** 2435 (1892).
3. R. Spence and W. Wild, *J. Chem. Soc.,* 506 (1935).
4. S. Ohe, *Computer Aided Data Book of Vapor Pressure,* Data Book Publishing Co., Tokyo, 1976, p. 109.
5. M. M. Brazhnikow and co-workers, *Zh. Prikl. Khim.* (*Leningrad*) **49,** 1041 (1976).
6. D. R. Stull and coworkers, *The Chemical Thermodynamics of Organic Compounds,* John Wiley & Sons, Inc., New York, 1969, p. 438.
7. H. von Wartenberg and B. Lerner-Steinberg, *Angew. Chem.* **38,** 591 (1925).
8. Ref. 6, p. 62.
9. *Handbook of Chemistry and Physics,* 56th Edition, CRC Press, Cleveland, Ohio, 1975, p. D276.
10. Yu. M. Blazhin and co-workers, *Zh. Prikl. Khim.* (*Leningrad*) **49**(1), 174 (1976).
11. *Fire Protection Guide on Hazardous Materials,* 5th ed., National Fire Protection Association, 1973, pp. 49–150.
12. *JANAF Interim Thermochemical Tables,* Thermal Laboratory, The Dow Chemical Co., Midland, Mich., Mar. 31, 1961.
13. Ref. 1, pp. 40, 42.
14. R. W. Gallant, *Hydrocarbon Process* **47**(5), 151 (1968).
15. R. Bieber and G. Trümpler, *Helv. Chim. Acta* **30,** 1860 (1947).
16. Ref. 1, pp. 60–61.
17. A. V. Rudnev and co-workers, *Zh. Fiz. Khim.* **51,** 2031 (1977).
18. Ref. 1, pp. 86–91, 500–501.
19. *Formaldehyde Product Bulletin,* Celanese Chemical Company, Inc., Dallas.
20. *Formaldehyde Product Bulletin,* E. I. du Pont de Nemours and Company, Inc., Wilmington, Del.
21. *Formaldehyde Product Bulletin,* Georgia-Pacific Corporation, Portland, Oregon.
22. G. Natta and M. Baccarredda, *G. Chim. Ind. Appl.* **15,** 273 (1933).
23. Ref. 1, p. 93.
24. B. Olsson and S.-G. Svensson, *Trans. Inst. Chem. Eng.* **53,** 97 (1975).
25. Ref. 1, pp. 115, 129.
26. A. E. Obraztsov and co-workers, *Zh. Prikl. Khim.* (*Leningrad*) **42,** 2393 (1969).
27. U.S. Pat. 2,527,654 (Oct. 31, 1950), C. Pyle and J. A. Lane (to E. I. du Pont de Nemours & Co., Inc.).
28. Yu. M. Blazhin and co-workers, *Zh. Prikl. Khim.* (*Leningrad*) **50**(1), 39 (1977).
29. S. J. Green and R. E. Vener, *Ind. Eng. Chem.* **47,** 103 (1955).
30. W. Dankelman and J. M. H. Daemen, *Anal. Chem.* **48,** 401 (1976).
31. Ya. Slonim and co-workers, *Vysokomol. Soedin. Ser. B* **XVII**(12), 919 (1975).
32. Z. Fiala and M. Kavrâtil, *Coll. Czech. Comm.* **39,** 2200 (1974).
33. D. A. Young, unpublished results, Celanese Research Company, Summit, N.J., 1978.
34. C. J. M. Fletcher, *Proc. Roy. Soc.* (*London*) **A146,** 357 (1934).
35. M. J. Marshall and D. F. Stedman, *Trans. Roy. Soc. Can.* **17** (Sect. III), 53 (1923).
36. Y. Miyazaki and I. Yasumori, *Bull. Chem. Soc. Jpn.* **40,** 2012 (1967).
37. H. Tropsch and O. Roehlen, *Abh. Kenntnis Kohle* **7,** 15 (1925).
38. H. C. Sutton and T. M. Downes, *Chem. Comm.,* 1 (1972).
39. P. R. Stapp, *J. Org. Chem.* **88,** 1435 (1978).
40. U.S. Pat. 2,246,037 (June 17, 1941), M. Gallagher and R. L. Hasche (to Eastman Kodak).
41. Jpn. Kokai 45 21,928 (July 24, 1970), (to Air Reduction).
42. Jpn. Kokai 71 16,728 (May 8, 1971), K. Kimura and I. Hiroo (to Toa Synthetic Chemical Co.).
43. U.S. Pat. 3,933,888 (Jan. 20, 1976), F. W. Schlaefer (to Rohm and Haas).
44. H. Hanna and S. Malinowski, *Chim. Ind.* (*Milan*) **51,** 1078 (1969).
45. T. Yashima and co-workers, *Nippon Kagaku Kaishi,* 325 (1974).
46. U.S. Pat. 2,335,691 (Nov. 30, 1943), H. O. Mottern (to Jasco, Inc.).
47. U.S. Pat. 2,412,762 (Dec. 17, 1946), A. R. Workman (to Cities Service).
48. L. F. Fieser and M. Fieser, *Reagents for Organic Synthesis,* John Wiley & Sons, Inc., New York, 1967, p. 132.

49. Belg. Pat. 817,848 (July 19, 1973), L. A. Ens (to Dow).
50. U.S. Pat. 2,232,867 (Feb. 25, 1941), E. Keyssner and W. Reppe (to General Anilin and Film).
51. L. Henry, *Bull. Acad. Roy. Belg.* **28,** 359, 366 (1894).
52. U.S. Pats. 2,744,904 (May 8, 1956), and 2,807,618 (Sept. 24, 1957), F. E. Cislak and W. R. Wheeler (to Reilly Tar and Chemical).
53. Ref. 1, p. 634.
54. Ger. Offen. 2,427,954 (Jan. 9, 1975), T. Yukawa, K. Kawasaki, and H. Wakamatsu (to Ajinomoto).
55. Belg. Pat. 858,628 (Jan. 2, 1978), R. G. Wall (to Chevron).
56. U.S. Pat. 4,016,209 (April 5, 1977), S. Suzuki (to Chevron).
57. Ref. 1, p. 7.
58. R. L. Mitchell, *Pet. Refiner* **35,** 179 (July 1956).
59. W. C. Hopkins and J. J. Fritsch, *Chem. Eng. Prog.* **51,** 361 (Aug. 1955).
60. Ref. 1, pp. 25–27.
61. N. C. Robertson in *The Chemistry of Petroleum Hydrocarbons,* Vol. 2, Reinhold Publishing Corporation, New York, 1955, pp. 365–74.
62. H. D. Medley and S. D. Cooley in *Advances in Petroleum Chemistry and Refining,* Vol. 3, Interscience Publishers, Inc., New York, 1960, pp. 325–29.
63. U. Gerloff, *Hydrocarbon Process.* **46,** 169 (June 1967).
64. B. H. McConkey and P. R. Wilkinson, *Ind. Eng. Chem. Process Des. Dev.* **6,** 436 (1967).
65. R. H. Newton and B. F. Dodge, *J. Am. Chem. Soc.* **55,** 4747 (1933).
66. H. Tadenuma and co-workers, *Hydrocarbon Process.* **45,** 195 (Nov. 1966).
67. *Hydrocarbon Process.* **46,** 180 (Nov. 1967).
68. *Chem. Week* **105,** 79 (Nov. 19, 1969).
69. A. R. Chauvel and co-workers, *Hydrocarbon Process.* **52,** 179 (Sept. 1973).
70. Ref. 1, p. 207.
71. M. Weimann, *Chem. Eng.* **77,** 102 (Mar. 9, 1970).
72. D. G. Sleeman, *Chem. Eng.* **75,** 42 (Jan. 1, 1968).
73. *Hydrocarbon Process.* **52,** 135 (Nov. 1973).
74. J. H. Marten and M. T. Butler, *Oil Gas J.* **72,** 71 (Mar. 11, 1974).
75. U.S. Pat. 4,072,717 (Feb. 2, 1978), G. Halbritter (to BASF).
76. U.S. Pat. 4,080,383 (Mar. 21, 1978), H. Diem and co-workers (to BASF).
77. H. Diem, *Chem. Eng.* **85,** 83 (Feb. 27, 1978).
78. Fr. Pat. 1,487,093 (June 30, 1967), (to E. I. du Pont de Nemours & Co., Inc.).
79. U.S. Pat. 2,519,788 (Aug. 22, 1950), W. A. Payne (to E. I. du Pont de Nemours & Co., Inc.).
80. U.S. Pat. 2,462,413 (Feb. 22, 1949), W. B. Meath (to Allied Chemical).
81. U.S. Pat. 3,959,383 (May 25, 1976), E. S. Northeimer (to E. I. du Pont de Nemours & Co., Inc.).
82. U.S. Pat. 4,076,754 (Feb. 28, 1978), G. L. Kiser and B. G. Hendricks (to E. I. du Pont de Nemours & Co., Inc.).
83. U.S. Pat. 3,629,997 (Dec. 28, 1971), C. W. DeMuth (to Borden).
84. U.S. Pat. 1,383,059 (June 28, 1921), G. C. Bailey and A. E. Craver (to Barrett Company).
85. U.S. Pat. 1,913,405 (June 13, 1933), V. E. Meharg and H. Adkins (to Bakelite Corp.).
86. U.S. Pat. 3,198,753 (Aug. 3, 1965), F. Traina (to Montecatini).
87. Brit. Pat. 1,080,508 (Aug. 23, 1967), (to Perstorp AB).
88. U.S. Pat. 3,408,309 (Oct. 29, 1968), A. W. Gessner (to Lummus).
89. U.S. Pat. 3,855,153 (Dec. 17, 1974), G. M. Chang (to Reichhold).
90. U.S. Pat. 2,813,309 (Nov. 5, 1957), C. L. Allyn and co-workers (to Reichhold).
91. *Chem. Eng.* **61,** 109 (Nov. 1954).
92. *Hydrocarbon Process.* **50,** 161 (Nov. 1971).
93. *Hydrocarbon Process.* **48,** 184 (Nov. 1969).
94. Ref. 92, p. 160.
95. *Hydrocarbon Process.* **55,** 150 (Nov. 1975).
96. U.S. Pat. 2,436,287 (Feb. 17, 1948), W. F. Brondyke and J. A. Monier (to E. I. du Pont de Nemours & Co., Inc.); Brit. Pat. 589,292 (June 17, 1947).
97. C. W. Horner, *Chem. Eng.* **84,** 108 (July 4, 1977).
98. *Chem. Mark. Rep.,* (Jan. 23, 1978).
99. *Chem. Mark. Rep.,* (Apr. 1, 1966); *Chem. Mark. Rep.,* (July 1, 1969); *Chem. Mark. Rep.,* (Oct. 1, 1972); *Chem. Mark. Rep.,* (July 1, 1975).

100. *Synthetic Organic Chemicals,* United States International Trade Commission.
101. J. L. Blackford, "Formaldehyde" in *Chemical Economics Handbook,* Stanford Research Institute, Menlo Park, Ca., 1977, p. 658.5032.
102. *Formaldehyde Product Bulletin,* Badische Anilin- und Soda-Fabrik AG, Ludwigshafen, Germany, Sept. 1975.
103. *Formaldehyde Product Bulletin,* Deutsche Gold- und Silber-Scheideanstalt AG, Frankfurt, Germany.
104. *Formaldehyde Product Bulletin,* Borden Chemical Co., Columbus, Ohio, Feb. 1978.
105. *Formox® Process for Producing Formaldehyde, Product Bulletin,* Reichhold Chemicals, Inc., White Plains, N.Y.
106. *Formaldehyde Product Bulletin,* Imperial Chemical Industries, Ltd., Billingham, UK, Feb. 1977.
107. ASTM Standard D 2378, American Society for Testing and Materials, Philadelphia, Pa.
108. Ref. 1, pp. 486–488.
109. Ref. 1, pp. 498–501.
110. U.S. Pat. 3,137,736 (June 16, 1964), R. H. Prinz and R. C. Kerr (to Celanese).
111. Ref. 1, pp. 93–100.
112. *Registry of Toxic Effects of Chemical Substances,* Vol. II, U.S. Dept. of Health, Education, and Welfare, 1977.
113. D. W. Fosset in *Industrial Hygiene and Toxicology,* 2nd ed., Vol. II, Interscience Publishers, Inc., New York, 1963.
114. N. F. Bartone, *J. Am. Water Works Assoc.* **50,** 21 (1958).
115. W. M. Grant, *Toxicology of the Eye,* 2nd ed., Charles C Thomas, Springfield, Ill., 1974.
116. F. Watanabe, *Jpn. J. Cancer Res.* **45,** 451 (1954).
117. H. S. Rosenkranz, *Bull. Environ. Contam. Toxicol.* **8,** 4 (1972).
118. P. Nettesheim, *Prog. Exp. Res.* **24,** 330 (1979).
119. *Suspected Carcinogens: A Subfile of the NIOSH Substances List,* 2nd ed., U.S. Dept. of Health, Education, and Welfare, NIOSH, Washington, D.C., 1976.
120. R. G. Solyanik and co-workers, *Genetika* **813,** 164 (1972).
121. G. A. Shevelava, *Toksikol. Nov. Prom. Kh. Veshchestv.* **12,** 78 (1971).
122. P. Propping and F. Vogel, *Triangle* **12**(2), 49 (1973).
123. H. Hurni and H. Ohder, *Food Cosmet. Toxicol.* **11,** 459 (1973).
124. R. G. Dickinson and A. L. Redmond, *J. Am. Chem. Soc.* **45,** 28 (1943).
125. Ref. 1, pp. 511–551.
126. Ref. 1, pp. 140–205.
127. U.S. Pat. 3,883,309 (May 13, 1975), K. Ishigawa and co-workers (to Kanebo Ltd.).
128. US. Pats. 2,568,016 and 2,568,017 (Sept. 18, 1951), A. F. MacLean and W. E. Heinz (to Celanese).
129. U.S. Pat. 4,036,891 (July 19, 1977), J. C. T. Moller and O. E. Hansen.
130. Ger. Offen. 2,037,759 (Feb. 10, 1972), F. V. Deisenroth (to Josef Meissner).
131. U.S. Pat. 3,492,357 (Jan. 27, 1970), L. B. Levy (to Celanese).
132. J. Mahieux, *Hydrocarbon Process.* **48**(5), 163 (1969).
133. U.S. Pat. 3,310,572 (Mar. 21, 1967), H. Delle and H. Mann (to Degussa).
134. U.S. Pat. 3,637,751 (Jan. 25, 1972), H. Fuchs and H. Sperber (to BASF).
135. U.S. Pat. 4,110,298 (Aug. 29, 1978), W. J. Wells and A. L. Stautzenberger (to Celanese).
136. Neth. Appl. 6,415,198 (June 30, 1966), (to Skanska Attikfabriken Aktiebolag).
137. Ref. 1, pp. 191–199.
138. U.S. Pat. 4,043,873 (Aug. 23, 1977), J. Ackermann and co-workers (to Societa Italiana Resine).
139. U.S. Pat. 4,125,540 (Nov. 14, 1978), A. Sugio and co-workers (to Mitsubishi).
140. U.S. Pat. 3,426,041 (Feb. 4, 1969), Y. Miyake and co-workers (to Toyo Koatsu Industries).

H. R. GERBERICH
A. L. STAUTZENBERGER
W. C. HOPKINS
Celanese Chemical Company, Inc.

FORMALS. See Formaldehyde.

FORMAMIDE, $HCONH_2$. See Formic acid and derivatives.

FORMIC ACID AND DERIVATIVES

FORMIC ACID

Formic acid [*64-18-6*] (methanoic acid) is a colorless, odorous acid, the first and by far the strongest of the unsubstituted series of carboxylic acids (qv). It was first described in 1671 by the British chemist Ray who noted that the air above redwood ant beds corroded lead and iron. Steam distillation of the ants (*Formica rufa*) gave a vinegarlike liquid that attacked lead. Dry distillation of the salt, however, gave a powerful acid entirely different from that obtained from vinegar. This acid occurs in fruits, vegetables, and leaves and roots of plants, and also in the secretions of numerous insects. For example, the bombardier beetle emits a mixture containing as much as 75% formic acid. Gay-Lussac developed a convenient laboratory preparation from oxalic acid and Liebig determined its constitution in 1836. Berthelot made it from potassium hydroxide and carbon monoxide in 1855 (1).

Formic acid is important in textile souring, leather preparation, cattle-fodder preservation, and chemical manufacture. It can be degraded biologically or chemically to innocuous substances in most environments. The anhydrous acid must be handled with the same care as concentrated sulfuric acid. It is a powerful dehydrating agent and serious burns can result from its action on the skin. It burns when ignited.

Physical Properties

Physical properties of formic acid are given in Table 1. Some resemble those of acetic acid but there are significant differences. Formic acid and water form maximum-boiling azeotropes that contain 77.5% acid [bp 107.3°C at 101.3 kPa (1 atm)] and 83.2% acid [bp 134.6°C at 2.4 MPa (24 atm)]. Inasmuch as most azeotropic mixtures boil below the boiling points of one or both constituents, this behavior is most unusual. Another distinctive feature is its increasing, rather than decreasing, latent heat as the temperature rises toward the boiling point. Dissociation of the vapor, ie, the dimer–monomer equilibrium, seems to be the cause. The dielectric constant of formic acid is approximately ten times that of acetic acid, and shows that there is an important contribution from the polar configuration of the molecules. Similarly, the dissociation constants for formic acid in water solutions are almost precisely tenfold those of acetic acid at the same temperature (12).

Chemical Properties

Formic acid may react (13) both as an acid and as an aldehyde because the carboxyl is bound to a hydrogen rather than an alkyl group. It is not a ready electron donor and its ion has far less tendency than acetate to enter the coordination spheres of various transition metal ions (see Coordination compounds). It decomposes readily (14) by dehydration, dehydrogenation, or through a bimolecular redox reaction:

$$HCOOH \rightarrow H_2O + CO$$

Table 1. Physical Properties of Formic Acid

Property	Value	Ref.
melting point, °C	8.4	2
boiling point, °C	100.7	3
density, g/cm^3		
at 20°C	1.220	4
at 25°C	1.213	5
refractive index, n_D^{25}	1.369	4
surface tension at 20°C, mN/m (= dyn/cm)	37.58	6
viscosity at 20°C, mPa·s (= cP)	1.784	7
heat of vaporization at 25°C, kJ/mol[a]	20.10	8
heat capacity at 17°C, C_p, J/(mol·K)[a]	98.78	2
molar entropy at 25°C, J/(mol·K)[a]	129.0	2
specific conductance, Ω·cm	6.08×10^{-5}	9
dielectric constant, 25°C	56.1	10
self-ionization constant	2.20×10^{-7}	9
dissociation constant in H_2O	3.739 ± 0.001	11

[a] To convert J to cal, divide by 4.184.

$$HCOOH \rightarrow H_2 + CO_2$$

$$2\ HCOOH \rightarrow H_2O + CO_2 + HCHO$$

Even at RT, formic acid decomposition rates are measurable. Traces of water, including the water formed in the decomposition, inhibit the reaction (15). The anhydrous acid catalyzes its own esterification with alcohols and polyols, but often also promotes dehydration to the ether or olefin. Dry formic acid adds easily to olefins to give formate esters. With hydrocarbons, even one as inert as adamantane, in the presence of sulfuric acid carboxylation occurs through the Koch–Haaf reaction (16):

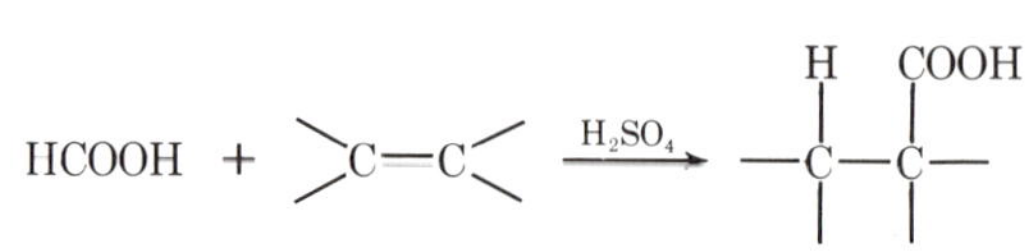

Alcohols and alkyl halides may be carboxylated in this way.

Formic acid or its aqueous solutions dissolve many of the more active, main group metals and their oxides to give the corresponding formates. It reduces peroxydisulfate, $(SO_3)_2O_2$, to sulfate ion, SO_4^{2-}, and sulfur dioxide to thiosulfate, $S_2O_3^{2-}$, and dithionite ions, $S_2O_4^{2-}$. Sodium formate and SO_2 in nonaqueous solutions give anhydrous sodium dithionite (17). Formic acid reduces various organic compounds such as imines, benzyl alcohols, and even certain diarylmethanes.

Manufacture

Direct synthesis of formic acid from water and CO has an unfavorable equilibrium, which can only be improved by operating at very high pressures and in the presence of copper halides or mineral acid catalysts. At 160°C the forward rate is low but at higher temperatures the equilibrium becomes less favorable (18). The concentration of formic acid present in equilibrium with any given pressure of CO is given in Table

Table 2. Thermodynamic Equilibrium of Formic Acid Synthesis

Formic acid, wt %	CO pressure, MPa[a] 25°C	100°C	218°C
1	0.03	0.63	6.95
10	0.30	7.14	79.1
20	0.72	17.0	188
50	3.65	2543	28,160

[a] To convert MPa to atm, divide by 0.101.

2. Operation of a direct process is not feasible under present technological limits. Reaction of CO with methanol yields methyl formate with a more favorable equilibrium (19). The ester may be decomposed to yield free acid in a second step (20).

Formic acid is mainly produced as a by-product of liquid-phase oxidation of hydrocarbons to acetic acid (21). There is a large patent literature on butane and naphtha oxidation to acetic acid and formic acid (see Hydrocarbon oxidation). Formic acid recovery from the crude products is complicated because it esterifies any alcohols present more readily than acetic acid. The formate esters thus formed are often lost when the low boiling fractions are recycled. Free formic acid is highly corrosive and must be separated as quickly as possible from acetic acid and the higher acids (22–23). Trichloroethylene, benzene, and gasoline are well known as formic acid entrainers (24). Pentane and hexane are inexpensive but effective azeotroping agents (25), as are toluene and ethylene dichloride (26) (see Azeotropic and extractive distillation).

About a third of the world formic acid production is based on methyl formate via formamide (27):

$$HCOOCH_3 + NH_3 \rightarrow HCONH_2 + CH_3OH$$

Formamide is easily hydrolyzed in the presence of H_2SO_4 to formic acid and ammonium sulfate (28). A yield of 93% formic acid was obtained in a pilot study of this process using industrial 72% H_2SO_4. No special concentration operation was required (29).

Another route to formic acid from methyl formate involves direct hydrolysis, the reverse of the esterification reaction (30). In principle, CO and methanol give methyl formate, and in a second step the ester is hydrolyzed to methanol and formic acid, which are separated. The net result is represented by the following equations:

$$CO + CH_3OH \rightarrow CH_3OOCH$$

$$CH_3OOCH + H_2O \rightarrow CH_3OH + HCOOH$$

$$CO + H_2O \rightarrow HCOOH$$

Efficient conversion requires hydrolysis under high pressure. Rapid reduction of the pressure, accompanied by cooling, permits recovery of the product (31–32). The methyl formate may also be prepared from sodium formate obtained as a by-product of pentaerythritol or disodium dithionite manufacture (33–34).

Formic acid is prepared from sodium formate using sulfuric acid or other cheap industrial source of acid in several small plants throughout the world. Stoichiometric amounts of sodium formate and H_2SO_4 are passed through an automatic balance pump at 35°C. The product is metered continuously into an evaporator where the formic acid is distilled and sodium sulfate recovered (35). Electrolysis of sodium formate in a cell fitted with an ion-selective permeable membrane can be used to produce sodium hydroxide and formic acid simultaneously. Addition of sodium sulfate suppresses formic acid decomposition during the electrolysis. With current efficiencies in excess of 85%, these processes are among the smallest consumers of energy per unit mass of formic acid recovered (36). Sodium formate is present in black liquor produced in pulping wood but its recovery as formic acid is significantly complicated by the presence of phenolic substances and higher fatty and resin acids (37).

In a new process, designed at the RCA Laboratories in Princeton, N.J., carbon dioxide in water is electrochemically reduced to formic acid liberating oxygen at the anode (38).

Shipping and Storage

Formic acid is shipped in tank cars, tank trucks, drums, and carboys; a DOT white label is mandatory. Provision must be made for the fact that the 98–99% acid has a higher thermal expansion than the 90% acid. When formic acid is held in storage for periods of six months or longer above 30°C, it slowly decomposes liberating CO; therefore, storage temperatures above 30°C should be avoided.

Stainless steel is the construction material of choice for handling formic acid; glass and ceramics are also satisfactory. Polyethylene and rubber furnish satisfactory liners for equipment exposed to formic acid. A number of proprietary acid-resistant enamel coatings can be used with formic acid under mild conditions.

Economic Aspects

Although formic acid was manufactured in Europe as early as 1896, it has been produced commercially in the United States only since the end of World War II. United States production is given in Table 3. Compared to acetic acid, it is a small-volume commodity. The annual growth rate from 1950 to 1974 was 5.9%. The wholesale price (1979) ranges $0.30–0.37/kg. About 30% of production is consumed by the textile industry and ca 15% by the leather industry.

Since hydrolytic processes starting either with formamide or methyl formate yield concentrations of 92–95%, it has been customary to dilute the acid to 90–85% for sale.

Table 3. United States Production of Formic Acid and Its Salts, Metric Tons

Year	90% Formic acid	Salts
1955	7,698	10,532[a]
1965	10,731	
1972	14,511	21,271[b]
1974	28,409	18,975[b]
1975		15,075[b]
1976	22,680	
1977	23,590	
1978	28,123	
1979	31,298	

[a] All formate salts.
[b] Sodium formate only.

The cost of obtaining 99–100% concentrations was considered greater than the increased transportation cost attendant upon shipping the lower concentrations. Since today most formic acid is recovered from oxidized hydrocarbons in essentially anhydrous form, the less concentrated grades may be expected to decline. Some of the smaller-volume producers, chiefly those who recover the acid from by-products, may be expected to have difficulty in competing with formic acid from hydrocarbon oxidation.

Specifications and Analytical Methods

Typical specifications for commercial grades of formic acid are shown in Table 4; other properties are given in Table 5. Formic acid may be identified by the reduction of mercury(II) chloride to mercury(I) chloride, a reaction also useful for quantitative analysis. The insoluble HgCl may be weighed directly or determined by various well-known volumetric procedures. Confirmation of the identity of formic acid can be obtained through the formation of a characteristic cyanine dye with *N*-methylquinaldinium *p*-toluenesulfonate in dimethyl sulfate solution. No other acid gives the reaction; the colored dyestuff with wavelength maxima at 565 and 610 mμ can be measured spectrophotometrically. Formate is usually determined by oxidation with standard potassium permanganate and back titration; for other analytical methods, see ref. 39.

Table 4. Specifications of Commercial Grades of Formic Acid

Assay	90%[a]	Technical[a]	Pharmaceutical[a]
formic acid, wt %, min	89–91	98.0	99.5
acetic acid, wt %, max		0.8	0.4
water, wt %, max			0.5
toluene, ppm, max			5
chlorides, ppm, max	20	20	5
heavy metals, ppm, max	5	5	5
iron, ppm, max	3	3	3
sulfates, ppm, max	10	10	10
color, Pt–Co units, max	20	20	15

[a] Substantially free of suspended matter.

Table 5. Physical Properties of Commercial Grades of Formic Acid

	Concentration, wt %			
Property	100	99	90	85
boiling point, °C	100.7	101.5	105	106
freezing point, °C	8.4	7.2	−5	−13.5
flash point, °C		59	50	
density, d_4^{25}	1.213	1.213	1.199	1.189
refractive index, n_D^{25}	1.369	1.369	1.368	1.368

Health and Safety Factors

When in contact with formic acid, its hazardous properties should be considered and safety precautions should be observed. Formic acid produces blisters on the skin almost instantly and it should be removed immediately and the skin rinsed thoroughly with water. Because of the heat of dilution of formic acid, it is not advisable to apply damp cloths or dilute $NaHCO_3$ solutions as the first treatment. Medical attention should be obtained for spills or splashes in the eyes, mouth, or nose, and in cases of swallowing.

Breathing the vapors causes intense irritation to the membranes of the nose and mouth. Mask, gloves, aprons, boots, and goggles are needed whenever the concentrated acid is handled. Safety shower, eye-washing facilities, and good ventilation of the area should be provided. The OSHA standard is 5 ppm.

A 1.5% dilution is scarcely toxic (LD_{50} for rats, ca 1210 mg/kg; for dogs, 4000 mg/kg). This solution, though it has about the same acidity as vinegar, is usually unacceptable because of its powerful odor and flavor, which differs strikingly from that of acetic acid.

Uses

Large quantities of formic acid are consumed by the textile and leather industries. Usually, it is employed as an inexpensive, volatile acid that is less corrosive than the mineral acids (see Leather; Textiles).

Formic acid neutralizes alkaline solutions and speeds rinsing in laundering. It is used for souring dye baths and as a dye-exhausting agent for cotton and cotton blends. It facilitates application of resin finishes to textiles for wash-and-wear properties. It is used in the acid fulling of wool (qv), particularly in blends with cotton and staple fiber.

In the leather industry, formic acid is used in dyeing as a substitute for sulfuric acid and for disinfection, swelling, deliming, and hair removal. It prevents mold growth on damp leather (qv).

A new and growing use is in silage and grain preservation where formic acid appears to inhibit or prevent mold growth. On silage, it changes the mode of natural fermentation. It is sometimes formulated with calcium formate, sugar, or sodium nitrate before applying to the fodder. Acetic or higher acids increase the effectiveness (40). Silage treated with formic acid can be used in dairy herds throughout the cold months of the year without milk decline. Neither mycotoxins, such as aflatoxin, nor their metabolites are present in the milk (41).

Small amounts of formic acid are used for coagulating rubber latex, in nickel-plating baths, in stripping the enamel from wire, and in the manufacture of drugs, dyes, flavors, and synthetic perfume ingredients. For example, 4,5-diamino-1,3-dimethyluracil boiled with formic acid furnishes the key formamido derivative that on cyclization gives sodium theophylline, precursor of theophylline and caffeine (see Alkaloids).

Traces of various formate esters are employed in the formulation of fragrances and flavors, especially fruit and honey mixtures. Methyl formate was formerly used as a grain fumigant but has been displaced by more effective agents. It has good insecticidal properties, coupled with low mammalian toxicity.

Formic acid and its salts have some application in the manufacture and regeneration of catalysts; eg, nickel catalysts (42), sulfur-poisoned catalysts (43), and lead-poisoned catalytic exhaust purifiers (44) (see Exhaust control).

BIBLIOGRAPHY

"Formic Acid" in *ECT* 1st ed., Vol. 6, pp. 875–883, by Benjamin Toubes, Victor Chemical Works; "Formic Acid" under "Formic Acid and Derivatives" in *ECT* 2nd ed., Vol. 10, pp. 99–103, by J. Frederic Walker, Consultant.

1. F. D. Chattaway, *Chem. News* **109,** 6103 (1914).
2. J. W. Stout and L. H. Fisher, *J. Chem. Phys.* **9,** 163 (1941).
3. W. R. Angus, G. T. W. Llewelyn, and G. Stott, *Trans. Faraday Soc.* **50,** 1311 (1954).
4. R. R. Dreisbach and R. A. Martin, *Ind. Eng. Chem.* **41,** 2875 (1949).
5. T. Ito and F. Yoshida, *J. Chem. Eng. Data* **8,** 315 (1963).
6. J. Timmermans, *Physicochemical Constants of Pure Organic Compounds,* Vol. 1, Elsevier, New York, 1950.
7. S. E. Sheppard and R. C. Houck, *J. Rheol.* **1,** 349 (1929).
8. J. Knoicek and I. Wadsö, *Acta Chem. Scand.* **24,** 2612 (1970).
9. T. C. Wehman and A. I. Popov, *J. Phys. Chem.* **72,** 4031 (1968).
10. J. F. Johnson and R. H. Cole, *J. Am. Chem. Soc.* **73,** 4536 (1951).
11. J. E. Prue and A. J. Read, *Trans. Faraday Soc.* **62,** 1271 (1966).
12. H. S. Harned and co-workers, *J. Am. Chem. Soc.* **55,** 652 (1933); **56,** 1042 (1934).
13. H. W. Gibson, *Chem. Rev.* **69,** 673 (1969).
14. P. Mars, J. J. F. Scholten, and P. Zwietering, *Adv. Catal.* **14,** 35 (1963).
15. H. N. Barham and L. W. Clark, *J. Am. Chem. Soc.* **73,** 4638 (1951).
16. H. Koch and W. Haaf, *Ann.* **618,** 251 (1958); G. A. Olah and J. A. Olah, *Friedel Crafts and Related Reactions,* Vol. 3, Interscience Publishers, a division of John Wiley & Sons, Inc., New York, 1964, p. 1284.
17. Jpn. Kokai 73 39,719 (Nov. 26, 1973), Y. Yoshikawa and T. Suzumori (to Mitsubishi Gas Chemical Co. Ltd.); Ger. Pat. 2,260,208 (Feb. 14, 1974), J. B. Heitman (to Pennwalt Corp.).
18. G. E. K. Brauch, *J. Am. Chem. Soc.* **37,** 2316 (1926); U.S. Pat. 1,606,394 (Nov. 9, 1926), W. C. Arsem.
19. W. Kotowski, *Chem. Tech. (Berlin)* **19,** 418 (1967).
20. Jpn. Kokai 75 11,887 (May 7, 1975), E. Yonemitsu, T. Tomita, and J. Kaida (to Mitsubishi Gas Chemical Co. Ltd.).
21. N. M. Emanuel, E. T. Denisov, and Z. K. Maizus, *Liquid Phase Oxidation of Hydrocarbons* (Engl. translation by B. J. Hazzard), Plenum Press, New York, 1967.
22. Br. Pat. 1,383,058 (Feb. 5, 1975), M. Wilkinson (to BP Chemicals International Ltd.).
23. Br. Pat. 3,859,346 (Jan. 7, 1975), R. J. Camm and H. Green (to BP Chemicals International Ltd.).
24. Ger. Pat. 513,024 (Feb. 8, 1928), I. Petersen and C. Engelbrecht (to I.G. Farbenind. A.G.).
25. Br. Pat. 727,078 (Mar. 30, 1955), (to Celanese Corporation of America).
26. Br. Pat. 1,049,901 (Nov. 30, 1966), T. Nakamura and co-workers (to Tekkosha).
27. U.S. Dept. of Commerce, *PB-952, PB-5077, PB-44962,* National Technical Information Service, Springfield, Va., 1947.
28. V. K. Krieble and K. A. Holst, *J. Am. Chem. Soc.* **60,** 2976 (1938).
29. K. Tokue, S. Okajima, H. Goto, and H. Tachibana, *Kogyo Kagaku Zasshi* **66,** 1814 (1963).
30. R. Linarte Lazcano and F. B. L. Rivera, *Rev. Inst. Mex. Pet.* **6**(3), 31 (1974).
31. Fr. Pat. 999,907 (Feb. 6, 1952), (to Soc. des produits chimiques de Clamecy).
32. Ger. Pat. 1,035,637 (Aug. 7, 1958), L. Hüter and M. Petzold (to Deutsche Gold- und Silberscheidenanstalt).
33. U.S. Pat. 3,578,703 (May 11, 1971), E. F. Schoenbrunn and E. H. Durell (to Escambia Chemical Corp.).
34. Jpn. Pat. 73 38,558 (Nov. 17, 1973), Y. Okumura and M. Masumori (to Nissan Chemical Industries Ltd.).
35. Fr. Pat. 1,429,084 (Feb. 18, 1966), (to Skanska Attikfabriken Aktiebolag).
36. Br. Pat. 787,976–7 (Dec. 18, 1957), (to Hooker Chemical Corp.).

37. U.S. Pat. 2,701,255 (Feb. 1, 1955), C. C. Heritage and W. G. Van Beckum (to Weyerhaeuser Timber Co.); U.S. Pat. 2,714,118 (July 26, 1955), J. E. Copenhaver, W. A. Biggs, Jr., W. H. Baxley, and J. T. Wise (to Sonoco Products Co.).
38. *Chem. Eng.* 17 (July 30, 1979).
39. R. M. Speights and A. J. Barnard, Jr., "Formic Acid" in F. D. Snell and L. S. Ettre, eds., *Encyclopedia of Industrial Chemical Analysis,* Vol. 13, Interscience Publishers, a division of John Wiley & Sons, Inc., New York, 1971, pp. 117–38.
40. U.S. Pat. 3,595,665 (Dec. 6, 1968), J. J. Huitson and P. W. J. Watts (to BP Chemicals (U.K.) Ltd.).
41. L. A. Goldblatt, ed., *Aflatoxin,* Academic Press, Inc., New York, 1969.
42. U.S. Pat. 1,390,684 (Sept. 13, 1921), C. E. Ellis.
43. Ger. Pat. 481,927 (Sept. 4, 1929), (to Gewerkschaft Kohlenbenzene).
44. U.S. Pat. 3,053,773 (Sept. 11, 1962), W. R. Calvert (to Oxy-Catalyst, Inc.).

FRANK S. WAGNER, JR.
Celanese Chemical Company, Inc.

FORMAMIDE

Formamide [*75-12-7*] (methanamide), mol wt 45.04, is a clear, hygroscopic liquid with viscosity similar to that of glycerol and a faint odor of ammonia. It is soluble in water, lower alcohols, and glycols, and insoluble in hydrocarbons, chlorinated solvents, and ethers. Formamide is a good solvent for proteins and salts owing to its high dielectric constant. Main applications are as solvent in the chemical industry, as softener for paper, as intermediate for the manufacture of formic acid and esters and hydrocyanic acid, and as a reaction medium.

Formamide was first synthesized in 1863 by Hofmann from ethyl formate and ammonia. On a large scale, it is produced from carbon monoxide and ammonia at elevated temperatures and high pressure by the two-stage process, with methyl formate as intermediate.

Formamide has been detected in space and is the only molecule found in interstellar clouds containing four different elements (1).

Properties

Table 1 lists the important physical properties of formamide. Association below 80°C is higher than that of water, decreasing rapidly with increasing temperature. Because of its high dielectric constant, formamide is an excellent solvent for many inorganic salts. It is freely miscible with water, lower alcohols, glycols, lower esters, and acetone, but insoluble in almost all aliphatic and aromatic hydrocarbons, chlorinated solvents, and ethers. Albumoses and peptones are soluble, but not albumins. Casein, corn starch adhesives, animal glue, and resins are soluble.

Formamide decomposes at the boiling point to NH_3, CO, HCN, and H_2O at a rate of about 0.5%/min. Hydrogen cyanide is obtained in high yields by heating formamide with a catalyst above 350°C. Formamide is hydrolyzed very slowly at room temperatures. Acids, bases, and elevated temperatures accelerate the hydrolysis.

Table 1. Physical Properties

Property	Value
boiling point, °C	210
boiling range, 5–95% at 2.4 kPa[a], °C	111–112
density at 20°C, g/cm^3	1.133
dielectric constant	109 ± 1.5
dipole moment at 30°C	3.37
freezing point, °C	2.55
heat of combustion, kJ/mol[b]	564.4
heat of vaporization, kJ/mol[b]	75.4
refractive index	
n_D^{15}	1.4491
n_D^{25}	1.4468
specific heat at 19°C, kJ/(kg·K)[b]	2.30
surface tension at 20°C, mN/m (= dyn/cm)	58.2
vapor pressure, kPa[a]	
at 70.5°C	0.13
at 122.5°C	2.67
at 157.5°C	13.33
viscosity at 30°C, mPa·s (= cP)	2.926
ignition temperature, °C	>500
flash point, °C	175

[a] To convert kPa to mm Hg, multiply by 7.5.
[b] To convert J to cal, divide by 4.184.

Because of its bifunctionality, formamide is a highly reactive intermediate reacting either at the carbonyl group or at the amide nitrogen to give a large number of nitrogen-containing heterocyclic compounds.

Reaction with inorganic acids gives formic acid and ammonium salts. Organic halides or alcohols, in the presence of hydrogen halides, zinc sulfate, or mercuric chloride, give the formate esters. In concentrated ammonia solution, formamide is oxidized by potassium permanganate to cyanuric acid (see Cyanuric acid). Reaction with β-diketones and β-iminoketones at 180–200°C gives 4,6-disubstituted pyrimidines. Good yields of 4,5-disubstituted imidazoles are obtained with aliphatic, aromatic, and heterocyclic acyloins at 150–180°C. Formamide forms complexes with cobalt, copper, and nickel salts. The reaction with P_2S_5 gives thioformamide; *N*-halogen derivatives are obtained with hypohalous acids.

Manufacture

In the first process (2) for the production of formamide, carbon monoxide and ammonia reacted under pressure and at high temperature. In addition to technical problems encountered, only low yields were obtained. An important advance was achieved when alkali alcoholates in alcoholic solution were employed as catalysts (3) (see Alkoxides, metal).

Today, formamide is produced by one of two continuous processes: in the direct synthesis, NH_3 and CO react at 10–30 MPa (100–300 atm) and 80–100°C in methanolic sodium methoxide. Water-cooled, slim, high-pressure cylinders are used as reactors. The catalyst can be separated from the product by adding water before distillation

(4), spraying the liquid phase as mist into the reactor (5), or vacuum distillation (6). The methanol is recovered by adding benzene (7) or water (8) before distillation. Other catalysts available (9) are not used for large-scale production.

The two-stage synthesis, developed by BASF, presents fewer problems.

$$CO + CH_3OH \rightarrow HCOOCH_3$$

$$HCOOCH_3 + NH_3 \rightarrow HCONH_2 + CH_3OH$$

In the first stage, carbon monoxide and methanol, in the presence of catalytic sodium methoxide, form methyl formate, which is distilled and thus separated from the catalyst. In the second stage, the methyl formate (97%) is treated with liquid or gaseous ammonia at 0.2–0.6 MPa (2–6 atm) and 80–100°C, and formamide is obtained. The thermal stability of methyl formate is higher than that of formamide. The methanol has to be of high purity for recycling to the methyl formate synthesis. The BASF process (10) produces pure formamide and pure methanol. In the first distillation, all light fractions are stripped from the crude product. The distillate is recycled and the bottoms are separated from the remaining methanol under vacuum. The product is of high purity and needs no further distillation. Instead of liquid ammonia, gaseous ammonia can be used to react with methyl formate at atmospheric pressure at 20–30°C (11).

Recently, a process using ammonia, hydrogen, carbon dioxide, and a special metal–organophosphorus catalyst has been proposed (12–13).

Shipment and Storage

For storage and transportation, tanks of stainless steel (eg, V2A) or aluminum are indicated. When small amounts of iron are permissible, mild steel can be used for tank cars and drums. Caffeine, adenine, and other purines stabilize formamide in storage and upon heating (14).

Copper, brass, and even lead are attacked. Most paints and finishes are unsatisfactory protection in contact with formamide because of its high solvency. Natural rubber but not neoprene is attacked. For gaskets and packing, white asbestos can be used.

Formamide is available in tank cars, tank containers, and in 60- and 220-kg drums either of polyethylene- or polyethylene-coated steel. No restrictions or special labelings are required for transportation.

Economic Aspects

The leading producer of formamide is BASF Aktiengesellschaft (Federal Republic of Germany), with main uses in captive production. Other manufacturers are Leuna Werke (German Democratic Republic), Nitto Chemical Industries (Japan). Formamide is also manufactured in Czechoslovakia with main uses in captive production. Prices range from \$0.70 to 0.90/kg (1979).

Specifications; Quality Control; Analytical Methods

Formamide as currently marketed by BASF has the specifications shown in Table 2.

Formamide can show slight color caused by the presence of iron or polymeric hydrocyanic acid. The determination of the freezing point is the best method for quality control. Nitrogen is determined by the Kjeldahl method and formic acid is titrated potentiometrically with caustic in aqueous solution. The water content is measured by the Karl Fischer method. Iron is determined colorimetrically as its thiocyanate complex, and ammonium formate is titrated as formic acid.

Health and Safety Factors

Formamide has a low acute toxicity. The oral LD_{50} for various species of animals is above 1250 mg/kg. A cumulative effect was established in mice after prolonged oral administration of 10% of the LD_{50}. Formamide leads to slight temporary irritation when applied undiluted to the skin of animals; it causes slight irritation without permanent damage when applied to the eyes. Embryo lethal and teratogenetic effects were found in rats after combined oral, intraperitoneal, and epicutaneous applications. No effect on the reproduction ability of male rats was obtained with concentrations of ca 6 mg/m^3 in air.

Based on the animal experiments, a TLV of 20 ppm (37 mg/m^3) has been established. Precautions that should be observed when handling formamide include avoidance of prolonged inhalation of vapors or contact of the liquid with the skin or eyes. Women of child-bearing age should not be employed in the production and processing of formamide (15–17).

Uses

Formamide has a wide variety of applications as an intermediate and solvent in the chemical industry. It is used in the large-scale production of formic acid by reaction with inorganic acids. Hydrogen cyanide is made by passing formamide over a catalyst at high temperatures (see Cyanide, hydrogen cyanide). It is used to synthesize imidazoles (18), pyrimidine (18), 1,3,5-triazines (19), and other chemicals. A complete synthesis of purine alkaloids such as caffeine, theobromine, and theophylline has been attempted by a simplified method using formamide (18) (see Alkaloids). A review of its application in chemical reactions is contained in the general references.

As a solvent, formamide has found a wide range of applications (see Solvents,

Table 2. Formamide Specifications

Assay	Value
formamide, % min	99.5
methanol, %	0.1
ammonium formate, %	0.1
water, %	0.1
color, APHA	10
freezing point, °C	2.1–2.3

industrial). Its excellent solvent properties are widely exploited in the manufacture and processing of plastics. It can be used for spinning acrylonitrile copolymers and as solvent in the polymerization of unsaturated amines for the production of ion-exchange resins and in antistatic finishing of plastics or in forming conductive coatings on plastic particles. Formamide has the ability to wet cellulose fibers and cause them to swell, enhancing the adhesion of polymers on cellulose substrates. It is therefore used in lignin–casein inks and as solvent in felt-tip pens.

To increase the load-bearing capacity of soil, formamide is applied as delayed gel-producing reactant for sodium silicate. Because of its high dielectric constant and solvency for inorganic salts, it is used as a nonaqueous electrolytic solvent. In the pharmaceutical industry, formamide is a solvent for the crystallization of penicillin and dihydrostreptomycin sulfate (20) (see Antibiotics). Other applications include as solvent for separating chlorosilanes, as a drilling-mud additive (see Petroleum), in the manufacture of deicing fluids for aircraft runways (21), as additive for hydraulic fluids (qv), and for the purification of oils and fats.

BIBLIOGRAPHY

"Formamide" in *ECT* 1st ed., under "Formic Acid," Vol. 6, pp. 881–882, by Benjamin Toubes, Victor Chemical Works; "Formamide" in *ECT* 2nd ed., under "Formic Acid and Derivatives," Vol. 10, pp. 103–108, by H. Louderback, E. I. du Pont de Nemours & Co., Inc.

1. *New Sci. Sci. J.* **50,** 191 (Apr. 22, 1971).
2. Ger. Pat. 390,798 (July 10, 1921), K. H. Meyer (to IG Farben/BASF AG).
3. Ger. Pat. 550,749 (Aug. 26, 1924), R. Weitzel (to IG Farben/BASF AG).
4. Ger. Pat. 61,536 (May 5, 1968), G. Peinze (to VEB-Leuna-Werke).
5. Ger. Pat. 1,142,163 (Jan. 10, 1963), E. Kottenmaier (to Knapsack-Griesheim AG).
6. Ger. Pat. 13,169 (Nov. 27, 1954), K. Smeykal and co-workers (to VEB-Leuna-Werke).
7. Ger. Pat. (Jan. 30, 1930), R. Roth (to IG Farben/BASF AG).
8. Ger. Pat. (Mar. 5, 1973), W. Rhierfelder and A. Peinze (to VEB-Leuna-Werke).
9. Jpn. Kokai 10,250–69 (Sept. 16, 1966), (to Toyo Rayon).
10. Ger. Pat. 1,215,130 (July 3, 1962), E. German and H. Friz (to BASF AG).
11. Ger. Pat. 924,928 (Mar. 28, 1937), W. B. Tanner (to E. I. du Pont de Nemours & Co., Inc.).
12. U.S. Pat. 3,530,182 (Sept. 22,1970), P. Haynes and co-workers (to Shell Oil).
13. Jpn. Kokai 52,087–112 (Jan. 14, 1976), (to Mitsui Petrochemicals).
14. U.S. Pat. 2,346,425 (Apr. 11, 1944), E. C. Kirkpatrick (to E. I. du Pont de Nemours & Co., Inc.).
15. G. Kimmerle, *Ullmanns Encyklopädie der Technischen Chemie,* Vol. 11, Verlag Chemie, GmbH, Weinheim, 1976, p. 709.
16. *Documentation of the Threshold Limit Values,* 3rd ed., American Conference of Governmental Industrial Hygienists, Cincinnati, Ohio, 1971, p. 322.
17. *Formamide Data Sheet,* E. I. du Pont de Nemours & Co., Inc., Wilmington, Del., 1976.
18. H. Bredereck and co-workers, *Angew. Chem.* **71,** 753 (1959).
19. H. Bredereck, D. Smerz, and R. Gompper, *Ber. Dtsch. Chem. Ges.* **94,** 1883 (1961).
20. U.S. Pat. 2,744,892 (May 8, 1956), L. Katz (to Schenley Ind.).
21. Belg. Pat. 647,597 (May 7, 1963), G. Nottes (to BASF AG).

General References

Formamide, Technical Leaflet, BASF Aktiengesellschaft, 67 Ludwigshafen, 1973.
H. Kiefer and H. Lang, *Ullmanns Encyklopädie der Technischen Chemie,* Vol. 11, Verlag Chemie GmbH, Weinheim, 1976, pp. 703–705.
H. Bredereck and co-workers, *Ber. Dtsch. Chem. Ges.* **97,** 3397 (1964).
H. W. Gibson, *Chem. Rev.* **69,** 686 (1969).

L. F. Fieser and M. Fieser, *Reagents for Organic Synthesis,* Vol. 3, John Wiley & Sons, Inc., New York, 1972, p. 147.

CLAUDIO L. EBERLING
BASF Aktiengesellschaft

DIMETHYLFORMAMIDE

Dimethylformamide [*68-12-2*], $HCON(CH_3)_2$, DMF, mol wt 73.09, is a colorless, high-boiling, polar liquid with a faint but typical odor. It can be distilled without decomposition and is completely miscible with water and many organic compounds but not with aliphatic hydrocarbons. As an aprotic liquid with a high dielectric constant, DMF is an excellent solvent. It was first synthesized in 1893; however, it gained technical importance only in the 1950s as solvent for the spinning of acrylic fibers. Since then, many new applications have been found; for example, as solvent for polyurethane and polyamide coatings and films; selective medium for extraction of aromatics from crude oil; reaction medium; selective solvent of gases; solvent for dyes; solvent for electrolytes in galvanization baths; and paint remover and cleaner. All these applications led to a world production capacity of about 225,000 metric tons with prices ranging from $0.90 to 1.10/kg (1979).

Properties

DMF is a colorless, hygroscopic liquid that forms azeotropes with *o*-, *m*-, and *p*-xylenes (1), formic acid, and tetrachloroethylene. The physical properties are listed in Table 1. DMF has a high flash point; however, care should be taken in its use since its vapors are flammable when mixed with air.

The high solvency power of DMF makes it eminently suitable as a solvent for salts, gases, and compounds with high molecular weight. This solvency power can be ascribed to the small molecule, the high dielectric constant, the electron donor properties, and the ability to form complexes. It forms solvates with many salts at room temperature. Gases such as acetylene, butadiene, isoprene, sulfur dioxide, and HCl are soluble in DMF; ammonia, carbon monoxide and dioxide, and oxygen and hydrogen are only sparingly soluble.

Aqueous solutions have little tendency to hydrolyze. Even after 120 h of reflux, a 50% aqueous solution was hydrolyzed only 0.17%. However, hydrolysis is accelerated by acids or bases. The characteristic amine odor of the commercial product is due to moisture pick-up and subsequent slight hydrolysis to dimethylamine. Only a few ppm of dimethylamine are required to produce this odor.

Solutions of elementary bromine in DMF decompose in an exothermic, explosive reaction. Mixtures with halogenated hydrocarbons such as carbon tetrachloride and hexachlorobenzene in the presence of iron decompose, and particularly at elevated temperatures. Methylene chloride, *o*-dichlorobenzene, and ethylene chloride do not have this effect.

Table 1. Physical Properties of DMF

Property	Value
boiling point, °C	153
critical constants	
temperature, °C	374
pressure, MPa[a]	4.48
density, g/cm^3	0.276
density at 25°C, g/cm^3	0.9445
dielectric constant at 25°C	36.7
dipole moment, C·m (D)	12.7×10^{-30} (3.82)
freezing point, °C	−61
heat of combustion, kJ/mol[b]	1915
heat of vaporization at 153°C and 101 kPa (1 atm), kJ/mol[b]	42.3
refractive index, n_D	
at 20°C	1.4301
at 25°C	1.4269
relative vapor density[c] at STP	2.51
specific heat, kJ/(kg·K)[b]	
20°C	2.03
surface tension at 25°C, mN/m (= dyn/cm)	35.2
vapor pressure, kPa[d]	
at 20°C	0.35
at 60°C	3.46
at 100°C	19.95
viscosity at 20°C, mPa·s (= cP)	0.82
ignition temperature, °C	410
flammable limits in air at 20°C, 101 kPa (1 atm), % vol	
lower limit	2.2 (70 g/m^3)
upper limit	16 (500 g/m^3)
flash point, °C	58

[a] To convert MPa to atm, divide by 0.101.
[b] To convert J to cal, divide by 4.184.
[c] Air = 1.
[d] To convert kPa to mm Hg, multiply by 7.5.

Caution should be taken to avoid violent reactions when DMF is exposed to nitrates, chromic acid, or triethylaluminum.

Manufacture; Shipping

The most commonly used processes are the one-stage reaction of carbon monoxide with dimethylamine (eq. 1) and the two-stage process with methyl formate and dimethylamine (eqs. 2–3).

$$CO + (CH_3)_2NH \xrightarrow{\text{cat.}} HCON(CH_3)_2 \tag{1}$$

$$CO + CH_3OH \xrightarrow{\text{cat.}} HCOOCH_3 \tag{2}$$

$$HCOOCH_3 + (CH_3)_2NH \longrightarrow HCON(CH_3)_2 + CH_3OH \tag{3}$$

In the one-stage process, a solution of dimethylamine in methanol reacts with carbon monoxide in the presence of sodium methylate (2) or metal carbonyls (3–4) at 110–150°C and pressures of 1.5–2.5 MPa (15–25 bar). The product is purified by distillation.

In the two-stage process, methyl formate is first produced from carbon monoxide and methanol under high pressure at 60–100°C in the presence of sodium methylate. The methyl formate is distilled and then reacts with dimethylamine at 80–100°C and low pressure. The DMF is purified by distillation (5). Capital investment and energy consumption are lower for the one-step process.

In other DMF processes, hydrocyanic acid reacts with methanol in the presence of water and a titanium catalyst (6–7), or with dimethylamine and a catalyst (8). DMF can also be manufactured from carbon dioxide, hydrogen, and dimethylamine in the presence of halogen-containing transition metal compounds (9–10).

DMF is recovered from air by scrubbing with water (11) and from aqueous solution of more than 20% by distillation. Dilute aqueous solutions can be extracted with methylene chloride in stainless steel equipment.

Recycled DMF must meet high quality standards before being reused in spinning processes. Traces of impurities affect the viscosity of the spinning solution and the degree of whiteness and strength of the fibers. Formic acid is corrosive, and with amines acts as an inhibitor in the polymerization of polyurethanes. DMF is separated from amines by distillation; catalytic decomposition (12) or reaction with dimethylamine is used for separation from formic acid (13).

DMF is shipped in tank trucks and tank containers, but is also available in 200-kg steel drums.

Economic Aspects

World production capacity of DMF is ca 225×10^3 t/yr with approximately half located in Europe (see Table 2). U.S. production for 1979 was ca 15,000 t, with prices ranging from \$0.90 to 1.10/kg. The main use (65–75%) of DMF is as solvent for acrylic fibers and polyurethanes; 15–20% is used for the production of pharmaceutical products, and the rest for pesticides and other uses.

Specifications; Analysis; Storage and Handling

DMF marketed by BASF has the specifications shown in Table 3. Purity is routinely determined by gas chromotography or by conductivity measurements (14). Traces of formic acid may be determined as its tetrabutylammonium salt, and traces of bases by potentiometric titration with perchloric acid solution.

DMF in air is determined by gas chromatography. Direct-reading analysis tubes are available in the range of 10–40 ppm (TLV 10 ppm) for quick measurements. For wastewater, chemical methods based on the hydrolysis of DMF with acids are used.

Pure DMF is essentially noncorrosive to metals. However, copper, tin, and their alloys should be avoided. Ideal materials for its handling and storage are nonalloy (carbon) steels, austenitic CrNi steels (eg, V2A), and aluminum. Seals and pipelines should be made of polytetrafluoroethylene, polyethylene, or polypropylene of high molecular weight. Ethylene–propylene rubber may also be used (see Elastomers, synthetic). Graphite can be used for lubricating moving parts in contact with DMF.

Table 2. DMF Manufacturers and Capacities[a]

Manufacturer	Location	Capacity t/yr	World production, %
Air Products	Pensacola, Fla., United States	7,000	3.2
BASF Aktiengesellschaft	Ludwigshafen, FRG	60,000	27
BASF Quimica da Bahia S.A.	Camaçari, Brazil	6,000[b]	2.7
Celenese Mexicana	Cosoleacaque, Veracruz, Mexico	6,000[c]	2.7
Chinook	Toronto, Canada	10,000	4.5
E. I. du Pont de Nemours & Co., Inc.	Belle, West Virginia, United States	41,000	18.5
Ertisa	Huelva, Spain	5,000	2.3
ICI-Petrochemicals Division	Billingham, UK	15,000	6.7
Korea Fertilizer Co.	Ulsan, Republic of Korea	8,000[b]	3.6
Leuna Werke	Leuna, GDR	10,000	4.5
Mitsubishi Gas Chemicals	Matsuhama, Niigata, Japan	10,000	4.5
Nitto Chemicals	Tsurumi, Yokohama, Japan	26,000	11.7
Union Chimique Bedrijven (UCB)	Gent, Belgium	18,000	8.1
Total		*222,000*	

[a] Including captive use.
[b] On stream 1980.
[c] Being planned (1979).

Table 3. Specifications for DMF

Assay	Method	Value
DMF[a], % min	gas chromatography	99.9
methanol, ppm max	gas chromatography	100
water, % max	Karl Fischer	0.03
formic acid, ppm max	titration	20
N-methylformamide, ppm max	gas chromatography	100
dimethylamine, ppm max	titration	20
color, max	ASTM D 1209-62	10

[a] Clear and free from suspended matter.

Oil and greases are not suitable. The high solvency power of DMF has to be considered.

Since DMF is hygroscopic, it should be kept under a blanket of dry nitrogen. High purity DMF, as required for acrylic fibers, is best stored in aluminum tanks.

Health and Safety Factors

Single exposures to DMF by inhalation, skin contact, or ingestion are not particularly hazardous. The oral LD_{50} is on the order of 2000 mg/kg or above for various species of animals; the dermal LD_{50} (rabbit) is ca 5000 mg/kg. When applied to the skin of animals, undiluted DMF acted only as a slight irritant, and contact with the eye produced temporary, moderate conjunctival inflammation and transient corneal damage. High, sublethal doses led to an embryotoxic and teratogenetic effect in animals after oral and intraperitoneal administration.

Animal experiments gave a TLV of 10 ppm. Excessive dermal contact and exposure to levels well exceeding the TLV led to loss of appetite, nausea, and abdominal complaints. In some cases, symptoms of liver and pancreatic disorders were observed. Finally, DMF may cause alcohol intolerance with flush reactions in exposed workers (15–18).

Uses

The main application of DMF is as solvent in industrial processes, especially for polar polymers with strong intermolecular forces such as PVC and polyacrylonitrile (see Solvents, industrial). The high solvent power of DMF can be increased further by adding small amounts of THF. DMF solutions of high molecular weight polymers are processed to fibers, films, and surface coatings.

DMF allows the economic production of the acrylic fibers, giving solutions that readily permit spinning. For the production of synthetic leather of high quality, DMF is the choice solvent for polyurethanes (see Leather-like materials). Wire enamels based on polyamides, polyurethanes, and others are also produced with DMF as solvent.

Since DMF dissolves or dissociates a large number of inorganic compounds as well as organic substances, it is used as solvent, catalyst, or intermediate in many reactions. In the pharmaceutical industry, it is used as crystallization medium for vitamins (qv), hormones (qv), and sulfonamides (see Antibacterial agents, synthetic).

Since salts can be dissolved and dissociated in DMF, the solutions are used in electrolytic capacitors and in several electrolytic processes, eg, electroplating baths.

DMF is a highly selective and powerful solvent for certain hydrocarbons and inorganic gases. It is therefore used as selective absorbent in chemical processes such as the extraction of acetylene from ethylene streams (19), butadiene from C_4-hydrocarbons (20), and H_2S and SO_2 from flue gas. In petroleum processing, DMF is used pure or in blend with other solvents for the extraction of aromatics from mixtures.

Other applications of DMF are as paint stripper and cleaner for machinery, moldings, and gas pipelines. Further uses are the separation of terephthalic from isophthalic acid, recovery of sulfonic acids from petroleum fractions, and separation of chlorosilanes.

BIBLIOGRAPHY

"Dimethylformamide" in *ECT* 2nd ed., under "Formic Acids and Derivatives," Vol. 10, pp. 109–114, by H. Louderback, E. I. du Pont de Nemours & Co., Inc.

1. L. H. Horsey, *Azeotropic Data III,* Advances in Chemistry Series 116, American Chemical Society, Washington, D.C., 1973.
2. Belg. Pat. 852,300 (Mar. 12, 1976), (to UCB SA).
3. Ger. Pat. 863,000 (Aug. 10, 1940), H. Krzikalla and E. Woldan (to BASF AG).
4. U.S. Pat. 3,446,842 (May 27, 1969), K. Nozaki (to Shell Oil Co.).
5. Ger. Pat. 714,311 (July 24, 1937), R. Seidler and H. Klein (to IG Farben/BASF AG).
6. Jpn. Pat. 72 08,050 (Mar. 8, 1972), S. Senoo and co-workers (to Asahi Chemical Industry Co., Ltd.).
7. Ger. Pat. 2,143,074 (Oct. 10, 1974), Kominami and co-workers (to Asahi Chemical Industry Co., Ltd.).
8. Jpn. Kokai 72 08,051 (Nov. 27, 1968), Takahashi and Fukuoka (to Asahi Chemical Industry Co., Ltd.).
9. U.S. Pat. 3,530,182 (Sept. 22, 1970), P. Haynes and co-workers (to Shell Oil Co.).
10. Jpn. Kokai 2,036–617 (Sept. 17, 1975), (to Mitsui Petrochemicals).

11. Ger. Pat. 2,634,173 (July 30, 1975), V. H. Kareeva and co-workers (to Synthetic Resins Plasti).
12. Ger. Pat. 1,055,524 (Oct. 25, 1957), A. Brandl and R. Hoffmeister (to Phrix-Werke AG).
13. Ger. Pat. 2,018,616 (Nov. 12, 1970), H. Matsuzawa, H. Itoh, and T. Yoshioka (to Mitubishi Rayon Co., Ltd.).
14. J. Juillard, *Pure Appl. Chem.* **49,** 885 (1977).
15. D. Henschler, ed., *Gesundheitsschädliche Arbeitsstoffe, Toxikologische-arbeitsmedizinische Begründung von MAK-Werten,* Verlag Chemie GmbH, Weinheim, 1975.
16. D. Henschler, ed., *Gesundheitsschädliche Arbeitsstoffe, Toxikologische-arbeitsmedizinische Begründung von MAK-Werten,* Verlag Chemie GmbH, Weinheim, 1978.
17. *Documentation of the Threshold Limit Values,* 3rd ed., American Conference of Governmental Industrial Hygienists, 1971.
18. A. Kutzsche, *Arzneim. Forsch.* **15,** 618 (1965).
19. F. Rottmayr and co-workers, *Chem. Anlagen Verfahren,* 69 (Oct. 1972).
20. *Chem. Eng. News* **44,** 54 (Mar. 7, 1966).

General References

Dimethylformamide Technical Leaflet, BASF Aktiengesellschaft, 67 Ludwigshafen, 1974.

Properties and Uses of Dimethylformamide, E. I. du Pont de Nemours & Co., Inc., Wilmington, Del., 1972.

R. S. Kittila, *DMF—Dimethylformamide, Chemical Uses,* E. I. du Pont de Nemours & Co., Inc., Wilmington, Del., 1967.

R. S. Kittila, *DMF—Supplement to Dimethylformamide, Chemical Uses,* E. I. du Pont de Nemours & Co., Inc., Wilmington, Del., 1973.

H. Kiefer and H. Lang, *Ullmanns Encyklopädie der Technischen Chemie,* Vol. 11, Verlag Chemie GmbH, Weinheim, 1976, pp. 705–708.

S. S. Pizey, *Synthetic Reagents,* Vol. 1, John Wiley & Sons, Inc., New York, 1974, pp. 1–99.

L. F. Fieser and M. Fieser, *Reagents for Organic Synthesis,* Vol. 5, John Wiley & Sons, Inc., New York, 1975, p. 247.

J. A. Riddick and W. B. Bunger, "Organic Solvents" in *Techniques of Chemistry,* 3rd ed., Vol. II, John Wiley & Sons, Inc., New York, 1970.

R. Keller and co-workers, *Properties of Nonaqueous Electrolytes,* NASA CR-1425, Washington, D.C., 1969.

H. W. Lang, *Chem. Ztg.* **84,** 239 (1960).

CLAUDIO L. EBERLING
BASF Aktiengesellschaft

FRICTION MATERIAL. See Brake linings and clutch facings.

FRIEDEL-CRAFTS REACTIONS

In 1877, at the Sorbonne in Paris, Charles Friedel and James Mason Crafts showed that anhydrous aluminum chloride could be used as a condensing agent in a general synthetic method for furnishing an infinite number of hydrocarbons. In the continuation of their work (stretching over fourteen years), they extended their studies of the catalytic effect of aluminum chloride in organic reactions to various fields in which they discovered: (*1*) reactions of organic halides and unsaturated compounds with aromatic and aliphatic hydrocarbons; (*2*) reactions of anhydrides of organic acids with aromatic hydrocarbons; (*3*) reactions of oxygen, sulfur, sulfur dioxide, carbon dioxide, and phosgene with aromatic hydrocarbons; (*4*) cracking of aliphatic and aromatic hydrocarbons; and (*5*) polymerization of unsaturated hydrocarbons.

Thus in the early stages of investigation of the catalytic activities of aluminum chloride in organic chemistry, Friedel and Crafts were able to show that it is a most versatile catalyst effecting alkylations, dealkylations, acylations, polymerizations, and an almost unlimited number of other reactions. The diversity of reactions is astounding. Aluminum chloride and related catalysts bring about reactions between aromatic and aliphatic compounds, and inorganic and organic compounds; they serve in the degradation of aromatic and aliphatic compounds, in substitution and addition reactions, and in various other reactions such as polymerizations. Most of these reactions are classified by chemists as reactions of the Friedel-Crafts type and, with time, it has become a generally accepted concept that any organic reaction brought about by the catalytic action of anhydrous aluminum chloride, or related catalysts, is a Friedel-Crafts reaction. The important industrial processes for production of high octane gasoline, ethylbenzene, synthetic rubber, plastics, and detergent alkylates are all based on the Friedel-Crafts principle. The scope of the reactions is extremely wide and thus they form a large part of the more general field of electrophilic reactions. The published literature is tremendous, and for more detailed information, monographs (1–2) and other comprehensive reviews (3–4) should be consulted (see also Alkylation; Aluminum compounds; Boron compounds; Catalysis).

The observation that aluminum chloride is not, by any means, the only specific catalyst in the reaction was included in the very first papers of Friedel and Crafts. Ferric and zinc chlorides, as well as the double salt sodium aluminum chloride, could also be employed, but these were less reactive.

To approach the problem of definition, it is necessary to come to a clear understanding that not one but a number of reactions bear the general name Friedel-Crafts and a large number of other reactions are related to this type (2). Today Friedel-Crafts-type reactions are considered to be any substitution, isomerization, elimination, cracking, polymerization, or addition reaction that takes place under the catalytic effect of Lewis acid-type acidic halides (with or without cocatalysts) or proton acids. Originally, the evolution of hydrogen halide in the course of the reaction was considered characteristic; however, this is by no means any longer a limiting condition. Using the term in a more general sense points to the relation of these reactions with the Friedel-Crafts reactions.

The scope of the Friedel-Crafts reaction is not limited to the formation of carbon—carbon bonds. The formation of carbon—oxygen, carbon—nitrogen, carbon—sulfur, carbon—halogen, carbon—phosphorus, carbon—deuterium, carbon—boron, and many other types of bonds all conform with the general Friedel-Crafts principle.

Although Friedel and Crafts made their original observation on a reaction involving replacement of hydrogen in an aliphatic compound, ie, amyl chloride, the main emphasis of the reaction was first of all on aromatic compounds. The preparation of aliphatic compounds involving Friedel-Crafts methods was of minor importance until World War II, when isomerization of paraffins and cycloparaffins and the polymerization of olefins achieved considerable importance. The development of the aliphatic chemistry of Friedel-Crafts reactions stems largely from advances made in the production of relatively pure aliphatic hydrocarbons and their utilization for motor fuels (Ipatieff and co-workers) and from the increasing importance of polyolefins. The need for ethylbenzene in the manufacture of styrene, detergent alkylates, and related products led to the utilization of Friedel-Crafts reactions in large-scale production.

Despite the seemingly wide variety of types, Friedel-Crafts reactions can be divided into two general categories: alkylations and acylations. Within these two broad areas there is considerable diversity.

ALKYLATION

Alkylation of Aromatic Compounds

In the usual Friedel-Crafts alkylation of aromatics, a hydrogen atom of the aromatic nucleus is replaced by an alkyl group through the interaction of an alkylating agent in the presence of Friedel-Crafts catalyst. The most frequently used alkylating agents are: alkyl halides, alkenes, alkynes, alcohols, esters (of carboxylic and inorganic acids), ethers, aldehydes and ketones, paraffins and cycloparaffins, thiols (mercaptans), sulfides, amines (via diazotization), and thiocyanates.

The overall reaction using alkyl halides, alcohols, and alkenes as alkylating agents in the presence of aluminum chloride may be written as follows:

$$C_6H_6 + RX \xrightarrow{AlCl_3} C_6H_5R + HX$$

$$C_6H_6 + ROH \xrightarrow{AlCl_3} C_6H_5R + H_2O$$

$$C_6H_6 + RCH{=}CH_2 \xrightarrow{AlCl_3} RCH(C_6H_5)CH_3$$

Mechanism. The mechanism of alkylation and of other related Friedel-Crafts reactions is best explained by the carbocations concept. The alkylation of benzene with isopropyl chloride may be used as a general example:

$$(CH_3)_2CHCl + AlCl_3 \rightleftharpoons [CH_3\overset{+}{C}HCH_3][AlCl_4^-]$$

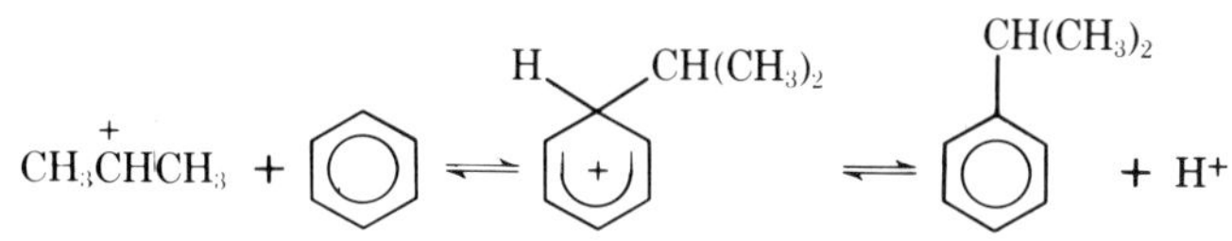

Lewis acid catalysts, such as $AlCl_3$ or BF_3, coordinate strongly with nonbonded electron pairs but they interact only weakly with bonded electron pairs. Therefore, n-donor reagents, such as alkyl halides, can react with Lewis acid catalysts under complete exclusion of moisture or any other proton source:

$$RX + MX_{3\text{ or }5} \rightleftharpoons R^{\delta+} \ldots X \ldots M^{\delta-}X_{3\text{ or }5} \rightleftharpoons [R^+]\, MX^-_{4\text{ or }6}$$

However, strong protic acid catalysts are needed when π- or σ-donor alkylating agents are used to produce carbocationic or highly polarized donor–acceptor-complex, reactive alkylating intermediates:

$$RCH{=}CH_2 + HMX_{4\text{ or }6} \rightleftharpoons R\overset{+}{C}HCH_3\, MX^-_{4\text{ or }6}$$

$$RC{\equiv}CH + HMX_{4\text{ or }6} \rightleftharpoons R\overset{+}{C}{=}CH_2\, MX^-_{4\text{ or }6}$$

$$R_2CH{-}CH_3 + HMX_{4\text{ or }6} \rightleftharpoons R_2\overset{+}{C}HMX^-_{4\text{ or }6} + CH_4$$

With superacidic media, the carbocationic intermediates, long postulated to exist during Friedel-Crafts type reactions (5–7), can be observed. Structural aspects of these carbocations were studied in high acidity–low nucleophilicity solvent systems using spectroscopic methods such as nmr, ir, Raman, esr, and x-ray crystallography (see Analytical methods).

Dealkylation, Transalkylation, and Disproportionation. The action of aluminum chloride is not restricted to the introduction of alkyl groups into aromatics. It also removes alkyl groups from alkylbenzenes (dealkylation, disproportionation) (8). Alkylbenzenes, heated with $AlCl_3$, form mixtures of benzene and polyalkylated benzenes:

$$C_6H_5R \xrightarrow[\Delta]{AlCl_3} C_6H_6 + C_6H_5R + C_6H_4R_2 + C_6H_3R_3 + \ldots$$

The tendency to disproportionate is on the order of the following (9):

$$CH_3 < C_2H_5 < n\text{-}C_3H_7 < i\text{-}C_3H_7 < tert\text{-}C_4H_9$$

In one of the most important commercial alkylations, the preparation of ethylbenzene from benzene and ethylene, polyethylbenzenes are also produced. The overall formation of polysubstituted products is minimized by recycling the higher ethylation products for the ethylation of fresh benzene (10). By adding to the benzene feed the calculated equilibrium amount of dialkylbenzene, a high conversion of ethylene to monoethylbenzene can be achieved (11–12) (see Xylenes and ethylbenzene).

Mono- and Polyalkylations. It had been assumed (13) that the tendency toward polysubstitution during Friedel-Crafts alkylation is owing to the greater reactivity of the initially produced alkylbenzene toward further substitution. This effect, however, has been shown to be limited since the alkyl groups on a benzene nucleus have only a small effect on the rate of further alkylation (14). The actual alkylation occurs in a heterogeneous reaction system, specifically in the catalyst layer, and the reason for polysubstitution is the preferential extraction of the early reaction product by this catalyst layer (14). The tendency toward polysubstitution may be minimized by the use of a mutual solvent for the hydrocarbon and catalyst layer, by high speed stirring, or by operating in the vapor phase or at a temperature sufficiently high for the aluminum chloride to be soluble in the hydrocarbon layer.

Orientation. An additional consideration is the orientation involved in the introduction of more than one alkyl group. Alkylation in the presence of aluminum chloride and other catalysts yields a considerable proportion of meta dialkylbenzenes, as well as the expected ortho and para isomers (13). The relative extent of so-called normal and abnormal orientation is a function of the alkylation conditions. In general, the more vigorous the conditions (reaction time and temperature, relative amount of catalyst, absence of solvent, etc), the greater is the tendency toward the formation of meta derivatives.

Isomerization. A characteristic of the Friedel-Crafts alkylations is the tendency of alkyl groups to rearrange during the alkylation process. As isomerization and disproportionation are always possible during alkylations, these side reactions should be considered wherever Friedel-Crafts conditions are employed. Generally, neither the configuration of the entering group nor the orientation in the nucleus can be accurately predicted. In fact, alkylation appears to be, in this respect, entirely different from acylation, which generally gives reliable orientation without isomerization phenomena.

In the course of alkylations two different types of isomerizations take place. The first involves the alkylating agent (alkyl halide, alcohol, etc) itself, which may undergo rearrangement and may generally be converted to the most highly branched isomer. Thus, the alkylation of benzene with *n*-propyl chloride in the presence of aluminum chloride yields *n*-propylbenzene and isopropylbenzene (cumene), and with *n*-butyl chloride *sec*-butylbenzene is formed. The rearrangement of the alkyl groups during alkylation usually proceeds in the direction primary → secondary → tertiary. Under the usual conditions of a Friedel-Crafts alkylation, the alkyl group rearrangement is initiated by migration of a hydrogen atom and not by direct carbon rearrangement. Under conditions favoring the formation of more stable carbocations, isomerization takes place more readily.

A second type of rearrangement involves the alkylated aromatic products. *n*-Butylbenzene is readily isomerized to isobutylbenzene and *sec*-butylbenzene under the catalytic effect of aluminum chloride or related Friedel-Crafts catalysts. As might be expected, the tendency toward rearrangement of alkyl groups during Friedel-Crafts alkylations depends on the alkylating agent and the reaction conditions (catalyst, solvent, temperature, etc).

Because of the second type of rearrangement, alkylation of benzene with tertiary alkyl halides yields secondary alkylbenzenes rather than tertiary alkylbenzenes as the principal products (15). For example, the *tert*-hexylbenzene, which is first formed by the reaction of benzene with 2-chloro-2,3-dimethylbutane isomerizes largely to 2,2-dimethyl-3-phenylbutane:

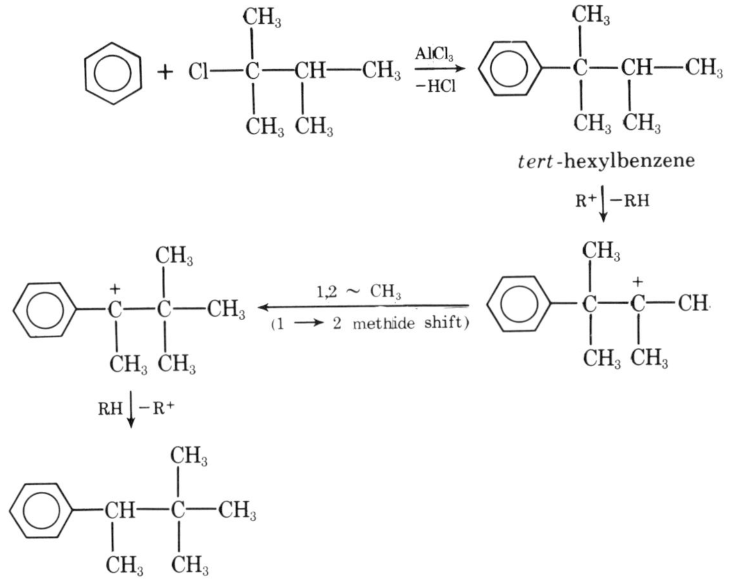

On the other hand, with ferric chloride as the catalyst, the *tert*-hexylbenzene does not undergo isomerization and is isolated as such.

The isomerization of dialkylbenzenes generally results in a considerable increase of the meta isomer, in accordance with the fact that a thermodynamically controlled equilibrium is approached in which the relative ratios of the ortho, meta, and para isomers can be theoretically predicted based on their thermodynamic stabilities (16).

Similar thermodynamically controlled equilibria are obtained in the treatment of dihalo- (17) and polyhalobenzenes, as well as haloalkylbenzenes (18) with Friedel-Crafts catalyst systems. Aryl groups migrate in a way similar to the migration of alkyl and cycloalkyl groups.

Alkylation with Alkyl Halides in the Presence of Isoparaffin. Alkylation of benzene with *tert*-butyl chloride at RT in the presence of $AlCl_3$ and an isoparaffin as the solvent, produces not only *tert*-butylbenzene but also alkylbenzene in which the alkyl group is formed from the isoparaffin (19). For example, the gradual addition of *tert*-butyl chloride to a stirred mixture of benzene, isopentane, and $AlCl_3$ yields mainly 2-methyl-2-phenylbutane during the early stage of the reaction. The relative quantity of the *tert*-butylbenzene increases as the addition proceeds owing to 2,3-methyl shift. The *tert*-butyl cation apparently reacts more rapidly with isopentane than with benzene and, therefore, induces the alkylation of benzene with the isoparaffin.

Stereoselective Alkylation. Stereoselective alkylation of benzene with optically active 4-valerolactone catalyzed by $AlCl_3$ with 50% net inversion of configuration was observed (20). The following mechanism was suggested:

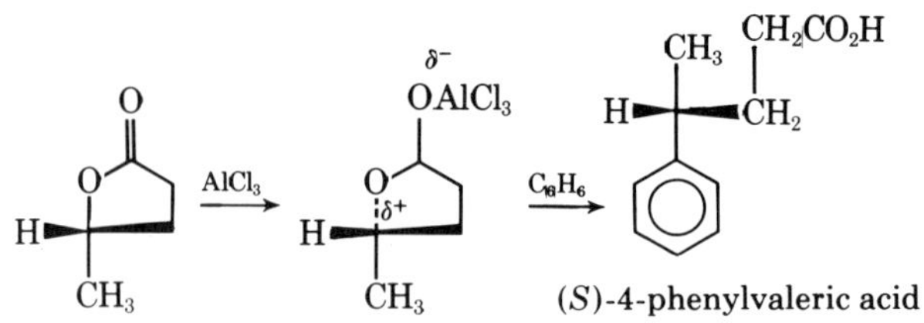

The internal ion-pair complex of the lactone with $AlCl_3$ can racemize by bond rotation and subsequently revert to starting material to give racemized lactone or it can attack benzene with retention of optical activity. Consistent with this mechanism was the observation that dilution with CS_2 increases racemization.

Other stereoselective alkylation reactions were also observed; eg, both 2-methyltetrahydrofuran (21) and 3-chlorobutanoic acid (22) alkylate benzene with inversion of configuration.

Haloalkylation. Haloalkyl groups can be introduced directly by processes similar to Friedel-Crafts alkylations into aromatic and, to a certain degree, into aliphatic compounds.

As haloalkylations involve bi- or polyfunctional alkylating agents, they must be carried out under conditions that promote the primary haloalkylation but not, to any substantial degree, secondary alkylations with the initially formed haloalkylated products.

The most frequently used haloalkylating agents are: aldehydes and hydrogen halides, haloalkyl ethers, haloalkyl sulfides, acetals and hydrogen halides, di- and polyhaloalkanes, haloalkenes, haloalcohols, haloalkyl sulfates, haloalkyl *p*-tosylates, and miscellaneous haloalkyl inorganic acid esters.

Haloalkylations include halomethylations, haloethylations, and miscellaneous higher haloalkylations. Under specific conditions, bis- and polyhaloalkylations can also be carried out.

Haloalkylations are accompanied by alkylation of the aromatic by the product, ie, the haloalkylated product reacts with the original reagent, yielding diarylalkanes or cyclialkylated products, eg:

$$C_6H_6 \xrightarrow[AlCl_3]{CH_3Cl} C_6H_5CH_2Cl \xrightarrow{AlCl_3} C_6H_5CH_2C_6H_5$$

Acidic halide catalysts used in chloromethylations include (23) $ZnCl_2$, $ZnCl_2$ + $AlCl_3$, $SnCl_4$, $SnCl_2$, $AlCl_3$ + ketones, $AlCl_3$ + pyridine, $AlCl_3$ + tertiary amines, $AsCl_3$, $FeCl_3$, BF_3, $TiCl_4$, TiF_4, $BiCl_3$, $SbCl_3$, and $SbCl_5$. The most widely used proton-acid catalysts are HCl, H_2SO_4, H_3PO_4, $H_2SnCl_6.6H_2O$, $ClSO_3H$ + CH_3OH, CH_3COOH, and benzene-, naphthalene, and *p*-toluenesulfonic acids (23).

Zinc chloride is used most frequently. Its activity is increased sometimes by fusion with a small amount of aluminum chloride. In other instances, however, sufficient catalytic effect is obtained with a mineral acid alone.

Reactions have been carried out under heterogeneous conditions with an excess of aromatic compound serving as the solvent, and under homogeneous conditions, as in glacial acetic acid solution (24). Other solvents are diethyl ether, dioxane, dimethoxymethane, carbon tetrachloride, chloroform, ethylene chloride, perchloroethylene, nitrobenzene, ligroin (light petroleum distillate), and carbon disulfide.

Great care must be exercised in chloromethylation reactions, since chloromethyl ethers are highly carcinogenic.

Cyclialkylations. As a rule, Friedel-Crafts alkylation of aromatics with difunctional compounds leads partly to open-chain alkylates but, if conditions are suitable, mainly to cyclialkylation products (25). These compounds react at each end of the difunctional open chain, and thus attach a new cycle to the aromatic nucleus. Such difunctional compounds probably first form a monoalkyl derivative with the aromatic compound yielding an aryl-substituted monofunctional compound. The latter then undergoes an intramolecular alkylation forming the new ring. Intermolecular reactions between a bifunctional alkylating agent and an aromatic compound can also take place.

Intermolecular Cyclialkylations. Ring formation readily occurs in the alkylation of aromatics with di- and polyhalides. Of particular interest is the reaction of di- and trihalomethanes with aromatics in the presence of aluminum chloride. In the reaction of dichloromethane with benzene, besides diarylmethanes, anthracene derivatives are also formed (26):

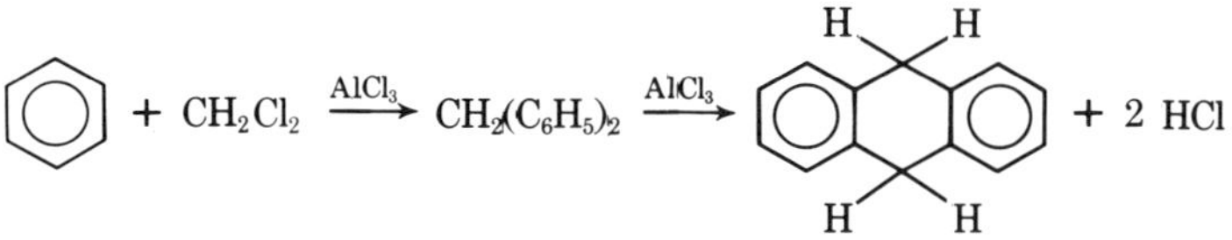

9,10-dihydroanthracene

Cyclialkylation of aromatic compounds with 1,4- or 1,3-glycols in the presence of aluminum chloride leads to the formation of derivatives of tetrahydronaphthalene or indane, respectively (27):

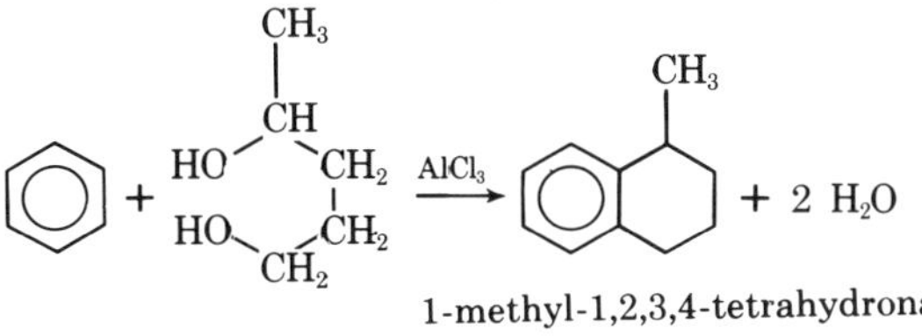

1-methyl-1,2,3,4-tetrahydronaphthalene

Dienes can also be used in Friedel-Crafts cyclialkylations. For example, treatment of phenol with 2,5-dimethyl-1,4-hexadiene gives 5,5,8,8-tetrahydro-2-naphthol (25):

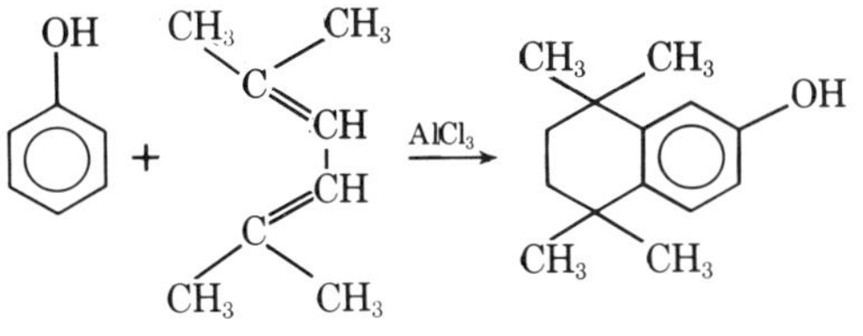

Intramolecular Cyclialkylation. Intramolecular alkylation of haloalkylbenzenes is perhaps the simplest intramolecular counterpart of the classical Friedel-Crafts alkylation reaction. Primary, secondary, or tertiary phenylalkyl chlorides undergo ring closure. Compounds of the general formula $C_6H_5(CH_2)_nCl$ cyclize preferentially to give six-membered ring compounds:

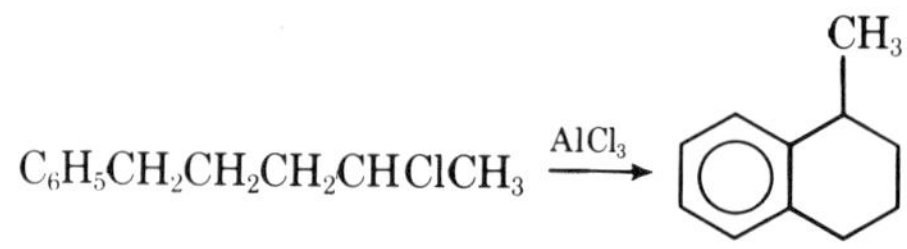

When a tertiary carbocation intermediate is involved, five-membered rings can be formed:

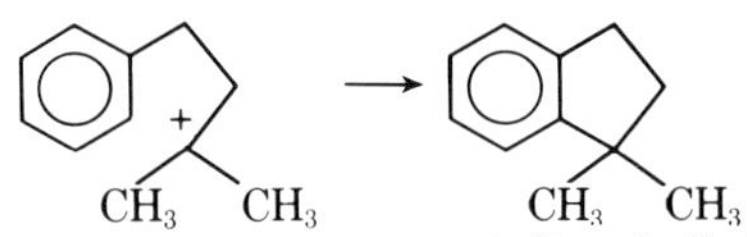

1,1-dimethylindane

Arylalkenes. Aromatic hydrocarbons with an unsaturated side chain undergo ring closure when heated with Lewis acid halides (28):

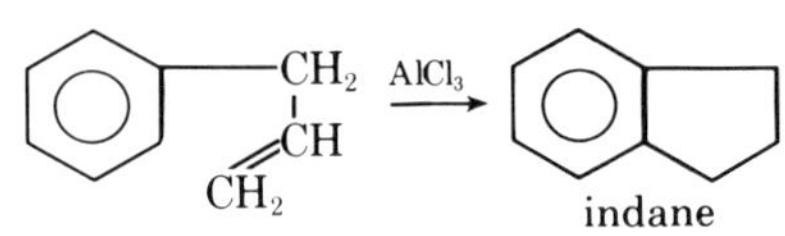

indane

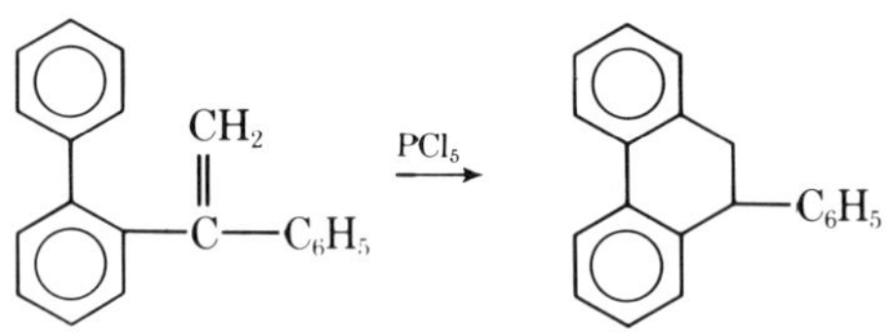

as-phenyl-*o*-biphenylethylene

9-phenyl-9,10-dihydrophenanthrene

Arylalkanes. Alkylbenzenes containing an α-tertiary hydrogen, such as in isopropyl and *sec*-butyl side chains, rapidly transfer this hydrogen as a hydride ion producing an α-carbocation analogous to that formed in the protonation of styrene compounds. Subsequent alkylation of an olefin followed by ring closure produces indanes by much the same process. *p*-Cymene, on treatment with a branched olefin, such as methylcyclohexene, or trimethylethylene in the presence of an acid catalyst, yielded 1,3,3,6-tetramethyl-1-*p*-tolylindane:

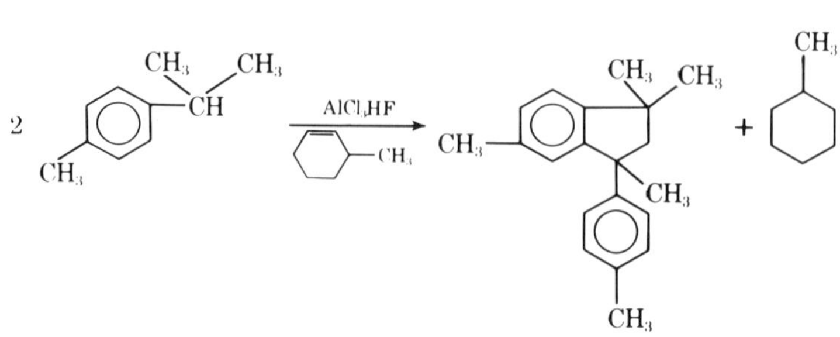

Arylaliphatic Alcohols. Tertiary 3- and 4-phenylalkanols undergo cyclialkylation readily with H_2SO_4. Primary and most secondary 3-phenylalkanols do not undergo cyclialkylation even at higher temperatures with H_3PO_4 (29). Cyclization of phenylalkyl alcohols to tetrahydronaphthalenes and indanes has been extensively investigated (30).

Aliphatic Compounds

The alkylation of olefins by branched-chain paraffins, first carried out by Ipatieff and co-workers (31) over aluminum chloride, made it possible to alkylate olefins with paraffins of normal and branched structures (except methane and ethane).

$$\text{isobutane} \xrightarrow{\text{ethylene}} \text{hexanes} \xrightarrow{\text{ethylene}} \text{octanes} \xrightarrow{\text{ethylene}} \text{decanes}$$

The first industrial aliphatic alkylation process, the sulfuric acid catalyzed alkylation of branched paraffins with olefins, was developed in 1938 (32). Because of its simplicity, low cost, the wide variety of raw materials, and the high-quality alkylate, this process is still widely used. In the presence of 96–97% sulfuric acid, isobutane ($(CH_3)_2CHCH_3$), isopentane ($(CH_3)_2CHCH_2CH_3$), and isohexane ($(CH_3)_2CHCH_2CH_2CH_3$) are easily alkylated by olefins (with the exception of ethylene), producing alkylates with high octane numbers.

The suggested reaction pathway is shown below:

$$R_2C{=}CH_2 \xrightarrow{H^+} R_2\overset{+}{C}CH_3 \xrightarrow{R'_3CH} R_2CHCH_3 + R'_3C^+$$

$$R'_3C^+ + R_2C{=}CH_2 \rightarrow R_2\overset{+}{C}{-}CH_2CR'_3 \xrightarrow{H^-} R_2CHCH_2CR'_3$$

Products do not contain 2,2,3-trimethylbutane or 2,2,3,3-tetramethylbutane, which would be expected as the primary alkylation products in direct alkylation of isobutane with propene and isobutylene, respectively. In fact, the process involves olefin alkylation by the carbocation.

On the other hand, under superacidic conditions, alkanes are readily alkylated via front-side σ-insertion by carbocationic alkylating agents. The direct alkylation of the tertiary C—H σ-bond of isobutane, eg, was demonstrated (33). Even in the sterically unfavorable reaction of *tert*-butylfluoroantimonate with isobutane, 2% of the C_8 fraction was 2,2,3,3-tetramethylbutane:

$$(CH_3)_3C^+ + HC(CH_3)_3 \rightleftharpoons (CH_3)_3CC(CH_3)_3 + H^+$$

No alkenes are formed (34). For example, when using $HF{:}TaF_5$ in a flow system for alkylation of excess ethane with ethylene (in a 9:1 molar ratio), only *n*-butane was obtained; isobutane was not detectable even by gas chromatography. Only the σ-insertion mechanism can account for these results:

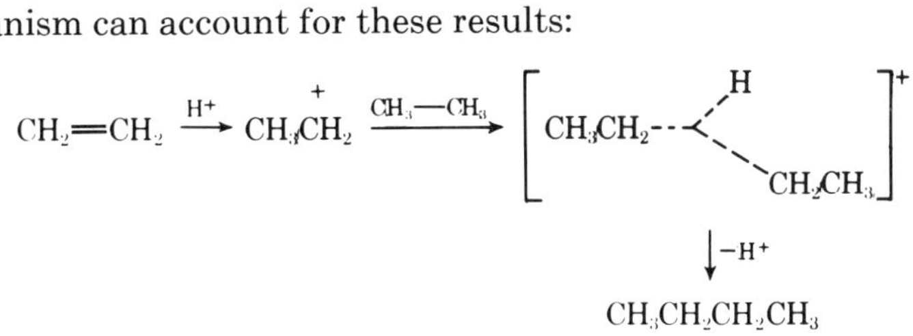

If the ethyl cation would alkylate ethylene, the reaction would proceed through butyl cations, inevitably leading also to the formation of isobutane:

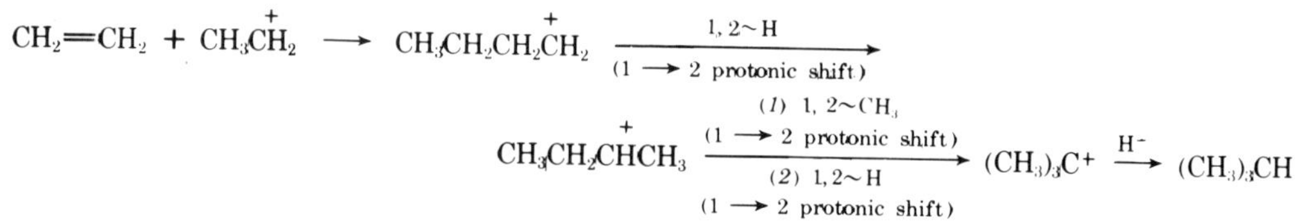

The condensation of alkyl halides containing at least three carbon atoms with olefinic hydrocarbons and haloolefins is catalyzed by Friedel-Crafts halide catalysts (35). The primary reaction consists of the addition of the alkyl group and the halogen atom to the double bond, the halogen adding to the carbon atom that holds the smaller number of hydrogen atoms. In general, the reaction takes place more readily with tertiary than primary and secondary halides.

Depending largely upon the particular reactants and the catalyst, the products are mainly alkyl halides containing quaternary carbon atoms, and are obtained in yields up to 75%. These compounds may further be dehydrohalogenated to the corresponding alkylated olefins and haloolefins.

Lower alkanes such as methane and ethane were polycondensed in superacid solutions at 50°C, yielding higher liquid alkanes (36). The proposed mechanism for the oligocondensation of methane requires the existence of protonated alkane (pentacoordinated carbonium ions):

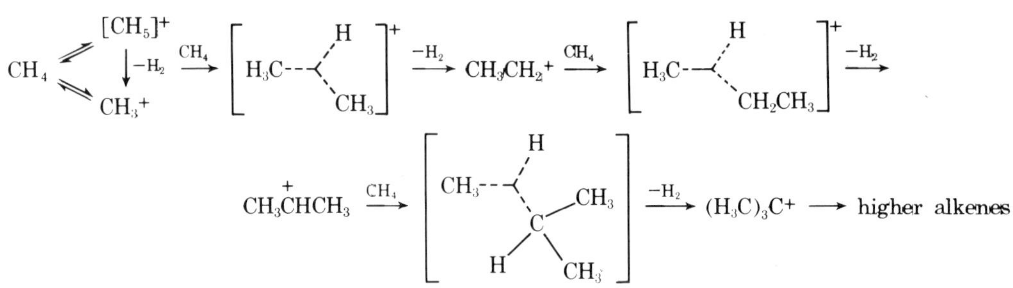

Arylation of Aromatic Compounds

In contrast to the Friedel-Crafts alkylations, arylations of aromatics are by no means as well known or as easy to carry out. They include: (*1*) dehydrogenating condensation (Scholl reactions); (*2*) arylation with aryl halides; and (*3*) arylation with diazonium halides.

Dehydrogenating Condensation. The elimination of two ring hydrogens accompanied by the formation of an aryl—aryl bond under Friedel-Crafts conditions is known as the Scholl reaction. The dehydrogenating condensations can take place in either an inter- or an intramolecular way. Intermolecular Scholl reactions are numerous and include such reactions as formation of biphenyl from benzene:

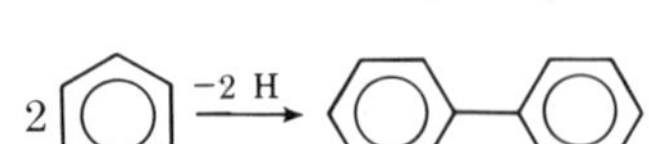

of perylene from naphthalene (through binaphthyl):

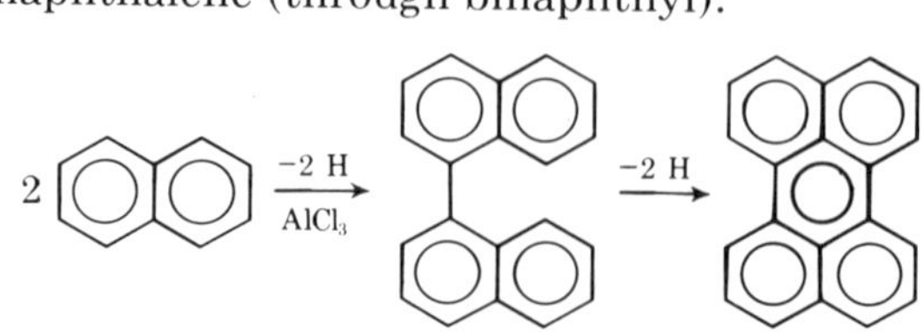

of 2,2′-dipyridyl from pyridine:

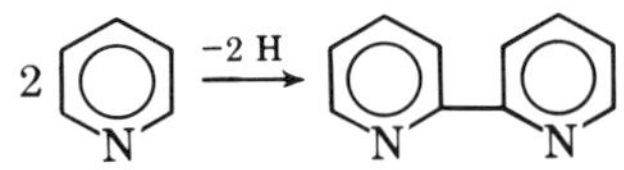

and the formation of high molecular weight polycondensate aromatics.

Aryl Halides. Because of their inertness, aryl halides are rarely used as electrophilic arylating agents in Friedel-Crafts-type reactions. However, under suitable conditions, simple halobenzenes may be used as Friedel-Crafts arylating agents (37). Their decreasing order of reactivity is F ≫ Cl > Br. Aromatic hydrocarbons are arylated with fluorobenzene in the presence of aluminum halides to the corresponding biphenyls. The directive effects are in agreement with an electrophilic aromatic substitution:

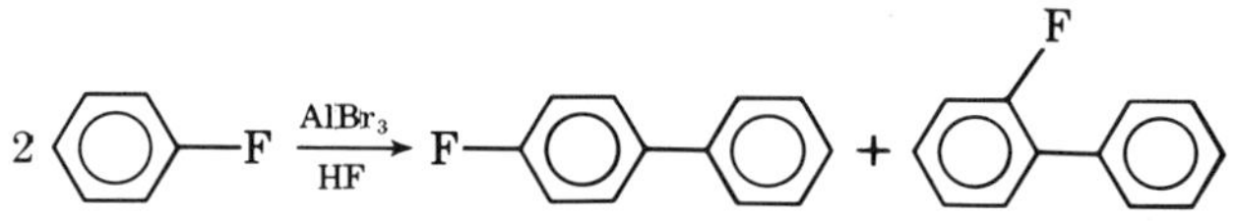

p-fluorobiphenyl *o*-fluorobiphenyl

Diazonium Halides. In the investigation of the decomposition reaction of aryldiazonium tetrafluoroborates in nitrobenzene, it was found that in addition to fluorobenzene, 3,3′-dinitrobiphenyl was formed (38). An ionic type of arylation reaction seems to take place.

Decomposition of aryldiazonium tetrafluoro-, tetrachloro-, and tetrabromoborates in aromatic solvents leads to electrophilic ring arylation (39).

Isomerization of Saturated Hydrocarbons

The ability of Friedel-Crafts-type superacids to protonate not only π-bonds but also the weaker σ-bonds of hydrocarbons at relatively low temperatures, allows the isomerization of saturated hydrocarbons which is of substantial practical importance. Straight-chain alkanes of the gasoline range (C_5–C_8 alkanes), have considerably lower octane numbers than their branched isomers. The isomerization of alkanes can be effected by catalysts such as $AlCl_3$, $AlBr_3$, HF–SbF_5, HF–TaF_5, HSO_3–SbF_5, and the like. The stronger superacids allow the reaction to be carried out at relatively low temperatures, which favor the branched alkanes. Recently, processes for the isomerization of straight-chain alkanes in the gas phase over acid or molecular-sieve catalysts were described (40) (see Molecular sieves).

Isomerization of straight-chain hydrocarbons is of particular importance in so-called lead-free gasoline (qv). Addition of higher octane aromatic hydrocarbons or olefins is less advisable because of pollution problems (41). Efficient additives are methyl or *tert*-butyl ethers.

Polymerization

The Friedel-Crafts polymerization of olefins did not attract much attention until the 1930s (see Polymerization). Since then a considerable amount of interest has been focused on olefin polymerizations catalyzed by $AlCl_3$ and BF_3 and a variety of other Lewis-acid-type halides and protic acids; three main types are recognized (42): (*1*) conversion of low molecular weight olefins to gasoline-range olefins (43); (*2*) conversion to intermediate molecular weight polymers for use as synthetic lubricant oils; and (*3*) conversion to high molecular weight polymers having 5,000–100,000 monomer units.

Under suitable conditions, the degree of polymerization may be controlled to produce any of the above types. Typical monomers that can readily be converted to high molecular weight polymers are isobutylene, styrene, α-methylstyrene, butadiene, isoprene, and vinyl alkyl ethers.

Even alkanes, when treated with superacids, can undergo oligocondensation (as described above). For example, highly branched polyalkanes, of molecular weight up to 700, were obtained by treating gaseous alkanes (C_1–C_4) with liquid superacids at RT (44) (see also Olefin polymers).

ACYLATION

Acylation of Aromatic Compounds

Ketone Synthesis. In the Friedel-Crafts ketone synthesis, an acyl group is introduced into the aromatic nucleus by an acylating agent such as an acyl halide, acid anhydride, ester, or the acid itself:

$$C_6H_6 + RCOX \xrightarrow{AlCl_3} C_6H_5COR + HX$$
$$X = \text{halide}, -OH, -OR, -OCOR$$

Ketenes, amides, and nitriles also may be used; aluminum chloride and boron trifluoride are the most common catalysts (see Ketones).

Friedel-Crafts acylation of aromatics is of considerable practical value owing to the importance of aryl ketones and aldehydes. Acylating agents in general are more reactive than alkylating agents, and the reaction of acyl halides with aromatic compounds in the presence of a Friedel-Crafts catalyst proceeds more readily than the corresponding alkylations with alkyl halides. These reactions are also successful with aromatic hydrocarbons and their substituted derivatives including halo, alkoxy, and even amino derivatives, although as a general rule, electronegative groups have an inhibiting effect. Usually, it is difficult to introduce more than one acyl group into an aromatic ring, although diacylation products are reported in the literature. A high yield of diacetylmesitylene was obtained when mesitylene was acetylated with a sixfold excess of catalyst and the Elbs method was used (44). In another study (45), 63% conversion to the diacylated product was obtained in the acetylation of fluoranthrene. Acetylation of benzene with 2,6-dimethoxybenzoyl chloride gave the following products (46):

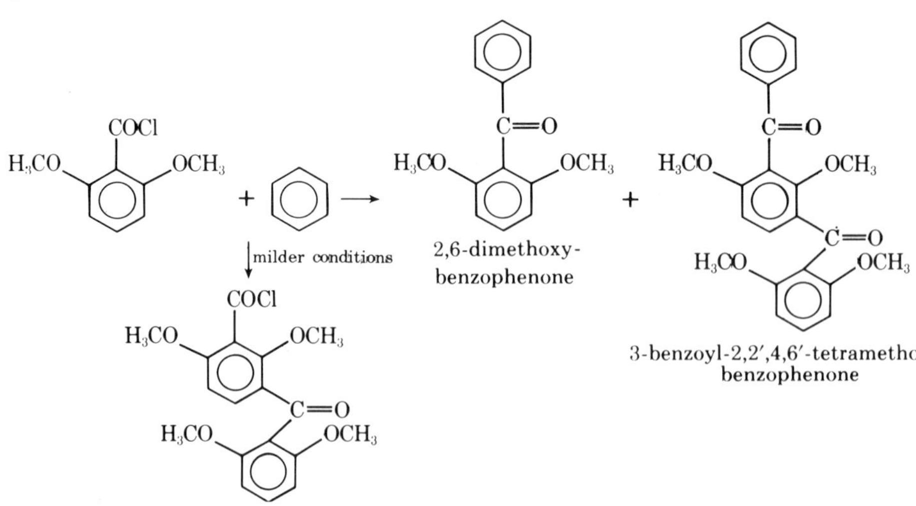

2,6-dimethoxy-benzophenone

3-benzoyl-2,2′,4,6′-tetramethoxy-benzophenone

3-(2,6-dimethoxybenzoyl)-2,6-dimethoxy-benzoyl chloride

It is characteristic of the Friedel-Crafts acylation of ortho–para directing aromatic compounds that the para derivatives are formed, in most cases, with almost complete exclusion of the ortho isomer. The reason for this behavior was long attributed to steric factors, although exceptions were known (47). In a recent systematic study of the effect of substituents in acylating agents on the selectivity of Friedel-Crafts acylation (48), the $AlCl_3$-catalyzed benzoylation of toluene with benzoyl chloride gave 8% 2- and 91% 4-methylbenzophenone. In contrast, acylation with 2,6-dinitrobenzoyl chloride gave 42% ortho isomer and 55% para isomer.

The basicity of the aromatic compound also affects the isomer distribution. This behavior is best explained by a change in the nature of the transition state of the highest energy intermediate ion, imparting more of the character of the starting hydrocarbon (49).

1-Phenylnorbornane is benzoylated faster than isopropylbenzene or toluene despite the bulkiness of the norbornyl group probably because of hyperconjugation (50). Moreover, hyperconjugation of the C—C bond is at least as or more important as that of the C—H bond since the fastest reaction is that with 1-phenylnorbornane which has no α-hydrogen atom.

Interaction of the aromatic derivative, eg, C_6H_6 acyl halide (RCOX), and aluminum halide (AlX_3) liberates hydrogen halide and produces a complex of aromatic ketone and aluminum halide from which the ketone is liberated by hydrolysis:

$$C_6H_6 + RCOX + AlX_3 \longrightarrow HX + \begin{bmatrix} C_6H_5\underset{\displaystyle R}{\underset{|}{C}}O\,AlX_3 \end{bmatrix} \xrightarrow{H_2O} C_6H_5COR$$

Elbs Method. The most widely used procedure for ketone synthesis is that of Friedel and Crafts themselves, further developed by Elbs (51), in which the catalyst is added as the last reactant.

Bouveault Method. In general, addition of the acyl halide to a mixture of the aromatic compound and aluminum halide [Bouveault method (52)] is not preferred, since the presence of hydrogen halide from impurities, moisture, or formed after the first addition of acyl halide may cause extensive isomerization or disproportionation.

Perrier Method. In the Perrier method (53) the catalyst (aluminum chloride) and the acyl halide are allowed to react before the addition of the aromatic compound. Thus the addition compound of the acyl halide and the catalyst is prepared first and then allowed to react with the hydrocarbon.

Stable acyl fluoride–Lewis acid fluoride complexes, found to be oxocarbonium (acylium) salts (54), can be used advantageously in a modification of the Perrier synthesis:

$$CH_3COF + BF_3 \rightleftharpoons CH_3CO^+BF_4^-$$

$$CH_3CH_2COF + SbF_5 \rightleftharpoons CH_3CH_2CO^+SbF_6^-$$

$$C_6H_5COF + AsF_5 \rightleftharpoons C_6H_5CO^+AsF_6^-$$

The reaction of acyl chlorides or bromides with anhydrous silver salts leads to the preparation of acylium salts that are highly reactive acylating agents. As these metathetic reactions are carried out in acid-free media (with the exception of the byproduct conjugate acid formed in the substitution reaction with aromatics), they are applicable to systems otherwise sensitive to Lewis or Brønsted–Lowry acid catalysts:

$$RCOCl + AgSbF_6 \rightleftharpoons RCO^+SbF_6^- + AgCl$$

Whereas alkylation with alkyl halides requires only catalytic amounts of catalysts (aluminum halide), acylation affords ketones in yields that are proportional to the amount of aluminum halide added. The reaction is complete when a little more than a molar proportion of catalyst is used.

Apparently one molar equivalent of catalyst combines with the acyl halide, giving a 1:1 addition compound, which then acts as the active acylating agent:

$$RCOX + AlX_3 \rightarrow RCO^+AlX_4^-$$

Evidence supporting the formation of 1:1 addition compounds is further substantiated by the actual isolation of stable acyl halide–Lewis acid complexes.

The conclusion (55) that the catalyst is firmly attached to the resulting ketone and is no longer able to activate more acyl halide is obvious. However, this conclusion is not wholly justified since halogen exchange between acyl and aluminum halides continues in the presence of excess ketone. Apparently, the presence of ketone does not prevent formation of acyl cation but lowers its efficiency as an acylating agent. This deactivation is probably effected by association:

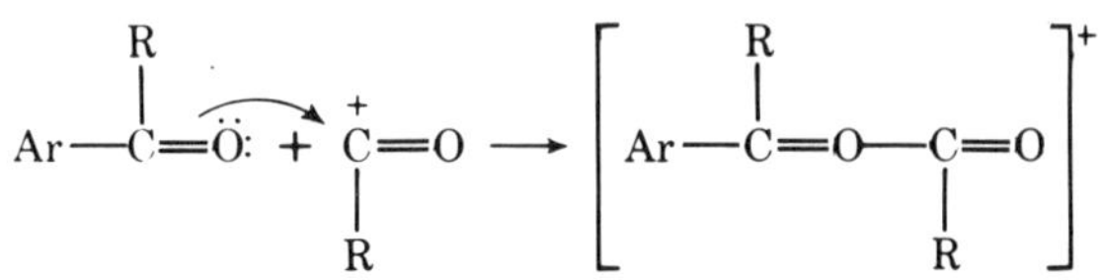

Similarly, the ability of solvents or added substances, eg, *m*-dinitrobenzene, acyl halides, nitrobenzene, nitromethane, diethyl ether, benzophenone, *p*-methoxybenzophenone, to retard or even to suppress Friedel-Crafts reactions is related to their ability to solvate acyl cations, and thus to reduce the electron demand and the acylating activity.

A further consequence of association of acylating agents with basic compounds is an increase in the bulk of the reagent, and greater resistance to attach at the more sterically hindered positions of aromatic compounds. Thus acylation of chrysene and phenanthrene in nitrobenzene or in carbon disulfide occurs to a considerable extent in an outer ring, whereas acylation of naphthalene leads to extensive reaction at the less reactive but sterically less hindered 2-position.

Cycliacylation. Cyclic ketones can be prepared by intramolecular Friedel-Crafts acylation of an aromatic ring that has an acyl halide group in an attached side chain. The method is used in the preparation of hydrindones, tetralones, chromones, xanthones, etc. Cyclizations of ketoacyl chlorides leading to the formation of coumarandiones, chromandiones, indandiones, and anthraquinones are further examples of intramolecular cycliacylations; five- and six-membered ring ketones are generally obtained in good yield by this method.

In an alternative procedure the acids themselves are cycliacylated, catalyzed by hydrogen fluoride. γ-Phenylbutyric acid can be cyclized to α-tetralone by this method:

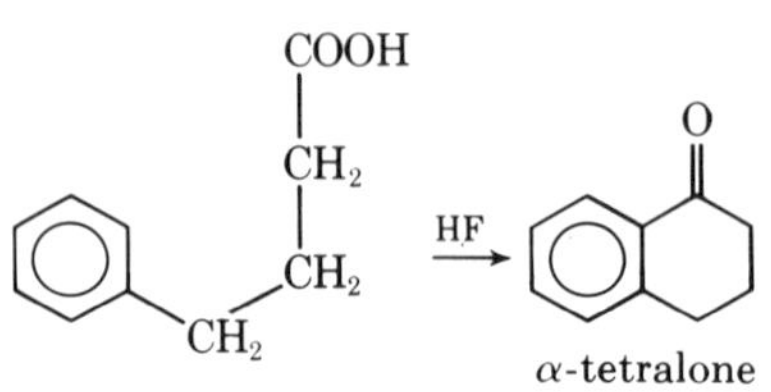

Polyphosphoric acid (PPA) is often employed as a catalyst for cycliacylations.

Cycliacylations readily take place in intermolecular acylations involving bifunctional acylating agents. Both functional groups may be acyl (as in the case of α,ω-diacyl halides) or one may be an alkylating group (as in unsaturated acyl halides or certain haloacyl halides) (56):

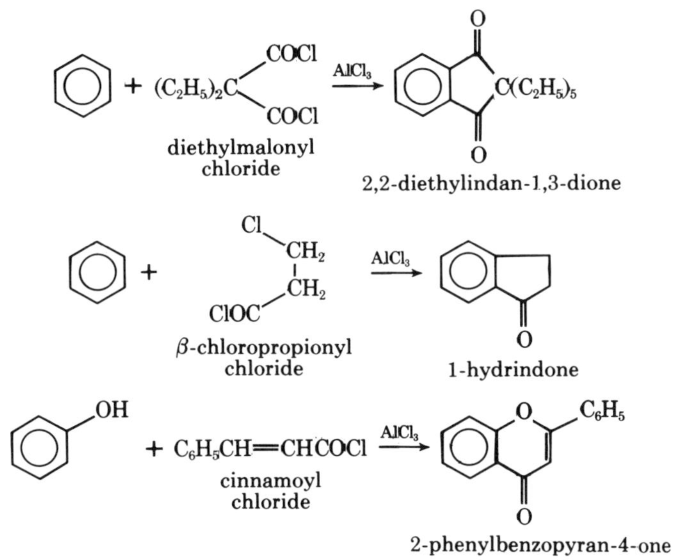

Polystyrene can be cross-linked by its acylation with bifunctional acylating agents such as adipoyl, sebacoyl, or malonyl chlorides in the presence of $AlCl_3$ in CS_2 solution at 0°C (57).

Aldehyde Synthesis. Formylation would be expected to take place when formyl chloride or formic anhydride reacts with an aromatic compound in the presence of aluminum chloride or other Friedel-Crafts catalysts. However, the acid chloride and anhydride of formic acid are both too unstable to be of preparative interest.

Gattermann-Koch Synthesis. In the presence of aluminum chloride and a small amount of cuprous halide, a mixture of hydrogen chloride and carbon monoxide serves as a formylating agent of aromatics (58):

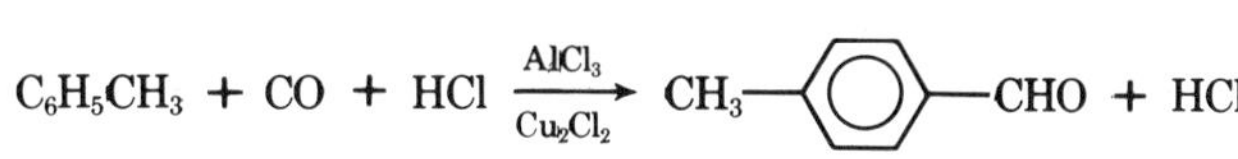

p-tolualdehyde

Current theories suggest that the presence of formyl chloride is not necessary since the actual alkylating agent, CHO^+, can be produced by protonation of carbon monoxide or its complexes. However, it is difficult to obtain an equimolar mixture of anhydrous hydrogen chloride and carbon monoxide in the laboratory. A suitable laboratory preparation involves the reaction of chlorosulfonic acid with formic acid:

$$HCO_2H + ClSO_3H \rightarrow H_2SO_4 + CO + HCl$$

or the reaction of benzoyl chloride with formic acid:

$$HCO_2H + C_6H_5COCl \rightarrow C_6H_5CO_2H + CO + HCl$$

The Gattermann-Koch synthesis is suitable for the preparation of simple aromatic aldehydes from benzene and its substituted derivatives, as well as from polycyclic aromatics. The para isomers are produced preferentially. Aromatics with meta-directing substituents cannot be formylated (59).

Formylation. The stability of formyl fluoride (60), HCOF, allows a direct, simple Friedel-Crafts formylation reaction (54):

$$C_6H_6 + FCHO \rightarrow C_6H_5CHO + HF$$

Attempts to use the stable acetic–formic anhydride with Lewis acid metal-halide catalysts result only in acetylation:

$$C_6H_6 + CH_3COOOCH \xrightarrow{AlCl_3} C_6H_5COCH_3 + H_2O + CO$$

However, using anhydrous HF as a catalyst, a small amount of aldehyde is also formed. This is in accordance with the fact that acetic–formic anhydride gives both acetyl and formyl fluoride with HF. By continuous removal of the low boiling HCOF, the reaction can be directed to the predominant formation of this compound (54).

Gattermann Reaction. The Gattermann-Koch reaction cannot generally be applied to formylation of phenolic compounds. To overcome this difficulty, Gattermann (61) allowed a mixture of hydrogen cyanide and hydrogen chloride to react with phenols in the presence of zinc or aluminum chloride:

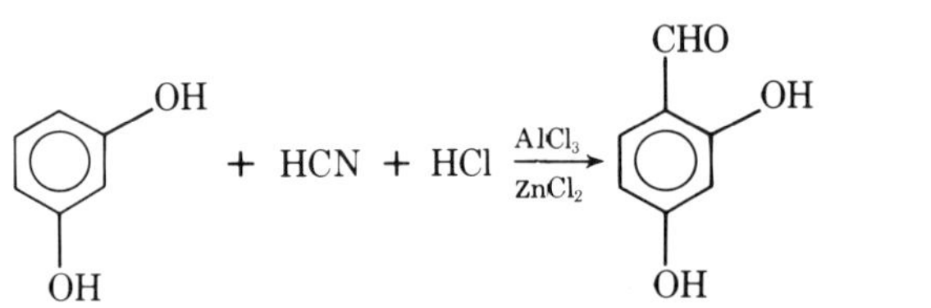

2,4-dihydroxybenzaldehyde

With zinc or sodium cyanides, hydrogen cyanide is generated *in situ* by addition of hydrogen chloride and at the same time zinc chloride, the required Friedel-Crafts catalyst is produced. This procedure has the great advantage that it eliminates the handling of hydrogen cyanide. If zinc chloride is not a sufficiently strong catalyst, aluminum chloride is added also. Curiously, very pure zinc chloride does not bring about the reaction; it is, however, activated by small amounts of metal halides.

In another modification (62) cyanogen bromide is used in place of hydrogen cyanide:

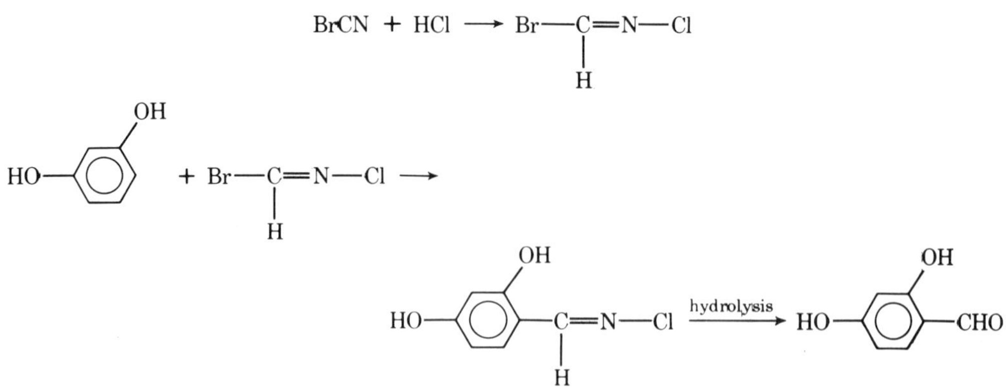

Vilsmeier Aldehyde Synthesis. Aromatic and heterocyclic compounds are formylated by reaction with dialkyl- or alkylarylformamides in the presence of phosphorus oxychloride or phosgene (63).

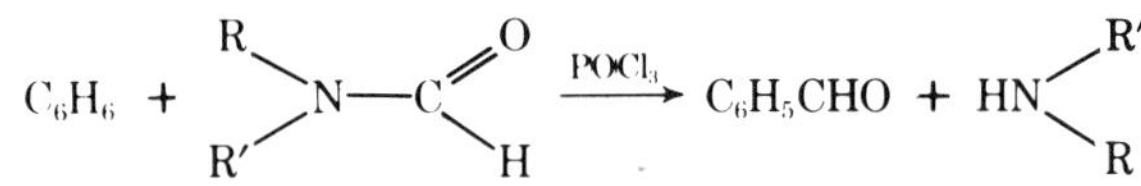

The Vilsmeier reaction is indeed an electrophilic substitution or a Friedel-Crafts-type formylation (64), since the intermediate cation formed by the interaction of phosphorus oxychloride with the formamide is the electrophilic reagent. Ionic addition compounds of the formamide with phosgene or phosphorus oxychloride are also possible (65).

Aromatic compounds are formylated also by dichloromethyl methyl ether or trialkyl orthoformate (66):

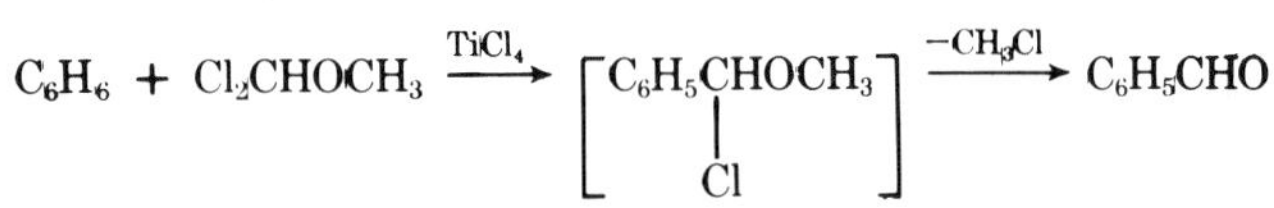

$$C_6H_5OH + HC(OR)_3 \xrightarrow{AlCl_3} p\text{-}HOC_6H_4CH(OR)_2 \xrightarrow[H^+]{H_2O} p\text{-}HOC_6H_4CHO$$

Syntheses of Aryl Nitriles and Aldoximes. Mercuric fulminate reacts with benzene and aluminum chloride containing some hydrated salt to give benzaldoxime. It is assumed that the reaction proceeds through primary formation of fulminic acid by the action of HCl on the fulminate:

$$C_6H_6 + C{=}NOH \rightarrow C_6H_5CH{=}NOH$$

Using freshly sublimed aluminum chloride, benzonitrile is formed (see Nitriles). The formation of benzonitrile is probably due to subsequent formation of cyanogen chloride, by the action of the catalyst on isocyanic (fulminic) acid, and condensation with the hydrocarbon (67):

$$C{=}NOH \xrightarrow{AlCl_3} C{=}NCl \rightarrow ClC{\equiv}N \xrightarrow[-HCl]{C_6H_6} C_6H_5CN$$

Cyanogen bromide condenses with toluene in the presence of $AlCl_3$ to give *p*-tolunitrile. The reaction appears to be more satisfactory with phenol ethers (68):

$$CH_3C_6H_5 + BrCN \xrightarrow{AlCl_3} p\text{-}CH_3C_6H_4CN + HBr$$

An alternative method (69) consists of the reaction of trichloroacetonitrile with a hydrocarbon and $AlCl_3$. A ketimine is formed which is hydrolyzed by treatment with alkali into the nitrile and chloroform. The reaction has been applied to toluene, some of its homologues, and phenols:

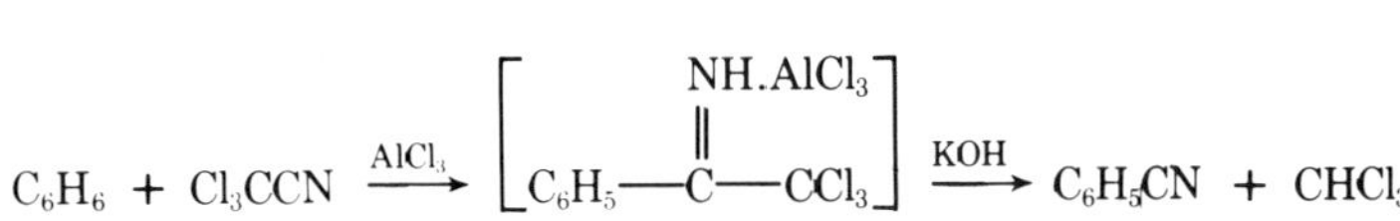

Syntheses of Arylcarboxylic Acids and Derivatives. The general Friedel-Crafts acylation principle can be successfully applied to the preparation of aromatic carboxylic acids. Carbonyl halides (eg, phosgene, carbonyl chloride fluoride, or carbonyl fluoride) are diacyl halides of carbonic acid. Phosgene or oxalyl chloride react with aromatic hydrocarbons in the presence of a Friedel-Crafts catalyst to give aroyl chlorides which yield acids on hydrolysis (70):

$$C_6H_6 + COCl_2 \xrightarrow[-HCl]{AlCl_3} C_6H_5COCl \xrightarrow{H_2O} C_6H_5COOH$$

However, excess aromatic gives the corresponding ketone:

$$C_6H_5COCl + C_6H_6 \xrightarrow[-HCl]{AlCl_3} C_6H_5COC_6H_5$$

Carbon disulfide as solvent favors the formation of the acid since the intermediate complex formed, $C_6H_5COCl.AlCl_3$, is insoluble in it and by precipitating avoids secondary ketone formation.

Ketone formation can also be avoided if one of the functional acyl halogens in phosgene is blocked. Carbamyl chlorides, readily obtainable by the reaction of phosgene with ammonia or amines, are very suitable reagents for the preparation of amides in direct Friedel-Crafts acylation of aromatics. The resulting amides can be hydrolyzed to the acids (71):

$$C_6H_6 + ClC(=O)NR_2 \xrightarrow{AlCl_3} C_6H_5CONR_2 \xrightarrow{hydrolysis} C_6H_5COOH$$

The practical applicability of this reaction has been demonstrated in an industrial preparation of terephthalic acid (72):

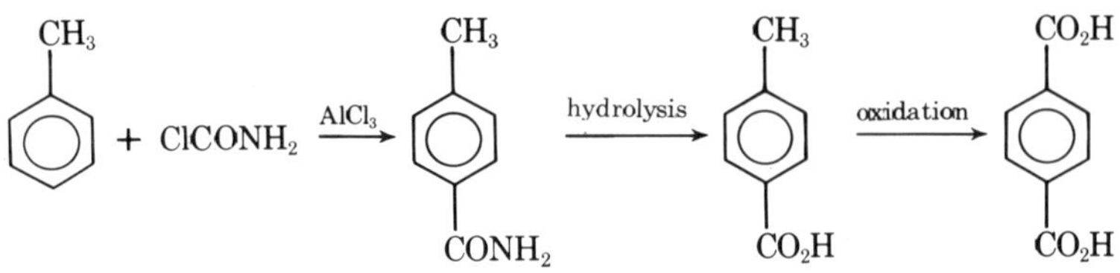

The reaction succeeds well also with substituted carbamyl chlorides such as $C_6H_5(C_2H_5)NCOCl$.

Amides result from the reaction of aromatic hydrocarbons with isocyanates, such as phenyl isocyanate, in the presence of aluminum chloride. Phenyl isothiocyanate similarly gives thioanilides (73):

$$C_6H_5NCO + C_6H_6 \xrightarrow{AlCl_3} C_6H_5NHCOC_6H_5$$

In these reactions the active acylating agent is, indeed, the carbamyl chloride, formed by interaction of isocyanate with hydrogen chloride (74):

$$C_6H_5NCO + HCl \rightarrow C_6H_5NHCOCl$$

Carbon dioxide can be looked upon as the acid anhydride of the dibasic carbonic acid. Accordingly, it is not unexpected that it reacts with benzene to give benzoic acid and benzophenone in the presence of aluminum chloride (75):

$$C_6H_6 + CO_2 \xrightarrow{AlCl_3} C_6H_5CO_2AlCl_2 + HCl \xrightarrow{H_2O} C_6H_5CO_2H$$

With more reactive substances, such as phenols, zinc or ferric chlorides may be substituted as catalysts (75). The reaction is, however, not easy to carry out in good yield and without side reactions. Elevated temperatures and high pressure are generally needed, and only reactive substrates (such as phenols) react readily.

Sulfonylation. Under Friedel-Crafts reaction conditions, sulfonyl halides and sulfonic acid anhydrides sulfonylate aromatics (76), a reaction that can be considered as the analogue of the related acylation with acyl halides and anhydrides. The products are sulfones (instead of ketones obtained in the acylation reactions). Sulfonyl chlorides are the most frequently used reagents, although the bromides and fluorides also react:

$$C_6H_6 + RSO_2Cl \xrightarrow{AlCl_3} C_6H_5SO_2R + HCl$$

Methanesulfonic and benzenesulfonic anhydrides are the anhydrides used in Friedel-Crafts sulfonylation reactions:

$$C_6H_6 + (CH_3SO_2)_2O \rightarrow C_6H_5SO_2CH_3 + CH_3SO_3H$$

Benzenesulfonic anhydride was claimed to be superior to benzenesulfonyl chloride in Friedel-Crafts sulfonylation (77). Catalysts used besides aluminum chloride are ferric chloride, antimony pentachloride, aluminum bromide, and boron trifluoride (78).

Sulfonation. The general Friedel-Crafts acylation principle can be applied in the case of sulfonation to the halides and anhydride of sulfuric acid (halosulfonic acids, sulfur trioxide). Aluminum chloride and boron trifluoride are effective catalysts in certain sulfonations with halosulfonic acids. In a general sense, sulfonations by sulfuric acid or oleum should also be considered in the scope of Friedel-Crafts acylations, the acids providing both sulfonating agent and catalyst.

Sulfonation of aromatic hydrocarbons with sulfuric acid is catalyzed by hydrogen fluoride or, at lower temperatures, by boron trifluoride (79). The products obtained are more uniform and considerably less sulfuric acid is needed, probably because the BF_3 forms complexes with the water formed in the reaction, and thus prevents dilution of the sulfuric acid.

The migration of alkyl or halogen atoms during the sulfonation of polyalkyl (polyhalo) benzenes containing at least four substituents is known as the Jacobsen rearrangement (80). Thus 2,3,5,6-tetramethylbenzenesulfonic acid is rearranged chiefly to 2,3,4,5-tetramethylbenzenesulfonic acid (see Sulfur compounds).

Sulfination. Benzene and its homologues react with SO_2 in the presence of $AlCl_3$ and HCl to form sulfinic acids (81):

$$C_6H_6 + SO_2 \xrightarrow[HCl]{AlCl_3} C_6H_5SO_2H$$

Nitration. The general Friedel-Crafts acylation principle can also be applied to nitrations involving nitryl halides and dinitrogen pentoxide (the halides and the anhydride of nitric acid).

In a more general sense, nitric acid nitrations catalyzed by proton acids (H_2SO_4, $HClO_4$, HF, etc) or by Lewis acid-type metal halides (BF_3, $AlCl_3$, etc) should also be considered Friedel-Crafts-type reactions in analogy to the ketone syntheses where carboxylic acids are used as acylating agents.

Friedel-Crafts nitration using nitryl chloride with aluminum chloride as catalyst has been reported (82):

$$C_6H_6 + NO_2Cl \xrightarrow{AlCl_3} C_6H_5NO_2 + HCl$$

Anhydrous silver tetrafluoroborate is also a suitable cation-forming agent in these reactions (83):

$$C_6H_6 + NO_2Cl + AgBF_4 \rightarrow C_6H_5NO_2 + AgCl + HF + BF_3$$

Dinitrogen tetroxide is an effective Friedel-Crafts nitrating agent (84) for aromatics in the presence of aluminum chloride, ferric chloride, or sulfuric acid (85).

Dinitrogen pentoxide is a powerful nitrating agent, even in the absence of catalysts, preferably in sulfuric acid solution (86). Solid dinitrogen pentoxide is known to be the nitronium nitrate, $(NO_2)^+(NO_3)^-$. The use of BF_3 as catalyst has been reported (87).

Stable nitronium salts such as $(NO_2)^+(BF_4)^-$, $(NO_2)^+(AsF_6)^-$, $(NO_2^+)_2(SiF_6^{2-})^-$, $(NO_2)^+(HS_2O_7)^-$, etc, are extremely powerful Friedel-Crafts nitrating agents for aromatics (88):

$$ArH + (NO_2)^+(BF_4)^- \rightarrow ArNO_2 + HF + BF_3$$

Nitronium tetrafluoroborate is easily obtained from N_2O_5, HF, and BF_3:

$$N_2O_5 + 2\ HF + 3\ BF_3 \rightleftharpoons 2(NO_2)^+(BF_4)^- + BF_3.H_2O$$

or from nitric acid, HF, and BF_3 (89).

The nitrations can be carried out in homogeneous media, using tetramethylene sulfone or nitromethane as solvent. A large variety of aromatic compounds have been nitrated with nitronium salts in excellent yields in nonaqueous media. Thus sensitive compounds, otherwise easily hydrolyzed or oxidized by nitric acid, can be nitrated without secondary effects. Nitration of aromatic compounds is considered an irreversible raction; however, recently the reversibility of the reaction had been demonstrated when 9-nitroanthracene, as well as pentamethylnitrobenzene transnitrated benzene, toluene, and mesitylene in the presence of superacids (90) (see Nitration):

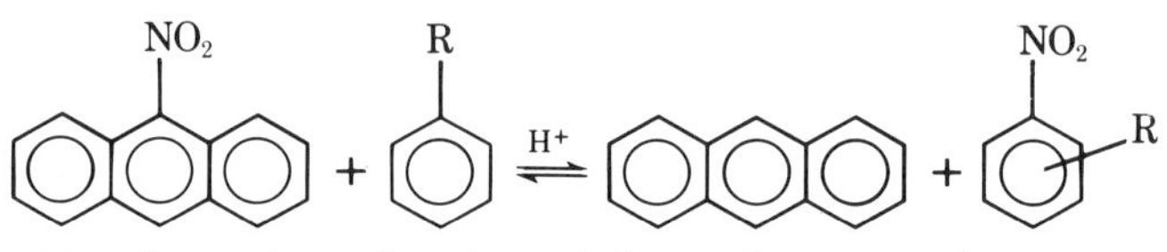

Amination. The direct introduction of the amino group into aromatic hydrocarbons was first reported by Graebe in 1901 (91). Hydroxylamine hydrochloride in the presence of aluminum or ferric chlorides reacts with benzene or toluene to give a low yield of aniline or toluidine:

$$C_6H_6 + NH_2OH.HCl \xrightarrow[-H_2O]{AlCl_3} C_6H_5NH_2.HCl$$

Hydroxylamine sulfate (92), *N*-alkylhydroxylamines (93), hydroxylamine-*O*-sulfonic acid and some of its derivatives were used in similar reactions:

$$C_6H_6 + H_2NOSO_3H \xrightarrow{AlCl_3} C_6H_5NH_2 + H_2SO_4$$

Data on the reaction of chloramine with aromatic hydrocarbons are insufficient, although in certain cases amination has been indicated.

$$C_6H_6 + ClNH_2 \rightarrow C_6H_5NH_2.HCl$$

An unusual predominately meta substitution takes place in the amination of toluene in the presence of trichloramine–$AlCl_3$ (94). The following *o*-substitution pathway was proposed (see Amines):

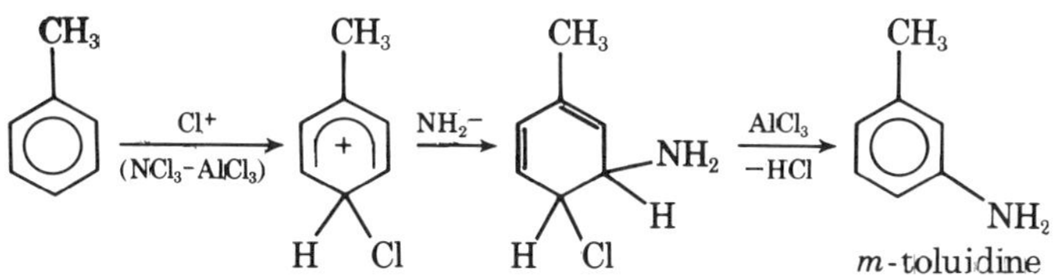

Perchlorylation. The stability of perchloryl fluoride, $FClO_3$, made possible the development of a new type of aromatic substitution reaction, perchlorylation, a reaction closely related to the Friedel-Crafts acylation (95):

$$C_6H_6 + FClO_3 \xrightarrow{AlCl_3} C_6H_5ClO_3 + HF$$

Aluminum bromide was found to be an equally active catalyst, whereas boron trifluoride is considerably less active probably because of its limited solubility in aromatic hydrocarbons. The perchloryl aromatics are interesting compounds but they must be handled with care because of their explosive nature and sensitivity to mechanical shock and local overheating.

Halogenation. The halogenation of a wide variety of aromatic compounds, a generalized acylation reaction, proceeds readily in the presence of ferric chloride, aluminum chloride, and related Friedel-Crafts catalysts. Halogenating agents used are elementary chlorine, bromine, or iodine, and interhalogens (such as iodine monochloride and bromine monochloride):

$$C_6H_6 + X_2 \xrightarrow{FeCl_3} C_6H_5X + HX \ (X = Cl, Br, I)$$

Although iodination can also be achieved, oxidative conditions must be provided in order to remove the HI formed, which otherwise tends to reduce the ring-iodinated compounds (96).

The extent of halogenation is regulated by the amount of halogenating agent used and by reaction conditions. Alkylated benzenes, phenols, phenol ethers, and polynuclear hydrocarbons undergo ring halogenation with ease.

Bromination can be conveniently effected by transfer of bromine from one nucleus to another. As the Friedel-Crafts isomerization of bromoaromatic compounds generally takes place through an intermolecular mechanism, the migrating bromine atom serves as a source of positive bromine, thus effecting ring brominations (18,97). For example, bromotoluene was prepared from toluene by bromination with the same brominating agents. 2,4,6-Tribromophenol was prepared by bromination of phenol with dibromobenzene.

Hydroxylation. The direct synthesis of phenol from benzene is of special interest since phenol is widely used in industry and benzene is a relatively cheap starting material for phenol. However, electrophilic hydroxylation of benzene with H_2O_2 or peracids, catalyzed by $AlCl_3$, BF_3, etc, usually results in a mixture containing also polyhydroxybenzenes and oxidized products.

Attempts to hydroxylate benzene with aqueous H_2O_2 in the presence of hydrogen fluoride also failed, and the only product was a high melting black, alkali-soluble resin when the reaction temperature was 0°C (98). This is owing to the polyhydroxylation of the phenol formed at the first stage, which is more reactive toward nucleophiles than benzene itself. This system, however, seems to be drastically influenced by the presence of carbon dioxide. When the same procedure was applied for the hydroxylation of benzene in the presence of carbon dioxide, 37% phenol was formed along with 16% catechol, and 37% hydroquinone (98). The rest was not characterized. Hydroxylation of benzene and substituted benzenes with H_2O_2 in superacid media (99) reduces polysubstitution significantly:

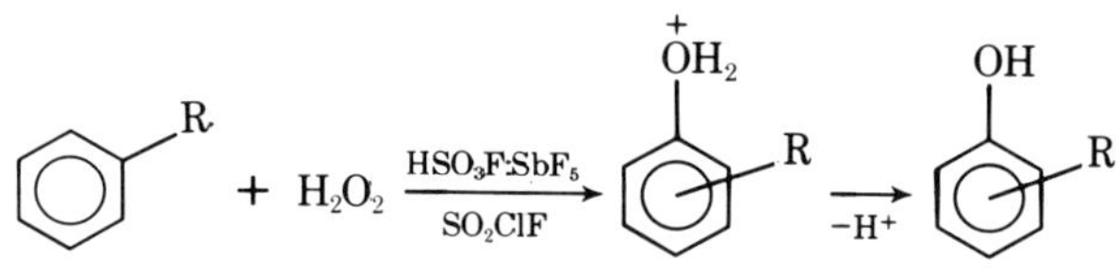

The phenol formed in the reaction is immediately protonated and the phenyloxonium ion is unreactive toward electrophiles.

Acylation of Aliphatic and Cycloaliphatic Compounds

Similar to alkylation, not only aromatic but also aliphatic and cycloaliphatic compounds undergo Friedel-Crafts acylation reactions.

Olefins and cycloolefins give unsaturated ketones with acyl halides (100). Saturated chloroketones are formed as intermediates, usually with elimination of HCl. Similar products may be obtained by using acid anhydrides:

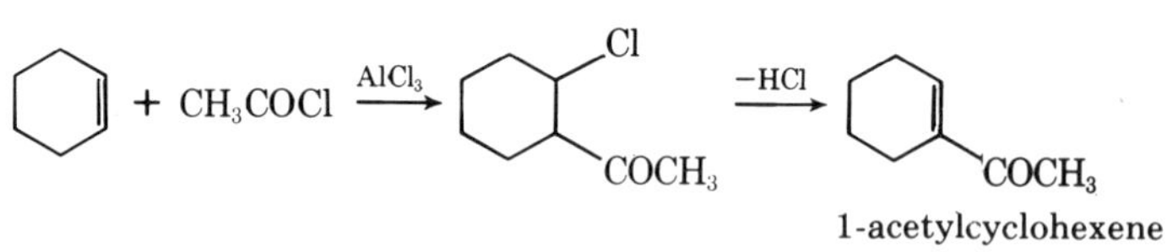

Acylation of acetylenic compounds produces β-chlorovinyl ketones:

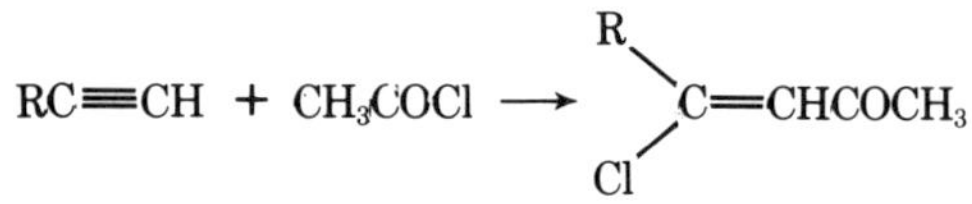

Even saturated hydrocarbons may give ketones with acyl chlorides (101). For example, cyclohexane and acetyl chloride react in the presence of aluminum chloride to give 1-acetyl-2-methylcyclopentane:

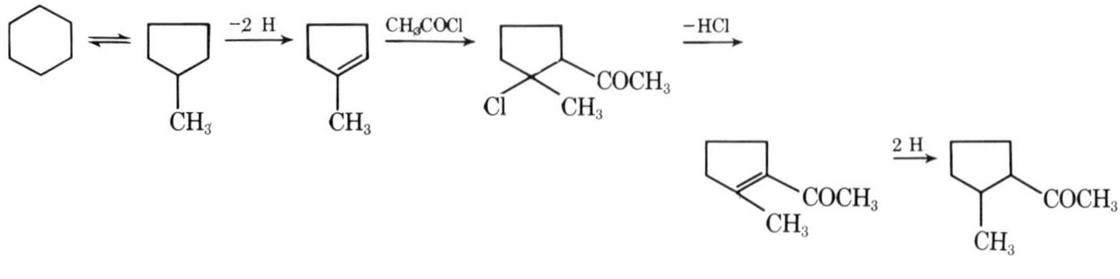

Aldehyde Syntheses. Friedel-Crafts-type aliphatic aldehyde syntheses are considerably rarer than the corresponding aromatic syntheses. However, it should be pointed out that the hydroformylation reaction of olefins (102) and the related oxo synthesis are effected by a catalyst which is a strong acid, eg, tetracarbonylhydrocobalt, $HCo(CO)_4$ (see Oxo reaction):

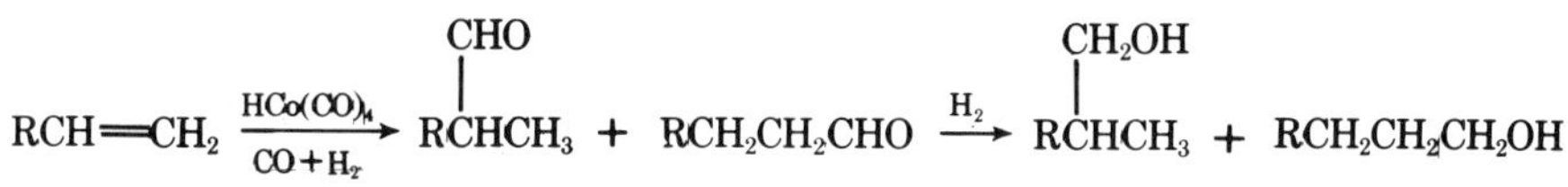

The Gattermann-Koch reaction when applied to alkenes or alkanes gives ketones or acids but not aldehydes. However, the Vilsmeier aldehyde synthesis can be applied to aliphatic compounds. For example, 1,2-dialkoxyethylenes react with *N*-methylformanilide and $POCl_3$ to give alkoxymalondialdehydes:

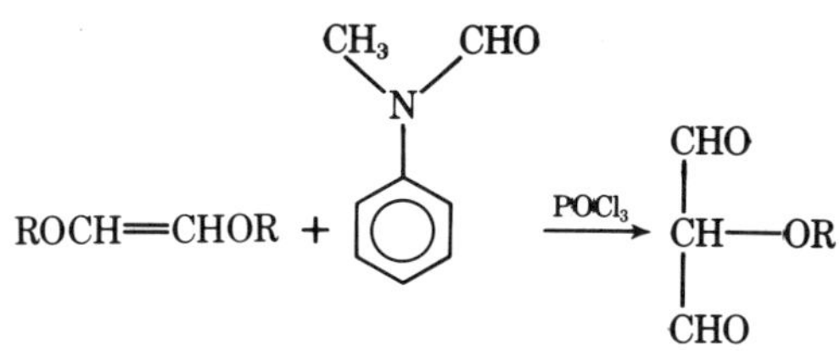

Syntheses of Carboxylic Acids and Derivatives. Alkenes are carbonylated in the presence of acid catalysts at 75–100°C and under pressures of 60–90 MPa (600–900 atm) to give carboxylic acids (103):

$$(CH_3)_2C{=}CH_2 + CO + H_2O \xrightarrow[BF_3.2H_2O]{H_2SO_4} (CH_3)_3CCOOH$$

Olefins are carbonylated in concentrated sulfuric acid at moderate temperatures (0–40°C) and low pressures with formic acid, which serves as the source of carbon monoxide (104):

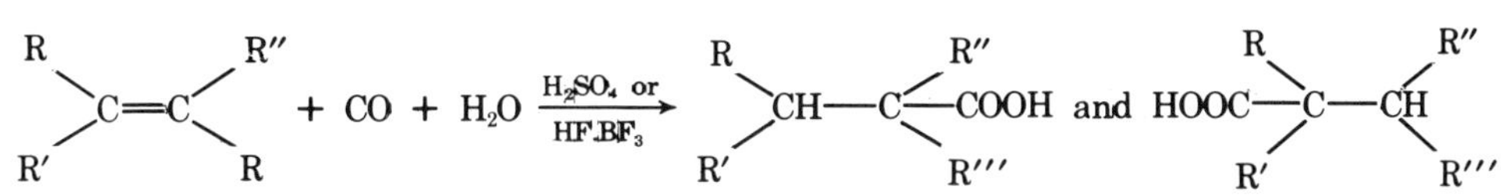

Liquid hydrogen fluoride, preferably in the presence of boron trifluoride, is an equally good catalyst and solvent system (see Carboxylic acids).

Nitration. Alkanes undergo electrophilic nitration with nitronium salts such as $(NO_2)^+(PF_6)^-$ in a protic solvent such as CH_2Cl_2 and sulfolane (105):

$$RH \xrightarrow[CH_2Cl_2\text{–sulfolane, 25°C, in darkness}]{(NO_2)^+(PF_6)^-} RNO_2 + H^+ + (PF_6)^-$$

CATALYSTS

Friedel-Crafts-type catalysts are electron acceptors, and fall into the general class of acids as defined by Lewis (see Catalysis).

Activity. In order to compare the activity of Friedel-Crafts catalysts, it is necessary to do activity studies in the same reaction system. Thus, the alkylating ability of benzyl chloride was selected to evaluate the relative catalytic activity of a large number of Lewis acid halides. The activity index is the lowest temperature at which reaction occurs (106). The results of this study were summarized by suggesting four categories of catalyst activity (107).

Group A, Very Active. Catalysts that lead to high yields in the benzylation reactions but at the same time cause extensive intra- and intermolecular isomerization: $AlCl_3$, $AlBr_3$, AlI_3, $GaCl_3$, $GaCl_2$, $GaBr_3$, GaI_3, $ZrCl_4$, $HfCl_4$, $HfBr_4$, HfI_4, SbF_5, NbF_5, $NbCl_5$, TaF_5, $TaCl_5$, $TaBr_5$, MoF_6, and $MoCl_5$.

Group B, Moderately Active. Catalysts that lead to high yields of benzylation without significant side reactions (isomerizations): $InCl_3$, $InBr_3$, InI_3, $SbCl_5$, WCl_6, $ReCl_5$, $FeCl_3$, $AlCl_3$–RNO_2, $AlBr_3$–RNO_2, $GaCl_3$–RNO_2, SbF_5–RNO_2, and $ZnCl_2$.

Group C, Weak. Catalysts that lead to low yields of benzylation without side reactions: BCl_3, BBr_3, BI_3, $SnCl_4$, $TiCl_4$, $TiBr_4$, $ReCl_3$, $FeCl_2$, and $PtCl_4$.

Group D, Very Weak or Inactive. This group comprises a great number of metal, alkaline earth, and rare earth element halides.

Acidic Halides (Lewis Acids). All metal halide-type Lewis catalysts, generally known as Friedel-Crafts catalysts, have an electron-deficient central metal atom capable of electron acceptance from the basic reagents. The most frequently used are aluminum chloride and bromide, followed by $BeCl_2$, $CdCl_2$, $ZnCl_2$, BF_3, BCl_3, BBr_3, $GaCl_3$, $GaBr_3$, $TiCl_4$, $ZrCl_4$, $SnCl_4$, $SnBr_4$, $SbCl_5$, $SbCl_3$, $BiCl_3$, $FeCl_3$, and UCl_4.

Easy availability and the low cost of aluminum chloride are partially responsible for its wide use in industry. Although aluminum chloride is frequently thought of as $AlCl_3$, at ordinary temperatures it is, in fact, the dimer Al_2Cl_6 (see Fig. 1), which prevails up to 440°C; between 440 and 880°C there is an equilibrium mixture of monomer and dimer. At higher temperatures, only the monomer exists, although above 1000°C some ionic dissociation takes place. However, under the usual Friedel-Crafts reaction conditions, the catalytically active species is always the monomeric Lewis acid $AlCl_3$ (see Aluminum compounds). Although Friedel and Crafts in their original work described investigations with anhydrous aluminum chloride, conditions demanding the application of absolutely anhydrous aluminum chloride are very rarely encountered.

It is rather difficult to obtain a Lewis acid-type metal halide, be it aluminum

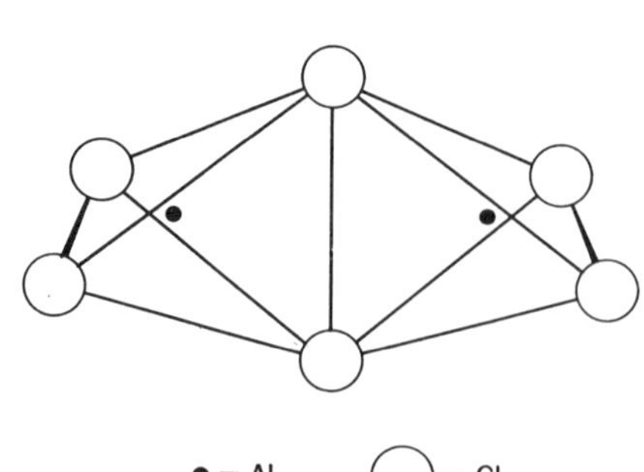

Figure 1. Aluminum chloride structure.

chloride, ferric chloride, boron trifluoride, or any other catalytic metal halide, in an absolutely anhydrous state. Even if they are anhydrous, utmost care and precaution must be exercised during their application to a reaction mixture if the claim is to be made that the reaction was carried out in the absence of moisture or other impurities.

In view of these facts, it is obvious that neither the original inventors, Friedel and Crafts, nor the thousands of subsequent researchers who carried out most successful work with aluminum chloride and related catalyst systems, worked under truly anhydrous conditions. Impurities such as water, oxygen, hydrogen halides, organic halides, etc (generally called promoters or cocatalysts) were present in almost all cases.

In spite of the fact that complete exclusion of moisture is usually advised in aluminum chloride reactions, the presence of traces of moisture has fortunately been found to accelerate rather than hinder the reactions. In many cases, the presence of these so-called initiators or cocatalysts is indeed essential (108). The beneficial action of traces of moisture has been observed especially in reactions involving the olefinic double bond (alkylation with olefins, polymerization, etc).

The inactivity of pure anhydrous Lewis acid halides in Friedel-Crafts polymerization of olefins was first demonstrated by Ipatieff and Grosse in 1936 (109), who found that pure, dry aluminum chloride does not react with ethylene. Subsequently it was shown (110) that boron trifluoride alone does not catalyze the polymerization of isobutylene when kept absolutely dry in a vacuum system. However, polymers form upon admission of traces of water. The active catalyst is boron trifluoride hydrate, $BF_3.H_2O$, ie, a conjugate protic acid $H^+(BF_3OH)^-$.

Cocatalysts of two types occur in Friedel-Crafts systems: (*1*) proton-donor substances, such as hydroxy compounds (water, alcohols) and proton acids (hydrogen chloride, sulfuric acid, phosphoric acid, organic acids, etc), and (*2*) cation-forming substances (other than proton), including cocatalysts such as alkyl and acyl halides, as well as a variety of other donor substances (oxygen, sulfur, nitrogen, halogen, etc, donors), leading to cations other than the proton. Besides carbocations formed with alkyl and acyl halides, oxonium, sulfonium, halonium, etc, complexes can act as activators in different Friedel-Crafts reactions.

Friedel-Crafts catalysis occurs in two ways: In the first, a positively polarized complex or carbocation is formed by the interaction of the catalyst with the reagent containing a functional group having a donor atom with nonbonded electron pairs (such as halogen, oxygen, nitrogen, sulfur, etc) thus acting as an n-donor. The carbocationic electrophile then reacts with the π-donor substrate (aromatic, olefinic, or acetylenic hydrocarbons). Such a reaction, of course, can, in principle, take place under absolutely anhydrous conditions and in the absence of any cocatalyst, since the Lewis acid catalysts themselves are suitable acceptors to interact with the aforesaid n-donors. Lewis acids, alone, however, are generally insufficient catalysts in reactions of olefins or acetylenes or in isomerization of hydrocarbons. Generally, a cocatalyst is required to form a strong conjugate acid or carbocation, which then initiates the reaction. In the second type, the catalyst formally extracts a hydride ion from a saturated hydrocarbon, thus causing carbocationic initiation of the reaction. This takes place in alkylations with alkanes, in isomerization of saturated hydrocarbons, and similar reactions. Neat Lewis acids do not abstract hydride ions to provide the active species for subsequent Friedel-Crafts-type reactions. A cocatalyst that forms a powerful conjugate acid with the Lewis acid is needed for initiation. Hydride abstraction is effected by the proton acid.

Impurities and the methods of preparation and the physical state of the catalyst may affect both yield and direction of a Friedel-Crafts reaction. Although traces of moisture in the aluminum chloride may accelerate the effect, other impurities, such as ferric chloride, can markedly decrease the yield.

Metal Alkyls and Alkoxides. Acidic halide catalysts effect Friedel-Crafts reactions through their Lewis acid activity. Consequently, Lewis acidic metal alkyls and metal alkoxides also possess catalytic activity in Friedel-Crafts reactions.

Metal alkyls (eg, aluminum, boron, zinc alkyls) are fairly active catalysts. Hyperconjugation with the electron-deficient metal atom, however, tends to decrease the electron deficiency. The effect is even stronger in alkoxides which are, therefore, fairly weak Lewis acids.

The present discussion does not encompass catalyst systems of the Ziegler-Natta type (such as $AlR_3 + TiCl_4$), although certain similarities with Friedel-Crafts systems are apparent. For example, alkylation of aromatics with olefins in the presence of a Ziegler-Natta catalyst results in longer chain alkylates and is obviously related to the general Friedel-Crafts alkylations (see Ziegler-Natta catalysts).

The most important application of metal alkoxides in reactions of the Friedel-Crafts type is that of aluminum phenoxide as a catalyst in phenol alkylation (111). Aluminum phenoxide is also the active catalyst when aluminum metal is used. Phenol is sufficiently acidic to react with aluminum with the formation of $(C_6H_5O)_3Al$. Aluminum phenoxide, when dissolved in phenol, greatly increases the acidic strength. It is believed that, similar to the alkoxoacids of Meerwein (112), an aluminum phenoxoacid is formed, which is a strong conjugate acid of the type $HAl(OC_6H_5)_4$. This acid is then the catalytically active species (see Alkoxides, metal).

Proton Acids (Brønsted Acids). Many Friedel-Crafts reactions carried out with metal halide catalysts are, indeed, proton catalyzed as shown by the very close relationships of these reactions to those carried out with proton acid (Brønsted acid) catalysis. Sulfuric acid is the most widely used example, followed by anhydrous hydrogen fluoride and many others.

Acidic Oxides and Sulfides (Acidic Chalcogenides), Modified Zeolites. Chalcogenide catalysts include a great variety of solid oxides and sulfides. The most widely used comprise alumina, silica, and mixtures of alumina and silica (either natural or synthetic), in which other oxides such as chromia, magnesia, molybdena, thoria, tungstic oxide, and zirconia may also be present, as well as certain sulfides such as sulfides of molybdenum. A large variety of compositions is possible. The composition and structure of different types of bauxites, floridin, Georgia clay, and other natural aluminosilicates are still not well known. Some synthetic catalysts, other than silica–alumina compositions, representative of the acidic chalcogenides are BeO, Cr_2O_3, P_2O_5, TiO_2, and $Al_2(SO_4)_3$ which may be regarded as Al_2O_3, $3SO_3$, $Al_2O_3.xCr_2O_3$, $Al_2O_3.Fe_2O_3$, $Al_2O_3.MnO$, $Al_2O_3.CoO$, $Al_2O_3.Mo_2O_3$, $Al_2O_3.V_2O_3$, $Cr_2O_3.Fe_2O_3$, MoS_2, and MoS_3. In contrast to sulfuric acid which may be regarded as a fully hydrated chalcogenide, the members of this group are seldom very highly hydrated under conditions of use.

Silica–alumina has been studied most extensively. The acidity could rest in Lewis acid or Brønsted acid sites. Dehydrated silica–alumina is inactive as isomerization catalyst but addition of water increases activity until a maximum is reached; additional water then decreases activity. The effect of water suggests that Brønsted acidity is responsible for catalyst activity (113). At any rate, silica–alumina is quantitatively at least as acidic as 90% sulfuric acid (114).

Acidic Cation-Exchange Resins. Brønsted acid catalytic activity has resulted in the successful use of so-called solid acids (acidic chalcogenides) in a variety of reactions. Based on a similar principle it is easily understandable that acidic cation-exchange resins, which are also solid acids, are active catalysts.

Cation-exchange catalysts are used in esterification, acetal synthesis, ester alcoholysis, acetal alcoholysis, alcohol dehydration, ester hydrolysis, and sucrose inversion. The solid acid type permits simplified procedures when high boiling and viscous compounds are involved because the catalyst can be separated from the products by simple filtration. Unsaturated acids and alcohols which can polymerize in the presence of proton acids can thus be esterified directly and without polymerization.

Sulfonated styrene–divinylbenzene cross-linked polymers have been applied in many of the previously mentioned reactions and also in the acylation of thiophene with acetic anhydride and acetyl chloride (115). Resins of this type (Dowex 50, Amberlite IR-112, and Permutit Q) are particularly effective catalysts in the alkylation of phenols with olefins (such as propylene, isobutylene, diisobutylene), alkyl halides, and alcohols (116) (see Ion exchange).

Superacids. In the 1960s a class of acids millions of times stronger than mineral acids was discovered; acids stronger than 100% sulfuric acid are called superacids (117).

The determination of acidity by measurement of H^+ activity ($pH = -\log a_{H^+}$) does not hold for very concentrated acid solution, where Hammett's acidity function is generally used (118):

$$H_o = pK_{BH^+} - \log \frac{BH^+}{B}$$

where pK_{BH^+} is the dissociation constant of the conjugate acid and BH^+/B is the ionization ratio. Typical H_o values are 12.6 for 100% H_2SO_4, and 11.0 for anhydrous HF.

Protic Superacids. Fluorosulfuric acid (HSO_3F) is one of the strongest Brønsted acids known with $H_0 = -15.1$. This acidity is somewhat lower than that of H_2SO_4, SO_3, ie, $H_2S_2O_7$. However, because of its stability, ease of purification, its wide liquid range (mp = −89°C, bp = 162°C) and relatively low viscosity (1.56 mPa·s or cP at 28°C), it is more convenient to use.

Perfluoroalkanesulfonic acids also show high acidity. The parent trifluoromethanesulfonic acid (triflic acid), CF_3SO_3H, is commercially prepared by electrochemical fluorination of methanesulfonic acid (119–120). It has an H_o value of −14.6 (121). The higher homologues show somewhat decreasing acidities (122).

Brønsted-Lewis Superacid Mixtures. Conjugate Friedel-Crafts acids prepared from protic and Lewis acids, such as $HCl–AlCl_3$ and $HCl–GaCl_3$ are, indeed, superacids with an estimated H_o value of −15 to −16. They have proved to be effective catalysts in hydrocarbon transformations (123). In the early 1960s acid systems were prepared (117) comprising a pentafluoride of group V elements, particularly SbF_5 and a strong Brønsted acid such as HF, FSO_3H, CF_3SO_3H, etc. Magic Acid ($HSO_3F–SbF_5$) is one of the most successful systems; fluoroantimonic acid, $HF–SbF_5$ is similarly used. The acidity of HF or HSO_3F is increased sharply by adding SbF_5 (124–125). A composition of 4 mol % SbF_5 in HF has an H_o value of −19 and with higher SbF_5 concentrations acidities can increase to −25. These very highly acidic systems are being utilized for petrochemical transformations such as isomerization of straight-chain alkanes (126), alkane–alkene alkylations (127), and the like (128).

Solid Superacids. Most large-scale petrochemical and chemical industrial processes are preferably carried out, whenever possible, over solid catalysts. Hence, the increased interest in solid superacidic catalysts. Acidic oxides, discussed earlier, are used extensively. However, their relatively low acidity necessitates higher operating temperatures. Even at high temperatures, they are inefficient to promote such reactions as the transethylation of benzene with di- and polyethylbenzenes, etc. Recently, new solid acid systems were developed with considerably higher acidity than those of acidic oxides. These include acidic zeolites as well as a variety of new catalysts. Graphite-intercalated $AlCl_3$ is an effective solid Friedel-Crafts catalyst but loses catalytic activity because of partial hydrolysis and leaching of the Lewis acid halide from the graphite. Aluminum chloride can also be complexed to sulfonated polystyrene resins but again the stability of the catalyst is limited.

More stable catalysts are obtained by using fluorinated graphite or fluorinated alumina as backbones, and Lewis acid halides, such as SbF_5, TaF_5, and NbF_5, which have a relatively low vapor pressure. These Lewis acids are attached to the fluorinated solid supports through fluorine bridging. They show high reactivity in Friedel-Crafts type reactions including the isomerization of straight-chain alkanes such as *n*-hexane.

Another type of solid superacid is based on perfluorinated resin sulfonic acid such as the acid form of DuPont's Nafion resin, a copolymer of a perfluorinated epoxide and vinylsulfonic acid or solid, high molecular weight perfluoroalkanesulfonic acids such as perfluorodecanesulfonic acid, $CF_3(CF_2)_9SO_3H$. Such solid catalysts were found very efficient in many alkylations of aromatic hydrocarbons (129) and other Friedel-Crafts reactions (130).

Metathetic Cation-Forming Agents. A number of effective Friedel-Crafts catalysts are capable of forming cations in a way other than protonation or hydride abstraction. Their activity is not strictly catalytic because they are generally consumed in the cation-forming reaction. Anhydrous silver salts such as $AgClO_4$, $AgBF_4$, $AgSbF_6$, $AgPF_6$, $AgAsF_6$, Ag_3PO_4, etc, are representatives of this class of compounds.

SOLVENTS

Generally, in Friedel-Crafts processes, an excess of the hydrocarbon that is undergoing substitution is used as solvent. Thus in alkylations or acylations of benzene, the latter compound is almost invariably used as solvent. If, however, neither reactant is a liquid under the reaction conditions, or a modification of the catalyst activity is necessary, an additional solvent may be employed. Solvents frequently used are carbon disulfide, petroleum ether, ethylene chloride, methylene chloride, and carbon tetrachloride. Nitrobenzene and nitroalkanes are also adequate not only because they possess good solvent powers but also because they form addition complexes with aluminum chloride (and other acidic metal halides) and thus ameliorate the action of the catalyst. Complex formation may lessen the tarring and disproportionation frequently caused by aluminum chloride, thereby making the introduction of alkyl and acyl groups a milder reaction applicable to sensitive compounds, such as phenols, amines, benzyl halides, or quinolines. When highly reactive substrates are to be substituted and mild condensing agents are employed, solvents such as o-dichlorobenzene or diethyl ether, or even benzene may be used. Under these conditions, the alkylation or acylation of the substrate in question is so rapid that the solvent under the same reaction conditions does not have an opportunity to react. For example, furan may be acetylated with acetic anhydride using benzene as the solvent:

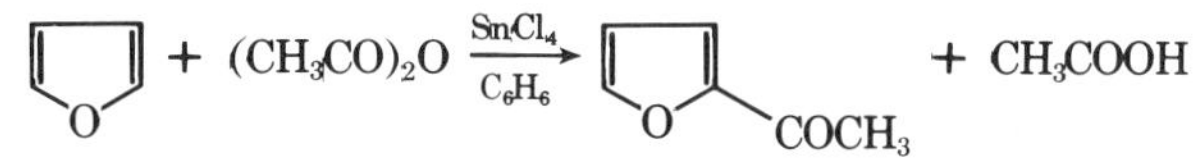

Other solvents acquiring considerable importance in Friedel-Crafts syntheses are liquid sulfur dioxide, dimethyl sulfoxide, and tetramethylene sulfone.

Sulfur dioxide particularly is used extensively in carbocationic chemistry as a solvent of low nucleophilicity. In recent years SO_2ClF has also been used because of its lower nucleophilicity, lower freezing point, and higher boiling point.

As noted previously, the course of Friedel-Crafts-type reactions is, in some instances, greatly affected by the solvent used in the reaction. For example, the acylation of naphthalene occurs in the α-position when carbon disulfide or tetrachloromethane is the solvent, whereas substitution occurs in the β-position with nitrobenzene or nitromesitylene.

BIBLIOGRAPHY

"Friedel-Crafts Reactions" in *ECT* 1st ed., Vol. 6, pp. 883–892, by G. W. Pedlow and P. E. Hoch, General Aniline & Film Corporation; "Friedel-Crafts Reactions" in *ECT* 2nd ed., Vol. 10, pp. 135–166, by George A. Olah and Chris A. Cupas, Western Reserve University.

1. G. A. Olah, ed., *Friedel-Craft and Related Reactions,* Vols. 1–4, Interscience Publishers, a division of John Wiley & Sons, Inc., New York, 1963–1965.
2. G. A. Olah, ed., *Friedel-Crafts Chemistry,* John Wiley & Sons, Inc., New York, 1973.
3. G. Kranzlein, *Aluminum Chlorid in der Organischen Chemie,* 3rd ed., Verlag Chemie, Berlin, Ger., 1939.
4. C. A. Thomas, *Anhydrous Aluminum Chloride in Organic Chemistry,* Reinhold Publishing Corp., New York, 1961.
5. G. A. Olah, ed., *Reactive Intermediates in Organic Chemistry,* John Wiley & Sons, Inc., New York, 1968–1972.
6. G. A. Olah and P. V. R. Schleyer, eds., *Carbonium Ions,* John Wiley & Sons, Inc., New York, 1968–1972.
7. D. Bethell and V. Gold, *Carbonium Ions, An Introduction,* Academic Press, Inc., New York, 1967.
8. O. Jacobsen, *Ber.* **18,** 338 (1885).
9. N. G. Bun-Hoi and P. Cagneunt, *Bull. Soc. Chim. Fr.* **9,** 887 (1942).
10. A. W. Francis and E. E. Reid, *Ind. Eng. Chem.* **38,** 1194 (1946).
11. Ger. Pat. 2,239,712 (Feb. 22, 1973), N. L. Carr and D. Y. C. Ko (to Gulf R&D Co.).
12. Ger. Pat. 2,239,675 (Feb. 22, 1973), N. L. Carr and D. Y. C. Ko (to Gulf R&D Co.).
13. C. C. Price in R. Adams, ed., *Organic Reactions,* Vol. 3, John Wiley & Sons, Inc., New York, 1946, p. 1.
14. A. W. Francis, *Chem. Rev.* **43,** 257 (1948).
15. L. Schmerling and J. P. West, *J. Am. Chem. Soc.* **76,** 1917 (1959).
16. W. J. Taylor, D. D. Wagman, M. G. Williams, K. S. Pitzer, and F. D. Rossini, *J. Res. Natl. Bur. Stand.* **37,** 95 (1946).

17. G. A. Olah, W. S. Tolgysei, and R. E. A. Dear, *J. Org. Chem.* **27,** 3445, 3449 (1962).
18. G. A. Olah and M. Meyer, *J. Org. Chem.* **27,** 3464 (1962).
19. L. Schmerling, *J. Am. Chem. Soc.* **97,** 6134 (1975).
20. J. I. Brauman and A. J. Pandell, *J. Am. Chem. Soc.* **89,** 5421 (1967).
21. J. I. Brauman and A. Solladie-Cavalo, *Chem. Commun.,* 1124 (1968).
22. S. Suga, T. Nakajima, Y. Nakamoto, and K. Matsumoto, *Tetrahedron Lett.,* 3283 (1969).
23. Ref. 1, Vol. 2, Part 2, Chapt. XXI.
24. H. F. Fieser and M. D. Gates, Jr., *J. Am. Chem. Soc.* **62,** 2335 (1940)
25. H. A. Bruson and J. W. Kroeger, *J. Am. Chem. Soc.* **62,** 36 (1940).
26. Y. Lavaux and M. Lombard, *Bull. Soc. Chim. Fr.* **7,** 913 (1910).
27. I. P. Tsukervanik and N. I. Bogdanova, *J. Gen. Chem. USSR* **22,** 410 (1953).
28. E. D. Bergmann, *Chem. Rev.* **29,** 529 (1941).
29. A. A. Khalaf and R. M. Roberts, *J. Org. Chem.* **34,** 3571 (1968).
30. R. O. Roblin, D. Davidson, and M. T. Bogert, *J. Am. Chem. Soc.* **57,** 151 (1935).
31. V. N. Ipatieff, A. V. Grosse, H. Pines, and V. I. Komarewski, *J. Am. Chem. Soc.* **58,** 913 (1936).
32. S. F. Birch, A. E. Dunstan, F. A. Pim, and T. Tait, *J. Inst. Pet. Technol.* **24,** 303 (1938).
33. G. A. Olah and J. A. Olah, *J. Am. Chem. Soc.* **93,** 1256 (1970).
34. M. Siskin, R. H. Schlosberg, and M. P. Kocsi, *Symposium on New Hydrocarbon Chemistry, A.C.S. San Francisco Meeting, Aug. 29–Sept. 3, 1976, Preprints,* p. 429; R. H. Schlosberg, M. Siskin, W. P. Kocsi, and F. J. Parker, *J. Am. Chem. Soc.* **98,** 7723 (1976).
35. L. Schmerling, *J. Am. Chem. Soc.* **67,** 1152 (1945); **68,** 1650 (1946).
36. G. A. Olah and R. H. Schlosberg, *J. Am. Chem. Soc.* **90,** 2726 (1968).
37. D. V. Nightingale, *Chem. Rev.* **25,** 329 (1939); G. A. Olah, W. S. Tolgyesi, and R. E. A. Dear, *J. Org. Chem.* **27,** 3441 (1962).
38. G. Makarova, M. K. Matveeva, and E. A. Gribehenko, *Bull. Acad. Sci. USSR Div. Chem. Sci.,* 1399 (1958); A. N. Nesmeyanov and L. G. Makrova, *Bull. Acad. Sci. USSR Div. Chem. Sci.,* 213 (1947).
39. G. A. Olah and W. S. Tolgysesi, *J. Org. Chem.* **26,** 2053 (1961).
40. U.S. Pat. 4,116,880 (Sept. 26, 1978), G. A. Olah.
41. *Occupation Safety and Health Administration Standards and Regulations of U.S. Government,* U.S. Government Printing Office, Washington, D.C.
42. A. G. Oblad, G. A. Mills, and H. Heinemann in P. H. Emmett, ed., *Catalysis,* Vol. 6, Reinhold Publishing Corp., New York, 1958, p. 341.
43. V. N. Ipatieff, B. B. Corson, and G. Egloff, *Ind. Eng. Chem.* **27,** 1077 (1935).
44. D. T. Roberts, Jr., and L. E. Calihan, *J. Macromol. Sci. Chem.,* 1629, 1641 (1973); P. H. Goore and J. A. Hiskins, *J. Chem. Soc. C,* 517 (1970).
45. N. P. Bun-Hoi, P. Mabille, and Do-Cao Thang, *Bull. Soc. Chim. Fr.,* 981 (1968).
46. P. Fletcher and W. Marlow, *J. Chem. Soc. C,* 937 (1970).
47. D. Papa, E. Schwenk, and A. Klingsberg, *J. Am. Chem. Soc.* **68,** 2133 (1946).
48. G. A. Olah and S. Kobayashi, *J. Am. Chem. Soc.* **93,** 6964 (1971).
49. Ref. 1, p. 581.
50. F. R. Jensen and B. E. Swart, *J. Am. Chem. Soc.* **91,** 5686 (1969).
51. K. Elbs, *J. Prakt. Chem.* **35,** 503 (1887).
52. L. Bouveault, *Bull. Soc. Chim. Fr.* **17,** 1014 (1897).
53. G. Perrier, *Bull. Soc. Chim. Fr.* **31,** 859 (1903).
54. G. A. Olah and S. J. Kuhn, *J. Org. Chem.* **26,** 237 (1961).
55. H. Ulich and G. Heyne, *Z. Elektrochem.* **41,** 509 (1935).
56. M. Freund and K. Gleischer, *Ann.* **373,** 291 (1910).
57. F. Wolf, K. Fredrich, and G. Schwachula, *Plaste Kaut.* **16,** 727 (1969).
58. L. Gattermann and J. A. Koch, *Ber.* **30,** 1622 (1897).
59. L. Bert, *Compt. Rend.* **221,** 77 (1945).
60. A. N. Nesmeyanov and E. J. Kahn, *Ber.* **67,** 370 (1934).
61. L. Gattermann, *Ber.* **31,** 1149 (1808); *Ann.* **357,** 313 (1907).
62. P. Karrer, *Helv. Chim. Acta* **2,** 89 (1919).
63. A. Vilsmeier, *Chem. Ztg.* **75,** 133 (1951).
64. H. Lorenz and R. Witzinger, *Helv. Chim. Acta* **28,** 600 (1945).
65. Z. Arnold and F. Sorm, *Chem. Listy* **51,** 1082 (1957); C. Jutz, *Ber.* **91,** 850 (1958).
66. A. Rieche, H. Gross, and E. Hoft, *Ber.* **95,** 88 (1960).
67. R. Scholl and F. Kacer, *Ber.* **36,** 322 (1903).

68. P. Karrer and E. Zeller, *Helv. Chim. Acta* **2,** 361 (1920).
69. J. Houben and W. Fischer, *Ber.* **66,** 339 (1933).
70. C. Friedel, J. M. Crafts, and E. Ador, *Compt. Rend.* **85,** 673 (1877).
71. H. Hopff, *Angew. Chem.* **60,** 245 (1948).
72. F. Runge, H. Reinhardt, and G. Kuhnhanss, *Chem. Tech. (Berlin),* 644 (1956).
73. U.S. Pat. 1,892,990 (Jan. 3, 1933), F. Linner (to Beck, Koller & Co.).
74. M. T. Bogert and M. Meyer, *J. Am. Chem. Soc.* **44,** 1568 (1922).
75. J. F. Norris and J. E. Wood, *J. Am. Chem. Soc.* **62,** 1428 (1940).
76. S. C. J. Olivier, *Rec. Trav. Chim.* **33,** 91, 163, 244 (1914).
77. L. Field, *J. Am. Chem. Soc.* **74,** 394 (1952).
78. W. E. Truce and C. W. Vriesen, *J. Am. Chem. Soc.* **75,** 5032 (1953).
79. R. J. Thomas, W. F. Anzilotti, and G. F. Hennion, *Ind. Eng. Chem.* **32,** 408 (1940).
80. L. I. Smith in R. Adams, ed., *Organic Reactions,* Vol. 1, John Wiley & Sons, Inc., New York, 1942, p. 370.
81. E. Knoevenagel and J. Kenner, *Ber.* **41,** 3315 (1908).
82. C. C. Price and C. A. Sears, *J. Am. Chem. Soc.* **75,** 3276 (1953).
83. G. A. Olah, A. E. Pavlath, and S. J. Kuhn, *Chem. Ind. (London),* 50 (1957).
84. A. Schaarschmidt, *Ber.* **57,** 2065 (1924); *Angew. Chem.* **39,** 1457 (1926).
85. L. A. Pinek, *J. Am. Chem. Soc.* **49,** 2536 (1927).
86. A. Klemenz and K. Scholler, *Z. Anorg. Allgem. Chem.* **141,** 231 (1927).
87. G. B. Bachman and J. L. Denver, *J. Am. Chem. Soc.* **80,** 5871 (1958).
88. G. A. Olah, S. J. Kuhn, and A. Mlinko, *J. Chem. Soc.,* 4257 (1956).
89. S. J. Kuhn and G. A. Olah, *J. Am. Chem. Soc.* **83,** 4564 (1961).
90. G. A. Olah, S. C. Narang, and J. A. Olah, *J. Am. Chem. Soc.* **101,** 1805 (1979).
91. C. Graebe, *Ber.* **34,** 1778 (1901).
92. U.S. Pat. 2,585,355 (Feb. 12, 1952), J. S. Turski.
93. P. Kovacic and G. L. Foote, *J. Am. Chem. Soc.* **83,** 743 (1961).
94. P. Kovacic and R. M. Lange, *J. Org. Chem.* **78,** 968 (1965); P. Kovacic and J. A. Levitsky, *J. Am. Chem. Soc.* **88,** 1000 (1966); P. Kovacic and J. F. Gormish, *J. Am. Chem. Soc.* **88,** 3819 (1966).
95. C. E. Inmann, R. E. Oesterling, and E. A. Tyezkowsky, *J. Am. Chem. Soc.* **80,** 5284 (1958).
96. E. Wertyporoch, *Ann.* **493,** 153 (1932).
97. G. A. Olah, W. S. Tolgyesi, and R. E. A. Dear, *J. Org. Chem.* **27,** 3455 (1962).
98. J. A. Vesely and L. Schmerling, *J. Org. Chem.* **35,** 4028 (1970).
99. G. A. Olah and R. Ohnishi, *J. Org. Chem.* **43,** 865 (1978).
100. H. Wieland, L. Bettag, *Ber.* **55,** 2246 (1922).
101. C. D. Nenitzescu and E. Cioranescu, *Ber.* **69,** 1820 (1936).
102. G. Salamon, *Rec. Trav. Chim.* **68,** 903 (1949).
103. T. A. Ford, H. W. Jacobsen, and F. C. McGrew, *J. Am. Chem. Soc.* **70,** 3743 (1948).
104. H. Koch and W. Haaf, *Angew. Chem.* **70,** 311 (1958); *Ann.* **618,** 251 (1958).
105. G. A. Olah and H. C. Lin, *J. Am. Chem. Soc.* **93,** 1259 (1971).
106. T. Asaoka, C. Shimasaki, K. Taki, M. Funayama, M. Sakano, and Y. Kamimura, *Yuki Gosei Kagaku Kyokai Shi* **27,** 783 (1968).
107. G. A. Olah, S. Kobayashi, and M. Tashiro, *J. Am. Chem. Soc.* **96,** 7448 (1972).
108. C. D. Nenitzescu and E. I. P. Cantuniari, *Ber.* **66,** 1097 (1933).
109. V. N. Ipatieff and A. V. Grosse, *J. Am. Chem. Soc.* **58,** 915 (1936).
110. A. G. Evans, D. Holden, P. H. Plesch, M. Polanyi, H. A. Skinner, and M. A. Weinberger, *Nature* **157,** 102 (1946).
111. A. J. Kolka, J. P. Napolitano, and G. G. Ecke, *J. Org. Chem.* **21,** 712 (1956).
112. H. Meerwein and T. Bersin, *Ann.* **476,** 113 (1920).
113. F. E. Condon in P. H. Emmett, ed., *Catalysis,* Vol. 6, Reinhold Publishing Corp., New York, 1953, p. 43.
114. A. Roe in R. Adams, ed., *Organic Reactions,* Vol. 5, John Wiley & Sons, Inc., New York, 1949, p. 200.
115. U.S. Pat. 2,711,414 (June 21, 1955), T. R. Norton (to The Dow Chemical Co.).
116. B. Loev and J. T. Massengele, *J. Org. Chem.* **22,** 988 (1957).
117. G. A. Olah, G. K. Suryaprakash, and J. Sommer, *Science* **206,** 13 (1979).
118. L. P. Hammett and A. J. Deyrup, *J. Am. Chem. Soc.* **54,** 2721 (1932).
119. R. N. Haszeldine and J. M. Kidd, *J. Chem. Soc.,* 4228 (1954).

120. T. Gramstad and R. N. Haszeldine, *J. Chem. Soc.*, 173 (1976).
121. R. D. Howells and J. D. McCowen, *Chem. Rev.*, 69 (1977).
122. J. Groudin, R. Sagnes, and A. Commeyras, *Bull. Soc. Chim. Fr.*, 1779 (1976).
123. D. A. McCowley and A. P. Line, *J. Am. Chem. Soc.* **74**, 6246 (1956).
124. M. Kilpatrick and F. E. Lubersky, *J. Am. Chem. Soc.* **76**, 5863 (1954).
125. R. J. Gillespie and T. E. Peel, *J. Am. Chem. Soc.* **95**, 5173 (1973).
126. U.S. Pat. 3,838,489 (Oct. 1, 1974), J. E. Mahan and J. R. Norell.
127. U.S. Pat. 3,708,553 (Jan. 2, 1973), G. A. Olah (to Esso Research and Engineering Co.).
128. G. A. Olah and J. Sommer, *Recherche* **10**, 625 (1979).
129. G. A. Olah and D. Meidar, *Nouv. J. Chim.* **3**, 269 (1979); G. A. Olah, D. Meidar, R. Malhotra, S. C. Narang, and J. A. Olah, *J. Catalysis* **61**, 96 (1980).
130. G. A. Olah, D. Meidar, and J. A. Olah, *Nouv. J. Chim.* **3**, 275 (1978); G. A. Olah and J. Kaspi, *Nouv. J. Chem.* **2**, 585 (1979).

GEORGE A. OLAH
DAVID MEIDAR
University of Southern California

FRUIT JUICES

The commercial production of fruit juices began in Europe shortly after the turn of the century when the Boehi process was developed for the storage of apple juice under CO_2 and refrigeration (1). This bulk storage permitted distribution of juice in returnable bottles over an extended period. Prior to that time fruit juices were available for only short periods after harvest and before the onset of fermentation.

In the United States, the sale of bottled or canned juices—tomato, grape, and citrus—started around 1929. During World War II, commercial juice production was greatly increased and the market continued to expand in the postwar years through the application of new technologies. Frozen single-strength juices were introduced, followed shortly by frozen concentrates containing add-back or recovered volatile flavoring materials, mixed drinks, and more recently the distribution of aseptically packaged chilled juices. Many of the steps in the flow diagram in Figure 1 are described in this article. The composition of a number of fruit juices and concentrates is shown in Table 1.

Raw Materials

In the early stages of the juice industry, when it was largely a salvage operation, the principal source of raw material was fruit that was misshapen, poorly colored, or had skin blemishes which made it unsuitable for the fresh market and as canned or frozen fruit. Currently, raw materials are selected for their suitability for juice production.

Variety and maturity are important factors affecting suitability for juice production (3). For some juices such as citrus (4) or grape (5), only a few varieties are used for a distinctive flavor or because of freedom from bitterness. When a range of varieties is available, for example apples, blends must be used to achieve continuity of flavor (6), or fruit grown in a warmer climate having a high sugar content may be mixed with apples from a cooler region to achieve a desired sugar–acid balance. This sugar–acid ratio is based on Brix readings (wt % sugar) obtained with a refractometer, and total acidity by titration (7) (see Sugar).

Juice factories frequently employ field persons to advise growers on the application of sprays to the growing crops so that residues on harvested fruit are within prescribed limits (see Trace and residue analysis). They also may sample the crop before harvest for analysis, and coordinate harvesting with factory-production schedules (8). Such coordination is essential where the fruit is mechanically harvested and the flesh is broken. Payment for raw materials is frequently based on specifications that are either official government grades or state market standards. Official graders may be employed to test each load.

Manufacture

Cleaning. On delivery to the plant, the fruit is passed over a roller conveyor or between revolving brushes to remove leaves or other debris before entering a tank containing a chlorine solution or detergent. Strong currents of wash solution direct the floating fruit toward a roller elevator over which sprays of fresh water remove the wash solution (9). It is important that chlorine be removed from fruit containing polyphenols since they may react with chlorine to form chlorophenols which give a medicinal flavor to the juice. In addition to removal of debris or dirt, washing reduces the number of microorganisms on the fruit, thereby making subsequent juice sterilization processes more effective. Reduction in microbial content by washing is difficult where the interior of the fruit is open to the atmosphere, eg, apples (10).

A second brushing is sometimes combined with fresh-water sprays. The speed of brush rotation, alignment, and stiffness of the bristles are varied with the rigidity of the fruit surface and the intent of the operation, as, for example, the removal of bruised tissues.

Sorting. The fruit receives final inspection on roller conveyors to ensure freedom from imperfections before it is segregated into different size grades. Spots of rot on a few apples can cause an adverse flavor in a large volume of juice. During the final inspection, samples are frequently analyzed.

Mechanically harvested grapes are dumped from collection bins into a hopper from which individual berries and partial clusters are elevated to a stemmer–crusher on a woven-wire belt. Foreign materials such as leaves, cluster stems, etc, are removed manually; juice draining to the base of the hopper is combined with that from the crushed pulp.

Several types of sizing equipment have been used, eg, inclined diverging belts or rolls, roller conveyors with varying space between the rolls, or revolving drums located at different heights above a feed conveyor. The fruit of each size category is collected on a cross-conveyor and delivered to a temporary storage bin or directly to the extraction units. Stationary surfaces on the grader are usually covered with foam rubber to minimize bruising.

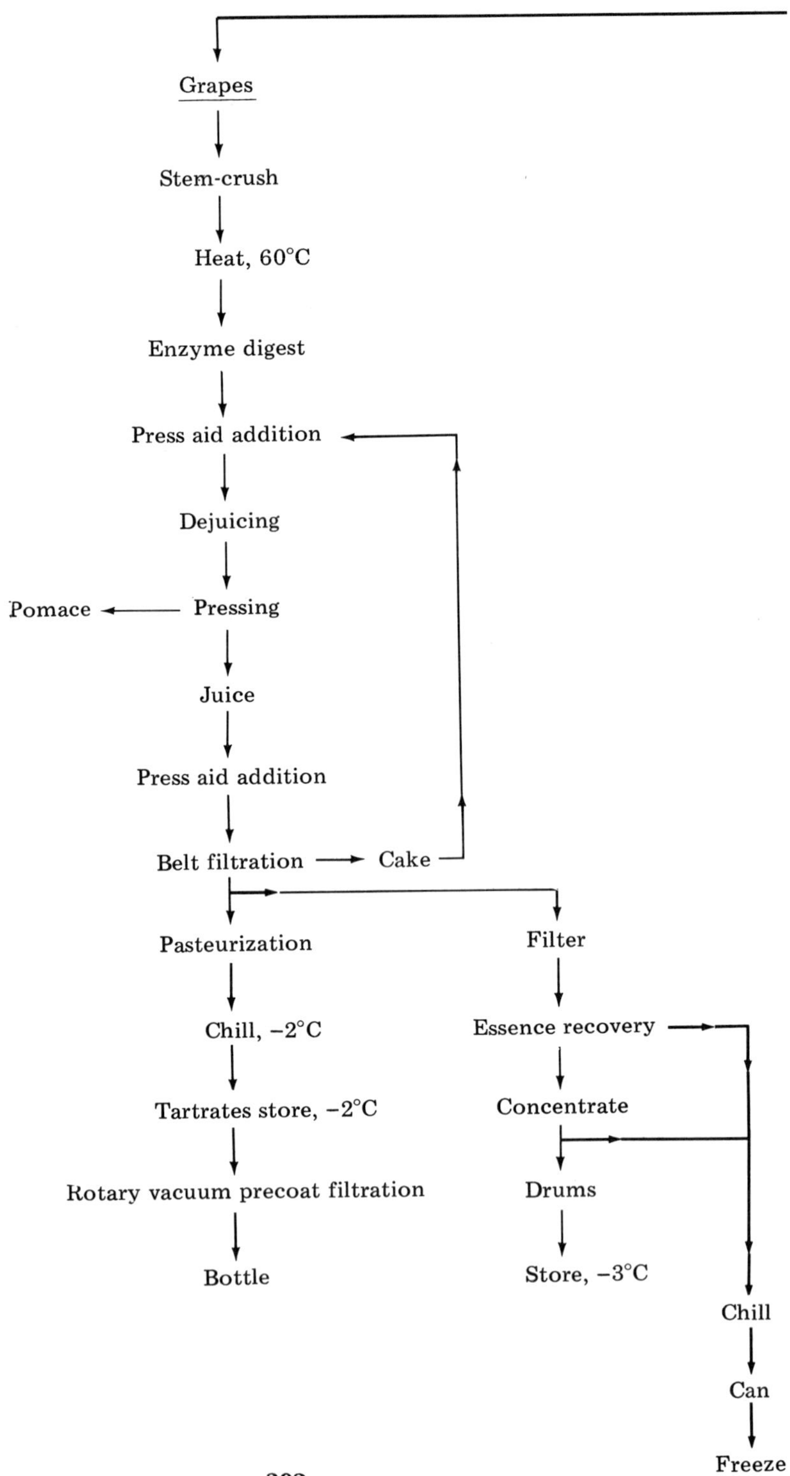

Grapes
Stem-crush
Heat, 60°C
Enzyme digest
Press aid addition
Dejuicing
Pomace
Pressing
Juice
Press aid addition
Belt filtration
Cake
Pasteurization
Chill, −2°C
Tartrates store, −2°C
Rotary vacuum precoat filtration
Bottle
Filter
Essence recovery
Concentrate
Drums
Store, −3°C
Chill
Can
Freeze

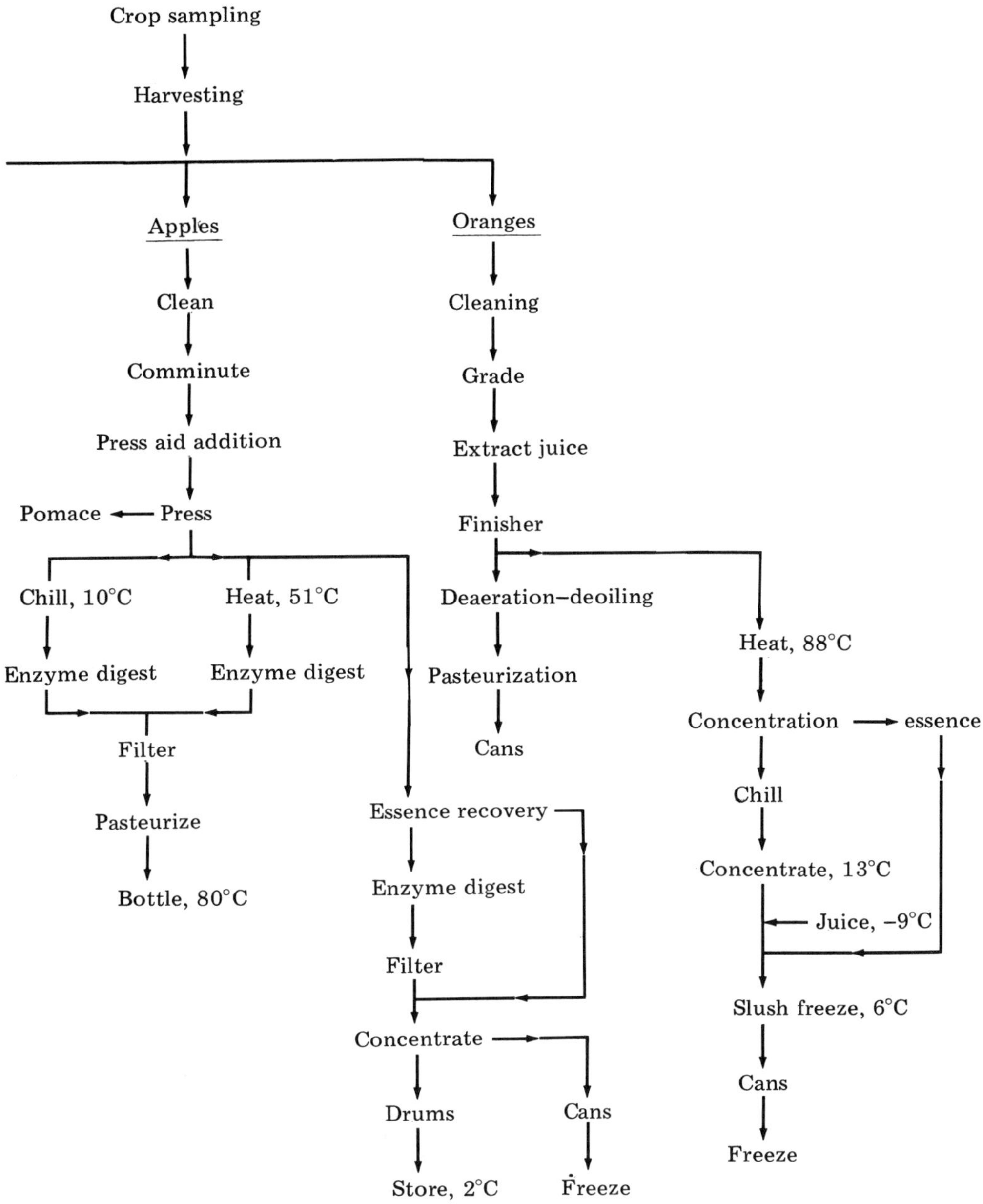

Figure 1. Fruit juice flow diagram.

Table 1. Composition of Fruit Juices, per 100 g[a]

Fruit juice	Water, %	Food energy, kJ[b]	Proteins, g	Fat, g	Carbohydrates, g	Ash, g	Ca, mg	P, mg	Fe, mg	Na, mg	K, mg	Vitamin A, IU[c]	Thiamin, mg	Riboflavin, mg	Niacin, mg	Ascorbic acid, mg
acerola	94.3	96	0.4	0.3	4.8	0.2	10	9	0.5	3			0.02	0.06	0.4	1600
apple	87.8	197	0.1	tr[e]	11.9	0.2	6	9	0.6	1	101		0.01	0.02	0.1	1
grapefruit, S-S[d]	89.2	172	0.5	0.1	9.8	0.4	8	14	0.4	1	162	10	0.03	0.02	0.2	34
sweetened	86.4	222	0.5	0.1	12.8	0.4	8	14	0.4	1	162	10	0.03	0.02	0.2	31
conc frozen	62	607	1.9	0.4	34.6	1.1	34	60	0.4	4	604	30	0.14	0.06	0.7	138
grapefruit–orange																
blend, S-S[d]	88.7	180	0.6	0.2	10.1	0.4	10	15	0.3	1	184	100	0.05	0.02	0.2	34
sweetened	86.9	209	0.5	0.1	12.2	0.4	10	15	0.3	1	184	100	0.05	0.02	0.2	34
conc[f]	59.1	657	2.1	0.5	37.1	1.2	29	47	0.4	2	623	380	0.23	0.03	1.1	144
grape, S-S[d]	82.9	276	0.2	tr[e]	16.6	0.3	11	12	0.3	2	116		0.04	0.02	0.2	tr[e]
conc[f]																
sweetened	52.8	766	0.6	tr[e]	46.3	0.3	10	15	0.4	3	118	20	0.06	0.10	0.7	15
grapejuice drink	86.0	226	0.1	tr[e]	13.8	0.1	3	4	0.1	1	35		0.01	0.01	0.1	16
lemon, S-S[d]	91.6	96	0.4	0.1	7.6	0.3	7	10	0.2	1	141	20	0.03	0.01	0.1	42
conc[g]	58.0	485	2.3	0.9	37.4	1.4	33	47	0.9	5	658	80	0.14	0.06	0.3	230
lemonade,																
conc[f]	48.5	816	0.2	0.1	51.1	0.1	4	6	0.2	2	70	20	0.02	0.03	0.3	30
lime	90.3	109	0.3	0.1	9.0	0.3	9	11	0.2	1	104	10	0.02	0.01	0.1	21
limeade, conc[e]	50.0	783	0.2	0.1	49.5	0.2	5	6	0.1	tr[e]	59	tr[e]	0.01	0.01	0.1	12
orange, S-S[d]	87.4	201	0.8	0.2	11.2	0.4	10	18	0.4	1	199[h]	200	0.07	0.02	0.3	40
sweetened	86.5	218	0.7	0.2	12.2	0.4	(10)[d]	18	0.4	1	(199)	200	0.07	0.02	0.3	40
conc frozen[f]	58.2	661	2.3	0.2	38.0	1.3	33[d]	55	0.4	2	657	710	0.30	0.05	1.2	158
dehydrated	1.0	1590	5.0	1.7	88.9	3.4	84	134	1.7	8	1728	1680	0.67	0.21	2.9	359
orange–apricot juice drink	86.7	209	0.3	0.1	12.7	0.2	5	8	0.1	tr[e]	94	580	0.02	0.01	0.2	16
pear nectar	86.2	218	0.3	0.2	13.2	0.1	3	5	0.1	1	39	tr[e]	tr[e]	0.02	tr[e]	tr[e]
pineapple, S-S[b]	85.6	230	0.4	0.1	13.5	0.4	15	9	0.3	1	149	50	0.05	0.02	0.2	9
conc frozen[f]	53.1	749	1.3	0.1	44.3	1.2	39	29	0.9	3	472	50	0.23	0.06	0.9	42
pineapple–grapefruit juice drink	86.0	226	0.2	tr[e]	13.6	0.2	5	5	0.2	tr[e]	62	10	0.02	0.01	0.1	16
prune	80	322	0.4	0.1	19.0	0.5	14	20	4.1	2	235		0.01	0.01	0.4	2
tangerine, S-S[b]	88.8	180	0.5	0.2	10.2	0.3	18	14	0.2	1	178	420	(0.06)[h]	(0.02)[h]	(0.01)[h]	22
conc frozen[f]	58	678	1.7	(0.7)[h]	38.2	1.3	62	48	0.7	2	613	1460	0.20	0.06	0.4	96
tomato	93.6	79	0.9	0.1	4.3	1.1	7	18	0.9	200	227	800	0.05	0.03	0.8	16

[a] Ref. 2.
[b] To convert kJ to kcal. divide by 4.184.
[c] IU = International units.
[d] Single-strength.
[e] Trace.
[f] Concentrated, must be diluted with water before use.
[g] Lemon juice concentrate is a product in itself.
[h] Imputed value.

Comminution. A variety of methods are used to break the fruit for juice release depending on the structure of raw material, the clarity desired in the final juice, and enzymatic discoloration and destruction of pectin. The most common disintegrator is the hammermill or a variation of it where fixed or free-swinging hammers force the fruit particles through a screen. The degree of disintegration is varied by the use of blunt hammers for impact disintegration or sharp hammers for chopping, the diameters of the screen holes beneath the hammers, and by hammer speed. A variation of the hammermill is the angle disintegrator in which the fruit is fed into the center of the mill and discharged through a circular screen into a collection annulus. In Europe, the fruit is fed into the center of a cylinder (a grater), and propelled by bars past stationary knives. Grated fruit contains less fines and permits the formation of a more porous press cake.

A modified hammermill consisting of curved knives revolving over a grating is used to chop tomatoes before heat treatment at 82°C for pectin esterase inactivation by pumping through a tubular heat exchanger in the manufacture of tomato juice (see Fig. 2). When greater conservation of pectin is desired for the production of tomato puree or concentrate, the chopped material is dropped into a tank of stirred hot pulp where pectic enzymes are immediately inactivated. A revolving steam coil at the bottom of the tank provides agitation and maintains a pulp temperature of ≥93°C (11).

When mechanical disintegration is not feasible because of pit breakage, whole peaches or plums are heated to ≥93°C in an enclosed screw or ribbon conveyor by direct injection of culinary-grade steam. Retention time may be 10–20 min until oxidizing enzymes are inactivated and the flesh separates easily from the whole pits in a pulper.

Stemmers are used to separate grape berries and other fruit harvested as clusters from the stems, vine, or other debris and reduce the berries to a pulp (5). The fruit clusters are fed into a slowly revolving cylinder having perforations of ≥1.9 cm, and they are subjected to the action of beaters revolving in the opposite direction at a much faster speed, thereby knocking the berries off the clusters, and partially crushing them

Figure 2. The Fitz-Mile Comminutor used for pulping and juicing apples, tomatoes, oranges, etc. Courtesy of The Fitzpatrick Co., Elmhurst, Ill.

before they are discharged through the drum perforations. The paddles are slightly twisted so that the stems are propelled towards the discharge end of the drum. The pulp is sufficiently liquid so that it can be pumped through a tubular heat exchanger to raise the temperature for pigment release and enzyme digestion.

Enzyme Treatment of Pulp. Many fruit pulps have a mucilaginous character that makes juice extraction difficult, especially with continuous presses. The nature of this mucilage is not known but it is commonly believed to be due to pectin (see Gums). The slipperiness of the pulp can be overcome somewhat by the addition of press aid. In the production of Concord grape juice, the crushed grapes are heated to 60°C, a pectic enzyme preparation added, and held at this temperature for 45 min. Digestion is complete when the juice of the pulp, which is squeezed in several layers of cheese cloth and mixed with two parts of ethanol or 2-propanol, gives no stringy-type precipitate after one minute. In Europe, ground apples have been treated with commercial pectic enzymes following aeration of the pulp to facilitate juice extraction on a belt press (12). As an alternative to aeration, the addition of 400–1000 ppm of poly(1-vinyl-2-pyrollidinone) to the pulp was found to protect pectic enzymes from inactivation by polyphenols (13). In the United States, finely comminuted apples have been digested with pectic enzymes before extraction on a rotary vacuum precoat filter (14). When a fruit is digested with pectic enzymes, residual enzyme activity is adsorbed on the pulp, and an additional enzyme treatment may be required for the clarification of the extracted juice (see Enzymes, industrial).

Press Aids. Fibrous materials are frequently added to the fruit pulp to facilitate juice liberation, or to provide a stable structure to withstand the application of pressure (6). Several types of press aid are used: diatomaceous earth (see Diatomite), coarse wood flour, rice hulls, and bleached and unbleached wood pulp. Diatomaceous earth and wood flour are sometimes sprinkled over press cloths, or are added to the pulp before the cloths are filled for hydraulic pressing, thereby making discharge of the spent cakes easier and also increasing the amount of juice (see Filtration). When continuous presses are used, more fibrous press aids are required to develop a cake structure that permits compression without slippage against a revolving screw. The most common of the latter materials are bleached and unbleached wood pulp which liberates a minimum of soluble materials that may affect the flavor or color of the juice adversely. The unbleached wood pulp is available in bale form and must be shredded, whereas the sheet form of bleached pulp is frequently defibrated in a hammermill. Both the shredded and the defibrated pulps should be free of clumps or short fibers for maximum effectiveness.

The press aid must be thoroughly dispersed throughout the fruit pulp before it enters the press; for this purpose, a mixing trough equipped with open-loop paddles made from square barstock has been used. The composition of the fruit entering the press by gravity is controlled by feeding the pulp to the mixer with a variable speed screw pump; the addition of the unbleached pulp, when it is used, is controlled by the rate of feed to the shredder, or when bleached pulp is used, by the rate of feeding to the hammermill. From the hammermill, the defibrated pulp is pneumatically conveyed to a cyclone from which it drops into the mixer. Rice hulls may be added to the mixing trough by a small screw conveyor or by a vibratory feeder. The amount of bleached or unbleached pulp added to the fruit may vary 0.5–5 wt %, depending on the type, variety, and maturity of the fruit, the degree of defibration of the press aid, and thoroughness of dispersion in the fruit pulp. If too little fiber is added, the juice yield is low and there is an exceptional amount of suspended solids in the juice. Too much press aid gives a decreased juice yield owing to absorption by the press aid and the cake may be so dry that it cannot be discharged from the press. Rice hulls are frequently added at the rate of 0.25–1% to prevent blinding of the screen holes in the press cage by fruit particles. Excessive addition of rice hulls can cause serious erosion of the press cage and screw.

When juice is to be extracted from a warm puree, such as crushed grapes or stone-fruit pulps, pieces of sheet pulp are added to the tanks near the end of the digestion period. The fluidity of the mass and agitation by a turbine-type stirrer permits dispersion of the cellulose fibers without prior defibration (5). Then the mixture is placed in a revolving reel so that 30–40% of the juice is removed by gravity, thereby increasing the capacity of a screw press. The sheet pulp aids dejuicing by reducing blinding of the reel screen, and facilitates extraction of juices from the screened solids in subsequent pressing.

Extraction. Owing to composition and structural differences in fruits, a variety of procedures are used to extract juice. They are divided into two classes: those used for (*1*) citrus or (*2*) deciduous fruits.

Citrus. Size-graded oranges, lemons, or grapefruit are automatically conveyed to one of several types of extractors (9). In one type of extractor (FMC Inc.), the fruit is located between two cups attached to metal strips having sharp-edged metal tubes at their base. When the upper cup descends, a hole is cut in the top and bottom of the fruit and, on further compression, the rag, seeds, and juice sacs are compressed in the bottom tube between the two plugs of skin. A piston moves up inside the bottom tube expressing the juice through perforations in the tube wall. In another type of extractor (Brown Co.), the individual fruits are wedged between tapered vertical revolving disks, and cut in half as they pass a stationary knife. A conveyor guides the halves, cut surface down, to a stationary turret where flexible fingers grip them as reamers move in and remove the juice sacs. Sometimes the juice stream is divided at the reamers according to the desired oil and pulp contents in the juice since these components increase with greater reaming pressure. In extraction of citrus juices it is desirable to have as gentle an extraction procedure as possible, and minimal contact time between juice and pulp to reduce the amount of limonin which has an intensely bitter flavor. Lime juice may be extracted by feeding the whole fruit directly to a screw press. The amount of suspended solids in citrus juices is controlled in a subsequent separation in a finisher, and the oil level is adjusted by centrifugation or vaporizing under a vacuum.

Deciduous. Until a decade ago, deciduous juices were extracted with hydraulic presses (6). Loosely woven nylon cloths draped over shallow forms were filled with comminuted fruit and the edges folded over to form cheeses that were separated by racks having square lattice openings. These racks permitted juice drainage and provided stability to a stack of "cheeses" when it was moved into position between press platens. Juice started to flow during cheese formation and reached a maximum flow rate with a pressure of 172 kPa (25 psi) on the cheese. Thereafter, the flow declined until the end of cake compression at 689–1030 kPa (100–150 psi). The juice separated during the initial stage of pressing (free-run juice) frequently is sweeter and has less acid and astringent materials. Usually a press cycle lasts 30 min. Juice yields vary from 70% for apples to 87% for hot Concord grapes. The suspended solids in the juice are usually less than 1 vol % as measured by centrifugation at 1200 G for 30 min. Hydraulic presses, available in a range of sizes, provide flexibility in extracting a variety of fruits.

Usually several presses are operated in sequence to maintain a steady flow of juice to succeeding clarification and canning operations. The chief disadvantages of hydraulic presses are their relatively low juice production rates and the labor in handling and cleaning large press cloths. They are operated usually without the addition of press aid to the ground fruit although the application of coarse wood flour or diatomaceous earth to the cloths before filling facilitates separation of the spent cake.

A variety of intermittently and continuously operated presses have been introduced to facilitate juice production and reduce labor costs. A horizontal basket press (6) (Bucher-Guyer) utilizes a perforated cylinder enclosed with stationary heads and a moving circular platen mounted on the piston of a hydraulic cylinder. The basket has a hinged door so that on rotation it is filled from above and the spent cake discharged below. During pressing, the platen moves in to compress the fruit mass and on retraction a series of perforated nylon hoses strung between the heads break up the cake before recompression. These hoses also act as drainage channels for the juice. Compression and breaking of the cake are repeated seveal times during a pressing cycle of about one hour. The addition of a press aid (eg, rice hulls) is frequently required for efficient juice extraction.

Pneumatic-drum presses have also been used in which a rubber tube inside a horizontal perforated cylinder is inflated, pressing the pulp radially against a cloth lining the cylinder (6). The cylinder is filled and emptied through a hinged section of the drum surface. When filled with fruit pulp, the closed cylinder is rotated to distribute the fruit mass uniformly on the cylinder surface and air pressure is applied to the rubber tube until the juice flow ceases. Then the air pressure is released and the cake is broken up by revolving the cylinder before air pressure is applied again. This process is repeated 4 or 5 times, gradually increasing air pressures to a maximum of 620 kPa (90 psi). Press aids in the form of rice hulls, wood fiber, or sheet pulp are usually added to the pulp to facilitate juice extraction.

A recently introduced horizontal hydraulic press (Stoll) has a sliding gate, and a mixture of fruit pulp and rice hulls (a press aid) is dropped from a hopper between two vertical platens mounted on hydraulic pistons. The recessed surface of each platen is covered with closely spaced bars to permit juice drainage. The bottom of the chamber consists of a similar platen. One platen moves in compressing the mass and finally the second platen recedes to allow discharge of the cake. This hydraulic press has a cycle time of 30–150 s and an hourly capacity of up to 9 metric tons.

A variety of methods has been proposed for the continuous extraction of juices such as screws, conveyor belts, drums, and countercurrent extraction with warm water. Of these, the screw type (Vincent, Jones, Ritz) is probably the most widely used (6) (see Fig. 3). A mixture of fruit pulp and press aid is propelled by a screw through a reinforced perforated cylinder having $\leq$1.6 mm ($\geq$12 mesh) dia or smaller. On the end of the screw shaft, a sliding cone positioned by a pneumatic cylinder resists the pulp discharge, thereby causing a compressed cake to be discharged. To reduce slippage of the pulp on the screw, resistor bars protrude in the path of the notched screw. In some presses, a lobed wheel protrudes through the cage and meshes with the screw flights to reduce pulp slippage. At the feed end of the screw, the discharged juice is essentially "free-run," and near the cone section the juice contains more extracted and suspended solids owing to the shearing action of the screw as it forces the pulp past the cone. The main factors affecting the operation of a screw press are: the mucilaginous nature of the fruit, amount and degree of dispersion of the press aids, screw

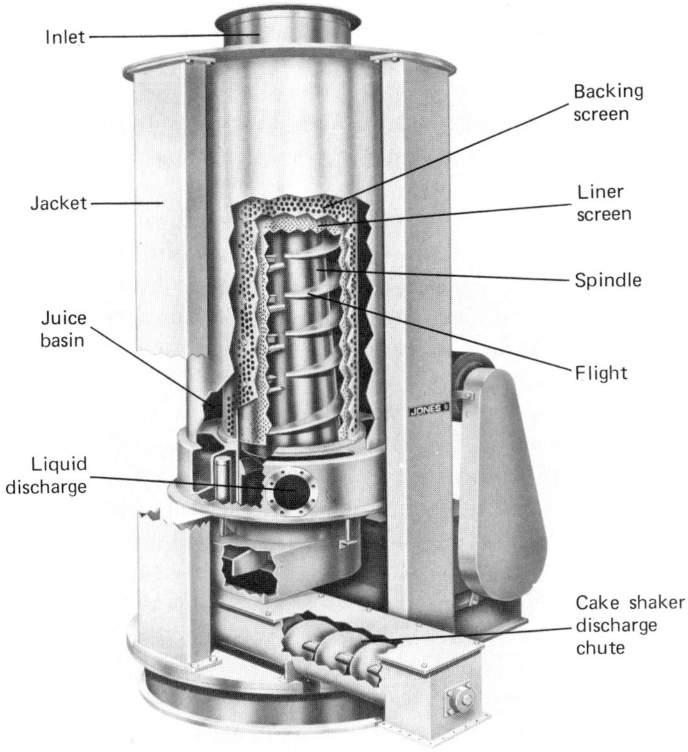

Figure 3. The Beloit-Jones Vertical Pressmaster for continuous extraction. Courtesy of Beloit Corp., Jones Division, Dalton, Mass.

speed, and counter pressure on the cone. The mucilaginous nature of the fruit pulp can be diluted by the addition of press aids, and in some cases, by enzyme digestion. Addition of too much press aid can reduce the juice yield and produce a cake so dry that the press will plug. An increase of screw speed over a limited range enlarges press capacity, but too high a screw speed and cone pressure increases the suspended solids content of the juice, and causes later problems in clarification. Both horizontal and vertical screw presses are employed, some having tapered cones, and others with screws of uniform or varying flight spacing. Under good operating conditions, it is possible to obtain apple juice yields of 85% containing 2% suspended solids by volume.

Continuous plate presses (Willmes) have been used in apple juice production (6). Ground apples mixed with press aid are spread on a cloth belt and the edges folded to form a pocket before the belt passes between two rows of vertical plates attached to roller chains with sprockets under tension. As the belt moves forward, the cake is compressed and the juice collected below in a pan. At the discharge end, the belt flattens before passing over return rolls and the cake is discharged. Accurate belt loading is essential and difficulties may be encountered from belt alignment and wear but juice yields of 600 L/t (175 gal/short ton) of apples have been obtained with 2% suspended solids in the juice.

Endless horizontal belts of cloth interwoven with stainless steel wires or hinged perforated plates have been used to convey fruit pulp while under compression from rollers (12). A variation of this principle is compression of the fruit between a belt and a revolving perforated drum. After compression, the belt is separated and the drum cleaned before reloading.

Both screening (6) and solid-bowl decanter types of centrifuges have been tried, but their capacity, juice clarity, and yield are not as high as that obtained with other processes (see Centrifugal separation).

Rotary–vacuum–precoat filtration has been used to separate juice from finely ground and enzyme-digested pulp (14). Usually the pulp is diluted with equal volumes of water and juice to reduce its consistency; as a result insoluble fruit particles adhere to the cake. The cake may be leached by atomized water sprays during the drying portion of the filter cycle. When water is used, utilization of the filtrate is restricted to manufacture of juice concentrate.

In a countercurrent extraction process similar to that used for sugar beets, apples are cut in crinkle-faced slices and fed to the base of a steam-jacketed inclined conveyor having flights that permit good mixing with water fed at 55°C to the upper end of the diffuser (15). Residence time is approximately one hour and the liquid drained from the base is within 0.5–1.5° Brix (0.5–1.5 wt % sugar) of juice pressed from the apples. The extracted spices are pressed between continuous belts and the juice returned to the diffuser. The advantages claimed for this extraction method are less oxidative browning, better extraction of the volatiles, and a pomace suitable for pectin manufacture (15). Government regulations may limit the use of this juice to concentration because of the addition of water.

Clarification. The clarities of juices vary widely. Citrus and tomato juice have high insoluble solids contents, whereas cranberry, grape, and many apple juices have little or no suspended solids and are nearly translucent. The presence of a cloud in some juices probably results from the adherence of the flavoring constituents to fruit solids. It was recognized early in the citrus industry that clarified juices were lacking in flavor. For this reason and the unsightly appearance of sediment in bottled juice, considerable research was devoted in the citrus industry to the causes of solids flocculation. Preservation of the initial pectin content by rapid heating and cooling to inactivate the natural enzymes was found to be very important (16). Size of the suspended particles is another factor influencing flocculation and this may be adjusted by screening, homogenization, colloid milling, or by selective filtration. In the production of opalescent or natural-type apple juice, ascorbic acid is added to the fruit pulp before pressing or to the juice as it comes from the press to retain more of the apple flavor (6). This juice is cloudy but in time the insoluble solids settle. Salt (NaCl) is sometimes added in conjunction with ascorbic acid to enhance its antioxidant capacity. This process has also been used in the production of white grape juices (see Antioxidants; Dispersants; Flocculating agents).

In the production of clear juices, the natural pectin is degraded with pectic enzymes to facilitate filtration, and prevent the subsequent precipitation of the pectin. Other sources of colloidal haze in filtered juices are polymerized tannins or compounds formed by combination of tannins with proteins. In other instances, the haze may be due to the denaturation of proteins. To overcome tannin precipitates, gelatin (qv) is sometimes added to the juice during enzyme digestion. The amount and type of gelatin added to form a floc must be carefully controlled since an excess of gelatin causes a haze. Laboratory tests are frequently conducted by adding graduated amounts of gelatin and tannin to samples of juice and noting the amounts required for greatest clarity after standing overnight (6). Such tests are frequently necessary because of the variation in the natural tannin and protein contents of the juice. If the haze is due solely to proteins, the juice is sometimes heated to 71°C and cooled before filtration. Bentonite is sometimes used to remove proteins but color may also be removed and the slimy nature of the bentonite precipitate makes it difficult to filter the juice.

Digestion of the juice with commercial pectic enzyme preparations may be carried out at 13°C for 15–18 h or at 49°C for 1 or 2 h (see Enzymes, industrial). Intermediate digestion temperatures are not used because of the possibility of alcoholic fermentation. If a higher digestion temperature is used, the juice may be clarified by rotary vacuum precoat filtration followed by a small-cartridge or leaf-type filter to remove any filter aid. If juices are digested at low temperatures, sufficient time usually elapses for flocculation of the solids, and the clear juice is decanted and passed through a pressure-leaf filter. The precipitated solids may be used for vinegar stock or the juice extracted in high pressure plate-and-frame filters. The filter aids used are diatomaceous earth of fine to medium grades depending on the clarity desired and the nature of the solids. Completeness of pectin destruction is of importance in maintaining good filtration rates and filtrate clarity, and this is commonly measured as noted above by diluting 1 part of juice with 2 parts of ethanol or 2-propanol. If pectin persists in the juice, a stringy precipitate usually forms within 2 min.

To avoid the formation of tartrate crystals in bottled grape juice, the filtered juice must be stored at −2°C, or just above the freezing point, for 6 to 8 wk to permit the deposition of potassium acid tartrate crystals (5). Prior to storage, the juice is pasteurized at 88°C and then cooled in storage at −2°C before it is pumped into clean storage tanks. Unless care is taken in the sanitation of the pasteurization and cooling equipment, piping, and tanks, yeast growth may cause alcohol production or generation of off-flavors.

Pasteurization. Fruit juices generally have a pH of 4 or less and spoilage is usually limited to yeast and mold growth (17). Under such an acid condition, commercial sterility is achieved by filling the containers at 88°C, followed by inversion for one minute or slow cooling under water sprays. In addition, this heat treatment is usually sufficient to inactivate pectolytic and oxidative enzymes. The ascospores of some molds such as *Byssochlamys* and certain *Penicillium* commonly found on fruit surfaces are more heat-tolerant and known to withstand 3 hours at 85°C (18). This mold contamination can be greatly reduced by rejection of fruit-bearing rots, brushing the fruit surface with water sprays, or immersion in chlorine water followed by a fresh-water rinse. Clarification of the juice by use of fining agents such as gelatin, followed by filtration also reduces the danger from mold spoilage. Tomato juice frequently has a pH higher than 4 and the juice must be held at 121°C for 0.7 min to destroy *Bacillus coagulans*, a flat-sour spoilage-type organism (19).

Both shell-and-tube and plate-type heat exchangers are used for pasteurizing fruit juices. The heating medium may be hot water under pressure, steam, or steam under vacuum. The heating surface should be sufficient that the temperature differential between the final product temperature and the heating medium is not greater than −12°C. A juice velocity of 0.9–1.5 m/s is desirable. In addition to controlling the heating medium through sensing of the temperature of the discharged juice, provision should be made for recycling the juice should the discharge temperature drop below the set point. Periodically, the heating surfaces are fouled by deposits of pectins or proteinaceous materials, as evidenced by a widening temperature differential, and provision should be made for diversion of juice flow as cleaning solutions are circulated through the pasteurizer. When heating unclarified juices, the openings in the heat exchanger should be large enough to pass fruit particles or shreds of press aid.

Chemical Preservatives. Many governments require a listing of ingredients, including preservatives, on the container label, and only permit use of those materials that are considered to be noninjurious to health (see Food additives). The most common preservatives used in fruit juices are the potassium and sodium salts of sorbic, benzoic, and sulfurous acids (see Sorbic acid). The amount of sodium benzoate permitted is 0.1 wt % and a common rate of addition to apple juice is 0.05–0.075 wt % (20). In an acidic juice, the benzoic acid (qv) may impart an objectionable flavor at concentrations greater than 0.08 wt %. The sodium benzoate should be dissolved in a minimal amount of water before addition to the juice to prevent the immediate precipitation resulting from its direct addition to the juice. Potassium sorbate is more soluble and less likely to impart a flavor, and when added to apple juice at 0.05 wt %, it allows freshness to be retained for 2 wk at 10°C. The storage life of juices treated with chemical preservatives is seriously curtailed if they are heavily contaminated during preparation, or not refrigerated as much as possible. The amount of SO_2 added as a gas or a sulfite salt varies widely with pH and solids content of the juice or concentrate. At low pH, free SO_2 is the most effective form. At high pH, dissociated SO_2 is bound by sugars, aldehydes, etc, and is not available as an antiseptic. Concentrates containing more solids require more SO_2 than single-strength juices, hence the amount of SO_2 added may vary 350–600 ppm (21). When SO_2 is added to a concentrate intended for canning as a single-strength juice, the SO_2 must be removed by flash distilling the heated juice in a vacuum, or the SO_2 will be reduced by the tin creating an H_2S odor and darkening the interior of the container. In some juices, as little as 50 ppm SO_2 may be detected by a sensitive palate (see Flavor characterization).

Freezing Preservation. Fruit juices can be preserved by freezing. The juice is filled into waxed paper or plastic containers approved for food use so that a 10 vol % headspace is left for expansion during freezing. The juice should be as cold as possible before filling and in some cases the juice or concentrate is chilled to a slush in a wiped-film heat exchanger. The containers are placed on refrigerated shelves with provision for forced-air circulation around them, or in the case of round containers, conveying with rotation through a liquid refrigerant. The frozen juice should be held at a uniform temperature of −18°C to reduce ice crystal growth and damage to the colloidal solids.

Concentration and Aroma Recovery. Large quantities of fruit juices are concentrated for economy of transport or storage, and sold to the consumer as frozen concentrates or in bulk for manufacture of various products as blends of single-strength juices, juice drinks, jellies, alcoholic beverages, wines, and liqueurs (cordials), or in soft drinks where clarity of the product is desirable. Concentration by evaporation of the water by heating is the most common method, although aromatic flavoring materials may be lost with the water removed, or altered by prolonged heating (see Flavors and spices). Early procedures for recoving the volatile components involved heating the juice so that 5–40 vol % was flashed off at atmospheric pressure. These vapors were fed to a distillation column having stripping and rectification sections. The vent gases were scrubbed with refrigerated condensate to reduce entrainment.

The distillate drawn from the reflux section was 1/150 the volume of the feed juice. The stripped juice was concentrated, and essences were added to the concentrate, or packaged separately (22). When the essence was concentrated to 300-fold, difficulties were encountered, for example, with Concord grape aroma because the ethanol concentration interfered with separation of methyl anthranilate. The trend to flash distill the juice prior to vacuum concentration (23), and to add this condensate back to an over-concentrated juice prevents exposure of the juice to high temperatures. Sometimes the over-concentrated juice is diluted with single-strength juice to restore some of the original flavor. The procedure used for making fruit concentrate and essence is shown in Figure 4.

Fruit essences are stable for an extended period, especially under refrigeration, but when returned to the juice they usually have limited stability.

The evaporators (effects) used for juice concentration are double- or triple-effect rising, falling, or rising–falling film units and may employ tubular or plate surfaces (24) (see Evaporation). The trend is toward the use of more effects for heat economy. Forced-circulation evaporators or vacuum pans are seldom used because of the long retention times and danger from carmelization. An evaporator commonly used in the citrus industry (TASTE) may have five or more effects in the form of tall vertical tube bundles in which the juices flow down the heated surface and are pumped to the top of the succeeding effect (23). Residence times as short as 5 min are claimed for this evaporator.

Fouling of the heated evaporator surfaces is commonly encountered due to deposition of hespiridin (citrus), pectin, proteins, or tartrates (grape) and provision should be made for cleaning in place every 6 h. Inspection of the evaporator interior should be made for microbial growth especially in the vapor–liquid separators (23).

Centrifugal plate evaporators have been used for limited production of high-quality concentrate. Wiped-film evaporators have been used to attain high degrees of concentration in juices that are very viscous, eg, tomato puree.

In freeze-concentration, the juice produces superior concentrates when chilled to a slush from which the concentrated solutes are removed by centrifugation, compression, or in a wash-column. Difficulties have been encountered in separating viscous concentrates from ice (25).

Reverse osmosis (qv) has also been considered as a means of concentrating fruit juices, but the loss of flavoring materials and the high pressures required to effect the desired degree of concentration of low molecular weight solids have restricted its use.

When concentrates are to be used in jelly manufacture, soft drinks, or in alcoholic beverages, wine, and liqueurs (cordials), the initial juice must be thoroughly depectinized and have a high degree of clarity.

Dehydration. Drying pure fruit juices is difficult because of the low melting points of the solids and their hygroscopicity. Under conditions of low water vapor pressure as in freeze-drying, the juice must be frozen in such a manner that the ice crystals are thoroughly distributed throughout the mass, or a concentrated syrup results rather than dry solids (26). However, by the addition of various ingredients, the hygroscopicity of the solids may be reduced and flavor retention increased (22,26). Such additions have a dilution effect so that there is less fruit solid in the final powdered form (see Drying).

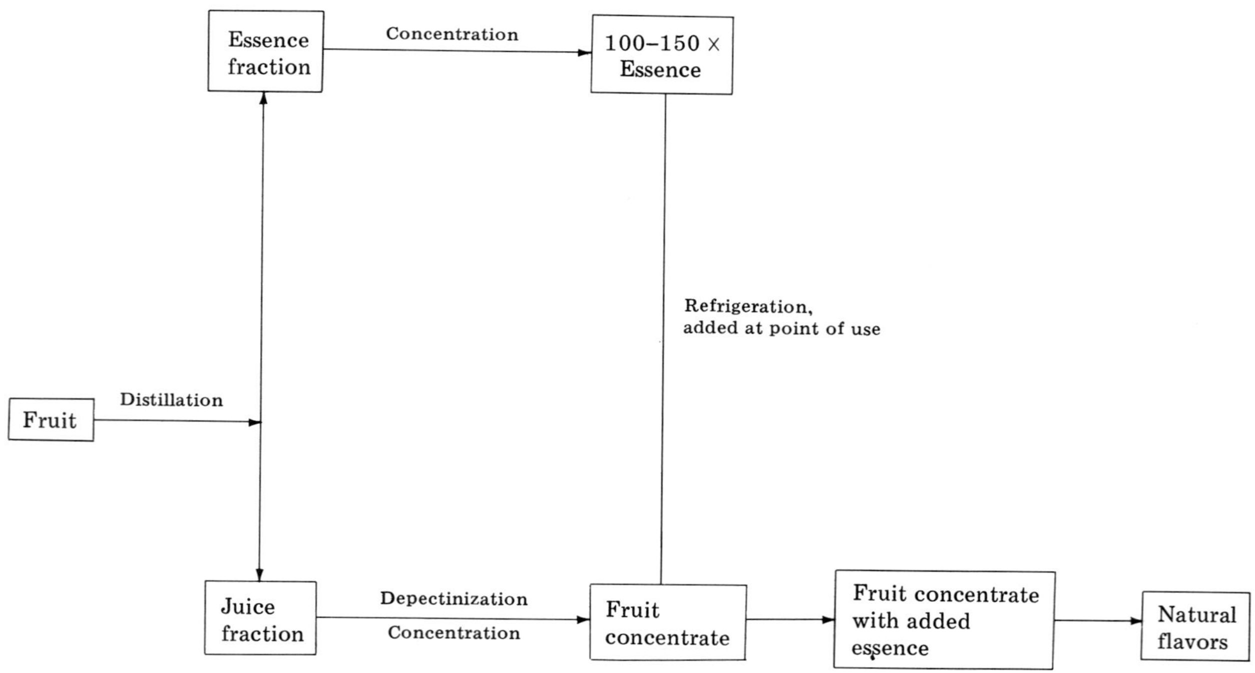

Figure 4. Production of fruit concentrate and essence.

Table 2. U.S. Per Capita Consumption of Fruit Juices[a], Kilograms Product Weight

Year	Canned							Chilled orange[c]	Citrus[d] concentrate	Frozen citrus
	Orange	Grapefruit	Blended	Apple	Grape	Prune	Pineapple[b]			
1966	0.69	0.78	0.15	0.53	0.28	0.50	1.65	1.37	0.44	6.67
1970	0.79	1.35	0.15	1.21	0.26	0.50	1.35	1.97	0.66	9.96
1974	0.67	1.61	0.10	1.66	0.30	0.33	1.06	2.12	1.11	12.97
1977	0.65	1.36	0.10	1.52	0.20	0.41	1.09	2.29	1.36	13.63

[a] Ref. 27.
[b] Includes concentrates on a single-strength basis.
[c] Chilled juice produced directly from fruit.
[d] Single-strength equivalent.

Consumption

Accurate statistics on current fruit juice production are difficult to obtain because juices are marketed in a variety of forms as fresh, chilled, frozen, canned, or concentrated products. The usage of concentrates of different densities has greatly increased in recent years because of the economy in shipping, storage, and ease of remanufacture as single-strength juice, juice-ades, carbonated beverages (qv), and jellies. For these reasons, data on the per capita consumption basis (Table 2) may approach most nearly the relative usage of various juices.

BIBLIOGRAPHY

"Fruit Juices" in *ECT* 2nd ed., Vol. 10, pp. 167–178, by M. K. Veldhuis, U.S. Fruit and Vegetable Products, USDA, Winter Haven, Fla.

1. H. Lüthi in D. K. Tressler and M. A. Joslyn, eds., *Fruit and Vegetable Juice Processing Technology,* AVI Publishing Co., Westport, Ct., 1961, pp. 667–695.
2. B. K. Watt, A. L. Merrill, R. K. Pecot, C. F. Adams, M. L. Orr, and D. F. Miller, "Composition of Foods" in *Agriculture Handbook No. 8,* USDA, Washington, D.C., Dec. 1963.
3. J. Shacklady in A. Wollan, ed., *Food Industries Manual,* Chemical Publishing Co., New York, 1970, p. 278.
4. W. C. Cooper and H. Chapot in S. Nagy, P. E. Shaw, and M. K. Veldhuis, eds., *Citrus Science and Technology,* Vol. 2, Avi Publishing Co., Westport, Ct., 1977, pp. 1–120.
5. C. S. Pederson in D. K. Tressler and M. A. Joslyn, eds., *Fruit and Vegetable Processing Technology,* Avi Publishing Co., Westport, Ct., 1971, pp. 234–271.
6. J. C. Moyer and H. C. Aitken in ref. 5, pp. 186–233.
7. P. Dupaigne in ref. 1, pp. 538–550.
8. J. G. Woodroff in J. G. Woodroff and B. S. Luh, eds., *Commercial Fruit Processing,* Westport, Ct., 1975, p. 58.
9. R. E. Berry and M. K. Veldhuis in ref. 4, pp. 177–248.
10. C. R. Marshall and V. T. Walkley, *Food Res.* **16,** 515 (1951).
11. A. Lopez, *A Complete Course in Canning,* The Canning Trade, Inc., Baltimore, Md., 1969, pp. 335–343.
12. L. deVos, *Proc. Int. Fruit Juice Union (Palmero)* **10,** 191 (1970).
13. A. P. Verspuy, W. Pilnik, and L. deVos, *Flussiges Obst.* **37,** 518 (1970).
14. J. A. Murch and N. F. Roger, *Food Eng.* **36**(2), 42 (1964).
15. H. Luthi and U. Glunk, *Flussiges Obst.* **41,** 498 (1974).
16. H. E. Swisher and L. E. Swisher in ref. 4, pp. 269–271.
17. H. Luthi in C. O. Chichester, E. M. Mrak, and G. F. Stewart, eds., *Advances in Food Research,* Vol. 9, Academic Press, New York, 1959, pp. 221–284.
18. D. F. Splittstoesser in L. R. Beuchat, ed., *Food and Beverage Mycology,* Avi Publishing Co., Westport, Ct., 1978, pp. 92–96.
19. S. Leonard in ref. 5, pp. 447–448.
20. G. Borgstrom in ref. 1, pp. 179–215.
21. E. J. Hsu in ref. 8, pp. 113–140.
22. J. L. Bomben, B. Solhe, H. A. C. Thijssen, and R. L. Merson in C. O. Chichester, E. M. Mrak, and G. F. Stewart, eds., *Advances in Food Research,* Vol. 20, Academic Press, New York, 1975, pp. 2–111.
23. R. E. Berry and M. K. Veldhuis in ref. 9, pp. 215–226.
24. C. H. Mannheim and N. Passey in A. Spicer ed., *Advances in Preconcentration and Dehydration of Foods,* John Wiley & Sons, Inc., New York, 1974, pp. 151–193.
25. H. A. C. Thijssen in ref. 24, pp. 115–147.
26. C. J. King in ref. 24, pp. 435–486.
27. *The Almanac of the Canning, Freezing, Preserving Industries,* E. Judge & Sons, Westminster, Md., 1979, p. 576.

JAMES C. MOYER
New York State Agricultural Experiment Station

FUEL CELLS. See Batteries and electric cells, primary.

FUELS, SURVEY

Until a few years ago, fuel generally referred only to those substances that could be ignited in air and continued to burn and to release heat that could be used to meet the needs of society. The principal fuels include the so-called noncommercial fuels (wood, wastes, agricultural products, and animal dung), and commercial fuels (peat, lignites, coal, oil, and natural gas). On reacting with oxygen in the air, these fuels produce heat and residual products that consist mainly of carbon dioxide and water vapor.

Recently, with advances in science and technology, the term fuels has been expanded to include other means of supplying heat and energy; these include reactions of a variety of types, eg, rocket fuels, and various forms of nuclear energy, eg, uranium and, possibly at some later date, thorium. The renewable energy resources include hydroelectric, solar, wind, geothermal, and tidal energy sources, which can also be used to provide heat and power.

Fuel Consumption and Productivity

Energy is important because of the many ways in which it is used and how pervasive it is in daily living. Although energy serves a variety of purposes, consumer willingness to alter consumption patterns differs from use to use, eg, priority of heating a residence over use of a snowmobile. Moreover, as a result of societal changes, energy is used differently than in the past, eg, a higher percentage of energy is used for transportation and a smaller percentage for space heating. It is almost certain that in the future, energy will be used differently than it is now, although the nature of the changes that will occur is difficult to predict.

At one time, it seemed that there were direct connections between economic growth and human welfare, and between energy use and economic growth; both of these views have been challenged in recent years. Those who consider economic growth and human welfare to be closely related point to the better health, educational services, and labor-saving appliances in economically advanced countries. Challengers of this view insist that the quality of life is reduced by economic growth through, eg, increased pollution, crowding, and crime that accompany this growth. No matter how the debate about the relationship between economic growth and human welfare is resolved, the less developed countries are striving to industrialize to improve their economic position, even if adverse consequences have to be tolerated.

The historical relationship between economic growth and energy use suggests that to increase economic growth, additional amounts of energy are required; however, there may be substantial flexibility in the amount of additional energy required per unit of increased gross domestic product (GDP); the aggregate value of domestically produced goods and services.

After the oil embargo of 1973–1974, which was accompanied by an abrupt increase in the price of oil and of other energy forms, a concern emerged regarding the availability of energy, and a fear that an energy scarcity could develop. This was intensified in 1979 with the unsettled political conditions in Iran and the sharp price increases promulgated by the Organization of Petroleum Exporting Countries (OPEC). These new concerns are in sharp contrast to the situation over the past 100 years when energy was both abundant and its real value declined steadily.

Energy and Gross Domestic Product. The view that levels of economic growth and energy consumption are closely linked seemed to be supported by the energy and GDP growth rates in the period prior to 1973. Another view has emerged: there is considerable flexibility in the energy required to achieve a set of economic goals.

The relationship between the price of energy and economic growth appears to be less important than one would expect. For example, in the United States, the energy sector is only a small component of the economy. In 1970, it was less than 4% of the GDP and even at the much higher energy prices in 1975, it was less than 8%. In most other countries, the percentage of GDP spent on energy is higher than in the United States, but in most industrialized countries it is still less than 15%.

The relationship between GDP and energy in the United States, although frequently used as a means to predict future energy demands, has changed substantially over the years. The GDP growth rate averaged 3.1% during the forty-year period from 1925–1965, whereas the consumption of energy was only about 80% of this value or 2.5% a year. As shown in Figure 1, the energy–GDP (in constant dollars) ratio in the United States increased from 1880–1920 by a factor of ca two, after which it declined steadily until 1950, when it leveled off at about 70% of its peak 1920 value.

Energy–GDP ratios for other industrialized countries vary widely from those in the United States and Canada. Table 1 gives the energy–GDP ratios (expressed in megajoules per dollar) for nine developed countries in 1972. It indicates that, even among the industrialized countries the ratio, when expressed in index numbers, varies widely: from 54–120% of that of the United States.

These data imply that some countries could reduce their energy–GDP ratio without adversely affecting economic growth; however, a comparison of the causes of the differences in energy use among countries is more complex than it first appears. At least 40% of the high energy–GDP value for the United States is caused by the large size of the country, dispersed population patterns, and the large number of single-

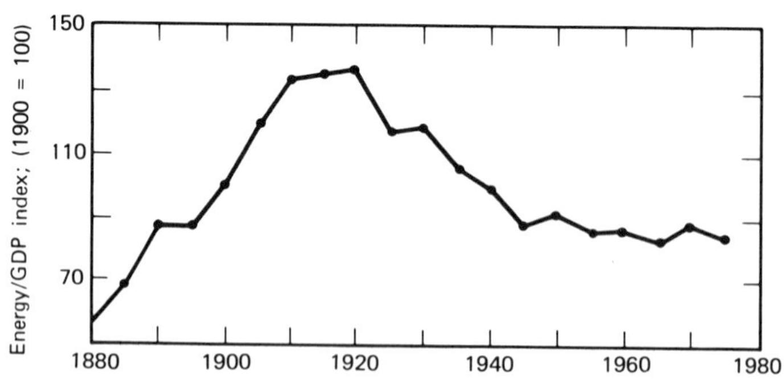

Figure 1. Index of energy/GDP ratios in the United States: 1880–1978 (1). Data for the 1880–1950 period are from ref. 2. Data for the 1955–1975 period are from ref. 3.

Table 1. Per Capita Energy Consumption and Per Capita Gross Domestic Product, Nine Developed Countries, 1972[a,b]

	Per capita		Energy/GDP	Index numbers (U.S. = 100)		
				Per capita		Energy/GDP
Country	GDP, $	Energy, GJ[c]	ratio, MJ/$[d]	GDP	Energy	ratio
United States	5643	353.6	62.7	10.0	1000	100
Canada	4728	354.8	74.9	8.4	1000	119
France	4168	140.3	33.6	7.4	400	54
FRG	3991	174.5	43.7	7.1	490	69
Italy	2612	100.9	38.6	4.6	290	63
Netherlands	3678	198.3	53.9	6.5	560	86
United Kingdom	3401	161.1	47.4	6.0	460	76
Sweden	5000	224.9	45.0	8.9	640	72
Japan	3423	122.9	19.1	6.1	350	57

[a] Ref. 4.
[b] The hydro and nuclear component of primary energy consumption is converted into joules on the basis of fuel inputs into fossil-fueled power plants, assuming 35% efficiency. Foreign GDPs are expressed in U.S. dollars, using a real purchasing power basis of comparison rather than market exchange rates.
[c] To convert GJ to 10^6 Btu, divide by 1.054.
[d] To convert MJ to 10^3 Btu, divide by 1.054.

family homes that are energy-intensive compared to apartments or multifamily homes common in most other countries. Canada's high energy use is the result of a large number of energy-intensive industries in that country, eg, the pulp and paper, metallurgical, and chemical industries.

The remaining 60% of the high energy–GDP value can be explained largely by differences in energy prices. The low efficiencies of U.S. automobiles compared to those in Western Europe are the result of much lower gasoline taxes in the United States. Lower cost for fuels in some countries also makes substitution of energy for labor and capital more economically attractive so that energy consumption patterns per unit of GDP in those countries tend to be higher than where fuel prices are high.

The variations of energy use per unit of GDP over time are related more to fundamental changes that have occurred in the composition of energy supply than they are to other patterns. This includes the growth rate in energy used as electricity (which requires energy inputs much greater than for other fuels), which was about twice the rate of total energy growth, and a dramatic shift from solid fuels use to liquids and at a later date to gas. At the same time, there were major increases in energy efficiency in the generation of electricity and in rail transportation. Despite these very large changes in energy supply patterns, energy consumption per unit of GDP changed only at a modest rate so that forecasts of drastic changes in this ratio in the future must be viewed cautiously.

Structure of Energy Demand. Historically, heat and power were supplied mainly by wood or vegetable and animal wastes and these noncommercial fuels are still used widely in underdeveloped countries. By the end of the nineteenth century when the industrial revolution had made significant headway, more convenient and higher quality fuels were used extensively in place of wood. In 1900, the total energy consumed worldwide was equivalent to 38.5 EJ (36.6×10^{15} Btu) per person: 58% was supplied by coal, 22% by wood, 17% by animal dung and farm waste, and 3% by oil and gas.

Both world energy consumption and population increased rapidly after the turn of the century, with an annual average energy growth rate of 2.6%/yr from 1900–1975. By 1975, average annual per capita consumption was ca 64.2 GJ (60.9 × 10^6 Btu). Table 2 shows the total and per capita energy consumption and population in selected years from 1925 to 1975.

The annual percentage rate of change in energy consumption, population, and energy use per capita also is shown at the bottom of Table 2. Annual per capita energy consumption has accelerated from 1.1% between 1925 and 1950 to 3.1% between 1950 and 1960 to 3.7% between 1960 and 1970. The sharply increased price of energy after 1973 reduced the growth rate of per capita energy consumption between 1970 and 1975 to 0.3%/yr.

The worldwide average per capita growth rate for energy from 1960–1970 ranged from 2.2% for the United States to 7.4% for Japan; a number of other industrialized countries had annual per capita growth rates in the range of 3.0–3.5% in the same period.

Energy Conservation. Energy conservation has become a new worldwide objective as energy prices increased sharply and concern about energy availability has grown. Conservation is the economic reduction in energy use per unit of output or of activity, not a reduction in energy use at any cost. Energy conservation meets society's objectives of the rational use of world resources. It is not an austerity situation that must be endured. In general, conservation measures are more difficult to implement than the alternative supply enhancement options because, except for the industrial sector, many opportunities occur in small units, with the majority of decisions required by individual consumers.

The degree of increased efficiency of energy use (owing to higher prices, better consumer information, improved technology) is highly uncertain because of the absence of a good data base on which to make estimates. The most optimistic proponents of energy conservation view it as a simpler, cleaner means of achieving economic goals and helping the economy through investments in conservation rather than in investments to enhance energy supply.

Conservation opportunities vary widely among countries and in the different

Table 2. World Energy Consumption and Population, Selected Years, 1925–1975[a]

Year	Total energy consumption, EJ[b]	Population, millions	Energy consumption per capita, GJ[c]
1925	46.64	1890	24.7
1950	80.97	2504	32.3
1960	130.7	2990	43.7
1970	266.1	3609	62.6
1975	254.7	3967	64.1
Average annual percentage rate of change			
1925–1950	2.2	1.1	1.1
1950–1960	4.9	1.8	3.1
1960–1970	5.5	2.0	3.7
1970–1975	2.5	2.0	0.3

[a] Updated from data in ref. 5.
[b] To convert EJ to 10^{15} Btu (quad), divide by 1.054.
[c] To convert GJ to 10^6 Btu, divide by 1.054.

energy-using sectors. Fuel efficiency can be greatly improved in the transportation sector, particularly in the United States where relatively low fuel taxes, geographical conditions, and personal preferences have resulted in large, inefficient automobiles. Recent regulations in the United States mandate greatly improved fuel efficiencies for transportation in the future.

The industrial sector is more price sensitive than other sectors and can be expected to reduce its energy consumption per unit of product as energy prices rise.

In the building sector, with its long turnover times for existing structures, the savings in energy use will occur more slowly. Builders of houses and commercial buildings, unlike industrial executives, are most interested in first cost and not in life-cycle energy cost of structures, so that investments in energy conservation are usually below the optimum level.

The Organization for Economic Cooperation and Development (OECD) countries, following the 1973–1974 embargo, embarked on a conservation program and have projected a reduction of primary energy consumption from 1.44 metric tons of crude oil equivalent (coe) per thousand U.S. dollars to 1.36 t of coe in 1985. A metric ton of coe is equal to 41.9 GJ (39.8×10^6 Btu). However, even with vigorous conservation efforts, the International Energy Agency (IEA) countries project increases in annual energy consumption of 3.7% from 1976–1985 and 3.5% from 1976–1990. Although use is still increasing, this is a significantly lower projection of energy consumption growth rates than was made in earlier years.

Energy Utilization, and Health and the Environment. The environmental aspects of energy utilization have become extremely important because both future fuel costs and the size of energy demand will depend to a large extent on how these issues are treated. Environmental questions affect both energy utilization and energy supply, but only the environmental issues related to energy utilization are discussed here.

The environmental movement emerged in the late 1960s, first in the United States and later in other industrialized countries, as an important political force that was considered in decisions concerning major economic activities that have the potential to create adverse effects on land, water and air. Although energy utilization generally does not create the hazardous or toxic substances that many industrial operations create, the massive amounts of fuels that are produced, transported, and consumed evolve into difficult environmental problems.

The type and magnitude of environmental impacts are different for each fuel form during its utilization. For example, natural gas is the most benign of the fossil fuels, creating a minimum of air pollutants, ie, sulfur and nitrogen oxides, particulates, unburned hydrocarbons, trace metals, and carbon monoxide. Oil combustion produces somewhat more air pollutants than gas, with the amount of sulfur oxide emissions depending on the sulfur content of the oil. Coal is the most polluting of the fossil fuels, usually producing more air pollutants of all kinds than either oil or gas (see Air pollution).

The major concern regarding the air pollutants emitted from the combustion of fossil fuels is the long-term chronic health effects that appear to be related to large concentrations of air pollutants. The questions involving health impacts of air pollution are controversial and there still is incomplete knowledge about which of the air pollutants or their transformation products are the cause of health damage. In addition, the sulfur or nitrogen oxides (or both) are believed to be the cause of acid rain which has harmful effects on foliage and soils and on aquatic flora and fauna (6).

These environmental considerations affect the level of fuel utilization, their cost, and the share that each fuel form contributes to the energy supply. These effects occur because of the costs involved in reducing emissions. Methods for pollution reduction include switching to other higher quality fuels, upgrading the fuels before use, finding new technology to burn the fuel cleanly, or applying control technology to remove pollutants from the combustion products (see Air pollution control methods).

Removal of air pollutants, particularly when coal is burned, is practiced widely in the United States by using equipment for scrubbing sulfur oxides from the flue gases. Since these control technologies add to the cost of fuel utilization, their use can affect fuel selection. The greatest penalty in preventing environmental damage is on coal and the least on natural gas. The annual cost (1977 dollars) of particulate control is the equivalent of about a dollar per metric ton for coal, whereas sulfur oxide scrubbing, using limestone and disposal of the sludge produced, is \$7 to \$12 per metric ton for coal. The control of nitrogen oxides usually is accomplished by modifying combustion techniques; costs are estimated at ca \$0.20 per metric ton of coal (see Burner technology).

Ash disposal from plants using coal can also create environmental problems, particularly if they are disposed of in such a way that soluble salts or toxic trace elements can be leached from the ash and can find their way into waterways where they can enter the food chain. Disposal of ash in lined ponds costs ca \$0.70/t coal used.

The use of nuclear energy creates only minor air-pollution problems, but thermal pollution is larger per unit of electricity produced than for a fossil-fuel plant, and disposal of spent nuclear-fuel elements in a safe manner is a difficult problem (see Nuclear reactors). In addition, accidents at nuclear plants pose a major potential threat to human health and the environment.

Energy Supply and Resources

The Changing Primary Fuel Supply. The historical change in mix of primary energy supplies, which started with a shift from wood to coal during the Industrial Revolution, is still undergoing modification. Table 3 shows the distribution of major energy sources by regions for 1950 and 1975. For the world as a whole, coal's share declined from nearly 56 to 33% of total energy use in this twenty-five year period, although the absolute tonnage consumed increased over 80% because of the rapidly expanding demand for energy (see Coal). Use of oil grew from 29 to 44% of total demand and gas increased from 9 to 20% (see Petroleum). The decline in coal's share of primary energy consumed occurred in every region, with the smallest decline in communist Asia and Eastern Europe, and the largest in Western Europe and Japan. The percentage of natural gas use increased sharply in North America and Western Europe, whereas oil consumption increased sharply in all but two regions, with the largest increases in Japan and Western Europe (see Gas, natural).

Not shown in this table is the amount of nuclear energy used for the generation of electricity. As recently as 1971, installed nuclear energy capacity was 25.2 GW out of a total world electrical capacity of 1200 GW, or ca 2.1%. But by 1975, this had increased to 98 GW out of a total of 1570 GW or about 6.3%.

There is worldwide controversy over the future of nuclear energy. Complex issues that have been raised include: the adequacy of the resource base, the need for breeder

Table 3. Distribution of Energy Sources for Major Regions, 1950 and 1975[a]

	1950						1975					
				Primary	Total[c]					Primary	Total	
	Coal, %	Oil, %	Natural gas, %	electricity, %[b]	%	EJ[d]	Coal, %	Oil, %	Natural gas, %	electricity, %[b]	%[c]	EJ[d]
North America	38.0	38.7	16.7	6.4	99.8	38.85	21.3	44.8	30.6	3.3	100.0	78.7
Canada	40.6	28.6	2.8	27.9	99.9	2.86	11.1	49.7	27.9	11.3	100.0	7.0
United States	37.8	39.5	18.0	4.7	100.0	35.99	22.3	44.3	30.9	2.5	100.0	71.8
Western Europe	77.4	14.3	0.3	8.0	100.0	18.42	25.1	54.7	15.9	4.2	99.9	44.9
Oceania	65.3	27.3		7.4	100.0	0.98	41.7	47.4	6.9	4.2	100.2	3.1
Latin America	9.8	72.9	8.3	9.0	100.0	2.53	5.6	72.5	16.6	5.4	100.1	9.5
Asia (exclusive of communist countries)	53.3	28.5	1.4	16.9	100.1	4.40	29.7	62.2	4.9	3.1	99.9	20.6
Japan	61.9	5.0	0.2	32.9	100.0	1.83	20.0	73.4	3.2	3.4	100.0	12.3
other Asia	46.0	48.3	2.4	3.3	100.0	2.17	44.2	45.7	7.4	2.7	100.0	8.3
Africa	61.4	36.9		1.7	100.0	1.37	47.6	43.8	5.7	2.9	100.0	4.8
USSR and Eastern Europe	81.4	14.6	2.3	1.7	100.0	13.53	43.2	32.0	23.8	1.0	100.0	60.0
USSR	75.6	19.7	2.5	2.3	100.1	8.88	34.6	36.8	27.5	1.1	100.0	43.1
Eastern Europe	92.5	4.8	2.0	0.7	100.0	4.65	65.1	20.0	14.1	0.8	100.0	17.0
communist Asia	92.8	0.9		6.3	100.0	1.32	82.3	15.9	0.8	1.1	100.1	19.1
world, %	55.7	28.9	8.9	6.5	100.0	80.97	32.8	44.1	20.4	2.8	100.1	244.6
world, EJ[d]	45.1	23.4	7.2	5.27			80.2	107.7	50.0	6.7		

[a] Ref. 5.
[b] Electricity from hydroelectric power and nuclear energy.
[c] Total may not add to 100 because of rounding.
[d] To convert EJ to 10^{15} Btu (quad), divide by 1.054.

reactors in place of lightwater reactors (breeder reactors would extend the resource base by a factor of 60–70), the safety of nuclear plants, proliferation and sabotage opportunities, recycling of plutonium, and the difficulties of long-term safe disposal of spent nuclear wastes. These concerns were highlighted by the accident at the Three Mile Island nuclear power plant at Harrisburg, Pa., in the spring of 1979. Because of the number and complexity of these issues, projecting the future contribution of nuclear fuels to energy supplies is virtually impossible. It is certain that by 1985 nuclear fuel will supply much larger amounts of electricity than it now does since a number of new plants are already under construction.

Changes in primary energy supply continue, but there is a growing consensus that oil and gas resources are such that, in the period 1990–2010, oil and (at a later date) gas will decline in importance so that current trends in use of fuel will be modified significantly. Under these circumstances, coal and, if the above issues with respect to nuclear power can be resolved, nuclear energy will supply a larger share of energy demand. The unconventional resources (eg, geopressured gas, heavy oils, and oil shales) and the essentially renewable resources (eg, solar (qv), geothermal (qv), and biomass (see Fuels from biomass)) should become a significant portion of world energy supplies.

Energy Resources in Fossil Fuels. Resources are the total amounts of minerals, including reserves, that are expected to be produced. Reserves include that portion of the resources that are known with reasonable certainty and that can be produced and sold at present prices using extraction technology.

Information on energy resources has been collected since 1924 when the first survey was conducted by the World Power Conference. The most recent information is contained in ref. 7.

Coal. The total hard coal (bituminous and anthracite) resources of the world are estimated to be 7724×10^9 tons of coal equivalent (tce), the reserves of which are estimated to be 492×10^9 tce. A metric ton of coal equivalent has a heat content of 29.98 GJ (27.78×10^6 Btu). The estimate of the total resources of brown coal (subbituminous and lignite) are 2401×10^9 tce, of which 143×10^9 tce are reserves. Total resources of coal are thus $10{,}125 \times 10^9$ tce, of which 635×10^9 tce are reserves.

The communist countries are estimated to have 73% of the hard coal resources and 42% of the reserves, whereas North America has 16.5% of the resources and 25% of the reserves (see Coal). North America has 58% of the total brown coal resources and more than 45% of the reserves; the communist countries have a major share of the balance: 38% of the resources and 31% of the reserves (see Lignite and brown coal).

In recent years, the World Energy Conference has attempted to collect information based on a common geological basis, but extensive deviations occur because the information for some countries exists only for a particular set of geological conditions. The major difference among reporting countries is the maximum depth of deposit and minimum seam thickness that are included in the estimates.

In 1975, world coal production was ca 2.6×10^9 tce. The major producers were the USSR with 614×10^6 tce (24%), the United States with 581×10^6 tce (22.5%), and China with 349×10^6 tce (13.5%). World coal trade, however, was comparatively small compared to oil, with only about 5.7 EJ (5.4×10^{15} Btu) of energy in the form of coal being traded in 1975, compared to 59.6 EJ (56.5×10^{15} Btu) for oil for just the OECD countries in 1976.

The importance of the coal resource base is that it represents, despite the greater difficulties of producing, transporting, and utilizing it compared to oil and gas, an almost unlimited source of energy supply for the future. At current world energy consumption rates, the world coal resource base could supply all the energy required for about 1100 yr.

Peat. The resources of peat have not been surveyed as thoroughly as most of the other solid fuels because peat is not used nearly as extensively. In 1971, worldwide production of peat that was to be used for fuel was only 62×10^6 metric tons, compared to the total solid fuel production of 3×10^9 t in the same year. The most recent estimate (World Energy Conference, 1974) places total peat reserves at 211×10^9 tce (see Lignite and brown coal).

Petroleum. Estimates of the reserves and resources of petroleum (qv) (and to a lesser extent, natural gas) have received increasing attention in recent years because of concern over their incipient depletion worldwide and their importance in international energy trade. Estimates of ultimately recoverable petroleum resources are $180–263 \times 10^9$ t, of which ca 89×10^9 t are proved reserves. Regional oil reserves and resources are listed in Table 4. Reserves discovered in the past several years in Mexico are included in these estimates but they may turn out to be low as additional reserve data are developed. The three geographic areas with major petroleum resources and reserves are the Middle East, North America (mostly the United States), and the USSR. The Middle East contains 56% of the reserves and about 55% of the resources. North America has ca 8% of the reserves and about 12% of the resources. The USSR has 11% of the reserves and ca 10% of the resources. Recent discoveries of large petroleum sources in Mexico are an additional indication that major changes in the resource estimates may occur in the future, as they have repeatedly in the past, making the most recent projections of petroleum resource depletion perhaps appear too pessimistic. However, every nonrenewable resource eventually will be depleted and the extremely large demands for petroleum fuels, the massive quantities already consumed, and the great dependence of world economies on oil resources require that the world community be aware of the most reliable estimates of the state of the petroleum resource base.

Worldwide oil production in 1976 was approximately 2.9×10^9 metric tons; the Middle East produced 38%, the communist countries produced 19%, and the United States produced 14%. Whereas the world reserves-to-production ratio for coal was

Table 4. Reserves and Resources of Oil, 10^9 Metric Tons[a]

	Reserves	Resources
North America	6.9	20.5–34.2
South America	3.6	11.0–16.4
Western Europe	3.7	6.8–9.5
Eastern Europe	0.4	
Africa	8.1	13.7–20.6
Middle East	50.2	97.0–136.0
China	2.7	
USSR	10.3	19.1–27.2
other Asia	2.5	12.4–19.1
Australia	0.3	
Total	*88.7*	*180.5–263.0*

[a] Refs. 8–9.

245, that for petroleum was only 30; this is one of the major reasons for concern about future global petroleum availability.

Enhanced Oil Recovery, Heavy Oils, Bituminous Sands, and Oil Shale. The average oil recovery rate from reservoirs is 25–30% and this could be increased to 45–50% if enhanced oil recovery techniques are used (see Petroleum). Such improvements would increase appreciably the estimate of ultimately recoverable oil. An increase from 25 to 40% in recovery would add 100–160 × 10^9 t of oil to the estimate of total recoverable resources. There are also large reserves of heavy oil resources; some of these are being exploited currently, but most are not. These untapped resources could be produced if either new technology is developed to reduce the cost of their production or the price of oil from conventional sources becomes sufficiently high to make them competitive.

World deposits of heavy oils are believed to be very large but very few countries have attempted to estimate the quantity, because under current conditions, most of them cannot (except in a few special instances) compete with conventional sources of liquid fuels. Very large deposits of heavy oils are known to exist; those in Venezuela alone are estimated to be 150 × 10^9 t.

Bituminous tar sands (qv) are being produced commercially in Canada and production in 1976 was reported to be 2.5 × 10^6 t with new and larger capacity plants being constructed. Tar sand deposits also have been reported in Venezuela, the United States, Zaire, and Trinidad and Tobago, but no commercial operations exist in those countries. Total reserves of tar sands have not been estimated systematically, but in Canada the potential recoverable reserves are estimated to be 27.4 × 10^9 t.

Oil shale (qv) deposits are much more dispersed geographically than heavy oils and tar sands, and shale oil was produced commercially in small quantities in a number of countries, and is still produced in China and the USSR. Deposits of significant sizes were reported by the World Energy Conference (1978) for Sweden, Yugoslavia, the Federal Republic of Germany, the United States, Australia, New Zealand, and Brazil.

The total amount of oil in all the oil shale deposits of every grade and under all geologic conditions is enormous. Estimates range from 94–474 × 10^{12} metric tons of oil, or from 500–2500 times the estimated total recoverable resources of petroleum. However, the amount of oil per ton of shale is a major determinant in the cost of producing shale oil and only the higher-grade deposits can be expected to compete with other sources of liquids unless some entirely new technology is developed. The resource base of the highest grade and easily mined oil shale deposits is estimated to be 40 × 10^9 t, ca one-half the size of world oil reserves. Its widespread geographic distribution could help in commercializing oil shale because it occurs in many countries with little or no petroleum resources. However, large-scale operations are needed and capital investments of several billion dollars are required for individual installations and this involves a high financial risk that not many companies, or even countries, can afford to take. Moreover, the environmental problems that would be created by a large oil shale industry can be severe, particularly with regard to the disposal of the solid wastes.

Natural Gas. Natural gas has been used extensively in countries with gas resources (see Gas, natural). However, unlike petroleum, which can be transported at low cost between continents by oil tankers, natural gas transportation costs are high unless the consumer is supplied from sources that can be reached economically by pipeline.

Transport by sea and then pipeline is technically feasible, but the costs are much higher than transportation of liquids. Consequently, natural gas resources are less accessible to world consumers than are liquids.

Regional natural gas reserves and resources are listed in Table 5. The areas with the largest resources are the USSR, China and other Asian countries (31%), the Middle East (19%), and North America (19%). The largest reserves of natural gas are in the USSR (37%), the Middle East (29%), and North America (11%).

Natural gas and petroleum frequently occur in the same deposit and are often produced simultaneously, so that the potential for new discoveries is similar to those for petroleum. The estimated world natural gas and petroleum resources each translate into about equal amounts of energy.

In 1976, world natural gas production was 1.4×10^{12} m^3 (4.9×10^{10} ft^3) (on a heating value basis, only 46% that of petroleum). Large-scale utilization of natural gas was confined to the United States, Western Europe, and the communist countries.

Unconventional Gas Resources. Knowledge of the extent of the resource base for unconventional natural gas resources is very limited. In the United States, where production of natural gas from conventional sources appears to have peaked in the early 1970s, rough estimates have been made of the amount of natural gas that might be recoverable from unconventional gas deposits. These are listed in Table 6.

In addition, average recovery of gas from conventional natural gas deposits, already much higher than petroleum, could be increased from its current 70–80% by an additional 10 or 15%.

Table 5. Reserves and Resources of Natural Gas, 10^{12} m^3 [a,b]

	Reserves	Resources
North America	7.58	50.9–56.6
South America	2.24	22.6–25.5
Western Europe	3.90	14.2 (Western and Eastern Europe)
Eastern Europe	0.28	
Africa	5.86	28.3
Middle East	20.3	49.5
USSR	26.0	80.6 (USSR, China, other Asia)
China	0.7	
other Asia	2.6	
Australia	0.9	14.2
Total	*70.36*	*260.3–268.9*

[a] Refs. 8 and 10.
[b] To convert m^3 to ft^3, divide by 0.0283.

Table 6. Estimated Unconventional Gas Resources [a]

	10^{12} m^3 [b]
coal-bed degasification	8.7–23.3
devonian shale	14.6–17.4
tight formations	17.4
geopressured gas	85–1458

[a] Ref. 11.
[b] To convert m^3 to ft^3, divide by 0.0283.

World Fossil-Fuel Resources and Reserves. A summary of the estimates of world conventional fossil-fuel resources and reserves is shown in Table 7. Not included in Table 7 are the large quantities of unconventional resources not being produced.

Nuclear Energy Resources. Uranium (qv) is an energy resource that is being used to produce heat and to generate steam and electricity by nuclear fission (see Nuclear reactors). The amount of electricity that is produced from uranium is increasing rapidly, but information regarding available resources is still at an early stage of development compared to that for fossil fuels. Thorium (qv) also could be used as a fuel in some fission systems but no current commercial technology exists for using it. Regional uranium resources and reserves are listed in Table 8. The major deposits of uranium resources are found in the United States, Canada, Africa, and Western Europe.

In conjunction with lightwater reactor technology, the world uranium reserves (not including the reserves of the communist countries) have approximately one-half

Table 7. World Conventional Fossil Fuel Reserves and Resources[a]

	Reserves		Resources	
	10^3 EJ	Percent of total	10^3 EJ	Percent of total
coal	19.7	59.4	314	94.0
peat	6.7	20.4		
oil	4.0	12.1	8.2–11.9	3.0
natural gas	2.7	8.1	10.0–10.3	3.0
Total	*33.1*	*100.0*	*332–336*	*100.0*

[a] To convert EJ to 10^{15} Btu (quad), divide by 1.054.

Table 8. Uranium Reserves and Resources, Thousand Metric Tons[a]

	Reserves	Resources
United States	643	1696
Canada	182	838
Mexico	5	7
South America	60	74
Western Europe	87	487
Eastern Europe	na[b]	na[b]
Africa	572	772
Middle East		
USSR	na[b]	na[b]
China	na[b]	na[b]
other Asia	45	69
Australia	296	345
Total	*1890*	*4288*
heat content,		
10^3 EJ[c]	7.8 lightwater reactors	17.6
10^3 EJ[c]	467 fast breeder reactors	1057

[a] Ref. 12. Excludes 300,000 t of Swedish uranium not considered as reserves.
[b] Not available.
[c] To convert EJ to 10^{15} Btu (quad), divide by 1.054.

the heat content of the world coal reserves; yet world uranium resources are less than one-tenth of the world coal resources. In conjunction with fast-breeder reactor technology, world uranium resources have a heat content more than four times that of world coal resources. Thus uranium resources could supply world energy demand for hundreds of years if controversial international issues can be resolved regarding utilization of fast-breeder reactors.

Nuclear fusion, which uses the fuel forms, deuterium and tritium (for which there are no resource constraints) also may reach commercial development (see Fusion energy).

Renewable Energy Resources. Some of the wide variety of renewable energy resources that have been used extensively still are being used in many parts of the world. These include noncommercial fuels (animal dung, agricultural and municipal wastes, and wood) and water and wind power and geothermal and solar energy. Interest in the use of renewable resources has increased sharply with the growing concern regarding the depletion of world oil and gas resources.

Water Power. In 1976, hydroelectric sources produced ca 21% of the total world electricity, which represented about 2.5% of the total world energy consumption. Major hydroelectric production is concentrated in the United States, Canada, the USSR, Japan, Brazil, China, Norway, and Sweden. Hydroelectric energy contributes a large share to the energy used in Canada, Norway, Iceland, Switzerland, New Zealand, and Sweden. The amount of electrical generating capacity that could be installed worldwide if all the available sites were developed is approximately 2,857 GW compared to the installed capacity in 1976 of 384 GW. However, the share of total energy consumed from hydroelectric resources will continue to decrease, as it has recently, because of the greatly expanded total demand for energy and the limited rate at which suitable hydroelectric sites can be developed.

As the cost of energy from nonrenewable resources increases, the use of other energy resources that utilize water could become more widespread. These could include low-head turbines that can generate electricity from the flow of river waters and from the energy contained in tides in estuaries where the tidal range is unusually high. These resources for capturing the energy in water power already are being applied in some situations.

Wind Power. Potential wind power resources of the world are very large; wind energy was used widely in many countries over the past several centuries. However, wind power requires a minimum wind speed for satisfactory operation. Because the wind does not blow either steadily or persistently, windmills require a back-up energy system that appreciably increases total costs. New wind-generating concepts are being studied with regard to improved efficiency and reduced capital costs. It is unlikely that wind power will make a major contribution to world energy supplies but it may be of importance under certain conditions, eg, where wind power is available a large part of the time, in remote locations, in less developed countries, and for special applications, eg, irrigation (see Solar energy).

Geothermal Energy. An enormous reservoir of heat exists in the earth and reaches the earth's surface in many locations in the form of high temperature water or of steam. This energy resource is used at a few installations in Iceland, Italy, New Zealand, and the United States. The main application for this essentially renewable energy resource is generation of electricity, but it also has been used occasionally to heat buildings.

An extensive research program on geothermal energy (qv) began in the United

States in the mid-1970s and was aimed at developing new technology to utilize the energy from some types of geothermal resources that have not been exploited commercially to date. The size of the total geothermal base is very large but its use will probably be confined to those regions where the geothermal energy is close to the surface.

Ocean Thermal Gradients. Thermal gradients as great as 20°C exist between the ocean's surface and depths. By using a low boiling point fluid as the working medium, it is possible to generate electricity from these differences in ocean temperatures. Research is underway, but costs are unknown. As with geothermal energy, the use of ocean thermal gradients will be confined to certain areas, ie, where ocean thermal gradients are large (see Solar energy).

Solar Energy. The primary source of energy available to the earth is the sun, which radiates energy at an enormous rate. Only a very small part of this energy reaches the earth, although it is equivalent to more than 2000 times the world energy consumption in 1978. Solar energy (qv) can be used by direct thermal methods (heating water, and space conditioning—heating and cooling—and generating electricity by conventional thermal generating techniques) or it can be converted directly into electricity (see also Photovoltaic cells). Solar energy has been used to heat water in a number of countries but the number of commercial opportunities has been expanded by the sharp increase in prices of conventional fuels. Capital costs of solar installations for space conditioning still are high and require extensive storage facilities so it can be used at night and during overcast periods. At present, most solar energy space conditioning installations require a conventional energy back-up system.

Solar thermal electric energy systems concentrate the heat from the sun so that steam can be generated and converted to electricity in a conventional steam generator. The land area required for collecting and concentrating the solar energy is large and capital costs of collectors are high. The contribution of solar thermal conversion to electricity production will not be large by the year 2000.

Solar energy can be converted to electricity directly through the use of photovoltaic materials, in which electrons flow when the material is exposed to solar energy. The principles of this process have been tested successfully but costs must be greatly reduced before it can compete commercially with conventional methods of generating electricity. Means must be found to solve the very serious storage problem of solar energy; the only practical method is expensive storage batteries (qv).

In order to accelerate the growth of a commercial photovoltaic industry and to increase the procurement of photovoltaic systems to a level that will permit the use of low-cost production techniques, a program began for using a number of photovoltaic devices in U.S. federal buildings. However, the average price quoted for photovoltaic units is $11 per watt, or 15–20 times more than the cost of a coal-fired or nuclear power plant. Thus major technological breakthroughs are needed before solar photovoltaic energy systems will be competitive, even if a low-cost solution is found for the storage of solar power (see Photovoltaic cells).

Wood, Biomass, Wastes. These energy sources are fuels derived from nonfossil plant material. They include wastes and residues and deliberately managed fuel crops—terrestrial and marine. Wastes are organic materials created by consumers, and the residues from agriculture and forestry. The largest consumer wastes are from municipal sources and sewage. Agricultural wastes include manure and harvested crop residues, and forestry wastes include residues from wood-product mills, pulping mills, and logging operations (see Fuels from waste). Portions of these wastes are used for heat and power generation. Production and use of deliberately managed fuel crops has not become widely commercial (see Fuels from biomass).

Projections of Energy Demand

Future energy demand levels will depend mainly on expectations of key demographic and economic variables and on estimates of improvements in energy efficiency over time that would reduce the amount of energy required to produce a unit quantity of output. Although flexibility in energy demand per unit of GDP can have a significant impact on the total amount of required energy, the level of GDP remains the most important determinant of energy demand. Important demographic factors are the total number of people that will consume energy, and the size of the labor force.

In general, energy projections made before the 1973–1974 oil embargo anticipated much greater growth rates in energy use than are now expected. One important factor contributing to these reduced expectations is a widely held belief that economic growth rates will not be as large as they had been in the past nor as great as they were expected to be even a few years ago. Projections of future energy growth in Organization for Economic Cooperation and Development (OECD) and International Energy Agency (IEA) countries, and the year in which they were made, are shown in Table 9.

The original estimate for energy use by OECD countries in 1985 (made in 1974) was 290 EJ (275×10^{15} Btu); by 1977, this first was reduced to 236 EJ (224×10^{15} Btu), and later to 206 EJ (195×10^{15} Btu), a reduction of expectations of energy consumption of ca 20 and 30%, respectively. The most recent OECD projection is slightly higher than the lowest made in 1977 but it still represents an expected reduction in energy use from the 1974 projection of 25%.

Other projections of world energy use are listed in Table 10. The most striking characteristics of these projections are that, despite the large increases in energy prices and the massive efforts to promote energy conservation, the amount of energy that is expected to be consumed continues to increase at a relatively rapid pace.

Growth in energy consumption in developing countries is expected to be at a faster rate than in industrialized countries, but because the former start at a much lower base, the total amount of energy consumed by them will be small. The developing countries consumed 15% of the world energy demand in 1972; this is expected to grow to about

Table 9. Energy Consumption by OECD Countries, 10^3 GJ[a]

	1985	1990	2000
base case[b]	290		
reference case[c] (1977)	236		
1976–1977 review[d] (1977)	213		
1977–1978 review[d] (1977)	206	237	
reference case[e] (1978)	220	253	313

[a] To convert GJ to 10^6 Btu, divide by 1.054.
[b] Ref. 13.
[c] Ref. 14.
[d] Ref. 15.
[e] Ref. 16.

Table 10. Projections of World Energy Demand, EJ[a]

Projector	Date	1975 actual	1985	2000	2020
WAES[b]	1977	276	289–321[d]	369–480	
WEC[c]	1977	276	349–392	519–689	839–1323

[a] To convert EJ to 10^{15} Btu (quad), divide by 1.054.
[b] Workshop on Alternative Energy Strategies (17).
[c] World Energy Conference (11).
[d] Does not include consumption by communist countries.

Table 11. Selected Forecasts for the United States: Total Primary Energy Demand, EJ[a]

Forecaster[b]	Year of forecast	1985	1990	2000	Reference
Project Independence	1974	114.3–117.1			18
Federal Energy Administration, *National Energy Outlook*	1976	95.5–111.3	120.1–128.3		19
A Time to Choose					
historical growth		122.4		196.8	20
technical fix		96.6		130.7	
zero energy growth		92.9		105.4	
U.S. Department of the Interior, *Energy Through the Year 2000*	1972	122.9		202.3	21
	1975	109.1		172.2	
Shell	1976	101.4	115.9		
	1978	95.3	106.6		
Exxon	1977	98.0	113.8		
National Energy Plan	1977	102.2			
Central Intelligence Agency	1977	103.9			
Congressional Research Service	1977	96.1–103.7	110.1–119.5		
Department of Energy (MOPPS)	1978	99.7			
Department of Energy (Energy Information Administration)	1978	96.1–102.1	106.1–115.3		
Energy in America's Future, The Choices Before Us	1979			121.2	22

[a] To convert EJ to 10^{15} Btu (quad), divide by 1.054.
[b] Refs. 18–22.

25% in the year 2000. Similar downward revisions of expected energy use have been made in studies examining the U.S. energy situation (Table 11). Projections made as recently as 1972–1975 imply expected energy consumption rates of 108.6–122.3 EJ (103–116 × 10^{15} Btu) of energy to be required in 1985. More recent estimates are about 105.5 EJ (100 × 10^{15} Btu), a decline of nearly 10%.

BIBLIOGRAPHY

"Fuels, Survey" in *ECT* 1st ed., Vol. 6, pp. 892–902, by R. A. Sherman, Battelle Memorial Institute; "Fuels, Survey" in *ECT* 2nd ed., Vol. 10, pp. 179–191, by Albert Parker, Consultant.

1. S. H. Schurr, *EPRI J.* **3**(4), 16 (May 1978).
2. B. Netshert, V. Eliasberg, J. Lerner, and H. H. Landsberg, *Energy in the American Economy, 1850–1975*, Johns Hopkins University Press for Resources for the Future, Baltimore, 1960.

3. J. Alterman, *The Energy/Real Gross Domestic Product Ratio,* Bureau of Economic Analysis Staff Report, U.S. Department of Commerce.
4. J. Darmstadter, J. Dunkerley, and J. Alterman, *How Industrial Societies Use Energy: A Comparative Analysis,* Johns Hopkins University Press for Resources for the Future, Baltimore, Md., 1977.
5. J. Darmstadter and S. Schurr, *Philos. Trans. Roy. Soc. London,* 276 (1974).
6. "Air Quality and Stationary Source Emission Control," *A Report to U.S. Senate Committee on Public Works,* Commission on Natural Resources, Serial 94-4, Mar. 1975, p. 623.
7. *Survey of Energy Resources 1978,* World Energy Conference (formerly World Power Conference).
8. L. Auldridge, *Oil Gas J.* **75**(53), 97 (Dec. 26, 1977).
9. R. Nehring, *Giant Oil Fields and World Resources,* Report R-2284-CIA, The Rand Corporation, Santa Monica, Calif., June 1978.
10. J. D. Parent and H. R. Linden, *A Survey of U.S. and Total World Proved Reserves and Remaining Recoverable Reserves of Fossil Fuels and Uranium,* Institute of Gas Technology, Chicago, Ill., Jan. 1977.
11. *World Energy Resources, 1985–2000,* World Energy Conference, IPC Science and Technology Press, Guildford, UK, 1978.
12. *Uranium: Resources, Production and Demand,* International Atomic Energy Agency/OECD Nuclear Energy Agency, Paris, 1973.
13. *Energy Prospects to 1985,* OECD, Paris, 1974, p. 56.
14. *World Energy Outlook,* OECD, Paris, 1977, p. 14.
15. *Energy Policies and Programs of IEA Countries,* OECD, Paris, 1977, p. 30.
16. *Steam Coal: Prospects to 2000,* OECD, Paris, 1978, p. 18.
17. C. L. Wilson and co-workers, *Energy—Global Prospects 1985–2000,* Workshop on Alternative Energy Strategies, McGraw-Hill Book Co., New York, 1977.
18. *Project Independence,* Federal Energy Administration, GPO, Washington, D.C., 1974.
19. *National Energy Outlook,* Federal Energy Administration, GPO, Washington, D.C., 1976.
20. *A Time to Choose: America's Energy Future,* Energy Policy Project of the Ford Foundation, Ballinger Publishing Co., Cambridge, Mass., 1974.
21. W. Dupree and J. Corsentino, *Energy Through the Year 2000 (Revised),* U.S. Department of the Interior, Bureau of Mines, Washington, D.C., GPO, 1975.
22. S. H. Schurr, J. Darmstadter, H. Perry, W. Ramsey, and M. Russell, *Energy in America's Future, The Choices Before Us,* Johns Hopkins University Press for Resources for the Future, Baltimore, Md., 1979.

General References

Refs. 1, 7, 9–17 are general references.

"Economic Growth and Energy Conservation: Historical and International Lessons," AAAS Selected Symposium #22, *AAAS (American Association for the Advancement of Science), 144th Meeting, Washington, D.C., 1978.*

The International Energy Situation: Outlook to 1985, ER-77-102400, Central Intelligence Agency, Washington, D.C., 1977.

J. Darmstadter and H. H. Landsberg, "The Economic Background," *Daedalus* **104**(4), (1975), special issue on "The Oil Crisis: In Perspective."

W. H. Hogan, *Future Energy Demand,* Ballinger Publishing Co., Cambridge, Mass., 1978.

Energy Demand and Supply Scenarios in 2000, Institute for Energy Analysis, Oak Ridge, Tenn., 1978.

Survey of Energy Resources, World Energy Conference, U.S. National Committee of the World Energy Conference, New York, 1974.

HARRY PERRY
Resources for the Future

FUELS FROM BIOMASS

Renewable carbon resources capable of conversion to synthetic fuels that are not depletable must ultimately be developed if the economy of the world continues to rely on organic fuels as a primary energy source. Technology for manufacturing large supplies of organic fuels and chemicals from nonfossil carbon must be commercialized. Land- and water-based vegetation and photosynthetic organisms, all of which are categorized as biomass, represent a practical source of this carbon.

Renewable carbon resources is actually a misnomer because the earth's carbon is in a perpetual state of flux. Carbon is not consumed in the sense that it is no longer available in any form. Many reversible and irreversible chemical reactions occur in such a manner that through the carbon cycle all forms of carbon, including fossil resources, are renewable. It is simply because of time that one carbon form is more renewable than another. If society could wait several million years so that natural processes can replenish depleted petroleum or natural gas deposits, presuming that replacement does in fact occur, then there would never be a shortage of organic fuels as they are distributed and accepted in the world's energy markets. Unfortunately, this cannot be done; therefore, fixed carbon-containing materials that renew themselves over a time span short enough to make them continuously available in large quantities should be sought. Land- and water-based biomass and organic wastes are effectively major sources of carbon that meet these requirements.

The capture of solar energy as fixed carbon in biomass via photosynthesis is the key initial step in the growth of biomass:

$$CO_2 + H_2O + \text{light} \xrightarrow{\text{chlorophyll}} (CH_2O) + O_2$$

Carbohydrate, represented by the building block (CH_2O) is the primary organic product. For each gram–mole of carbon fixed, ca 470 kJ (112 kcal) are absorbed. According to radioactive tracer experiments, the oxygen liberated in the process comes exclusively from the water. Although there are still many unanswered questions regarding the detailed molecular mechanisms of photosynthesis, the prerequisites for plant biomass production are well established: carbon dioxide, light in the visible region of the electromagnetic spectrum, the sensitizing catalyst chlorophyll, and a living plant are essential. The upper limit of the capture efficiency of the incident solar radiation in biomass has been variously estimated at 8–15% but in most actual situations, it is generally in the 1% range or less (1).

The main features of biomass-to-energy technology are illustrated in Figure 1. Conventionally, biomass is harvested for feed, food, and materials-of-construction applications or is left in the growth areas where natural decomposition occurs. The decomposing biomass or the waste products from the harvesting and processing of biomass, if disposed of on or in land, can in theory be partially recovered after a long period of time as fossil fuels. This is indicated by the dashed lines in Figure 1. Alternatively, the biomass and any wastes that result from its processing or consumption could be directly converted into synthetic organic fuels if suitable conversion processes were available. Although not shown in the figure, the energy content of biomass and wastes could be diverted instead to direct heating applications by combustion. Another route to energy products is to grow certain species of hydrocarbon-producing biomass, such as the rubber tree (*Hevea braziliensis*), in which high energy hydrocarbons are formed within the tree by natural biochemical mechanisms. In this case, biomass serves the dual role of a carbon-fixing mechanism and a continuous source of hydrocarbons without being consumed in the process. Other plants, such as the guayule bush, produce hydrocarbons also but must be harvested to recover them. Thus, conceptually, it can be seen from Figure 1 that there are several different pathways by which energy products and synthetic fuels might be manufactured.

Another approach to the development of fixed-carbon supplies from renewable carbon sources is to convert carbon dioxide outside the biomass species into synthetic fuels and organic intermediates. The ambient air, which contains an average of about 320 ppm by volume of carbon dioxide, the dissolved carbon dioxide and carbonates in the oceans, and the earth's large terrestrial carbonate deposits could serve as renewable carbon sources. But because carbon dioxide is the final oxidation state of fixed carbon, it contains no chemical energy. Energy must be supplied in a reduction step. A convenient method of supplying the required energy and simultaneously reducing the oxidation state is reduction of carbon dioxide with elemental hydrogen. The end product can be methane, CH_4, the dominant component of natural gas:

$$CO_2 + 4\,H_2 \rightarrow CH_4 + 2\,H_2O$$

With all components in the ideal gas state, the standard enthalpy of the process is exothermic by −165 kJ (−39.4 kcal) per gram–mole of methane formed. Biomass can serve as the original source of hydrogen which then effectively acts as an energy transfer agent from the biomass to the carbon dioxide.

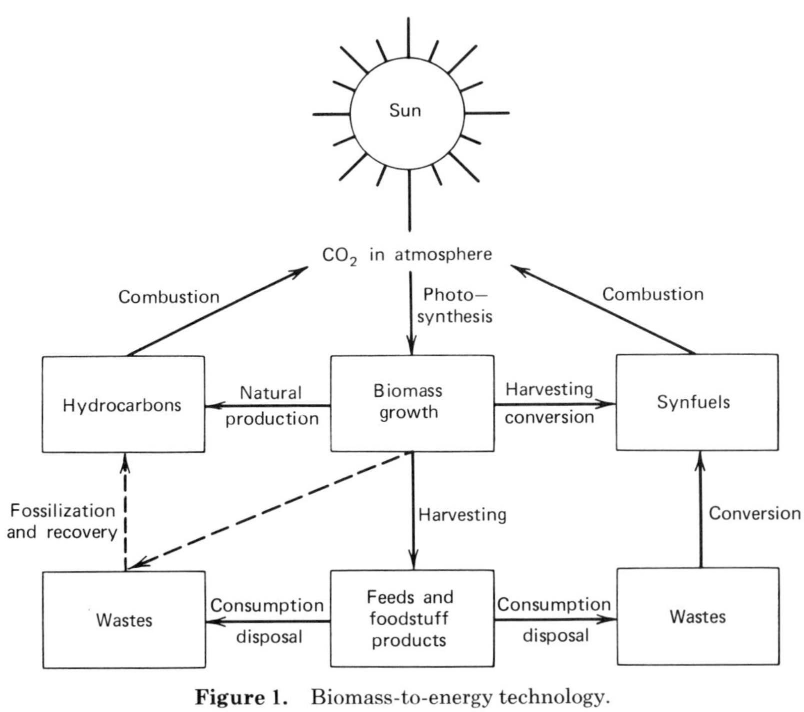

Figure 1. Biomass-to-energy technology.

Distribution of Carbon

The estimation of the amounts of biomass carbon on the earth's surface is the ultimate problem in global statistical analysis. Yet reasonable projections have been made using the best available data, maps, surveys, and a host of assumptions. The validity of the results in their entirety is impossible to support with hard data because of the scope of the analysis. Nevertheless, such analyses must be performed to assess the feasibility of biomass energy systems and the gross types of biomass that might be available for energy applications.

The results of one such study are summarized in Table 1; this is perhaps one of the better attempts to conduct an analysis of biomass carbon distribution (2). Each ecosystem on the earth is considered in terms of area, mean net carbon production per year, and standing biomass carbon. It is immediately apparent that of the total net carbon fixed on earth each year, forest biomass, which is produced on only 9.5% of the earth's surface, contributes more than any other source. Marine sources of net fixed carbon are high too, as might be expected, because of the large area of the earth occupied by water but it is evident that the high turnover rates of carbon in the marine environment result in relatively small steady-state quantities of standing carbon. The

Table 1. Estimates of Net Photosynthetic Production of Dry Biomass Carbon and Standing Biomass Carbon for World Biosphere[a,b]

Ecosystem	Area, 10^6 km^2 [c]	Mean net carbon production		Standing biomass carbon	
		$t/(hm^2 \cdot yr)$[c]	10^9 t/yr[c]	t/hm^2	10^9 t
tropical rain forest	17.0	9.90	16.83	202.5	344
boreal forest	12.0	3.60	4.32	90.0	108
tropical seasonal forest	7.5	7.20	5.40	157.5	118
temperate deciduous forest	7.0	5.40	3.78	135.0	95
temperate evergreen forest	5.0	5.85	2.93	157.5	79
Total	*48.5*		*33.26*		*744*
extreme desert-rock, sand, ice	24.0	0.01	0.02	0.1	0.2
desert and semidesert scrub	18.0	0.41	0.74	3.2	5.8
savanna	15.0	4.05	6.08	18.0	27.0
cultivated land	14.0	2.93	4.10	4.5	6.3
temperate grassland	9.0	2.70	2.43	7.2	6.5
woodland and shrubland	8.5	3.15	2.68	27.0	23.0
tundra and alpine	8.0	0.63	0.50	2.7	2.2
swamp and marsh	2.0	13.50	2.70	67.5	14.0
lake and stream	2.0	1.80	0.36	0.1	0.02
Total	*100.5*		*19.61*		*85*
Total continental	*149.0*		*52.87*		*829*
open ocean	332.0	0.56	18.59	0.1	3.3
continental shelf	36.6	1.62	4.31	0.004	0.1
estuaries excluding marsh	1.4	6.75	0.95	4.5	0.6
algae beds and reefs	0.6	11.25	0.68	9.0	0.5
upwelling zones	0.4	2.25	0.09	0.9	0.04
Total marine	*361.0*		*24.62*		*4.5*
Grand total	*510.0*		*77.49*		*833.5*

[a] Ref. 2.
[b] Dry biomass is assumed to contain 45% carbon.
[c] $1\ km^2 = 1 \times 10^6\ m^2$ (0.3861 sq mi); to convert $t/(hm^2 \cdot yr)$ to short ton/(acre·yr), divide by 2.24.

low turnover rates of forest biomass make it again the largest contributor to standing carbon reserves. The forests produce about 43% of the net carbon fixed each year and contain over 89% of the standing biomass carbon of the earth. Tropical forests are the largest sources of these carbon reserves, followed by temperate deciduous and evergreen forests and the savanna and grasslands. Cultivated land is one of the smaller producers of fixed carbon and is only about 9% of the total terrestrial area of the earth.

The remaining carbon transport mechanisms on earth are primarily physical mechanisms, such as the solution of carbonate sediments in the sea and the release of dissolved carbon dioxide to the atmosphere by the hydrosphere. The earth's total carbon mass amounts to about 20×10^{16} metric tons of which 99.75% is in sediments (lithosphere) and the remainder distributed in the sea, atmosphere, and in living matter (3). Because of the relatively short lifetimes of phytoplankton compared to those of land biomass, there is a much larger amount of carbon in usable land biomass at any given time. The great bulk of carbon, however, is contained in the lithosphere as carbonates in rock. The carbon deposits that contain little or no stored chemical energy—although some high temperature deposits could provide considerable thermal energy—consist of lithospheric sediments and atmospheric and hydrospheric carbon dioxide. Together, these carbon sources comprise 99.9% of the total carbon estimated to exist on the earth. The fossil fuel deposits are ca 0.05% of the total, and the nonfossil energy-containing deposits make up the remainder, or ca 0.02%.

Biomass carbon is thus a very small fraction of the total carbon inventory on earth but it is a very important fraction. It helps to maintain the delicate balance between the atmosphere, hydrosphere, and biosphere necessary to support all life forms. At the same time, it serves as a perpetual source of food and materials. In the past, biomass carbon has served also as a primary energy source for the industrialized nations of the world. It still does today for developing countries and may again become a dominant source of energy products throughout the world because of fossil fuel depletion and the steadily increasing costs of petroleum and natural gas. Overshadowing many of these factors is the potentially damaging effects that large-scale fossil fuel combustion and the resultant increases in atmospheric carbon dioxide may have on the earth's life support systems (4). The utilization of biomass carbon as a primary energy source would not add any new carbon dioxide to the atmosphere because carbon would simply be recycled between the surface of the earth and the air.

Energy Impact

By examining the potential amounts of synfuels that might be produced from renewable carbon sources and comparing these amounts with fossil-fuel demands, the percentage of demand can be estimated that might be satisfied by particular nonfossil energy resources.

Presuming the existence of suitable conversion technology, Table 2 summarizes the quantities of synthetic crude oil and natural gas that might be manufactured at a thermal efficiency of conversion of 60% from the organic wastes estimated to be produced in the United States in 1971 and 1980. The waste types, essentially all derived from biomass, range from agricultural crop and food wastes, which are generated in the largest quantity, to municipal sewage solids, and are all renewable. The information in Table 2 indicates that if all organic wastes generated in 1971 in the United States

Table 2. Estimates of Synfuels by Conversion of Organic Wastes Generated in the United States in 1971 and 1980, 10^6 t/yr[a]

Source	1971	1980
agricultural crop and food wastes[b]	354	354
manure	181	241
urban refuse	117	201
logging and wood manufacturing residues	50	54
miscellaneous organic wastes	45	54
industrial wastes[c]	40	45
municipal sewage solids	11	13
Total	*798*	*962*
net oil potential, L/yr[d]	188×10^9	227×10^9
net gas potential, m^3/yr[e]	182×10^9	219×10^9

[a] Unless otherwise stated.
[b] Assumes 70% dry organic solids in major crop-waste solids.
[c] Assumes 100 million t/yr of industrial wastes in 1971.
[d] Based on an oil yield of 236 L/dry t of organic wastes (1.35 petroleum bbl/short ton) and 60% thermal efficiency. To convert bbl to L, multiply by 159.0.
[e] Based on a gas yield of 228 m^3/dry t of organic waste (3.65 ft^3/lb) and 60% thermal efficiency.

were converted to synthetic crude oil (syncrude), about 188 GL (1.18×10^9 bbl) would have resulted. This corresponds to ca 16% of the oil consumed in the United States in 1971 and is over 82% of the United States oil imports for 1971. Similarly, the substitute natural gas (SNG) potential in Table 2 is 182×10^9 m^3 (6.43×10^{12} ft^3); this corresponds to over 27% of the natural gas consumed in the United States in 1971. The estimates for 1980 indicate a larger syncrude or SNG potential because the larger population in 1980 produces more waste materials. Thus the apparent contribution that organic wastes could make to the United States energy markets is a significant fraction of the demand. However, the gas and oil potentials indicated in Table 2 cannot actually be realized in practice because most of this waste is not collected and is thus not available for conversion to synfuels.

For example, Table 3 compares the gas and oil potentials from the organic wastes generated and those collected in the United States in 1971. Substantially less waste is collected than generated. This reduces the gas or oil that might be produced from wastes to about 15% of the total synfuel potentially available if all wastes were converted. Collection could of course be changed by legislation or other methods to increase the amounts of wastes collected and that would be available for conversion. But even with legal mandates to collect and process as much of the wastes as possible, considerable quantities would still not be capable of collection, such as manures that are not generated in central feedlot locations or dairy barns, and agricultural residues left in the fields. Collection costs would be excessive in many of these cases (see Wastes).

Table 3 shows that urban refuse is not generated in the largest amount but is collected in the largest quantity. Municipal collection systems are responsible for this apparent anomaly. This has prompted many studies and process developments leading to the recovery and recycling of urban refuse components such as ferrous scrap and the utilization of combustible fractions as refuse-derived fuel (RDF) for direct combustion applications. Certain recyclable components from urban wastes, such as

Table 3. Estimates of Synfuel Potential from Organic Wastes Generated and Collected in the United States in 1971, 10^6 t/yr[a]

Source	Generated	Collected
urban refuse	117	64.4
manure	181	23.6
agricultural crop and food wastes	354	20.5
industrial wastes	40	4.7
logging and wood-manufacturing residues	50	4.5
miscellaneous organic wastes	45	4.5
municipal sewage solids	11	1.4
Total	*798*	*123.6*
net oil potential, L/yr[b]	188×10^9	29.2×10^9
net gas potential, m^3/yr[c]	182×10^9	28.2×10^9

[a] Unless otherwise stated.
[b] To convert L to bbl, divide by 159.
[c] To convert m^3 to ft^3, multiply by 35.3.

aluminum, are energy-intensive materials. Hence, recycling is equivalent to a considerable incremental savings in energy that would otherwise be needed to manufacture virgin materials. As for conversion to synfuels, the fact that the main markets for fuels are in and near urban areas should continue to stimulate the development and commercialization of processes for manufacturing synfuels from urban wastes. This reduces the transmission or transportation costs of the particular synfuel and can be an important benefit for on-site manufacturing plants. The proximity of the raw material and specific industries in urban areas may also provide the opportunity to have a captive renewable fuel source. The waste survey data shown in Table 3 suggests that only small fractions of certain wastes and residues are collected and, therefore, modest improvements in the collection methods could supply disproportionately larger amounts of waste for energy applications. Unfortunately, the energy recovered from wastes is minimal. It is estimated that of the total energy consumed in the United States in 1977, about 82×10^9 GJ (77.7×10^{15} Btu) less than 2×10^8 GJ (1.9×10^{14} Btu) were derived from solid wastes excluding forestry residues (5). Thus wastes are sources of energy that remain largely untapped. It should be emphasized, however, that even if all the wastes and residues generated were converted to substitute organic fuels, only a small fraction of the total fossil fuels needs could be satisfied. Less than 10% of the United States fossil fuel demand, which was about 70.5×10^9 GJ (66.9×10^{18} Btu) in 1977, could be met by wastes if a conversion efficiency of 60% to energy products is assumed. Nevertheless, organic wastes can be a valuable supplemental source of synfuels (see also Fuels from waste).

Virgin biomass, in contrast, represents an excellent source of synfuels. As already pointed out, extremely large quantities of carbon are fixed each year in the form of land- and water-based biomass. Using a nominal biomass heating value of 16 GJ per dry metric ton (13.8×10^6 Btu/short ton) and assuming that the total world gas and oil consumption in 1980 is ca 200×10^9 GJ (190×10^{18} Btu), the solar energy trapped in 12.5 Gt of biomass would be equivalent to the world's consumption of these fossil fuels. Since it is estimated that about 77×10^9 t of carbon, or 173×10^9 t of biomass equivalent, most of which is wild and not under human control, is fixed on the earth each year (2), biomass can certainly be considered as a raw material for conversion to large supplies of renewable synfuels. Under controlled conditions, biomass might be grown specifically for energy applications.

A more realistic assessment of biomass as an energy resource can be made by calculating the average surface areas needed to produce sufficient biomass at different annual yields to meet certain percentages of fuel demand for a particular country, and then to compare these required areas with those that might be made available (see Table 4). It is assumed that biomass, whether it be trees, plants, grasses, algae, or waterplants, has a heating value of 15.1 GJ/dry t (12.9×10^6 Btu/short ton), is grown at yields of 25, 50, and 100 dry t/(hm^2·yr) and is converted in integrated biomass planting, harvesting, and conversion systems to SNG at an overall thermal efficiency of 50%. These assumed conditions of biomass production and conversion are either within the range of present technology and agricultural practice, or are believed to be attainable in the future. The average areas were then calculated to meet 1.58, 10, 50, and 100% of the projected United States demand for natural gas in 1980. A percentage of 1.58 is equal to a daily production of 26.9×10^6 m^3 (normal cubic meter: 15°C, dry, 101.3 kPa or 1 atm) (1×10^9 SCF) of SNG. The range of areas required at the low-yield level is between $(2.04–129.1) \times 10^4$ km^2, whereas at the high-yield level, the areas are between $(5.1–32) \times 10^3$ km^2. Figure 2 shows three yield levels along with the percentage area of the United States needed at any selected gas demand supplied by SNG from biomass. Relatively large areas are required but not so much as to make the use of land- or fresh water-based biomass for energy applications impractical. When compared with the area distribution pattern of the United States (Table 5) (6) it is seen that selected areas or combinations of areas might be utilized for biomass energy applications. Areas that are not used for productive purposes might be suitable, or possibly biomass for both energy and foodstuffs or energy and forest products applications can be grown simultaneously or sequentially in ways that would benefit both. In addition, relatively small portions of the bordering oceans might supply the needed biomass growth areas, in which case marine plants would be grown and harvested.

This approach to synfuel production presumes that chemical or biological conversion processes can be used to convert biomass to either a reduced form, such as methane, identical with the natural hydrocarbon, or to some intermediate partially oxygenated form, such as methanol, both of which can be used as synthetic fuels. The direct route, alluded to in Figure 1 as natural production of hydrocarbons, can possibly bypass the harvesting–conversion routes. Some biomass species as already mentioned produce hydrocarbons as metabolic products. Natural rubber and terpene production from selected biomass species, eg, as well as other reduced compounds could be refined to conventional fossil fuels.

Table 4. Potential Substitute Natural Gas in United States from Biomass at Different Crop Yields

1980	Average area required, km^2 [b]		
Demand, % [a]	25 t/(hm^2·yr) [c]	50 t/(hm^2·yr) [c]	100 t/(hm^2·yr) [c]
1.58	20,400	10,200	5,100
10	129,000	64,500	32,300
50	645,500	323,000	161,000
100	1,291,000	645,500	323,000

[a] U.S. demand in 1980 estimated to be 244×10^8 GJ or 653×10^9 m^3 (231×10^{11} ft^3).
[b] 1 km^2 = 1×10^6 m^2 (0.3861 sq mi).
[c] Yields expressed as dry t; to convert t/(hm^2·yr) to short ton/(acre·yr) divide by 2.24.

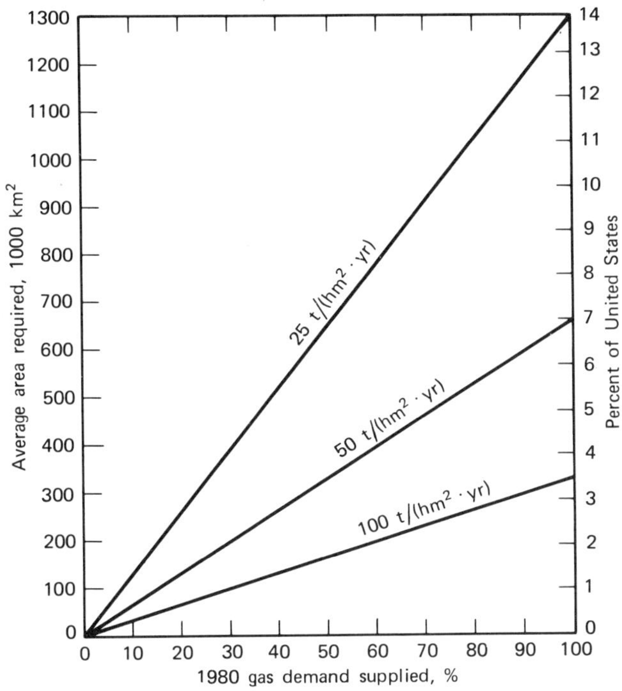

Figure 2. Average area required and percent of United States area vs 1980 gas demand. To convert t/(hm^2·yr) to short ton/(acre·yr), divide by 2.24.

Table 5. Area Distribution of United States, 1974[a]

Source	10^6 km^2 [b]	%
cropland		
for crops	1.46	15.6
for pasture	0.336	3.6
idle	0.085	0.9
grassland pasture	2.42	25.8
forest land	2.91	31.1
special use land	0.595	6.3
water	0.203	2.2
urban area	0.142	1.5
other land	1.22	13.0
Total	*9.37*	*100.0*

[a] Ref. 6.
[b] 1 km^2 = 1 × 10^6 m^2 (0.3861 sq mi).

The third source of renewable carbon consists of the deposits and reservoirs of essentially nonenergy carbon forms, ie, carbon dioxide and the carbonates. The availability of such raw materials cannot be questioned, although low cost separation and energy-efficient recovery of very small concentrations of carbon dioxide present technological challenges. Another basic problem resides in the fact that all the energy must be supplied by a second raw material such as elemental hydrogen. Hydrogen (qv) would have to be made available in large quantities and from a nonfossil source, or the purpose of the synfuel system would be defeated. Conceptually, there is no difficulty in developing such hydrogen sources. Hydrogen can be produced by water electrolysis and thermochemical and photolytic splitting of water. Electrical power and thermal energy can be supplied by nonfossil-powered nuclear reactors, and by means of wind, ocean thermal gradients, wave action, and solar-actuated devices. Hydrogen can also be manufactured from biomass and by direct action of solar energy on certain catalytic surfaces.

As already pointed out, ca 12.5×10^9 t of biomass, or ca 5.6 Gt of carbon equivalent, would have approximately the same energy content as all the gas and oil consumed in the world in 1980. This amount of carbon corresponds to ca 0.7% of the total standing biomass carbon of the earth, or to ca 7% of one year's net fixed-carbon production. Under present conditions of human- and nature-controlled production of fixed-carbon supplies, the utilization of some of this carbon for energy applications seems to be a logical use of a renewable raw material. Forest biomass is especially interesting for these applications because of its abundance. It is evident that the expansion of controlled biomass carbon production should also be considered because this would result in new additions to biomass carbon supplies. For example, the steady-state carbon supplies in marine ecosystems might conceivably be increased under controlled conditions over its current low levels by means of marine biomass plantations in areas of the ocean that are dedicated to this objective.

Thus sufficient supplies of renewable carbon are available to make biomass carbon a basic raw material of a substitute fossil fuel industry. Presuming that suitable conversion processes are available, and that the demand for established organic fuels and intermediates continues, commercialization of a synfuels industry based on renewable carbon to supply a significant portion of energy demand is, at the very least, a technically feasible concept.

Chemical Characteristics

Because of many different types and species, the chemical characteristics of biomass vary over a broad range as illustrated in Table 6. In this table, typical analyses and the energy contents of land- and water-based biomass (wood, grass, kelp) and wastes (manure, urban refuse, primary sewage sludge) are compared with those of cellulose, peat, and bituminous coal.

Cellulose is the chief structural element and main constituent of many biomass forms. In trees, eg, the concentration of cellulose is generally ca 40–50% of the dry weight, whereas lignin and compounds related to cellulose comprise most of the remaining organic components. However, in giant brown kelp cellulose is only a minor component. Mannitol, a hexahydric alcohol formed by reduction of the aldehyde group of D-glucose to a methylol group, and alginic acid, a polymer of mannuronic and glucuronic acids, are the main carbohydrates (see Gums).

Plant biomass contains much less fat and protein than carbohydrates. Fatty constituents are usually present in the smallest amounts, proteins in larger amounts, and carbohydrates the largest. Approximate crude protein values are obtained by multiplying the organic nitrogen analyses by 6.25. This factor is used because the average weight percentage of nitrogen in pure dry protein is ca 16%, although the protein content of each biomass species can best be determined by amino acid assay. The calculated crude protein values of the biomass species in Table 6 range from almost

Table 6. Typical Compositions and Heating Values of Wastes, Biomass, Peat, and Coal

Analysis	Pure cellulose	Pine wood	Kentucky blue grass	Giant brown kelp[a]	Feedlot manure	Urban refuse[b]	Primary sewage sludge	Reed sedge peat	Illinois bituminous coal
elemental, wt %									
C	44.44	51.8	45.8	27.65	35.1	41.2	43.75	52.8	69.0
H	6.22	6.3	5.9	3.73	5.3	5.5	6.24	5.45	5.4
O	49.34	41.3	29.6	28.16	33.2	38.7	19.35	31.24	14.3
N		0.1	4.8	1.22	2.5	0.5	3.16	2.54	1.6
S		0.0	0.4	0.34	0.4	0.2	0.97	0.23	1.0
C (MAF)[c]	44.44	52.1	52.9	45.3	45.9	47.9	59.5	57.2	75.6
proximate, wt %									
moisture		5–50	10–70	85–95	20–70	18.4	90–97	84.0	7.3
volatile matter		99.5	86.5	61.1	76.5	86.1	73.47	92.26	91.3
ash		0.5	13.5	38.9	23.5	13.9	26.53	7.74	8.7
high heating value,									
MJ/kg[d] (dry)	17.51	21.24	18.73	10.01	13.37	12.67[e]	19.86	20.79	28.28
MJ/kg[d] (MAF)[c]	17.51	21.35	21.65	16.38	17.48		27.03	22.53	30.97
MJ/kg[d] carbon	39.40	41.00	40.90	36.20	38.09		45.39	39.38	40.99

[a] *Macrocystis pyrifera.*
[b] Combustible fraction.
[c] Moisture and ash free.
[d] To convert MJ/kg to Btu/lb, multiply by 430.
[e] As received with metals.

nothing for pine wood to a high of ca 30% for Kentucky blue grass. The protein content of grasses is strongly dependent on the growing procedures used before harvest, particularly the fertilization methods. Some biomass species, however, such as legumes, fix nitrogen from the atmosphere and often contain high protein concentrations.

The energy content of biomass is a very important parameter in the conversion to energy products and synfuels. The different components in biomass would be expected to have different heats of combustion because of varying chemical structures and carbon content, as illustrated by the fuel values listed in Table 7 (7). The more reduced the state of carbon, the higher the fuel value. Monosaccharides have the lowest carbon content, highest degree of oxygenation, and lowest heating value. As the carbon content increases and the degree of oxygenation is reduced, the structures become more hydrocarbonlike and the heating value increases. Fatty components thus have the highest heating values. Cellulose, the dominant component in most biomass, has a high heating value (product water in liquid state) of 17.51 MJ/kg (7.53×10^3 Btu/lb).

Typical low heating values (product water in vapor state) of selected biomass are shown in Table 8 (8–9). Woody and fibrous materials appear to have energy contents between 19 and 21 MJ/kg (8.2 and 9×10^3 Btu/lb, respectively); the water-based algae, *Chlorella,* has a higher value, probably because of its higher protein content. Oils derived from plant seeds are much higher in energy content and approach the heating value of paraffinic hydrocarbons. High concentrations of inorganic components of a given biomass species can greatly affect its heating value because inorganic materials generally do not contribute to the heat of combustion. This is illustrated by the heating values for giant brown kelp, which leaves an ash residue equivalent to ca 40 wt % of the dry weight, as shown in Table 6. On a dry basis, the high heating value is about 10 MJ/kg (4.3×10^3 Btu/lb), and on a dry, ash-free basis the heating value is ca 16 MJ/kg (6.9×10^3 Btu/lb).

Conversion

Various methods can be used to produce energy or gaseous, liquid, and solid fuels from biomass. In addition, chemicals can be produced by a wide range of processing

Table 7. Approximate Fuel Values of Selected Biomass Components[a]

Component	Carbon, %[b]	MJ/kg[c,d]
monosaccharides	40	15.6
disaccharides	42	16.7
polysaccharides	44	17.5
lignin	63	25.1
crude protein	53	24.0
fat	75	39.8
carbohydrate	41–44	16.7–17.7
crude fiber[e]	47–50	18.8–19.8

[a] Ref. 7.
[b] Approximate values.
[c] Dry.
[d] To convert MJ/kg to Btu/lb, multiply by 430.
[e] Contains ca 15–30% lignin.

Table 8. Typical Fuel Values of Selected Biomass and Fossil Materials[a]

Material	Low heating value, MJ/kg[b] (dry)
wood	
pine	21.03
beech	20.07
birch	20.03
oak	19.20
oak bark	20.36
bamboo	19.23
fiber	
coconut shells	20.21
buckwheat hulls	19.63
bagasse	19.25
green algae	
Chlorella	26.98
seed oils	
cottonseed	39.77
rapeseed	39.77
linseed	39.50
amorphous carbon	33.8
paraffinic hydrocarbon	43.3
crude oil	48.2

[a] Refs. 8–9.
[b] To convert MJ/kg to Btu/lb, multiply by 430.

techniques. Table 9 summarizes the important feed, process, and product variables considered for the development of a synfuel-from-biomass process. An almost infinite number of feed, process, and product combinations exist but not all are feasible. For example, the separation of the small amounts of metals present in biomass and the

Table 9. Feeds, Processes, and Products

Biomass feed	Primary conversion process	Primary energy products	
land-based	separation	energy	thermal
trees	combustion		steam
plants	pyrolysis		electric
grasses	hydrogenation		
	anaerobic fermentation	solids fuels	char
water-based	aerobic fermentation		combustibles
single-cell algae	biophotolysis		
multicell algae	partial oxidation	gaseous fuels	methane (SNG)
water plants	steam reforming		hydrogen
	chemical hydrolysis		low-thermal-value gas
	enzyme hydrolysis		medium-thermal-value gas
	other chemical conversions		light hydrocarbons
	natural processes		
		liquid fuels	methanol
			ethanol
			higher hydrocarbons
			oils
		chemicals	

direct combustion of high moisture content algae are technically possible, but energetically unfavorable.

The moisture content of the biomass is especially important in the selection of a suitable conversion process. The giant brown kelp, *Macrocystis pyrifera,* which is under development as a source of SNG, contains as high as 95% intracellular water; consequently, thermal gasification techniques such as pyrolysis and hydrogasification cannot be used directly without first drying the algae. Anaerobic digestion methods are preferred because the water does not have to be removed. Wood, on the other hand, can often be processed by several different thermal conversion techniques without drying. Figure 3 illustrates the effects of thermal drying on biomass used for synfuel production as SNG presuming the feed has a heating value of 11.6 MJ/dry kg (5000 Btu/lb). Reduction of an initial moisture content of 60 wt % by thermal drying to 30% moisture content requires the equivalent of ca 20% of the feed energy content and leaves ca 80% of the feed energy available for SNG production. Clearly, a main portion of the feed's equivalent energy content can be expended for drying, and the properties of the feed such as water and energy content, composition, size distribution, noncombustibles, and organisms must be carefully considered.

Within a given process type, a particular process may also have specific requirements. For example, anaerobic digestion and alcoholic fermentation are both biological

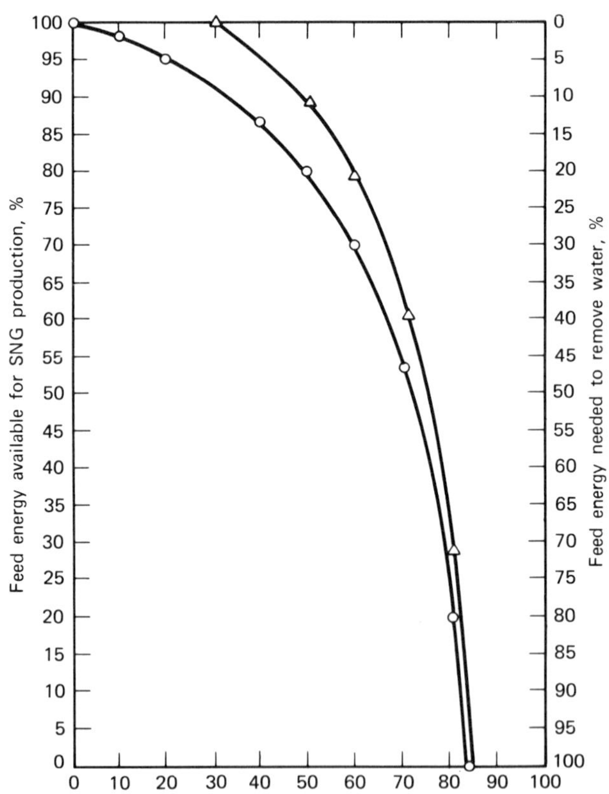

Figure 3. Effect of feed moisture content on energy available for synfuel production. Feed: 11.63 MJ/kg (5000 Btu/lb) dry. ○, 0% moisture in dried feed; △, 30% moisture in dried feed.

conversion processes, but animal manure, which has a relatively high biodegradability, is not equally applicable as a feedstock for both processes. The degree of complexity of the process design is another factor affecting its practicality. Some processes, such as combustion, are intrinsically simple in design, whereas others, such as alcoholic fermentation, consist of many different unit operations and are complex. Capital and operating costs are dictated by the particular process design, the logistics of raw material supply, and plant size. Generally, the more complex the process, the higher the costs.

Biomass conversion processes can conveniently be divided into four groups: physical, biological–biochemical, thermal, and chemical.

Physical Processes. ***Particle Size Reduction.*** Particle size reduction prepares the biomass feed for direct fuel use, fabrication into pellets, conversion processes, storage, and shipment.

The final particle size required depends upon the conversion process. For thermal gasification processes, eg, the size of the material influences the rate at which the gasification process occurs. In biological processes such as anaerobic digestion to produce methane, the smaller the particle, the higher the reaction rate because more surface area is exposed to the organisms. Particle size reduction consists of one or more of the unit operations that make up the front end of the total processing system. Currently, two basic types are in commercial use, wet shredders and dry shredders. Wet shredders employ a hydropulping mechanism in which a high-speed cutting blade pulverizes a water slurry of the feed over a perforated plate. The pulped material passes through the plate and the nonpulping materials are ejected. Dry shredders are either vertical- or horizontal-shaft hammermills. Rotating metal hammers on a shaft reduce the particle size of the feed material until the particles are small enough to drop through the grate openings. Maintenance costs for dry shredders are generally higher than those for wet shredders. Particle size reduction units such as agricultural choppers and tree chippers are usually hammermills or are equipped with knife blades that reduce particle size by cutting or shearing action (see Size reduction).

Separation. It is sometimes desirable to separate biomass into two or more components for different applications. Examples are the separation of agricultural biomass into foodstuffs and residues that may serve as fuel or as a raw material for synfuel manufacture, the separation of forest biomass into the darker bark-containing fractions and the pulpable components, and the separation of marine biomass to isolate various chemicals. Common operations such as extraction, distillation, filtration, and crystallization are often used as well as industry-specific methods characteristic of farming, forest products, and specialized biomass industries.

Drying. Drying refers to the vaporization of all or part of the water in the biomass material. When the biomass is thermally processed directly for energy recovery, it may be necessary to partially dry the raw biomass before conversion, otherwise more energy might be consumed to operate the process than that produced in the form of recovered energy or fuels. Open-air solar drying is probably the cheapest drying method. Raw materials that are not sufficiently stable to be dried by solar methods can be dried more rapidly using spray dryers, drum dryers, and convection ovens. For large-scale applications, forced-air type furnaces and dryers designed to use stack gases are probably more efficient (see Drying).

Fabrication. Solid fuel pellets from a variety of biomass, particularly wood and wood residues, are manufactured by extrusion and other techniques and in some cases, a binding agent such as a thermoplastic resin is incorporated in the biomass. The fabricated products are stated to be more uniform in combustion characteristics than the raw biomass. Depending on the composition of the additive, such as the binding agents in the pelletized fuel, the heat of combustion can be higher or lower than that of the unpelletized biomass (see Pelleting and briquetting).

Biological–Biochemical Processes. Fermentation (qv) is a biological process in which a water slurry or solution of the raw material interacts with microorganisms and is enzymatically converted to other products. Biomass can be subjected to fermentation conditions to form a variety of products; two of the most common fermentations yield methane and ethanol. Biochemical processes include those that occur naturally within the biomass.

Anaerobic Digestion. Methane can be produced from water slurries of biomass by anaerobic digestion in the presence of mixed populations of anaerobes. This process has been used for many years to stabilize municipal sewage sludges for purposes of disposal. Presuming the biomass is all cellulose, the chemistry can be represented in simplified form:

$$(C_6H_{10}O_5)_x + x\,H_2O \xrightarrow{\text{hydrolysis}} x\,C_6H_{12}O_6 \xrightarrow{\text{acidification}} 3x\,CH_3COOH \xrightarrow{\text{methanation}} 3x\,CH_4 + 3x\,CO_2$$

The complex organic compounds are first converted to lower molecular weight, soluble products, primarily carboxylic acids, in the water slurry by the acidogenic bacteria present in the digestor. Methanogenic bacteria then convert these intermediates to a gas with heating values ranging from ca 19.6 to 29.4 MJ/m^3 (500–750 Btu/SCF). The main gas components are methane and carbon dioxide. Residual ungasified solids containing more nitrogen, phosphorus, and potassium than the feed solids are also formed. In some systems, these solids have application as animal feeds and fertilizers. The conventional high rate digestion process is conducted under nonsterile conditions in large, mixed, anaerobic fermentation vessels at near atmospheric pressure at ca 35°C (mesophilic range) or 55°C (thermophilic range) and at reactor residence times of 10–20 d. The pH is maintained in the 6.8–7.2 range.

Raw digestor gas has been used for many years as a fuel. It can be upgraded to pure methane or SNG by scrubbing the carbon dioxide or by other known processes.

The overall thermal efficiency of the anaerobic digestion of biomass to methane is a function of the process design and the raw material characteristics. Without pretreatment of the feed or the recycled ungasified solids to increase biodegradability, the feed can generally be gasified at overall thermal efficiencies of ca 30–60%. Yields of methane would be expected to range up to 0.30 m^3/kg (5 SCF/lb) of volatile solids added to the digestors. A typical biological gasification plant can contain hydrolysis units, anaerobic digestors, gas clean-up and dehydration units, and liquid effluent treatment units (see Fig. 4). Advanced designs under development include two-phase, plug-flow, packed-bed, and fluidized-bed digestors.

Alcoholic Fermentation. Certain types of starchy biomass such as corn are readily converted to ethanol under anaerobic fermentation conditions in the presence of specific yeasts (*Saccharomyces cerevisiae*) and other organisms (Fig. 5). However, alcoholic fermentation of biomass containing high cellulose concentrations can only

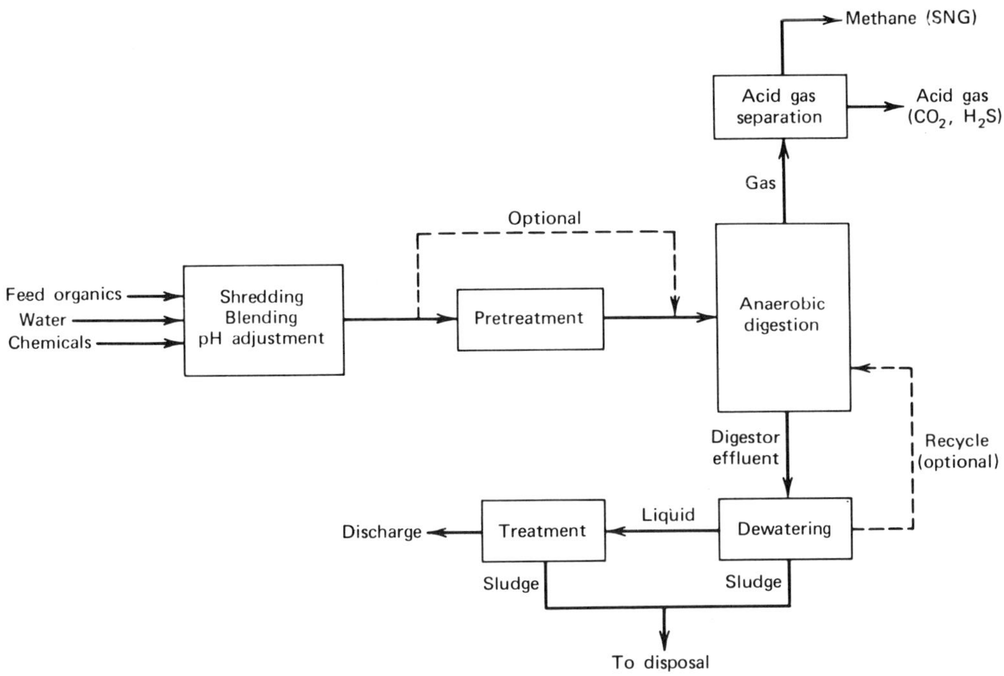

Figure 4. Methane production by anaerobic digestion.

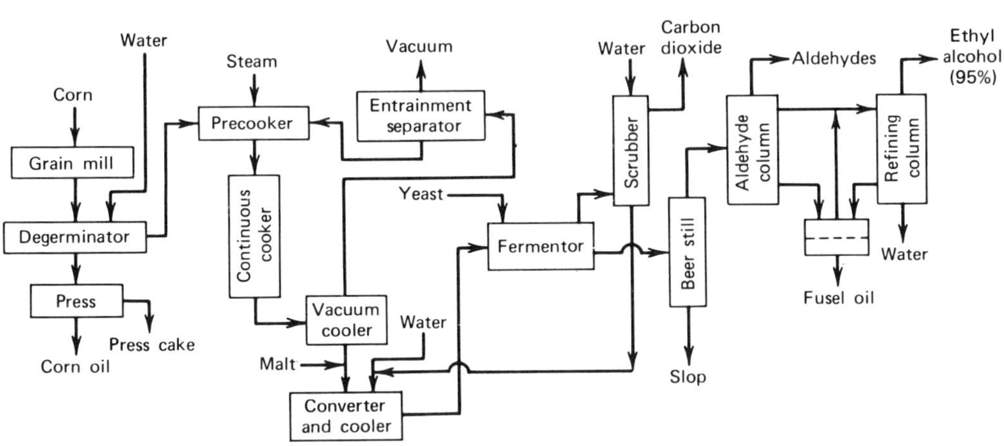

Figure 5. Typical flow scheme for manufacture of ethyl alcohol from corn.

be performed in high yield after the polysaccharide components are converted to monosaccharides by acid- or enzyme-catalyzed hydrolysis:

$$\left(C_6H_{10}O_5\right)_x + x\ H_2O \xrightarrow{\text{hydrolysis}} x\ C_6H_{12}O_6 \xrightarrow{\text{fermentation}} 2x\ C_2H_5OH + 2x\ CO_2$$

A commercial alcohol fermentation plant for biomass would include units to shred and separate the combustible fermentable organic fraction from the nonfermentable components, hydrolysis units to produce glucose concentrates if the feed were high in cellulose components of low degradability, fermentors, distillation towers, and dehydration units (Fig. 6). The degradable organic material is converted at high efficiencies and residence times of 1–2 d to a "beer" that contains about 10% ethanol. This broth is heated to remove the product alcohol as overhead by distillation, and the resulting 50–55% alcohol distillate is distilled again to yield 95% alcohol and by-product aldehydes and fusel oil. Bottoms from the beer still contain low volatility components from the fermentation called stillage which is normally processed further to yield high protein animal feeds.

The thermal efficiency of ethanol production in a fermentor is relatively high but the overall thermal efficiency of the process is low because of the many energy-consuming steps and by-products formed. Alcohol yields are about 40–50 wt % of the weight of the fermentable fraction in the feed. Substantial improvements in the overall thermal efficiency of alcoholic fermentation should be possible by improving the thermal efficiencies of the auxiliary unit operations. Alcoholic fermentation under reduced pressure and systems in which polysaccharide hydrolysis and fermentation occur together are being developed to provide improved process performance.

Biophotolysis. The decomposition of water to hydrogen and oxygen using the radiant energy of visible light and the photosynthetic apparatus of green plants, certain bacteria, or blue–green algae is called biophotolysis. The concept has been studied in the laboratory for several years but a practical process has not yet been developed. In biophotolysis the water is first oxidized to liberate molecular oxygen and electrons which are raised from the level of the water–oxygen couple, +0.8 to 0.0 V by Photosystem II:

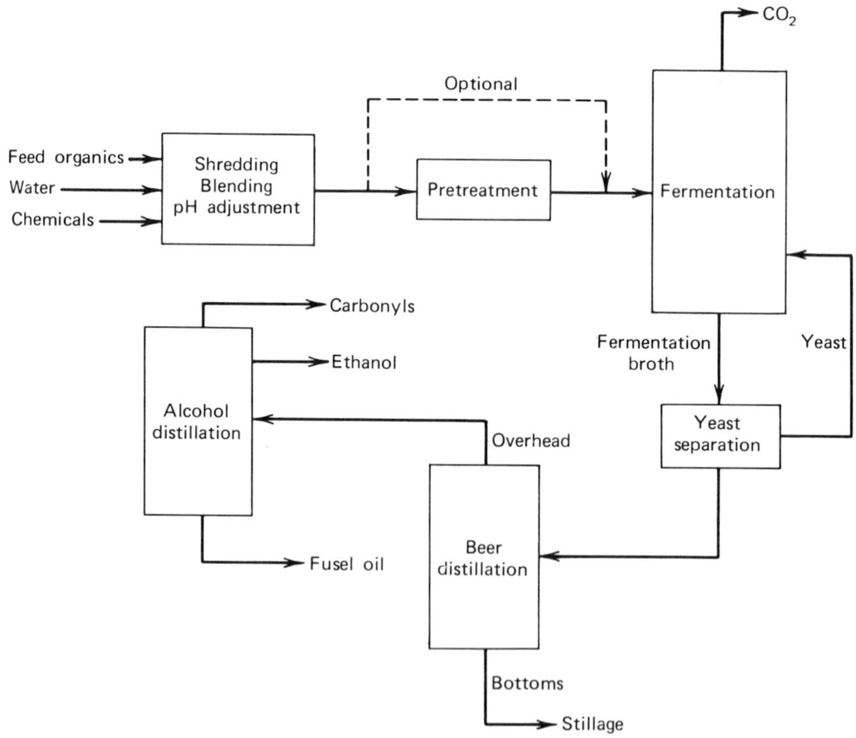

Figure 6. Ethanol production by alcoholic fermentation.

$$H_2O \xrightarrow{\text{light}} \tfrac{1}{2}\,O_2 + 2\,H^+ + 2\,e$$

The electrons undergo the equivalent of a partial oxidation process in a dark reaction to a positive potential of +0.4 V, and then Photosystem I raises the potential of the electrons to as high as −0.7 V. Under normal photosynthesis conditions, these electrons reduce triphosphopyridine nucleotide (TPN) to TPNH which reduces carbon dioxide to organic plant material. In the biophotolysis of water, these electrons are diverted from carbon dioxide to a microbial hydrogenase for reduction of protons to hydrogen:

$$2\,H^+ + 2\,e \rightarrow H_2$$

Thus the overall chemistry is simply the photolysis of water to hydrogen and oxygen.

Various single-stage schemes, in which hydrogen and oxygen are produced separately, have been under study, and the thermodynamic feasibility of the chemistry has been experimentally demonstrated. The upper limit of efficiency of the biophotolysis of water has been projected to be 3% for well-controlled systems. This limits the capital cost of useful systems to low-cost materials and designs. However, the concept of water biophotolysis to afford a continuous, renewable source of hydrogen is quite attractive and may ultimately lead to a practical hydrogen-generating system (see Hydrogen energy).

Natural Processes. Hydrocarbon production in land-based biomass by natural biochemical mechanisms is a well-known phenomenon. Commercial production of natural rubber, the highly stereospecific polymer of *cis*-1,4-polyisoprene, is an established technology. Natural rubber has a molecular weight range of ca 500,000–2,000,000 and is tapped as a latex from the hevea rubber tree (*Hevea braziliensis*). The desert shrub guayule, which grows in the southwestern United States and in northern Mexico, is another biomass species being developed as a source of natural rubber almost identical with hevea rubber (10). The idea of growing guayule and extracting the rubber latex from the whole plant was tested in full-scale plantations during the rubber shortage in World War II and found technically feasible. Terpene extraction from pine trees and other biomass species is also established technology.

Recently, many plants native to or grown in North America have been tested as sources of hydrocarbons (11–12) to identify biomass species that produce isoprenoid hydrocarbons, especially those of lower molecular weight than natural rubber that would be more amenable to standard petroleum refining methods (13). Desert plants have been investigated that might be grown in arid or semiarid areas without competition from biomass grown for foodstuffs, whereas other work has concentrated on perennial species adapted to wide areas of North America. Table 10 lists biomass species that contain isoprenoid hydrocarbons. Some species in the Euphorbiaceae family, which includes *Hevea braziliensis,* form hydrocarbons having molecular weights considerably less than that of natural rubber at yields as high as 10 wt % of the plant. This corresponds to hydrocarbon yields of about 4 $m^3/(hm^2{\cdot}yr)$ (25 bbl/$(m^2{\cdot}yr)$). "Gasoline plantations" are thus not totally in the realm of science fiction.

Figure 7 illustrates a scheme that might be used for separating the various hydrocarbon components (11). Acetone extraction removes the polyphenols, glycerides, and sterols, whereas benzene extraction removes the hydrocarbons. If the biomass species in question contained low concentrations of the nonhydrocarbon components,

Table 10. Hydrocarbon-Producing Biomass Species Suitable for North America [a]

Family	Genus–species	Common name
Aceraceae	*Acer saccharinum*	silver maple
Anacardiaceae	*Rhus glabra*	smooth sumac
Asclepiadiaceae	*Asclepias incarnata*	swamp milkweed
	sublata	desert milkweed
	syriaca	common milkweed
	Cryptostegia grandiflora	Madagascar rubber vine
Caprifoliaceae	*Lonicera tartarica*	red tarterium honeysuckle
	Sambucus canadensis	common elder
	Symphoricarpos orbiculatus	coralberry
Companulaceae	*Companula americana*	tall bellflower
Compositae	*Ambrosia trifida*	giant ragweed
	Cacalia atriplicifolia	pake Indian plantain
	Chrysathamnus nauseosus	rabbitbrush
	Cirsium discolor	field thistle
	Eupatorium altissimum	tall boneset
	Parthenium argentatum	guayule
	Silphium integrifolium	rosin weed
	laciniatum	compass plant
	terbinthinaceum	prairie dock
	Solidago graminifolia	grass-leaved goldenrod
	leavenworthii	Edison's goldenrod
	rigida	stiff goldenrod
	Sonchus arvensis	sow thistle
	Vernonia fasciculata	ironweed
Euphorbiaceae	*Euphorbia denta*	
	lathyrus	mole plant, gopher plant
	pulcherima	poinsetta
	tirucalli	African milk bush
Gramineae	*Agropyron repens*	quack grass
	Elymus canadensis	wild rye
	Phalaris canariensis	canary grass
Labiatae	*Pycnanthemum incanum*	western mountain mint
	Teucrium canadensis	American germander
Lauraceae	*Sassafras albidium*	sassafras
Rhamnaceae	*Ceanothus americanus*	New Jersey tea
Rosaceae	*Prunus americanus*	wild plum
Phytolaccaceae	*Phytolacea americana*	pokeweed

[a] Ref. 11.

exclusive of the carbohydrate and protein fractions, direct extraction of the hydrocarbons with benzene or a similar solvent might be preferred.

The main steps in the mechanism of polyisoprene formation in plants are known. Mevalonic acid, a key intermediate derived from plant carbohydrate via acetyl coenzyme A, is converted into isopentenyl pyrophosphate (IPP) via phosphorylation, dehydration, and decarboxylation. IPP then rearranges to dimethylallyl pyrophosphate (DMAPP). DMAPP reacts with IPP, releasing pyrophosphate to form another allyl pyrophosphate containing ten carbon atoms. The chain can successively build up by five-carbon units to yield polyisoprenes by head-to-tail condensations, or alternatively, tail-to-tail condensations of two C_{15} units can yield squalene, a precursor of sterols. Similar condensation of two C_{20} units yields phytoene, a precursor of carotenoids. This

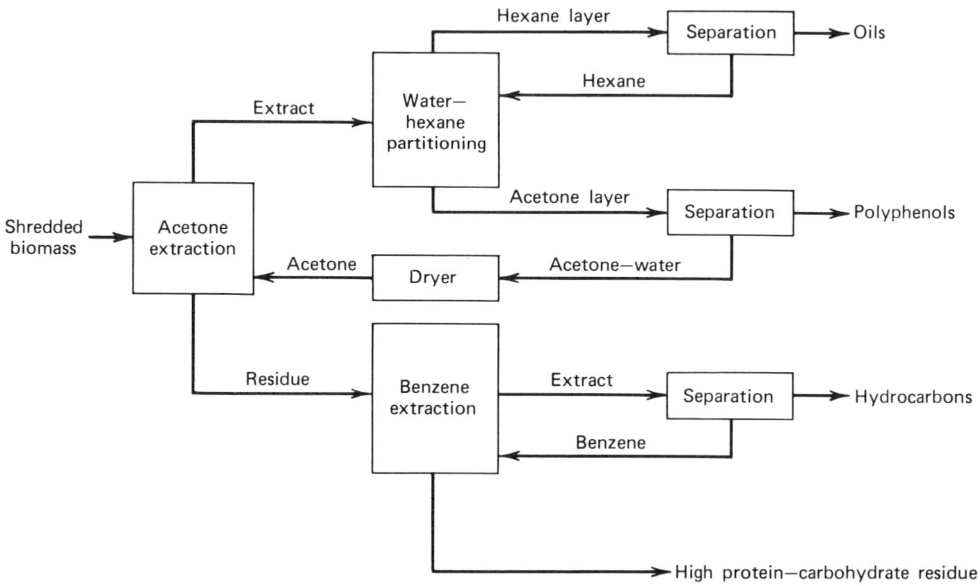

Figure 7. Operations for processing hydrocarbon-containing biomass.

information is expected to be of use to control of hydrocarbon structures and yields obtained from biomass.

Thermal Processes. Thermal processes for the production of energy and fuels from biomass usually involve irreversible chemical reactions, heat, and the transfer of chemical energy from reactants to products. The two largest classes of thermal processes are combustion and pyrolysis. A third class can be described either as a combination of combustion and pyrolysis reactions or a thermochemical process in which conversion of the feed is facilitated by a reactant such as water or hydrogen. For convenience, these processes are grouped together as miscellaneous processes. Gaseous, liquid, and solid fuels can be produced by pyrolysis processes and several processes in the miscellaneous category.

Combustion. Complete combustion (incineration, direct firing, burning) is the rapid chemical reaction of feed and oxygen to form carbon dioxide, water, and heat. The heat released is a function of the enthalpy of combustion of the biomass. Agricultural products, such as bagasse (qv) generated in sugar cane plantations, forestry residues, and wood chips have been used as fuels in combustion systems for many years. The recovered heat is used for the production of steam, electric power in a steam-electric plant, and drying.

Many types of combustion equipment are commercially available. They differ mainly in the design of the combustion chambers, operating temperature, and heat transfer mechanism. Refractory-lined furnaces operating at about 1000°C were standard until the introduction a few years ago of waterwall incinerators (qv). Ash build-up occurs rapidly in refractory-lined furnaces, and excess air must be introduced to control the wall temperature. The waterwall incinerator has combustion chamber walls containing banks of tubes through which water is circulated thereby reducing the amount of cooling air needed. Heat is transferred directly to the tubes to produce steam.

Another type of combustion unit operates at about 1600°C to produce a molten slag which on quenching forms a granular frit rather than the usual ash. The higher operating temperature is obtained by preheating the combustion air or by burning auxiliary fuel. In fluidized-bed systems, air is dispersed through an orifice plate at the bottom of the combustion unit. The dispersed air passes through a bed of sand or residual inorganic particles recovered from combustion which causes the effective volume of the bed to increase and the bed to become fluidized. The feed is introduced to this rapidly mixed bed where flameless combustion occurs at about 650°C. This temperature is substantially below flame temperature and, because of the lower heat input requirements, many high moisture biomass feeds can be combusted without supplemental fuel. Many other variations, such as stationary and rotating-shaft furnaces, suspension firing systems, and stationary and moving grates, are available (see Furnaces, fuel-fired).

The specific design most appropriate for biomass combustion and energy recovery depends on the kinds, amounts, and characteristics of the biomass; the final energy form desired (heat, steam, electric); the relationship of the system to other units in the plant (independent, integrated); whether recycling or co-combustion is practiced; the disposal method for residues; and environmental factors.

Pyrolysis. Pyrolysis (retorting, destructive distillation, carbonization) is the thermal decomposition of an organic material in the absence of oxygen. For biomass, pyrolysis generally starts near 300–375°C. Chars, organic liquids, gases, and water are formed in varying amounts depending particularly on the biomass composition, heating rate, pyrolysis temperature, and residence times in the pyrolysis reactor. As might be expected, higher temperatures and longer residence times promote gas production, whereas higher liquid and char yields result from lower temperatures and shorter residence times. No matter what the pyrolysis conditions are, with the exception of extremely high temperatures, the product mixture has a complex composition, and selectivity for specific products is low even with a single-feed component.

Depending on the pyrolysis temperature, inorganic materials ashed to varying degrees, unconverted organic solids, and fixed carbon residues produced by thermal decomposition of the organics remain in the char fraction. The liquid fraction contains a complex mixture of organic chemicals having much lower average molecular weights than the feed. The liquid fraction obtained from highly cellulosic feeds usually contains acids, alcohols, aldehydes, ketones, esters, heterocyclic derivatives, and phenolic compounds. The pyrolysis gas is a low-thermal-value gas with a heating value of ca 3.9–15.7 MJ/m^3 (100–400 Btu/SCF). It contains carbon dioxide, carbon monoxide, methane, hydrogen, ethane, ethylene, minor amounts of higher gaseous organics, and water vapor. Separation is the main problem, and further refining may be necessary.

Pyrolysis processes may be endothermic or exothermic depending on the temperature of the reacting system. For most biomass feeds containing mostly highly oxygenated cellulosic fractions, pyrolysis is endothermic at low temperatures and exothermic at high temperatures. Energy to drive the process is often obtained from a portion of the feed or the pyrolysis products such as the char. At low temperatures, pyrolysis is generally reaction-rate controlled, and at high temperatures, the process becomes mass-transfer controlled.

The data in Table 11 show how temperature affects yields, and gas and char compositions obtained by pyrolysis of cellulose (14). Gas yield increases with tem-

Table 11. Effect of Temperature on Pyrolysis of Cellulose[a]

	Temperature, °C			
	500	650	800	900
components, wt %				
carbon dioxide	44.8	31.8	20.6	18.3
carbon monoxide	33.5	30.5	34.1	35.3
methane	12.4	15.9	13.7	10.5
hydrogen	5.56	16.6	28.6	32.5
ethane	3.03	3.06	0.77	1.07
ethylene	0.45	2.18	2.24	2.43
char composition, wt %				
volatile matter	21.8	15.1	8.13	8.30
fixed carbon	70.5	70.7	79.1	77.2
ash	7.71	14.3	12.8	14.5
product yield, wt %				
gases	12.3	18.6	23.7	24.4
liquids	61.1	59.2	59.7	58.7
char	24.7	21.8	17.2	17.7
gas yield				
m^3/kg[b] combustibles	0.114	0.166	0.216	0.202
MJ/kg[c] combustibles	1.39	2.63	3.33	3.05
high heating value				
MJ/m^3 [d]	12.3	15.8	15.4	15.1
MJ/kg[c]	28.1	28.6	26.7	26.5

[a] Ref. 14.
[b] To convert m^3/kg to ft^3/lb, multiply by 16.0.
[c] To convert MJ/kg to Btu/lb, multiply by 430.
[d] To convert MJ/m^3 to Btu/ft^3, multiply by 26.9.

perature. Although the heating value of the product gas remains about the same, significant increases in hydrogen concentration and energy yield occur, and carbon dioxide concentration decreases. As the temperature increases from 500 to 900°C, the char yields decrease along with the volatile matter concentration but the energy value of the char does not.

Pyrolysis reactor designs include fixed beds, moving beds, suspended beds, fluidized beds, entrained-feed solids reactors, stationary vertical-shaft reactors, inclined rotating kilns, horizontal shaft kilns, high temperature (1000–3000°C) electrically heated reactors with gas-blanketed walls, single- and multihearth reactors, and a host of other designs.

A fluidized two-bed system in the development stage uses two fluidized-bed reactors containing sand as a heat transfer medium (15–16). Combustion of char from the pyrolysis reactor takes place within the combustion reactor. The heat released supplies the energy for pyrolysis of the combustible fraction in the pyrolysis reactor. Heat transfer is accomplished by sand flow from the combustion reactor at 950°C to the pyrolysis reactor at 800°C and return of the sand to the combustion reactor (Fig. 8). This configuration separates the combustion and pyrolysis reactions and keeps the nitrogen in air separated from the pyrolysis gas. It yields a pyrolysis gas that can be readily upgraded to SNG by shifting, scrubbing, and methanating without regard to nitrogen separation. The initial pyrolysis gas from high-cellulose feeds contains about 37 mol % hydrogen, 35 mol % carbon monoxide, 16 mol % carbon dioxide, and 11 mol % methane. This is a low-thermal-value gas having a high heating value of about

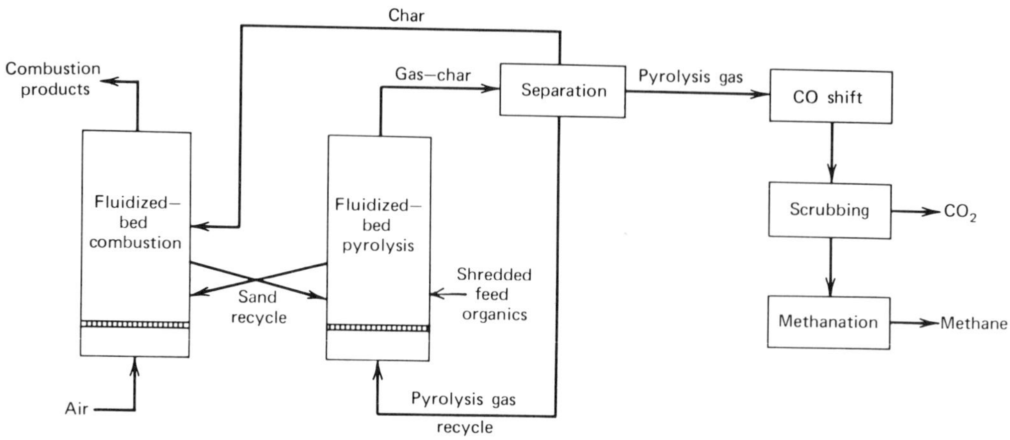

Figure 8. Methane production by pyrolysis using fluidized sand and char recycle in two-bed system.

13.5 MJ/m³ (344 Btu/SCF). The projected gas yields are about 667 m³ (17,000 SCF) of pyrolysis gas and about 196 m³ (5000 SCF) of methane per dry ton of feed (15).

An example of a liquid-fuel production system under development in a short-residence time pyrolysis process is shown in Figure 9 (17). In this process, the high-cellulose feed is dried in a rotary kiln to about 4 wt % moisture content and finely divided to a 200-μm (14-mesh) particle size. The feed is rapidly passed through the pyrolysis reactor. The raw product mixture leaves the reactor at ca 510°C. Separation of the gas from the char and rapid quenching to ca 80°C yields the liquid fuel. The

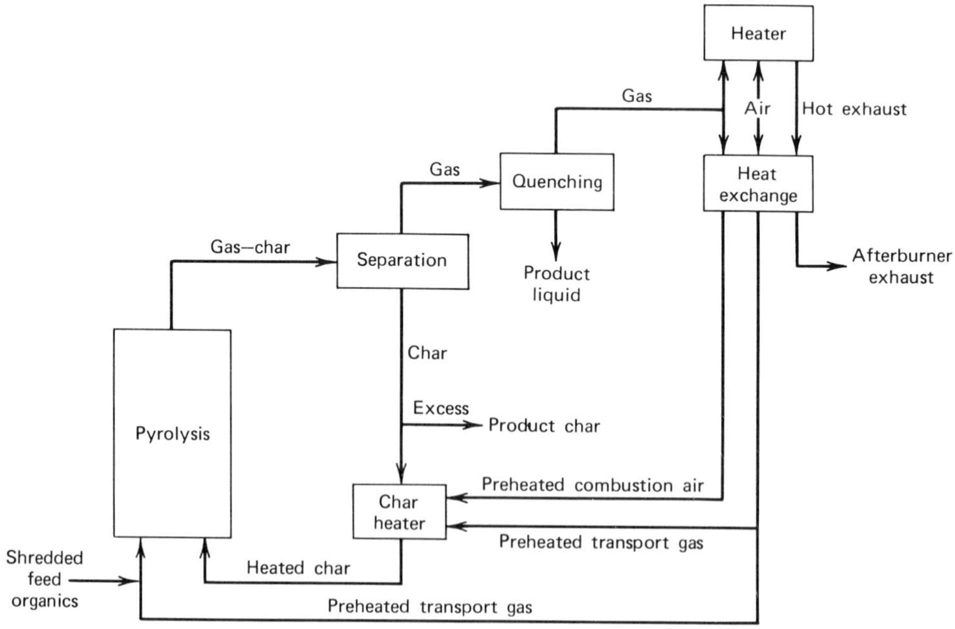

Figure 9. Liquid-fuel production by flash pyrolysis using char recycle.

remaining gas goes through a series of clean-up steps for in-plant use. The raw product yields are ca 10 wt % water, 20 wt % char, 30 wt % gas, and 40 wt % liquid fuel. The product char has a heating value of ca 20.9 MJ/kg (9000 Btu/lb), contains ca 30 wt % ash, and is produced at an overall yield of ca 7.5 wt % of the dry feed. The corresponding values of the liquid fuel are about 24.4 MJ/kg (10,500 Btu/lb), 0.2–0.4% ash, and 22.5 wt % of dry feed as received or ca 0.16 m^3/t (1 bbl/t). This product has been proposed for use as a heating oil; its properties are compared with a typical No. 6 fuel oil in Table 12. Some important differences exist but successful combustion trials in a utility boiler with the liquid fuel have been carried out.

Miscellaneous Thermal Processes. Many thermal conversion processes can be classified as partial oxidation processes in which the biomass is supplied with less than the stoichiometric amount of oxygen needed for complete combustion. Under these conditions, low-thermal-value gases, similar to pyrolysis gases are formed that can contain high concentrations of hydrogen and carbon monoxide. Such gaseous mixtures, termed synthesis gases, can be converted to a large number of chemicals and synthetic fuels by established processes (18). In some partial oxidation processes, the various chemical reactions may occur simultaneously in the same reactor zone. In others, the reactor may be divided into a combustion zone, which supplies the heat to promote pyrolysis in a second zone, and perhaps a third zone for drying, the overall result of which is partial oxidation. Both air and pure oxygen have been utilized for such systems.

A three-zoned vertical shaft reactor furnace is shown in Figure 10 (19). Coarsely shredded feed is fed to the top of the furnace, and as it descends through the first zone, is dried by the ascending hot gases, which are partially cleaned by the feed. The temperature is reduced from ca 315°C to 40–200°C. The dried feed then enters the pyrolysis zone in which the temperature ranges from 315 to 1100°C. The resulting char and ash descend to the hearth zone where the char is partially oxidized with pure

Table 12. Typical Properties and Analysis of Liquid Fuel and No. 6 Fuel Oil[a]

Properties	Liquid fuel	No. 6 fuel oil
heating value, MJ/kg[b]	24.6	42.3
density, g/cm^3	1.3	0.98
pour point, °C	32[c]	15–30
flash point, °C	56[c]	65
viscosity at 87.8°C (SUs)	1150[c]	90–250
pumping temperature, °C	71	46
atomization temperature, °C	116	104
analysis[d], wt %		
carbon	57.5	85.7
hydrogen	7.6	10.5
sulfur	0.1–0.3	0.5–3.5
chlorine	0.3	
nitrogen	0.9	2.0
oxygen	33.4	
ash	0.2–0.4	0.5

[a] Ref. 17.
[b] To convert MJ/kg to Btu/lb, multiply by 430.
[c] Containing 14% water as produced.
[d] Dry basis.

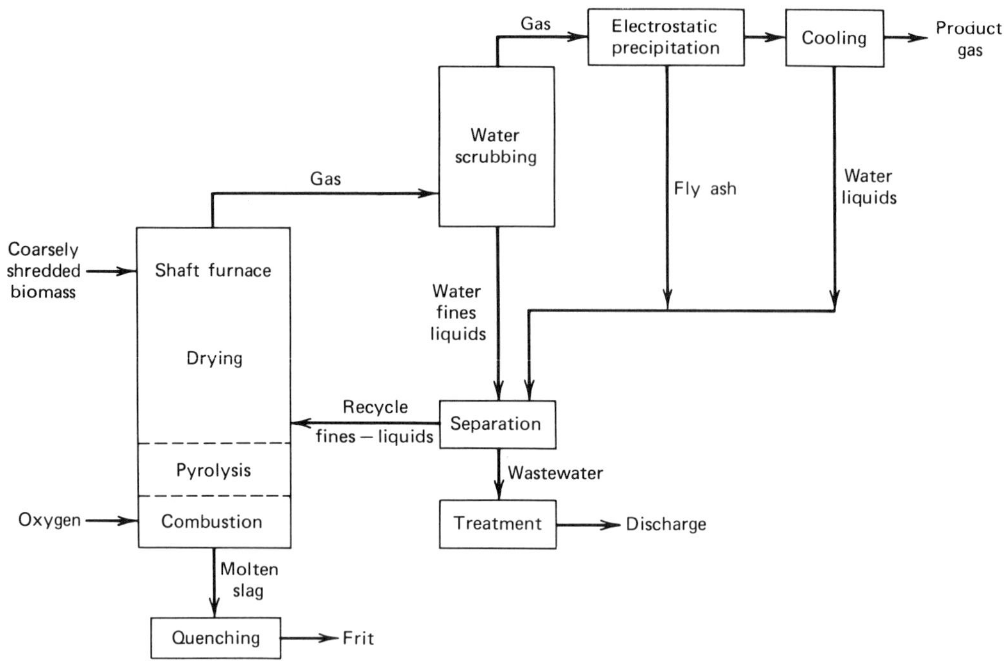

Figure 10. Production of synthesis gas in three-zone shaft furnace.

oxygen. Slagging temperatures near 1650°C occur in this zone and the resulting molten slag of metal oxides forms a liquid pool at the bottom of the hearth. Continuous withdrawal of the pool and quenching forms a sterile granular frit. The product gas is processed to remove fly ash and liquids, which are recycled to the reactor. A typical gas analysis is 40 mol % carbon monoxide, 20 mol % hydrogen, 23 mol % carbon dioxide, 5 mol % methane, and 5 mol % carbon; it has a high heating value of ca 14.5 MJ/m^3 (370 Btu/SCF).

An example of the partial oxidation of biomass in which air and the biomass are supplied without zone separation in the gasifier is the molten salt process (20). The shredded biomass and air are continuously introduced beneath the surface of a sodium carbonate-containing melt which is maintained at ca 1000°C. The resulting gas passes through the melt. Acid gases are absorbed by the alkaline salts and the ash is also retained in the melt. The melt is continuously withdrawn for processing to remove the ash and returned to the gasifier; tars or liquid products are not formed. The heating value of the product gases depends on the amount of air supplied, and is essentially independent of the type of organic feed. The greater the deficiency of air needed to achieve complete combustion, the higher the fuel value of the product gas. Thus, with about 20, 50, and 75% of the theoretical air, the respective high heating values of the gas are about 9.0, 4.3, and 2.2 MJ/m^3 (230, 110, and 55 Btu/SCF).

Steam is also blended with air in some gasification units to promote the overall process via the endothermic steam–carbon reaction to form carbon monoxide and hydrogen. This was common practice at the turn of the last century when so-called producer gasifiers were employed to manufacture low-thermal-value gas from different types of biomass and wastes. The producer gas from biomass generally had heating values ca 5.9 MJ/m^3 (150 Btu/SCF), and the energy yields as gas ranged to ca 70% of the energy contained in the feed. Many gasifier designs were offered for the manufacture of producer gas from biomass, and several types of units are still available for purchase. Thousands of producer gasifiers operating on air and wood were used during World War II, particularly in Sweden, to power automobiles, trucks, and buses. The engines needed only slight modification to operate on low-thermal-value producer gas.

Another approach to obtain energy from biomass is based on hydrogenation. Hydrogen, which can either be generated from the feed or the conversion products, or be obtained from an independent source, reacts directly with the feed or intermediate process streams at elevated pressures and temperatures to yield substitute fuels. In theory, highly oxygenated biomass should be capable of reduction to liquid and gaseous fuels at any level between the initial oxidation state of the feed and methane. For a cellulosic material containing hydroxyl groups, the reactions might consist of dehydroxylation and depolymerization by hydrogenolysis, during which there is a transition from solid to liquid to gas:

$$R(OH)_x + y\,H_2 \rightarrow RH_y(OH)_{x-y} + yH_2O$$

$$R—R' + H_2 \rightarrow RH + R'H$$

Hydrogenation is mainly employed for the production of methane by hydrogasification. In one process, synthesis gas is produced and then the carbon monoxide is methanated with hydrogen to yield methane. In another route (see Fig. 11), the feed reacts directly with hydrogen (21) and the shredded biomass is converted with hydrogen-containing gas to a gas containing relatively high methane concentrations in the first-stage reactor. The product char from the first stage is used in a second-stage reactor to generate the hydrogen-rich synthesis gas for the first stage. From experi-

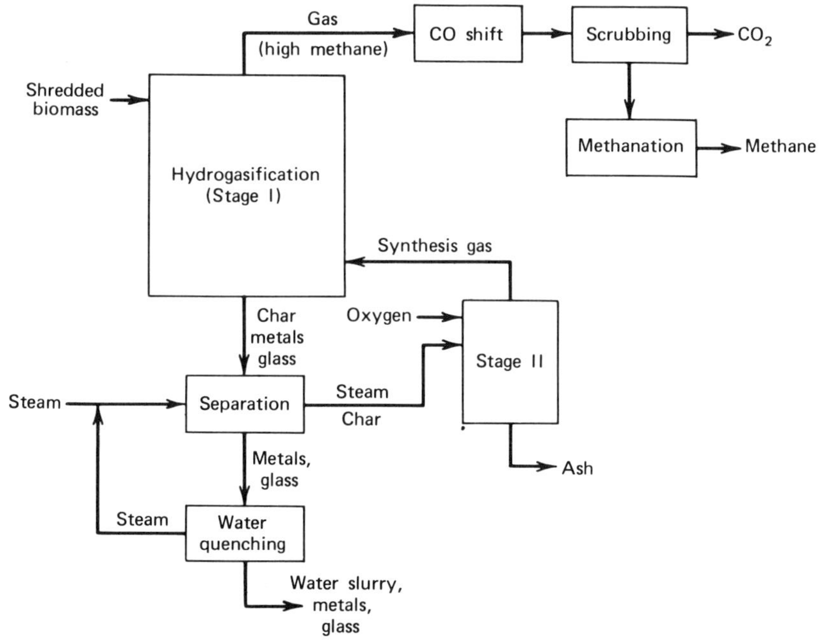

Figure 11. Methane production by hydrogasification.

mental results with the first-stage hydrogasifier operated in the free-fall and moving-bed modes at 1.72 MPa (17 atm) and 870°C with pure hydrogen, calculations shown in Table 13 were made to estimate the composition and yield of the high methane gas produced when the first stage is integrated with an entrained-char gasifier as the second stage. Although the methane content of the raw product gas is projected to be higher in the moving-bed reactor than in the falling-bed reactor, gas from the first stage must still be adjusted in H_2:CO ratio in a shift converter, scrubbed to remove carbon dioxide, and methanated to obtain SNG.

A conversion process is being developed in which biomass reacts in a water solution of sodium carbonate and carbon monoxide gas at elevated temperature and pressure to form a heavy liquid fuel (22). Biomass and the combustible fraction of wastes have been converted at weight yields of 40–60% at 250–425°C and 10–28 MPa (101–282 atm). Lower viscosity products are generally obtained at higher reaction temperatures and solid or semisolid products are obtained when the reaction temperature is below 300°C. However, because of the high nitrogen and oxygen content and the boiling characteristics and high viscosity range of the liquid products, these products cannot be classified as synthetic crude oils. Conventional refining methods could not be used to upgrade this type of material to standard petroleum derivatives but the product might be improved by process modification. It has been suggested that part of the mechanism involves formate production from carbonate, dehydration of the vicinal hydroxyl groups in the cellulosic feed to carbonyl compounds via enols, reduction of the carbonyl group to an alcohol by formate and water, and regeneration of formate (22). The liquefaction mechanism probably involves catalytic hydrogenation because carbon monoxide would be expected to form at least some hydrogen by the water–gas shift reaction.

Chemical Processes. Chemicals have long been manufactured from biomass, especially from wood (silvichemicals), by many different fermentation and thermochemical methods. For example, continuous pyrolysis of wood was used by the Ford Motor Company in 1929 for the manufacture of various chemicals as shown schematically in Figure 12; the product distribution is shown in Table 14 (23). Indeed, wood alcohol (methanol) was manufactured on a large scale by destructive distillation of wood for many years until the 1930s and early 1940s when the economics became more favorable for methanol manufacture from fossil-fuel-derived synthesis gas. As the

Table 13. Estimated Gas Compositions and Yields from Integrated Hydrogasification Process at Stage I

	Free fall	Moving bed
composition, mol %		
carbon monoxide	45.9	51.9
hydrogen	31.9	13.3
methane	10.4	17.2
carbon dioxide	10.1	16.1
ethane	1.2	1.1
benzene	0.5	0.4
yield, m^3/kg^a dry feed	1.1	0.95
fraction of total methane produced in Stage I after methanation	0.26	0.52

[a] To convert m^3/kg to ft^3/lb, multiply by 16.0.

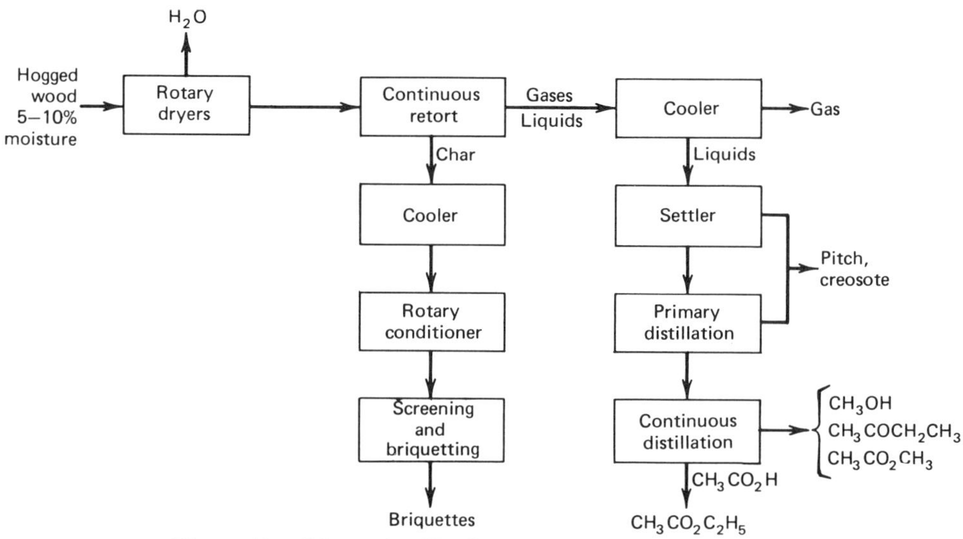

Figure 12. Schematic of Ford Motor Company process (1929).

Table 14. Product Yields from Wood Pyrolysis[a,b]

Product	Yield per metric ton dry wood
gas[c], m^3	156
charcoal, kg	300
ethyl acetate, L	61.1
creosote oil, L	13.6
methanol, L	13.0
ethyl formate, L	5.3
methyl acetate, L	3.9
methyl ethyl ketone, L	2.7
other ketones, L	0.9
allyl alcohol, L	0.2
soluble tar, L	91.8
pitch, kg	33.0

[a] Ref. 23.
[b] Feed: 70% maple, 25% birch, 5% ash, elm, and oak; av temperature, 515°C.
[c] CO_2, 37.9 mol %; CO, 23.4 mol %; CH_4, 16.8 mol %; N_2, 16.0 mol %; O_2, 2.4 mol %; H_2, 2.2 mol %; hydrocarbons, 1.2 mol %. To convert m^3 to ft^3, multiply by 35.3.

shortage of fossil fuels becomes more severe and the prices of natural gas and petroleum increase further, it is expected that some of the older biomass-based chemical manufacturing processes will become popular again (see Chemurgy).

In the production of chemicals from biomass, wood is still the raw material of choice for the manufacture of certain chemicals, although many of them cannot currently compete with fossil-based products. The chemistry of silvichemical production is related directly to the chemical composition of trees (ca 50% cellulose, 25% hemicelluloses, and 25% lignins). However, specialty chemicals are often manufactured from nonwoody biomass because they occur naturally in certain plant species or can easily be derived from these plants. Examples are the alginic acids from giant brown kelp (*Macrocystis pyrifera*) and physiologically active alkaloids from particular plants. Ethanol (qv) has of course been manufactured for chemical applications from starchy biomass by fermentation for many years.

The availability of C_6 sugars, such as glucose, is an important factor in the development of a biomass chemicals industry. Unfortunately, most biomass is high in cellulose rather than C_6 sugars. Cellulose, because of its relatively low biodegradability, cannot easily be converted to fermentation alcohol and other products without first liberating the monosaccharides. Trees and fibrous biomass, such as plant stalks and reedy plants, contain cellulose in partially crystalline form sometimes complexed with other materials. The low degradability of wood cellulose, which can exist as lignin–cellulose complexes, is attributed to these factors. Such complexes can be degraded to glucose concentrates by several hydrolytic methods, the most common of which is hydrolysis with dilute sulfuric acid. However, at the temperatures necessary to form glucose, a large portion of the product is converted to by-products such as 5-hydroxymethyl furfural (see Furan derivatives). Glucose yields are usually near 50% of theory because of by-product formation. Enzyme-catalyzed hydrolysis of cellulose has afforded much higher glucose yields but the particle size of the cellulosic material subjected to hydrolysis must be reduced to the micron range to facilitate depolymerization. The cost of particle size reduction tends to outweigh the advantages of higher glucose yields.

The yield of monosaccharides can be improved by dissolving the cellulose, separating the insolubles, and then subjecting the solution to hydrolysis or precipitating the cellulose before hydrolysis. This method destroys crystallinity and any cellulose complex that may be present, and gives high yields of glucose even with highly fibrous biomass. Special solvents such as Cadoxen and 65% sulfuric acid have been suggested for this application (24).

Another approach to improve alcoholic fermentation combines saccharification and fermentation. Enzyme-catalyzed cellulose hydrolysis and fermentation to alcohol takes place in the same vessel in the presence of enzyme and yeast (25). Reduced fermentor pressures and enzyme and yeast recycling result in 70–80% ethanol yields. These process modifications coupled with more energy-efficient distillation and heat exchanger improvements are projected to make fermentation ethanol and low value biomass competitive with ethanol from ethylene.

Multiproduct processes for biomass chemical plants in which the operating conditions can be manipulated to vary the product distribution as a function of demand or other factors, or in which an optimum mix of products is chosen based on feedstock characteristics, appear to have some merit even though no full-scale systems have yet been commercialized. The process depicted in Figure 13 (26) illustrates how such a

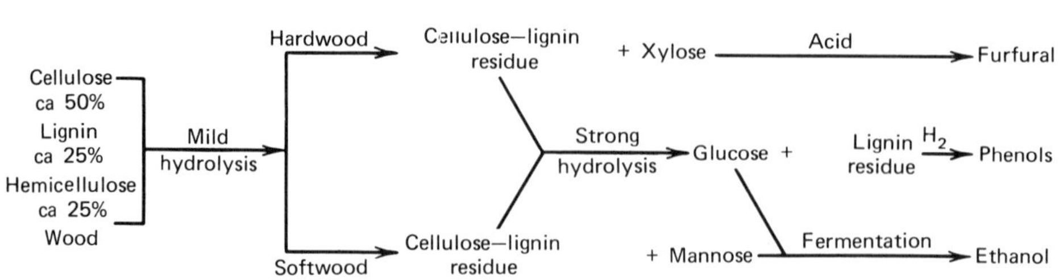

Figure 13. Routes to ethanol, phenols, and furfural from wood.

plant might function. Mild acid hydrolysis of the hemicelluloses in wood affords either a predominantly xylose or mannose solution, depending on the type of wood feed, and a cellulose–lignin residue. Strong acid treatment of this residue yields a glucose solution, which can be combined with mannose for alcoholic fermentation, and a lignin residue. Phenols can be made from this residue by hydrogenation, and furfural can be made from xylose by strong acid treatment. The products are thus ethanol, furfural, and phenols.

Obviously, there are many different routes to organic chemicals from biomass because of its high polysaccharide content and reactivity. The practical value of the conversion processes selected for commercial use with biomass depends on the availability and price of the same chemicals produced from petroleum and natural gas.

Production

The manufacture of synfuels and energy products from biomass requires that suitable quantities of biomass be grown, harvested, and transported to the conversion plant site. Many questions must be studied in great detail to select the proper species or mixture of species for operation of the system. These questions concern such matters as growth cycle; fertilization; insolation; temperature; precipitation; propagation and planting procedures; soil and water needs; harvesting methods; disease resistance; growth area competition from biomass for food, feed, and fiber; growth area availability; the possibilities for simultaneous or sequential growth of biomass for synfuels and foodstuff or other applications; and nutrient depletion. Since at least 250,000 botanical species, of which only about 300 are cash crops, are known in the world, it would seem that biomass selection for energy applications could easily be achieved. This does not necessarily follow simply from the multiplicity of biomass species that can be considered for energy usage. Indeed, compared to the total known botanical species, a relatively small number will be used for the manufacture of synfuels and energy products.

In the ideal case, biomass chosen for energy applications should be high-yield, low-cash-value species that have short growth cycles and that grow well in the area and climate chosen for the biomass energy system. Fertilization requirements should be low and possibly nil if the species selected fix ambient nitrogen, thereby minimizing the amount of external nutrients that have to be supplied to the growth areas. In areas having low annual rainfall, the species grown should have low water needs and be able to utilize efficiently available precipitation. For land-based biomass, the requirements should be such that the crops can grow well on low-grade soils and do not need the best classes of agricultural land. After harvesting, growth should commence again without the need for replanting. Surprisingly, several biomass species meet many of these idealized characteristics and appear to be quite suitable for energy applications.

Photosynthesis. The biochemical pathways in atmospheric carbon dioxide fixation involve the decomposition of water to form oxygen, protons, and electrons; the transport of these electrons to a higher energy level via Photosystems II, I, and several electron transfer agents; the concomitant generation of reduced nicotinamide adenine dinucleotide phosphate (NADPH) and adenosine triphosphate (ATP); and reductive assimilation of carbon dioxide to carbohydrate. The initial process is believed to be the absorption of light by chlorophyll which promotes the decomposition of water.

The ejected electrons are accepted by ferredoxin (Fd), a nonheme iron protein. The reduced Fd starts a series of electron transfers to generate ATP from adenosine diphosphate (ADP) and inorganic phosphate, and NADPH. The overall stoichiometry including the reduction of carbon dioxide to carbohydrate can be represented in simplified form:

$$2\ H_2O(l) \rightarrow 4\ H^+ + 4\ e + O_2(g)$$

$$4\ Fd^{3+} + 4\ e \rightarrow 4\ Fd^{2+}$$

$$3\ ADP + 3\ Pi \rightarrow 3\ ATP$$

$$4\ Fd^{2+} \rightarrow 4\ Fd^{3+} + 4\ e$$

$$2\ NADP^+ + 4\ H^+ + 4\ e \rightarrow 2\ NADPH + 2\ H^+$$

$$CO_2(g) + 3\ ATP + 2\ NADPH + 2\ H^+ \rightarrow (CH_2O) + 3\ ADP + 2\ NADP^+ + 3\ Pi + H_2O(l)$$

$$\text{Overall:}\quad CO_2(g) + H_2O(l) \rightarrow (CH_2O) + O_2(g)$$

For each of the two light reactions, one photon is required to transfer each electron; a total of eight photons is thus required to fix one molecule of carbon dioxide. Assuming that the carbon dioxide is in the gaseous phase and that the initial product is glucose, the standard Gibbs free energy change at 25°C is 480 kJ (114 kcal) per mole of carbon dioxide assimilated and the corresponding enthalpy change is 470 kJ (112 kcal).

The maximum efficiency with which photosynthesis can occur has been estimated by several methods. The upper limit has been projected at ca 8–15%, depending on the assumptions made; ie, the maximum amount of solar energy trapped as chemical energy in the biomass is 8–15% of the energy of the incident solar radiation.

The relationship of the energy and frequency of a photon is given by:

$$e = \frac{hc}{\lambda}$$

where e = energy content of one photon, J; h = Planck's constant, 6.624×10^{-34} J·s; c = velocity of light, 3×10^8 m/s; and λ = wavelength of light, m.

Assume that the wavelength of the light absorbed is 575 nm and is equivalent to the light absorbed between the blue (400 nm) and red (700 nm) ends of the visible spectrum. This assumption has been made by several investigators for green plants to calculate the upper limit of photosynthesis efficiency. The energy absorbed in the fixation of one mole of carbon dioxide is then given by:

$$\text{energy absorbed} = (6.624 \times 10^{-34})\ (3.00 \times 10^8)\ (575 \times 10^{-9})^{-1}\ (8)\ (6.023 \times 10^{23})$$
$$= 1.67\ \text{MJ (399 kcal)}$$

Since 470 kJ (112 kcal) of solar energy is trapped as chemical energy in this process, the maximum efficiency for total white light absorption is 28.1%. Further adjustments are usually made to account for the percentages of photosynthetically active radiation in white light, the light that can actually be absorbed, and respiration. The amount of photosynthetically active radiation in solar radiation that reaches the earth is estimated to be about 43%. The fraction of the incident light absorbed is a function of many factors such as leaf size, canopy shape, and reflectance of the plant; it is estimated to have an upper limit of 80%. This effectively corresponds to the utilization of 8 photons out of every 10 in the active incident radiation. The third factor results from biomass respiration. A portion of the stored energy is used by the plant, the amount of which depends on the properties of the biomass species and the environment. For purposes of calculation, assume that about 25% of the trapped solar energy is used by the plant, thereby resulting in an upper limit for retention of the nonrespired energy of 75%. The upper limit for the efficiency of photosynthetic fixation of biomass can now be estimated to be 7.2% (0.281 × 0.43 × 0.80 × 0.75). For the case where little or no energy is lost by respiration, the upper limit is estimated to be 9.7% (0.281 × 0.43 × 0.80). The low efficiency limit might correspond to land-based biomass whereas the higher efficiency limit might be closer to water-based biomass such as unicellular algae. These figures can be transformed into dry biomass yields by assuming that all the fixed carbon dioxide is contained in the biomass as cellulose, $\leftarrow C_6H_{10}O_5 \rightarrow_x$, from the equation:

$$Y + \frac{CIE}{F}$$

where Y = yield of dry biomass, t/(m^2·yr); C = constant, 3.1536; I = average insolation, W/m^2; E = solar energy capture efficiency, %; and F = energy content of dry biomass, MJ/kg.

Thus for high cellulose dry biomass, an average insolation of 184 W/m^2 [1404 Btu/(ft^2·d)], which is the average insolation for the continental United States, a solar energy capture efficiency of 7.2%, and a high heat of combustion of 17.51 MJ/kg (7529 Btu/lb) for cellulose, the yield of dry biomass is 239 t/(hm^2·yr) [107 short ton/(acre·yr)]. The corresponding value for an energy capture efficiency of 9.7% is 321 t/(hm^2·yr) [143 short ton/(acre·yr)]. These yields of organic matter can be viewed as an approximation of the theoretical yield limits for land- and water-based biomass. Some estimates of maximum yield are higher and some are lower than these figures depending upon the values used for I, E, and F, but they serve as a guideline to indicate the highest yields that a biomass production system could be expected to achieve under normal environmental conditions. Unfortunately, real biomass yields rarely approach these limits. Sugarcane, eg, which is one of the high yielding species of biomass, typically produces total dry plant matter at yields of about 80 t/(hm^2·yr) [36 short ton/(acre·yr)] (see Sugar).

Yield is plotted against solar energy capture efficiency in Figure 14 for insolation values of 150 and 250 W/m^2 [1142 and 1904 Btu/(ft^2·d)] which spans the range commonly encountered in the United States, and for biomass energy values of 12 and 19 MJ/dry kg (5160 and 8170 Btu/lb). The higher the efficiency of photosynthesis, the higher the biomass yield. However, for a given solar energy capture efficiency and incident solar radiation, the yield is projected to be lower at the higher biomass energy values (curves A and B, curves C and D). From an energy production standpoint, this simply means that a higher-energy-content biomass could be harvested at lower yield levels and still compete with higher yielding but lower-energy-content biomass species. It is also apparent that for a given solar-energy-capture efficiency, yields similar to those obtained with higher energy content species should be possible with a lower energy content species even when it is grown at a lower insolation (curves B and C). Finally, at the solar-energy-capture efficiency usually encountered in the field, about 1% or less, the spread in yields is much less than at the higher energy capture efficiencies. It is important to emphasize that this interpretation of biomass yields as functions of insolation, energy content and energy capture, although based on sound

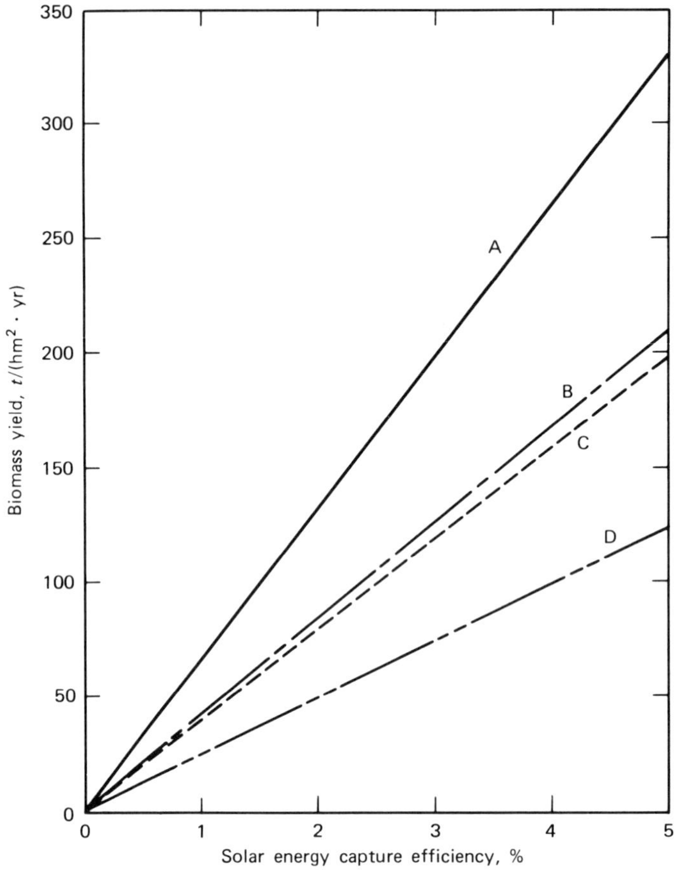

Figure 14. Effect of solar energy capture efficiency on biomass yield. To convert W/m^2 to Btu/(ft^2·d), multiply by 7.616; to convert MJ/kg to Btu/lb, multiply by 430. To convert t/(hm^2·yr) to short ton/(acre·yr), divide by 2.24.

	I, W/m²	*F, MJ/kg (dry)*
A	250	12
B	250	19
C	150	12
D	150	19

principles, is still a theoretical analysis of living systems. Because of the many uncontrollable factors in the field, such as the changes in climate and seasonal changes in biomass composition, departures from the norm can be expected.

The biochemical pathways involved in the conversion of carbon dioxide to carbohydrate play an important role in our understanding of the molecular events of biomass growth. Carbon dioxide fixation proceeds via three different biochemical energy transfer pathways (27–28). The first is called the Calvin three-carbon cycle, and involves an initial three-carbon intermediate of 2-phosphoglyceric acid (PGA). This cycle, also referred to as the reductive pentose phosphate cycle, is used by autotrophic photochemolithotrophic bacteria, algae, and green plants. For every three molecules of carbon dioxide converted to glucose in a dark reaction, nine ATP, six NADPH, and twelve Fd molecules are required. From three molecules each of carbon dioxide and ribulose-1,5-diphosphate (RDP) in the chloroplasts, six molecules of PGA are formed. After these carboxylation reactions, a reductive phase occurs in which six molecules of PGA are successively transformed into six molecules of diphosphoglyceric acid (DPGA) and six molecules of 3-phosphoglyceraldehyde, a triosephosphate. The three molecules of RDP are regenerated by five molecules of the triose phosphate and the cycle is initiated again. The other triose phosphate molecule is used to generate glucose-6-phosphate and energy and electron carriers. Plant biomass species are called C_3 plants when they use the Calvin cycle which is common in many fruits, legumes, grains, and vegetables. In general, C_3 plants exhibit low rates of photosynthesis at light saturation, low light saturation points, sensitivity to oxygen concentration, rapid photorespiration, and high carbon dioxide compensation points (ca 50 ppm). The carbon dioxide compensation point is the carbon dioxide concentration in the surrounding environment below which more carbon dioxide is respired by the plant than is photosynthetically fixed. Typical C_3 biomass species are peas, sugar beet, spinach, alfalfa, *Chlorella, Eucalyptus,* potato, soybean, tobacco, oats, barley, wheat, tall fescue, sunflower, rice, and cotton.

The second pathway is called the C_4 cycle because the carbon dioxide is initially fixed as the four-carbon dicarboxylic acids, malic or aspartic acids. Phosphoenolpyruvate (PEP) reacts with one molecule of carbon dioxide to form oxaloacetate (OAA) in the mesophyll of the biomass, and then malate or aspartate is formed. The C_4 acid is transported to the bundle sheath cells where decarboxylation occurs to regenerate pyruvate which is returned to the mesophyll cells to initiate another cycle. The carbon dioxide liberated in the bundle sheath cells enters the C_3 cycle in the usual manner. Thus no net carbon dioxide is fixed in this portion of the C_4 cycle, and it is the combination with the C_3 cycle which ultimately results in carbon dioxide fixation. The subtle differences between the C_4 and C_3 cycles are believed responsible for the wide variations in biomass properties. In contrast to C_3 biomass, C_4 biomass is usually produced at higher yields and has higher rates of photosynthesis, high light saturation points, insensitivity to oxygen concentrations below 21 mol %, low levels of respiration, low carbon dioxide compensation points, and greater efficiency of water usage. It often occurs in areas of high insolation, hot daytime temperatures, and seasonal dry periods. Typical C_4 biomass includes important crops such as corn, sugarcane, and sorghum, and forage species and tropical grasses such as bermuda grass. Even crabgrass is a C_4 biomass. At least 100 genera in 10 plant families are known to exhibit the C_4 cycle.

The third pathway is called crassulacean acid metabolism or CAM. It refers to the capacity of chloroplast-containing tissues to fix carbon dioxide in the dark via phosphoenolpyruvate carboxylase leading to the synthesis of free malic acid. The mechanism involves the β-carboxylation of PEP by this enzyme and the subsequent reduction of OAA by malate dehydrogenase. CAM has been documented in at least 18 families including the family Crassulaceae, and 109 genera of the *Angiospermae*. Biomass species in the CAM category are typically adapted to arid environments and have low photosynthesis rates and high water usage efficiencies. Examples are cactus plants and the succulents such as pineapple. The initial carbon dioxide fixation can take place in the dark with much less water loss than the fully light-dependent C_3 and C_4 pathways. CAM biomass also conserves carbon by recycling of endogenously formed carbon dioxide. Several CAM species show temperature optima in the 12–17°C range for carbon dioxide fixation in the dark. The stomates in CAM plants open at night to allow entry of carbon dioxide and then close by day to minimize water loss. The carboxylic acids formed in the dark are converted to carbohydrates when the radiant energy is available during the day. Relatively few CAM plants have been exploited commercially.

Appreciable differences in net photosynthetic assimilation of carbon dioxide are apparent between C_3, C_4, and CAM biomass species. The generally lower yields of C_3 biomass are caused by its higher photorespiration rate. If it could be reduced, the net yield of biomass would increase. Considerable research is in progress to achieve this rate reduction by chemical and genetic methods but, as yet, only limited yield improvements have been made. Such an achievement with C_3 biomass would be expected to be very beneficial for foodstuff production and biomass energy applications.

The specific carbon dioxide-fixing mechanism used by plants affects the efficiency of photosynthesis. Therefore, it is desirable from an energy utilization viewpoint to choose plants that exhibit high photosynthesis rates to maximize the yields of biomass in the shortest possible time. Obviously, however, numerous factors other than the carbon dioxide-fixing mechanism affect the efficiency of photosynthesis. Insolation; the amounts of available water, nutrients, and carbon dioxide; temperature; and the transmission, reflection, and biochemical energy losses within or near the plant affect the efficiency. For lower plants such as the green algae, many of these parameters can be controlled by man. But for conventional biomass growth that is subjected to the natural elements, it is not feasible to control all of them.

Climate and Environment. The biomass species selected for energy applications must be compatible with the climate to facilitate operation of fuel farms. Insolation, rainfall, and temperature have the most influence on the productivity of indigenous or transplanted species. Information compiled over the years in meteorological records and from agricultural practice supplies a valuable data bank from which to develop biomass energy applications. Atmospheric carbon dioxide concentration and the availability of nutrients are also important.

Insolation. The intensity of the incident solar radiation at the earth's surface is a key factor in photosynthesis; natural biomass growth does not take place without solar energy. Insolation varies with location and time, and as is well known, is high in the tropics and near the equator (see Table 15) (29). Insolation values over the continental United States are shown in Table 16 (30). To place the amount of energy that strikes the earth in the proper perspective, the annual insolation on about 0.1 to 0.2% of the surface of the continental United States is equivalent to all the energy consumed by the United States in one year.

Figure 15 shows how the mean annual insolation varies at the earth's surface with

Table 15. Insolation at Various Latitudes for Clear Atmospheres[a]

Location	Latitude	Maximum, W/m^2 [b]	Minimum, W/m^2 [b]	Average[c], W/m^2 [b]
equator	0°	315	236	263
tropics	23.5°	341	171	263
mid-earth	45°	355	70.9	210
polar circle	66.5°	328	0	158

[a] Ref. 29.
[b] To convert W/m^2 to $Btu/(ft^2 \cdot d)$, multiply by 7.616.
[c] Yearly total divided by 365.

Table 16. Daily Solar Radiation at Selected Locations in the United States During 1970[a]

Location	Average total daily insolation, W/m² [b] January	April	July	October	Annual average, W/m² [b]
Tucson, Arizona	146	289	288	208	229
Fresno, California	93.2	290	338	187	229
Lakeland, Florida	135	260	247	189	210
Indianapolis, Indiana	90.0	188	242	120	157
Lake Charles, Louisiana	109	215	236	175	191
Saint Cloud, Minnesota	75.9	178	275	104	157
Glasgow, Montana	72.0	190	299	118	175
Ely, Nevada	108	257	288	176	210
Oklahoma City, Oklahoma	80.8	212	264	155	183
San Antonio, Texas	113	198	286	182	199
Burlington, Vermont	76.3	182	208	99.7	146
Sterling, Virginia	90.9	173	233	113	159
Seattle–Tacoma, Washington	36.5	179	276	98.1	151

[a] Ref. 30.
[b] To convert W/m^2 to Btu/(ft^2·d), multiply by 7.616.

location (31). At a given latitude, the incident energy is not constant and often exhibits large changes over relatively short distances. The insolation values in Table 16 and Figure 15 were used to calculate the solar energy capture efficiencies for the biomass listed in Table 17 assuming that all of the organic matter is cellulose. The production figures shown in Table 17 represent annual yields obtained under good growth conditions (7–8,27,32–33). The solar energy capture efficiency estimates indicate that C_4 plants are usually better photosynthesizers than C_3 plants, and that high insolation alone does not always correlate with high biomass yield and capture efficiency. Although there are a few exceptions, Table 17 shows a trend in this direction.

Precipitation. Precipitation as rain, snow, sleet, or hail is governed by movement of air and is generally abundant wherever air currents are predominately upward. The greatest precipitation should therefore occur near the equator. The average annual precipitation in the continental United States (without Alaska) is shown in Figure 16 (34). Table 18 (35) summarizes precipitation at different locations in the United States; the average annual rainfall is about 79 cm.

The moisture needs of aquatic biomass are presumably met in full because growth occurs in liquid water but the growth of land biomass can often be water-limited. The water requirements for good growth for many biomass species are in the range of 50–76 cm (36). A few crops such as wheat exhibit good growth with much less moisture. Without irrigation, water is supplied during the growing season by the water in the soil at the beginning of the season and by rainfall. Biomass productivity is generally related to precipitation, though not to rainfall alone because of the differences in soil characteristics, water evaporation rates, and infiltration. Some low rainfall areas can be made productive through irrigation.

Temperature. Most biomass species such as corn, kenaf, and napier grass, grow well in the United States between 15.6 and 32.2°C. Tropical grasses and certain warm-season biomass grow best at 35–40°C (37). Cool-weather biomass such as wheat may show good growth below 15°C, and certain marine biomass such as the giant brown kelp only survives in water below 20–22°C (38). Below 6.1°C biomass growth does not

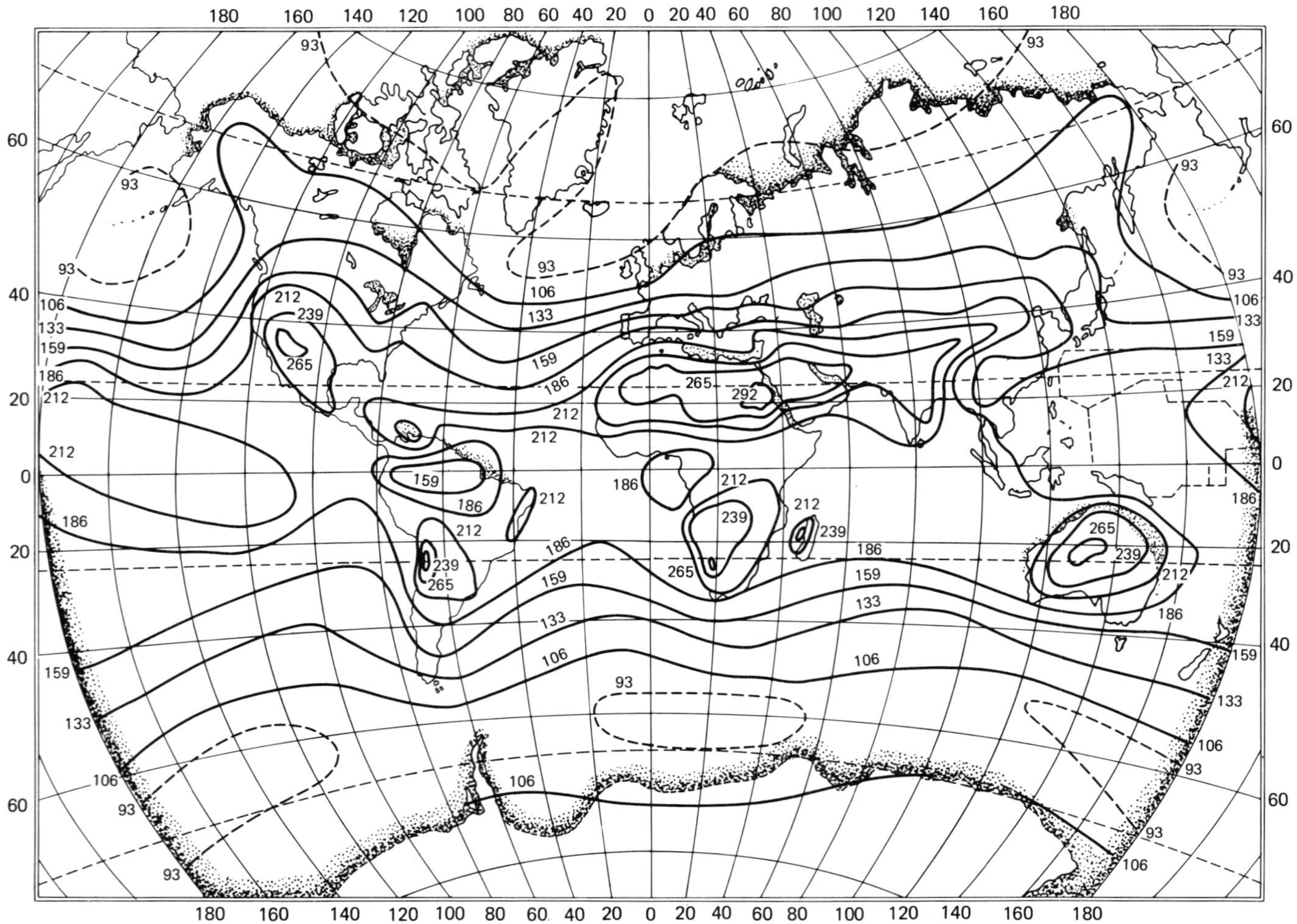

Figure 15. Mean annual insolation at earth's surface, W/m² (adapted from ref. 31).

Table 17. Biomass Productivity and Estimated Solar Energy Capture Efficiency[a]

Location	Biomass community	Annual production of dry matter, t/(hm²·yr)[b]	Average insolation, W/m²[c]	Solar energy capture efficiency[d], %
Sweden	enthrophic lake angiosperm	7.2	106	0.38
Denmark	phytoplankton	8.6	133	0.36
Mississippi	water hyacinth	11.0–33.0	194	0.31–0.94
Minnesota	maize	24.0	169	0.79
New Zealand	temperate grassland	29.1	159	1.02
West Indies	tropical marine angiosperm	30.3	212	0.79
Nova Scotia	sublittoral seaweed	32.1	133	1.34
Georgia	subtropical saltmarsh	32.1	194	0.92
England	coniferous forest, 0–21 yr	34.1	106	1.79
Israel	maize	34.1	239	0.79
New South Wales	rice	35.0	186	1.04
Congo	tree plantation	36.1	212	0.95
Holland	maize, rye, two harvests	37.0	106	1.94
Marshall Islands	green algae	39.0	212	1.02
FRG	temperate reedswamp	46.0	133	1.92
Puerto Rico	*Panicum maximum*	48.9	212	1.28
California	algae, sewage pond	49.3–74.2	218	1.26–1.89
Colombia	pangola grass	50.2	186	1.50
West Indies	tropical forest, mixed ages	59.0	212	1.55
Hawaii	sugarcane	74.9	186	2.24
Puerto Rico	*Pennisetum purpurcum*	84.5	212	2.21
Java	sugarcane	86.8	186	2.59
Puerto Rico	napier grass	106	212	2.78
Thailand	green algae	164	186	4.90

[a] Refs. 7–8, 27, 32–33.
[b] To convert t/(hm²·yr) to short ton/(acre·yr), divide by 2.24.
[c] To convert W/m² to Btu/(ft²·d), multiply by 7.616.
[d] Approximate estimates only; probably too high.

occur. Table 19 (35) summarizes seasonal temperature fluctuations in the United States. The growing season is clearly longer in the southern portion of the country. In some areas such as Hawaii, the Gulf states, southern California, and the southeastern Atlantic states, the temperature is usually conducive to biomass growth for most of the year.

The effect of temperature fluctuations on net carbon dioxide uptake is illustrated by Figure 17 (39). As the temperature increases, net photosynthesis increases for cotton and sorghum to a maximum value and then rapidly declines. Ideally, the biomass species grown in an area should have a maximum rate of net photosynthesis as close as possible to the average temperature during the growing season in that area.

Atmospheric Carbon Dioxide Concentration. Higher concentrations of carbon dioxide than are normally present in air promote carbon fixation and increase biomass yields. Although in confined, environmentally controlled enclosures such as hothouses, carbon dioxide-enriched air can be used to stimulate growth; in large-scale open systems, such as those envisaged for biomass energy farms, this is not practical. For

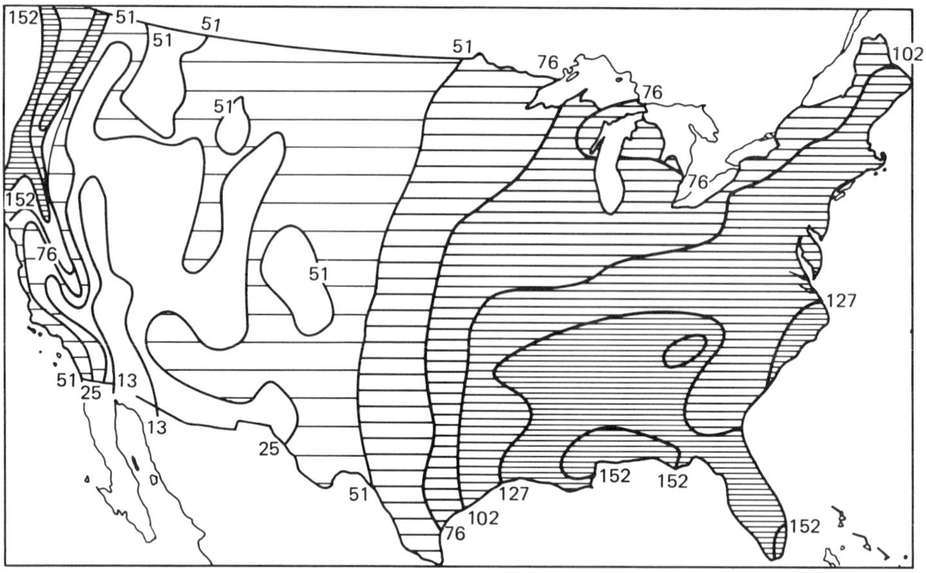

Figure 16. Normal annual precipitation in the U.S., cm (adapted from ref. 34).

Table 18. Average Monthly and Annual Precipitation in the United States, 1941 to 1970, cm[a]

Location	January	April	July	October	Annual
Juneau, Alaska	10.0	7.6	11.9	19.9	139
Phoenix, Arizona	1.8	0.8	1.9	1.2	17.9
Los Angeles, California	6.4	2.8	0.03	0.6	29.4
San Francisco, California	11.1	4.0	0.03	2.5	49.6
Miami, Florida	5.5	9.1	17.6	20.8	152
Honolulu, Hawaii	11.2	3.5	1.5	3.8	58.2
Indianapolis, Indiana	7.3	9.8	9.3	6.4	98.4
New Orleans, Louisiana	11.5	10.5	17.1	5.7	144
Minneapolis, Minnesota	1.9	5.2	9.4	4.5	65.9
Great Falls, Montana	2.2	3.0	3.2	1.7	38.1
Reno, Nevada	3.1	1.2	0.7	1.1	18.3
Oklahoma City, Oklahoma	2.8	8.8	6.8	6.5	79.7
Burlington, Vermont	4.4	6.7	9.0	7.0	82.7
Norfolk, Virginia	8.5	6.9	14.5	7.8	114
Seattle, Washington	14.7	6.2	1.8	9.9	98.5

[a] Ref. 35.

aquatic biomass production, carbon dioxide enrichment of the water phase may promote biomass growth if carbon dioxide concentration is a limiting factor since the growth often occurs by uptake of carbon dioxide from both the air and liquid phase near the surface.

For some high-growth-rate biomass species, the carbon dioxide concentration in the air among the leaves of the plant is often considerably less than that in the surrounding atmosphere. Photosynthesis may be limited by the carbon dioxide concentrations under these conditions when wind velocities are low and insolation is high.

Table 19. Average Monthly and Annual Temperature in the United States, 1941 to 1970, °C[a]

Location	January	April	July	October	Annual
Juneau, Alaska	−4.7	3.8	13.2	5.4	4.6
Phoenix, Arizona	10.7	19.8	32.9	22.3	21.3
Los Angeles, California	12.5	14.9	20.3	18.4	16.5
San Francisco, California	9.1	12.9	16.9	16.1	13.8
Miami, Florida	19.6	23.9	27.9	25.4	24.2
Honolulu, Hawaii	22.4	23.8	26.7	26.1	24.8
Indianapolis, Indiana	−2.3	11.3	23.9	13.2	11.3
New Orleans, Louisiana	11.6	20.3	27.7	21.0	20.2
Minneapolis, Minnesota	−11.0	7.3	22.2	10.0	6.7
Great Falls, Montana	−6.4	6.3	20.7	9.1	7.2
Reno, Nevada	−0.1	8.2	20.7	10.2	9.7
Oklahoma City, Oklahoma	2.7	15.8	27.5	16.9	15.5
Burlington, Vermont	−8.4	6.1	21.0	9.3	6.9
Norfolk, Virginia	4.7	14.3	25.7	16.5	15.2
Seattle, Washington	3.4	9.3	18.1	11.2	10.6

[a] Ref. 35.

Nutrients. All living biomass requires nutrients other than carbon, hydrogen, and oxygen to synthesize cellular material. The macronutrients are nitrogen, phosphorus, and potassium; nutrients required in lesser amounts are sulfur, sodium, magnesium, calcium, and iron, and the micronutrients manganese, cobalt, copper, zinc, and molybdenum of which only trace quantities are needed. For land-based biomass, all these elements are usually supplied by the soil and eventually, if not replaced through fertilization, the nutrients are depleted. Some biomass, such as the legumes, meet all or part of their nitrogen requirements through fixation of atmospheric nitrogen. Water-based biomass, eg, marine kelp, use the natural nutrients in ocean waters. Fresh-water biomass, such as water hyacinth, is often grown in water enriched with nutrients in the form of wastewater. The growth of the plant is stimulated and at the same time, the influent waste is stabilized because its components are taken up by the plant as nutrients. So-called luxuriant growth of water hyacinth on sewage in which more than the needed nutrients are removed from the waste can be used as a substitute wastewater treatment method.

Whole plants typically contain 2 wt % N, 1 wt % K, and 0.5 wt % P; therefore, at a yield of 20 t/(hm^2·yr) [8.9 short ton/(acre·yr)], harvesting of the whole plant without return of any of the plant parts to the soil corresponds to the annual removal of 400 kg N, 200 kg K, and 100 kg P per square hectometer (hectare). This illustrates the importance of fertilization, especially of these macronutrients, to maintain fertility of the soil. Indeed, biomass growth is often nutrient limited and the yield correlates with fertilizer rates. For example, for United States corn production from 1945 to 1970, nitrogen fertilizer applications increased from 8 to 125 kg/hm^2, while corn yields increased from 2132 to 5080 kg/hm^2 (40). Average nitrogen fertilizer applications for production of wheat, rice, potato, and brussel sprouts are currently about 73, 134, 148, and 180 kg/hm^2, respectively, in the United States (40). Much of the success of the green revolution is claimed to be the result of increased fertilizer applications. Estimates of balanced fertilizers needed to produce various land biomass species are shown in Table 20 (36). Note that alfalfa does not require added nitrogen because of its ni-

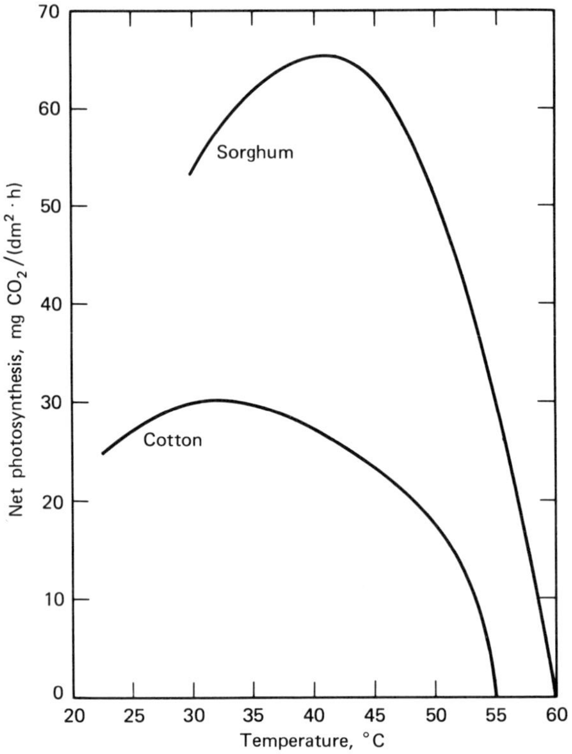

Figure 17. Effect of temperature on net photosynthesis for sorghum and cotton leaves. To convert mg/(dm^2·h) to lb/(ft^2·h), multiply by 2.373×10^{-3}.

Table 20. Estimated Fertilizer Requirements of Selected Biomass[a], Kilogram Required Per Whole Dry Plant

Biomass	N	P_2O_5	K_2O	CaO
alfalfa	0.0	12.3	34.0	20.7
corn	11.8	5.7	10.0	0.0
kenaf	13.9	5.0	10.0	16.1
napier grass	9.6	9.3	15.8	8.5
slash pine[b]	3.8	0.9	1.6	2.3
potato	16.8	5.3	28.3	0.0
sugar beet	18.0	5.4	31.2	6.1
sycamore	7.3	2.8	4.7	0.0
wheat	12.9	5.3	8.4	0.0

[a] Ref. 36.
[b] 5 years.

trogen-fixing ability. It is estimated that it can fix in one year ca 130–600 kg of elemental nitrogen per square hectometer (41).

Normal weathering processes release nutrients from soils but not in amounts large enough to promote maximum biomass yields. Fertilization is usually essential. Since nitrogenous fertilizers are currently manufactured from fossil fuels, mainly natural gas, and since fertilizer needs are usually the most energy-intensive inputs in a biomass production system, a careful analysis of the integrated biomass production–conversion system is necessary to ensure that net energy production is positive. Tradeoffs between synfuel outputs, nonsolar energy inputs, and biomass yields are required to operate a system that produces only energy products (see Fertilizers).

Land Availability. The availability of land suitable for production of land-based biomass is estimated by the land capabilities classification scheme developed by the U.S. Department of Agriculture (42) in which land is divided into eight classes (see Table 21). Classes I–III are suited for cultivation of many kinds of crops; Class IV is suited only for limited production; and Classes V–VIII are useful only for permanent vegetation such as grasses and trees. The actual usage of this land at the time of the survey is shown in Table 22 (43), illustrating that of all the land judged suitable for cultivation in Classes I–III, only about one-half was used as cropland.

This survey suggests that there is ample opportunity to produce biomass for en-

Table 21. Land Capability Classification by U.S. Department of Agriculture[a]

Class	Description	$km^2 \times 10^2$	%[b]
I	few limitations that restrict use	18.22	3.13
II	moderate limitations that reduce the choice of plants or require moderate conservation practice	121.15	20.79
III	severe limitations that reduce the choice of plants or require special conservation practices, or both	117.97	20.25
IV	very severe limitations that reduce the choice of plants or require very careful management, or both	72.90	12.51
V	not likely to erode but other limitations, impractical to remove, that limit use largely to pasture, range, woodland, or wildlife habitat	12.56	2.16
VI	severe limitations that make soils generally unsuited to cultivation and limit their use largely to pasture or range, woodland, or wildlife habitat	110.95	19.04
VII	severe limitations that make soils unsuited to cultivation and that restrict use largely to pasture or range, woodland, or wildlife habitat	128.96	22.13
VIII	limitations that preclude use for communical plants and restrict use largely to recreation, wildlife habitat, water supply, or to esthetic purposes		

[a] Ref. 42.
[b] Total 582.71×10^2 km^2 (about 60% of total U.S. area).

Table 22. Summary of Nonfederal Land Use in the United States in 1975 by Use Type[a]**, 10^2 km^2**

Land class	Cropland	Pasture and range	Forest	Urban	Water	Other land
I	13.51	2.49	0.88	0.46	0.06	0.81
II	76.03	22.98	15.21	1.99	0.28	4.66
III	49.71	35.60	24.69	1.45	0.70	5.81
IV	16.15	28.62	23.37	1.15	0.19	3.43
V	0.62	3.49	7.18	0.10	0.27	0.90
VI	5.35	72.13	30.37	0.72	0.14	2.25
VII–VIII	0.67	65.72	50.25	0.86	1.07	10.40
Total	*162.04*	*231.03*[b]	*151.95*[b]	*6.73*	*2.71*	*28.26*
Percent	*27.81*	*39.65*	*26.08*	*1.15*	*0.47*	*4.85*

[a] Ref. 43.
[b] The combined areas constitute about 66% of nonfederal land.

ergy applications on nonfederal land that is not used for foodstuffs production. Large areas of land in Classes V–VIII not suited for cultivation would appear to be available for biomass energy applications, and sizable areas in Classes I–IV that are not being used for crop production seem to be available. Land now used for crop production could also be considered for simultaneous or sequential growth of biomass for foodstuffs and energy. Portions of the federally owned lands, which are not included in this survey, might also be dedicated to biomass energy applications. Careful design and management of land-based biomass production areas could very well result in improvement or upgrading of lands to higher land capability classifications.

Water Availability. The production of marine biomass in the ocean even on the largest scale envisaged for energy applications would require only a very small fraction of the available ocean areas. For example, the U.S. Navy estimated that a square 753 km on each edge off the coast of California may be sufficient to produce enough giant brown kelp for conversion to methane to supply all of the natural gas needs of the U.S. Furthermore, the benefits to other marine life from a large kelp plantation have been well documented. Any conflicts that might arise would be concerned primarily with ocean traffic. With the proper plantation design for marine biomass and precautionary measures to warn approaching ships, it is expected that marine biomass growth could be sustained over long periods.

Fresh-water biomass could in theory be grown on the 20×10^6 km^2 of fresh water in the United States. However, several difficulties are encountered in large-scale, fresh-water biomass energy systems. About 80% of the fresh water in the United States is located in the northern states but several of the fresh-water biomass species considered for energy applications require a warm climate such as that found in Gulf states. The fresh-water areas suitable for biomass production in the southern states, however, are much smaller than those in the north, and the density of usage is higher in southern inland waters. Overall, these characteristics make small-scale aquatic biomass production systems more feasible for energy applications. In the future, the practicability of constructing large artificial lakes for this purpose might be examined.

Land-Based Biomass. Evaluation of land-based biomass energy applications has been aimed at selecting high yield biomass species, characterizing their physical and chemical properties, defining their growth requirements, and rating their energy use potential. Several species have been proposed specifically for energy usage, and others have been recommended for multiple uses, one of which may be as an energy resource. For example, sugarcane bagasse (qv) the fibrous material remaining after sugar extraction, is used in several sugar factories as a boiler fuel. Most land-based biomass plantations operated for energy production or synfuel manufacture will probably also yield products for nonenergy markets; production of single-energy products will be the exception rather than the rule.

Forest Biomass. About one-third of the world's land area is forestland (44). Silviculture, or the growth of trees, is practiced by five methods: exploitative, conventional extensive, conventional intensive, naturalistic, and short rotation (44). The exploitative method is simply the harvesting of trees without regard to regeneration, whereas in the conventional extensive method natural regeneration is encouraged. Conventional intensive silviculture is the growing and harvesting of commercial tree species in essentially pure stands such as Douglas fir and pine on tree farms. The naturalistic method is the growth of selected mixed tree species including hardwoods in which the species are selected to match the ecology of the site. In short-rotation silviculture, suggested as the most suitable method for energy applications, fast-growing trees are harvested every few years, in contrast to once every 20 or more years. Fast-growing trees such as cottonwood, red alder, and aspen are intensively cultivated and mechanically harvested every 2–5 yr when they are 3–6 m high and only a few centimeters in diameter. The young trees are converted into chips for further processing or direct fuel use and the small remaining stems or stumps form new sprouts which are intensively cultivated again. Short-rotation silviculture produces an annual dry yield of several tons of biomass per hm^2, apparently without large energy inputs for fertilization, irrigation, cultivation, and harvesting.

Historically, trees are important resources and still serve as important energy resources in many developing countries. Several studies of temperate forests indicate productivities of ca 9–28 t/(hm^2·yr) [4–12.5 short ton/(acre·yr)]; corresponding yields of tropical forests are higher, ranging ca 20–50 t/(hm^2·yr) [9–22 short ton/(acre·yr)] (45). These yields are obtained using conventional forestry methods over long periods of time, 20–50 yr or more. Productivity is initially low, increasing slowly for about 20 yr, and then declining. Coniferous forests grow even in the winter months in moderate temperatures; they do not exhibit the yield fluctuations characteristic of deciduous forests.

The eucalyptus tree, of the myrtle family *Myrtaceae,* has been studied in great detail as a renewable energy source (46). There are approximately 450 to over 700 identifiable species of eucalyptus. The eucalyptus is a rapidly growing tree native to Australia and is a prime candidate for energy use because it reaches a size suitable for harvesting in about 7 yr. Several species have the ability to coppice after harvesting, and frequently as many as four harvests can be obtained from a single stump before replanting is necessary. In several South American countries, eucalyptus trees are converted to charcoal and used as fuel. Eucalyptus wood has also been used to power an integrated sawmill–wood distillation, and charcoal–iron plant in Western Australia. Several large areas of marginal land in the western and central regions of California and the southeastern United States may be suitable for establishing eucalyptus energy farms.

Other studies have demonstrated that deciduous trees are preferred over conifers for the production of such fuels as methane (47). Several species can be started readily from clones, resprout copiously and vigorously from their stumps at least five or six times without loss of vigor, and exhibit rapid initial growth. They can also be grown on sites with slopes as steep as 25% where precipitation is 50 cm or more per year. It has been estimated that yields of ca 18–22 dry t/(hm^2·yr) [8–10 short ton/(acre·yr)] are possible on a sustained basis almost anywhere in the United States eastern and central time zones from deciduous trees grown in dense plantings.

Grasses. Grasses are very abundant forms of biomass (48). About 400 genera and 6000 species are distributed all over the world growing in all land habitats capable of supporting higher forms of plant life. Grasslands cover over one-half the continental United States; about two-thirds of this land is privately owned. Grass, as a family (Gramineae), includes the great fruit crops, wheat, rice, corn, sugarcane, sorghum, millet, barley, and oats, and many species of sod crops which provide forage or pasturage for all types of farm animals (see Wheat and other cereal grains). In the concept of grassland agriculture, grass-related species are also included such as the legume family, the clovers, alfalfas, and many others. Grasses are grown as farm crops and for decorative purposes, preserving the balance of productive capacity of lands by crop rotation, controlling erosion on sloping lands, the protection of water sheds, and the stabilization of arid areas. Many advances in grassland agriculture have been made over the last 30–40 yr through breeding and the use of improved species of grass, alone or in seeding mixtures; cultural practices, including amending the soil to promote herbage growth best suited for specific purposes; and the adoption of better harvesting and storage techniques. Until recently, very little of this effort has been directed to energy applications. A few examples of energy applications of grasses can be found such as the combustion of bagasse for steam and electric power but many other opportunities exist that have not yet been developed.

Perennial grasses have been suggested as raw materials for conversion to synfuels (47). Most perennial grasses can be grown vegetatively, and reestablish themselves rapidly after harvesting. More than one harvest can usually be obtained per year. The warm-season grasses are preferred over the cool-season grasses because their growth increases rather than declines as the temperature rises to its maximum in the summer months. In certain areas, rainfall is adequate to permit harvesting every 3 to 4 weeks from late February into November, and yields of ca 18–24 t/(hm^3·yr) [8–11 short ton/(acre·yr)] of dry equivalent grasses are believed to be obtainable in managed grasslands. Table 23 (47) lists warm-season grasses that have been proposed for the manufacture of synfuels.

Experimental work has shown that cool-season grasses such as Kentucky blue grass and warm-season grasses such as coastal Bermuda grass (*Cynodon dactylon*) can be converted to methane by conventional anaerobic digestion techniques (49–50). Certain nutrients can be incorporated into some harvested grass to facilitate their conversion by biological means without the addition of excessive chemicals (50).

As already mentioned, sugarcane is used commercially as a combination food and fuel crop, and it might be used as a model for other biomass energy systems. It grows rapidly, produces high yields, the fibrous bagasse is used as boiler fuel, and cane-derived ethanol is used as a motor fuel in gasoline blends (gasohol). Sugarcane plantations and the associated sugar processing and ethanol plants are actually biomass

Table 23. Promising Warm-Season Grass Species for Methane Plantations [a]

Species	Localities [b]	Annual yields [c]	Comments
perennial sorghums and their hybrids	plains, south	high	Sudan grasses, Johnson grasses and other warm-season hybrids are promising for localities with alkaline soils; several harvests per year
Bermuda grasses coastal midland Suwanne	south and south central states	high	most promising of all warm-season grasses, especially for localities with acid soils; can be harvested several times per year
related to sugarcane	Louisiana and Florida	very high	limited suitable sites
related to bamboo	south central United States	untested	
Bahia grasses	Florida and southern coastal plains	high	competes with Bermuda grasses when fertilized; effect on overall yield is in dispute

[a] Ref. 47.

[b] Regions in which species grow naturally, or have been successfully introduced, or have been extensively tested.

[c] High is in the range of 18–22 dry t/(hm^2·yr) [8–10 short ton/(acre·yr)]; very high may be as much as 45 dry t/(hm^2·yr) [20 short ton/(acre·yr)] in specially suitable sites.

fuel farms. About one-half of the organic material in sugarcane is sugar, the rest is fiber. Dry cane yields per year have been reported to range as high as 80–85 t/hm^2 (36–38 short ton/acre). Normal cultivation of sugarcane provides dry annual sugarcane yields of ca 50–59 t/hm^2 (22–26 short ton/acre) (7) (see Sugar).

Other productive grasses that have been given serious consideration as raw materials for the production of energy and synfuels include sorghum and their hybrids. Tropical grasses are very productive and normally give an annual yield of 50–60 t of organic matter per hm^2 on good sites (7). The tropical fodder grass *Digitaria decumbens* has been grown at yields of organic matter as high as 85 t/(hm^2·yr) [38 short ton/(acre·yr)] (7).

It is apparent that there are many grasses and related plants that can be considered for energy applications. They seem to have many of the desirable characteristics needed for land-based biomass energy systems.

Other Cultivated Crops. Other high yielding land biomass species have also been proposed as renewable energy sources (51), including kenaf (*Hibiscus cannabinus*), an annual plant reproducing by seed only, sunflower (*Helianthus annuus* L.), which is an annual oil seed crop grown in several parts of North America, and a few others such as the polyisoprene-containing plant species described previously. Kenaf is highly fibrous and exhibits rapid growth, high yields, and high cellulose content. It is a potential pulp crop and is several times more productive than the traditional pulpwood trees. Maximum economic growth usually occurs in less than 6 months and consequently two croppings may be possible in certain regions of the United States. Without irrigation, heights of 4–5 m are average in Florida and Louisiana but 6-m plants have been observed under near optimum growing conditions. Yields as high as 45 t/(hm^2·yr) [20 short ton/(acre·yr)] have been obtained (see Fibers, vegetable). The sunflower is a prime candidate for biomass energy applications too because of its rapid growth, wide adaptability, drought tolerance, short growing season, massive vegetative production, and adaptability to root harvesting. Dry yields have been projected to be as high as 34 t/hm^2 (15 short ton/acre) per growing season.

Water-Based Biomass. The average net annual productivities of dry organic matter on good growth sites for some land- and water-based biomass are shown in Table 24 (7). With the exception of phytoplankton, aquatic biomass seems to exhibit higher net organic yields than land biomass. Water-based biomass considered to be the most suitable for energy applications include the unicellular and multicellular algae and water plants.

Algae. Unicellular algae such as the species *Chlorella* and *Scenedesmus* have been produced by continuous processes in outdoor light at high photosynthesis efficiencies. *Chlorella*, for example, has been reported to be produced at rates as high as 1.10 t/(hm^2·d) [0.5 short ton/(acre·d)] (dry) (52); this corresponds to an annual rate of 401 t/(hm^2·yr) [179 short tons/(acre·yr)]. These figures are probably in error but there is no reason why very high yields cannot be achieved because the process of producing algae can be almost totally controlled. Algae are also produced as slurries in lakes and ponds where the depth of the biomass-producing area as well as plant yield per unit volume of water are important parameters. The nutrients for algae production can be supplied by sewage and other wastewaters. Most unicellular algae are grown in fresh water which tends to limit their energy applications to small-scale algae farms. The high water content of unicellular algae also tends to limit the conversion processes to biological processes.

Table 24. Average Net Annual Biomass Yields on Fertile Sites[a]

Dry organics, t/(hm²·yr)	Climate	Ecosystem type	Remarks
1	arid	desert	better yield if hot and irrigated
2		ocean phytoplankton	
2	temperate	lake phytoplankton	little influence by man
3		coastal phytoplankton	probably higher in some polluted estuaries
6	temperate	polluted lake phytoplankton	in agricultural and sewage runoffs
6	temperate	freshwater submerged macrophytes	
12	temperate	deciduous forest	
17	tropical	freshwater submerged macrophytes	
20	temperate	terrestrial herbs	possibly higher yields if grazed
22	temperate	agriculture, annuals	
28	temperate	coniferous forests	
29	temperate	marine submerged macrophytes	
30	temperate	agriculture, perennials	
30		salt marsh	
30	tropical	agriculture, annuals	including perennials in continental climates
35	tropical	marine submerged macrophytes	including coral reefs
38	temperate	reedswamp	
40	subtropical	cultivated algae	better yield if CO_2 supplied
50	tropical	rain forest	
75	tropical	agriculture, perennials, reedswamp	

[a] Ref. 7.

Macroscopic multicellular algae, or seaweeds, are under serious consideration as renewable energy sources, including the giant brown kelp *Macrocystis pyrifera* (53–55), the red algae (*Rhodophyta*), and the floating algae *Sargassum*. Giant brown kelp has been studied in detail and harvested commercially off the California coast for many years (38). Because of its high potassium content, giant brown kelp was a commercial potash source during World War I. At present, organic gums and thickening agents and alginic acid derivatives are manufactured from it (see Gums). Off the east coast of the United States, *Laminaria* seaweed is harvested for the manufacture of alginic acid derivatives. In tropical seas not cooled by upwelled water, species of the *Sargassum* variety of algae may be suitable as renewable energy sources. Several species of *Sargassum* grow naturally at the reefs surrounding the Hawaiian Islands. The very high water content of macroscopic algae suggests that biological conversion processes rather than thermochemical conversion processes should be used for synfuel manufacture. The manufacture of coproducts from macroscopic algae, such as polysaccharide derivatives, along with synfuel might make it feasible to use thermochemical processing techniques on intermediate process streams.

Water Plants. The productivity of some salt marshes is similar to that of seaweeds. *Spartina alterniflora* has been grown as net annual dry yields of about 33 t/(hm²·yr) [14.7 short tons/(acre·yr)] including underground material on optimum sites (7). Other emergent communities in brackish water, including mangrove swamps, appear to have annual organic productivities of up to 35 t/(hm^2·yr) [156 short tons/(acre·yr)] (7) but insufficient information is available to judge their value in biomass energy systems. Fresh water swamps are believed to be highly productive and offer opportunities for energy production. Both the reed *Arundo Donax,* and bulrush, *Scirpus Lacustris,* appear to produce 57–59 t/(hm^2·yr) [25.4–26.3 short tons/(acre·yr)] (7).

A very strong candidate for energy applications is the water hyacinth (*Eichhornia crassipes*). This species of aquatic biomass is highly productive as might be expected from the fact that it grows in warm climates and has submerged roots and aerial leaves like reedswamp plants. It has been estimated that water hyacinth could be produced at rates up to about 150 t/(hm^2·yr) [67 short tons/(acre·yr)] if the plant were grown in a good climate, the young plants always predominated, and the water surface was always completely covered (7,56–57). Currently, water hyacinth has no competitive uses and is considered to be an undesirable species on inland waterways; it is a very hardy, disease-resistant species (58).

Systems Analysis

The overall design of a biomass-to-synfuel system is very important to its successful operation because the system will be large and require integration of many different operations such as planting, growing, harvesting, transporting, and converting biomass to gaseous and liquid synfuels. The detailed design depends on the type, size, number, and location of the biomass growth and processing areas. In the ideal case, the synfuel production plants are located in or near the biomass growth areas to minimize the cost of transporting the harvested biomass to the plant. The nonfuel effluents are recycled to the growth areas, as shown in Figure 18. If such a synfuel plantation can be developed, it would be equivalent to an independent system with inputs of solar radiation, air, carbon dioxide, and minimal water, and one output, synfuel. The nutrients are kept within the ideal system eliminating addition of external fertilizers and chemicals. Environmental and disposal problems are minimized.

Various modifications of the idealized design in Figure 18 can be conceived for large-scale usage; eg, addition of wastewater influent into the biomass growth area and the growth of water hyacinth for the treatment of wastewater by luxuriant uptake of nutrients and for conversion to synfuels. In this case, inorganic material would build up in the biomass growth area and residual material from the conversion plant could

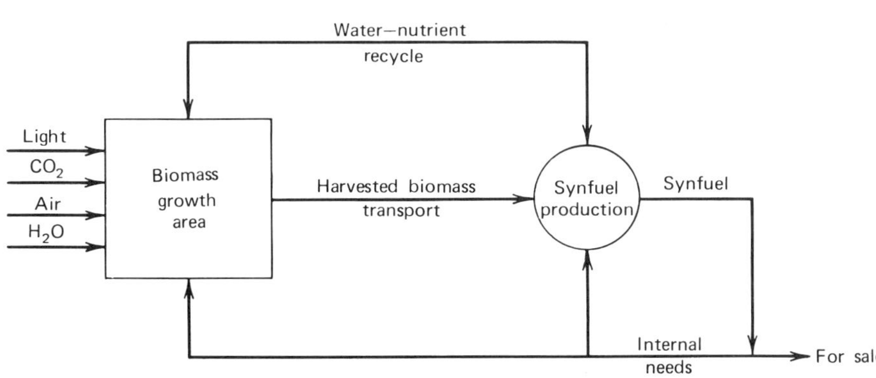

Figure 18. Idealized synfuel plantation.

be partially removed or bled from the system as the synfuel is produced. This product might be considered to be a coproduct along with the synfuel.

Marine biomass plantations such as those envisaged for giant brown kelp and the production of methane might be constructed after a design similar to that shown in Figure 18. The SNG plant could be located on a floating platform near the kelp growth area or on shore.

Economics. The practical value of a biomass synfuel plantation depends on the cost of the salable energy products and many economic analyses have been performed on biomass conversion processes, production, and integrated biomass–synfuel systems (59).

In a recent study (60), the synfuel conversion plant was assumed to be located in the center of the biomass-producing area, and the product was delivered to the plant gate. The transmission or transport costs of the products were not included in the economic calculations. The analyses were based on regulated utility financing and calculated to yield a 15% discounted cash flow return on equity and a 9% return on debt with a debt-to-equity ratio of 65:35. Income tax was assumed to be 52% (see Table 25).

With few exceptions, comparison of the projected synfuel costs in 1985 (as 1977 dollars) in Table 25 with the projected costs of conventional fuels on the same basis (Table 26) shows that synfuels from biomass are more expensive than conventional fuels. Furthermore, a few of the systems listed in Table 25 that do not appear to be economically feasible have already been commercialized or are about to be (SNG from manure, electric power from wood, alcohol from sugarcane). Thus a very careful study of comparative economic analyses is essential and continual improvement and revisions must be made. As the price of crude oil and natural gas continues to increase, it is expected that biomass energy products will become more competitive with fossil fuels.

Energetics. The net energy production efficiency of an integrated biomass energy system is extremely important to its development since the goal is to design and operate environmentally acceptable systems to produce new supplies of salable energy from biomass at the lowest possible cost and with the minimum consumption of energy inputs. It is necessary to quantify how much energy is expended and how much salable energy is produced in each fully integrated system. An energy budget similar to an economic budget should be prepared because the capital, operating, and salable energy cost projections and the conversion process efficiency alone are insufficient to choose and design the best systems. These values do not necessarily correlate with net energy production. Furthermore, the capital energy investment consumed during construction of the system should be recovered during its operation. Comparative analyses of similar systems for production of synthetic liquid and gaseous fuels from the same feedstock or of different systems that yield the same fuels from different biomass should be performed by consideration of the economics and the net energetics.

The following method of analyzing the net energetics of the biomass energy system can be used: let E_f, E_x, and E_p represent the energy content of the dry biomass feed, the sum of the external nonsolar energy inputs into the total system, and the energy content of the salable fuel products, respectively. The system can be represented by:

$$\underset{E_f}{\text{biomass}} \xrightarrow{\;E_x\downarrow\;} \underset{E_p}{\text{salable fuel}}$$

Table 25. Summary of Comparative Economic Analyses, 1977 Dollars in 1985[a]

Feed	Feed cost and amount: \$/dry t	Feed cost and amount: \$/GJ	Feed cost and amount: dry t/d	Energy products: Type	Energy products: GJ × 10^3/d	Energy products: Units/d	Conversion process: Type[b]	Conversion process: Efficiency, %[c]	Plant cost: Capital, \$ 10^6	Plant cost: Operating, \$ 10^6/yr[d]	Product price, \$/GJ[e]: With feed cost	Product price, \$/GJ[e]: Ex feed cost	Product price, \$/GJ[e]: Feed cost as % of product cost
manure	5.46	0.35	408	intermediate-thermal-value gas	2.38	ca 0.127 hm^3	AD	33.5[f]	10.8	1.87	4.62	3.67	21
manure	5.46	0.35	408	SNG	2.15	0.058 hm^3	AD	30.2[f]	13.6	2.6	6.65	5.67	15
manure	5.46	0.35	41	SNG	0.215	0.0058 hm^3	AD	30.2[f]	2.4	0.6	13.69	12.70	7
wood	21.16	1.06	2722	SNG	38.0	1.082 hm^3	OBG	63[f]	238.5	25.2	6.08	4.55	25
kelp	110.23[g]	6.59	5443[g]	SNG	35.4	0.951 hm^3	AD	35[f]	115.7	16.0	19.62	2.65	86
wheat straw	27.56	1.76	2722	gas[h]	8.9	0.476 hm^3	AD	34[f]	65.7	27.4	20.92	9.38	55
wheat straw	27.56	1.76	454	gas[h]	1.5	0.0793 hm^3	AD	34[f]	12.6	5.1	22.52	10.52	53
wood	21.16	1.06	2722	methanol	35.0	2300 m^3	OBG	58	268.7	29.4	7.36	5.70	23
algae	82.67[g]	4.95	1021[g]	sugar[i]		454 t	H		27.9	5.2	0.24[j]		
				ethanol	6.0	288 m^3	F	32	28.0	7.0	25.50	10.43	59
sugar cane	71.65	4.57	2500	sugar		454 t		45	49.2	6.8			
				ethanol	5.8	284 m^3	F	90	21.0	5.0	30.52	10.11	67
wheat straw	16.53	1.06	2966	ethanol	5.8	287 m^3	H, F	11	127.8	62.6	49.86	41.22	17
wood	21.16	1.06	2722	fuel oil	32.3	837 m^3	CL	53	144.9	15.5	5.09	3.29	35
wood	21.16	1.06	2722	fuel oil	19.9	590 m^3	P	33.1	61.4	10.3	4.27	1.23	71
				char	24.3	41.9 t	P	40.5			1.18		
wood	21.16	1.06	2722	steam	46.8	15,600 t[k]	DC	77	94.1	11.8	2.84	1.59	44
wood	21.16	1.06	2722	electricity	12.9	150 MW	DC	21	165.6	19.3	15.53	11.01	29
wood	21.16	1.06	2722	ammonia	29.7	1399 t	OBG	49	267.3	20.6	181[l]	139[l]	23

[a] Adapted from ref. 60; To convert GJ to 10^6 Btu, divide by 1.054.

[b] AD, anaerobic digestion; OBG, oxygen-blown gasifier; H, hydrolysis; F, alcoholic fermentation; CL, catalytic liquefaction; P, pyrolysis; DC, direct combustion.

[c] 100 (Product energy content ÷ Feed energy content).

[d] Excludes feedstock cost and depreciation.

[e] Conversion plant in center of feed-producing area, product at plant gate, calculated to yield 15% discounted cash flow return on equity and 9% return on debt (65% debt-35% equity), 52% income tax, 1977 dollars in 1985, all figures on regulated utility basis.

[f] These values are too low and should be at about 50% if the best techniques are employed.

[g] Ash free.

[h] 18.6 MJ/m^3 (500 Btu/ft^3).

[i] 50% of sugar converted to ethanol.

[j] Cost per kg.

[k] 3100 kPa (30.6 atm), 432°C.

[l] Cost per ton, corresponding figure for nonregulated utility basis is \$274 per ton including feed cost,

Table 26. Midpoint of Fuel Market Price Range in 1985 (1977 Dollars)[a]

Fuel	\$/GJ[b]
electric power (av)	9.29
gasoline	3.60
coal-derived steam	3.51
ammonia	3.13
natural gas	2.84
low-S residual oil	2.56
low-thermal gas	2.56
intermediate-thermal gas	2.56
crude oil	2.46
coal	1.33

[a] Ref. 60.
[b] To convert to $\$/10^6$ Btu, multiply by 1.054.

The ratio, $(E_p - E_x)/E_x$, termed the net energy production ratio, indicates how much more (or less) salable fuel energy is produced than that consumed in the integrated system if the external energy consumed is replaced and it is assumed that the biomass feed energy is zero. This is a reasonable assumption because the energy value of biomass is derived essentially 100% from solar radiation. Net energy production ratios greater than zero indicate that an amount of energy equivalent to the sum of the external nonsolar energy inputs and an additional energy increment of salable fuel are produced; the larger the ratio, the larger the increment. The ratio $100\ E_p/(E_f + E_x)$ is the overall fuel production efficiency of the system or simply the net energy output divided by the gross energy input. Finally, another useful ratio is the value of net energy output divided by the gross energy output, $100\ (E_p - E_x)/E_p$. This ratio expresses the percentage of the energy output which is new energy added to the economy.

Examples of simple models for biomass energy production excluding conversion to synfuels are illustrated in Tables 27 and 28. Table 27 presents an analysis of a short-rotation tree plantation biomass production (61). The biggest energy consumer is fertilization; the energy cost of ammonia production and biomass transportation is included. However, net energy production efficiency is high and about fourteen times more energy is produced in the form of wood biomass than the external nonsolar energy inputs needed to operate the system.

Similarly, the results for a mixed biomass plantation, including species of short-rotation hardwoods, sunflower, and kenaf, is shown in Table 28 (62). Again, fertilization consumes the largest amount of external energy, followed by the energy cost for irrigation. The net energy production efficiency is high; about twenty times more energy is produced as wood biomass energy than the external nonsolar energy inputs.

These systems analyses suggest that biomass plantations can be designed to operate at high net energy efficiencies, and that improved fertilization, such as nitrogen fixation or the recycling of nutrients, may offer additional benefits.

For the total integrated biomass production–conversion system, the arithmetic product of the efficiencies of biomass production and conversion is the efficiency of the overall system. An overall conversion efficiency near 45% would thus be produced by integrating the biomass plantation illustrated in Table 27 or 28 with a conversion process that operated at an overall efficiency of 50%. Every operation in the series is thus equally important.

Table 27. Net Energy Analysis of Short-Rotation Wood Biomass Production[a,b]

Operation	Energy consumed per metric ton dry wood produced, MJ/t[c]	Total consumption, %
cultivation and planting	9.3	0.8
fertilization	604	52.1
harvesting	87	7.5
transport to trucks	115	9.9
load trucks	55	4.7
80.5 km transport to user	221	19.0
unload	nil	
auxiliary	69	5.9
Total	*1160*	*99.9*
biomass energy produced, GJ/(hm²·yr)[d]	196[e]	
energy production efficiency, %	93.4[f]	
net energy production ratio	14.1[g]	

[a] Adapted from ref. 61.
[b] Yield: 11.2 t/(hm²·yr) [5 short ton/(acre·yr)] dry; planting cycle: 20 yr; fertilization: 224 kg N/(hm²·yr) (200 lbs N/(acre·yr)).
[c] To convert MJ/t to Btu/1000 lb, multiply by 430.
[d] To convert GJ to 10^6 Btu, divide by 1.054.
[e] Assumes heating value of biomass is 17.5 MJ/dry kg.
[f] Net energy output ÷ gross energy output × 100, 100 $(E_p - E_x)E_p$.
[g] Net energy output ÷ energy consumed, $E_p - E_x/E_x$.

These simplified treatments correspond to net energy analyses using the first law of thermodynamics. Some energy analysts feel that only an analysis based on the second law can provide the ultimate answers in terms of where more available energy, in the thermodynamic sense, can be found to permit efficiency maximization. Others believe that the conventional energy balance is optimal because it is more realistic and easier to use. Indeed, for integrated synfuel production systems, entropic losses may not always be definable for all segments of the system, and a rigorous second-law analysis may not be possible.

The location of the system boundaries is of paramount importance in the net energy analysis of integrated biomass energy systems. Thus, tractors may be used to plant and harvest biomass. The fuel requirements of the tractors are certainly part of E_x but is the energy expended in manufacturing the tractors also part of E_x? Some analysts believe that a complete study should trace all materials of construction and fossil fuels used back to their origin.

Research, Development, and Commercialization

Biomass energy research projects expanded rapidly in the 1970s (4). The federal program, Fuels From Biomass, sponsored by the U.S. Department of Energy, has grown from $600,000 in fiscal 1975 to 26.9 million dollars in fiscal 1979, and is projected to be 118.9 million dollars in fiscal 1981. In July, 1978, Canada's Minister for Energy, Mines, and Resources announced a 380 million dollar 5-year program to develop renewable energy resources, a large portion of which has been allocated to biomass energy development. Projects include wood gasification and combustion systems, the conversion of agricultural biomass to energy products, the use of methanol fuels for diesel-driven equipment, and the development of equipment for chipping, hogging, and

Table 28. Net Energy Analysis of Mixed Biomass Production[a,b]

Operation	Energy consumed per metric ton dry biomass produced, MJ/t[c]	Total consumption, %
planting and fertilization	584	68.8
harvesting	32	3.8
fresh hauling on plantation	53	6.2
solar drying	1	0.1
dry hauling on plantation	40	4.7
irrigation	138	16.3
Total	*848*	*100.0*
biomass energy produced, GJ/(hm²·yr)[d]	1176[e]	
energy production efficiency, %	95.1[f]	
net energy production ratio	19.6[g]	

[a] Adapted from ref. 62.
[b] Yield: 67.3 t/(hm²·yr) (30 short t/acre·yr) dry; species: annuals and perennials including short rotation hardwoods; fertilization: 672 kg N/(hm²·yr) (600 lb N/(acre·yr)).
[c] To convert MJ/t to Btu/1000 lb, multiply by 430.
[d] To convert GJ to 10^6 Btu, divide by 1.054.
[e] Assumes heating value of biomass is 17.5 MJ/dry kg.
[f] Net energy output ÷ gross energy output × 100, 100 $(E_p - E_x)E_p$.
[g] Net energy output ÷ energy consumed, $E_p - E_x/E_x$.

crushing forest biomass. It appears that Canada is pursuing biomass energy at an even higher rate than the United States. Brazil is producing ethanol from sugarcane for use in gasohol and is conducting several research programs on biomass energy. Many other countries are developing biomass energy in the laboratory and in the field. It would appear that the energy market is beginning to close the cycle and return to its original position in which nonfossil renewable fuels were dominant.

Biomass Production. ***Forests.*** Improved tree growth via short-rotation forestry is under intensive study and investigation by major timber producers. Advanced designs of tree harvesters, logging residue collection and chipping units, and rapid planting machinery are being evaluated.

Chemical injections into pine trees have been reported to stimulate natural production of resins and terpenes. Combined oleoresin–timber production in mixed stands of pine and timber trees shows promise of increased timber and energy product yields.

Hybridizing techniques seem to be leading to super trees that have short growth cycles and yield larger quantities of wood. Fast-growing clones of hybrid poplars are being developed for energy farms in which the trees are ready for harvest in as little as 10 years and yield up to 30 m³/(hm²·yr) [429 ft³/(acre·yr)]. Genetic and environmental manipulation has also led to valuable techniques for the fast growth of saplings in artificial light and controlled atmospheres, humidity, and nutrition.

Test plots aimed at energy applications in the midwestern, northwestern, and southeastern United States are underway with improved species of *Populus* and jack pine to obtain optimum spacings, rotations, and culture conditions. Short-rotation intensive culture has already demonstrated improved yields.

In the Amazon jungle of Brazil, perhaps the largest tree plantation in the world is being integrated with energy, pulp, and chemical production facilities. Fast-growing Caribbean pine and Gmelina trees are used.

Grasses and Cultivated Plants. Land-based plant biomass production techniques are being optimized for warm season grasses, sugarcane, sweet sorghum, and other species. Higher yields and lower costs are being developed through closer plant spacing, improved fertilization and irrigation methods, and more frequent harvesting. Species improvements via genetic manipulation and nitrogen fixation (qv) are being investigated.

The feasibility of producing high yields of polyisoprene-containing plants such as the guayule plant is being studied. Full-scale tests are expected to take place in northern Mexico and the southwestern United States in the 1980s if the work currently in progress shows promise. Growth of the jojoba bush on lands irrigated by seawater is being evaluated. This approach presents interesting possibilities for the production of drought-resistant hydrocarbon-containing species that need little or no fertilization in arid climates.

Water-Based Biomass. Work on marine biomass has concentrated primarily on California giant brown kelp. A test farm module containing a submerged polypropylene grid to which kelp plants are attached has been designed for deployment in the deep waters of the Pacific Ocean off the California coast.

Commercial-scale plants for the growth of unicellular algae in special salt-water ponds are under construction in Israel. The algae will be used to manufacture glycerol and an animal feed containing 60% protein.

Several organizations and local governments in Florida, Louisiana, and the western states are planning to treat wastewater by utilizing the growth of water hyacinth in the sewage influent. The water hyacinth will be harvested and anaerobically digested to produce methane or dried in the desert sun and then burned as fuel in a steam–electric plant.

Conversion. ***Physical Processes.*** Fuel pellets fabricated from cellulose are now commercially available. These solid fuels have heating values of 18.6–20.9 MJ/kg (8000–9000 Btu/lb), and are reported to cost about the same as coal with a heating value of 29.1 MJ/kg (12,500 Btu/lb) on an equivalent energy content basis. A thermoplastic binder under development improves mechanical and combustion properties.

A mechanical cud has been developed for processing hydrocarbon-containing biomass. This device squeezes and shears plant leaves, stems, and roots into a pliable mass of fibers bound by the hydrocarbons which can be extracted with solvent. Another device separates the epidermis, rind fibers, and sugar-containing pith of sugarcane. High-fiber cane can easily be processed; the sugar is cleaner, and the rind fibers have a higher value than bagasse.

Anaerobic Digestion. Research on anaerobic digestion for the gasification of high moisture content biomass to form intermediate-thermal-value gases (18.6–29.7 MJ/m^3, 500–800 Btu/SCF) high in methane emphasizes feedstock evaluation, fermentation rates and methane yields, and digestor designs. Certain biomass such as giant brown kelp, water hyacinth, coastal Bermuda grass, and other species have good biodegradabilities, and afford relatively high methane yields under conventional digestion conditions. Species such as peat and woody biomass give lower yields of methane, which can be improved by pretreatment. Overall, conversion efficiencies to methane seem to have leveled around 50% volatile solids destruction efficiencies and 50–60% energy recovery efficiencies as methane.

Alcoholic Fermentation and Alcohol-Derived Fuels. An important pretreatment step has been devised that dissolves cellulose, thereby eliminating hydrolytic problems caused by lignocellulose complexes and crystallinity. The treated cellulose can be converted to glucose in high yields (90+%). This technique practiced on woody biomass opens routes to many high efficiency biomass applications such as alcohol production. It appears to have much greater potential than enzyme-catalyzed hydrolysis methods which usually require a feed particle size of the order of micrometers.

One of the most interesting developments in ethanol production is the first commercial plant by Petrobras, the Brazilian oil monopoly, to manufacture ethanol from the cassava plant (manioc, tapioca). The plant, located in Curvelo, Minas Gerais, has a capacity of 60,000 L alcohol/d, and will produce carbon dioxide and stillage by-product. The net energy production ratio of this type of plant including the cassava production facilities has been estimated to be +3.2. However, there is considerable disagreement in the United States as to whether the net energy ratio of grain alcohol plants is positive (see Gasoline).

The Gulf Oil approach to ethanol from biomass is a combination enzyme-catalyzed hydrolysis–alcoholic fermentation process. The hydrolytic enzyme and yeast for fermentation are present in the same vessel and the reactions occur simultaneously. An 8000-L per day pilot plant is under construction in Kansas. Battelle is developing the conversion of biomass to ethanol and utilization of the residue from the fermentors to manufacture hydrogen in a photoelectrolysis plant.

Mobil Oil's gasoline-from-methanol process has the potential of eliminating many of the problems encountered in gasohol fuel production. Methanol is converted directly into high octane gasoline over a synthetic zeolite catalyst at a cost of 1.3–2.6¢/L (5–10¢/gal) of gasoline produced and no diesel fuel or residue of any kind is formed. Higher alcohols such as ethanol can also be converted to gasoline by this process. The key factor in the commercial use of this process is the cost of the feed alcohol; present prices preclude commercialization.

Thermal Processes. *Combustion.* Many wood combustion facilities are in commercial use in the United States, Canada, and elsewhere for the production of heat, steam, and the cogeneration of steam and electric power. Plans have been announced by a number of companies including public utilities to build new wood-powered generation plants in the Midwest, on the West Coast, and in New England. Burlington, Vermont, eg, is proceeding with a 50-MW plant scheduled for completion in 1982. A larger 100-MW wood-powered steam–electric plant in Maine is in the design stage under sponsorship of the U.S. Department of Energy. At the present time, major users of wood and other biomass combustion systems for power generation are the sugar, pulp and paper, and wood products industries. Small-scale units operated by home owners, farmers, and small businesses are receiving renewed interest.

Steam-electric production from agricultural and forest biomass such as bagasse and hogged wood is thus established commercial technology in the United States. Many of the existing plants are designed to burn fossil and nonfossil fuels either simultaneously or separately. According to a recent survey (63) of installations that generate their own power, pulp and paper mills burn 46% wood-derived fuels and 54% fossil fuels, whereas forest-product mills burn 98% wood fuels and 2% fossil fuels.

Pyrolysis and Other Thermochemical Processes. Thermochemical gasification is being developed by many research groups. These systems include partial oxidation, steam reforming, steam-oxygen reforming, hydrogasification, and pyrolysis reactions. Most of this work is aimed at maximizing the production of synthesis gas. A few systems studied on a laboratory scale have produced significant quantities of olefins which have a higher heating value than synthesis gas. Some reports indicate that thermochemical gasification methods that convert the moist feed without drying operate at higher efficiencies, as might be expected, and provide higher synthesis-gas yields. Alkaline salts, lime, and wood ash mixed with the biomass feed have been reported to be effective catalysts. Most thermochemical gasification systems also yield char and liquid products, depending on reaction conditions and feed characteristics.

Fluidized-bed gasifiers, multiple-hearth gasifiers for high-moisture-content biomass, hydrogasification units, and small-scale producer gasifiers are being evaluated. Small-scale producer gasifiers for biomass delivering up to 10.5 GJ/h (10×10^6 Btu/h) of low-thermal-value gas to gas-fired boilers are under test and several manufacturers offer small commercial units. In European countries, producer gasifier designs developed during World War II for wood are being modernized for possible stationary and mobile applications (see Fuels, synthetic—gaseous fuels).

Thermochemical liquefaction is much more limited in scope than gasification. Currently, only two direct liquefaction processes are under development in the United States: flash pyrolysis (Occidental Research Corp.) and the Perc process in which carbon monoxide reacts with biomass slurries in an aqueous carbonate solution. Both processes yield heavy fuel oils. No hydroliquefaction research seems to be in progress.

Large-scale thermal processing plants to manufacture low-thermal-value gas and liquid fuels from nonfossil feeds do not seem to be faring well in the United States. Start-up and technical problems have delayed operation of the plants under development. In Europe, two plants have already been built using the Andco-Torrax process, and two are under construction, but none has been announced for the United States.

Analysis of Large Areas. There have been many regional studies of biomass energy potential. One was a comprehensive analysis of biomass energy possibilities for the state of Hawaii. Hawaii is ideal for such a study because there are no indigenous petroleum resources. In 1977 about 92% of Hawaii's energy came from imported petroleum, 7% from burning bagasse, and 1% from hydroelectric facilities. The study recommended improved usage of bagasse, energy recovery from municipal solid wastes (MSW), establishment of short-rotation eucalyptus energy farms and wood-powered electric plants, and marine energy farms for the kelp *Sargassum,* five species of which are native to Hawaii.

Prospects. Available fossil fuels are still sufficiently low in cost to make the economics of producing substitute fuels from biomass unattractive if the system is used only as an energy resource. Large-scale integrated biomass synfuel systems are therefore not expected to be in operation until the late 1980s and early 1990s.

The large research effort on biomass is expected to lead to slow evolution of commercial synfuel processes and specialty and heavy organic chemicals. Small-scale systems for the individual farmer are expected to supply the farm's energy requirements. These systems will be supplied with captive sources of biomass and waste material generated on the farm. It is also likely that large building complexes such as schools and apartments will be able to incorporate similar systems using captive wastes and delivered biomass in urban areas.

Based on available technology and the research and development work in progress, biomass energy applications are expected to grow and contribute increasingly to satisfy energy demand.

BIBLIOGRAPHY

1. D. L. Klass, *Chemtech* **4,** 161 (1974).
2. R. H. Whittaker and G. E. Likens, "The Biosphere and Man" in H. Lieth and R. H. Whittaker, eds., *Primary Productivity of the Biosphere,* Springer Verlag, New York, 1975.
3. B. Bolin, "The Carbon Cycle" in *The Biosphere,* W. H. Freeman and Co., San Francisco, Calif., 1970.
4. D. L. Klass, "Energy From Biomass and Wastes—1978 Update" in *Symposium Papers Energy From Biomass and Wastes, Sponsored by Institute of Gas Technology, Washington, D.C., Aug. 14–18, 1978,* pp. 1–27.
5. E. T. Hayes, *Science* **203,** 233 (1979).
6. *Statistical Abstracts of the United States,* U.S. Department of Commerce, U.S. Government Printing Office, Washington, D.C., 1978.
7. D. F. Westlake, *Biol. Rev.* **38,** 385 (1963).
8. J. S. Burlew, ed., *Algae Culture From Laboratory to Pilot Plant, Publication 600,* Carnegie Institute of Washington, Washington, D.C., 1953, pp. 55–62.
9. C. D. Hodgman, ed., *Handbook of Chemistry and Physics,* 31st ed., Chemical Rubber Publishing Co., Cleveland, Ohio, 1949, p. 1537.
10. *Guayule: An Alternative Source of Natural Rubber,* National Academy of Sciences, Washington, D.C., 1977.
11. R. A. Buchannan and F. O. Otey, "Multi-Use Oil- and Hydrocarbon-Producing Crops in Adaptive Systems for Food, Material, and Energy Production," *paper presented at 19th Annual Meeting, Society for Economic Botany, St. Louis, Mo., June 11–14, 1978.*
12. M. Calvin, *Chemtech* **7,** 352 (1977).
13. P. B. Weisz, W. O. Haag, and P. G. Rodewald, *Science* **206,** 57 (1979).
14. D. A. Hoffman and R. A. Fitz, *Environ. Sci. Technol.* **2,** 1023 (1968).
15. D. L. Klass and S. Ghosh, *Chemtech* **3,** 689 (1973).
16. S. B. Alpert and co-workers, *Pyrolysis of Solid Wastes: A Technical and Economic Assessment, NTIS PB 218-231,* SRI, Menlo Park, Calif., Sept., 1972.
17. G. T. Preston in *Symposium Papers Clean Fuels From Biomass, Sewage, Urban Refuse, Agricultural Wastes, Sponsored by Institute of Gas Technology, Orlando, Fla., Jan. 27–30, 1976,* pp. 89–114.
18. D. L. Klass in ref. 17, pp. 21–58.
19. T. F. Fisher, M. L. Kasbohm, and J. R. Rivero in ref. 17, pp. 447–459.
20. S. J. Yosim and K. M. Barclay, *Preprints of Papers, 171st National Meeting ACS, Div. of Fuel Chem.* **21**(1), 73 (Apr. 5–9, 1976).
21. H. F. Feldman and co-workers, *Hydrocarbon Process.* **55**(11), 201 (1976).
22. H. R. Appell and co-workers, *U.S. Bur. Mines Rep. Invest.* **8013,** (1975).
23. E. R. Riegel, *Industrial Chemistry,* 2nd ed., The Chemical Catalog Co., New York, 1933, Chapt. 16, p. 257.
24. M. R. Ladisch, C. M. Ladisch, and G. T. Tsao, *Science* **201,** 743 (1978).
25. U.S. Pat. 4,009,075 (Feb. 22, 1977), W. H. Hoge (to Bio-Industries, Inc.).
26. R. Katzen Associates, *Chemicals From Wood Wastes,* USDA Forest Products Laboratory, Madison, Wisc., Dec. 14, 1975.
27. R. S. Loomis, W. A. Williams, and A. E. Hall, *Ann. Rev. Plant Physiol.* **22,** 431 (1971); E. I. Rabinowitch, *Photosynthesis,* Vols. 1–2, Interscience, New York, 1956.
28. C. B. Osmond, *Ann. Rev. Plant Physiol.* **29,** 379 (1978).
29. B. J. Brinkworth, *Solar Energy for Man,* John Wiley & Sons, Inc., New York, 1973.
30. U.S. Dept. of Commerce, *Climatological Data, National Summary,* Vol. 21, Nos. 1–12, U.S. Government Printing Office, Washington, D.C., 1970.

31. H. J. Critchfield, *General Climatology,* 3rd ed., Prentice-Hall, Inc., Englewood Cliffs, N.J., 1974, p. 22.
32. R. S. Loomis and W. A. Williams, *Crop Sci.* **3,** 67 (1963).
33. T. R. Schneider, *Energy Convers.* **13,** 77 (1973).
34. S. S. Visher, *Climatic Atlas of the United States,* Harvard University Press, Cambridge, Mass., 1954.
35. *Statistical Abstracts of the United States,* U.S. Department of Commerce, U.S. Government Printing Office, Washington, D.C., 1976.
36. W. L. Roller and co-workers, *Grown Organic Matter as a Fuel Raw Material Source, NASA Report CR-2608,* Ohio Agricultural Research and Development Center, Washington, D.C., Oct. 1975.
37. M. M. Ludlow and G. L. Wilson, *J. Aust. Inst. Agric. Sci.* **36,** 43 (Mar. 1970).
38. W. J. North, ed., *The Biology of Giant Kelp Beds (Macrocystis) in California,* Cramer, Lehre, FRG, 1971, p. 12.
39. M. El-Sharkawy and J. D. Hesketh, *Crop Sci.* **4,** 514 (1964).
40. J. Krummel in ref. 17, pp. 359–370.
41. H. J. Evans and L. E. Barber, *Science* **197,** 332 (1977).
42. *Land Capability Classification, Agricultural Handbook 210,* Soil Conservation Service, U.S. Department of Agriculture, 1966, 21 pp.
43. R. I. Dideriksen, A. R. Hidlebaugh, and K. O. Schmude, *Potential Cropland Study, Statistical Bulletin No. 578,* Soil Conservation Service, U.S. Department of Agriculture, 1977.
44. S. H. Spurr, *Sci. Am.* **240,** 76 (1979).
45. A. A. Nichiporovich in *Photosynthesis of Productive Systems,* Israel Program for Scientific Translations, Jerusalem, Israel, 1967.
46. E.O. Mariani in ref. 4, pp. 29–38.
47. *Solar SNG, Final Report American Gas Association Project IU-114-1,* Prepared by InterTechnology Corp., American Gas Association, Washington, D.C., Oct. 1975.
48. *Grass The Yearbook of Agriculture 1948,* U.S. Department of Agriculture, U.S. Government Printing Office, Washington, D.C., 1948.
49. D. L. Klass, S. Ghosh, and J. R. Conrad in ref. 17, pp. 229–252.
50. D. L. Klass and S. Ghosh, "Methane Production by Anaerobic Digestion of Bermuda Grass," *paper presented at Symposium Biomass as a Nonfossil Fuel Source, ACS/CSJ Joint Chemical Congress, Honolulu, Hawaii, Apr. 1–6, 1979.*
51. J. A. Alich, Jr., and R. E. Inman, *Effective Utilization of Solar Energy to Produce Clean Fuel, Grant No. GI 38723,* Final Report for National Science Foundation, Stanford Research Institute, Palo Alto, Calif., June, 1974.
52. R. Retovsky, *Continuous Cultivation of Algae, Theoretical and Methodological Bases of Continuous Culture of Microorganisms,* Academic Press, Inc., New York, 1966.
53. D. L. Klass and S. Ghosh in *Symposium Papers Clean Fuels From Biomass and Wastes, Sponsored by Institute of Gas Technology, Orlando, Fla., Jan. 25–28, 1977,* pp. 323–351.
54. D. L. Klass, S. Ghosh, and D. P. Chynoweth, "Methane Production From Aquatic Biomass by Anaerobic Digestion of Giant Brown Kelp," *paper presented at 175th National Meeting, ACS, Anaheim, Calif., Mar. 15, 1978.*
55. D. P. Chynoweth, D. L. Klass, and S. Ghosh in ref. 4, pp. 229–251.
56. M. G. McGarry, *Process Biochem.* **6,** 50 (1971).
57. J. L. Yount and R. A. Grossman, *J. Water Pollut. Control Fed.* **42,** 173 (1970).
58. E. S. Del Fosse in ref. 53, pp. 73–99.
59. P. B. Weisz and J. F. Marshall, *Science* **206,** 24 (1979).
60. F. A. Schooley and co-workers, *Mission Analysis for the Federal Fuels From Biomass Program, Quarterly Progress Report, Contract EY-76-C-03-011PA131,* Stanford Research Institute, Palo Alto, Calif., June 12, 1978.
61. N. Smith and T. J. Corcoran, "The Energy Analysis of Wood Production for Fuel Applications," *Preprints of Papers Presented at 171st National Meeting, ACS, Fuel Chemistry Division, Symposium on Net Energetics of Integrated Synfuel Systems* **21**(2), 9 (Apr. 1976).
62. R. E. Inman, "An Energy Budget for a Hypothetical Biomass Plantation" in ref. 61, pp. 21–27.
63. D. W. Pingrey, "Wood Fueled Electric Generating Plants," *2nd Annual Fuels From Biomass Symposium, Sponsored by U.S. Department of Energy, Rensselear Polytechnic Inst., June 21, 1978, Troy, New York.*

General References

Symposium Papers Proceedings: Clean Fuels From Biomass, Sewage, Urban Refuse, Agricultural Wastes, Jan. 27–30, 1976, Orlando, Fla., Sponsored by Institute of Gas Technology, Chicago, Ill.
Symposium Papers Proceedings: Clean Fuels From Biomass and Wastes, Jan. 25–28, 1977, Orlando, Fla., Sponsored by Institute of Gas Technology, Chicago, Ill.
Symposium Papers Proceedings: Energy From Biomass and Wastes, Aug. 14–18, 1978, Washington, D.C., Sponsored by Institute of Gas Technology, Chicago, Ill.
Proceedings From Executive Conference: Energy Production From Biomass and Wastes, Jan. 21–23, 1979, Orlando, Fl., Sponsored by Institute of Gas Technology, Chicago, Ill.
Bio-Energy Directory, The Bio-Energy Council, Washington, D.C., June 1978.
Bio-Energy Directory, The Bio-Energy Council, Washington, D.C., May 1979.
Proceedings: Fuels From Biomass Symposium, Apr. 18–19, 1977, University of Illinois, Urbana, Ill., Sponsored by Department of Energy.
Proceedings: The Second Annual Symposium on Fuels From Biomass, Vols. I and II, June 20–22, 1978, Troy, N.Y., Sponsored by U.S. Department of Energy.
Proceedings: Forest and Field Fuels Symposium, Oct. 12–13, 1977, Sponsored by The Biomass Energy Institute Inc.; Winnipeg, Can.
Proceedings: 1977 TAPPI Forest Biology Wood Chemistry Conference, June 20–22, 1977, Madison, Wisc., Sponsored by The Technical Association of the Pulp & Paper Industry.
Proceedings: A Conference on Capturing the Sun Through Bioconversion, Mar. 10–12, 1976, Washington, D.C., Coordinated by Washington Center for Metropolitan Studies.
Proceedings: 3rd Annual Biomass Energy Systems Conference Abstracts, SERI/TP-33-285, June 5–7, 1979, Golden, Colorado, Sponsored by U.S. Department of Energy.
Chemical Congress ACS/CSJ Preprints, Biomass As a Non-Fossil Fuel Source, Symposium Sponsored by Div. of Petroleum Chemistry, Inc., ACS **23**(2), 408–503 (Mar. 1979).
Proceedings: Mid-American Biomass Energy Workshop, May 21–23, 1979; West Lafayette, Ind., Sponsored by U.S. DOE, USDA, Mid-American Solar Energy Complex, and Purdue University.

DONALD L. KLASS
Institute of Gas Technology

FUELS FROM WASTE

Wastes from which energy can be recovered are many and diverse; the following types are most important: wood wastes from pulp mills, sawmills, and plywood mills; municipal solid waste (MSW); crop residues such as bagasse, corn cobs, and straw; and animal manure, principally from cattle feedlots (see Wastes, industrial and municipal).

The rates at which such wastes are generated (1–3), and utilization rates have been estimated (4–5). The annual generation and utilization rates for the various waste-based fuels given in Table 1 show that wood wastes are the dominant energy feedstock. MSW is rapidly emerging as a feedstock for fuels and energy production. Thus most of this chapter will focus on those two fuels, treating other waste sources primarily by comparison.

Few, if any, sources of fuel and energy exhibit more heterogeneity, both between and within specific sources, than do organic wastes. Wood waste varies in bulk density, moisture content, particle size, percent ash, and other critical fuel parameters. Moisture content, eg, may vary 8–65%. Municipal waste in any city varies in moisture content as a function of weather, and in combustibles content as a function of season. Crop residues and animal fecal matter present analogous problems of variability. However, many companies are using these wastes for fuel, as Table 1 shows.

Characteristics of Wastes for Fuels

Waste fuels share the following common characteristics: they are solid, hygroscopic, modest in carbon content compared to coal or oil, and they are low in bulk density. Thus they are fuels for which markets are local, and the economics of their use is based upon fossil-fuel replacement rather than on established market pricing mechanisms. The driving characteristic is that the fuel value of each type of waste is determined by its carbon content on a dry weight basis (8–9). Thus the following regression holds with an $r = 0.982$:

$$HHV_D = 0.475\ C - 2.38 \tag{1}$$

where HHV_D = higher heating value (HHV), MJ/kg dry matter; C = % carbon. The English equivalent regression is:

$$HHV_D = 188.0\ C - 131.5 \tag{2}$$

where HHV_D is the heating value in Btu/lb for oven dry wood.

From equation 1 the net heat content of all waste fuels can be estimated by the following equation:

$$NHV_G = HHV_D\ (1 - MC) \tag{3}$$

where NHV_G = net heating value (green) and MC = moisture content, decimal basis.

Wood-Waste Characteristics. Wood wastes exhibit the most potential among wastes for an increase in their energy contribution (8) and are already frequently used. Their present and potential sources include sawmills, plywood mills, logging operations, and unmerchantable stand conversions. The problem of heterogeneity in higher

Table 1. Waste Generation and Utilization[a]

Waste type	Dry waste generated annually, t × 10^6	Dry waste used annually, t × 10^6	Waste used annually, %
primary manufactured wood residue[b]	57	45	79
logging residue	82	negl.	negl.
crop residue[c]	363	2	0.5
feedlot manure	190	negl.	negl.
municipal solid waste	150	6	4.0

[a] Refs. 1, 4–7.
[b] Does not include coarse wood residues used for manufacturing pulp or particleboard.
[c] Mostly bagasse, cotton gin waste, and corn seed burned in one Indiana power plant.

heating value (*HHV*), moisture content (*MC*), net heating value (*NHV*), particle size distribution, and ash content is inherent in these fuels.

Heterogeneity of wood is caused by differences among log size to be processed, wood species, tree age, and forest management regime. Differences in manufacturing residues, or hog fuels, are more severe; they result from the process configuration in the sawmill or plywood mill, and alternative markets for residues. Table 2 gives representative analyses for softwoods and hardwoods.

Among the manufacturing operations generating wood wastes, sawmills are the most important, and their machine configuration significantly influences fuel quality. Logs may be stored on dry paved or unpaved decks, or in fresh or saltwater ponds. Logs may be debarked mechanically or hydraulically. Initial breakdown may be performed in chipping or sawing headrigs. Products may be kiln dried before being planed. Pond storage increases fuel moisture content by 10–15% (11–13); headrigs may increase the quantity of fine fuel produced by ca 30% (14). Kiln drying may change the moisture content of the total fuel produced by 5% (15). Table 3 presents typical values for components of hog fuel.

Residues from forestry operations vary by species, age, and management regime

Table 2. Proximate and Ultimate Analysis of Softwoods and Hardwoods[a]

Analysis	Softwoods, %	Hardwoods, %
proximate		
volatiles	40.6	52.4
fixed carbon	12.4	12.9
ash	1.0	2.5
moisture	46.0	32.2
Total	*100.0*	*100.0*
ultimate (dry basis)	wt %	wt %
H	6.1	6.2
C	52.1	51.0
O	40.0	39.7
N	0.1	0.4
S	0.0	0.0
ash	1.7	2.7
Total	*100.0*	*100.0*

[a] Refs. 9–10.

Table 3. Particle Size and Moisture Ranges for Hog-Fuel Components From a Sawmill[a]

Component	Hog fuel, average %	Particle size, mm	Moisture content, wt %	Ash content, wt %
bark	36.6	0.8–102	25–75	1.0–20.0
sawdust	37.1	0.8–9.5	30–60	0.5–2.0
planer shavings	27.2	0.8–12.7	16–40	0.1–1.0

[a] Refs. 15–16.

but they are less heterogeneous than process residues. Percent fines and moisture content are not influenced by pond storage or processing.

Gross bulk density for wood residues ranges 240–400 kg/m^3 (15–25 lb/ft^3) about a mean of 320 kg/m^3 (20 lb/ft^3). Thus, on a net heating-value basis, cubic volume wood residues are quite modest, eg, 3.3 GJ/m^3 (90×10^3 Btu/ft^3). This low concentration of heat content/unit volume in terms of fuels availability means that wood residues rarely can be transported more than 32 kilometers economically. Thus most are used by firms generating them, and relatively small volumes are sold.

Municipal Waste Characteristics. Municipal solid waste (MSW) varies in composition by city, by season of year, and by weather conditions. The fuel fraction comes primarily from paper, wood debris, and to a lesser extent plastics. Table 4 lists the average composition of MSW by locality showing that paper waste ranges 21.5–53.3% about a mean of 37.4%. Wood, leather, and rubber waste ranges 0.3–2.1% about a mean of 1.2%. There are many proximate and ultimate analyses of MSW resulting from the variations shown in Table 4. A few are presented in Table 5.

The U.S. Bureau of Mines (19) analyzed ten MSW samples. The as-received mean HHV was 11.2 MJ/kg (4.83×10^3 Btu/lb) with a standard deviation (σ) of 0.9 MJ/kg (385 Btu/lb). The range was 9.9–12.2 MJ/kg (4.26–5.26 $\times 10^3$ Btu/lb). Ash content ranged 5.8–15.2% about a mean value of 10.3% ($\sigma = 3.1\%$). Moisture content ranged 12.6–32.0% about a mean of 20.8% ($\sigma = 5.6\%$). Bulk density for MSW fuels, as represented by the St. Louis, Missouri, MSW processing system, average ca 96 kg/m^3 (6 lb/ft^3) (6).

Again the heat content–unit volume data indicate transportation problems for waste-based fuels. However, 62% of the MSW is generated in 150 major metropolitan areas (20). The demand for municipal trash services preconcentrates the materials effectively.

Table 4. Municipal Waste Composition in Selected Cities, Wt %[a]

Waste component	Oceanside, N.Y.	Berkeley, Calif.	Raleigh, N.C.	Flint, Mich.	Tampa, Fla.	Madison, Wisc.
food waste	16.7	12.5	31.8	36.0	9.1	15.3
yard waste	0.3	12.5	8.4	0.3	41.5	13.8
glass	11.9	11.3	11.9	23.2	6.0	10.1
metal	10.6	8.7	9.2	14.5	5.9	6.7
paper	53.3	44.6	38.9	21.1	24.1	42.4
plastics and textiles	2.2	3.0	0.0	0.8	5.2	1.6
wood, leather, and rubber	1.5	0.3	0.0	0.8	2.1	1.1
other	3.5	7.1	0.0	3.3	6.1	9.0
Total	*100.0*	*100.0*	*100.2*	*100.0*	*100.0*	*100.0*

[a] Ref. 17.

Table 5. Representative Proximate and Ultimate Analysis for MSW, Wt %

Analysis	Source[a]	Source[b]
proximate		
volatile matter	43.0	48.6
fixed carbon	6.7	9.0
ash	7.0	25.9
moisture	43.3	16.5
Total	*100*	*100*
ultimate (dry basis)		
H	8.2	3.8
C	27.2	30.9
O	56.8	21.7
N	0.7	0.6
S	0.1	0.2
Cl	0.0	0.3
ash	7.0	25.9

[a] Ref. 18.
[b] Ref. 6.

Characteristics of Agricultural Residues. Agricultural and crop processing residues include bagasse (qv), rice straw, rice hulls, peanut shells, cotton gin trash, corn cobs, grasses, and a host of grain stalks. The range in values for some of these sources is shown in Table 6. Variations in plant material come from type of plant and type of processing. Similar variations in manures are a function of animal type (eg, cattle, hogs, chicken) and age of manure.

For the crop wastes, a bulk density of 40–256 kg/m^3 (2.5–16 lb/ft^3) is reported (22). A bulk density value of 200 kg/m^3 (12.5 lb/ft^3) has been ascribed to bagasse, the pulp remaining after extraction of sugar from cane (7).

Wastes as Fuel

Wastes can be used as fuel directly or upgraded into more concentrated solid, liquid, and gaseous sources of energy. Most are burned directly with little, if any, upgrading; hence the primary consideration must be the use of wastes as fuel.

Waste Fuel Combustion Pathways. The basic combustion equations are stated as:

$$C + O_2 \rightarrow CO_2 \quad \Delta H = 33.9 \text{ MJ/kg } (14.6 \text{ Btu/lb}) \tag{4}$$

$$H_2 + 2\,O_2 \rightarrow 2\,H_2O \quad \Delta H = 142.9 \text{ MJ/kg } (61.5 \text{ Btu/lb}) \tag{5}$$

Table 6. Representative Ultimate Analyses for Agricultural Materials (Dry Basis), Wt %[a]

Analysis	Bagasse	Feedlot manure	Rice hulls	Grass straw
H	6.1	5.7	5.5	6.0
C	47.3	41.2	39.4	45.0
O	35.3	33.3	36.0	41.9
N	0.0	2.3	0.5	0.5
S	0.0	0.3	0.2	0.5
ash	11.3	17.2	18.4	6.1

[a] Refs. 1, 18, 21.

However, the pathways of combustion are critical to the conversion and energy recovery process. The process follows the following sequential steps for any given fuel particle: initial heat up, drying, pyrolysis to volatiles and char, and then combustion. Pyrolysis is the most critical step. Temperatures of >420 K are required for drying; 550 K is a minimum for effective pyrolysis (23). The volatiles produced burn in flaming combustion with rapid rates of heat release. The oxidation of char occurs as glowing combustion, with slower heat release.

Influences on Wood Pyrolysis. Several variables influence the distribution of precombustion or pyrolysis products. The dry fuel material influences the distribution. The components that are higher in energy content (eg, lignin) tend to promote char formation (Table 7). The temperature of the reactor also influences char and volatile formation. Lower temperatures favor orderly degradation to char, and higher temperatures favor gasification reactions and volatiles production (9).

The third variable affecting char and volatile formation is particle size and shape, hence surface-to-mass ratio. Because the thermal conductivity of solid fuels is low, the surface layer of a large particle may undergo initial pyrolytic degradation rapidly while the interior regions of the particle are heating slowly and forming char rather than gaseous products. The expression governing this variable is the thermal conductivity of solid wood [k_t, W/(cm·K)]

$$k_t = \left[(5.18 + 0.096\ MC_a)S_{o,a} + 0.57\left(1 - \frac{S_{o,a}}{1.46}\right)\right] \times 4.184 \times 10^{-4} \tag{6}$$

where MC_a = % moisture at 0–40%; and $S_{o,a}$ = specific gravity, oven-dry volume. For fuel particles with low surface–mass ratios the results are slow heating rates, particularly at the center of the particle, hence a higher percentage char from pyrolysis; eg, $k_t = 11.8 \times 10^{-4}$ W/(cm·K) for dry Douglas fir.

The fourth variable affecting precombustion products is moisture content, which exerts several influences simultaneously. Moisture cools the combustor by acting as a heat sink. Levelton (24) reports measured flame temperatures of 1920 K in combustors burning dry wood, and 1200 K in combustors burning 50% MC wood. This cooling influence promotes char formation. The second, reinforcing influence is effectively putting a temperature ceiling on the fuel particle while moisture is present. The presence of moisture promotes heat transfer into the center of the particle. For wood and lignocellulosic materials with MC $\geqslant$ 40%, the k_t value W/(cm·K) is calculated as follows:

$$k_t = \left[(5.18 + 0.131\ MC_a)S_{o,a} + 0.57\left(1 - \frac{S_{o,a}}{1.46}\right)\right] \times 4.184 \times 10^{-4} \tag{7}$$

The influence of moisture is noted by the change in term from 0.096 to 0.131 MC_a. The final influence is the steam–carbon reaction:

Table 7. Heat of Combustion and Char Formation for Cellulose and Lignin[a]

Source	ΔH^{25} combustion, kJ/g[b]	Char yield on pyrolysis, wt %
cellulose	−17.36	14.9
lignin (klason)	−26.69	59.0

[a] Ref. 9.
[b] To convert kJ to kcal, divide by 4.184.

$$C + H_2O \rightarrow CO + H_2 \quad K_{eq} = 1 \text{ at } 923 \text{ K} \tag{8}$$

This reaction increases volatiles production at virtually all temperatures experienced in a combustor but it occurs only on the surface of the larger fuel particles and is of minor importance.

Ash and other inerts also serve as a heat sink, cooling the combustor (25), again increasing char formation and decreasing volatile production. Thus, although proximate analyses of as-received lignocellulosics may show ca 7.5% fixed carbon, ca 42% volatiles, ca 1% ash, and ca 50% H_2O, the percent of fuel burning in the solid phase is substantially higher when wet material is burned.

Precombustion Considerations for MSW. The influences on pyrolysis products are somewhat different for MSW. Newsprint, which is made from groundwood pulp, contains holocellulose and lignin but much of the paper fraction of MSW is delignified. It, therefore, yields more volatiles and less char. Similarly, the petrochemical products of MSW favor volatile production over char formation.

Particle size is of consequence in terms of complete fuel burnout but is less significant in governing the distribution of pyrolysis products. Solid particles in the >65 cm^3 size are far less common in MSW than in wood waste. Moisture performs the same functions in pyrolysis of MSW. Ash plays a more prominent role as MSW contains such materials as glass and iron which serve as significant heat sinks in the firebox. Paper, particularly magazine stock, carries substantial quantities of clay which is counted as ash. Combustion of either wood or MSW, therefore, becomes oxidation of pyrolysis products by a series of free-radical reactions.

Wood-Waste Combustion Systems

Wood-combustion systems include pile burning and modified pile burning systems (such as Dutch ovens and cell furnaces), semisuspension or spreader-stoker boilers, and suspension burners for dry material. Of these, spreader-stokers are by far the most commonly employed as they can handle varying fuel quality and changes in steam load.

Fuel Handling and Preparation. The extensive fuel-handling systems required for wood-fired systems often cost up to 25% of the total installation cost. Some residue handling systems keep bark and sawdust separate from planer shavings in order to segregate the dry fuels. All systems involve some particle-size reduction for bark, screening, and magnetic removal of tramp iron. If the waste is dry (eg, planer shavings) and is to be burned in suspension, careful attention is paid to reducing particle size to <4 mm and to covered storage (26).

Fuel drying has been tried periodically but problems of fines generation in the dryer have often made it unsuccessful. Fines tend to cling to the large, wet particles. In dryers these particles are liberated from their piggy-back position, causing operating problems. The boiler itself is the most cost-effective drying mechanism despite penalties in efficiency. Densification of wood and crop wastes is also frequently proposed, as pelletization produces a more uniform, low moisture fuel. However, it is of limited use at best, and is generally not cost effective for large systems (24).

Wood-Combustion Systems. Most of the wood that is burned is wet, heterogeneous material, necessitating the use of a spreader-stoker or a pile-burning system. The modern spreader-stoker dominates; it is a waterwall furnace with traveling grates, regenerative air heaters, and extensive pollution control devices. The influence of moisture on its performance becomes apparent. As moisture content increases, the percentage of excess air rises. Figure 1 is an excess air–fuel moisture curve. The use of such quantities of excess air results in wood combustors that are limited by their volume of flue gas. As moisture content increases from 30 to 50%, the energy recovery capacity of the boiler drops by 13% for the same mass of dry wood (Table 8) (see Furnaces, fuel-fired).

The presence of ash and dirt mandates use of multiple cyclones, precipitators, and/or scrubbers. Normal combustion conditions do not produce large amounts of slagging with wet material as the ash softening temperature is reported to be 1520 K and the ash fusion temperature is estimated to be 1550 K (29). Some clinkering does occur but it is not a serious problem (30).

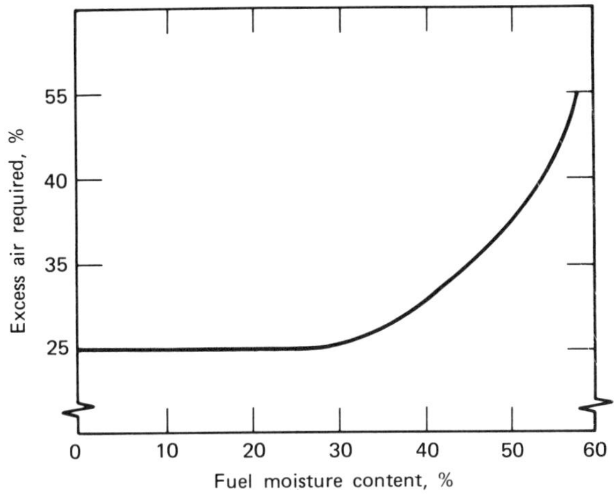

Figure 1. Excess air required as a function of fuel moisture content (28).

Table 8. Energy Recovery from Burning 30% and 50% Moisture Wood[a]

Input streams	30% MC wood Mass, t	30% MC wood Energy, GJ[b]	50% MC wood Mass, t	50% MC wood Energy, GJ[b]
wood fuel				
wood (dry)	25.0	475.2	25.0	475.2
water in wood fuel	10.7		25.0	
Total	*35.7*	*475.2*	*50.0*	*475.2*
combustion air				
stoichiometric dry air	160.3		160.3	
excess dry air	40.1		80.2	
H_2O in air	1.5	0.1	1.8	0.1
output streams at 477 K				
ash	0.25	negl.[c]	0.25	negl.[c]
dry gas and excess air	210.9	0.12	250.8	0.3
water vapor in stack gas	26.4	79.0	41.0	122.9
misc. losses		23.1		23.1
heat available for use as steam		373.1		329.0

[a] Basis: 25 oven-dry t wood/h.
[b] To convert GJ to 10^6 Btu, divide by 1.054.
[c] 1.45×10^{-4} GJ.

Sulfur poses no problems in wood combustion, since these fuels are virtually sulfur-free. NO_x formation from N_2 in the air is similarly inconsequential because of the relatively low flame temperatures associated with wood combustion. However, as Table 1 shows, small fractions of N exist in hardwoods. Because they are present in highly reduced forms, some NO can be produced during combustion. Table 9 presents recent estimates of emissions from wood burning (see Air pollution).

Dry wood reduced to <4 mm size may be segregated and burned in suspension under conditions causing all ash to be reported as fly ash. Either dry heat from combustion gases or steam may be produced; the former is most common. Such systems maximize production of volatiles and minimize production of chars by the synergistic effects of small particles, dry fuels, and higher combustor temperature (eg, 1370 K).

For all wood- and crop-waste fired systems, a series of equations to estimate operating efficiencies has been developed (32):

$$\text{moisture heat loss} = MC \times [538.9 + (373 - T_1) + (0.253 \times T_2)] \times 4.184 \quad (9)$$

$$\text{hydrogen heat loss} = 0.54 \times [538.9 + (373 - T_1) + (0.253 \times T_2)] \times 4.184 \quad (10)$$

$$\text{dry gas and excess air losses} = [(T_2 - T_1)(0.796\,A) + 0.843] \times 4.184 \quad (11)$$

These losses are calculated in kJ/kg of as-received fuel where A = fraction excess air; MC = moisture content expressed as a fraction; T_1 = incoming air, K; and T_2 = stack temperature, K. The operating efficiency of the system can be estimated as:

$$CE\% = \frac{NHV - \left[\left(\sum_{i=1}^{3} L\right) + 0.05\,NHV\right]}{NHV} \times 100 \quad (12)$$

where CE = combustion efficiency; NHV = net heat value of the fuel; and L = losses calculated in equations 9–11. The 0.05 NHV term estimates otherwise unaccounted losses. Typically, wet-wood-fired combustors operate at 65–70% thermal efficiency. Dry-wood-fired combustors operate at 80–84% thermal efficiency.

Several other approaches to burning wood are being commercialized: fluidized-bed combustion, co-combustion with coal, and two-stage pyrolytic combustion. Fluidized-bed systems can handle unusually wet and dirty materials, co-combustion can minimize excess air energy losses while diluting sulfur emissions from coal, and two-stage pyrolytic combustion can handle wet wastes efficiently and cleanly, with a minimum of pollution control equipment (see Fluidization).

Table 9. Air Pollution from Wood Combustion Compared to Residual Oil[a]

Pollutant	Fuel: Wood waste[b], kg/h	Fuel: Residual oil[c], kg/h
particulates	0.95	6.90
SO_2	8.07	96.16
NO_x	10.43–49.90	18.14
hydrocarbons	19.96	0.32

[a] Basis: 10^5 GJ/h or ca 10^{11} Btu/h of fuel. Ref. 31.

[b] Assumes baghouse control.

[c] Assumes 2% S in oil as fired, no stack gas clean-up system.

The Uses of Wood-Combustion Systems. Over 90% of all wood waste burned is used to raise steam for process purposes, to generate electricity, or both. Process steam generated is typically 345 kPa to 1.72 MPa (50–250 psi) and 475–560 K.

Cogeneration is also possible by use of a back-pressure or extraction turbine. From 45,000 kg/h steam at 4.14 MPa (600 psi), 670 K, 2 MW/h can be generated. For systems producing 227,000 kg steam/h at 8.62 MPa (1250 psi) and 755 K, 33 MW/h can be generated. Minimum conditions appear to be internal demand for 35,000 kg process steam and electricity demand of 1.5–2.0 MW (27,33). The advantage of this system over condensing turbines is better efficiency. The heat rate for wood-fired, back-pressure cogeneration systems is 6.34 MJ/(kW·h) (ca 6×10^3 Btu/kW·h) (34).

The efficiency gain by cogeneration can be seen from Figure 2 which shows the incremental quantities of fuel required and energy produced assuming raising steam at 5.86 MPa (850 psi) and generating 5 MW of electricity. The heat rate for the power generated is 6.4 GJ/(kW·h) (6.1×10^6 Btu/kW·h).

A third option for wood-waste combustion systems is generation of electricity in condensing turbine–generator systems. Burlington, Vermont, has one 5-MW system operating and the Eugene Water and Electric Board of Eugene, Oregon, has three such units on line.

Two limitations are associated with wood-fired condensing turbine systems: scale and heat rates. The maximum practical size generally is ca 50 MW–75 MW (35–36). Beyond that size, materials handling difficulties overwhelm the system. The heat rate for wood-fired systems depends upon scale as Figure 3 shows. These rates are substantially higher than those associated with coal.

Economics of Wood Waste Combustion Systems. Wood-fueled systems have high capital and operating costs. For field-created boilers of 5.96–8.72 MPa (850–1250 psi) at 45,000 kg/h capacity, installed costs, in 1978 dollars, are ca \$154/kg. This cost de-

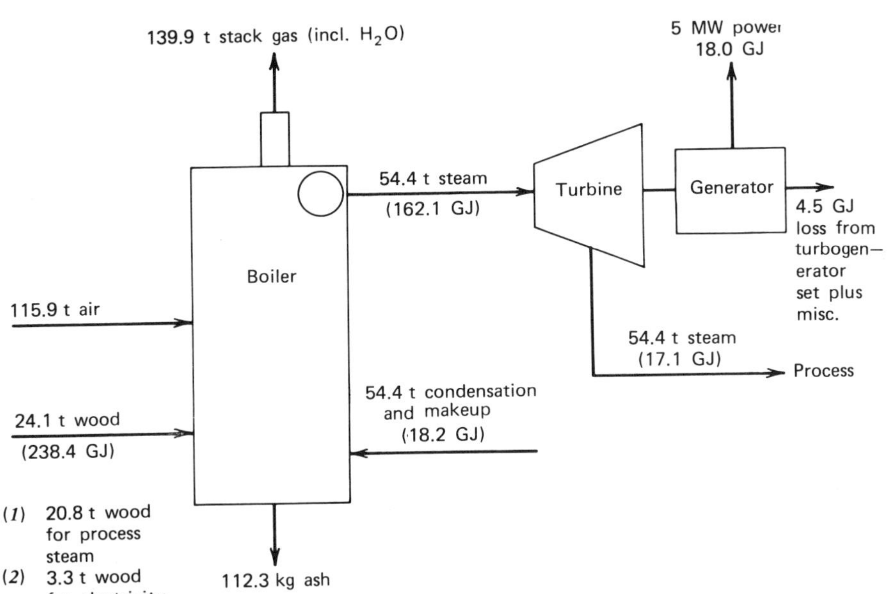

Figure 2. Simplified materials and energy balance for cogeneration system showing fuel required for steam and power (basis of calculation: 1 h). To convert GJ to 10^6 Btu, divide by 1.054.

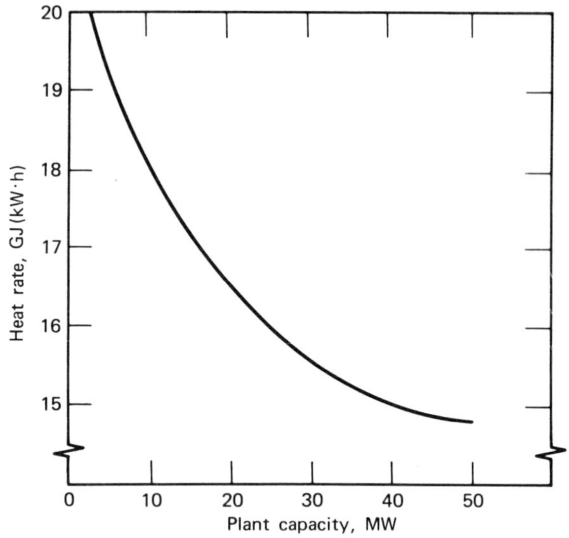

Figure 3. Electric power plant heat rates as a function of scale (33). To convert GJ to 10^6 Btu, divide by 1.054.

clines to \$132/kg at 113,400 kg/h of capacity and then remains essentially flat (35). Back pressure turbines in the 5–25 MW size exhibit capital costs conforming to the function:

$$C = \$74{,}500\ \text{MW} + \$949{,}500 \qquad (13)$$

where C = purchase price in 1978 dollars and MW = megawatt of capacity (37). These prices exclude engineering or installation costs.

The capital cost curve for total boiler–condensing turbine systems expressed in 1978 \$/kW installed oscillates about \$1100/kW in the 5–50 MW range (35–36,38–39).

The most useful overall economic comparison places a value on the wood fuel as a function of oil not burned or electricity not consumed. Tables 10 and 11 compare wood and oil systems for both process steam and cogeneration assuming \$100/m^3 (\$16.00/bbl) oil and 3.5¢ kW·h electricity; 50% M fuel is assumed. Based upon these data, a series of wood price predictors have been developed for various systems. In the case of process steam raising, the equation is based on oil saved; for cogeneration it is based on both oil and electricity saved, and for condensing turbines it is based on 3.5¢/(kW·h) generation costs. These functions are presented below:

$$WV_s = 0.196\,P - \$10.53 \qquad (14)$$

$$WV_{cg} = 0.172 + 1.64\,E - \$12.27 \qquad (15)$$

$$WV_e = 4.45\,E - \$12.34 \qquad (16)$$

where WV_s = wood value in process steam system (\$/t); P = cost of petroleum (\$/m^3) [$WV_s = 1.23\,P - \$10.53$ if P is in \$/bbl and $WV_{cg} = 1.08\,P + 1.64\,E - \$12.27$]; WV_{cg} = wood value in cogeneration system (\$/t); WV_e = wood value in electric-fired system (\$/t); and E = marginal cost of electricity at the bus bar (¢/(kW·h)).

Table 10. Comparative Costs of Wood-Waste-Fired Systems and Oil-Fired Systems[a], 10^3 (1978) Dollars

Cost	Wood system	Oil system
capital cost		
site preparation	106	93
fuel handling	939	100
boiler and related	3692	943
spares	150	50
Total direct cost	*4887*	*1186*
engineering	733	178
commissioning and contingencies	843	205
working capital	305	78
less investment tax credit	(1354)	(0)
Total	*5414*	*1647*
annual capital cost[b]	838	255
annual operating cost	250	125
annual maintenance	406	99
local taxes and insurance	169	41
fuel		4229[c]
Total annual cost	*1663*	*4749*

[a] Basis: 54,400 kg/h, 2.07 MPa (20.5 atm) steam. Ref. 40.
[b] Assumes 15% after tax ROI.
[c] Assumes 8000 h/yr operation; $100/m^3 ($16/bbl) residual oil.

Table 11. Comparative Costs of Wood and Oil-Fired Systems Assuming Cogeneration[a], 10^3 (1978) Dollars

Cost	Wood system	Oil system
capital cost		
boiler as installed	5414	1647
turbogenerator as installed (less ITC)	1687	1687
Total	*7101*	*3344*
annual capital cost[b]	1100	517
annual operating cost	50	50
annual maintenance	426	201
local taxes and insurance	178	84
fuel		4820[c]
Total	*1754*	*5672*

[a] Basis: 54,400 kg/h, 5.86 MPa (58 atm), 5 MW. Ref. 40.
[b] 15% after tax ROI.
[c] 8000 h/yr operation, $100/m^3 ($16/bbl) residual oil.

Similar price functions can be developed with moisture content as the independent variable. If dry fuels are burned separately in direct heat applications, estimated values are $45.50/t (26).

When wood wastes are not used as fuel, they must either be incinerated without heat recovery or disposed of as landfill. Because the latter practice is becoming increasingly expensive, a disposal credit of $2.00–5.00/t (1978 dollars) can be added to the fuel value of wood fuels (40). Table 12 gives wood values as fuel, as a function of use technology, with and without disposal credits.

Table 12. Current Values of Wood-Waste Fuels with Current Technology and Price[a], (1978) Dollars

	Wood value	
Technology	No disposal credit	With \$2.50/t disposal credit
steam only[b]	13.10	15.60
cogeneration[b]	13.80	16.30
electricity generation[c]	7.75	10.25

[a] Assumes 50% moisture wood.
[b] Oil at \$100/m^3 (\$16/bbl).
[c] Electricity at 3.5¢/(kW·h).

Municipal-Waste Combustion

Fuel Preparation and Handling. For municipal waste processing ranges from virtually nothing (in water-wall incineration) to systems for MSW shredding, ferrous metals recovery, glass and nonferrous metals removal, and possibly reshredding the light MSW fraction to <25 mm size. The light MSW may be reduced to dust size and embrittled chemically for suspension burning (6). The vendors developing such systems include the American Can Co. and Raytheon Co. A typical simplified flow sheet for refuse-derived fuel (RDF) is shown in Figure 4.

Many add-on options to the Figure 4 flow sheet are possible. These include two and three stages of air classification for maximum paper and plastics recovery, trommel screening for isolation of glass and heavy organics, electrostatic separation of aluminum, and secondary processing of fuels to various particle sizes. All of these systems produce a distinct class of solid fuels referred to as RDF. The production of RDF is the most common technology, other than incineration, for recovery of energy from MSW. Table 13 shows the influence of processing on heat content of solid fuels.

The Black Clawson Co. has developed a second approach to MSW processing. In the Black Clawson system particle size reduction and large inorganic materials removal are accomplished in a hydropulper, in a slurry of 97% water, 3% solids. The

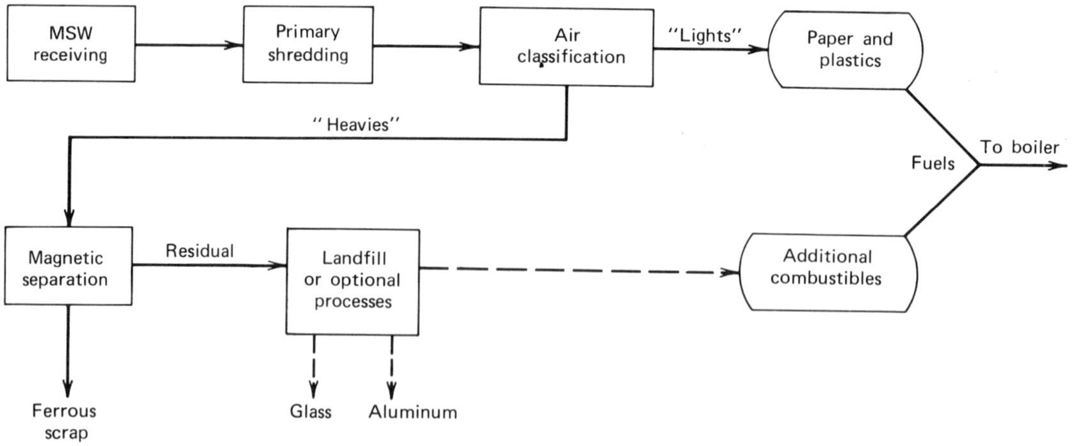

Figure 4. Simplified schematic of municipal solid waste (MSW) processing.

Table 13. The Influence of Processing on MSW Fuels[a]

MSW fuel	Heating value, MJ/kg[b]	Moisture, wt %	Ash, wt %
unprocessed municipal waste	9.37–10.54	ca 25	ca 25
RDF (eg, St. Louis, Milwaukee)	11.71–16.40	ca 25	ca 20
Eco Fuel II	18.74–19.91	ca 12.5	ca 10

[a] Refs. 6, 17, 41.
[b] To convert MJ to 10^3 Btu, divide by 1.054.

water not only serves as a carrying medium but also reduces the fines generation during particle size reduction. Following hydropulping the slurry is passed through a series of classifiers where metal and glass particles are removed by capitalizing on density differences. The slurry is then screened and dried to ca 50% MC. The resulting fuel is relatively homogeneous and has combustion characteristics similar to those of sawdust or bark.

Combustion of MSW Fuels. Combustion of municipal wastes as fuel must be considered in three segments: combustion of unprocessed refuse; combustion of RDF as the sole fuel; and cocombustion of RDF with coal or oil.

Incineration with heat recovery, perhaps the simplest technology, is widely used in Europe (42). Several European designs have been licensed for use in the United States, among them the von Roll system licensed by Wheelabrator-Frye and installed in Saugus, Massachusetts. In this technology, MSW enters the furnace and is tumbled down a series of grates where drying, ignition, and combustion occur. Combustion occurs at 810–1255 K. Combustion gases pass through boiler tubes for steam raising.

In the Saugus, Massachusetts installation, MSW is dumped into a 6080-metric-ton capacity pit where large objects (eg, refrigerators) are removed by crane, and the refuse is mixed. The system then consumes 45.4 t/h to produce 136,100 kg of steam at 700 K/4.31 MPa (800°F/625 psi) conditions (43).

Incineration has two inherent difficulties. The heterogeneity of the material makes distribution of combustion air difficult (25) and 100% excess air, or more, is often used. Typically incinerators produce 520–580 kJ/MJ of input (44) (see Incinerators).

Combustion of RDF is a more efficient approach to recovering energy from MSW. Standard systems employed are usually spreader-stokers unless dustlike particles (d-RDF) are prepared. In that case, suspension burning may be used. Because the preparation systems improve the homogeneity of the fuel, and because most of the combustion-resistant PVC has been removed from the fuel fraction, less excess air and higher combustion temperatures can be used. Thermal efficiencies of 78% have been reported for such systems (6).

Cofiring of 10–20% RDF with 90–80% coal is also popular, and is currently being used in Ames, Iowa; Milwaukee, Wisconsin; Rochester, New York; Chicago, Illinois, and elsewhere. Thermal efficiencies in the >65% range are common (45). Principal problems are that burnout may be ca 87% complete, as experienced in St. Louis, when the wastes are discharged from the boiler grates to sluice water (6). Cofiring with coal requires minimal boiler modification, and RDF is compatible with coal as a fuel. Certain solid fuels under development such as Eco Fuel II may be slurried with oil;

under research conditions the combination has burned much like a coal–oil slurry (6).

Uses of MSW Combustion Systems. Incineration systems are used to raise industrial steam as in Saugus, Massachusetts, where the incinerator supplies steam to the Lynn Works of General Electric; and to raise district heat and circulate cooled water, as in Nashville, Tennessee. In the Saugus installation, cogeneration is employed.

For the RDF fuels the overwhelming market is utilities—hence the generation of electricity. Some chemical plants and cement kilns have also considered the use of RDF fuels. At this writing, Hooker Chemical of Niagara Falls, New York, has such a system under construction.

Economics of MSW Combustion. The economics of MSW processing, based primarily upon energy recovery, are complicated by the following factors: (*1*) recovery of ferrous metals, (*2*) periodic recovery of nonferrous metals and/or glass; (*3*) periodic skimming of high-grade paper fractions; (*4*) avoidance of landfilling costs; (*5*) access to low cost pollution-control revenue bonds; and (*6*) interrelationships between municipal and private industry organizations. These systems are economically feasible only when credit for not landfilling is included in the calculation.

The capital cost of a 910 t/d waterwall incinerator is ca 50×10^6 (1978) dollars (46). For the RDF systems, assuming modifications of an existing boiler, costs are about 20×10^6 dollars for the same capacity system. Typically fuels and energy products are sold at or slightly below the marginal cost of competing fuels (eg, self-raised steam; coal) with the differential revenue required being provided by user charges to the government in the area. Typical user charges are in the $9–13.50/t range.

Other Wastes as Fuel

Among other wastes burned as fuel are bagasse (qv), which supplies 66% of the cane sugar industry's energy needs (7); coffee grounds, used to fuel instant-coffee manufacturing operations (47) (see Coffee); corn seed used to fuel a utility boiler in Logansport, Indiana (48); cotton gin trash (49); and a host of other agriculturally based products. Economics and efficiencies associated with these processes are similar to those detailed for wood. The primary difference is in the acceptable scale of operations, as commodities of lower bulk density place more severe constraints on the materials handling system.

The seasonal nature of crop harvesting combined with the bulk density problem make the economics less attractive for crop wastes than for wood. Storage is required, and the storage systems must inhibit low-grade oxidation and fuel degradation, spontaneous combustion, rodent and pest infestation, and other similar problems (22).

Conversion Systems Applied to Wastes

Numerous pyrolysis, gasification, and liquefaction systems developed for coal conversion have been adapted to wastes. These systems are in various stages of research, development, and demonstration. Additional research is applying anaerobic digestion, acid hydrolysis, and enzymatic hydrolysis of wastes to production of methane-rich gases and alcohol fuels. Among these systems gasification is considered most promising since it capitalizes on the high ratio of volatile-to-fixed carbon associated with wastes, the higher atomic hydrogen-to-carbon ratios, and the virtual absence of sulfur (50). Gasification also permits continued use of many existing natural gas fired boilers threatened by uncertainties in existing fuel supplies (see Fuels, synthetic).

The basic principles of gasification include drying of the waste in the reactor, pyrolysis of the materials to produce volatiles, and partial or total oxidation of the char formed in pyrolysis. The oxidation process supplies heat to drive the reactions and, in some cases, adds CO to the product gas.

Several small commercial residue gasifiers exist including the inclined grate Conifer of American Fyr Feeder. In the Conifer, complete combustion of fixed carbon on the grate drives off the volatiles as gaseous fuels, and close-coupled combustion of the gaseous fuel follows. This form of gasification capitalizes on the high volatile-to-fixed carbon ratio (ca 5.7:1) and produces a gaseous fuel consisting of CO, CO_2, H_2, CH_4, and C_2H_x. Constant fuel feeding on the grate ensures a continuing supply of fixed carbon for combustion to drive the gasification reactions. The air supply controls the gasification rate.

New systems being developed separate gasification and combustion functions more completely, and carry out the reaction of a portion of the char with steam to enhance gas production by the following reaction:

$$H_2O + C \rightarrow CO + H_2 \tag{17}$$

This endothermic reaction is used in some cases both to increase gas production and to keep the reaction below ash-softening and fusion temperatures.

The primary distinction among such gasifiers is the source of O_2 used to drive the gasification reactions. Air is used in the Andco-Torrax and Moore-Canada systems. Pure oxygen is used in the Union Carbide Purox gasifier. In air-blown units, reaction temperatures are in the 1225–1245 K range. In the oxygen-blown system, hearth temperatures of 1921 K are reached. The ash forms a slag removed in molten form. Table 14 compares the gas composition of air and oxygen-blown systems.

Benemann (5) reports thermal efficiencies for low and intermediate heat content gas systems of 44 and 62%, respectively, when combustion of the gaseous fuel is included in the calculation. The estimated fuel prices are \$4.46/GJ (\$4.70/10^6 Btu) for low heat content gas and \$2.80/GJ (\$2.95/10^6 Btu) for synthesis gas (5). Although these

Table 14. Analysis of Air-Blown and Oxygen-Blown Gasifier Products, Vol % [a]

Gas component	Air blown (Moore-Canada)	Oxygen blown (Purox)
H_2	18.0	26.0
CO	22.8	40.0
CO_2	9.2	23.0
CH_4	2.5	5.0
C_xH_y	0.9	5.0
O_2	0.5	0.5
N_2	45.8	0.5
Total	*99.7*	*100*
heat content, MJ/m^3 (Btu/lb)	ca 6.73 (ca 150)	15.26 (ca 350)

[a] Ref. 51.

values are reported for municipal waste, they are comparable to costs for other residues such as wood or crop waste. Environmental emissions of gasification systems, resulting from tars condensed during gas cleaning and purification, require extensive wastewater cleanup before discharge. The Purox process, eg, has been estimated to produce 60,000–70,000 mg/L of BOD loadings (44).

Owing to demand for liquid fuels, research on the thermochemical liquefaction of organic wastes is proceeding at a pilot plant in Albany, Oregon. Its capability is the processing of 3.6 t/d of wood chips into ca 1 m^3 of heavy oil (2.72 t/d to 5 bbl). The general chemistry of this pilot plant is described by the following formula (52):

$$0.61\ (C_6H_{9.13}O_{4.33}) + 0.23\ CO + 0.08\ H_2 \xrightarrow[\text{catalyst}]{H_2O} H_2O + 0.64\ CO_2 + 0.53\ (C_6H_{6.93}O_{1.22}) \qquad (18)$$

The fuel oil has the following composition: C, 77%; H, 10.7%; N, 2.8%; O, 8.8%; and S, 0.3% (53). Its higher heating value is 34.9 MJ/kg (15,000 Btu/lb). It has a specific gravity of 1.03 and a pumping temperature of 333 K (53). Although wood is the currently employed feedstock, municipal waste and crop wastes give the highest rates of carbon conversion and oil production, and wood has the highest rate of CO consumption (52). The present scale of operation precludes precise economic evaluation of the commercial prospects for this process (54), but initial estimates indicate a production cost substantially higher than that associated with coal liquefaction.

Conclusions

Wastes are a widely heterogeneous source of fuels and energy—a heterogeneity caused by generating processes. The consequence of this heterogeneity is that the direct use of wastes as fuel, with a minimum of processing, in combustion processes used to raise heat, steam, and electricity, is most practical. Gasification to industrial-grade low and medium heat value gas is also significant since it permits use of existing gas-fired boilers. In both cases, the economics are based upon industrial use with minimum transportation, and are influenced by alternative disposal costs.

BIBLIOGRAPHY

1. L. L. Anderson in L. L. Anderson and D. A. Tillman, eds., *Fuels From Waste,* Academic Press, Inc., New York, 1977, pp. 2–15.
2. L. L. Anderson, *U.S. Bur. Mines Inf. Circ.,* **8549,** (1972).
3. K. Howlett and A. Gamache, *Silvicultural Biomass Farms Vol. VI. Forest and Mill Residues as Potential Sources of Biomass, ERDA Contract E(49-18)-2081,* The Mitre Corp., McLean, Va., 1977, 124 pp.
4. D. A. Tillman, *Chemtech* **7,** 611 (1977).
5. J. R. Benemann, *Biofuels: A Survey,* Electric Power Research Institute, Palo Alto, Calif., 1978.
6. J. G. Abert and H. Alter in K. V. Sarkanen and D. A. Tillman, eds., *Conversion of Biomass to Fuels and Chemicals,* Vol. I, Academic Press, Inc., New York, 1979, pp. 146–205.
7. W. Arlington in D. A. Tillman, K. V. Sarkane, and L. L. Anderson, eds., *Fuels and Energy from Renewable Resources,* Academic Press, Inc., New York, 1977, pp. 249–256.
8. D. A. Tillman, *Wood As An Energy Resource,* Academic Press, Inc., New York, 1978.
9. F. Shafizadeh and W. F. DeGroot in F. Shafizadeh, K. V. Sarkanen, and D. A. Tillman, eds., *Thermal Uses and Properties of Carbohydrates and Lignins,* Academic Press, Inc., New York, 1976, pp. 1–18.
10. R. Arola, *Energy and the Wood Products Industry,* Forest Products Research Society, Madison, Wisc., 1976, pp. 34–45.
11. R. A. Currier, "Converting Bark into Opportunities," paper presented at *Proc. Oregon State University, Corvallis, Oregon, Mar. 8–9, 1971.*

12. H. J. Hardy, *Modern Sawmill Techniques,* Vol. 3, Miller Freeman, San Francisco, Calif., 1974.
13. M. Leman, "Wood Residues as An Energy Source," paper presented at *Proc. Forest Products Research Society (No. P-75-13), Madison, Wisc., 1976,* pp. 60–65.
14. D. A. Hartman, W. A. Atkinson, B. S. Bryant, and R. O. Woodfin, *Conversion Factors for the Pacific Northwest Forest Industry,* Institute of Forest Products, College of Forest Resources, University of Washington, Seattle, Wash., 1978.
15. H. Resch, "Energy Recovery from Wood Residues," paper presented at the *Eighth World Forestry Congress, Jakarta, Indonesia, 1978.*
16. D. C. Junge, *Boilers Fired with Wood and Bark Residues,* Forest Research Laboratory, Oregon State University, Corvallis, Oregon, 1975.
17. *NCRR Bull.* **3**(2), 3 (1973).
18. I. Wender, F. W. Steffgen, and P. Yavorsky in T. F. Yen, ed., *Recycling and Disposal of Solid Wastes,* Ann Arbor Science, Ann Arbor, Mich., 1974, pp. 43–99.
19. H. Schultz, P. M. Sullivan, and F. E. Walker, *U.S. Bur. Mines Rep. Invest.* **8044,** (1975).
20. W. E. Franklin, *Potential for Resource Recovery in the United States,* Franklin Associates, Prairie View, Kan., 1975.
21. K. W. Pober and H. F. Bauer in ref. 1, pp. 57–72.
22. T. R. Miles in ref. 7, pp. 225–248.
23. F. Shafizadeh and P. S. Chin in I. S. Goldstein, ed., *Wood Technology: Chemical Aspects,* ACS Press, Washington, D.C., 1977.
24. B. H. Levelton and Assoc., Ltd., *An Evaluation of Wood Waste Energy Conversion Systems,* Environment Canada, Vancouver, B.C., Can., 1978.
25. J. B. Edwards, *Combustion: The Formation and Emission of Trace Species,* Ann Arbor Science, Ann Arbor, Mich., 1974.
26. L. H. Furman and L. G. Desmon, *For. Prod. J.* **26**(9), (1976).
27. R. L. Jamison in ref. 6, pp. 28–50.
28. G. Villesvik, personal communication, Apr. 1, 1979.
29. J. H. White, "Wood Waste and Bark as Fuel in a Particle Board Plant," paper presented at the *Joint Power Generation Conference, ASME, Dallas, Tex., Sept. 10–13, 1978.*
30. *Steam: Its Generation and Use,* 39th ed., Babcock and Wilcox, New York, 1978.
31. P. Boys, EPA, communication to P. Hofer, Nov. 3, 1978.
32. P. J. Ince, *USDA Forest Service General Technical Report FPL 13,* USDA, Madison, Wisc., 1977.
33. W. B. Archibald and J. A. Gabriel, "Hardware for Energy Generation in the Forest Products Industry," paper presented at *Proc. Forest Products Research Society, Madison, Wisc., 1979.*
34. N. Methven, "Increased Energy from Biomass: 1985 Possibilities and Problems," paper presented at *Proc. Pacific Northwest Bioconversion Workshop, Oct. 24–26, 1978, Portland, Oregon.*
35. J. S. Bethel and co-workers, Energy from Wood, *Report to the U.S. Congress,* Office of Technology Assessment, Washington, D.C., 1979.
36. H. C. Mason, D. H. Edel, and G. O. Tolman, *Study and Report on the Direct Conversion to Electrical Power of Forest Residues,* H. C. Mason and Assoc., Gladstone, Oregon, 1972.
37. C. Bruya, personal communication, Apr. 1, 1979.
38. C. Bliss and D. O. Blake, *Silvicultural Biomass Farms, Vol. V.: Conversion Processes and Costs, Tech. Report 7347,* The Mitre Corp., McLean, Virginia, 1977.
39. *General Facts: Wood Waste Steam—Electric Generating Plant,* The Washington Water Power Company, Spokane, Wash., 1979.
40. D. A. Tillman in ref. 6, pp. 54–86.
41. N. L. Treat, *Net Energy From Municipal Solid Waste,* Institute for Energy Analysis, Oak Ridge, Tenn., 1977.
42. Resource Planning Associates, *European Waste to Energy Systems: An Overview,* For Energy Research and Development Administration, Washington, D.C., 1977.
43. W. K. MacAdams, *Mineral Resources and the Environment,* Suppl. Report, National Research Council, National Academy of Sciences, Washington, D.C., 1975.
44. D. A. Tillman in ref. 1, pp. 17–39.
45. D. A. Tillman, *Environ. Sci. Technol.* **9,** 418 (1975).
46. D. A. Tillman in ref. 1, pp. 211–226.
47. R. G. Schweiger, *Power,* S1 (Feb. 1975).
48. *Energy Data Reports,* U.S. Department of Energy, Washington, D.C., Mar. 1979.
49. W. F. Lalor in ref. 7, pp. 257–274.
50. J. S. Bethel and co-workers, *Renewable Resources for Industrial Materials,* National Research Council, National Academy of Sciences, Washington, D.C., 1976.

51. R. M. Rowell and A. E. Kokanson in ref. 6, pp. 118–144.
52. H. R. Appell in ref. 1, pp. 121–139.
53. A. L. Hammond, W. D. Metz, and T. H. Maugh, *Energy and the Future,* AAAS, Washington, D.C., 1973, p. 75.
54. *Chem. Eng.,* 41 (Aug. 27, 1979).

DAVID A. TILLMAN
Consultant

FUELS, SYNTHETIC

GASEOUS FUELS

In the United States, the conversion of fossil fuels to gas for more convenient transportation, storage, and utilization dates back to about 1800 (1). This "manufactured gas" was produced in coke ovens or similar types of retorts by simply heating coal to vaporize its volatile constituents (see Coal, carbonization). The resulting gas typically had a volumetric analysis of 52% hydrogen, 32% methane, 8% carbon monoxide, 6% nitrogen, and 2% carbon dioxide. Its heating value was about 20.5 MJ/m^3 (550 Btu/SCF). Only ca 20% of the coal was converted to gas; the remainder was comprised of several by-products ranging from hydrocarbon liquids to tar and coke. This gas was widely used until the advent of natural gas, and is still used in steel mills where coke has to be produced for the blast furnace.

A variation of manufactured gas appeared in the late 1800s when the "blue water-gas" generators were introduced. In these cyclic generators, coke was burned with air to heat it to incandescence. The air was turned off, and steam blown over the hot coke to produce carbon monoxide and hydrogen. When the coke cooled, the steam was turned off and the combustion process repeated. Two chambers were used so that gas could be produced continuously, first in one chamber and then the other. The heating value of this gas was low, about 10.2 MJ/m^3 (275 Btu/SCF). Liquid fuels were gasified for the first time when oil was added to the hot blue water gas. The heat decomposed the oil into methane and heavier hydrocarbons, which increased the heating value of the blue water gas above that of manufactured gas. This new product was called carbureted water gas. Carburetion was originally introduced to add luminosity to gas lights (2).

The water–gas machines used almost all the coal to produce gas; the char was burned for heat to operate the process. Producer-gas generators, which appeared about the same time, also converted essentially all the coal to gas. Both types were continuous gasifiers. Coal, fed at the top, descended by gravity; ash was discharged through grates at the bottom. Air was introduced at the bottom and the coal burned to produce heat and products of combustion. As these hot products rose through the descending coal bed, the coal reacted to give carbon monoxide and hydrogen; some methane was also formed. The producer gas, which contained 50% nitrogen from the air, had a low heating value of ca 5.6 MJ/m^3 (150 Btu/SCF). A much-refined form of the producer has been adapted to use oxygen instead of air and to operate at moderately high pressure. The leading example, the Lurgi pressure gasifier, is currently one of the primary candidates for production of pipeline-quality gas from coal in the United States.

Thermal cracking (pyrolysis) of oil was an important method for producing gas in the years following its use for carburetion of water gas. Many water–gas sets were converted (3) into straight oil-gasification units. A few of these were used for base-load city gas supply; most were used for peak-load situations in the winter. A few new plants were also built to furnish gas the year round, eg, Honolulu.

In the 1940s, the British Gas Council, which had been working on coal hydrogasification, experimented with the hydrogasification of oil (4). In the ensuing years, the Institute of Gas Technology (IGT) also did much work in this field to supplement the Council's extensive work in thermal cracking (5–6). In 1964, the Council found that light distillates (7) having end boiling points below 182°C and containing no sulfur could be catalytically reformed by an autothermic process into essentially pure methane. This method was extensively used in England until North Sea gas came into use. Several plants were built in the United States in the early 1970s. No further construction has occurred because of supply problems and high cost of the feedstock (see Feedstocks).

Only a trivial amount of fossil fuels have been used for gasification since natural gas became a dominant energy source. Plentiful supplies of coal (qv), oil shale (qv), peat, and biomass (see Fuels from biomass), however, have led to extensive research and development of gasification processes (see also Lignite and brown coal). The following processes could become economically competitive with natural gas because of changes in the price structure and may be needed in future on short notice in view of U.S. dependence on overseas oil and gas supplies (see Fuels, survey).

Coal

Coal is gasified either by partial combustion or by hydrogasification. Very often a process is a combination of both modes, depending on the product desired: low, medium, or high heat-value gas. Table 1 lists many of the existing commercial gasification processes together with many that are under development.

Low Heat-Value Gas. Low heat-value (low-Btu) gas usually has a heating value of 3.3–5.6 MJ/m^3 (90–150 Btu/SCF), and is formed by partial combustion of coal with air, usually in the presence of steam. The primary reactions are:

$$C + \frac{1}{2}\,O_2 \rightarrow CO \qquad (1)$$

$$C + H_2O \rightarrow CO + H_2\ (\text{steam}) \qquad (2)$$

$$CO + H_2O \rightarrow CO_2 + H_2 \qquad (3)$$

The combustible components of the gas are carbon monoxide and hydrogen, diluted with carbon dioxide and nitrogen. The gas has a low flame temperature unless the combustion air is strongly preheated. Its use has been limited essentially to steel mills, where it is produced as a by-product of blast furnaces. It is now being widely proposed for use in electric power for small requirements of manufacturing concerns. The DOE is sponsoring several such installations. Only one is described here. Most of the others may be found in the references.

A popular choice for these "backyard" gas producers is the Wellman-Galusha

Table 1. Tabulation of Some Gasifiers and Types of Systems Existing or in Development[a]

Gasifier or process designation	Type of contact	Heat source	Ash removal	Pressure[b], MPa	Stage of development
ATGAS	iron molten bath	O_2	as slag	7	laboratory—currently inactive
Battelle Agglomerating	fluidized bed	circulating hot ash	agglomerated ash	7	pilot plant—inactive
BI-GAS	entrained	O_2	as slag from cyclone combustor	7	pilot plant under reassessment
CO_2 acceptor	fluidized bed	circulating hot dolomite	as fines off regulator bed	1	pilot plant—currently inactive
COGAS	fluidized bed	circulating char	as slag	0.1	demo plant design
Combustion Engineering	entrained	air	as slag	0.1	pilot plant—in start-up
electrothermal	fluidized bed	electricity	char removal	7	pilot plant—inactive
Exxon Catalytic Gasification	fluidized bed	H_2 preheat through heat exchange	withdrawal of bed material		laboratory
Foster Wheeler	entrained	air	as slag	2	conceptual design
gasification using nuclear heat	fluidized bed	nuclear by hot helium	withdrawal of bed material	4	pilot plant
Hydrane	entrained + fluidized bed	O_2	depends on H_2 production method	7	laboratory
HYGAS	entrained + fluidized bed	O_2	withdrawal of 3rd-stage material	7	demo plant design
Koppers-Totzek	entrained	O_2	as slag	0.1	commercial
Lurgi	fixed bed	O_2	dry ash through revolving grate	3	commercial
molten salt	molten-salt bath	O_2	withdrawal and refining of molten salt	8.5	pilot plant—under construction
Riley-Morgan[c]	fixed bed	air	dry ash through grate	0.1	commercial design in development
Ruhr 100	fixed bed	O_2	dry ash through revolving grate	10	pilot plant—under design
Saarberg-OTTO	entrained molten-slag bath	O_2	as slag	2.5	pilot plant
Shell-Koppers	entrained	O_2	as slag	3–4	pilot plant
Slagging Lurgi	fixed bed	O_2	as slag	3	demo plant design
steam-iron	entrained + fluidized	circulating iron–iron oxide	withdrawal of producer bed material	7	pilot plant—in start-up

Table 1. *(continued)*

Gasifier or process designation	Type of contact	Heat source	Ash removal	Pressure[b], MPa	Stage of development
Synthane	fluidized bed	O_2	withdrawal of bed material as char	7	pilot plant—in start-up
Texaco	entrained	O_2	as slag	4	pilot plant operational (Germany); demo plant design
U-GAS	fluidized bed	air or O_2	as ash agglomerates	0.6	pilot plant operational; demo plant design
Wellman-Galusha[c]	fixed bed	air	dry ash through revolving grate	0.1	commercial
Westinghouse[c]	fluidized bed	air	as ash agglomerates	1.5	pilot plant
Winkler	fluidized bed	air or O_2	as percentage of cycloned fines	0.1	commercial
Winkler	fluidized	air or O_2	as percentage of cycloned fines	1.1	pilot plant—in start-up

[a] Ref. 8.
[b] To convert MPa to atm, divide by 0.101.
[c] Low heat-value gas only.

unit (9) because of its long history of successful operation. This atmospheric pressure, air-blown gasifier (Fig. 1) is fed by a two-compartment lockhopper system. The upper coal-storage compartment feeds intermittently into the feeding compartment, which feeds the gasifier section almost continuously except for brief periods when the feeding compartment is being loaded from storage. In small units up to 1.5 m ID, ash is removed by shaker grates. In larger units, ash is removed continuously from the bottom of the gasifier into an ash-hopper section by a revolving grate. The grate is constructed of flat, circular steel plates set one above the other with edges overlapping. The grate, revolving eccentrically within the gasifier, causes ash to fall from the coal bed as the space between the grate and the shell increases, and then pushes the ash down into the ash-hopper as the space decreases. The smaller size units are brick-lined, although the larger sizes are unlined and water-jacketed. Combustion air, provided by a fan, passes over the warm 82°C jacket water causing the water to vaporize to provide the necessary steam for gasification. Gas leaves from above the fuel bed at 428–538°C for bituminous coal.

The Wellman-Galusha unit has two configurations: the standard type for gasification of noncaking coals and the agitator type for gasification of any type of coal. The latter uses a slowly revolving horizontal arm, not shown in Figure 1, which also spirals vertically below the surface of the fuel bed to retard channeling and maintain a uniform bed. Use of the agitator not only allows operation with caking coals but also can increase the capacity of the gasifier by about 25% for use with other coals. Capacities of various standard size gasifiers range from as low as 3.4 g/s of anthracite to as high as 882 g/s of bituminous coal. One kilogram of good bituminous coal will make

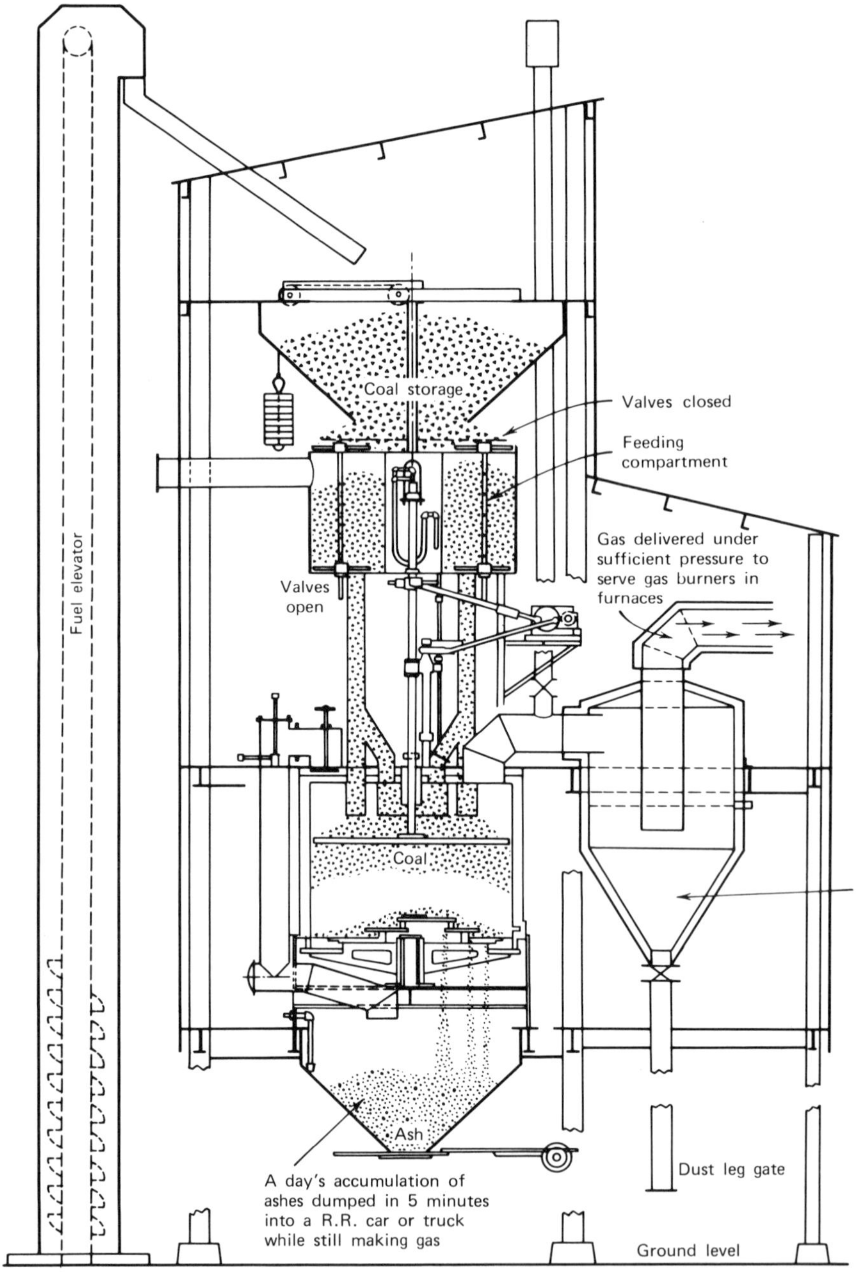

Figure 1. The Wellman-Galusha agitator-type gas producer makes possible the gasification of either bituminous coal, anthracite, or coke without mechanical adjustment. Courtesy of McDowell-Wellman Gas Products (9).

4.0 m^3 of the gas shown in Table 2 (64.1 ft^3/lb). This requires 2.65 m^3 (42.5 ft^3/lb) of air for gasification.

Medium Heat-Value Gas. Medium heat-value gas has a heating value between 9.3 and 20.5 MJ/m^3 (250 and 550 Btu/SCF). At the lower end of this range, the gas is produced like low heat-value gas, except that an air separation plant is added and relatively pure oxygen is used instead of air to partially oxidize the coal. This eliminates the nitrogen and doubles the heating value from 5.2 to 10.6 MJ/m^3 (140–285 Btu/SCF).

The Winkler gasifier is a commercial example of medium heat-value gas processes. When blown with oxygen, a gas is produced which is largely carbon monoxide and hydrogen. In the Winkler Process (10) (Fig. 2), finely crushed coal is gasified in a fluidized state; oxygen and steam are introduced in the bottom. The coal is fed by lockhoppers and a screw feeder into the bottom of the fuel bed. Sintered ash particles settle on a grate, where they are cooled by the incoming oxygen and steam; a rotating, cooled rabble moves the ash toward a discharge port. The ash is then conveyed pneumatically to a disposal hopper. The gasifier is a cylindrical vessel, 5.5 m dia and 21.3 m high.

The gas along with entrained ash and char particles, which are subjected to further gasification in the large space above the fluid bed, exit the gasifier at 954–1010°C. The hot gas is passed through a waste-heat boiler to recover the sensible heat, and then through a dry cyclone. Solid particles are removed in both units. The gas is further cooled and cleaned by wet scrubbing, and if required, an electrostatic precipitator is included in the gas-treatment stream (see Air pollution control methods). Ash collection and disposal is somewhat of a problem.

Best performance is obtained with the use of lignites or noncaking subbituminous coals. Careful consideration is given to ash fusion data, reactivity, and size consistency. At present, units are operated at atmospheric pressure.

Davy Powergas, Inc., of Lakeland, Fla., has considered a joint venture to build an air-blown, 2 kg/s (173 metric tons per day) Winkler gasifier, which will be designed to operate at pressures up to 205 kPa (15 psig) and will allow the use of a wide range of feedstocks including high-sulfur coke, crude oil, residual oil, and char.

The Rheinbraun process is another development of the Winkler process. Rheinische Braunkohlenwerke plans to use this high pressure modification as the basis for a 21.8 t/d pilot plant that will operate at 1.13 MPa (150 psig). This plant will be

Table 2. Typical Gas Compositions[a]

Components	Bituminous	Anthracite
CO	25	27.1
CH_4	2.3[b]	0.5[b]
illuminants	0.6	
H_2	14.5	16.6
CO_2	4.7	5.0
N_2	52.8	50.8
MJ/m^3 [c] HHV (higher heating value)	6.2	5.4

[a] Ref. 9.

[b] Methane increases with increasing volatile content.

[c] To convert MJ/m^3 to Btu/ft^3, multiply by 26.87.

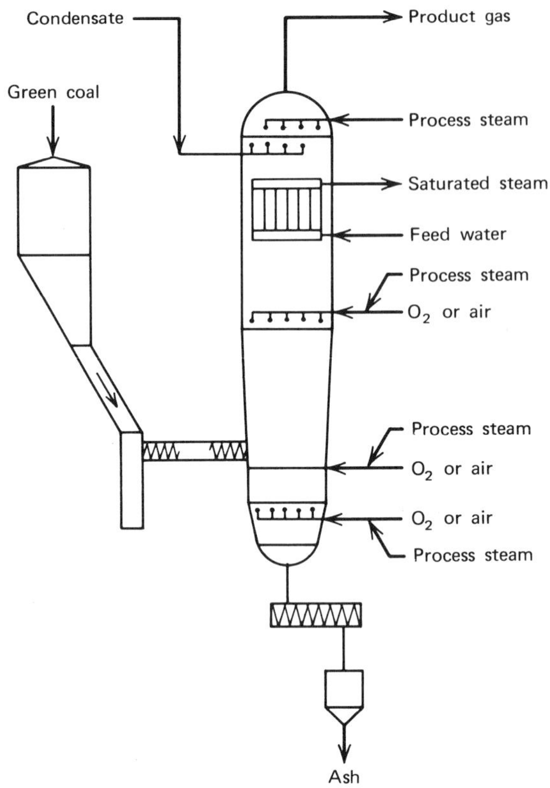

Figure 2. Winkler generator (10).

used to gasify lignite primarily to produce a carbon monoxide–hydrogen mixture as a reducing gas for direct reduction of ores. The company also is looking at the possibility of producing fuel gas.

The process is used for the commercial production of atmospheric pressure synthesis gas. The first commercial Winkler gasifier was put into operation at Leuna, Germany in 1926. Since then, 36 producers in 16 installations have been designed and constructed by Bamag Verfahrenstechnik GmbH, a German affiliate of Davy Powergas.

The Koppers-Totzek (K-T) process (11–12) is a second example of a fully commercial medium heat-value gas process. Whereas the Winkler Process employs a fluidized bed, the K-T process uses an entrained-flow system. Other systems using entrained flow such as Texaco, Shell-Koppers, BI-GAS, Combustion Engineering, and Foster-Wheeler are in various stages of development.

In the K-T process, dried and pulverized coal is conveyed continuously by a screw into a mixing nozzle. From there a high velocity stream of steam and oxygen entrains the coal into a gasifier. The gasifier (Fig. 3) is a refractory-lined, cylindrical vessel. The refractory lining is designed to conduct a selected amount of heat to a surrounding water jacket in which low pressure process steam is generated. The lining is thin (about 5 cm), and made of a high alumina cast material. In a two-headed gasifier two burner heads are placed 180° apart at either end of the vessel. Four burner heads, 90° apart, are used in a four-headed gasifier. The largest gasifiers are 3 to 4 m dia at the middle,

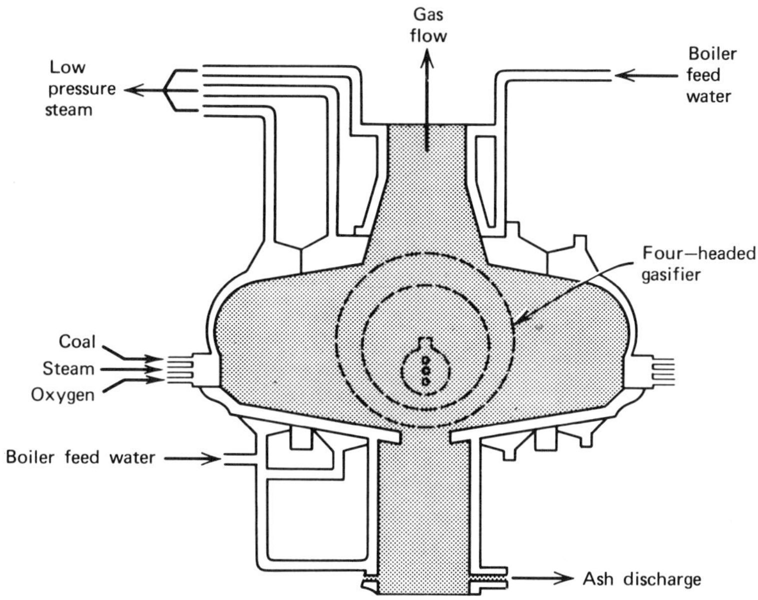

Figure 3. K-T gasifier (12).

tapering to 2–3 m at the burner ends and are about 19 m long. The reactor volume is about 30 m^3 (1060 ft^3) for the two-headed design, and 64 m^3 (2260 ft^3) for the larger, four-headed models.

The process is carried at moderate pressures (slightly above atmospheric) but at very high temperatures that reach a maximum of 1900°C. Even though the reaction time is short, 0.6–0.8 s, the high temperature prevents the occurrence of any tar, condensable hydrocarbons, or phenols in the product gas. The absence of liquid simplifies the subsequent gas clean-up steps.

Normally about half of the coal ash is removed from the bottom of the gasifier as a quenched slag. The balance is carried overhead in the gas as droplets which are solidified when the gas is cooled with a water spray. A fluxing agent is added, if required, to the coal to lower the ash fusion temperature and increase the molten slag viscosity.

Conversion of carbon in the coal to gas is very high. With low-rank materials such as lignite and young subbituminous coal, the conversion borders on 100%, and for high-volatile A coals, it is 90–95%. Unconverted carbon appears mainly in the overhead material.

The composition of the gas from the gasifier as well as composition of the feed coal and quantities of products obtained for gasification of an Illinois coal are shown in Table 3. Sulfur removal is facilitated in the process because nearly all of it appears in the gas. Distribution of sulfur in the gas is typically 90% H_2S and 10% COS with only a trace of SO_2. No CS_2 or CH_3SH is formed.

Table 4 lists the K-T gasifiers that have been constructed. The smaller equipment gasifies about 305 metric tons per day and the larger four-headed type converts up to 780 t/d.

Two medium heat-value gas processes under development employ a potentially

Table 3. K-T Gasifier Data for Illinois Coal[a]

Gasifier feed	
dried coal to gasifier analysis, wt %	
C	61.94
H_2	4.36
N_2	0.97
S	4.88
O_2	6.73
ash	19.12
moisture	2.00
Total	*100.00*
gross heating value, kJ/g[b]	26.5
process steam: dried coal, wt:wt	0.271
oxygen: dried coal, wt:wt	0.704
purity, %	98.0
Gasifier products	
jacket steam: dried coal, wt:wt	0.202
high press steam: dried coal, wt:wt	1.146
raw gas analysis (dry basis), vol %	
CO	55.38
CO_2[c]	7.04
H_2	34.62
N_2	1.01
H_2S	1.83
COS	0.12
Total	*100.00*
gross heating value, MJ/m^3 [d]	10.8
gas make: dried coal, m^3/t[e]	1857
slag make: dried coal, wt:wt	0.190
process efficiency, %	85.0
coal to gas efficiency, %	75.8

[a] Ref. 11.
[b] To convert kJ/g to Btu/lb, multiply by 430.3.
[c] Depending on the particular coal and the intended use of the gas, the CO_2 content may be as high as 12–13%.
[d] To convert MJ/m^3 to Btu/ft^3, multiply by 26.87.
[e] To convert m^3/t to ft^3/short ton, multiply by 32.0.

improved method of collecting the ash by forming rather large, agglomerated spheres. These are IGT's U-GAS (13), and the Westinghouse fluidized-bed process (14). The latter has been aimed at low heat-value gas production. U-GAS, a similar but mechanically simpler version, has been developed to produce medium heat-value gas by a single-stage ash-agglomerating gasification system. Ash is discharged as a dry, nonslagged material with low carbon content by selectively heating a small conical volume in the fluidized bed in such a manner that the ash is heated to its incipient softening point. The ash will selectively stick to itself as the carbon is burned out (15). As the ground coal is gasified, the resultant ash particles stick together to form agglomerates that leave the gasifier by discharging through the oxygen–steam inlet when the agglomerates are sufficiently large and dense (Fig. 4).

In the experimental program to date, coke-oven char and Kentucky No. 9 coal (both washed and unwashed) have been successfully gasified at rates up to 22.7 t/d at atmospheric pressure.

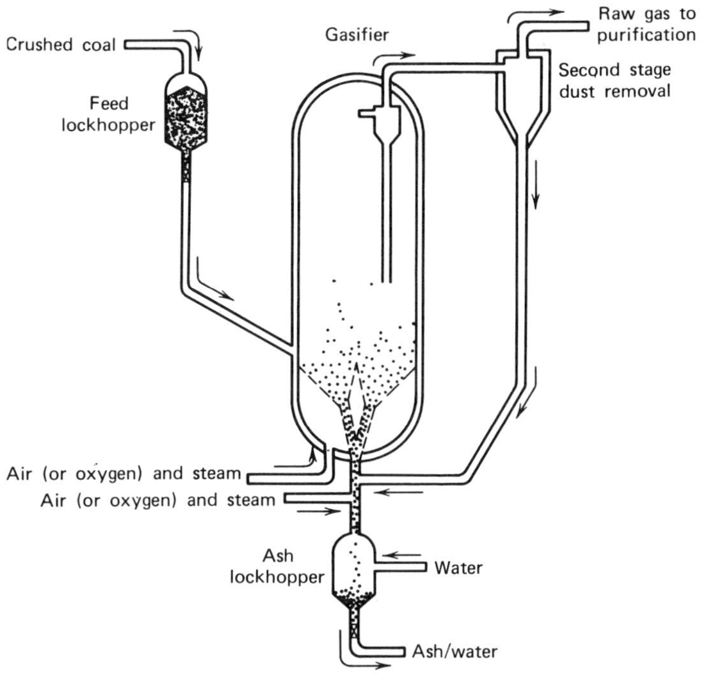

Figure 4. Schematic of the U-GAS gasifier (14).

In Figure 4, crushed raw coal is introduced to the reactor at the desired pressure through a conventional lockhopper system. The coal is gasified into a carbon monoxide–hydrogen mixture by oxidation with oxygen and steam in a fluidized bed. Mixtures of oxygen and steam enter the reactor in two places. The first feed mixture has a low concentration of oxygen in steam and enters through a fluidization deck, resulting in a uniformly fluidized bed maintained at 954–1010°C. The second mixture has a high oxygen concentration and enters at the center of the fluidization deck at high velocity through a specially developed nozzle. This high velocity stream "blows a bubble" above the nozzle. Locally, the temperature is $\geqslant$38°C hotter than the bed because of the high oxygen content of the stream. Ash agglomerates are formed in this zone and drop through the nozzle into a residue hopper where they are cooled with water and discharged. The medium heat-value gas discharges from the vessel through an internal cyclone where most of the entrained coal fines are removed and returned to the bed. The gas is further cleaned of solids in an external second-stage cyclone. The dust from that cyclone is returned to the nozzle and reenters the bed where it is burned. The raw gas is purified in conventional ways.

Although U-GAS was developed to furnish a low or medium heat-value gas for combined cycle, electric power generation (U means utility), or for an industrial fuel, it also can be used as a synthesis gas for chemical production or methanated to SNG (substitute natural gas). Current development is directed toward the production of medium heat-value gas for industrial use by customers of Memphis Light, Gas and Water Co. The program, which is cosponsored by DOE, aims at the design and construction of a plant to gasify 2540 metric tons of Kentucky coal per day into 52.7 T J/d (50×10^9 Btu/d) of gas.

Table 4. Tabulation of Existing Koppers-Totzek Plants[a]

Company	Feedstock	Number of gasifiers	Capacity, m^3 CO + H_2/d[b]	Synthesis	Ordered (year)
Charbonnages de France, Mazingarbe Plant, Paris, France	bituminous coal dust, coke-oven gas, residual gas	1	75,000–150,000	methanol and ammonia synthesis	1949
Typpi Oy, Oulu, Finland[c]	bituminous dust, oil, peat	3[d]	140,000	ammonia synthesis	1950
Nihon Suiso Kogyo Kaisha, Ltd., Tokyo, Japan	bituminous coal dust	3	210,000	ammonia synthesis	1954
a nitrogen plant in Puentes de Garcia Rodriguez (Coruna), Spain[c]	lignite dust	3	242,000	ammonia synthesis	1954
Typpi Oy, Oulu, Finland[c]	bituminous coal dust, oil, peat	2[d]	140,000	ammonia synthesis	1955
S.A. Union Chimique Belge, Zandvoorde Plant, Brussels, Belgium[c]	bunker fuel oil, plant can be extended to include coal dust gasification system	2[d]	176,000	ammonia synthesis	1955
Amoniaco Portugues S.A.R.L., Estarreja Plant[c]	heavy naphtha; plant can be extended to include lignite dust and anthracite dust gasification system	2	169,000	ammonia synthesis	1956
Ptolemais Nitrogen Plant, Greece[c]	lignite dust, bunker fuel oil	4	629,000	ammonia synthesis	1959
a nitrogen plant in Puentes de Garcia Rodriguez, Spain (reconstruction)[c]	lignite dust or naphtha	1	175,000	ammonia synthesis	1961
Attaka Nitrogen Plant, Suez, Egypt (ARE)[c]	refinery residual gases, L.P.G. and naphtha	3	778,000	ammonia synthesis	1963
Chemical Fertilizer Company Ltd. of Thailand, Fertilizer Plant, Mae Moh, Lampang, Thailand[c]	lignite dust	1	217,000	ammonia synthesis	1963
Azot Sanayii T.A.S., Kütahya Plant, Ankara, Turkey[c]	lignite dust	4	775,000	ammonia synthesis	1966
VEB Hydrierwerk Zietz, German Democratic Republic[c]	distillation residues and fuel oil	2	360,000	raw gas (hydrogen production)	1966
Kobe Steel Ltd., Kobe, Japan, for Industrial Development Corp. of Zambia, Kafue, Lusaka, Zambia	bituminous coal dust	1	214,320	ammonia synthesis	1967
Nitrogenous Fertilizers Industry S.A., Ptolemais Nitrogen Plant, Athens, Greece[c]	lignite dust	1	165,000	ammonia synthesis	1969

Table 4 (*continued*)

Company	Feedstock	Number of gasifiers	Capacity, m^3 CO + H_2/d[b]	Synthesis	Ordered (year)
The Fertilizer Corp. of India Ltd., Ramagundam Plant, New Delhi, India[e]	bituminous coal dust	3	2,000,000	ammonia synthesis	1969
The Fertilizer Corp. of India, Talcher Plant, New Delhi, India[e]	bituminous	3	2,000,000	ammonia synthesis	1970
Nitrogenous Fertilizers Industry S.A., Ptolemais Nitrogen Plant, Athens, Greece[c]	lignite dust	1	242,000	ammonia synthesis	1970
The Fertilizer Corp. of India, Korba Plant, New Delhi, India[e]	bituminous coal dust	3	2,000,000	ammonia synthesis	1972
AE & CI Ltd., Modderfontain Plant, Johannesburg, South Africa[c]	bituminous coal dust	6	2,150,000	ammonia synthesis	1972
Indeco Chemicals Ltd., Lusaka, Kafue, Zambia[c]	bituminous coal dust	1	220,800	ammonia synthesis	1974
Indeco Chemicals Ltd., Lusaka, Kafue, Zambia[f]	bituminous coal dust	2	441,600	ammonia synthesis	1975
Petrobras San Jeronimo, Brazil[g]	bituminous coal dust	2	1,300,000	ammonia synthesis	1979

[a] Ref. 12. All plants have been built by Krupp-Koppers (formerly H. Koppers) of Essen, Federal Republic of Germany, process licensors.
[b] To convert m^3 to ft^3, multiply by 35.3.
[c] Plants in daily operation.
[d] The plants in Finland and Belgium have converted to an oil feedstock using Shell's partial oxidation process and continue daily operation.
[e] The Indian plants in Ramagundam and Talcher were in the start-up phase during the first quarter of 1979. The process engineering for the Korba plant is complete but not construction; however, further work has been suspended.
[f] The latest gasification expansion of the plant in Zambia is complete. However, the expanded utilities support is incomplete preventing operation.
[g] Krupp-Koppers announced the latest contract signed in March, 1979. Start-up is scheduled in 1983.

High Heat-Value Gas. High heat-value (high-Btu) gas is generally referred to as SNG or pipeline-quality gas. Its composition is as close to pure methane as possible (see Hydrocarbons, C_1–C_6). Any of the medium heat-value gases that consist of carbon monoxide and hydrogen (often called synthesis gas) can be converted to SNG by methanation, a low-temperature catalytic process that combines carbon monoxide and hydrogen to form methane and water.

$$CO + 3\,H_2 \rightarrow CH_4 + H_2O \qquad (4)$$

Upon drying, the methanated gas is nearly pure methane. Prior to methanation, the gas from the gasifier must be thoroughly purified, especially from sulfur compounds. The composition of the gas must be adjusted, if required, to contain 3 parts hydrogen to 1 part carbon monoxide to fit the stoichiometry of equation 4. This is accomplished by a catalytic water–gas shift reaction:

$$CO + H_2O \rightarrow CO_2 + H_2 \qquad (5)$$

The ratio of hydrogen to carbon monoxide is controlled by shifting only part of the gas stream. After the shift, the carbon dioxide (formed in the gasifier and in the water-gas reaction) and the sulfur compounds formed during gasification are removed from the gas by any of a number of processes. These scrubbing, shifting, and methanating steps considerably add to the complexity and cost of making SNG as compared to medium heat-value gas.

The processes that have been developed specifically for SNG often make as much methane in the gasification step as possible to minimize the methanation step. In addition, methane formation is highly exothermic, and the more that can be formed in the gasifier where the heat can be used for the endothermic steam–carbon reaction to produce hydrogen and carbon monoxide, the more efficient the process.

$$C + H_2O \rightarrow CO + H_2 \qquad (6)$$

The production of SNG can be illustrated by describing the fully commercial Lurgi process. The Lurgi (16) gasifier (Fig. 5) is similar in principle to the Wellman-Galusha unit. It is designed for operation at pressures up to 3.2 MPa (450 psig). Coal is fed through a lockhopper mounted on top of the gasifier. A rotating distributor provides uniform coal feed across the bed. When processing caking coals, blades attached to the distributor (not shown in Fig. 5) rotate within the bed to break up agglomerates. A revolving grate supports the bed at the bottom and serves as a distributor for steam and oxygen. Solid residue is removed at the bottom of the gasifier through an ash lockhopper. The entire gasifier vessel is surrounded by a water jacket in which high pressure steam is generated.

Crude gas leaves from the top of the gasifier at 288–593°C, depending on the type of coal fed. The composition of gas also depends on the type of coal feed and is notable for the relatively high methane content when contrasted to gases produced at lower pressures or higher temperatures. Typical compositions of raw gas from various types of coals are shown in Table 5. These can be used as produced for electric power production or can be treated to remove carbon dioxide and hydrocarbons to provide synthesis gas for ammonia, methanol, and synthetic oil production. The gas is made suitable for methanation to SNG by a partial shift and carbon dioxide and sulfur removal. In any of these cases, synthesis takes place at high pressure, which gives the Lurgi gasification process an advantage over other processes using low pressure gasifiers.

As in most of the high heat-value processes, the raw gas is in the medium heat-value-gas range and can be employed directly in that form. Removing the carbon dioxide raises the heating value, but not enough to be worthwhile without methanation.

Table 6 is a listing of all the Lurgi gasifiers that have been installed as of late 1978. The only units currently operating are the 13 at Sasolburg, and the 5 at Essen (Lunen). The former produce synthesis gas for oil production and the latter a fuel gas which fires a combined-cycle electric generating plant. The Essen plant is the first Lurgi unit to use air rather than oxygen.

Recent attention has been directed toward the use of Lurgi gasifiers for plants to produce SNG in the United States. In these plants, the gas would undergo the cleanup, shift, and methanation steps described. In addition, a new version of the process is under development in which the ash is removed as molten, rather than dry, slag. This method increases the reaction capacity and reduces the steam requirement.

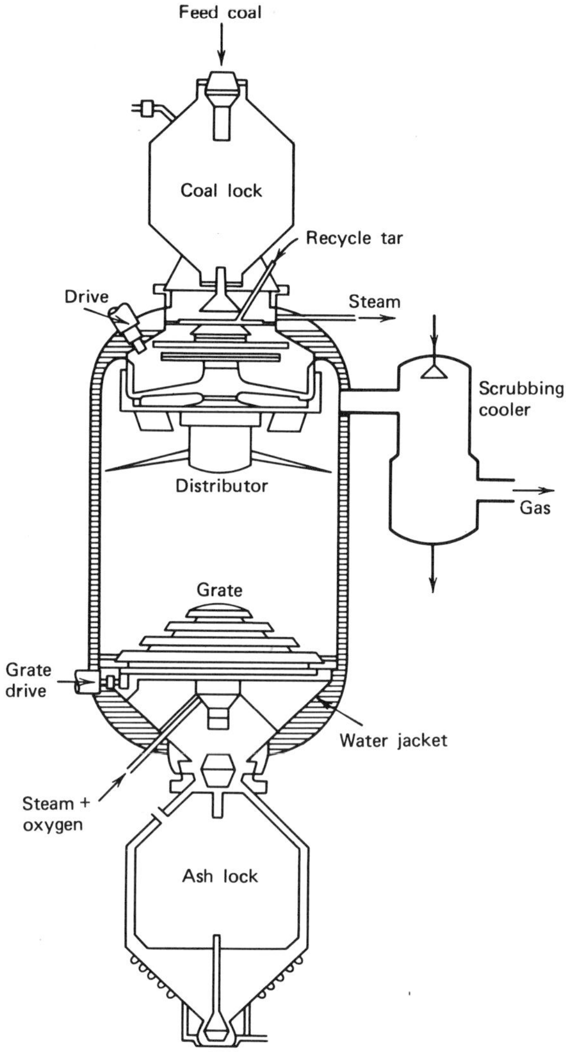

Figure 5. Lurgi pressure gasifier (17).

Additional information on this extensive subject can be obtained from refs. 18–42.

Peat

The United States, as well as the rest of the world, contains immense resources of peat, a cellulosic plant residue (see Lignite and brown coal). The peat is easily obtained because it lies above ground. However, it contains 90–95% water. In Europe, it is commonly air-dried to 30% water, and is used as a fuel similar to coal. In the United States, it has only been used in small quantities for agricultural purposes.

IGT recently joined the Minneapolis Gas Company in an effort to make the extensive deposits in northern Minnesota available to the residents of that state in the

Table 5. Lurgi Product Gas Analyses [a]

	Lignite	Subbituminous coal	Low volatile coal
Components, vol %			
CO_2	31.9	28.2	26.5
C_nH_m	0.5	0.3	0.1
CO	17.4	20.6	21.4
H_2	36.4	39.6	43.5
CH_4	13.5	10.5	8.0
N_2	0.3	0.8	0.5
Total	*100.0*	*100.0*	*100.0*
heating value as produced, MJ/m^3 [b]	12.1	11.3	10.8
heating value, CO_2 removed, MJ/m^3 [b]	17.7	15.8	14.7

[a] Ref. 16.
[b] To convert MJ/m^3 to Btu/ft^3, multiply by 26.87.

Table 6. Lurgi Coal Gasification Plants [a]

Year	Location	Number	Diameter, m	Service	Capacity, 10^3 m^3/d [b]
1936	Hirschfelde, FRG	2		town gas	28
1940	Bohlen, FRG	5	2.5	town gas	250
1943	Bohlen, FRG	5	2.5	town gas	280
1944	Brun, Czechoslovakia	3	2.5	town gas	210
1949	Brun, Czechoslovakia	3	2.5	town gas	250
1954	Sasolburg, South Africa	10	3.7	synthesis gas	3700
1955	Dorsten, FRG	6	2.7	town gas	1600
1956	Melbourne, Australia	6	2.7	town gas	620
1957	Daud Khel', Pakistan	2	2.7	synthesis gas	120
1960	Westfield, Scotland	4	2.7	town gas	710
1963	Coleshill, England	5	2.7	town gas	1300
1963	Seoul, Republic of Korea	3	3.2	synthesis gas	420
1966	Sasolburg, South Africa	3	3.7	synthesis gas	1700
1969	Essen, FRG	5		fuel gas	4500
1978	Sasolburg, South Africa	3 [c]	4.0	synthesis gas	
		1 [c]	4.9		
1981	Sasolburg, South Africa	36	4.0	synthesis gas	42,000

[a] Refs. 16, 18–20.
[b] To convert m^3/d to ft^3/d, multiply by 35.3.
[c] Under construction.

form of SNG (43). Funding was furnished by Minneapolis Gas and DOE to develop the PEATGAS process, a two-stage gasification process, which is now ready for large pilot-plant operation.

Peat, dried to 25–50 wt % moisture, is crushed into particles smaller than 1.6 mm, and fed to the top of a reactor through lockhoppers (see Fig. 6). At the top of the reactor, a fluidized-bed stage completes the drying of the incoming peat by exchanging heat with gases from the hydrogasification stage below. Dried peat flows down through a standpipe into a lift-line where it is entrained by the gas from the peat–char gasification stage. The mixture of peat and synthesis gas flows up the lift-line in a dilute phase. This dilute-phase, cocurrent, short residence time hydrogasification section produces as much as 80% of the total methane made in the process. At the top of the lift-line, the gas and the solids (char) are disengaged. Gas flows up to the drying stage, and char flows down to the char gasification stage where it reacts with steam and oxygen in a fluidized bed to make synthesis gas. The feed carbon is essentially completely converted. The hot ash is discharged from the system and cooled by recirculating water. The resultant ash–water slurry is filtered, and the filtrate recycled.

After fines are removed, the gas from the top of the PEATGAS reactor flows through a CO-conversion reactor where the hydrogen:carbon monoxide ratio is adjusted to 3.7. An oil- and sulfur-resistant catalyst is used in the CO-conversion reactor. After recovering the liquid products downstream of the CO-shift reactor, the gas is purified to remove hydrogen sulfide and carbon dioxide before entering the methanation reactor. About 84% of the total methane made in the plant is formed in the reactor. The remainder is made in the catalytic methanation reactor. The dried product gas has a heating value of 35.7 MJ/m^3 (960 Btu/ft^3).

In addition to making SNG, the process also makes by-product benzene, oil, anhydrous ammonia, and sulfur. Energy required for producing oxygen is supplied internally from the peat. For making a medium heat-value gas, the CO-conversion and catalytic methanation steps downstream of the PEATGAS reactor are not required. Operating conditions are 3.55 MPa (500 psig) and 232°C in the dryer, 802°C in the hydrogasifier lift line, and 371°C in the steam–oxygen combustion zone.

A large plant has an overall thermal efficiency of about 67%. The following yields of various products and by-products per metric ton of dry Minnesota reed-sedge peat (received at the plant with 50% moisture) can be expected:

SNG	262 m^3 (9250 ft^3)
benzene	17.8 L (4.7 gal)
oil	35.6 L (9.4 gal)
anhydrous ammonia	18 kg
sulfur	1.8 kg

Economic studies indicate that SNG produced from peat can be competitive with other SNG sources and a cheaper and more secure source of energy than imported natural gas or oil.

Oil Shale

Oil shale (qv) is another immense natural resource that is not being exploited for a variety of reasons. Much work has been done on converting it to oil. Essentially, the only efforts to produce SNG from it have been done by IGT under American Gas Association sponsorship (44).

This process is capable of recovering almost all of the organic carbon in the shale as useful gaseous and liquid products. This has been demonstrated in equipment ranging in size from a laboratory unit using only a few grams of shale to a large adiabatic process development unit (PDU) with a 0.25 kg/s shale-feeding capacity. This high organic carbon conversion is obtained by controlled shale heating in hydrogen at

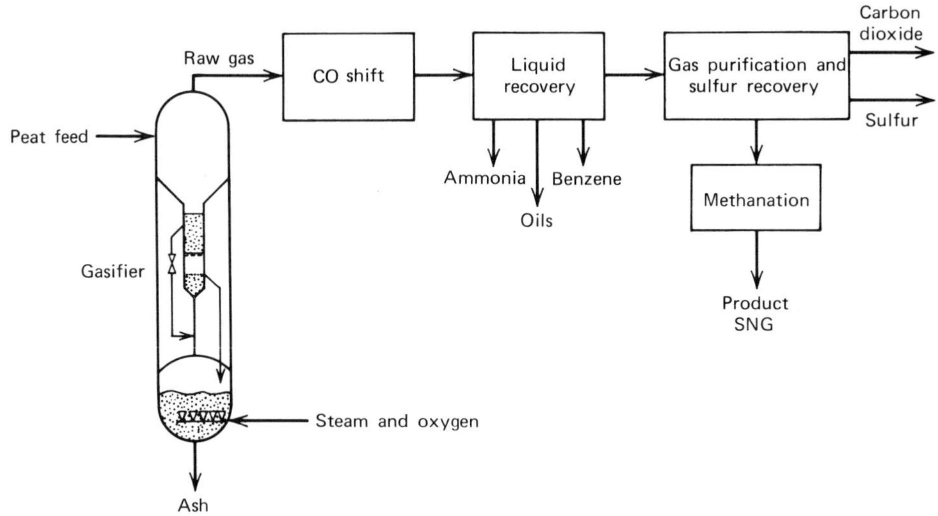

Figure 6. PEATGAS Process (43).

moderate pressure. When coupled with countercurrent heating and cooling of the shale, not only is the organic carbon conversion high, but most of the thermal energy in the spent shale is recovered for use in the process. By proper selection of operating conditions, the process can be used to produce primarily gas or oil. Although most of the work done so far in developing this process has been directed toward aboveground processing, the process appears to be applicable to modified *in situ* processing as well.

This process has been demonstrated with western (Colorado) oil shales and with eastern (New Albany) shales. Total organic carbon conversions are somewhat lower with the eastern shales than with the western shales. However, when compared with oil yields by conventional retorting, the process is quite attractive with eastern shales, converting 2.5–3 times more kerogen into a useful product than simple retorting, which typically recovers only about 30% of the organic carbon content.

Figure 7 is a schematic diagram showing the principal steps of one variation of the IGT process. Raw, crushed shale in a moving bed passes downward through a reactor divided internally into three continuous-operation zones. The feed shale is prehydrogenated and preheated in the upper zone and hydroretorted or hydrogasified, or both, in the middle zone. The spent shale is cooled in the bottom zone. Hydrogen feed to the bottom zone recovers essentially all the sensible heat in the spent shale, then uses it to preheat the feed shale in the upper zone. This hydrogen stream bypasses the middle zone, where most of the retorting occurs.

A separate hydrogen stream passes through the middle zone and hydroretorts and/or hydrogasifies the prehydrogenated kerogen to produce gaseous and vaporized hydrocarbons. These effluent gases and vapors are cooled, and the resulting oils are collected and fractionated. The heavy-oil fraction can be used for hydrogen production; the high-quality, low-sulfur fraction can be a coproduct. Alternatively, hydrogen can be produced by catalytic steam reforming of the lighter fractions. The effluent gases are further processed to pipeline-quality gas. If desired, any suitable portion of the light or heavy oil fractions can be hydrogasified and upgraded to produce additional quantities of high heat-value gas.

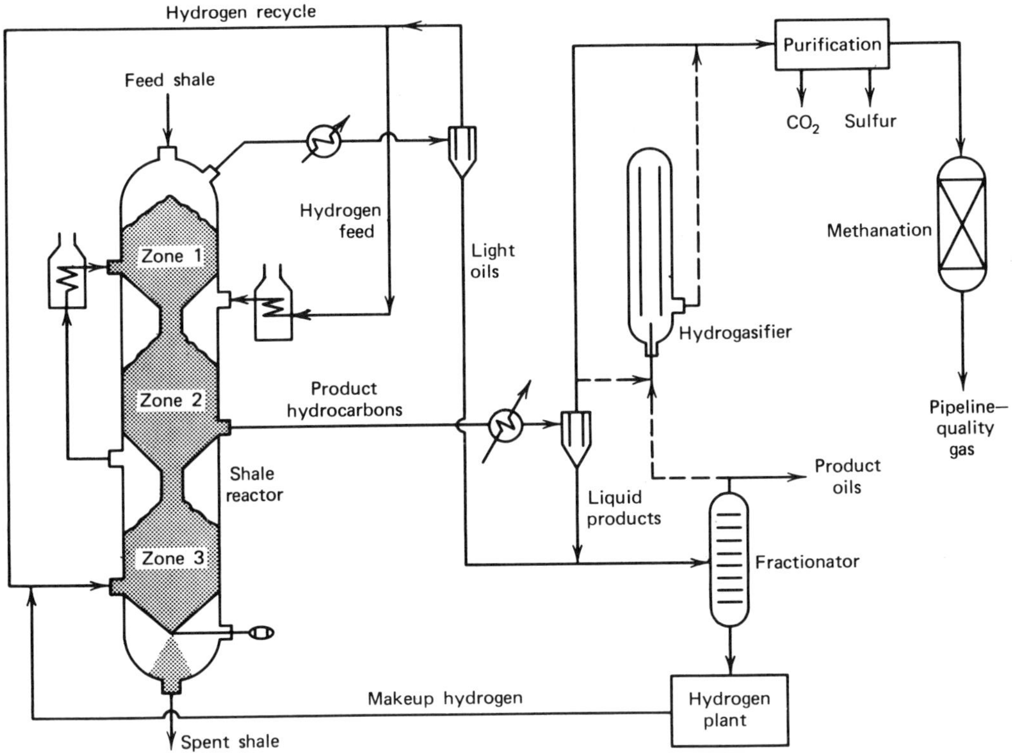

Figure 7. IGT/A.G.A. oil–shale process (45).

As mentioned above, this process can be adapted for either high gas or high oil yields. This process flexibility is achieved by varying the way the middle zone is operated. If the temperature is kept below 649°C, the principal product will be oil boiling below ca 388°C, and a relatively small amount of gas.

The process developers are currently awaiting a decision to build a large-scale pilot plant in the eastern part of the United States.

Hydrocarbon Liquids

The Middle East war of 1973 and the subsequent Arab oil embargo cast considerable doubt on oil supplies. Although extensive work has been done in the United States, the United Kingdom, and Japan on the production of SNG from hydrocarbon liquid feedstocks, many naphtha-reforming plants had been constructed (13 U.S. plants with a total capacity of 37.55×10^6 m^3/d (1326×10^6 ft^3/d) of SNG; six plants in the United Kingdom with a total capacity of 7.50×10^6 m^3/d (265×10^6 ft^3/d); and several plants with a total capacity of about 6.91×10^6 m^3/d (244×10^6 ft^3/d) of SNG in Japan). Most plans for future construction have been shelved, and little interest has been shown in oil gasification since that time. Until then, oil gasification for SNG production was a well-developed art.

Thermal Cracking Processes. U.S. commercial production of high heat-value oil gas by thermal cracking increased during the 1940s and reached a peak in 1952 when the equivalent of 368×10^6 m^3 (13×10^9 ft^3) of natural gas was produced by this method (4). Most of this gas was used for peakshaving and to ensure against interrupted deliveries.

All the high heat-value oil gas sets used in the United States were operated on a cyclic basis. Because thermal cracking oil is endothermic, heat must be supplied on an overall basis to operate the process. The heat was furnished by burning oil and previously deposited carbon. Cracking the oil produced a mixture of hydrogen, methane, high molecular weight saturated and unsaturated hydrocarbons, light aromatic oils, tar, and carbon.

The chemistry of oil–gas production was established in the laboratory about 1950. On the basis of "true oil gas" composition and properties, data were correlated to exclude, as far as possible, the vagaries of individual unit operations. True oil–gas heating values range from 26.1–59.6 MJ/m^3 (700–1600 Btu/ft^3) depending on the severity of cracking conditions, ie, temperature and residence time. The composition of the oil gases produced (5) from a wide variety of feedstocks depends on the severity of cracking much more than on the feedstock type (Fig. 8). However, the yield of true oil gas, compared with those of tar and deposited carbon, is strongly dependent on feedstock type, as shown in Figure 9 (44,46–49).

Most thermal cracking processes are operated without a catalyst. The Petrogas process and the Hall process are typical examples.

A typical improved Petrogas oil–gas set arrangement, designed by the Gas Machinery Co., is shown in Figure 10 (50). In this setup, modern technology has been applied to increase process thermal efficiency by improved waste-heat recovery techniques. Heavy-oil feedstocks, such as the sp gr $^{15}_{15}$ 0.934 (20° API) Alaskan reduced crude containing 7.6% Conradson carbon residue(CCR)-forming materials, which is used commercially in the Petrogas process by Gasco, Inc. (formerly Honolulu Gas Co.), produces about 50% liquid products plus carbon, by weight (Fig. 9). The sensible heat of this liquid fuel oil or tar product can be used as shown in Figure 10 to reheat incoming makeup oil. In addition, the high pressure waste-heat boiler shown in Figure 9 is used to produce steam from the sensible heat of the blow gases.

The object of Hall-type processes is to produce additives that are the closest possible substitutes for natural gas. Cracking conditions are not as vigorous as those sometimes employed in other processes. Distillates and crude or residual oils are used for feedstock, although gasification efficiency and output decrease because of carbon deposition as heavier oils are used.

CC content of oil used, %	0.2	3.2	6.0	13.0
gas output per unit, 10^3 m^3/d (10^3 ft^3/d)	168 (5950)	144 (5100)	144 (5100)	101 (3575)

Heating values are 36–47 MJ/m^3 (950–1250 Btu/ft^3), depending on the type of feedstock used and its Conradson carbon (CC) content. Typical operating results are shown in Table 7.

Cyclic Catalytic Processes. The three most highly developed cyclic catalytic processes are the Segas process developed by Britain's Southeastern Gas Board, the O.N.I.A.-G.E.G.I. process, a French process, and the UGI Cyclic Catalytic Reforming process. The conformation of equipment and operating procedures used in these

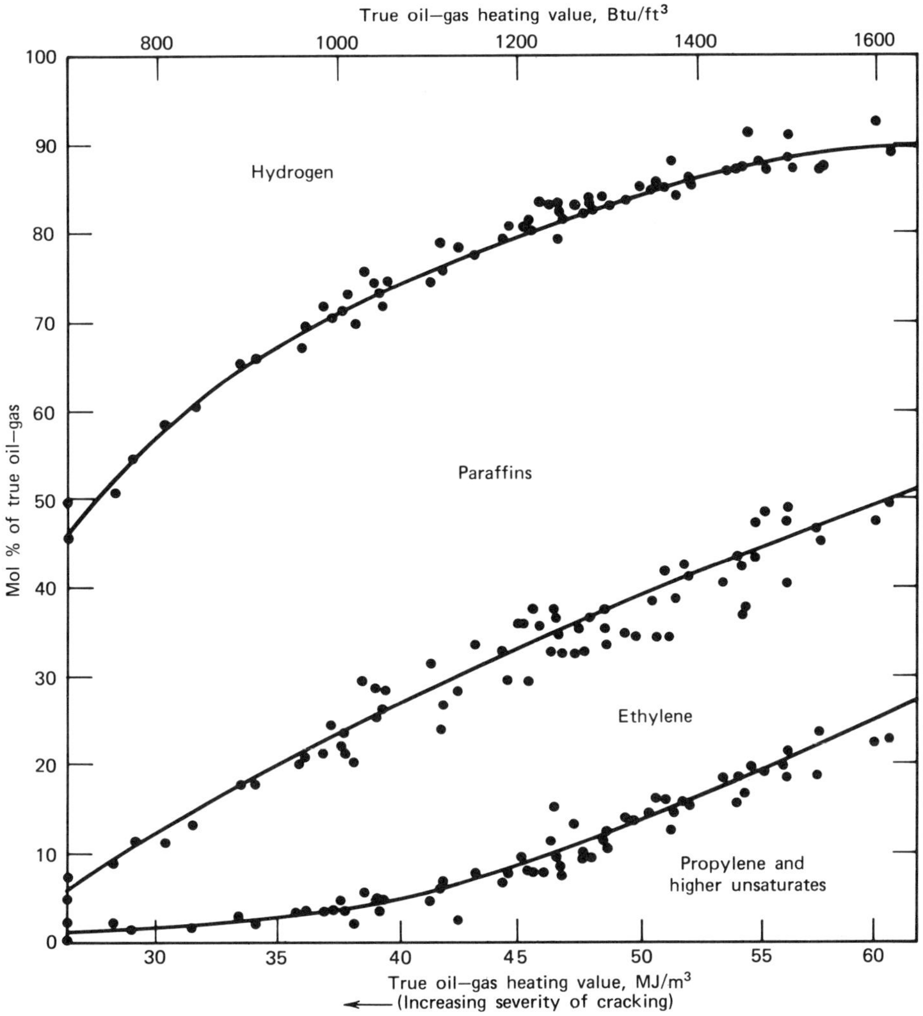

Figure 8. Relationship between true oil-gas heating value and composition for all types of feedstocks over normal range of cracking conditions (44).

processes is very similar to that used for straight thermal cracking, if the packed bed of catalyst is visualized as taking the place of the checker-brick chamber used for the initial ("make") phase of the operation.

Table 7 (49) shows that the catalyst used in these sets tends to promote steam-reforming reactions that lead to a product gas containing more hydrogen and carbon monoxide and fewer unsaturates than the product from a straight thermal cracking process. Cracking severities are higher than those from straight thermal cracking, and the resulting gas is more suitable for use as a medium heat-value gas than the rich gas produced by straight thermal cracking. The catalyst also influences rates in the basic thermal cracking reactions, leading to higher gas yields and lower tar and carbon yields, as shown in Figure 9, at a given feedstock weight. However, because of the endothermic characteristic of the reactions involving steam, more tar and supplemental heating

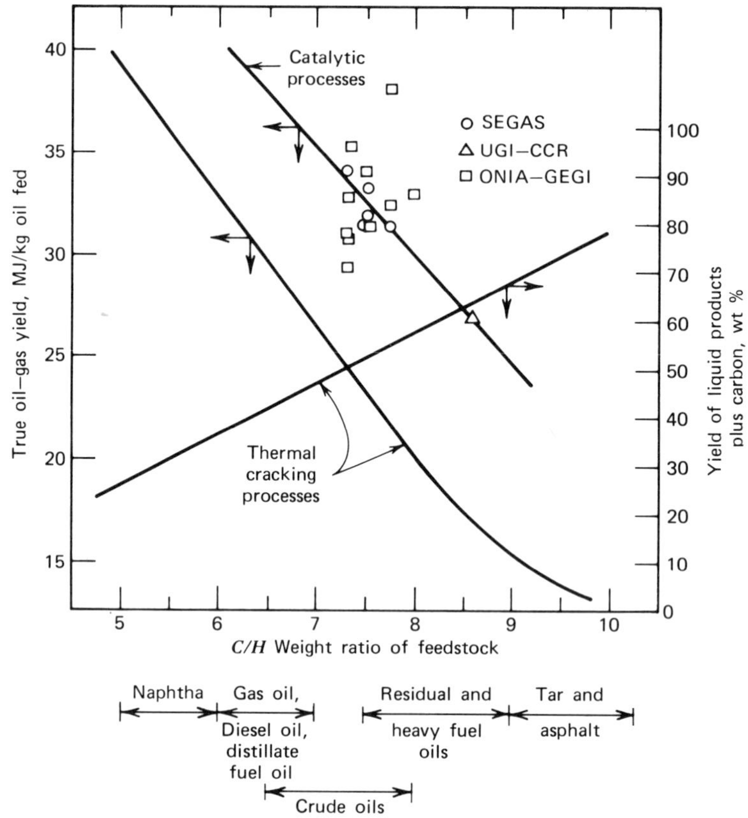

Figure 9. *C/H* ratio of feedstock vs oil–gas yields(44,46–49). To convert MJ/kg to Btu/lb, multiply by 430.2.

oil must be burned during the heating phase of a catalytic operation than in straight thermal cracking; therefore, the spread between fuel and conversion efficiencies is more narrow for the catalytic processes.

Continuous Thermal Cracking Processes. The cyclic processes have the following disadvantages: they are limited to atmospheric pressure operation, produce gases that are not interchangeable with natural gas, and are environmentally unacceptable. Continuous thermal cracking processes help to overcome these disadvantages. The equipment used is designed to accommodate the deposition of carbon.

The Fluid-Bed Thermal Cracking (FTC) process, developed by IGT, successfully gasified distillate and residual oils containing 6–7 wt % Conradson carbon residue. The Thermofor Pyrolytic Cracking (TPC) process, developed by Surface Combustion Division of Midland Ross Corp., differed from the FTC process because cracking was carried out on a moving bed of coke pebbles instead of on the fluidized bed of finely divided solids used in the FTC process. The TPC process also satisfactorily gasified both distillate and residual oils. Both of these processes produced high heat-value oil gas that was free of combustion product dilution, and as a result, the SNG produced was interchangeable with natural gas. In addition, closer control was maintained over the cracking conditions.

The basic problem in making SNG from heavy oils is the addition of hydrogen

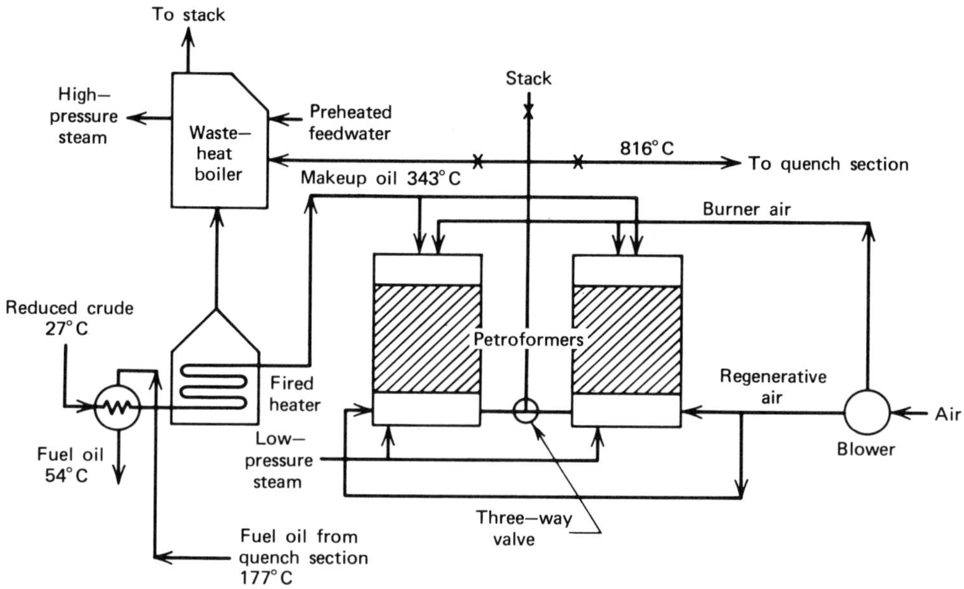

Figure 10. Improved heat recovery system for the Petrogas Process.

to make methane from oil with the approximate empirical formula, $(CH_2)_n$. This can be accomplished by direct reaction with hydrogen. Generally, as the feedstock weight is increased, similar operating conditions of hydrogasification lead to decreasing hydrocarbon gas yields and increasing yields of aromatic liquids, with solid carbon appearing in the crude and residual oil weight range. The best single variable to express this dependence is the C/H weight ratio of the feed oil. Figure 11 (51) shows the distribution of feed carbon resulting from similar operating conditions (760°C, 7 MPa (1000 psig), and ca 1500 m^3 H_2/m^3 oil (200 ft^3/gal)) in tests with a wide variety of petroleum feedstocks from a number of different sources. Reactor types used were the empty tube, fluidized-bed hydrogenator (FBH), gas recycle hydrogenator (GRH), and checker-brick chambers.

The major secondary variable that influences yields from oils of various types is their aromatic content. A feedstock of a given C/H ratio that contains a large proportion of aromatic constituents can produce a larger proportion of liquid products, and is more likely to produce solid carbon than a predominantly paraffinic feedstock.

Operating conditions, as well as feedstock type, have a profound effect on the distribution of feed carbon in hydrogasification. The most important variables affecting product distribution are hydrogen-to-oil ratio and gasification temperature. Figure 12 summarizes the effects of stoichiometry with diesel oil at a constant operating temperature, 760°C. The hydrogen-to-oil ratio is expressed as a percentage of the stoichiometric hydrogen required to completely convert the feed oil to methane, 1575 m^3 H_2/m^3 oil (ca 210 ft^3/gal). The C/H ratio of the oil used in these tests was 6.29, indicating low aromaticity. The amount of carbon deposited was not influenced by system pressure in the 3.50–10.5 MPa (500–1500 psig) range studied, but was significantly affected by the hydrogen-to-oil feed ratio. When 50% of the stoichiometric hydrogen was used, 12% of the carbon in the feed was deposited as a solid. When 100% of the stoichiometric hydrogen was used, no carbon was deposited.

Table 7. Operating Results of Thermal Cracking Processes[a]

Process	Hall	Segas[b]
feedstock, °API[c]	14.3	15.5
process feedstock, g/m³ (lb/10³ ft³) of SNG	1485 (92.8)	552 (34.5)
process steam, g/m³ (lb/10³ ft³) of SNG	207.8 (13.0)	1522 (95.0)
dry gas composition, vol %		
H_2	20.1	49
$CO_2 + CO$	9.4	26
N_2	13.4	3.0
paraffins	32.6	15.3
$CH_2{=}CH_2$	15.1	6.7
higher unsaturates	9.4	
Total	*100.0*	*100.0*
HHV[d], MJ/m³ (Btu/ft³)	36.6 (982)	18.7 (502)
total recuperated steam, g/m³ (lb/10³ ft³)		1.52 (95.0)
supplemental fuel, kJ/m³ (Btu/10³ ft³)	2390 (64.1)	975 (26.2)
HHV of net tar residue, kJ/m³ (Btu/10³ ft³) product gas	16530 (444)	1410 (38)
cold gas efficiency, %		
conversion[e]	57.7	78.9
fuel[f]	67.3	81.5

[a] Ref. 49.
[b] Southeastern Gas Board, UK.
[c] $°\text{API} = \dfrac{141.5}{\text{sp gr } 15/15°\text{C}} - 131.5.$
[d] HHV = higher heating value.
[e] $\text{Conversion efficiency} = \dfrac{\text{HHV of gas}}{\text{HHV of process fuel feedstock.}}$
[f] $\text{Fuel efficiency} = \dfrac{\text{HHV gas + HHV tar + steam fed}}{\text{HHV total fuel.}}$

The relative quantities of gaseous and liquid products at a given stoichiometric ratio depend primarily on temperature. Increasing the operating temperature from 593 to 760°C at the exact stoichiometric ratio, for example, decreases the liquid product yield from 25 to 8 wt % of the diesel-oil feed.

Two SNG processes based on direct hydrogasification have been commercialized; others have been developed to the semiworks pilot stage. These processes generally are based on one of the two following possibilities: (*1*) the use of process feedstocks with a C/H ratio of about 6.8 or less (No. 2 fuel oil or lighter) and operation of the hydrogasification reactor without carbon formation; and (*2*) hydrogasification of any weight of process feedstock and the removal or burn-off of the solid carbon deposited in the gasification zone.

A hybrid process also is possible in which a heavy-oil feedstock is hydrocracked prior to hydrogasification to reduce its C/H ratio to the range required for carbon-free operation.

Processes based on carbon-free operation are typified by the British Gas Corporation's GRH process (52). Thirty-six commercial installations based on the GRH reactor are currently in use, mostly for enrichment of lean gas for town gas production. The GRH reactor is shown in Figure 13 (53).

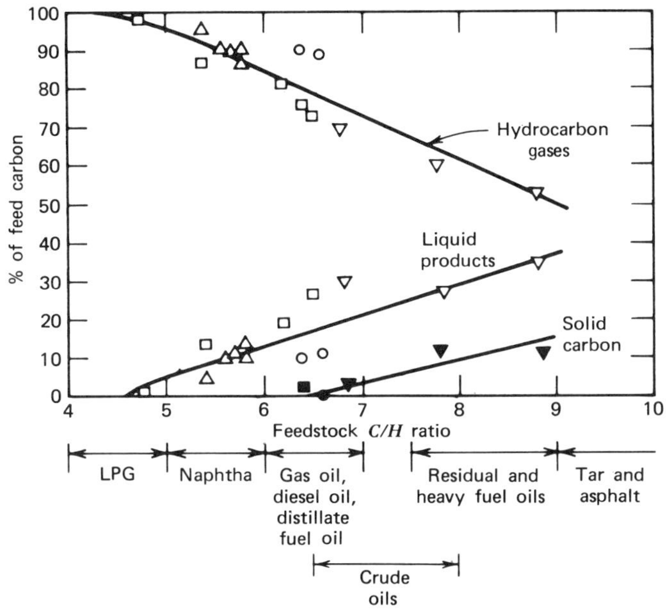

Figure 11. Distribution of feed carbon in high pressure hydrogasification (51). Data source—●, ○; *IGT Research Bulletin No. 29* (6); ■, □, △: British Gas Council; and ▼, ▽: unpublished IGT data. Operating conditions: 732–788°C, 3.5–10.4 MPa (500–1500 psig), and 85–100% stoichiometric hydrogen.

In operation, process oil and hydrogen are introduced through the center tube, and product gas is aspirated around the annulus; mixing supplies the preheat required to keep the unit in heat balance. Product gas is withdrawn from the side of the unit at the top. Thus, the product gas is continuously recycled around the inner tube of the unit, inside of which the reaction actually occurs.

Although a 37.22 MJ/m^3 (1000 Btu/ft^3) gas comprised of 44 vol % methane, 25 vol % ethane, 24 vol % hydrogen, and 7 vol % other constituents can be produced easily, directly making 95% methane in the GRH reactor is difficult because carbon is deposited in the hydrogen-poor atmosphere of a gasifier operated at such a high hydrogen conversion rate. To make SNG consisting almost entirely of methane by the GRH process, some type of methanator or hydrogen–methane separator must follow the GRH reactor to remove unreacted hydrogen.

Figure 14 shows an IGT method for processing crude oils too heavy to be hydrogasified directly. The crude is first hydrocracked to yield a light fraction, which has a *C/H* ratio suitable for hydrogasification, and a heavy fraction, which is used for hydrogen production by partial oxidation (54). In this scheme, which is a hybrid of the GRH process mentioned above, excess hydrogen in the product gas from the hydrogasifier is methanated by using by-product carbon dioxide from the hydrogen plant.

An alternative available for hydrogasifying feedstocks with a high *C/H* ratio is to accept some carbon deposition during hydrogasification and to use the deposited carbon for heat or for hydrogen production. Both IGT (54) and the Hydrocarbon Research Institute (55) did pilot-plant work based on this concept, which is an extension of the British Gas Corporation's FBH process (56). The FBH reactor, shown

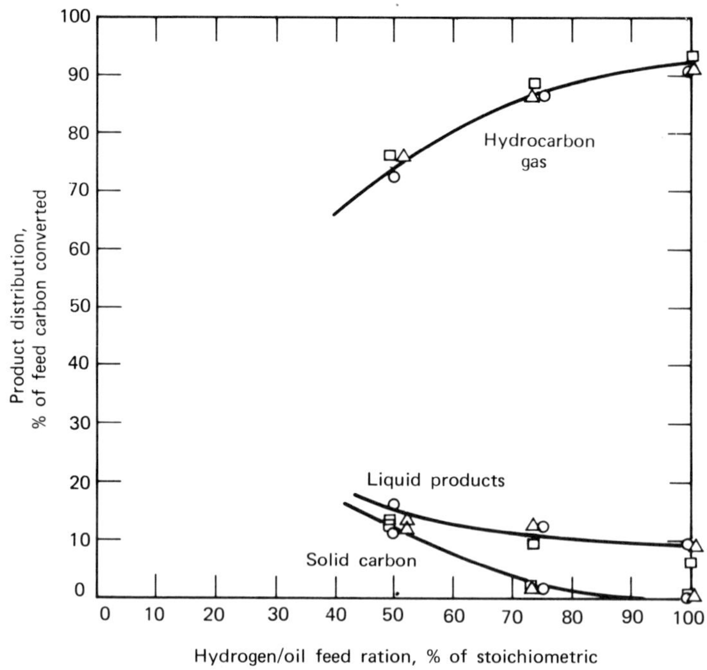

Figure 12. Increased hydrogen-to-oil ratio eliminates carbon deposition (51). Pressure, MPa (psig): ○, 3.5 (500); △, 7 (1000); □, 10.4 (1500). Temperature, 760°C.

in Figure 15 and described in numerous publications of the British Gas Corporation, has been operated on a 1.4×10^5 m^3/d (5×10^6 ft^3/d) scale in Japan.

A typical simplified diagram of the IGT heavy-oil hydrogasification process is shown in Figure 16. The feed oil is heated and charged to the bottom of the hydrogasifier, where it is contacted with hydrogen-rich gas at 800°C and 10.1 MPa (1450 psig). The product stream from the hydrogasifier passes through a steam generator, a water scrubber, a monoethanolamine (MEA) acid-gas removal system (see Alkanolamines; Carbon dioxide), and a drying step, yielding high heat-value gas.

Solids containing carbonaceous residues from the hydrogasification step flow from the hydrogasifier to the gasifier. Oxygen and steam are added to the gasifier to convert the carbonaceous residues to a synthesis gas. Because of the hydrogen requirements in the hydrogasifier stage, some of the feed oil usually must be added to the gasifier, and the rest to the hydrogasifier for conversion into gas.

The product gas from the gasifier is cooled and shifted, and the carbon dioxide is removed by hot carbonate scrubbing. The hydrogen-rich gas from the carbonate-scrubbing step is compressed, reheated to 93°C, and returned to the hydrogasifier lift leg, where it is used to lift and fluidize the solids returning to the hydrogasifier from the gasifier stage.

Partial Oxidation. As in the case of coal, SNG from heavy oil can be produced by partially oxidizing the oil to a mixture of carbon monoxide and hydrogen, which is methanated catalytically to produce methane of any required purity. The initial partial oxidation step consists of the reaction of liquid feedstock of any weight with a quantity of oxygen insufficient to burn it completely, making a mixture consisting of carbon monoxide, carbon dioxide, hydrogen, and steam.

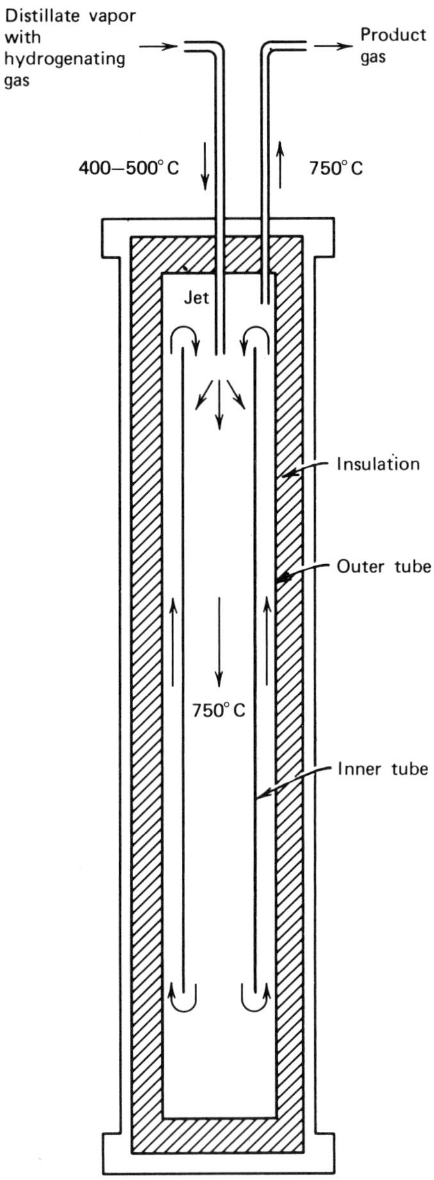

Figure 13. Gas-recycle hydrogenator (54).

Success in partially oxidizing heavy fuel oil feedstocks depends mainly on details of the burner design (58–59). Some solid carbon, usually 1–4% of the original feedstock weight, is formed. The relationship between the theoretical gas composition and the experimental results as a function of the amount of carbon deposition is shown in Figure 17 (60). This curve also shows that the conversion efficiency of the process is maximized at about 2% unconverted carbon, which for heavy oil corresponds to a specific oxygen feed rate of about 0.26 m^3 O_2/m^3 CO + H_2. The ratio of hydrogen to carbon monoxide in the product gas is a function of reaction temperature and stoi-

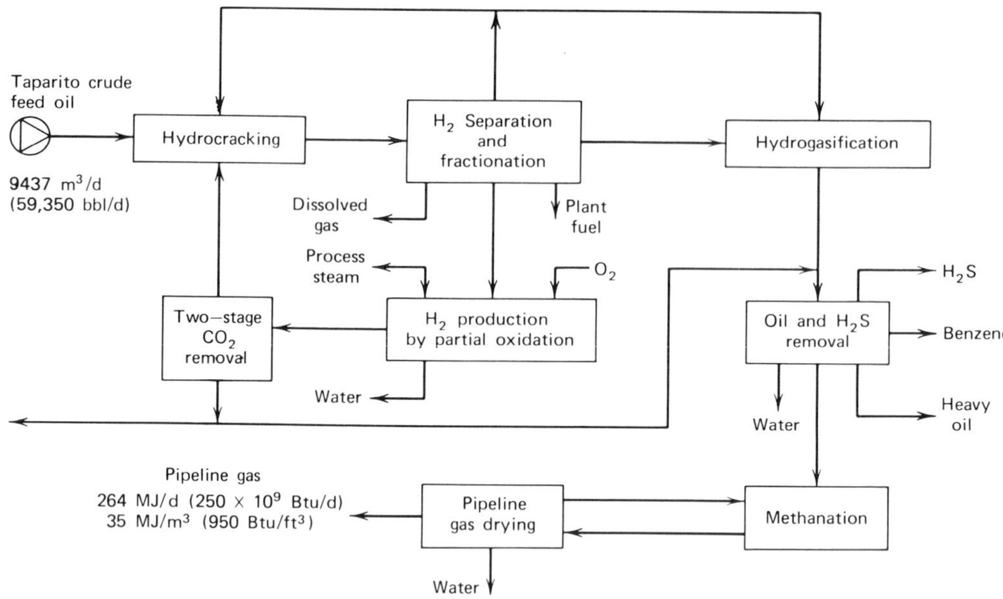

Figure 14. Production of SNG by combined hydrocracking–hydrogasification.

chiometry and can be adjusted, if desired, by varying the ratio of carrier steam to oil fed to the unit (61).

To make SNG by partial oxidation, using commercial processes, virtually all of the methane in the product gas must be produced by catalytic methanation of carbon monoxide and hydrogen. A generalized process based on this concept is shown in Figure 18 (54).

The feed is mixed with recycled carbon and fed, together with steam and oxygen, to a reactor in which partial combustion takes place. Heat from the reaction gasifies the rest of the feed, and by-product coke is formed. The heavier feedstocks tend to produce more carbon than can be consumed through recycling; thus some must be withdrawn. After carbon separation by water scrubbing, the synthesis gas that is available can be converted into hydrogen or into SNG by the steps described in the coal section.

The partial oxidation process is well adapted to the production of hydrogen for other SNG processes because it can use residual oil or similar stocks of little value for other purposes.

Several companies, including Shell Oil Co. (58), Texaco, Inc. (60–61), and Koppers Co. (59), currently offer partial oxidation processes capable of handling liquid feedstocks of any weight. The Koppers-Totzek process was developed for coal and is an atmospheric pressure operation, whereas the Shell and Texaco processes can be operated at any desired pressure. The exothermicity involved in the methanation step reduces the thermal efficiency compared with other schemes. On the other hand, any weight feedstock can be used, plants can be built over a wide range of capacities (reactors with capacities up to 1.05×10^6 m^3/d (37×10^6 ft^3/d) of hydrogen and carbon monoxide have been built), and the gasification step has been thoroughly tested commercially.

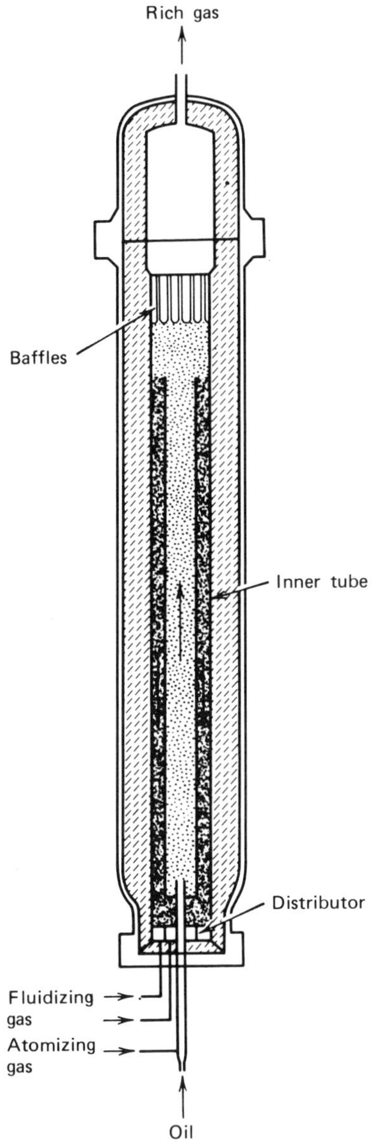

Figure 15. Fluid-bed hydrogenation (FBH) reactor developed by the British Gas Corporation.

A Texaco patent (62) claims that synthesis gases containing 10–26% methane (dry basis) can be made directly through partial oxidation by limiting the reaction temperatures. An SNG process based on this type of partial oxidation reactor would be more attractive than those based on traditional partial oxidation methods.

Steam-Reforming Processes. When relatively light feedstocks (naphthas with *C/H* ratios of less than 6.0, about 182°C end boiling point, and limited aromatic content) are available, high-nickel-content catalysts can be used to simultaneously conduct a variety of near-autothermic reactions, which result in the essentially complete conversions of the feedstocks to methane. The approximate overall net reaction is as follows:

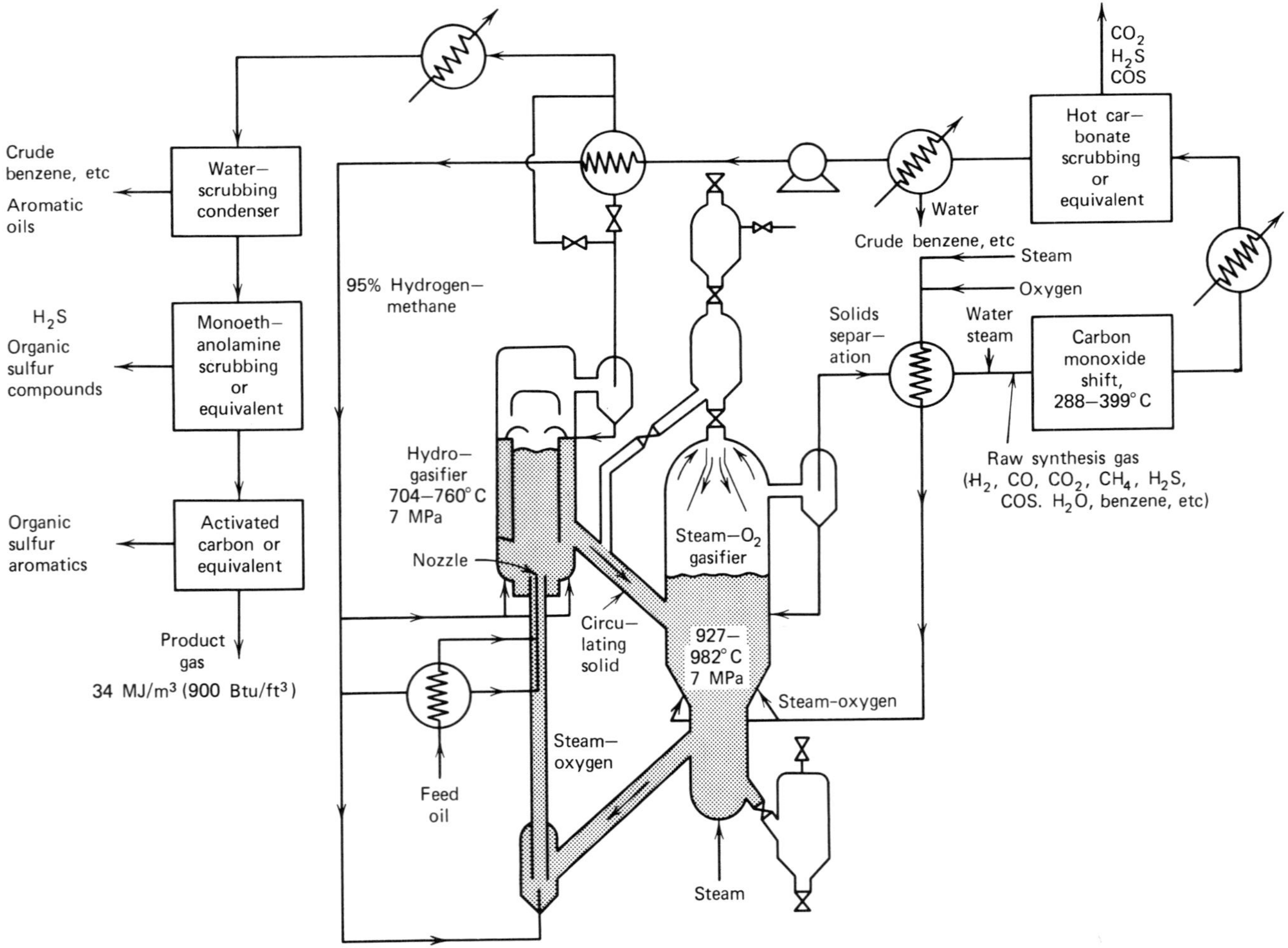

Figure 16. IGT oil-gasification process (57). To convert MPa to atm, divide by 0.101.

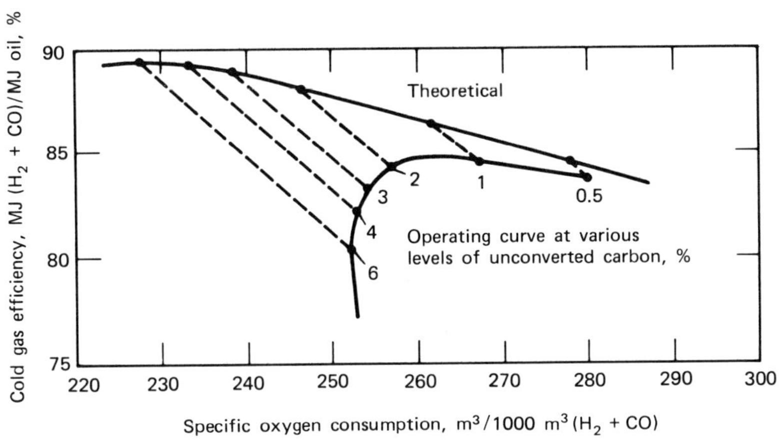

Figure 17. Oxygen consumption of partial-oxidation reactors (60). Oil properties: specific gravity, sp gr $^{15}_{15}$ 1.00 (10° API); viscosity, 200 SFs (Saybolt seconds furol) at 50°C, s^{-1}; sulfur, 2.0 wt %; nitrogen, 0.25 wt %; ash, 0.05 wt %. Operating conditions: steam/oil ratio, 0.5 kg/kg; oil–steam temp, 399°C; oxygen temp, 149°C; purity, 95%; pressure, 3.5 MPa (500 psig).

$$CH_3(CH_2)_3CH_3 + 2\ H_2O \rightarrow 4\ CH_4 + CO_2 \tag{7}$$

Because of limitations on the activity of practical catalysts, this reaction must be carried out in stages, the first of which is carried out at 427–482°C and 1.5–2.9 MPa (200 to 400 psig), and amounts approximately to the following reaction:

$$C_5H_{12} + 2.4\ H_2O \rightarrow 3.7\ CH_4 + 1.1\ CO_2 + 1.0\ H_2 + 0.2\ CO \tag{8}$$

In ensuing catalytic stages, usually termed hydrogasification and methanation (not to be confused with the noncatalytic, direct hydrogasification processes described above), the remaining carbon monoxide and hydrogen are shifted and react to produce additional methane.

Typical methane yields from feedstocks of varying weight are shown in Figure 19 (63). Overall plant thermal efficiency is usually quite high, ranging from 88–98% depending on feedstock weight and heat recovery arrangements.

There are now 13 commercial SNG plants in the United States using A steam-reforming technology, primarily that of the British Gas Corporation's Catalytic Rich Gas (CRG) process (Table 8). These plants have a combined capacity of about 36.8 $\times 10^6$ m^3 (1300 $\times 10^6$ ft^3) of SNG per day. Additionally, three more plants are in the planning stages, and five projects have been suspended for various reasons (63).

A block flow diagram of the British Gas Corporation's CRG process is shown in Figure 20. Feedstock naphtha, containing up to 500 ppm of sulfur, is hydrodesulfurized (see Petroleum), reducing the sulfur level to less than 5 ppm. The naphtha is then treated in a secondary desulfurizer in which the sulfur level is reduced to less than 0.2 ppm. The naphtha is mixed with steam, heated, and introduced into the CRG reactor where, in the presence of an active nickel catalyst, reforming proceeds at 2.86–4.24 MPa (400–600 psig) and 454°C. The reforming reactor effluent is cooled and introduced into the primary methanator, where a major portion (but not all) of the hydrogen reacts with carbon monoxide and carbon dioxide to form methane.

After cooling to remove the heat of exothermic reaction, a portion of the excess water vapor is removed, and the gas is introduced into a secondary methanator where

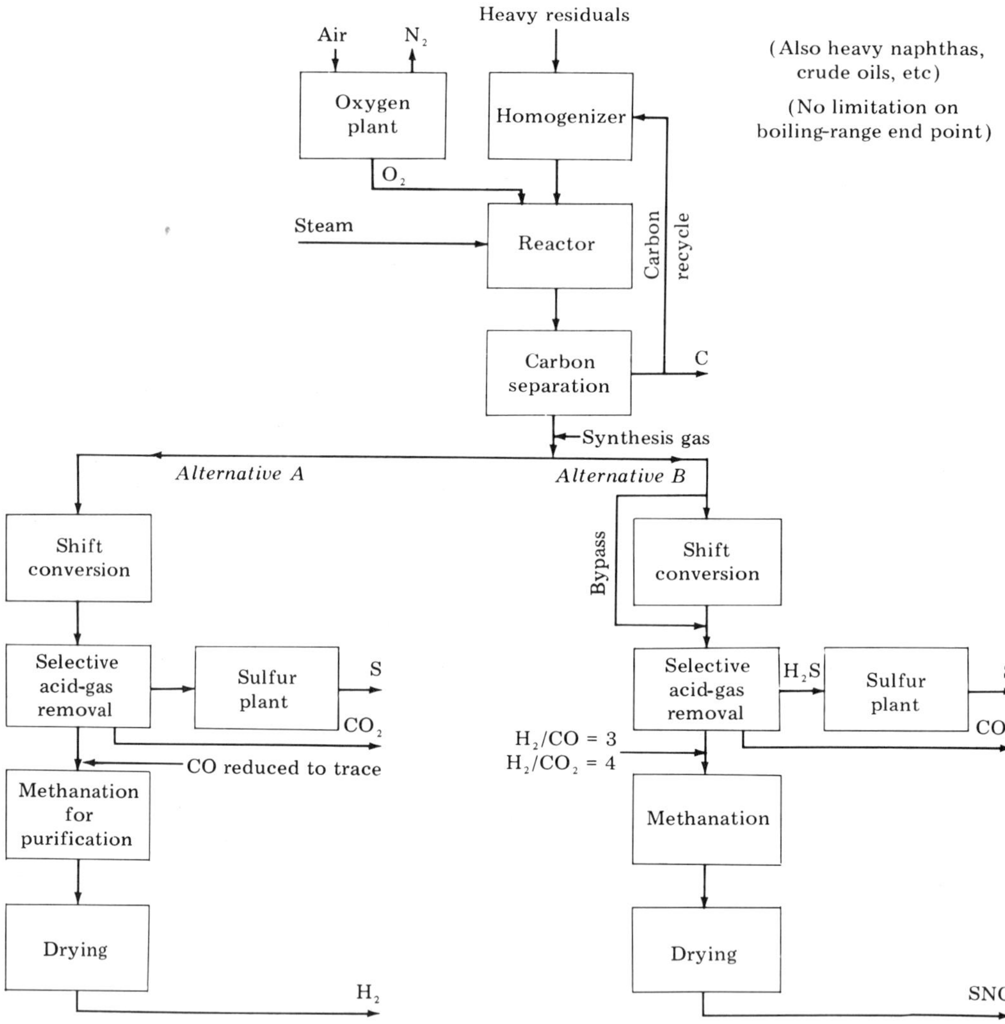

Figure 18. Partial-oxidation process (47). Alternative A: hydrogen production. Alternative B: SNG production.

all but a small portion of the remaining hydrogen is consumed. The gas is scrubbed to remove excess carbon dioxide, dried, and introduced into the pipeline system.

Operation of the Methane-Rich Gas (MRG) process (64), developed by the Japan Gasoline Co., and the Lurgi Gasynthan process (65) are essentially similar. Various process configurations are used, depending on feedstock properties and product gas specifications.

Economic Aspects

The economics of SNG production are ill-defined because these technologies have not been widely realized commercially, and owing to the current rapid inflation rate and the difficulty of predicting future costs. The cost of SNG from light-hydrocarbon

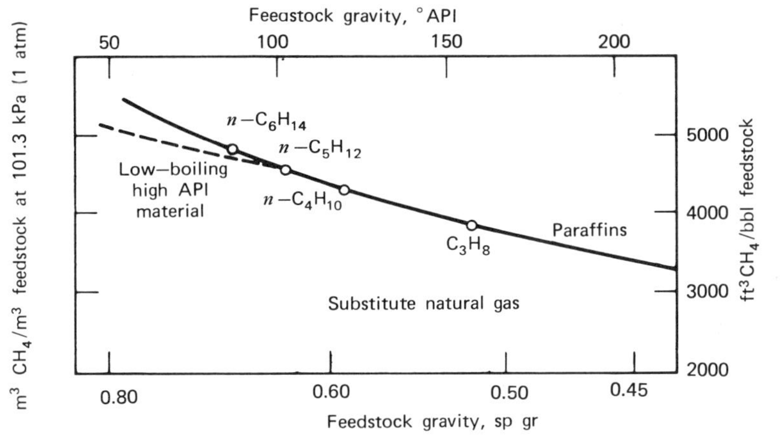

Figure 19. SNG yields from the CRG Process (62).

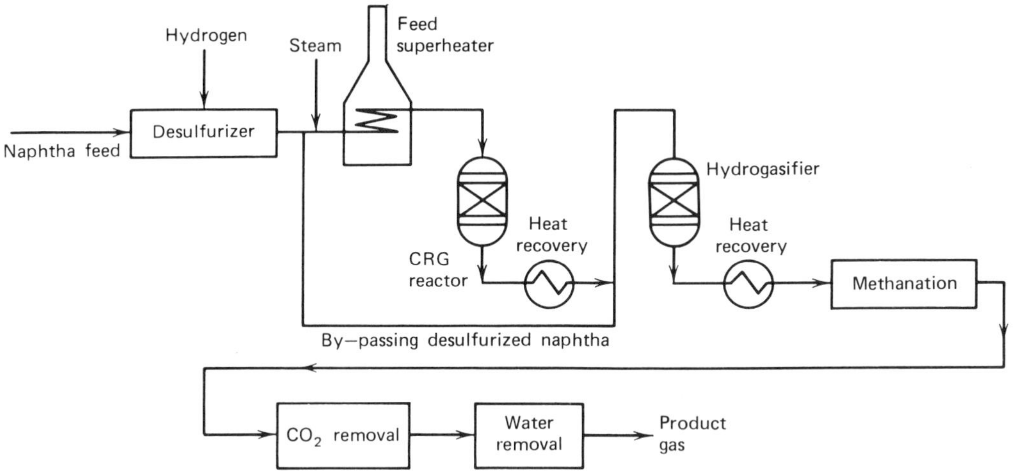

Figure 20. British Gas Council CRG Process.

liquids is primarily a function of feedstock cost. For base-load operation, the cost of SNG (1979 dollars) from commercial U.S. plants was estimated to be \$3.80–4.70/GJ (\$4.5/10^6 Btu) or higher for those plants operated on a seasonal or peakload basis. High heat-value gas from coal from large baseload plants would be \$2.80–5.70/GJ (\$3–6/10^6 Btu), depending on the process, feedstock cost, and location. For relatively large plants (47–95 J/d or 50–100 × 10^9 Btu/d), the cost of medium heat-value gas should be \$3.30–4.70/GJ (\$3.50–5.00/10^6 Btu). For large low heat-value gas plants, costs would be \$2.40–3.30/GJ (\$2.50–3.50/10^6 Btu). Cost from small, low, and medium heat-value plants would be about the same, or higher, than the cost from large high heat-value plants because of the diseconomics of small-scale production.

Table 8. Status of SNG from Petroleum Plants (as of Aug. 1977)[a]

Company operational plants	Feedstock: Design volume, 10^3 m^3/d (10^3 bbl/d)	Feedstock: Type	Feedstock: Source	SNG design capacity, 10^6 m^3/d (10^6 ft^3/d)	Investment, $\$10^6$	Process for SNG	1976 annual production, 10^9 m^3/d (10^9 ft^3/d)	Location of plant	Remarks
Algonquin SNG, Inc.	4.3 (26.8)	naphtha	domestic	3.4 (120)	60	CRG	0.53 (18.9)	Freetown, Mass.	Designed for seasonal use. Start-up was in Dec. 1973. Operates at design capacity.
Ashland Oil, Inc.	1.7 (11)	naphtha	imported, crude	1.7 (60)	21.3	CRG	0.51 (18.1)	Tonawanda, N.Y.	National Fuel Gas Distribution Corporation will buy 495×10^6 m^3/yr (17.5×10^9 ft^3/yr) for 10 yr at 1.7×10^6 m^3/d (60×10^6 ft^3/d) during the winter and at 1.02×10^6 m^3/d (36×10^6 ft^3/d) during the summer. Prices are subject to the average monthly cost to the refinery. Recent cost was about \$3.32/GJ (\$3.50/10^6 Btu).
Boston Gas Co.	1.6 (10)	propane	imported, Exxon	1.1 (40)	8	MRG	0.09 (3.2)	Everett, Mass.	Designed for seasonal use. Start-up was in May 1974. Operates at full capacity.
The Brooklyn Union Gas Co.	2.2 (13.7)	naphtha	domestic, Exxon	1.7 (60)	46	CRG	0.17 (5.9)	Brooklyn, N.Y.	Designed for seasonal use. Start-up was in May 1974. Operates at full capacity.
Columbia LNG Corp.	11.1 (70)	propane, and butane, natural gas liquids	offshore butane and propane and domestic natural gas liquids	7.1 (250) (2 trains)	48.2	CRG	1.66 (58.8)	Green Spring, Ohio	First train start-up was in Mar. 1974 and second train in mid-1974. Plant designed for base load operations.
CNG Transmission Company	1.0 (6.45)	propane, butane, and natural gas liquids	mostly imported; some domestic	0.8 (29.5)	7	CRG	0.06 (2.1)	Chesapeake, Va.	Designed for seasonal use. Start-up was in Nov. 1974. Operates at full design capacity. Recent average prices were about \$0.134/m^3 (\$3.80/

									10^3 ft^3) with feedstock prices of \$104–127/t (\$115–140 per short ton). Plant was modified to use propane rather than plant output as plant fuel.
Consumers Power Co.	8.0 (50)	natural gas liquids	imported, Canada, some domestic	5.7 (200) (2 trains)	155	CRG	1.69 (59.9)	Marysville, Mich.	First train start-up was in Aug. 1973, with a second train in Apr. 1974. Plant is capable of exceeding design capacity up to 7×10^3 m^3/d (250×10^6 ft^3/d). Recent price was \$0.121/m^3 (\$3.43/10^3 ft^3).
Gasco, Inc.	477 (3000)	naphtha	imported	0.5 (16.6)	7.6	Lurgi Gasyn-than	0.09 (3.2)	Barbers Point, Oahu, Hawaii	Start-up was in Sept. 1974. Operating at about 283×10^3 m^3/d (10×10^6 ft^3/d).
Northern Illinois Gas Company	5.1 (32)	natural gas liquids	domestic	5.2 (185)	80	CRG	1.74 (61.4)	Aux Sable, Ill.	Start-up was in Fall 1974. Plant can exceed design capacity up to 283×10^3 m^3/d (225×10^6 ft^3/d).
	2.5 (16)	naphtha	domestic						
Public Service Electric and Gas Co.	0.6 (4)	naphtha	domestic	0.6 (20)	12	CRG		Harrison, N.J.	Designed for seasonal use. Start-up was in Mar. 1973.
							0.09 (3.1)		
Public Service Electric and Gas Co.	4 (25)	naphtha	imported	3.5 (125)	50	CRG		Linden, N.J.	Designed for seasonal use. Start-up was in Dec. 1974.
Baltimore Gas and Electric Co.	1.9 (12)	naphtha	domestic	1.7 (60)	40	Lurgi Gasyn-than		Sollers Point, Md.	Designed for seasonal use. Plant testing began in Dec. 1976. Full-scale production was anticipated for the 1979–1980 heating season.
Peoples Gas Light & Coke Co.	5.2 (33)	naphtha	domestic	4.5 (160)	100	CRG	1.15 (40.7)	Elwood, Ill.	Full-scale, base-load operations began in Mar. 1976.
Total	*499.5 (3141.5)*	naphtha		*37.5 (1326.1)*	*635.1*		*7.78 (275.3)*		
	26.8 (168.45)	propane, butane, and natural gas liquids							

[a] Ref. 63.

BIBLIOGRAPHY

"Manufactured Gas" in *ECT* 1st ed., Vol. 8, pp. 765–800, by W. H. Fulweiler, Consultant, "Gas, Manufactured," in *ECT* 2nd ed., Vol. 10, pp. 353–442, by Martin A. Elliot, Illinois Institute of Technology and Henry R. Linden, Institute of Gas Technology; "Carbon Monoxide–Hydrogen Reactions," in *ECT* 2nd ed., Vol. 4, pp. 446–489, by Helmut Pichler and Annemarie Hector, Carl-Engler- und Hans-Bunte-Institut für Mineralöl und Kohleforschung der Technischen Hochschule Karlsruhe

1. "Synthesized Gaseous Hydrocarbon Fuels," *National Gas Survey, FERC-0008,* Department of Energy, Washington, D.C., 1978.
2. S. Clegg, *A Practical Treatise on the Manufacture and Distribution of Coal-Gas,* 50th ed., Triibner and Co., London, 1868.
3. J. M. Reid, "Early Development of SNG Processes," *paper presented at SNG Symposium I, Institute of Gas Technology, Chicago, Mar. 12–16, 1973.*
4. F. J. Dent and co-workers, *Gas. Counc. GB Res. Commun.,* (GC37), (Nov. 1956); *Gas. J. (GB)* **288,** 600, 606, 610 (Dec. 1956); *Gas World (GB)* **144,** 1078, 1080 (Nov. 1956).
5. H. R. Linden and E. S. Pettyjohn, "Selection of Oil for High Btu Gas," *IGT Res. Bull. No. 12,* Chicago, 1952.
6. G. B. Schultz and H. R. Linden, "High Pressure Hydrogasification of Petroleum Oils," *IGT Res. Bull. No. 29,* Chicago, 1960.
7. F. J. Dent and co-workers, "The Use of Light Distillate for the Production of 500 Btu (4525 kcal/m_η^3, 18.95 MJ/m_η^3) Town Gas," *paper presented at the Ninth International Gas Conference, The Hague (Scheveningen), The Netherlands, Sept. 1–4, 1964.*
8. F. C. Schora, *Pipe Line Ind.* **49**(3), 35 (Sept. 1978).
9. "Wellman Galusha Gas Producers," *Res. Bull. No. 576A,* McDowell-Wellman Engineering Co., Cleveland, Ohio, May, 1976.
10. J. H. Martin, I. N. Banchik, and T. K. Subramaniam, "Winkler Technology for Clean Fuels From Coal," *presented at the Symposium Comparative Economics of Synthetic Fuels Processing, New York, Apr. 4–9, 1976; Report of the Committee on Production of Manufactured Gases,* Report No. IGU/B-76, London, 1976.
11. J. F. Farnsworth, *Ind. Heat.* **41**(11), 38 (Nov. 1974).
12. J. F. Farnsworth, "The Production of Gas from Coal Through a Commercially Proven Process," *Coal Gas Fundamentals Symposium, IGT, Chicago, 1973.*
13. J. W. Loeding and I. G. Stanfill, "Application of the U-GAS Process for the Production of Intermediate Calorific Value," *Conference preprint presented at the 14th World Gas Conference, Toronto, Can., May 27–June 1, 1979.*
14. D. H. Archer, L. A. Salvador, and E. S. Vidt, "The Westinghouse Fluidized Bed Process," *paper presented at the 10th Synthetic Pipeline Gas Symposium, Chicago, Ill., Oct. 30–Nov. 1, 1978.*
15. L. Jéquier, L. Longchambon, and G. Van de Putts, *J. Inst. Fuel* **33,** 584 (Dec. 1960).
16. J. C. Hoogendoorn, "Gas From Coal for Synthesis of Hydrocarbons, Status of SASOL II," *paper presented at the Ninth Pipeline Gas Symposium, Chicago, Ill., Oct. 31–Nov. 2, 1977.*
17. *Report of the Committee on Production of Manufactured Gases,* Rep. No. IGU/B-76, International Gas Union, London, 1976.
18. J. C. Hoogendoorn, "Gas From Coal With Lurgi Gasification at SASOL," *paper presented at the Clean Fuels From Coal Symposium, Institute of Gas Technology, Chicago, Ill., Sept. 10, 1973.*
19. P. F. H. Rudolph, "The Lurgi Process–The Route to SNG From Coal," *paper presented at the Fourth Pipeline Gas Symposium, Chicago, Ill., Oct. 30–31, 1972.*
20. "Lurgi Pressure Gasification Performance Record," *Lurgi Express Information 0 1018/12.71,* Lurgi Mineralötechnik, GmbH, Frankfurt, FRG, Dec. 1971.
21. "Fixed-Bed Coal Gasification for Production of Industrial Fuel Gas," *DOE Rep. No. FE-2220-26 for DOE Contract No. EX-76-C-01-2220,* Gilbert/Commonwealth, Energy Research Division, Reading, Pa., 1977.
22. K. H. Krieb, "KDV Process (High-Pressure Coal Gasification) for Power Generation," *paper presented at the 13th World Gas Conference, London, June 7–11, 1976.*
23. *New Coal Gasifier To Be Tested,* Special Pub. 78-1, Foster Wheeler Energy Corp., Turnpton, N.J., Dec. 1971.
24. R. P. Lewis, "Installation of a FW-Stoic Gasifier for the Generation of a Clean Boiler Fuel," *paper presented at the 85th National Meeting of A.I.Ch.E., Univ. of Minnesota, Minneapolis, Minn., June 4–8, 1978.*

25. R. J. Rutherford and A. H. Rawdon, "The Riley-Morgan Gasifier," *EPRI Conference Proceedings, Special Report EPRI-SR-1, EPRI Conference, Palo Alto, Calif., Apr. 8–10, 1974.*
26. A. H. Rawdon, R. A. Lisauskas, and S. A. Johnson, "Operation of a Commercial Size Riley-Morgan Coal Gasifier," *paper in Proceedings of the American Power Conference, Institute of Gas Technology, Chicago, Ill., 1976.*
27. D. E. Woodmansee and P. M. Palmer, "Gasification of Highly Caking Coal in the GEGAS Pressurized Gas Producer," *paper presented at the ACS Symposium on Gasification Kinetics, New Orleans, La., Mar. 24–25, 1977.*
28. "Wilputte Low-Btu Fuel Gas Process," *Res. Bull. No. 7862,* Wilputte Corp., Murray Hill, N.J., May 1, 1978.
29. G. R. Cooper, "Operating Overview of a Producer Gas Plant (12 Machines) at Kingsport, Tennessee," *paper presented at the Fifth Annual International Conference on Coal Gasification, Liquefaction, and Conversion to Electricity, University of Pittsburgh, Pittsburgh, Pa., Aug. 1–3, 1978.*
30. R. V. Rahfuse, A. J. Liberatore, and G. R. Friggens, "Gasification of Caking-Type Bituminous Coal at 75 to 150 psig in a Stirred-Bed Gas Producer," *ERDA Rep. No. MERC/TPR-75/3,* Morgantown, W. Va., 1975.
31. "Produce Your Own Gas for Your Own Business," *Brochure No. 3M 177 CA 4104,* Applied Technology Corp., Houston, 1977.
32. D. Hebden, "High-Pressure Gasification Under Slagging Conditions," *paper presented at the Seventh Synthetic Pipeline Gas Symposium, Chicago, Ill., Oct. 27–29, 1975.*
33. J. D. Sudbury, G. F. Curran, and J. A. Phinney, "Gasification of Caking Coals in the BGC/Lurgi Slagging Gasifier—A Status Report," *paper presented at the Fifth Annual International Conference on Coal Gasification, Liquefaction, and Conversion to Electricity, University of Pittsburgh, Pittsburgh, Pa., Aug. 1–3, 1978.*
34. R. C. Ellman, B. C. Johnson, and M. M. Fegley, "Studies in Slagging Fixed-Bed Gasification at the Grand Forks Energy Technology Center," *paper presented at the 10th Synthetic Pipeline Gas Symposium, Chicago, Ill., Oct. 30–Nov. 1, 1978.*
35. C. Fink, G. Curran, and J. Sudbury, "CO_2 Acceptor Process Pilot Plant—1973," *paper presented at the Fifth Synthetic Pipeline Gas Symposium, Chicago, Ill., Oct. 29–31, 1973.*
36. C. E. Fink, "CO_2 Acceptor Process Pilot Plant—1975," *paper presented at the Seventh Synthetic Pipeline Gas Symposium, Chicago, Ill., Oct. 27–29, 1975.*
37. M. D. Mitsak and co-workers, "Gas Supply: Consider an Old Approach With Perhaps a Twist," *paper presented at the Fifth Annual International Conference on Coal Gasification, Liquefaction and Conversion to Electricity, University of Pittsburgh, Pittsburgh, Pa., Aug. 1–3, 1978.*
38. J. Huebler, F. C. Schora, and B. S. Lee, "The HYGAS Coal-Gasification Process" in D. M. Considine, ed., *Energy Technology Handbook,* McGraw-Hill, New York, 1977, pp. 1-201–1-211.
39. R. Lewis and co-workers, "Update of Synthane Pilot Plant Status," *paper presented at the 10th Synthetic Pipeline Gas Symposium, Chicago, Ill., Oct. 30–Nov. 1, 1978.*
40. L. E. Furlong, and N. C. Nahas, "Catalytic Coal Gasification Process Research and Development," *paper presented at the 10th Synthetic Pipeline Gas Symposium, Chicago, Ill., Oct. 30–Nov. 1, 1978.*
41. K. H. van Heek, H. Juntgen, and W. Peters, "Gasification of Coal Using Process Heat From HTR," *paper presented at the 10th Synthetic Pipeline Gas Symposium, Chicago, Ill., Oct. 30–Nov. 1, 1978.*
42. R. J. Eby, N. McClintock, and R. Bloom, Jr., "The Illinois Coal Gasification Group Project—COGAS Process," *paper presented at the 10th Synthetic Pipeline Gas Symposium, Chicago, Ill., Oct. 30–Nov. 1, 1978.*
43. W. W. Bodle, D. V. Punwani, and M. C. Mensinger, *Chem. Technol.* **8,** 559 (Sept. 1978).
44. B. B. Bennett, *J. Inst. Fuel* **35,** 338 (Aug. 1962).
45. F. C. Schora, P. B. Tarman, H. L. Feldkirchner, and S. A. Weil, *Hydrocarbon Process.* **56**(4), 107–110 (Apr. 1977).
46. G. L. Calderwood, "The 'ONIA-GEGI' Process for Cyclic Catalytic Cracking of Liquid Hydrocarbons," *Proc. Oper. Sect. Am. Gas Assoc.* 89, (May, 1958).
47. H. R. Linden and M. A. Elliot, *Am. Gas J.* **186,** 22 (Feb. 1959).
48. C. G. Milbourne, "Production of Catalytic Oil Gas From Bunker C Oil in the U.G.I. Cyclic Catalytic Reforming Sets," *Am. Gas Assoc. Proc.* 533 (Oct. 1956).
49. E. R. Ward, *Gasmaking,* British Petroleum Co., Ltd., London, 1965; *Gasmaking and Natural Gas,* British Petroleum Co., Ltd., 1972.

50. C. A. Stokes and A. J. Middleton, "SNG From Heavy Distillates and Reduced Crudes by the Petrogas Process," *paper presented at the Project S-128 Symposium on Heavy Oil Gasification, Institute of Gas Technology, Chicago, Ill., Aug. 1973.*
51. J. C. Janka and P. B. Tarman, "SNG From Heavy Oil," *paper presented at the Fifth Synthetic Pipeline Gas Symposium, Chicago, Ill., Oct. 29–31, 1973.*
52. R. W. Tittle and W. Hartley, "The Potential of the Gas Recycle Hydrogenator in the Production of SNG," *paper presented at the SNG Symposium I, Institute of Gas Technology, Chicago, Ill., Mar. 12, 1973.*
53. B. H. Thompson and H. L. Conway, "The Hydrogenation of Hydrocarbons in Relation to the Manufacture of Substitute Natural Gas," *paper presented at the 65th Annual Meeting A.I.Ch.E., New York, Nov. 26–30, 1972.*
54. J. Huebler, J. Janka, G. Seay, and P. Tarman, "Pipeline Gas From Crude Oil—Combined Hydrocracking–Hydrogasification," *paper presented at A.I.Ch.E. Synthetic Gas Symposium, New York, Nov. 26, 1972.*
55. *Oil Gas J.* **71**(15), 32, 33 (Apr. 9, 1973).
56. *Oil Gas J.* **71**(7), 36, 37 (Feb. 12, 1973).
57. U.S. Pat. 3,708,269 (Jan. 2, 1973), H. R. Linden (to Institute of Gas Technology).
58. C. J. Kuhre and C. J. Shearer, *Oil Gas J.* **71,** 85 (Sept. 6, 1971).
59. R. Wintrell and co-workers, "Production of Gas From Coal by the Koppers-Totzek Process," *paper presented at Clean Fuels From Coal Symposium, Institute of Gas Technology, Chicago, Ill., Sept. 10, 1973.*
60. W. L. Slater and R. M. Dille, "The Chemistry of Synthesis Gas Generation by Partial Oxidation at Various Pressures," *paper presented at the International Congress of Industrial Chemistry, Brussels, Sept. 20, 1966.*
61. C. P. Marion and W. L. Slater, "Manufacture of Tonnage Hydrogen by Partial Combustion—The Texaco Process," *Section 3 Paper 22-PD9 presented at the Sixth World Petroleum Conference, Frankfurt, FRG June 19–26, 1963.*
62. E. J. Nobles, M. Van Sickles, and S. Crosslands, *Chem. Eng. Prog.* **68,** 39 (Dec. 1972).
63. "Update on Status of SNG-From-Petroleum Plants," *Gas Supply Rev.* **5**(12), 9 (Sept. 1977).
64. R. A. Erickson, "JGC Methane Rich Gas Process," *paper presented at SNG Symposium I, Institute of Gas Technology, Chicago, Ill., Mar. 12–16, 1973.*
65. G. A. White, "The Gasynthane Process: A BASF/Lurgi Development," *paper presented at SNG Symposium I, Institute of Gas Technology, Mar. 12–16, 1973.*

JACK HUEBLER
JOHN C. JANKA
Institute of Gas Technology

LIQUID FUELS

Synthesis of liquids that may serve as fuels may be taken strictly to imply the creation of such liquids, as in the Fischer-Tropsch synthesis, or in the synthesis of methanol, from gases. Recently, however, the term synthetic liquid fuels is being applied to all potential fuel liquids, regardless of source, not normally produced from natural crude petroleum. Hence, fuel liquids prepared from naturally occurring bitumen deposits qualify as synthetics, even though the sources are natural liquids. More properly, the term has come to refer to prepared liquids with characteristics approaching those of the fuel liquids in current commerce, specifically gasolines, kerosenes, or jet fuels, and fuel oils. Synthesis may include liquefaction of coals, destructive distillation of kerogen or coal, and extraction or hydrogenation of organic matter in coke liquids, coal tars, tar sands, or bitumen deposits, in addition to processes that fit the classical interpretation of the organic chemist.

The first liquids commonly combusted were animal renderings burned to produce light. Such liquids were too valuable to be consumed solely for heat production, and devices for deriving power from their combustion did not exist. Whale oil, eg, was always a luxury commodity; its price on the Eastern seaboard of the United States in 1850 was in excess of 53¢/L ($2/gal). The search for low cost light was the driving force behind the production of combustible liquids before 1900, and was responsible for the creation of the petroleum industry.

The large-scale use of coal, which began in England in the 19th century, first led to significant reductions in the costs of hydrocarbon liquids, and to their wider availability. The gas distilled from coal was used to light homes at the end of the 18th century (1), leading to the establishment in 1812 in London of the first company to generate gas from coal for lighting purposes. The production of coal tar, and the separation therefrom of various coal liquids, as a concomitant of the production of illuminating gas, thus probably predates its production from the coking operations associated with iron production. The coal tars produced in gas works may have been the first synthetic liquids turned to fuel use in quantity.

Hofmann's proof of the existence of benzene in the light oil derived from coal tar (2) first established coal tar and coal as chemical raw materials. Soon thereafter the separation of coal-tar light oil into substantially pure fractions produced a number of the aromatic components now known to be present in significant quantities in petroleum-derived liquid fuels. These separation procedures are still used for the recovery of BTX (benzene–toluene–xylene) and related substances (benzol, motor benzol) from coke-oven operations (2) (see BTX processes).

Coal gas for home lighting was also an expensive luxury until the late 19th century. By the middle of the 19th century it was realized, both in England and in the United States, that kerosene, or coal oil, distilled from coal, could produce a luminous combustion flame, and its commercialization for that purpose was rapid. By the time of the United States Civil War some 87 m^3/yr (23,000 gal/yr) of lamp oil was produced in the United States from the distillation not only of coal but also of oil shale and natural bitumen.

Rock oil or natural petroleum frequently was found by early settlers in the United States to be associated with subsurface brine deposits. American Indians had used such crude oils in a variety of medicine and paint formulations, and some entrepreneurs among the settlers bottled and sold oil residues as cure-alls. In 1859 high gravity, low sulfur crude oil was discovered which produced high quality kerosene with minimal processing, and the world's first oil boom erupted. Until the end of the century, kerosene was the only substance of value extracted from natural crude oil; its cost was too high for heating purposes but it was used widely for lighting until it was replaced by electricity.

Refiners slowly learned to use the residues from kerosene production. Brick kilns were fired with crude oil residues, as well as with the natural gas which had been discovered in the search for oil. Pennsylvania crudes, in particular, were shown to produce superior lubricating oils and greases, and these products began to displace traditional vegetable-derived lubricants about the turn of the century.

As the market for lamp oil collapsed, gasoline increased in value. The widespread use of liquids as fuels dates from that time. Of the total estimated world fuel consumption of 1.16×10^9 metric tons of bituminous coal equivalent in 1900, something less than 3% was in the form of liquid (oil) or gas (natural gas). By 1960 total world fuel consumption had increased to an estimated 4.6×10^9 t of coal equivalent, of which 29% was in the form of petroleum oils, and 14% natural gas.

Carbonaceous fuel substances, other than natural petroleum, are widely distributed throughout the world (see also Coal; Oil shale; Asphalt; Lignite; Tar sands).

In the Western hemisphere, Canada has large tar sand, bitumen (or very heavy crude oil), and coal deposits. The United States has very large reserves of coal and shale. Venezuela has an enormous bitumen deposit and Brazil has significant oil shale reserves. Coal is also found in Brazil, Columbia, Mexico, and Peru.

The total resource base is immense by any estimate. Unfortunately, estimates of the quantities of in-place reserves which may be recoverable with current technology range no higher than ca 10%. Hence, considerable pressure for new recovery technology will develop in the next century, even if fuel usage is obviated by other developments.

Most carbonaceous substances may be combusted more or less directly in suitable equipment to produce heat or power. Direct combustion may represent the means for deriving maximum benefit of the heating value of particular fuel substances, although the necessity to limit emissions of noxious compounds, especially of sulfur or nitrogen, and to control disposition of noncombustible residues, may significantly reduce the overall thermal efficiency of the combustion process and add to the cost of work so produced.

Liquid fuels possess inherent advantageous characteristics. They are much more readily stored, transported, and metered than solids or tars, and are much easier to process or clean by chemical means. Noncombustibles may be reduced to very low levels by treatment, and sulfur and nitrogen compounds may be removed. The energy densities of clean hydrocarbon liquids may be very high relative to solid or semisolid fuel substances, which may contain mineral matter and moisture. Liquid fuels are now essential to fuel-powered conveyances, which, with few exceptions, are designed to function only with relatively clean, low viscosity liquids.

In general, liquid hydrocarbon fuels produced from natural petroleum, as well as the crude petroleum itself, are chemically distinguished from natural solid fuel substances (coal, peat, lignite, bitumen, kerogen) by the higher hydrogen-to-carbon content of the liquids. The hydrogen-to-carbon atomic ratio in a high volatile bituminous coal is ca 0.8, whereas a Texas crude oil may analyze ca 1.6, and a highly paraffinic Pennsylvania crude oil may show a ratio of ca 2.0. Isooctane has a ratio of 2.25.

If paraffinic liquids are extracted from solid fuel substances, as by the destructive distillation of bitumen or oil shale, the hydrogen content of the residual material is reduced even further, and the residues become more refractory. The yields of liquids so derivable are generally low, even when a significant fraction of the hydrogen is extractable. Thus production of fuel liquids from nonliquid carbonaceous fuel substances may be enhanced only by the introduction of additional hydrogen in a synthesis process.

Coal Pyrolysis

As the temperature of carbonaceous matter is increased, decomposition ultimately occurs. Melting and dehydration may also occur. Coals exhibit more or less definite decomposition temperatures, as indicated by melting and rapid evolution of volatile components, including potential fuel liquids, during destructive distillation (3). Composition is determined by the Fischer assay (4). Table 1 summarizes an extensive survey of North American coals subjected to laboratory pyrolysis. The yields of light oils so derived average no more than ca 8.3 L/t (2 gal/short ton), and tar yields of ca 125 L/t (30 gal/short ton) are optimum for high volatile bituminous coals (5).

Coal pyrolysis has been studied at both reduced and elevated pressures (6), and in the presence of a variety of agents and atmospheres (7) (see Coal). Although important to the study of coal structure and reactions, coal pyrolysis, as a means to generate liquids, has thus far proved to have limited commercial value. Several modern coal conversion processes, however, enhance the recovery of potential fuel liquids.

COED Process. Sponsored by the Office of Coal Research of the U.S. Department of the Interior, the COED process was developed by FMC Corporation as Project COED (Char–Oil–Energy Development) from 1962 through 1975 (8–10). Bench-scale experiments led the way to construction in 1965 of a process development unit employing multistaged, fluidized-bed pyrolysis to process 45 kg/h (11). Correlated studies included hydrotreating of COED oil (12), high temperature hydrodesulfurization of COED char, and investigations of char-oil and char-water slurry pipelining economics (13).

A pilot plant capable of processing up to 33 t/d and hydrotreating 4.7 m^3/d (30 bbl/d) was started up in 1970 (14), and was operated successfully for a number of years (15). In 1975 federal funding was terminated by the Energy Research and Development Administration, which had succeeded the Office of Coal Research, and the pilot plant was disassembled. Development was continued, however, as part of the proprietary COGAS development (FMC Corporation) which employs COED technology (16).

The COED concept, shown in Figure 1, is designed to recover liquid, gaseous, and solid fuel components. Heat is generated by the reaction of oxygen with a portion of the char in the last pyrolysis stage, and is carried countercurrently through the train by the circulation of hot gases and char. Heat is also introduced by the air combustion of gas to dry feed coal and to heat fluidizing gas for the first stage. The number of stages in the pyrolysis, and the operating temperatures in each, may be varied to accommodate high volatile bituminous and subbituminous feed coals with widely ranging caking or agglomerating properties.

Table 1. Averaging Yields and Range of Yields of Fischer Assay of Various Coals[a,b]

Rank of coal	Coke, %		Tar, L/t[c]		Light oil, L/t[c]		Gas, m^3/t[d]		Water, %	
	Average	Range	Average	Range	Average	Range	Average	Range	Average	Range
semianthracite			3.2		0.14					
low volatile bituminous	89.7	85.8–93.3	39.6	29.0–58.4	4.69	3.36–7.41	59.8	54.4–66.6	3.2	1.1–6.6
medium volatile bituminous	83.3	77.4–90.4	86.9	44.6–117.8	7.68	4.92–10.58	66.0	47.3–76.2	4.1	2.8–7.0
high volatile A bituminous	75.5	68.8–81.4	142.1	105.3–187.2	10.53	6.81–15.09	67.0	57.5–80.2	6.0	3.0–9.2
high volatile B bituminous	70.4	66.0–73.2	139.4	111.8–198.3	10.03	7.13–15.82	68.3	56.4–82.3	11.1	10.2–13.1
high volatile C bituminous	67.1	65.4–68.6	124.2	85.1–178.5	8.65	5.93–12.47	61.2	53.0–70.4	15.9	12.0–19.1
high volatile C bituminous or subbituminous A	59.1		94.3	84.6–112.2	7.59	6.26–8.88	90.4		23.4	
subbituminous A			81.9	81.0–82.8	6.21	6.12–6.26				
subbituminous B	57.6	54.8–59.9	70.8	60.7–76.8	6.12	5.24–7.13	90.4	62.2–93.8	27.8	23.3–30.4
lignite	36.5		69.9	30.8–124.2	5.47	2.90–8.69	71.4		44.0	
cannel	58.8	44.1–69.0	338.1	247.0–498.2	23.28	16.84–34.13	61.5	51.0–72.1	3.7	2.0–4.8

[a] Ref. 5.
[b] As-received basis; maximum temperature, 500°C.
[c] To convert L/t to gal/short ton, divide by 4.6.
[d] To convert m^3/t to ft^3/short ton, multiply by 29.4.

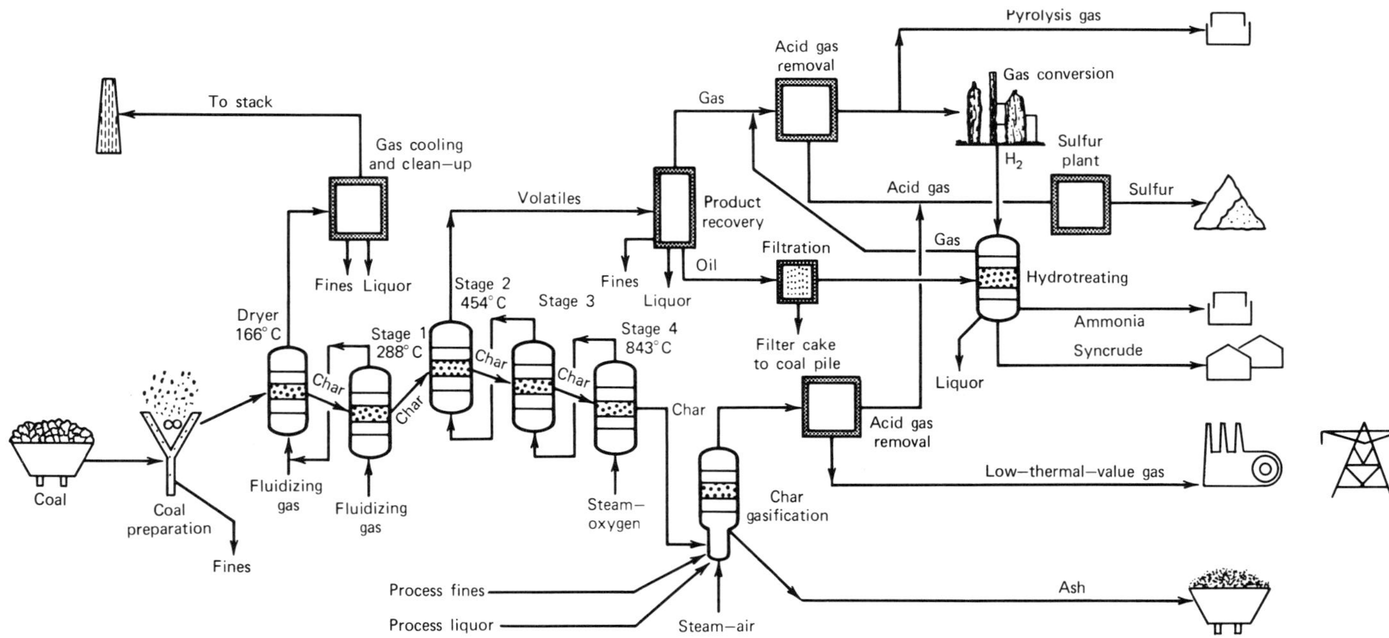

Figure 1. COED process with added char gasification (9).

Oil condensed from the released volatiles from the second stage is filtered on a rotary precoat pressure filter and catalytically hydrotreated at high pressure to produce a synthetic crude oil. Medium heat-content gas produced after the removal of H_2S and CO_2, is suitable as clean fuel. The pyrolysis gas produced, however, is insufficient to provide the fuel requirement for the total plant. Residual char, 50–60% of the feed coal, has a heating value and sulfur content about the same as feed coal, and its utilization may thus largely dictate process utility. In Figure 1, the char that is discharged from the final pyrolyzer vessel is gasified along with process residues in a fluidized gasifier to produce additional low-heat-content product gas.

Properties. Estimated product yields from two possible coal feeds, a low sulfur Western coal, and a high sulfur Midwestern coal, are shown in Table 2. Note that utility requirements are derived from external sources, and that syncrude yields are higher than those reported in other estimates (17).

The properties of the char products are shown in Table 3. The char derived from the low sulfur Western coal may be directly suitable as plant fuel, with only minor addition of clean process gas to stabilize its combustion. Flue gas desulfurization may not be required. Flue gas from the combustion of the char derived from the high sulfur Illinois coal, however, requires desulfurization before it may be discharged into the atmosphere (per current United States New Source Performance Standards for large coal-fired boilers) (see also Sulfur recovery).

Typical COED syncrude properties are shown in Table 4. The properties of the

Table 2. Product Yields for a Commercial COED Plant[a]

	Utah coal	Illinois no. 6 coal
feeds		
coal, t/d, dry	21,770	21,770
oxygen, t/d	3,450	3,450
products		
syncrude, m^3/d[b]	5,040	4,134
net pyrolysis gas[c], m^3/s[d]	37	36
high heating value (HHV), kJ/L[e]	1.9	1.9
char, t/d	11,972	11,746
sulfur, t/d	65	444
ammonia, t/d	50	44
products with char gasification added		
syncrude, m^3/d	5,040	4,134
net pyrolysis gas[f], m^3/s[d]	66	66
heating value (HHV), kJ/L[e]	1.9	1.9
net low-thermo-value gas, m^3/s	454	416
heating value (HHV), kJ/L[e]	0.6	0.6
sulfur, t/d	144	785
ammonia, t/d	50	44

[a] Ref. 9.
[b] To convert m^3/d to ft^3/h, multiply by 1.47.
[c] Gas yield after supplying plant fuel gas and feedstock to the hydrogen plant. Assumes purchased utilities.
[d] To convert m^3/s to ft^3/s, multiply by 35.3.
[e] To convert kJ/L to Btu/SCF, multiply by 268.
[f] Gas yield after supplying feedstock to the hydrogen plant.

Table 3. Properties of COED Char Product[a]

Property	Utah coal	Illinois no. 6
volatile matter, wt %	6.1	2.7
fixed carbon, wt %	80.2	77.0
ash, wt %	13.7	20.3
higher heating value, MJ/kg[b], dry	28.6	25.6
elemental analysis, wt %, dry		
carbon	81.5	73.4
hydrogen	1.3	0.8
nitrogen	1.5	1.0
sulfur	0.5	3.4
oxygen	1.5	1.0
chlorine	0.006	0.1
iron[c]	0.28	

[a] Ref. 9.
[b] To convert MJ/kg to Btu/lb, multiply by 430.
[c] Included in ash above.

oil products depend heavily on the severity of hydroprocessing. The degree of severity also markedly affects costs associated with hydrogen production and compression. Syncrudes derived from Western coals have much higher paraffin and lower aromatic content than those produced from Illinois coal. It is estimated (18) that ca 34 vol % of the COED syncrude produced from Illinois coal might be converted to acceptable, low sulfur, high octane gasoline using typical oil refinery processing.

Independent studies have been conducted with syncrude and char samples produced by the COED pilot plant (19–21). In general, properties of the COED products have been found compatible with expected industrial requirements.

Occidental Petroleum Coal Conversion Process (Garrett Coal Pyrolysis, Oxy Coal Conversion). A 1968 survey by the Garrett R&D Company (now the Occidental Research Co.) pointed to pyrolysis or thermal decomposition as the most direct procedure to produce liquid fuels from coal. The liquid yields from pyrolysis were inversely related to the temperature and to the residence time. Although high rates of heat transfer to coal were desirable, the presence of oxygen or the use of direct combustion within the system to supply process heat adversely affected liquid yields.

The desired heating process for coal particles in an ideal system was mathematically simulated and it was estimated that coal particles of 100-μm diameter could be heated throughout their volumes to decomposition temperature (450–540°C) within 0.1 s.

Operation of a small pilot plant was sufficiently encouraging to lead to the construction of a large pilot facility at LaVerne, California, in 1971. This unit, shown schematically in Figure 2, was reported to operate successfully at feed rates up to 136 kg/h (3.2 t/d).

Hot product char carries heat into the entrained bed to obtain the high heat transfer rates required. Feed coal must be dried and pulverized. A portion of the char recovered from the reactor product stream is cooled and discharged as product. The remainder is reheated to 650–870°C in a char heater blown with air. Gases from the reactor are cooled and scrubbed free of product tar. Hydrogen sulfide is removed from the gas, and a portion is recycled to serve as the entrainment medium.

Table 4. Typical COED Syncrude Properties[a]

Property	Utah A-seam	Illinois no. 6 seam
specific gravity, (°API)	0.934 (20)	0.929 (22)
pour point, °C	16	−18
flash point, closed cup, °C	24	16
viscosity, mm^2/s at 38°C (= cSt)	8	5
ash, wt %	0.01	0.01
moisture, wt %	0.1	0.1
metals, ppm	10	10
elemental analysis, wt %		
C	87.2	87.1
H	11.0	10.9
N	0.2	0.3
O	1.4	1.6
S	0.1	0.1
ASTM distillation		
initial bp, °C	138	88
10%	221	134
30%	277	199
50%	349	270
70%	416	316
90%	493	362
end point (95%)	510	397
hydrocarbon type analysis, liquid vol %		
paraffins	23.7	10.4
olefins	0	0
naphthenes	42.2	41.4
aromatics	34.1	48.2

[a] Ref. 9.

Scale-up of this technique, especially for overcoming coal agglomeration in pyrolytic systems, remains to be demonstrated. Product gas may be used to generate hydrogen for use in a tar hydrotreater to produce either synthetic crude oil or a low sulfur liquid fuel.

Properties. Only data from a laboratory unit have been reported (22–24). A high volatile western Kentucky bituminous coal, whose tar yield by Fischer assay was ca 16%, gave a tar yield of ca 26% at a pyrolysis temperature of 537°C. Tar yield peaked at ca 35% at 577°C and dropped off to 22% at 617°C.

The char heating value is essentially equal to that of the starting coal, and the tar has a lower hydrogen content than other pyrolysis tars. The product char is not suitable for direct combustion because of its sulfur content (2.6%).

The TOSCOAL Process. The Oil Shale Corporation (TOSCO) piloted the low temperature carbonization of Wyoming subbituminous coals over a two-year period in its 23 t/d pilot plant at Rocky Flats, Colorado (25). The principal objective was the upgrading of the heating value in order to reduce transportation costs (on a heating value basis). Hence, the solid char product from the process represented 50 wt % of the starting coal but had 80% of its heating value.

Furthermore, 60–100 L (14–24 gal) oil, with sulfur content below 0.4 wt %, could be recovered, per metric ton coal, from pyrolysis at 427–517°C. The recovered oil was suitable as low sulfur fuel.

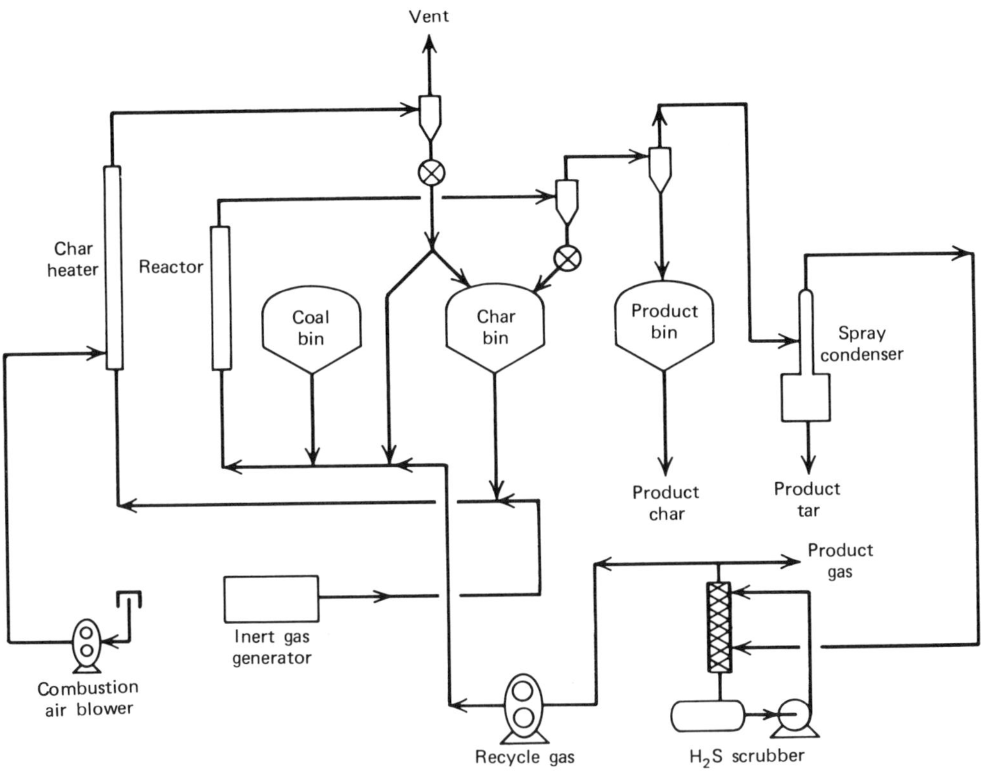

Figure 2. Occidental (Garrett) coal conversion pilot plant (22).

Figure 3 is a flow sheet of the Rocky Flats pilot plant. Coal is fed from hoppers to a dilute-phase, fluid-bed preheater. Preheated feed is transported to a pyrolysis drum, where it is contacted by hot ceramic balls. Pyrolysis drum effluent is passed over a trommel screen that permits char product to fall through. Product char is thereafter cooled and sent to storage.

The ceramic balls pass over the screen, and are recycled to a ball heater via a mechanical elevator. Pyrolysis vapors are condensed and fractionated. Separated gas is used to heat the ceramic balls.

Properties. Results for the operation with subbituminous coal from the Wyodak mine near Gillette, Wyoming, are shown in Table 5. Char yields decreased with increasing temperature, and oil yields increased. The Fischer assay laboratory method closely approximated the yields and product assays that were obtained with the TOSCOAL process.

The volatiles contents of product chars decreased from ca 25%, at a retort temperature of 427°C, to ca 16%, at 517°C. Char (lower) heating values, on the other hand, increased from ca 26.75 MJ/kg (11,500 Btu/lb) at 427°C to 29.5 MJ/kg (12,700 Btu/lb) at 517°C. Chars in this range of heating values are suitable for boiler fuel application, and the low sulfur content (about equal to that of the starting coal) permits direct combustion without need for additional desulfurization. These char products, however, are pyrophoric and require special handling in storage and transportation systems.

Properties of the tar oil products are given in Table 6. The oils change only slightly

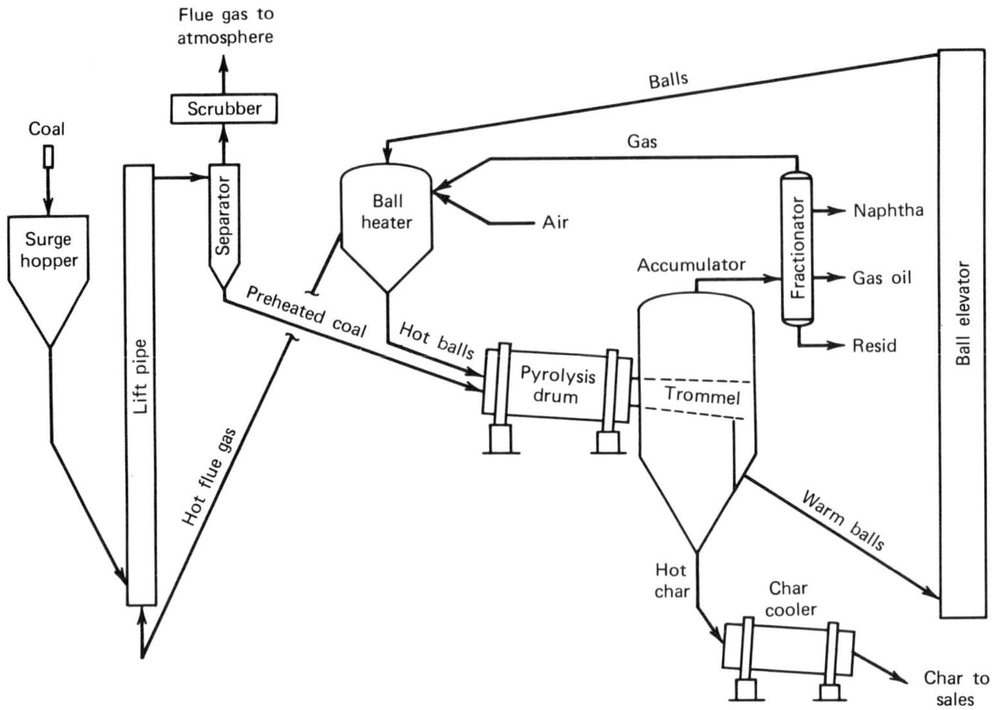

Figure 3. TOSCOAL process (25).

Table 5. TOSCOAL Retorting of Wyodak Coal[a]

	Product yields, kg/t of as-mined coal		
	Retort temperature		
Component	427°C	482°C	521°C
char	524.5	505.8	484.4
gas (C_3 and lighter)	59.5	78.4	63.0
oil (C_4 and heavier)	57.0	71.5	93.1
water[b]	351.0	351.0	351.0
Total	*992.0*	*1006.7*	*991.5*
Recovery, %	*99.2*	*100.7*	*99.1*

[a] Ref. 26.
[b] Value assumed from Fischer assay and moisture content. The addition of steam to the process prevented accurate measurement of water produced in retorting.

with change in the retorting temperature; sulfur levels are low. The fraction boiling up to 230°C contains 65 wt % of phenols, cresols, and cresylic acids.

TOSCO tar oils have high viscosity and may not be transported by conventional pipelines. Core-flow techniques may prove applicable. Alternatively, heated tank cars might be used, or the oils might be processed, as by visbreaking, to reduce viscosity.

Heating values of product gas on a dry, acid gas-free basis are in the natural gas range if butanes and heavier components are included. The production of gas undiluted with nitrogen or excess carbon dioxide is a considerable advantage.

Table 6. TOSCOAL Oil Properties[a]

Properties	Retort temperature 427°C[b]	482°C	521°C
analysis, wt %			
carbon	81.4	80.7	80.9
hydrogen	9.3	9.1	8.7
oxygen	8.3	9.4	9.3
nitrogen	0.48	0.7	0.7
sulfur	0.43	0.2	0.2
chlorine	0.0	0.0	0.0
ash	0.0	0.2	0.1
Total	*99.91*	*100.3*	*99.9*
heating values			
gross, MJ/kg[c]	38.59	37.72	37.13
net, MJ/kg[c]	36.61	35.75	35.26
specific gravity, (API)			
primary oil	1.015 (7.9)	1.040 (4.5)	1.061 (1.9)
calculated, with C_4 and heavier components of gas added	0.978 (13.2)	0.985 (12.1)	1.027 (6.2)
pour point, °C	32	38	35
Conradson carbon, wt %	7.6	9.9	11.4
distillation[d], °C			
2.5 vol %	212	216	199
20 vol %	302	288	235
50 vol %	407	413	385
viscosity (SUs)			
at 82°C	122	123	128
at 90°C	63	66	69

[a] Ref. 26.
[b] Feed coal was different from that used for 482 and 521°C.
[c] To convert MJ/kg to Btu/lb, multiply by 430.
[d] Combination of true boiling point and D 1160 distillations.

Coal Carbonization

In the by-product recovery of a modern coke oven, coal tar is removed first by cooling the gases emanating, and light oil is removed last by scrubbing the gas with solvents. Other products, ammonia, phenols, pyridine, naphthalene, may be recovered between these operations. The constituents of coal tar, light oil, and gas usually overlap considerably, ie, the fractional condensation does not effectively separate individual components. Assuming the lowest boiling coal tar constituent to be benzene (bp 80.09°C), and the highest boiling to be naphthalene, the overlapping compositions of gas, light oil, and tar may be as shown graphically in Figure 4. Many chemical compounds have been identified (2) in these substances. Included are most of the significant constituents of petroleum-derived fuel liquids, although only a few components are present in sufficient quantity to make their commercial recovery feasible (see Coal conversion process, carbonization).

The precise compositions of the light oil and coal tar recovered from coke-oven gas is a distinct function of the design of the recovery system, as well as of the properties of the starting coal. In general, 12.5–16.7 L/t (3–4 gal light oil/short ton) of coal carbonized is recovered from high temperature coke-oven operations. Light oil may

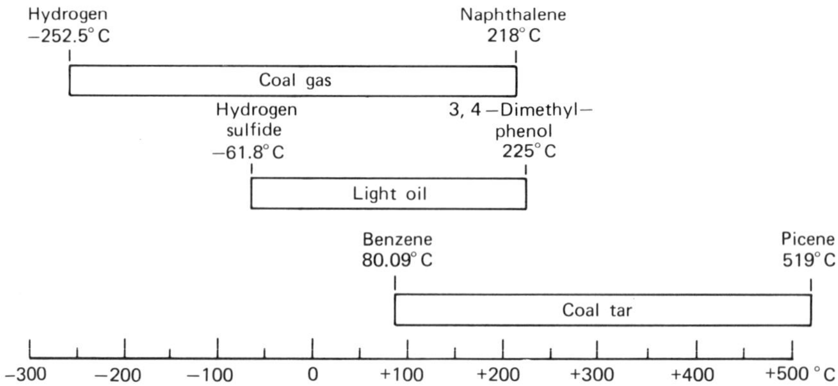

Figure 4. Boiling ranges of carbonization products (2).

contain 55–70% benzene, 12–20% toluene, and 4–7% xylene. Unrecovered light oil appearing in the effluent coal gas may comprise ca 1 vol % and contribute ca 5% of the gas's heating value. Refining of light oil consists mainly of sulfuric acid washing, followed by fractional distillation.

Large-scale recovery of light oil was commercialized in England, Germany, and the United States toward the end of the 19th century (27). Industrial coal-tar production dates from the earliest operation of coal–gas facilities. The principal bulk commodities derived from coal tar are wood-preserving oils, road tars, industrial pitches, and coke. Naphthalene is obtained from tar oils by crystallization, tar acids are derived by extraction of tar oils with caustic, and tar bases by extraction with sulfuric acid. Coal tar generally contains less than 1% benzene and toluene, and may contain up to 1% xylene.

The total United States production of BTX from coke-oven operations is barely significant compared to current petroleum product consumptions. Thus a total of ca 16,700 m^3 (105,000 bbl) benzene, 1845 m^3 (11,600 bbl) toluene, and 477 m^3 (3000 bbl) xylene was reported to have been produced during January, 1979, which compares with the current total United States petroleum consumption of 2.7×10^6 m^3/d (1.7×10^7 bbl/d), or with the current total gasoline consumption of 1.1×10^6 m^3/d (7×10^6 bbl/d) (see also Benzene; BTX processes Toluene; Xylene).

Coal and Coal-Tar Hydrogenation

Hydrogenation of coal and other carbonaceous matter with high pressure hydrogen was patented in 1913 (28), and a 1 t/d plant was operating in 1922, using hydrogen at 10 MPa (100 atm) and 400°C to treat brown coal tar, to give a liquid that comprised 25 wt % gasoline (boiling at 75–210°C) and 40 wt % middle oil (210–300°C). The pitch residue had a specific gravity of 1.04, and a solidification point of 15°C. Fischer and Tropsch (29) showed that the degree of liquefaction increased with decrease in coal rank. Liquid products were of low quality, however, high in oxygen, nitrogen, and sulfur content, owing to low hydrogenation rates and polymerization of primary products (30).

I. G. Farbenindustrie A.G. improved hydrogenation catalysts and procedures in the 1920s, and by 1936, 272,000 t/yr of motor fuel were produced by the hydrogenation of brown coal and coal tar at a facility constructed at Leuna. Some 363,000 t/yr was being produced at that time in three other German plants (31). The total German output from these facilities was ca 1.4×10^6 t/yr two years later (32). The total number of coal hydrogenation plants in Germany increased during World War II to twelve, with total capacity of about 4×10^6 t/yr (100,000 bbl/d) of aviation and motor gasolines.

In 1935 an ICI coal hydrogenation plant at Billingham, England produced ca 136,000 t motor fuel per year from bituminous coal and coal tar.

Experimental plants for hydrogenating coal or coal tar were operated in Japan, France, Canada, and in the United States before, or during, World War II. Much of the technology has remained proprietary. In general, coal-in-oil slurries containing iodine or stannous oxalate catalyst were subjected to liquid-phase hydrogenation at pressures of 25–70 MPa (250–700 atm). Liquids produced were fractionated, and the middle oils were then subjected to vapor-phase hydrogenation over molybdenum-, cobalt-, or tungsten sulfide-on-alumina catalysts (33). About 1 t of crude motor fuel was recovered from 4.5 t of coal, from which all necessary hydrogen and power requirements for the production were also obtained.

Reference 34 provides a critical review of the physical and chemical aspects of coal hydrogenation, and the whole field is reviewed in reference 35.

Developments in the United States. A large number of proprietary coal hydrogenation process variants have been proposed in the United States and elsewhere during this century. Much of the technology originally directed to the catalytic hydrogenation of coals and coal tars in Germany was later applied to the hydrorefining of petroleum fractions in the United States but commercial interest in coal hydrogenation in this country was offset by the relative abundance of domestic petroleum up to World War II. The U.S. Bureau of Mines, however, did follow the European developments, dating its interest in the production of coal liquids from about 1924 (36).

The huge demand for liquid fuels during World War II prompted the passage of the *Synthetic Liquid Fuels Act of 1944,* whereby the Secretary of the Interior was authorized to construct, maintain, and operate demonstration plants to produce liquid fuels from coal, oil shale, and other substances, including agricultural and forestry products. An Office of Synthetic Liquid Fuels was established to administer the various programs.

The Bureau of Mines began work on coal liquefaction in 1936, at which time a 45 kg/d experimental coal hydrogenation unit was constructed. The expanded program which followed on the passage of *Public Law 290* in April 1944, culminated in the construction and operation of a 45 t/d coal hydrogenation demonstration plant at Louisiana, Missouri, in 1949 (37), where a variety of problems and processing variations were investigated (38–39). However, cost studies (40) showed that production of gasoline via coal hydrogenation could not compete with natural petroleum, and demonstration plant operations were terminated in 1953.

Work on coal hydrogenation was thereafter continued by the Bureau of Mines on a laboratory scale (41–43). In one of these variants (44) coal–oil pastes admixed with catalyst in tubular reactors were hydrogenated at high pressure and low residence times to give improved yields of liquid products. The original thrust of the work was to hydrodesulfurize coal economically to produce environmentally acceptable boiler fuel (45). This process was named Synthoil (46).

In the 1960s the Office of Coal Research (OCR) funded the development of a number of coal liquefaction process variants incorporating hydrogenation which are discussed below.

Synthoil Process. Figure 5 presents a simplified flow diagram of the Synthoil process (U.S. Bureau of Mines, now U.S. Department of Energy). The reactor is packed with a commercial silica-stabilized cobalt molybdate-on-alumina catalyst in the form of 0.3 by 0.3 cm cylindrical pellets, leaving ca 50% void space in the reactor.

Feed coal is ground to pass 100% through a 149-μm (100-mesh) sieve and 70% through a 74-μm (200-mesh) sieve. Coal is kept suspended in tar or in recycled product oil in a stirred feed system. Coal slurries containing up to 50 wt % coal have been fed, although most data are reported for 30% coal feeds. Compressed hydrogen is admixed with the feed coal slurry ahead of a preheater that serves to bring the combined stream to reaction temperature. It is estimated that only slight reaction occurs ahead of the reactor because of the absence of catalyst and the short residence time.

The product issuing from the reactor is cooled regeneratively and separated into gaseous, liquid, and solid phases in a series of receivers and coolers. A portion of the recovered oil may be centrifuged, filtered, or otherwise treated to remove solids, and then may be recycled to coal slurry preparation.

The fraction of product that is soluble in pentane or benzene is termed oil, usually comprising low molecular weight liquids. Asphaltene is a higher molecular weight fraction not soluble in pentane but soluble in benzene. The fraction insoluble in benzene or pentane is the residue, comprising unreacted or polymerized portions of coal or tar.

The highly turbulent flow condition is employed in such a way that potentially deactivating species erode from the catalyst surfaces or are otherwise prevented from deposition. Similarly, the high turbulence apparently maintains unconverted coal and ash in suspension with the result that reactor plugging or hot-spot carbonization is avoided. However, if the ash content of feed exceeds ca 5%, excessive pressure drop results.

The process removes all inorganic sulfur (pyritic or sulfate) in coal, producing H_2S in the reactor off-gas. Only organic sulfur has been found in the products (45). Processing conditions vary with total pressures of 13.8–27.6 MPa (2000–4000 psi), and temperatures of 400–450°C. Residence time for liquid in this system could not be precisely determined. A variety of coals have been processed, and each has yielded acceptably desulfurized oil in one to three passes through the reactor. At the high system pressure, water could be used to absorb H_2S and NH_3 from reactor off-gas, permitting recycle of this hydrogen-rich stream to the reactor without adverse affect on desulfurization reactions.

The Bureau of Mines Process Evaluation Group published (47–48) an economic study of a projected commercial Synthoil facility, designed to produce ca 17,170 m^3/d (108,000 bbl/d) of low sulfur, heavy fuel oil from high sulfur, western Kentucky coal, in 1974. By that time a 23 kg/h experimental system was being operated, and a 9–14 t/d pilot unit had been designed. Construction of the pilot unit began in 1975.

The DOE announced in June, 1978, that active development of the Synthoil process had been terminated, however, since efforts to extend activity of the pelletized catalyst beyond ca 400 h had not succeeded. Although the system containing inactive catalyst provided adequate sulfur removal, hydrodesulfurization rates were not sufficiently high to warrant large-scale development.

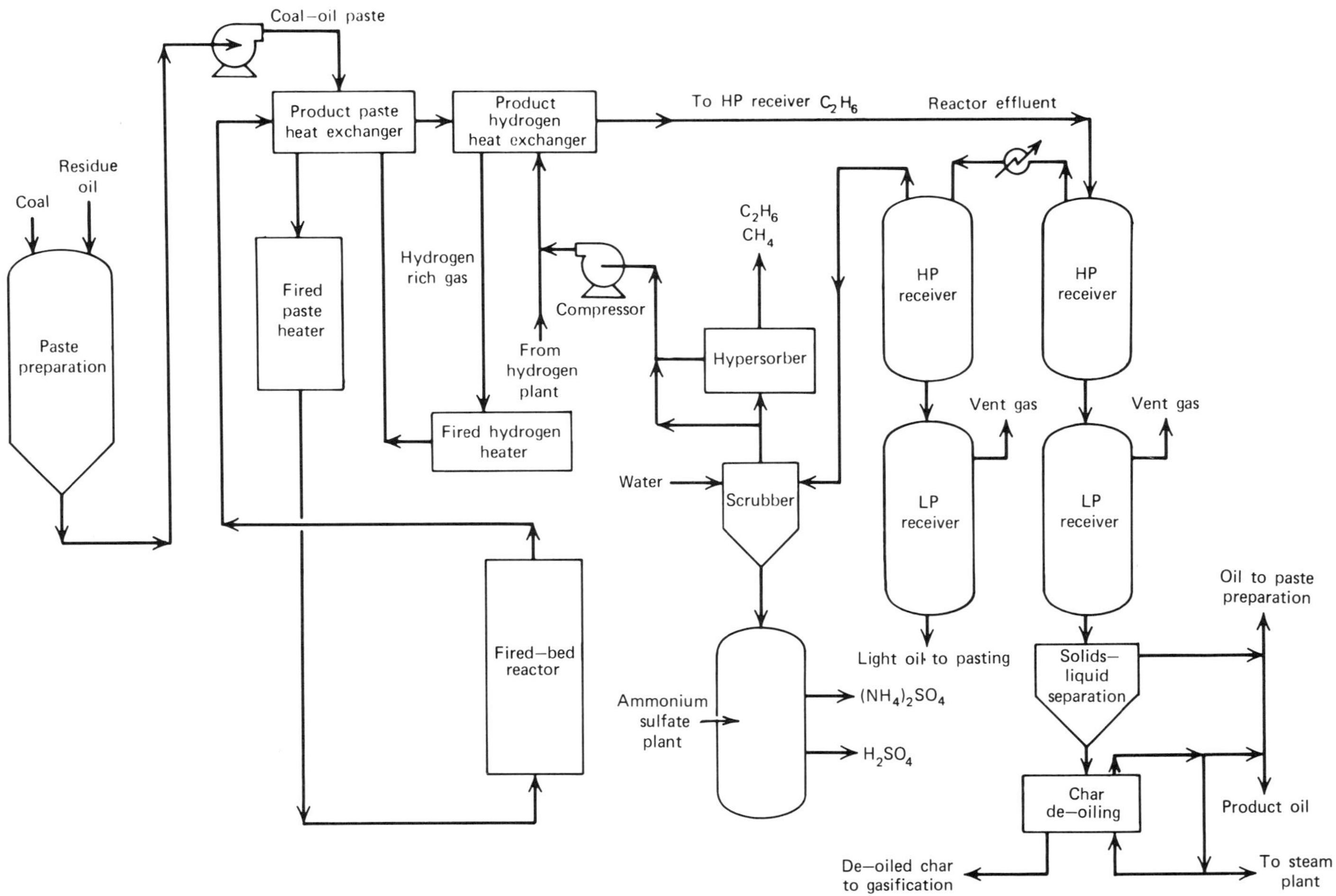

Figure 5. Synthoil process flow sheet (47). HP = high pressure; LP = low pressure.

Properties. Table 7 summarizes processing conditions and yields from the three-pass hydrodesulfurization of three different coals in a laboratory (2.25 kg/h) Synthoil reactor. The bulk of the sulfur remains in the benzene-insoluble residues.

A variety of catalysts have been evaluated (46,49). In general, cobalt molybdate was preferred to tin, molybdenum, and nickel catalysts for sulfur removal but tin catalysts were preferred for oil production. The Synthoil process and products are reviewed in references 50 and 51, respectively.

H-Coal Process. The H-Coal process (Hydrocarbon Research, Inc., HRI, subsidiary of Dynalectron Corp.) for the conversion of coal to liquid products is an application of HRI's ebullated-bed technology for the conversion of heavy oil residues into lighter fractions (see Fluidization). In 1965 HRI obtained a contract from the Office of Coal Research to develop a process to convert coal to liquid fuels. This contract entailed bench-scale and process development unit (PDU) operations, and preparation of a conceptual process design.

As shown schematically in Figure 6, coal is dried, pulverized, and slurried with coal-derived oil. The coal–oil slurry is charged continuously with hydrogen to a reactor of unique design (53) containing a bed of ebullated catalyst, where the coal is hydrogenated and converted to liquid and gaseous products; the upward passage of the

Table 7. Results from Three-Pass Hydrodesulfurization of Coal Slurries in Tar[a,b]

Results	Pittsburgh seam coal	Middle Kittaning no. 6 seam coal	Indiana no. 5 seam coal
sulfur in feed, wt %	0.8	1.3	1.41
sulfur in whole product, wt %	0.3	0.37	0.42
yield of whole product, wt % of feed	93.0	94.4	92.9
viscosity of whole product, SUs at 82°C	54.0	42.0	20.1
yield of fuel oils[c], wt % of product	84.0	86.8	87.1
sulfur in fuel oils fraction, wt %		0.18	0.14
benzene insoluble residue[d], wt % of product	16.0	13.2	12.9
sulfur in insolubles, wt %		1.54	1.88
solvent analysis of product, wt %			
oil (pentane-soluble)	59.5	65.1	69.2
asphaltene	24.6	21.7	17.9
benzene-insoluble residue	16.0	13.2	12.9
Total	*100.1*	*100*	*100*
elemental analysis of product[e], wt %			
carbon	91.9	91.9	92.3
hydrogen	6.2	6.5	6.5
nitrogen	1.0	0.8	0.8
sulfur	0.3	0.37	0.42
oxygen (by difference)	0.6	0.4	0.0

[a] Ref. 45.

[b] Operating conditions: reactor, 0.79 cm ID × 20 m long; catalyst, silica-promoted cobalt molybdate; temperature, 450°C; initial liquid feed, 30 wt % coal + 70 wt % tar; liquid feed rate, 2.26 kg/h; liquid feed throughput, 2268 kg/(h·m^3 reactor volume) [4000 lb/(ft^3·d)]; and operating pressure, 13.8 MPa (2000 psi).

[c] Fuel oils defined here as the pentane-soluble oil fraction plus the asphaltene fraction as determined by solvent analysis; equivalent to a product after quantitative removal of ash and organic insolubles.

[d] Includes ash.

[e] Ash-free basis.

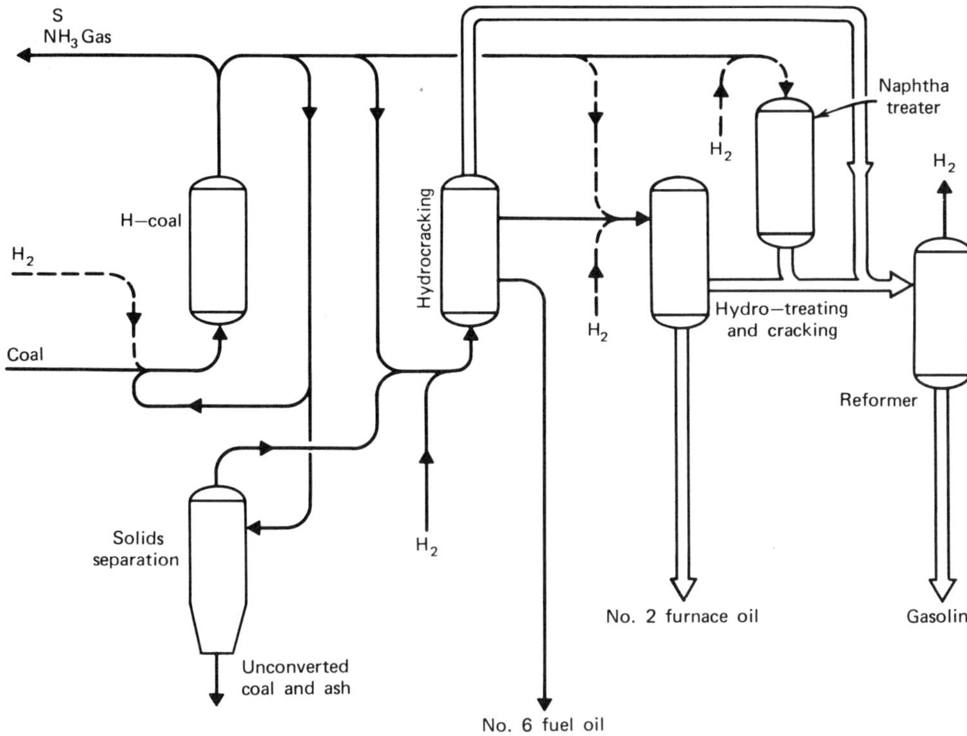

Figure 6. HRI's H-Coal process (52).

reactants maintains the catalyst in a fluidized state. Unconverted coal, ash, liquid, and gaseous products leave the reactor as overhead products, whereas the coarser catalyst is retained in the reactor. Catalyst can be added and withdrawn continuously to maintain a constant activity level. The reactor controls the temperature and contact with the catalyst; consequently, the reaction occurs at moderate pressure. The liquid product from the reactor is a synthetic crude oil that can be converted to gasoline or heating oil by conventional refining processes. Alternatively, under milder operating conditions, a clean fuel gas and low sulfur fuel oils may be produced. The relative yields of these products depend on the desired sulfur level in the heavy fuel oil.

The processing sequence shown in Figure 6 is only one approach for obtaining a desired product spectrum. In general, reaction products are separated by fractionation and absorption. Unreacted coal may be fed into a fluid coker that produces gas, gas oil, and dry char. The coker gas oil, along with gas oils separated from the main reactor effluent, may be subjected to hydrocracking for conversion to lighter products.

Reactor off-gas may be processed to remove hydrogen sulfide and ammonia and to produce salable sulfur and ammonia products. The naphtha and lighter gas-oil components collected from the reactor and hydrocracker effluents constitute the synthetic crude oil product.

Overall hydrogen requirements are estimated to exceed the potential quantity of hydrogen that may be supplied by steam reforming the total gas product. Hence, in general, additional coal is gasified in a steam–oxygen atmosphere, and the gaseous product treated, as in an oxygen-blown gasification train, to produce the necessary hydrogen. Additional coal is needed to supply power and steam for the installation which includes an oxygen plant and other auxiliaries. Estimated material balances and some product properties have been disclosed (53–54).

In 1976, Ashland Oil (Ashland Synthetic Fuels, Inc.) was awarded the prime contract to construct a 540 t/d (600 short tons of coal-per-day) H-Coal pilot plant adjacent to its refinery at Catlettsburg, Kentucky, by an industry-government underwriting consortium. Construction was completed in 1979, and the unit is expected to be fully operational by mid-1980.

Properties. The properties of a blend of naphtha, gas oil, and H-oil liquids, which constitute an expected synthetic crude product from an H-Coal operation, are given in Table 8. Properties of liquid products are a function of feed coal (53) and may be tailored significantly by manipulation of processing conditions.

Solvent-Refined Coal Process. In Germany in the 1920s the anthracene oil fraction recovered from pyrolysis, or coking, of coal was utilized to extract 35–40% of bituminous coals at low pressures for the purpose of manufacturing low cost newspaper inks (55).

Tetralin was found to have higher solvent power for coals, and the I.G. Farben Pott-Broche process (56) was developed, wherein a mixture of cresol and tetralin was used to dissolve ca 75% of brown coals at 13.8 MPa (2000 psi) and 427°C. The extract was filtered, and the filtrate vacuum distilled. The overhead was distilled a second time at atmospheric pressure to separate solvent, which was recycled to extraction, and a heavier liquid, which was sent to hydrogenation. The bottoms product from vacuum distillation, or solvent-extracted coal, was carbonized to produce electrode carbon. Filter cake from the filters was coked in rotary kilns for tar and oil recovery. A variety of liquid products was obtained from the solvent extraction–hydrogenation system (55). A similar process was employed in Japan during World War II to produce electrode coke, asphalt, and carbonized fuel briquettes (57).

There was little interest in solvent processing of coals in the United States, although the action of solvents on coals was used extensively in coal analysis and characterization. A method to reduce the sulfur content of coal extracts by heating with sodium hydroxide and zinc oxide was patented (58).

In the 1960s the Office of Coal Research (OCR) contracted with the Spencer Chemical Company to study the technical feasibility of a coal deashing process, and

Table 8. Properties of Syncrude from H-Coal Process[a]

Property	Boiling range			Total
	Initial to 190°C	190–343°C	343–524°C	
specific gravity, (°API)	0.767 (53.0)	0.915 (23.2)	1.05 (3.5)	0.863 (32.4)
vol % on total	40.0	54.2	5.8	100.0
analysis, wt %				
carbon	84.5	88.8	89.4	87.3
hydrogen	13.6	11.0	10.2	11.9
oxygen	1.7			0.6
nitrogen	0.1	0.1	0.1	0.1
sulfur	0.1	0.1	0.3	0.1
Total	*100*	*100*	*100*	*100*

[a] Ref. 53.

a process in a 45 kg/h flow unit (59) was demonstrated. Development was continued by The Pittsburgh & Midway Coal Mining Co. (P&M), a Gulf Oil Corp. subsidiary, and a pilot plant able to process 44–45 t/d was completed in late 1974 (60).

The process has a number of designations, including solvent deashing, solvent processing, the PAMCO process, the P&M process, the Solvent-Refined Coal (SRC) process, and most recently, the Pittsburgh & Midway Solvent-Refined Coal process.

A flow diagram of the SRC process is shown in Figure 7 (61). Coal is pulverized and mixed with a solvent to form a slurry containing 25–35 wt % coal. The slurry is pressurized to ca 7 MPa (1000 psig), mixed with hydrogen, and heated to ca 425°C. The solution reactions are completed in ca 20 min, during which time nearly all of the organic matter in the coal is dissolved. The reaction product is flashed to separate gases, and the liquid is filtered to remove the mineral residue (ash and undissolved coal). The liquid is then fractionated to recover the solvent, which is recycled to slurry more feed coal.

During each pass through the reactor a small amount of solvent is converted to light liquids and gases; however, this loss is more than balanced by the amount of coal that is converted to solvent. Separate solvent feed is required for initial start-up. The liquid remaining after the solvent has been recovered is a heavy residual fuel called solvent-refined coal, containing less than 0.8% sulfur and 0.1% ash. It melts at ca 177°C and has a heating value of ca 37 MJ/kg (16,000 Btu/lb), regardless of the quality of the coal feedstock. Smaller quantities of hydrocarbon gases and light distillate liquids are also produced. The solvent refining process removes all inorganic sulfur and 60–70% of the organic sulfur in the coal.

The action of solvents on coal is complex and not completely understood (62–63). In the process patent (64), it is indicated that the dissolving time is critical to the success of the process. The viscosity of the solubilized slurry in the dissolver first increases to a high value, with the result that gels may be formed if the solvent-to-coal ratio is too low. The viscosity then drops to some minimum value but thereafter rises again. It is essential to separate undissolved residue at the minimum viscosity.

The activity of the solvent is apparently more important than the action of gaseous hydrogen in this type of uncatalyzed hydrogenation. At first anthracene oil was used, a coal-derived material that was employed in the earliest coal extraction plants. However, it is not available in the quantities required to start-up a commercial SRC operation. Hence, research has been directed to use of petroleum-derived aromatic oils as start-up solvents (60).

In the early 1970s production of low sulfur, ashless (solid) boiler fuel was the preferred commercial application (61). This basic process (SRC-I) yielded small amounts of liquid oil products with additional processing. Liquid output was significantly increased by the coal–oil–gas (COG) refinery concept (65–66). In this design, a high degree of hydroconversion was incorporated to maximize distillate yields, and sulfur, nitrogen, and oxygen were removed from the liquid products by hydrotreatment. The primary liquid product was a light refinery liquid or syncrude, with a specific gravity of 0.85 (35° API), and a 347°C endpoint. About one quarter of the total coal feed to the design was directly gasified, and the gas was treated to produce the plant's hydrogen requirement. Cryogenic separation was employed to isolate the hydrogen from pipeline-quality gas, and to recover propane and butane (LPG) by-products.

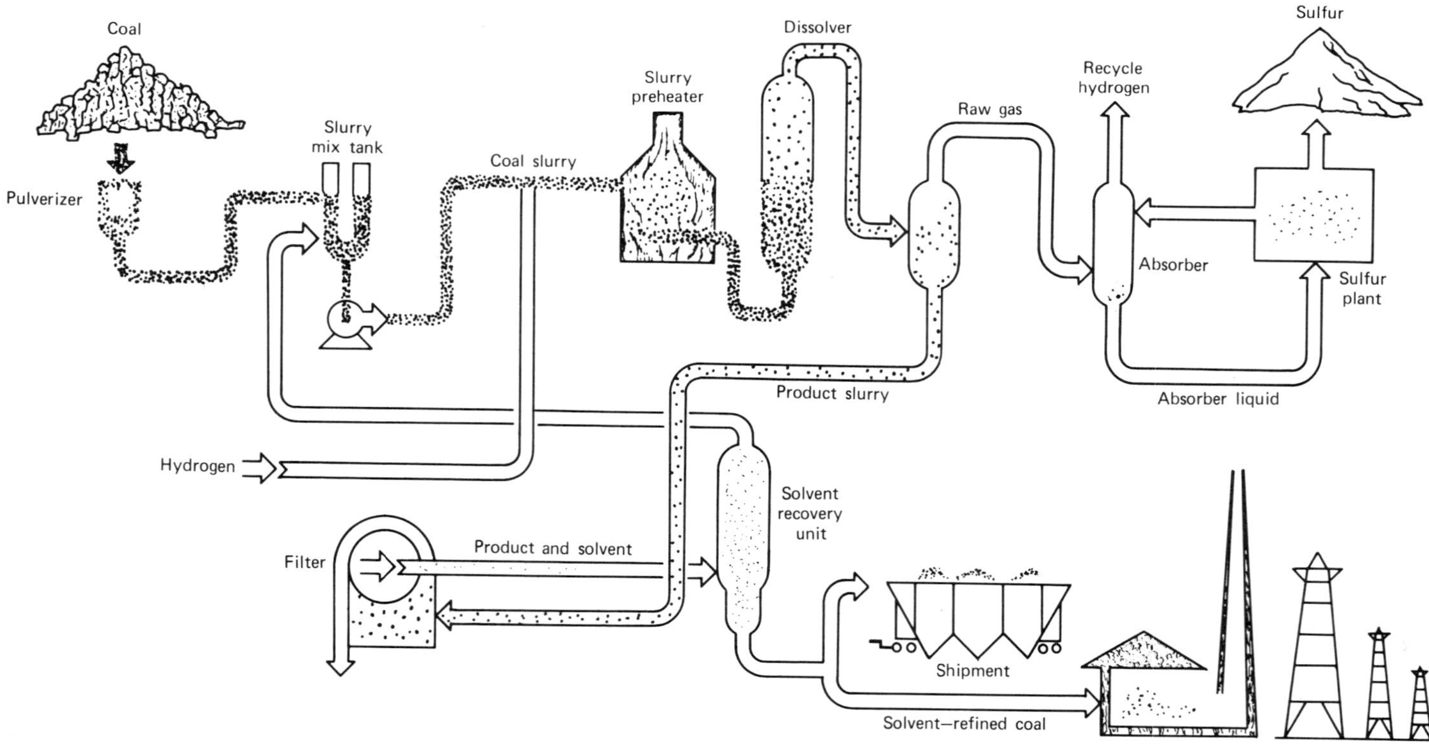

Figure 7. Solvent-refined coal process (61).

More recently, the SCR-II process has been developed (Fig. 8), in which hydrocracking occurs in the solution (hydrogenation) vessels (67). A low viscosity fuel oil in the primary distillate product in this case, although naphtha and LPG are also recovered.

The DOE has recently entered into a joint project with Gulf Oil which may lead to the construction of a 5400 t/d demonstration SRC-II plant at Morgantown, West Virginia. Present schedules call for operation by 1984. Partial funding for this project may come from the government of the Federal Republic of Germany and private German interests.

Southern Company Services, Inc. has operated a 5 t/d SRC-I pilot plant at Wilsonville, Alabama, since 1975. A joint project with DOE may result in the construction of a 5400 t/d SRC-I demonstration unit near Newman, Kentucky.

Properties. Table 9 shows properties of solvent-refined coal (SRC-I) derived from Kentucky no. 11 coal feed.

The properties of the liquid fuel oil produced by the SRC-II process are influenced by the particular processing configuration. However, in general, it is an oil boiling between 177 and 487°C, with a specific gravity of 0.99–1.00, and a viscosity at 38°C of 40 SUs (67). Pipeline gas, propane, and butane (LPG), and naphtha (C_5 to 176°C bp) are also recovered as fuel products from an SRC-II complex.

Consol Synthetic Fuel Process (CSF). Consolidation Coal Company (Consol) initiated coal liquefaction studies in the early 1950s. Subsequently, the OCR contracted the Ralph M. Parsons Company to assess the Consohio process, as it was then known, for converting coal to gasoline (Project Gasoline) (68). This process included coal extraction, low temperature carbonization of the residue from extraction, and hydroconversion of the coal extract.

Molten zinc chloride was investigated as a catalytic medium for coal hydrogenation in the early 1960s (69–70). Catalyst regeneration and metallurgical problems were never resolved, however.

A 900 kg/h pilot plant was constructed at Cresap, West Virginia, in 1967 (71–75). However, serious mechanical problems were encountered (71,74), in addition to declining catalyst activity, and the inability to demonstrate extract hydrogenation. Nevertheless, granulated extract produced at Cresap was successfully treated via the H-Oil process by Hydrocarbon Research, Inc. in 1970.

The CSF process is similar to the P&M SRC process; the main difference is that hydrogen gas is not introduced directly to the coal extraction reactor but is used to rehydrogenate the extraction donor solvent after the extract has been removed.

Figure 9 shows the four processing steps that are combined to convert coal into extract, char, and gas. Coal preparation includes drying and crushing. Coal–solvent slurry is heated to 385–410°C at 2.86 MPa (400 psig) in a coil-furnace preheater. Furnace product is discharged into a down-flow, agitated, staged extractor. The fundamental reaction is the depolymerization and solution of a major portion of the coal in an aromatic solvent, as a result of hydrogen transfer to the coal from a donor solvent.

Originally, three methods of separating undissolved coal from the solvent–extract solution were installed at Cresap. All operated at ca 315°C and 1.1 MPa (150 psig) to avoid precipitation and gelation of extract below ca 290°C. The most successful separation method used hydroclones (liquid cyclones). A two-stage system recovered 90% of the extract with ash content of one percent or less. Continuous precoat filtration

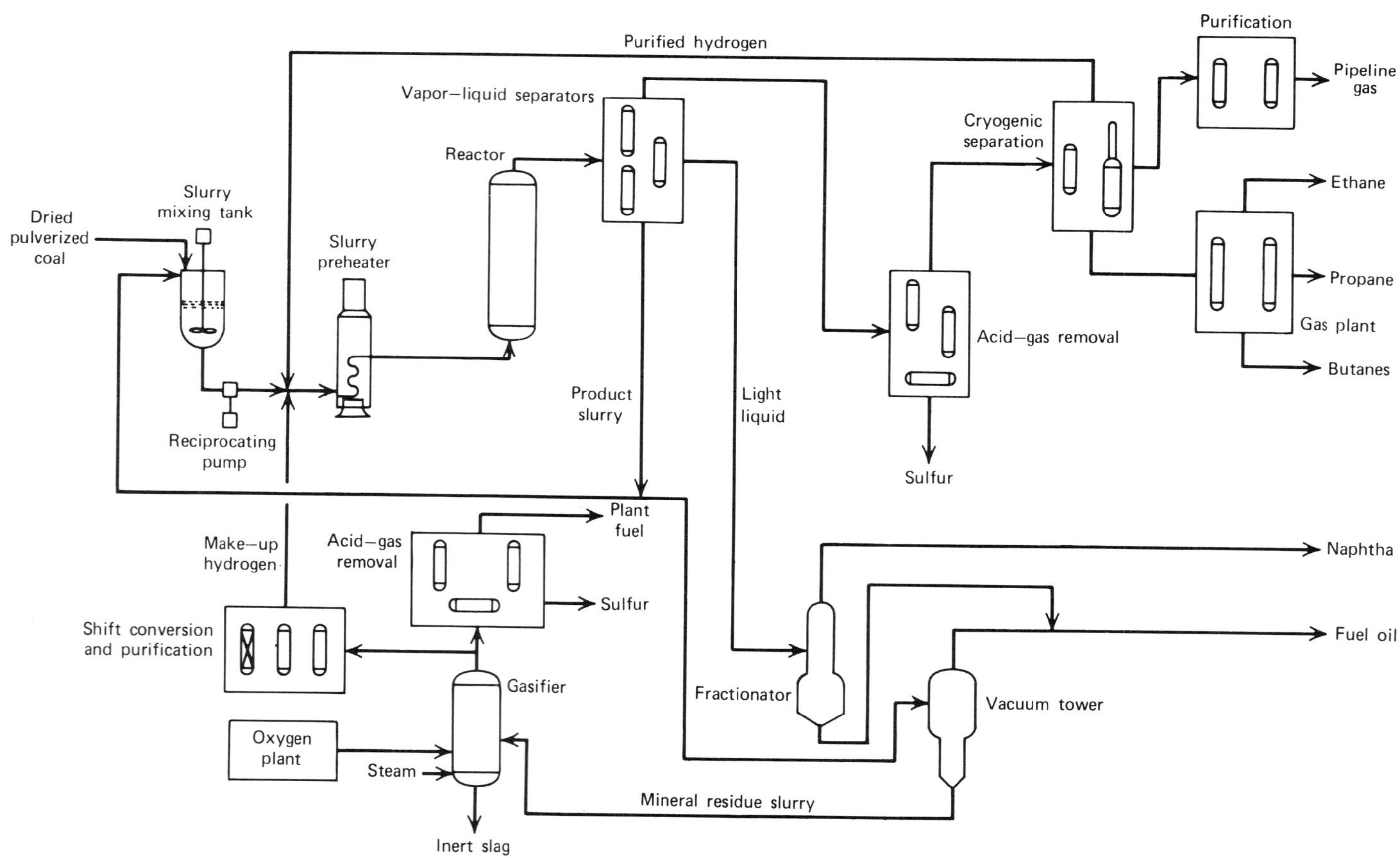

Figure 8. SCR-II process (67).

Table 9. Properties of Solvent-Refined Coal (SRC-I)[a]

Property	Kentucky no. 11 coal feed	SRC-I
carbon, wt %	78.45	88.16
hydrogen, wt %	5.20	5.23
nitrogen, wt %	1.19	1.54
sulfur, wt %	3.74	1.17
oxygen, wt %	11.40	3.42
Total	*99.98*	*99.52*
ash, wt %	7.33	0.48
volatile matter, wt %	42.88	36.6
melting point, °C		220
density, g/cm^3	1.33	1.24
heating value, MJ/kg[b]	32.5	36.6

[a] Ref. 64.
[b] To convert MJ/kg to Btu/lb, multiply by 430.

was not successful, presumably because of mechanical problems. Filtration was, in any event, considerably complicated by the necessity to maintain temperatures above 315°C throughout the system.

The extraction residue leaves the hydroclone separation system as a concentrated (40 wt % solid) slurry with solvent and small amounts of extract. The slurry is pyrolyzed (carbonized) at 430–480°C and low pressure to recover the solvent and to produce additional distillate.

Pyrolysis was accomplished by injecting the hydroclone underflow into a fluidized bed of char, using a nonclogging spray nozzle. The carbonizer was fluidized by a mixture of air and recycle gas; partial combustion of char with air provided the necessary heat for the system. Overhead vapors were quenched and oil-scrubbed to recover condensables. Control of char particle size in the fluid carbonizer was a major problem at Cresap.

The hydroclone extract is vacuum distilled to recover solvent and syncrude distillate. Coal extract is thermally degraded above 340°C but must be maintained above ca 290°C to prevent gelation. It is otherwise abrasive and nonlubricating.

Hydrogenation of extract at Cresap was planned in a staged, ebullated-bed reactor system at 24–29 MPa (3500–4200 psi) operating at 430°C, but was prevented by mechanical problems.

Finally, distillation of the hydrogenated extract separates low sulfur syncrude from donor solvent; the latter is recycled to the extraction step.

Properties. Only estimates of potential yields could be derived from Consol's operation of the Cresap pilot plant. Approximately 70 wt % of the coal feed was estimated to be recoverable as liquid (extract) product but additional hydrogen was required equal to ca 3% of the coal feed weight. As already indicated, H-Oil processing was successfully applied to Cresap extract, and converted some 70% of the extract to a distillate boiling at 200–440°C (74). Wheeler presented estimates of yields for two conceptual design options arising from its evaluation of Project Gasoline (75).

Exxon Donor Solvent–Coal Liquefaction Process. Exxon's research into coal liquefaction (76) resulted in the EDS process designed to produce high yields of liquid products.

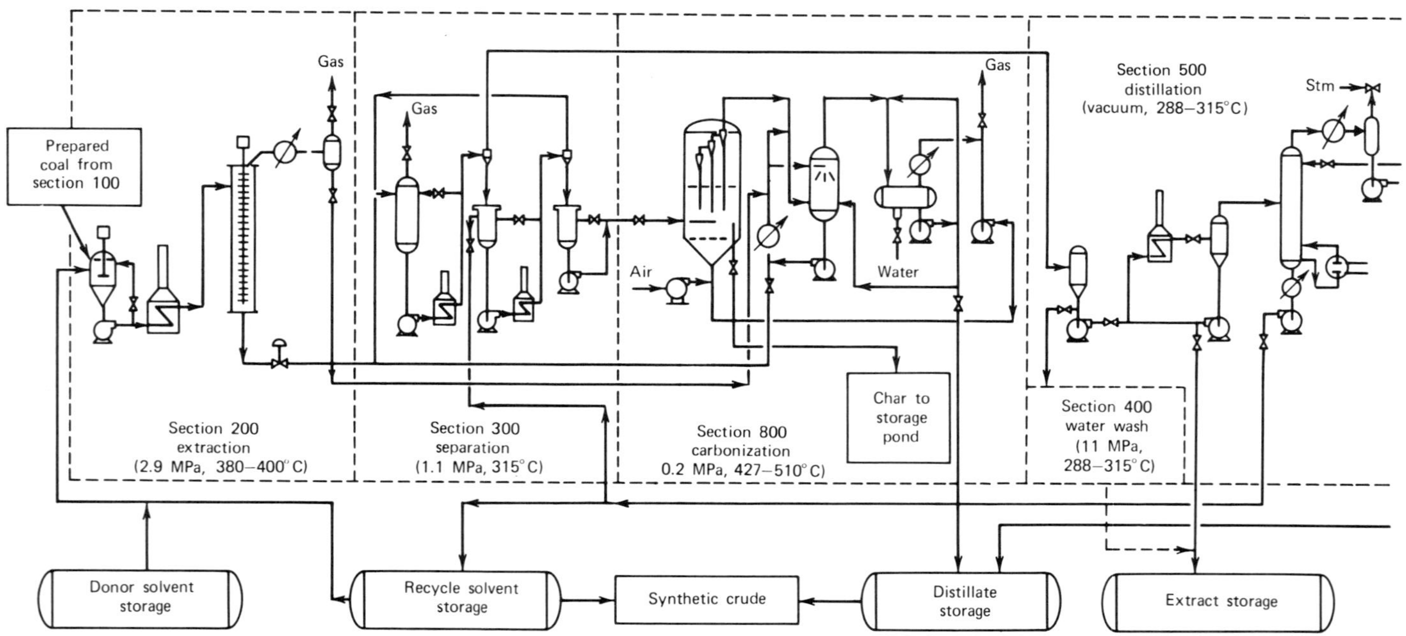

Figure 9. Extract production at coal liquefaction pilot plant, Cresap, W.Va. (73). To convert MPa to atm, divide by 0.101.

In the process sequence, shown in Figure 10, the feed coal is crushed, dried, and slurried with hydrogenated recycle solvent (the donor solvent) and fed to the liquefaction reactor in admixture with gaseous hydrogen. The reactor is an upward plug–flow design operating at 430–480°C and at ca 14 MPa (2000 psi) total pressure.

The reactor effluent is separated by conventional distillation into recycle solvent, light hydrocarbon gases, C_4 to 537°C bp distillate, and a heavy vacuum bottoms stream containing unconverted coal and ash. The recycle solvent is hydrogenated in a conventional fixed-bed catalytic reactor, and recycled to the liquefaction reactor.

The heavy vacuum-bottoms stream is fed to a Flexicoking unit. This is a commercial (77) petroleum process that employs circulating fluidized beds at low pressures [0.3 MPa (50 psi)] and intermediate temperatures (480–650°C in the coker and 815–980°*C* in the gasifier) to produce high yields of liquids or gases from organic material present in the feed. Residual carbon is rejected with the ash from the gasifier fluidized bed. The total liquid product is a blend of streams from liquefaction and the Flexicoker.

The process's hydrogen requirement may be obtained by steam reforming the light hydrocarbon gases, from partial oxidation of the heavy vacuum bottoms stream, or from additional coal.

In 1976 Exxon Research & Engineering Company entered into a six-year joint research project with the DOE, which is supplying half the funding, to develop the EDS Process. The Electric Power Research Institute (EPRI), Japan Coal Liquefaction Development Company (JCLD), Phillips Petroleum Company, Atlantic Richfield Company, and Ruhrkohle A.G. are participating with Exxon Co. USA in funding the industrial sector of the project.

Under the direction of Exxon Co. USA, a 230 t/d pilot plant has been constructed at Baytown, Texas. Operations designed to demonstrate operability of the EDS liquefaction section are in progress.

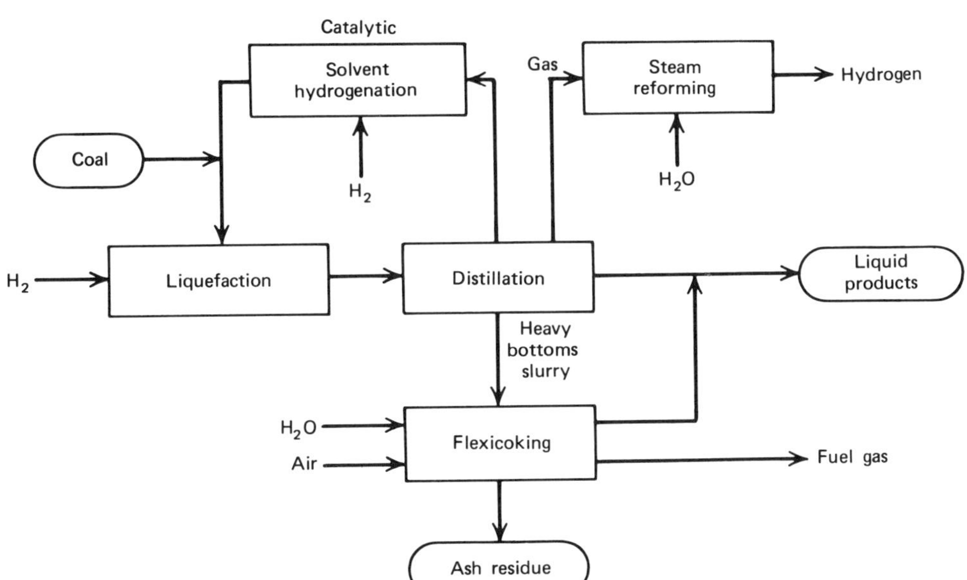

Figure 10. Exxon donor solvent process (76).

Properties. Pilot-unit data indicate that the EDS process may accommodate a wide variety of coal types, including bituminous, subbituminous, and lignites. Overall process yields, which include liquids from both liquefaction and Flexicoking, are shown in Figure 11. The bituminous coals give total liquid yields of 43–45 wt % of the dry feed coal rate, the subbituminous coal produces ca 40% liquid, and the lignites yield 33–35% liquid. These liquids have higher nitrogen contents than are found in similar petroleum fractions. Sulfur contents reflect the sulfur levels of the starting coals (ca 4.0 wt % sulfur in the dry, bituminous coal, 0.5 wt % in the subbituminous, and 1.2 wt % sulfur in the dry lignite).

As shown in Table 10, naphthas (C_5 to 204°C bp) produced by the EDS process have higher concentrations of cycloparaffins and phenols than do petroleum-derived naphthas, whereas *n*-paraffins are present in much lower concentrations. The sulfur and nitrogen concentrations in coal naphthas are high compared to those in petroleum naphthas.

Gas-oil fractions (204–565°C) from coal liquefaction show even greater differences in composition compared to petroleum-derived counterparts than do the naphtha fractions (78). The coal-gas oils consist mostly of aromatics (60%), polar heteroaromatics (25%), asphaltenes (8–15%), and saturated compounds (<10%). Petroleum-gas oils, on the other hand, contain more than 50% saturated compounds, less than 5% polar heteroaromatics, and no asphaltenes. Furthermore, the aromatics of petroleum-gas oils have longer side chains.

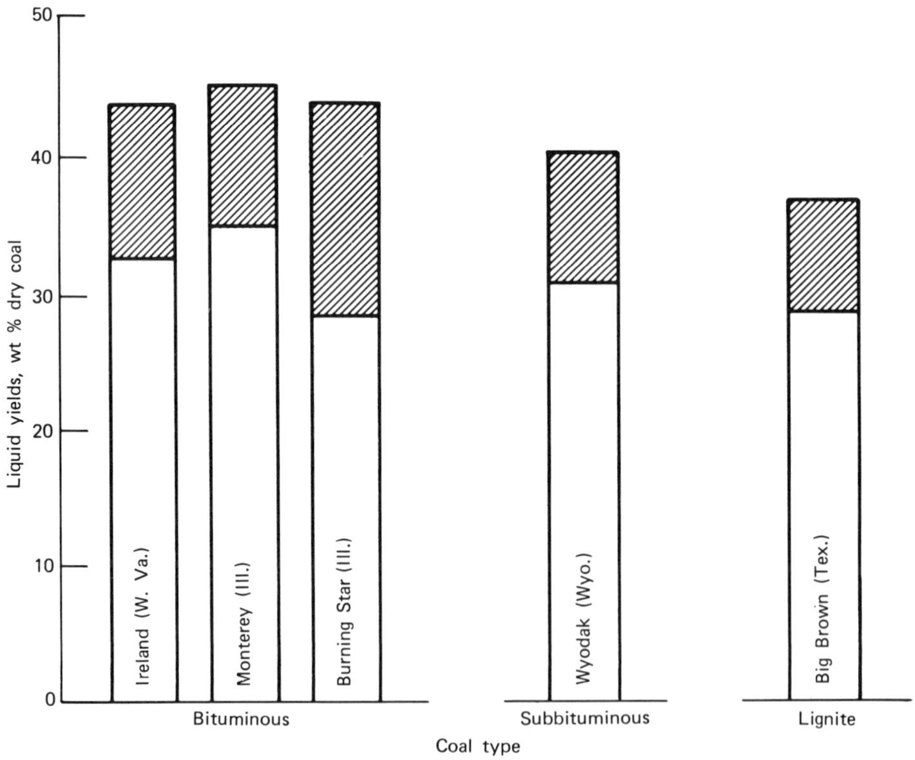

Figure 11. Preferred liquefaction–coking liquid yields in EDS process (76). ▨, Flexicoking liquids; □, liquefaction liquids.

Table 10. Composition of Naphthas (C_5 to 204°C bp) from Various Sources[a]

	EDS coal liquefaction		Petroleum naphthas	
Component, wt %	Illinois coal	Wyodak coal	Cycloparaffinic[b]	Paraffinic[c]
saturated compounds	69.9	60.8	77.9	81.3
paraffins	13.4	19.4	36.0	58.2
cycloparaffins	56.5	41.4	41.9	23.1
olefins (approx)	0.5	3.7		
aromatics	17.0	28.6	22.1	18.7
benzenes	11.7	25.1	20.8	17.2
indans, tetralins	5.1	3.5	1.0	1.2
indenes	0.2	0.0		
naphthalenes	0.01	0.0	0.3	0.3
phenols	12.6	6.9	traces	traces
Total	*100.0*	*100.0*	*100.0*	*100.0*
sulfur	0.57	0.10	0.035	0.049
nitrogen	0.15	0.18	0.0001	0.0001
oxygen	1.60	2.49		

[a] Ref. 78.
[b] Prudhoe Bay.
[c] Arab Light.
[d] wt % of total sample.

Fischer-Tropsch Synthesis

In 1902 Sabatier and Senderens reported (79) the synthesis of methane by the catalytic reduction of carbon monoxide and hydrogen over nickel and cobalt catalysts at atmospheric pressure:

$$CO + 3\ H_2 \rightarrow CH_4 + H_2O$$

$$2\ CO + 2\ H_2 \rightarrow CH_4 + CO_2$$

In the early 1920s Badische Anilin- und Soda-Fabrik announced the specific catalytic conversion of carbon monoxide and hydrogen at 20–30 MPa (200–300 atm) and 300–400°C to methanol (80–81), a process subsequently widely industrialized. At the same time Fischer and Tropsch announced the Synthine process (82–83), in which an iron catalyst effects the reaction of carbon monoxide and hydrogen to produce a mixture of alcohols, aldehydes, ketones, and fatty acids at atmospheric pressure.

In the classical normal pressure synthesis (84), higher hydrocarbons are produced by net reactions similar to those observed by Sabatier but at temperatures below the level at which methane is formed:

$$n\ CO + 2\ n\ H_2 \rightarrow C_nH_{2n} + n\ H_2O$$

$$2\ n\ CO + n\ H_2 \rightarrow C_nH_{2n} + n\ CO_2$$

In the mid 1930s improvements in catalysts and techniques (85–87) culminated in the licensing of the process to Ruhrchemie to produce liquid hydrocarbons and paraffin waxes using precipitated cobalt-on-kieselguhr catalysts.

Subsequently, Fischer and Pichler developed a medium pressure synthesis (88) at 0.5–2 MPa (5–20 atm) with dispersed cobalt catalysts which improved hydrocarbon yields by 10–15%. The yields of paraffin wax, in particular, could be increased to 45% of the total liquid product, hydrotreating of catalyst (required in the normal pressure process) was avoided, and catalyst life was extended (89–91).

Figure 12 shows the marked influence of pressure on yields. Beyond the optimum

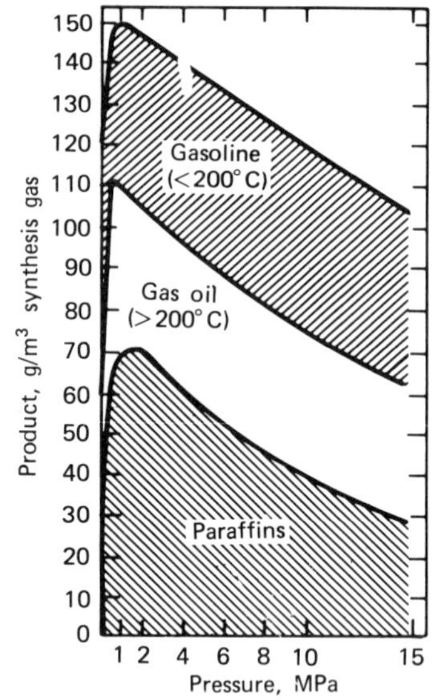

Figure 12. Influence of pressure on yields of gasoline, gas oil, and paraffin (cobalt catalyst). To convert MPa to atm, divide by 0.101. To convert g/m^3 to lb/ft^3, multiply by 6.24×10^{-5}.

pressure of about 2 MPa (20 atm), paraffin yield decreases but little change is found in the ratio of gasoline and gas-oil yields.

Fischer and Pichler furthermore discovered that reasonable yields could be obtained with precipitated iron catalysts at 1–3 MPa (10–30 atm), and that very high melting waxes could be synthesized at 10–100 MPa (100–1000 atm) over ruthenium catalysts. At the same time a related process, the Oxo synthesis (see Oxo process), was announced (92). Early in World War II, Fischer and co-workers developed the Isosynthesis process for the production of low molecular weight isoparaffins at high temperatures and pressures over thoria and mixtures of alumina and zinc oxide (93–96). In the early 1960s polymethylenes were synthesized with activated ruthenium catalysts at high pressures.

Industrial operation of the Fischer-Tropsch synthesis involves five steps: (*1*) synthesis gas manufacture; (*2*) gas purification by removal of water and dust, and hydrogen sulfide and organic sulfur compounds; (*3*) synthesis of hydrocarbons; (*4*) condensation of liquid products and recovery of gasoline from product gas; and (*5*) fractionation of synthetic products. Only the synthesis reactor and its method of operation was unique to the process. For low pressure synthesis it incorporated elaborate bundles of cooling tubes immersed in the catalyst, whereby circulating water removed the heat of reaction, limiting the conversion to methane which has produced high temperatures. In the pressure process, bundles of concentric tubes, with catalyst arranged in the annuli, through and around which cooling water flowed, served as conversion units. In both systems, the conversion units each contained about 10 m^3 (ca 350 ft^3) of catalyst, and were rated at a capacity of ca 4.8 m^3 (30 bbl) of liquid product per day.

During World War II nine commercial plants were operated in Germany, five using the normal pressure synthesis, two the medium pressure process, and two having converters of both types. The largest plants had capacities of ca 400 m^3/d (2500 bbl/d) of liquid products. Combined production reached ca 2800 m^3/d (18,000 bbl/d) during the war. Cobalt catalysts were used exclusively, although the shortage of cobalt in 1943 led to pilot-plant experiments in which the feasibility of hydrocarbon production with iron catalysts was proved. The best yields, however, were 15% less than with cobalt.

Developments Outside Germany. In the late 1930s experimental work in England (97–99) led to the erection of large pilot facilities for Fischer-Tropsch studies (100). In France, the Societe Courrieres-Kuhlmann licensed the process from Ruhrchemie, and constructed a commercial facility near Calais which produced about 150 m^3 (ca 940 bbl) of liquid hydrocarbons per day. In Japan, two full-scale plants were operating under Ruhrchemie license, with combined capacity of ca 400 m^3 (2500 bbl) of liquids per day.

In the mid 1930s Universal Oil Products reported (101–102) that gasoline of improved quality could be produced by cracking the high boiling fractions of Fischer liquids. A consortium (known as Hydrocarbon Synthesis, Inc.) entered into an agreement with Ruhrchemie to license the Fischer synthesis outside Germany. Experimental work performed before World War II has remained largely proprietary, and the group was dissolved with the advent of war.

In the United States, as in other countries, the scarcities of war revived interest in the synthesis of fuel substances. A study of the economics of Fischer synthesis led to the conclusion that the large-scale production of gasoline from natural gas offered hope for commercial utility. In the Hydrocol process (Hydrocarbon Research, Inc.) natural gas was treated with high-purity oxygen to produce the synthesis gas which was converted in fluidized beds of iron catalysts (103). A large facility was erected for Carthage Hydrocol, Inc., at Brownsville, Texas, in 1950, and mechanical problems, which had caused erratic operation initially, were solved within a few years. However, the economic utility of the process was undermined by the rapid development of low cost Middle East crude-oil resources.

Table 11 compares the synthesis conditions for the German process with data from a Hydrocol pilot plant.

The U.S. Bureau of Mines has tested other techniques, including direct heat exchange with recycling oil in fixed or moving beds of granular catalysts, and conversion in suspensions or slurries of finely powdered catalysts in oil (105).

Table 11. Comparison Between Fixed-Bed and Fluidized Catalysts[a]

operating data	German fixed-bed catalyst	Hydrocol fluidized catalyst
temperature, °C	180–220	300–350
throughput vol gas/(vol catalyst·h)	100	3000[b]
heat of reaction, MJ/(m^3 catalyst·h)[c]	209	6279
yield of C_3 and higher, g/m^3 synthesis gas[d]	165	165
yield of C_3 and higher, kg/(m^3 catalyst·h)	16.5	495
synthesis converters required for production of 1×10^6 t/yr	635–907	5

[a] Ref. 104.
[b] The expanded volume of the catalyst is considered.
[c] To convert MJ/m^3 to Btu/ft^3, multiply by 26.87.
[d] To convert g/m^3 to lb/ft^3, multiply by 6.243×10^{-5}.

In the Union of South Africa in 1947 the establishment of an oil-from-coal industry was encouraged by legislation. Soon thereafter the South African Coal, Oil, and Gas Corporation Ltd. (Sasol) was founded with share capital provided by the Industrial Development Corporation on the government's behalf. A large synthesis facility was constructed at the site of a vast coal field utilizing a combination of fluidized Synthol process technology (106) with Ruhrchemie-Lurgi fixed-bed Arge synthesis units. The system started up in 1955.

In 1974 Sasol's coal mine had a production capacity of 16,000 t coal/d. About 6,000 t/d was being fed to a total of 13 Lurgi gasifiers (see Coal). Average gasifier load was ca 39,000 m^3/h (14×10^5 ft^3/h). About 5400 t coal/d was consumed additionally in the direct production of electricity and steam for the synthesis complex. The oxygen plant, the largest in the world at that time, produced 2360 t/d of 98% oxygen. A vast array of chemical and fuel products, including 1.5×10^6 t gasoline/yr were produced from a combination of fixed- and fluidized-bed reactors arranged to maximize syngas usage.

Currently, a major expansion of the Sasol facility is underway. Sasol II, so-called, was scheduled to consume an average of 40,000 t coal/d, including the outputs of two new coal mines being constructed as part of the expansion.

Properties. Fischer-Tropsch liquid obtained with cobalt catalysts is roughly equivalent to a very paraffinic natural petroleum oil (see Petroleum–petroleum composition) but is not so complex a mixture. Straight-chain, saturated aliphatic molecules predominate but monoolefins may be present in an appreciable concentration. Alcohols, fatty acids, and other oxygenated compounds may represent less than 1% of the total liquid product. The normal pressure synthesis yields ca 60% gasoline, 30% gas oil, and 10% paraffin (mp 20–100°C). The medium pressure synthesis yields 35% gasoline, 35% gas oil, and 30% paraffin. The octane rating of the gasoline is too low for direct use as motor fuel (107).

Most of the German gasoline production was blended into motor fuels with benzene derived from coking. The gas oil could be used directly as a superior Diesel fuel; some was also used for soap (qv). The paraffin (referred to as gatsch) was used primarily for the synthesis of fatty acids and hard soaps. The propane and butane gases were also used as motor fuels; some propylenes and butylenes were polymerized over phosphoric acid to high octane gasoline, and some olefins to lubricating oils.

Table 12 presents typical values for the composition of the technical-scale reaction products of the normal and medium pressure syntheses.

Recent Developments. Extensive research continues to interpret the mechanism of the Fischer-Tropsch synthesis; eg, investigation of the role of alkali (109–110). A review of recent theoretical studies can be found in ref. 111.

The most significant recent achievement in a related field is the methanol conversion process announced by Mobil in 1976 (112). A wide variety of heteroorganic compounds, including methanol and dimethyl ether, may be efficiently transformed to C_2–C_{10} hydrocarbons using a new class of shape-selective zeolites (113) (see Molecular sieves). A potential mechanism has been postulated (114) whereby methanol, eg, is converted to olefins, predominantly in the C_2–C_5 range. Finally the C_2–C_5 olefins are converted to paraffins, cycloparaffins, and olefins containing more than 6 carbon atoms which fall predominantly in the gasoline boiling range (C_4–C_{10}). Gasoline so produced is chemically similar to petroleum-derived gasoline, and has unleaded research octane numbers of 90 to 95. This Mobil gasoline is superior in both yield and quality to that produced by the traditional Fischer-Tropsch processes (113).

Table 12. Properties of Normal and Medium Pressure Syntheses Products [a]

Constituent	Wt % of total products listed[b]		d_4^{15}	Olefins, vol %	bp index[c]	Mol wt (av)	No. C atoms	Cloud pt, °C	Pour pt, °C	Octane no., research method	Cetane No.	
											Motor	Research
normal pressure synthesis[d]												
gasol ($C_3 + C_4$)	12	12		50			$C_3 + C_4$					
gasoline, to 185°C	49		0.689	37	108	100	C_4–C_{10}			52		
gasoline, to 200°C		54	0.693	34	115	115	C_4–C_{11}			49		
diesel oil, 185–320°C	29		0.760	15	245	190	C_{11}–C_{18}	−13	−18		87	100
diesel oil, 200–320°C		24	0.766	13	260	205	C_{12}–C_{19}	−5	−9		92	105
soft paraffins, 320–450°C	7	7	0.900	iodine value			$>C_{18}$					
hard paraffins, >450°C	3	3	0.930	2								
Total	*100*	*100*										
medium pressure synthesis[e]												
gasol ($C_3 + C_4$)				30			66% C_4 33% C_3					
gasoline, to 185°C	35		0.685	20	110	100	C_4–C_{10}			28		
gasoline, to 200°C		40	0.689	18	120	115	C_4–C_{11}			25		
diesel oil, 185–320°C	35		0.760	10	240	190	C_{11}–C_{18}	−7	−11		87	100
diesel oil, 200–330°C		35	0.766	8	255	205	C_{12}–C_{19}	−2	−5		92	105
soft paraffins												
>320°C	30		0.90	iodine value			$>C_{18}$		80[f]			
>330°C		25	0.90	2			$>C_{19}$		80[f]			
Total	*100*	*100*										

[a] Ref. 108.

[b] Total yield per cubic meter of synthesis gas: normal pressure synthesis, 148 g; medium pressure synthesis, 145 g of liquid products and 10 g of gasol.

[c] The boiling point index is the sum of the temperatures at which 5, 15, 25 . . . 95 vol % distill over, divided by 10.

[d] At 101 kPa (1 atm); 180–95°C; catalyst, 100 Co:5 ThO_2:7.5 MgO:200 kieselguhr; 1 CO:2 H_2 (18–20% inert components); throughput, 1 m^3 synthesis gas/(kg Co·h) (3.5 ft/lb Co·h); two stage; no recycle.

[e] At 707 kPa (7 atm abs); 175–95°C; catalyst, 100 Co:5 ThO_2:7.5 MgO:200 kieselguhr; 1 CO:2 H_2 (18–20% inert components; throughput, 1 m^3 synthesis gas/(kg Co·h); two stage; no recycle.

[f] Approximate.

A fixed-bed version (115) is considered ready for commercialization. Mobil Oil has recently proposed a joint project to the government of New Zealand which may lead to the construction of a fixed-bed facility which could supply up to half of New Zealand's gasoline requirement, currently ca 6360 m^3/d (40,000 bbl/d), by 1984. In that case, available natural gas would be converted to methanol, which would in turn be converted to gasoline.

A fluidized-bed version has been piloted at a level of ca 0.63 m^3/d (4 bbl/d) (116). The DOE has recently indicated interest in a joint project of the FRG government, and private German interests in the construction of a 16 m^3/d (100 bbl/d) fluidized-bed pilot unit in the FRG.

Shale Oil

In the United States, shale oil, or oil derivable from oil shale, represents the largest potential source of liquid hydrocarbons that can be readily processed to fuel liquids similar to those derived from natural petroleum (see Oil shale).

A number of foreign countries produce liquid fuels from oil shale, but there is no such industry in the United States at present, although more than fifty companies were producing oil from coal and shale in the United States in 1860 (117–118).

Private oil companies resumed interest in the vast domestic oil-shale deposits during and immediately after World War II. This industrial activity prompted the government to terminate large-scale shale-oil research which had been instituted by the *Synthetic Liquid Fuels Act of 1944.* The very rapid increase in Middle East crude-oil production following the war, however, drove worldwide crude prices down and shale development was again curtailed. After the Arab oil embargo of 1973 several companies reactivated shale-oil process development programs for the third or fourth time (119–120).

The significance of oil-shale resources may be exaggerated. No more than 3–5% of the total domestic resource is estimated to be economically extractable with present-day technology. Moreover, no process has yet been reported to produce crude shale oils at costs that may compete with those of natural petroleum. Development of a significant domestic shale-oil industry using today's technology requires enormous capital outlays. In the current framework of high inflation, spurred by the rapid increase in the price of natural petroleum, and unclear environmental constraints (121), establishment of a shale-oil industry requires direction by the federal government.

Heavy Oil

The definitions used to distinguish among naturally occurring heavy petroleum oils, bitumens, asphalts, and tars, among other classifications, are subject to broad variations. More than 10% of the world's current crude oil production has an API gravity below 20°, or a specific gravity greater than 0.934^{15}_{15}. Oils with sp gr^{15}_{15} >0.904–0.934 (20–25° API) are considered heavy oils in most classifications. However, the context must be clear. Safaniyah crude oil produced in Saudi Arabia with API gravity of sp gr^{15}_{15} 0.893 (27° API) carries the designation Arabian heavy, and, in petroleum parlance, is generally referred to as heavy crude oil. Yet its production method does not differ from that of Arabian medium or Arabian light crude oils.

Energy in the form of injected water or CO_2 may be supplied to increase the rate of production of light crude oils. Application of heat to the reservoirs, eg, with hot water, steam, heated CO_2, fireflood, or *in-situ* combustion, however, is generally associated with the production of heavier, viscid crudes.

Heavy crude oil is widely distributed, and it is difficult to estimate reserves separate from normal crude oil reserves or from tar sands deposits. Current estimates of petroleum reserves frequently include a large heavy oil component, which can only be produced at significantly higher cost than light oil.

Most heavy oil production is currently concentrated in California, Canada, and Venezuela. About half of California's current crude oil production of 143,000 m^3/d (900,000 bbl/d), eg, is heavy oil, much of it coming from the Kern River field, near Bakersfield, where oil gravity averages sp gr_{15}^{15} 0.993–0.972 (11–14° API). Production generally involves steam drives, or the injection of steam into reservoirs through special wells in prescribed sequences. Oil–water mixtures are recovered and often separated water is treated and reinjected.

Tar Sands

Tar sands are considered to be sedimentary rocks with natural porosity, whose pore volume is occupied by viscous, petroleumlike hydrocarbons. It has been pointed out that tar is a manufactured article of commerce, and that sand is a term applied to particulate mineral matter, so that tar sands may be a misnomer. Similar semantic problems may be incurred with the terms oil sands, rock asphalts, asphaltic sandstones, and malthas, or malthites, all of which have been applied to the same resource. The hydrocarbon component of tar sands is properly termed bitumen (see Tar sands).

Distinctions between tar sands' bitumens and heavy oils are based largely on differences in viscosities. Whereas heavy oils might be produced by the same techniques used for the lighter crude oils, the bitumens in tar sands are too viscous for these techniques. Consequently these oil-bearing stones have to be mined and specially processed to recover contained hydrocarbon.

Tar sands have been reported on every continent except Australia and Antarctica. The best known deposits are the Athabasca of Canada, where more than 50,000 km^2 in northeastern Alberta is underlain with an estimated $(47–63) \times 10^9$ m^3 $[(3–4) \times 10^{11}$ bbl] of recoverable bitumen. By many estimates, Canada may have a reserve of 160×10^9 m^3 $(1 \times 10^{12}$ bbl) bitumen. Venezuela may have the largest accumulations in the world; the Orinoco heavy-oil belt has been estimated to contain 635×10^9 m^3 $(4 \times 10^{12}$ bbl). The USSR may have a reserve of ca 160×10^9 m^3 $(1 \times 10^{12}$ bbl).

Until recently, the plant of the Great Canadian Oil Sands Ltd. (GCOS) (Sun Oil Company) at Fort McMurray, Alberta, Canada, which has been operating since 1967, was the only commercial tar sands enterprise of any consequence in the Western world. Typically, some 8050 t/d (55,000 bbl/d) of a synthetic crude oil were being produced from 110,000 t/d of tar sands (122).

The plant of Syncrude Canada, Ltd. with an allowable production rate of 20,000 m^3/d (129,000 bbl/d), has recently come on stream with an initial production capability about equal to that of GCOS.

Alcohol Fuels

The lower alcohols, especially methanol and ethanol, have been widely used separately, and in blends with gasolines and other hydrocarbons, to fuel internal combustion engines. Since these compounds, as a class, may be considered to be already partially oxidized, incorporating a mole of oxidized hydrogen in each molecule, they differ from the hydrocarbons that make up gasoline principally in their lower heating values and in their higher vaporization heat requirements (Table 13). Their solubilities in gasolines (124–125) and their effects on gasoline vapor pressure (126) are of special importance in the formulation of fuel blends, eg, when alcohols are used to extend volumes of gasolines. The sensitivity of alcohol–gasoline blends to the presence of water is a complex subject (125,127) which must be considered in such application.

Although the alcohols, as particular compounds, may be obtained in extremely pure form, those materials that may be used as fuels are apt to be more or less contaminated with other substances, depending on the alcohol source. This, along with the wide variability in the chemical constitution of gasolines, makes the formulations of blends extremely difficult to define precisely, and results obtained from testing such blends in fuel applications may be limited. The presence of small amounts of particular compounds may markedly affect mutual solubilities, water tolerances, volatilities, or octane ratings, among other properties, of alcohol–gasoline fuel blends.

Claims relating to improvements in fuel economy, exhaust emissions, and driveability for alcohol–gasoline blends must be critically examined. The fuel economy, in km/L (mi/gal), must decrease proportionally to increases in the alcohol content of the blend. Exhaust emissions should be changed to the same extent as with gasolines alone, as carburetting is modified to accommodate the blended fuel (123). Driveability, starting, vapor lock, and corrosion problems must also increase with increasing alcohol content when blends are introduced to current systems designed to operate with straight gasolines. There is little doubt that particular engine systems may be designed

Table 13. Thermal Properties of Alcohol and Hydrocarbon Fuels[a]

Property	Methanol	Ethanol	Isoctane	Typical gasoline
heating value				
higher, liq fuel–liq water, MJ/kg[b]	22.7	29.8	47.8	47.2
lower, liq fuel–water vapor, MJ/kg[b]	19.9	26.7	44.3	44.0
liq fuel–water vapor, MJ/L[c] at 20°C	15.8	21.1	30.6	32.1
gas fuel–water vapor, MJ/kg[b] at 20°C,	21.1	27.7	44.6	44.3
stoichiometric mixture				
in vapor state, MJ/m^3 [d] at 20°C	3.44	3.46	3.50	3.55
fuel in liq state, MJ/kg[b] of air	3.1	3.0	2.9	3.0
latent heat of vaporization at 20°C				
MJ/kg[b]	1.2	0.9	0.3	ca 0.4
MJ/L[c]	0.9	0.7	0.2	ca 0.3
MJ/kg of air[b,e]	0.2	0.1	0.02	ca 0.02

[a] Ref. 123.
[b] To convert MJ/kg to Btu/lb, multiply by 430.
[c] To convert MJ/L to Btu/gal, multiply by 3589.
[d] To convert MJ/m^3 to Btu/ft^3, multiply by 26.85.
[e] For a stoichiometric mixture.

or modified to optimize performance for a given blend. However, such modified systems may have limited tolerance for deviations from design fuel specifications.

To date, no process has been shown to produce alcohols at lower cost per volume than gasolines derived from natural petroleum. Moreover, the energy equivalent required to produce alcohols is generally greater than the heat derivable from their combustion (128) (see Gasoline).

Methanol. Methanol has been produced by the destructive distillation of wood but is now largely obtained from carbon monoxide and hydrogen gases (synthesis gas) which are produced by the partial oxidation of methane (natural gas). Synthesis gas may be produced, however, by the partial oxidation, or gasification, of virtually any carbonaceous fuel resource, including coals, shale oils, and petroleum.

Depending on the synthesis process (see Methanol) employed, more or fewer impurities, including water, may be present in the product. The term methyl fuel describes certain methanol products that may also contain significant quantities of water and higher alcohols and other oxygenated compounds. Water removal from such mixtures by distillative processes is generally complicated by the formation of azeotropes. The associated energy requirements may outweigh the advantages afforded by the production of impure alcohol, or the inclusion of higher alcohols in methanol fuels.

Straight methanol has long been a preferred fuel for racing engines because of the much higher compression ratios at which methanol may be combusted relative to gasolines. This is translatable for racing purposes at equivalent power outputs to engines of considerably reduced weight. However, fuel consumptions roughly three times that of gasoline, on a km/L (mi/gal) basis, and extremely high emissions of unburned fuel and carbon monoxide, can result (129).

The engine performance of methanol compared to that of gasolines has been subjected to various assessments (130–132). Much of the complexity relates to the multidimensional nature of the test parameters, including the inability to truly equilibrate the engine systems for fuels of widely differing properties. Other studies (133–137) have shown that methanol combustion may be optimized in existing automotive engines to take advantage of its leaner misfire limit, providing increased efficiency compared to gasoline, and with significantly lower nitrogen oxide emissions. Fuel consumptions are roughly double those for gasoline on a volumetric basis. Moreover, emission of unburned methanol may be high. Formaldehyde has also been reported in the exhausts of cars fueled with straight methanol or methanol–gasoline blends.

Because of its high latent heat of vaporization, intake manifolds of normally carburetted, methanol-fueled engines must be heated. Methanol cannot vaporize sufficiently to enable a cold engine to start below ca 10°C. Therefore, a more volatile fuel component, such as propane or butane, either must be blended into methanol fuel, or must be injected separately into the manifold for starting, if a separate heating system is not provided.

The ASTM octane rating methods for hydrocarbons, which are determined at the fuel–air ratio for maximum knock intensity, lose much of their significance in the rating of alcohols because pure alcohols may be used at leaner carburetor settings than gasoline. A recent rating (138) gives a research octane number of 109.6 and a motor octane number of 87.4 for pure methanol.

Methanol is a polar liquid that corrodes metals, and causes deterioration in plastic and elastomeric materials not affected by gasolines. Lead–tin (Terneplate)-coated steel used for automotive fuel tanks, and fittings and parts constructed of copper, brass, carbon steels, aluminum, magnesium, methacrylates, and Viton are attacked by straight methanol. Plastic carburetor and fuel-gauge elements, fuel-pump seals and diaphragms, and fiber gaskets have also failed in methanol service. Engine failure may result from prolonged use of straight methanol (129).

Refiners have added methanol to gasolines, especially to aircraft fuels, as antiicing or antistalling agent, at concentrations generally less than 1 but as high as ca 2%. When tested in airplane engines, methanol increases power output more than isopropanol, another common gasoline antiicing additive (see Antifreezes and deicing fluids; Aviation and other gas turbine fuels).

Depending on the gasoline composition and temperature, as little as 0.05% water may cause phase separation of a blend of 10% methanol in gasoline (123). At least 5% of all domestic gasoline distribution facilities have been estimated to contain at least 0.05% water.

In Germany, in the early 1950s, a 50:50 mixture of methanol and isopropanol was blended with gasoline, first at a level of 7.5% and later at 1.5% (123). Complaints about stalling, power loss, and phase separation caused the ratio to be changed to 60:40 methanol–isopropanol but this apparently aggravated the problems. The practice was discontinued in 1970 when a tax was placed on alcohol.

In the United States, the Oxirane Chemical Co. has recently introduced (139) a gasoline additive, Oxinol, comprising equal parts of methanol and *tert*-butyl alcohol, which is purported to improve the performance of engines using unleaded gasolines, when incorporated into such gasolines at levels of 3.0–5.5%.

Methanol has been combusted in stationary engines and gas turbines (140), and may offer advantages in future central-station, combined-cycle electricity generation systems (141). It shows promise of emitting considerably less nitrogen oxides (NO_x) than more conventional fuels (142). Conversion of the natural gas coproduced with Alaskan North Slope crude oil, eg, to methanol, has been proposed to facilitate shipment as an alternative to a separate gas pipeline.

Although methanol may be used also in nonmotive service for space and process heating, its toxicity, low flash point, and high cost generally militate against such usage, except by large consumers in special circumstances.

The Energy Co. of Sao Paulo (CESP), Brazil has recently proposed (143) a controversial project calling for the construction of 65 methanol plants, which would be fueled by eucalyptus trees growing on a 52,000 km^2 area. Methanol would be produced by the fixed-bed gasification of wood. The main thrust of CESP's program is directed toward replacement of imported fuel oil needed for the generation of electricity.

Ethanol. Ethanol (qv) is produced both from ethylene derived from the cracking of petroleum fractions (35%) and by the fermentation (qv) of sugars derived from grains or other biomass. Many of its relevant properties are similar to those of methanol. Although it may be a more desirable fuel or fuel component than methanol, its significantly higher cost (volume basis) outweighs these advantages.

The term gasohol has recently come into wide usage to identify, generally, a blend of gasoline and ethanol, with the latter derived from grain. The term may also be applied to blends of methanol or other alcohols in gasolines or other hydrocarbons, without regard to the sources of components (see Gasoline).

Brazil's Alcohol Program (Proalcool). In Brazil, the sugar industry was instrumental in the enactment of legislation, in 1931, making ethanol addition to gasoline compulsory at a level of 5% (144). Excess molasses and sugar were converted to alcohol in distilleries attached to sugar mills as a means to stabilize sugar prices.

Most Brazilian automobile engines are set to operate fuel-rich, because of the low octane ratings of available gasolines; consequently, few gasohol problems have been experienced. Moreover, start-up and phase-separation problems are not incurred because ambient temperatures are high. Car manufacturers in Brazil supply systems optimized for the available fuels.

Starting in the city of Sao Paulo in 1977, and extending to the entire state of Sao Paulo in 1978, a gasohol incorporating 20% ethanol was mandated. Brazil's National Alcohol Program (Proalcool) has set an initial goal of providing the 20% fuel mixture nationwide by 1980–1981, which represents an estimated annual requirement of 3.6 $\times 10^9$ L absolute (anhydrous) ethanol, in addition to the 1 $\times 10^9$ L required in Brazil for nonfuel uses.

A system of special tax, warranty, and price considerations has been enacted to advance the aims of Proalcool. By 1981 some 243 distilling plants are expected to be operating.

About 2.2 $\times 10^6$ hm^2 (5.4 $\times 10^6$ acres) was planted in sugarcane in Brazil in 1976, producing ca 90 $\times 10^6$ t cane. Brazilian experience has been that one metric ton of sugarcane yields 71 L ethanol. An increase in cane production of ca 42 $\times 10^6$ t/yr, requiring an increase in the planted area of 1 $\times 10^6$ hm^2 (2.5 $\times 10^6$ acres), and the generation of some 314,000 additional direct jobs in cane cultivation, have been estimated for the attainment of Proalcool's initial goals.

Sugars in juices extractable from sugarcane can be directly fermented by yeasts to ethanol whereas the starches in grains or other biomass must first be saccharified, generally by controlled enzymatic action. Hence, in Brazil, ethanol is also produced from cassava (manioc or mandioca) root and from the babacu palm. In all cases, the energy requirements for production of fuel-quality (anhydrous) ethanol by distillation are high but the bagasse from juice extraction may be burned to supply this requirement when sugarcane is the starting material. This is not possible if cassava root is the raw material.

By far the largest product of cane alcohol distilleries is stillage (vinhoto), or bottom slops from distillation, which is generated at a rate twelve to thirteen times in volume that of the ethanol produced. Total stillage output from Brazilian distilleries in 1977 was estimated at 20 $\times 10^6$ m^3 (ca 50 $\times 10^8$ gal). Because of its high content of soluble impurities, stillage presents a formidable pollution problem such that Brazil's 1977 output was equivalent in biological oxygen demand (BOD) to the sewage associated with 50 million people.

In Brazil, in 1978, the cost of production of ethanol from sugarcane was estimated to be almost exactly double the cost of gasoline derived from imported petroleum (by volume).

Gasohol in the United States. In the United States, and especially in the Midwest, gasohol is associated with the production of ethanol from grain. Corn is generally the grain crop under consideration, following the thrust of legislation of the state of Nebraska, which provided a 1.3¢/L (5¢/gal) reduction in the state gasoline tax for a 90% gasoline/10% ethanol gasohol blend ostensibly to help stabilize the corn market. Proponents of gasohol frequently argue that "distress" corn, or corn which is below grade or unnecessary for food purposes is the fuel ethanol feedstock of preference. Such claims must be critically examined, since ethanol may be produced from a very wide range of cellulosic biomass, as well as from extractable sugars.

Nebraska also established the Agricultural Products Industrial Utilization Committee to administer that state's grain alcohol program, which in turn sponsored a 3.2×10^6 km (2×10^6 mi) gasohol road test, completed in 1977, resulting in a number of claims relating to improved fuel economy and reduced exhaust emissions for gasohol. In this much-publicized experiment, the gasohol was produced from ethanol derived as an adjunct to the manufacture of wood pulp in a unique facility operated by the Georgia-Pacific Corporation in Bellingham, Washington.

The Georgia-Pacific facility was originally a part of the federal synthetic fuel program instituted during World War II. It currently produces ca 20.8×10^3 m^3 (5.5×10^6 gal) annually of 190-proof ethanol from the yeast fermentation of the six-carbon sugar produced by its calcium sulfite pulping process. The manufacturing cost of ethanol produced by this plant may be even lower than ethanol produced from ethylene or by other fermentation routes. This may be owing in part to low associated capital charges, and from the marketability of by-products. It is not clear whether the cost advantages for this operation would be realized in new, expanded facilities.

Other food crops, especially milo or sweet sorghum, have been proposed as specific starting points for agricultural fuel ethanol production in the United States. As in other countries, proponents of fuel ethanol production generally seek to stabilize some segment of the agricultural market but require governmental financing, loan guarantees, or tax benefits to bridge the gap between the cost of ethanol and petroleum-derived fuels (128).

In addition, a number of federally sponsored research programs seek to effect the microbiological degradation of cellulosic biomasses, and the direct conversion to ethanol, or to soluble sugars (see Fuels from biomass).

BIBLIOGRAPHY

"Fuels, Synthetic Liquid" in *ECT* 1st ed., Vol. 6, pp. 960–983, by J. H. Arnold and Helmut Pichler, Hydrocarbon Research, Incorporated; "Carbon Monoxide–Hydrogen Reactions" in *ECT* 2nd ed., Vol. 4, pp. 446–489, by Helmut Pichler and Annemarie Hector, Carl-Engler und Hans-Bunte-Institut für Mineralöl- und Kohleforschung der Technischen Hochschule Karlsruhe.

1. L. Shnidman, "Utilization of Coal Gas" in H. H. Lowry, ed., *Chemistry of Coal Utilization,* Vol. 2, John Wiley & Sons, Inc., New York, 1945.
2. W. L. Glowacki, "Light Oil From Coke-Oven Gas" and J. O'Brochta, S. E. Wooldridge, and E. O. Rhodes, "The Chemical Nature of Coal Tar" in ref. 1.
3. M. J. Burges and R. V. Wheeler, *Fuel* **5,** 65 (1926).
4. F. Fischer and H. Schroder, *Z. Angew. Chem.* **33,** 172 (1920); J. B. Goodman and co-workers, "Low Temperature Carbonization Assay of Coal in a Precision Laboratory Apparatus," *U.S. Bur. Mines Bull.,* 530 (1953).
5. W. A. Selvig and W. H. Ode, *U.S. Bur. Mines Bull.* **571,** (1957).
6. H. C. Howard in ref. 1, Vol. 1, pp. 761–773.
7. *Ibid.,* Suppl. Vol., pp. 340–394.
8. J. F. Jones and co-workers, *Chem. Eng. Prog.* **62**(2), 73 (1966).
9. J. A. Hamshar, H. D. Terzian, and L. J. Scotti, "Clean Fuels From Coal by the COED Process," *paper presented at EPA Symposium on Environmental Aspects of Fuel Conversion Technology, St. Louis, Missouri, May 1974.*
10. C. D. Kalfadelis and E. M. Magee, *Evaluation of Pollution Control in Fossil Fuel Conversion Processes, Liquefaction, Section 1, COED Process, EPA-650/2-74-009e,* EPA, Washington, D.C., 1975.

11. R. T. Eddinger and co-workers, *Char Oil Energy Development, Office of Coal Research R&D Report No. 11,* Vol. I (PB 169, 562) and Vol. II (PB 169, 563), U.S. Government Printing Office, Washington, D.C., Mar. 1966.
12. J. F. Jones and co-workers, *Char Oil Energy Development, Office of Coal Research R&D Report No. 11,* Vol. I (PB 173, 916) and Vol. II (PB 173, 917), U.S. Government Printing Office, Washington, D.C., Feb. 1967.
13. M. E. Sacks and co-workers, *Char Oil Energy Development, Office of Coal Research Report 56, Interim Report No. 2, GPO Cat. No. 163.10:56/Int. 2,* U.S. Government Printing Office, Washington, D.C., Jan. 1971.
14. J. F. Jones and co-workers, *Char Oil Energy Development, Office of Coal Research R&D Report No. 56, Final Report, GPO Cat. No. 163.10:56,* U.S. Government Printing Office, Washington, D.C., May 1972.
15. J. F. Jones and co-workers, *Char Oil Energy Development, Office of Coal Research R&D Report No. 73, Interim Report No. 1, GPO Cat. No. 163.10:73/Int. 1,* U.S. Government Printing Office, Washington, D.C., Dec. 1972.
16. L. H. Dierdoff, Jr., and R. Bloom, Jr., "The CoGas Project," *paper presented at SAE West Coast Meeting, Portland, Oregon, Aug. 1973.*
17. *Char Oil Energy Development, Office of Coal Research R&D Report No. 73-Interim Report No. 2, GPO Cat. No. 163.10:73/Int. 2,* U.S. Government Printing Office, Washington, D.C., July 1974.
18. M. F. Greene, L. J. Scotti, and J. F. Jones, "Low Sulfur Synthetic Crude Oil From Coal," *paper presented at ACS National Meeting, Div. of Fuel Chemistry, Los Angeles, Calif., Apr. 1974.*
19. J. J. Demeter and co-workers, "Further Studies of the Combustion of Pulverized Char," *paper presented at ASME Winter Annual Meeting, Detroit, Mich., Nov. 11, 1973.*
20. *Gasification of COED Chars in a Koppers-Totzek Gasifier, EPRI Report AF-615 Final Report, Project RP 264-1, 1978.*
21. J. P. Gallagher and co-workers, *Chem. Eng. Prog.* **75,** 56 (1979).
22. A. Sass, "The Garrett Research and Development Company Process for the Conversion of Coal into Liquid Fuels," *paper presented at 65th Annual AIChE Meeting, New York, Nov. 29, 1972.*
23. *Oil Gas J.,* 78 (Aug. 26, 1974).
24. A. Sass, *Chem. Eng. Prog.* **70**(1), 72 (1974).
25. F. B. Carlson and co-workers, *Chem. Eng. Prog.* **69**(3), 50 (1973).
26. F. B. Carlson, L. H. Yardumian, and M. T. Atwood, "The TOSCOAL Process for Low Temperature Pyrolysis of Coal," *paper presented at the American Institute of Mining, Metallurgical and Petroleum Engineers, San Francisco, Calif., Feb. 22, 1972, and to the American Institute of Chemical Engineers, New York, Nov. 29, 1972.*
27. W. Tiddy and M. J. Miller, *Am. Gas J.* **153**(3), 7 (1940).
28. Ger. Pat. 301,231 (Nov. 26, 1919), F. Bergius and J. Billwiller.
29. F. Fischer and H. Tropsch, *Ges. Abhandl. Kenntis Kohle* **2,** 154 (1918).
30. F. Bergius, *Pet. Z.* **22,** 1275 (1926).
31. *Gas World* **104,** 421 (1936).
32. *Pet. Times* **42,** 641 (1939).
33. H. H. Storch, L. L. Hirst, C. H. Fischer, and G. C. Sprunk, *U.S. Bur. Mines Tech. Pap.* **622,** (1941).
34. H. H. Storch, "Hydrogenation of Coal and Tar" in ref. 1.
35. W. R. K. Wu and H. H. Storch, *U.S. Bur. Mines Bull.* **633,** (1968).
36. A. C. Fieldner and co-workers, *U.S. Bur. Mines Tech. Pap.* **666,** (1944).
37. M. L. Kastens and co-workers, *Ind. Eng. Chem.* **41,** 870 (1949).
38. J. L. Wiley and H. C. Anderson, *U.S. Bur. Mines Bull.* **485,** I (1950), II (1951), III (1952).
39. C. C. Chaffee and L. L. Hirst, *Ind. Eng. Chem.* **45,** 822 (1953).
40. Bituminous Coal Staff, *U.S. Bur. Mines Rep. Invest.* **5506,** (1959).
41. E. L. Clark and co-workers, *Ind. Eng. Chem.* **42,** 861 (1950).
42. L. L. Newman and A. P. Pipilen, *Gas Age* **119**(10), 16 (1957); *ibid.,* **119**(11), 18 (1957).
43. U.S. Pat. 2,860,101 (Nov. 11, 1958), M. G. Pelipetz (to the United States of America).
44. S. Akhtar, S. Friedman, and P. M. Yavorsky, *U.S. Bur. Mines Tech. Prog. Rep.* **35,** (1971).
45. S. Akhtar and co-workers, "Process for Hydrodesulfurization of Coal," *paper presented at 71st National AIChE Meeting, Dallas, Tex., Feb. 20, 1972.*
46. B. Linville and J. D. Spencer, *U.S. Bur. Mines Inf. Cir.* **8612,** (1973).

47. P. Wellman and S. Katell, "The Economics of Coal Conversion Systems," *Preprint No. SPE 5097, AIME Meeting, Houston, Tex., Oct. 6, 1974.*
48. "Economic Analysis of Synthoil Plant Producing 50,000 Barrels Per Day of Liquid Fuels From Two Coal Seams: Wyodak and Western Kentucky," *ERDA Report 76-35, Process Evaluation Group, Morgantown, West Virginia, Nov. 1975.*
49. P. M. Yavorsky and co-workers, "Process Developments: Fixed-Bed Catalysis of Coal to Fuel Oil," *paper presented at 65th Annual AIChE Meeting, New York, Nov. 26, 1972.*
50. R. Salmon and co-workers, "Evaluation of the Synthoil Process," Vols. I and II, *ORNL-520 and ORNL-5210, ERDA Contract No. W-7405-eng-26,* Oak Ridge National Laboratory, Union Carbide Corporation, New York, June 1977.
51. T. Aczel, R. B. Williams, R. J. Pancirov, and J. H. Karchmer, "Chemical Properties of Synthoil Products and Feeds," Vols. I–V and Summary Report, *ERDA Contract No. E(46-1)-8007,* Exxon Research and Engineering Company, Linden, N.J., Feb. 1977.
52. *Coal Age,* 101 (May 1976).
53. G. A. Johnson and co-workers, "Present Status of the H-Coal Process," *Paper 30, IGT Coal Symposium, Chicago, Ill., 1973.*
54. C. Hoertz, *Energy Technol.* **5,** 347 (1978).
55. "High Pressure Hydrogenation at Ludwigshafen-Heidelberg," Vol. IA, *FIAT Final Report No. 1317, ATI No. 92,762,* Central Air Documents Office, Dayton, Ohio, 1951.
56. H. H. Lowry and H. J. Rose, *U.S. Bur. Mines Inf. Cir.* **7420,** (1947).
57. A. Baba and co-workers, *Rep. Resources Res. Inst. Jpn.* (22), (1955).
58. U.S. Pat. 2,221,866 (Nov. 19, 1940), H. Dreyfus.
59. D. L. Kloepper and co-workers, *Solvent Processing of Coal to Produce a De-ashed Product, Contract 14-01-0001-275, OCR Report No. 9,* U.S. Government Printing Office, Washington, D.C., 1965.
60. V. L. Brant and B. K. Schmid, *Chem. Eng. Prog.* **68**(12), 55 (1969).
61. B. K. Schmid and W. C. Bull, "Production of Ashless, Low-Sulfur Boiler Fuels From Coal," *paper presented at ACS Division of Fuel Chem. Symp. on Pollution Control, New York, Sept. 12, 1971.*
62. J. F. Cudmore and R. E. Guyot, *Solvent Extraction of Coal-Batch Autoclave Studies, P.R. 67-6,* Australian Coal Industry Research Laboratories Ltd., Sydney, Australia, 1967.
63. R. M. Baldwin and co-workers, "Mathematical Modeling of the Selectivity of the Solvent-Refining Process in the Removal of Sulfur From Coal," *paper presented at 78th National Meeting AIChE, New York, Aug. 1974.*
64. U.S. Pat. 3,341,447 (Sept. 12, 1967), W. C. Bull, L. G. Stevenson, D. L. Kloepper, and T. F. Rogers (to the United States of American and Gulf Oil Corporation).
65. *Demonstration Plant, Clean Boiler Fuels From Coal, OCR R&D Report No. 82, Interim Report No. 1,* Vol. I (1973), Vol. II (1974), Vol. III (1975), Ralph M. Parsons Company, Los Angeles, Calif.
66. M. E. Frank and B. K. Schmid, "Economic Evaluation and Process Design of a Coal-Oil-Gas (COG) Refinery," *paper presented at Symp. on Conceptual Plants for the Production of Synthetic Fuels From Coal, AIChE 65th Annual Meeting, New York, Nov. 26, 1972.*
67. B. K. Schmid and D. M. Jackson, "The SRC-II Process," *paper presented at the Third Annual International Conference on Coal Gasification and Liquefaction, University of Pittsburgh, Pittsburgh, Pa., Aug. 3–5, 1976;* and D. M. Jackson and B. K. Schmid, "Production of Distillate Fuels by SRC-II," *paper presented at ACS Div. of Ind. and Eng. Chem. Symposium, Colorado Springs, Col., Feb. 12, 1979.*
68. *1970 Final Report—Consol Synthetic Fuel Process, OCR R&D Report No. 45,* Ralph M. Parsons Company, Los Angeles, Calif., 1970.
69. *Research on Zinc Chloride Catalyst for Converting Coal to Gasoline, OCR R&D Report No. 39,* Vol. III, Book 1, Consolidation Coal Company, New York, 1968.
70. *Pre-Pilot Plant Research on the CSF Process—Phase I, Regeneration of Zinc Chloride Catalyst, OCR R&D Report No. 39,* Vol. III, Book 2, Consolidation Coal Company, New York, 1968.
71. *Engineering Evaluation of Project Gasoline, OCR R&D Report No. 59,* Foster Wheeler Corp., Livingston, N.J., July 28, 1970.
72. W. W. Bodle and K. C. Vyas, *Oil Gas J.,* 76 (Aug. 26, 1974).
73. J. A. Phinney, "Coal Liquefaction at the Cresap, West Virginia Pilot Plant," *paper presented at AIChE Coal Conversion Pilot Plant Symposium, Pittsburgh, Pa., June 2, 1974.*
74. *Pilot-Scale Development of the CSF Process, OCR R&D Report No. 39,* Vol. IV, Book 3, Consolidation Coal Company, New York, Aug. 1971.

75. *Engineering Evaluation and Review of Consol Synthetic Fuel Process, OCR R&D Report No. 70,* Foster Wheeler Corp., Livingston, N.J., Feb. 1972.
76. W. R. Epperly and J. W. Taunton, "Exxon Donor Solvent Coal Liquefaction Process Development," *paper presented at Coal Dilemma II ACS Meeting, Colorado Springs, Colo., Feb. 12, 1979.*
77. D. E. Blaser and A. M. Edelman, "Flexicoking for Improved Utilization of Hydrocarbon Resources," *paper presented at API 43rd Mid-Year Meeting, Can., May 8, 1978.*
78. L. E. Swabb, Jr., G. K. Vick, and T. Aczel, "The Liquefaction of Solid Carbonaceous Materials," *paper presented at The World Conference on Future Sources of Organic Raw Materials, Toronto, Can., July 10, 1978.*
79. P. Sabatier, *Catalysis, Then and Now,* Part II, Franklin Publishing Co., Englewood, N.J., 1965.
80. Fr. Pat. 571,356 (May 16, 1924), (to Badische Anilin-und Soda-Fabrik).
81. Fr. Pat. 580,905 (Nov. 19, 1924), (to Badische Anilin-und Soda-Fabrik).
82. F. Fischer and H. Tropsch, *Ber.* **56,** 2428 (1923).
83. F. Fischer, *Die Umwandlung der Kohle in Ole,* Borntraeger, Berlin, Ger., 1923, p. 320.
84. F. Fischer and H. Tropsch, *Ber.* **59,** 830, 832, 923 (1926).
85. F. Fischer, *Brennstoff-Chem.* **11,** 492 (1930).
86. F. Fischer and H. Koch, *Brennstoff-Chem.* **13,** 61 (1932).
87. F. Fischer and K. Meyer, *Brennstoff-Chem.* **12,** 225 (1931).
88. F. Fischer and H. Pichler, *Brennstoff-Chem.* **20,** 41, 221 (1939).
89. K. Peters, *Comparison of I.G. Work on Fischer Synthesis, Technical Oil Mission Report, Reel 13,* Library of Congress, Washington, D.C., July 1941.
90. H. Pichler, *Medium Pressure Synthesis on Iron Catalyst, (Pat. Appl.), Technical Oil Mission Report, Reel 100,* Library of Congress, Washington, D.C., 1937–1943.
91. H. Pichler, *Medium Pressure Synthesis on Iron Catalyst, Technical Oil Mission Report, Reel 101,* Library of Congress, Washington, D.C., June 1940.
92. U.S. Pat. 2,327,066 (Aug. 17, 1943), O. Roelen.
93. F. Fischer, *Ole Kohle* **39,** 517 (1943).
94. H. H. Storch, N. Golumbic, and R. B. Anderson, *The Fischer-Tropsch and Related Synthesis,* John Wiley & Sons, Inc., New York, 1951.
95. "The Fischer-Tropsch and Related Processes for Synthesis of Hydrocarbons by Hydrogenation of Carbon Monoxide" in W. G. Frankenburg, V. I. Komarewsky, and E. K. Rideal, *Advances in Catalysis,* Vol. I, Academic Press, Inc., New York, 1948, pp. 115–156.
96. H. H. Storch, "Synthesis of Hydrocarbons From Water Gas" in ref. 1, p. 1797.
97. O. C. Elvins and A. W. Nash, *Fuel* **5,** 263 (1926).
98. O. C. Elvins, *J. Soc. Chem. Ind. (London)* **56,** 473T (1927).
99. A. Erdeley and A. W. Nash, *J. Soc. Chem. Ind. (London)* **47,** 219T (1928).
100. W. W. Myddleton, *Chim. Ind.* **37,** 863 (1937); *J. Inst. Fuel,* **11,** 477 (1938); *Colliery Guardian* **157,** 286 (1938).
101. G. Egloff, *Brennst. Chem.* **18,** 115 (1937).
102. G. Egloff, E. F. Nelson, and J. C. Morrell, *Ind. Eng. Chem.* **29,** 555 (1937).
103. P. C. Keith, *Oil Gas J.* **45**(6), 102 (1946).
104. H. Pichler, *Brennst. Chem.* **30,** 105 (1959).
105. J. H. Crowell, H. E. Benzon, J. H. Field, and H. H. Storch, *Ind. Eng. Chem.* **42,** 2376 (1950).
106. W. B. Johnson, *Pet. Ref.,* **35** (Dec. 1956).
107. A. E. Sands, H. W. Wainwright, and L. D. Schmidt, *Ind. Eng. Chem.* **40,** 607 (1948); A. E. Sands and L. D. Schmidt, *Ind. Eng. Chem.* **42,** 2277 (1950).
108. H. Pichler, "Synthesis of Hydrocarbons from Carbon Monoxide and Hydrogen," *Technical Oil Mission Report, Reel 259,* Library of Congress, Washington, D.C., 1947, frames 467–654.
109. M. E. Dry and G. J. Oosthuizen, *J. Catal.* **11,** 18 (1968).
110. M. E. Dry, T. Shingles, L. J. Boshoff, and G. J. Oosthuizen, *J. Catal.* **15,** 190 (1969).
111. P. J. Denny and D. A. Whan, "The Heterogeneously Catalyzed Hydrogenation of Carbon Monoxide" in C. Kemball, ed., *Catalysis,* Vol. II, Chemical Society, London, Eng., 1978.
112. S. L. Meisel, J. P. McCullough, C. H. Lechthaler, and P. B. Weisz, *Chemtech* **6,** 86 (1976).
113. U.S. Pats. 3,894,106 (July 8, 1975), C. D. Chang, W. H. Lang, and A. J. Silvestri (to Mobil Oil Corporation); 3,894,107 (July 8, 1975), S. A. Butter, W. W. Kaeding, and A. T. Jurewicz (to Mobil Oil Corporation); 3,928,483 (Dec. 23, 1975), C. D. Chang, A. J. Silvestri, and R. L. Smith (to Mobil Oil Corporation).
114. C. D. Chang and A. J. Silvestri, *J. Catal.* **47,** 249 (1977).

115. J. J. Wise and A. J. Silvestri, *Oil Gas J.*, 140 (Nov. 22, 1976).
116. J. E. Penick, S. L. Meisel, W. Lee, and A. J. Silvestri, "Mobil Process for Conversion of Coal and Natural Gas to Gasoline," *paper presented at Institute of Chemical Engineers Alcohol Fuels Conference, Sydney, Australia, Aug. 9, 1978.*
117. M. J. Gavin, *U.S. Bur. Mines Bull.* **210,** (1922).
118. M. J. Gavin and J. S. Desmond, *U.S. Bur. Mines Bull.* **315,** (1930).
119. H. Shaw, C. D. Kalfadelis, and C. E. Jahnig, *Evaluation of Methods to Produce Aviation Turbine Fuels From Synthetic Crude Oils—Phase I, Technical Report AFAPL-TR-75-10,* Vol. I, Air Force Aero Propulsion Laboratory, Wright-Patterson Air Force Base, Ohio, Mar. 1975.
120. C. D. Kalfadelis, *Evaluation of Methods to Produce Aviation Turbine Fuels From Synthetic Crude Oils—Phase II, Technical Report AFAPL-TR-75-10,* Vol. II, Air Force Aero Propulsion Laboratory, Wright-Patterson Air Force Base, Ohio, May 1976.
121. Colony Development Operation, *An Environmental Impact Analysis for a Shale Oil Complex at Parachute Creek, Colorado,* Vols. I, II, III, Denver, Colo., 1974.
122. R. D. Hynphreys, F. K. Spragins, and D. R. Craig, "Oil Sands—Canada's First Answer to the Energy Shortage," *Proc. 9th World Petroleum Congress, Tokyo, Japan, May 11, 1975,* Vol. V, p. 17.
123. American Petroleum Institute, Task Force EF-18 of the Committee on Mobile Source Emissions, *Alcohols—A Technical Assessment of Their Application as Fuels, Publication No. 4261,* API, New York, July 1976.
124. A. W. Nash and D. A. Howes, *Alcohol Fuels,* John Wiley & Sons, Inc., New York, 1935.
125. A. W. Crowley and co-workers, "Methanol-Gasoline Blends—Performance in Laboratory Tests and in Vehicles," *Paper 750419, Automotive Engineering Congress and Exposition, SAE, Detroit, Mich., Feb. 1975.*
126. E. E. Wigg and R. S. Lunt, "Methanol as a Gasoline Extender—Fuel Economy, Emissions, and High Temperature Driveability," *Paper 741008, Automobile Engineering Meeting, Society of Automotive Engineers, Toronto, Can., Oct. 1974.*
127. G. Egloff and J. C. Morrell, *Ind. Eng. Chem.* **28,** 1080 (1936).
128. E. V. Anderson, *Chem. Eng. News,* 8 (July 31, 1978).
129. T. Powell, "Racing Experiences with Methanol and Ethanol Based Motor-Fuel Blends," *Paper 750124, Automotive Engineering Congress and Exposition, Society of Automotive Engineers, Detroit, Mich., Feb. 1975.*
130. G. D. Ebersole and F. S. Manning, "Engine Performance and Exhaust Emissions: Methanol Versus Isooctane," *Paper 720692, Society of Automotive Engineers, Detroit, Mich., Jan. 1972.*
131. E. S. Starkman, H. K. Newhall, and R. D. Sutton, *Comparative Performance of Alcohol and Hydrocarbon Fuels, Special Publication SP-254,* Society of Automotive Engineers, Warrendale, Pa., June 1964.
132. R. K. Pefley, M. A. Saad, M. A. Sweeney, J. D. Kilgroe, and R. E. Fitch, *Study of Decomposed Methanol as a Low Emission Fuel—Final Report, Environmental Protection Agency Contract No. EHS 70-118 PB 202732,* EPA, Washington, D.C., Apr. 30, 1971.
133. J. A. Harrington and R. M. Pilot, "Combustion and Emission Characteristics of Methanol," *Paper 750450, Automotive Engineering Congress and Exposition, Society of Automotive Engineers, Detroit, Mich., Feb. 1975.*
134. W. W. Bernhardt and W. Lee, "Combustion of Methyl Alcohol in Spark Ignition Engines," *Paper 136, 15th International Symposium on Combustion, Tokyo, Japan, Aug. 1974.*
135. H. G. Adelman, D. G. Andres, and R. S. Devoto, "Exhaust Emissions From a Methanol-Fueled Automobile," *Paper 720693, West Coast Meeting, San Francisco, Calif., Society of Automotive Engineers, August 1972.*
136. R. M. Tillman, O. L. Spilman, and J. M. Beach, "Potential for Methanol as an Automotive Fuel," *Paper 750118, Automotive Engineering Congress and Exposition, Society of Automotive Engineers, Detroit, Mich., Feb. 1975.*
137. R. D. Fleming and T. W. Chamberlain, "Methanol as an Automotive Fuel, Part I—Straight Methanol," *Paper 750121, Automotive Engineering Congress and Exposition, Society of Automotive Engineers, Detroit, Mich., Feb. 1975.*
138. W. J. Most and J. P. Longwell, "Single Cylinder Engine Evaluation of Methanol—Improved Economy and Reduced NO_x," *Paper 750119, Automotive Engineering Congress and Exposition, Society of Automotive Engineers, Detroit, Mich., Feb. 1975.*
139. *Chem. Week,* 30 (Jan. 2, 1980).

140. P. M. Jarvis, "Methanol as a Gas Turbine Fuel" in *Methanol as an Alternate Fuel,* Vol. II, Engineering Foundation Conference, Henniker, N.H., July 1974.
141. S. C. Stinson, *Chem. Eng. News,* 28 (Apr. 2, 1979).
142. H. Shaw, "The Effect of Water on Nitric Oxide Production in Gas Turbine Combustors," *paper presented at 20th Annual International Gas Turbine Conference, Houston, Tex., Paper No. 75-GT-70, March 2, 1975.*
143. *Chem. Week,* 38 (Mar. 14, 1979).
144. V. Yand and S. C. Trindade, *Chem. Eng. Prog.,* 11 (Apr. 1979).

General Reference

H. H. Lowry, ed., *Chemistry of Coal Utilization,* Vols. I and II, John Wiley & Sons, Inc., New York, 1945; Suppl. Vol., 1963.

CHARLES D. KALFADELIS
HENRY SHAW
Exxon Research and Engineering Company

FULLER'S EARTH. See Clays (uses).

FULMINATES. See Explosives.

FUMARIC ACID, *trans*-HOOCCH═CHCOOH. See Maleic acid, fumaric acid, and maleic anhydride.

FUMIGANTS. See Insect control technology.

FUNCTIONAL FLUIDS. See Hydraulic fluids.

FUNGICIDES, AGRICULTURAL

Agricultural fungicides are described in this article. An alphabetical list of the fungicides mentioned in the text with their CAS Registry Numbers is given in Table 1. Industrial fungicides are described in other Encyclopedia articles (see Chemotherapeutics, antimycotic; Industrial antimicrobial agents).

The use of fungicides in agriculture is necessitated by the great losses caused by a wide variety of plant-pathogenic fungi. The USDA estimates the annual losses from plant diseases in the United States exceed 3×10^9 dollars. Many of these losses are minor and cannot be economically controlled by fungicides. To be economic, the cost of controlling plant diseases by the application of fungicides must be offset by potential gains of severalfold. The largest tonnages of fungicides are required by apples, pears, bananas, cereals, cocoa, coffee, cotton, potatoes, tobacco, and grapes. Crops such as apples, potatoes, bananas, and grapes cannot be grown without fungicides in current production areas.

Although worldwide production figures of agricultural fungicides are not readily available, those of the United States serve to illustrate the magnitude of the quantities. Because of the proprietary nature of most of the fungicides, exact figures are difficult to obtain but annual sales of organic fungicides by U.S. producers exceeds 14,000 metric tons. The inorganic fungicides, sulfur and copper sulfate, bring the total to over 82,000 t. The value of the organic fungicides produced in the United States exceeds 1×10^8 dollars. In addition to annual increases in tonnages used, there has been a significant shift during the past twenty years toward the organic fungicides. This shift results from extensive industrial research aimed at finding more and better fungicides. In spite of research and development costs estimated to be as high as 15×10^6 dollars per compound, this research continues in many laboratories throughout the world.

Thousands of compounds are evaluated each year in multiple bioassay tests employing fungi and host plants under greenhouse and laboratory conditions. Compounds not eliminated by this procedure are evaluated in the field under varying environmental conditions. If the compound shows promise in the field, if the results of toxicological and residue studies are satisfactory, if the various regulatory agencies of government approve the compound, and if the economics of production are favorable, suitable formulations may be developed and the compound may be marketed as an agricultural fungicide. The variety of the fungicides available today shows that a number of compounds have survived this lengthy elimination procedure.

Agricultural fungicides are applied not only to foliage but also to seeds, to soil, and to harvested plant parts for control of fungus diseases. For the latter, various treatments are used, eg, dipping the produce in a wash tank containing fungicide; impregnating fungicide into the wrapper, crate liner, carton, or finishing wax; and fumigating the produce with fungicide in a special room, railroad car, or tank.

Fungicides are usually applied in a water suspension with hydraulic sprayers but are sometimes used as dusts, granules, or fumigants. Adjuvants such as solvents, emulsifiers, spreaders, stickers, and dispersants, and extenders such as clay and talc are frequently used in the preparation of fungicide formulations. These give the spray or dust characteristics that allow easy and efficacious application. Combinations of fungicides may also be used to increase the spectrum of organisms controlled.

Table 1. Alphabetical List of Fungicides Referred to in the Text

Compound	Structure number	CAS Registry number
Actidione		[*66-81-9*]
benomyl	**(12)**	[*17804-35-2*]
blasticidin S		[*2079-00-7*]
Bordeaux mixture		
captan	**(7)**	[*133-06-2*]
carboxin	**(10)**	[*5234-68-4*]
chloranil	**(2)**	[*118-75-2*]
chloro(2-hydroxyphenyl)mercury	**(1)**	[*90-03-9*]
chloroneb		[*2675-77-6*]
chlorothalonil	**(9)**	[*1897-45-6*]
copper carbonate		[*7492-68-4*]
copper, 8-hydroxyquinoline complex		[*13014-03-4*]
copper oxychloride		[*1332-40-7*]
copper silicate		[*1344-72-5*]
copper sulfate, tribasic, monohydrate		[*1332-03-2*]
copper zeolites		
cuprous oxide		[*1317-39-1*]
dichlone		[*117-80-6*]
dichloran		[*99-30-9*]
dimethirimol	**(14)**	[*5221-53-4*]
dinocap		[*39300-45-3*]
dodine	**(6)**	[*2439-10-3*]
ethazole		[*2593-15-9*]
ethirimol	**(15)**	[*23947-60-6*]
fenaminosulf		[*140-56-7*]
ferbam		[*14484-64-1*]
folpet	**(8)**	[*133-07-3*]
fuberidazol		[*3878-19-1*]
glyodin	**(5)**	[*556-22-9*]
hexachlorobenzene		[*118-74-1*]
iprodione		[*36734-19-7*]
kasugamycin		[*6980-18-3*]
lime–sulfur (calcium polysulfide)		[*1344-81-6*]
MBC (methyl 2-benzimidazolyl carbamate)		[*10605-21-7*]
maneb		[*12427-38-2*]
nabam	**(4)**	[*142-59-6*]
nickel chloride		[*7718-54-9*]
nickel sulfate		[*7786-81-4*]
PCNB (quintozene), pentachloronitrobenzene		[*82-68-8*]
polyoxins		
sulfur		[*7704-34-9*]
2,3,5,6-tetrachloronitrobenzene		[*117-18-0*]
tetramethylthiuram disulfide	**(3)**	[*137-26-8*]
thiabendazole	**(11)**	[*148-79-8*]
thiophanate	**(13)**	[*23564-06-9*]
thiarimol		[*26766-27-8*]
1,2,4-trichloro-3,5-dinitrobenzene		[*2678-21-9*]
1,3,5-trichloro-2,4,6-trinitrobenzene		[*2631-68-7*]
triphenyltin acetate		[*900-95-8*]
triphenyltin hydroxide		[*76-87-9*]
zinc bis(dimethyl dithiocarbonate)		[*137-30-4*]
zineb		[*12122-67-7*]

Early Fungicides

Sulfur and Polysulfides. Sulfur, probably used as a fungicide since the time of Homer, did not become established as an agricultural fungicide until the nineteenth century when it came into use for the control of powdery mildew on grapes, fruit trees and hops, and leaf curl of peach. Elemental sulfur may be applied by dusting, spraying, or in the greenhouse, by sublimation. As a spray or dust, sulfur must be finely divided, and as a spray, it must be used with a surface-active agent. A number of "wettable" sulfurs are available. Lime–sulfur, prepared by boiling sulfur and calcium oxide together to form calcium polysulfides, has been widely used for the control of fungus diseases, particularly those of fruit trees. Lime–sulfur, although more fungitoxic than elemental sulfur, has a greater tendency to be phytotoxic (damaging to leaves), a serious problem with any inorganic sulfur fungicide (see Sulfur compounds). Sulfur is also incompatible with many other pesticides. Toxicologically one of the safest fungicides, sulfur has been largely replaced by less phytotoxic, more specific, effective organic compounds.

Heavy-Metal Fungicides. In 1882, Millardet observed that grapevines daubed with a mixture of copper sulfate and lime to deter pilferers were not defoliated by downy mildew. Following this lead, he developed a new fungicide called Bordeaux mixture after the area of its discovery. The fungicidal properties of copper sulfate had been previously demonstrated, but phytotoxicity precluded widespread use. Millardet's discovery was the ability of lime to make the mixture safe (note: this process is correctly called safening) but not destroy its fungitoxicity. The use of Bordeaux mixture spread rapidly, and until the advent of organic fungicides following World War II, it was known as the universal fungicide. Despite its long tenure, the exact composition of the mixture is elusive. Initially the reaction was thought to be: $CuSO_4 + Ca(OH)_2 \rightarrow Cu(OH)_2 + CaSO_4$ (see Calcium compounds; Copper compounds). The product is much more complex and now is thought to be a mixture of cupric calcium sulfates that varies with the ratio of copper sulfate to lime and the age of the preparation (1).

Despite the safening effect of lime, Bordeaux mixture was too phytoxic for some plants. This shortcoming led to the development of other fixed copper fungicides such as copper silicate, copper zeolite, cuprous oxide, copper oxychloride, copper carbonate, and tribasic copper sulfate. Copper carbonate and cuprous oxide, used extensively as seed treatments, were replaced first by organic mercurials, then other organic compounds. The copper complex of 8-hydroxyquinoline is an effective foliar fungicide but its high cost has limited its use. Copper has also been used in complexes with other metals, eg, $15CuO.10ZnO.6CrO_3.24H_2O$, which is used as a potato fungicide (2).

Early in the century, Reihm, a German plant pathologist, following a lead from Ehrlich's work on the use of organic arsenic for syphilis therapy, developed chloro(2-hydroxyphenyl)mercury (1), as a seed-treatment fungicide. Subsequently, a number of organic mercury derivatives were synthesized and evaluated, leading to the development of several important agricultural fungicides of the general structure, RHgX. These compounds are no longer used as agricultural fungicides because they are too poisonous.

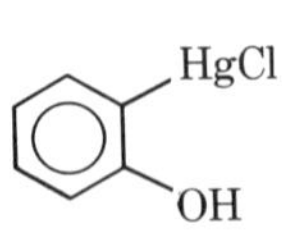

(1)

The stannous and stannic ions are relatively inactive biologically, but organotin derivatives of the type R_3SnX are highly fungitoxic (3). The tripropyl and tributyl derivatives are the most fungitoxic but are too phytotoxic for use on foliage. The less fungitoxic but safer triphenyltin acetate, $(C_6H_5)_3SnOOCCH_3$, and hydroxide, $(C_6H_5)_3SnOH$, have been developed.

Nickel chloride and nickel sulfate have been used successfully for control of rust and rust-like diseases. These compounds control cereal rusts, but they have limited use owing to economic considerations.

Organic Fungicides

Quinones. Based upon his knowledge of their reactions in rubber chemistry, ter Horst (5) first proposed the quinones (qv) as fungicides in 1938. Chloranil (**2**) (2,3,5,6-tetrachloro-1,4-benzoquinone) was reported as a seed protectant in 1940 (4). This introduction is viewed by many as the beginning of the organic fungicide era. Shortly thereafter, 2,3-dichloro-1,4-naphthoquinone (dichlone) was introduced as a foliar fungicide. Hundreds of other quinones were evaluated, but only chloranil and dichlone have been developed commercially.

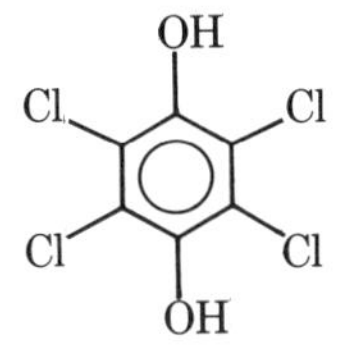

(**2**)

Organic Sulfur Compounds. The first dithiocarbamates were patented in 1934 but were not developed as fungicides until the early 1940s. These compounds are derivatives of dithiocarbamic acid, H_2NCSSH, and were known to the rubber industry as accelerators (see Rubber chemicals). The first derivative developed was tetramethylthiuram disulfide (**3**), still used widely as a vegetable and corn seed treatment. The first dithiocarbamate widely used on foliage was ferric dimethyldithiocarbamate, $[(CH_3)_2NCSS]_3Fe$, known as ferbam.

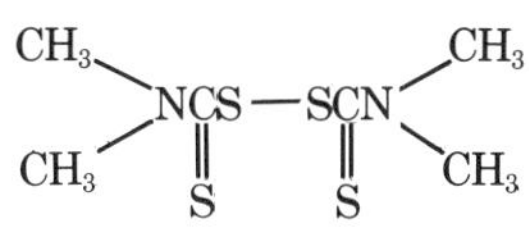

(**3**)

The zinc salt was used commonly when the black deposit of ferbam was undesirable. These materials would probably have replaced Bordeaux mixture except for the development of the ethylenebisdithiocarbamates.

The first of this new group of carbamates to be introduced was disodium ethylenebisdithiocarbamate, $NaSSCNH(CH_2)_2NHCSSNa$ (**4**), or nabam.

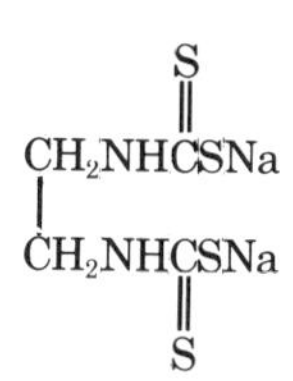

(**4**)

Although highly effective as a fungicide, nabam was unstable and would probably not have become successful but for the discovery that it was stabilized by adding zinc sulfate to the spray tank to produce zineb (5). The manganese-stabilized material (maneb) was later found to be even more effective and is the dominant form used today.

Imidazolines and Guanidines. Heptadecyl-2-imidazolinium acetate (**5**) is effective for the control of apple scab and cherry leaf spot (6). This compound, known as glyodin, is synergistic with another fungicide, dodecylguanidinium acetate (**6**), known as dodine. Dodine found wide use because of its ability to destroy established infections of apple scab, a property shown previously only by the organic mercurials (see also Cyanamides).

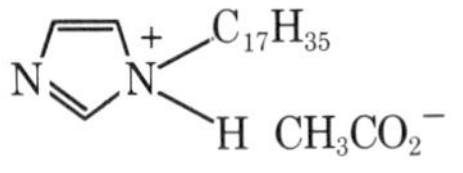

(**5**)

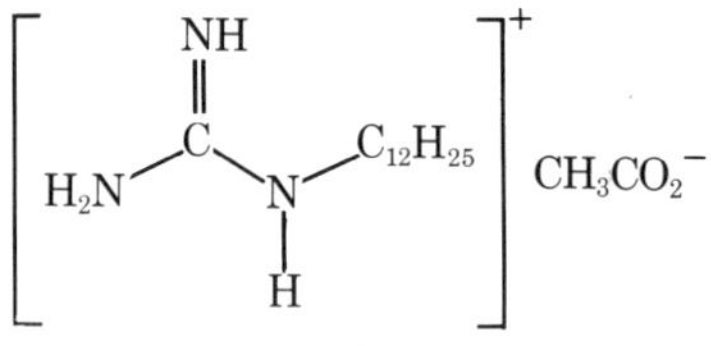

(**6**)

Trichloromethylthiocarboximides. Two of the trichloromethylthiocarboximides, *N*-(trichloromethylthio)-4-cyclohexene-1,2-dicarboximide (captan) (**7**), and *N*-(trichloromethylthio)phthalimide (folpet) (**8**), are important fungicides. Captan is one of the least phytotoxic agricultural fungicides and is widely used as a foliage fungicide on fruit, as well as a seed treatment and soil fungicide. Folpet controls powdery mildew, against which captan is ineffective.

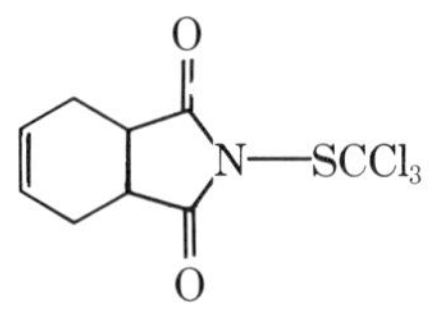

(**7**)

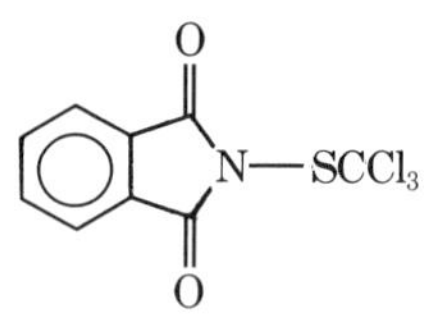

(**8**)

Chlorinated and Nitrated Benzenes. Among the chlorinated and nitrated benzenes, 2,3,5,6-tetrachloronitrobenzene, pentachloronitrobenzene (PCNB), 1,2,4-trichloro-3,5-dinitrobenzene, 1,3,5-trichloro-2,4,6-trinitrobenzene, and hexachlorobenzene have been used as soil or seed fungicides (see Nitrobenzenes). PCNB (quintozene) controls Botrytis diseases, club root of crucifers, scab of potato, and Rhizoctonia damping-off of seedlings. A related material, 2,6-dichloro-4-nitroaniline (dichloran), is an effective soil and foliage fungicide against Botrytis and Sclerotinia diseases. In 1965, 1,4-dichloro-2,5-dimethoxybenzene was introduced to control Rhizoctonia disease of cotton seedlings (7). This material, called chloroneb, has some systemic activity but is used only as a seed treatment. Also at this time, a new, broad spectrum foliar fungicide, tetrachloroisophthalonitrile (**9**), was introduced.

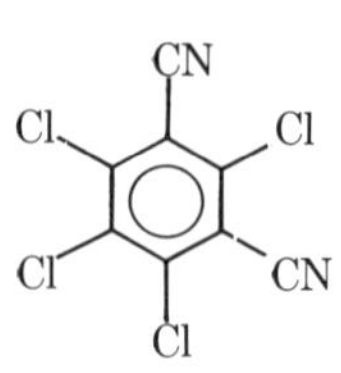

(**9**)

This safe, effective material, called chlorothalonil, has become one of the most important protectant agricultural fungicides.

Systemic Fungicides

Oxathiins. Before the report of the systemic fungicidal activity of some 1,4-oxathiins, protection from commercial fungicides was restricted to activity at or near the point of application to the plant where frequent applications of material were necessary to protect new growth. The oxathiins permitted systemic protection of new growth (8). This systemic activity marked the beginning of the third era of agricultural fungicides.

This group is mainly active against the class Basidiomycetes although some of the imperfect fungi are also sensitive. One of the more effective of the series is 2,3-dihydro-5-carboxanilido-6-methyl-1,4-oxathiin, known as carboxin (**10**).

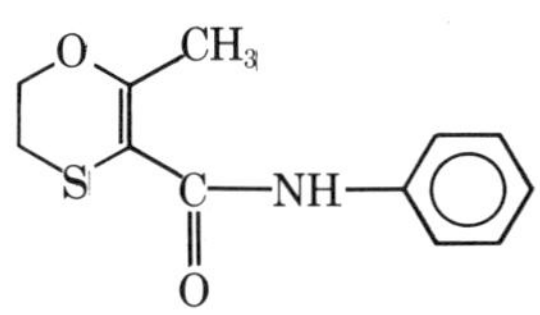

(**10**)

Carboxin is used as a seed treatment for smuts of cereals and for Rhizoctonia diseases of seedlings.

Benzimidazoles. Almost simultaneous to the announcement of the systemic activity of the oxathiins, a second group of systemic fungicides appeared. The first of the series was 2-(4-thiazoyl)-benzimidazole (**11**), which was originally developed as an anthelmintic (9) (see Chemotherapeutics, anthelmintic). This material, called thiabendazole, is primarily transported upward in the plant but downward movement has been reported. Methyl (1-butylcarbamoyl)-2-benzimidazolylcarbamate (benomyl) (**12**) is an excellent systemic fungicide with protectant and eradicant activity. Benomyl rapidly hydrolyzes to methyl 2-benzimidazolylcarbamate (MBC), which may be the fungitoxic principle. Subsequent to the development of benomyl, another group of systemic fungicides (thiophanates) appeared that had activity similar to benomyl. It appears that thiophanate (**13**) is metabolized to MBC, which is responsible for at least a portion of the observed fungicidal activity (10).

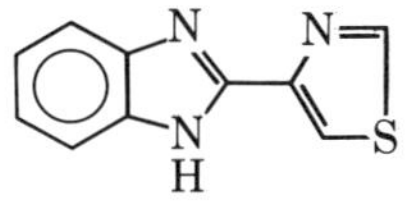

(**11**)

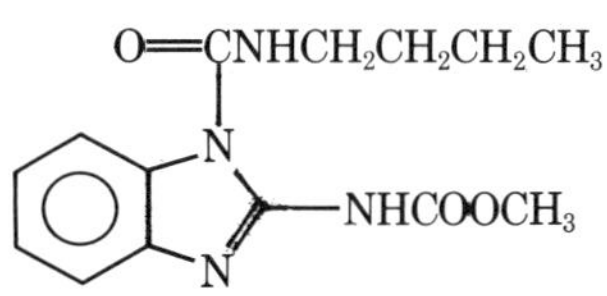

(**12**)

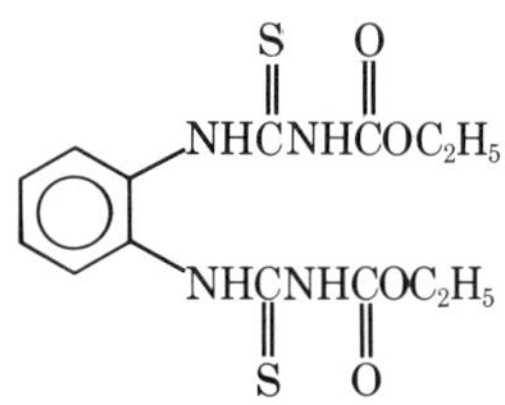

(**13**)

The benzimidazoles are extensively used and are probably the most successful of the systemic fungicides currently in use.

Pyrimidines. The pyrimidine nucleus has long been associated with biological activity including protectant fungicidal and insecticidal action. The key to development of systemic fungicidal activity for the hydroxypyrimidines was the oxygen function placed at position 4. Both 5-butyl-2-dimethylamino-6-methyl-4(1*H*)-pyrimidinone (dimethirimol) (**14**) and the 2-ethylamino analogue (ethirimol) (**15**) have achieved commercial use as systemic agents against powdery mildew (11). When barley seed is treated with ethirimol, the plants are protected against powdery mildew for the entire season without another application. Another analogue, α-(2,4-dichlorophenyl)-α-phenyl-5-pyrimidinemethanol (triarimol) (**16**), is a more broad spectrum fungicide, valued for its ability to arrest further development of fungal lesions.

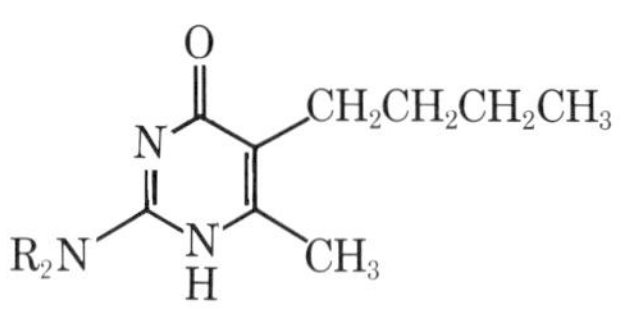

(**14**) R = CH_3
(**15**) R = C_2H_5

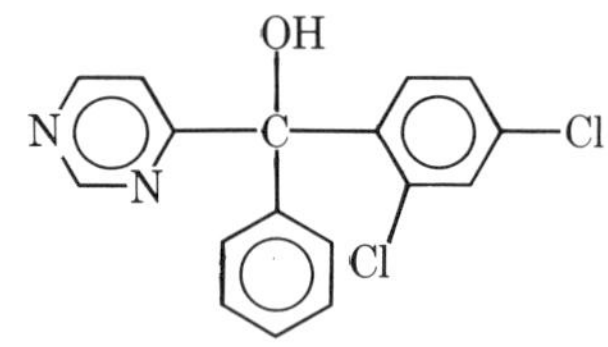

(**16**)

Miscellaneous Fungicides

Fungicides other than those previously described continue to contribute significantly to the control of fungal diseases. Dinocap, a mixture of dinitrooctylphenylcrotonates developed as an acaricide and mite ovicide, proved to be active against powdery mildews. It is an effective protectant as well as an eradicant for established lesions. Iprodione, 1-isopropylcarbamoyl-3-(3,5-dichlorophenyl)-hydantoin, is a new contact fungicide active against a broad spectrum of foliar diseases. Fenaminosulf, sodium *p*-(dimethylamino)benzenediazosulfonate, and ethazole, 5-ethoxy-3-trichloromethyl-1,2,4-thiadiazole, are narrow spectrum fungicides active against Pythiacious species of fungi in soil. They are both commonly used as seed treatments for field crops and soil drenches for nursery crops.

Antibiotics. The use of antifungal antibiotics in agriculture has been limited by high cost and short residual life. Cycloheximide (Actidione) is the only antibiotic to receive significant usage in the United States, but its remarkable activity against many fungi is counteracted by its general phytotoxicity. The antibiotics have enjoyed the greatest success in Japan where they are used primarily on rice. The antibiotics blasticidin S, kasugamycin, and the polyoxins are used extensively to control rice blast, a disease that defies economic control by synthetic fungicides (see Antibiotics, nucleosides). These antibiotics possess systemic activity, which contributes greatly to their success.

Resistance

Prior to the appearance of the highly specific systemic fungicides, rarely did phytopathogenic fungi develop tolerance to fungicides. Tolerance to some systemic fungicides such as dimethirimol was noticed within one year of introduction (12). Tolerance to benomyl developed rapidly in *Botrytis cinerea*, somewhat slower in

Cercospora sp., and much slower in the powdery mildews (13–16). As might be expected, fungal strains with developed tolerance to benomyl also have cross-tolerance to thiabendazole, fuberidazol, and thiophanate (17). With the recognition of acquired tolerance came the adoption of strategies to avoid selection for tolerance, such as alternation of materials.

Mode of Action

The mode of action of agricultural fungicides has been actively studied, but in most cases the actual biochemical basis for activity remains elusive. In general, the broad-spectrum, protectant-type materials apparently have several sites of action that cause deleterious effects in many enzyme and membrane systems. This multiple-site action decreases the likelihood of the fungus genetically adapting to the fungicide. In contrast, the site of action of the highly-specific, systemic compounds seems to be confined to one system. These fungicides exert their action through one of three different methods: (*1*) inhibition of energy production, (*2*) interference with biosynthesis, or (*3*) disruption of cell structure.

The oxathiins are an example of the first type of action. Carboxin blocks mitochondrial respiration in sensitive fungi by blocking the action of succinodehydrogenase as well as the substrate side of cytochrome b and ubiquinone (18–19). The pyrimidine, triarimol, is an example of a compound that interferes with biosynthesis. Triarimol and related materials inhibit advanced stages of sterol synthesis (20–21) by interfering with the activity of several synthetic enzymes. The benzimidazoles are materials that cause disruption of cell structure. The active material binds with microtubular protein preventing assembly of the spindle fibers that are necessary for mitosis (22).

Although many fungitoxic molecules are available, it is obvious from the preceding discussion that relatively few are suitable for agricultural use. The physiochemical characteristics required for a chemical to be useful in agriculture have ruled out the use of many very active fungicides. This results from the narrow margin of selectivity available when trying to kill or inhibit the germination or growth of one plant, the fungus, on or in a higher plant, as well as the type of stability required to function in the field. Much has been published on various aspects of fungicidal action such as structure–activity relations, detoxification, the influence of environment, and mode of action. Although much has been learned about the influence of these and other factors, no satisfactory system has been devised for predicting what chemicals are fungitoxic or even what modifications can be made on a newly discovered toxic nucleus to make it even more effective. The empirical approach initially yielded compounds of low, nonspecific activity, but now is yielding materials with increasing specificity and activity. The empirical approach will probably continue to be used in the discovery of new fungicides but it is hoped that as more basic information becomes available, more rational and effective approaches can be used.

BIBLIOGRAPHY

"Fungicides, Agricultural" in *ECT* 1st ed., Vol. 6, pp. 984–991, by J. G. Horsfall, The Connecticut Agricultural Experiment Station; "Fungicides, Agricultural" in *ECT* 2nd ed., Vol. 10, pp. 220–228, by D. C. Torgeson, Boyce Thompson Institute for Plant Research.

1. B. S. Magdoff, H. P. Burchfield, and J. Schechtman, *Contrib. Boyce Thompson Inst.* **19**, 267 (1958).

2. J. B. Harry, R. H. Wellman, F. R. Whaley, H. W. Thurston, Jr., and W. A. Chandler, *Contrib. Boyce Thompson Inst.* **15,** 195 (1948).
3. G. J. M. van der Kerk and J. G. A. Luijten, *J. Appl. Chem.* **4,** 314 (1954).
4. H. S. Cunningham and E. G. Sharvelle, *Phytopathology* **30,** 4 (1940).
5. J. W. Henberger and T. F. Manns, *Phytopathology* **33,** 1113 (1943).
6. R. H. Wellman and S. E. A. McCallan, *Contrib. Boyce Thompson Inst.* **14,** 151 (1946).
7. T. C. Ryker, *Proc. Cotton Disease Council* **25,** 137 (1965).
8. B. von Schmeling and M. Kulka, *Science* **152,** 659 (1966).
9. T. Staron and C. Allard, *Phytiatr. Phytopharm.* **13,** 163 (1964).
10. J. W. Vonk and A. K. Sijpesteijn, *Pestic. Sci.* **2,** 160 (1971).
11. R. S. Elias and co-workers, *Nature* **219,** 1160 (1968).
12. K. J. Bent and co-workers, *Proc. 6th Br. Insectic. Fungic. Conf. Brighton* **1,** 274 (1971).
13. G. J. Bollen and G. Scholton, *Neth. J. Plant Pathol.* **77,** 83 (1971).
14. S. G. Georgopoulos and C. Dovas, *Plant Dis. Rep.* **57,** 321 (1973).
15. W. T. Nishijima and E. B. Smalley, *Phytopathology* **69,** 69 (1979).
16. E. Shabi and T. Katan, *Phytopathology* **69,** 267 (1979).
17. S. Kosaka, *J. Pestic Sci.* **2,** 323 (1977).
18. H. Lyr, G. Ritter, and G. Casperson, *Z. Allg. Mikrobiol.* **12,** 275 (1972).
19. H. Lyr, T. Shewe, D. Zanke, and W. Müller in *Mechanisms of Action of Systemic Fungicides,* Akademie-Verlag, Berlin, 1975, p. 153.
20. J. L. Sherald and H. D. Sisler, *Pestic. Biochem. Physiol.* **5,** 477 (1975).
21. M. R. Siegel and N. N. Ragsdale, *Pestic. Biochem Physiol.* **9,** 48 (1978).
22. L. C. Davidse in *Mechanisms of Action of Systemic Fungicides,* Akademie-Verlag, Berlin, 1975, pp. 137–143.

General References

M. R. Siegel and H. D. Sisler, eds., *Antifungal Compounds,* 2 vols., Marcel Dekker, Inc., New York, 1977.

R. W. Marsh, ed., *Systemic Fungicides,* John Wiley & Sons, Inc., New York, 1972.

E. J. BUTTERFIELD
D. C. TORGESON
Boyce Thompson Institute for Plant Research, Inc.

FUNGICIDES, INDUSTRIAL. See Industrial antimicrobial agents.

FURAN DERIVATIVES

Furfural and its derivatives comprise an industrially significant class of heterocyclic compounds, the furans (1). Furan derivatives may be simple furans, ie, when the nucleus occurs as a free monocycle, or condensed furans, when the nucleus is fused to another ring.

Radicals derived from furan are named in the same manner as the analogous radical in the benzene series. Typical furan radicals are 2 (or α)-furyl (**1**), 2-furfuryl (**2**), 2-furoyl (**3**), and 2-furfurylidene (**4**):

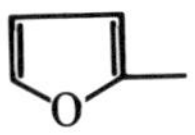

(1)

(2)

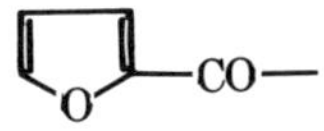

(3)

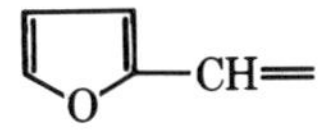

(4)

The furan nucleus is a cyclic, dienic ether with some aromatic character (2). A stabilization energy of 96 ± 4 kJ/mol (23 ± 1 kcal/mol) is observed for the furan ring, whereas for the benzene ring it is at 155 ± 4 kJ/mol (37 ± 1 kcal/mol) (3). Furan compounds are less aromatic (hence more reactive) than their benzene analogues and undergo a variety of addition reactions (4).

Reactivity of the furan nucleus is directly related to the electron density at particular ring atoms. Canonical forms (**5**) through (**8**) represent the overall electron distribution (5–6).

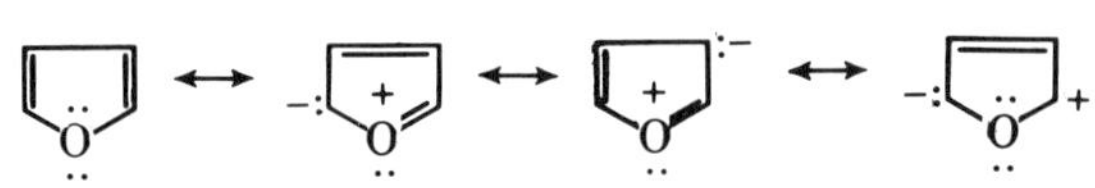

(5) **(6)** **(7)** **(8)**

Centers of high electron density at the α positions of the ring are readily supported by the behavior of furans in electrophilic substitution reactions. Such reactions proceed with great facility at an α position, if available. Orientation of incoming groups is similar to the benzene series; eg, on nitration, 3-methylfuran [*930-27-8*] yields 2-nitro-3-methylfuran [*38829-39-9*] (ortho) and 3-furancarboxaldehyde [*498-60-2*] yields 2-nitro-4-furancarboxaldehyde [*72918-24-2*] (meta). For comprehensive reviews on furan chemistry, see refs. 1, 7; for furan polymers, see refs. 8–9.

The balance between aromatic and aliphatic character is markedly affected by substituents. Furan and its homologues, for example, function as dienes in the Diels-Alder reaction. With maleic anhydride, furan readily forms 7-oxabicyclo[2.2.1]hept-5-ene-2,3-dicarboxylic anhydride [*5426-09-5*] (**9**) in excellent yield.

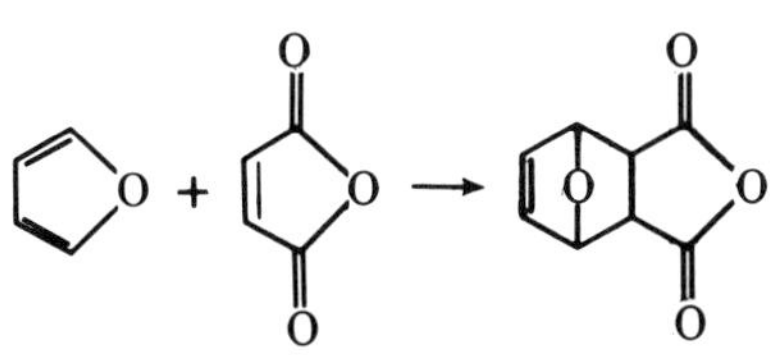

(9)

Halofurans, alkoxyfurans, furfuryl esters and ethers, and furfural diacetate [*613-75-2*] behave similarly. On the other hand, furans with electron-withdrawing substituents (eg, furfural [*98-01-1*], 2-furoic acid [*26447-28-9*], and nitrofurans) fail as dienes even with the strongest dienophiles.

Furan and its homologues are readily cleaved on acid hydrolysis to open-chain dicarbonyl compounds. 2,5-Dimethylfuran [*625-86-5*], for example, is hydrolyzed in excellent yield to 2,5-hexanedione. Certain substituents alter the course of the cleavage reaction giving rise to γ-ketocarboxylic compounds; furfuryl alcohol, for example, yields levulinic acid (**10**) (1).

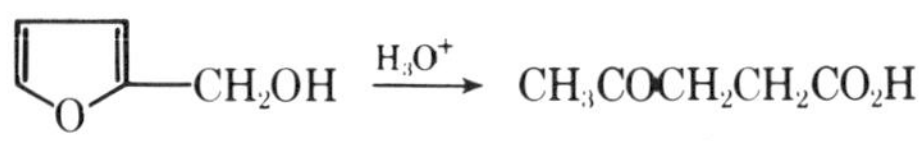

(**10**)

Negatively substituted furans, however, are quite stable to cleavage by acids because the electron-withdrawing substituent deactivates the ring but renders it rather vulnerable to nucleophilic attack. Thus 2-nitrofuran [*609-39-2*], which is remarkably resistant to nonoxidizing acids, is easily cleaved by strong alkali yielding undefined aldehydic resinous products.

In view of the susceptibility of the furan nucleus to cleavage and/or polymerization, considerable care is required in the selection of reagents and reaction conditions to achieve desired products in reasonable yield.

The lesser aromatic character of the furan series is further demonstrated by the paucity of phenol or aniline analogues. Free furanols or furylamines do not exist and any reaction that could result in them invariably affords the tautomeric dihydro derivative or its decomposition product. Hydrolysis of 2-acetoxyfuran [*25414-23-7*], for example, does not yield the desired 2-furanol (**11**); instead, cleavage occurs resulting in β-formylpropionic acid, presumably by hydrolysis of 3-butenolide, the tautomer of 2-furanol.

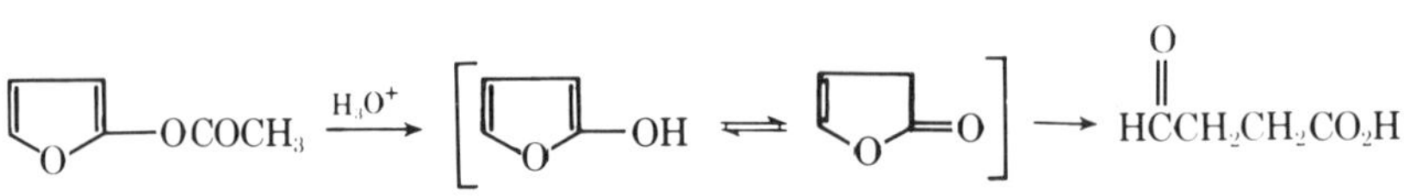

(**11**)

There are two classes of partially saturated monocyclic furans, the 2,3- (**12**) and the 2,5-dihydrofurans (**13**) [*1191-99-7, 1708-29-8*].

(**12**)

(**13**)

2,3-Dihydrofurans are cyclic enol ethers of γ-hydroxycarbonyl compounds to which they revert on mild acid hydrolysis. The 2,5-dihydrofurans, prepared by acid-catalyzed cyclodehydration of appropriate *cis*-2-butene-1,4-diols, are less sensitive to acid cleavage than 2,3-dihydrofurans. A particular group of 2,5-dihydrofurans is prepared by controlled oxidative addition of methanol or acetic acid to furans that have no electronegative substituents (10). These 2,5-dihydrofurans are cyclic acetals or acylals of 1,4-dicarbonyl compounds to which they are readily converted by hydrolysis under mildly acidic conditions. For example, 2,5-dimethoxy-2,5-dihydrofuran [*332-77-4*] (**14**), prepared from furan and methanol, hydrolyzes to the commercially utile malealdehyde (**15**).

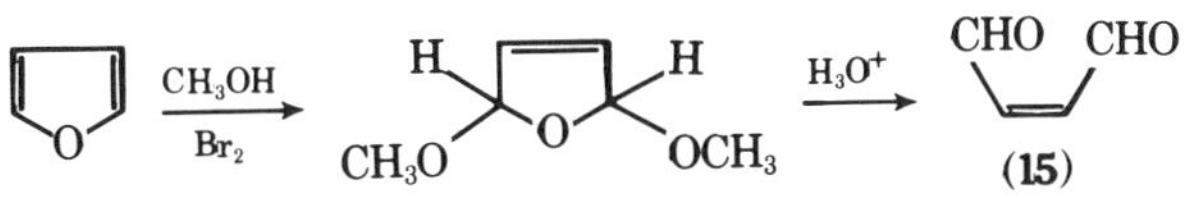

(14)

Tetrahydrofuran [*109-99-9*] is a fully saturated cyclic ether. It is manufactured by cyclodehydration of 1,4-butanediol and by catalytic hydrogenation of furan (see Ethers).

The chemical properties of the industrially significant furan derivatives are characteristic of the functional groups affixed to the furan nucleus. In many instances, however, those properties are modified because of the highly reactive furan ring.

For descriptions of all of the published chemical reactions of furan derivatives see refs. 1, 7, 9, and 11–12.

Furfural

Physical Properties. Furfural [*98-01-1*] (2-furancarboxaldehyde), a colorless liquid aldehyde with a pungent, aromatic odor reminiscent of almonds when freshly distilled, darkens appreciably on exposure to air. It is miscible with most of the common organic solvents but only slightly miscible with saturated aliphatic hydrocarbons. For the solubilities of representative organic compounds in furfural see refs. 13–14. Inorganic compounds, generally, are quite insoluble in furfural.

In Table 1 are collected the most important physical properties of furfural. These are supplemented by plots (Fig. 1) of the vapor–liquid compositions for the system: furfural–water (15–16).

Chemical Properties. The chemical properties of furfural are characteristic of aromatic aldehydes but with unique variances attributable to the furan ring.

Furfural is very thermally stable in the absence of oxygen. At temperatures as high as 230°C, exposure for many hours is required to produce detectable changes in the physical properties of furfural, with the exception of color (17).

Furfural can be oxidized to furoic acid [*26447-28-9*], reduced to furfuryl alcohol [*98-00-0*], and converted to furan by decarbonylation over selected catalysts. With concentrated alkalies, furfural undergoes the Cannizzaro reaction yielding furfuryl alcohol and sodium furoate [*17256-62-1*].

Just as other aldehydes, furfural condenses with compounds possessing active methylene groups such as aliphatic carboxylic esters and anhydrides, ketones, aldehydes, nitriles, and nitroparaffins; with furfuryl alcohol and phenolic compounds, resins are formed.

The electron-withdrawing character of the aldehyde group makes furfural quite resistant to hydrolytic fission. Even at high temperatures, long exposure is required to effect extensive destruction of furfural by dilute acids (18). Black polymer results from acidic treatment of this type, and the rate of its formation is dependent upon the hydrogen ion concentration and temperature.

Table 1. Physical Properties of Furan Derivatives

	Furfural [*98-01-1*]	FA[a] [*98-00-0*]	Furan [*110-00-9*]	THFA[b] [*97-99-4*]
General properties				
molecular weight	96.09	98.10	68.08	102.13
boiling point at 101.3 kPa (1 atm), °C	161.7	170	31.36	178
freezing point, °C	−36.5		−85.6	<−80
metastable crystalline form		−29		
stable crystalline form		−14.63		
refractive index, n_D				
20°C	1.5261	1.4868	1.4214	1.4250
25°C	1.5235			1.4499
density, d_4, at 20°C	1.1598	1.1285	0.9378	1.0511
vapor pressure, 100 Pa (0.75 mm Hg)				
−15°C			130	
0°C			277	
20°C			658	
50°C			1980	
60°C	21	8.5		10
80°C	56	25		28
100°C	132	68		68
120°C	280	164		156
140°C	567	361		324
vapor density (air = 1)	3.3	3.4	2.36	3.5
critical pressure, P_c, MPa[c]	5.502		5.32	
critical temperature, T_c, °C	397		214	
solubility, wt %, in water				
20°C	8.3	∞		∞
25°C			1	
alcohol; ether	∞	∞	∞	∞
Thermodynamic properties				
heat of vaporization, kJ/mol[d]	38.6		27.1	
specific heat (liq), J/(g·K)[d]				
20°C			1.699	
20–27°C				1.774
20–100°C	1.741			
25°C		2.100		
specific heat (vap), J/(g·K)[d]				
31.36°C			1.021	
98.99°C			1.251	
heat of combustion (liq), kJ/mol[d]	2344	2548	2092	2965
Fluid properties				
viscosity, mPa·s (= cP)				
20°C			0.38	6.24
25°C	1.49	4.62		
surface tension, mN/m (= dyn/cm)				
25°C		ca 38		37
29.9°C	40.7			

Table 1. *(continued)*

	Furfural [98-01-1]	FA[a] [98-00-0]	Furan [110-00-9]	THFA[b] [97-99-4]
Electrical properties				
dielectric constant				
20°C	41.9			
23°C				13.6
Flammability properties				
explosion limits (in air), vol %	2.1–19.3	1.8–16.3	2.3–14.3	1.5–9.7
flash point, °C				
Tag closed cup	61.7	65	−35.5	
Tag open cup				83.9
ignition temperature, °C	315	391		282

[a] Furfuryl alcohol.
[b] Tetrahydrofurfuryl alcohol.
[c] To convert MPa to atm, divide by 0.101.
[d] To convert J to cal, divide by 4.184.

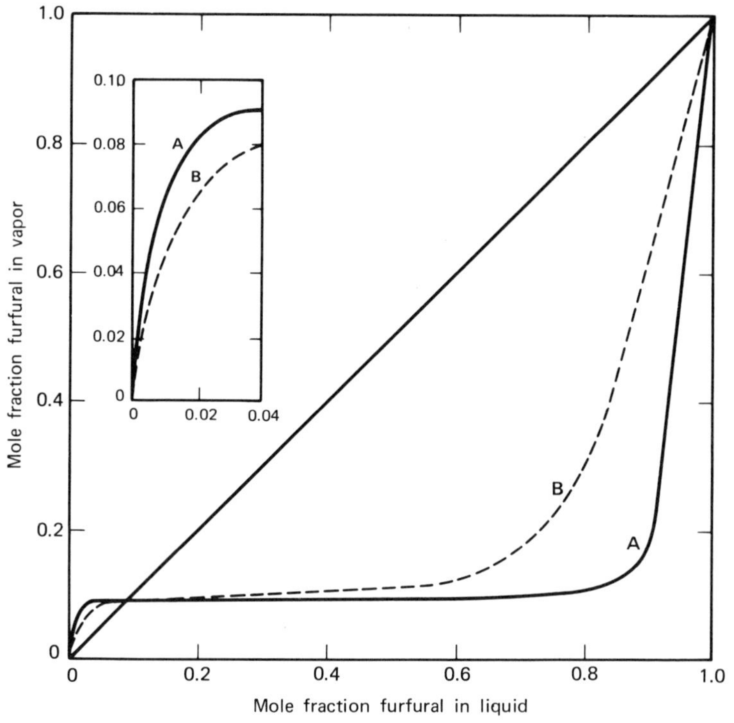

Figure 1. Vapor–liquid equilibria in the furfural–water system. A at 101 kPa (1 atm) (15). B at 598 kPa (5.92 atm) (16).

The resinification of furfural initiated by acids or heat has been observed for years and many interpretations of the polymerization mechanism have been proposed (1,17,19–21). Because of the complexity of this polymerization, key intermediates have not been isolated. Conditions of polymerization, whether aqueous or anhydrous, inert or oxygen atmosphere, all affect the composition of the polymer. Recent work on the resinification of furfural (22–23) has introduced new insight on the polymerization mechanism, particularly with respect to thermal reaction at 100–250°C in the absence of air. Based on the isolation and characterization of two intermediate products (16–17),

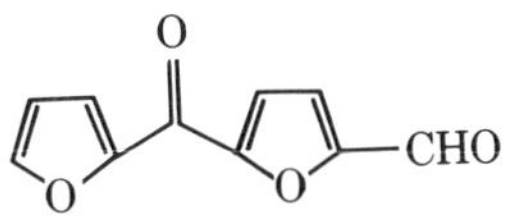

(16) [*57583-51-4*]

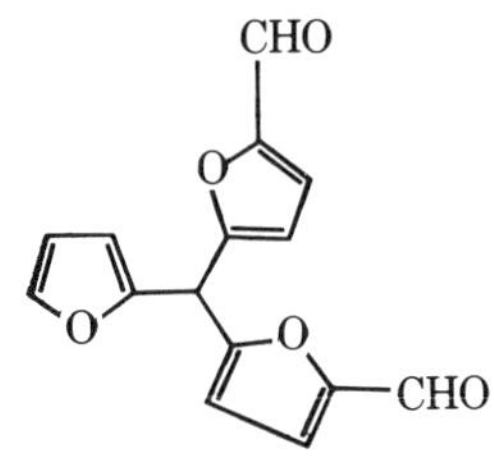

(17) [*72918-23-1*]

(18) was proposed for the final resin (9).

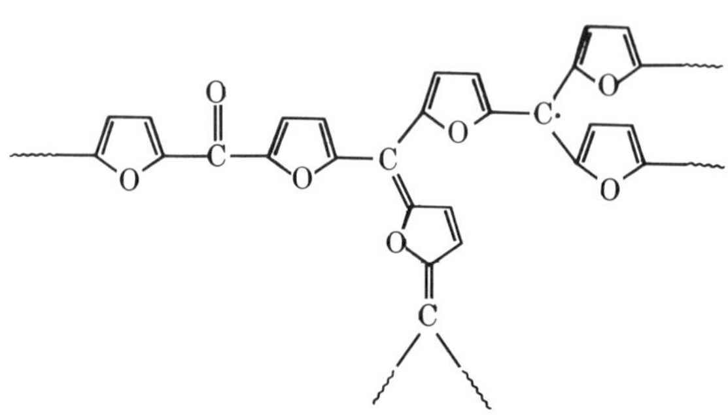

(18)

The presence of stable free radicals in the final polycondensate is supported by the observation that traces of (18) have a strong inhibiting effect on the thermal polymerization of styrene.

The photopolymerization of furfural by uv has not received much attention. Although the products of furfural polymerized by heat are branched polycondensates with highly conjugated structures, the photopolymer of furfural is a linear polyaddition product (22,24). The gas-phase photolysis of furfural in the $n \rightarrow \pi^*$ and $\pi \rightarrow \pi^*$ transitions (25) proceeds with fragmentation to CO, furan, and C_3-hydrocarbon, but a certain amount of resinification is also noted (about 5% quantum yield with excitation of the $n \rightarrow \pi^*$ transition) (see Photochemical technology).

Vacuum liquid-phase photolysis by uv at room temperature produced resin (22–23). The $\overline{DP}_n$ of these polymers was only about 5 with predominance of the following unit:

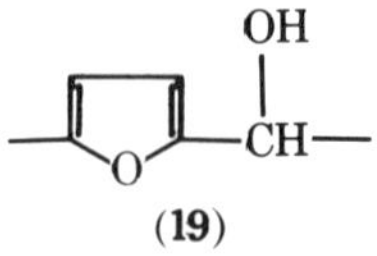

(19)

The presence of one carbonyl group per oligomer molecule was also observed. The fact that furoin was isolated is supportive of the proposed mechanism.

Copolymerization of furfural with dimethylketene at low temperature using anionic initiators occurs through the carbonyl bond (26). The polyester formed has a regular structure indicating an alternating addition of the two monomers. The polymer repeating unit was suggested to be:

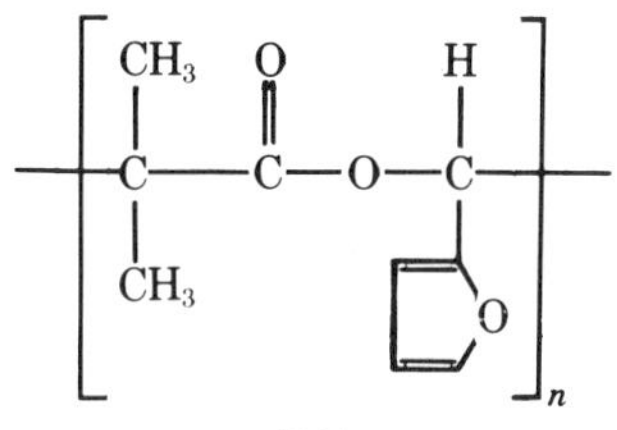

(20)

Furfural reacts with poly(vinyl alcohol) under acid catalysis to effect acetalization of the hydroxyl groups (27–28). An addition product of the polymer with maleic anhydride has been reported (29–30). This "poly(vinyl furfural)" has a high molecular weight and is prepared in good yield.

Nitration and halogenation of furfural occurs under carefully controlled conditions with introduction of the substituent at the open 5-position (31–32).

Manufacture. Furfural is produced from annually renewable agricultural sources such as nonfood residues of food crops and wood wastes. Of the various components of these vegetable materials, the pentosan polysaccharides (xylan, arabinan) are the major precursors of furfural and are almost as widely distributed in nature as cellulose (qv) (1) (see Alcohols, polyhydric).

Cereal grasses and by-products such as corncobs, cottonseed hulls, and rice hulls constitute one large natural storehouse; bagasse (qv), a by-product of sugarcane harvesting, another; and wood and wood products yet another. The chief constituent of the pentosan fraction of these materials is xylan, a polysaccharide having backbone chains of (1→4)-linked β-D-xylopyranosyl residues (33). According to published data (34), xylan with a small amount of arabinan, a highly branched polysaccharide of (1→3) and (1→5)-linked α-L-arabinofuranosyl units (35), accounts for 25–30% of cereal straws and grains, approximately 15–25% of deciduous woods and 5–15% of coniferous woods, based on the dry matter of these plants.

Theoretically, all pentosan-containing substances are potentially usable but only a relatively few (corncobs, oat hulls, rice hulls, and bagasse), available in large tonnages within a limited radius of the furfural-producing plant, are commercially significant (1,36–37). The mechanism for the formation of furfural from C_5-sugars or their precursors has not been unequivocally established. The potential for side reactions complicates the recovery of furfural and limits its practical yield which is far from quantitative under any commercial mode of operation. Thus the economics of furfural production is highly dependent upon pentosan content, availability, and cost of raw material, as well as the expense of collecting, shipping, and handling the material.

Furfural is produced commercially in batch or continuous digesters (37) where the pentosans are hydrolyzed to pentoses and the pentoses subsequently cyclodehydrated to furfural:

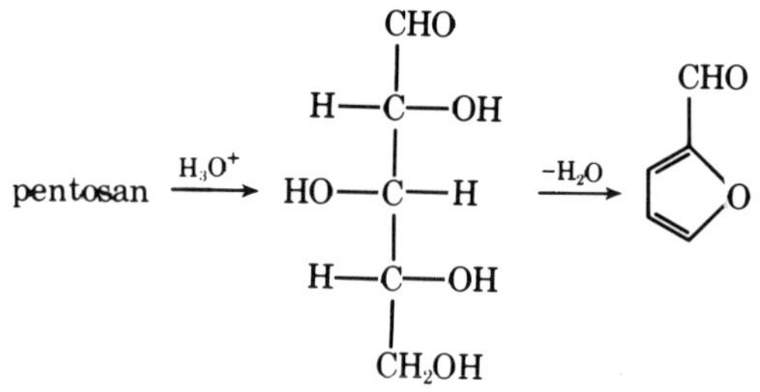

(21)

In all processes, raw material is charged to the digester and treated with strong inorganic acid. High pressure steam is introduced through the mass and, after attaining operating temperature and pressure, furfural is steam distilled. The condensed reactor vapors are fed to a stripping column from which an enriched furfural–water distillate is withdrawn and condensed in a decanter where it separates into two layers. The furfural-rich layer containing about 6% water is processed further to obtain the furfural of commerce, and the water-rich layer containing ca 8% furfural is recycled to the columns as reflux. The furfural recovery process is illustrated in Figure 2; flow charts for the Escher-Wyss and Rosenlew continuous processes appear in ref. 37. In general, regardless of the process employed, the cellulose in the raw material undergoes only partial degradation and remains with the lignin and ash as the by-product residue.

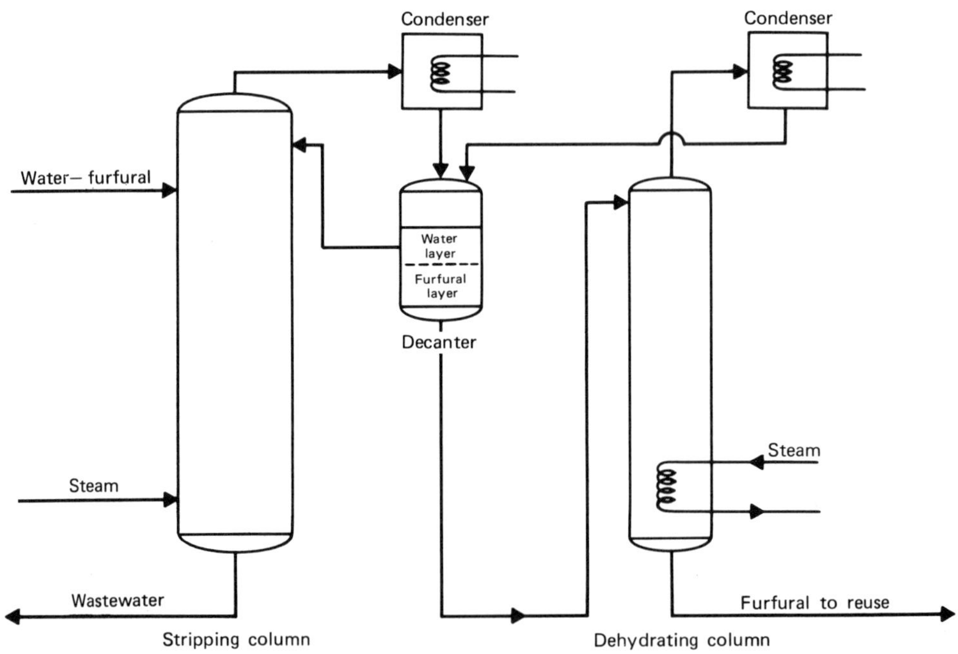

Figure 2. Furfural recovery from aqueous solutions.

A number of procedures for quantitative estimation of furfural have been reported (1), yet not one of these has universal acceptance. For the most part, they can be divided into general categories: those involving reactions of the aldehyde group, and those that are dependent upon behavior characteristic of the furan ring. Accordingly, the determination of furfural is generally subject to interference by carbonyl compounds or other unsaturated substances.

Chemical methods leave much to be desired from the standpoint of accuracy and reproducibility. A quantitative measurement of furfural in water may be carried out by uv spectroscopy; absorbance is measured at 276 nm (38). However, uv-absorbing impurities interfere with the analysis. A gas chromatographic method to measure furfural in the presence of water has been developed and has been adopted as the standard analytical procedure in the laboratories of The Quaker Oats Company. Using this technique, 2-methylfurfural is also analyzed in the presence of furfural.

Furfural is shipped with few problems in steel tank cars, aluminum tank trucks, and steel cans. Storage in either aboveground or underground installations is satisfactory. Because furfural is an excellent solvent and penetrant, care must be taken that all joints are secure and that the pump and valve packings are in good condition. Drum lots may be stored for months without appreciable change in physical properties.

Flammability of furfural is comparable to that of kerosene or No. 1 fuel oil. Because of its chemical reactivity, furfural should not be mixed with strong acids or alkalies, nor stored near strong oxidizing chemicals. When furfural is stored for long periods in contact with air, there is a gradual darkening of color, increase in acidity, and polymer formation.

According to published data (39) as of October, 1978, estimated noncommunist world capacity for furfural production was believed to total about 180,000 metric tons with annual demand estimated at 136,000 t. The worldwide capacity, production and consumption of furfural for 1965, 1973, and 1978 are listed in Table 2. Also shown for those years are the then prevailing U.S. domestic bulk furfural prices. In contrast to the shortage years of 1974 and 1975, domestic supplies of furfural are now more than ample, and although world consumption has slipped between 8 and 10% from 1975 levels, new capacity continues to be announced or brought on stream to create a supply–demand imbalance. The January, 1980, selling price was \$1.14/kg.

Uses. Furan derivatives including furfural, furfuryl alcohol, and tetrahydrofurfuryl alcohol are used as solvents in a variety of industrial applications.

Furfural and furfuryl alcohol are specialty solvents. They also are reactive solvents and contribute low viscosity to resin formulations. Thermosetting resins containing

Table 2. Worldwide Furfural Supply/Demand[a], 1000 Metric Tons

	1965	1973	1978
capacity	100	141	204[b]
production	91	136	159
consumption	91	136	136
U.S. bulk price, ¢/kg	27.6	41.3	114

[a] Refs. 36–37.
[b] Includes 22,700 t estimate for Eastern Europe.

furfural and/or furfuryl alcohol (furan resins) demonstrate specialty properties including corrosion resistance, high carbon yield, stability at elevated temperature, low fire hazard, and excellent physical strength. These properties are of industrial importance in foundry sand core/mold making, fiberglass composites, mortars, cements, plastic insulation foams, refractory mixes, high-carbon composites, and aggregate binders, among others.

Furfural serves as a source of furfuryl alcohol, tetrahydrofurfuryl alcohol, furan, tetrahydrofuran, poly(oxytetramethylene) glycol, and a variety of synthetic resins. Other chemicals that have been derived, on a lesser scale, include methylfuran, methyltetrahydrofuran, di- and tetrahydropyran, levulinic acid, γ-valerolactone, pyrrole and pyrrolidine, pyridine and piperidine, and lysine.

The principal usage of furfural is as selective solvent for separating saturated from unsaturated compounds in petroleum refining, gas oils, and diesel fuel; for the extractive distillation of butadiene and other C_4-hydrocarbons for the manufacture of synthetic rubber; and for the production of light-colored wood (qv) rosins (see Extraction). It is both a solvent and a processing aid for refining anthracene. Furfural is also an ingredient in several furfuryl alcohol resins and in phenol-aldehyde resins. It is also used as a resin solvent and wetting agent in the manufacture of abrasive wheels and brake linings (qv).

In its use as a solvent, furfural is consumed in lube (lubricating) oil refining and butadiene extraction. Little expansion is seen in the U.S. and European lube-oil industries. However, important new lube-oil capacity is under development in several Middle East countries including Iran, Iraq, and Saudi Arabia, and in Africa. Furfural refining has been used in this application for over thirty years and is a time-tested process. The lighter oils, where furfural processing appears to excel, are growing at the expense of the heavier oils in an otherwise stagnant market.

Phenol and furfural are the most common extraction solvents. From a technical/economic standpoint, each has advantages as well as disadvantages as reported in a recent article on the production of lube oil (40). However, there is a movement away from phenol, motivated primarily by its negative safety, health, and environmental aspects.

In extraction processes using furfural, recently designed plants have been replacing packed columns with rotating disk contactors (RDCs) (40). The ability to control mixing intensity by adjusting rotor speed permits optimum mixing for each stock, resulting in higher efficiencies and increased yields. In addition to improving efficiencies, RDCs give lower transition times because the majority of holdup in the column is solvent. It has also been reported that coking is practically eliminated by using RDCs.

Optimization of solvent recovery, control of impurity buildup in the circulating solvent, reduction in solvent loss, and control of corrosion have led to reliable, lower cost, base-stock processing.

Furfural's use as an extractive solvent in the recovery–purification of butadiene is declining. Co-product butadiene from heavy-liquid olefin crackers is displacing dehydrogenation butadiene and it is projected that between 1985 and 1990 all U.S. production of dehydrogenation butadiene will cease (see Butadiene).

Furfural, used as a reactive solvent in cold-blending of pitch (see Tar and pitch), forms resinous impregnants that interact at carbonization temperatures. These "resin pitch" combinations have curing and carbonization characteristics better than those of either component alone. The cold-blending process significantly improves the environmental acceptance of pitch and offers industry an alternative to hot processing.

High quality carbon fibers have been produced (41) by pyrolyzing, in a nonoxidizing atmosphere, resinous fibers prepared from a viscous copolymer of furfural and pyrrole. Yield of carbon fiber having a glossy surface and only 3% voids was 66%. Compared with other graphite-type fibers, these carbon fibers had improved mechanical strength, higher oxidation resistance, and lower density (see Carbon).

Important thermosetting resins are obtained by interaction of furfural with phenol. The reaction occurs under both acidic and basic catalysis. Other large uses of furfural together with phenol are in the manufacture of resin-bonded grinding wheels and coated abrasives (8).

Furfural reacts with ketones to form strong, cross-linked resins of considerable technical interest in the USSR and to a lesser degree by the U.S. Air Force (42–43). The so-called furfurylidene acetone monomer, a mixture of 2-furfurylidene methyl ketone [*623-15-4*] (**22**), bis-(2-furfurylidene) ketone [*886-77-1*] (**23**), mesityl oxide, and other oligomers, is obtained by condensation of furfural and acetone in an alkaline medium (44). Treatment of the "monomer" with an acidic catalyst leads initially to polymer of low molecular weight and ultimately to cross-linked, black, insoluble, heat-resistant resin (45).

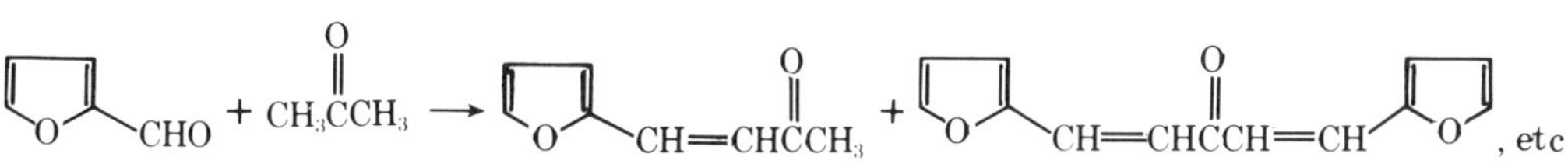

(**22**) (**23**)

Furfural–acetone resins have been used to form resin–aggregate mixtures referred to as organic concretes. Despite the reportedly excellent properties, there has been virtually no commercial use of such resins outside the USSR. The structures and polymerization mechanisms of these furfural–aldehyde–ketone polymers are discussed in a recent review (9).

Resins prepared from the condensation products of furfural with urea, formaldehyde, phenols, etc, modified by appropriate binders and fillers are described in the technical literature (1). These cross-linked systems are applied in specific thermosetting applications described in broad patent coverage.

Toxicology. Furfural, the key member of the industrial furan chemicals, is known as a moderately toxic substance. Apparently because of its broad natural occurrence in foods and beverage, it was selected by the National Cancer Institute for bioassay, using lifetime rat and mouse studies. Recent mutagenicity studies (*in vitro*) in five standard strains of *Salmonella typhimurium* show no evidence of activity by furfural or derived members of the furan and tetrahydrofuran series.

Furfural is known to penetrate intact skin but with no fatal effect (rabbits) up to 500 mg/kg. Several drops of furfural in standard (rabbit) tests show significant irritation but no irreversible damage to the eye. Specific toxicological data are presented in Table 3.

Table 3. Toxicology of Furan Derivatives[a]

	Furfural	FA[b]	THFA[c]
LD_{50}, oral, mg/kg,			
rat	149	149	5250[d]
dog	950		
guinea pig			3000
LC_{50}, inhalation, ppm,			
rat	96–260	233	
mouse	260–370		
occupational exposure limit,			
8 h-TWA, ppm	5	50	na[e]

[a] Refs. 46–50.
[b] Furfuryl alcohol.
[c] Tetrahydrofurfuryl alcohol.
[d] See ref. 50 for additional values.
[e] Not available.

Furfuryl Alcohol

Physical Properties. Furfuryl alcohol (2-furanmethanol) is a colorless, liquid, primary alcohol with mild odor. On exposure to air, it gradually darkens in color. It is completely miscible with water, alcohol, ether, acetone, and ethyl acetate, and is soluble in most organic solvents with the exception of the paraffinic hydrocarbons.

The physical constants of furfuryl alcohol are listed in Table 1. The density and refractive index of furfuryl alcohol change when it is exposed to heat, acid or air (51), and the rate of change in these properties is a function of temperature and time of exposure.

Chemical Properties. The chemical properties of furfuryl alcohol are characteristic of the hydroxymethyl group; however, the reactivity of the furan ring, intensified by carbonium ion formation under acid catalysis, introduces complexities. Furfuryl alcohol undergoes the reactions expected of a primary alcohol such as oxidation, esterification, and etherification. Although stable to strong alkali, furfuryl alcohol is sensitive to acid; thus it imposes limitations on the conditions used in many of the typical alcohol reactions. For example, esterifications must be performed under neutral or basic conditions, or via base-catalyzed transesterification. Likewise, etherifications are best done under basic condensation conditions to suppress polymerization of the furan ring.

Under acidic conditions furfuryl alcohol polymerizes to black resinous polymers. Co-condensations also occur with phenolic compounds, formaldehyde and/or other aldehydes. In dilute aqueous acid, the predominant reaction is one of cleavage-rearrangement to form levulinic acid (**10**) (levulinic esters in alcoholic media).

Reduction of furfuryl alcohol can yield tetrahydrofurfuryl alcohol [*97-99-4*], methylfuran [*534-22-5*], methyltetrahydrofuran [*96-47-9*], or straight chain compounds by hydrogenolysis of the ring. Ethoxylation and propoxylation of furfuryl alcohol provide ether alcohols useful in coatings, printing ink, and paint stripper formulations.

Acid-catalyzed resinification is the most important industrial reaction of furfuryl alcohol. The chemistry of furfuryl alcohol polymerization initiated by heat, acidic catalysts, and alumina has been studied at various laboratories. The concepts available up to the early 1950s are summarized in ref. 1. Polymerization intermediates have been identified more positively since then with the advent of more advanced spectroscopic and chromatographic analytical techniques (52). Kinetics of polymerization (53–54) and elucidation of polymer structure (55) have resulted in improved understanding of the mechanism. Details of the mechanism have been described in a review of furan polymers (9). Based on the accumulated data, furfuryl alcohol has to be considered a bifunctional monomer in the initial stage and its "normal" reactions give linear chains or oligomers containing essentially two repeating units (**24–25**) with (**25**) predominating.

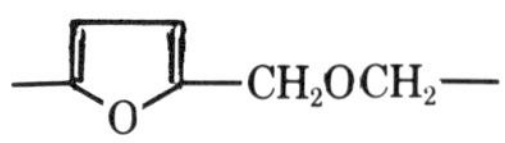

(**24**)

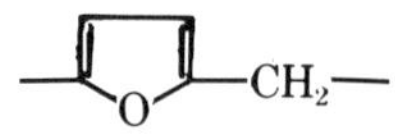

(**25**)

The polymerization or resinification of furfuryl alcohol is highly exothermic and, depending upon the activity and concentration of the catalyst used, requires careful control of the reaction temperature. Control is generally accomplished by cooling the reaction mixture with refluxing solvent and/or an external cooling fluid; in extreme instances, to avoid loss of control, emergency neutralization of the catalyst is essential.

The degree of polymerization is carried to the desired point as measured by viscosity. When the proper viscosity has been reached, the reaction is terminated by adjusting the pH of the system to 5–8. Such liquid resins can be stored for six months or longer.

The liquid resins have linear structures formed by intermolecular dehydration of the hydroxyl group of one molecule and the active α-hydrogen atom of another. Furfuryl alcohol, its dimer and higher homologues (**26**) [*25212-86-6*] are major constituents in the product mixture; however, considerable amounts of other species (**27–29**) are present as demonstrated by thin-layer chromatographic (tlc) densitometry analysis (see Table 4).

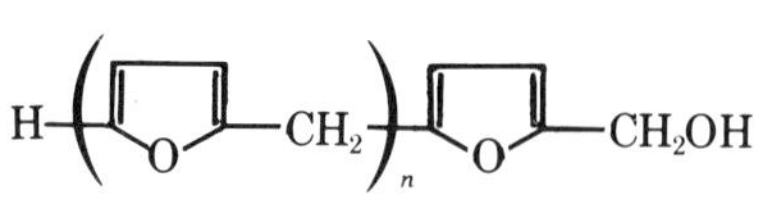

(**26**)

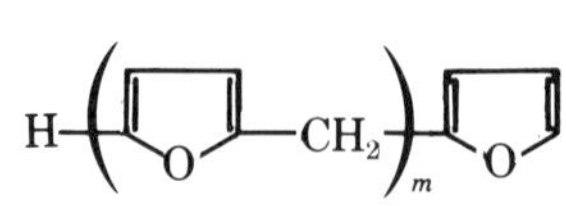

(**27**)

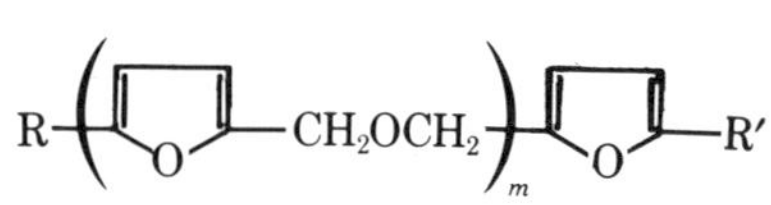

(**28**) R = H; R′ = H or $HOCH_2$

(**29**) R, R′ = $HOCH_2$ [*73246-91-0*]

As shown, the polymer end groups of these oligomers are —CH_2OH and unsubstituted furan rings. Identification of 2,2′-difurylmethane [*1197-40-6*] and higher homologues (**27**) indicates that formaldehyde, detected as a gaseous by-product of the reaction, is released by terminal carbinol groups and/or from internal ether bridges. An adjunct to the understanding of this mechanism is the evidence obtained that polymerization

Table 4. Compositions of Liquid Furan Resins[a]

	Furfuryl alcohol and homologues (26)	Difurylmethane and homologues (27)	Difurfuryl ether and homologues (28)	Difunctional (29)	wt % OH
FA[b] resin	25%	12%	35%	28%	4–7
BHMF[c] resin			5%	95%	16–18

[a] The analyses were performed with Safety Kotes self-charring silica gel plates (Applied Science Laboratories, Inc., Ann Arbor, Mich.). Values shown are based on thin-layer chromatographic densitometry.
[b] Furfuryl alcohol.
[c] 2,5-Bis(hydroxymethyl)furan.

of 2,5-bis(hydroxymethyl)furan [*1883-75-6*] produces no 2,2′-difurylmethane or higher homologues (**27**) (56). Analysis by tlc shows such resin having greater than 95% difunctional components (terminal hydroxymethyl groups, **29**) as compared to 28% for the resin from a conventional furfuryl alcohol polymerization.

In a proprietary procedure, furfuryl alcohol has been successfully hydroxymethylated to produce 2,5-bis(hydroxymethyl)furan. Bis(hydroxymethyl)furan has also been prepared by hydrogenation of 5-hydroxymethylfurfural [*67-47-0*].

Under the reaction conditions employed, hydroxymethylation is favored kinetically over polymerization of furfuryl alcohol. Good yields of monomer can be obtained. If resins derived from bis(hydroxymethyl)furan are desired, a change in reaction conditions yields difunctional oligomers (resins) with hydroxyl content of 16–18%. Composition of such resins is shown in Table 4; the higher hydroxyl content results in enhanced resin reactivity.

Manufacture. Furfuryl alcohol has been manufactured on an industrial scale by employing both liquid-phase and vapor-phase hydrogenation of furfural (57–58). Copper catalysts are preferred because they are selective and do not promote hydrogenation of the ring.

Furfuryl alcohol is shipped in bulk or drums. It is a powerful solvent and penetrant; containers, tanks, lines, and valves used to transport the product should be in good condition to avoid potential leakage. Furfuryl alcohol can be stored in containers lined with baked phenolic coatings; however, it should not be put in containers that are coated with lacquers, varnishes, or epoxy resins because it is an excellent solvent for many such coatings.

Furfuryl alcohol, on long storage, becomes progressively darker and less water-soluble, a change that is also caused by heat, acidity, and exposure to air. The reactions responsible for this change in water-solubility may be retarded by the addition in small quantity of an organic or inorganic base. Commercial furfuryl alcohol, FA (Quaker Oats), however, does not contain inhibitors.

Furfuryl alcohol is comparable to kerosene or No. 1 fuel oil in flammability. In the presence of concentrated mineral acids or strong organic acids, furfuryl alcohol reacts with explosive violence. Therefore, precautions should be taken to avoid contact of such materials with the alcohol. Caution is also recommended to avoid over-catalysis in the manufacture of furfuryl alcohol resins.

Furfuryl alcohol capacity is in place throughout the world with annual production (as of Oct. 1978) estimated to be in the range of 112,000–116,000 metric tons (36). Consumption of furfuryl alcohol and furfural is concentrated in the industrialized countries where demand primarily resides in the foundry industry, because furfuryl alcohol resins are the basis of significant industrial binder technology servicing this field (36–37). The March 1980 selling price was $1.23/kg.

Uses. Furfuryl alcohol, because of its chemical reactivity with acids, is used as a monomer in manufacturing furfuryl alcohol resins. The acid-catalyzed polycondensation of furfuryl alcohol is described as essentially a bifunctional, linear aggregation process accompanied by side reactions which introduce occasional branching points and polyconjugated sequences in the polymer chains leading to ring cleavage, structural rearrangement and ultimately to cross-linked, black resins (9). These furan resins are heat-stable and remarkably resistant to acid, alkali, and solvent. Resins of variant structure and properties have been prepared by polymerization of furfuryl alcohol and other co-reactants such as furfural, formaldehyde, glyoxal, resorcinol, phenolics, epoxies, and urea.

Industrially, furan resins derived from furfuryl alcohol are the most important in both utility and volume. The final cross-linked product displays outstanding chemical, thermal, and mechanical properties.

Furfuryl alcohol is used alone or in combination with other solvents for various cleaning and paint removing operations.

Furfuryl alcohol is a reactive plasticizer (qv) for phenolic resin in the manufacture of cold-molded grinding wheels. In this application, it improves adhesion between the abrasive grain and the phenolic resin, and reacts with the phenolic resin as the wheels cure (see Abrasives).

Major markets for furfuryl alcohol resins comprise the production of cores and molds for casting metals, corrosion-resistant fiber-reinforced plastics (FRPs) and corrosion-resistant cements and mortars. Under development are applications for low fire-hazard foams. The industrial value of furfuryl alcohol is a consequence of its low viscosity, high reactivity, and the outstanding chemical, mechanical, and thermal properties of its polymers (eg, corrosion resistance, nonburning, low smoke emission, and excellent char formation).

Metal-Casting Cores and Molds. In the foundry industry there is a highly significant binder technology based on furfuryl alcohol resins. Furan No-Bake (FNB) resins are accepted on a worldwide basis along with a number of competitive chemical binders. Total chemical binder consumption in the foundry industry is ca 136,000 t in North America.

FNB binders are very prominent in the production of sand cores and molds that cure without heat (59–61). Recently, ingot-mold foundries have converted to the FNBs (62). The FNB process uses reactive furan resin binder with strong organic acid or phosphoric acid as catalyst and eliminates long oven-baking cycles. Other money-saving benefits that accrue include: fewer or no reinforcing rods; less pattern wear; dimensional accuracy of the castings; easy shakeout of core sand after pouring; and minimal clean-up of the finished casting. Most important, the process is simple, requiring fewer steps to produce cores, and can be operated by semiskilled personnel.

Furan hot-box resins are used in both ferrous and nonferrous foundries (63–64). In this process, resin and catalyst are intimately mixed with dry sand and then blown into heated metal boxes containing a cavity the shape of the desired core. In seconds, the surface of the sand mass hardens and, as soon as the core has cured sufficiently to be dimensionally stable, the box is opened and the core removed.

Automotive cores with excellent dimensional accuracy and high strengths are made via this twenty-year-old process. Availability and cost of natural gas are causing concern in parts of the world and lower energy curing techniques are under development. However, to date, the hot-box process yields cores at the lowest cost per ton of mixed sand.

A warm-box resin based on 2,5-bis(hydroxymethyl)furan (BHMF) oligomers has recently been reported and is under test in U.S. and European foundries (65). Lower binder levels are reported effective at core-curing temperatures of 149–177°C. Core production rate using warm-box resins is equivalent to the rate with hot-box resins and with significant improvement in working environment (eg, reduced temperature and lowered emissions).

A furan resin cold-box curing system is presently under commercialization (66). Cold-box curing of cores offers many advantages: lower energy utilization, highly reproducible dimensional accuracy of the cores, use of less expensive patterns, and rapid core production. The cold-box process uses sulfur dioxide gas to trigger hardening of sand mix in a box at ambient temperature, as compared to the hot-box process which requires expensive iron core boxes, and energy to heat them and subsequently cure the resin–sand mix. The cold-box sand mix consists of furan resin (1.0–1.5 wt %) and peroxide (20–40 wt %, based on resin). Because it does not harden until gassed, this sand mix can be prepared advantageously in a conventional muller. The process works well with core boxes made of wood, plastic or metal. Sulfur dioxide gasing takes only seconds and cores are instantly ready for stripping. Castings made via this process are of high quality.

Corrosion-Resistant Fiber-Reinforced Plastic (FRP). Fiberglass reinforcement bonded with furfuryl alcohol thermosetting resins provides plastics with unique properties. Excellent resistance to corrosion and heat distortion coupled with low flame spread and low smoke emission are characteristics that make them valuable as laminating resins with fiberglass (67–68). Hand-lay-up, spray-up, and filament-winding techniques are employed to produce an array of corrosion-resistant equipment, pipes, tanks, vats, ducts, scrubbers, stacks, and reaction vessels for industrial applications throughout the world (see Laminated and reinforced plastics).

Reinforced furan resins have been used for many years in process piping and in underground sewer or waste-disposal systems. With a wide range in pH acceptability and good solvent resistance, furan piping has been a logical choice for many services, eg, kilometers of Permanite (Maurice A. Knight Co.) furan pipe are installed in Monsanto plants (69) (see Piping systems).

The highly cross-linked thermoset character of cured furan resin provides substantially greater corrosion resistance than other resins used in fiberglass-reinforced plastics. Hence, furan FRP is recommended (70) for high performance chemical process equipment in service with chlorinated aromatics, highly oxygenated organic solvents, and other admixtures of organics and aggressive aqueous media of a nonoxidizing nature.

Commercially manufactured corrosion-resistant off-the-shelf pipe using furan resins and fiberglass has been available since 1977 (71).

Another characteristic of furan FRP is its strength at elevated temperature as demonstrated in the comparative thermal gravimetric analysis (tga) curves (Fig. 3).

The inherent fire resistance and low smoke properties of furan resins appear to be related to the high carbon yields and the low molecular weight volatiles evolved on pyrolysis.

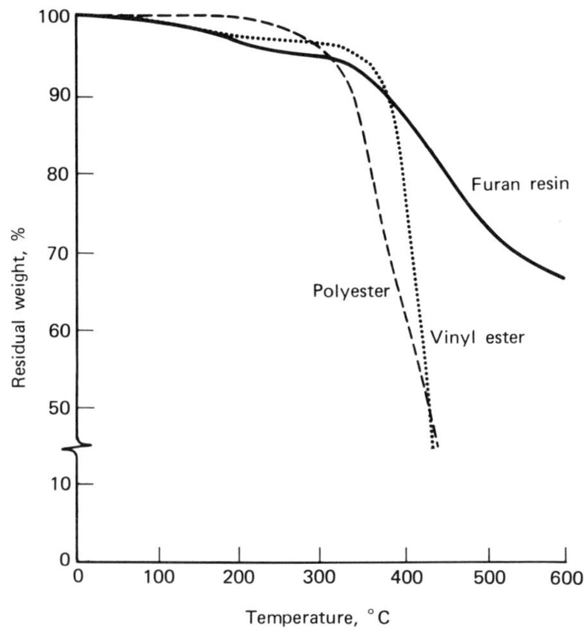

Figure 3. Thermal gravimetric analysis curves for typical resins.

Furan FRP is used by the Montreal Metro in the construction of the End Cap (the exterior front end of the driven car). It was selected because the low flame spread and low smoke evolution properties would minimize the risk of tunnel or chimney effects in case of end-to-end conflagration of cars (72).

A new, all liquid furan resin–catalyst system has been reported (73–75) which provides low flame spread and low smoke development without need for synergists or filler. Properly cured furan–fiberglass laminates tested in the ASTM E 84 Tunnel test consistently meet the <25 flame spread and <50 smoke development indexes. A corner-room burn test clearly portrayed the nonburning characteristics of furan FRP (76–78).

Low Fire-Hazard Foams. Combustibility characteristics of foamed plastic insulation are receiving widespread attention today. Considerable effort is devoted to develop low fire-hazard foams. The nonburning properties of furan composites also is evidenced in foam products (see Foamed plastics).

In a substantial advance of the state of the art, a novel low hazard thermal-insulation foam board, NF2 (Koninklijk Emballage Industrie), has been commercialized (79–80). NF2 insulation is a modified furfuryl alcohol foamed resin with high compressive strength and low friability formed into a dry, rigid, cellular insulating board with integral glass-mat facings (see Insulation, thermal).

The distinctive characteristic of NF2 insulation is low hazard behavior under the influence of fire. It is reported to be not easily ignitable (ie, no flame retardants need to be added). It exhibits essentially no surface spread of flame and no flash-over (vapor ignition) tendency; further diminishment in ignitability is achieved with aluminum foil or other facings. At fire temperatures it is converted to carbon without melting or generating significant amounts of combustible gases. It emits negligible amounts of smoke and significantly diminishes obnoxious gas evolution during the critical phase of fire. It has a markedly lower calorific value compared with all known foamed plastics and shows remarkably high burn-through resistance for its weight (80).

Rigid cellular isocyanate-modified foams with improved friability and flame-retardance properties have been prepared (81–82) by catalytically condensing an organic polyisocyanate in the presence of a compound that promotes carbodiimide linkages, a trimerization catalyst, and a furfuryl alcohol polymer (see Isocyanates, organic).

Diminishment in the friability of polyisocyanurate foams with retention or improvement in the flame retardance and char-forming properties has been reported by incorporating furfuryl alcohol (83).

Polyurethane cellular products modified with furfuryl alcohol polymer have also been reported to yield structurally stable char on carbonization. Without the furfuryl alcohol polymer a very weak char is obtained (84).

Cements and Mortars. One of the original applications for furan resins was in the jointing of bricks and masonry. Furan resins have been extensively used for formulating mortars, grouts, and "setting beds" for brick linings in structures exposed to corrosive media, especially concentrated acids, and for setting tile in floors exposed to alkaline cleaning solutions or other corrosive chemical media (see Chemical grouts). Mortars and grouts for these applications are usually formulated as two-package systems and mixed just prior to use. Silica fillers are satisfactory for many applications, but carbon flour (ground coke or other nongraphitic, nonactivated carbon) is used where resistance to hydrogen fluoride, fluoride salts, or hot, concentrated alkaline cleaning solutions is required.

High Carbon Yield. Furfuryl alcohol (FA) and/or furfural are reactive monomers and effective in producing high carbon yield; FA requires acidic catalysis and furfural basic catalysis. Mixtures of FA and furfural are generally catalyzed with acid although some systems may be catalyzed with base.

FA alone or in combination with other cross-linkers such as phenolic resins, chemical by-products and pitch, catalyzed with acid, gives carbon yields of 35–56%. Furfural together with cyclohexanone, pitch, or phenolic resins gives, under acid catalysis, yields of 35–55% carbon; under basic catalysis yields of 5–50% are achieved. Furfurylideneacetone resins (**22–23**), catalyzed by acid or base, give carbon yields of 48–56 and 30–35%, respectively (85) (see Carbon).

Anisotropic and isotropic carbon are produced from furfural-modified systems; glassy carbon is produced primarily from FA or BHMF resins (85).

Toxicology. Furfuryl alcohol, classed as moderately toxic, is less toxic than furfural. Like furfural, this compound shows a tendency to penetrate intact skin (rabbit test). See Table 3 for specific toxicological information.

Furan

Physical Properties. Furan, a colorless liquid with strong ethereal odor, is both low-boiling and highly flammable. It is miscible with most common organic solvents but only slightly soluble in water. The physical properties of furan are listed in Table 1.

Chemical Properties. Furan is a heat-stable compound although, at 670°C in the absence of catalyst, or at 360°C in the presence of nickel, it decomposes to form a mixture consisting mainly of carbon monoxide, hydrogen, and hydrocarbons (86). Substitution and addition reactions can be effected under controlled conditions, with reaction occurring first in the 2,5 positions.

Gas-phase reaction of ammonia (87) or primary amines (88) with furan over alumina at 400–450°C gives rise to pyrrole or *N*-substituted pyrroles in low yield. Another useful furan reaction is acylation under Friedel-Crafts (qv) conditions.

Catalytic hydrogenation of furan to tetrahydrofuran is accomplished in either liquid or vapor phase. Hydrogenation is essentially quantitative over nickel catalysts but is generally accompanied by hydrogenolysis over the noble metals. Furan can be catalytically oxidized in the vapor phase with oxygen-containing gases to maleic anhydride (89).

Temperature-resistant aromatic polymers have been synthesized from maleic anhydride double adducts of bis-furans by dehydration to the corresponding aromatic dianhydride and subsequent reaction with aromatic di- or tetraamines (90). The role of the furan derivative is purely synthetic because literally no trace of furan ring structure is found in the final products (see Heat-resistant polymers).

Furan reacts with carbon atoms via a novel rearrangement which does not involve deoxygenation. A recent investigation (91) reports the reaction of carbon atoms from tetrazylazide with furan to form *cis*-2-penten-4-ynal (**30**).

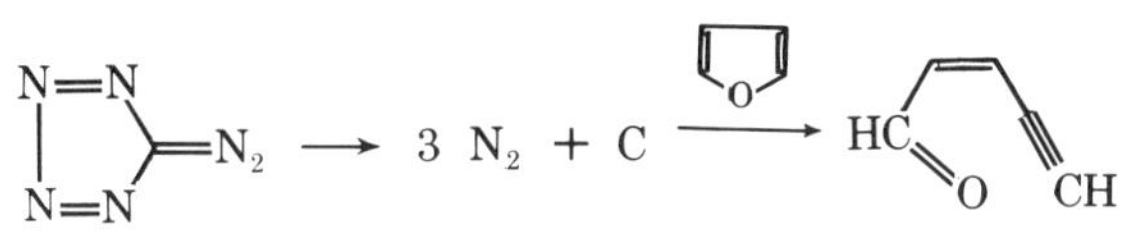

(**30**)

Furan and maleic anhydride undergo the Diels-Alder reaction to form the tricyclic 1:1 adduct (**9**). Other strong dienophiles also add to furan (92). Although both endo and exo isomers are formed initially, the former rapidly isomerizes to the latter in solution, even at room temperature. The existence of a charge–transfer complex in the system has been demonstrated (93–94).

The radical-catalyzed polymerization of furan and maleic anhydride has been reported to yield a 1:1 furan–maleic anhydride copolymer (93,95). The structure of the equimolar product, as shown by nmr analyses, is that of an unsaturated alternating copolymer (**26**) arising through homopolymerization of the excited donor–acceptor complex (95–96).

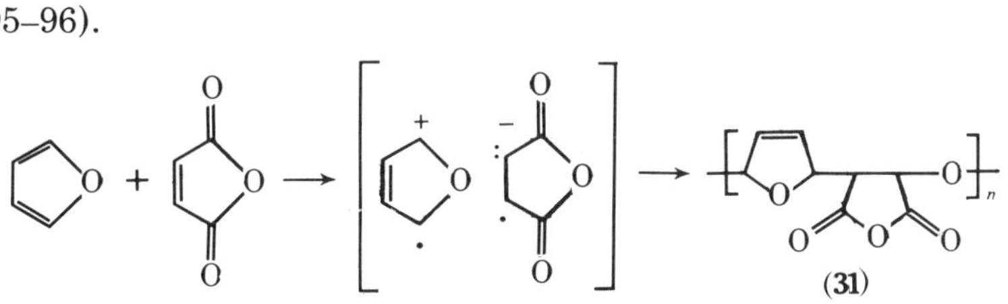

(**31**)

Manufacture. Furan is produced commercially by decarbonylation of furfural. Noncatalytic pyrolysis of furfural at 700°C, in a quartz reactor with vapor residence time of 5 s, generates low yields of furan.

Vapor-phase contact of furfural with steam over catalysts such as lime (97), or a mixed chromite of zinc and manganese (98) provides high yields of furan.

Furfural passed over nickel or cobalt contact surfaces at 280°C in the presence of hydrogen yields furan and carbon monoxide (99–101). Inclusion of hydrogen with the feed vapors increases both the yield of furan and the effective life of the catalyst. 2-Methylfuran [*534-22-5*], *n*-butyraldehyde, and propylene are by-products.

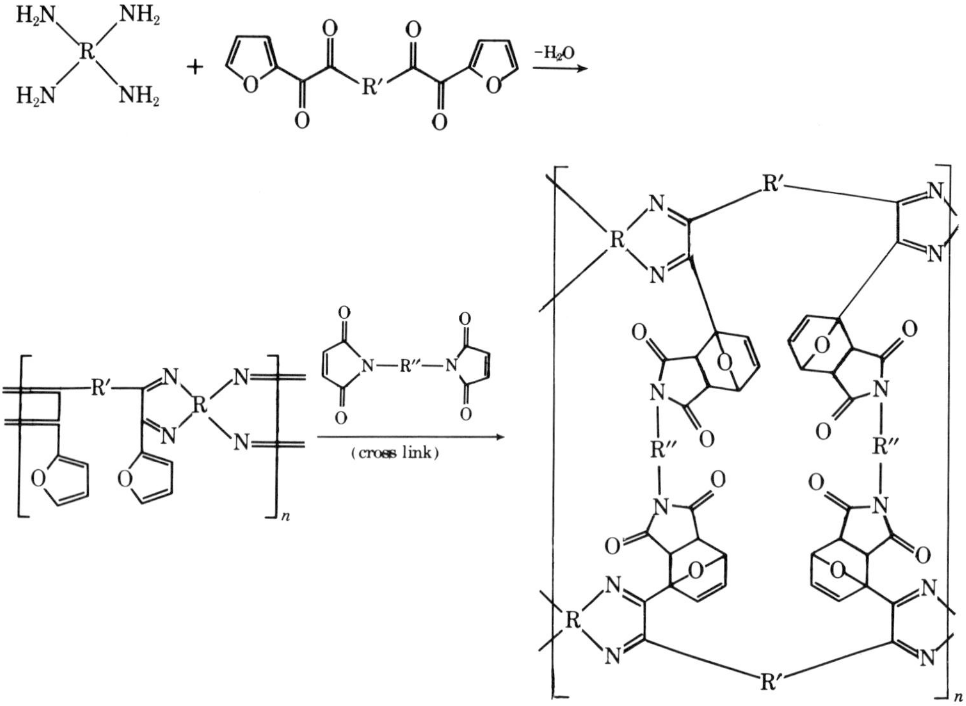

(**32**)
[*3006-93-7*]

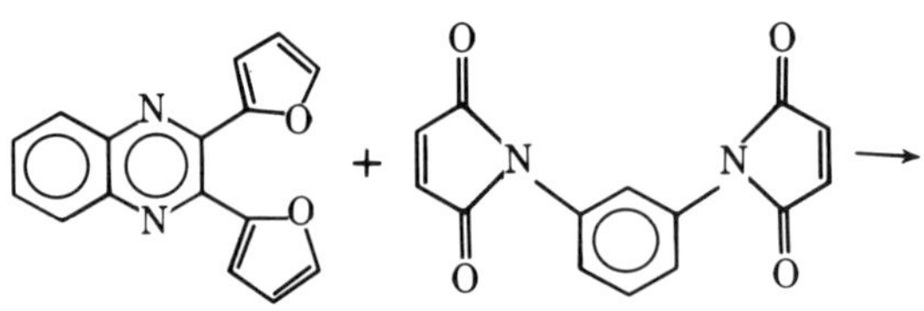

[*57490-73-0*]

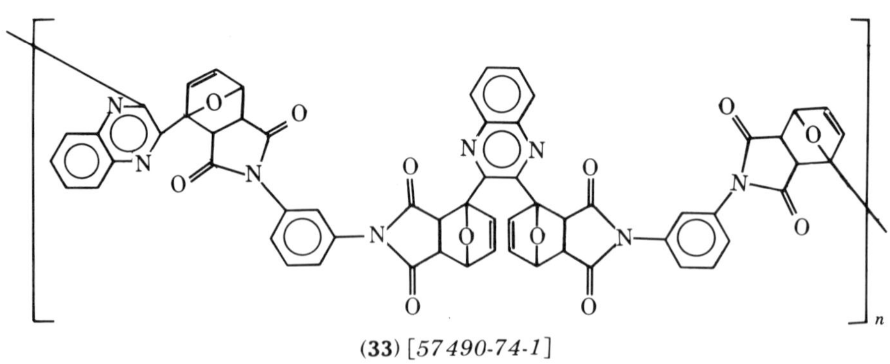

(**33**) [*57490-74-1*]

Other decarbonylation patents (102–103) describe the contact of furfural with a basic salt of an alkali metal in the presence of a noble metal catalyst at elevated temperatures. Liquid-phase decarbonylation of furfural in the presence of a noble metal catalyst and a basic salt of an alkali metal is also described (104).

A high temperature catalytic process for producing furan from butadiene (qv) has been described (105) using a molybdenum–bismuth catalyst with reported high selectivity. More recently (106), the coproduction of furan and maleic anhydride by oxidation of butadiene has been reported. The industrial significance of these methods has not been established. Furan has also been prepared by decarboxylation of 2-furoic acid.

Furan should be kept from heat and flame because of its low boiling point, low flash point, and high flammability. Unstabilized furan slowly forms an unstable peroxide on exposure to the air and, therefore, care should be taken in distilling furan to dryness.

The selling price of furan (March 1980) was \$2.09/kg.

Uses. Furan is utilized as a chemical building block in the production of other industrial chemicals for use as pharmaceuticals, herbicides, stabilizers, and fine chemicals.

Furan is readily hydrogenated in essentially quantitative yield over nickel catalysts; hence it is a source of commercial tetrahydrofuran (THF). Reaction with hydrogen sulfide over alumina at ca 400°C produces thiophene (107). Furan undergoes the Diels-Alder reaction with strong dienophiles. Thus cross-linked polymers have been obtained from reaction of a linear polyimidazole or polyquinoxaline, containing pendent furan groups, with an olefinic end-capped aromatic compound (108–109). These polyimidazole and polyquinoxaline polymers (**32, 33,** respectively) have alicyclic oxy cross-links repeated throughout the polymer where the furan constituent has undergone a Diels-Alder reaction with the aromatic olefin group. The alicyclic oxy linkage is subsequently aromatized by application of heat to produce high performance polymers suitable for use at temperatures up to 400°C.

Other linear polymers suitable for high temperature structural laminates are obtained by the reaction of bis(furfuryl)imide via Diels-Alder reactions (110–111). Stability in air up to 500°C has been reported for these polymers. The prepolymers have pendent phenyl substituents and are soluble in organic solvents, an important processing improvement.

Polymeric addition compounds (**34**) of furan and ethylene, where n is a whole number from 1 to 50, have been prepared (112) by addition of >2 mol furan to 1 mol ethylene at 120–250°C and high pressures followed by fractional extraction of the solid product formed. These polymers have interesting physical and chemical properties. They appear to be highly crystalline (x ray) and show good thermal stability which increases directly with molecular weight. The presence of ether bridges makes it possible to convert them into other compounds that retain the double hydrocarbon chain (ladder) structure and may bear various functional groups, such as hydroxyl or halogen, and isolated or conjugated double bonds (**35**),

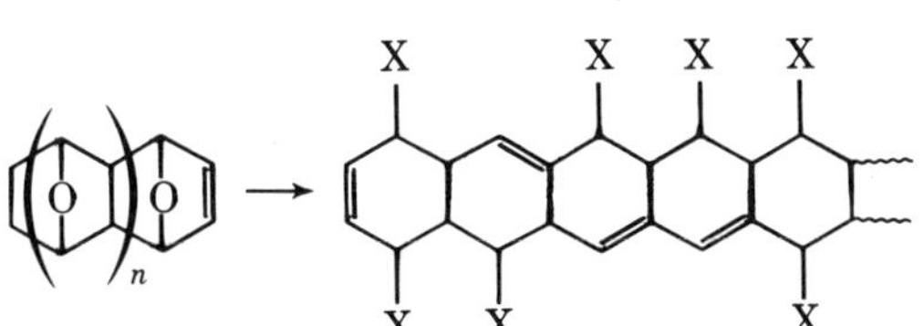

(**34**) (**35**)

where X is a functional group. Proposed applications for such compounds include usage as flocculants, pesticides, and pharmaceuticals.

Washing and cleaning agents containing salts of maleic acid–furan copolymers (113) form complexes with alkaline earth ions. These cleaning compositions do not contain phosphorus or nitrogen and find use in metal, foodstuff, and machine dishwashing products.

Toxicology. Furan is much more toxic than the other furan chemicals (114), and with its high volatility, constitutes a more serious hazard. Although no LD_{50} or LC_{50} results have been determined, the minimum fatal dose (rats) is 300 mg/kg. Recurring symptoms of toxicity were seen (dogs) at 39 ppm following nine six-hour exposures. Human exposures have been controlled routinely to less than 10 ppm (115). Because furan is one of the volatile constituents in tobacco smoke, its toxicity (and metabolism) in the lung is now being studied.

Tetrahydrofurfuryl Alcohol

Physical Properties. Tetrahydrofurfuryl alcohol (2-tetrahydrofuranmethanol) is a colorless, high-boiling liquid with a mild, pleasant odor. It is completely miscible with water and the common organic solvents. Tetrahydrofurfuryl alcohol is a moderately hydrogen-bonded solvent, essentially nontoxic, biodegradable, and with a low photochemical oxidation potential. The more important physical properties of tetrahydrofurfuryl alcohol are listed in Table 1.

Chemical Properties. Without inhibitors, tetrahydrofurfuryl alcohol is susceptible to autoxidation but remains colorless and with low carbonyl buildup after protracted periods of storage if a stabilizer such as Polygard (Uniroyal Chemical; EMIC) is added. Peroxide accumulation is low and readily dischargeable on redistillation.

The reactions of tetrahydrofurfuryl alcohol are characteristic of its bifunctional structure involving the carbinol and/or ether moieties. As a primary alcohol, it undergoes normal displacement or condensation reactions affording new functional groups (eg, halides, esters, alkoxylates, ethers, glycidyl ethers, cyanoethyl ethers, amines, etc). As a cyclic ether, it is susceptible to hydrolytic or hydrogenolytic cleavage to a variety of open-chain compounds, some of which may be recyclized to different heterocyclic species.

All the common monobasic (116) and dibasic esters (117) of tetrahydrofurfuryl alcohol have been prepared by conventional techniques; the dibasic esters exhibit distinctive properties as primary and secondary plasticizers for vinyl polymers. Tetrahydrofurfuryl acrylate [*2399-48-6*], a specialty monomer, has been produced by carbonylation (nickel carbonyl and acetylene) of the alcohol (118) as well as by ester interchange (119) (see Acrylic acid).

Tetrahydrofurfuryl alcohol reacts with ammonia to give a variety of nitrogen-containing compounds depending on the conditions employed: over a barium hydroxide-promoted skeletal nickel–aluminum catalyst, 2-tetrahydrofurfurylamine [*4795-29-3*] is produced (120); with palladium on alumina catalyst in the vapor phase (250–300°C), pyridine is the principal product (121); and at low pressure and 200°C over a reduced nickel catalyst, piperidine is obtained in good yield (122).

Tetrahydrofurfuryl alcohol is oxidized to 2-tetrahydrofurfural [*7681-84-7*] in good yield by passage, with oxygen, over silver gauze at 500°C (123). Reaction of the aldehyde with an alkali metal cyanide and ammonium carbonate gives 5-(2-tetrahydrofurfuryl)hydantoin [72918-25-3] (**36**), a useful intermediate for the production of lysine (124).

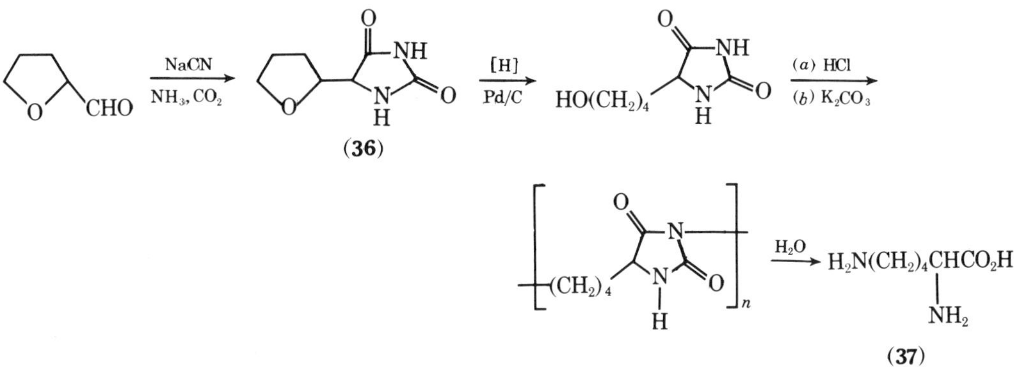

Tetrahydrofurfuryl alcohol is surprisingly resistant to hydrogenolysis; under vigorous conditions, however, cleavage of the ring or side chain occurs (125–130).

When tetrahydrofurfuryl alcohol is passed over a variety of alumina-containing catalysts at temperatures ranging from 200–500°C, it undergoes dehydration and ring expansion to 2,3-dihydropyran, a major chemical intermediate for 1,5-difunctional open-chain aliphatic compounds (87,131–133). The analogous 2,3-dihydrothiopyran is similarly prepared by reaction of tetrahydrofurfuryl alcohol with hydrogen sulfide over alumina at 300°C (134–135).

Manufacture. Tetrahydrofurfuryl alcohol is produced commercially by the hydrogenation of furfuryl alcohol.

Normal precautions for chemicals of mild toxicity are applicable to the safe handling and storage of commercial tetrahydrofurfuryl alcohol.

Discoloration in storage rarely occurs if the proper precautions are observed.

The selling price (March 1980) of THFA (Quaker Oats) was $1.34/kg.

Uses. Tetrahydrofurfuryl alcohol is of interest in chemical and related industries where low toxicity and minimal environmental impact are important (12). For many years THFA has been used as a specialty organic solvent. The fastest growing applications are in formulations for crop sprays, cleaners, paint strippers, water-based paints, and the dyeing and finishing of textiles and leathers.

A major use of THFA is as an ingredient in proprietary stripping formulations. These formulations are used broadly in the automotive industry to remove protective coatings, paint, and grime prior to final painting and finishing. Stripping formulations containing THFA are also available for the lifting and removal of epoxy coatings.

Because tetrahydrofurfuryl alcohol is virtually colorless, it is used in lacquer formulations for all colors as well as water-white clear products. More specifically, THFA is a wetting dispersant for most pigments. It has a high boiling point, high toluene dilution ratio, and good miscibility with oils, eg, linseed and soya, and is an excellent solvent for a wide range of resins.

Tetrahydrofurfuryl alcohol is used as solvent and carrier for industrial and commercial cleaning formulations. Product applications include the commercial cleaning of aircraft, ships, and trucks as well as the industrial cleaning requirements for machine tools and dies.

The solvent properties of tetrahydrofurfuryl alcohol and its low rate of evaporation are utilized in biocides and pesticides where it serves as solvent/carrier. THFA has been cleared by the EPA for use in formulations sprayed on growing food crops. The fact that THFA is totally miscible with water and completely biodegradable enhances its use in agricultural sprays used for pest control, weed control, and growth retardation.

THFA also has been cleared for use in California under Rule 66. It meets the rigid specifications for uv degradation in environmentally sensitive areas. In addition to domestic use, a substantial amount of tetrahydrofurfuryl alcohol is used as a solvent for agricultural chemicals in Europe.

THFA is a solvent and coupling agent for a phosphate-type insecticide used to control the gypsy moth. Esters of THFA are used in preparations employed as insect repellents. THFA is also used as a solvent/carrier for an EPA-approved paper slimicide formulation. In this application, the exceptional solvent action of THFA prevents separation of the active ingredient, especially at low operating temperatures.

Tetrahydrofurfuryl acrylate and methacrylate [*2455-24-5*], reactive unsaturated monomers, are readily polymerized and easily cross-linked by exposure to heat, peroxide catalysts, or uv radiation.

Tetrahydrofurfuryl acrylate is used as a co-reactive viscosity depressant for vinyl-terminated epoxy systems. It is also used in the formulation of uv-curable printing inks, coatings, paints, and adhesives. Tetrahydrofurfuryl methacrylate serves as a reactive curing agent in the peroxide-catalyzed production of nitrile rubber. The monomer functions as plasticizer during processing and polymerizes during cure to yield hard vulcanizates useful in electrical cable coatings.

Tetrahydrofurfuryl alcohol is used in elastomer production. As a catalyst solvent, it finds application in the manufacture of chlorohydrin rubber. Additionally, THFA is used as a catalyst solvent–activator and reactive diluent in epoxy formulations for a variety of applications. Where exceptional moisture resistance is needed, as for outdoor applications, furfuryl alcohol (FA) is used jointly with THFA in epoxy adhesive formulations.

Esters of tetrahydrofurfuryl alcohol, especially tetrahydrofurfuryl oleate [*5420-17-7*], are used as plasticizers (qv). The oleate is a high-boiling, stable, light-colored plasticizer that is compatible with many cellulose derivatives, poly(vinyl chloride), polystyrene, polyacrylates, synthetic and natural resins, rubber, and chlorinated derivatives. It improves the low temperature flexibility of vinyl films, provides internal lubrication, and imparts excellent draping properties to films and sheeting.

In dyeing applications, tetrahydrofurfuryl alcohol permits higher dye concentrations and wider latitude in solution temperature because of its high solvency for dyes. Through swelling action on synthetic fibers and polarity with respect to leather, THFA facilitates deeper penetration of dye into the substrate circumventing mere surface tinting. As a dye leveling agent, tetrahydrofurfuryl alcohol aids in the attainment of evenness in color shade.

Toxicology. Completely saturated, tetrahydrofurfuryl alcohol shows low toxicity. It is slightly irritating to skin (rabbit) and mildly irritating to the eye (rabbit). In a 90-d feeding test (dogs), the no-effect level occurred at a dietary concentration of 800 parts per million. Other toxicological data are given in Table 3.

2,5-Bis(hydroxymethyl)furan

Physical Properties. 2,5-Bis(hydroxymethyl)furan (BHMF) [*1883-75-6*] is a colorless, crystalline monomer. Properties are listed in Table 5.

Table 5. Physical Properties of 2,5-Bis(hydroxymethyl) furan

Property	Value
molecular weight	128.13
boiling point at 1.1 kPa (8.3 mm Hg), °C	162
melting point, °C	76.5–77
solubility in	
water, ethanol, acetone, pyridine, tetrahydrofuran	soluble
hexane, toluene, methylene chloride	insoluble

Chemical Properties. BHMF monomer behaves as a diol and undergoes esterification (base-catalyzed transesterification), alkoxylation, glycidyl ether formation, cyanoethylation, etherification, urethane formation, carbonate formation, etc (136–139).

Resinification occurs in the presence of acid catalysts. Resins (**29**) produced are difunctional in hydroxymethyl moiety. These resins are characterized by enhanced reactivity compared to acid-catalyzed furfuryl alcohol resins. They convert to highly cross-linked thermosetting polymers that demonstrate unique characteristics including high char yield on carbonization, low smoke evolution, low flame-spread, corrosion resistance, and superior high temperature stability.

Manufacture. Monomeric 2,5-bis(hydroxymethyl)furan and its oligomeric resins are produced from furfuryl alcohol and formaldehyde. The monomer has also been prepared by chemical or catalytic reduction of 5-hydroxymethylfurfural (140–141).

Uses. 2,5-Bis(hydroxymethyl)furan may be used to prepare linear polymers (142) that can subsequently be cross-linked under a variety of conditions as, eg, acid catalysis, heat, Diels-Alder condensation, etc.

Linear polyesters, polyurethanes, and polyamides containing furan rings in the polymer backbone have been made using BHMF and 2,5-furandicarboxylic acid [*3238-40-2*] (136–139).

BHMF and its saturated derivative, 2,5-bis(hydroxymethyl)tetrahydrofuran, are used as diol extenders in cast polyurethane elastomers and as components of epoxy resin systems (143).

Oligomers of BHMF are used in the foundry industry as warm box resins (65) (see Furfuryl Alcohol, Uses, Metal-Casting Cores and Molds, above). Because BHMF oligomers contain two or more reactive hydroxymethyl groups on the polymer chain, they are especially useful in the modification of urethane and isocyanurate foams to provide low friability, high compressive strength, good insulating properties, and as determined by the ASTM E 84 Tunnel Test, low flame spread and low smoke emission.

Toxicology. 2,5-Bis(hydroxymethyl)furan exhibits low toxicity. The oral LD_{50} is 2500–5000 mg/kg (male rats) and 5000–10,000 mg/kg (female rats). Skin and eye irritation (rabbits) are in the slight–mild ranges. In recent tests, no mutagenic effects were detected.

BIBLIOGRAPHY

"Furfural and Other Furan Compounds" in *ECT* 1st ed., Vol. 6, pp. 995–1008, by H. R. Duffey, The Quaker Oats Company, and H. J. Barrett, E. I. du Pont de Nemours & Co., Inc.; "Furfural and Other Furan Compounds" in *ECT* 2nd ed., Vol. 10, pp. 237–250, by A. P. Dunlop, The Quaker Oats Company.

1. A. P. Dunlop and F. N. Peters, *The Furans,* ACS Monograph 119, Reinhold Publishing Corp., New York, 1953.
2. M. J. Cook, A. R. Katritzky, and P. Linda, *Adv. Heterocycl. Chem.* **17,** 255 (1974).
3. V. Schomaker and L. Pauling, *J. Am. Chem. Soc.* **61,** 1769 (1939).
4. A. Albert, *Heterocyclic Chemistry: Introduction,* 2nd ed., Oxford University Press, New York, 1968.
5. L. Pauling, *The Nature of the Chemical Bond and the Structure of Molecules and Crystals,* 2nd ed., Cornell University Press, Ithaca, N.Y., 1940.
6. G. W. Wheland, *The Theory of Resonance and Its Application to Organic Chemistry,* John Wiley & Sons, Inc., New York, 1944.
7. M. V. Sargent and T. M. Cresp "Furans" in D. Barton and W. D. Ollis, eds., *Comprehensive Organic Chemistry; The Synthesis and Reaction of Organic Compounds,* Vol. 4, Pergamon Press Ltd., Oxford, Eng., 1979, pp. 693–744.
8. K. J. Siegfried "Furan Polymers" in N. M. Bikales, ed., *Encyclopedia of Polymer Science and Technology,* Vol. 7, John Wiley & Sons, Inc., New York, 1967, pp. 432–445.
9. A. Gandini, *Adv. Polym. Sci.* **25,** 47 (1977).
10. N. Clauson-Kaas, F. Limborg, and J. Fakstorp, *Acta Chem. Scand.* **2,** 109 (1950).
11. *QO Furfural,* Bulletin 203D, 1980; *QO Furfuryl Alcohol,* Bulletin 205C, 1980; *QO Furan,* Technical Bulletin No. 149, 1965, The Quaker Oats Company, Chicago, Ill.
12. *QO Tetrahydrofurfuryl Alcohol,* Bulletin 206, The Quaker Oats Company, Chicago, Ill.
13. A. J. Hammer, *Chem. Ind.,* 608 (1933).
14. F. Trimble, *Ind. Eng. Chem.* **33,** 660 (1941).
15. G. H. Mains, *Chem. Met. Eng.* **26,** 779 (1922).
16. R. G. Curtis and H. H. Hatt, *Aust. J. Sci. Res.* **A1,** 213 (1948).
17. A. P. Dunlop and F. N. Peters, Jr., *Ind. Eng. Chem.* **32,** 1639 (1940).
18. D. L. Williams and A. P. Dunlop, *Ind. Eng. Chem.* **40,** 239 (1948).
19. J. Z. Marcusson, *Z. Angew. Chem.* **32,** 113 (1919).
20. G. Illari, *Gazz. Chim. Ital.* **77,** 389 (1947).
21. Y. Nakamura and M. Saito, *Kogyo Kagaku Zasshi* **62,** 1173 (1959).
22. N. Galego, Ph.D. Thesis, University of Havana, 1975.
23. N. Galego and A. Gandini, *Rev. CENIC Cent. Nac. Invest. Cient. 1 Cienc. Fis.* **6,** 163 (1975); *Chem. Abstr.* **84,** 18211p (1976).
24. N. Galego and A. Gandini, *J. Polym. Sci. Polym. Chem. Ed.* **15,** 1027 (1977).
25. A. Gandini, J. M. Parsons, and R. A. Back, *Can. J. Chem.* **54,** 3095 (1976).
26. G. Natta, G. Mazzanti, G. F. Pregaglia, and G. Pozzi, *J. Polym. Sci.* **58,** 1201 (1962).
27. V. V. Girdyuk, Yu. K. Kirilenko, L. A. Vol'f, and A. I. Meos, *Zh. Prikl. Khim.* **39,** 2601 (1966).
28. I. V. Kamenskii, S. M. Filimonova, Ngo Van Lam, B. Ya. Eryshev, and N. B. Krovyakova, *Plast. Massy* (5), 5 (1973); *Chem. Abstr.* **79,** 79683n (1973).
29. V. V. Girdyuk, Yu. K. Kirilenko, L. A. Vol'f, and A. I. Meos, *Zh. Prikl. Khim.* **40,** 1386 (1967).
30. N. G. Shelkunov, I. B. Klimenko, V. V. Girdyuk, and L. A. Vol'f, *Khim. Geterotsikl. Soedin.* (5), 755 (1969); *Chem. Abstr.* **72,** 67586a (1970).
31. W. J. Chute and G. F. Wright, *J. Org. Chem.* **10,** 541 (1945).
32. H. Gilman and G. F. Wright, *J. Am. Chem. Soc.* **52,** 2550 (1930).
33. R. L. Whistler and E. L. Richards "Hemicelluloses" in W. Pigman and D. Horton, eds., *The Carbohydrates,* Vol. IIA, Academic Press, Inc., New York, 1970, pp. 447–467.
34. R. I. Browne, *Handbook of Sugar Analyses,* John Wiley & Sons, Inc., New York, 1912.
35. G. O. Aspinall, *Polysaccharides,* Pergamon Press Ltd., Oxford, Eng., 1970, pp. 103–115.
36. R. F. Bradley, *Furfural,* Chemical Economics Handbook, Menlo Park, Cal., Dec. 1978.
37. *Making and Marketing Furfural: Added Value for Agro-Industrial Wastes,* International Trade Center UNCTAD/GATT, Geneva, Switz., 1979.
38. J. E. Stone and M. J. Blundell, *Can. J. Res.* **28B,** 676 (1950).
39. *Chem. Mark. Rep.,* **214**(14), 16 (Oct. 2, 1978).
40. B. Gillespie, L. W. Manley, and C. J. DiPerna, *Chem. Technol.* **8,** 750 (Dec. 1978).

41. Jpn. Pat. 78 19,693 (June 22, 1978), M. Futamoto, U. Kawabe, and S. Hosoki (to Hitachi Ltd.).
42. *AID (Air Information Division) Report 61-19,* Armed Services Technical Information Agency, NTIS (National Technical Information Service), Springfield, Va., Feb. 1961.
43. *AID Report 61-5,* Armed Services Technical Information Agency, NTIS, Springfield, Va., Jan. 1961.
44. D. A. Isacescu, I. Gavat, C. Stoicescu, C. Vass, and I. Petrus, *Rev. Roum. Chim.* **10,** 219 (1965); *Chem. Abstr.* **63,** 16284 (1965).
45. I. V. Kamenskii, N. V. Ungurean, B. M. Kovarskaya, and I. V. Itinskii, *Plast. Massy* (12), 9 (1960); *Chem. Abstr.* **55,** 15985 (1961).
46. L. A. Woods and M. H. Seevers, unpublished reports to The Quaker Oats Company, 1953–1956.
47. *Code of Federal Regulations,* Vol. 29, CFR 1910.1000, Table Z-1, 1979.
48. *Criteria For a Recommended Standard: Occupational Exposure to Furfuryl Alcohol,* DHEW (NIOSH), *Publ. No. 79-142,* Government Printing Office, Washington, D.C., 1979.
49. K. H. Jacobson and co-workers, *Am. Ind. Hyg. Assoc. J.* **19,** 91 (1958).
50. *Registry of Toxic Effects of Chemical Substances,* Vol. II, NIOSH, 1977, p. 434.
51. A. P. Dunlop and F. N. Peters, Jr., *Ind. Eng. Chem.* **34,** 814 (1942).
52. J. B. Barr and S. B. Wallon, *J. Appl. Polym. Sci.* **15,** 1079 (1971).
53. A. K. A. Rathi and M. Chanda, *J. Appl. Polym. Sci.* **18,** 1541 (1974).
54. T. A. Krishnan and M. Chanda, *Angew. Makromol. Chem.* **43,** 145 (1975).
55. E. M. Wewerka, E. D. Loughran, and K. L. Walters, *J. Appl. Polym. Sci.* **15,** 1437 (1971).
56. R. H. Leitheiser, M. E. Londrigan, and C. A. Rude, *Proceedings ACS Meeting, April 2–6, 1979, Honolulu, Hawaii.*
57. R. H. Wojcik, *Ind. Eng. Chem.* **40,** 210 (1948).
58. U.S. Pat. 2,754,304 (July 10, 1956), S. Swadesh (to The Quaker Oats Company).
59. W. D. Wakefield, *Production* (Oct. 1976).
60. R. F. Frankenberg, *Foundry Met. Treat.,* 87 (Jan. 1979).
61. *Int. Mod. Foundry,* (2), 2 (1978).
62. A. Thomas and D. V. Allen, *Met. Australas.,* 8 (July 10, 1978).
63. W. H. Buell, *Foundry* **89**(2), 64 (1961).
64. A. Dorfmuller, Jr., *Foundry* **90**(4), 54 (1962).
65. R. H. Kottke and A. E. Bloomquist, *AFS Trans.* **86,** 215 (1978).
66. J. H. Schaum, *Mod. Cast.* **68,** 50 (Dec. 1978).
67. M. B. Launikitis, *Manag. Corros. Plast.* **Series 3,** 190 (1977).
68. R. H. Leitheiser, K. B. Bozer, and D. D. Watson, *Furan Resins for FRP Composites, National Association of Corrosion Engineers, Annual Meeting, San Francisco, Calif., Mar. 14–18, 1977.*
69. O. H. Fenner and J. E. Callahan, *Mater. Perform.* **16**(12), 37 (1977).
70. R. M. Webster, *National Association of Corrosion Engineers, Annual Meeting, San Francisco, Calif., Mar. 14–18, 1977.*
71. E. D. Edmisten, K. D. McGuire, and R. H. Currieo, *Symposium Sponsored by NACE, SPI, SPE, AIChE, TAPPI, and SPI Canada, New Orleans, Oct. 1977.*
72. G. Belanger, *30th Annual Conference, Reinforced Plastics/Composites Institute, SPI, Feb. 4–6, 1975.*
73. R. H. Leitheiser, M. E. Londrigan, D. W. Akerberg, and K. B. Bozer, *Effectiveness of Flame Retardant Additives in Furan Resins, European Conference on Flammability and Fire Retardants, Copenhagen, Denmark, July 13–14, 1978.*
74. K. B. Bozer, M. E. Londrigan, and D. W. Akerberg, *Effectiveness of Flame Retardant Additives in Furan Resins, 33rd Annual Conference of the Reinforced Plastics/Composites Institute, SPI, Section 6-B, Washington, D.C., Feb. 7–10, 1978.*
75. K. B. Bozer, M. E. Londrigan, D. W. Akerberg, and B. R. Bogner, *Effectiveness of Flame Retardant Resins—Large Scale Test Results, 34th Annual Technical Conference, Reinforced Plastics/Composites Institute, SPI, New Orleans, La., Jan. 29–Feb. 2, 1979.*
76. K. B. Bozer, M. E. Londrigan, R. H. Leitheiser, and D. W. Akerberg, *Flame Retardant Furan Resins, National Fire Protection Association Meeting, Montreal, Quebec, Can., Nov. 13–16, 1978.*
77. R. H. Leitheiser, *Flame Retardant Furan Resins, Flame-Retardant Chemicals Association Meeting, San Francisco, Calif., Oct. 29–Nov. 1, 1978.*
78. R. H. Leitheiser, M. E. Londrigan, D. W. Akerberg, and K. B. Bozer, *Fire Prevention Science and Technology* **22,** 21 (1979).
79. U.S. Pat. 4,016,111 (Apr. 5, 1977), and Ger. Offen. 2,514,542 (Oct. 23, 1975), P. L. Wolff and J. A. G. Gent (to Koninlijke Emballage Industrie van Leer B. V.).

80. *The Comparative Fire Characteristics of Cellular Plastics and NF_2 Low Hazard Thermal Insulation, Technical Data No. 1,* Van Leer Ltd., Environmental Products Division, UK, 1979.
81. U.S. Pat. 4,029,611 (June 14, 1977), M. Cenker, T. Narayan, and J. T. Patton, Jr. (to BASF Wyandotte Corp.).
82. U.S. Pat. 3,922,238 (Nov. 25, 1975), and Ger. Offen. 2,523,029 (Nov. 27, 1975), T. L. Narayan and M. Cenker (to BASF Wyandotte Corp.).
83. Ger. Offen. 2,546,183 (July 1, 1976), and Belg. Pat. 834,468 (Aug. 20, 1976), M. Cenker and T. L. Narayan (to BASF Wyandotte Corp.).
84. A. K. Thoeni, G. K. Baker, and C. H. Smith, *J. Cell. Plastics* **7,** 294 (1971).
85. *Impregnating Formulations Based on Furan Chemicals,* Technical Bulletin No. 190, Chemicals Division, The Quaker Oats Company, Chicago, Ill., 1979.
86. C. D. Hurd and A. R. Goldsby, *J. Am. Chem. Soc.* **54,** 2558 (1932).
87. O. W. Cass, *Chemurgic Dig.* **8**(6), 6 (1949).
88. Yu. K. Yur'ev, *J. Gen. Chem. USSR* **8,** 1934 (1938).
89. N. A. Milas and W. L. Walsh, *J. Am. Chem. Soc.* **57,** 1389 (1935).
90. A. A. Berlin, B. I. Liogon'kii, B. I. Zapadinskii, E. A. Kazantseva and A. O. Stankevich, *Vysokomol. Soedin. Ser. A,* **18,** 926 (1976); *Chem. Abstr.* **85,** 94712p (1976).
91. S. F. Dyer and P. B. Shevlin, *J. Am. Chem. Soc.* **101,** 1303 (1979).
92. A. Norton, *Chem. Rev.* **31,** 319 (1942).
93. G. B. Butler, J. T. Badgett, and M. Sharabash, *J. Macromol. Sci. Chem.* **4**(1), 51 (1970).
94. B. Kamo, I. Morita, S. Horie, and S. Furusawa, *Polym. J.* **6**(2), 121 (1974).
95. N. G. Gaylord, S. Maiti, B. K. Patnaik, and A. Takahashi, *J. Macromol. Sci. Chem.* **6,** 1459 (1972).
96. N. G. Gaylord, M. Martan, and A. B. Deshpande, *J. Polym. Sci. Polym. Chem. Ed.* **16,** 1527 (1978).
97. O. W. Cass, *Chem. Ind.* **60,** 612 (1947).
98. U.S. Pat. 2,374,149 (Apr. 17, 1945), G. M. Whitman (to E. I. du Pont de Nemours & Co., Inc.).
99. U.S. Pat. 2,846,449 (Aug. 5, 1958), W. H. Banford and M. M. Manes (to E. I. du Pont de Nemours & Co., Inc.).
100. U.S. Pat. 3,021,342 (Feb. 13, 1962), D. G. Manly (to The Quaker Oats Company).
101. R. Paul, *Compt. Rend.* **206,** 1028 (1938).
102. U.S. Pat. 3,223,714 (Dec. 14, 1965), D. G. Manly and J. P. O'Halloran (to The Quaker Oats Company).
103. U.S. Pat. 3,007,941 (Nov. 7, 1961), H. B. Copelin and D. I. Garnett (to E. I. du Pont de Nemours & Co., Inc.).
104. U.S. Pat. 3,257,417 (June 21, 1966), A. P. Dunlop and G. W. Huffman (to The Quaker Oats Company).
105. U.S. Pat. 3,238,225 (Mar. 1, 1966), W. F. Brill and A. J. Besozzi (to Petro-Tex Chemical Corp.).
106. Jpn. Kokai 73 38,292 (June 5, 1973), H. Inoue, K. Mizutani, and H. Ito (to Toa Gosei Chemical Industry Co., Ltd.).
107. Yu. K. Yur'ev and V. A. Tranova, *J. Gen. Chem. USSR* **10,** 31 (1940).
108. U.S. Pat. 3,904,584 (Sept. 9, 1975), R. J. Jones and M. K. O'Rell (to TRW Inc.).
109. U.S. Pat. 3,905,941 (Sept. 16, 1975), R. J. Jones (to TRW Inc.).
110. U.S. Pat. 3,927,027 (Dec. 16, 1975), R. J. Jones and H. E. Green (to TRW Inc.).
111. U.S. Pat. 3,951,902 (Apr. 20, 1976), R. J. Jones, H. N. Cassey, and H. E. Green (to TRW Inc.).
112. Brit. Pat. 1,402,705 (Aug. 13, 1975), and Ger. Offen. 2,420 875 (Nov. 7, 1974), G. Stigliani, R. Manetti, A. Carbonaro, and G. Dall'Asta (to Montedison S.p.A).
113. U.S. Pat. 3,850,832 (Nov. 26, 1974), and Ger. Offen. 2,162,673 (July 19, 1973), B. Wegemund, H. J. Lehmann, and E. Schmadel (to Henkel & Cie GmbH).
114. E. M. Koch and M. H. Cahan, *J. Pharmacol.* **26,** 281 (1925).
115. *Furan Products,* DuPont Technical Information Bulletin, FC 1-560, ca 1959.
116. R. T. Wragg, *J. Chem. Soc.,* 7162 (1965).
117. L. H. Brown and J. W. Hill, *J. Chem. Eng. Data* **5**(1), 56 (1960).
118. W. Reppe and co-workers, *Ann.* **582,** 1 (1953).
119. Yu. V. Mironov, V. V. Korshak, I. V. Kamenskii and A. P. Yusova, *Tr. Mosk. Khim. Tekhnol. Inst.* (52), 181 (1967); *Chem. Abstr.* **68,** 68798a (1968).
120. U.S. Pat. 2,636,902 (Apr. 28, 1953), A. W. C. Taylor, P. Davies, and P. W. Reynolds (to Imperial Chemical Industries Ltd.).
121. U.S. Pat. 3,238,214 (Mar. 1, 1966), D. G. Manly, J. P. O'Halloran, and F. J. Rice (to The Quaker Oats Company).

122. U.S. Pat. 3,163,652 (Dec. 29, 1964), D. G. Manly, J. P. O'Halloran, and F. J. Rice (to The Quaker Oats Company).
123. Brit. Pat. 593,617 (Oct. 21, 1947), J. G. M. Bremner and R. R. Coats (to Imperial Chemical Industries Ltd.).
124. U.S. Pat. 2,557,904 (June 19, 1951), E. C. Britton, Y. J. Dickert, and H. C. White (to Dow Chemical Company).
125. E. P. Goodings and C. L. Wilson, *J. Am. Chem. Soc.* **73,** 4801 (1951).
126. Brit. Pat. 571,418 (Aug. 23, 1945), C. L. Wilson and W. H. Bagnall (to Revertex Ltd.).
127. C. L. Wilson, *J. Chem. Soc.,* 52 (1945).
128. E. P. Goodings and C. L. Wilson, *J. Am. Chem. Soc.* **73,** 4798 (1951).
129. U.S. Pat. 2,768,978 (Oct. 31, 1956), J. A. Robertson (to E. I. du Pont de Nemours & Co., Inc.).
130. S. Landa and J. Mostecky, *Collect. Czech. Chem. Commun.* **20,** 430 (1955).
131. R. Paul, *Bull. Soc. Chim. France* **14,** 158 (1947).
132. O. W. Cass, *Ind. Eng. Chem.* **40,** 21C (1948).
133. U.S. Pat. 3,518,281 (June 30, 1970), D. C. Holtman (to Emery Industries, Inc.).
134. R. F. Naylor, *J. Chem. Soc.,* 2749 (1949).
135. Yu. K. Yur'ev and E. G. Vendel'shtein, *J. Gen. Chem. USSR* **22,** 751 (1952).
136. J. A. Moore and J. E. Kelly, *Macromolecules* **11,** 568 (1978).
137. J. A. Moore and J. E. Kelly, *J. Polym. Sci. Polym. Chem. Ed.* **16,** 2407 (1978).
138. Jpn. Pat. 64 2096 (Feb. 29, 1964), and 64 24,730 (Nov. 7, 1964), M. Izumi, S. Matsumura, Y. Hachihama, and T. Shono (to Sumitomo Electric Industries, Ltd.).
139. Jpn. Kokai 73 40,763 (June 15, 1973), I. Matsumoto, K. Nakagawa, and K. Horiuchi (to Banyu Pharmaceutical Co., Ltd.).
140. F. H. Newth and L. F. Wiggins, *Research(London)* **3,** Suppl. 50 (1950).
141. U.S. Pat. 3,083,236 (Mar. 26, 1963), U. Torleif, J. D. Garber, and R. E. Jones (to Merck & Co., Inc.).
142. *QO Farez B-260, Technical Bulletin No. 186,* The Quaker Oats Company, Chicago, Ill., 1978.
143. *BHMF: 2,5-Bis(hydroxymethyl)furan, Technical Bulletin No. 194,* The Quaker Oats Company, Chicago, Ill., 1979.

General References

Refs. 1, 8–9.
M. V. Sargent and T. M. Cresp "Furans" in D. Barton and W. D. Ollis, eds., *Comprehensive Organic Chemistry: The Synthesis and Reactions of Organic Compounds,* Vol. 4, Pergamon Press Ltd., Oxford, Eng., 1979, pp. 693–744.

W. J. McKillip
E. Sherman
The Quaker Oats Company

FURFURAL AND OTHER FURAN COMPOUNDS. See Furan derivatives.

FURNACES, ELECTRIC

INTRODUCTION

The term, electric furnace, applies to all furnaces that use electrical energy as their sole source of heat. The definition distinguishes such apparatus from traditional fuel-fired furnaces (see Furnaces, fuel-fired) in which heat is produced directly by combustion of fossil fuels (eg, coal, oil, or gas). Electric furnaces are used mainly for heating solid materials to desired temperatures below their melting points for subsequent processing, or melting materials for subsequent casting into desired shapes, ie, electric heating furnaces or electric melting furnaces; the latter includes so-called holding furnaces, which store a molten charge received from separate melting furnaces. Important to a discussion of electric furnaces is the classification of the manner in which the electrical energy is converted into heat. Thus three distinct types of widely used industrial furnaces can be distinguished; they are: electric resistance furnaces (qv), electric arc furnaces (qv), and electric induction furnaces (qv).

The conversion of electrical energy into heat in each of the type of furnaces above is schematically illustrated in Figure 1. Common to all is a charge that is to be heated or melted, A, a refractory furnace lining, B, and an electric power supply, C. The refractory lining separates the hot furnace interior from the work area. It must withstand high temperatures and provide thermal insulation to conserve energy. The lining is often contained in a steel structure. The power supply is normally ac (dc is rarely used) at standard power line frequency (60 Hz in North America) and it usually includes a transformer so that the most suitable furnace voltage is obtained. However, certain induction furnaces require a higher frequency.

Electric Resistance Furnaces

In all electric resistance furnaces, the power supply is connected directly to an electric resistance in the interior of the furnace lining in which all electrical energy is converted to heat. Widely used and best known is the indirect-heat resistance furnace (Fig. 1**a**) (also known as electric resistor furnace). It resembles an electric toaster or kitchen oven. Resistors, D, are attached to the furnace roof or walls, or both, and are connected electrically to the power supply. Conventional metallic resistors serve to ca ≤1150°C; special resistor materials are needed for higher temperatures. Electrical energy is converted to heat in the resistors, and then the heat is transmitted to the charge by radiation and convection. This furnace type is suited for heating or melting, and there is no inherent dependence on the electrical conductivity of the charge; however, temperatures above 1700°C are not easily attained.

The other form of electric resistance furnace is the direct-heat type (Fig. 1**b**). Here the resistance is the charge, A, itself which may be a solid mass, an agglomeration of solid particles, or a pool of molten material. The power supply is connected to this

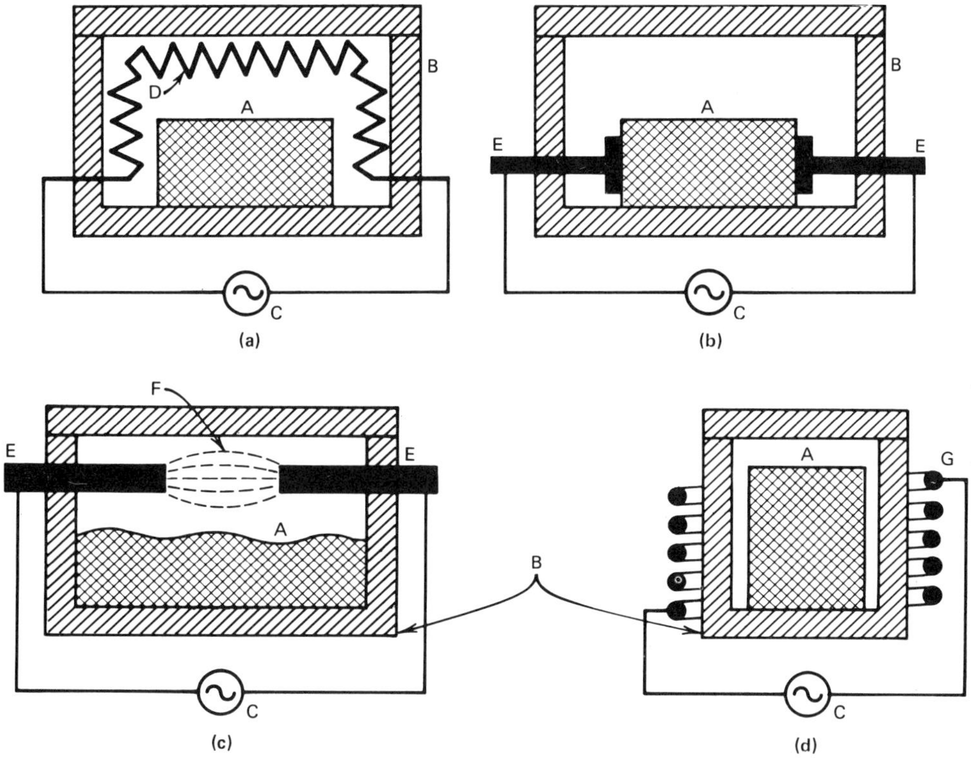

Figure 1. Main types of electric furnaces: (**a**) resistance furnace, indirect heat (resistor furnace); (**b**) resistance furnace, direct heat; (**c**) arc furnace; (**d**) induction furnace. A, charge to be heated or melted; B, refractory furnace lining; C, electric power supply; D, resistors; E, electrodes; F, electric arc; G, induction coil.

resistance through so-called electrodes, E, that are hot on the inside end and cold on the outside end. Use of direct-heat resistance furnaces is limited by the electrical conductivity of the charge and by the electrode system. It is well-suited for carbon, molten glass, and molten salts. It cannot be used for nonconducting charges and generally it is not practical for metals because of their high conductivity.

Electric Arc Furnaces

In electric arc furnaces (Fig. 1**c**), the conversion of electrical energy to heat occurs primarily within the electric arc, F, formed between spaced electrodes to which the power supply is connected. The arc has a high temperature (3700°C), and heat is transmitted to the charge primarily by radiation. A widely used electrode material is graphite, which can withstand high temperatures. Figure 1**c** illustrates the pure arc-furnace principle; however, in most industrial furnaces, the arcs are formed between one or more electrodes and an electrically conducting charge. Most arc furnaces are large. They are used primarily for melting steel, ferro alloys, and nonmetallic substances. Operating temperatures are 1000–3000°C.

Electric Induction Furnaces

In electric induction furnaces (Fig. **1d**), the conversion of electrical energy to heat takes place in the charge; however, no electrodes are used. Instead, the power supply is connected to a cold induction primary coil, G, on the exterior of the furnace lining. Electrical energy then is transmitted by transformer effect, causing a high secondary current to flow in the charge. Therefore, functioning of this furnace does not require the presence of any material other than the charge in the hot zone enclosed by the lining. This is a rather unique advantage. When necessary, extremely high temperatures can be attained. Induction furnaces require a charge of relatively high electrical conductivity. Thus they are best suited for heating and melting all metals and for heating graphite.

Economic Aspects

Selection of an industrial furnace for a given purpose usually involves deciding between electric and fuel-fired furnaces. Only electric furnaces are practical above 1700°C, but both types generally are considered for lower temperatures. Electric furnaces frequently have a higher capital (amortization) cost. Generally, there is not a decisive difference in energy cost as long as the cost of electricity is based on its generation from fossil fuels. However, in addition to amortization and energy, the following potentially significant furnace operating-cost factors should always be examined: operating labor cost, charge-material loss (oxidation and vaporization), furnace-related raw material cost differences, cost of rejects (effect of process control), environmental-control costs (investment and operation), and furnace-maintenance costs (including relining and electrodes). Electric furnaces often are selected in preference to fuel-fired furnaces because the former are characterized by a significantly lower total operating cost resulting from savings in the six areas outlined above. The same total operating cost analysis should be used when making the selection between various electric furnace types suitable for the same job.

The increasing importance of environmental considerations tends to favor selection of electric furnaces. Fuel-fired furnaces emit large volumes of hot products of combustion that contain objectionable gases and particles, which may include vapors or fumes from certain molten charges. The cost of treating these products of combustion to meet clean air standards can be very high, and it will increase as standards are tightened. In electric furnaces (particularly resistance and induction furnaces), air pollution control is required only where pollutants emanate from the charge, but cost is much lower in the absence of a large hot gas volume.

Electric furnaces have a much higher efficiency and, therefore, release considerably less heat into their surroundings, thereby minimizing the need to cool the work area. Induction furnaces are best in this respect because the charge is heated internally and there are no elements operating above charge temperature. Noise control also is increasingly important. Resistance furnaces are quiet. Coil vibration noise may be a problem in induction furnaces, but can be minimized by coil design. Arc furnaces and fuel-fired furnaces generate noise which may have to be controlled (see Noise pollution; Insulation, acoustic).

BIBLIOGRAPHY

"Furnaces, Electric" in *ECT* 1st ed., Vol. 7, pp. 1–23, by Victor Paschkis, Columbia University; "Furnaces, Electric" in *ECT* 2nd ed., Vol. 10, pp. 252–278, by Victor Paschkis, Columbia University.

General References

W. Trinks, *Industrial Furnaces, Fuels, Furnace Types and Furnace Equipment—Their Selection and Influence Upon Furnace Operation,* 4th ed., Vol. II, John Wiley & Sons, Inc., New York, 1967, 358 pp., emphasis placed on heating furnaces (fuel-fired and electric) rather than melting furnaces.

A. G. E. Robiette, *Electric Melting Practice,* John Wiley & Sons, Inc., New York, 1972, 412 pp., arc, induction, resistor, and special melting furnaces are discussed.

K. Kegel, ed., *Elektrowaerme, Theorie und Praxis,* Union Internationale d'Electrothermie, Verlag W. Girardet, Essen, FRG, 1974, 902 pp., a very complete and up-to-date reference for all electric furnaces.

C. F. Haecker, "Iron Melting and Pollution Control: An Opportunity to Progress," *Foundry* **97,** 56 (Aug. 1969), economic comparison: cupola, arc, and induction furnaces for iron foundry melting, including pollution control.

E. Mittmann, "The Effect of Metal Loss on the Economics of Aluminum Melting Furnaces," *Giesserei* **63**(1), 4 (1976), economic comparison: fuel-fired and electric aluminum melting furnaces, including metal loss.

Foundry Yearbook, 1977, Part III, Foundry Trade Journal, Redhill, Surrey, Eng., p. 254 (Table 21 is a detailed application chart for fuel-fired and electric foundry melting furnaces).

E. Doetsch and H. Doliwa, "Economical and Process Technology Aspects of Cast Iron Melting," *Elektrowaerme Int.* **37**(B3), B157 (1979), contains an economic comparison: fuel-fired and electric iron foundry melting furnaces.

MARIO TAMA
Ajax Magnethermic Corporation

ARC FURNACES

Arc furnaces used in electric melting, smelting, and electrochemical operations are of two basic designs: the indirect and the direct arc. The arc of the indirect-arc furnace is maintained between two electrodes and radiates heat to the charge. The arcs of the direct-arc furnace are maintained between the charge and the electrodes, making the charge a part of the electrical power circuits. Not only is heat radiated to the charge, but the charge is heated directly by the arc and the current passing through the charge.

Indirect-Arc Furnaces

Indirect-arc furnaces are used primarily in foundries for melting copper, copper alloys (eg, bronze, brass, etc), and other nonferrous metals having a low melting point (see Copper). They have also been used for producing molten iron and, occasionally, molten steel (see Iron; Steel). The typical indirect-arc furnace is a single-phase furnace utilizing two horizontally mounted graphite electrodes, each of which project into an end of a refractory-lined horizontally mounted cylindrical steel shell. One electrode is set manually and the other electrode is automatically adjusted to maintain the preset voltage and current of the arc. Each electrode is connected to a multivoltage tap transformer and reactor through flexible copper cables and bus bars.

After the furnace has been charged and the charge has been melted, the furnace is oscillated through an increasing angle until ninety percent of the refractory has been covered by molten metal on each oscillation. This intermittent metal covering of the refractories not only protects the refractories from the high temperature of the arc, but also cools the refractories by absorbing any excessive heat (1). This oscillating or rocking action improves refractory life, increases melting efficiency, and reduces thermal losses.

Direct-Arc Furnaces

Open-Arc Furnaces. Most of the open-arc three-phase furnaces (Fig. 1) are used in melting and refining operations for steel and iron. They are also used in melting operations for copper, slag, refractories, and other materials having melting points above 1000°C.

Today's standard furnace consists of: a refractory-lined steel shell to contain the charge and/or molten metal; a set of graphite electrodes; electrical equipment, bus bars, and flexible conductors to energize the electrode (see Fig. 2); equipment to regulate the position of the electrodes and thereby control the energy input; a means to slag the furnace and inspect the refractory; a method to tilt the furnace to empty it; and a means to allow the furnace to be recharged. Practically all furnace shells today are short vertical cylinders made of welded steel plate and are reinforced and water-

Figure 1. An open-arc furnace after pouring its "heat."

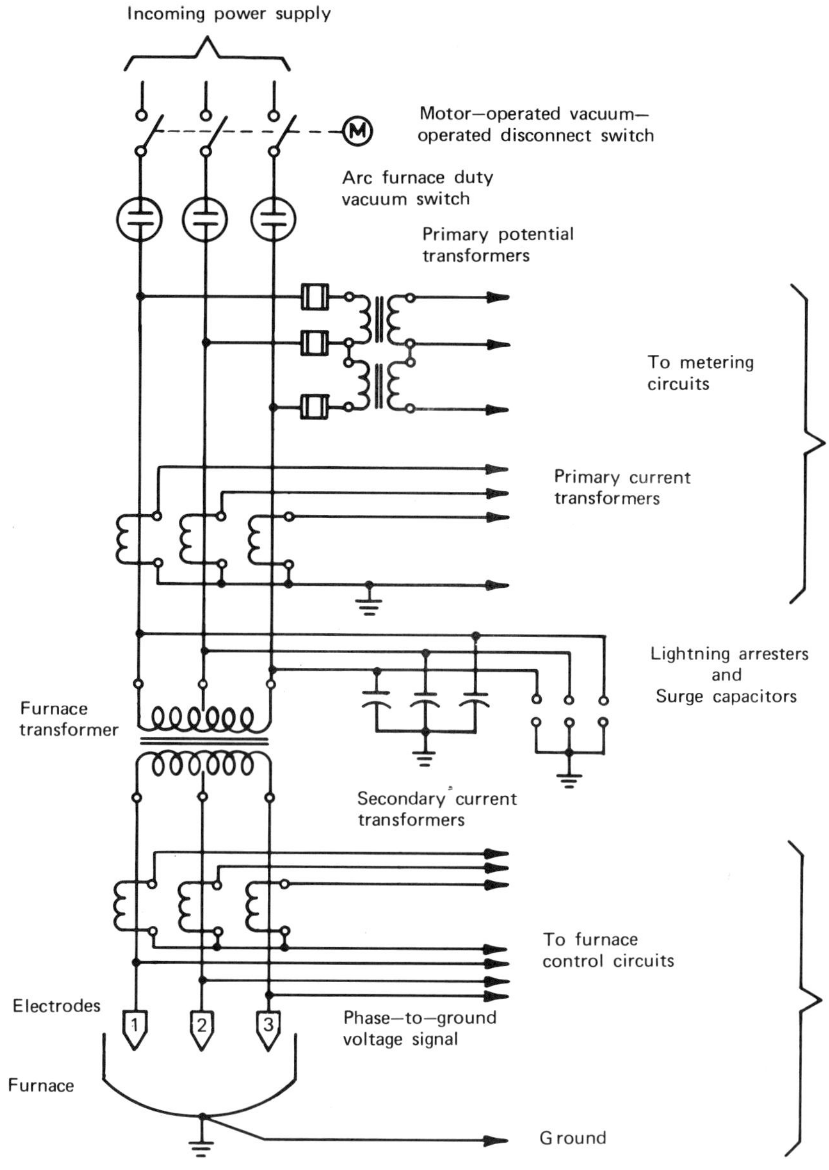

Figure 2. Three-phase direct-arc furnace schematic.

cooled around the openings. The bottom usually has a dished steel shape so that the refractory bottom lining can form an inverted arch to better withstand the molten metal pressure. The dished bottom also helps prevent formation of overinsulated areas and allows the hearth to expand and contract freely with temperature changes with no overstressing of the refractories.

Some shells may be horizontally split so that one or two sections, containing worn refractories, are easily removed and replaced with prelined shell sections so that the nonproductive furnace time required to replace these refractories is minimized (2).

Tapered shells are also used to increase the charge capacity or hot-metal capacity of existing furnaces. The furnace shell contains a tapping spout to direct the molten contents when the furnace is emptied. A water-cooled door is located diametrically across from the spout for the additions of alloys, fluxes, oxygen injection, etc, and allows the removal of slag. A second door is used on some furnaces to allow for further sampling, taking metal temperatures, "gunning" (spraying the eroded refractory lining with refractory material to rebuild the lining).

The roof, in the form of a dome, is composed of refractory brick held in place by a water-cooled steel roof ring. Water-cooled electrode glands are used on small foundry furnaces to maintain the openings in the roof for the electrodes. On large steel-mill furnaces, refractory glands are used because they require less maintenance than water-cooled electrode glands.

Refractory Linings. The refractory linings (3–4) for furnaces designed for melting ferrous materials may be acidic, basic, or neutral (see Refractories). Silica is a major acid refractory used in iron and steel foundry operations. Alumina, a neutral refractory, is used mainly for furnace roofs for all ferrous operations and in the walls for iron foundry operations. Magnesite, a basic refractory, is used primarily in steel foundries and steel mills where the sulfur or phosphorus content of the metal, or both, must be reduced or where alloy loss must be minimized. Recently, magnesite has been used in high powered basic furnace roofs (5). Usually, the bottom is lined with two or more courses of refractory brick to serve as a safety lining. Over the compatible safety hearth (acidic, neutral, or basic) granular material is rammed in approximately 5-cm layers until the desired bottom contour is attained. The wall of the furnace is lined with compatible brick, eg, magnesite for basic operation, silica for acid operation, and alumina for neutral operation. Insulating brick is not used because it increases the rate of refractory consumption.

Acid linings are the least expensive linings and are used wherever the melting process allows it. However, they are subject to spalling and, therefore, a more expensive lining may be more economical in an intermittent operation. Fire clay and linings containing a higher percentage of alumina are used when spalling or high temperatures require their use, or when a basic lining cannot be used because of chromium contamination of the molten metal. Graphite also is used in the walls nearest the electrode in iron foundry furnaces. Water-cooled cast or fabricated panels (6–7) are becoming very popular and are replacing up to 75% of the wall refractories above the sill or slag line. Energy consumption with the panels is equal to or less than with refractory linings unless there are extended holding periods. The average life of the refractory walls and roofs in ferrous melting furnaces is approximately 200 cycles (heats) and has attained greater than 400 cycles, especially with patching or gunning, or both. In ultrahigh power furnaces (2000 kVA/m^2 hearth area), the refractory life may be reduced to 100 cycles or less. The use of water-cooled panels has extended wall life to 1500 cycles, and to as much as 3000 cycles or more in the colder areas of the furnace. The refractory hearth usually lasts 5 yr or more because it can be patched easily between cycles.

Uses. The standard three-phase arc furnaces are available in sizes of 200 kg to 500 t and shell diameters of 1–12 m. Furnace transformer ratings are available from 200 to >150,000 kVA. Nearly all open-arc furnaces used in foundries and steel mills are three-phase and contain individually controlled jib-type electrode arms, each supporting a vertical column of graphite electrodes. The electrode arms are raised or lowered to maintain the desired arc characteristics, arc voltage, and arc current. This action takes place within a fraction of a second after the error signals are generated; the degree of movement depends upon the strengths of these signals. Each electrode arm is continually moving because its arc characteristics are changing continually as scrap falls away from or against the electrodes, as the electrodes erode, as the atmosphere in the furnace changes, etc. Each electrode arm's electrical conductors are connected through flexible cables to the bus bars or tubes of the delta closure extensions and onto a multivoltage tap transformer. Furnace transformers of ≤7500 kVA also contain a multitap reactor to provide sufficient inductive reactance to offset the negative characteristic of the arc.

Electrodes. Almost all the electrodes (8–9) used in most open-arc furnaces are prefabricated and are made either of dense graphite or of regular graphite. Carbon electrodes seldom are used in melting furnaces and those that are used are being replaced by graphite electrodes because of the latter's higher conductivity, smaller diameter, lower weight, and smaller diameter electrode circle for a given size transformer. Dense graphite electrodes are used instead of regular density electrodes whenever greater mechanical strength or slightly higher conductivity is required and are available in diameters of 32–711 mm. Regular density electrodes are available in sizes of 178–610 mm diameter. Carbon electrodes are usually rated at electric current or current densities of 4.5–9 A/cm^2 and graphite electrodes at 15.5–46.5 A/cm^2.

The electrode diameter normally is selected on the basis of its current-carrying capability and its mechanical strength. The major cause of electrode consumption usually is oxidation because of the high furnace temperatures, the furnace atmosphere, and the high gas velocities. This oxidation rate is further accelerated by the electrode's surface temperature being increased by passage of current. Another factor is the stress that is imposed on the electrode by tilting the furnace, swinging the electrodes aside, scrap falling against the electrode, arc forces, etc.

Electrode consumption for ferrous melting furnaces usually averages 4–6 kg/t of molten metal for single-slag practice and may be increased by one third or more for a double-slag operation. In the event the molten metal must be held in the furnace after it has been processed, the electrode consumption is increased further. A typical energy consumption for a single-slag steel practice is 550 kW·h/t and as much as 725 kW·h/t for a double-slag process.

Voltage. The voltage chosen for open-arc furnaces must be high enough to compensate for the voltage drops caused by the resistance and inductance of the primary and secondary electrical circuits and still have the required power input available at the arcs (10). In the smaller furnaces, the voltage also must be high enough to penetrate any thin oxide coatings on the scrap. Also, it must provide a sufficient area of meltdown; otherwise, the electrodes bore through the scrap, melting insufficient metal to cover the hearth and arc on the bottom refractories. The highest phase-to-phase no-load voltage for a 200 kVA production furnace usually is ca 200 V and may be over 800 V for a 100,000 kVA furnace. Lower voltages are also available for the operator to use for the melting, superheating, refining, and holding periods during a furnace heat cycle; the lowest voltage is approximately one third of the highest voltage.

For a given voltage tap, the operating electrode current can be <20 to >100% of the rated current depending upon the quality of the electrode regulator and positioner. The current also will reach zero when there is no arc (because the electrode is too far from the charge) and maximum when the scrap falls against two electrodes and causes a short circuit. Figure 3 shows a typical arc-furnace power curve for a given voltage

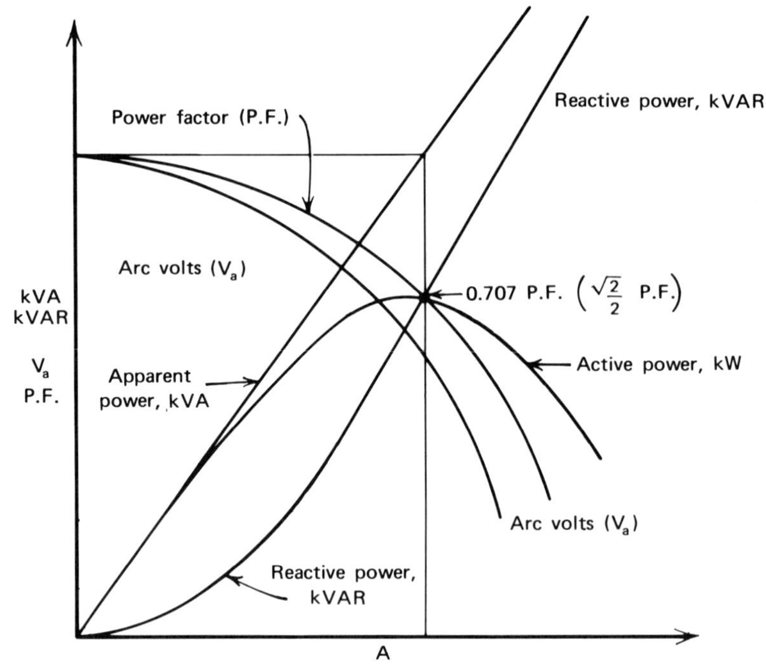

Figure 3. Typical arc-furnace power curve.

input. As indicated, the power factor and arc voltage are highest at very low currents, whereas the maximum power input is attained at a power factor (P.F.) slightly higher than 0.707 ($\sqrt{2}/2$) depending on the electrical characteristics of the primary and secondary circuits.

Vacuum-Arc Furnace. Another type of open-arc furnace is the vacuum-arc furnace (11) which is used for melting metals that have high temperature melting points, eg, titanium, molybdenum, and tungsten, or for upgrading alloy steels (see Fig. 4). An electrode of the material to be melted is cast or formed from metal powder. An arc is formed between the starter button, B, and the electrode, E. As the electrode is melted, an ingot is formed in the water-cooled copper or steel mold, M. The furnace or mold cavity usually is under a vacuum but it may also be filled with an inert gas. In some instances, a water-cooled tungsten-tip electrode is used and the material to be melted is dropped into the melting chamber and the rate of feed is coordinated with the power input.

Plasma-Arc Furnace. The plasma-arc furnace, sometimes used in the production of castings of high alloy steels and special alloys, usually has a furnace shell similar to that of the three-phase conventional open-arc furnace used in the production of iron and steel. However, water-cooled nonconsumable electrodes are used to conduct direct current and argon (to serve as the plasma base) to the sealed furnace interior. There may be one (13) or more (14) fixed electrodes in the furnace roof or sidewall and a water-cooled bottom electrode extending through the refractory hearth and in contact with the molten metal serving as the anode. It is said that the increased operating costs for argon and refractories are compensated for by the savings in alloy and graphite electrodes.

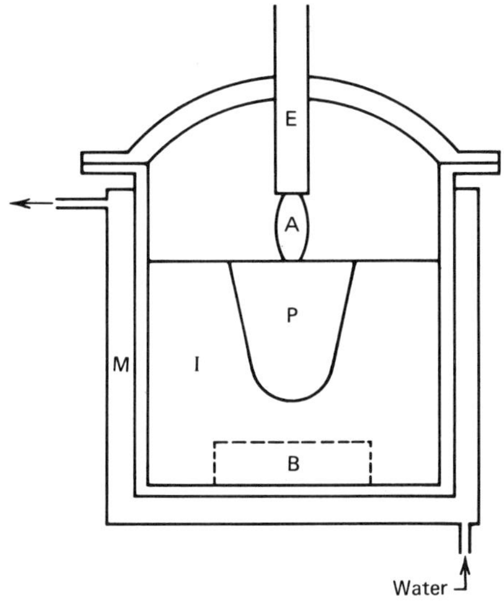

Figure 4. Consumable-electrode melting in the vacuum-arc furnace (12). A = arc; B = button; E = electrode; I = ingot; L = lid; M = water-cooled mold; P = pool of molten metal.

Argon and other inert gases have been added to the other types of open arc furnaces to decrease alloy loss and to decrease the noise level but with little commercial success.

Submerged-Arc Furnaces. Furnaces used for smelting and for certain electrochemical operations are similar in general design to the open-arc furnace in that they are usually three-phase, have three vertical electrode columns and a shell to contain the charge. They are used in the production of pig iron, phosphorus, calcium carbide, copper, other metals and compounds (15), and some types of high temperature refractories.

When a smelting campaign is started, carbonaceous material, eg, coke or a conductive material, is placed on the hearth. The electrodes are lowered and arc on this material at a low power input. The charge materials (ores, reductants, etc) are added slowly. As the material becomes molten and conductive, the power input and charging rate are correspondingly increased to the desired production rate when the furnace is filled with the charge materials. The electrodes continue to arc on the molten material, usually a layer of slag on top of the molten metal in a ferroalloy operation. Submerged arc describes the operation in that the arc area is surrounded by the charge material in different stages of melting. The space immediately beneath the electrode is filled with ionized gases through which the arcs travel, as shown in Figure 5. These gases travel upward through the burden, preheating it, and burn at the top with open flames. If the gases contain toxic or desired volatile materials, they are collected for further processing. An open-arc furnace also can be considered a submerged-arc furnace when the electrodes, which arc on the molten metal, are surrounded by scrap and other charge materials. However, this period of time is only a small portion of the total heating time.

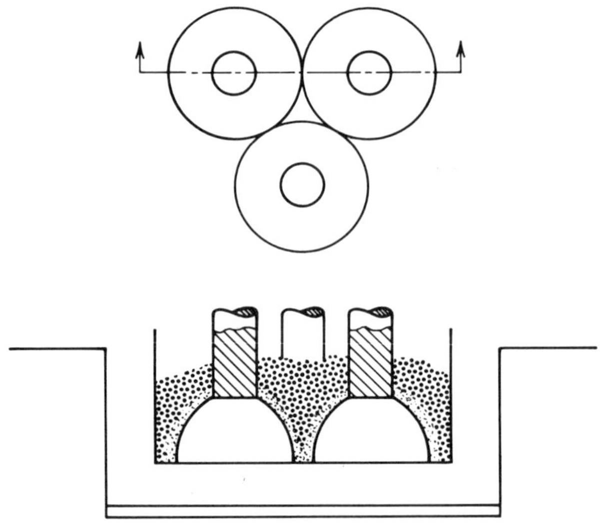

Figure 5. Reaction zones in a three-phase submerged-arc furnace (12).

Arc-Resistance Furnace. The arc-resistance furnace is very similar to the submerged-arc furnace except the electrodes of the former are immersed in the molten material—usually slag or a nonmetallic material. There are minute arcs between the bottom and sides of the electrode, but the major heat source is the passage of current through the resistive, yet conductive molten material. The furnace interior may be filled with unmelted charge material, as in a submerged-arc furnace, or may contain only the molten material. Most of the submerged-arc and arc-resistance furnaces do not tilt and usually do not have roofs. Where the volatilized materials and gases are toxic or cannot be exposed to air, the shell is covered with a sealed monolithic roof so that the gases and vapors can be ducted away from the furnace for subsequent collection.

Most furnace shells are short vertical cylinders but may also be triangular, elliptical, or rectangular in plan view. Single-phase furnaces may have one or two movable electrodes; the furnace with one movable electrode has a permanent electrode in the bottom. Three-phase furnaces usually have three movable electrodes but there is at least one furnace that has six movable electrode columns. A few of the smaller furnaces must tilt to expedite emptying a portion of the molten material since the transfer must be extremely rapid to keep the high melting point material from freezing. In cases of materials having melting points over 2200°C, the vertical portion of the furnace shell is removed to allow cooling of the materials for a day or two. After cooling, the fused material is broken into smaller pieces for further processing and the nonfused material is added to the next furnace charge. Where there is no appropriate refractory material that will withstand the high temperature and the chemical reaction with the material to be processed, the unmelted material is used for lining the furnace. In other instances, the shell is cooled by water sprays.

Electrodes. Because of the numerous different processes, there are many different types of electrodes (9) in use. There are prefabricated graphite, prefabricated carbon, self-baking, and composite electrodes (see Carbon). Graphite electrodes are used primarily in small furnaces or in sealed furnaces and are available in diameters of ≤152 cm. Carbon electrodes, made in diameters of ≤152 cm or 76 by 61 cm rectangular, are used primarily in processes where the oxidation rate is rather high.

The self-baking electrode (16–17) consists of a cylindrical steel casing that has internal radial fins and is filled with a carbonaceous paste during operation of the furnace. As the electrode is consumed and automatically lowered, the heat from the furnace and the electrical current passing through the electrode bake the carbonaceous paste and drive off the volatiles. New sections of casing also are welded on as the electrode is consumed. Self-baking electrodes, having diameters of ≤250 cm, are used mainly on calcium carbide, pig iron, and different types of ferroalloy furnaces (see also Aluminum and aluminum alloys).

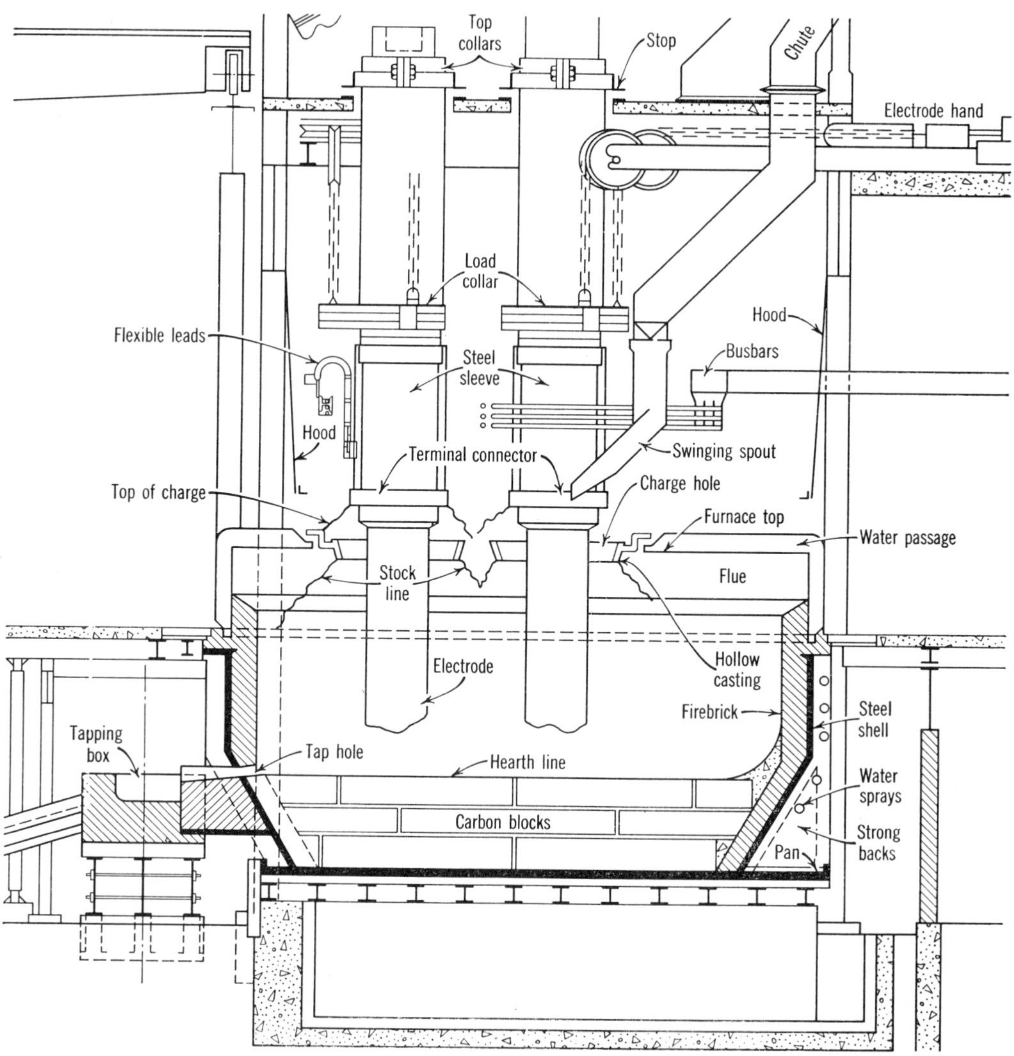

Figure 6. Design of a ferroalloy furnace. Courtesy of Union Carbide Corporation, Mining and Metals Division.

Table 1. Electrode Consumption, g/kWh

standard ferromanganese	5.2
pig iron	3.7
75% ferrosilicon	5.2
silicon	1.2
calcium carbide	6.8
phosphorus	1.5
nickel matte	6.2–10

Furnaces requiring larger electrodes use a composite electrode which consists of carbon or graphite blocks held on an electrically insulated steel frame.

A typical large three-phase ferroalloy furnace using prefabricated carbon electrodes is shown in Figure 6. The hearth, where molten materials come in contact with refractories, is usually composed of carbon blocks backed by safety courses of firebrick. In the upper section, where the refractories are not exposed to the higher temperatures, superduty or regular firebrick may be used. The walls of the shell also may be water-cooled for extended life. Usually, the furnace shell is elevated and supported on beams or on concrete piers to allow ventilation of the bottom. When normal ventilation is insufficient, blowers are added to remove the heat more rapidly. The shell also may rest on a turntable so that it may be oscillated slightly more than 120° at a speed equivalent to 0.25–1 revolution per day in order to equalize refractory erosion or buildup.

The larger electrodes are usually supported from overhead by chains or steel bands which are, in turn, connected to winches or hydraulic cylinders that position the electrode in proportion to the error signals. The electrodes are held in alignment by insulated vertical guides. The arcs are quite stable because of the carbonaceous atmospheres in the arcing zone which allows use of slow and less responsive electrode regulators and positioners (18).

Table 1 shows some of the typical electrode consumption figures for baked electrodes in a submerged-arc furnace operation.

Energy consumptions for submerged-arc and arc-resistance furnaces are higher than for the direct-arc furnace because the latter is used primarily for melting and refining metals. In contrast, the submerged-arc and arc-resistance furnaces are used in melting of compounds, eg, ores, slag, etc, that have much higher specific heats. Energy also must be furnished to allow the endothermic reactions to proceed to completion. Therefore, in a reduction operation, energy consumption figures can vary from a low of 2000 to over 20,000 kW·h/t of product, depending on the materials to be processed and the product desired. In submerged-arc and arc-resistance furnaces, the electrode size, electrode spacing, diameter–height ratio of the lined shell, voltage, and current are critical factors that must be closely coordinated to optimize energy consumption, electrode consumption, and furnace production (19–20).

BIBLIOGRAPHY

"Arc Furnaces" under "Furnaces, Electric" in *ECT* 1st ed., Vol. 7, pp. 1–23, by Victor Paschkis, Columbia University; "Arc Furnaces" under "Furnaces, Electric" in *ECT* 2nd ed., Vol. 10, pp. 252–278, by Victor Paschkis, Columbia University.

1. W. H. Smith, "Refractory Ramming Mixes for Electric Furnaces," *First A.F.S. Electric Ironmelting Conference Proceedings, Michigan State University, East Lansing, Mich., Nov. 1970,* pp. 1-1–1-18.

2. J. W. Wild, "The Split Shell: A Case Evaluation," *AIME Electric Furnace Proceedings, St. Louis, Mo., 1976,* Vol. 34, pp. 301–303.
3. *AIME Electric Furnace Steelmaking,* Vol. 1, Wiley-Interscience, Inc., New York, 1962, pp. 153–174.
4. P. T. Troell and A. L. Renkey, "Refractory Brick for Electric Arc and Induction Furnaces in the Iron Foundry," in ref. 1, pp. 1-17–1-20.
5. C. R. Davis and B. H. Baker, "Basic Roofs for Electric Furnaces," *AIME Electric Furnace Proceedings, Houston, Texas, 1975,* Vol. 33, pp. 136–144.
6. K. S. Roberts and O. E. Prenn, "Hot Spot Protection for U.H.P. Electric Arc Furnaces Progress Report," in ref. 5, pp. 147–151.
7. R. Assenmacher, E. Eisner, and D. Ameling, "Water Cooling Systems in Electric Arc Furnaces," *Proceedings of the Third International Iron and Steel Congress, Chicago, Ill., Apr. 16–20, 1978,* pp. 570–579.
8. *Electrode Arc Furnace Electrode Digest,* Union Carbide Corp., Carbon Products Division, 1975.
9. J. A. Persson, "Carbon and Graphite Electrodes Thermal Stresses and Current Carrying Capacities," *AIME Electric Furnaces Proceedings, Chicago, Ill., 1977,* Vol. 35, pp. 115–126.
10. J. A. Ciotti, "A New Era in Melting," *AIME Electric Furnace Proceedings, Pittsburgh, Pa., 1970,* Vol. 28, pp. 130–139.
11. D. E. Cooper, K. S. Snow, J. C. Fulton, and J. W. Tommeney, "Feasibility Experiments in Vacuum Arc Remelting of 28 in. Diameter Ingots in Three-Electrode AC Furnaces," *AIME Electric Furnace Proceedings, Pittsburgh, Pa., 1965,* pp. 158–164.
12. V. Paschkis and J. Persson, *Industrial Electric Furnaces and Appliances,* 2nd ed., Interscience Publishers, Inc., New York, 1960.
13. R. J. McCullough, "The Plasmarc Furnace—A New Concept in Melting Metals," *AIME Electric Furnace Proceedings, Cincinnati, Ohio, 1962,* Vol. 20, pp. 319–332.
14. C. McCombe, *Foundry Trade J. Int.,* 60 (Dec. 1978).
15. W. Fettweis, G. Rath, and W. Haiduk, "Present Technology and Development Trends in the Production of Ferroalloys," in ref. 9, pp. 21–26.
16. R. P. Van Ness, "Comparative Value of Soderberg and Prebaked Electrodes," *AIME Electric Furnace Proceedings, Pittsburgh, Pa., 1961,* Vol. 19, pp. 179–192.
17. R. E. Scherrer and M. L. Stott, "Recent Developments in Soderberg Electrode Practice in Carbide Operation," *AIME Electric Furnace Proceedings, Buffalo, N.Y., 1964,* Vol. 22, pp. 136–141.
18. O. D. Jordan, "Automatic Load and Electrode Position Control of a Submerged Arc Furnace," *AIME Electric Furnace Proceedings, Chicago, Ill., 1963,* Vol. 21, pp. 311–319.
19. W. M. Kelly, "Design and Construction of the Submerged-Arc Furnace," *Carbon and Graphite News,* Vol. 5, No. 1, Union Carbide Corp., Carbide Division, April/May 1958.
20. J. A. Persson, "The Significance of Electrode-to-Hearth Voltage in Electric Smelting Furnaces," in ref. 10, pp. 168–169.

DAN OAKLAND
Whiting Corporation

INDUCTION FURNACES

Induction furnaces are transformers in which the electrically conducting furnace charge constitutes the secondary and load resistance. They are suitable for heating and melting all metals; principal industrial uses are for iron, copper, aluminum, zinc, and their alloys. Furnace capacities range from a few kilograms to several hundred metric tons, and power ratings range from 20–20,000 kW. There are three basic types: channel induction-melting furnaces, coreless induction-melting furnaces, and induction heating furnaces.

Induction furnace efficiency is the ratio of net useful power (P_n) in the charge and power input (P_o) to the furnace. Electrical losses (P_e) (mostly resistive losses in the primary coil) are usually a constant percentage of P_o. Thermal losses (P_t) by heat flow through furnace walls, etc, for the most part are constant during operation and are independent of P_o. Figure 1 shows net power ($P_n = P_o - P_e - P_t$) and efficiency (P_n/P_o) as a function of P_o. It also shows induced power ($P_i = P_o - P_e$). Production rate is proportional to P_n. Relative magnitude of P_e and P_t differs substantially for the various induction furnace types, as described below. Efficiency always improves with increasing utilization.

Power control equipment is similar for all types of induction furnaces that operate at the same frequency. Equipment for furnaces that operate at line frequency (60 Hz in North America) includes variable output voltage transformers and associated switching means to regulate furnace power. Power controls for high frequency furnaces

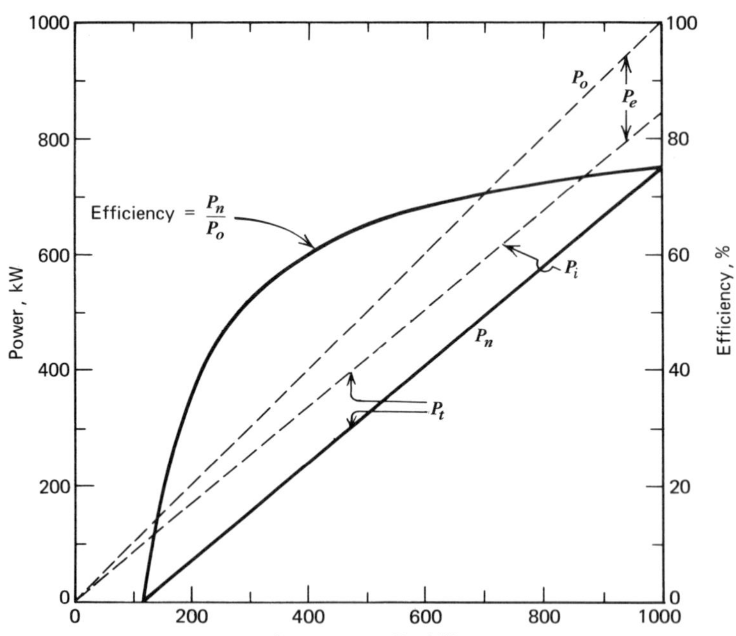

Figure 1. Induction furnace efficiency. Typical characteristic of a 1000 kW furnace. Example: P_e = 15% of P_o and P_t = 100 kW. P_n = useful power; P_o = power input; P_e = electrical loss; P_i = induced power; P_t = thermal loss.

(180–10,000 Hz) include frequency converters, predominantly of the thyristor type, which convert three-phase line frequency to single-phase high frequency power either directly (1) or through intermediate rectification (2); thyristor gate control regulates furnace power. All power controls include substantial capacitor banks which compensate for the inductive component of the furnace current.

Channel Induction-Melting Furnaces

Design. A channel induction-melting furnace (previously known as a core-type induction furnace) operates at line frequency. It has the same electrical elements as a power transformer; however, the secondary coil and load resistance are formed by a single closed turn of molten metal (3). The magnetic iron core and primary coil assembly thread a steel-encased refractory block which contains the annular secondary cavity or channel for the melt. Very high currents (up to 150 kA) are induced in the molten secondary at low voltage. The channel communicates with the lower portion of a pool of molten metal contained in a refractory-lined hearth. High electromagnetic pressure in the channel tends to constrict or pinch off the molten secondary. The hydrostatic head from the adjacent pool prevents such constriction. Electromagnetic pressure gradients also create a vigorous molten metal circulation (3–4), which distributes energy by forced convection from the channel throughout the melt in the hearth. A uniform temperature and alloy results.

The assembly of steel casing, refractory block, transformer core, and primary coil is the inductor of a channel induction furnace. Coils are cooled either by forced air or by water. High-powered inductors usually have two primary coils on a common magnetic core (3); the molten secondary then consists of two parallel single turns with a common center channel (Fig. 2). Inductors are attached to the furnace hearth structure with a bolted flange. When an inductor lining wears out, the inductor is detached and replaced with a lined spare, without relining the hearth.

Inductors have power ratings of up to 2500 kW. When higher melting rates are needed, several inductors are attached to the same hearth to form a multiple-inductor furnace (Fig. 3). These hold charges of up to several hundred metric tons. Most furnaces tilt or rotate to discharge a stream of molten metal; others are stationary, discharging metal by overflow, pumping, or tapping. Channel induction-melting furnaces have low electrical losses (P_e = ca 5% of P_o) resulting from the close coupling by magnetic core and the high secondary resistance of the confined channel. Thermal losses typically range from 10–30% of full power, depending on the operating temperature and the desired ratio of furnace capacity and hourly melting rate. Therefore, the efficiency at full power ranges from 65–85%.

Channel induction-holding furnaces with relatively low power ratings are widely used. They store molten metal received from separate melting furnaces. Some are gas tight and control the gas pressure above the melt surface to discharge metal through a sloped tubular spout in the side wall (5). This enables either accurate repetitive pouring of castings or a controlled continuous casting process. Neutral or reducing gas atmospheres are sometimes used to prevent oxidation.

Operation. The furnace requires an uninterrupted molten secondary in order to function. It must be primed by pouring molten metal into the preheated channel to start operation. Priming normally occurs only after relining. The furnace is then continuously maintained with a pool of molten metal at temperature, often for years.

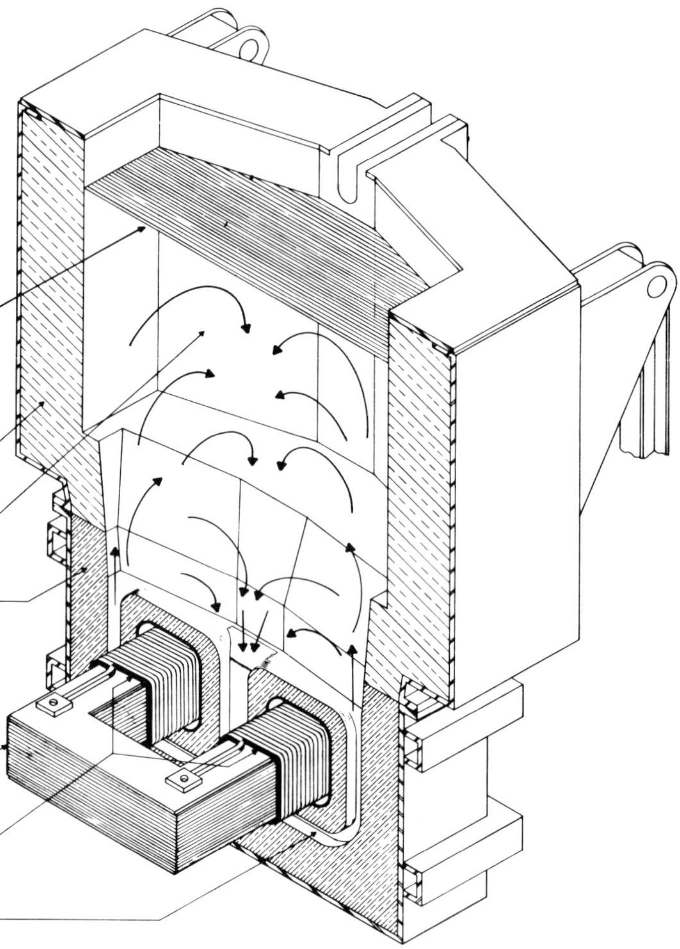

Figure 2. Channel induction-melting furnace. Tilting furnace with one high-powered twin-coil inductor. Courtesy of Ajax Magnethermic Corporation.

Minimal thermal cycling stresses result in long refractory life. Melting is carried out by dropping charge material into the molten pool. After melting, the charge is poured off by tilting (leaving a residual pool) and the cycle repeats. In some applications, melting and pouring are simultaneous and continuous. Temperature is often controlled from a thermocouple placed in the melt through a power-regulating pyrometer. Close control is achieved because all heat is generated directly in the circulating melt.

Applications. Channel induction-melting furnaces are preferred for steady production (at least two shifts per day) where major alloy changes are not required. They are not suitable for operations with frequent shutdowns or major alloy changes. Tilting or rotary furnaces are used to melt cast iron (qv) and to store molten iron at temperature (duplexing) in foundries (6). They are used almost exclusively for melting and holding copper alloys in brass mills (3) to minimize oxidation and vaporization loss of zinc. They produce oxygen-free copper by melting and casting under a controlled atmosphere (7). Furnaces with automatic gas-pressure pouring are used to cast copper-alloy plumbing fittings (8).

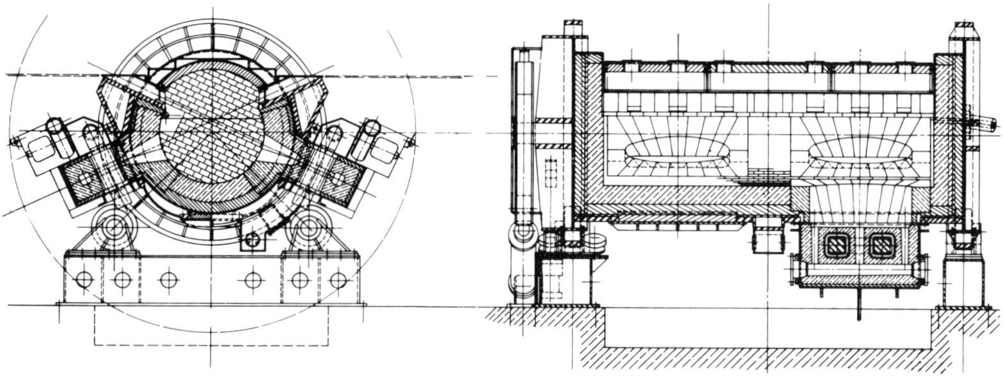

Figure 3. Multiple inductor furnace. Rotary channel induction furnace with four twin-coil inductors, hydraulic rotation. Courtesy of Ajax Magnethermic Corporation.

Large stationary multiple-inductor furnaces melt all zinc cathodes in electrolytic zinc refineries (9) for subsequent casting into slabs and other primary shapes (see Zinc). Large stationary furnaces are used for continuous galvanizing in steel mills (10), ie, coating sheet steel with zinc for rust prevention; submerged rolls guide the steel strip through the molten zinc (see Steel). Small channel induction-melting and holding furnaces are used for aluminum die-casting. A number of large aluminum melting furnaces are in use (see Aluminum).

Coreless Induction-Melting Furnaces

Design. Coreless furnaces lack a magnetic iron core to link the primary and secondary. The charge is placed in a crucible-shaped refractory container surrounded by a cylindrical primary coil. Coreless induction melting is made possible by the electromagnetic phenomenon known as the skin effect. Alternating current induced in a metallic secondary tends to concentrate at the surfaces facing the primary coil (Fig. 4) and to rapidly decline exponentially towards the inside. Practically, the current is concentrated in a layer of the thickness d, equivalent to an annular single-turn secondary with a relatively good coupling to the coil; d is known as depth of penetration or reference depth (11).

The diameter of the melt should be at least 6 d to obtain adequate electrical efficiency. In nonmagnetic conductors d is proportional to the square root of the ratio of resistivity and frequency (11). The order of magnitude of d for most molten metals is 30–80 mm at line frequency and 7–20 mm at 1000 Hz. When the charge consists of many metal pieces, a secondary current flows in each piece. Energy is efficiently transmitted to such a charge if the pieces have a smallest dimension of at least 3 d.

High currents in a molten charge create electromagnetic pressures and result in a convex shape of the melt surface, termed meniscus. Pressure gradients also cause a vigorous melt circulation (12) in a pattern as shown in Figures 4 and 5. The use of a higher frequency decreases d and thereby increases the secondary resistance. A lower secondary current then is required to attain a desired power level and the meniscus and turbulence are reduced. Therefore, high frequency (180–3000 Hz) is preferred for small furnaces (50–2000 kg). High power levels and melting rates are attained

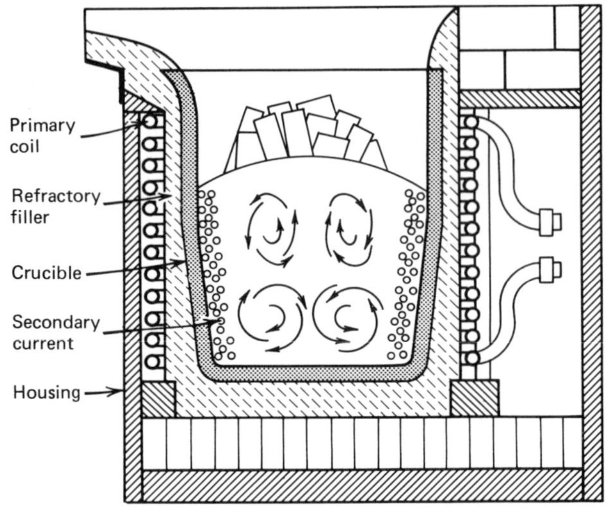

Figure 4. Small coreless-induction furnace. 500 kg high-frequency furnace with insulating board housing and crucible.

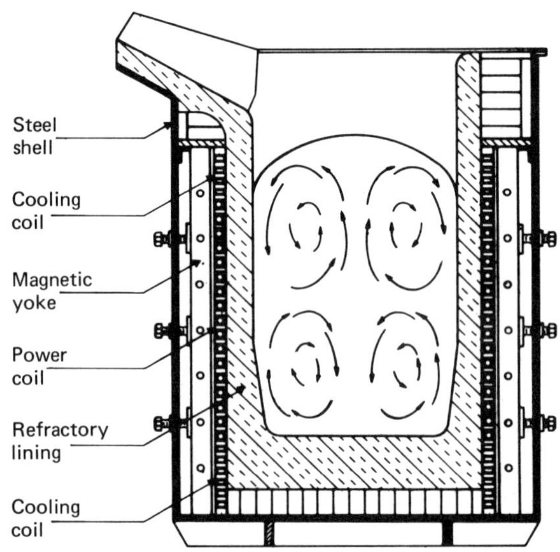

Figure 5. Large coreless-induction furnace. 20-t line-frequency furnace with steel shell, yoke structure, and rammed monolithic lining.

without excessive turbulence and meniscus, and cold charges that consist of small pieces melt down quickly (13). A typical high-frequency furnace is shown in Figure 4; it has a housing of structural insulating board to minimize stray inductive power losses.

Coreless furnaces ranging from 3–60 t are predominantly powered at line frequency. The resulting greater surface turbulence is advantageous for melting finely divided scrap, causing the scrap to submerge quickly. Most furnaces of more than a

2-t capacity have a steel shell structure (Fig. 5). Vertical magnetic iron laminations (yokes) are positioned between the coil and shell to minimize stray inductive losses. Coreless furnaces always have water-cooled coils made of copper tubing. The melt is discharged by tilting around an axis close to the spout. Electrical losses are high, owing to the inherent low secondary resistance and a skin effect limitation on primary coil resistance. P_e is about 20% of P_o for iron melting and 30% of P_o for copper or aluminum alloys. In high-frequency installations, there is an additional 6–10% loss in the frequency converter; however, thermal losses are only ca 10%. Therefore, the efficiency of coreless furnaces at full power ranges from 50–70%.

Operation. A high-frequency furnace melting cycle is started by placing a partial cold charge of new metal, scrap, and alloy additions into the empty furnace. After this is melted (Fig. 4), the remainder of the charge is added, alloy composition is adjusted, and the entire charge is poured off; then the cycle is repeated. Alloys can change from cycle to cycle. The furnace is cooled down when not melting. The lining or crucible life is normally in the range of 50–200 cycles (heats), depending on temperature and application. Relining cost is low. In most large line-frequency coreless furnace operations, a molten pool of at least 70% of full capacity is maintained. Each charge is 30% or less, dropped onto the turbulent surface of the molten pool. Large furnaces are maintained with molten metal or with auxiliary heat during off-shifts to minimize thermal cycling stresses on the monolithic refractory.

Applications. High-frequency coreless furnaces are preferred for operations with frequent alloy changes and those with less than two shifts per day. They are widely used in foundries producing castings of alloy steel (14) and copper alloys. They are suited for precious metals and for all special alloys that are cast above 1500°C. Coreless induction furnaces also are preferred for melting under vacuum (15). In some installations the furnace and casting mold are placed in a common vacuum-tight enclosure. In others, the steel shell forms the vacuum-tight structure and includes means for sealing the cover and spout. Large line-frequency coreless furnaces are widely used in high-production iron foundries (16), especially for charges including steel scrap, and in aluminum mills for recovering finely divided scrap (foil, scalpings, and turnings).

Induction Heating Furnaces

Induction heating furnaces are generally known as induction heaters or induction coils. They rely on the skin effect (described above for coreless furnaces) for the efficient transfer of energy to the charge. The charge usually is a finished or semifinished metal product of specific shape and dimensions. A cylindrical coil is used for a round product; if the product has a square, rectangular, or other shaped cross section, the coil may be shaped to conform to the product. Most induction heaters are for steel, copper, or aluminum (qv). There are three types that are extensively used: static induction heaters, continuous induction heaters, and induction heaters for surface-hardening. The first two are used for through-heating, ie, raising the temperature throughout the mass of the product, usually for subsequent hot-forming. The latter apply localized heat only to specific surface areas of steel products.

Induction heaters are used primarily for through-heating steel, to 1100–1300°C. Efficiency is high while the product is magnetic (up to the Curie temperature, 768°C) and much lower in the nonmagnetic temperature range (17). Line frequency normally is suitable if the smallest product dimension is at least 15 cm. A higher frequency is required for smaller sizes.

Energy is generated in a surface layer of the thickness d. Heat transfer to the center region of the product is by conduction only. A significant temperature differential may occur between the surface and the center when a high power density is used. Power density is often decreased upon approach to the final temperature to attain a satisfactory temperature uniformity for hot-forming (18). Induction through-heating is faster than any other heating method and minimizes work in process and oxidation loss.

Static Induction Heaters. In static heaters, the inserted product remains in place (static) within the coil while power is applied. When final temperature is reached, the power is shut off, the product is transferred to the next operation, and the cycle is repeated. Static heating permits individual control of final temperature. In small static induction heaters, the coil axis is horizontal. Large heaters have a coil with a vertical axis and an elevator mechanism disposed underneath for insertion and retraction of billets or slabs of steel, copper, or aluminum. Figure 6 shows a 20 MW static induction heater for heating 27-t steel slabs to 1300°C in one hour (19–20). Each rectangular coil turn surrounds the entire slab lengthwise. A group of such heaters is needed to supply a steel-rolling mill.

A special form of static induction heating is the indirect heating of materials of poor electrical conductivity in an induction-heated container. Substances such as polymers, resins, and dyes are processed at 300–350°C in large magnetic steel vessels which are efficiently induction-heated at line frequency (21). High-frequency induction-heated graphite containers often serve as heat sources for processing above 1800°C.

Continuous Induction Heaters. Continuous heaters are widely used to deliver billets of steel, copper, or aluminum to forging or extrusion presses at a steady rate. Several

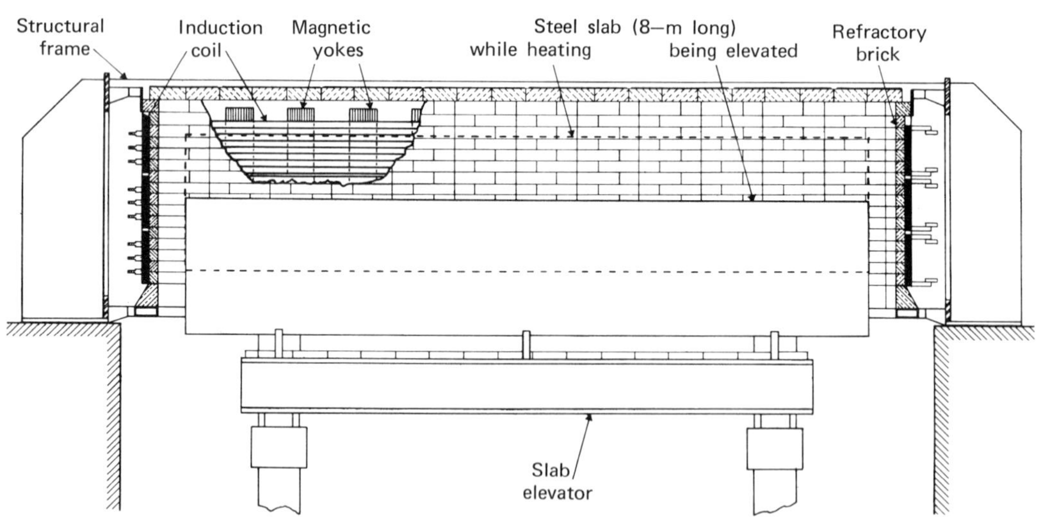

Figure 6. Static induction heater. 20 MW induction heater for slabs 8-m long, weighing 27 t. Courtesy of Ajax Magnethermic Corporation.

induction coils are arranged in line with a common horizontal axis. Billets form a continuous column, usually sliding on alloy steel guide rails within each coil and moved by a pusher mechanism (22). Power density is lower in the final coils. Continuous heaters are also used to heat long steel bars for rolling. Bars are transported by rolls located at the ends of the heater and between coils. Heat-treating of heavy-duty pipe (23) is shown schematically in Figure 7. The pipe is first induction-heated to about 900°C, then quenched continuously by way of a water spray ring as it leaves the final induction coil. The pipe subsequently is tempered in another continuous induction heater.

Induction Heaters for Surface Hardening. Hardening of only the surface (or a selected surface region) of a steel product is achieved by applying a high frequency, eg, 10 kHz, to an induction coil at a very high power density. Heat is generated only in a very thin skin in which the temperature rises rapidly. As soon as the surface attains the proper temperature (850–950°C), power is shut off and a high velocity quench spray is applied. This results in hardening to a predetermined depth beneath the surface; however, the core of the product remains soft (24). Hardening is achieved only in the current-carrying surface regions, ie, those directly facing the coil. Therefore, the coil can be shaped to harden only a selected region. Induction-hardening coils usually have just one turn or a few turns; they are supplied with low voltage through a transformer. Controlled motion of the product relative to the coil and quench spray (scanning) is used widely for shafts and similar products. A great variety of shaped coil configurations and associated mechanical handling arrangements are in use (24–25). Surface hardening improves wear resistance and fatigue strength. Many parts of automobiles, trucks, tractors, etc, are surface-hardened by induction.

Special induction heaters are widely used for brazing and annealing and for many other applications not described here. Dielectric heating furnaces are not induction furnaces; in principle they are capacitors energized at very high frequency to heat an electrically nonconducting charge by dielectric loss (26).

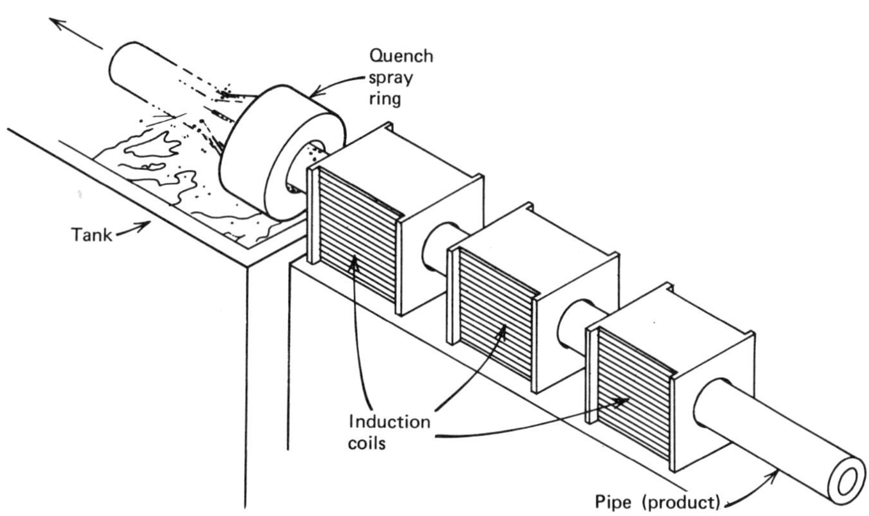

Figure 7. Continuous induction heater. A line of induction coils followed by a quench spray for pipe heat treating.

BIBLIOGRAPHY

"Induction Furnaces" under "Furnaces, Electric" in *ECT* 1st ed., Vol. 7, pp. 1–23, by Victor Paschkis, Columbia University; "Induction Furnaces" under "Furnaces, Electric" in *ECT* 2nd ed., Vol. 10, pp. 252–278, by Victor Paschkis, Columbia University.

1. G. Havas and R. A. Sommer, *IEEE Trans. Ind. Electron. Control Instrum.* **17,** 321 (1970).
2. A. P. Baines and M. Garaway, "Solid State Induction Melting for Investment Casting," *paper 7 presented at the 13th Annual Conference, British Investment Casters' Technical Association, 1975.*
3. M. Tama, *J. Metals* **26**(1), 18 (1974).
4. U.S. Pat. 3,595,979 (July 27, 1971), W. E. Shearman (to Ajax Magnethermic Corporation).
5. U.S. Pat. 2,936,326 (May 10, 1960), M. Tama (to Ajax Magnethermic Corporation).
6. A. S. Leary, "High-Powered Iron Melting Channel Furnaces," *paper 27 presented at Assoc. Ing. Electr. Inst. Electrotechnique Montefiore Meeting, Liege, Belgium, Oct. 2–6, 1978.*
7. D. Janicijevic, *Met. Prog.* **79,** 112 (Feb. 1961).
8. F. L. Riddell, "Automatic Pouring Practices in American Brass Founderies," *paper 10 presented at 39th International Foundry Congress, May 1972.*
9. A. C. Jephson, A. Y. Bethune, and R. C. Kelahan, *J. Metals* **18,** 947 (1966).
10. M. Tama, "Ajax Induction Furnaces for Galvanizing," *42nd Meeting, The Galvanizers Committee, Canton, Ohio, Oct. 28, 1960.*
11. C. A. Tudbury, *Basics of Induction Heating,* Vol. 1, J. F. Rider, New York, 1960, pp. 1-44–1-47.
12. *Ibid.,* pp. 1-124–1-125.
13. H. M. Rowan, *Foundry* **84,** 56 (Sept. 1971).
14. M. W. Hubbard, *Steelmaking for Steel Founders—Induction Furnace Practice,* Monograph 1, Steel Castings Research & Trade Association, Sheffield, England, 1972.
15. R. C. Leinach and H. J. Hamijian, *J. Metals,* **18,** 219 (1966).
16. H. G. Heine, "Coreless Induction Furnaces," *1st AFS Electric Ironmelting Conference,* pp. 5-7–5-20.
17. Ref. 11, pp. 1-66–1-67.
18. N. V. Ross, *IEEE Trans. Ind. Gen. Appl.* **6,** 449 (1970).
19. B. E. McArthur, *Iron Steel Eng.* **45,** 640 (1968).
20. U.S. Pat. 3,424,886 (Jan. 28, 1969), N. V. Ross (to Ajax Magnethermic Corporation).
21. K. G. Webley, "Induction Heating of Steel Reactor Vessels," *paper presented at AIChE Chemical Process Industry Symposium, Philadelphia, Pa., June 5–8, 1978.*
22. C. N. Howell, "Heavy Steel Production with Induction Heating," *4th Biennial AIEE Conference on Electric Heating, Philadelphia, Pa., 1959,* pp. 117–123.
23. R. M. Storey, *Met. Prog.* **101,** 95 (Apr. 1972).
24. T. H. Spencer and co-workers, in *Metals Handbook,* 8th ed., Vol. 2, ASM Committee on Induction Hardening, American Society Metals, Metals Park, Ohio, 1964, pp. 167–188.
25. A. M. Guthrie and K. C. Archer, *Heat Treat. Met.* **1,** 15 (1975).
26. C. DeDieri and W. A. G. Voss in K. Kegel, ed., *Elektrowaerme,* W. Girardet, Essen, 1974, pp. 631–666.

General References

Electric Melting and Holding Furnaces in the Ironfounding Industry (*Conference Record*), University of Technology, Loughborough, Sept. 20–22, 1967, British Cast Iron Research Association, Birmingham, Eng., 1968, 464 pp., an extensive symposium of worldwide operating experience with channel and coreless induction melting in iron foundries.

K. H. Brokmeier, *Induktives Schmelzen,* W. Girardet, Essen, 1967, 543 pp., a very complete treatment of European induction melting technology.

C. A. Tudbury, *Basics of Induction Heating,* 2 Vols., J. F. Rider, New York, 1960, an excellent, well-illustrated introduction to induction heating for nonspecialists.

MARIO TAMA
Ajax Magnethermic Corporation

RESISTANCE FURNACES

Resistance furnaces are categorized by a combination of four factors: batch or continuous; protective atmosphere or air atmosphere; method of heat transfer (radiation, convection, and conduction); and operating temperature. The primary method of heat transfer in an electric furnace is usually a function of the operating temperature range. The three methods of heat transfer are radiation, convection, and conduction. Of the three, radiation and convection apply to all of the furnaces described above. Conductive heat transfer is limited to special types of furnaces covered later.

Operating temperature ranges are classed: low, medium, and high. Unfortunately, there is no standard or precise definition of these ranges. Generally, a low temperature furnace operates below 760°C, medium temperature ranges from 760–1150°C, and furnaces operating above 1150°C are high temperature furnaces. There is also some confusion in the use of the terms, furnace and oven. The term, oven, is used when temperatures are below 760°C and furnace is applied for higher temperatures; however, indiscriminate use of the words occur. The term furnace is used here regardless of operating temperature.

Batch and Continuous Furnaces

Furnaces must be either batch or continuous. The determination of the need for either type of furnace is dependent on production rate and the physical size and weight of the work to be processed.

Batch Furnaces. Batch furnaces carry out the desired time–temperature cycle for the work to be processed by subjecting the entire furnace and charge of work to the particular cycle. Batch furnaces are most often used for one or more of the following applications: very large and/or heavy charges; low production rates; infrequent operation; variable time–temperature cycle; and processing material that must be in batches because of previous or subsequent operations. Larger batch furnaces are often of the elevator or car-bottom type. A typical electric elevator furnace is shown in Figure 1. In this furnace, the charge of one or many pieces is loaded onto the hearth of a car. The car is moved under the furnace and is hoisted into the furnace by way of an elevator mechanism which is part of the furnace. Very large or heavy loads are often processed in a car-bottom furnace which is similar to the elevator furnace except that the furnace is not elevated and the car carries the work into the furnace through a door at one end.

Medium-sized loads are often processed in a bell furnace, as shown in Figure 2. The operation of this furnace is opposite to that of an elevator furnace: the work load is placed on a stationary hearth and the furnace is lowered over the hearth. Bell furnaces are often arranged with two or more bases (hearths) which permits more efficient use of the furnace because one base can be unloaded/loaded as the furnace carries out a heating cycle on another base.

Small loads are commonly processed in a box furnace (see Fig. 3). The load is placed on the furnace hearth through a door. Other versions include the pit furnace which is a box furnace with the door on top and which is often installed in a pit with the top of the furnace near floor level.

Figure 1. Elevator furnace. Courtesy of Wellman Thermal Systems Corp.

Continuous Furnaces. Continuous furnaces are applicable for: uniform charges of work that arrive at the furnace continuously; moderate to high production rates; constant time–temperature cycle; and continuous operation over at least one and preferably two or three shifts per day. The desired time–temperature cycle is designed into the continuous furnace. The charge is subjected to this cycle by moving it through chambers operating at different temperatures. Although temperatures may be varied, a sequence, eg, heat, hold, and cool, is a part of the furnace design and it may be difficult and expensive to change this cycle once the furnace is built. Although multicycle continuous furnaces are possible, they are expensive and are subject to design limitations. Continuous furnaces are usually named for the method used to convey the material through the furnace.

The roller-hearth furnace shown in Figure 4 is used to process both large and heavy, and small and light parts. Unless the configuration of the work will permit it to roll on a roller conveyor, the charge is carried on a tray or—in the case of small parts—in a basket which is in turn carried by a tray. For any given operating temperature, the weight that can be carried per unit length of the furnace is limited by the strength of the rolls.

Light loads are often processed in a mesh-belt furnace shown in Figure 5. Mesh-belt furnaces usually carry the work load directly on the mesh belt. At a given operating temperature, loading per unit area of the belt is limited by its tensile strength. Cast-link-belt furnaces function in the same manner as mesh-belt furnaces except that the former carry heavier loads because the belt is made from suitable alloy castings instead of woven wire. The belt is normally contained in the furnace on both the working and return sides, whereas the mesh belt usually exits the furnace with the work load and

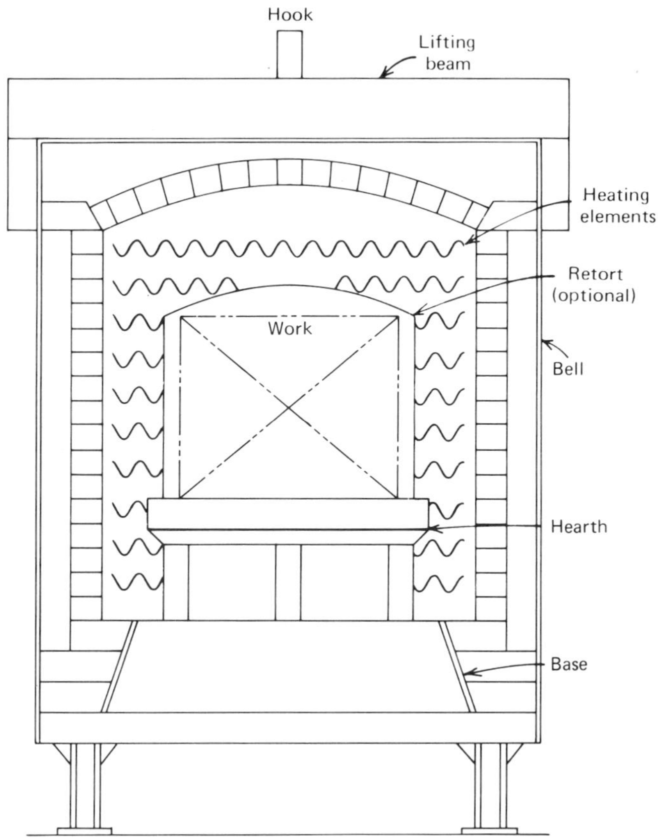

Figure 2. Typical electric bell furnace.

returns outside the furnace. Because of its large weight, it is uneconomical to let the cast-link belt cool on the return and reheat it with the work.

In pusher furnaces, the work is pushed through the furnace in steps by a hydraulic or electro-mechanical mechanism which pushes each load into the furnace, thus pushing all work in the furnace ahead one work space. The walking-beam furnace lifts the work load on a walking beam, advances the load a step, and returns the work to the hearth. The walking beam then returns to its original position in preparation for the next step.

Furnace Atmospheres

Electric furnaces can operate either with air in the interior of the furnace or with a protective atmosphere; the choice is dictated by the process requirements of the work. The furnace must be designed for the atmosphere to be used, because the combination of temperature and atmosphere are significant factors in selecting internal materials used in the furnace construction; this applies particularly to the selection of heating element (resistor) material. It is feasible and common to design an electric furnace that can operate in both air and protective atmospheres although, in general, shortened element life results from frequent alternating between reducing atmospheres and

Figure 3. Box furnace. Courtesy of Wellman Thermal Systems Corp.

oxidizing atmospheres. There are exceptions to this rule as some resistor materials must be periodically oxidized if used in a reducing atmosphere. Others are limited to a particular atmosphere.

Air-Atmosphere Furnaces. Air-atmosphere furnaces are applied to processes where the work load can tolerate the oxidation that occurs at elevated temperatures in air. In some special applications, the oxidation is not only tolerable but is desired. Some furnaces heat the work solely to promote oxidation. Furnaces designed for air operation are not completely gas-tight which results in somewhat lower construction costs. There are no particular problems encountered in selecting the insulation systems because almost all refractory insulations are made up of oxides. Heating-element materials are readily available for the common temperature ranges used with air atmospheres.

Protective-Atmosphere Furnaces. Protective-atmosphere furnaces are used where the work cannot tolerate oxidation or where the atmosphere must provide a chemical or metallurgical reaction with the work. In some cases, mainly in high temperature applications, the atmosphere is required to protect the electric heating element from oxidation.

Protective-atmosphere furnaces are of two general types. In one type, the work is inside a muffle (retort) and the protective atmosphere is inside the muffle. The

Figure 4. Roller-hearth furnace. Courtesy of Wellman Thermal Systems Corp.

outside of the muffle and the interior of the furnace operate in air and are designed accordingly. The other type is gas-tight and the atmosphere is introduced directly into the furnace, obviating the expensive and expendable retort or muffle. It does require careful selection of the internal furnace parts which must not be adversely affected by the atmosphere. The selection of electric heating elements must be carefully made with respect to operating temperature and atmosphere. The best material to use for a given application is a function of the combination of temperature and atmosphere.

The true operating temperature range and atmosphere must be specified in a description of an electric controlled-atmosphere furnace. Frequently, a higher temperature is specified than required for the contemplated operations, apparently with the thought that the higher temperature construction results in a safety factor. In the case of heating elements it can result in more expensive materials which, at the true operating temperature, are actually inferior to less expensive element materials. In addition to heating elements, there are other materials used in furnace construction that are satisfactory for one temperature range but that are not suitable for a lower temperature operation.

Figure 5. Mesh-belt furnace. Courtesy of Wellman Thermal System Corp.

Low Temperature Convection Furnaces

Low temperature convection furnaces are designed to transfer the heat from the heating elements by forced convection. Convection is normally used in furnaces operating below 760°C because it is the most effective means of heat transfer that can maintain good uniformity of temperature on various configurations of work load. Convection furnaces also are used in this range of temperatures where it is important that no part of the work load exceed the controlled temperature. This is accomplished by shielding the work load from any view of the heating elements and by controlling the temperature of the air or atmosphere (which carries the heat from the heating elements to the work) at the desired maximum temperature.

One design for a low temperature convection furnace shown in Figure 6 utilizes an external circulating fan, heating chamber, and duct system. The fan draws air from the furnace and passes it through the external heating chamber and back into the furnace past the work. This system minimizes the chance that the work receives any direct heat radiation. It is—at least in theory—less efficient because the external blower, heating chamber, and ductwork constitute added external surfaces that are subject to heat losses.

The other design, shown in Figure 7, functions similarly but all components are inside the furnace. An internal fan moves air (or a protective atmosphere) from the bottom of the furnace and up a passage between a baffle and the heating elements (which are mounted on the furnace wall) and forces the air down over the work. Depending on the specific application, the flow direction may be reversed. This design eliminates floor-space requirements and added heat losses of the external system but requires careful design to prevent radiant-heat transfer to the work.

Heating elements operating <760°C are almost always of a chrome–nickel re-

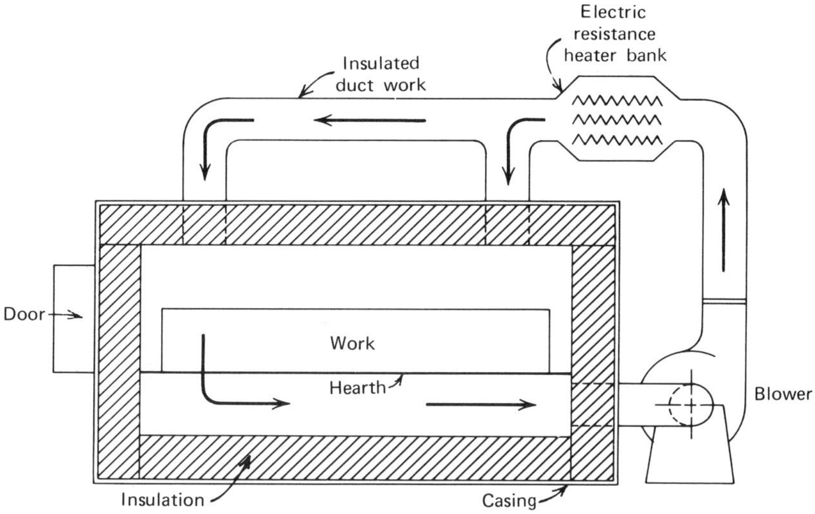

Figure 6. External heating chamber convection furnace.

sistance alloy and are in the form of ribbon, cast alloy, open wire coils, or sheathed construction. Several alloys are suitable in this temperature range and all are satisfactory if properly applied. In general, the more expensive alloys are used when physical space limitations dictate high unit power dissipation from the element.

Low Temperature Radiation Furnaces

Low temperature radiation furnaces are of the ir-heater type. Heat transfer is by direct radiation from a high temperature heating element. Control of the heat is obtained by controlling the time of exposure to the heat radiation. This type of furnace is normally used for such applications as drying of paint films. Heating elements are nickel–chrome resistance wire which is wound on ceramic supports or contained in sheaths. Other materials include tungsten resistors in glass or quartz envelopes which exclude oxygen from the resistors.

Medium Temperature Radiation Furnaces

Medium temperature radiation furnaces generally are used at 760–1150°C. The major part of the heat is transferred directly to the work by radiation from the heating elements and by radiation to the furnace refractory which reradiates the energy to the work. Heating elements may be located in the sidewalls, roof, or floor of the furnace. Locations of heating elements must be selected to assure uniform temperatures in the work zone. Elements located below the hearth level require protection from falling work or other contaminants. Figure 8 shows a typical cross section through a radiation furnace.

From 760 to 960°C, circulating fans (normally without baffles) are used to improve temperature uniformity and overall heat transfer by adding some convection heat transfer. They create a directional movement of the air or atmosphere but not the positive flow past the heating elements to the work as in a convection furnace. Heating

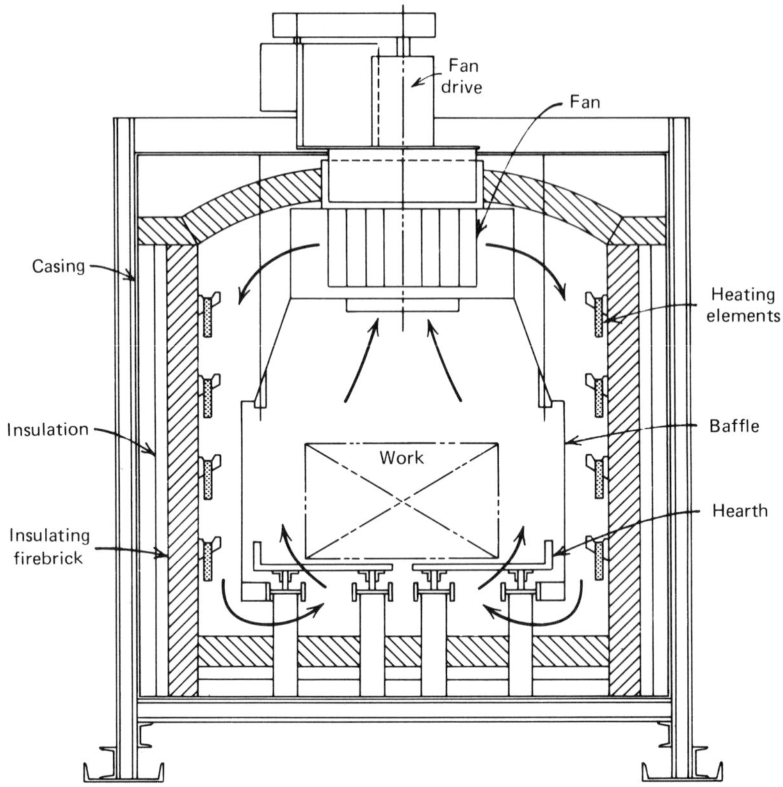

Figure 7. Internal heating-element convection furnace cross section.

elements are commonly chrome–nickel alloys in the forms described previously. Sheathed elements are limited to the very low end of the temperature range, whereas—at the upper end—silicon carbide resistors may be used. It is in this temperature range that the selection of heating element materials—based on the combination of temperature and atmosphere—becomes critical (1).

High Temperature Radiation Furnaces

High temperature radiation furnaces operating above 1150°C are similar in construction to medium temperature radiation furnaces. The insulation system must be designed to withstand the high temperatures, and internal structural parts become critical.

At temperatures above 1150°C, alloys used for the hearth or material handling systems in low and medium temperature furnaces lose strength rapidly (2) and temperatures are reached where ceramic refractories are required to support the work. This results in less use of roller hearth and belt-type hearths and more use of pushers or walking-beam designs for continuous furnaces.

Chrome–nickel-alloy heating elements that are commonly used in low temperature furnaces are not suitable above the very low end of the range. Elements commonly used as resistors are either silicon carbide, carbon, or high temperature metals, eg,

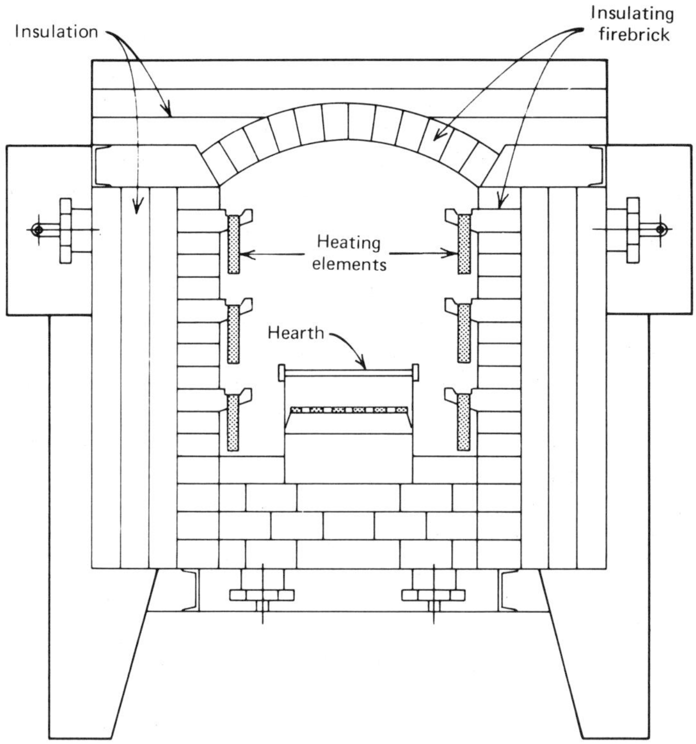

Figure 8. Typical medium temperature furnace cross section.

molybdenum and tungsten. The latter impose stringent limitations on the atmosphere that must be maintained around the heating elements to prevent rapid element failure (3).

Refractory selection becomes critical with high temperature radiation furnaces that have molybdenum or tungsten heating elements. Although these elements are stable in vacuum or inert atmospheres, many applications require a reducing atmosphere, often high in hydrogen content and very dry. In these cases, there is the possibility of reducing the oxides that make up refractory insulations. Published data are available relating temperature and dew point at which hydrogen will reduce the various oxides present in insulations (4). If this point is reached, the reduction process begins to destroy the insulation system of the furnace, thus limiting the maximum practical operating temperature to less than the capability of the heating element material (see Refractories).

Vacuum Radiation Furnaces

Vacuum furnaces are used where the work can be satisfactorily processed only in a vacuum or in a protective atmosphere. Most vacuum furnaces use molybdenum heating elements. Because all heat transfer is by radiation, metal radiation shields are used to reduce heat transfer to the furnace casing. The casing is water-cooled and a sufficient number of radiation shields between the inner cavity and the casing reduce the heat flow to the casing to a reasonable level. These shields are substitutes for the insulating refractories used in other furnaces.

Conduction Furnaces

Conduction furnaces utilize a liquid at the operating temperature to transfer the heat from the heating elements to the work being processed. Some furnaces have a pot filled with a low melting metal, eg, lead, or a salt mixture (eg, sodium chloride and potassium chloride) with a radiation-type furnace surrounding the pot. Although final heat transfer to the work is by conduction from the hot lead or salt to the work, the initial transfer of heat from the resistors to the pot is by radiation.

Conduction furnaces are of two general types. One has a pot or crucible with suitable exterior insulation. Sheathed resistance elements are inside the pot which contains molten lead or another low melting metal. The molten metal can be the conductive medium which transfers heat to the work which is immersed in it, or the molten metal may be the work. Such furnaces are often used to supply molten lead, type metal, etc. As the molten metal is removed, bars of the metal are added for melting. The initial charge of solid metal does not provide good surface contact with the heating elements and, because of this, the metal around the heating elements is often melted initially with a torch or other auxiliary heat during startup. The salt-bath furnace is the other type of conduction furnace. A molten salt not only provides the medium for conductive heat transfer, the salt is the heating resistor. These furnaces commonly are applied for temperatures ranging from slightly above the melting point of the salt to ca 1260°C.

The furnace shown in Figure 9 consists of a metal (or for higher temperatures, ceramic) crucible that is surrounded by a suitable insulating refractory and the outer casing. Metal electrodes that are immersed in the salt are connected to a low voltage power source. Placement of electrodes is important because the mean path through the salt determines the resistance of the salt which serves as heating resistor. The current should flow from electrode to electrode with a negligible amount of current passing through the work when it is immersed in the bath.

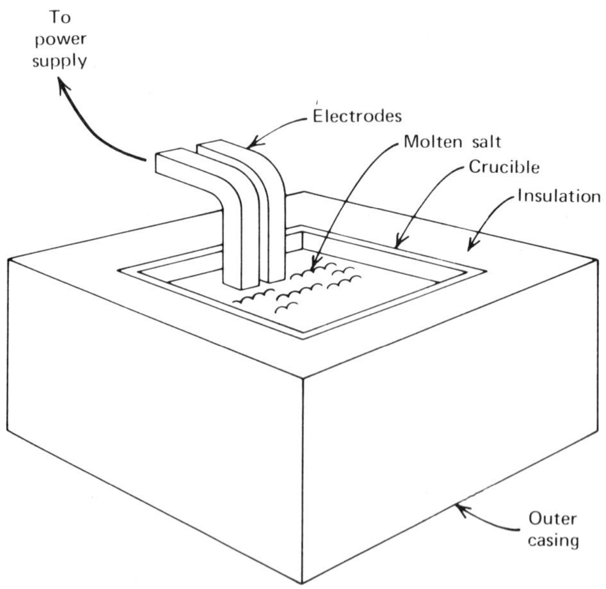

Figure 9. Salt-bath furnace.

Cold sodium chloride, either granulated or solidified after melting, has a high electrical resistance and low heat conductance (see Sodium compounds). It is, therefore, necessary to make provisions for melting the initial change of salt. This can be done by melting it in a radiation furnace and pouring the molten salt into the salt-bath furnace or by melting the salt in the salt-bath furnace with a torch. The torch method is necessary when a salt bath solidifies as a result of a planned shutdown or a prolonged power failure. If the shutdown is planned, heat-conducting metal tubes can be placed in the bath to improve heat transfer when remelting the salt.

Direct-Heat Electric-Resistance Furnaces

Direct-heat electric furnaces use the material to be heated as the resistor, and the furnace consists of an insulated enclosure to return the heat, a power source of suitable voltage, and means of attaching the power leads to the work (see Fig. 10). This type of furnace has several limitations which have prevented widespread use to date. Since the work is the resistor, it must have a uniform cross section between power connection points, and the material must be homogeneous. Varying sections or nonuniformities in the material would produce hot or cold spots in proportion to the change in electrical resistance. Also, a given furnace must be designed for work in which each piece to be heated has about the same resistance and power requirements. Although voltage and power can be controlled, a furnace designed to heat a part with a given cross section and length would probably not have the voltage required to heat a part of twice the length and half the cross section or have the current capacity to heat a part of half the length and twice the cross section.

There are additional problems in making the electrical connections to the work to be heated. The connection must have low electrical resistance to prevent overheating at the point of contact but such a connection has a low resistance to heat transfer, thus conducting heat away from the work. These problems have limited the use of direct-resistance heating mainly to heating of pipe, tubing, bars, or small identical parts. This heating is a one-piece-at-a-time batch operation and often does not use a furnace (insulated housing), instead it is used as a preheater for a forming operation that takes place as soon as the work reaches the desired temperature.

There are large-scale operations using direct-heat resistance furnaces. These are mainly in melting bulk materials where the liquid material serves as a uniform resistor.

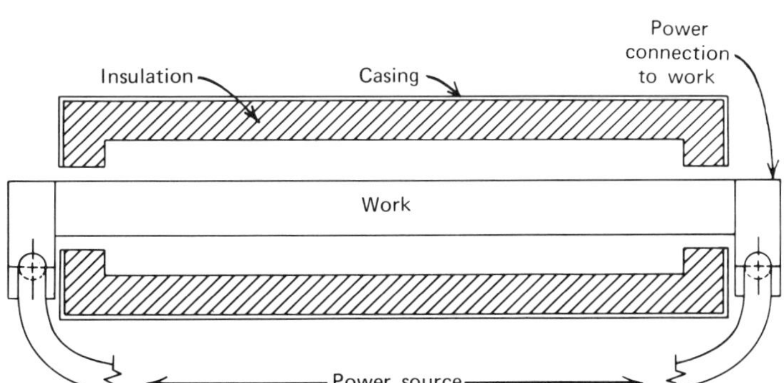

Figure 10. Direct-heat resistance furnace.

The material is contained in a crucible of fixed dimensions which, coupled with a given resistivity of the material, fixes the total resistance within reasonable limits. The most common application for this type of direct-heat electric resistance furnace is the melting of glass (qv).

BIBLIOGRAPHY

"Resistance Furnaces" under "Furnaces, Electric" in *ECT* 1st ed., Vol. 7, pp. 16–23, by Victor Paschkis, Columbia University; "Resistance Furnaces" under "Furnaces, Electric" in *ECT* 2nd ed., Vol. 10, pp. 269–278, by Victor Paschkis, Columbia University.

1. C. D. Starr and F. M. Cole, *Ind. Heat. Mag.,* (Oct. 1963).
2. *High Alloy Data Sheets, Heat Series,* Alloy Casting Institute Division, Steel Founders' Society of America, New York, 1973.
3. *How To Make Out With Moly,* Schwarzkopf Development Corp., Holliston, Mass.
4. *Hydrogen Atmospheres, Their Influence on Serviceability of Refractories, Paper 2,* Babcox & Wilcox Co., Refractories Division, Augusta, Georgia, 1971.

General References

W. Trinks and M. Mawhinney, *Industrial Furnaces, Principles of Design and Operation,* 5th ed., Vol. I, John Wiley & Sons, Inc., New York, 1961, 486 pp.

W. Trinks and M. Mawhinney, *Industrial Furnaces, Fuels, Furnace Types and Furnace Equipment—Their Selection and Influence upon Furnace Operation,* 4th ed., Vol. II, John Wiley & Sons, Inc., New York, 1967, 358 pp.

V. Paschkis and J. Persson, *Industrial Electric Furnaces and Appliances,* 2nd ed., Interscience Publishers, a division of John Wiley & Sons, Inc., New York, 1960.

RICHARD M. MILLER
Wellman Thermal Systems Corp.

FURNACES, FUEL-FIRED

A furnace is a device (enclosure) for generating controlled heat with the objective of performing work. In fossil-fuel furnaces, the work application may be direct (eg, rotary kilns) or indirect (eg, plants for electric power generation). The furnace chamber is either cooled (waterwall enclosure) or not cooled (refractory lining). In this article, furnaces related to metallurgy such as blast furnaces are excluded as they are covered under the associated topic; see, eg, Iron (see also Burner technology; Coal; Furnaces, electric).

Among the technologies in existence by ca 4000 BC, which included the manufacture of synthetic lapis lazuli, the development of the first true pottery kilns must rate as a major achievement (1). For polychrome pottery to be successfully manufactured, it was essential to separate the fire (fuel) from the work (clay pottery). The excavations performed in the near east (Mesopotamia in antiquity) indicate that these early kilns were probably constructed in accordance with Figure 1. With the smelting of copper in pit hearths predating by perhaps a millenium the start of the bronze age at ca 3000 BC, another important advance was the invention of the bellows at ca 2000 BC. Bellows supply combustion air where it is needed and are used as a means of raising furnace temperature.

Waterwall furnaces were employed by the ancient Greeks and Romans for household services. A water boiler, shown in Figure 2, found in Pompeii, was constructed of cast bronze and incorporated the water-tube principle (2). The earliest recorded instance of boilers performing mechanical work (130 BC) was Hero's engine (3) which is the earliest known reaction turbine. Furnaces, in general, and waterwall furnaces, in particular, were neglected for about the next 1600 years. In part, this may be ascribed to the fact that steam as a working fluid had no application until the invention of the first commercially successful steam engine at the end of the 17th century, followed by Newcomen's engine in 1705 with self-regulating steam valves. These machines were generally used for pumping water from coal mines (2). From Figure 3 the sequence furnace-boiler-work is very apparent and foreshadows the eventual emergence of the utility power plant which, in essence, is not different in concept.

In fire-tube furnaces developed in the 19th century, such as typified by the Scotch-Marine boiler (Fig. 4), thin currents of water contact a multiplicity of tubes; thus, the hot gases transmit heat simultaneously to all regions of the bulk of the water. Therefore, this boiler–furnace combination steams readily and responds promptly to load changes, and is, for a given amount of heating surface, the least expensive of all furnace–boiler installations (4). Furnaces of this type, such as the steam locomotive furnace–boiler design (see Fig. 5), had the obvious disadvantage that pressure was limited to ca 1 MPa (150 psi). The development of seamless, thick-wall tubing for stationary power plants (ie, water-tube furnaces) and other engines for motive power, such as diesel-electric, has in many cases eclipsed the fire-tube boiler. For applications calling for moderate amounts of lower pressure steam, however, the modern fire-tube boiler continues to be the indicated choice (5).

A key development in water-tube furnace design was the Babcock and Wilcox boiler of 1877 (Fig. 6) (3). This can be considered the direct evolutionary ancestor of the 1000 MW, steam power plants a century later (see Steam).

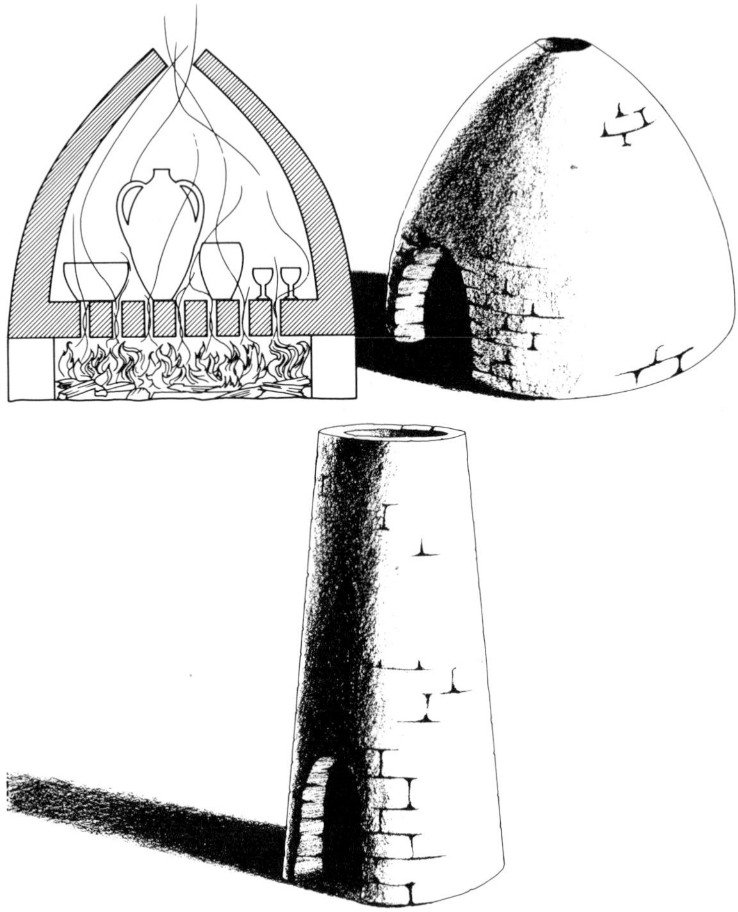

Figure 1. Mesopotamian pottery kiln of beehive construction 4000–3500 BC. Subsequent Egyptian pottery kiln of the period ca 3000 BC shown at bottom is of the familiar chimney shape.

Classification

Furnaces are either cooled (waterwall-enclosed chamber) or not cooled (refractory-lined chamber). In general, the basic structure roughly resembles either a rectangular box or a cylinder with variations for generally good reasons (6). For example, if the material being processed is a liquid, the furnace bottom must provide a bath. Hence, the open-hearth furnace is used for steel melting and refining, the reverbatory furnace for copper, and glass tanks for various materials. If the material tends to be lumpy, as in smelting, cupola melting, or lime burning, the furnace is constructed vertically to make gravity feeding possible.

The roof of the typical furnace collapses unless it is built as a sprung arch (self-supporting arch), or (a relatively modern development) is flat but suspended (along with the rest of the furnace) from an enclosing structural framework. All modern power-plant waterwall furnaces of any significant megawatt rating are of suspended construction.

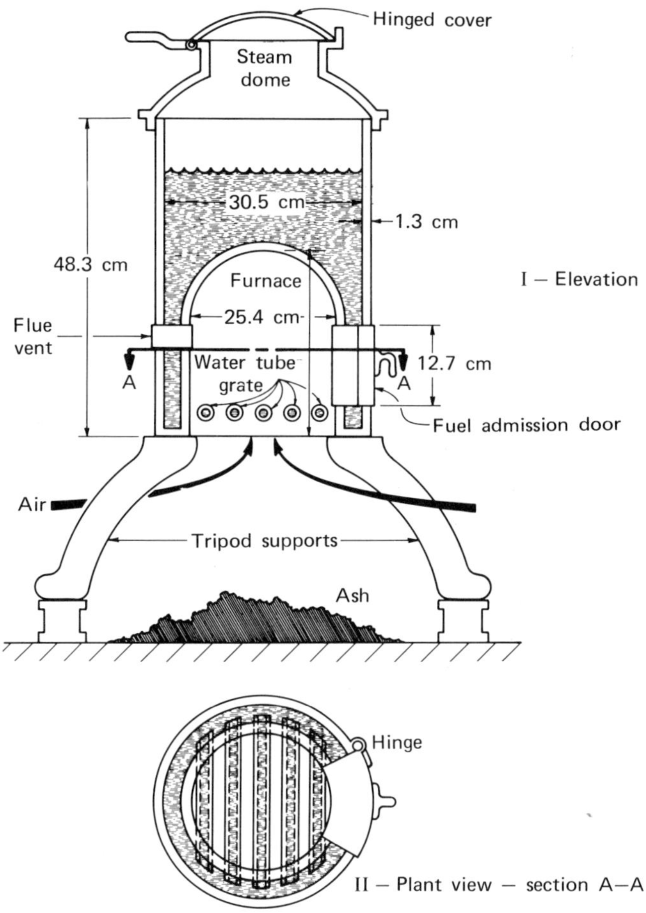

Figure 2. Pompeiian water boiler, earliest known example of waterwall furnace.

If a refractory-wall furnace is very large, the walls must be relatively thick to reduce wall losses and provide the necessary structural strength required to withstand the weight of the roof and thermal stresses. Since bricks are very expensive, large furnaces are built with two or even three different types of brick, with the outer brick generally the more highly insulating and the inside brick able to withstand the highest temperatures. The middle bricks are, therefore, often of lower quality and carefully selected since their softening temperature must not be exceeded.

Wall losses through most refractory walls are ca 10% of the heat supplied by the fuel. Losses increase with rising operating temperature. In special cases, eg, in glass tanks, losses can be as high as 30–35%. In these instances, very high values are required to maintain the refractory at a temperature below which it does not melt or collapse.

A furnace may be direct-fired or indirect-fired. The indirect-fired is known as a muffle furnace, and in such furnaces the combustion gases are separated from the stock being heated to prevent contamination.

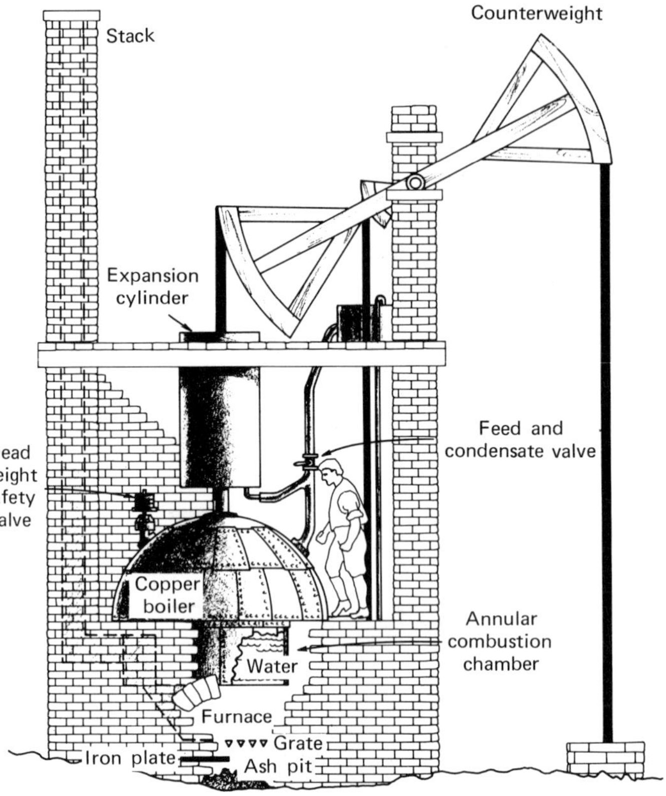

Figure 3. Newcomen's engine, 1705.

Fuels

The carbon–hydrogen ratio of oil, gas, and coal is a variable used widely in fuel technology as shown, eg, in Figure 7. However, its effect on flame temperature is usually misrepresented (7). It is commonly believed that flame temperature is a significant function of heat of combustion. But this is only true at very low values of heat of combustion, below ca 13.9 MJ/kg (6000 Btu/lb), because the largest weight of material moving through most furnaces is atmospheric nitrogen (7). This usually outweighs the contribution from the fuel by 10 to 1 or more (see Table 1). Coupled with a statistical requirement of roughly 0.0283 m^3 (1 ft^3) of cold air for every 105 kJ (100 Btu) released, fossil-fuel combustion is roughly equivalent to a release of 3.7 MJ/m^3 (100 Btu/ft^3) of combustion mixture at STP (or 5.4 MJ/m^3 (145 Btu/ft^3) for the combustion of CO and H_2). Consequently, most adiabatic flame temperatures are ca 2000 ± 100°C (6). This holds for the heat of combustion range of ca 14–140 MJ/kg (6,000–60,000 Btu/lb). The range includes CO and H_2 although they are not strictly fossil fuels.

Figure 8 illustrates this trend in adiabatic flame temperatures. Also indicated is the consequence of another statistical result, ie, flames extinguish at a roughly common low limit. This corresponds to a heat-release density of ca 1.9 MJ/m^3 (50 Btu/ft^3) of fuel–air mixtures, or half that for the stoichiometric ratio. It also corresponds to a flame temperature, as indicated, of ca 1220°C. As these are statistical

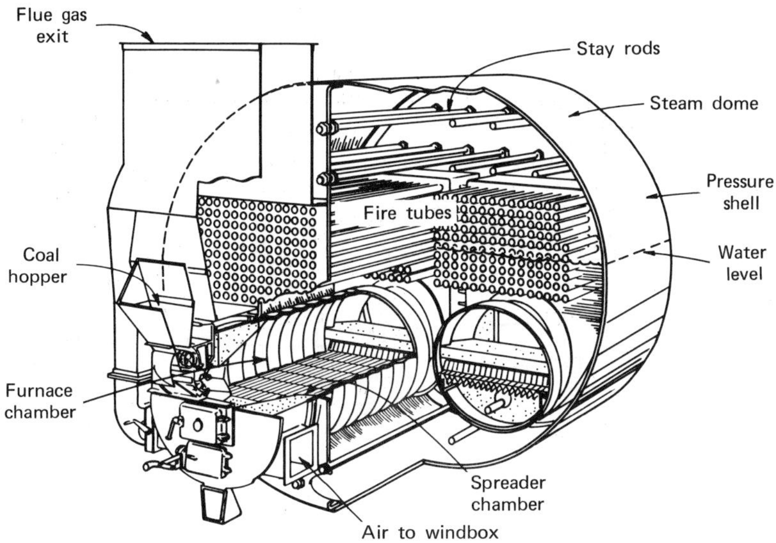

Figure 4. Scotch-Marine boiler, ca 1930.

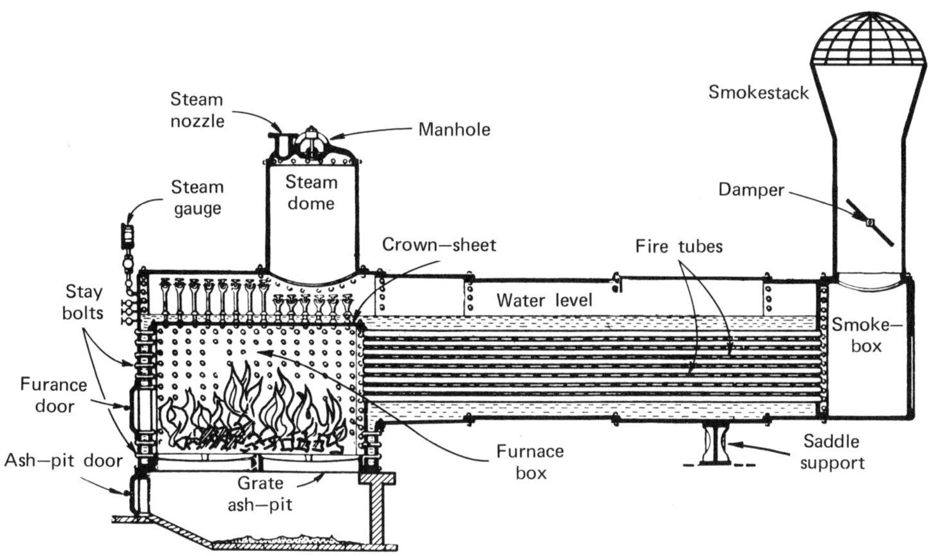

Figure 5. Locomotive furnace, ca 1870.

quantities, the same numerical values of flame temperature, low-limit excess air, and so forth, can be expected to apply to coal–air mixtures and to fuels derived from coal (see Fuels, synthetic).

The insensitivity of adiabatic flame temperature to heat of combustion does not necessarily apply to the operational flame temperature, which is the flame temperature found in an actual furnace (remembering that this refers to some average temperature). The higher excess air requirements at higher C/H ratios coupled with greater thermal loads on longer flames generally results in markedly lower operational temperatures as the C/H ratio increases.

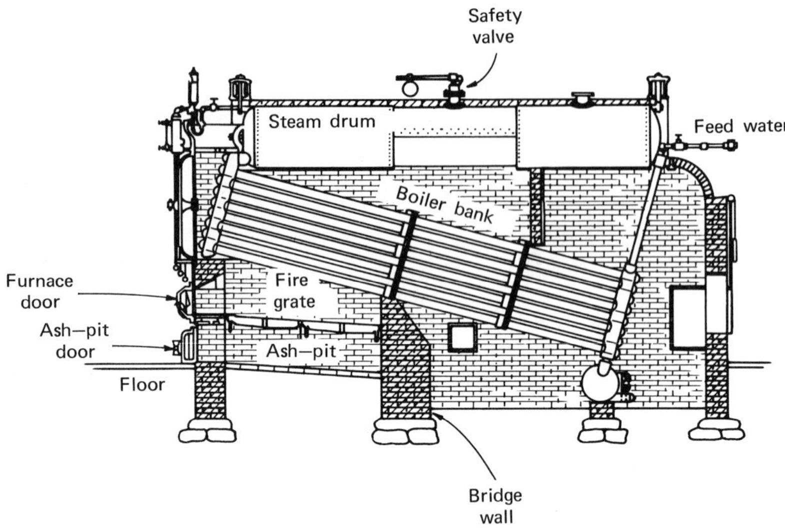

Figure 6. Early Babcock and Wilcox boiler, 1877.

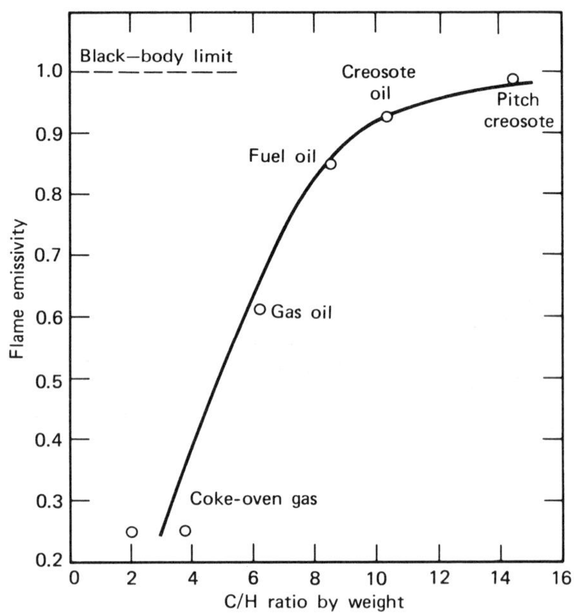

Figure 7. Influence of C/H ratio on flame emissivity.

The fuels listed in Table 2 are generally representative of fuels to be encountered over the range of industrial furnaces and, depending on the type (waterwall or refractory wall), exhibit operating temperatures considerably different from adiabatic values.

Oil and coal are currently the most important fuels in industrial use. Although the last few decades witnessed an upsurge in oil- and gas-fired power-plant furnaces,

Table 1. Theoretical Air Requirement

Fuel	Stoichiometric combustion air (dry), wt/wt fuel
gas	
blast furnace	1.54
average natural	15.86
coke oven	11.36
oil, bunker C	13.83
coal	
coke by-product	10.03
anthracite	10.64
bituminous, Eastern low vol	10.80
bituminous, Eastern high vol	10.38
bituminous, Midwest	8.54
subbituminous, Wyo.	7.03
lignite, N. Dakota	5.33
wood, 30% moisture	4.06
50% moisture	2.90
bagasse, 50% moisture	2.61

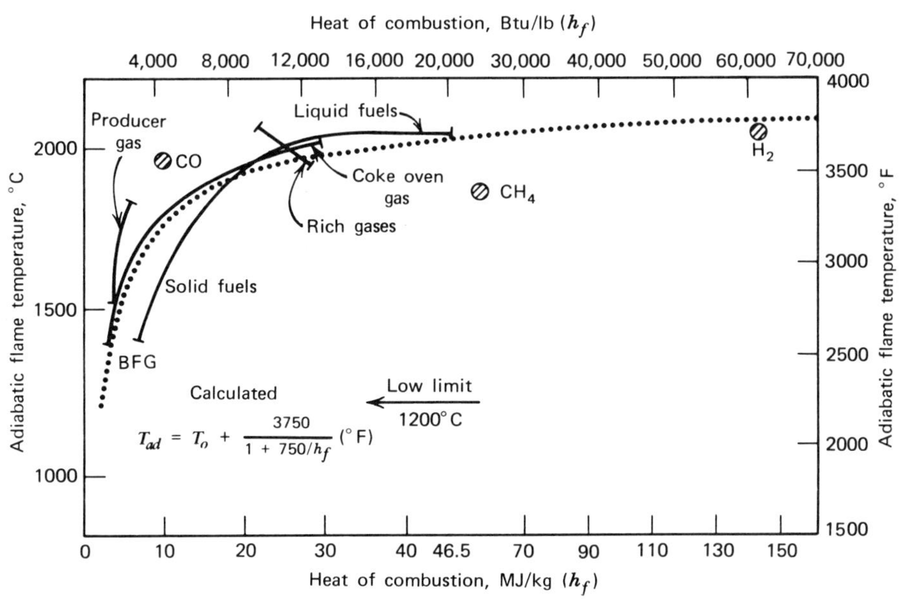

Figure 8. Variation of adiabatic flame temperature with heat of combustion. BFG = blast furnace gas. Note change of scale at 46.5 MJ/kg (20,000 Btu/lb).

the increasing worldwide shortage of these fuels in the 1970s has resulted in new power-plant furnaces being almost exclusively coal fired (8). In fact, most industrial furnaces are either oil-fired kilns or coal-fired power plants.

The various grades of oil marketed as fuels are designated by number and summarized in Table 3.

Table 2. Fuel Analyses, Wt %

Fuel	C	H_2	S	N_2	O_2	Ash	Moisture	HHV[a], MJ/kg[b]
gas								
blast furnace	15.86	0.20		53.97	28.91		1.06	2.62
natural	69.20	23.09	0.35	5.81	1.55			51.02
coke oven	48.62	20.03		4.91	26.44			39.95
oil (fuel)	85.5	11.5	1.6	0.7	0.7			43.03
coal								
coke by-product	85.0	0.7	1.0	1.3	0.5	10.7	0.8	29.51
anthracite	83.9	2.9	0.7	1.3	0.7	8.0	2.5	31.91
bituminous low vol	80.7	4.5	1.8	1.1	2.4	6.2	3.3	33.28
bituminous high vol A	76.6	4.9	1.3	1.6	3.9	9.1	2.6	31.65
bituminous high vol B	67.6	4.3	1.8	1.4	5.4	13.5	6.0	28.37
bituminous midwest	63.4	4.3	2.3	1.3	7.6	8.7	12.4	26.56
bituminous west	66.7	5.2	0.5	1.2	9.7	8.7	8.0	27.91
subbituminous A	60.6	4.6	0.5	1.3	13.3	7.3	12.4	24.65
subbituminous B	54.6	3.8	0.4	1.0	13.2	3.8	23.2	21.91
subbituminous C	47.3	3.3	0.7	0.6	12.1	9.5	26.5	19.12
lignite, North Dakota	42.4	2.8	0.7	0.7	12.4	6.2	34.8	16.77
lignite, Texas	41.0	3.1	0.7	0.7	12.3	9.9	32.2	16.86
wood, 30% moisture	34.3	4.2			30.8	0.7	30.0	15.39
50% moisture	24.5	3.0			22.0	0.5	50.0	11.00
bagasse	22.5	3.0			23.5	1.0	50.0	9.76

[a] High heating value.
[b] To convert MJ/kg to Btu/lb, multiply by 430.

The transition from a choice of multiple fossil fuels to various ranks of coal, with the subbituminous varieties a common choice, does in effect, entail a fuel-dependent size aspect in furnace design. A controlling factor on power-plant furnace design is, in fact, the ash content and composition of the coal. If the waterwall deposition thereof (slagging) is not properly allowed for or controlled, the furnace may not perform as predicted. Furnace size varies with the ash content and composition of the coals used. The ash composition for various coals of industrial importance is shown in Table 4.

Power-Plant Furnaces

In 1977, 13% of the electric power consumed in the United States was generated by nuclear energy (see Nuclear reactors). It is expected that before the year 2000 this will have increased to 25% with the remainder supplied by fossil-fuel and hydroelectric power (8–9). The bulk of electric power generation will probably be supplied by coal-fired power-plant furnaces supplying steam to turbogenerators. In terms of megawatts supplied, the coal-fired power plant is, therefore, the foremost component of importance in the energy supply system. As stated previously, power-plant furnaces are of waterwall type and are generally designed for steam pressures in the range of 12.4–24.1 MPa (1800–3500 psi); the latter value is referred to as supercritical, ie, >22.1 MPa abs (3208 psia).

An example of a modern, tangentially fired, supercritical, lignite-fuel furnace is shown in Figure 9. This unit, at maximum continuous ratings, supplies 2450 metric tons per hour superheat steam at 26.6 MPa (3850 psi) and 544°C, and 2160 t/h reheat

Table 3. Properties of Fuel Oils

Grade[a]	Color	Nominal analysis, wt %					Description	Specific gravity, 15/15°C	Viscosity at 37.8°C, mm^2/s (= cSt)
		S	$O_2 + N_2$	H_2	C	ash			
no. 1 (kerosene)	light	0.1	0.2	13.2	86.5	trace	distillate oil for vaporizing pot-type burners	0.8251	160
no. 2	amber	0.7	0.2	12.7	86.4	trace	distillate oil most commonly used for domestic heating; over the years this has become an increasingly complex blend (used also for power-plant warmup)	0.8654	268
no. 4	black	1.50	0.48	11.9	86.1	0.02	residual oil; does not require preheat; industrial application (alternative fuel in power plants)	0.9279	1500
no. 5	black	2.00	0.7	11.7	85.55	0.05	residual oil; requires preheat; for industrial installations and also fired in power plants	0.9529	5000
no. 6 (bunker C)	black	2.80	0.92	10.5	85.70	0.08	residual oil; requires preheat; the major fuel used in oil-fired power plants	0.9861	36,000

[a] No. 3 fuel oil was discontinued in 1948.

Table 4. Ash Properties of Typical Coals

	Eastern bituminous	Midwest bituminous	Western subbituminous	N. Dakota lignite
Ash composition, wt %				
acid SiO_2	44.8	44.7	33.5	13.8
acid Al_2O_3	28.0	18.4	18.0	10.7
base Fe_2O_3	17.8	18.3	7.3	8.4
base CaO	2.0	5.3	18.4	24.2
base MgO	0.6	1.1	4.7	7.6
base Na_2O	0.3	1.8	0.2	9.0
base K_2O	1.7	2.1	0.2	0.4
acid TiO_2	1.4	0.8	0.7	0.4
acid P_2O_5	0.3	0.2	0.2	0.6
acid SO_3	0.8	5.8	15.3	20.3
other	2.3	1.5	1.5	4.6
base–acid ratio	0.30	0.45	0.59	2.0
Fe_2O_3–CaO ratio	8.9	3.5	0.4	0.35
Ash fusibility temperatures, °C				
initial deformation	1210	1120	1130	1130
softening	1350	1150	1200	1200
fluid	1420	1240	1270	1350

steam at 5.32 MPa (772 psi) and 541°C. These are the values at the superheater and reheater outlet, respectively. Supercritical fluid-pressure installations are, however, only rarely needed. Most power plants operate at subcritical pressures in the range of 12.4–19.3 MPa (1800–2800 psi).

Although furnace sizes (dimensions for a given MW production) do not vary too widely between principal manufacturers, the type of firing employed by each is generally quite distinctive. This indicates that the furnace size is not strongly controlled by the type of firing system, particularly for pulverized-coal firing (10) (70% through 74-μm (200 mesh) screen for a mean size of ca 40–50 μm). The furnace needs to be sufficiently large to permit the oxygen enough time to penetrate (diffuse through) the blanketing CO_2 layer evolving from the burning coal particle. The residual ash particles are, of course, considerably smaller than the parent coal particle, on the order of a mean size of ca 10 μm before postcombustion agglomeration. Although flame temperatures should be high for combustion efficiency in order to minimize CO formation and combustible carbon loss, it is further required that the combustion products (gases) are sufficiently cooled to enter the convection banks below the temperature at which slagging occurs. These contradictory conditions (aside from pollution control requirements) influence the furnace size and have led to solutions such as tangential firing. Another popular and widely applied solution was the introduction of separate combustion chambers auxiliary to the main furnace. This is generally referred to as cyclone firing or the cyclone furnace (9). This furnace (see Fig. 10) is a horizontal, refractory-lined, water-cooled cylinder firing crushed coal (95% through a 4.76-mm (4-mesh) screen). Fuel and air are mixed by a swirling (centrifugal) motion which promotes turbulence. The fuel is burned at high heat-release rates in the range of 16.7–30.0 GJ/(m^3·h) [(4.5–8.0) × 10^5 Btu/(ft^3·h)] at combustion temperatures in excess of 1649°C. At these temperatures the coal ash content forms a liquid slag film on the

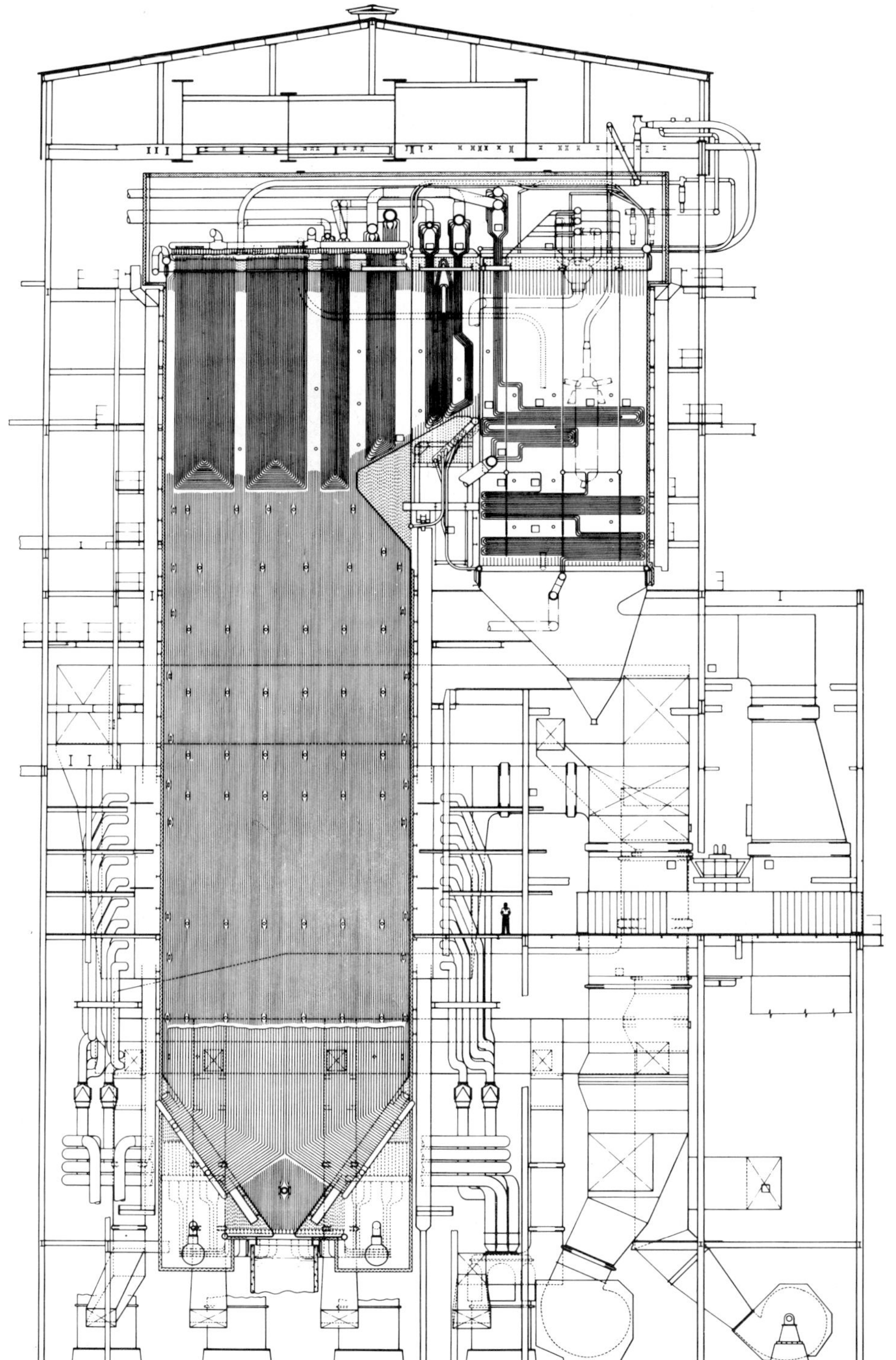

Figure 9. Martin Lake unit of Texas Utilities Services, Inc., 750 MW; a representative of modern supercritical design. Courtesy of Combustion Engineering, Inc.

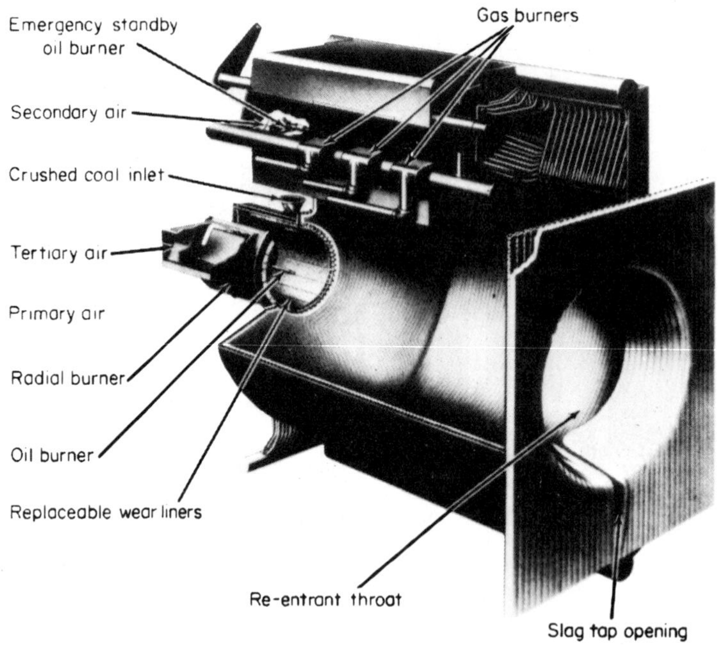

Figure 10. Cyclone furnace; in effect a high temperature, slag removal combustion chamber auxiliary to the gas-cooling boiler furnace. Courtesy of Babcock & Wilcox, Inc.

water-cooled refractory-lined wall because of the centrifugal force imparted to the coal particles. The gaseous products of combustion are discharged into the main gas-cooling boiler furnace and the film of molten slag on the walls continually drains away from the burner end and discharges through the tap opening into the main boiler furnace from which it drains into a tank for further disposition. In tangential firing systems, a cyclone (ie, vortex flame) is created in the main furnace, ie, the fuel and air admission system is corner-installed. For opposed firing, the fuel–air system is generally designed to impart an initial swirl to the reactants before furnace entry.

Industrial Furnaces

Generally speaking, industrial furnaces are an order of magnitude smaller than power-plant furnaces since the applications are usually on an individual basis (hospital complex, chemical plant, paper mill, etc) rather than feeding power to a regional electric grid (4). Like the power-plant furnace, the function of the industrial furnace usually is to generate steam, generally for a chemical process, mechanical power, or heating application, rather than electric power generation.

Package Boilers. These boilers are also known as shop-assembled boilers. Low-capacity units are shipped complete with fuel-burning equipment, safety and combustion controls, and boiler trim. Large-capacity units, because of tunnel and overpass shipping clearances and railroad flatcar limitations, are designed as multiple integrated-component packages such as the complete furnace–burner assembly, forced-draft fan assembly, and the like. In this category of steam generators, the fire-tube furnace competes with waterwall furnaces, particularly for duty in the capacity range of 205 kPa to 1.8 MPa (15–250 psig) and 1800–9100 kg steam/h. Figure 11 shows a modern fire-tube furnace of four-pass design. Like the fire-tube furnace, the waterwall package furnace, because of the temperature quenching effect of the immediate proximity of the cold sink (steam and water), requires a long flame path (ie, residence time) to permit the combustion reaction to go to completion.

Paper Mill (Recovery) Furnaces. The conversion of wood chips into cellulose and lignins by chemical pulping requires considerable amounts of steam. Paper is made from the cellulose fibers, which amount to ca 50% of the decomposition products, with the remainder (including bark) being burned to generate the process steam (see Pulp). In terms of power generated, the paper industry ranks second (after electric utilities), all of the generated power being consumed in-plant. The pulp is produced by either the kraft or the soda process. The main chemical constituent is sodium sulfate (kraft process, on the order of 80% of the production) with the sodium carbonate process (soda ash) roughly accounting for the remaining production. In the kraft process, wood chips are cooked under pressure (by steam heating) in an aqueous solution of NaOH and Na_2S, known as white liquor. In this operation the lignin binder is dissolved, freeing the cellulose fibers. The spent cooking liquor (containing the lignin), called black liquor, is successively steam and stack-gas dried for concentration; sodium sulfate (Na_2SO_4) is added to make up chemical losses. The main function of the recovery furnace is to reduce the sodium sulfate content of the black liquor to sodium sulfide (Na_2S). The heat obtained from the combustion of the organic black liquor constituent as well as from the supplementary fuel produce steam and molten smelt, largely composed of

Figure 11. Four-pass fire-tube boiler. Courtesy of Cleaver Brooks, a division of Aqua-Chem Corp.

Na_2CO_3 and Na_2S. It is tapped from the furnace bottom and dissolved in water to form green liquor for further processing. In the recovery furnace shown in Figure 12, liquor spray nozzles are used to inject the heavy black liquor into the furnace of the recovery unit.

The burning of the ligneous portion of the black liquor produces sufficient heat in the furnace to sustain flash drying of residual moisture, salt-cake reduction, and chemical smelting. The heat in the gas passing through the furnace, boiler, and economizer produces steam for power and process.

Large Industrial Furnaces (Waterwall). In contrast to the shop-assembled, waterwall furnace (package boilers), the large industrial furnaces are erected on site. Steam capacities are in the range of 45–450 t of superheated steam/h and generally serve industries such as steel, oil refining, automotive, sugar, and chemical. In distinction

Figure 12. Black liquor recovery furnace. Courtesy of Combustion Engineering, Inc.

to the power-plant utility furnace, the industrial furnace is of considerably smaller capacity and is likely to be of modular design and operationally flexible. It generally is designed to burn two or more fuels of the industrial range comprised of coal, oil, gas, refuse, wood products, bagasse, sewage sludge, and other solid, liquid, or gaseous waste products. Consequently, the furnace is often equipped with several firing methods in order to accommodate different fuels. For example, it may be arranged for tangential, horizontal, or stoker firing of single or multiple fuels.

Refractory-Wall Furnaces. Many types of refractory-wall furnaces are used for melting or heat treatment of metals, at temperatures ranging from less than 540°C to as high as 1760°C, such as ceramic and glass kilns and furnaces, ovens and dryers operating from 50°C to 600°C, and stills and retorts (7).

These furnaces may operate batchwise or continuous. In the batch, intermittent, or periodic types, the content is heated at the desired temperature for the stipulated time and then removed. In the continuous type, the charge moves at a predetermined rate through one or more heating zones to emerge in most cases at the end opposite the point of entry. Figures 13 and 14 are representative examples of typical, industrial refractory-wall furnaces.

Analysis

The rising demand for fuel efficiency and performance guarantees are imposing increasing requirements for analytical complexity. Furnace orientation in this area has developed only over the past 50 years. Starting with the rosin analysis of the 1920s (11), subsequent development of furnace and combustion analysis proceeded along distinctively dissimilar lines. The USSR approach (12–13) in particular includes considerable theoretical (mathematical) emphasis, whereas the European–British–American approach concentrates on a fuel and flame chemistry orientation (14–16). Today the analytical emphasis has shifted toward a heat-transfer oriented view of predicting furnace performance; (*1*) wall absorption rate and gas temperature profiles, and (*2*) pollutant formation. Among the various models (17), the Hottel zoning method (18) is gaining in popularity (19). Another analytical model that lends itself to engi-

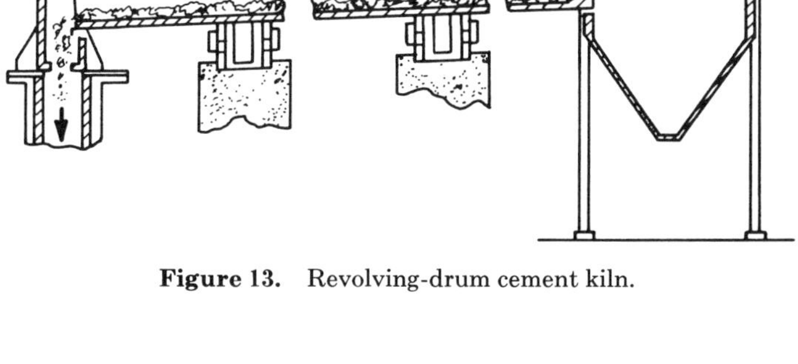

Figure 13. Revolving-drum cement kiln.

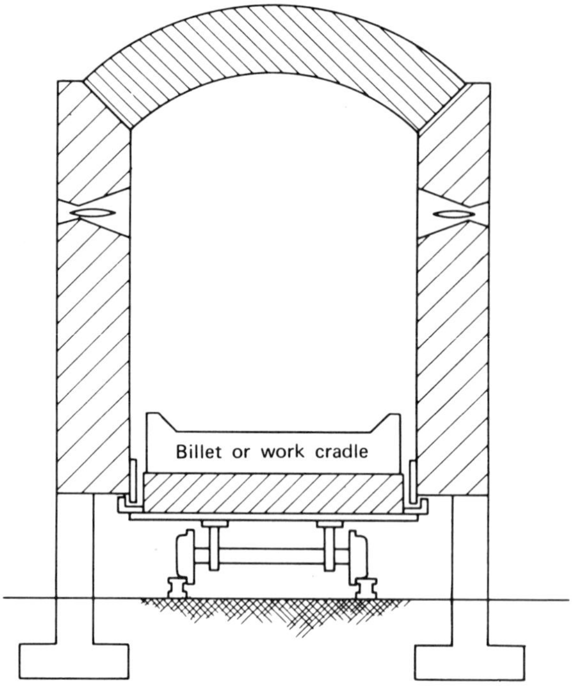

Figure 14. Car-type annealing furnace.

neering analysis of furnaces is generally referred to as PSR theory (perfectly stirred reactor theory) (20) and is, in fact, applied on an industrial basis (21) where it is important to obtain a quantitative evaluation of the emissivity of the combustion products (22).

The analytical mechanisms for predicting the corresponding pollutant formation associated with fossil-fuel-fired furnaces lag the thermal performance prediction capability by a fair margin. The most firmly established mechanism at this time is the prediction of thermal NO_x formation (23). The chemical kinetics of pollutant formation is, in fact, a subject of intense research (see Burner technology).

New Developments

New furnace concepts in an evolutionary stage include fluidized-bed furnaces (24), coal gasification furnaces (25), and MHD furnaces (26–27). The major attraction for the first two is the intrinsic low pollution potential. This is not the case for MHD where high cycle efficiency and the direct conversion of fluid energy in an ionized gas to electrical energy (see Coal, magnetohydrodynamics) is the motivation. These developments exist on an operational pilot-plant scale but, as yet, have no commercial importance.

BIBLIOGRAPHY

"Furnaces, Fuel-Fired" in *ECT* 1st ed., Vol. 7, pp. 23–37, by Matthew H. Mawhinney, Consulting Engineer; "Furnaces, Fuel-Fired" in *ECT* 2nd ed., Vol. 10, pp. 279–294, by Matthew H. Mawhinney, Consulting Engineer.

1. H. Hodges, *Technology in the Ancient World,* Alfred A. Knopf Publ. Co., New York, 1970.
2. T. Croft and R. B. Purdy, *Steam Boilers,* 2nd ed., McGraw-Hill Publ. Co., New York, 1937.
3. *Steam—Its Generation and Use,* 38th ed., Babcock & Wilcox Publishing, New York, 1972.
4. O. de Lorenzi, ed., *Combustion Engineering—A Reference Book on Fuel Burning and Steam Generation,* Combustion Publ. Co., New York, 1947.
5. J. Griswold, *Fuels, Combustion, and Furnaces,* 1st ed., McGraw-Hill Publ. Co., New York, 1946.
6. R. H. Essenhigh, "Energy Use, Efficiency, and Conservation in Industry," *Proceedings of the EPA/ERDA Symposium on Environment and Energy Conservation, Denver, Colorado, Nov. 1975, EPA-600/2-76-212, ERDA-47.*
7. R. H. Essenhigh, *Future Fuel Supplies for Industry: Bases for Choice, ASHRAE Transactions,* Vol. 84, American Society of Heating, Refrigerating, and Air-Conditioning Engineers, New York, 1978, Part 1.
8. J. C. Fisher, *Energy Crises in Perspective,* 1st ed., Wiley-Interscience, New York, 1974.
9. D. M. Considine, ed., *Energy Technology Handbook,* 1st ed., McGraw-Hill Book Company, 1977.
10. R. H. Essenhigh, *J. Inst. Fuel,* (Jan. 1961).
11. R. H. Essenhigh, *Ind. Eng. Chem.* **59**(7), 52 (1967).
12. L. N. Khitrin, *Physics of Combustion and Explosion,* University of Moscow Press, Moscow, USSR, 1957, trans. Israel Scientific Translations Program, Office of Technical Services, *OTS 61-31205,* U.S. Dept. of Commerce, Washington, D.C., 1962.
13. L. A. Vulis, *Thermal Regimes of Combustion,* Kazakh State University, USSR, G. C. Williams, trans. ed., McGraw-Hill Book Co., Inc., New York, 1961.
14. M. A. Field, D. W. Gill, B. B. Morgan, and P. G. W. Hawksley, *Combustion of Pulverized Coal,* BCURA, Leatherhead, Surrey, Eng., 1967.
15. M. W. Thring, *The Science of Flames and Furnaces,* John Wiley & Sons, Inc., New York, 1952.
16. B. Lewis and G. von Elbe, *Combustion, Flames and Explosions of Gases,* Academic Press, Inc., New York, 1961.
17. T. J. Tyson, "The Mathematical Modeling of Combustion Devices," paper presented at *Proceedings of the Stationary Source Combustion Symposium, Vol. I, U.S.E.P.A., Washington, D.C., EPA-600/2-76-1522, June 1976.*
18. H. C. Hottel and E. S. Cohen, "Radiant Heat Exchange in a Gas-Filled Enclosure," paper presented at *ASME/AIChE Joint Heat Transfer Conference, University Park, Pa., Aug. 1957, AIChE Paper No. 57-HT-23.*
19. K. K. Boon, "A Flexible Mathematical Model for Analyzing Industrial P. F. Furnaces," M.S. Thesis, University of Newcastle, Newcastle, NSW, Australia, Sept. 1978.
20. R. H. Essenhigh, "A New Application of Perfectly Stirred Reactor (P.S.R.) Theory to Design of Combustion Chambers," *Technical Report FS67-1(u),* Pennsylvania State University, Dept. of Fuel Science, Mar. 1967.
21. K. A. Bueters, J. G. Cogoli, and W. W. Habelt, "Performance Prediction of Tangentially Fired Utility Furnaces by Computer Model," paper presented at the *Fifteenth Symposium on Combustion, Tokyo, Japan, Aug. 25–31, 1974,* The Combustion Institute, Pittsburgh, Pa., 1975.
22. K. A. Bueters, *Combustion* **45**(9), (1974).
23. K. A. Bueters and W. W. Habelt, *NO_x Emissions from Tangentially Fired Utility Boilers,* AIChE Symposium Series, Vol. 71, American Institute of Chemical Engineers, New York, 1975, Parts I–II.
24. J. B. Anderson and W. R. Norcross, *Combustion* **50**(8), 9 (1979).
25. J. T. Stewart and T. D. Pay, "Coal Gasification Processes and Equipment Available for Small Industrial Applications," paper presented at the *Fifth Annual International Conference on Coal Gasification, Liquefaction and Conversion to Electricity, University of Pittsburgh, Pittsburgh, Pa., Aug. 1–3, 1978.*
26. H. B. Palmer and J. M. Beer, ed., *Combustion Technology—Some Modern Developments,* Academic Press, Inc., New York, 1974.
27. L. P. Harris, *Combustion* **50**(9), 21 (1979).

KEES A. BUETERS
Combustion Engineering, Inc.

ROBERT H. ESSENHIGH
The Ohio State University

FURS, SYNTHETIC

Fur is the fine, soft, thick, hairy covering or coat of a mammal, usually consisting of a double coating of hair that includes a layer of comparatively short, soft, curly, barbed hairs next to the skin, protected by longer, smoother, stiffer hairs that grow up through the shorter layer (1).

The more desirable animal furs, such as that of the mink, the fox or the sable, normally contain the two distinct layers of hair alluded to above, a soft, dense undercoat designated as the base or ground hair, underhair or underwool, and a longer protective covering commonly referred to as the guard hair (2). Not all furs contain both layers, eg, Persian lamb contains only the underwool portion, and monkey fur is made of guard hairs but no underhair. Seal and beaver furs, as commonly sold, contain only the underhair portion of the natural pelt since the guard hairs are removed in processing the pelt. Simulating natural furs in the broadest sense can encompass a wide spectrum of constructions from the simplest fleeces and pile fabrics in natural fur colorations to complex synthetic furs, duplicating to the degree possible the ombré-contoured multilayered appearance of a coat made from natural mink pelts.

Animal skins and their derivative, leather (qv), undoubtedly were the first materials used by prehistoric man to clothe himself. As mankind progressed, the art of fabric forming, first by the weaving process and later by the knitting process, was developed to provide a broad range of essentially two-dimensional fabrics which could be incorporated into wearing apparel (3–4). Fabrics designed for outerwear use in the colder regions of the world are normally quite bulky and heavy, and oftentimes uncomfortable to wear, although the construction, weight, bulk and fiber content of two-dimensional fabrics can be widely varied. Thus natural furs continued to be highly prized, not only for their esthetic qualities but also because they combined superior insulating qualities and suppleness in relationship to their weight (see also Insulation, thermal).

As a result, fur grew to become an important article of trade (5), both along the caravan routes of central Asia and between the various Baltic and North Sea ports of Europe. Much of Canada and northwestern United States was opened up as a direct result of the North American fur trade which peaked during the early and mid 19th century to supply the fur needs of Europe. As furs became a scarce commodity, owing to the rapid growth of the world's human population, fur substitutes have been produced from the materials and methods at hand to supplement the natural supply.

Yarn-insertion techniques in the weaving process were first developed in the Middle East to produce three-dimensional pile fabrics. These improvements spread west with the Arab conquest of Sicily in the ninth century, reaching Italy following the French occupation of that island in 1266. By 1315, Sicilian weavers were producing low pile velvets in Florence; and in the mid 16th century, the woven velvet trade was also well established in Genoa and Venice. From there, pile-fabric weaving spread to Arras in France, the low countries, Germany and ultimately England.

Commonly, in both woven and knit piles, a typical ground fabric was formed into which an extra set of yarns was locked to produce a surface structure consisting of loops. Once the loops were cut and the yarn forming the loops unraveled into its constituent fibers, a velvet or plush fabric resulted, depending upon the final height of the sheared pile. From here it was but a short step to producing a reasonably good imitation of an artificial fur by controlling the coloration, content and height of the yarn loops extending from the basic woven or knit structure.

As the Industrial Revolution developed during the 18th and 19th centuries, mechanical looms, knitting machines and finishing equipment (6) were progressively introduced that for the first time made practical the mass production of pile fabrics including velvets, plushes, velours, terrys, pile carpeting and true artificial furs. The first knitting machine was invented in England in 1589 by Lee, and Arkwright developed the first successful cotton-spinning frame in 1769. In the United States, Tompkins introduced spring-needle knitting in 1846; Heusch developed the modern cloth-shearing blade in 1850; and Barfuss in 1859 built the first model warp-knit or Raschel machine, as the tempo of technical development in fabric forming steadily accelerated.

The one missing ingredient for the production of a truly synthetic fur was the lack of suitable chemical fiber to simulate the hairs of an actual animal pelt (see Fibers, chemical). The 19th century production of artificial fur was limited to the natural fibers then available, eg, cotton (qv), flax, silk (qv), wool (qv), and similar natural animal hairs from the cashmere goat, alpaca, camel and vicuna (7–10) (see also Fibers, vegetable). Chemical fibers began with the cellulosic fibers, commercial manufacture of rayon (qv) in 1884, cuprammonium rayon in 1890, then viscose rayon in 1892, and acetate in 1918 (see Cellulose acetate and triacetate fibers). These were followed by the long-chain linear polymers, nylon (see Polyamides) in 1938, polyesters (qv), in 1939–1941, modacrylics and acrylics in 1950 (see Acrylic and modacrylic fibers) and polypropylene (see Olefin polymers) in 1951 (11–14), giving the synthetic fur technologist a full range of fibers from which to produce imitation furs.

Of the twelve to fifteen main categories of chemical fiber commercially available today, only three (acrylic, polyester, and modacrylic categories) are important for the manufacture of synthetic fur, although individual fibers from other categories (and even from some natural-fiber categories) continue to be incorporated (15).

Methods of Manufacture

Historically, the two methods used to produce artificial fur have been based upon the weaving and the knitting processes. In each instance, insertion of an extra set of yarns into the basic woven or knit structure results in a fabric face made up of yarn loops, which with cutting or shearing and the unraveling of the individual fibers making up the face yarn tufts remaining, produces the effect of a pile or furlike surface on one side of the fabric. Additional methods of forming furlike fabrics have been adopted, ie, there are other manufacturing techniques that can be used, depending upon the esthetic, economic, and performance characteristics desired in the fur fabric.

Direct material costs commonly constitute more than 50% of the total costs associated with such a fabric at the manufacturer's level. A much smaller fraction is attributable to direct labor costs, packaging, and distribution expenses. The direct material cost in turn can be broken down into the fiber content of the pile face and its backing structure. Although substantial differences exist between the labor costs of producing synthetic fur by the various competing manufacturing methods, the key determinants of a method's cost competitiveness remain the amount of value that must be added to its raw material content relative to the esthetics and performance of the resulting fabric.

From the standpoint of value added, the fibrous materials commonly used to manufacture synthetic fur fabrics can be categorized in ascending order as fiber tow, staple, flock, sliver, slit-film yarn, textured filament yarn, and spun yarn. Fabric substrate structures in ascending order of value added are nonwovens, knit and woven fabrics (see Nonwoven textiles). The 13 methods most commonly used to produce synthetic furs are given in Table 1, roughly ranked in their order of cost competitiveness based upon the value added to the materials employed.

Pile-Fabric Weaving and Spring–Needle Knitting Processes. The most widely-used methods for producing artificial fur were the pile-fabric weaving process and the spring–needle knitting process. Pile-fabric weaving (28–29), best epitomized today by the modern shuttleless double-pile-fabric weaving machines supplied by such European loom manufacturers as Güsken, Iwer, and Günne, continues to enjoy an enviable reputation for producing premium-grade synthetic furs, most particularly of the multilayered type, combining both base and guard hair (see Fig. 1). The spring–needle knitting machine (23), developed in the United States by Tompkins in 1846, traditionally dominated the manufacture of less expensive, single-layered furlike fabrics, fleeces, and shearlings.

Sliver-Knitting Processes. During the past 20 years, both processes have been displaced by the sliver-knitting process as the leading synthetic fur production method. Although patents granted in England and France in the early 1900s involved the sliver knitting process, G. W. Borg of the United States developed sliver knitting (17–19) into a commercial process to produce knit-pile fabrics in the late 1940s (33). Extremely flexible, this process can produce a wide range of both low and high pile fabrics, and with the introduction of electronic needle selection in the latest Morat and Wildman pattern sliver-knitting machines, the process can approach pile-fabric weaving in the variety and quality of the synthetic fur produced. Boasting high productivity, moderate capital costs and the use of less expensive sliver in place of yarn to produce its pile structure, this process enjoys over 50% of the United States market for synthetic fur, and its market share continues to grow. Of lesser importance in that market is fine-gauge tufting (31) which has become an important factor in the production of shearlings and double-needle–bar Raschel (warp knit) (26), which is being adopted for high volume production of low pile height, furlike fabrics.

Synthetic-Fur Finishing

Just as in the preparation of natural furs for outerwear use, finishing technique is as important or even more important to the final synthetic fur than the fabric-forming process itself. All of the forming methods listed in Table 1 produce, with few exceptions, a fabric with either a looped surface structure or a dull, often matted and shaggy, pile fabric that is not salable in its initial state.

Depending upon the final fur appearance being simulated, various finishing processes are employed to create the desired appearance. In all cases where a surface loop of yarn is first produced, the loop must be cut either on the forming equipment, by napping (30) or by a later shearing process (30,34–35). The formed fabric produced in practically all cases also must be sheared to evenly crop the pile prior to further processing. Where the face fiber is composed of yarn tufts, these tufts need to be thoroughly tigered (36–37), napped, or otherwise brushed to unravel the individual fibers from the tufts and thus create a pile rather than a tufted surface.

In most cases, particularly where a knit structure is involved, the fabric itself must be stabilized, usually by tentering it (35,38) and either heat-setting or applying a

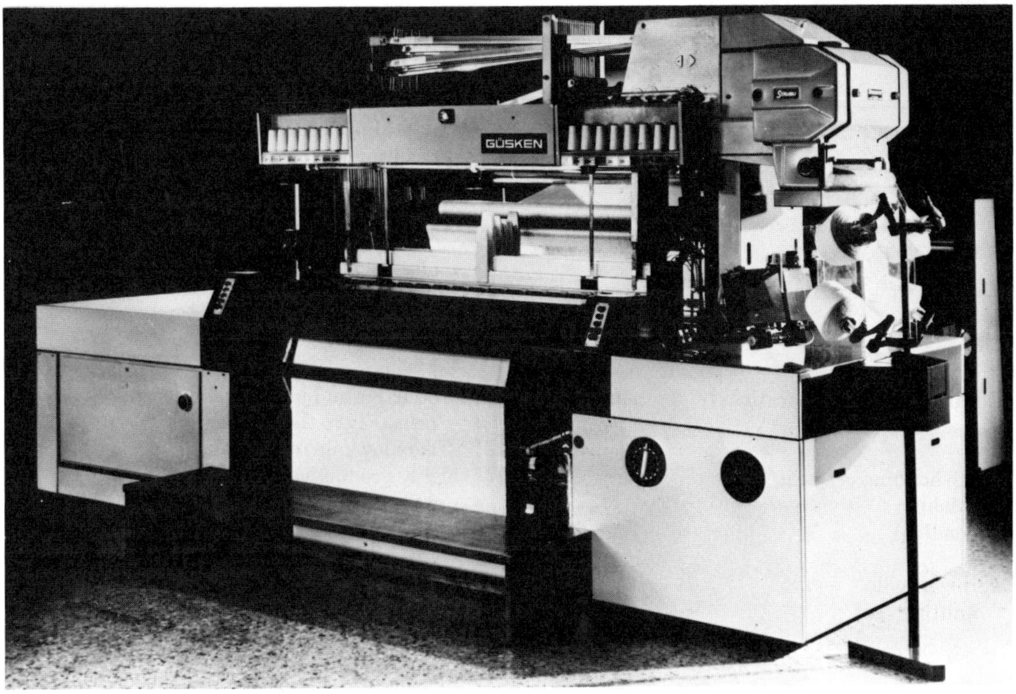

Figure 1. The Güsken GMV-80 double-layer velvet weaving machine, Jean Güsken G.m.b.H. & Co., KG, Viersen, FRG. Courtesy of Bolliger Corp.

polymer-based backcoating to the nonpile side of the backing structure. Even after these preliminary finishing steps have been completed, the resulting fabric exhibits a dull matte appearance best described as woolly. Although some synthetic furs are marketed with that appearance, most are finished further to resemble more closely the natural furs after which they are patterned.

Simulated shearling fabrics, designed to resemble a sheared sheepskin, require the simplest of finishing techniques. Such fabrics, normally knit or tufted from special polyester fiber (although acrylics also are used sometimes for the purpose), may be rough-sheared before being moistened and tumbled (39) to develop the characteristic clumped appearance of a natural shearling. Then they require only tentering and backcoating, or heat-setting to complete the finishing process. Where a free-flowing pile with a high luster is desired, synthetic-fur fabrics are commonly subjected to alternating electrifications and shearings in final finishing.

Electrification. The fiber used in the face of a pile fabric usually has been crimped in prior converting operations, for greater resiliency, bulk and ease of handling, resulting in a fabric with very diffused light-refraction characteristics and a generally matte appearance prior to final finishing. After its possible napping or tigering and initial rough shearing, the crimped face fiber is subjected to electrification (36–37), a combing and polishing technique that at the first pass removes most of the crimp from the fiber above the finish line by bringing the pile surface in contact with a revolving heated, grooved metal cylinder. By removing most face-fiber crimp, the first electrification irregularly elongates the pile fibers and thus is usually followed by another shearing action.

Table 1. Synthetic-Fur Manufacturing Methods

Method	Material content		Characteristics	Refs.
	Pile	Backing		
needle punch	fiber (staple)	fiber (staple)	low cost but limited as to resulting fabric esthetics and performance; probably best suited to the production of inexpensive fleecelike fur fabrics	16
sliver knitting	fiber (sliver)	yarn	relatively low cost but flexible method, capable of producing a broad range of both low and high pile synthetic furs	17–19
flocking	fiber (flock)	nonwovens, woven-fabric substrates	productive and low cost, limited primarily to the production of very low pile furlike fabrics	20–21
stitch bonding (Malipol bonding)	yarn	yarn	highly productive, low cost; best suited to the volume production of inexpensive fleeces	22
spring–needle knitting	yarn	yarn	productive, low cost; can produce a fairly broad range of synthetic furs in the low to medium pile height range	23
loop knit single-jersey plush knitting (circular)	yarn	yarn	productive, relatively low cost; commonly used to produce terrys or velours, this method is capable of producing a fairly broad range of low pile synthetic furs	19, 24–25
pile-warp knitting	yarn	yarn	highly productive, reasonably low cost; suited to the volume production of lower pile height fur fabrics	26
modified double knitting (circular)	yarn	yarn	not a particularly productive or flexible method but currently attractive because of the availability of underutilized double-knitting equipment; again generally limited to lower pile height synthetic furs	27
weaving	yarn	yarn	the historic method to produce synthetic furs, now limited primarily to the production of premium-grade fabric owing to the economics involved; capable of producing the multilayer hair effects of the true fur	28–29
surface pile raising (napping, teaseling, etc)	yarn	yarn	another historic method of producing fleeces and similar low pile, furlike fabrics by raising surface fibers from two-dimensional woven or knit goods by napping, teaseling, or similar fiber-raising methods	30

Table 1 (*continued*)

Method	Material content		Characteristics	Refs.
	Pile	Backing		
fine-gauge tufting	yarn	nonwovens, woven-fabric substrates	highly productive, low cost, best suited to the production of shearlings and similar fabrics	31
I-tuft bonding	yarn, fiber	nonwovens, woven-fabric substrates	productive, capable of producing a wide range of synthetic-fur fabrics	32
U-tuft bonding	yarn	nonwovens, woven-fabric substrates	although capable of producing synthetic-fur fabrics, most developmental efforts to date have been directed toward the manufacture of pile carpeting by this method	33

Subsequent electrifications remove most remaining fiber crimp but the primary objective of these further electrifications is not to straighten the upper portions of the face fibers as much as it is to polish them, ideally evenly around their circumference to achieve the desired final fiber luster and flow. Each electrification may be followed by a shearing until the desired surface appearance is achieved, although most or all shearings may be omitted in the production of shaggy pile fabrics such as simulated monkey fur. However, typical highly finished synthetic furs, such as simulated seal, beaver, or rabbit, may be napped or rough-sheared, or both, then tentered and stabilized, tigered, and finally alternately electrified and sheared to achieve the desired luster and hand. The shearing operation today is performed by rotating helical knife arbors, not unlike a reel-type lawn mower (see Fig. 2).

Shearing. When producing highly finished, multilayered synthetic furs simulating those of mink, sable, or fox, similar final finishing techniques are employed but the face-fiber blend is altered to contain both heat-shrinkable and nonheat-shrinkable fibers (19). An initial rough-shearing operation shears both the shrinkable and nonshrinkable fiber to a common height, before tentering and stabilizing the fabric. Then when heat is applied to stabilize the fabric, the heat-shrinkable fibers are heat-shrunk below the level of the nonheat-shrinkable fibers, thereby simulating the lower underhair portion of the natural fur being copied. The nonheat-shrinkable fibers, often of a heavier tex (denier), remain above this base and become the guard hairs of the finished synthetic fur.

Despite subsequent electrifications and shearings, two (or more) distinct layers remain visible in the final fabric. The final height relationship of these multiple layers can be established by the amount of shrinkage built into the heat-shrinkable fiber portion. In this manufacturing process, various chemicals or water (36), or both, may be applied to the face fibers at different points within the process to further influence the resulting smoothness and luster of the face pile.

Finishing. Finishing agents can be introduced during face-fiber or yarn preparation stages, such as picking, carding, or spinning, at napping or immediately prior to an electrification. These chemicals can include various waxes, resins, silicones, and fluorochemical-based topical finishes designed not only to enhance the appearance of the final fabric but also to provide soil release, water repellency, and other improved performance characteristics (see Textiles).

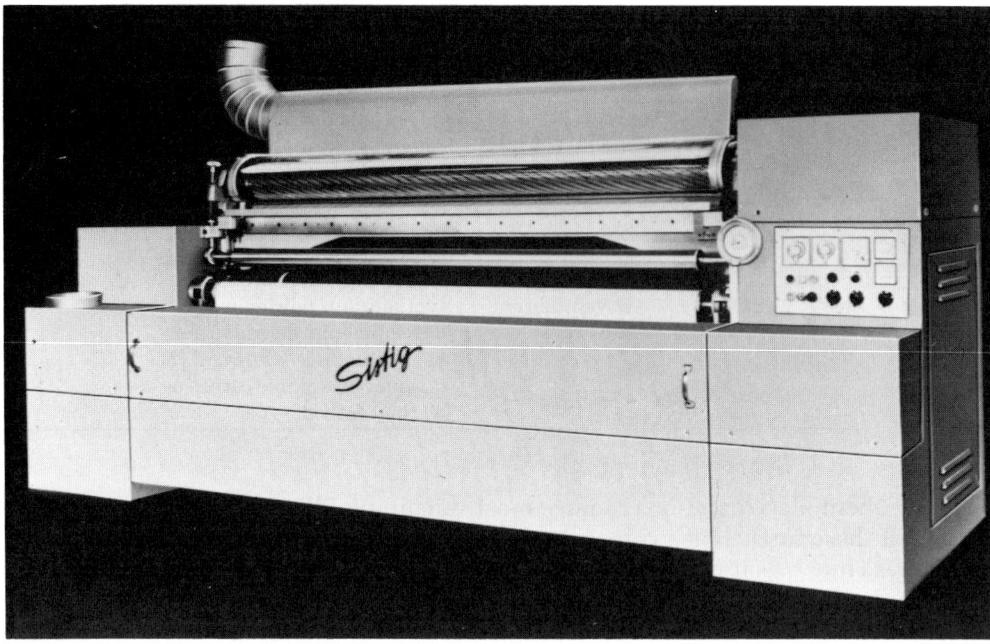

Figure 2. The Sistig polishing shear electrifier, Leo Sistig KG, Krefeld, FRG. Courtesy of Bolliger Corp.

Just as in the finishing of natural furs, many special effects can be created through the use of the special finishing equipment available to the fur trade. For example, embossing (36) using a heated engraved embossing roll, can be used to simulate the curl of such furs as karakul, Persian lamb, or broadtail. A wide variety of surface effect, swirling, and patterning machines are offered that permanently rearrange the lay of the face fibers and thus create attractive designs in the pile. Colorants can also be used to produce distinctive furlike effects.

Manufacture of the more complex synthetic furs usually requires the use of stock-dyed or yarn-dyed materials woven or knit into the pile face in some form of pattern through the use of jacquard- or electronic-patterning devices. Monochromatic furlike fabrics, such as those produced by the spring–needle process, commonly are piece-dyed from the greige after the fabric is formed but before it is finished. Usually piece dyeing is limited to lower pile constructions because of problems relating to pile-matting and even dye penetration.

Another popular approach to synthetic fur decoration with colorants is printing (40). Historically, roller and rotary-screen printing and spray painting have been used to imprint on pile fabrics typical animal fur patterns such as the spots of a cheetah or the stripes of a tiger. In recent years heat-transfer printing, because of its economics and ability to provide greater pattern fidelity, has begun to replace the older printing methods, particularly where polyester-faced synthetic furs are involved. As in the piece dyeing of pile fabrics, dye penetration can be critical, and the most successful heat-transfer printing of synthetic furs has been limited to lower pile heights up to ca 1 cm.

Energy Requirements

Energy requirements for the production of synthetic furs are relatively moderate; the main energy commitments occur in the prior processes of fiber manufacture, and fiber-or yarn-dyeing. Besides electricity for the fabric-forming and finishing equipment, and the compressed-air, vacuum and air-exhaust systems common to most textile mills, the other energy source usually required is the natural gas used to heat-set, or dry and cure the backcoating applied to the fur fabric during the tentering operation to stabilize it.

Natural gas also may be used to heat the electrifier cylinders and embossing rollers employed in finishing, although such rolls usually are heated electrically. It is probable, if the supply of natural gas is ever sharply curtailed, that its use in heating and drying will be replaced by more efficient electricity-based technology such as radio-frequency or electron-beam drying machines. Because of high capital costs, these methods are not economical currently; however, as energy costs rise, this situation will change. For those synthetic fur manufacturers whose processes permit them to piece-dye from the greige, their dyehouse represents a major consumer of energy.

Recently, the DOE subsidized the design and installation of a solar-collecting field and distribution system to meet part of the process hot-water requirements of a southern pile-fabric manufacturer's dyehouse (41) (see Solar energy).

Despite the recent large increases in the price of all forms of energy, the current cost of energy used in the actual manufacture of synthetic furs is relatively small, normally falling below 3% of the total cost of sales.

Health and Safety Factors

Toxicology. Toxicology problems relating to the manufacture of synthetic furs appear to be minor. Owing to the importance of the acrylic and modacrylic fibers (qv) to the synthetic-fur industry, concern was recently expressed about the possible release of residual monomeric acrylonitrile, considered to be carcinogenic, from such fibers during their processing or use. The level of possible acrylonitrile release is at or below the present limit of detection under normal temperature conditions, and approximates 0.0005% over an extended period at 180°C, an extreme condition not likely to be approached in either the processing or use of acrylic or modacrylic fibers.

The Toxic Substances Control Act (TSCA) (42–43) calls into question all chemicals presently used in the manufacture of synthetic furs. Probably any impact will be slight since the synthetic-fur industry has not adopted widely the formaldehyde-based or urea–formaldehyde-based topical treatments commonly used in processing permanent-press fabrics; therefore, it is not threatened by the possible ban on the future use of these compounds. In addition, flame-retardant topical coatings such as tris(2,3-dibromopropyl) phosphate, also have not been used widely on synthetic-fur fabrics (see Flame retardants, phosphorus compounds).

It is likely that any impact of the TSCA will be felt in two areas: (*1*) the choice of dyestuffs and auxiliaries used in imparting color to the fibers, and yarns used as raw materials by the industry, or to the resulting fabrics, where piece-dyeing is used; and (*2*) the choice of backcoatings used to stabilize the resulting synthetic furs. Benzidine-based dyestuffs already are being replaced, and unquestionably the use of other specific dyestuff families may eventually be eliminated.

Fortunately, the synthetic-fur industry will be spared the cost of compliance with cotton (qv) dust standards required by OSHA that could be imposed upon that segment of the United States textile industry that uses cotton fiber in significant volume. Very little cotton fiber is currently used in the manufacture of synthetic furs; the small amount used, primarily in backing yarns, will probably be eliminated to minimize compliance costs.

Flammability. Of all the problems facing the synthetic-pile industry, product liability claims, based upon the potential flammability (44–45) of its product, is probably receiving the greatest industry attention. Only fabrics made from asbestos, glass, or certain metallic fibers are completely fireproof, although certain high performance organic fibers made from selected polyamides (qv), chlorotrifluoroethylene–tetrafluoroethylene (CTFE–TFE) copolymers (see Fluorine compounds, organic), and aromatic polyamides are highly flame resistant (see also Aramid fibers). The objective of current flammable-fabric legislation is to provide clothing or household textiles that will not burn too rapidly when exposed to reasonably foreseeable flame sources (see Flame-retardant textiles).

Most synthetic furs are used in apparel applications and, therefore, fall under Part 1610 of the *Flammable Fabrics Act* (formerly *Commercial Standards 191-53*). This standard first took effect in July 1954, and provides for the use of the 45° angle tester to measure ease of ignition and speed of flame spread. Fabrics tested under this standard are classified into three categories, 1, 2 and 3, with those fabrics falling in category 3 deemed to be highly flammable and their use in wearing apparel prohibited. Most common textile fabrics, regardless of fiber type, do not fall into class 3 and the main impact of this law has been to restrict the manufacture of garments from extremely lightweight fabrics or pile fabrics.

The use of truly fireproof or flame-resistant fibers does not meet customer mandated performance or esthetic considerations. Pile fabrics also do not lend themselves readily to wet processing; therefore, most flame-resistant, topical finishes are not used. Of the three common fibers used in synthetic-fur manufacture, the modacrylics are inherently flame resistant; the polyesters commonly ignite with difficulty and burn slowly; and the acrylics ignite and burn readily. Flammability characteristics of a pile fabric normally vary with pile height, density, and fiber finish. The backing structure, the materials used in it and the backcoating applied to it all can appreciably affect a fabric's flammability rating.

The synthetic-fur-fabric designer, therefore, must not only take into consideration the esthetics and performance characteristics desired in the fabric but also its flammability rating. Most pile-fabric construction composed primarily of modacrylic fibers or a vinyl polymer such as the Cordelan (Japan) should be naturally flame resistant (ie, self-extinguishing) and readily meet the requirements of the apparel flammability standard (see Acrylic and modacrylic fibers; Vinyl polymers). Many pile fabrics made from polyester fibers (qv) also pass, although here more care must be taken to see that fiber content, pile height, fabric structure, and backcoating are balanced carefully to produce a satisfactory fabric.

In the cases of fabrics made predominantly from the acrylic family, a certain portion of a flame-resistant fiber, most commonly a modacrylic, may have to be blended into the fabric face in order to achieve an acceptable flammability rating. Certain fibers, such as the nonflame-resistant cellulosics, rayon, acetate, and cotton, are usually to be avoided at all costs, even if they meet other performance specifications.

Although most synthetic furs are used in wearing apparel, a small proportion is used as furniture upholstery or in the form of small carpets or rugs. Different federal flammability standards apply to these categories, eg, FF-1-70 in the case of small rugs or carpets, and the proposed new upholstered-furniture flammability standard for the upholstery fabric category. It may be necessary to modify design criteria for such synthetic fur items if the specific flammability requirements of these standards are to be met.

BIBLIOGRAPHY

1. *Webster's Third New International Dictionary, Unabridged,* G. & C. Merriam Company, Springfield, Mass., 1976, p. 923.
2. M. Bachrach, *Fur—a Practical Treatise: Geography of the Fur World,* Gordon and Breach, Science Publishers, Inc., New York, 1977.
3. "The History of Textiles from Prehistoric Times to the 20th Century," *Encyclopedia of Textiles*, 2nd ed., Prentice-Hall, Inc., Englewood Cliffs, N.J., 1972, pp. 153–192.
4. *The New Encyclopaedia Britannica,* 15th ed., Vol. 18, Encyclopaedia Britannica, Inc., Chicago, Ill., 1979, pp. 171–172.
5. Ref. 4, vol. 7, p. 811.
6. "Inventors and Inventions" in ref. 3, pp. 193–196.
7. M. Hawkins, ed., *Textile Handbook,* 4th ed., American Home Economics Association, Washington, D.C., 1970, pp. 7–12, 54–55.
8. M. Kapper, *Fabric Almanac,* 2nd ed., Fairchild Publications, Inc., New York, 1971, pp. 153–156.
9. M. D. Potter and B. P. Corbman, *Textiles, Fiber to Fabric,* 4th ed., Gregg Division, McGraw-Hill Book Company, New York, 1967, pp. 1–7, 197–276.
10. R. R. Kaswell, "Classification and Preparation of Natural Fibers" in *Wellington Sears Handbook of Industrial Textiles,* Wellington Sears Company, Inc., New York, 1963, pp. 5–35.
11. "Classification and Manufacture of Man-Made Fibers" in ref. 10, pp. 37–68.
12. Ref. 9, pp. 1–15, 277–398, 411–417.
13. *Man-Made Fibers Fact Book,* Man-Made Fiber Producers Association, Inc., Washington, D.C., 1978, pp. 4–24.
14. *Man-Made Fiber and Textile Dictionary,* 4th ed., Celanese Corporation, New York, 1978.
15. I. B. Wingate, *Textile Fabrics and their Selection,* 6th ed., Prentice-Hall, Inc., Englewood Cliffs, N.J., 1970, pp. 114–115.
16. Ref. 3, p. 492.
17. N. Hollen and J. Saddler, *Textiles,* 4th ed., MacMillan Company, New York, 1973, pp. 180–181.
18. Ref. 9, pp. 97–98.
19. Ref. 3, p. 373.
20. Ref. 3, p. 392, 492.
21. Ref. 17, pp. 160–161.
22. Ref. 9, p. 122.
23. *The Tompkins Knitting Machine,* Tompkins Brothers, Co., Inc., Syracuse, N.Y.
24. Ref. 17, p. 180.
25. J. C. Stevens, *Knitting Times* (National Knitted Outerwear Association, New York), 142–146, 230 (Apr. 18, 1977).
26. *Double Plush Raschel Machine,* Karl Mayer Textilmaschinenfabrik G.m.b.H., Obertshausen, FRG, 1973.
27. J. Getgood, "Double-Knit Pile Technology," *paper presented at Knitted Pile Fabrics Seminar, North Carolina State University, Raleigh, N.C., May 23, 1978.*
28. Ref. 9, pp. 76–77.
29. Ref. 3, pp. 372.
30. Ref. 9, p. 142.
31. Ref. 17, pp. 158–160.
32. D. Ward, *Text. Mon.* (Sept., Oct. 1978).
33. R. J. Casey, *Mr. Clutch, The Story of George William Borg,* The Bobbs-Merrill Company, New York, 1948, pp. 213–217.

34. Ref. 10, pp. 125–127.
35. Ref. 17, p. 221.
36. Ref. 9, p. 98.
37. Ref. 17, p. 180.
38. Ref. 9, pp. 134–135.
39. B. Magin, *Melliand Textilber.* **55,** 285 (1974).
40. "The Printing of Textiles" in ref. 3, pp. 451–473.
41. A. T. Pedersen, *Daily News Record* (Fairchild Publications, New York), (Feb. 23, 1977); (Oct. 30, 1978).
42. *Bus. Week,* 44B (Jan. 29, 1979).
43. S. M. Suchecki, *Text. Ind.,* 60 (Mar. 1979).
44. Ref. 17, pp. 241–242.
45. K. A. Howry in cooperation with C. A. Baker, *Mod. Text. Mag.* **59**(4), 19–21, 46 (1978); **59**(5), (1978); **59**(7), 34–36 (1978); **59**(8), 22–25, 30 (1978).

General References

References 2, 3, 7, 8, 9, 10, 13, 14, 15, and 17 are general references.

G. G. Denny, *Fabrics*, Eighth Edition, J. P. Lippincott Company, Philadelphia, Pa., 1962.

P. Perry, *Fur and Fur-Like Fabrics, Everything about Sewing Fur and Fur-Like Fabrics from Vogue Patterns,* Butterick Fashion Marketing Company, New York, 1971.

GUY KIECKHEFER
NORMAN C. ABLER
Borg Textile Corporation

FUSION ENERGY

Fusion is generally acknowledged to be the ultimate resource for mankind's long-term energy needs. Its attractiveness stems primarily from the fact that the fuel reserves for fusion are virtually limitless. Even conservative estimates lead to the astounding conclusion that there is sufficient fusion fuel to supply our growing projected requirements for a billion years, a period that exceeds the remaining lifetime of the solar system (see also Power generation).

The first realization of fusion's potential as a vast source of energy dates back to the 1930s when the nuclear burning mechanism of the sun was elucidated (1). In a complex sequence of reactions starting with hydrogen, atomic nuclei are fused to form heavier species. Because of a mass defect exhibited by the reaction products, large amounts of energy are released, as dictated by the well-known Einstein equivalence relating the energy E to the mass defect Δm: $E = \Delta mc^2$ Large-scale fusion energy production on earth was demonstrated dramatically in the early 1950s with the explosion of thermonuclear fusion bombs. Subsequently, a major international research effort was undertaken in order to harness this awesome power on a controllable scale for peaceful purposes. Several impressive advances over the past ten years have led to a well-founded feeling of optimism that fusion energy will become practical during the early 21st century.

In order to effect a fusion reaction between two atomic nuclei, it is necessary that they be brought together closely enough to experience each other's nuclear forces. Because all nuclei are positively charged, they repel one another via the electrostatic law of the repulsion of like charges. This electrostatic barrier can be surmounted by imparting sufficient kinetic energy to the reacting species. These repulsive forces increase rapidly with the magnitude of the nuclear charge; therefore, nuclear fusion is always limited to the lightest elements and their isotopes with the lowest atomic numbers.

Even with this constraint, close to twenty exothermic fusion reactions have been identified. These involve hydrogen and its isotopes deuterium (^2H or D) and tritium (^3H or T) and isotopes of helium (^3He), beryllium (^9Be), lithium (^6Li), and boron (^{11}B). The reactions of deuterium, tritium, and ^3He are the easiest to initiate, ie, they have the highest reaction cross section at the lowest reactant energies.

Deuterium–Tritium Fusion

The so-called D–T reaction, $D + T \rightarrow {}^4He + n$, is especially attractive because of its relative ease of ignition. The products of this reaction are an alpha particle (the helium nucleus, ^4He) and a free neutron (n) carrying kinetic energies of 3.5 and 14.1 MeV, respectively. In an electric power-generating facility the neutrons would be absorbed in a blanket surrounding the fusion region, and their kinetic energy would be converted to heat. Conventional power conversion systems would be used to transform this heat into electrical energy. Note that fusion reactions are extremely energetic, and yields are measured in units of MeV (millions of electron volts), where 1 MeV = 1.6×10^{-13} J (3.8×10^{-14} cal or 1.51×10^{-16} Btu).

Another set of reactions of practical interest involves only deuterons. The D–D reaction can proceed along either of the two following pathways with roughly equal probabilities: $D + D \rightarrow T$ + proton (4.0 MeV), or $D + D \rightarrow {}^3He + n$ (3.2 MeV). Finally, the D–He^3 reaction, $D + {}^3He \rightarrow {}^4He$ + proton, is noteworthy not only because of its high energy release (18.3 MeV) but because its reaction products are charged particles, offering the possibility of high-efficiency, direct energy conversion. A variation of the above reaction schemes is the so-called catalyzed D–D reaction whereby the external feedstock is deuterium but the ^3He and T produced in the D–D reactions are recycled and burned *in situ* to enhance the net energy yield.

Because of its relatively high reactivity (2), the D–T fusion-fuel cycle will be employed in the first generation of fusion reactors. Deuterium is heavy hydrogen (^2H) and occurs naturally in the ratio of 1:6700 relative to ordinary hydrogen: 30,000 kg water contains one kilogram deuterium. The separation of deuterium from water is a relatively simple and inexpensive process. Tritium, on the other hand, is a short-lived radioactive isotope of hydrogen and does not exist in nature since its half-life is only 12.4 yr. The initial inventory of tritium for a power-producing D–T fusion reactor (ca 10 kg) could be supplied, eg, from fission reactors where it is a by-product. Further tritium refueling would be accomplished by breeding additional tritium in the fusion reactor itself by absorbing the fusion-produced neutron in a blanket of lithium and exploiting the reaction: $n + {}^6Li \rightarrow T + {}^4He + 4.8$ MeV. Fortunately, ^6Li accounts for ca 7.5% of natural lithium and is abundantly available in the earth's crust and oceans. Detailed fusion-blanket designs must incorporate additional isotopes, such as ^{7}Li and ^{9}Be, which provide neutron-multiplying reactions, to compensate for the leakage and absorption losses of neutrons. A D–T fusion reactor, then, is in reality a consumer of deuterium and lithium. The estimated reserves (3) of lithium should provide sufficient resources for at least several hundred years of D–T fusion reactor operation, even allowing for a significant increase in the world demand for energy. A common fusion evolution scenario claims that D–T fusion will fulfill our needs until the more difficult fuel cycle burning pure deuterium can be implemented. From that point onward, deuterium alone will be the only fuel required for the fusion energy economy. Since each liter of seawater contains enough deuterium to supply the energy equivalent of 300 L of gasoline, the long-term energy needs of mankind would be assured (see Deuterium and tritium).

Although it has been stated that the D–T reaction is the easiest route to fusion-power production, it must be emphasized that it is no easy task to impart the required energy to the reactants to achieve a significant reaction rate. Relative kinetic energies between the deuterons and tritons of at least 10 keV are necessary to overcome the coulomb repulsion barrier, corresponding to relative particle velocities on the order of 10^6 m/s. Fusion-produced neutrons have, in fact, been created by impinging a beam of accelerated deuterons onto a solid target containing tritium. Unfortunately, a fusion reactor cannot be built around this concept because most of the incident beam energy is dissipated nonproductively through scattering and collision events in the target, and only a relatively small number of energy-producing fusion reactions occur. Other approaches, involving colliding beams of particles, have been proposed; however, such schemes are inherently of very low power density and are not likely to yield practical fusion devices.

Plasma Condition Required for Fusion Reaction

The most promising approach to attaining a significant reaction rate is to heat the reacting species to a high temperature, thereby imparting sufficient kinetic energy to the nuclei in the form of thermal motions. By doing so, the particles (eg, deuterons and tritons) may scatter amongst each other several times, keeping their energy within the system, before undergoing fusion reactions. At any given temperature, a system of particles in thermal equilibrium is characterized by a distribution of kinetic energies. The particles at the upper end of this distribution account for most of the fusion reactions in the system.

At these temperatures all matter exists in the plasma state, ie, it consists of totally ionized, positively charged atomic nuclei (ions) moving energetically throughout a cloud of negatively charged electrons, the system as a whole being electrically neutral. Generally speaking, the reactivity, or the fusion-power output rate, increases with increasing temperature. However, as the temperature of the plasma is raised, the radiation losses are also increased, primarily because of bremsstrahlung, or continuum radiation from the electrons. For any fusion-fuel system there exists a unique temperature at which the fusion-power production is precisely balanced by the radiation losses. This temperature is called the ideal ignition temperature (2), and equals ca 46 $\times 10^6$ K for a D–T plasma. Note that a net power-producing fusion system must be operated above the ideal ignition temperature. For a D–D plasma, this temperature is ca 4×10^8 K, a fact which reflects the difficulty in using pure deuterium fuel (see also Plasma technology).

Besides having to satisfy a minimum temperature requirement, a plasma must be sufficiently dense and must be contained for a long enough time to yield net power. If the plasma burns above the ideal ignition temperature for some time period τ, the fusion energy released must at least equal the energy required to heat the plasma to that temperature plus the energy radiated during that period. It can be shown that this condition is met providing that the product of plasma density N and confinement time τ exceeds a characteristic value which depends only on temperature. The minimum value of the product $N\tau$ represents the least stringent condition for the plasma to be a net producer of fusion energy. For D–T plasmas, this minimum occurs at a temperature of ca 1×10^8 K; for which $N\tau \cong 10^{14}$ s/cm^3. For D–D, the corresponding product is $N\tau \cong 10^{16}$ s/cm^3, again indicating that a pure deuterium system requires a higher quality of confinement. The minimum $N\tau$ product is called the Lawson criterion for net power production, and can be generally defined for any fusion-fuel system or confinement scheme.

Plasmas at these temperatures cannot be kept in ordinary containers because the energetic ions and electrons would rapidly collide with the walls and dissipate their energy. A major loss mechanism results from enhanced radiation by the electrons in the presence of elements sputtered off the container walls by the plasma ions. Therefore, some method must be found to contain the plasma at elevated temperatures without using material containers.

Magnetic Confinement

The two diverse approaches, magnetic and inertial confinement, are pursued vigorously, and form the basis of a large number of fusion research programs throughout the world. Magnetic confinement has been studied for the past twenty-five years. It is based upon the well-known fact that charged particles, such as electrons and ions (eg, deuterons, tritons), tend to be fixed to lines of magnetic force. Thus the essence of the magnetic confinement approach to fusion is to trap a hot plasma in a suitably chosen magnetic field configuration in the hope of exceeding the Lawson criterion (4).

Electrons and ions in magnetic fields spiral in tight circles around the field lines but translate freely along the direction of the magnetic field (5). Thus, the magnetic field of a long solenoid, eg, confines the plasma in two directions but does not prevent the particles from streaming from either end of the system. Furthermore, collisions between particles displace them from one field line to another, producing a net diffusion of plasma across the field toward the walls of the container. By employing more complex magnetic field configurations, fusion researchers have made significant progress toward solving the problem of magnetic confinement of plasmas.

A possible approach is to construct a configuration whereby the field lines close on themselves with the consequence that a particle that follows such a line remains within the system. The simplest way of producing such a configuration is to employ a torus, or doughnut-shaped container, with current-carrying coils wrapped around the minor diameter as shown in Figure 1. In this geometry, the magnetic lines of force are circles that traverse the torus and provide endless paths for the plasma ions and electrons to spiral about. Unfortunately, such a simple toroidal configuration suffers

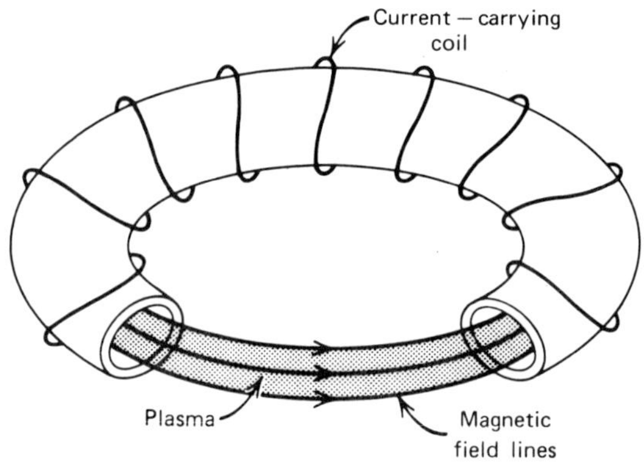

Figure 1. Cutaway view of toroidal field configuration.

a serious inherent weakness because the magnetic field strength is not constant across the plasma but rather stronger at the inner wall and weaker at the outer wall of the toroidal chamber (6). As a result, the ions and electrons drift in opposite directions across the field lines and establish an electric field within the plasma. This electric field, coupled with the applied magnetic field, then causes the plasma as a whole to move to the container wall and dissipate its energy.

This deleterious effect can be obviated by introducing an additional component of magnetic field, causing the field lines to circumscribe the torus without ever closing on themselves. The net magnetic field is then composed of a major, or toroidal field component produced by the current coils, plus a smaller poloidal component which gives the desired twist to the lines. Particle drifts weaken or nullify the harmful electric field and the plasma no longer tends to move to the walls.

Devices. The poloidal field component can be induced by adding another set of current-carrying coils that wind helically around the outside surface of the torus. Such a device is called a stellarator and was first developed at Princeton University during the 1950s. Subsequently, the USSR introduced an alternative means of producing the required magnetic field lines in a torus. A new class of plasma devices, called tokamaks, derive their poloidal field components from a current in the plasma itself, induced by a pulse from an external transformer. The tokamak is a simpler device than a stellarator, since fewer current coils are required to produce the necessary magnetic field. Two simple coils above and below the torus are required, however, to produce a vertical field component which interacts with the plasma current to position the plasma within the torus.

Tokamaks. Much of today's optimism concerning the scientific feasibility of fusion by magnetic confinement is owing to a series of noteworthy achievements in tokamak research (7). For example, at the Princeton large torus (PLT) a temperature of 75×10^6 K was reached without any serious instabilities occurring. Neutral beam heating power in excess of 2 MW was used, and the experiments yielded $N\tau \approx 10^{12}$ s/cm^3. Equally significant is the growing body of data that supports the theoretical models which predict that tokamak performance improves with increased device size (the so-called economy of scale). A main component of the PLT is a set of neutral-beam injectors that produce intense beams of deuterium for the purpose of heating the plasma. Initial heating of the plasma is accomplished by ohmic (ie, resistive) losses associated with the plasma current. However, as the plasma temperature increases, its resistivity decreases, and consequently resistive heating is capable of attaining a maximum temperature of ca 5×10^7 K. Additional heating means are required to reach ignition temperatures, and neutral-beam injection is considered a leading candidate for this function. The particle beams must be neutral in order to cross the magnetic field lines and enter the plasma where they deposit their kinetic energy via collisions in the form of heat.

Other tokamak programs have also made advances during the past few years. The Alcator at MIT has attained very high densities through the use of strong magnetic fields. Supplemental heating is provided by radio-frequency power, a promising alternative to neutral beams. Doublet II at General Atomics Company employs a noncircular plasma cross section, a shape which theoretically offers the potential for higher density operation with a given magnetic field strength. Large tokamak research programs in the USSR (T-10), Japan (JT-60), France (TFR), United Kingdom (Cleo), Italy, and Germany (Pulsator) indicate the international scope of this effort.

Several highly ambitious tokamak programs are already in progress for completion during the next 5 years. The tokamak fusion test reactor (TFTR) at Princeton marks a major advance as the first tokamak device to operate with D–T fuel by 1983. This machine is planned to operate at a scientific break-even level by producing 10 MW of fusion power with the injection of a similar amount of beam-injection power. Doublet III at General Atomics should come on line sooner with plasma operating parameters approaching reactor conditions. The joint European torus (JET) is being planned for Culham, United Kingdom, and will also employ a noncircularly shaped plasma. Finally, the USSR is planning the T-20 which would be the largest device of all, with a plasma minor radius of 2 m and a major radius of 5 m.

The tokamak offers great promise for demonstrating scientific break-even and the attainment of reactor operating conditions. However, its ultimate utility may be hampered by the fact that tokamaks tend to be very large devices characterized by relatively low power densities. A parameter β, which quantifies this concept, is defined as the ratio of the plasma pressure to the pressure exerted by the confining magnetic field B. It can be shown quite generally that the power density of a fusion device varies as $\beta^2 B^4$. Since B is limited by available magnet technology and allowable stresses in supporting structures, it is highly desirable to maximize β. Tokamaks may be inherently limited to a β of around 0.05–0.1, and could consequently exhibit unfavorable economic characteristics. Furthermore, they seem to operate best with small aspect ratios (defined as major radius divided by minor radius), leading to fat doughnuts which are difficult to access and maintain. Finally, it should be pointed out that tokamaks are usually referred to as pulsed devices, since their poloidal field component derives from the current pulse through the external transformer which couples to the plasma. For practical applications, a steady-state mode of operation would be preferable. In this respect tokamaks are less desirable than stellarators since the latter devices receive all their magnetic fields from independently controllable current-carrying coils.

Magnetic Mirrors. Another magnetic-confinement approach that may be well-suited to reactor applications is the magnetic mirror (8), a leading example of an open magnetic-field-line configuration. A simple mirror system is depicted in Figure 2. Particles gyrating about magnetic field lines move freely along these lines until they enter regions of increased field strength at either end of the device. Theoretical con-

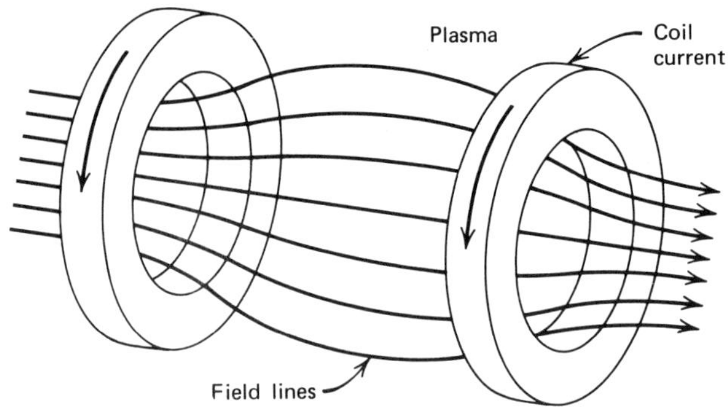

Figure 2. Simple magnetic mirror.

siderations dictate that, as the particles approach these end regions, they gyrate more energetically about the field lines and slow down in the direction of the lines. Ultimately, their kinetic energy is completely converted into gyration energy at which point they are reflected from these mirror points and return to the central, weaker field region. Particles whose motion is initially along the axis of the device (ie, parallel to the lines of force) are not capable of reflection and are lost through the ends. Nevertheless, the mirror approach offers an attractive alternative to toroidal devices since it is predicted that mirror-based fusion machines can operate in steady state with high values of β (approaching unity, or higher).

The center for mirror research is Lawrence Livermore Laboratories (LLL), Livermore, California. To date, the most successful experiment operated at LLL has been the 2XIIB device. Significant progress has been demonstrated on 2XIIB toward eliminating many destructive instabilities that have historically plagued the mirror concept. Indeed, stabilized operation yielded $\beta \cong 2$ and temperatures in excess of 10^8 K with the use of 6 MW neutral-beam heating power. However, primarily because of end losses, the confinement quality factor $N\tau$ was considerably less than those values attained in tokamak devices.

Improved variations of the mirror machine are being planned for implementation at LLL. Their common objective is to reduce leakage by using more effective end-plugging mechanisms. Concurrently, the amount of injected power required to maintain the plasma should decrease, thereby increasing the overall efficiency of the device. Another improvement is called the tandem mirror experiment (TMX), which began operation in 1979. This device consists of a central confinement chamber within a solenoid coil with higher-field mirror coils on both ends. The end cells are fed by high energy neutral beams injected at each end of the device. It is envisioned that plasma end losses are impeded by electrostatic potentials developed by the plasma as the electrons and ions attempt to leave the device at different rates.

Yet another possibility is the field-reversed mirror concept in which the diamagnetic nature of a plasma is exploited to cause the interior magnetic field lines within the central region to reverse and close on themselves. The plasma current responsible for this field modification is at right angles to the original field lines. It is predicted that high β and enhanced plasma confinement should be possible using the field-reversed mirror approach.

A large-scale mirror fusion test facility (MFTF) is being planned for operation at LLL in the early 1980s. Because of its significantly larger volume, it is hoped the MFTF will operate beyond the regime where loss mechanisms and instabilities have hampered the operation of smaller devices. Advances in magnet technology and beam injection also will be incorporated in MFTF.

Other Magnetic Confinement Approaches. Several other fusion experiments embodying alternative concepts of magnetic confinement are currently operational; others are being planned for the future. For example, the Elmo Bumpy torus (EBT) at Oak Ridge National Laboratory is a high β, large-aspect-ratio device designed to operate in an essentially steady-state mode. In essence, EBT is a chain of magnetic mirrors connected to form a torus. A unique feature is the use of high power microwaves to provide heating for the plasma (see Microwave technology). The latest version of an older approach to magnetic confinement fusion is embodied in the reversed-field toroidal pinch planned for operation soon at the Los Alamos Scientific Laboratory. Named ZT-40, this device utilizes a strong plasma current to produce a magnetic field that pinches the plasma, pulling it from the material walls and heating it by compression. A unique feature of ZT-40 is the reversal of the longitudinal component of the magnetic field near the outside of the plasma. The purpose of this reversed-field pinch is to suppress localized instabilities of the plasma column. Experimental results from ZT-40 are expected to add to a general understanding of how plasmas are contained and heated.

Inertial Confinement

Because the maximum density that can be confined is determined by the field strength of available magnets, fusion plasmas at reactor conditions are very diffuse media. Typical plasma densities are on the order of one-thousandth that of air at STP. The Lawson criterion is met by confining these plasmas for periods of ca 1 s. A totally different approach to controlled fusion attempts to create a much denser reacting plasma which, therefore, needs to be confined for a correspondingly shorter time. This concept is the basis of a rapidly growing and relatively new research area known as inertial confinement. In this approach, small pellets containing fusion fuel are compressed to extremely high densities by intense, focused beams of photons or energetic charged particles. Because of the substantially higher densities involved, the confinement times for this method can be much shorter. In fact, no external means are applied to effect the confinement; the inertia of the pellet mass is sufficient to provide ample time for net fusion energy release to occur before the pellet flies apart. Typical burn times and fuel densities are 10^{-10} s and 10^{25}–10^{26} per cm^3, respectively. These densities correspond to a few hundred to several thousand times that of ordinary condensed matter.

Laser-Driven Fusion. The possibility of heating small pellets of fusile fuel to thermonuclear temperatures was raised shortly after the development of the laser in the early 1960s (see Lasers). The unique characteristic of the laser, which makes it a potentially ideal energy source for inertial confinement fusion, is its ability to focus large amounts of radiant energy into small volumes in very short time intervals. In other words, extremely high peak power densities are now available which, in theory, can heat small amounts of matter to the temperatures required for fusion. In order to attain such temperatures, the mass of the fuel pellets must be kept quite low. As a result, the pellets are small objects whose typical dimensions in present experiments are much less than 1 mm. Fuel pellets in reactors would be considerably larger (ca 1 cm) because of the increased laser energy involved.

The basic processes relevant to laser-driven fusion (9–10) are illustrated in Figure 3. Energy from the laser is absorbed in the outer regions of the pellet, producing a hot, outwardly expanding plasma. The reaction to this radial expansion causes inward compression of the pellet core. As the core compresses, its temperature rises (preferably adiabatically) until the center ignites. A propagating nuclear burn front proceeds outwardly from the center until the pellet mass disassembles, quenching further reactions. The goal, of course, is to extract more energy from these fusion reactions than was supplied by the laser to ignite the fuel pellet.

A hypothetical laser-fusion reactor is shown in Figure 4. Fuel pellets are injected into a cavity and are ignited by a series of laser beams that are transmitted through optical windows. Neutrons emitted from each microexplosion are absorbed in a surrounding blanket. The blanket contains lithium and serves as a heat transfer medium and a breeding material for tritium production. Heat in the blanket is converted to electricity by conventional means. High average power output is obtained by repetitively injecting fuel pellets into the system and igniting them at a rate of one to several per second (11–12).

Theoretical considerations of the physical processes in laser-driven fusion, combined with energy-flow arguments can be used to estimate many requirements for a fusion reactor laser. Because pellet disassembly times are quite rapid, the laser pulse must deliver its energy in several nanoseconds (10^{-9} s) or less. The required energy content of the laser pulse probably needs to range from several hundred kJ to 1 MJ (ca 50–240 kcal). A more exact statement of the energy requirement and the necessary net laser efficiency depends upon a parameter Q, the pellet gain factor. It is defined as the ratio of the fusion energy yield to the laser energy incident upon the

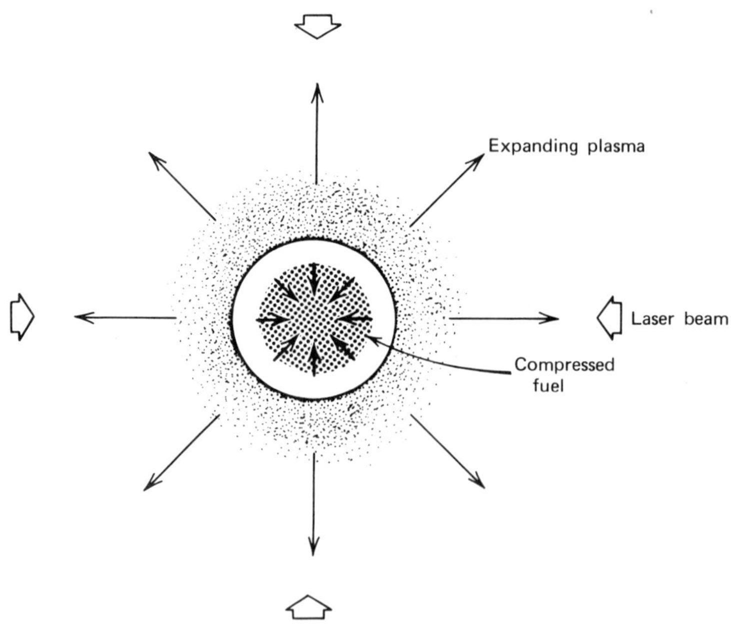

Figure 3. Laser-driven pellet compression.

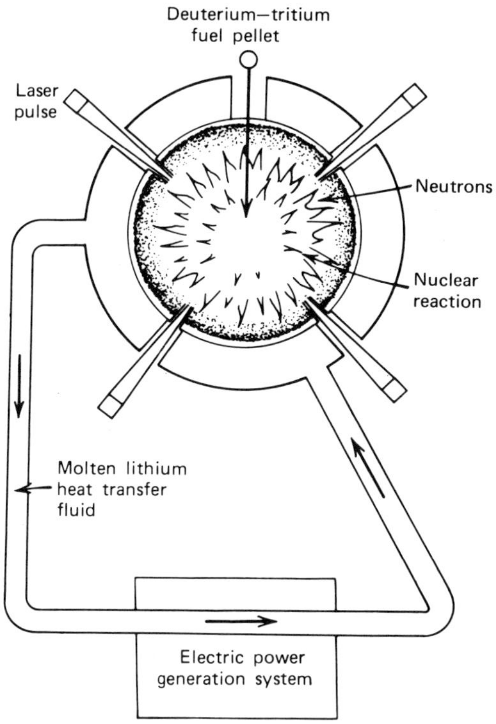

Figure 4. Laser-fusion, electric power-generation system.

pellet. Computer models of pellet designs for laser fusion lead to Q values of a few hundred to a thousand. Correspondingly, the laser efficiency must be at least 1–5% if the system is to be a net electrical energy producer with a reasonable amount of internal recirculating power. To obtain a high average power output, the laser should be capable of repetitive operation at the rate of 1–10 Hz. Finally, the wavelength of the laser should be optimized for efficient energy coupling to the fuel pellet. Visible or shorter wavelength light is generally thought to be preferable; however, long wavelength radiation (ie, ir) may be suitable with properly designed fuel pellets.

The neodymium-doped glass laser is a useful research tool for exploring the scientific feasibility of laser fusion. Although its intrinsic efficiency and repetition rate are too low for commercial applications, the glass laser is capable of extensive scale-up to power regimes where many of the fundamental physical issues of laser fusion can be explored. A large glass laser system called Shiva was recently completed at LLL. Shiva is designed to deliver pulse energies of 10 kJ (2.4 kcal), with peak power in excess of 20 TW. Experiments with targets filled with D–T gas mixtures have yielded more than 10^{10} fusion neutrons. In addition, the experiments indicate that pellet cores can be compressed to densities well above normal liquid values. Similar experiments are being conducted at the University of Rochester's Laboratory for Laser Energetics. This program, plus others at KMS Fusion and the Naval Research Laboratories, conduct fundamental experiments to elucidate the physics of the laser fusion process. Other glass laser systems are operated in the USSR, Japan, England, and France.

An alternative approach to laser fusion is being pursued at the Los Alamos Scientific Laboratories (LASL). The main laser system at LASL is an 8-beam CO_2 laser called Helios, which has an output energy of 10 kJ (2.4 kcal) and a peak power of 20 TW. Compared with the neodymium-glass laser, the CO_2 laser has a longer wavelength (10.6 μm vs 1.06 μm) and a higher theoretical efficiency (several % vs <1%). Because of its higher efficiency and potential for rapid pulse-repetition rate, the CO_2 laser could offer several advantages if the wavelength issue is favorably resolved.

An expanded version of Shiva (termed Nova) is being planned at Livermore for operation in 1983. With a projected power output of 200 TW, Nova should reach or exceed the scientific break-even level with 200 kJ pulses (ca 50 kcal). At Rochester a 24-beam system, Omega, will produce 15 TW by 1980 and perhaps up to 30 TW by 1982. Finally, LASL is constructing a 72-beam CO_2 laser system called Antares, which is designed to operate by 1983 at a pulse energy of 50 kJ (12 kcal) with a peak power of 200 TW.

During the next few years several uncertainties concerning laser fusion should be resolved. Foremost among scientific questions is whether pellets can be compressed and heated stably to the densities and temperatures required for producing a significant fusion yield. The main technological issue is whether a suitable laser can be developed with the required characteristics. Advanced laser research programs involving gas lasers with shorter wavelengths than CO_2 (10.6 μm) are underway at LLL, the Sandia Laboratories at Albuquerque, and elsewhere. Additional considerations of future practical concern relate to the economic fabrication of fuel pellets and the many materials and optics problems associated with reactor design and operation.

Particle-Beam Fusion. Inertial-confinement fusion may also be possible by substituting energetic charged-particle beams for lasers (13). The chief virtue of particle beams is that they have been proven to be far more efficient than lasers in converting electrical energy to useful beam energy. Proven efficiencies in high power systems are as high as 50%. Special techniques are required, however, to focus the beams onto small targets because beams of like charge tend to spread owing to electrostatic forces between the particles. Transport of the beams to the fuel pellet is not as straightforward as in the case of lasers, and therefore, reactors may be more difficult to design.

Electron and light ion-beam-driven fusion is being studied intensively at Sandia Laboratories. The largest *e*-beam device currently operating, Proto II, delivers 8 TW of peak power with a pulse energy of 100 kJ. Targets for *e*-beam experiments are larger than laser fusion targets since the electrons penetrate considerably deeper into the target. Pulse lengths for this approach are also correspondingly longer, and may be on the order of tens of nanoseconds. Successful heating and compression of fusion targets has been accomplished with the release of more than 10^6 neutrons from deuterium fusion. Large scale *e*-beam fusion experiments are also being conducted at the Kurchatov Institute in the USSR.

The latest facility under development at Sandia is a powerful electron and light ion-beam device, originally called electron-beam fusion accelerator (EBFA) and recently renamed particle-beam fusion accelerator (PBFA). The first phase (PBFA-I) is scheduled for operation mid-1980 and will be capable of delivering 1 MJ at a peak power of 30 TW. A second generation accelerator (PBFA-II) is planned, using the same tank structure of PBFA-I, but delivering over three times the power in 72 beams.

The use of heavy ions for imploding fuel pellets has also been proposed as an alternative approach to inertial confinement fusion. Compared with electrons, the more massive and highly charged ions should be more effective in coupling their energy to the pellet. It has been claimed that the required ion-beam systems can be readily extrapolated from technology already developed at existing storage rings and accelerators. Preliminary designs for reactors based upon this concept have been made at Brookhaven and Argonne National Laboratories. These designs employ heavy ions such as xenon and mercury, although some researchers believe that still heavier ions such as uranium might be best. Because of the large size of ion accelerators, it is difficult to perform ion-driven fusion experiments without a major programmatic commitment. Proposals for such experiments will probably emerge in the near future. In the meantime, some experimentation has been reported with light ions (eg, protons, deuterons) at Sandia, Cornell University, and the Naval Research Laboratories.

Health and Safety Factors; Environmental Aspects

Fusion offers environmental and safety advantages (15) as compared to fission that make it an attractive alternative (see Nuclear reactors). When operating with the D–T fuel cycle, fusion is by no means free from radiological hazards. The principal concerns relate to the tritium inventory in the power plant, ca 10 kg or 3.7×10^{18} Bq (10^8 Ci), and the activation of the structural materials of the reactor by the 14 MeV neutrons from the D–T reaction. Safety systems must be engineered to deal with unexpected tritium releases, and remotely operated mechanisms are required to perform routine periodic replacement of the activated structures. Compared with fission reactors, however, the radiological hazards of these materials are far less severe; in contrast to fission, the fusion process yields no radioactive reaction products or long-lived radioactive actinide wastes. The hazard potential of the radioactive substances associated with D–T fusion are less severe, resulting in a substantial mitigation of the handling and long-term storage problems associated with fission. Furthermore, a fusion reactor is inherently safer since the reactions are easily quenched, and runaway accidents are not possible. Finally, even though tritium is a government-licensed material, there are no current concerns regarding proliferation issues with respect to the D–T fuel cycle. The radiation hazards of tritium handling and structural activation have been quantified, and it has been concluded that, overall, fusion is substantially preferable with respect to accident risk and environmental impact. Fusion operation with the D–D cycle or advanced fuels (eg, D-^{3}He, p-^{11}B) would even strengthen these claims, especially if direct conversion of charged particle product energy is employed (16).

Long-Range Developments and Applications

The long-range role for fusion in the energy economy is in central station electrical power generation. When operating with the D–T fuel cycle, power is extracted from the thermalization of the neutron kinetic energy deposited in the blanket. Pulsed systems require storage techniques to provide a continuous output. High neutron fluxes may cause damage to reactor components and several materials-related problems are being studied. Neutron activation of structural materials is a serious concern, and provisions must be made for the periodic replacement and maintenance of components by remote handling techniques. Major technological challenges in the magnetic and inertial confinement approaches involve the development of large superconducting magnets and efficient high power drivers, respectively (see Superconducting materials).

Common to both approaches are problems in handling and recycling tritium and the engineering aspects of the blanket.

Despite these nontrivial issues, the pursuit of fusion power is well-justified because of its many advantages. Foremost, of course, is its unlimited supply of fuel, a factor dramatically highlighted by today's awareness of energy shortages. Studies of fusion power economics conclude that electricity from fusion may not be competitive with alternative sources such as coal and fission at their present costs. Even though fuel charges are relatively negligible, the capital, operational, and maintenance expenses prevent fusion from ever being a really inexpensive option.

Alternative uses of the neutrons from D–T fusion are also being considered. The most promising concept involves capturing the neutron in a fertile blanket material for the purpose of breeding fissile fuel (17–18) for a separate fission reactor. Using the abundantly available supply of depleted uranium, the fissile material ^{239}Pu can be bred and subsequently burned in a conventional nuclear reactor. The attractiveness of this scheme results from the fact that fusion is a neutron-rich process, and the 14.1 MeV neutrons from D–T fusion are eminently well-suited to this breeding application. Results of studies of fusion–fission hybrid schemes show that a single fusion reactor can supply enough fuel to support five or more fission reactors. In this mode, the hybrid approach competes with the fast breeder reactor with the advantages of potentially much higher breeding rates and the ability to operate in a subcritical regime. In a fusion–fission hybrid system, the worth of a fusion-produced neutron is greatly augmented because the fissile fuel it produces is highly energy intensive. Because of this effect, the fusion component of a hybrid system can operate under less stringent constraints with regard to energy output and efficiency. In addition, because of its off-line role as a fuel producer, the issues of steady-state operation and reliability are not nearly as crucial as for a pure fusion electrical power producer. These mitigating factors could allow fusion to make an earlier impact on the energy scene. For example, a fusion reactor operating with parameters below the Lawson criterion could still be useful in the hybrid context. Although it might be argued that the coupling of fusion to fission compromises many of its advantages, it is clear that hybrid implementation could serve a useful interim purpose until the ultimate goal of pure fusion is reached. A number of systems studies for hybrid applications are being carried out at various national laboratories and universities and in industry, both for laser-driven and magnetic-confinement fusion. The large proposed USSR tokamak, T-20, may be the first experimental device to incorporate this approach.

Finally, there are a number of other potential fusion applications which have received some study and analysis (19). For example, radiation and high temperature heat from a fusion reactor could be used to produce hydrogen by the electrolysis or radiolysis of water. This hydrogen would be employed in the synthesis of portable chemical fuels for transportation or industrial use (see Hydrogen energy). Several variations have been proposed such as the direct radiolysis of water by converting the neutron energy to ionizing radiation. Other multistep thermochemical cycles have also been analyzed, all of which operate best at the high temperatures that are available from a fusion reactor. The economics of the production of synthetic fuels such as hydrogen, methanol, and methane from fusion energy are highly uncertain. More creative schemes with greater efficiencies may be required to make this approach competitive. The transmutation of radioactive actinide wastes from fission reactors may also be feasible (see Actinides and transactinides). This idea would utilize the neutrons from a fusion reactor to convert these hazardous isotopes into more benign and easier to handle species. The practicality of this concept appears to be, at best, marginal; however, further analysis is continuing.

Other applications of fusion energy may evolve as researchers continue in their quest to harness this rich and powerful energy source. Although the technological challenges are substantial, the potential rewards are so great that the investment in fusion research is worthwhile. Considering the impressive progress achieved to date, and the ambitious programs under development, the future for fusion becomes progressively more optimistic. The next decade will provide a clearer indication of whether, or when, practical fusion energy will be realized.

BIBLIOGRAPHY

1. H. A. Bethe, *Phys. Rev.* **55,** 103 (1939).
2. S. Glasstone and R. H. Lovberg, *Controlled Thermonuclear Reactions,* D. Van Nostrand, New York, 1960.
3. W. C. Gough and B. J. Eastlund, *Sci. Am.* **224,** 50 (Feb. 1971).
4. J. D. Lawson, *Proc. Phys. Soc. B* **70,** 6 (1957).
5. G. Schmidt, *Physics of High Temperature Plasma,* Academic Press, Inc., New York, 1966.
6. F. F. Chen, *Introduction to Plasma Physics,* Plenum Press, New York, 1974.
7. H. P. Furth, *Bull. Am. Phys. Soc.* **24,** 51 (1979).
8. F. L. Ribe, *Rev. Mod. Phys.* **47,** 7 (1975).
9. J. S. Clarke, H. N. Fisher, and R. J. Mason, *Phys. Rev. Lett.* **30,** 89 (1973).
10. K. A. Brueckner and S. Jorna, *Rev. Mod. Phys.* **46,** 325 (1974).
11. M. J. Lubin and A. P. Fraas, *Sci. Am.* **224,** 21 (June 1971).
12. J. Nuckolls, L. Wood, A. Thiessen, and G. Zimmerman, *Nature (London)* **239,** 139 (1972).
13. G. Yonas, *Sci. Am.* **239,** 50 (Nov. 1978).
14. R. C. Arnold, *Nature (London)* **276,** 19 (1978).
15. J. P. Holdren, *Science* **200,** 168 (1978).
16. J. Dawson, *Bull. Am. Phys. Soc.* **24,** 42 (1979).
17. L. D. Lidsky, *Nucl. Fusion* **15,** 151 (1975).
18. H. A. Bethe, *Nucl. News,* 41 (May 1978).
19. L. Booth, *Dept. of Energy Report CONF-770593,* U.S. Government Printing Office, Washington, D.C., 1977.

GERALD M. HALPERN
Exxon Research and Engineering Company

G

G ACID, $(OH)C_{12}H_5(SO_3H)_2$. See Naphthalene derivatives.

GADOLINITE, $Be_2FeY_2Si_2O_{11}$, GADOLINIUM. See Rare-earth metals.

GALENA, GALENITE. See Lead.

GALLIUM AND GALLIUM COMPOUNDS

Gallium [*7440-55-3*], atomic number 31, was discovered through a study of its spectral properties in 1875 by P. É. Lecoq de Boisbaudran. It was the first element to be discovered after the publication of Mendeleev's periodic table; its discovery constituted a confirmation of the table which was reinforced shortly after by the discoveries of scandium and germanium. Gallium was named from *Gallia* in honor of its discoverer's homeland.

Gallium is a scarce element and is found most commonly in association with its immediate neighbors in the periodic table: zinc, germanium, and aluminum; it was also discovered in a sphalerite from the Hautes-Pyrénées. However, gallium is not a rare element: its concentration in the earth's crust, 10–20 g/t (10–20 ppm), is comparable to that of lead and arsenic. Up to 1% of gallium was found in a few rare ores, eg, South African germanite. Table 1 gives some data on gallium geochemistry (1); note that the ratio Ga:Al appears quite constant.

Because of the abundance of aluminum (qv) ores and alumina plants (and less importantly zinc ores and plants), these two are the main sources of gallium. Gallium is found also in the production of germanium from germanium ores and from certain coals; the ashes of their flue dust may contain 0.001–0.05%, and sometimes 1%, Ga. Micas also are gallium-rich (up to 0.1%) and gallium extraction from a combined spodumene and lepidolite deposit (Bernic Lake, Ontario, Canada) was considered.

Table 1. Gallium in Rocks and Minerals, Mean Values

Material	Number of studies	Ga, ppm	10^4 Ga:Al
igneous rocks	323	22	3.3
metamorphic rocks	90	19	2.2
sedimentary rocks	166	16	2.8
bauxites	76	55	2.2
sphalerites	31	35[a]	

[a] Observed to be quite variable, eg, 3–100 or more.

Physical Properties

Gallium, atomic weight 69.717, has two stable isotopes, ^{69}Ga [*14391-02-7*] (60.4%) and ^{71}Ga [*14391-03-8*] (39.6%). Twelve unstable isotopes (mass 63–76) are known. The radius of the atom is 0.138 nm, and the ionic radius of Ga^{3+} is 0.062 nm and that of Ga^{+} is 0.133 nm. Solid gallium has a metallic, slightly bluish appearance; the liquid is more white than silver.

The physical properties of gallium—especially its thermal properties—are exceptional (2). It has a low mp (29.78 ± 0.01°C) and it vaporizes above 2200°C; it has the longest liquid interval of the elements. Also it is easily supercooled. However, it expands during solidification (3.2%)—a property shared by only two other elements, germanium and bismuth. Its crystal structure is unusual for a metal: it crystallizes in the orthorhombic system, and it is very anisotropic; the latter property is attributed to the existence of Ga–Ga covalent bonds along its 001 axis. Principal physical properties of normal gallium are listed in Table 2. There are other isomorphic phases of gallium. Some, which are metastable at atmospheric pressure, are obtained from supercooled gallium; others are obtained at high pressures (3–4). Some of the physical properties of the metastable and high pressure forms are listed in Table 3. Only normal gallium (α or I) expands upon solidification.

Gallium wets almost all surfaces, especially in the presence of oxygen which promotes formation of a gallium suboxide film. Gallium diffuses very quickly into the crystal lattice of certain metals, particularly aluminum. If a line is traced with a piece of solid gallium on a sheet of aluminum, the aluminum immediately becomes brittle on that line as a result of the diffusion of gallium to the grain boundaries of the aluminum. Because of this property of embrittling aluminum, the DOT Office of Hazardous Materials has classified gallium in Hazard Class ORM-B and has placed restrictions on air shipment (5).

Chemical Properties

In accordance with its normal potential, gallium is chemically similar to zinc and is somewhat less reactive than aluminum. Just as is aluminum, it is protected from air-oxidation at ambient temperature by a fine film of oxide. It is entirely oxidized by air and by pure oxygen at ca 1000°C. Normally, gallium is trivalent; however, it also may be monovalent. Gallium forms some compounds of these two valences, eg, $Ga^{+}(Ga^{3+}Cl_4)$ [*13498-12-9*]. Gallium cannot have a valence of +2.

Gallium scarcely reacts with water below 100°C but it is completely oxidized by

Table 2. Physical Properties of Normal (I or α) Gallium

Property	Value
melting point, °C	29.77–29.78
boiling point, °C	ca 2200
density at melting point, g/cm³	
solid	5.904
liquid	6.095
heat of fusion, J/g[a]	79.8
heat capacity, J/(kg·K)[a]	
30°C	381.5
−17°C	361.4
−36°C	352.8
thermal conductivity, W/(m·K)	
solid at 20°C	axes { a 88.4; b 16.0; c 40.8
liquid at 77°C	28.7
cubic coefficient of expansion, °C⁻¹	
solid, at 0–20°C	5.95×10^{-5}
liquid, at 103°C	1.20×10^{-4}
at 600°C	1.03×10^{-4}
vapor pressure, Pa[b]	
1198 K	0.14
1249 K	0.44
1321 K	1.74
1478 K	16.5
crystallographic properties	orthorhombic Cmca space group; axes { a = 0.45186 nm; b = 0.76602 nm; c = 0.45258 nm; Z = 8
electrical resistivity, μΩ·cm	
solid at 20°C	axes { a 8.1; b 54.3; c 17.4; polycrystal 15.05
liquid at 30°C	25.795
Debye temperature, °C	51.6
viscosity, dynamic, mPa·s (= cP)	
32°C	1.810
50°C	1.739
surface tension, mN/m (= dyn/cm)	
30°C	709
400°C	733

[a] To convert J to cal, divide by 4.184.
[b] To convert Pa to mm Hg, divide by 133.3.

water in an autoclave at 200°C. At ambient temperature, concentrated inorganic acids react slowly with gallium. Oxidizing agents, such as aqua regia, H_2SO_4, H_2O_2 and perchloric acid, attack gallium effectively, especially when hot. The halogens readily react with gallium with heating, as do sulfur, selenium, tellurium, phosphorus, arsenic, and antimony. Nitrogen reacts with gallium only under particular conditions (see below), whereas boron seems not to react with it.

Table 3. Physical Properties of Metastable and High Pressure Forms of Ga

Property	Metastable forms at normal pressure			
	β	δ	ϵ	γ
melting point, °C	−16.3	−19.4	−28.6	−35.6
heat of fusion, J/g[a]	38.0	37.0	36.2	34.9
density at melting point, g/cm^3	6.22	6.21		6.20
crystal structure	monoclinic C2/c	trigonal $R\bar{3}m$		orthorhombic Cmcm
	High pressure form II		High pressure form III	
pressure, GPa[b]	2.8		2.6	
melting point, °C			48.7	
crystal structure	cubic $I\bar{4}3d$		tetragonal I4/mm	

[a] To convert J to cal, divide by 4.184.
[b] To convert GPa to kbar, multiply by 10.

Amphoteric character is one of gallium's principal properties. Gallium oxide is more acidic than alumina, and thus solutions of gallates(III) are not as unstable as those of aluminates. Solutions of alkali hydroxides readily attack gallium at 100°C. In alkaline solutions (eg, of $NaGaO_2$ [*12063-93-3*])—as in acidic solutions (eg, of $GaCl_3$)—the solubility can be very high, on the order of 1000 g Ga/L of solution. At temperatures of 500–1000°C or more, gallium is very corrosive to most metals. Only tungsten is unaltered; niobium, tantalum, and molybdenum are less resistant.

Extraction

Gallium as a By-Product of the Aluminum Industry. The sodium aluminate liquor in the Bayer-process plants that produce alumina is the best source of gallium. There is a considerable quantity of alumina produced in the world. The high concentration of gallium that is readily available from this solution obviates the preliminary extraction.

Gallium can be extracted by fractional carbonation which consists of treating the aluminate solution with carbon dioxide in several controlled stages (Fig. 1). Gallium also can be extracted by direct electrolysis of the aluminate solution at a strongly agitating mercury cathode (Fig. 2) or at a strongly agitating sodium amalgam. Gallium can be extracted from aluminate solution by reaction with Lewis-based organic complexes containing oxygen or, preferably, nitrogen: several processes of this type

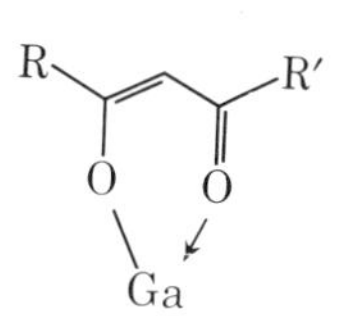

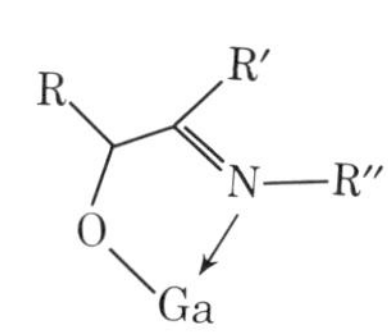

are being developed (see Chelating agents).

Gallium as a By-Product of the Zinc Industry. The gallium content of sphalerites generally is concentrated in the residues of zinc distillation and in the iron mud resulting from purification of zinc sulfate solutions. Gallium is extracted from these media with acidic solutions and $GaCl_3$ is extracted with an organic solvent, eg, ethyl or isopropyl ether.

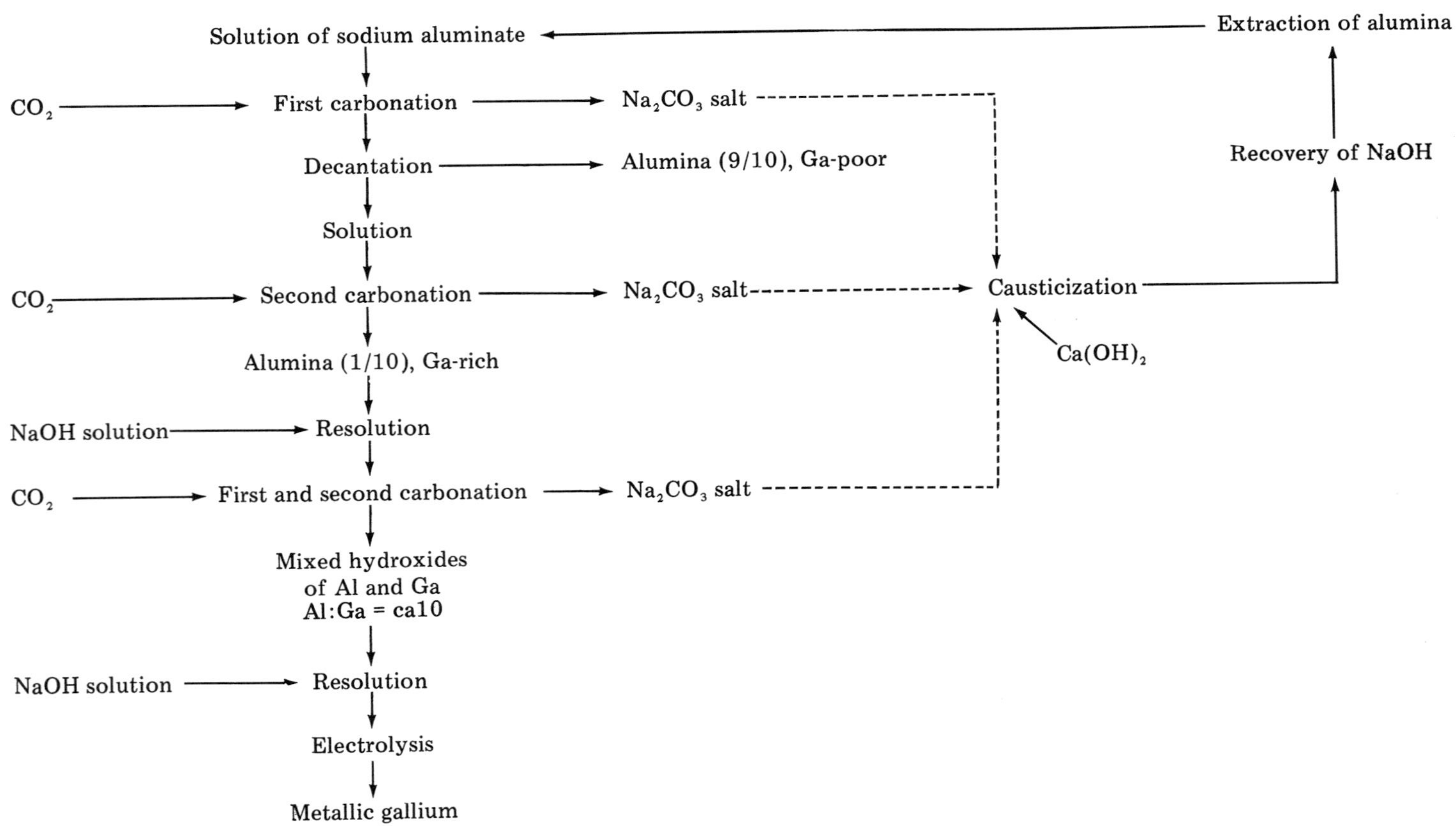

Figure 1. The carbonation process for gallium extraction (Beja process, Pechiney, 1946).

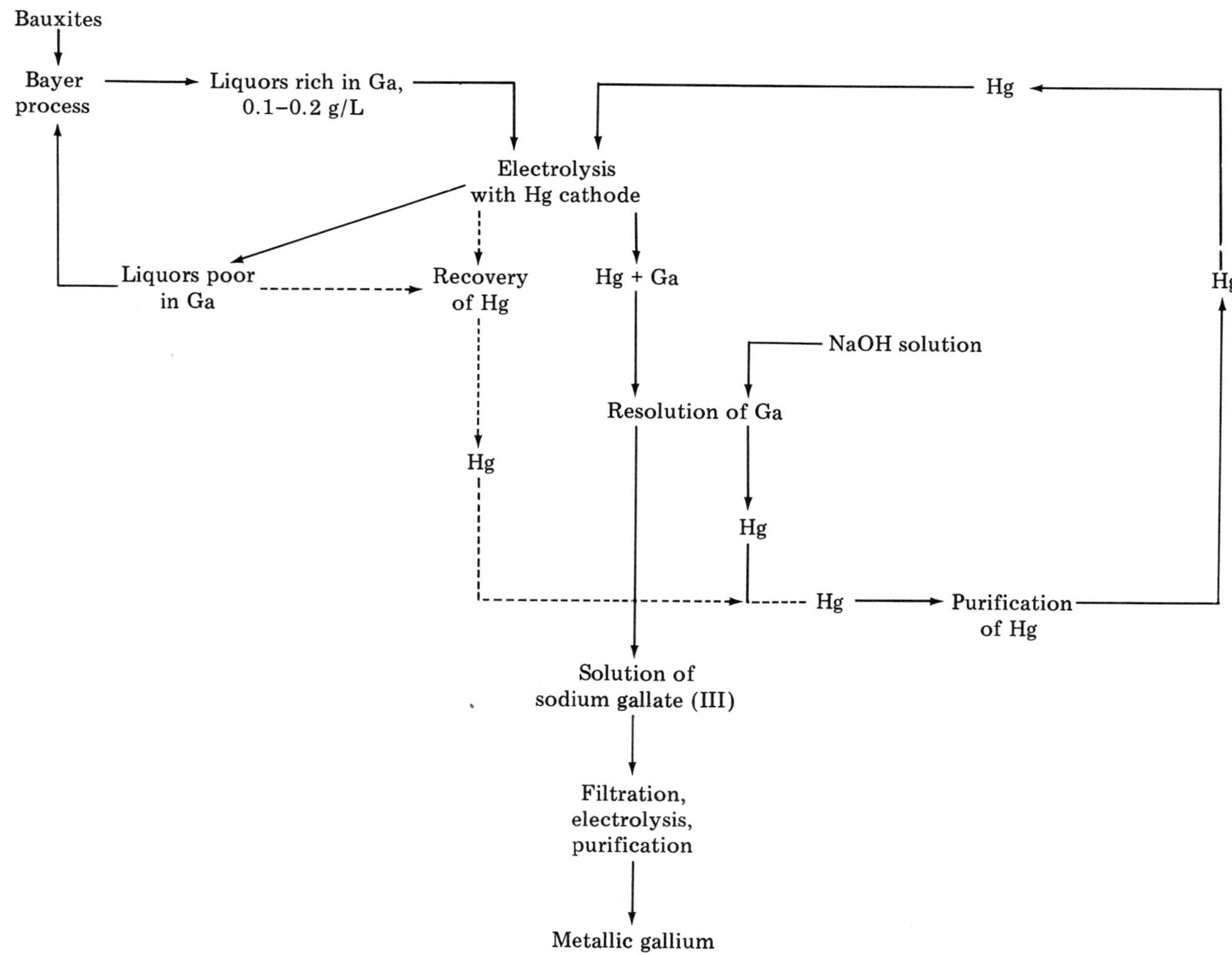

Figure 2. The mercury process (electrolysis) for gallium extraction (de la Bretèque process, Alusuisse, 1955).

Electrodeposition of Metallic Gallium. Extraction from aluminum or zinc ores produces gallium concentrates. Usually concentrates are transformed to sodium gallate(III) solutions by reaction with caustic soda solutions. Metallic gallium then is obtained by electrolysis between stainless steel electrodes. Conditions such as concentration, current density, and temperature vary with amounts of impurities. Gallium deposits at ca 40–60°C as a liquid, and is collected at the bottom of an apparatus of appropriate form (eg, a cylindrical polypropylene container with a conical bottom with a tap for gallium).

Purification. After hydrochloric and nitric acid washing, the gallium is filtered through porous plates (eg, sintered glass) to produce 99.9–99.99% pure Ga. Many gallium applications, particularly in electronics, require high purity (>99.9999%). This is achieved by operations such as electrochemical purification (anodic dissolution and cathodic deposition), heating under vacuum which allows the elimination of volatile elements, or fractional crystallization (6) (see also Fine chemicals).

Economic Aspects

The world consumption of gallium in the late 1950s was below 100 kg/yr and is estimated to be ca 14–16 t in 1979. World resources of 1.1×10^5 metric tons have been estimated as potentially economically and technologically recoverable. Of this, 8,400 t has been estimated as the United States share. Considering the foreseeable market development of III/V semiconductors (qv) and bubble memories (see Magnetic materials), one can expect a substantial growth during the next decade.

The price of gallium has fallen significantly in the course of this development: in 1920 ca $50/g, in the mid 1950s ca $4/g, and presently (1979) ca $0.51/g for large orders of the semiconductor-grade quality. Current (1979) prices, which have been stable over recent years, are ca $750/kg for semiconductor-grade (99.9999%) gallium metal, and $600/kg for the sesquioxide of >99.99% purity. A further remarkable drop in the future is not very likely, unless a new and much more economical process is developed for its production. As a consequence of its dissemination and the very dilute solutions available for its production, gallium will always be an expensive element. Methods have been developed to upgrade the substantial amounts of gallium scrap generated in the production of III/V semiconductors and GGG (gadolinium gallium garnet [*12024-36-1*], $Gd_3Ga_5O_{12}$).

Because of the inherent value of gallium waste and scrap from the electronics industry, some attempt has been made to recycle it. However, the difficulty in extracting the gallium from the very refractory garnets, and the environmental problem presented by the arsenic from GaAs have made recycling unattractive.

The principal producers of primary gallium are: Swiss Aluminium Ltd., Aluminum Company of America, and Vereinigte Aluminium-Werke (Federal Republic of Germany). The metal also is produced in some countries of Eastern Europe and of the Far East. Also, there are companies working on a scrap basis (see Recycling).

Analysis

Gallium is easily identified spectrographically, eg, using the lines 287–294 and 403–417 nm. It is somewhat difficult to separate gallium from aluminum and zinc by chemical means. The best method is the extraction of its chloride in acidic medium with an organic solvent, eg, diethyl or diisopropyl ether. Gallium can be identified also by colorimetry (with rhodamine B), fluorometry (with 8-hydroxyquinoline), gravimetrically (with 5,7-dibromo-8-hydroxyquinoline), or by complexometry and titration (with EDTA). The best modern method for the identification of gallium in aqueous media is by atomic absorption, which ensures precise determination and obviates the previously described separation.

Alloys and Intermetallic Compounds

Alloys. Gallium has complete miscibility in the liquid state with aluminum, indium, tin, and zinc, without formation of compounds. These binary systems are simple eutectics having the following properties:

Metal	*Ga, %*	*Temperature, °C*
Al	96	26.4
In	76	15.7
Sn	91.5	20.6
Zn	96.3	25.0

The systems obtained with gallium and bismuth, cadmium, germanium, mercury, lead, silicon, and thallium present miscibility gaps, and they give no intermetallic compounds.

Intermetallic Compounds. Numerous compounds formed with the transition elements have been reported (8). The principal compounds are listed in Table 4 (9–15). There are probably several Cs and Rb compounds, however, none is well known.

Lanthanides (and Yttrium). The following compounds are known: M = lanthanide: Dy, La, Nd, Sm, and Tb, MGa_2, hexagonal; MGa, orthorhombic; M_5Ga_3, tetragonal; Eu and Yb, MGa_2, hexagonal; Lu, Tm, MGa_3, cubic; MGa, orthorhombic; M_5Ga_3, hexagonal; $ErGa_3$ [*12785-97-6*], cubic; $ErGa_2$ [*60874-25-1*], hexagonal; Er_5Ga_3 [*12324-45-7*], hexagonal; $GdGa_2$ [*12690-00-5*], hexagonal; GdGa [*12310-95-1*], orthorhombic; Gd_3Ga_2 [*12683-05-5*], tetragonal; Gd_5Ga_3 [*12324-83-3*], tetragonal; $HoGa_3$ [*67352-19-6*], cubic; $HoGa_2$ [*12599-74-5*], hexagonal; HoGa [*12310-96-2*], orthorhombic; Ho_5Ga_3 [*12324-84-4*], hexagonal; $PrGa_2$ [*12599-76-7*], hexagonal; PrGa [*12310-99-5*], orthorhombic; Pr_5Ga_3 [*12361-85-2*], orthorhombic; YGa_2 [*12435-21-1*], hexagonal; YGa [*12160-65-5*], orthorhombic; and Y_5Ga_3 [*12024-34-9*], hexagonal.

Actinides. The following compounds are known: U_2Ga_3 [*37381-84-3*], orthorhombic; UGa_2 [*12064-23-2*], hexagonal; UGa_3 [*12024-33-8*], cubic (16); $PuGa_6$ [*12160-80-4*], tetragonal; $PuGa_4$ [*12160-77-9*], orthorhombic; $PuGa_3$ [*12024-31-6*], hexagonal; $PuGa_2$ [*12160-71-3*], hexagonal; PuGa [*12160-58-6*], tetragonal; Pu_5Ga_3 [*12160-74-6*], tetragonal; and Pu_3Ga [*12160-59-7*], tetragonal. Gallium also is used at low concentrations (ca 1%) to stabilize the δ phase of plutonium.

Compounds Other Than Intermetallic

Compounds with monovalent elements (17) are described below.

Hydrides. The hydride of Ga(I), GaH [*13572-92-4*], is detected only in the vapor phase. The hydride of Ga(III) or gallane, GaH_3 [*13572-93-5*], can be obtained by reaction of trimethylamine-gallane [*19528-13-3*], $(CH_3)_3N.GaH_3$, with boron fluoride,

Table 4. Gallium Intermetallic Compounds

Compound	CAS Registry number	Crystal form	Compound	CAS Registry number	Crystal form
Ag_3Ga	[57673-45-7]	hexagonal	Mn_2Ga	[12160-55-3]	hexagonal
AgGa	[12260-08-1]	hexagonal	Mn_3Ga	[12186-69-5]	hexagonal
$AgGa_2$	[12044-77-8]	cubic	Mo_8Ga_{41}	[55071-07-3]	tetragonal
AuGa	[12006-53-0]	orthorhombic	Mo_6Ga_{31}	[52015-37-9]	monoclinic
Au_2Ga	[54327-67-2]	orthorhombic	Mo_3Ga	[37381-83-2]	cubic
Au_7Ga_2	[12536-48-6]	hexagonal	$NaGa_4$	[12435-23-2]	tetragonal
$BaGa_4$	[12230-66-9]	tetragonal	Na_5Ga_8	[39297-66-0]	
$BaGa_2$	[12258-56-9]	hexagonal	$NbGa_3$	[12160-73-5]	tetragonal
BaGa	[12258-55-8]	hexagonal	Nb_5Ga_{13}	[12160-83-7]	orthorhombic
Ba_3Ga_2	[12258-59-2]	trigonal	Nb_5Ga_4	[12024-35-0]	hexagonal
Be_3Ga	[73156-95-3]		Nb_3Ga_2	[12306-31-9]	triagonal
$CaGa_4$	[12177-60-5]	tetragonal	Nb_5Ga_3	[12024-30-5]	triagonal
$CaGa_2$	[12258-68-3]	hexagonal	Nb_3Ga	[12024-05-4]	cubic
CaGa	[12258-67-2]	orthorhombic	$NiGa_4$	[56627-21-5]	cubic
$CeGa_2$	[12157-58-3]	hexagonal	Ni_2Ga_3	[12629-62-8]	tetragonal
CeGa	[20328-37-4]	orthorhombic	Ni_3Ga_4	[12435-24-4]	cubic
$CeGa_3$	[12360-78-0]	tetragonal	$Ni_{13}Ga_9$	[12500-13-9]	monoclinic
Ce_5Ga_3	[12360-78-0]	tetragonal	Ni_3Ga_2	[56541-04-8]	hexagonal
Ce_3Ga	[73623-49-1]	cubic	Ni_5Ga_3	[12401-05-7]	orthorhombic
$CuGa_2$	[12443-57-1]	tetragonal	Ni_3Ga	[12063-96-6]	cubic
CuGa	[12191-11-6]		O_5Ga_3	[60862-21-7]	tetragonal
Cu_2Ga	[69847-96-7]	cubic	$PdGa_5$	[52935-27-0]	tetragonal
Cu_9Ga_4	[12395-15-2]	cubic	Pd_3Ga_7	[73157-02-5]	cubic
Cu_3Ga	[68985-62-6]	cubic	PdGa	[59125-32-5]	cubic
$FeGa_2$	[12062-72-5]	tetragonal	Pd_5Ga_4	[73157-03-6]	cubic
Fe_3Ga_4	[53237-41-5]	monoclinic	Pd_5Ga_3	[52935-26-9]	orthorhombic
FeGa	[71771-30-7]	cubic	Pd_5Ga_2	[53095-62-8]	orthorhombic
Fe_6Ga_5	[53262-40-1]	monoclinic	Pd_2Ga	[12529-00-9]	orthorhombic
Fe_2Ga	[12160-14-4]	hexagonal	$PtGa_6$	[12411-32-4]	orthorhombic
Fe_3Ga	[12063-30-8]	cubic	Pt_3Ga_7	[12411-33-5]	cubic
$HfGa_3$	[73156-96-4]	tetragonal	$PtGa_2$	[12786-51-5]	cubic
$HfGa_2$	[12186-72-0]	tetragonal	Pt_2Ga_3	[12411-26-6]	tetragonal
Hf_2Ga_3	[73156-97-5]	orthorhombic	PtGa	[12411-22-2]	cubic
Hf_5Ga_3	[73156-98-6]	hexagonal	Pt_5Ga_3	[12411-27-2]	orthorhombic
Hf_2Ga	[12786-47-9]	tetragonal	Pt_2Ga	[12401-01-3]	cubic
$IrGa_3$	[60682-20-6]	tetragonal	Pt_3Ga	[12411-23-3]	cubic
Ir_3Ga_5	[12064-32-3]	tetragonal	Rh_2Ga_9	[66703-44-4]	monoclinic
IrGa	[60921-95-1]	cubic	$RhGa_3$	[60862-22-8]	tetragonal
KGa_4	[12435-22-2]	orthorhombic	$Rh_{10}Ga_{17}$	[12064-38-9]	tetragonal
K_5Ga_8	[73156-99-7]		RhGa	[73157-04-7]	cubic
LiGa	[12519-03-8]	cubic	$RuGa_3$	[60862-23-9]	tetragonal
Li_3Ga_2	[63705-92-0]		$RuGa_2$	[12064-21-0]	orthorhombic
Li_2Ga	[39343-49-2]		RuGa	[39388-93-7]	cubic
$LiGa_2$	[39343-50-5]		$SrGa_4$	[12160-78-0]	tetragonal
$LiGa_3$	[53570-25-5]		$SrGa_2$	[12259-26-6]	hexagonal
Mg_2Ga_5	[12411-31-3]	tetragonal	Sr_3Ga_2	[12259-27-7]	cubic
$MgGa_2$	[12411-25-5]	orthorhombic	Ta_3Ga_2	[60862-19-3]	tetragonal
MgGa	[12063-92-2]	tetragonal	Ta_5Ga_3	[12160-75-7]	tetragonal
Mg_2Ga	[12422-88-7]	hexagonal	$TiGa_3$	[12064-30-1]	tetragonal
Mg_5Ga_2	[12064-14-1]	orthorhombic	$TiGa_2$	[12186-73-1]	tetragonal
$MnGa_4$	[12160-76-8]	cubic	TiGa	[12398-41-3]	tetragonal
Mn_2Ga_5	[12763-96-1]	tetragonal	Ti_5Ga_4	[65453-84-1]	hexagonal
Mn_7Ga_6	[73157-00-3]	tetragonal	Ti_5Ga_3	[68565-51-5]	tetragonal
Mn_3Ga_2	[12160-70-2]	tetragonal	Ti_2Ga	[12500-08-2]	hexagonal
Mn_8Ga_5	[73157-01-4]	cubic	Ti_3Ga	[12183-38-9]	hexagonal

Table 4 (*continued*)

Compound	CAS Registry number	Crystal form	Compound	CAS Registry number	Crystal form
V_8Ga_{41}	[*55071-08-4*]	tetragonal	Zr_3Ga_5	[*73156-91-9*]	
V_2Ga_5	[*12024-48-5*]	tetragonal	Zr_2Ga_3	[*73156-92-0*]	
V_6Ga_7	[*12024-50-9*]	cubic	ZrGa	[*73156-93-1*]	tetragonal
V_6Ga_5	[*12024-49-6*]	tetragonal	Zr_5Ga_4	[*73156-94-2*]	hexagonal
V_3Ga	[*12024-15-6*]	cubic	Zr_3Ga_2	[*66703-42-2*]	tetragonal
$ZrGa_3$	[*12064-31-2*]	tetragonal	Zr_5Ga_3	[*66103-56-8*]	hexagonal
$ZrGa_2$	[*12186-74-2*]		Zr_2Ga	[*12786-50-4*]	tetragonal

BF_3, at −20°C (the hydride decomposes above −15°C). Halogen derivatives of gallane, especially $GaHCl_2$ [*13886-65-2*], and $GaHBr_2$ [*13886-66-3*], are obtained by reaction of trimethylsilane with $GaCl_3$ [*13450-90-3*] or $GaBr_3$ [*13450-88-9*] below −20°C. The dichlorogallane, $GaHCl_2$ (colorless crystals), is stable up to 29°C; the dibromogallane decomposes at −5°C.

Mixed hydrides, $MGaH_4$, also are known (M = Li [*17836-90-7*], Na [*32106-51-7*], K [*32106-52-8*], Rb [*32104-62-4*], and Cs [*32104-63-4*]). The lithium–gallium hydride, which is the most stable, has been studied as a reagent for organic reactions, similar to the aluminum compound, $LiAlH_4$.

A number of organic derivatives of gallane (addition or substitution compounds) are known. Some are much more stable than gallane, eg, $(CH_3)_3N.GaH_3$, $(CH_3)_3P.GaH_3$ [*15279-03-5*], and $(CH_3)_2N.GaH_2$ [*32342-80-6*], and are used for the preparation of other organic compounds of gallium.

Halides. ***Gallium(I).*** Halides of Ga(I) are known only in the vapor phase. Coordination complexes of GaCl and GaBr have been obtained with dioxane, morpholine, and acetylacetone.

Gallium(I–III). Compounds having the formula, Ga_2X_4 (X = Cl, Br, or I) are known; their structure, $Ga^+(Ga^{3+}X_4)$, has been better understood in recent years (18–19). Some of their properties are listed in Table 5. Other compounds ($Ga^+(Ga_2X_7)$ where X = Cl, Br) were recently discovered and their structure was determined (20–21).

Table 5. Properties of Gallium Halides

Compound	CAS Registry No.	Melting point, °C	Density, g/cm^3	Crystal structure	
GaF_3	[*7783-51-9*]	subl 800	4.47	trigonal	$R\bar{3}c$
$Ga(GaCl_4)$	[*24597-12-4*]	170.4		orthorhombic	Pnma
$Ga(Ga_2Cl_7)$	[*24688-86-6*]	(dec 77.3)	2.74	orthorhombic	$Pna2_1$
$GaCl_3$	[*13450-90-3*]	77.8	2.47	triclinic	$P\bar{1}$
$Ga(GaBr_4)$	[*18897-61-5*]	164.5	3.471	(2 forms; transition at 152°C)	
$Ga(Ga_2Br_7)$	[*50647-37-5*]	(dec 81.2)	3.631	orthorhombic	$P2_12_12_1$
$GaBr_3$	[*13450-88-9*]	124.1	3.74	orthorhombic	Pbca
$Ga(GaI_4)$	[*17845-89-5*]	211			
GaI_3	[*13450-91-4*]	212	4.15	monoclinic	$P2_1/c$

Gallium(III). The anhydrous fluoride, GaF_3, is obtained as colorless needles, which dissolve slowly in water, by thermal decomposition of the ammonium hexafluorogallate(III) [*14639-94-2*] in a stream of argon. The hydrate, $GaF_3.3H_2O$ [*22886-66-4*], is obtained by dissolving gallium or its hydroxide in hydrofluoric acid; the hydrate is more soluble in water. Thermal decomposition of $GaF_3.3H_2O$ yields the hydroxyfluoride, $16[Ga(OH,F)_3].6H_2O$, which is fcc at 200°C (22). Between 350 and 500 MPa (3450–4940 atm) at 300°C (7), the reaction of HF and Ga_2O_3 yields another hydroxyfluoride, $Ga_8[(OH)_{0.45}F_{0.55}]_{24}.3H_2O$, which is of the pyrochlore cubic type.

Gallium trichloride, $GaCl_3$, is obtained as colorless needles by reaction of chlorine or HCl on gallium at ca 200°C in a stream of nitrogen in an airtight apparatus to exclude traces of moisture. It can be purified by distillation in a stream of chlorine or nitrogen. Its structure seems to be a bridged dimer molecular lattice. A hydrate, $GaCl_3.H_2O$ [*23306-52-7*], mp 44.4°C, is known. The gallium trichloride is extremely soluble in water, ie, >800 g $GaCl_3$/L.

Gallium hydroxychloride, $GaCl_2OH$ [*73157-09-2*], can be obtained (23) by decomposition at 130°C of the complex, $GaCl_3.C_2H_5OH$ [*19379-37-4*]; the hydroxychloride melts at 120°C and is a cyclic tetramer.

Gallium tribromide, $GaBr_3$, is obtained by reaction of a stream of bromine vapor and nitrogen on gallium. The tribromide has seven hydrates, with 1, 2, 2.5, 3, 4, 6, and 15 H_2O (24).

Gallium hydroxybromide, $GaBr_2OH$ [*73157-08-1*], can be obtained by the same method as with the hydroxychloride; it melts at 175°C.

Gallium triiodide, GaI_3, is obtained by direct reaction of the elements or by reaction of iodine solution in carbon disulfide on gallium.

Gallium Oxyhalides. The compounds, GaOCl [*15588-51-9*] and GaOBr [*15605-97-7*], are obtained by reaction of Ga_2O_3 on $GaCl_3$ or $GaBr_3$ in a sealed tube at ca 300°C; they crystallize in the orthorhombic system.

Gallium Halogenates. The anhydrous perchlorate, $Ga(ClO_4)_3$ [*19854-31-0*], was obtained by reaction of the chlorine peroxide, Cl_2O_6, on $GaCl_3$ at −180°C (25): it is a white solid and is stable at ambient temperature.

Sulfohalides and Sulfohalogenates. The fluorosulfide GaSF [*73157-10-5*], is obtained by heating Ga_2S_3 and GaF_3 under pressure at 310°C. The sulfohalides, $Ga_9S_8Cl_{11}$ [*12268-03-0*] and $Ga_9S_8Br_{11}$ [*12049-26-2*] (monoclinic solids), were obtained by heating Ga_2S_3 with $GaCl_3$ in a sealed tube at 250°C or with $GaBr_3$ at 330°C (26). The chlorosulfate, $Ga(SO_3Cl)_3$ [*30338-20-6*], and its addition compounds with LiCl and NaCl were also obtained (27).

Compounds with Ammonia. Gallium halides give the following derivatives with ammonia: $GaF_3.NH_3$ [*73157-06-9*] and $GaF_3.3NH_3$ [*36171-86-5*]; $GaCl_3.NH_3$ [*50599-24-1*]; $GaBr_3.NH_3$ [*54955-92-9*]; $GaI_3.NH_3$ [*58384-90-0*]; and some poorly defined compounds of gallium chloride, bromide, and iodide that contain up to 14 NH_3 groups.

Mixed Halides. The following fluorogallates (III) were obtained with univalent elements: M_3GaF_6 where M = Cs [*16027-80-8*], K [*16061-04-4*], Li [*16061-07-7*], Na [*16061-62-4*], Rb [*16061-06-6*], Tl [*26846-24-2*], and NH_4 [*14639-24-2*], and $MGaF_4$ where M = Cs [*15002-96-7*], K [*15002-94-5*], Li [*39210-67-8*], Na [*15002-93-4*], Rb [*15002-95-6*], Tl [*26667-61-8*], and NH_4 [*18532-60-0*]. Among other fluorogallates(III), some typical examples are the compounds of $MGaF_5$, where M is Ca [*27658-92-0*], Cd [*58984-51-3*], Mn [*35745-87-0*], Pb [*27658-94-2*], or Sr [*27658-93-1*].

The following chlorogallates(III) have been described (19,21): $MGaCl_4$ where M = Cs [*21646-31-1*], K [*18154-89-7*], or Rb [*21646-29-7*]; and MGa_2Cl_7, where M = Cs [*12331-25-8*], K [*30617-31-1*], or Rb [*56531-09-0*], analogous to the compounds where M = Ga^+. The following chlorogallates(III) are known also but less precisely: $MGaCl_4$, where M = In [*12432-57-4*], Li [*15955-98-3*], Na [*15007-28-0*], or Tl [*26490-70-0*]; and several $M(GaCl_4)_2$ and $MGaCl_5$ compounds. The chlorogallate(III) $GaSbCl_6$ [*73157-05-8*] was studied (18). Chlorogallates(III) are also obtained with $AsCl_3$, $AsCl_5$, PCl_3, PCl_5, $POCl_3$, $SeCl_4$, $TeCl_4$, $MoCl_5$, $NbCl_5$, and $TaCl_5$.

The following bromogallates(III) were described (19,21): $MGaBr_4$ where M = Cs [*52582-09-9*], K [*50328-30-8*], or Rb [*52582-08-8*]; MGa_2Br_7 where M = Cs [*52724-44-4*], K [*39681-16-8*], or Rb [*52582-11-3*]; and MGa_3Br_{10} where M = Cs [*62974-40-7*] or Rb [*62974-41-8*]. The bromogallate(III) $SbGaBr_6$ [*12514-56-6*] also was studied (18).

The following iodogallates(III) have been described (28): $MGaI_4$ where M = Cs [*57146-51-7*] or Rb [*57146-50-6*]; and MGa_2I_7 where M = Cs [*57143-24-5*] or Rb [*57143-25-6*]; and $SbGaI_6$ [*12519-02-7*] (18). Gallium trihalides, particularly $GaCl_3$ and $GaBr_3$, form a large number of complexes with organic Lewis bases; these halides are strong donors. Complexes also are formed with compounds, especially those containing O, S, N, P, or As, eg, $LGaCl_3$, where L = $(C_2H_5)_2O$ or pyridine. Generally, because of its acceptor strength, the reactivity of trivalent Ga is ranked below aluminum and above indium and boron: B < Al > Ga > In > Tl, and B < Ga.

Gallium Oxides. The gallium oxides and hydroxides, preparation conditions, and their structures are listed in Table 6.

Gallium(I) Oxide. Ga_2O can be prepared by reduction of gallium(III) oxide with gallium under vacuum between 500 and 700°C, and subsequent purification by sublimation under vacuum at 500°C. It is a dark brown powder and is stable in cold, dry air. Above 700°C, it decomposes into gallium and Ga_2O_3.

Gallium(II) Oxide. GaO (unknown structure) was detected only in the vapor state by spectroscopy.

Gallium(III) Oxide. Ga_2O_3 is the single gallium oxide that is stable under normal conditions. Like alumina, it exists in several crystalline forms; the conditions of their formation are summarized in Table 6. The most stable form is the β oxide, mp ca 1725°C, which is obtained by calcination at ≥600°C of the hydroxides or certain salts, eg, nitrate, acetate, and oxalate. The α oxide can be obtained by heating the monohydroxide, GaOOH, at 300–500°C.

Gallium(III) oxide is very stable. It can be reduced at ca 600°C by hydrogen, carbon monoxide, or metallic gallium, It is more reactive than alumina and is strongly amphoteric (see Chemical Properties), forming gallium salts with acids and gallates(III) with bases. There is only one well-defined hydroxide, the monohydrate, GaOOH (or $Ga_2O_3.H_2O$). It can be prepared by reaction of water on metallic pure gallium in an autoclave at ca 200°C. The hydroxide, $Ga(OH)_3$, which probably has an amorphous structure, is obtained by neutralization of aqueous solutions of gallium salts or of alkaline gallates(III).

Gallates(III). The oxidation number (III) was recommended by the IUPAC to distinguish gallates(III) from the salts of gallic acid, 3,4,5-trihydroxybenzoic acid, which often are wrongly named gallates. The alkali metal gallates(III) are the only gallates(III) that are soluble in water; the former gives stable solutions at all concentrations, unlike aluminate solutions. They are obtained easily by direct reaction of metallic

Table 6. Gallium Oxides and Hydroxides

Compound	CAS Registry No.	Means of preparation	Crystal structure (similar compounds)	Density, g/cm^3
Ga_2O	[*12024-20-3*]	by reduction of Ga_2O_3 with Ga at 500–600°C		4.77
α-Ga_2O_3	[*12024-21-4*]	see text	trigonal, R$\bar{3}$C (corundum, α-Al_2O_3)	6.44
β-Ga_2O_3	[*12024-21-4*]	see text	monoclinic, C2/m (θ-Al_2O_3)	5.88
γ-Ga_2O_3	[*12024-21-4*]	by rapid dehydration of gels at ca 400°C	cubic, Fd3m (spinel)	
δ-Ga_2O_3	[*12024-21-4*]	by decomposition of $Ga(NO_3)_3$ at ca 250°C	cubic, Ia3 (MnO_3)	5.18
ϵ-Ga_2O_3	[*12024-21-4*]	by brief heating of δ-Ga_2O_3 at ca 550°C	orthorhombic, (κ-Al_2O_3)	
GaOOH	[*20665-52-5*]	by oxidation of Ga with H_2O at ca 200°C under pressure	orthorhombic, Pbnm (diaspore)	
$Ga(OH)_3$	[*12023-99-3*]	unstable gel		

gallium with solutions of the hydroxides of Na, K, Rb, or Cs. The ratio M:Ga can be as low as unity; the solutions can be obtained in very high concentration by evaporation, eg, 800 g Ga/L solution for the sodium gallate(III). Lithium gallate(III) is less soluble than $NaGaO_2$.

The solid gallates(III) of numerous elements have been studied and include gallates(III) of alkaline and other metals. The former group includes: Li_5GaO_4, $LiGaO_2$, and $LiGa_5O_8$; $MGaO_2$, where M = Cs, K, Na, or Rb; and $Na_2O.5.4Ga_2O_3$ and $Na_2O.$-$7Ga_2O_3$ (ionic conductors, of the type of β-alumina). The latter group includes: $MGaO_2$ (M = Cu or Ag); MGa_2O_4 (M = Ba, Sr, Co, Cu, Fe, or Zn); $CaO.nGa_2O_3$ (n = 0.33–1 or 2); $AlGaO_3$; and $LnGaO_3$ and $Ln_3Ga_5O_{12}$ with the lanthanides. Among these compounds, a number are studied for their useful magnetic or electric properties, especially spinels, perovskites, and above all, garnets, eg, $Ln_3Ga_5O_{12}$ (Ln = Dy, Sm, or Gd—see Uses).

Gallium Chalcogenides. It is very doubtful that gallium(I) sulfide and gallium(I) selenide exist. Gallium forms two types of compounds by reaction with sulfur, selenium, and tellurium, and these are listed in Table 7. One type is represented by GaS, GaSe and GaTe, in which the gallium is trivalent and the form is a lamellar crystal containing atomic groups such as S—Ga—Ga—S. In such a group, each gallium atom is surrounded by a tetrahedron containing 3 S and 1 Ga. This lamellar structure gives two-dimensional properties to these semiconductor compounds. The other compound type includes Ga_2S_3 (29), Ga_2Se_3 (30), and Ga_2Te_3, which have vacancies in their structure that also give particular electronic properties.

In the binary system, Ga–Te, another compound exists, Ga_2Te_5 (31). It is formed only at high temperatures but can be obtained at ambient temperature by quenching.

Mixed Chalcogenides. The mixed chalcogenides include: $MGaS_2$ (or Se_2 or Te_2), where M = Ag, Cu, In, or Tl; among these compounds, some are semiconductors, eg, $AgGaS_2$ [*12002-67-4*] or $CuGaS_2$ [*12018-83-6*]. Other mixed chalcogenides are: MGa_2S_4 (or Se_4 or Te_4) where M = Cd, Hg, or Zn; $MGaS_3$, where M = In [*12398-40-2*] or Y [*12018-83-6*]; and $M_{18}Ga_{10}S_{42}$ (or Se_{42}), where M = lanthanides.

Table 7. Gallium Chalcogenides

Compound	CAS Registry No.	Melting point, °C	Crystal structure	Density, g/cm^3
GaS	[*12024-10-1*]	962	hexagonal (lamellar structure)	3.86
Ga_2S_3	[*12024-22-5*]	1090	monoclinic, Bb (superstructure of wurtzite type)	3.77
GaSe	[*12024-11-2*]	960–965	hexagonal (lamellar structure)	5.01
Ga_2Se_3	[*12024-24-7*]	1005–1010	monoclinic, B2/m (superstructure of deformed blend type) cubic by quenching of liquid compound (disordered sphalerite type)	4.95
GaTe	[*12024-14-5*]	825	monoclinic or hexagonal (lamellar structure)	5.44
Ga_2Te_3	[*12024-27-0*]	792	cubic (sphalerite type)	5.57
Ga_2Te_5	[*73623-48-0*]	stable only at 400–495°C	tetragonal	5.85

Sulfates and Selenates. The anhydrous sulfate, $Ga_2(SO_4)_3$ [*13494-91-2*], is obtained by dissolving GaOOH in 50% H_2SO_4, evaporating, filtering, and drying to 360°C. It crystallizes from aqueous solutions as $Ga_2(SO_4)_3.18H_2O$. With alkali metal or ammonium sulfates, it gives alums, $MGa(SO_4)_2.12H_2O$, where M is Cs, K, Rb, or NH_4. Gallium selenate is obtained under the same conditions as the sulfate. Like the sulfate, it gives double selenates that are isomorphous with alums.

Gallium Compounds with Nitrogen and Boron. Direct combination of gallium and nitrogen (32) involves a gas discharge at 750°C for the dissociation of N_2 into atomic N which combines with Ga vapor; a thin film of GaN is deposited. The gallium nitride [*25617-97-4*], GaN, can be obtained by the reaction of ammonia on gallium at ca 1000°C, or on gallium compounds (oxide, chloride, sulfide, etc). GaN also is prepared by thermal decomposition of triammonium hexafluorogallate(III) in a stream of ammonia at ca 600°C. Gallium nitride (hexagonal, wurtzite type; density, 6.10 g/cm^3) is very stable in air to ca 900°C; it can be sublimed in a stream of ammonia at 1150°C. It has a colorless crystal form when pure. It is a semiconductor, and as such, promises practical applications, especially as an electroluminescent compound (see Chromogenic materials).

Double nitrides with metals, particularly with alkali metals, are known; eg, Li_3GaN_2.

The hydrated nitrate, $Ga(NO_3)_3.9H_2O$, is formed by dissolving gallium in nitric acid.

The azide, $Ga(N_3)_3$ [*73157-11-6*], is prepared by decomposition, at ca 250°C, of the addition compound, $GaF_3.NH_3$ [*73157-06-9*].

Contrary to previous indications there is apparently no gallium boride.

Gallium Compounds with Phosphorus, Arsenic, and Antimony. These 1:1 compounds, GaP [*12063-98-8*], GaAs [*12064-03-8*], and GaSb [*1303-00-0*], can be obtained by direct combination of the elements at high temperature; GaP and GaAs formation requires combination under pressure. The three compounds also can be prepared, mainly as thin films, by numerous exchange reactions in the vapor phase.

Gallium phosphide (density, 4.13 g/cm^3) melts at 1465°C under pressure; gallium arsenide (density, 5.40 g/cm^3) melts at 1238°C under pressure; gallium antimonide (density 6.096 g/cm^3) melts at 712°C. All three crystallize in the cubic system (sphalerite type) and possess semiconductor properties which make them (especially the arsenide) the most important in terms of practical application of gallium. Many mixed compounds also are used, eg, $GaAs_{1-x}P_x$ or $Ga_{1-x}Al_xAs$ (see Digital displays; Semiconductors).

The following phosphorus oxoacids are obtained by the reaction of gallium hydroxide on the oxoacids. The anhydrous gallium phosphate is prepared by heating the dihydrate or by the reaction of gallium on phosphoric acid in a sealed tube at 200°C: anhydrous phosphate, $GaPO_4$ [*14014-97-2*], hexagonal (β quartz type); hydrated phosphate, $GaPO_4.2H_2O$ [*23653-37-4*], orthorhombic; and pyrophosphates, $GaHP_2O_7$ [*34641-69-5*] and $Ga_4(P_2O_7)_3.xH_2O$ [*19584-43-1*] ($x \cong 23$).

The anhydrous gallium arsenate [*13811-89-7*], $GaAsO_4$, hexagonal (β-quartz type) is slightly soluble in water and can be obtained by the reaction of $GaCl_3$ on As_2O_3 at 800°C. It forms a hydrate, $GaAsO_4.2H_2O$ [*23653-38-5*] (orthorhombic) and it can form acidic salts, eg, $Ga(H_2AsO_4)_3.xH_2O$ with $x = 1$ [*38296-94-5*] or 5 [*38296-97-8*].

The gallium antimonate, $GaSbO_4$ [*28888-33-7*], is tetragonal (rutile type) and has a denisty of 6.540 g/cm^3.

Carbon Compounds of Gallium. There is no binary gallium carbide; however, some ternary carbides have been studied, principally with regard to their structure and magnetic properties: Mn_3GaC [*12069-65-7*], Nd_3GaC [*12127-26-3*], and Mo_2GaC [*12286-91-8*] (see Magnetic materials).

More than 1000 organic gallium compounds have been described in addition to the coordination compounds of the gallium hydride or halides already mentioned. These compounds include salts of organic acids, eg, acetate, oxalate, citrate and numerous complexes, eg, oxalates, $Ga(C_2O_4)^+$, $Ga(C_2O_4)_2^-$, $Ga(C_2O_4)_3^{3-}$; alkoxides or their derivatives by substitution, eg, $Ga(OCH_3)_3$ [*2746-72-7*], $Ga(OC_2H_5)_3$ [*2572-25-0*], $Ga(OC_6H_5)_3$ [*2572-17-0*], and $GaCl(OCH_3)_2$ [*21907-51-7*], and $GaCl_2(OCH_3)$ [*22381-97-1*]. Another group is the organogallium compounds, R_3Ga, which contain Ga—C bonds, where R is: methyl [*1445-79-0*] (mp −15.8°C), ethyl [*1115-99-7*] (mp 82.3°C), *n*-propyl [*29868-77-7*], isopropyl [*54514-59-9*], propenyl (cis or trans), butyl [*15677-44-8*], pentyl [*15677-45-9*], hexyl, decyl, phenyl [*55321-79-4*], and α-naphthyl. In addition, a number of their substitution or coordination compounds have been described.

Toxicology

The toxicity of metallic gallium or gallium salts is very low. The corrosive, poisonous, or irritating nature of some of its compounds is attributable to the anions or radicals with which it is associated. Gallium metal-organics, such as $Ga(CH_3)_3$, react vigorously with air, and can be explosive. The gallium halides, except the fluoride, hydrolyze in water to form the corresponding halogen acids. Gallium phosphide, arsenide, selenide, and telluride react slowly with water, and more vigorously with acids and bases, to liberate toxic compounds. The LD_{50} of the solution for mice is ca 3–4 g/kg $Ga(NO_3)_3$ [*13494-90-1*]. The ^{72}Ga and ^{67}Ga isotopes were studied, eg, as citrate salts, for detection of tumors: ^{72}Ga concentrates in bone tissues and ^{67}Ga seems to have a tumor-specific affinity.

Uses

Gallium is used primarily in compound form. Most important are gallium arsenide and gallium phosphide (so-called III/V compounds), which are semiconductors. Both compounds show electroluminescence, and are used for the manufacture of light-emitting diodes (qv) (LEDs). The emitted wavelength, depending on the width of the band gap, is in the ir region for GaAs (band gap 1.4 eV), and changes to the red and orange regions for mixed crystals, and to the yellow and green regions for suitably doped GaP (2.4 eV). GaAs in combination with (Ga,Al)As-layers is a material used in solid-state lasers, especially in telecommunication systems with optical fibers. There is increasing interest in its use in microwave devices and, based on its high electron mobility (8800 $cm^2/(V{\cdot}s)$) compared with 1900 $cm^2/(V{\cdot}s)$ for silicon at RT, for fast switching ICs (see Lasers; Fiber optics; Integrated circuits). Seldom has an element been channeled so completely into a single industry. This application seems ready-made for gallium since usage will never be large, ultra-high purity is required, and the cost of the multitude of processing steps that are required for device fabrication overshadows the high price of the starting material. GaAs also is a promising material for solar cells.

Of growing importance is the use of gallium oxide for growing highly perfect (<1 dislocation per cm^2) single crystals of gadolinium gallium garnet (GGG), the substrate material for magnetic garnet layers in bubble domain memories. This technology was developed during the last 10 years, and the new generation of computer memories having densities of ca 10^6 $bits/cm^2$ will be marketed soon (see Magnetic materials).

To a smaller extent gallium oxide is used—and has been used since ca 1965—for the preparation of Mg–Ga–spinel. When doped with Mn, the compound is used as a green-emitting phosphor for luminescence tubes in copying machines.

The compounds, V_3Ga and recently Nb_3Ga, which are superconductors with superior properties only in extremely high magnetic fields, are being studied but it is too early to predict the extent of their future applications (see Superconducting materials).

In the early 1980s gallium, in the form of a chloride solution, may be used in large amounts to measure the neutrino radiation formed by the fusion processes inside the sun.

However, none of the aforementioned applications has reached any large-scale commercial importance because of the rather peculiar physical properties of the element.

A new use of gallium is as a thermometric standard. Its melting point of 29.7714°C (I PTS 68) is used as a thermometric fixed point for the calibration of thermometers in clinical and biological applications. It has the only thermometric fixed point between 0 and 100°C (34) (see Temperature measurement).

BIBLIOGRAPHY

"Gallium" in *ECT* 1st ed., Vol. 7, pp. 53–58, by L. M. Foster, Aluminum Company of America; "Gallium and Gallium Compounds" in *ECT* 2nd ed., Vol. 10, pp. 311–328, by Pierre de la Bretèque, Soc. Française pour l'Industrie de l'Aluminium, Div. of Swiss Aluminium, Ltd.

1. P. de la Bretèque, *Gallium, Bulletin d'Information et de Bibliographie, No. 12,* Alusuisse France S.A., Marseille, 1974, pp. 11–101.

2. Ref. 1, *No. 8,* 1970, pp. 13–89.
3. L. Bosio, A. Defrain, and I. Epelboin, *Colloq. Int. C.N.R.S.* **201,** 325 (1971).
4. L. Bosio, *J. Chem. Phys.* **68**(3), 1221 (1978).
5. CFR **49,** 172.101 and 173.861–173.862.
6. P. de la Bretèque, *Mem. Sci. Rev. Met.* **67**(1), 57 (1970).
7. M. Rault, G. Demazeau, J. Portier, and J. Grannec, *Bull. Soc. Chim. Fr.,* 74 (1970).
8. K. Yvon and P. Feschotte, *J. Less Common Met.* **63,** 1 (1979).
9. Ref. 1, Nos. 5–13, 1966–1975.
10. M. Puselj and K. Schubert, *J. Less Common Met.* **38,** 83 (1974).
11. N. E. Alekseevskii and V. M. Zakosarenko, *Dokl. Akad. Nauk SSSR* **208**(2), 303 (1973).
12. M. J. Philippe, B. Malaman, and B. Roques, *C.R. Acad. Sci. Ser. C* **278**(17), 1093 (1974).
13. S. P. Yatsenko and co-workers, *Izv. Akad. Nauk SSSR Metal.* (1), 185 (1973).
14. K. Yvon, *Acta Crystallogr. Sect B* **30**(Pt. 4), 853 (1974).
15. P. Guex and P. Feschotte, *J. Less Common Met.* **46,** 101 (1976).
16. K. H. Buschow, *J. Less Common Met.* **31,** 165 (1973).
17. Ref. 1, *No. 14,* 1978, pp. 11–71.
18. E. Chemouni, *J. Inorg. Nucl. Chem.* **33,** 2317 (1971).
19. D. Mascherpa-Corral, Thesis, University of Montpellier, Fr., 1975.
20. Y. Dumas and A. Potier, *Bull. Soc. Chim. Fr.,* 2634 (1975).
21. D. Mascherpa-Corral and A. Potier, *Bull. Soc. Chim. Fr.,* 1912 (1973).
22. A. Baumer, R. Caruba, and G. Turco, *C.R. Acad. Sci. Ser. D* **271**(1), 1 (1970).
23. L. Moegele, *Z. Naturforsch. B,* **23,** 1013 (1968).
24. M. T. Roziere-Bories, A. Calmel, J. Roziere, and A. Potier, *Bull. Soc. Chim. Fr.,* 1285 (1974).
25. M. Chaabouni, J. L. Pascal, A. Pavia, A. Potier, and J. Potier, *J. Chem. Res.,* 80 (1977).
26. A. Hardy and D. Cottreau, *C.R. Acad. Sci. Ser. C* **262,** 739 (1966).
27. B. Vandorpe, M. Drache, and B. Dubois, *C.R. Acad. Sci. Ser. C* **276,** 73 (1973).
28. D. Mascherpa-Corral and A. Potier, *J. Inorg. Nucl. Chem.* **39,** 1519 (1977).
29. G. Collin, J. Flahaut, M. Guittard, and A. M. Loireau, *Mat. Res. Bull.* **11,** 285 (1976).
30. G. Ghemard, R. Ollitrault, and J. Flahaut, *C.R. Acad. Sci. Ser. C* **282,** 831 (1976).
31. F. Alapini, M. Guittard, M. Julien-Pouzol, and J. Flahaut, *C.R. Acad. Sci. Ser. C* **282,** 543 (1976).
32. B. B. Kosicki and D. Kahng, *J. Vac. Sci. Technol.* **6**(4), 593 (1969).
33. S. A. Kutolin and co-workers, *Metody Poluch. Khim. Reakt. Prep.* (24), 22 (1972).
34. H. E. Sostman, *Rev. Sci. Instrum.* **48,** 2 (Feb. 1977); L. Crovini and co-workers, *Metrologia* **13**(4), 197 (1977).

General References

K. Wade and A. J. Banister, *The Chemistry of Al, Ga, In, and Tl,* Pergamon Press, New York, 1974.

PIERRE DE LA BRETÈQUE
Alusuisse France S.A.

GAS CLEANING

Gas cleaning is an important feature of many industrial processes (1–2). Within the scope of this article, gas cleaning is understood to comprise those steps that are necessary for removal of impurities from a gas stream for process reasons. In many processes, cleaning of atmospheric effluents is required for compliance with regulations, and protection of the environment and the health and safety of the community (see Air pollution control methods). Typical requirements may involve the removal of a gaseous impurity that could poison a reaction catalyst; produce unwanted side reactions or serve as a diluent, reducing driving force concentration, in either case, reducing yields; or could reduce the purity of the process product. Similarly, particulates (either solid or liquid) may have to be removed from a gas stream for much the same reasons. Solid and tarry liquid particles may also cause operating problems in downstream equipment which result in plugging the system or abrasive wear of mechanical equipment. Sometimes, an impurity is salable and is separated for marketing.

Cleaning of coke oven and coal carbonization gases are examples of gas cleaning needs. The coal gas itself has value as a fuel for industrial combustion processes or can be mixed with other combustible gases for domestic use. Condensible tars that would plug lines and burners must be removed; they may have value for chemical uses. Ammonia is separated for its chemical or fertilizer value. Benzene, toluene, and naphthalene are other fractions sufficiently valuable for recovery. Removal of sulfur compounds prevents pollution when the coal gas is burned. In the synthesis of anhydrous ammonia, hydrogen is produced by steam reforming of natural gas. The resulting CO reacts further with steam over a shift catalyst producing more H_2 and converting the CO to CO_2. The CO_2 is then removed in a cleaning step (absorber–stripper) to produce process-quality hydrogen (see Alkanolamines; Carbon dioxide). The gases may require further cleaning to remove sulfur compounds (carbon adsorption) and other substances (chlorine, arsenic) that could poison the various catalysts (see Ammonia, Vol. 2, pp. 480–494). Absorption-type processes are widely practiced to remove H_2S and organic sulfur compounds from refinery gas and even natural gas. Table 1 in Absorption (Vol. 1, p. 53) lists a number of gas-cleaning processes using both water-based and organic absorbents. In the production of acetylene from natural gas and petroleum liquids by any of several flame and partial combustion processes (see Acetylene, Vol. 1, pp. 213–230), some soot (or acetylene black) is produced which must be removed from the gas stream before carrying out other petrochemical reactions. Electrostatic precipitators, fabric and granular-bed filters, and liquid scrubbers have been used.

In catalytic fluid-bed gas reactors, separation and recovery of entrained catalyst fines are required to reduce catalyst losses and maintain bed-fluidization properties. In the manufacture of elemental phosphorus, efficient removal of entrained dust particles from the electric furnace off-gas purifies the phosphorus and CO gas stream and prevents contamination of the product when it is separated by condensation from the carrier CO gas. In many of the coal gasification processes being developed for energy conversion, cleaning of the gas produced is extremely important. For instance, the gases resulting from underground coal gasification must be completely cleaned of condensed tars and solid particulates before being burned in a gas turbine to prevent turbine blade erosion. Finally, atmospheric air for process use may need to be cleaned by various methods for specific processes. Removal of dust may be necessary to prevent contamination of a high-purity chemical or pharmaceutical which is to be dried with air. Removal of bacteria and spores may be necessary when air is used in fermentation processes and pharmaceuticals (see Sterile techniques). Air for cryogenic use must be treated to remove CO_2, water vapor, and hydrocarbons in addition to dust (see Cryogenics).

Application Requirements

In order to design and select gas-cleaning equipment, the flow rate, composition, temperature, and pressure of the gas to be cleaned must be known, in addition to the amount of the contaminant and its chemical species and physical state. All equipment used for air pollution effluent abatement can generally be employed for gas cleaning. When present as particulates, the particle size distribution and apparent density of the particle are important. Figure 1 in Air pollution (Vol. 1, pp. 636–637) gives relative particle sizes and measuring methods, and equipment for particle collection for many common contaminants. However, it is best to ascertain the contaminant loading and size distribution by actual measurements. The method of process sampling is extremely critical for particulates. The loading on one side of a duct can be severalfold greater than on the other side. Therefore, a traversing-type sample should be obtained; 12 samples on each of 2 diameters at right angles are suggested. The flow should not be obstructed for 8–10 pipe diameters upstream or 2 pipe diameters downstream. Sampling should be isokinetic (probe-inlet flow equal and parallel to the local duct velocity). When particle size distribution is needed, *in situ* classification is preferable to laboratory particle size analysis. Cascade impactors are widely used in which the aerosol particles are fractionated according to their aerodynamic diameter. Sampling is complicated by the fact that an impactor cannot be used for traverse isokinetic sampling. For accurate work, individual impactor samples should be taken isokinetically at each traverse location.

Principles of sampling are discussed in refs. 3–8, particle size analysis in refs. 9–10 (see Sampling). Particle sampling and sizing methods are reviewed in ref. 11, and testing of gas cleaning equipment in ref. 12.

Removal of Gaseous Impurities

The five methods applicable for removing gaseous impurities are absorption, adsorption, chemical reaction, condensation, and incineration. The last item is valid only for combustible impurities. Absorption may be water-based or employ an organic liquid. Typically, packed and plate-type columns are used (see Absorption). In older industrial processes, towers are packed with quartz or coke, large ceramic shapes, or slat packing. Towers subject to plugging may occasionally use baffle trays or disk and doughnut trays. Open spray chambers are also used but are generally limited to 5–7 transfer units because of spray entrainment and loss of countercurrency. Rotating centrifugal atomizing gas absorbers are used in Europe (1) but only seldom in the United States.

Under certain circumstances, pollutant gases can be removed with solid absorbents, eg, the removal of H_2S with iron oxide (Fe_2O_3). Organic sulfur compounds can be removed by hydrogenation with catalysts composed of cobalt, nickel, and molybdenum sulfides, and sulfur with zinc oxide. Most of these processes are old (1), and are generally in poor repute among engineers. Static beds are generally utilized and reaction efficiencies are low because solids with high surface areas have not been developed. However, with the need to develop processes for synthetic natural gas and its purification, this area may be developed, and high surface area solid absorbents suitable for use in fluid and traveling beds may be employed. Lime and limestone (qv), in solid form, remove 50–60% SO_2 from flue gases. The higher removal efficiency achieved with liquid slurries has discouraged further research with dry solids.

Adsorption as a means of gas purification is discussed in Vol. 1 under Adsorptive separation, and Air pollution control methods (pp. 655–659). Carbon adsorption is especially useful to remove organic compounds almost completely from large volumes of gas (see Carbon). The Purasiv HR System (13) is an interesting dual fluid-bed unit in which adsorption and regeneration occur simultaneously and continuously in a common vessel. Molecular sieves (qv) have also been found highly selective for certain gaseous impurities in dry gas streams. When impurities are present in the carrier gas at concentrations of 10–100 ppmv, use of adsorption as a pollutant concentrator may be economically attractive before destroying the impurities by chemical reaction, oxidation, or incineration. A 40-fold increase in concentration can be achieved while simultaneously transferring the pollutant to air or an inert gas. The economics of incineration of trace-contaminated gases are compared with adsorption concentration in ref. 14 (see Exhaust control; Incineration).

Chemical reaction includes absorption methods wherein a reactive material is added to the absorbing liquid. Apart from this, gas-phase and gas-solid chemical reactions can be employed to enhance the separability. For instance, a reaction may be employed (such as chlorination) to increase the molecular weight and reduce the vapor pressure of the undesirable species. The result may be condensation of the pollutant as a liquid or solid. Again, a reaction may be carried out with the help of catalysts to change the molecular structure of the pollutant to make it more readily absorbable.

Pollutants with high vapor pressures can be condensed by cooling and thus their concentration reduced (see Air pollution control methods, Vol. 1, p. 659). This method is satisfactory where 80–90% removal is sufficient. Refrigerated cooling by itself or in conjunction with gas compression reduces pollutant concentrations but is expensive unless cooling and compression is needed for other process reasons.

Combustible pollutants can be destroyed by incineration provided the carrier gas itself is not harmed by the operation (see Incinerators). Usual designs can be adapted to reduce combustible concentrations to 40–50 ppmv, but longer residence times (0.7–2.0 s) and higher temperature (1200 K and above) are required to achieve combustible material concentrations below 10 ppmv. Economics and methods of removal of volatile organics from gas streams to levels of 5 ppmv are reviewed in ref. 15.

Removal of Particulates

Particles are generally removed by gravity and its enhancement with centrifugal force, inertial impaction and direct interception, diffusional deposition, and flux-force interaction, including electrophoresis, thermophoresis, diffusiophoresis, and Stephan flow. All of these are discussed in Vol. 1, pp. 660–664, 685, and 701–704. Selection of particle removal devices for gas cleaning depends on the particle size and the required removal efficiency.

Common treatment devices for solid particles include cyclones, fabric and granular-bed filters, impaction devices, electrostatic precipitators, and both low- and high-energy wet scrubbers (see Air pollution control methods). Except for fabric and granular-bed filters, these devices can also be used for removal of liquid particulates. Cyclones and electrostatic precipitators may require a somewhat different design for liquids to prevent reentrainment. Mixtures of liquid and solid particulates can be handled in devices suitable for liquids if the ratio of liquid to solid is high enough to prevent formation of sticky pastes within the collector. The addition of sprayed liquid can be helpful where insufficient mist is present in the gas stream. Devices circulating a wash liquid such as wet scrubbers and liquid-washed electric precipitators are most useful in such situations. Primarily suitable for liquids are centrifugal and zig-zag (or chevron) entrainment separators, knitted fiber-pad mist eliminators, and a variety of depth-filtration fiber-bed mist eliminators. These devices are discussed in Vol. 1, pp. 694–695 and 701. Efficiency models for cyclones, zig-zag baffles, tube banks, packed beds, sieve plates, and knitted-mesh entrainment separators have been developed (16).

The Petersen separator (17–18) is an interesting device for the efficient removal of fine particles (Fig. 1). Originally, the collector consisted of a coiled spring with specially shaped top and bottom sections and extremely small gaps (typically 300–1200 μm) between successive faces. The gap can be adjusted by tightening the tension on the spring. The gas enters the inside of the spring and is forced to travel out through the wave-shaped opening between successive coils. Liquid particulates impact on the surface of the gap, coalesce, and drain from the spring. Collection efficiencies exceeding 99% have been reported on submicrometer size particles. Pressure drops are 3.6–6.0 kPa (15–25 in. water). In a modification of this device offered by Union Carbide, the

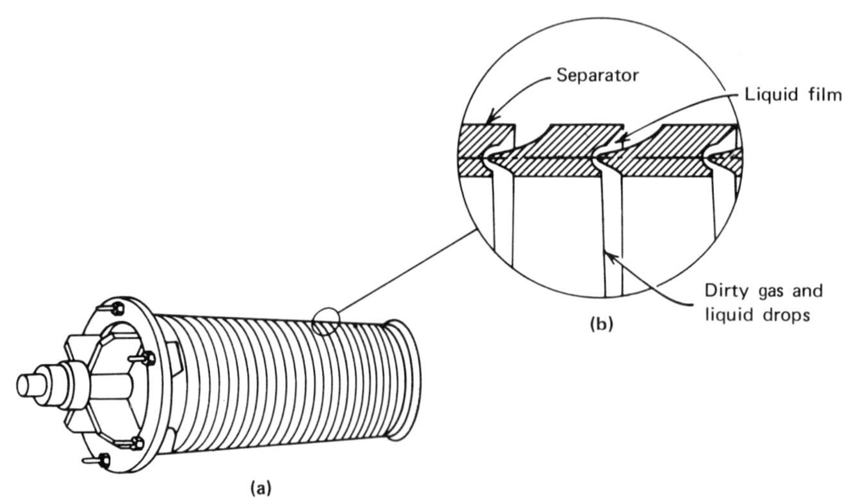

Figure 1. Petersen separator. (**a**) Assembly of a Petersen candle. One or more are bottled in a tube sheet at the flange, which requires the dirty gas to enter the inside of the candle. (**b**) Cross section through a portion of the separator rings or spiral spring. Liquid droplets or solid particulate impact on the recessed face of the ring as the gas passes through the gap at high velocity.

spiral spring has been replaced by molded rings of polysulfone, glass-filled polypropylene, or 304 stainless steel. Developed originally to collect submicrometer liquid mists of phosphoric and sulfuric acids and oil mists, it has been applied more recently to removal of fine solids such as pigment dusts, ammonium nitrate, urea fumes, and iron sulfide particles from natural gas. Water sprays are added at the entrance to the spring to provide a continuous flush when fine solids are handled.

When fine particle removal is needed for gas cleaning and purification, as in elemental phosphorus production, or for product collection, as in carbon black manufacture, electric precipitators are frequently employed (19). The utilization of wet and corrosion-resistant electrostatic precipitators (Fluid-ionics), augmenting the collection efficiency of other types of both wet and dry particle collectors (see Air pollution control methods, Vol. 1, pp. 683–686) is increasing. Electrically-augmented bag filters have been tested (20). Wet electrically-augmented collectors are particularly attractive for efficient removal of mixtures of solid and liquid fine particles. A low-pressure-drop (0.5–1.0 kPa (2–4 in. of water)) electrically-augmented venturi scrubber with 95–98% collection efficiency on submicrometer particles is available (21). Union Carbide has developed a high-intensity ionizer that can be used in conjunction with water-washed collecting tubes. Other promising methods for the removal of fine particulates are discussed in ref. 22 (particle collection in fluidized beds), ref. 23 (high-gradient magnetic separation of airborne particulate), and ref. 24 (a cleanable reticulated-foam filter).

Air Filters

Dust is removed from gas or air with filters referred to as air cleaners or air filters (25–28) (see also Filtration). They are designed for lower concentrations of particulate matter in the carrier gas such as that encountered in outdoor air or air recirculated within a building. Although they are typically employed in ventilation, heating, and air conditioning, they are very important for processes requiring clean air. In the former, removal of coarser dusts (>1–2 μm) may be satisfactory, but for some processes such as the assembly of electronic devices in clean surroundings, efficient removal of submicrometer material may be needed. In some processes involving fermentation, pharmaceuticals, and food, removal of microorganisms (2) may also be needed (see Table 1) (see Air conditioning; Sterile techniques).

Filters of many different types are available ranging in efficiencies for 0.3 μm dioctyl phthalate (DOP) smoke from practically zero to 99.99%. Filters are often classified as viscous, dry, or automatic. The latter classification includes both viscous and dry types which are provided with some means for automatic cleaning or media replacement as well as electrostatic collectors. Air filters may also be classified as

Table 1. Filter Penetration of Microorganisms, %

Medium	DOP[a]	1.0 μm bacteria	0.03 μm virus
glass paper	0.01	0.0001	0.0036
asbestos–cellulose paper	5.0	0.4	3.5

[a] DOP = Dioctyl phthalate.

cleanable or disposable. In the latter group, the entire filter including frame may be discarded, or a permanent frame with replaceable media may be used. Low pressure drop is essential for air-cleaning filters to minimize power requirements when handling large volumes of air.

Viscous filters consist of a filter mat that is coated with an oil or other chemicals of high viscosity and flash point which acts as an adhesive for the impacting dust. In stationary filter panels, the filter medium is usually retained between coarse screens contained in a metal frame. The thickness of the medium may vary from 2.5–10 cm. It can consist of a variety of materials including glass and chemical fibers, animal hairs, wood shavings, metal turnings, and layered wire woven into gauze or crimped. Filters with finer fiber elements are more efficient. Dimensions of the containing frame vary but typical sizes are 0.5–0.6 m on a side. Face velocities are typically 150–200 cm/s. When more than one filter is needed, they are normally fastened into an angle-supporting frame. When dirty, metallic media are generally removed, washed and steamed, reoiled, and returned to service.

Dry filters are supplied in units of similar size to viscous filters but are deeper. The air-flow passages are smaller, so that lower velocities must be used through the medium. To compensate for this lower velocity, the medium is usually arranged either in the form of parallel pockets or an accordian pleat in order to maintain a high frontal velocity approaching the filter. The filter medium frequently consists of sheets of cellulose pulp, cotton, felt, or spun glass. With cheaper media, the entire filter or the medium is discarded when dirty. Mechanical loading devices are sometimes used to replace cellulose media in permanent metallic frames. More expensive media such as synthetic felts are sometimes reclaimed by dry cleaning (see Felts). Some newer washable dry filters use reticulated foam (see Foamed plastics) or self-charging mats of woven plastic fibers which retain dust particles electrostatically.

Automatic filters of viscous type employ screen curtains or moveable leaves of mesh or gauze that are slowly moved either continuously or at intervals on a vertical

Table 2. Efficiency of Air Filter Media

Media	Air flow rate, mm/s at filter surface	Pressure drop, kPa[a]	Collection efficiency on 0.3 μm DOP[b]
glass fiber pads			
loose 3-μm fibers, 12-mm thick	170	0.01	63
loose 1.3-μm fibers, 12-mm thick	150	0.45	91
loose 1.3-μm fibers, 25-mm thick	140	0.38	99.4
12-mm of 3-μm fibers followed by 12-mm of 1.3-μm fibers	150	0.23	94
woven glass fibers			
fine weave fabric	26	0.02	48
coarse weave fabric	26	0.005	22
wool felt (reverse jet cleaning)			
at start	100	0.20	30
at 840 continuous hours	100	0.70	92
commercial glass fiber absolute filter	20	0.28	99.99
vacuum cleaner bag paper	70	0.30	30

[a] To convert kPa to mm Hg, multiply by 7.5.
[b] DOP = Dioctyl phthalate.

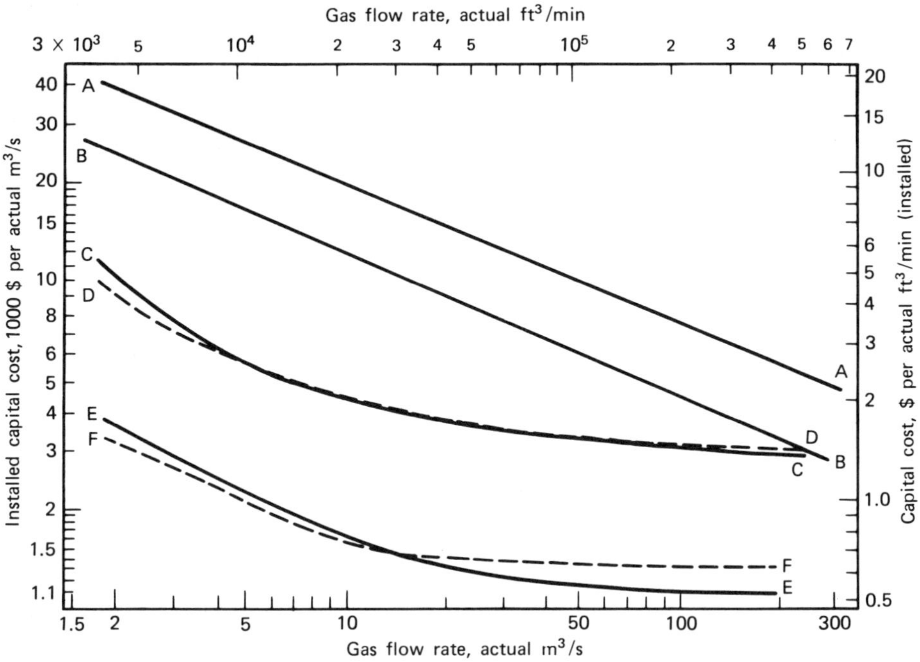

Figure 2. Installed capital cost of gas-cleaning particulate-removal equipment in fourth quarter, 1977 dollars (Marshall and Swift Equipment Cost Index: 518.4. A, high efficiency precipitator: 98–99% on 2–3 μm particles, 93–95% on submicrometer particles. B, average efficiency precipitator: 90% on 2–3 μm particles. C, fabric filters with moderate temperature synthetic bags, woven or felt. D, high energy wet scrubber. E, low to moderate energy wet scrubber. F, high efficiency cyclone separator. Costs are based on carbon-steel construction. Water treatment and disposal costs where required, are not included.

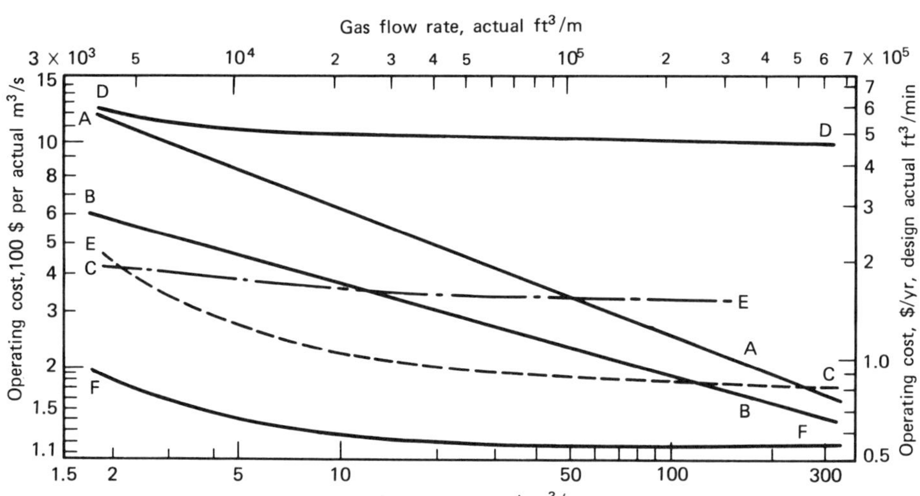

Figure 3. Operating cost of particulate gas-cleaning equipment in 1977 dollars. Curve designation is the same as Figure 2 (when operated at capacity). Costs include capital charges: 31.3%/yr (repairs, 10%, depreciation, 8.3%, taxes and insurance, 2%, capital overhead costs, 2.5%, interest, 8.5%); electricity: 3.5¢/kW·h; make-up water: 10¢/m^3 (40¢/1000 gal). Operating labor, supervision, and payroll charges are not included. Water treatment and disposal costs, where applicable, are also not included.

track. At the bottom, the medium dips into an oil bath where the collected dust settles to the bottom and is removed periodically as a sludge. Automatic filters of the dry type expose a layer of close-packed glass or chemical fibers (bonded together) to the gas stream for a predetermined time or pressure differential. The clean medium is unrolled from a fresh spool and the spent medium is rerolled on another spool. When high removal efficiency is required, two-stage electrostatic precipitators (see Air pollution control, Vol. 1, pp. 684–685) are also used for air cleaning.

Ref. 28 discusses empirical efficiency test methods for air filters, the performance of various filter types based on these test methods, and general principles for filter selection and maintenance. The test methods are useful in comparing the relative efficiency of one filter vs another on specific dusts but give no measure of the efficiency of the collector as a function of particle size. Table 2 compares the collection efficiency of various media for 0.3 μm DOP smoke.

Economic Aspects

Figure 2 presents the relative purchase costs of various devices for removing particulates from gases adjusted to 1977 dollars. Figure 3 presents comparative operating costs based on typical capital charges. Labor costs are expected to be essentially the same for all devices. Such means of comparing costs are not wholly equitable since some devices (electric precipitators, bag filters, and high-energy wet scrubbers) would be more efficient, especially on fine particulate, than cyclones and low energy scrubbers. Refs. 29–30 give charts for adjusting collector size for equivalent efficiency and give a basis for more equitable cost comparison.

BIBLIOGRAPHY

"Gas Cleaning and Purification" in *ECT* 1st ed., Vol. 7, pp. 83–106, by H. F. Johnstone, University of Illinois; "Gas Cleaning" in *ECT* 2nd ed., Vol. 10, pp. 329–352, by S. Sylvan and R. F. Logsdon, American Air Filter Co., Inc.

1. G. Nonhebel, ed., *Gas Purification Processes,* Geo. Newnes Ltd., London, 1964.
2. W. Strauss, *Industrial Gas Cleaning,* 2nd ed., Pergamon Press, Oxford, 1975.
3. H. B. H. Cooper, Jr. and A. T. Rossano, Jr., *Source Testing for Air Pollution Control,* Environmental Research and Applications, Inc., Wilton, Conn., 1971.
4. *Fed. Regist.* **41,** 23059 (June 1976).
5. *Industrial Guide for Air Pollution Control, Publ. EPA-625/6-78-004,* EPA, 1978.
6. H. J. Paulus and R. W. Thron in A. C. Stern, ed., *Air Pollution,* 3rd ed., Vol. III, Academic Press, New York, 1976, Chapt. 14, pp. 525–587.
7. *Methods of Air Sampling and Analysis,* 2nd ed., American Public Health Assn., Washington, D.C., 1977.
8. J. W. Leatherdale in P. N. Cheremisinoff and R. A. Young, eds., *Air Pollution Control and Design Handbook,* Part I, Marcel Dekker, Inc., New York, 1977.
9. J. D. Stockham and E. G. Fochtman, *Partical Size Analysis,* Ann Arbor Science, Ann Arbor, Michigan, 1977.
10. H. E. Hesketh, *Fine Particle in Gaseous Media,* Ann Arbor Science, Ann Arbor, Mich., 1977.
11. W. B. Smith, P. R. Cavanaugh, and R. R. Wilson, *Technical Manual: A Survey of Equipment and Methods for Particulate Sampling in Industrial Process Streams, Publ. EPA-600/7-78-043,* EPA, Washington, D.C., 1978.
12. P. W. Spaite and J. O. Buckle in A. C. Stern, ed., *Air Pollution,* 3rd ed., Vol. IV, Academic Press, New York, 1977, pp. 77–86.
13. *Chem. Eng.* **84,** 39 (Aug. 29, 1977).
14. B. Grandjacques, *Pollut. Eng.* **9,** 28 (Aug. 1977).

15. B. B. Crocker, "Removal of Hazardous Organic Vapors from Vent Gases," *Proceedings of the APCA Specialty Conference on Control of Specific (Toxic) Pollutants, Gainesville, Fla., Feb. 13–16, 1979, APCA, Pittsburgh, Pa., 1979.*
16. S. Calvert, S. Yung and J. Leung, *Entrainment Separators for Scrubbers—Final Report, Publ. EPA 650/2-74-119-b,* EPA, Washington, D.C., available as NTIS Report PB-248050, Springfield, Va., Aug. 1975.
17. H. Petersen, *Achema Jahrbuch,* Vol. II, Dechema, Frankfurt, FRG, 1971/73 (special reprint).
18. Ref. 2, pp. 211–213; W. Strauss in ref. 12, pp. 306–307.
19. S. Oglesby, Jr. and G. B. Nichols, *A Manual of Electrostatic Precipitator Technology—Part II, Application Areas,* NTIS Report PB-196381, Southern Research Institute, Birmingham, Ala., 1970.
20. D. J. Helfritch, *Chem. Eng. Prog.* **73,** 54 (Aug. 1977).
21. T. Dembinsky and J. F. Vicard, "A Low Energy, High Velocity Electrostatic Precipitator—The Electrodynamic Venturi Scrubber," *preprint 78-17.6 presented at APCA 71st Annual Meeting, Houston, Texas, June 1978.*
22. Y. Doganoglu, V. Jog, K. V. Thambimuthu, and R. Clift, *Trans. Inst. Chem. Eng.* **56,** 239 (1978).
23. D. C. Drehmel and C. H. Gooding, *J. APCA* **29,** 534 (1979).
24. J. D. Brady, D. W. Cooper, and M. T. Rei, *Chem. Eng. Prog.* **73,** 45 (Aug. 1977).
25. Ref. 2, pp. 353–366.
26. R. L. Lucas in R. H. Perry and C. H. Chilton, eds., *Chemical Engineers' Handbook,* 5th ed., McGraw-Hill, New York, 1973, pp. 20-115–20-120.
27. *ASHRAE Handbook and Product Directory—1975 Equipment,* American Society of Heating, Refrigeration and Air Conditioning Engineers, Inc., New York, 1975, Chapt. 10.
28. C. N. Davies, *Air Filtration,* Academic Press, New York, 1973.
29. M. L. Kinkley and R. B. Neveril, *Capital and Operating Costs of Selected Air Pollution Control Systems, Publ. EPA-450/3-76-014,* EPA, Washington, D.C., May 1976.
30. R. B. Neveril, J. W. Price, and K. L. Engdahl, *J. APCA* **28,** 829, 963, 1069, 1171, 1253 (Aug.–Dec. 1978).

B. B. CROCKER
Monsanto Company

GAS HYDRATES. See Hydrocarbons, C_1–C_6; Water, supply and desalination.

GASKETS. See Packing materials.

GAS, MANUFACTURED. See Fuels, synthetic.

GAS, NATURAL

Natural gas is defined as a naturally occurring mixture of hydrocarbon and nonhydrocarbon gases found in the porous geologic formations beneath the earth's surface, often in association with petroleum (1). To obtain a marketable product, the raw natural gas flowing from gas or oil wells must be processed to remove water vapor, inert or poisonous constituents, and condensable hydrocarbons. The processed gas is principally methane, with small amounts of ethane, propane, butane, pentane, carbon dioxide, and nitrogen. This gas can easily be transported from the producing areas to the market in underground pipelines under pressure or liquefied at low temperatures and transported in specially designed ocean-going tankers (see Pipelines; Transportation).

Natural gas is used principally as a source of heat in residential, commercial, and industrial service. Because of its clean burning quality, the convenience of utilization, low cost, and abundance, natural gas supplies approximately one third of the total energy requirements of the United States.

Seepage of natural gas from surface rocks and springs and its combustion properties have been known since the earliest recorded history. There is evidence that many of the early temples of worship were located near known gas seeps. The records of travels of Julius Caesar noted that a famous *fontaine ardente* or burning spring was located near Grenoble, France. It is known that as early as 615 AD gas wells had been drilled in Japan; however, it is not clear how the gas was utilized. In 900 AD the Chinese used bamboo tubes to transport natural gas to their salt works. The heat generated by the burning of the gas was used to evaporate the water from the brine, leaving crystallized salt. Gas seeps and columns of fire were reported in the Ohio Valley and the Appalachian area during the early settlement of the United States. Early salt producers in the United States faced large financial losses when their drill hit a gas pocket, because it usually resulted in an explosion or fire and loss of their equipment. In 1840 in Butler County, Pennsylvania, natural gas was again used to evaporate salt brine, much as the Chinese had done 1000 years earlier.

The first well drilled in the United States, with the expectation of finding natural gas, occurred in 1821 near Fredonia, New York. The well was 8 m deep and gas was transported in log pipes to nearby houses and lighted 30 burners. General Lafayette, during his visit to the United States in 1825, arrived in Fredonia by stagecoach at 2:00 AM to find the town brilliantly illuminated with natural gas in his honor. The Fredonia Gas, Light, and Waterworks Company was formed in 1865 as the first natural-gas company in the country.

As the exploration for oil increased, so did the discovery of natural gas, which was wasted by flaring (burning it in a stack close to the well). In 1872 the gas that had been venting from an oil well near Titusville, Pennsylvania, was transported to the town in an iron pipeline (8.9 cm dia, 8.8 km long) to serve 250 domestic and industrial customers, marking the first use of iron pipe to transport natural gas. The gas industry had its beginning serving local areas near known gas supplies in the Appalachian area. The first compressor station was placed in operation in Rexford, Pennsylvania, in 1880. The first long distance pipeline consisted of 193 km of parallel 20-cm wrought-iron screwed pipes running from Greenstown, Indiana, to Chicago, Illinois. With the invention of the Welsbach mantle in 1885, the sale of natural gas for lighting increased; however, industrial sales provided the largest gas markets.

As natural gas became more readily available, it was mixed with, or replaced manufactured gas. The industry continued to develop locally around gas reserves as they were discovered in the Appalachian area, Oklahoma, Kansas, Louisiana, Texas, California, and the Rocky Mountain areas. In 1931 a 1600 km high pressure pipeline (61 cm dia) was constructed from the panhandle of Texas to Chicago. This was the first modern, long distance transmission line built to deliver gas from the producing states of the southwest to the large markets located in the north and east. The development of high strength pipe and electric welding, which made the pipe joints stronger than the pipe itself, spurred the construction of other long distance pipelines, until today there is a network of ca 418,400 km of transmission lines carrying gas to 45 million customers in every state, except Hawaii, where manufactured gas is sold.

Natural gas is found in areas of the earth that are covered with sedimentary rocks. These sediments were first laid down during the Cambrian period, ca 500 million years ago, and this process continued until the end of the Tertiary period ca 100 million years ago. These sediments contain the organic source materials from which natural gas and petroleum were produced. Gas and petroleum, being less dense than the water present in the rocks, tended to migrate upward until contained under impervious rock barriers. Variation in deposition, together with the folding and faulting that occurred in the sedimentary rock layers over millions of years, formed anticlines, stratigraphic traps, and other types of reservoirs where petroleum and gas accumulated. Deposits are found in thirty-four states of the United States and in at least 68 other countries.

Reserves

Natural gas is a nonrenewable resource and, because of its value as an energy source, reports of available reserves are of great national interest. The best estimates of the total proven gas reserves of the United States and Canada are provided annually by industry associations (2). These estimates are prepared by regional committees made up of geologists and petroleum engineers, and are based on technical studies of the natural gas that, with reasonable certainty, will be produced from all known oil and gas reservoirs under existing economic and operating conditions. Natural-gas reserves in the United States are classified into two categories based on the reservoir occurrence: (*1*) nonassociated gas is defined as free natural gas not in contact with crude oil in the reservoir, and (*2*) associated, dissolved gas is the combined natural gas that occurs in crude oil reservoirs, either as free gas or as gas in solution with crude oil.

Total proved gas reserve estimates are the sum of all nonassociated gas, associated, dissolved gas, plus all of the recoverable gas volumes available in underground storage reservoirs. Table 1 shows the 1977 estimates for the United States, together with the preliminary annual production volumes. Approximately 55% of the gas reserves are located in the Gulf Coast states of Louisiana and Texas. The last large addition was made in 1970 when the gas available in the North Slope area of Alaska was included. Similar data for Canada (Table 2) shows that 78% of those proved gas reserves are located in the province of Alberta. Proved gas reserves in the United States increased steadily each year until 1967, when they peaked at ca 8.3×10^{12} m^3 (293×10^{12} ft^3) and have declined since that time to ca 6×10^{12} m^3 (209×10^{12} ft^3) by 1977 (see Fig. 1), whereas the Canadian reserves continued to increase. This indicates the maturity of the natural-gas industry in the United States, where ca 1.6×10^6 km of pipelines and mains gather, transport, and distribute gas. The decline in proved gas reserves

Table 1. Estimated Total Proved United States Reserves of Natural Gas[a,b], 10^6 m^3[c]

State	1977 Production[d]	Reserves as of December 31, 1977 Total gas	Non-associated	Associated–dissolved	Under-ground storage[e]
Alabama	1,064	21,106	20,288	818	0
Alaska	5,305	901,181	155,471	745,710	0
Arkansas	3,032	47,570	42,634	4,218	718
California[f]	9,395	147,165	55,504	80,427	11,234
Colorado	5,288	57,356	49,660	6,644	1,052
Florida	1,234	6,096	0	6,096	0
Illinois	44	11,827	29	30	11,768
Indiana	6	1,508	26	1	1,509
Kansas	21,831	337,630	330,372	4,068	3,190
Kentucky	1,606	21,092	16,042	1,191	3,859
Louisiana[f,g]	199,001	1,491,540	1,216,456	266,383	8,701
Michigan	3,707	50,709	17,388	16,633	16,688
Mississippi	2,532	36,977	31,876	2,681	2,448
Montana	1,524	29,555	23,097	1,909	4,549
Nebraska	92	1,932	157	426	1,346
New Mexico	32,607	337,770	267,262	69,416	837
New York	283	7,001	3,910	4	3,087
North Dakota	725	11,099	130	10,942	0
Ohio	2,821	41,316	26,725	4,222	10,369
Oklahoma	47,068	331,576	259,639	64,903	7,035
Pennsylvania	2,613	53,362	37,291	324	15,748
Texas[f]	193,281	1,759,688	1,315,412	438,126	6,151
Utah	1,780	21,202	11,463	9,672	67
Virginia	232	1,955	1,955	0	0
West Virginia	4,171	67,260	55,258	1,374	10,628
Wyoming	9,571	112,188	94,111	16,549	1,529
miscellaneous[h]	21	5,644	640	93	4,912
Total United States	*550,834*	*5,913,305*	*4,032,796*	*1,752,860*	*127,425*
Gulf of Mexico detail					
Louisiana	117,409	915,119	781,988	133,130	0
Texas	6,621	148,078	146,876	1,202	0
Total Gulf of Mexico	*124,030*	*1,063,197*	*928,864*	*134,332*	*0*

[a] Courtesy of the American Gas Association (2).
[b] At 101.3 kPa (1 atm) and 16°C.
[c] To convert m^3 to ft^3, multiply by 35.3.
[d] Preliminary net production.
[e] Proven recoverable gas contained in underground storage reservoirs, including native and net injected gas. (First reported on a recoverable basis in 1973).
[f] Includes offshore.
[g] Reported quantities include reserves estimated to be recoverable from some reservoirs considered natural gas bearing, based on electrical logs, core data, and other available engineering and geological data.
[h] Includes Arizona, Iowa, Maryland, Minnesota, Missouri, South Dakota, Tennessee, and Washington.

Table 2. Remaining Marketable Natural-Gas Reserves in Canada[a,b], 10^6 m^3[c]

		Remaining reserves at December 31, 1977				
Provinces[d]	1977 Net production[e]	Non-associated	Associated	Dissolved	Under-ground storage	Total
Territories	628	152,104				152,104
British Columbia	8,605	180,246	7,286	1,794		189,326
Alberta[f]	59,874	1,086,736	164,553	57,140	1,286	1,309,721
Saskatchewan	1,223	20,250	1,203	1,530	282	23,266
Ontario	239	3,507			5,407	8,915
other eastern Canada	3	283	40			323
Total	*70,572*	*1,443,126*	*173,082*	*60,464*	*6,975*	*1,683,655*

[a] Courtesy of the American Gas Association (2).
[b] At 101.3 kPa (1 atm) and 16°C.
[c] To convert m^3 to ft^3, multiply by 35.3.
[d] Mackenzie Delta gas reserves included for the first time in 1974. Arctic islands gas reserves included for the first time in 1975.
[e] Preliminary estimate.
[f] After adjustment for straddle-plant processing shrinkage and fuel.

does not signal an early end to this industry, since it relies on a continuing exploration program to maintain a long-term gas supply. The present estimates range from 5.6 $\times 10^{12}$ m^3 (198×10^{12} ft^3) probable supplies to 11.3×10^{12} m^3 (399×10^{12} ft^3) possible supplies and these will sustain the industry for many years.

The world reserves (Table 3) continue to increase, spurred by improved price and technological advances in the developing countries. Large gas deposits are now being discovered in Mexico, South America, and in the Far East (see also Fuels, survey; Petroleum, resources).

Numerous studies have been made in attempting to arrive at the total remaining recoverable reserves of natural gas from conventional sources; however, such studies are, at best, only a calculated guess. These studies (4) indicate that the total remaining recoverable reserves of natural gas from conventional sources in the world amount to ca $(2.6–2.7) \times 10^{14}$ m^3 [$(9.18–9.53) \times 10^{15}$ ft^3]. Based on these estimates and present proved reserves, ca 2.0×10^{14} m^3 (7.0×10^{15} ft^3) remain to be discovered and proved in the world by future exploration.

Properties and Composition

Methane, the principal constituent of natural gas, is colorless and odorless. It is a simple, nonpoisonous asphyxiant. When properly mixed with air, it burns with a faint, luminous flame; however, when mixed with air in concentrations ranging between the lower and higher limits of flammability, natural gas explodes. Table 4 (5) shows some of the more important physical properties of the common hydrocarbon compounds and impurities found in gas streams (see Hydrocarbons, C_1–C_6). The relationship of mass, pressure, volume, and temperature of natural gas follow the ideal gas laws when including the compressibility factor.

$$PV = ZNRT$$

where P = absolute pressure, kPa (psia); V = volume, L (ft^3); Z = compressibility

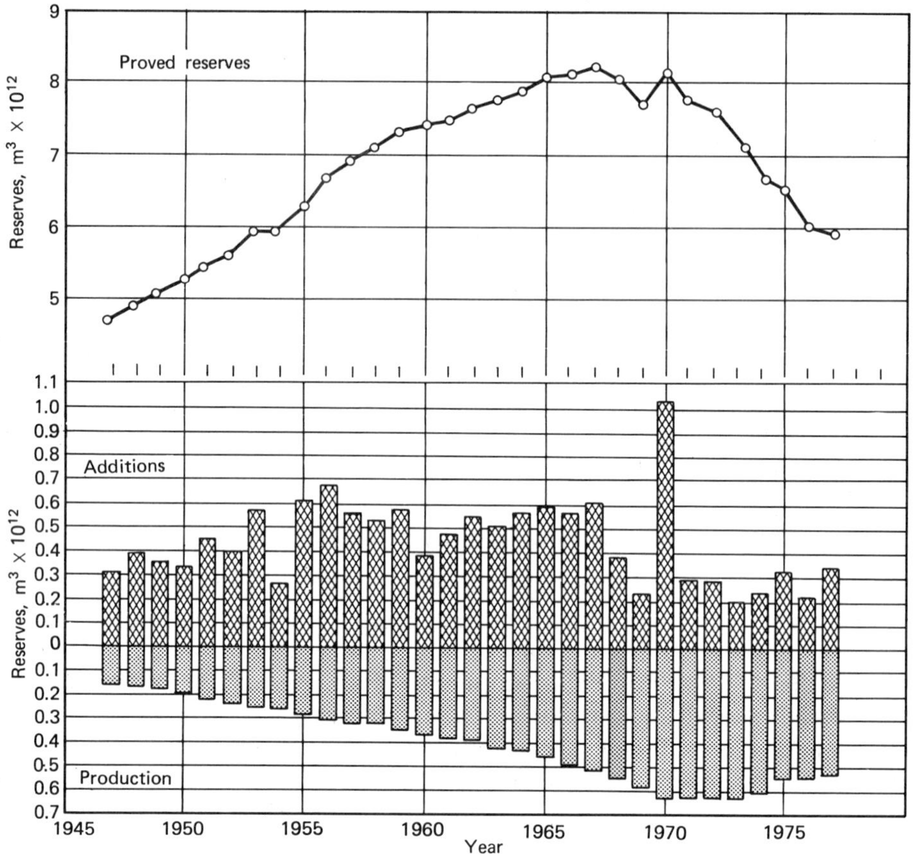

Figure 1. United States natural gas reserves. To convert m^3 to ft^3, multiply by 35.3.

factor, dimensionless; N = mass, mol; R = universal gas constant, 8.314 J/(mol·K); T = absolute temperature, K. The compressibility factor depends on pressure, temperature, and composition of the gas and can be calculated with reasonable accuracy (6) or obtained by laboratory tests.

Natural gas is classified in several broad categories based on the chemical composition: (*1*) wet gas contains condensable hydrocarbons such as propane, butane, and pentane; (*2*) lean gas denotes an absence of condensable hydrocarbons; (*3*) dry gas is a gas whose water content has been reduced by dehydration process; (*4*) sour gas contains hydrogen sulfide and other sulfur compounds; and (*5*) sweet gas denotes an absence of hydrogen sulfide and other sulfur compounds. Natural gas sold to the public is described as lean, dry, and sweet as illustrated in Table 5.

The composition of natural gas at the wellhead varies widely from field to field (Table 6). Many undesirable components may be present that must be removed by processing before delivery to the pipeline. Natural-gas liquids must be extracted from the wet gas in sufficient quantity to eliminate the possibility of condensation in the pipeline. Accumulations of these liquids are a safety hazard and restrict the gas flow. Natural-gas liquids are usually propane and higher hydrocarbons; ethane may also be included. Propane and butane are known as liquefied petroleum gas (LPG), whereas

Table 3. Estimated Proved World Reserves of Natural Gas in 1978[a]

Location	Natural gas reserves, 10^9 m^3 [b]	Location	Natural gas reserves, 10^9 m^3 [b]
North America	9,286	Africa, Northern	4,585
Canada	1,775	Algeria	3,455
Mexico	1,699	Egypt	82
United States	5,672	Libya	866
Trinidad and Tobago	170	Morocco	0.2
others	0.2	Tunisia	182
Central America	0.06	Africa, Central and Southern	1,545
South America	2,151	Angola	47
Argentina	432	Cameroun	1
Bolivia	176	Congo	28
Brazil	44	Gabon	0.9
Chile	133	Nigeria	1,466
Columbia	127	Zaire	2
Ecuador	8	Middle East	17,454
Peru	38	Abu Dhabi	531
Venezuela	1,193	Bahrain	260
Western Europe	3,364	Dubai	54
Austria	14	Iran	10,533
Denmark	156	Iraq	755
France	70	Israel	0.3
FRG	177	Kuwait	1,055
Greece	1	Divided Neutral Zone	538
Italy	180	Oman	75
Netherlands	1,649	Qatar	1,657
Norway	406	Saudi Arabia	1,917
Spain	13	Syria	50
Sweden		Turkey	0.8
United Kingdom	670	others	27
others	28	Asia, Far East	2,217
Eastern Europe	23,793	Burma	5
Albania	11	People's Republic of China	736
Bulgaria	13	India	239
Czechoslovakia	13	Japan	19
DDR	81	Pakistan	623
Hungary	113	Taiwan	26
Poland	127	Thailand	187
Romania	348	others	382
USSR	23,040	Oceania	2,684
Yugoslavia	47	Australia	798
		Brunei	194
		Indonesia	685
		Malaysia	827
		New Zealand	179
		Philippines	0.5
Total			*67,079*[c]

[a] Ref. 3.
[b] To convert m^3 to ft^3, multiply by 35.3.
[c] Rounded.

Table 4. Physical Constants[a] of Common Natural Gas Constituents[b]

Component	Chemical formula	Molecular weight	Boiling point, °C	Specific gravity: Liquid at 16°C/16°C	Specific gravity: Gas[a] (air = 1.00)	Gross heating value[a], MJ/m^3 [c]
methane	CH_4	16.043	−161.5	0.3[d]	0.5539	37.56
ethane	C_2H_6	30.070	−88.6	0.3564[e]	1.0382	65.80
propane	C_3H_8	44.097	−42.0	0.5077[e]	1.5225	93.65
n-butane	C_4H_{10}	58.124	−0.5	0.5844[e]	2.0068	121.36
isobutane	C_4H_{10}	58.124	−11.7	0.5631[e]	2.0068	121.01
n-pentane	C_5H_{12}	72.151	36.1	0.6310	2.4911	149.16
isopentane	C_5H_{12}	72.151	27.8	0.6247	2.4911	148.82
carbon dioxide	CO_2	44.010	−78.5	0.827[e]	1.5195	
hydrogen sulfide	H_2S	34.076	−60.3	0.79[e]	1.1765	23.70
nitrogen	N_2	28.013	−195.8	0.808[f]	0.9672	
water	H_2O	18.015	100.0	1.000	0.6220	

[a] At 101.35 kPa (1 atm) and 16°C.
[b] Ref. 5.
[c] To convert MJ/m^3 to Btu/ft^3, multiply by 26.88.
[d] Apparent value at 16°C.
[e] At saturation pressure and 16°C.
[f] Density of liquid, g/mL, at normal boiling point.

Table 5. Typical Gas Specifications[a]

Assay	Value
methane, vol %	89.5–92.5
ethane, vol %	5.1–2.0
propane, vol %	2.1–0.7
butanes and heavier, vol %	1.6–0.5
nitrogen, vol %	0.6–1.9
carbon dioxide, vol %	1.1–2.4
odorant sulfur and H_2S, mg/m^3 [b]	0.0044–0.0874
specific gravity, g/mL	0.60–0.65
heating value range, MJ/m^3 [c]	38.33–34.05

[a] Ref. 7.
[b] To convert mg/m^3 to $lb/10^6\ ft^3$, divide by 16.
[c] To convert MJ/m^3 to Btu/ft^3, multiply by 26.88.

pentane and higher hydrocarbons are known as natural gasoline (see Gasoline; Liquefied petroleum gas). There is a ready market for all of these natural gas liquids which helps offset the cost of processing.

Water vapor must be removed because accumulation of water at pipeline operating pressures results in solid gas hydrates at temperatures well above 0°C. The buildup of these icelike crystals restricts or stops the gas flow (see Clathration). Pipeline specifications are based on obtaining a water dew-point temperature below the lowest temperatures expected in the flowing gas. In the Rocky Mountain area the water content is usually limited to 80 mg/m^3 (5 $lb/10^6\ ft^3$). Carbon dioxide and nitrogen are inert gases and can be removed to improve the heating value of the remaining gas stream. It is common practice to limit the inert gases to 3% of the total gas volume.

Table 6. Composition of Various Natural Gases

	Rio Arriba County, N. Mex.	Terrell County, Texas	Stanton County, Kansas	San Juan County, N. Mex.	Olds Field, Alberta, Canada	Cliffside Field, Amarillo, Texas[a]
component, mol %						
methane	96.91	45.64	67.56	77.28	52.34	65.8
ethane	1.33	0.21	6.23	11.18	0.41	3.8
propane	0.19		3.18	5.83	0.14	1.7
butanes	0.05		1.42	2.34	0.16	0.8
pentanes and heavier	0.02		0.40	1.18	0.41	0.5
carbon dioxide	0.82	53.93	0.07	0.80	8.22	0.0
hydrogen sulfide		0.01			35.79	0.0
nitrogen	0.68	0.21	21.14	1.39	2.53	25.6
helium						1.8
Total	*100.00*	*100.00*	*100.00*	*100.00*	*100.00*	*100.0*
total sulfur, mg/m^3 [b]	0	0.27	0	0	984	
classification						
wet				x		
lean	x	x	x		x	x
sweet	x		x	x		x
sour		x			x	
gross heating value, MJ/m^3 [c]	37.6	17.3	34.9	46.8	30.0	30.7
specific gravity	0.574	1.077	0.733	0.741	0.882	0.711

[a] Ref. 8.
[b] To convert mg/m^3 to $lb/10^6\ ft^3$, divide by 16
[c] To convert MJ/m^3 to Btu/ft^3, multiply by 26.88.

Hydrogen sulfide must be removed from the natural gas since it is extremely dangerous and poisonous, and in the presence of air or water, corrosive. Many pipeline specifications limit the amount of hydrogen sulfide to less than 5.7 mg/m^3 gas (0.36 $lb/10^6\ ft^3$). Even at these low concentrations, the possibility of corrosion over a long period of time makes it necessary to avoid use of copper or brass tubing in kitchen ranges or other home heating appliances. Helium is present in some natural gas produced from fields in the middle of the continent and Rocky Mountain areas, and is regarded as a valuable component if present in concentrations sufficiently large for economic recovery (ca 0.3–0.7% or higher) (see Helium-group gases).

Natural gas is odorized before distribution to provide a distinctive odor which alerts customers of possible gas leaks in their equipment. Commercial odorant, containing mercaptans, aliphatic sulfides, or cyclic sulfur compounds, is added at rates varying from 4 to 24 mg/m^3 (0.25–1.5 $lb/10^6\ ft^3$).

In some instances the presence of appreciable concentrations of impurities has been turned to economic advantage. The hydrogen sulfide removed from natural gas can be reduced to elemental sulfur which is a basic chemical raw material. In 1976, 57 natural-gas treatment plants recovered ca 1.3×10^6 metric tons elemental sulfur, representing 12% of the total United States sulfur production (see Sulfur). Although the value of elemental sulfur varied widely from location to location, the average for the United States was $37.02/t (9). The only raw material from which helium can be extracted economically in large quantities is helium-bearing natural gas. The helium is recovered by utilizing a cryogenic process, where all other components of the gas stream are condensed to a liquid and removed. There are eight plants in the United States producing helium, two are operated by the Bureau of Mines and six by private industry. The Bureau of Mines also operates a helium conservation system consisting of a pipeline and an underground storage reservoir at the Cliffside Field near Amarillo, Texas, where crude helium (50–60% helium and 50–40% nitrogen) is stored for future use. High-purity helium (min 99.995% purity) is marketed by the Bureau of Mines at $1.24/m^3$ ($35/1000 ft^3), a price set in 1961 at the inception of the conservation program to recover the cost of the project. The average price received by private producers declined to $0.78/m^3$ ($22/1000 ft^3) in 1976 (9).

Gas Conditioning

Gas conditioning is required to alter the composition of the wet natural gas flowing from the wellhead to provide uniform lean, dry, sweet gas for consumer use. Processing facilities are usually located in the fields and vary in size from small wellhead installations to large, centrally located plants. The conditioning of natural gas involves a large variety of processes, each designed to remove a specific impurity.

Water vapor, generally present in gas produced from underground reservoirs, is removed either by absorption by hygroscopic liquids or adsorption on activated solid desiccants (see Drying agents). Diethylene and triethylene glycol have gained the widest acceptance because of their high affinity for water, low cost, chemical stability, low solubility for natural gas, and low foaming tendencies (see Glycols). The glycol dehydration system consists of an absorber tower, where the gas stream comes in close contact with a countercurrent flow of glycol, and a regenerator, wherein the water is distilled from the glycol. The dehydration effect is based on the concentration of the glycol leaving the regenerator. In field operation triethylene glycol can be concentrated to 99%, resulting in processed gas that contains less than 48 mg water vapor/m^3 (3 lb/10^6 ft^3). Because of the relatively low cost and ease of operation, glycol dehydration units are predominantly employed. Dehydration units utilizing solid desiccants, such as silica gel, activated aluminum, or molecular sieves (qv), are of the fixed-bed type. These dehydration units consist of three towers filled with the solid desiccant, together with the necessary equipment to provide regeneration. The first tower is on the line, processing the gas stream, while the second tower is being regenerated with hot gas, and the third tower is being cooled after regeneration. This process is used when extremely dry gas is required (see Adsorptive separation).

To remove acid gases such as carbon dioxide, hydrogen sulfide, and other sulfur compounds from natural gas, many processes are available, such as (*1*) chemical solvent processes which employ a solvent in aqueous solution; the solvent reacts with the impurity at contact temperature and pressure, forming a chemical complex that later decomposes in the regeneration equipment; (*2*) physical solvent processes which employ liquid organic solvents that absorb the impurity when in contact with natural gas; and (*3*) dry-bed processes which employ fixed beds of solid granular material that serve to promote chemical reaction of the impurities or remove the impurity by adsorption. Monoethanolamine, diethanolamine, and diglycolamine are typical solvents used (see Alkanolamines). Monoethanolamine (MEA) is the oldest and most widely used solvent to remove hydrogen sulfide and carbon dioxide from natural gas (see Carbon dioxide). The chemical reaction between carbon dioxide, hydrogen sulfide, and the amine takes place in a countercurrent gas–liquid absorber. The rich solution is regenerated in a steam-stripping still, operating at ca 115°C and 143 kPa (6 psig). Monoethanolamine is cheap, highly reactive, and stable. However, vaporization losses are high, hydrogen sulfide cannot be removed selectively in the presence of carbon dioxide, and MEA is attacked by other sulfur compounds, such as carbonyl sulfide, that may be present. Diethanolamine is usually preferred to treat gas containing sulfur compounds such as carbonyl sulfide, because it is resistant to attack. Since diglycolamine was developed in 1965, it has received wide acceptance in the industry because of lower plant costs. The circulation rate of diglycolamine is approximately one half that of monoethanolamine, which results in a reduction in the size of the necessary plant equipment. The products of degradation caused by carbonyl sulfide can be controlled by a simple batch reclaimer operating at 193°C. The treated gas leaving a diglycolamine plant contains less than 5.7 mg hydrogen sulfide/m^3 (0.36 lb/10^6 ft^3) and 0.01 vol % carbon dioxide. Furthermore, diglycolamine withstands freezing to −40°C in very cold climates. This is a very important consideration.

Fluor solvent, Sulfinal, and Selexol are typical of the physical solvent processes, all of which are relatively new and of a proprietory nature. Generally, these processes are most applicable for treating high pressure 2.17–7.00 MPa (300–1000 psig) natural gas. They remove carbon dioxide, hydrogen sulfide, and are not subject to degradation in the presence of carbonyl sulfide. The main advantage of these solvents over MEA is their ability to remove undesirable acid gas at lower circulation rates, thus reducing the plant investment and operating costs. The disadvantage of the physical solvents is their affinity for absorbing pentane and higher hydrocarbons. The Fluor solvent process, which utilizes propylene carbonate, has found wide application in treating high pressure natural gas containing a large percentage (20% or higher) of carbon dioxide and low concentrations of hydrogen sulfide. The solvent is regenerated by merely flashing the gas to a vacuum through a hydraulic turbine. The Selexol process utilizes poly(ethylene glycol) dimethyl ether as the solvent. It is employed primarily for the bulk removal of the acid gases from natural gas that contains high concentrations of carbon dioxide and low concentrations of hydrogen sulfide. The solubility of hydrogen sulfides in the solvent is seven to ten times greater than that of carbon dioxide, which permits preferential removal of the hydrogen sulfide. Sulfinal is unique in this class of processes because the solvent is a mixture of sulfolane (tetrahydrothiophene-1,1-dioxide) and diisopropanolamine. The sulfolane acts as a physical solvent, whereas the diisopropanolamine acts as a chemical solvent giving the process the ability to remove hydrogen sulfide to pipeline specifications while leaving part of the carbon dioxide in the gas stream (10).

A typical example of the dry-bed process is the iron-sponge process, which is one of the oldest and simplest processes available for the selective removal of hydrogen sulfide from natural gas. The iron sponge consists of wooden chips coated with a hydrated iron oxide. The hydrogen sulfide reacts with the sponge to form iron sulfide. The bed can be partially regenerated with air but must be replaced periodically. Today the use of this process is limited to small volumes of gas containing less than 45 mg hydrogen sulfide/m^3 (2.8 lb/10^6).

These acid-gas-treating processes represent only a few of the numerous processes available. The proper selection of a process is based on a knowledge of the inlet gas composition, available pressure, flow rate, and desired composition of the treated gas stream.

Consideration for the environment and the economic value of sulfur result in the current practice of converting hydrogen sulfide to elemental sulfur. The Claus process is the most widely used method of producing elemental sulfur. In this process the acid gases are combined with a stoichiometric quantity of air to burn one third of the total hydrogen sulfide to sulfur dioxide and water:

$$2\,H_2S + 3\,O_2 \rightarrow 2\,H_2O + 2\,SO_2$$

A shift reaction then occurs between the sulfur dioxide and the remaining hydrogen sulfide; a catalyst is used to reduce the operating temperature:

$$2\,H_2S + SO_2 \rightleftharpoons 2\,H_2O + 3/2\,S_2$$

The EPA has issued stringent sulfur emission standards which limit the amount of sulfur dioxide that may be released into the atmosphere. Hence, practically all hydrogen sulfide removed from natural gas must be converted to elemental sulfur. To meet these standards, the tail gas from even the most efficiently designed Claus plants requires further processing. No process has so far gained wide acceptance. The ASR Sulfoxide process (Alberta Sulfur Research, Ltd.) employs an organic sulfoxide as a liquid catalyst medium to carry out the traditional Claus sulfur reaction. A typical modern Claus plant tail gas contains 2% hydrogen sulfide and 1% sulfur dioxide, and can be processed to a concentration of 200 ppm hydrogen sulfide and sulfur dioxide. The cost for tail-gas cleanup cannot be offset by the value of the recovery of the additional elemental sulfur and must be considered as an additional cost for sour gas.

Natural-gas liquids are recovered from natural gas because of their economic value and because they pose a safety hazard if allowed to condense and accumulate in transmission lines and distribution systems. The extent to which gas is processed depends on the available gas flow rate, liquefiable content, market availability, and product demand and price. Natural-gas liquids can be sold as an ethane-free mixture of propane, butane, pentanes, and higher hydrocarbons. In larger plants the components can be separated into specification products. In areas where the petrochemical industries are concentrated, ethane is recovered; elsewhere, the ethane adds to the value of the natural gas. Wet gas is usually processed in an oil-absorption or natural-gasoline plant. Until World War II the oil-absorption process operated at ambient temperature and ca 30% propanes, 70% butanes, and all pentane and heavier liquids were recovered. Plants constructed since that time apply refrigeration to an oil-absorption process and operate at temperatures as low as −45°C, recovering up to 50% ethane and heavier liquids. Basically, an oil-absorption plant consists of an absorber column for absorbing the hydrocarbons in a high boiling gas oil, a stripping column to remove hydrocarbon liquids from the absorption oil, and several fractionation towers to separate and grade the liquids for sale.

In many recent applications, the cryogenic turboexpander process has replaced oil absorption as the least expensive plant to build and operate for the extraction of ethane and higher hydrocarbons (11). The gas is refrigerated by reducing the pressure of the gas stream through an expansion engine. A turboexpander plant consists of large surface, low pressure-drop heat exchangers, the turboexpander, and a separation system to remove the condensed hydrocarbon liquids. The horsepower developed by the turboexpander is used to partially recompress the lean gas stream. Additional compressors are added, when required, to minimize overall pressure drop across the plant. A turboexpander plant designed to operate at −118°C can recover up to 90% ethane, 99% propane, and 100% butanes and higher hydrocarbons. The turboexpander process may provide the first economical method of extracting nitrogen (qv) from natural gas. Previously, gases with high nitrogen content (10–75 vol %) could not be delivered to pipelines because of low heating value. The sale of natural gas liquids is necessary for economic operation of a natural gasoline plant. During 1976, 763 plants treated 5×10^{11} m^3 (17.7×10^{12} ft^3) natural gas to 89% of the total United States marketed production, to recover 9.3×10^7 m^3 (24.7×10^9 gal) liquids valued at \$3.28 $\times 10^9$.

Production

During the early stages of development of the oil and gas industry in the United States, enormous amounts of natural gas were flared in conjunction with the production of crude oil. Several factors contributed to this waste of a valuable resource, including a lack of conservation and proration laws, inadequate markets, and a lack of transmission lines and inadequate technical knowledge of how to cope fully with the many problems arising in the production of oil and gas. Today, the widespread gas transmission network provides ready access to large markets and most states have rigid conservation and proration laws which prohibit waste of natural gas. The total gas vented or flared in the fields amounts to ca 0.6% of the gross production; ten years ago it was ca 2%.

The advanced technology available today has greatly reduced the underground waste of both oil and gas. Improved well materials and techniques have reduced leakage from the reservoirs. Advanced formation-fracturing provides better drainage to the well bore, thus increasing the recovery of both oil and gas (see Petroleum). Reservoir engineering studies, taking into account all known characteristics of the porous formation and the fluids involved, provide better information to control the field delivery volumes, thus reducing the possibility of the formation water trapping valuable production. Oil fields are repressured to increase the recovery of crude oil (see Petroleum). In such operations, a portion of the gas produced with the oil is compressed and reinjected into the reservoir. When the crude oil has been exhausted, the gas so employed becomes available for sale. To obtain the maximum recovery of all hydrocarbons in certain gas condensate reservoirs, gas must be recycled to maintain the formation pressure above the hydrocarbon dew point of the reservoir fluid. The hydrocarbons in this type of reservoir exist as a single-phase system at the reservoir temperature and pressure. When the single-phase system is reduced in pressure, the higher hydrocarbons are condensed. This phenomenon is known as retrograde condensation. If the pressure reduction occurs in the reservoir, the higher hydrocarbons condense and are lost. If, however, the formation pressure is maintained by recycling gas, the pressure reduction occurs in a separator located at the wellhead and the higher hydrocarbons can be recovered and sold. After all the liquids have been produced from the reservoir, the recycled gas is then available for sale.

In a number of other countries possessing substantial proved reserves, conservation practices are not employed to the same extent as in the United States because of the lack of transportation facilities and the availability of large markets but in many areas extensive projects are underway to utilize this valuable resource. Algeria has raised its total marketed production to 1.05×10^{10} m^3 (37.1×10^{10} ft^3), almost all of which was exported as liquefied natural gas (LNG). In addition, plans are under consideration to construct additional LNG facilities that would increase the present

Table 7. World Production of Natural Gas[a], 10^6 m^3 [b]

Country	Marketed production[c] in 1974	Marketed production[c] in 1976
North America and Caribbean Islands		
Barbados	2	4[d]
Canada	86,218	86,837
Cuba	19	25[d]
Mexico	15,879	16,361
Trinidad and Tobago	1,938[e]	2,224
United States	611,511	564,854
South America		
Argentina	7,240	7,332[d]
Bolivia	1,714	1,749
Brazil	498	547[d]
Chile	3,610	4,016
Colombia	1,863	1,889
Ecuador	28[d]	31[d]
Peru	901[e]	991[d]
Venezuela	13,475	13,583
Europe		
Albania[d]	203	204
Austria[f]	2,094	1,956
Belgium[g]	64	34
Bulgaria	180	110[d]
Czechoslovakia[f]	975	937
Denmark[d,h]		
France	7,627	7,090
DDR	7,730	7,927[d]
FRG[f]	20,191	18,629
Hungary[i]	5,100	6,082
Italy	15,298	15,637
Netherlands	83,703	97,278
Norway[d,h]		
Poland[f]	5,738	6,696
Romania	28,636	29,827
Spain	1	1
USSR	260,489	320,921
United Kingdom[f]	34,822	37,266
Yugoslavia	1,447	1,730
Africa		
Algeria	5,620	9,930
Angola[d]	68	57
Congo (Brazzaville)	19	15
Egypt	43[d,e]	380
Gabon	637	169
Libya[i]	9,773	14,311
Morocco	59	43[d]
Nigeria	404	524[d]
Rwanda	1	0.2
Tunisia	201	214
Asia		
Afghanistan	3,200	2,543
Bahrain	1,932	2,179
Bangledesh	488	991[d]
Brunei	5,006	8,460
Burma[j]	139[d]	153[d]

Table 7 (*continued*)

Country	Marketed production[c] in 1974	Marketed production[c] in 1976
People's Republic of China[d]	33,972	45,296
India	717	1,523
Indonesia	1,133[d]	3,580
Iran	22,290	22,471
Iraq[h]	1,189	1,654
Israel	66	58
Japan[f]	2,847	2,493
Kuwait[k]	5,291	5,580
Malaysia (Sarawak)	481[d,e]	640
Oman	43[d]	57
Pakistan	4,955	5,200
Qatar[i]	1,300	1,476
Saudi Arabia[k]	6,200[d]	6,794[d]
Syria	180	221
Taiwan	1,568	1,841
Turkey[d]	69	74
United Arab Emirates		
Abu Dhabi[i]	1,200	1,133[d]
Dubai[d]	538	566
Sharjah[d]	0.2	0.2
Oceania		
Australia	4,511	5,928
New Zealand	300	876
Total	*1,335,674*[e]	*1,400,198*[l]

[a] Ref. 9.
[b] To convert m^3 to ft^3, multiply by 35.3.
[c] Comprises all gas collected and utilized as a fuel or as a chemical industry raw material as well as that used for gas left in fields, including gas used in oil fields and/or gas fields as a fuel by producers, even though it is not actually sold.
[d] Estimate.
[e] Revised.
[f] Includes output from coal mines.
[g] Total production obtained from coal mines.
[h] No marketed production reported; there probably was some small field use in Denmark, Norway, and Sharjah; and in Norway natural-gas liquids were extracted in each year 1974–1976 but available information is inadequate to estimate output levels.
[i] Includes gas reinjected to reservoirs, if any.
[j] Data are for year ending June 30 of that stated.
[k] Includes one half of production reported for the former Kuwait–Saudi Arabia neutral zone.
[l] Rounded.

marketed production by at least three times. Algerian LNG is transported in ocean going tankers at atmospheric pressure and −162°C to markets in Europe and the United States. Indonesia has expanded its marketed production nearly three times in the last two years (9), and two LNG plants are under construction with a total capacity to liquefy ca 1.6×10^{10} m^3 (56×10^{10} ft^3) natural gas annually. The worldwide marketed production of natural gas for recent years as reported by the Bureau of Mines is shown on Table 7.

Stimulated by higher prices for natural gas and crude oil and the desire to be less dependent on foreign oil supplies, the exploration for oil and gas in the United States is continuing at a rapid pace. The magnitude of the oil and gas production industry is evidenced by the operation of almost 138,000 gas wells and 503,000 oil wells. In 1977, 46,000 additional wells were drilled, of which almost 18,000 were completed as oil wells and 10,000 as gas wells.

Economic Aspects

The United States is a net energy consuming nation. The 1976 estimated consumption of all forms of energy resources in the United States was 78 EJ (74×10^{15} Btu), of which natural gas supplied 21.3 EJ (20.2×10^{15} Btu) and natural gas liquids supplied 2.7 EJ (2.56×10^{15} Btu). Natural gas and natural-gas liquids supplied ca 31% of the nation's total energy requirements; the balance was supplied by crude petroleum (44%), coal and lignite (18%), hydropower (4%), and nuclear power (3%). Annual production of energy resources during the same period is estimated at 63 EJ (60×10^{15} Btu), of which wet natural gas supplied 36.4%. The balance is provided by crude petroleum (28.8%), coal and lignite (26.4%), hydropower (5.0%), and nuclear power (3.4%) (1). The net energy deficit is primarily made up by importation of crude oil (see Fuels, survey; Petroleum, resources).

The growth of the natural-gas utility industry since 1965 in terms of customers, revenues, and energy sales is shown in Table 8.

The price of natural gas is increasing rapidly in the United States. In 1976 residential customers paid an average of \$1.88/GJ (\$1.98/10^6 Btu), up 17.2% over 1975; commercial customers \$1.59 (\$1.68), up 20.9%; and industrial customers \$1.25 (\$1.32), up 33.3%. The variation in prices reflects service costs related to providing service to the low volume residential customer, having a wide variation in demand and an intricate distribution system, in contrast to the high, steady volume industrial customer, requiring little in the way of distribution systems and operating under interruptible service contracts. The average price of \$1.88/GJ (\$1.98/10^6 Btu) paid by the residential consumer in 1976 was, for the most part, the least expensive energy service available when compared with the average equivalent prices of electricity at \$9.60/GJ (\$10.11/10^6 Btu) and oil at \$2.86/GJ (\$3.01/10^6 Btu). Prices for natural gas vary widely from area to area as a result of the distance between the point of production and the market. For example, although the average cost of natural gas paid by all consumers in the United States was \$1.52/GJ (\$1.60/10^6 Btu) in 1976, it varied from a low of \$1.04 (\$1.09) in New Mexico to \$3.23 (\$3.40) in Connecticut.

The growth of utility-industry customers has reached a plateau. In 1976 the total number of customers increased 0.9%, ie, residential 1.3% and commercial 1.1%, whereas industrial declined by 2.5%. The decline in industrial customers is the result of industries that experienced continuing curtailment for long periods of time during the winter and have permanently switched to other fuels. Total gas sales in 1976 declined 0.3% from the 1975 level as a result of greater curtailment of gas to industry because of supply limitations, conservation efforts being made by all classes of customers, and the loss of industrial customers. Electric power generation represented the largest decline in gas sales during 1976 since in many areas coal provides the least expensive year-round supply of energy for this purpose. The industries using the largest volumes of gas are: electric power generation; chemicals and allied products; primary iron and steel industries; fuel for petroleum refining; food and related products; primary nonferrous industries; and fabricated metal products.

Table 8. United States Natural Gas Utility [a]

Service	1965	1970	1975	1976
residential				
customer, thousands	34,340.8	38,097.0	40,950.3	41,337.6
revenues, millions of dollars	4,029.6	5,207.3	8,445.5	9,941.0
sales, EJ [b]	4.2	5.2	5.3	5.3
commercial				
customers, thousands	2,789.7	3,130.9	3,366.9	3,371.5
revenues, millions of dollars	1,053.6	1,620.4	3,302.6	4,075.0
sales [c], EJ [b]	1.4	2.1	2.5	2.6
industrial [d] and other				
customers, thousands	207.0	254.2	237.3	232.3
revenues, millions of dollars	2,323.8	3,454.8	7,326.3	9,685.3
sales, EJ	7.0	9.6	7.9	7.8
Total				
customers, thousands	*37,337.5*	*41,482.1*	*44,554.5*	*44,941.4*
revenues, millions of dollars	*7,407.0*	*10,282.5*	*19,074.4*	*23,701.3*
sales, EJ	*112.6*	*16.9*	*15.7*	*15.7*

[a] From ref. 1.
[b] To convert EJ to quad (10^{15} Btu), divide by 1.054.
[c] Primarily for heating.
[d] Chemical, steel, and power plants, and various types of industry that use large volumes of gas for fuel or processing operations.

The average wellhead price of natural gas in 1976 was 2.0¢/m^3 (58.0¢/1000 ft^3), representing an increase of ca 240% since 1970 and contributing substantially to the increased price to the consumer. Higher field prices have stimulated the search for new natural gas supplies.

Transportation and Storage

Production wells in the United States are connected to the consumers by a complex network of buried pipelines, consisting of 11.3×10^4 km of field gathering lines, 42×10^4 km of transmission lines, and 106×10^4 km of distribution mains. Transmission lines, the backbone of the network, consist of pipelines ranging in diameter up to 107 cm and operating at pressures up to 8.38 MPa (1200 psig), together with pressure-boosting compressor stations with a total of ca 9 GW (12×10^6 hp). Since 1930 the interstate transmission system has continued to grow in response to the discovery of new gas supplies and an ever-increasing market demand. Construction of ca 7900 km of transmission line (91–142 cm dia) is planned for the new gas supplies available in the Alaskan North Slope area to markets in the midwest and Pacific coast (12). The growth of the gas transmission industry has been assisted by the technological advancements in the development of high pressure, thin walled pipe, electric arc-welding methods and equipment; mechanical equipment for efficient pipelaying; and high capacity, low maintenance, fuel efficient, reciprocating and gas turbine compressors; all of these reduce investment costs (see also Pipelines). In 1976 the gas utility industry transmission plant was valued at 22×10^9 dollars.

Gas transmission costs vary widely. In general, for a specific transmission system, they are the sum of operating expenses and costs associated with the investment in facilities, including a fair return. These costs relate more to the cost of the facilities than to how much or how little gas is being delivered; therefore, the load factor (the ratio of the daily average delivery to the maximum daily delivery in a year's time) is important in the determination of the unit transportation cost. A pipeline operating at a 50% load factor has nearly twice the unit transportation cost as the same line operating at full capacity.

Careful studies of the field potentials and market demands have been used to determine the proper size for high load factor operation of most pipeline systems constructed in the United States since World War II. It is essential that pipelines be operated at a high load factor to keep transmission costs low. Today the market load factor is nearly 70%, as a result of the low load-factor, space-heating sales that require high deliveries during cold winter periods and low deliveries during the summer, combined with the high load-factor industrial sales. This includes many industrial customers that operate on interruptible-service-type contracts, requiring the use of an alternative fuel such as coal or fuel oil during periods of peak demand.

Underground storage near market areas provides the most economical means of balancing variations in market demand with delivery capability of transmission lines. Storage service is provided through use of depleted gas and oil fields and underground water-bearing strata with the proper structural configuration to hold gas. Gas is compressed and injected into such reservoirs during the summer months, when market demand is low, and is withdrawn in the winter months when the market demand exceeds the maximum production capability of fields connected to the transmission lines. In 1976 there were 386 underground reservoirs in the United States that contained 1.95×10^{11} m^3 (6.9×10^{12} ft^3) of gas; 5.83×10^{10} m^3 (2×10^{12} ft^3) was withdrawn from storage, and 5.42×10^{10} m^3 (1.9×10^{12} ft^3) was injected.

Recent technological advancement in the use of liquefied natural gas (LNG) provided the gas industry with new methods to solve the problems of storage and transportation. Natural gas can be reduced to $1/600$ of the volume occupied in the gaseous state by cryogenic processing, safely stored or transported in double-walled insulated metal containers at near atmospheric pressure and −162°C, and when required, can be regasified (see Cryogenics). In the United States, 102 LNG plants furnish gas for distribution during periods of maximum market demand, when gas supply is not adequate. (This is known as peak shaving service.) In 1976 these plants had a combined storage capacity equivalent of ca 1.7×10^9 m^3 (61×10^9 ft^3) with a short-term daily delivery rate of 2.1×10^8 m^3 (7.5×10^9 ft^3). The new LNG technology is being used to develop a worldwide natural gas transportation system. At present, eleven LNG baseload liquefication plants are in operation or under construction in Algeria, Libya, Indonesia, Alaska, and Borneo, to serve as export terminals for natural gas. In addition, seventeen baseload receiving terminals are in service in England, France, Italy, Japan, and the United States, to import LNG and regasify it for delivery into the gas distribution network. These facilities are connected by a fleet of large ocean-going carriers capable of delivering up to the equivalent of 7×10^7 m^3 (2.5×10^9 ft^3) of gas per trip. Of these, nine large carriers are in service between Arzew, Algeria, and the United States, delivering the equivalent of 2.83×10^7 m^3 (10^9 ft^3) of natural gas per day to import terminals located at Cove Point, Maryland, and Elba Island, Georgia (13–14).

Applications

Natural gas is used mainly as fuel to provide heat for homes, commercial buildings, and industrial processing. Combustion of natural gas is a chemical reaction in which the hydrocarbons are rapidly oxidized with the evolution of heat and light, involving a complicated series of chemical reactions with the point of first attack being through the hydrogen atoms. The general chemical equation is shown below:

$$CH_4 + 2\,O_2 \rightarrow CO_2 + 2\,H_2O$$

Combustion is said to be complete when all hydrocarbons have been converted to carbon dioxide and water. Perfect combustion is the complete burning of the gas with the theoretical amount of air or oxygen required for the reaction. Combustion is initiated by applying sufficient heat to raise the temperature of the gas–air mixture to the ignition point. At this point, the local rate of heat generation is sufficient to propagate the flame through the combustible gas (see Burner technology; Furnaces, fuel-fired).

Gas-burning appliances used today are the result of continuing technological development. They consist of a burner for combustion of gas and a heat transfer system to conduct heat away by conduction, convection, or radiation. Most burners used for space heating operate at atmospheric pressure. Burners vary in configuration but must be capable of completely burning the gas, operate quietly and over a wide heating range without flashback or lifting of the flame, and must be made of materials that withstand many heating and cooling cycles. To obtain greater efficiency or higher temperatures, industrial systems employ high pressure mixing of gas and either air or oxygen. Industrial equipment is specially designed for each application. The efficiency of gas-burning appliances is usually considered as the ratio of the heat output to total heat available from the complete combustion of the gas. Most gas heating systems have an operating efficiency of 50–80%. Advancing technology has developed new and unusual burner designs such as the pulse combustion processes, which are used in residential heating units that operate at efficiencies as high as 95%.

Chemical Uses. Natural gas and natural-gas liquids furnish a portion of the fuel and feedstock requirements of the chemical industry. During 1976 chemicals and allied products represented ca 6% of the total deliveries made by the natural-gas utility industry in the United States. In addition, substantial sales of gas are made by producers directly to the chemical manufacturers that are not reflected in the utility company statistics. The chemical industry also utilizes ca 21% of the natural gas liquids produced in the United States. It is estimated that approximately 30% of the natural gas and all natural-gas liquids provided to the chemical industry are utilized as feedstock or raw material from which a variety of chemicals are manufactured (15). The largest chemical use of methane is through its reaction with steam at elevated temperatures to form synthesis gas used for the production of ammonia and methanol, most of which is utilized in fertilizers, plastics, and adhesives (see Fuels, synthetic). Ethylene (qv) is the largest volume chemical building block obtained from natural-gas liquids. It is produced by the steam cracking of ethane or propane. In 1977 it was estimated that ca 65% of the 1.12×10^3 t ethylene produced in the United States would be derived from liquid petroleum gases used as feedstocks. The remainder is supplied by the steam cracking of heavier liquids derived from crude oil, a process that coproduces propylene, butadiene, and benzene. 1,3-Butadiene can also be produced by catalytic dehydrogenation of butane obtained from natural gas. In 1977 ca 45% of the 1.9×10^3 t butadiene produced in the United States was derived from the dehydrogenation process (see Butadiene).

Other chemicals, such as propylene, benzene, and xylene, can be derived from natural gas liquids. However, the petrochemical industry relies on crude oil as its major feedstock for these products because of the relatively large available supply (see BTX processes; Xylenes; Petroleum).

Governmental Regulation

The natural gas industry in the United States is subject to regulation as a substitute for direct competition. Regulation is accomplished through various state and federal agencies that derive their powers and duties from specific legislation. These agencies are typically empowered to interpret their legislative authority by issuing rules and regulations that define the law and compliance with the law. Normally such agencies are considered as part of the executive branch of the government but their duties include a combination of legislative, judicial, and administrative functions. Through the years the patterns of regulation have been continually modified as a result of agency actions, new legislation, and court decisions in response to changing economic and political conditions, and technological advancements. The policies adopted by these agencies determine the future of the industry.

Distribution and sale of gas to consumers is normally regulated by states or municipalities where the gas is sold. Most states have utility commissions with broad quasijudicial powers to regulate distribution companies in matters pertaining to tariffs, service standards, records of accounts, reports, and financing. Because of the constitutional right of the federal government to regulate commerce among the states, Congress has given the Federal Energy Regulatory Commission (FERC) jurisdiction over the interstate transmission of natural gas. The FERC has jurisdiction over the wholesale price at which an interstate pipeline company can sell gas to a distribution company for resale. It also regulates the field price producers can receive for their gas. In addition, it has jurisdiction over the import and export of natural gas, the construction or abandonment of interstate pipeline facilities, and the system of accounts and reports of the interstate pipeline companies.

The production of natural gas is regulated by the states in which it is produced. The applicable laws may vary from state to state but are generally designed to protect the public from above-ground or underground waste of this valuable resource and protect the property rights of those who own an interest in the underground reservoirs. State conservation commissions usually have jurisdiction over the drilling and abandonment of wells, field-development proposals, allowable production volumes, and allocation of production between various lease areas.

The 95th Congress passed the *Natural Gas Policy Act of 1978* as a part of the country's first National Energy Plan. This act is a substantial revision of past policy in gas pricing and regulation. Generally, it brings intrastate gas sales at the wellhead under FERC control for the first time, and provides for a specified monthly wellhead price increase for all natural gas, for the deregulation of newly discovered natural gas, under certain conditions, in 1985, and for an incremental pricing scheme designed to make large industrial consumers absorb a portion of the higher natural-gas prices. It was hoped that the increased gas prices would encourage conservation through more efficient use and spur new exploration and production activity. Other parts of the National Energy Plan allow tax incentives for the installation of home insulation, and require new industrial and electric power plants to use coal or a fuel other than oil or gas. The entire plan is a significant and far-reaching program supposedly designed to change the country's pattern of energy consumption and production. The plan is complex and presents a major administrative challenge for the federal agencies because of changing political, economic, and supply conditions.

Concern for the environment, such as air quality or land use planning, have spawned additional federal regulations that have decisive effects on the operations of the natural gas industry.

Outlook

The future of natural gas must be viewed in the context of the overall world energy scene. At present, the commercial forms of primary energy used in the world are petroleum, natural gas, coal, hydroelectricity, and nuclear electricity. Approximately 80% of the total world energy is consumed in the large industrial regions of the USSR, western Europe, Japan, and North America. The demand for primary energy in a particular country or region is determined by the relative importance of industry, agriculture, transportation, etc, in the economic structure and the preference of each demand sector for a particular form of energy. The choice of supply to meet these demands is determined by the relationship of many factors such as: location and size of resources; basic cost structures; governmental intervention; environmental protection; and others. Some consumers have the ability to switch from one primary energy resource to another as cost or other factors may require. Other consumer demands, such as the feedstock requirements of the petrochemical industry, must continue to rely only on oil or natural gas. The relative efficiency of the different forms of energy to the consumer is an important factor in the choice of a resource. For example, the average efficiency at which thermal electricity is currently being generated is ca 32%, compared with direct fuel uses of the various hydrocarbons of 50–90%. To conserve natural resources and reduce costs to the consumer, it is necessary to remove this conversion loss and provide for the direct use of fossil fuels where possible.

Numerous in-depth studies have been made by various organizations, such as the Workshop on Alternative Energy Strategies (WAES) (16), in an effort to predict future world and national energy needs, and to suggest actions that would provide continued economic growth from the available world resources. The main WAES finding foresees that resource and production limitations could restrict the world oil supply as early as the late 1980s, and increase rapidly to a gaping deficit before the year 2000. These deficits must be satisfied by other energy sources. In order to reduce the demand for oil it is necessary to (*1*) vigorously develop replacement fuels from conventional sources such as natural gas, coal, and nuclear electricity; (*2*) develop new technology that will provide conservation of all energy sources; and (*3*) develop renewable energy supplies.

An assessment of the worldwide gas resource base indicates that it is capable of supporting a production growth rate of 4.4%/yr through the year 2000. This would increase the present production rate from 1.4×10^{12} m^3/yr (50×10^{12} ft^3/yr) to ca 3.7×10^{12} m^3/yr (1.3×10^{14} ft^3/yr). The actual growth rate depends on favorable economic and political factors necessary to develop the required market. The areas with the greatest potential for future natural gas development include the OPEC countries and the USSR (17).

The future prospects for the United States gas industry depend on maintaining production from conventional sources in the 48 contiguous states and development of several higher cost supplemental gas sources shown in Table 9 (18). The level of conventional gas production in the 48 contiguous states depends on continued proved reserve additions of ca 4.24×10^{11} m^3/yr (15.0×10^{12} ft^3/yr). This requires a vigorous exploration and development program that can only be encouraged by a continuing federal regulatory policy that allows the field price of gas to rise approximately to the price equivalent of alternative fuels. The gas reserves located under the North Slope of Alaska were added to the proved reserves of the United States in 1970, and now represent over 10% of the total. After extended federal regulatory hearings, the pipelines required to connect these supplies to the 48 contiguous states are projected to be completed in 1985 at a cost of about 14×10^9 dollars.

Nonconventional gas sources include methane in gaseous coal seams; natural gas in tight sands located principally in the Rocky Mountain area; methane, either free or dissolved in water and trapped in sedimentary rock at the higher than normal pressure (geopressure) that underlies large portions of Texas and Louisiana Gulf Coast; and natural gas in the organically rich clay formations known as Devonian shale, found principally in the eastern United States. These sources could be exploited with new technology. Present reasonable estimates (19) indicate that the ultimately recoverable gas from these sources are $(9\text{–}48) \times 10^{12}$ m^3 ($(3.2\text{–}17) \times 10^{14}$ ft^3). Evaluation of these resources is uncertain. Other areas where new technologies could make a significant contribution to future gas supplies (20) include production of gas from marine and land biomass, thus providing a renewable gas supply; supplementing the natural gas by the production of hyrogen by thermochemical decomposition of water; and development of *in situ* gasification of unminable coal seams.

In light of the energy outlook the chemical industry will gradually move away from its heavy dependence on natural gas for fuel toward the use of oil and eventually coal (15). Natural gas as a chemical feedstock will be principally confined to the production

Table 9. Estimated Total Future United States Gas Supply [a]

Sources	Annual gas volumes, 10^{10} m^3 [b] 1985	1990
contiguous 48 states production from conventional sources	51.5	49.5
supplemental sources		
Canadian imports	4.5	4.5
Mexican imports	2.5	4.2
LNG imports	3.68	5.66
North Slope Alaska gas	2.5	3.68
coal gasification	0.14	1.13
SNG[c] from petroleum	1.13	1.13
new technologies	1.4	2.83
Total	*67.5*	*72.7*

[a] Ref. 18.
[b] To convert m^3 to ft^3, multiply by 35.3.
[c] Substitute natural gas.

of ammonia and methanol. The use of natural-gas liquids as a chemical feedstock will decline in the future as a result of a gradually declining supply, coupled with increasing use of these liquids as a fuel in rural areas.

Adequate worldwide gas reserves are available for long-range development and the natural gas industry should continue to grow in spite of many political, financial, and marketing problems facing such development. No one can really guarantee the future of the highly developed, mature industry in the United States; however, a rational federal policy would go a long way toward setting priorities and removing uncertainties that would allow development of the necessary supplemental gas supplies and new technologies necessary to sustain the industry in the long term.

BIBLIOGRAPHY

"Gas, Natural" in *ECT* 1st ed., Vol. 7, pp. 59–73, by F. S. Lott, Bureau of Mines, U.S. Department of the Interior; "Gas, Natural" in *ECT* 2nd ed., Vol. 10, pp. 443–462, by Dysart E. Holcomb, United Gas Corporation.

1. *Gas Facts,* American Gas Association, Arlington, Va., 1976.
2. *Reserves of Crude Oil, Natural Gas Liquids and Natural Gas in the United States and Canada as of December 31, 1977,* Vol. 32, published jointly by the American Gas Association, American Petroleum Institute, and Canadian Petroleum Association, 1978.
3. *World Oil* **189**(3), 64 (Aug. 15, 1979).
4. J. D. Parent and H. R. Linden, *A Survey of United States and Total World Production, Proved Reserves and Remaining Recoverable Resources of Fossil Fuels and Uranium as of December 31, 1975,* Institute of Gas Technology, Chicago, Ill., Jan. 1977.
5. *Engineering Data Book,* 9th ed., Gas Processors Suppliers Association, Tulsa, Oklahoma, 1977.
6. D. L. Katz and co-workers, *Handbook of Natural Gas Engineering,* McGraw-Hill Book Co., Inc., New York, 1959.
7. *General Information on Natural Gas Utilization for Gas Appliance Manufacturers, Architects, and Design Engineers,* Mountain Fuel Supply Company, Salt Lake City, Utah, Aug. 1971.
8. L. W. Brandt and L. Stroud, *Open File Information and Data Relating to the Extraction of Helium from Natural Gas by Low Temperature Processes,* Bureau of Mines, Amarillo, Texas, May 1, 1959.
9. *Mineral Yearbook,* Vol. I, U.S. Bureau of Mines, Washington, D.C., 1976.
10. *Proceedings of the 1973 Gas Conditioning Conference, University of Oklahoma,* 104.
11. C. H. Jordan, *Chem. Eng. Prog.* **68**(9), 53 (1972).
12. D. Hale, *Pipeline Gas J.* **207**(11), 19 (1979).
13. D. Hale, *Pipeline Gas J.* **203**(7), 18 (1976).
14. D. Hale, *Pipeline Gas J.* **203**(7), 23 (1976).
15. K. L. Mai, *Chem. Eng.* **84**(12), 122 (1977).
16. S. Basile, ed., *Energy Supply—Demand Interactions to the Year 2000,* Themit Press, Cambridge, Mass., and London, Eng.
17. W. T. McCormick, Jr., R. B. Kalisch, and T. J. Wander, *Oil Gas J.* **76**(7), 103 (1978).
18. W. T. McCormick, Jr., *Outlook for Conventional U.S. Gas Supplies and Implications for Development of Supplemental Sources,* June 1979.
19. *Potential Supply of Natural Gas in the United States (as of December 31, 1978),* Potential Gas Agency-Colorado School of Mines, Golden, Colo.
20. *New Technologies for Gas Energy Supply and Efficient Use,* American Gas Association, Arlington, Va., Apr. 1979.

General References

References 1, 6, 9, and 10 are general references.

R. N. Maddox, *Gas and Liquid Sweeting,* 2nd ed., John M. Campbell, Norman, Oklahoma, 1974.

J. M. Campbell, *Gas Conditioning and Processing,* 2nd ed., John M. Campbell, Norman, Oklahoma, 1970.

JOHN H. HILLARD
Mountain Fuel Supply Company

GASOHOL. See Ethanol; Fuels, synthetic, liquid fuels; Gasoline and other motor fuels.

GASOLINE AND OTHER MOTOR FUELS

Practicable internal-combustion engines date back little more than 100 years. Nicolaus Otto built the first gasoline-type engine in 1876. And Rudolf Diesel's compression-ignition engine first ran reasonably successfully in 1897. Motor vehicles using such engines did not appear until the late 1880s, and they remained a toy of the rich until well into the 1900s. For a review of gas-turbine fuels, see the article Aviation and other gas-turbine fuels.

Gasoline

Gasoline, the preferred fuel for the remarkable Otto engine, evolved from an unwanted by-product to an indispensable mainstay of modern life in less than one normal human lifetime. As recently as 1900, gasoline was little more than a disposal problem for the infant petroleum industry in the production of stove and lamp oils, lubricating oils, and greases. Today, the gasoline engines in U.S. passenger cars alone produce more than fifty times the power generated by all of the nation's central power plants. In so doing, they consume more than 400 GL (1×10^{11} gal) of gasoline every year. The modern private motor car transports its owners and their families more than 1.6×10^{12} km (a trillion miles) each year in the U.S. alone, and more than twice that in the world as a whole. Trucks, which account for only about 20% of total U.S. motor-vehicle registrations, travel almost a third as far annually in carrying the nation's products to consumers. Today, almost everything we eat, wear, and use is moved at some stage by truck. Approximately four-fifths of all Americans who commute to work do so in privately owned automobiles.

During the pleasure-driving summer months of 1978, U.S. gasoline consumption averaged 1.25 GL/d (330×10^6 gal/d), and it failed to slacken much in the following fall and winter. Experts believe consumption will peak in the next few years, however, and will be somewhat less by 1990 than it was in 1980. One contribution to this decrease is an expected slowdown in growth of the U.S. driving population. Improved fuel economy is another major reason for leveling off of U.S. gasoline consumption. As a result of government-mandated standards for new-car production, the average fuel economy of all cars on the road is expected to increase from 1977's 5.7 km/L or 17.4 L/100 km (13.5 mpg) to 9.4 km/L or 10.7 L/100 km (32 mpg) in 1990. These two trends, plus growth of diesel-powered cars to more than 6% of the total U.S. car population, will more than offset a projected increase in average distance driven per driver—from 12,000 km/yr (7500 mi/yr) in 1978 to 16,000 km/yr (10,000 mi/yr) in 1990 (1).

Requirements for Good Gasoline. To satisfy modern high-performance automotive engines, today's gasolines must meet exacting specifications. Moreover, some of their characteristics must be tailored to fit prevailing local conditions. For example, gasolines designed for winter use are blended to vaporize at lower temperatures than do summer gasolines. Other characteristics, such as knock resistance (indicated by octane number), must be adjusted to meet the needs of the particular engines in which the gasoline will be used. Thus, some discussion of the requirements for a good gasoline is essential before explaining the role of each process used in making such a gasoline.

Vapor pressure. In order for an engine to start at the prevailing atmospheric temperature, its gasoline must form a substantial amount of combustible vapors at that temperature. The conventional U.S. measure of this vaporizing ability is the Reid vapor-pressure test (ASTM Method D 323). By measuring the vapor pressure exerted by the lowest-boiling components of a gasoline in a standardized apparatus under standard conditions, this test indicates whether the gasoline will give off enough vapors to start an engine at a given temperature. If the vapor pressure is too high, there is danger of "vapor lock," in which premature vaporization in fuel lines or the fuel pump partially or completely stops the flow of liquid fuel.

Distillation characteristics. For fast warmup, smooth acceleration, and proper distribution of the fuel among an engine's cylinders, the gasoline must vaporize increasingly as carburetor and manifold temperatures rise. In the United States, this characteristic is measured by ASTM Method D 86, which consists of distilling a 100-mL sample of gasoline, condensing the distillate, and recording the temperatures at which various percentages of the sample have distilled. The temperature at which the first drop of gasoline evaporates and is condensed is called the initial boiling point (IBP). Normally, temperature readings are taken each time the level of condensed gasoline passes a 10-mL mark on the 100-mL receiving graduate. The data are reported as the temperatures at which 10, 20, 30%, etc, of the sample has evaporated.

Volatility Requirement. Taken together, the Reid vapor pressure and the ASTM distillation curve tell much about a gasoline's volatility, ie, whether it has the proper vaporizing characteristics for the climate and altitude where it will be used. This is essential; otherwise, the gasoline may not provide proper warmup characteristics on the one hand or, on the other hand, may cause vapor lock. Partial vapor lock reduces the top speed and power of an engine by restricting the flow of liquid fuel and, by making the fuel–air mixture "lean," may cause knocking combustion. Complete vapor lock chokes off the engine's fuel supply, stalling it and making it difficult to restart until the fuel system has cooled for some time. Table 1 shows average volatility data for U.S. regular, unleaded, and premium-grade gasolines, both in summer and winter.

Reid vapor pressure and the 10% point of the ASTM distillation are the main indicators of the "front-end volatility" which determines the quick-starting and vapor-lock characteristics of a gasoline. The slope of the ASTM distillation curve from the 10% point to the 50% point indicates the ability of the gasoline to supply a proper fuel mixture during the warmup period, particularly during periods of acceleration. Rapid warmup characteristics are desirable both for good vehicle performance and to minimize the fuel waste that takes place as long as the choke is in operation. Recent industry studies show that the 90% point also is an indicator of driveability during the warmup period. The 90% point, or the slope of the 50–90% curve, indicates the proportion of higher-boiling components in the fuel. Once the engine is up to normal

Table 1. Average Volatility Data for U.S. Motor Gasolines[a]

Property	Unleaded		Regular		Premium	
	Winter	Summer	Winter	Summer	Winter	Summer
initial boiling point, °C	28	32	28	32	28	32
10% point, °C	42	49	42	48	42	48
50% point, °C	100	104	94	99	99	102
90% point, °C	165	167	168	172	164	166
end point, °C	207	209	210	213	208	209
Reid vapor pressure, kPa[b]	86	67	85	67	85	68

[a] Data from ref. 2.
[b] To convert kPa to psi, multiply by 0.145.

operating temperature, these components contribute to fuel economy. But until then, they tend to distribute unevenly in the manifold and cylinders and, by remaining partially unvaporized, to wash lubricating oil off cylinder walls and dilute the crankcase oil. The end point, or highest temperature reached during the ASTM distillation, indicates the extent to which "heavy ends" (highest-boiling fractions) are present in the gasoline.

For accurately tailoring gasolines to the vaporization characteristics required for a given climate and altitude, refiners either measure the amounts of vapor formed at various fuel temperatures or employ a Coordinating Research Council (CRC) method based on Reid vapor pressure and the ASTM distillation temperatures at the 10, 20, and 50% points. The latter method (3) yields a curve, such as that in Figure 1, showing the relationship between fuel temperature and the volume of vapor formed per unit volume of gasoline (the vapor–liquid ratio, or V/L). The vapor-forming characteristics of two gasolines can differ markedly even though they have the same

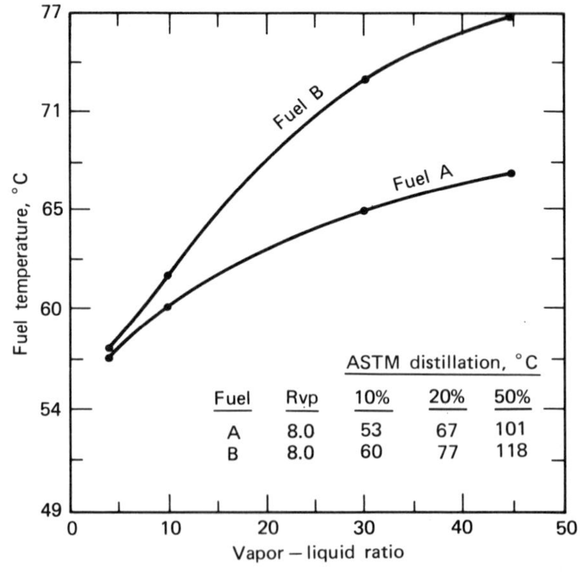

Figure 1. Vapor-forming characteristics of gasolines. Courtesy of Ethyl Corporation.

Reid vapor pressure (Rvp), as is true of the two summer gasolines in Figure 1. Because fuel A is considerably more volatile than fuel B throughout the 10 to 50% range of the ASTM distillation, it produces a given V/L ratio at a lower fuel temperature than does fuel B, although both fuels have an Rvp of 55 kPa (8.0 psi). In addition to indicating the vapor-forming tendencies of fuels, such V/L curves can be related to the V/L tolerances of specific automotive fuel systems. These vary among makes and models, but most recent-model cars can tolerate a V/L of 20–25. Older cars and most trucks and buses require lower V/L ratios, ca 15–20.

Octane-Number Requirement. To prevent annoying, fuel-wasting, potentially damaging engine knock at all engine speeds and loads, gasoline must have high antiknock quality (octane number) throughout its entire distillation range. Ideally, the fuel–air mixture in each engine cylinder should burn smoothly and evenly following timed ignition by the spark plug. But sometimes, as the spreading flame front sweeps across the combustion chamber, the yet-unburned portion of the fuel (end gas) is so heated and compressed that it autoignites, detonating throughout in an instant. Instead of being pushed down smoothly in the power stroke, the piston is given a hard instantaneous rap to which it cannot respond because it is connected to a crankshaft that is, in turn, connected to other pistons going through different phases of the engine cycle at the engine speed then prevailing. This abrupt energy release causes high-frequency pressure fluctuations throughout the combustion chamber, which register on the ear as the sharp metallic noise called knock. Fuel energy that could have been converted to useful power is dissipated in the form of pressure waves and increased heat radiation to surrounding engine parts and the cooling water. In addition to producing an objectionable sound and wasting fuel energy, prolonged knocking overheats valves, spark plugs, and pistons, potentially shortening their life substantially and causing engine-damaging preignition (ignition of the fuel–air mixture before the timed spark, ie, considerably before beginning of the power stroke). Knock is believed to be the result of chemical reactions that begin in the fuel–air mixture almost as soon as it is drawn into the cylinder during the intake stroke (4). As the mixture is progressively heated by the cylinder walls, by compression prior to the power stroke, and by burning gases after the spark plug fires. some of the fuel hydrocarbons undergo a series of cracking and oxidation reactions that convert them to unstable compounds prone to autoignition and instantaneous detonation (5–8). In essence, the presence or absence of knock in an engine depends on the outcome of a race between the advancing flame front and certain precombustion reactions in the gas.

Fairly early in the study of knock, it was recognized that the chemical structure of fuel hydrocarbons largely determines their tendency to cause knock and that straight-chain paraffins are more prone to knocking than branched-chain paraffins, olefins, or cyclic hydrocarbons. Likewise cycloparaffins (naphthenes) are more knock-prone than aromatics (unsaturated cyclics). This knowledge has led to development of several octane-improvement processes. Knock also is combated by negative catalysts, the antiknock compounds used to head off or to slow precombustion reactions that might otherwise convert fuel hydrocarbons to autoigniting, detonating compounds.

In 1919, C. F. Kettering and T. Midgley, Jr. undertook lengthy, systematic research into the cause and prevention of knock. They showed that knock could be suppressed by chemical fuel additives. This culminated in discovery of the remarkably effective antiknock action of tetraethyllead (TEL) and other alkyllead compounds.

As many as 143 other knock suppressors were tested along the way, but none could compare with the alkyllead compounds (see Lead compounds). And to date, extensive research has failed to turn up any other antiknock with the cost-effectiveness of the alkylleads.

Soon after refiners began adding tetraethyllead to gasoline, it became evident that some yardstick was needed for measuring the antiknock quality of motor fuels. In 1926, Graham Edgar of Ethyl Corporation developed the octane scale, which has become the worldwide standard for that purpose. For zero on the scale, he selected a straight-chain hydrocarbon, *n*-heptane, which burns with considerable knock. For 100, he chose a nonknocking branched-chain hydrocarbon, 2,2,4-trimethylpentane, which is often incorrectly referred to as "isooctane." By blending these two hydrocarbons in varying percentages, Edgar was able to produce primary reference fuels to match the knock resistance of any gasoline sample. Octane number was defined as the percentage of "isooctane" required in a blend with *n*-heptane to match the knocking behavior of the gasoline being tested. Thus, if a blend of 67% 2,2,4-trimethylpentane and 33% *n*-heptane was required to match the knock resistance of a particular gasoline sample when both were run in a test engine under specified conditions, the sample was said to have an octane number of 67. In order to rate fuels of more than 100 octane number, a method has since been devised to match the antiknock quality of the test fuel with that of reference fuels consisting of "isooctane" plus known amounts of tetraethyllead. Thus, the knock value of a fuel having the same antiknock quality as 2,2,4-trimethylpentane plus 1.5 mL of TEL is expressed as 100 + 1.5. A curve similar to Figure 2 is used to convert such knock values to octane numbers above 100.

The CFR (Cooperative Fuel Research) knock-test engine has been adopted as the standard for determining octane number. Basically, it is a single-cylinder, four-stroke engine in which the compression ratio can be varied at will. Auxiliary equipment includes a pickup unit to detect pressure impulses from detonation, an electronic amplifier, and a meter to record knock intensity. To determine a fuel's antiknock

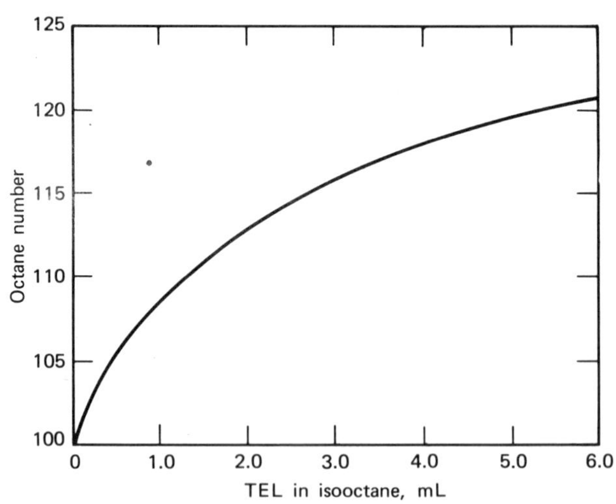

Figure 2. Relationship between "knock value" and octane numbers above 100. Courtesy of Ethyl Corporation.

quality, the CFR engine is operated on the fuel under a standard set of conditions and its compression ratio is adjusted to give a standard level of knock intensity. This knock level is then bracketed by two blends of the reference fuels, one of which blends knocks a little more than the test fuel, the other of which knocks a little less. The knock rating of the test fuel is determined by interpolation between the knock readings of the reference fuels to find the reference-fuel composition that just matches the knockmeter reading of the test sample.

The antiknock quality of today's gasolines is rated by two laboratory knock-test procedures, both of which employ the CFR engine in the manner just described. The Motor method (ASTM D 2700) employs a set of standard operating conditions that correlated well with the fuels and automobile engines of the 1930s. The Research method (ASTM D 2699) employs a different, less severe set of operating conditions adopted when it became apparent that newer refining processes and engine improvements gave gasolines much better road performance than their Motor-method ratings would indicate. Both methods continue in use, however, because together they predict a gasoline's road performance better than either alone. If two fuels have the same Motor-method octane number (MON), the one with the greater Research-method rating (RON) will usually satisfy a greater percentage of the cars on the road. Likewise, if the Research ratings are equal, the fuel with the higher Motor rating will usually perform better on the road. The difference between a gasoline's Research rating and its Motor rating (RON minus MON) is called its sensitivity; this difference indicates how sensitive the gasoline is, in terms of antiknock performance, to more severe engine operating conditions. Among fuels of equal Research octane number, the fuel having the least sensitivity generally will give the best road antiknock performance. A gasoline's sensitivity depends on its hydrocarbon composition: paraffins contribute little or no sensitivity, as might be expected from the fact that the reference fuels themselves are paraffins and, by definition, have the same octane numbers by either method. Olefins and aromatics are quite sensitive and, therefore, impart sensitivity to the cracked and reformed gasolines that contain them.

The importance of octane number in evaluating gasoline stems from its role in permitting increases in engine compression ratio. The power obtainable from combustion of a gasoline–air mixture in an engine can be increased by more highly compressing the mixture before igniting it. True, such compression requires work, the energy for which must be supplied by the fuel, but the extra power produced exceeds the work required for compression. A substantial net gain in efficiency and power results from less heat rejection to the cooling system and from putting more of the heat energy of the fuel to useful work. How great the gain can be is readily seen by comparing the maximum cylinder pressures, before and during combustion, in an average 1929–1930 automobile engine of 5:1 compression ratio (CR) with those of a 1960s engine having a compression ratio of 10:1. In the old lower-compression engine, the fuel–air mixture was compressed to a pressure of about 690 kPa (100 psi) prior to ignition. During combustion, it exerted a power-producing maximum pressure of 2.76 MPa (400 psi). But in higher-compression engines, compressing the mixture to 1.38 MPa (200 psi) prior to ignition raises the power-producing push to 6.20 MPa (900 psi). In the old 5:1 CR engine, the pressure increase due to combustion was only 2.07 MPa (300 psi), compared with 4.83 MPa (700 psi) in the 10:1 CR engine. Thus, engines of the 1960s derived 30–35% more power from the same amount of gasoline thanks to the increases in fuel antiknock quality that have made possible corresponding increases in engine compression ratio. In recent years, the compression ratios of U.S. cars have been considerably reduced to permit their operation on the unleaded gasoline required by their catalytic emission-control systems (see Exhaust control, automotive).

Figure 3 shows the approximate gasoline antiknock quality required to permit a given engine compression ratio. As can be seen, 1929–1930 cars of 5:1 CR operated satisfactorily on gasoline of 70–75 octane number, but high-performance engines of 10:1 CR require a fuel of about 100 octane number.

As may be seen, the octane-number scale at the right of Figure 3 is not linear; instead, each unit of increase in gasoline octane number below the 100-ON level permits a larger increase in engine compression ratio than did the 1-ON increase just before it. In other words, octane numbers become "larger" in their effect on knock-limited compression ratio as fuel antiknock quality approaches 100 ON. To provide a linear relationship between the antiknock quality of aviation gasolines and their performance in the supercharged aircraft engines of World War II, another fuel-rating system was developed: the Army–Navy performance number, or PN, system (see Aviation and other gas turbine fuels). For octane numbers above 100, the relationship between the two scales is expressed by the following formula:

$$ON = 100 + \frac{1}{3}(PN - 100)$$

For octane numbers below 100, the relationship is:

$$ON = 128 - (2800/PN)$$

During the course of fuel-engine research, other methods of expressing gasoline antiknock quality have been developed. Among them are highest useful compression

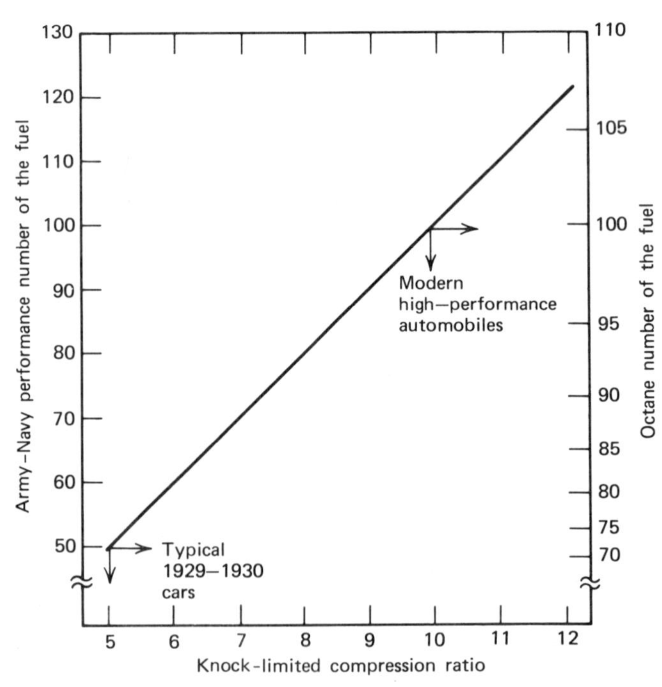

Figure 3. Fuel antiknock quality required as compression ratio is increased.

ratio (HUCR), aniline equivalent, benzene equivalent, and knock-limited imep (indicated mean effective pressure). But regardless of how antiknock quality is expressed, adequate antiknock resistance remains one of the most important requirements of a good gasoline.

Low Gum Requirement. A good gasoline must be substantially free of gum-forming materials in order to avoid clogging of engine fuel systems, carburetor malfunction, sticking of engine intake valves, and manifold deposits that would cause exhaust smoke and loss of engine power. Gum formation in gasoline is the result of oxidation and polymerization of unsaturated fuel components, particularly dienes or more highly unsaturated compounds, the end products being resinous materials. Modern catalytic refining processes produce very little or none of the worst gum-forming unsaturates once present in thermally cracked and thermally reformed gasoline fractions. And as is explained later, certain processes in the manufacture of today's gasolines are aimed at converting any potential gum formers into unreactive forms prior to final blending of the finished gasoline.

The amount of existent gum in a gasoline (ie, the amount that has already been formed and is dissolved in the gasoline) is determined by ASTM Method D 381. This method involves evaporating a 50-mL sample to near dryness by a jet of heated air, washing the residue with a liquid hydrocarbon (*n*-heptane) to remove any high-boiling additives that may be present, and then weighing the washed residue of gummy material. The result of the test is usually expressed as mg of air-jet (or existent) gum per 100 mL of sample. Most motor gasolines contain less than 3 mg of such gum per 100 mL.

Although the test just described measures the gum currently present in a gasoline, it does not indicate how much more might form during storage at the refinery, during transportation, in service-station tanks, or in the motorist's fuel tank. Some years ago, when refinery processes produced considerably more gum formers than they do today, the need for a rapid method of predicting ultimate gum formation led to development of a so-called oxidation stability test, ASTM Method D 525. This method, which is still used to identify unstable gasolines, consists of placing a sample of the gasoline in a pressure cylinder, pressurizing the bomb to 690 kPa (100 psi) with oxygen, immersing the pressurized bomb in boiling water, and continuously measuring the pressure inside the bomb. Eventually, the pressure drops sharply, about 14 kPa (2 psi) or more within a 15-min interval, indicating the break point. The time from start of the test to the break point indicates the oxidation stability of the gasoline. A good modern gasoline should exceed 240 min in this test.

Another gum-stability test, ASTM Method D 873, in effect combines the air-jet existent gum test with the oxidation stability test to determine the potential residue in aviation gasolines. A 100-mL sample of the fuel is exposed to oxygen in a heated bomb for a specified time (in h) and then is evaporated to dryness by the air-jet method. The result is expressed as mg of x-hour-aging gum per 100 mL where x is the time of exposure in the oxygen bomb. A U.S. military specification for aviation gasolines, MIL-G-5572C, calls for no more than 6.0 mL of 16-h-aging gum per 100 mL.

Although they are reasonably good predictors, none of the quick gum tests described above is always accurate in predicting gasoline stability in actual storage. The best laboratory method for predicting stability involves storage at 38°C for a protracted period of time, eg, 12 weeks, followed by determination of actual gum formed.

Low Sulfur Requirement. Gasolines containing hydrogen sulfide or the thiols (mercaptans) naturally present in crude oil would be extremely offensive to the motorist. These and other sulfur compounds are also corrosive to metals and could cause serious damage both to refinery equipment and to the engine. Over and above this, many of the sulfur compounds present in crude oils poison refining catalysts and are antagonists toward antiknock compounds. For these reasons, a number of refinery processes have been developed either to remove sulfur compounds from petroleum products or to convert them to inoffensive forms. Actual removal is preferred (see below).

The total sulfur content of a gasoline is measured by ASTM Method D 1266. This involves burning a measured amount of the gasoline in a wick-type lamp and recovering the products of combustion for chemical analysis by titration with standardized alkali solution. The result is expressed as wt % sulfur. The total sulfur content of modern gasolines is low. For example, U.S. averages in the summer of 1978 were 0.026 wt % for unleaded, 0.040 wt % for leaded regular, and 0.032 wt % for leaded premium gasolines.

Another ASTM Method (D 130) is used to determine a gasoline's corrosiveness to copper. In this test, a sample of the gasoline and a polished copper strip are placed in a test tube, which is then heated in a water bath at 50°C for 3 h. If the strip shows more than a very slight discoloration or shows any trace of corrosion, the gasoline is considered to have failed this test.

Manufacture. Shortly after 1900, as motor vehicles began to appear in growing numbers, gasoline came to have some marketable value as a refinery product. Around 1912, U.S. demand for motor fuels began to exceed the amount resulting from refinery crude-oil runs geared to the demand for kerosene, fuel oils, etc. Since then, the U.S. petroleum industry's biggest single problem has been devising new ways to meet the tremendous demand for gasoline without, at the same time, overproducing other petroleum products. Thus, it has devoted great time and effort to the development of processes for increasing the gasoline yield obtainable from crude oil. Simultaneously, in order to satisfy the fuel-quality requirements resulting from continual progress in engine design, the industry also has had to develop other processes for improving gasoline properties, even, if necessary, at the expense of yield. In achieving both of these somewhat conflicting goals, refiners have advanced gasoline manufacture from the status of a relatively simple art to one of the world's most complex and efficient technologies. The processes by which they produce today's high-octane gasolines are described below.

Distillation. Petroleum refining begins with the distillation of crude oil into fractions of different boiling ranges (see Distillation; Petroleum). The crude oil is heated to 370–430°C, pumped into a fractionating tower, and separated into the following major fractions (Table 2). Because they are naturally occurring fractions of crude oil, the naphtha fractions obtained by distillation are called virgin naphtha, or straightrun gasoline. Both the amounts of these naphthas and their hydrocarbon compositions depend on the type of crude oil being distilled. Thus, straightrun gasolines differ widely in such properties as specific gravity, vaporization characteristics, and antiknock quality. The light naphtha fraction is usually of sufficiently high ON to be used as a component of finished gasoline without any more refining than is needed to remove undesirable impurities. The heavy naphtha is catalytically reformed to higher-octane blending stock. The kerosene and light gas–oil fractions, referred to

Table 2. Boiling Range of Principal Crude-Oil Fractions

Principal crude-oil fractions	Approximate boiling range, °C
light naphtha	0–100
heavy naphtha	100–200
kerosene	200–250
light gas oil	200–300
heavy gas oil	300–400
reduced crude	400 to >600

collectively as middle distillates, are used in the production of kerosene, jet fuel, diesel fuel, and furnace oils. The heavy gas–oil fraction may be used in heavy diesel fuel, industrial fuel oil, and bunker oil. However, in the United States, gasoline demand is so high that much of the heavy gas oil and other heavy oils recovered from the reduced crude are cracked into gasolines by processes described below.

Recovery of Natural Gasoline. Around 1912, it became apparent that distillation of crude oil alone could not satisfy the rapidly growing demand for gasoline. If the required amount of crude oil had been distilled, oversupplies of kerosene, fuel oils, and other products would have been staggering. Therefore, plants were built to recover the gasoline-range hydrocarbons present in "wet" natural gas as it comes from the well (see Gas, natural). The first natural-gasoline plants used compression, followed by cooling to the original temperature, to condense the liquid hydrocarbons from wet gas. However, most modern plants recover these liquid hydrocarbons by absorbing them in petroleum fractions somewhat heavier than straightrun gasoline. The wet gas is passed through a column in contact with the absorption oil, and the absorbed natural gasoline is recovered from the oil by a form of distillation, using steam to increase the separation of lighter from heavier hydrocarbons.

Because natural gasoline (also called casinghead gasoline) is a highly volatile mixture of light hydrocarbons, only a limited amount of it can be included in finished gasoline without exceeding desired volatility characteristics. Moreover, the antiknock quality of natural gasoline is relatively low since it contains a large amount of straight-chain paraffinic hydrocarbons.

Cracking. Early in the history of petroleum refining, it was found that higher-boiling hydrocarbons could be broken down, or cracked, into lower-boiling ones by subjecting them to high temperatures for an appreciable length of time. In fact, a form of cracking was used as early as 1869 to convert heavier crude-oil fractions into kerosene. Use of cracking to produce gasoline began in 1913 and grew rapidly as the nation's motor-vehicle population multiplied. Because temperatures as high as 540°C are required for such thermal breakdown, pressures of 2.4–6.9 MPa (350–1000 psi) are used to keep the feedstock largely in the liquid phase while it undergoes the cracking reactions. Otherwise, individual vaporized hydrocarbons would be over-decomposed, forming light gases and laying down cokelike solid deposits inside the cracking unit. Such high pressures are not required, however, when cracking is conducted in the presence of a catalyst. The catalyst, which may consist of naturally occurring clays or synthetic compounds containing silica and alumina, offers a large surface area on which the cracking reactions take place. Modern catalytic-cracking units employ catalysts in the form of small beads or finely divided, fluidized powder. The carbon deposits formed on the surface of these catalysts during cracking are burned off (thereby regenerating the catalyst) by subjecting the hot catalyst to a stream of air in a separate catalyst–regeneration step. Continuous operation is attained by constantly circulating a portion of the catalyst from the reaction zone, through the regenerator, and back to the reaction zone.

Gasoline-range fractions produced by catalytic cracking make highly desirable blending components for high-octane gasoline. The lower-octane paraffinic and cycloparaffinic hydrocarbons in the feedstock are largely converted to higher-octane olefins and aromatics. Thus, in addition to its prime function of being a gasoline-production process, cracking also serves, in a sense, as an octane-improvement process. Catalytically cracked gasolines have octane numbers ranging from about 92 to 94 (Research method) before addition of antiknocks. And they respond reasonably well to the addition of antiknocks, gaining about 5–7 octane numbers from the addition of 0.8 mL/L (3 mL/gal) of tetraethyllead.

Hydrocracking. This process, which consists of cracking in the presence of added hydrogen, has recently been gaining favor as a highly flexible refining process because it permits wide variations in yields of gasoline and furnace oils to meet seasonal demand changes. With hydrocracking, refiners can convert heavy fractions into maximum amounts of gasoline and only small amounts of furnace oil, or *vice versa,* as needed. Moreover, hydrocracking is particularly effective in processing highly refractory (hard-to-crack) stocks. Thus, the process enables refiners to convert practically worthless heavy cracked stocks into gasoline, fuel oil, or other highly salable products. However, if hydrocracked stocks are to constitute a fairly large proportion of the finished gasoline, they must be subjected to an octane-improvement process called reforming, since they lack the high-octane olefins present in catalytically cracked blending stocks.

Reforming. All reforming processes have the same general purpose: to convert low-octane gasoline-range hydrocarbons into higher-octane ones. Reforming is largely limited to the upgrading of heavier gasoline fractions, such as straightrun stocks boiling from 90 to 200°C, because lighter fractions do not contain substantial amounts of hydrocarbons suitable for reforming. Thermal (noncatalytic) reforming, which now has been almost completely replaced by catalytic reforming, uses equipment and operating conditions similar to those of thermal cracking. The resulting octane-number improvement stems mainly from cracking of low-octane heavy paraffins into lighter, higher-octane paraffins and olefins. Only about 80% of the feedstock volume processed is recovered as full-boiling-range gasoline, because the cracking reactions produce substantial amounts of gases. In catalytic reforming, however, the octane-improvement reactions include molecular rearrangements as well as molecule splitting. Depending on the catalyst and the operating conditions, the following reactions occur to greater or lesser extent: (*1*) cycloparaffins lose hydrogen to become the corresponding aromatics, which are higher in octane number (see also BTX processing); (*2*) straight-chain paraffins rearrange into their branched-chain isomers; (*3*) heavy paraffins of low octane number are hydrocracked to lighter paraffins of higher octane number; and (*4*) heavy paraffins lose hydrogen and form rings to become aromatics of high octane number.

Some of the hydrogen liberated by these reactions is recycled through the reforming unit to minimize carbon deposition on the catalyst. This hydrogen also reacts with sulfur-containing impurities to form easily removable hydrogen sulfide gas.

For many years, most reforming catalysts consisted of platinum on a high-surface-area support such as alumina or silica–alumina. Bimetallic catalysts combining platinum with another promoting metal, such as rhenium, were introduced in the late 1960s. Their higher activity and greater selectivity permitted higher output and/or greater octane upgrading in existing units, accounting in part for the increase in U.S. reforming capacity that occurred in the early 1970s.

The octane number of gasoline components subjected to reforming increases as temperature and reaction time are increased. However, increasing the severity of reforming conditions decreases the yield of gasoline-range product, partly because more of the feed is converted to gases and partly because conversion of paraffins and cycloparaffins to the more dense aromatic structure shrinks the volume of the liquid product. Thus, refiners usually limit reforming severity to the point at which reforming cost plus the value of lost product begins to exceed the cost of supplying additional octane numbers with antiknock additives. Reformed gasolines respond well to alkyllead antiknock fluids. For example, an 86-octane unleaded reformate would be expected to rate 96–97 octane when lead is added.

Alkylation. Unlike cracking, alkylation makes larger hydrocarbons from smaller ones; it produces gasoline-range liquids from refinery gases. Because of its cost, however, its major value stems from the exceptionally high antiknock quality of its product, rather than from the fact that it offers another way to make gasoline. Extensive use of alkylation began during World War II to produce the high-octane (100+) gasolines required by military aircraft. After the war, some alkylation units not required to meet the demand for aviation gasoline were operated to make highly desirable blending stocks for passenger-car gasoline. And more recently, a number of refiners have installed new units exclusively for such use.

Alkylation processes employ either sulfuric acid or hydrogen fluoride as a catalyst. The reaction, which consists of combining an olefinic gas with a paraffinic gas to produce branched-chain liquid paraffinic hydrocarbons, is conducted at low temperatures, 0–38°C, and pressures ranging from atmospheric to 1.0 MPa (145 psi). Isobutane is used as the paraffinic feed gas, and the preferred olefinic feed consists of butylenes, although propylene and pentylenes can also be used. The reaction of isobutane (C_4H_{10}) with butylenes (C_4H_8) produces mainly branched-chain octanes (C_8H_{18}). Similarly, with propylene as the olefinic gas, the product consists primarily of branched-chain heptanes (C_7H_{16}). The products of alkylation have high octane numbers, ca 95 RON unleaded and 106–107 RON when lead is added. Because their actual performance in engines on the road often exceeds their laboratory octane numbers, alkylates are very desirable components for blending into premium-grade gasolines (see Alkylation).

Polymerization. Like alkylation, this is a way of making gasoline from refinery gases. But in polymerization only the olefinic gases in the feed react, linking together to form olefinic liquids. Any paraffinic gases in the feed pass through the process unchanged. A typical polymerization reaction is that of two molecules of isobutylene (C_4H_8) combining into one molecule of a branched-chain octylene (C_8H_{16}). This reaction was used during World War II to produce 2,4,4-trimethyl-2-pentene, which was subsequently hydrogenated to make "isooctane" for aviation gasoline. In polymerization processes for making motor-gasoline components, refinery gases rich in propylene and butylenes are subjected to temperatures of 150–230°C and pressures of 1.0–8.0 MPa (150–1200 psi) in the presence of a catalyst (usually phosphoric acid). The olefinic gasolines made by polymerization rate high in octane number by the

Research method but relatively low by the Motor method of laboratory engine testing. Compared with alkylates, they are markedly inferior in actual antiknock performance on the road. Therefore, refiners who have ample isobutane and the required alkylation capacity usually prefer to alkylate rather than polymerize their available olefinic gases.

Isomerization. To date, petroleum-refinery use of isomerization to convert straight-chain hydrocarbons into branched-chain hydrocarbons of the same molecular weight has been confined mainly to the conversion of normal butane to isobutane needed for alkylation feed. Isomerization has been used to a lesser extent to upgrade the antiknock quality of natural gasolines and light straightrun fractions by converting their *n*-pentane and *n*-hexane contents to their higher-octane branched-chain isomers. Such use is growing in countries where cracking is not needed to meet gasoline demand and, therefore, light olefins from cracking are not available to contribute high antiknock quality to the lower-boiling gasoline fractions. Isomerization catalysts include aluminum chloride, antimony chloride, platinum, and other metals and metal compounds. Reaction conditions vary according to the catalyst and the feedstock, ranging from 40–480°C and 0.7–5.0 MPa (100–750 psi). Depending on feedstock composition, isomerized light fractions rate 93–107 octane by the Research method after addition of antiknock compound. Because they are highly branched, they contribute excellent road antiknock quality to finished gasolines.

Hydrogen Treating. As mentioned earlier, foul-smelling, corrosive sulfur compounds in gasoline can either be removed or be converted to less objectionable forms, removal being preferred. Early removal methods included use of strong aqueous solutions of sodium hydroxide, along with other chemicals, to dissolve the sulfur compounds, followed by separation of the aqueous layer. Later processes used clay (bauxite or fuller's earth) as a catalyst for cracking thiols at the carbon–sulfur bond, breaking them down into readily removed gases. Today however, sulfur removal usually is accomplished by hydrogen treating, wherein hydrogen reacts with the sulfur in objectionable compounds, forming easily removable hydrogen sulfide. Hydrogen treating is conducted at moderate temperatures, 250–450°C, and pressures, 2.0–5.5 MPa (300–800 psi), using oxides of cobalt and molybdenum as catalyst. The range of operating conditions is wide, because the purposes vary from mild treatment of reformer feedstock to severe hydrogenation of heavy fractions (see Petroleum).

Antiknock Compounds. The original, and still least expensive, method of increasing gasoline octane number is use of an antiknock compound. Each type of gasoline component has its own response to antiknock fluid, but octane-number increases of 10 to 15 are obtainable when adding 0.8–1.0 g Pb/L (3–4 mL/gal) of tetraethyllead to a quite responsive stock. Although a number of chemical compounds have been found to have antiknock effectiveness, none of the thousands investigated to date are as effective at as little cost as the alkyllead compounds. The original and still most widely used antiknock agent is tetraethyllead (TEL), $Pb(C_2H_5)_4$. However, tetramethyllead (TML), $Pb(CH_3)_4$, its physical mixtures with TEL, and chemically redistributed mixtures consisting of ethyltrimethyllead, diethyldimethyllead, and triethylmethyllead in addition to TEL and TML are also used commercially both in the United States and abroad. Because TML and TEL–TML mixtures are more volatile than TEL, they provide more even distribution of antiknock quality throughout the gasoline boiling range. They are particularly desirable where the lighter components of a gasoline are of relatively low octane number. By distributing their antiknock action more evenly throughout the vaporizing fuel as it goes to each of an engine's cylinders, TML and TEL–TML mixtures often provide superior antiknock performance in modern automobiles. Because of differing thermal stability, which affects the timing of their release of lead's antiknock action, TML and its mixtures with TEL may also show superior effectiveness per unit lead content in certain fuels under certain engine operating conditions (see Lead compounds).

In addition to the active ingredients (TEL, TML, or both), antiknock fluids contain chemical scavengers to remove the small amount of lead compounds that might remain in the engine after combustion of the fuel. These scavengers contain chlorine and/or bromine to convert the lead compounds to lead chloride and lead bromide, which are gaseous at the temperatures prevailing inside an engine. Ethylene dibromide and ethylene dichloride are the lead scavengers used commercially. A typical motor mix for automotive gasolines consists of about 62% TEL, 18% ethylene dibromide, 18% ethylene dichloride, and 2% of other ingredients, such as dye, petroleum solvent, and stability improver. For overall best performance of aviation piston engines, the scavenger consists entirely of ethylene dibromide, and a typical aviation mix includes about 61–62% TEL, 35–36% ethylene dibromide, and 3% of dye, solvent, inhibitor, etc.

In 1957, Ethyl Corporation announced a new antiknock agent, methylcyclopentadienylmanganese tricarbonyl(MMT), $CH_3C_5H_4Mn(CO)_3$ (see Carbonyls). When used alone, this compound is even more effective than the same amount of TEL or TML in many gasolines. However, MMT also costs somewhat more to manufacture than do the alkyllead additives. Thus, its main economic value originally lay in the powerful promoting effect it exerts on the antiknock action of the alkylleads. Only a tiny amount of the manganese compound added to leaded gasoline gives a marked synergistic effect, raising octane number far more than would be expected from the fuel's response to either antiknock alone. The mechanism of this promoting action still is not completely understood, but the manganese compound is believed to increase the effective catalytic surface area of the lead antiknock particles. Certain organic chemicals (eg, *tert*-butyl acetate) also have been shown to have some promoting effect on the antiknock action of TEL, the effect becoming appreciable at fuel octane-number levels of 100 and above. Thus, the manganese antiknock and organic TEL appreciators offer a route to even higher gasoline octane numbers than can be attained with the alkyllead antiknock compounds alone.

From 1975 to 1978, MMT's main use was as a replacement for lead antiknocks in production of U.S. unleaded gasolines. Since 1975, most passenger cars sold in the United States have required catalytic converters to meet government-imposed standards on exhaust-gas emissions. The catalysts have, in turn, required use of unleaded gasoline, because even a small amount of lead deposited on their active surfaces decreases their emission-controlling effectiveness. Thus, U.S. refiners faced the problem of producing ever increasing quantities of gasoline without the octane help, blending flexibility, and processing economy provided by lead antiknock compounds. To compensate, in part at least, for the loss of lead, they began using MMT in their unleaded grades. This allowed them to reduce somewhat the high energy consumption and yield losses involved in the more severe catalytic reforming required to meet octane-number specifications without lead. In 1978, however, MMT also was banned from use in U.S. unleaded gasolines on suspicion that it may increase exhaust-gas emissions.

To avoid gasoline shortages and minimize the amount of crude oil that the United

States must import, the EPA in mid-1979 temporarily suspended its ban on use of MMT in unleaded gasolines and relaxed its schedule for phasedown of lead content in leaded gasolines. For the same reasons, it also legalized the use of certain octane-improving blending agents in unleaded gasolines.

Blending Agents. Strictly speaking, TBA (*tert*-butyl alcohol), MTBE (methyl *tert*-butyl ether), and other alcohols and ethers receiving attention as possible replacements for organometallic antiknock compounds are not additives, but rather blending agents. The difference lies in how much is needed to achieve the desired effect. With additives such as antiknocks, antioxidants, metal deactivators, etc, required concentrations are very small, usually measured in tenths of a g/L. With the alcohols and ethers, however, concentrations of several percent are required for any significant improvement of a gasoline's octane number. In essence, they are other fuels of such superior octane number that blending them with gasoline in substantial amounts results in a composite gasohol or "gasether" of greater antiknock quality than the base gasoline. In a sense, the benzene, toluene, xylenes, and other aromatic hydrocarbons produced *in situ* by reforming gasoline are also blending agents or would be if they were produced separately and then added.

Until recently, deliberate production of nonhydrocarbon blending agents for gasoline ranged from nonexistent to minimal. Where such practice did exist, it was based on desire to dispose of a surplus commodity (see Gasohol and Other Alcohol Fuels below), and alcohol blending was found to be economically advantageous only with direct or indirect government subsidy (9). Other studies suggest that although ethanol may be at best close to the energy breakeven point, the ability to use nonpetroleum sources in its manufacture and further crude oil price increases may tip the balance in ethanol's favor as an energy producer (10). In addition to ambiguous economics, certain technical difficulties described below discouraged use of light alcohols as gasoline blending agents. Tertiary butyl alcohol (TBA) does not pose such problems, however, and it is a reasonably good octane improver. It is the primary ingredient of Atlantic Richfield Co.'s Arconol blending agent, and it has been used by at least one other major U.S. refiner in premium-grade unleaded gasoline. Methyl *tert*-butyl ether (MTBE) has been used as a blending agent to a minor extent in Europe for some time. Both MTBE and TBA were approved by the EPA early in 1979 for use in unleaded gasolines, via waivers of the right to ban them if they interfered with exhaust-emission control.

Tertiary butyl alcohol has a Research octane number of 108, compared with 106 for benzene, 110 for toluene, and 115 for methyl *tert*-butyl ether. At the time of the EPA's waiver action, the only U.S. manufacturer of gasoline-grade TBA was Oxirane Corp., a joint venture of Atlantic Richfield (Arco) and Halcon International. Mobil Oil Corp. reportedly also used TBA in a premium-grade unleaded gasoline it introduced in the winter of 1978–1979 (11).

Plans to market TBA in Europe have been announced by Oxirane Chemie (Netherlands), which is 50% owned by Atlantic Richfield Co. It has a 260,000 metric-ton-per-year plant scheduled for 1980 completion at Botlek, near Rotterdam (12).

MTBE's high blending value (a Research octane number of 118 and a Motor octane number of 101) has gained it considerable attention from U.S. refiners seeking ways to increase unleaded gasoline output and antiknock quality. As recently as the summer of 1978, world capacity to produce MTBE was only about 150,000 t/yr, limited largely to one plant in Italy and another in the Federal Republic of Germany. Less than a year later, plans had been announced for more than 900,000 t/yr of U.S. capacity in 1980 (13). The Suntech Division of Sun Petroleum Products Co. reportedly is studying ways to make MBTE and other alkyl ether blending agents from coal, rather than from the petroleum stocks required by existing processes. Suntech researchers point out that new routes to MTBE and similar blending agents are essential if they are to be of much real benefit in extending gasoline supply (14).

Current processes for making MTBE all require reaction of isobutylene and methanol in the presence of an ion-exchange resin catalyst. In analyzing the economics of gasoline octane upgrading with MTBE, Arco Technology Inc. has compared MTBE with motor alkylate, since the two octane improvers compete for the same butylene feedstock. To produce 0.8 ML/d (5000 bbl/d) of blending stock, total plant investment would be \$9.2 10^6 for MTBE and \$13.6 10^6 for motor alkylate. Selling price required to cover material and manufacturing costs would have to be 13.0¢/L for MTBE and 13.1¢/L for motor alkylate. Arco reports, "A comparison of octane values shows that MTBE has an average blending value 15 octane numbers above motor alkylate and 20 octane numbers above a typical gasoline pool. Depending on the individual refiner, additional octane may be worth 0.08–0.26¢/L (0.3–1.0¢/gal). Using 0.13¢/L (0.5¢/gal) as the expected octane value to an average refiner in the 1980s, there is a 2¢/L (7.5¢/gal) premium for MTBE over motor alkylate. Since this premium significantly exceeds the capital recovery cost of 1.5¢/L (5.5¢/gal) for alkylate, the MTBE process should compete with alkylation units already in use (15)."

Another analysis of MTBE's economics sees its future brighter in Europe than in the U.S. Chemical Technomics points out that the United States has much more catalytic cracking capacity than does Europe and that the hydrocarbon composition of its gasoline differs markedly from that in Europe. Thus, MTBE's blending value in typical U.S. gasolines is only about 119 ON, compared with 133 ON in European gasolines. Assuming an increase of 0.26 ON/L (one octane per gallon) is worth 0.4¢, MTBE would be worth 11.6¢/L (44¢/gal) in the United States, compared with 13.2¢/L (50¢/gal) in Europe (16).

A 1979 study of octane-improvement economics (17) predicts that, despite recent interest shown in them, blending agents will not play a significant role in future gasoline production because their energy-saving potential is nil. As much or more energy is required to make a blending agent as to produce high-octane catalytic reformate by severe refinery processing. Quite large amounts of blending agent are required for even modest octane improvement, eg, several thousand liters for just a 1-RON improvement of a 100,000-liter blend. Thus, transportation costs would be prohibitive except for short distances.

Evaluation of a blending agent's octane-improving effectiveness requires a correction for the amount of pressurizing butane that the blending agent's vapor pressure allows to be added or requires to be excluded from the blend. Because butane is a high-octane blending component in its own right, the correction can be expressed as a butane credit. For illustrative purposes, Table 3 shows butane credits (or debits) and typical octane-number gains that would be expected from adding 5000 liters (5%) of various blending agents to 100,000 liters of unleaded and leaded regular-grade gasolines (11).

***Antioxidants* (qv).** Widespread adoption of catalytic cracking and catalytic reforming (which produce very little unstable hydrocarbons) has greatly increased the inherent oxidation stability of modern gasolines. Because unstable components, eg,

Table 3. Net Octane Gains for 5% Addition of Blending Agents

Blending agent	Butane credit, %	ON gain due to 5% addition[a]			
		Unleaded regular		Leaded regular	
		RON	MON	RON	MON
toluene	1.071	1.07	0.69	1.34	1.13
MTBE[b]	0.347	1.16	0.84	1.44	1.17
TBA[c]	none	0.78	0.63	0.70	0.85
ethyl alcohol	−2.347	0.80	0.17	0.63	negative

[a] ON = octane number; RON = Research-method octane number; MON = Motor-method octane number.
[b] MTBE = methyl *tert*-butyl ether.
[c] TBA= *tert*-butyl alcohol.

thermally cracked stocks and polymer gasolines, are disappearing from use, relatively low antioxidant concentrations now are sufficient to protect against gum formation that would cause fuel-system problems and against the formation of fuel peroxides that could act as knock inducers. However, antioxidants still play an important role in overall fuel stability. Until recently, *p*-phenylenediamine, aminophenols, and 2,6-di-*tert*-butyl-*p*-cresol were among the most widely used gasoline antioxidants. With them, the required effective dosages were relatively large, ranging from 40–60 g/t (10–15 lb/1000 bbl) of gasoline. In 1958, however, a new family of highly effective antioxidants was introduced: the orthoalkylated phenols (see Alkylphenols). Moreover, combinations of these new compounds with phenylenediamine have been found to be at least twice as effective as either ingredient alone. With such combinations, average dosage is only about 10 g/t (2–3 lb/1000 bbl).

Metal Deactivators. Because copper accelerates gum formation by acting as a catalyst in gasoline oxidation, metal deactivators (eg, the Schiff's base, *N*,*N'*-disalicylidene-1,2-diaminopropane) are often added along with an antioxidant. Trace amounts of copper, picked up from piping or engine fuel systems, so powerfully promote oxidation that even the most effective antioxidant might not be able to provide adequate fuel stability without the help of a metal deactivator. The tiny amount of chelating agent added for this purpose functions by tying up trace copper in stable compounds that have no catalytic effect on oxidation (see Chelating agents).

Antirust Agents. Because rusting and corrosion can cause problems in storage tanks, pipelines, tankers, and the fuel systems of engines, several types of hydrocarbon-soluble compounds are used in gasolines as rust inhibitors (see Corrosion and corrosion inhibitors). Most of these compounds, which include various fatty acid amines, sulfonates, alkyl phosphates, and amine phosphates, function by coating metal surfaces with a very thin protective film that keeps water from contacting the surfaces. This surface-active property also helps in preventing buildup of gummy deposits in the carburetor and combats carburetor iceup during cold weather.

Anti-icing Agents. To prevent ice formation in engine fuel lines or the carburetor, one or the other of two types of anti-icing agent is added to gasolines that will be used in cool to cold weather. To avoid possible carburetor malfunction, such an additive is needed even when outdoor temperature is as high as 10°C because the cooling effect of fuel vaporization in the carburetor is great enough to condense and freeze moisture from the air onto chilled carburetor surfaces. The anti-icer may be either a freezing-point depressant or a surface-active agent. The freezing-point depressants, which include alcohols and glycols, act in the same manner as does antifreeze in an engine's cooling system. The surface-active agents, which include amides, amines, and amine or ammonium salts of organophosphates, have a different mode of action: they provide a coating that tends to prevent ice particles from sticking to each other or from building up on metal surfaces. Thus, ice particles pass harmlessly through the carburetor into the intake manifold, where engine warmth converts them to water vapor that eventually exits with the exhaust gases (see Antifreezes and deicing fluids; Surfactants).

Detergents. When an engine is idling, high-boiling fuel components, together with contaminants from exhaust and crankcase fumes drawn in through the air cleaner, may accumulate on the interior walls of the carburetor just below the throttle blade. By interfering with air flow and altering the air–fuel ratio, the resulting deposits can cause rough idling, frequent stalls, poor performance, and increased fuel consumption. To prevent formation of such deposits and to remove them from dirty carburetors, many refiners include a detergent additive in their gasolines. The effectiveness of these detergents, which include amides and alkylammonium dialkyl phosphates, stems from their surface-active properties.

Multipurpose Additives. Many refiners now use a multipurpose additive that combines detergent, deicing, and anticorrosion properties in a single formulation. Most of the action of such an additive comes from its surface-active ingredients, typically amino hydroxy amides. However, isopropyl alcohol, which also is included, contributes to anti-icing effectiveness.

Upper-Cylinder Lubricants. To help lubricate cylinders and top piston rings, to prevent valve and ring sticking, and to reduce intake-system deposits, refiners often include small amounts of a light lubricant in their gasolines. Through solvent action, these lubricants tend to rinse away deposits in the intake system from the carburetor and manifold to the underside of the intake valve. They usually consist of light mineral oils or low-viscosity cycloparaffinic distillates (see Lubrication).

Dyes. Hydrocarbon-soluble organic dyes have long been added to gasolines to indicate the presence of antiknock compound, to increase sales appeal, and to identify various brands or grades of gasoline. To facilitate blending into finished gasoline, particularly where automated in-line blending is practiced, liquid gasoline dyes have recently been introduced. Dye concentrations, always quite small, depend on the intensity of color desired by the particular refiner to meet a given color standard for the product.

Proprietary Additives. A number of refiners have developed gasoline additives for their own exclusive use. Some of these additives perform a single function; others are of the multipurpose type. For example, one large refiner some time ago announced a nickel-containing additive, nickel isodecyl orthophosphate, described as functioning as a deposit modifier, carburetor deicer, antirust agent, and wear inhibitor (12–13). Another major oil company announced that inclusion of a titanium-containing compound in its premium gasoline gave the fuel greatly increased ability to keep carburetors clean and improved the overall performance of the engine (14).

Blending of Components. Every refiner's aim, quite naturally, is to meet gasoline demand and quality requirements as economically as possible with the available processes and the aid of the additives just described. Since refiners normally produce both regular-grade and premium-grade gasolines, they must decide, from day to day, on the optimum split of each blending stock between those grades in order to make the amounts required of each grade and, simultaneously, achieve the desired product specifications at lowest cost. This is like shooting at a moving target, because the amounts of the different blending stocks used at a given time must be balanced against overall refinery operations and product inventories at that time, not just on current gasoline requirements. Long-range planning aids refiners in making the necessary day-to-day decisions. For example, in balancing refinery operations optimally to meet changing seasonal demand for heating oils, a refiner normally cracks less feedstock (or cracks less severely) in winter than in summer. Reduced cracking is based on two reasons: because savings are required in middle distillates for use in furnace oils and because somewhat less gasoline is needed in winter than in summer. Similarly, in the summer a refiner often has more refinery gases available than can be blended directly into gasoline without exceeding desired summer vapor pressure. In this case, it may be economical to convert the gases to alkylate or polymer gasoline. Such fluctuations in the supply of the various possible components make optimum gasoline blending an ever-changing problem. But another phenomenon complicates the problem still more: the fact that environment affects the blending behavior of the components. Gasoline blending is not linear in respect to the octane-number, volatility, and vapor-pressure contributions of the components; each component may contribute either more or less antiknock quality, volatility, or vapor pressure to a blend than would be expected from its own laboratory-measured properties and the amount added, the difference depending on the types and proportions of the other components.

Despite these complications, rapid yet quite accurate methods for predicting proper blending proportions have been developed over the years. A key concept in these methods is that each component can be assigned a set of blending numbers, different from its actual laboratory properties, that represent its behavior in the environment of the other components and additives with which it will be blended in a particular gasoline (18). Although they are not intrinsic to the component under all blending conditions, these assigned blending numbers prove to be reasonably valid over the range of component variations involved in a given refinery situation. Use of blending numbers permits linear calculation of each component's contribution on the basis of its proportion in the blend.

For long-range planning of refinery operations, the modern trend is toward use of electronic digital computers capable of solving complex sets of simultaneous equations. Typically, a linear program is set up with sufficient flexibility to represent the range of processing and blending variations possible during the period of time under consideration (see Computers). Using well-established mathematical procedures, the computer produces a series of operating plans for the mathematical refinery model. Various processing and blending alternatives are considered, until the most economical solution is determined. For gasoline manufacture, this optimal solution is one for which the cost of a very small additional improvement in octane number via adjustments in refinery processing is just equal to the cost of the same improvement via addition of antiknock fluid. Short-term or day-to-day blending decisions usually are made by a skilled chief chemist, who formulates each blend on the basis of a balance between factors existing at the time and guidance provided from long-range linear-program solutions.

In-Line Blending. Until the 1960s, almost all gasoline blending was done in batches, the various components being stored in separate tanks until the time to make up a new batch of finished gasoline. To save on tank requirements, vapor losses during storage and transfer, and blending manpower, many refiners have now adopted in-line blending systems that automatically and continuously combine a multiplicity of gasoline components in a blending manifold, from which finished gasoline exits in a constant stream. Under computer control, turbine meters measure the flow rate of each component, regulating blending proportions to the predetermined ratios. Antiknock and other additives are also added to the blend in this manner. Such systems can blend very large amounts of finished gasoline in a very short time. For example, each of the two in-line blenders at a large Texas refinery has a capacity of almost 12 ML/h (10,000 bbl/h) (15). A number of ingenious stream analyzers have been developed to monitor or control antiknock quality, Reid vapor pressure, volatility at the 20, 50, and 90% distillation points, etc. These monitors make it possible for refiners to deliver in-line blended gasolines directly into pipelines with confidence that they have met specifications (16–17).

Do-It-Yourself Blending. As premium-grade leaded gasoline becomes harder to find at U.S. service stations (having been replaced by one and sometimes two unleaded grades), owners of older, high-performance cars have become increasingly worried over how to obtain fuel that will satisfy them. Antiknock suppliers must turn down requests for private supplies of TEL or MMT, because the toxicity of the undiluted additives limits their blending into gasoline to specially trained personnel using specially designed equipment. Fortunately, there is a relatively little known trick for obtaining higher octane gasoline that has been proved over the years by U.S. tourists in Mexico. Mexico has for years had an unleaded premium-grade gasoline and a leaded regular grade, neither of which had an octane number high enough to satisfy many U.S.-built automobiles. Today's drivers can solve this problem the same way those touring Mexico did—by mixing the best unleaded gasoline available with a good leaded regular-grade gasoline. The result is a surprisingly good low-lead gasoline, in some cases one that is considerably higher in octane number than either of its components. The best proportions of leaded to unleaded can be determined by trial and error. In 1979, a mixture of 20% Shell Regular leaded gasoline (2 g TEL/gal or 0.53 g/L) (89 octane) with 80% Shell Super Unleaded (92 octane) gave a mixture of ca 94 octane—about the same as premium-grade leaded gasoline (19). At present (1980), a 25% Shell leaded (1.1 g TEL/gal or 0.29 g/L) to 75% Shell unleaded mixture raises the octane rating by only 0.1. However, the present (1980) level of TEL for all gasoline produced by a petroleum company is 0.8 g/gal (0.21 g/L). Thus allowed, if a petroleum company should produce as much leaded as unleaded gasoline, the level of TEL could be 1.6 g/gal (0.42 g/L).

Why leaded/unleaded blending works so unexpectedly well is explained by two facts not widely known outside the petroleum-refining industry. One of them is that refiners must use premium-grade blending stocks just to reach the minimum legal octane requirement for unleaded gasoline, ie, unleaded regular-grade gasoline is equivalent to leaded premium before lead is added. The other fact is that a gasoline's response to antiknock additive is not linear with additive concentration. The first quarter gram per liter added gives a big boost—as much as 10 octane numbers, the magnitude of the response depending on the gasoline's hydrocarbon composition. A second equivalent amount of additive gives a smaller additional boost. A third less than the second, etc. This explains the apparently magic octane-number improvement when an unleaded gasoline is blended with a leaded regular. The unleaded fuel's premium-grade blending stocks show a big response to the lead they share with the original leaded fuel. And the leaded fuel does not lose nearly as much by sharing, because its lead loss is, in effect, from the top of the response curve, where added concentration has much less effect.

Tests to Determine Road Antiknock Performance. ***Road Octane Numbers.*** As might be expected, laboratory octane ratings made in a single-cylinder engine, under fixed standard conditions, do not always predict just how a gasoline will behave in a multicylinder automobile engine operated over a wide range of speeds, loads, and weather conditions. Therefore, methods have been developed for rating gasolines in actual vehicles on the road or on indoor chassis dynamometers that duplicate road conditions. By providing temperature and humidity control in addition to means of simulating vehicle inertia, road loads, and air resistance, all-weather chassis dynamometers permit indoor road testing under weather conditions other than those prevailing outdoors at the particular time.

The precursor of all current road-rating procedures, and one still used today, was developed from road tests performed during 1932 and 1934 in Uniontown, Pa., under the auspices of the Cooperative Fuel Research Committee (now the Coordinating Research Council, or CRC). Both the petroleum and the automotive industries cooperated to develop a uniform and meaningful road test for fuel antiknock quality. Sometimes called the maximum-knock procedure, the Uniontown procedure compares the test fuel with reference fuels on the basis of the maximum-knock intensity occurring during full-throttle acceleration from about 16 km/h (10 mph). In this procedure, the rater is interested only in the maximum knock level obtained during the entire acceleration. The speed at which the same maximum-knock intensity is encountered with the matching reference fuel may be different from that at which it occurred with the test gasoline. Although the Uniontown procedure has the advantage of simplicity, it also has a number of disadvantages, perhaps the most serious being that the human ear cannot accurately detect small differences in intensity at high knock levels. Therefore, the precision and reproducibility of Uniontown ratings are often rather poor.

The modified Uniontown procedure, developed to attain better precision and reproducibility in road ratings, takes advantage of the fact that the ear is quite sensitive to small differences in knock at low levels of knock intensity. As in the original Uniontown method, full-throttle accelerations are made from low speed. During these accelerations, however, the basic ignition timing is varied just enough to produce trace knock on each of a series of primary reference blends. These resulting data on required spark advance for trace knock are plotted into a reference-fuel curve, usually at increments of two octane numbers between the reference fuels (Fig. 4). Then all that remains for determining the road octane number of a test fuel is to determine the amount of spark advance required to produce trace knock with the fuel. Locating that point on the reference-fuel curve and reading down to the octane-number scale indicates the test fuel's road antiknock rating. As shown in Figure 4, test fuel A developed trace knock at a spark advance of 14° before top center (btc) and therefore rated 98.3 octane number; test fuel B developed trace knock at a lesser spark advance (9 btc) and rated 95.8 ON. Different test fuels can be rated very rapidly by this procedure, once the reference-fuel curve has been established, because there is no need for bracketing

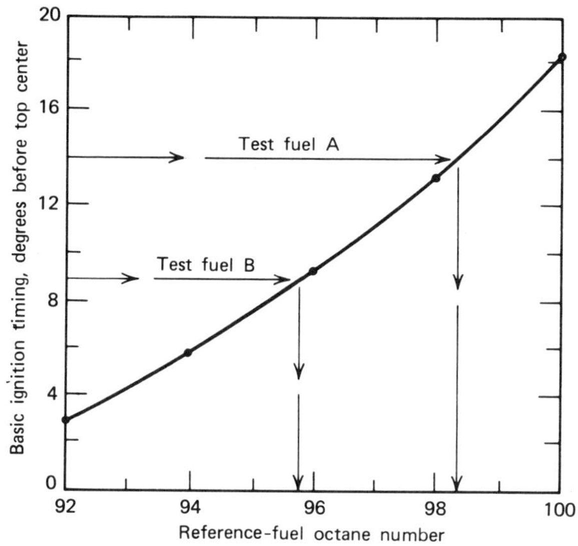

Figure 4. Reference-fuel curve for rating gasolines by the modified Uniontown procedure. Courtesy of Ethyl Corporation.

each test fuel individually between two reference blends. However, the reference-fuel curve must be checked fairly frequently, especially after the test car's engine has been shut down or when atmospheric conditions have changed.

Neither of the Uniontown techniques evaluates the effect of engine speed on a fuel's knocking characteristics. Therefore, in 1940, the CRC developed a procedure for determining the knock-limited spark advances for a fuel during full-throttle operation throughout an engine's speed range. Called the borderline technique (CRC Designation F-84), this involves observations during a series of accelerations, between each of which ignition timing is advanced slightly. During each acceleration, the speed at which knock dies out (can no longer be heard) is recorded, and the resulting data (spark advance and engine speed) locate one point on the borderline curve.

Today, a modified borderline procedure (CRC Designation F-27), in which spark advance is adjusted continuously during each acceleration to maintain trace knock throughout the speed range, has largely supplanted the original method. Because a curve of spark advance vs engine speed is obtained during a single acceleration, this procedure is much faster than the older one. A series of runs with primary reference blends establishes a family of curves like that in Figure 5. Thus, when a test fuel is run, plotting its trace-knock spark advances at the various engine speeds indicates its antiknock quality at those speeds. For example, in Figure 5, the modified borderline ratings of test fuels A and B at various speeds are shown in Table 4.

The ability of a test fuel to satisfy a car's octane-number requirements can be determined also by plotting the engine's standard spark-advance curve on the family of reference-fuel curves. Its intersection with each speed curve determines the engine's octane-number requirement at that speed. Thus, the engine of Figure 5 has its maximum requirement (97.4 ON) at 1500 rpm, and its requirements decrease with increasing engine speed to only 90.6 ON at 3500 rpm. The examples have purposely been chosen so that neither fuel A nor fuel B completely satisfies the engine over the entire

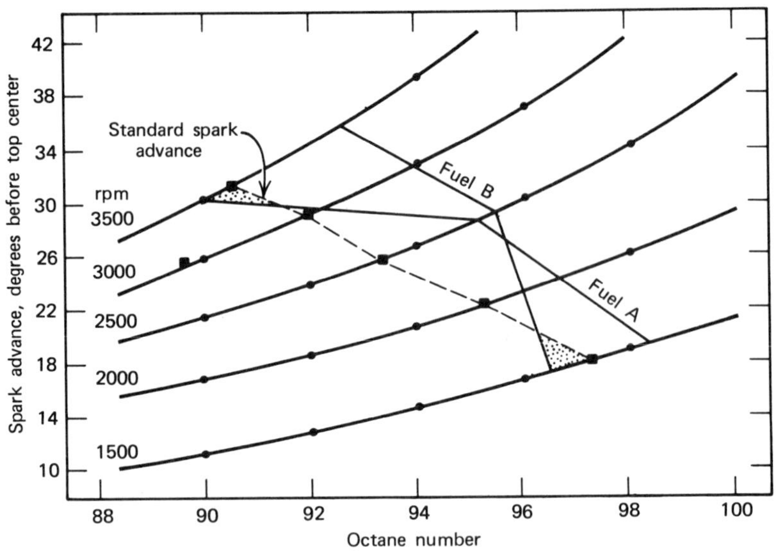

Figure 5. Modified borderline curves for two test fuels. Courtesy of Ethyl Corporation.

Table 4. Rating Tests of Fuels A and B

Engine rpm	Road octane number Fuel A	Fuel B
1500	98.4	96.7
2000	96.6	96.1
2500	95.0	95.4
3000	92.3	93.8
3500	90.0	92.5

speed range investigated. Note that fuel A, which gives good low-speed antiknock performance, will knock at speeds above about 3200 rpm (upper shaded area in Fig. 5). And fuel B, which is superior to fuel A at high speeds, will knock at speeds below about 1700 rpm (lower shaded area). Through choice of blending components and antiknock fluid, refiners tailor their gasolines to satisfy car requirements at all operating speeds. To obtain complete data on a car's octane-number requirements, they use a procedure (CRC Designation E-15) that determines its maximum requirement, its full-throttle requirements over the entire speed range, and its part-throttle requirements. Because weather conditions, particularly humidity, affect octane-number requirements, such tests are usually made on warm days of fairly low humidity or in a controlled-weather chassis dynamometer room constructed especially for the purpose.

Common Fallacies Concerning Gasolines. Over the years, a number of misconceptions about gasoline have cropped up again and again. Some are the products of wishful thinking; others stem from a lack of understanding or a misunderstanding of gasoline behavior in engines. Rumors of some suppressed invention, such as a special carburetor or miraculous fuel additive, that would give fuel economies of 43 to 85 km/L (or 100–200 mi/gal) fall in the wishful-thinking category. The following calculations were worked out some years ago in English units of measurement. To avoid the confusion of fractional conversions of base units, they have been left in their original form, with the results translated into metric units parenthetically. By definition, one British thermal unit (Btu) equals 778.26 ft·lbf of work (1.054 kJ), and one horsepower (hp) equals 33,000 ft·lbf of work per minute (745.7 W); thus 2545 Btu are required to produce 1 hp for 1 h. Because all gasolines, whether regular grade or premium, have almost identical energy content, about 115,000 Btu/gal (ca 32,000 kJ/L), 1 gal of gasoline could produce as much as 45 hp·h of work if every bit of its energy could be converted into useful work by the engine. However, the engine must reject still-hot combustion gases to admit a fresh fuel–air charge, and some 45% of the fuel's heat energy is lost for this reason. Another 20% is dissipated to the cooling water and/or air required to keep engine parts at temperatures they can stand. Friction losses in reciprocating and rotating engine parts consume about 5% of the fuel's energy. Thus, even modern high-compression engines convert only about 30% of a fuel's available energy into useful work. In other words, of the 45 hp·h theoretically available in 1 gal of fuel, only about 13.5 hp·h is converted to useful work at the engine's flywheel. And because some 5.5 hp is required just to keep a 2-ton automobile moving at a constant speed of 20 mph on a level road, the 13.5 hp·h of flywheel work could run the car for 13.5/5.5 or less than 2.5 h at that speed. Thus, maximum steady-speed fuel economy of the car would be around $2.5 \times 20 = 50$ miles per gallon or 21 km/L, and that could be achieved only at a constant speed of 32 km/h (20 mph) on a level road. Furthermore, because wind resistance and power losses due to tire flexing increase very rapidly with vehicle speed, the power required at highway speeds is much greater, ca 20 hp at 45 mph (14.9 kW at 72 km/h), 40 hp at 65 mph (30 kW at 105 km/h), and 80 hp at 80 mph (60 kW at 129 km/h). Therefore, if all of the flywheel output were available just to move the vehicle at constant speed on a level road, maximum attainable fuel economies would be about 13 km/L at 70 km/h (30 mpg at 45 mph), 9 km/L at 100 km/h (22 mpg at 65 mph), and 6 km/L at 130 km/h (13.5 mpg at 80 mph). And these figures do not take into consideration fuel consumption to power such accessories as power steering, power brakes, air conditioning, etc. Nor do they take into account fuel consumed in idling, hill climbing, accelerations, decelerations, etc. And finally, they do not take account of the fact that, for the acceleration performance demanded by motorists, cars are equipped with engines considerably larger and more powerful than needed just to maintain a given highway speed. Obviously then, dreams of a 100-km/L (200-mpg) automobile of the size and performance customary in the recent past are a myth beyond the bounds of possibility.

Another recurrent fallacy, that "premium fuel is just too hot for my engine—it would burn it up," results from lack of knowledge that all premium and regular-grade gasolines have practically the same heating value, around the 32 GJ/m^3 (115,000 Btu/gal) previously mentioned. The only truth in such statements is just the opposite: that a gasoline too low in octane number for the engine can cause a rise in engine temperature because of knocking. Closely allied to the "hot" gasoline fallacy is the erroneous belief that high-octane gasolines are more dangerous than low-octane ones. This belief probably stems from misunderstanding of what is meant when it is said that a high-octane gasoline is more powerful than a lower-octane gasoline. In heat content, any two gasolines are almost exactly equal, but because the higher-octane fuel permits higher compression of the fuel–air charge without encountering knock, it is capable of producing more power in an engine designed to make use of its superior antiknock quality. Because they contain volatile hydrocarbons, all gasolines should be handled with extreme care to avoid explosions resulting from accidental ignition; an "empty" gasoline can containing explosive vapors is equally as dangerous as a full can, if not more so. However, octane number has no effect on fire or explosion hazards, one way or the other. Gasoline put to work properly provides mankind with an extremely versatile source of power. Used wisely and handled correctly, it is safe as well as powerful.

Gasoline Substitutes

The Arab oil embargo of 1973–1974 and the gasoline shortages of the summer 1979, demonstrated the industrialized world's extreme and dangerous dependence on petroleum products, gasoline in particular. Some nations have faced this problem before, and a few, notably Germany and South Africa, developed operating synthetic petroleum industries. Both Germany and South Africa had little oil but ample coal—so they made their oil from coal. Other nations have other options: oil shale, tar sands, agricultural by-products, etc. Some, like the United States, have all of the mentioned options. Others, like Japan, have few, if any. A brief look at what can be done, starting with the various possible raw materials, seems appropriate in light of the world crude-oil situation.

Never, to date, have oil substitutes been able to compete in price with their oil-based counterparts, provided the latter were freely available. Germany turned to coal because adequate oil could not be obtained in wartime. South Africa turned to coal largely because oil-producing nations would not trade with it. Both nations had to build synthetic fuel industries that would be energy-effective, ie, would produce at least as much usable energy as is consumed in making it. According to most reports, this is technologically impossible, at present at least, by making ethanol or gasohol from agricultural products or by-products. Making fuel-grade alcohol via fermentation and distillation consumes more energy than it produces. Since the energy involved must come from other fuels, eg, oil, gas, or coal, the process is not energy-effective. However, the ability to use nonpetroleum sources in the manufacture of ethanol may eventually tip the balance in ethanol's favor (10). In fact, despite questionable current economics, gasohol is being used in Brazil and it is being test-marketed in the United States and elsewhere. Each new rise in OPEC oil prices increases the feasibility of all substitute fuels, including ethanol (see Chemurgy; Ethanol; Fuels, synthetic).

Germany built its synthetic-fuel industry on two processes: coal hydrogenation and catalytic synthesis. Under the aegis of I. G. Farbenindustrie, coal hydrogenation found particular favor. A 1945 report intimated that only intercompany rivalry prevented the catalytic synthesis (Fischer-Tropsch) process from playing as important a role as coal hydrogenation (20). Coal hydrogenation, which requires very high pressures at very high temperatures, is not easily adaptable to continuous-flow, electronically controlled processing such as practiced in petroleum refineries today. However, the Fischer-Tropsch process is adaptable to such processing, and the South Africans have adapted it.

Fischer-Tropsch Process. A detailed description of the Fischer-Tropsch process as operated by Germany during World War II appeared in 1946 (21). Basically, the process consists of the hydrogenation of carbon monoxide over such catalysts and under such conditions that the reaction product consists of hydrocarbons of various chain lengths and degrees of saturation. Depending upon a number of factors, mainly the ratio of hydrogen to carbon monoxide in the synthesis gas and the type of catalyst employed, one or the other of the following reactions predominates in the synthesis (although both almost always take place to some extent):

$$n\ CO + (2n + 1)\ H_2 \rightarrow C_nH_{2n+2} + n\ H_2O$$

$$n\ CO + 2n\ H_2 \rightarrow C_nH_{2n} + n\ H_2O$$

The first of these reactions (producing paraffinic hydrocarbons) is favored by relatively high ratios of hydrogen to carbon monoxide and catalysts of strong hydrogenating power. The second (producing olefinic hydrocarbons) is favored by low H/CO ratios and catalysts of less strong hydrogenating ability.

Gasoline made by the conventional Fischer-Tropsch process is notoriously low in octane number as a result of its straight-chain, largely paraffinic structure. The process developer, Franz Fischer, has stated (22) that optimum yield from single-pass atmospheric-pressure synthesis over cobalt catalyst was about 0.155 kg/m^3 (9.7 lb/1000 ft^3) of ideal synthesis gas. Of that total, gasoline fractions (C_5–200°C) accounted for 60% by weight. To maximize the octane-improving effect of the more olefinic volatile fractions, production Fischer-Tropsch gasoline was usually limited to a low final boiling point. Even so, the C_5–150°C fraction of the gasoline discussed here, which contained 45% olefins by volume, rated only 55 octane by the Motor method. Its susceptibility to upgrading by tetraethyllead was high, however, with moderate additions raising its antiknock quality 17 octane numbers.

To make more branched-chain hydrocarbons and thereby raise gasoline octane number, development of an "Iso-Synthesis" version of the Fischer-Tropsch process was begun by the Kaiser Wilhelm Institute of Coal Research in 1941. As described by the institute's acting director, Helmut Pichler, upon interrogation by the U.S. Technical Oil Mission (23), it combined three reactions in a single stage: (*1*) formation of higher alcohols (mainly isobutyl alcohol), (*2*) dehydration of the alcohols to olefins, and (*3*) hydrogenation of the olefins to the corresponding isoparaffins. Process conditions and catalysts are described in ref. 21. Typically, the butane fraction accounted for 60–70% of gasoline-range product and contained 90% isobutane. The pentane fraction, which accounted for 20–30% of total gasoline-range product, was 96–98% isopentane.

Efforts to develop a process for producing aromatics via a modified Fischer-Tropsch process also were undertaken by the Kaiser Wilhelm Institute. The same on-the-spot interrogation of Pichler (23) supplied what details are known. A 1:1 ratio of hydrogen to carbon monoxide was used. The yield of liquid hydrocarbons was very small. However, the liquid product contained about 50% aromatics (toluene and xylenes), the remainder consisting largely of cycloparaffins. Most of the synthesis gas was converted to methane and carbon. The Germans did not consider this synthesis very promising (24).

Theoretically at least, any carbonaceous material that can be decomposed to hydrogen and carbon monoxide can be used to make Fischer-Tropsch synthesis gas. Thus, the process offers a route to liquid fuels to nations that lack tar sands, oil shale, coal, or other fossil-fuel resources. Wood, for example, could be converted to usable Fischer-Tropsch synthesis gas if national priorities so dictated. To date, however, only coal and natural gas have appeared of commercial interest as raw materials for F-T synthesis gas. Processes by which they can be converted to suitable synthesis gas, and the H/CO ratio adjusted as desired, are described in ref. 21. As might be expected, decreasing the H/CO ratio in the synthesis gas results in a more olefinic product. A high H/CO ratio produces a saturated (paraffinic) product and favors high formation of methane. The amount of olefins in the product (conducive to higher octane rating) has been found to be inversely proportional to the hydrogen content of the synthesis gas. With conventional catalysts, a ratio of 2 H_2 to 1 CO seems to be about optimum for maximum yield of hydrocarbons (see also Hydrogen; Carbon monoxide).

All of the preceding comments about the Fischer-Tropsch process have been based on its World War II application or earlier research and development in Germany, Japan, and elsewhere. Currently, the South Africans have not publicly divulged much of what they have learned during the quarter century they have been operating their version of the process. In 1978, Sasol I (South African Coal, Oil and Gas Corp.'s first oil-from-coal plant) reportedly produced about 1.4 GL (8.5 million bbl) of synthetic crude oil (25). About the time of the Arab oil embargo in 1974, the government began funding construction of Sasol II, a $2.9 × 10^9 plant ten times as large as Sasol I. That plant is scheduled to go on stream in 1982. In 1979, the government announced plans for Sasol III, an additional $4 × 10^9 of Fischer-Tropsch facilities intended to double the complex capacity by the late 1980s. Sasol I and Sasol II reportedly should supply 13.5 GL/yr (230,000 bbl/d), 60% of the nation's oil needs, by 1982 (25) (see Fuels, synthetic—liquid fuels).

Mobil's Methanol-to-Gasoline Process. In 1975, Mobil Oil Corp. announced development of a process for making gasoline from methanol or other oxygenated hydrocarbons. At that time, it was believed to involve passing oxygenated hydrocarbons over a proprietary zeolite catalyst (see Molecular sieves) to form a product rich in aromatics that had a clear (unleaded) octane rating of 95 (26). Gasoline so produced was predicted to cost about two and a half times the cost of the methanol plus 1¢ per liter. With methanol then selling for 9.5–10.8¢/L (36–41¢/gal), that would make gasoline produced in this way cost 28–31¢/L ($0.94–1.07/gal). Since methanol can be produced from coal or natural gas, this meant that gasoline could be produced from gas or coal at those prices.

A 1976 report by Mobil on this work (27) revealed that the yield of hydrocarbons is practically 100% based on methanol, with about 25% being gaseous, ie, about 3% methane and ethane, 6% C_3s, and 16% C_4s. The other 75% consists of hydrocarbon liquids almost all of which were below C_{12}, the end point of conventional gasoline. The structure of the particular zeolite catalyst used prevents synthesis of molecules above gasoline range. In formulating a gasoline of desired volatility characteristics, addition of part of the butane produced to the 75% of the yield consisting of C_5-plus liquids raises total gasoline yield to 80%. "Furthermore, among the remaining C_3 and C_4 gases, olefins and isobutane are in proper balance for conversion by alkylation to additional gasoline of high octane quality," Mobil's vice president of R&D reported (27). "By introducing this step, we can build our yield to nearly 90% gasoline. . . . Variations in yield can be achieved by varying both operating conditions and catalyst properties. However, in all cases, the gasoline octane number is high—in the range of 90–100 unleaded."

More recent reports from Mobil's pilot plant indicate that yields from 636 L (4 bbl) of methanol a day averaged 238 L (1.5 bbl) of gasoline, 397 L (2.5 bbl) of water, and the rest gases that were not being utilized to increase liquid yield. Conversion cost in a commercial-scale plant was estimated at 1.3–2.6¢/L (5–10 ¢/gal) of gasoline produced. Gasoline made from coal via the process could not compete at the time with gasoline from crude oil, but as the supply of crude oil dwindles and its price rises, the process is predicted to become a major source of fuels (28).

Indications that this time may be at hand began appearing in mid-1979. The U.S. Department of Energy's Oak Ridge National Laboratory told the DOE's Fifth Underground Coal Conversion Symposium that synthetic gasoline selling for 21–34¢/L ($0.80–1.30/gal) before taxes and dealer costs may be possible in the 1980s by *in situ* coal gasification, methanol synthesis from the gas, and conversion of the methanol by the Mobil process (29). A 16,000-L/d (100-bbl/d) pilot plant in the Federal Republic of Germany reportedly is planned (30), the site being chosen because of German experience in producing methanol from coal. Badger Energy Inc. reportedly told the DOE that a large plant producing 92-octane unleaded gasoline from coal at a plant-gate price of 28.8¢/L ($1.09/gal) could be in full operation as early as 1989 (31). It would convert 57,000 t/d of southern Appalachian coal to methanol, then to 26.6×10^6 L/d (167,500 bbl/d) of gasoline using the Mobil process. By-products would include 2.3×10^6 L/d (14,500 bbl/d) of isobutane and 600 t/d of sulfur (31). Meanwhile, Mobil Oil New Zealand Ltd. reportedly has proposed a joint venture with the New Zealand government to convert surplus natural gas from the prolific Maui field into gasoline by the process. Estimates are that more than half of New Zealand's gasoline requirement could be so met by 1984 at a cost no greater than that for present imports (32).

Coal Liquefaction. Direct, one-step conversion of coal to high-octane gasoline is the goal of a process that Conoco Coal Development Co. and Shell Development Co. are jointly developing in a $\$2.5 \times 10^6$ pilot plant at the DOE, Library, Pa., research center (33). Basically, the process is catalytic hydrogenation conducted in a molten-salt reactor. Zinc chloride catalyst, powdered coal, and hydrogen are fed into a molten-salt reactor at a temperature of 340–440°C and a pressure of 10.0–30.0 MPa (1450–4350 psi). The ratio of coal to catalyst is about 1:1, and the pilot plant is sized to process 0.9 t/d. Early bench tests on coal extracts showed about 78% conversion in one hour to distillate boiling below 400°C, with 90% of the distillate boiling in the gasoline range. Besides being low in sulfur and nitrogen, the naphtha product from molten-salt hydrogenation has a high ratio of isobutane to normal butane (between 6:1 and 10:1), which contributes to high octane rating. Presumably, the isomer formation stems from zinc chloride's ability, as a Lewis acid, to promote formation of tertiary carbonium ions as reaction intermediates. Without further processing, the gasoline-range product reportedly rates about 90 Research octane number clear.

The putative economics of the molten zinc chloride process are encouraging. For a plant processing 28,000 t/d of Montana subbituminous coal to make 8.4 ML/d (53,000 bbl/d) of gasoline, the capital cost has been estimated at $\$1.9 \times 10^9$ in January, 1979, dollars. With coal charged at $7.39/t and a 10% discounted cash flow on 100% equity financing, the cost of 90–92 RON unleaded gasoline is 20.1¢/L (76¢/U.S. gal) (34).

Oil Shale and Tar Sands. Two major potential sources of gasoline, oil shale (qv) and tar sands (qv), have not yet been exploited to any appreciable extent. In the case of tar sands, extremely viscous bitumen must be removed from the surface of sand grains, to which it sticks tenaciously. In the case of oil shale, the solid kerogen content of the shale must be thermally cracked into liquid hydrocarbons. In both cases, the cost of recovery has, until recently, been too great for oil shale and tar sands to compete with petroleum, even when it must be imported. Now, however, these fossil-fuel resources look better with each new OPEC increase in the price of crude oil.

Regardless of how unrefined oil is obtained from oil shale or tar sands (by mining and retorting, by extraction processes, *in situ* methods, etc), the resulting crude can be processed into satisfactory gasoline by conventional refining methods. The surface mining operation conducted by Great Canadian Oil Sands Co. in the Athabasca tar-sands field employs delayed coking and hydrogenation to obtain stable distillates from the high molecular weight polymeric material, largely asphaltenes, present in the recovered crude. Syncrude Canada Ltd. employs fluid coking and hydrogenation. The processes crack the asphaltenes and help remove oxygen, nitrogen, and sulfur that could interfere in refinery processing (35).

Judging from estimated costs of crude recovery, tar sands promise to be more economical sources of gasoline than oil shales. The University of Utah has estimated costs of 6–9¢/L ($10–14/bbl) for recovery of crude from Uintah Basin tar sands by hot-water extraction or a thermal flashing method (36–37). British Petroleum Co. has estimated U.S. shale–oil costs at 12.5–16¢/L ($20–25/bbl), with an added cost of about 15¢/L ($2.30/bbl) if environmental considerations necessitate *in situ* recovery methods (35).

Gasohol and Other Alcohol Fuels. Blending of ethanol into gasoline has been tried many times in many countries, either to utilize surplus agricultural products or to create a market for expanded production. Alcohol use in motor fuel reached its height between World Wars I and II because of overdevelopment of alcohol production for munitions in World War I, rapid expansion of the European beet-sugar industry, and declines in the consumption of drinking alcohol.

During this period, alcohol–gasoline blends were marketed (usually only locally or seasonally) in 11 European countries and about 12 non-European areas. In 15 of those 23 countries, alcohol blending was compulsory, either nationwide or in certain areas. In the others, it was encouraged by tax incentives. In all cases, the cost in subsidies and lost taxes was greater at that time than the economic benefit from increased crop production (38). Therefore, most nations gave up alcohol blending after World War II. A 1952 British Institute of Petroleum study found that alcohol blending still was compulsory in three European countries: France, Switzerland, and the Irish Republic. Use of alcohol blending was appreciable in only three other countries: Brazil, Cuba, and Sweden. Of these, alcohol blending was compulsory in the two cane sugar-producing countries: Brazil and Cuba. Sweden repealed its compulsory-use law in 1947, but retained the tax incentive, which was enough to encourage paper producers to convert their sulfite wood wastes to alcohol.

The average penalty in subsidies and lost taxes for alcohol blending was $184 per metric ton of alcohol used (38). Calculations based on total tons used and the percent alcohol in the fuel blend show that, on overall average, the fuel blends contained 5.7% alcohol. Since the alcohol cost figures are based on excess cost above that of the gasoline replaced, each metric ton of gasoline containing alcohol bore an average of $10.50 penalty for alcohol blending. At about 160 L/t (8.5 bbl/t) of gasoline, the penalty averaged 0.8¢/L ($1.24/bbl).

Alcohol vs Hydrocarbons as Fuel. Comparing the properties of ethyl alcohol with those of a gasoline-range hydrocarbon (Table 5) suggests some of the advantages and disadvantages of alcohol. First, on a weight-to-weight basis, ethyl alcohols has only about two thirds the heating value of gasoline-range hydrocarbons, all of which have a heating value of about 46,500 J/g (21,100 Btu/lb), compared with alcohol's 29,800 J/g (13,500 Btu/lb). However, because ethyl alcohol's specific gravity is higher than that of most full-boiling gasolines (0.79 vs 0.70), the difference in heat content per unit

Table 5. Ethyl Alcohol vs Hydrocarbons[a]

	Ethyl alcohol	Isooctane
heating value, J/g[b]	29,800	46,500
heating value, MJ/cm^3 [c]	28,430	33,445
specific gravity	0.794	0.702
stoichiometric A/F ratio[d]	9	15
latent heat of evaporation, J/g[d]	840	328
Research octane number	106	100
Motor octane number	89	100

[a] Ref. 39.
[b] To convert J/g to Btu/lb, multiply by 0.430.
[c] To convert J to cal, divide by 4.184.
[d] A/F = air–fuel ratio.

volume is not as great as the difference per unit weight. Compared with isooctane, one of the lighter gasoline hydrocarbons, ethyl alcohol's heat-content deficiency is only about 27% on a volume basis. However, alcohol's deficit becomes greater when it is compared with full-boiling gasolines, especially those containing the relatively dense aromatic hydrocarbons present in catalytic reformates.

Because of ethyl alcohol's lower stoichiometric air–fuel ratio and its higher latent heat of evaporation, its use as engine fuel has a supercharging effect, increasing the power output obtainable from an engine of a given size. This effect and its high octane number explain its use in racing cars. However, this is of little or no advantage in passenger cars, and it is more than offset by higher fuel consumption with alcohol.

Adding a light alcohol (methanol, ethanol, or isopropanol) to gasoline increases its Reid vapor pressure and distorts or flattens the front half of the distillation curve. This may cause hard starting and vapor lock in hot weather. The distorted distillation curve and higher heats of vaporization may also have a combined deleterious effect on warmup in cold weather (40).

Alcohol's solubility in water creates another problem: possible separation of an alcohol–gasoline blend into an aqueous alcohol phase and a gasoline–hydrocarbon phase (ie, "splitting" of the blend). This not only means that the alcohol blended must be anhydrous, but also that great care must be taken to see that all storage, transport, and service-station tankage is completely water-free, which would be a major undertaking in most cases. A small amount of water is enough to cause phase separation, even in blends containing as much as 25% alcohol.

The best precautions possible at the refinery, in transit, and at service stations cannot prevent the possibility of water-phase separation in automobile fuel tanks. Such separation can cause corrosion, rough engine operation, starting problems, and fuel-line plugging.

Before antiknock compounds and modern refinery processing, ethyl alcohol's high octane rating made it attractive to increase fuel antiknock quality. In combination with the permissible increase in engine compression ratio, the increase in engine efficiency partially offsets the loss in fuel economy owing to ethyl alcohol's low heat content. However, today's gasoline base stocks and blending values are high enough, even unleaded, that alcohol addition cannot improve them appreciably.

As might be expected from its own octane ratings, 106 RON and 89 MON, ethanol has more effect on Research rating than on Motor and road ratings. Its effect is greatest with low-octane straightrun gasoline fractions and becomes less and less as base-stock antiknock quality increases.

Because unleaded gasoline is required by U.S. cars having catalytic exhaust-emission controls, the question arose whether ethanol could restore the antiknock quality lost by lead removal. An industrywide study of 1970 premium and regular gasolines showed that they exhibited almost none of the response to ethanol exhibited by the lower-quality gasolines of 20 years earlier.

In appraising alcohol blending, past and present, the July 1, 1979 issue of *International Petroleum Times* recalled England's experiment with Cleveland Discol (35). When introduced years ago, it contained 5–10% ethanol. By the time it was discontinued in 1976, it contained only 1% alcohol, and the ethanol had been changed to isobutyl alcohol (which illustrates ethanol's relative octane ineffectiveness in gasolines produced by modern refining processes). National circumstances may alter cases, however. When produced from waste molasses and bagasse from cane sugar refining, the alcohol blended into Brazilian gasohol in 1979 costs nearly twice as much as the gasoline. "But, in the context of the Brazilian economy, the substitution makes sense," the British study concluded.

Because of its huge demand for motor fuel, the United States poses an interesting challenge for gasohol proponents. Gasoline demand exceeds 380 GL (10^{11} gal) annually. Meeting that demand with gasohol containing 10% ethanol would require more than 38×10^9 L (10^{10} gal) of ethanol every year. This is roughly two and a half times current world output from all sources, including synthetic industrial alcohol (41).

Diesel Fuel

As a fuel for internal-combustion engines, diesel fuel ranks second only to gasoline. In 1977, total U.S. demand for diesel fuel was about 111 GL (30×10^9 gal) compared with about 400 GL (105×10^9 gal) for gasoline. Of the diesel fuel consumed, use in heavy-duty trucks, buses, tractors, and heavy construction equipment accounted for ca 46%, use in railroad locomotives for 15%, use in ships for 5%, industrial use for 15%, military use for 3%, oil-company use for another 3%, and miscellaneous uses for the remainder (42–43). Judged on the basis of the heavy work it accomplishes, diesel fuel's importance in the engine-fuel hierarchy is greater than the diesel fuel/gasoline consumption ratio indicates. Moreover, that ratio is likely to grow considerably, largely as the result of government-mandated standards for new-car fuel economy—standards that car makers say gasoline engines can attain only in subcompact-size vehicles. Auto industry executives believe that full-size cars can meet this standard only by use of diesel engines (44). Because diesel-powered cars of comparable size deliver considerably better fuel economy, knowledgeable prognosticators expect the U.S. diesel-powered car population to reach 13 million by 1985. By that time, diesel passenger cars (of negligible impact in the 1970s) will probably account for about 5% of all the motor fuel consumed in the United States, and more than 40,000 U.S. service stations will have diesel-fuel pumps (45), compared with about 11,000 such stations in 1977 and 5000 in 1974 (46).

Combustion in Diesel Engines. Unlike the spark-ignited gasoline engine, the diesel engine employs compression ignition. Through action of the pistons, air alone is drawn into the cylinders and compressed. Then, toward the end of the compression stroke, the fuel is injected into the compressed air and ignited by the heat of compression.

During injection, the fuel must be very finely broken up (atomized) and intimately mixed with the hot compressed air so that the maximum number of ignition points are created throughout the charge, thereby providing early and uniform ignition. Contrary to what might be expected, ignition does not occur the instant the fuel particles contact the hot compressed air. The length of the ignition delay depends largely on the hydrocarbon composition of the fuel; it is shorter for paraffinic fuels than for olefinic, cycloparaffinic, and aromatic fuels.

Long ignition delays result in excessively rapid pressure rise once combustion does occur and, therefore, cause rough engine operation. For many years, this sudden increase in pressure was believed to result from accumulation of fuel in the combustion chamber, but recent evidence indicates that a long delay allows time for certain chemical reactions to take place in the charge prior to its ignition. Presumably, the products of those reactions burn exceedingly rapidly once combustion begins, causing the undesirably rapid pressure rise. The role of cetane number in measuring a fuel's ignition-delay characteristics is described below; the role of additives in improving them will be discussed later.

Requirements for Good Diesel Fuel. ***Cetane Number.*** In addition to causing rough engine operation, long ignition delay causes misfiring and uneven or incomplete combustion in a cold engine at low temperatures, the result being loss of power and smoke in the exhaust gases. The procedure used in rating the ignition quality of diesel fuel (ASTM D 613) is similar to that used to measure gasoline octane number. Cetane (n-hexadecane), which has very high ignition quality (ie, low ignition delay), represents 100 on the cetane-number scale. Originally, α-methylnaphthalene was used to represent zero on the scale. More recently, 2,2,4,4,6,8,8-heptamethylnonane (HMN), which has a cetane number of 15, was made the low-quality reference fuel. It can be produced in very high purity by dimerization of diisobutylene (catalyzed by sulfuric and acetic acids) and subsequent hydrogenation of the olefinic mixture, the three possible isomers all hydrogenating to the 2,2,4,4,6,8,8-paraffin (see Alkylation).

Blends of the two hydrocarbons (cetane and HMN) represent intermediate ignition qualities, their cetane numbers being calculated, to the nearest whole number, from the percentages of cetane and HMN in the blend. The equation is as follows: cetane number = % cetane + 0.15 (% HMN). Thus, a reference blend containing 35% cetane and 65% HMN has a cetane number (CN) of 45.

Like laboratory octane-number determinations, cetane-number ratings are made in a single-cylinder engine having a variable compression ratio. The engine is run at 900 rpm, and injection is timed to begin at 13 crank-angle degrees before top center. Then, with the engine operating on the test fuel, its compression ratio is adjusted until combustion begins exactly at top center, thus producing an ignition delay of 13 crank-angle degrees, or 2.4 ms at 900 rpm. This procedure is repeated with reference-fuel blends until the test fuel is bracketed between two reference blends that differ by not more than 5 CN. The cetane number of the test fuel is then determined by interpolation between the two bracketing values.

As implied earlier, the cetane number of a fuel depends primarily on its hydrocarbon composition. In general, the paraffins have high cetane numbers, the aromatics have low cetane numbers, and the olefins and cycloparaffins fall in between. The relationship between the cetane number of a fuel and its performance in a diesel engine should not be confused with that of a gasoline's octane number and its performance in a spark-ignition engine. Raising octane number permits an increase in the compression ratio to increase power and improve fuel economy, and the maximum octane-number requirement of an engine usually occurs at full load. In the case of the diesel engine, however, the desirable level of cetane number is that which gives good ignition at light loads and low atmospheric temperature.

In general, high-cetane fuels eliminate engine roughness or diesel knock, permit an engine to be started at lower air temperatures, provide faster engine warmup without misfiring or producing white smoke, and reduce the rate of formation of engine varnish and carbon deposits. However, fuels having cetane numbers above the commercially available range can cause incomplete combustion and exhaust smoke if their ignition delay is too brief to allow proper mixing of fuel and air in the combustion space. Diesel fuels marketed in the U.S. have cetane numbers ranging from about 35 to 65. Most manufacturers of truck and bus engines and of diesel locomotives specify use of fuels of 45 CN and above. Tractor manufacturers and makers of heavy construction equipment vary in their specifications, but most call for 40 CN or better.

Volatility. Although some reasonably volatile light fractions are needed in diesel fuel to provide easy engine starting and more complete combustion under a variety of operating conditions, the light fractions generally are low in cetane number. Moreover, they have lower heat content per unit volume than the heavier fractions. On the other hand, a preponderance of heavy fractions might improve fuel economy, but too much could cause excessive deposit formation in the engine and difficult starting. Volatility characteristics also influence the amount and kind of exhaust smoke and odor. Therefore, desirable diesel-fuel volatility characteristics are obtained by careful balancing of light and heavy components available from distillation and cracking at the refinery. The temperatures at which 10, 50, and 90% of a fuel evaporate in the ASTM D 86 distillation test are important control points in achieving balanced diesel-fuel volatility. In general, most diesel fuels have an initial distillation temperature of around 160°C and a 90% point of 290–360°C, the latter being determined by the fuel grade and intended use.

Pour Point. Obviously, a diesel fuel must be able to flow at the lowest atmospheric temperature that may be encountered in its use. The temperature at which a fuel ceases to flow is called its pour point. As fuel temperature decreases toward this point, the fuel becomes progressively more difficult to pump through fuel-supply lines, filters, and the injection system. The pour point of a fuel is closely related to its component hydrocarbons: cycloparaffins have low pour points, but also have relatively low cetane numbers; conversely, paraffins have high pour points, but are desirable because of their high cetane numbers. Because low pour points often can be obtained only with lower cetane numbers or higher fuel volatility, the pour-point specification should not be set lower than necessary for the fuel's use under the climatic conditions anticipated. Most diesel-engine manufacturers recommend that fuels have a pour point 6°C below the lowest atmospheric temperature at which the engine will be operated.

Cloud Point. When cooled to a temperature somewhat above the pour point, diesel fuel becomes cloudy, forming wax crystals and other solid materials that can clog fuel filters and supply lines. The temperature at which these solids begin to form is the cloud point. The cloud-point specification may actually be more important than that of pour point. Both cloud point and pour point are determined by ASTM Method D 97. Although most diesel fuels have cloud points 4–6°C above their pour points, some have them as much as 8–11°C higher. In general, engine manufacturers recommend that cloud point be below the temperature of use and not more than 6°C above the pour point. Development of dispersant-type additives that keep wax crystals in a form that will flow through fuel-system filters is tending to diminish the significance of the cloud-point test. In a number of oil companies, work is progressing on "flowability" tests that give a better indication of a distillate fuel's behavior under cold conditions.

Viscosity. For optimum performance of diesel-engine injector pumps, the fuel should have the proper "body" or viscosity (see Rheological measurements). Too low viscosity may make necessary more frequent maintenance of the injection system; too high viscosity may cause excessively high pressures in the injection system. Moreover, viscosity affects the degree of atomization of the fuel upon injection into the cylinders. Two methods are used to measure diesel-fuel viscosity: ASTM D 88, which measures viscosity directly in Saybolt universal seconds, and ASTM D 445, which measures kinematic viscosity (viscosity divided by density) and is reported in mm^2/s (= centistokes). For use in high-speed diesel engines subject to frequent and fairly wide variations in load and speed, the recommended range of kinematic viscosity is 1.4 to 2.5 mm^2/s at 38°C. For similar engines in service involving relatively heavy engine load and uniform speed, kinematic viscosities in the range of 2.0 to 5.8 mm^2/s at 38°C are recommended. Grade 4-D diesel fuel, which consists of more viscous distillates and their blends with residual fuel oil and is intended for use in low-to-medium-speed engines operating under sustained loads at substantially constant speed, may have kinematic viscosity between 5.8 and 26.4 mm^2/s at 38°C.

Gravity. The greater the density of a diesel fuel, the greater is its heat content per unit volume. Because diesel fuel is purchased on a volume basis, density is stipulated in purchasing specifications and often is measured during delivery inspections. In the United States, the common measurement of density is API gravity (ASTM Method D 287). The relationship between API gravity and specific gravity is:

$$°\text{API at } 15.6°\text{C} = (141.5/\text{sp gr at } 15.6°\text{C}) - 131.5$$

Thus "degrees API" is inversely proportional to density; 0°API is equal to a density of 1.074 kg/L (8.962 lb/gal) and a specific gravity of 1.0760, whereas 100°API equals a density of 0.609 kg/L (5.086 lb/gal) and a specific gravity of 0.6112. Because the specific gravity of a diesel fuel depends largely on the refiner's choice of blending stocks to attain desired volatility characteristics and cetane number, relatively few engine manufacturers specify the gravity of fuel to be used in their engines. Of those who do, most set a range of about 30–40°API (or a specific gravity range of 0.8762 max to 0.8251 min).

Flash Point. The flash point of a fuel indicates the temperature below which it can be handled without danger of fire. This is the temperature to which the fuel must be heated to create sufficient fuel vapors above the surface of the liquid fuel for ignition to occur in the presence of an open flame. The flash point of diesel fuel is determined by ASTM Method D 93, using a closed Pensky-Martens tester in which the fuel is heated at a rate of 5.6°C per minute. A test flame is introduced into the vapor space at 30-second intervals, and the fuel temperature reached when the flame ignites the vapors is called the flash point. Specifications on flash point vary with the grade of diesel fuel, 38°C being as low as ever specified. Many states and insurance companies have mandatory flash-point requirements, which must be taken into account when setting the specification for a given application.

Carbon Residue. The tendency of a diesel fuel to form carbon deposits in an engine can be roughly predicted by one or the other of two carbon-residue tests: the Ramsbottom coking method (ASTM D 524) or the Conradson carbon test (ASTM D 189). Basically, both methods determine the amount of carbon residue remaining after evaporation and chemical decomposition of the fuel at elevated temperature for a specified length of time. The Ramsbottom method involves heating the last 10% residue of the ASTM distillation at 549°C for 20 min; the result is reported as percent carbon residue on 10% residuum. For use in high-speed diesel engines operating over a range of loads and speeds, specifications usually call for no more than 0.15% Ramsbottom carbon residue. Because ignition-control additives interfere with the carbon-residue tests, without affecting actual deposit formation in an engine, these laboratory tests should be made on the base fuel before addition of the ignition-control agent.

Sulfur Content. During intermittent engine operation or operation at low temperatures, when condensation of moisture is appreciable, sulfur compounds in the fuel may cause cold corrosion and increased engine wear. Specifications on the sulfur content of diesel fuel usually are based on a bomb test (ASTM D 129). For high-speed engines operating over a range of speeds and loads, 0.50% sulfur in the fuel generally is considered maximum, whereas 1.0% is the generally accepted maximum for use in similar engines operated at high loads and fairly uniform speeds. The copper-strip corrosion test (ASTM D 130) described above also is applied to diesel fuels, ratings of 2 to 3 generally being considered acceptable.

Ash Content. Because the fuel injectors of diesel engines are designed to very close tolerances, they are sensitive to any abrasive material in the fuel. Therefore, the maximum permissible ash content of the fuel usually is specified. This is measured by ASTM D 482, which consists of burning a small sample of the fuel in a weighed container until all the carbonaceous material has been consumed. The amount of unburned residue or ash should not exceed 0.01 to 0.02% of the weight of the sample in high-speed engines, but it may be as high as 0.10% for low-speed and medium-speed engines operated under sustained loads at nearly constant speed.

Grades of Diesel Fuel. Throughout the preceding discussion of diesel-fuel requirements, mention has repeatedly been made of the fact that these requirements vary with engine design and type of service. To facilitate choice of an economical fuel that meets given engine requirements, the ASTM has established a classification of diesel fuels by grades to reflect their suitability for engine types and operating conditions. This classification (ASTM D 975) defines the grades as follows:

Grade No. 1-D includes volatile fuel oils from kerosene to the intermediate distillates. Fuels of this grade are recommended for high-speed engines whose operating conditions involve frequent and fairly wide variations in engine load and speed. Such fuel is required for use at abnormally low temperatures.

Grade No. 2-D includes distillate gas oils of lower volatility. This grade is suitable for use in high-speed engines not subject to wide variations in load and speed, ie, those operated under relatively high loads and uniform speeds. Grade No. 2-D fuels can be used in engines not requiring fuels having the greater volatility and other properties specified for grade No. 1-D fuels.

Grade No. 4-D consists of more viscous distillates and their blends with residual fuel oils. Normally lower in cost than the other two grades, this grade is usually satisfactory only for low-speed and medium-speed engines operated under sustained load at nearly constant speed.

The limiting requirements for these three grades of diesel fuel, as established by the ASTM, are shown in Table 6. There is no longer a No. 3-D specification because it somewhat overlapped the 2-D and 4-D specifications and was not commercially available. Because of the wide variation in diesel-engine design and application, operators of diesel-powered equipment should follow the manufacturer's recommendations in selecting the proper fuel for a given combination of engine design and operating conditions.

Diesel Fuel Manufacture. In order of decreasing ignition quality, the hydrocarbon types rank in the following order: normal paraffins, olefins, cycloparaffins, branched paraffins, aromatics—just the reverse of their ranking for gasoline antiknock quality. Because straightrun distillates contain the greatest amounts of normal paraffins and cycloparaffins and the least amounts of branched paraffins and aromatics, they are the preferred stocks for diesel-fuel blending. Cracked fractions, which are relatively rich in branched paraffins and aromatics, are less desirable from the standpoint of ignition quality. However, they do offer some advantages: because of their greater density, they offer more heat energy per liter, and their pour and cloud points generally are lower than those of straightrun distillates. Therefore, high-quality diesel fuels usually are made by blending a high proportion of straightrun distillates with lesser amounts of cracked fractions. The amount of cracked material desirable in diesel fuel depends largely on the cetane-number specification that must be met. In the United States, where a high level of cracking is necessary to meet gasoline demand, the large supply of cracked fractions and the relatively small supply of straightrun distillates make substantial use of cracked material in diesel fuel economically desirable. This has been made possible without sacrifice in cetane number or storage stability through commercial availability of an ignition-improving chemical additive (see below) and through use of hydrogenation to convert unstable unsaturated hydrocarbons to stable,

Table 6. ASTM Limiting Requirements for Diesel Fuel Oils (ASTM D 975)

Diesel fuel oil	Flash pt, °C, min	Pour pt, °C, max	Water and sediment, vol %, max	C residue on 10% residuum, %, max	Ash, wt %, max
No. 1-D	38 or legal	[a]	trace	0.15	0.01
No. 2-D	52 or legal	[a]	0.10	0.35	0.02
No. 4-D	54 or legal	[a]	0.50		0.10

Diesel fuel oil	90% distillation temp, °C		Kinematic viscosity at 38°C, mm^2/s (= cSt)		S, wt %, max	Cu strip corrosion, max	Cetane number, min
	Min	Max	Min	Max			
No. 1-D		288	1.4	2.5	0.50	No. 3	40
No. 2-D	282[b]	357	2.0[b]	5.8	1.0	No. 3	40
No. 4-D			5.8	26.4	2.0		30

[a] For cold-weather operation, the pour point should be specified 6°C below the ambient temperature at which the engine is to be operated except where fuel-oil heating facilities are provided.

[b] When pour point less than −18°C is specified, the minimum viscosity shall be 1.8 mm^2/s and the minimum 90% point shall be waived.

high-cetane ones. The boiling range of the fractions included in diesel fuel extends from a low of about 150°C to a high of about 370°C, depending on the grade of fuel being blended.

Diesel Ignition Improver. During the postwar period, U.S. demand for diesel fuel increased sixfold, and demand for other petroleum products little more than doubled. To meet this demand economically required some means of utilizing those distillates in the diesel-fuel range that lacked only ignition quality to make them suitable engine fuels. Research by Ethyl Corporation led to the discovery that primary amyl nitrates, $C_4H_9CH_2ONO_2$, not only are highly effective in increasing cetane number, but also, unlike most other reasonably effective materials, retain effectiveness during fuel storage and have no adverse effect on engine cleanliness, wear, smoke, and odor. The first commercial ignition improver resulting from this discovery (a mixture of nitrates of primary amyl alcohols) increased the cetane rating of typical diesel fuels an average of four cetane numbers when used in a concentration of 0.1% by volume and seven cetane numbers at a concentration of 0.2% by volume. Ethyl Corporation subsequently introduced ignition improvers based on *n*-hexyl nitrate (in 1966) and on octyl nitrate (in 1978). Use of these cetane improvers allows refiners to utilize more cracked fractions in diesel fuel, thereby producing large quantities of high-cetane diesel fuel economically. By allowing refiners to upgrade stored heating oils to diesel-fuel quality quickly and economically, either at the refinery or at bulk terminals, cetane improvers help them meet sudden local increases in demand for diesel fuel.

Stability Improvers. Like gasoline, diesel fuel may form gums during storage. Some types of diesel-fuel gum are soluble in the fuel and can be detected only as a residue after the fuel has been evaporated. These can cause buildup of gum and varnish on injection-system parts. Moreover, when two or more diesel-fuel stocks containing dissolved gum are blended, insoluble solid particles sometimes form. One explanation is that, although the gums are soluble in their original stocks, some of them at least are insoluble in the resulting blend. These precipitated gums and sludgelike particles can clog diesel-engine filters. The chemistry involved in formation of the sludgelike insoluble materials is not completely understood, but oxidation and thermal reactions appear to be involved (47). Ordinary antioxidants are ineffective in preventing this sludge formation and may even be deleterious. Therefore, the stabilizers usually employed in diesel fuels are peptizing agents or dispersants. Typical dosage is about 100 ppm by weight, roughly five times the amount of antioxidant used in gasoline.

Corrosion Inhibitors. The corrosion inhibitors (qv) used in diesel fuels are generically similar to those used in gasoline and, like them, produce their protective effect primarily by a physical surface action. The typical dosage is about 50 ppm by weight.

Multipurpose Additive. Since the early 1960s, Ethyl Corporation has marketed a multifunctional additive for diesel fuels (MPAD). A mixture of nonionic and anionic surface-active agents, MPAD has both detergent properties to prevent buildup of gum and varnish on critical injection-system parts and surface-active properties to protect injectors against rusting during idle periods and to prevent rust in fuel-handling systems. By performing these functions, it also maintains diesel exhaust smoke levels at those normal for a clean engine and reduces smoke levels from dirty engines. The smoke-control effect is the result of preventing injector malfunctioning, such as plunger sticking, plugging of the nozzle, and rusting, which would distort fuel-spray patterns, causing exhaust smoke and loss of power. Typically, this additive is used in a concentration of about 200 ppm by weight.

Conductivity Improvers. Like home-heating oil and other middle distillates, diesel fuel can build up a powerful charge of static electricity, especially during the pumping involved in loading operations. To increase the electrical conductivity of distillate fuels, thereby causing static to dissipate rather than build up, refiners may add small amounts of a polymeric amine salt or other conductivity improver to their diesel fuels (see also Polymers, conducting).

Broad-Spectrum Distillate Additives. Ashless (metal-free) additives recently developed for multifunctional use in distillate fuels offer broad-spectrum protection for diesel fuels. Complex proprietary mixtures of organic compounds are designed to disperse particulates, retard oxidation, deactivate trace copper, improve conductivity, provide antirust protection, and improve water tolerance.

Ethyl Alcohol. During winter in very cold areas, ethanol is sometimes added to diesel fuel in the field (not at the refinery) to prevent ice from plugging fuel lines and filters. Care must be taken, however, not to add too much because alcohol lowers diesel-fuel flash point and can increase corrosive wear on engine fuel pumps and injectors. The rule of thumb is never to use more than one pint of alcohol per 100 gallons (1.25 mL/L) of fuel (48).

Liquefied Petroleum Gases

Liquefied petroleum gases (LPG), which consist primarily of propane, butanes, and their mixtures, have long been of interest as engine fuels because of their relatively low cost and their high octane numbers (see Liquefied petroleum gas). In 1977, use as engine fuel accounted for about 4.5 GL (1.2×10^9 gal), or about 7% of the LPG sold in the United States. Use in LPG-powered farm tractors accounts for almost half of the LPG consumed as engine fuel. Another major engine use is in large stationary engines, such as those in pumping stations and power-generation plants. The necessity for special sources of fuel supply, ie, service stations equipped with LPG storage tanks and dispensing pumps, limits vehicle use of LPG largely to city buses, large truck fleets, and construction equipment.

To understand the requirements for a good LPG engine fuel, some knowledge of the history of LPG use in engines is desirable. For many years, the octane numbers of butanes and butane-propane mixtures were more than adequate for the engines in which they were used. However, as engine compression ratios and gasoline octane numbers were progressively raised, the antiknock quality of commercial LPG mixtures became first marginal and then inadequate for use in heavy-duty trucks. Fuels consisting primarily of propane still enjoyed a substantial octane-number advantage over available commercial gasolines, but improvement in gasoline antiknock quality and in the compression ratio of gasoline engines narrowed the differences in efficiency and fuel economy obtainable with gasoline and LPG engines. Finally, in the late 1950s, the differences became so small that LPG-engine manufacturers indicated reluctance to design higher-compression, more efficient engines unless users could be assured of an adequate supply of fuel of uniformly high octane number. Consequently, in 1962, the Natural Gas Processors Association offered propane HD-5, a new grade of propane-engine fuel, distinct from the LPG sold as commercial propane. The major change in specifications is a limitation of propylene content to 5%, whereas commercial-grade propane may contain much larger amounts of propylene. Of the commercial-grade propane obtained from natural gas, about 87% meets the NGPA limitation on propylene content. However, because commerical propane obtained from refinery gases may contain considerably more than 5% propylene, petroleum refiners favor an alternative specification for motor-grade propane that would stipulate a minimum octane number of 94–95 as measured by a modified Motor method.

Production of Propane. Natural gas, the main source of propane, varies in composition, but a typical analysis might show 80–90% by volume of methane, 5–8% ethane, 4–6% propane, 1–3% butanes, and 1–5% pentanes and heavier hydrocarbons (see Gas, natural). Generally, propane and heavier hydrocarbons are separated from the methane and ethane by passing the natural gas at a pressure of about 3.45 MPa (500 psi) upward through an absorber tower, countercurrent to a flow of oil. The oil absorbs the propane and heavier hydrocarbons, while the methane–ethane mixture exits from the top of the tower as dry natural gas. The absorbent oil is then distilled to liberate the propane and heavier hydrocarbons, and the overhead product is subjected to a series of fractional distillations to recover and separate them. Variations in separation methods include omission of the oil-absorption step, its replacement by adsorption on charcoal, or low-temperature (refrigerated) fractionation. Because the boiling points of pure propane and pure propylene are only a few degrees apart (−42.5°C for propane and −47.0°C for propylene), separating substantially pure propane from propylene-rich refinery gases requires exacting and costly processing.

Upgrading LPG Antiknock Quality. The antiknock performance of commercial liquefied petroleum gases can be made to equal or exceed that of pure propane by addition of tetramethyllead (TML) antiknock compound (49). Under severe operating conditions in an engine that could be operated knock-free on 99% pure propane, the amounts of TML found to give satisfactory performance with 99% pure butane, 99% pure propylene, and high-propylene commercial LPG are shown in Table 7.

By modification of the ASTM knock-rating methods only to the extent necessary to rate gaseous fuels, Felt and Kerley (49) determined the amounts of TML necessary in various blends of the primary components of commercial LPG to make those blends at least equal to pure propane in both Research- and Motor-method ratings (49). The results are shown in Table 8.

This work has shown that the blending characteristics of propane with propylene, butane, or isobutane are not linear as regards octane number. It has also demonstrated that alkyllead antiknocks are particularly effective under severe engine operating conditions and that leaded LPG fuels are much more resistant to knock under some actual operating conditions than would be predicted from laboratory rating by either the modified Motor or the Research method.

Table 7. Performance of TML

Gaseous fuel	g Pb/L[a]
butane	0.26
propylene	0.14
commercial LPG[b]	0.13

[a] To convert to g Pb/gal, multiply by 3.8. Additions of TML or TEL are also often reported as mL/L. For TML, 1 mL/L = 0.405 g Pb/L. For TEL, 1 mL/L = 0.278 g Pb/L.

[b] Composed of 46% propane, 52.6% propylene, and traces of ethane and butane.

Table 8. TML Required in LPG to Equal Propane

Fuel composition	TML required, g Pb/L[a]	Octane number	
		Research	Motor
pure propane	0	112	96
80% propane, 20% butane	0.01	112	98
60% propane, 40% butane	0.08	112	101
40% propane, 60% butane	0.29	112	105
80% propane, 20% propylene	0.04	113	96
60% propane, 40% propylene	0.16	114	96
40% propane, 60% propylene	0.48	114	96

[a] To convert to g Pb/gal, multiply by 3.8.

LPG Locomotive Fuel. Because LPG is considerably cheaper than diesel fuel in certain sections of the United States, some U.S. railroads have considered converting diesel locomotives to operate on propane-rich LPG. Diesel-fuel shortages, possible allocations, and rising prices are new factors that could affect their decisions. At least one railroad has tested a spark-ignition propane locomotive converted from the diesel compression-ignition cycle. However, for maximum savings in fuel cost, a locomotive should be able to alternate between fuels at will, depending on which is cheaper in the area where the locomotive happens to be at the time. Indications are that dual-fuel operation (burning two fuels simultaneously) may prove to be the key to successful use of LPG in locomotives (50). In such operation, the primary fuel is LPG, but a small amount of diesel fuel is also injected into the engine to provide the compression ignition necessary for diesel-engine operation. The LPG is admitted to the engine via the air manifold or a special intake valve and is mixed with air either during the intake stroke or early in the compression stroke. Then, near the end of the compression stroke, a small quantity of diesel fuel (usually less than 10% of the total fuel consumed at full power) is added through the diesel-fuel injector. Compression ignition of the diesel-fuel charge ignites the LPG–air mixture, which then provides most of the engine's power.

For dual-fuel operation in diesel locomotives, the antiknock quality of the LPG fuel must be high enough for knock-free combustion at the compression ratio required for ignition of the diesel-fuel pilot charge. Even a good grade of LPG consisting mostly of propane is too low in octane number for knock-free operation of a locomotive engine at outputs near the full-load power it can develop on diesel fuel alone. Until recently, use of LPG as the primary fuel in diesel engines necessitated either reducing power output by 30–40% in order to avoid destructive knock or markedly lowering the engine's compression ratio and switching to ignition by spark plugs. The latter alternative was about as undesirable as the former, because it would reduce engine efficiency and prevent operation of the modified engine on diesel fuel when desired. However, addition of tetramethyllead to high-propane LPG at a concentration of only 0.26 mL/L (1 mL/gal) permitted knock-free operation of a small diesel engine at 100% of the full-load output obtainable on diesel fuel alone, as compared with only 67% of full-load diesel output when using the same LPG fuel without TML (51). Moreover, larger amounts of TML permitted dual-fuel operation at outputs above full-load diesel power.

The practicability of dual-fuel operations of diesel locomotives has since been shown (52). In dual-fuel operation of a two-cylinder test version of the diesel locomotive engine developed by the Electro-Motive Division of General Motors, addition of TML to high-propane LPG increased knock-limited power as much as 44%. The knock-limited power was not quite equal to rated full-power output on diesel fuel alone, but it could be made so by somewhat lowering engine compression ratio. When the compression ratio was reduced from 13.6:1 to 11.9:1, rated output was obtained with as little as 0.17 mL/L LPG (0.65 mL TML/gal). With 0.53 mL/L of TML (2 mL/gal), knock-limited power exceeded rated diesel power by about 20%. In addition to showing that leaded high-propane LPG can serve as an economical, practical fuel for locomotives, these tests have indicated its potential for dual-fuel operation of many industrial diesel engines (51–52).

Natural Gas

In many ways, natural gas is a highly desirable fuel for internal-combustion piston engines. On the basis of cost per joule of heat energy, it is often the least expensive engine fuel available in a given area, and its high antiknock quality permits design of highly efficient high-compression engines. Moreover, in the dry, sulfur-free form readily available from pipelines, it is an extremely clean fuel that burns cleanly in engines, producing very little engine deposits and little or no corrosion of engine parts. However, because natural gas cannot be stored in a vehicle's fuel tank, its use as engine fuel is limited to stationary installations, such as engines that power heavy-duty compressors or drive generators in central power stations (see Gas, natural).

At one time, very large piston engines were the primary power source for pumping natural gas through the United States' immense transcontinental pipelines. In the 1960s, U.S. installations of natural-gas engines totaled about 12.3×10^6 kW (16.5×10^6 hp). Large slow-speed engines used to power pipeline compressors accounted for about 42% of that total (53). Other large slow-speed engines driving electrical-power generators, municipal and industrial water pumps, and other heavy-duty stationary equipment accounted for about 24%. The remaining 34% of installed natural-gas engine power consisted of smaller higher-speed engines engaged in a variety of lighter-duty jobs, eg, powering smaller generator installations and driving pumps at oil wells and water wells. Today, gas turbines increasingly are being used for many of the applications in which gas-fueled piston engines formerly were the preferred source of power. A typical gas-fueled engine of the 1930s converted about 26% of its fuel's heat energy into useful power (54). A more modern engine of 15:1 compression ratio has an efficiency of ca 36%.

Engine-Fuel Properties of Natural Gas. Because the heating value of hydrogen is about four times that of carbon, the higher a hydrocarbon's hydrogen-to-carbon ratio, the greater its heating value. Thus as can be seen in Table 9, methane has the highest heating value of any hydrocarbon. It also has very high antiknock quality: 133 RON and 130 MON, as measured by the modified ASTM methods for rating gaseous fuels. Although the composition of natural gas varies somewhat with its source, the main constituent is methane. The "dry" natural gas supplied by pipelines ranges in heat content from ca 37 to 45 MJ/m^3 (1000–1200 Btu/ft^3). Because of its high methane content, it has excellent antiknock quality. For example, the natural gas whose composition is given in Table 9 rated 132 RON and 125 MON by the modified ASTM methods, not much below the antiknock quality of methane itself. With large interstate

Table 9. Composition and Properties of Typical Natural Gas

Components	Heating value (lower)[a] MJ/kg	Heating value (lower)[a] Btu/lb	Octane number[b] Research	Octane number[b] Motor	Vol % in gas[c]
methane	50.2	21,597	133	130	87.1
ethane	47.9	20,597	115	103	3.4
propane	46.6	20,015	110	99.6	1.0
n-butane	46.0	19,795	93.4	91.6	0.49
isobutane	46.0	19,795	100	98.4	0.03
pentanes	45.9	19.745			0.06
hexanes	45.2	19,432			0.09
heptanes	45.1	19,408			0.03
nitrogen	nil	nil			7.1
carbon dioxide	nil	nil			0.31
helium	nil	nil			0.35

[a] Ref. 55. Note: lower heating values do not include any of the heat that would be liberated by condensation of the water vapor formed during combustion of the fuel. Their use here is appropriate because internal-combustion engines reject this water as vapor in the hot exhaust gases.
[b] As determined by Felt and Kerley (49) by modified ASTM methods.
[c] Detroit area natural gas used by Felt and Kerley.

pipelines now delivering natural gas to every state in the continental United States, high-compression engines can be designed to take full advantage of this fuel's superior antiknock quality.

Dual-Fuel Operation on Natural Gas. A number of large stationary diesel engines are operated on the dual-fuel principle, using natural gas as the primary fuel and a small pilot charge of diesel fuel for compression ignition. Conversion of diesel engines to such an operation permits use of natural gas when it is available and rapid changeover to diesel fuel when home-heating loads make natural gas supplies scarce for industrial uses. Industrial consumers who contract for natural gas supplies on such an interruptible basis usually are granted a substantial reduction in the price they pay for gas when it is readily available. Thus, there is considerable economic incentive for conversion of large diesel installations to dual-fuel operation on natural gas.

The Total-Energy Concept

As central air conditioning has become almost as prevalent as central heating in public, commercial, and industrial buildings, interest has grown in systems that can supply all the energy requirements of a large building or complex of buildings, ie, total-energy systems. Until recently, however, residential total-energy systems have been considered impracticable, because of size, cost, noise, and exhaust emissions from the internal-combustion engines that would power them. Thus, application of the total-energy concept has been limited to relatively large power consumers: those needing generating capacity of 1000–20,000 kw. In addition, engine-powered residential systems did not attract much interest while utility rates for electricity were low and savings from home power-generation systems too small to justify their cost. Now, however, greater home use of electricity and heating fuels, combined with the sharply higher prices charged for them, has altered the economic picture. A recent market-research study (56) predicts an eight-fold increase in dollar sales of total-energy systems from 1975 to 1980, with home installations growing from nil to 40% of the dollar volume over that brief period. By 1990, the study predicts, total-energy systems will be a $3 billion-per-year business, and home installations will account for almost two-thirds of it.

Total-energy systems vary in components employed, but all have one thing in common: an engine to provide the power required for heating, cooling, lighting, etc. As early as the mid-1960s quite efficient systems had been devised. One of them (57) used a diesel-type engine fueled on then-cheap natural gas to drive a generator. Waste heat recovered from the engine's cooling system and exhaust gases supplied the energy for heating and air conditioning. Highly efficient, high-compression, spark-ignition engines designed for operation on natural gas also competed for the role of prime mover in total-energy systems, as did liquid-fueled piston engines and gas turbines. The cited 1977 survey (56) considers Rankine (steam-type) engines feasible for the residential market today and the Stirling engine "capable of providing an alternative within the next few years." Compared with gasoline or diesel engines, the Rankine and Stirling engines offer very low noise and pollution, the study notes, adding that the Stirling is potentially highly efficient as well. Employing external combustion, Rankine and Stirling engines could use a wide range of fuels, some of which might prove less expensive than gasoline or diesel fuel. Regardless of the type of engine and fuel used, chances are that a substantial number of private homes soon will be lighted, heated, and air-conditioned by total-energy systems deriving energy from motor fuels.

BIBLIOGRAPHY

"Gasoline" in *ECT* 1st ed., Vol. 7, pp. 113–117, by Mildred M. Hicks-Bruun, Consultant; "Gasoline and Other Motor Fuels" in *ECT* 2nd ed., Vol. 10, pp. 463–498, by John C. Lane, Ethyl Corporation.

1. *Oil Daily,* 1 (Apr. 5, 1978).
2. *Motor-Gasoline Survey, Winter 1977–78; Motor-Gasoline Survey, Summer 1978,* U.S. Department of Energy.
3. *CRC Handbook,* Coordinating Research Council, Inc., New York, 1946, pp. 153–155.
4. C. Walcutt and E. B. Rifkin, *Ind. Eng. Chem.* **43,** 2844 (1951).
5. G. A. Ball, "A Quick Look at Engine Combustion," *SAE National Fuels and Lubricants Meeting, Tulsa, Okla., Nov. 4–5, 1954.*
6. W. G. Lovell, "Combustion Problems in Internal Combustion Engines," *Fifth International Symposium on Combustion, University of Pittsburgh, Pittsburgh, Pa., Aug. 30–Sept. 3, 1954.*
7. E. B. Rifkin, *Proc. Am. Pet. Inst. Sec. III* **38,** 60 (1958).
8. W. G. Lovell, "Some Chemistry of Future High-Compression Engines," *James Clayton Lecture, Automobile Division, Institution of Mechanical Engineers, London, Eng., Mar. 1, 1960.*
9. T. C. Owtram, *J. Inst. Pet.* **38,** 820 (1952).
10. R. S. Chambers and co-workers, *Science* **206,** 789 (Nov. 16, 1979).
11. *Chem. Eng. News* **57,** 7 (Feb. 19, 1979).
12. *Eur. Chem. News,* 20 (Oct. 14, 1977).
13. *Chem. Eng.* **86,** 91 (May 21, 1979).
14. *Chem. Eng. News* **57,** 26, 36 (June 25, 1979).
15. *Oil Gas J.* **76,** 62 (June 26, 1978).
16. *Chem. Week,* 36 (June 28, 1978).
17. G. H. Unzelman and G. W. Michalski, "Octane Improvement Economics—Antiknocks and Alternatives," *National Petroleum Refiners Association Annual Meeting, San Antonio, Tex., Mar. 25–27, 1979.*
18. *Eur. Chem. News,* 28 (May 28, 1979).
19. *Chicago Tribune,* (Action Line), 1 (Apr. 7, 1979).
20. *Nat. Pet. News* **37,** R-861 (1945).
21. J. C. Lane, *Pet. Refiner* **25,** 8, 355, 423, 493, 587 (1946).

22. F. Fischer, *Berichte* **71A,** 56 (1938).
23. V. Haensel, *Tech. Oil Mission Rept. No. 9,* June 15, 1945; Office of Tech. Services Rept. PB 284, Dept. of Commerce, 1945.
24. D. R. Dewey, *Office of Tech. Services Rept. No. PB 890,* Dept. of Commerce, 1945.
25. *Bus. Week* (2594), 58 (July 16, 1979).
26. *Chem. Week,* 39 (June 18, 1975).
27. *Chemtech* **8,** 86 (Feb. 1976).
28. *Chem. Market. Rep.,* 5 (Jan. 16, 1978).
29. *Oil Daily,* 3 (July 9, 1979).
30. *Chem. Eng.* **86,** 69 (June 18, 1979).
31. *Oil Gas J.* **77,** 76 (July 2, 1979).
32. *Chem. Eng. News* **57,** 19 (July 9, 1979).
33. *Chem. Week,* 37 (Dec. 21, 1977).
34. *Chem. Week,* 32 (Aug. 8, 1979).
35. *Int. Pet. Times,* 11 (July 1, 1979).
36. *Platt's Oilgram News* **57,** 1 (June 22, 1979).
37. *Oil Daily,* 3 (July 12, 1979).
38. T. C. Owtram, *J. Inst. Pet.* **38**(346), 820 (1952).
39. *Alcohols—A Technical Assessment of Their Application as Fuels,* American Petroleum Institute, *Publ. No. 4261,* July 1976.
40. G. H. Unzelman, E. G. Forster, and A. M. Burns, "Are There Substitutes for Lead Antiknocks," *preprint No. 47-71 presented at American Petroleum Institute Division of Refining Meeting, San Francisco, Ca., May 14, 1971.*
41. *Chem. Market. Rep.* **215**(26), 11 (June 25, 1979).
42. *Diesel Fuel Oils, 1978,* BETC/PPS-78/5, U.S. Dept. of Energy Technical Information Center, Bartlesville, Okla., Nov. 1978.
43. *Nat. Pet. News* (Factbook Issue) **70**(6A), 80 (June 1978).
44. *Detroit News,* 4-D (May 19, 1977); *Wall Street J.,* 11 (May 20, 1977).
45. *Nat. Pet. News.* **70**(5), 14 (May 1978).
46. *Ward's Automotive News,* 394 (Dec. 12, 1977).
47. G. Edgar and H. A. Beatty, *Ind. Eng. Chem.* **48,** 1853 (1956).
48. *Diesel Fuel Questions and Answers,* American Petroleum Institute, 1400-6/77-30M.
49. A. E. Felt and R. V. Kerley, "Antiknock Compounds Applied to LP-Gas," *65th Annual Meeting, ASTM, New York, 1962.*
50. W. T. Steele, Jr., *Ethyl News* **30**(6), 14 (1963).
51. A. E. Felt and W. A. Steele, Jr., *SAE Trans.* **70,** 644 (1962).
52. A. V. Cabal, A. E. Felt, and H. A. Williams, Jr., "Performance of Leaded LP-Gas in an EMD-567 Dual Fuel Engine," *SAE preprint 763A presented at National Powerplant Meeting, Chicago, Ill., Oct. 14–17, 1963.*
53. R. S. Ridgway, *J. Eng. Power* **87**(2), 169 (1965).
54. R. I. Streeter and L. C. Lichty, *Internal Combustion Engines,* McGraw-Hill Book Co., New York, 1933.
55. A. P. Fraas, *Combustion Engines,* McGraw-Hill Book Co., New York, 1948.
56. *Nat. Pet. News* **69**(7), 42 (July 1977).
57. *Pet. Chem. Eng.* **36**(11), 64 (1964).

JOHN C. LANE
Ethyl Corporation

GASTROINTESTINAL AGENTS

The compounds currently used as gastrointestinal agents may be divided into several classes according to their general activity: agents for peptic ulcer therapy, laxatives, antidiarrheals, and antiemetics. Most of the compounds described in this article are discussed in the *United States Pharmacopeia* (USP–NF) (1) and approved for use by the FDA.

Antipeptic Ulcer Therapy

The primary therapeutic aim in antiulcer therapy is either prevention of the secretion of acid by the parietal cells of the stomach or neutralization of the acid before it comes in contact with the ulcerated area in the duodenum.

Established therapy for duodenal ulcers has been the utilization of antacids and anticholinergics. Antacids have been shown recently to be effective in healing duodenal ulcers, but only if given in very large and frequent doses. Anticholinergics can also be effective, but usually side effects are limiting (see Cholinesterase inhibitors).

The therapy for duodenal ulcers has changed dramatically with the introduction of cimetidine [*51481-61-9*] (1) (Tagamet) into the United States in 1977 (2). This compound, which apparently works through its histamine (H-2) blocking activity, is an effective antisecretory in humans and experimental animals and at this time has FDA approval for short-term treatment of duodenal ulcers (see Histamine and histamine antagonists). Approval for long-term maintenance therapy or for treatment of esophageal reflux and gastric ulcers is pending FDA approval.

The market for antiulcer agents is very large and is comprised of both prescription-only products, such as cimetidine and anticholinergics, and over-the-counter (OTC) products that are mainly antacids. The estimated market for cimetidine in 1980 is $\$5 \times 10^8$. The rest of the market for use as antiulcer therapy is more difficult to estimate but probably is close to $\$10^8$/yr.

The following contains information on cimetidine and a sampling of the more important commercially available antacids.

Cimetidine. ***Properties.*** Cimetidine (1) (Tagamet, SKF 92334), *N*-cyano-*N′*-methyl-*N″*-[2-[[5-methyl-1*H*-imidazol-4-yl)methyl]thio]ethyl]guanidine, has the properties shown in Table 1.

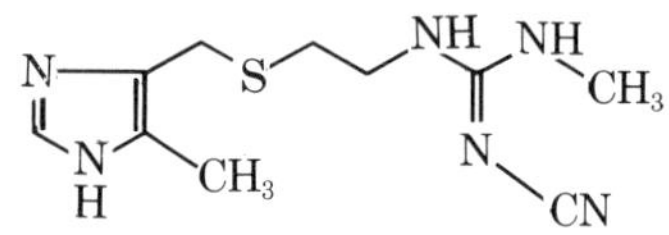

(1)

Table 1. Properties of Cimetidine

Property	Value
mol wt	252.33
mp, °C	141–143
soly in H_2O at 37°C, %	1.14[a]
pK_a (imidazole)	6.8

[a] Solubility increases in dilute HCl.

Preparation. Cimetidine is prepared by any of the three following ways (3): (*1*) by reflux of *N*-cyano-*s*-dimethylguanidine and 5-methyl-4-[(2-aminoethyl)-thiomethyl]imidazole in acetonitrile (the latter is made by the reaction of cysteamine and 5-methyl-4-(hydroxymethyl)imidazole (from 5-methylimidazole and formalin) in acetic acid, in hydrochloric acid, or hydrobromic acid); (*2*) by reaction of 5-methyl-4-[(2-aminoethyl)thiomethyl]imidazole with cyanodithioimidocarbonic acid dimethyl ester to give the isothiourea, followed by reaction with methylamine; and (*3*) by refluxing *N*-methyl-*N′*-[2-[[(5-methyl-imidazol-4-yl)-methyl] thio]ethyl]-2-thiourea (metiamide) and lead cyanamide (the first is prepared by reaction of 5-methyl-4-[(2-aminoethyl)thiomethyl]imidazole with methyl isocyanate).

Toxicology. This compound has not shown significant toxicity in humans on chronic dosing. Adverse reactions noted with cimetidine have been mild gynecomastia, presumably due to prolactin release; mild and transient diarrhea; muscular pain; dizziness; and rash.

Use. The first and only approved histamine (H-2) antagonist currently is used widely for therapy of duodenal ulcers (see Histamines and histamine antagonists). Claims for effectiveness in gastric ulcer and esophageal reflux have also been made, and are currently being tested in clinical trials. This compound has effective, antigastric secretory activity in laboratory animals and man by both oral and parenteral administration. Its mechanism of action is believed to be owing to the ability to block histamine stimulation of acid secretion. It is also effective against other stimulants of gastric acid such as gastrin, insulin, and feeding.

Aluminum Hydroxide Gel. Colloidal aluminum hydroxide [*21645-51-2*], Amphojel, is a suspension in which each 100 g contains the equivalent of 3.6–4.4 g aluminum oxide. Amphojel may contain a variety of flavoring agents. It is a white viscous suspension, from which small amounts of water may separate on standing. It is translucent in thin layers.

Preparation. Aluminum hydroxide gel, $Al(OH)_3$, may be prepared by a number of methods. The products vary widely in viscosity and particle size, also in the rate of solution. Such factors as degree of supersaturation, pH during precipitation, temperature, and nature and concentration of by-products present affect the physical properties of the gel.

In one manufacturing process, aluminum chloride is treated with a solution containing sodium carbonate and sodium bicarbonate. The product of this reaction is mixed with the precipitate obtained by reaction of a solution of aluminum chloride and ammonia. The mixed magma is dialyzed, the product is mixed with glycerol (qv), and sodium benzoate is added into the mixture which is then passed through a colloid mill.

Analytical and Test Methods. See the USP for official assay methods (1). One gram of Amphojel of suspension raises the pH of 12.5–25 mL of simulated gastric juice (0.1 *N* HCl) to pH 3.5. The rate of neutralization is slow, 5–15 min.

Toxicity. Aluminum hydroxide gel is nontoxic, but has the following side effects: constipation; nausea or vomiting owing to astringent action or taste; hypophosphatemia and osteomalacia owing to interference with calcium absorption; and complexation with tetracycline and interference with absorption of this and other classes of drugs.

Uses. Aluminum hydroxide gel is used as an antacid, and for treatment of phosphate nephrolithiasis.

Precipitated Calcium Carbonate. Calcium carbonate [*471-34-1*], $CaCO_3$, is a fine white, microcrystalline powder without odor or taste. It is stable in air. An aqueous suspension is close to neutrality. It is practically insoluble in water, insoluble in alcohol, and it dissolves with effervescence in dilute acetic, hydrochloric, and nitric acids.

Preparation. This substance can be prepared by the double decomposition of calcium chloride and sodium carbonate in aqueous solution. Its density and fineness are governed by the concentration of the solutions. Heavy and light forms are available.

Analytical and Test Methods. See the USP for official assay methods (1). One gram of $CaCO_3$ neutralizes 110 mL of 0.1 *N* HCl in 10 min, and 162 mL within 2 h.

Toxicology. Toxic effects include: long-term therapy with large doses taken with milk or other sources of phosphate may cause renal pathology (milk–alkali syndrome) and systemic alkalosis (7–19% of calcium is absorbed); increased urinary calcium may favor calcific renal stones; calcium carbonate may be constipating; and rebound gastric acid secretion may occur with high doses which neutralize gastric contents (greater than 2 g).

Uses. Calcium carbonate is an effective gastric antacid for use in peptic ulcer disease in a dose of 2.5 g/h. Usual dose for OTC usage is 1–2 g as needed. It can be used also as a dentifrice (qv).

Magnesia and Alumina Oral Suspension. Maalox is a mixture containing magnesium hydroxide and variable amounts of aluminum oxide in the form of aluminum hydroxide and hydrated aluminum oxide: 2.9–4.2% magnesium hydroxide [*1309-42-8*], $Mg(OH)_2$ and 2.0–2.4% aluminum oxide [*1344-28-1*], Al_2O_3. A mixture of 4.9–6.6% combined magnesium hydroxide and aluminum oxide may contain a flavoring agent and antimicrobial agents in a total amount not exceeding 0.5%.

Preparation. The magnesia and alumina suspension is prepared by treatment of an aqueous solution containing aluminum and magnesium salts in the desired proportion with sodium hydroxide. The coprecipitated aluminum and magnesium hydroxides are collected by filtration, washed free of soluble salts, and stabilized by the addition of a suitable hexitol.

Use. Maalox is an antacid. It is also available in tablet formulations.

Magnesium Oxide. ***Properties.*** Magnesium oxide [*1309-48-4*], MgO, magnesia, is available as a very bulky white powder known as light magnesium oxide, or a relatively dense white powder known as heavy magnesium oxide. It absorbs moisture and carbon dioxide when exposed to air. It is practically insoluble in water, insoluble in alcohol, and soluble in dilute acids.

Preparation. Light or heavy magnesium carbonate is exposed to a red heat and carbon dioxide and water are expelled, leaving light or heavy magnesium oxide. The density is also influenced by the calcining temperature; higher temperatures yield more compact forms.

Analytical and Test Methods. See the USP for official assay methods (1). One gram of magnesium oxide neutralizes 87 mL of 1 *N* HCl in 10 min and 305 mL in 2 h.

Use. Magnesium oxide is an effective nonsystemic antacid in that it is converted to the hydroxide. It does not neutralize gastric acids excessively and it does not liberate carbon dioxide. The light form is preferable to the heavy for administration in liquids since it is suspended more readily.

Magnesium Trisilicate. ***Properties.*** Magnesium trisilicate [*39365-87-2*], $2MgO.3SiO_2.xH_2O$, is a compound of magnesium oxide and silicon dioxide with varying proportions of water containing not less than 20% magnesium oxide and not less than 45% silicon dioxide (magnesium silicate hydrate). It is a fine white odorless, tasteless powder free from grittiness. Its suspension is neutral or only slightly alkaline. It is insoluble in water and in alcohol, and is decomposed readily by mineral acids with the liberation of silicic acid.

Preparation. Magnesium trisilicate is prepared by precipitation of a solution of sodium silicate of the proper composition (MgO to SiO_2 ratio equal to 1:1.5) with a solution of magnesium chloride or sulfate. The precipitate of the magnesium trisilicate is filtered, washed, and dried at a low temperature.

Analytical and Test Methods. See the USP for official assay methods (1). One gram of magnesium trisilicate neutralizes 10 mL of 0.1 *N* HCl in 10 min and 15 mL in 2 h.

Use. Magnesium trisilicate is a nonsystemic antacid and an adsorbent. It has a slow onset of activity, and it is a relatively weak antacid. It may cause diarrhea in large doses owing to soluble magnesium salts.

Magaldrate. ***Properties.*** Magaldrate, tetrakis(hydroxymagnesium) decahydroxydialuminate dihydrate [*39366-43-3*], $Al_2Mg_4(OH)_{14}.2H_2O$, also called aluminum magnesium hydroxide and Riopan, contains the equivalent of 28–39% magnesium oxide and 17–25% aluminum oxide. It is a white odorless, crystalline powder insoluble in water and alcohol, and soluble in dilute solutions of mineral acids.

Preparation. Magaldrate is prepared by precipitation from aqueous solutions of sodium or potassium aluminate and a magnesium salt under controlled conditions of concentration and temperature. The precipitated product is collected by filtration, washed to remove soluble by-products, and dried (4).

Test Methods. See the USP for assay methods (1).

Use. Magaldrate is a gastric antacid, with somewhat more efficacy than aluminum hydroxide. It does not appear to disturb electrolyte balance or bowel function.

Sodium Bicarbonate. ***Properties.*** Sodium bicarbonate [*144-55-8*], $NaHCO_3$, is a white crystalline powder. It is odorless and has a saline and slightly alkaline taste. It is stable in dry air, but slowly decomposes in moist air. Its solubility is 1 g in 10 mL water, and in hot water it is converted into carbonate. Sodium bicarbonate is insoluble in alcohol.

Preparation. Sodium bicarbonate may be prepared by the ammonia-salt (Solvay) process. Carbon dioxide is passed through a solution of sodium chloride in ammonia water. Sodium bicarbonate is precipitated and ammonium chloride remains in solution. The ammonium chloride is heated with lime to regenerate ammonia (see Alkali and chlorine products).

Test Methods. See the USP for official assay methods (1). One gram of $NaHCO_3$ neutralizes 115 mL 0.1 *N* HCl.

Uses. Sodium bicarbonate is a gastric antacid that may cause systemic alkalosis on overdose and may contribute to edema due to sodium retention. It is useful for systemic acidosis since both deficient ions are present in the same molecule and can be used topically as a moist paste or solution as an effective antipruritic. It is also an ingredient of many effervescent mixtures, alkaline solutions, etc.

Laxatives

Laxatives are agents that facilitate the passage and elimination of feces. These agents are used most commonly as self-treatment and are rarely prescribed by physicians. The use of laxatives should be limited to those few conditions in which they are warranted and should not be used chronically. Some of the legitimate uses are the presence of hardened and dry stools causing partial or complete intestinal obstruction, for softening stools in patients for whom straining could be harmful and as preparation for diagnostic tests. Laxatives have traditionally been classified into four categories: irritant (stimulant) cathartics, saline, bulking agents, emollients (lubricants) and fecal softeners. Bulking agents containing indigestible fiber appear to be the most physiological mechanism of action and should be the agents of choice when constipation is a problem, except for cases in which purgation is needed (see Dietary fiber). Over 2×10^8/yr is currently being spent on OTC laxatives.

Danthron. ***Properties.*** Danthron (**2**), 1,8-dihydroxyanthraquinone [*117-10-2*], also known as Dorbane and Chrysazin, is an orange crystalline powder that begins to sublime at ca 75°C and melts at 190–197°C. It is practically insoluble in water, very slightly soluble in alcohol, and slightly soluble in ether, benzene, and solutions of sodium hydroxide. It is sparingly soluble in chloroform.

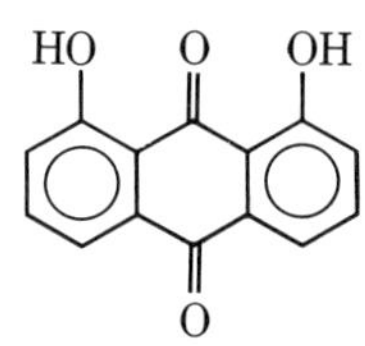

(**2**)

Preparation. Danthron may be prepared by sulfonation, with HgO as a catalyst, of anthraquinone to 1,8-anthraquinonedisulfonic acid, followed by heating with lime.

Uses. Danthron is a cathartic. It is eliminated in the urine as a red dye. It is also an intermediate in the manufacture of dyes and forms lakes with calcium, barium, and lead.

Bisacodyl. ***Properties.*** Bisacodyl (**3**), 4,4′-(2-pyridylmethylene)diphenol diacetate [*603-50-9*], also known as Dulcolax, is a white to off-white crystalline powder in which particles of 50 μm dia predominate. It is very soluble in water, freely soluble in chloroform and alcohol, soluble in methanol and benzene, and slightly soluble in diethyl ether.

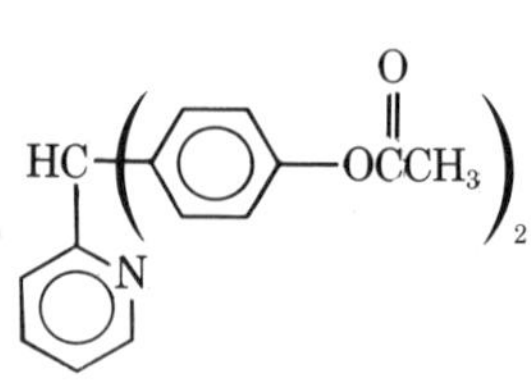

(**3**)

Preparation. Bisacodyl may be prepared from 2-pyridinecarboxaldehyde by condensation with phenol with the aid of a dehydrant such as sulfuric acid (5). The resulting 4,4′-(2-pyridylmethylene)diphenol is esterified by treatment with acetic anhydride and anhydrous sodium acetate. Crystallization is from ethanol.

Safety. Inhalation of bisacodyl and contact with eyes, skin, and mucous membranes should be avoided.

Uses. Bisacodyl is a contact laxative that may be given orally or rectally. It is often used for evacuation of the bowel prior to surgery or diagnostic examinations. It may obviate the need for a cleansing enema.

Plantago Seed. Plantago seed, also called psyllium seed and plantain seed, is the cleaned, dried, ripe seed of *plantaso psyllium* Linne or, *plantago indica* Linne or *plantago ovata* Forskal (blond or Indian psyllium). It is used as a bulking agent.

Phenolphthalein. ***Properties.*** Phenolphthalein (4), 3,3-bis(*p*-hydroxyphenyl) phthalide [*77-09-8*], is a white or faintly yellowish-white crystalline powder. It is odorless, stable in air, and it melts not lower than 258°C. Phenolphthalein is practically insoluble in water; one gram is soluble in 15 mL alcohol and 100 mL diethyl ether.

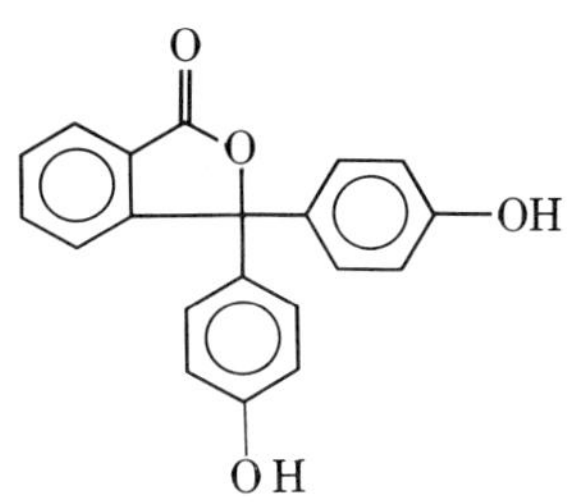

(4)

Preparation. Phenolphthalein may be prepared by mixing phenol, phthalic anhydride, and sulfuric acid, and heating at 120°C for 10–12 h. The product is extracted with boiling water, then the residue dissolved in dilute sodium hydroxide solution, filtered, and precipitated with acid.

Analysis. Official assay methods are reported in the USP (1).

Uses. Phenolphthalein is a cathartic drug, and the basis of many laxatives. It may cause red urine if alkaline, and may cause rash.

Dioctyl Calcium Sulfosuccinate. ***Properties.*** Dioctyl calcium sulfosuccinate (5), calcium salt [*128-49-4*] of 1,4-bis(2-ethylhexyl)sulfosuccinate, also known as Surfak, is a white amorphous solid having the characteristic odor of octyl alcohol. It is very slightly soluble in water, and very soluble in alcohol, polyethylene glycol 400, and corn oil.

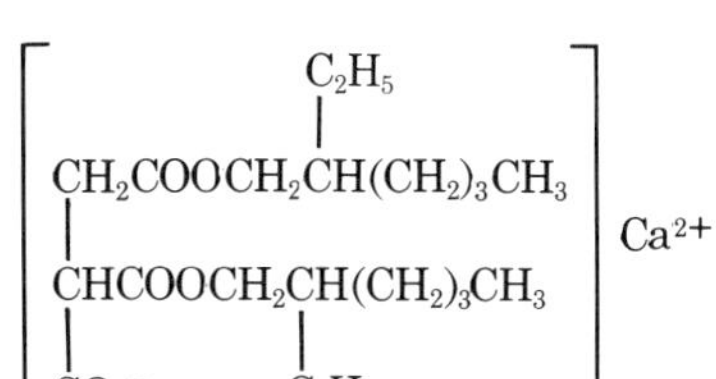

(5)

Preparation. It may be prepared directly from dioctyl sodium sulfosuccinate, dissolved in isopropanol, by reaction with a methanolic solution of calcium chloride (6).

Uses. It is used both as a fecal softening agent, and an emulsifier, wetting, or dispersing agent for external preparations (see Emulsions).

Dioctyl Sodium Sulfosuccinate. ***Properties.*** Dioctyl sodium sulfosuccinate (**6**), sodium 1,4-bis(2-ethylhexyl) sulfosuccinate [*577-11-7*], also called Aerosol OT Dry, Colace, and Doxinate, is a white wax-like, plastic solid with a characteristic odor suggestive of octyl alcohol, usually available in the form of pellets.

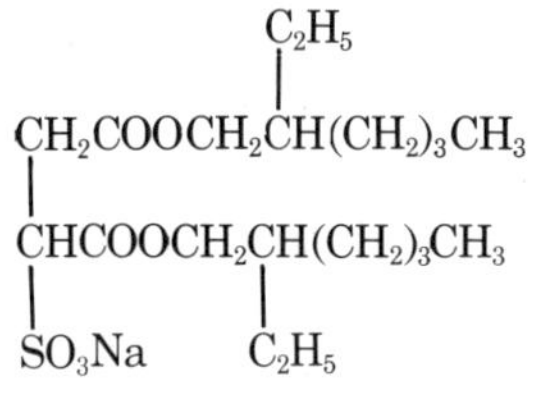

(**6**)

One gram of the sodium salt slowly dissolves in about 70 mL water. It is freely soluble in alcohol and glycerol, and very soluble in hexane. It decomposes in the presence of strong alkali.

Preparation. It may be prepared by treatment of maleic anhydride with 2-ethylhexanol to produce dioctyl maleate, which is allowed to react with sodium bisulfite under conditions conducive to saturation of the olefinic bond with simultaneous rearrangement of the bisulfite to the sulfonate (7).

Uses. It is both a surface-active agent for use as a fecal softener, and a wetting agent in industrial, pharmaceutical, cosmetic, and food applications (see Surfactants).

Magnesium Citrate Solution. ***Properties.*** Magnesium citrate [*3344-18-1*] (3:2), citrate of magnesia, $C_{12}H_{10}Mg_3O_{14}$, contains in each 100 mL an amount of magnesium citrate corresponding to 1.55–1.9 g of MgO (see Citric acid). The composition of magnesium citrate solution is shown in Table 2.

It is a colorless to slightly yellow, clear effervescent liquid with a sweet, acidulous taste and a lemon flavor.

Preparation. In the preparation of a kilogram of the magnesium citrate solution, anhydrous citric acid (49.4 g, see Table 2) is dissolved in 330 mL of hot purified water. Magnesium carbonate in 220 mL purified water is added slowly and stirred until dissolved. The syrup is added and the mixed liquids heated to the boiling point. The

Table 2. Contents of Magnesium Citrate Solution, 1 kg

magnesium carbonate	33 g
citric acid	49.4 g
syrup	132 mL
talc	11 g
lemon oil	0.22 mL
potassium bicarbonate	5.5 g
purified water	770 mL

lemon oil is triturated with talc and added to the hot mixture, which is filtered into a strong, sterile bottle of appropriate size. Boiled purified water is added to make 770 mL. The bottle is stoppered with cotton and allowed to cool, the potassium bicarbonate is added, and the bottle is stoppered securely. The mixture should be shaken occasionally and the bottle kept on its side in a cool place. Sodium bicarbonate (4.6 g) in tablet form may be substituted for potassium bicarbonate. The solution may also be further carbonated using carbon dioxide. The solution may be stabilized by pasteurization, sterilization or by the addition of 66 g citric acid and a quantity of magnesium carbonate equivalent to 13.2 g MgO. Precipitation on standing is increased by the presence of sucrose and CO_2 and decreased by sterilization.

Use. Magnesium citrate solution is a saline cathartic, usually packaged in a single-use container. It is used often as a preoperative or prediagnostic cathartic.

Magnesium Sulfate. ***Properties.*** Magnesium sulfate heptahydrate [*10034-99-8*], $MgSO_4.7H_2O$, also known as bitter salts and epsom salts, are small, colorless crystals, usually needle-like, have a cooling, saline, and bitter taste that effloresce in warm, dry air. The heptahydrate loses 5 molecules of water at 100°C. Its aqueous solution is neutral. One gram of epsom salts is soluble in 1 mL water, 0.2 mL boiling water, or ca 1 mL glycerol. The heptahydrate is sparingly soluble in ethanol.

Preparation. Magnesium sulfate heptahydrate may be prepared by neutralization of sulfuric acid with magnesium carbonate or oxide, or it can be obtained directly from natural sources. It occurs abundantly as a double salt. It also can be obtained from the magnesium salts that occur in brines used for the extraction of bromine (qv). The brine is treated with calcium hydroxide to precipitate magnesium hydroxide. Sulfur dioxide and air are passed through the aqueous suspension to yield magnesium sulfate (see Magnesium).

Analysis. See the USP for assay methods (1).

Uses. Magnesium sulfate is a saline cathartic. It is used topically as a saturated magnesium sulfate solution in water as therapy for skin disorders, and intravenously as an anticonvulsant.

Sodium Phosphate. ***Properties.*** Disodium phosphate heptahydrate [*7782-85-6*], $Na_2HPO_4.7H_2O$, is a colorless or white granular salt. It effloresces in warm dry air. One gram of the heptahydrate dissolves in 4 mL H_2O, and it is only very slightly soluble in alcohol.

Preparation. Bone phosphate or bone ash, obtained by heating bones to whiteness consists mainly of tribasic calcium phosphate. Phosphorite (phosphate rock) may also be used. Finely ground phosphatic material is mixed with a quantity of sulfuric acid, a little in excess of the amount required to transform the phosphate into monobasic calcium phosphate. The mixture is leached with hot water and a concentrated solution of sodium carbonate sufficient to convert half of the phosphate into the dibasic sodium salt is added. Then the mixture is boiled.

$$CaH_4(PO_4)_2 + Na_2CO_3 \rightarrow CaHPO_4 + Na_2HPO_4 + H_2O + CO_2$$

After filtering, the solution is concentrated and the sodium phosphate allowed to crystallize.

Analysis. See the official assay in the USP (1).

Use. Disodium phosphate heptahydrate is used as a saline cathartic.

Cascara Sagrada. Cascara sagrada, also known as sacred bark, chitten, dogwood, coffeeberry, bearberry, bitter bark, and bearwood, is the dried bark of *Rhamnus purshiana* DeCandolle. It is in the form of flattened or transversely curved pieces, 1 to 5-mm thick. It is brown, purplish-brown or brownish-red, and has a characteristic odor and bitter taste. It should be collected at least 1 yr prior to use. The active constituents are aloe-emodin [*481-72-1*], iso-emodin [*476-62-0*], purshianin [*1393-00-6*], and several resins. See the official assay in the USP (1).

Use. Cascara sagrada is used as a cathartic. It is most useful when prepared as a fluid extract or aromatic fluid extract. It tends to be a mild laxative causing little discomfort. However, on prolonged administration, it may result in characteristic melanotic pigmentation of the rectal mucosa.

The bitter taste can be lessened if the ground cascara sagrada is moistened and mixed with magnesium or calcium hydroxide, owing to neutralization of the acid constituents. However, this treatment may lessen its activity.

Psyllium Hydrophilic Mucciloid. Psyllium hydrophilic mucciloid, Metamucil, is a white to cream-colored, slightly granular powder with little or no odor and a slightly acid taste made from the mucilagenous portion (outer epidermis) of blond psyllium seeds (*plantago ovata*).

Use. Metamucil is a very popular bulking agent for chronic treatment of many intestinal disorders in which a high-fiber diet is indicated (see Dietary fiber).

Castor Oil. ***Properties.*** Castor oil [*8001-79-4*] is the fixed oil from the seeds of *Ricinus communis* Linné. Pale yellowish or almost colorless, it is a transparent viscid liquid with a faint, mild odor and a bland taste followed by a slightly acrid and usually nauseating taste. Its specific gravity is between 0.945 and 0.965. Castor oil is soluble in alcohol, miscible with anhydrous alcohol, glacial acetic acid, chloroform, and diethyl ether.

Castor oil consists chiefly of the glycerides of ricinoleic and isoricinoleic acids. The cathartic activity is due to the hydrolysis of ricinolein to ricinoleic acid in the small intestine. The seed contains a poisonous albumin (ricin) and base (ricinine).

Preparation. The oil may be obtained by cold expression of the decorticated seeds. The oil is steamed under vacuum to eliminate odors and coagulate the toxic albumin. Fuller's earth or activated charcoal may be used for further purification.

Uses. Castor oil is a cathartic, and is used topically as an emollient. It has been used to initiate labor at term; however, this is dangerous and should not be done.

It is used also as a hair tonic, and a solvent for removing foreign substances from the eye. Castor oil is now used primarily as an industrial material (see Castor oil).

Mineral Oil. ***Properties.*** Mineral oil, liquid petrolatum, is a mixture of liquid hydrocarbons obtained from petroleum (qv). It may contain a suitable stabilizer.

Mineral oil is a colorless, transparent, oily liquid, free or nearly free from fluorescence. It is tasteless and odorless when cold and develops not more than a faint odor of petroleum when heated. Its specific gravity is 0.860–0.905, and the kinematic viscosity is not less than 38.1 mm^2/s (= cSt) at 37.8°C.

Mineral oil is insoluble in water or alcohol, and miscible with most fixed oils, but not castor oil. It is soluble in volatile oils.

Preparation. Mineral oil may be prepared by distillation of the residuary liquid, boiling at 330–390°C, remaining after removing lighter hydrocarbons from petroleum. It is purified and decolorized with sulfuric acid and sodium carbonate, and allowed to pass through animal charcoal or fuller's earth while hot. On cooling, some solid paraffins separate. The liquid is redistilled and the portion boiling below 360°C rejected. American mineral oil consists mainly of aliphatic hydrocarbons containing 15–20 carbon atoms, with lower molecular weight hydrocarbons predominating in light mineral oil which produces a lower specific gravity and viscosity than heavy oils. Russian or Caucasian oils contain cyclic polymethylene (naphthene) hydrocarbons with hydrogen atoms of the ring carbons often replaced by low molecular weight alkyl groups. Both kinds of mineral oil can slowly undergo various reactions when exposed to heat and light. Tocopherol (Vitamin E) can be added as a retardant.

Analysis. See the USP for official assay methods (1).

Uses. Mineral oil is used as a laxative. It is used also as a vehicle for drugs. When applied to nasal mucosa care must be taken to prevent aspiration into the lung which may cause lipid pneumonia. Mineral oil can prevent absorption of lipid-soluble vitamins or it can be absorbed when formulated in a fine emulsion and cause changes in the liver and mesenteric lymph nodes.

Antidiarrheals

Diarrhea is a commonly occurring problem. It is usually self-limiting and of short duration. Diarrhea may be induced by a variety of causes. The following outlines some of these factors. *Increased accumulation of small intestinal contents;* this includes: (*1*) increased intestinal secretion which may be enterotoxin-induced (eg, cholera, *E. coli,* etc), hormone-induced, or drug-induced (eg, caffeine, prostaglandins, laxatives, etc); (*2*) decreased intestinal absorption caused by decreased small intestinal surface area, mucosal disease (eg, tropical sprue), or osmotic deficiency (disaccharidase or lactase deficiency); and (*3*) rapid transit of contents. *Increased accumulation of colonic contents;* this includes: (*1*) increased colonic secretion owing to hydroxy fatty acids, bile acids; exudation (eg, inflammatory bowel disease, amebiasis, etc); and (*2*) decreased colonic absorption caused by decreased surface area, mucosal disease, and osmotic factors. A final factor is *rapid transit* (eg, irritable bowel syndrome, etc).

It should be apparent from the list of possible causes that diagnosis and alleviation of the cause, if possible, is of prime importance. Often this is not possible and therapy is used to alleviate the inconvenience and pain of diarrhea. These compounds only mask the underlying factors producing the problem. Diarrhea in infants is a particularly serious problem owing to the loss of body fluids. Antidiarrheals do not prevent the loss of fluid and electrolytes into the large intestine and, although they may prevent frequent defecation, the serious imbalance of body electrolytes and fluids is not affected.

Commonly used antidiarrheals work by one of two mechanisms. Narcotic analgesics are constipating and are antidiarrheal due to the effect on intestinal propulsion (see Analgesics). This is not due to a smooth muscle-relaxing activity but results from increased nonpropulsive-phasic smooth muscle contractions. This is apparently the mechanism of such commonly used antidiarrheals as codeine [*76-57-3*], morphine [*57-27-2*], diphenoxylate [*915-30-2*], and loperamide [*53179-11-6*]. The other types of agents are much less effective and appear to act by binding water in the intestine or by adsorption of toxins.

Iodochlorhydroxyquin. Iodochlorhydroxyquin (**7**), 5-chloro-7-indo-8-quinolinol [*130-26-7*], also called Vioform and Clioquinol, is a spongy, yellowish-white or brownish-yellow powder. It is almost insoluble in water, cold ethanol, and ether, and soluble in 43 parts boiling ethanol and 128 parts chloroform. It can be prepared by chlorination with sulfuryl chloride of 8-quinolinol dissolved in chloroform, followed by iodination of the 5-chloro-8-hydroxyquinoline with iodine dissolved in alkali.

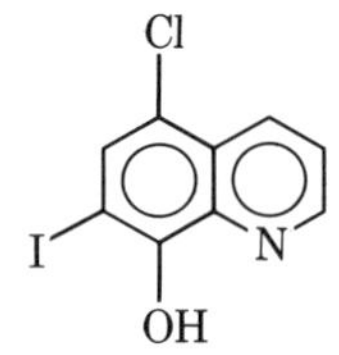

(7)

Use. Iodochlorhydroxyquin has been used prophylactically against so-called traveler's diarrhea on the basis of its systemic amebicidal activity. It was easily available as an over-the-counter drug in most countries in which traveler's diarrhea is a problem. In recent years, it has been shown to be ineffective for this use since the cause of the problem has been shown to be bacterial (toxigenic *E. coli*). Severe toxicity has also been noted, especially in Japan. It has been implicated in SMON, a crippling nervous disorder that has affected thousands of Japanese. In children, after 2 to 3 weeks of therapy, this compound and a close analogue, diodoquin [*83-73-8*], can cause optic nerve atrophy and peripheral neuropathy; and in adults it can cause transverse myelitis. These compounds have been taken off the market in Japan and many other countries; however, they are still widely prescribed in developing countries under a variety of names (8).

Diphenoxylate Hydrochloride. ***Properties.*** Diphenoxylate hydrochloride (8), ethyl 1-(3-cyano-3,3-diphenylpropyl)-4-phenylpiperidinecarboxylate hydrochloride [*3810-80-8*], also called Lomotil, is a white, odorless, crystalline powder that melts at 220–226°C. It is soluble in methanol, sparingly soluble in ethanol and acetone, slightly soluble in water and isopropyl alcohol, freely soluble in chloroform, and practically insoluble in ether and hexane.

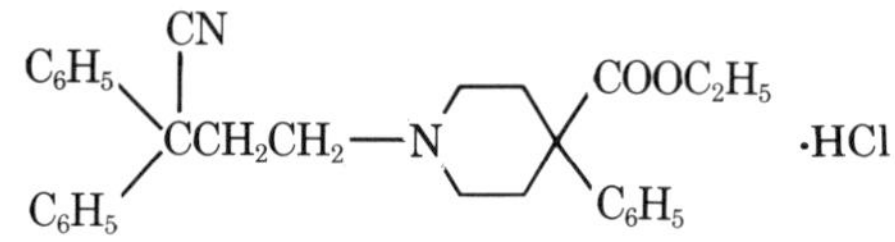

(8)

Preparation. Ethyl 4-phenylisonipecotate is condensed with 2,2-diphenyl-4-bromobutyronitrile by refluxing in toluene with either an excess of the ester or another suitable dehydrobrominating agent. Extraction of the filtered reaction product with HCl followed by alkalinization of the extract precipitates the diphenoxylate free base which is converted to the hydrochloride by extraction with ether and treatment of the extract with hydrogen chloride (9).

Use. Diphenoxylate hydrochloride is an antidiarrheal. Commercial forms are only available mixed with atropine, which is added to discourage usage of excessive amounts and abuse.

Loperamide. Loperamide (**9**), 4-(4-chlorophenyl)-4-hydroxy-*N*,*N*-dimethyl-α,α-diphenyl-1-piperidine butanamide [*53179-11-6*], also called Immodium, is practically insoluble in water (0.002%) at physiological pH. Its crystals are not affected by light, and it is not hygroscopic.

Loperamide may be synthesized by the method described in ref. 10.

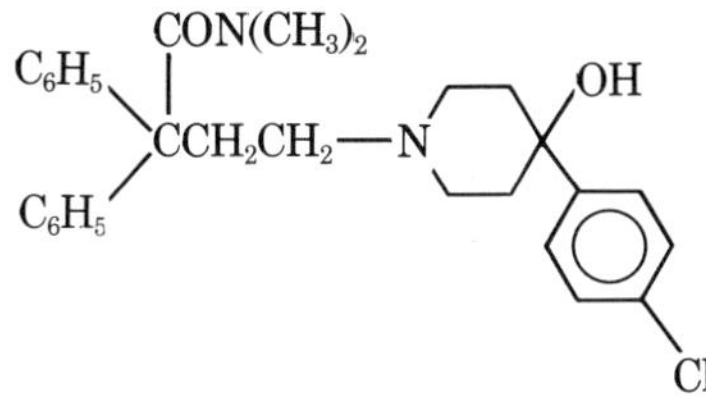

(**9**)

It is a colorless to slightly yellow, clear effervescent liquid with a sweet, acidulous taste and a lemon flavor.

Use. Loperamide is similar in use to diphenoxylate; however, it is formulated without atropine sulfate. It is reported to have less central nervous system side effects than diphenoxylate.

Polycarbophil. Polycarbophil [*73038-24-1*], a copolymer of acrylic acid and divinyl glycol (1,5-hexadiene-3,4-diol), consists of white to creamy white granules having a slight, characteristic esterlike odor. It contains a maximum of 1.5% water. It swells, but it is insoluble in water and most organic solvents.

Polycarbophil's calcium salt is the antidiarrheal Carbofil [*9003-97-8*], also called Sorboquel.

Polycarbophil is prepared by copolymerization of acrylic acid and divinyl glycol in a hot salt slurry using azobisisobutyronitrile as the initiator (11) (see Initiators).

Use. Polycarbophil binds free fecal water and therefore decreases the fluidity of stools. It is most active in the slightly acid or alkaline medium of the small bowel and colon.

Sulfasalazine. Sulfasalazine (**10**), 5-[[*p*-(2-pyridylsulfanoyl)phenyl]azo]salicylic acid [*599-79-1*], also called Azulfidine and salicylazosulfapyridine, is a light brownish-yellow to bright yellow fine powder that is practically tasteless and odorless. It melts at ca 255°C with decomposition.

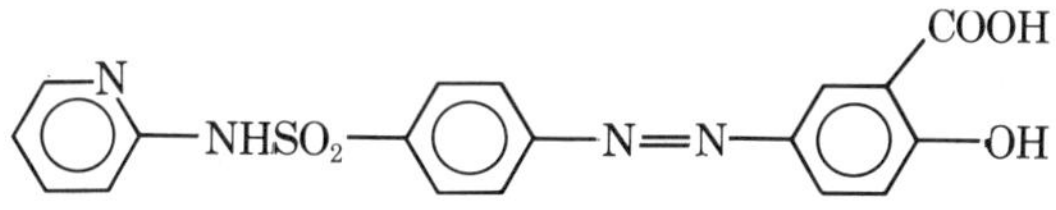

(**10**)

It is very slightly soluble in ethanol, and practically insoluble in water, diethyl ether, chloroform, and benzene. It is soluble in aqueous solutions of alkali hydroxides. Sulfasalazine may be prepared from *N*′-(2-pyridyl)sulfanilamide by diazotization and coupling with salicylic acid.

Use. Sulfasalazine is not used as a antidiarrheal as such, but it is useful for the treatment of inflammatory bowel diseases such as ulcerative colitis and Crohn's disease. Its action is purported to be due to the breakdown in the colon to 5-aminosalicylic acid and sulfapyridine. It has also been shown to be an effective blocker of prostaglandin synthesis in the colon (see Prostaglandins).

Antiemetics

Nausea and vomiting are frequent symptoms of disease and can occur from a variety of causes. Some of the more common causes are pregnancy, chemotherapy, motion sickness, and gastrointestinal infections. In most cases, it is self-limiting. However, in those cases in which fluid and electrolyte loss are important, such as in infants or debilitated patients, the control of vomiting is very important. Drugs described here all appear to act at the chemoreceptor trigger zone or on the vestibular apparatus (see also Histamine and antihistamine antagonists).

Cyclizine Hydrochloride. ***Properties.*** Cyclizine hydrochloride (**11**), 1-(diphenylmethyl)-4-methylpiperazine hydrochloride [*303-25-3*], Marezine, is a white crystalline powder or small colorless crystals. It is odorless or nearly so and has a bitter taste. It melts indistinctly and with decomposition at ca 285°C.

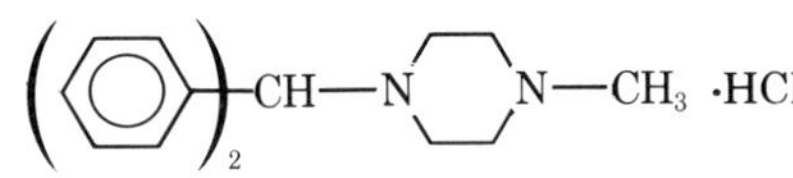

(11)

One gram of cyclizine hydrochloride is soluble in 115 mL water, 115 mL ethanol, and 75 mL chloroform. It is insoluble in diethyl ether.

Preparation. Benzhydryl chloride is condensed with *N*-carbethoxypiperazine and the carbethoxy group is then split off by heating with potassium hydroxide. The *N*-benzhydrylpiperazine thus formed is isolated from the reaction product and methylated at the N^4-position by treatment with formic acid and formaldehyde to give the free base. The purified base may then be converted to the hydrochloride by passing hydrogen chloride into a solution of base in an appropriate solvent (12).

Use. This compound has antihistaminic activity and is useful in the treatment of motion sickness. It may also be effective in the control of postoperative nausea and vomiting. It should not be used by pregnant women. Large doses may cause drowsiness and dryness of the mouth.

Dimenhydrinate. ***Properties.*** Dimenhydrinate (**12**), 8-chlorotheophylline compound [*523-87-5*] with 2-(diphenylmethoxy)-*N*,*N* dimethylethylamine (1:1), Dramamine, is a white crystalline, odorless powder that melts between 102 and 107°C. It is slightly soluble in water, freely soluble in ethanol and chloroform, and sparingly soluble in diethyl ether.

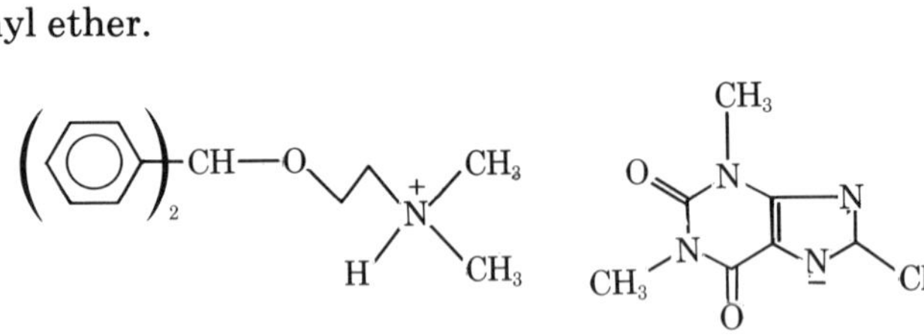

(12)

Preparation. Dimenhydrinate is prepared by combining dimethylaminoethyl benzhydryl ether with 8-chlorotheophylline and refluxing in an isopropyl alcohol solution. The crystalline precipitate of dimenhydrinate that forms on cooling is collected by filtration, washed with cold ethyl acetate, and dried (13).

Use. Dimenhydrinate is an antiemetic especially useful as an antinauseant in motion sickness, and for syndromes associated with vertigo such as Menières syndrome, radiation sickness, and vestibular dysfunction.

Meclizine Hydrochloride. ***Properties.*** Meclizine hydrochloride (**13**), 1-(*p*-chloro-α-phenylbenzyl)-4-(*m*-methylbenzyl)piperazine dihydrochloride [*31884-77-2*], also called Piperazine, Antivert, and Bonine, is a white or slightly yellowish crystalline powder. It has a slight odor, is tasteless, and it melts at 217–224°C.

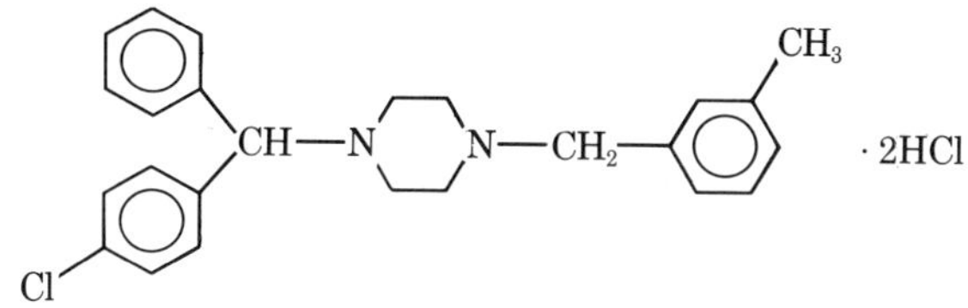

(**13**)

The hydrochloride is practically insoluble in water and ether. It is freely soluble in chloroform, pyridine, methylacetamide, and acid alcohol–water mixtures, and slightly soluble in dilute acids or alcohol.

Preparation. Meclizine hydrochloride is prepared by condensing *N*-(*m*-methylbenzyl)piperazine with *p*-chlorobenzhydryl chloride in the presence of triethylamine. The purified base is dissolved in a suitable solvent and converted to the dihydrochloride by a stream of hydrogen chloride (14).

Procloperazine Maleate. ***Properties.*** Procloperazine maleate (**14**), 2-chloro-10-[3-(4-methyl-1-piperazinyl)propyl]phenothiazine maleate [*84-02-6*] (1:2), or Compazine, is a white or pale-yellow, crystalline powder. It is almost completely odorless, and its saturated solution is acidic to litmus. It is practically insoluble in water and ethanol, and slightly soluble in warm chloroform.

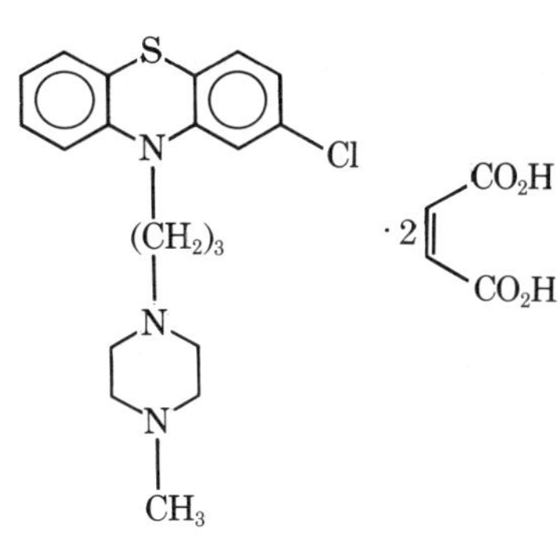

(**14**)

Preparation. A toluene solution of 1-(3-chloropropyl)-4-methylpiperazine and 2-chlorophenothiazine is refluxed with sodamide for several hours. After filtration, and distillation of the toluene, the proclorperazine is obtained by short-path distillation under high vacuum. The base is dissolved in a suitable solvent and treated with a double-equimolar quantity of maleic acid. The salt (**14**) precipitates from solution (15).

Use. Proclorperazine maleate is an effective antiemetic and tranquilizing agent. It is not particularly effective for motion sickness. Adverse reactions may include extrapyramidal reactions, motor restlessness, dystonias, tardive dyskinesia, and contact dermatitis.

BIBLIOGRAPHY

"Antacids, Gastric" in *ECT* 1st ed., Vol. 1, pp. 930–934, by J. C. Krantz, Jr., University of Maryland; "Antacids, Gastric" in *ECT* 2nd ed., Vol. 2, pp. 427–431, by John C. Krantz, Jr., University of Maryland.

1. *The United States Pharmacopeia XX* (*USP XX–NFXV*), The United States Pharmacopeial Convention, Inc., Rockville, Md., 1980.
2. *Chem. Week.*, 23 (Aug. 24, 1977).
3. U.S. Pat. 3,876,647 (Apr. 8, 1975), and Ger. Offen. 2,344,779 (Mar. 14, 1974), G. J. Durant and co-workers (to Smith Kline and French Laboratories).
4. U.S. Pat. 2,923,660 (Feb. 2, 1960), G. Hallman (to BykGulden Lomber Chemische Fabrik GBBH).
5. U.S. Pat. 2,764,590 (Sept. 25, 1956), A. Kottler and E. Seeger (to Dr. Karl Thomae GmbH).
6. U.S. Pat. 3,035,973 (May 22, 1962), L. J. Klotz (to Lloyd Brothers, Inc.).
7. U.S. Pat. 2,028,091 (Jan. 14, 1936), A. O. Jaeger (to American Cyanamid and Chemical Corp.).
8. *The New York Times*, 18 (June 19, 1977).
9. U.S. Pat. 2,898,340 (Aug. 4, 1954), P. A. J. Janssen.
10. R. A. Stokbroekx and co-workers, *J. Med. Chem.* **16,** 782 (1973); U.S. Pat. 3,714,159 (Jan. 30, 1975), P. A. J. Janssen, C. J. E. J. Niemegeers, R. A. Stokbroekx, and J. Vandenberk (to Janssen Pharmaceutica).
11. U.S. Pat. 3,202,577 (Aug. 24, 1965), R. L. Markus (to White Laboratories).
12. U.S. Pat. 2,630,435 (Mar. 3, 1953), R. Baltzly and J. C. Castillo (to Burroughs Wellcome and Co.).
13. U.S. Pats. 2,499,058 and 2,534,813 (Feb. 28 and Dec. 19, 1950), J. W. Cusic (to G. D. Searle and Co.).
14. U.S. Pat. 2,709,169 (May 24, 1955), H. Morren (to Union Chemique Belge Société Anonyme).
15. U.S. Pat. 2,902,484 (Sept. 1, 1959), R. J. Horclois (to Société des Unines Chemique Rhone-Poulenc).

General References

F. P. Brooks, *Gastrointestinal Pathophysiology,* 2nd ed., Oxford University Press, New York, 1978.
A. Osol and J. E. Hoover, *Remington's Pharmaceutical Sciences,* 15th ed., Mack Publishing Co., Easton, Pa., 1975.

HENRY I. JACOBY
McNeil Laboratories

GELATIN

Gelatin [*9000-70-8*], Type A and Type B, is obtained by the partial hydrolysis of collagen, the chief protein component in skins, bones, hides, and white connective tissues of the animal body (see also Glue; Proteins). Type A is produced by acid processing of collagenous raw materials and has an isoelectric point between pH 7 and 9, whereas Type B is produced by alkaline or lime processing and has an isoelectric point between pH 4.6 and 5.2. Mixtures of Types A and B as well as gelatins produced by modifications of these processes may exhibit isoelectric points outside of the stated ranges (1). Gelatin is a hydrolysis product obtained by hot water extraction and does not exist in nature.

Gelatin is used in the food, pharmaceutical, and photographic industries which take advantage of its unique properties such as reversible gel-to-sol transition of aqueous solutions; viscosity of warm aqueous solutions; capability to act as a protective colloid; water permeability; and insolubility in cold water but complete solubility in hot water.

Animal glue and hydrolysate (liquid protein) are lower molecular weight protein products obtained by a more complete hydrolysis of collagen and thus can be considered as containing lower molecular weight fractions of gelatin. Colloidal gels, such as bentonite clay, which gel reversibly because of their thixotropic properties, bear obviously no chemical or physical relationship to gelatin from collagen.

Physical and Chemical Properties

Commercial gelatin produced in the United States ranges from coarse granules to fine powder. Dry commercial gelatin contains about 9–12% moisture and is an essentially tasteless, odorless, brittle solid with specific gravity between 1.3 and 1.4. The physical and chemical properties of gelatin are measured on aqueous gelatin solutions and are functions of (*1*) the source of collagen, (*2*) the method of manufacture, (*3*) conditions during extraction and concentration, (*4*) thermal history, (*5*) pH, and (*6*) the chemical nature of impurities or additives. Gelatin is classified as a derived protein since it is obtained from collagen by a controlled partial hydrolysis. This makes gelatin a heterogeneous mixture of polypeptides; properties desired are obtained by blending products from selected extracts.

Solubility. In commercial applications (food, pharmaceuticals, and photography), gelatin is used as a solution. Gelatin is practically insoluble in absolute alcohol, acetone, carbon tetrachloride, ether, benzene, petroleum ether, and most other nonpolar organic solvents, but soluble in water, acetic acid, and aqueous solutions of polyhydric alcohols such as glycerol, propylene glycol, sorbitol, and mannitol. The viscosity of aqueous gelatin solutions increases with increasing concentrations and decreasing temperatures.

Dry gelatin absorbs water exothermally. The rate and degree of swelling is a characteristic of the particular gelatin. Swelled gelatin granules dissolve rapidly in water above 35°C. Gelatin freeze-dried with water-soluble sugar dissolves rapidly in cold water but then reverts quickly to the stable gel form. For more detailed discussion of gelatin solutions and solvation, see ref. 2. Many water-soluble organic solvents are compatible with gelatin but interfere with gelling properties (3). The cross-linking of gelatin matrix by chemical means is used extensively in photographic products, and this so-called hardening permanently changes the solubility of gelatin (4–5).

Stability. Dry gelatin stored in airtight containers at room temperature has a shelf life of many years. However, it decomposes above 100°C. For complete combustion, temperatures above 500°C are required. Aqueous solutions or gels of gelatin are highly susceptible to bacterial growth and breakdown by proteolytic enzymes. Stability is a function of pH and electrolytes and decreases with increasing temperature because of hydrolysis.

Swelling. The swelling property of gelatin is not only important in its solvation but also in photographic film processing and the dissolution of pharmaceutical capsules. The swelling of gelatin does not fit a simplified equation derived for polymers (6) where the swelling depends primarily on the number of cross-links and the interaction constant of polymer and solvent. In gelatin, the cross-links are believed to involve multiple interactions between chains that form fibrils; this reduces swelling (7). The stability of these cross-links depends on pH, temperature, time, and electrolytes. The pH and electrolytes affect swelling which can be explained by the simple Donnan equilibrium theory, treating gelatin as a semipermeable membrane. This explains why gelatin exhibits the lowest swelling at its isoelectric point. Electrolytes affecting swelling of gelatin have been arranged in the order of Hofmeister or lyotropic series:

$$Ba^{2+}, Ca^{2+}, Mg^{2+}, Cs^{+}, Rb^{+}, NH_4^{+}, K^{+}, Na^{+}, Li^{+},$$
$$SO_4^{2-}, CH_3COO^{-}, Cl^{-}, Br^{-}, NO_3^{-}, I^{-}, SCN^{-}$$

At pHs below the isoelectric point, the choice of anions can control swelling, whereas above the isoelectric point, proper choice of cations reduces swelling. The anions break hydrogen bonds which probably account for increased swelling.

The rate of swelling follows approximately a second-order equation (8). In photographic products, the swelling of the gelatin layer is controlled by coating conditions, drying conditions, chemical cross-linking, and the composition of the processing solutions (see Photography). Conditioning at 90% rh and 20°C for 24 h greatly reduces swelling of hot dried film coatings (9). The ratio of lateral to vertical swelling is of great concern in the photographic industry since it can cause severe curling of photographic papers or films when changes in humidity or general moisture content take place.

Chemical Composition and Structure

Gelatin is not a single chemical substance. The main constituents of gelatin are large and complex polypeptide molecules of the same amino acid composition as the parent collagen, covering a broad molecular weight distribution range. Most commercial gelatin contains molecular species from 15,000 to above 250,000 (10), with the average between 50,000 and 70,000 mol wt. Molecular weight distribution studies have been carried out by fractional precipitation with ethanol or isopropanol and by complexing with anionic detergent molecules. The coacervates are isolated and recovered as gelatin fractions (11–12).

Analysis shows the presence of amino acids from 0.2% tyrosine to 30.5% glycine (see Amino acids). The five most common amino acids are glycine, 26.4–30.5%; proline, 14.8–18%; hydroxyproline, 13.3–14.5%; glutamic acid, 11.1–11.7%; and alanine, 8.6–11.3%. The remaining amino acids in decreasing order are: arginine, aspartic acid, lysine, serine, leucine, valine, phenylalanine, threonine, isoleucine, hydroxylysine, histidine, methionine, and tyrosine (13). Warm gelatin solutions are more levorotatory than expected on the basis of the amino acid composition, indicating Gly–Pro–Pro and Gly–Pro–Hypro sequences (14). The α-chain form of gelatin behaves in a solution like a random-coil polymer, whereas the gel form may contain as much as 70% helical conformation. The remaining molecules in nonhelical conformation are linking helical regions together in the matrix. Structures have been studied with the aid of an electron microscope (7). The structure of the gel is a combination of fine and coarse interchain networks; the ratio depends on the temperature during the polymer–polymer and polymer–solvent interaction leading to bond formation. The rigidity of the gel is approximately proportional to the square of the gelatin concentration. Presence of crystallites, indicated by x-ray diffraction pattern, are believed to be at the junctions of the polypeptide chains (15). Homogeneous α-chain gelatin has been prepared by pretreating collagen with pronase in the presence of 0.4 *M* $CaCl_2$, and extracting the gelatin with hot water at 80°C and pH 7.0 after inactivating the enzyme and removing the salts. The molecular weight of a single α-chain is about 95,000.

Amphoteric Character. The amphoteric character of gelatin is due to the functional groups of the amino acids and the terminal amino and carboxyl groups created during hydrolysis. The distribution of these groups and the resulting isoionic point is determined by the manufacturing process. It can be determined conveniently by mixed-bed ion-exchange resin and is expressed in pH units. Titrating the free acidic or basic groups gives an indication of the amphoteric characteristics of gelatin. The amphoteric character and isoionic point of Type A gelatin depend on the acid processing time; a typical isoionic point is between pH 8.0 and 9.0. Type B gelatin exhibits more reproducible amphoteric characteristics, reaching an isoionic point of 5.2 after 4 weeks of liming which drops to 4.8 after a prolonged or more vigorous liming process. The isoionic point is reproducible, whereas the isoelectric point depends on the salts present. The isoelectric point can be found by determining a pH value at which gelatin solution exhibits maximum turbidity (16). Many isoionic point references are recorded as isoelectric points even though the latter is defined as a pH at which gelatin has net charge of zero and thus shows no movement in the electric field (17).

Gelation. A most useful and unique property of gelatin solution is its capability to form reversible gel–sols. A gelatin solution gels at temperatures below 35°C. This conversion temperature is determined as setting point (sol to gel) or melting point (gel to sol). Commercial gelatins melt between 23 and 30°C, with the setting point being lower by 2–5°C. Melting point determination, as described in *British Standards* (18), uses as end point,when colored carbon tetrachloride slides halfway down the gelatin–test tube interface. Several methods have been used to determine the setting point of gelatin but all are time consuming (19).

Manufacture and Processing

Technology for gelatin manufacture was developed in the early 1920s. Acid and lime processes have separate facilities and are not interchangeable. In the past, bones and ossein (decalcified bone) have been supplied by India and South America. Today, bones from slaughter houses and meat packing plants are an important source. The supply of bones has been greatly increased since the meat-packing industry introduced packaged and fabricated meats, assisted by the growth of fast-food restaurants. Dried and rendered bones yield about 14–18% gelatin, whereas pork skins yield about 18–22% (see also Meat products).

Most Type A gelatin is made from pork skins, yielding grease as a by-product which is also marketable. The process includes macerating the skins, washing to remove extraneous matter, swelling for 10–30 h in 1–5% hydrochloric, phosphoric, or sulfuric acid, reduction of acidity by washing, adjustment of pH, and pumping to extraction tanks. There, four to five extractions are made at temperatures increasing from 55–65°C for the first extract to 95–100°C for the last extract. Each extraction lasts about 4–8 h. The grease is then removed and the gelatin solution is filtered and for most applications deionized. Concentration to 30–40% solid is carried out in two stages by continuous vacuum concentration. The viscous solution is chilled, noodled, and dried at 30–60°C on a continuous wire-mesh belt. The dry gelatin is then ground and blended to specification.

Type B gelatin is made mostly from bones, but also from hides and skins. The bones for Type B gelatin are crushed and degreased at the rendering facilities, which are usually located at a meat-packing plant. Rendered bone pieces (0.5–4 cm) with less than 1% fat are demineralized by 4–7% hydrochloric acid in 7 to 14 days. Dibasic calcium phosphate is precipitated and recovered from the spent liquor. The demineralized bones (ossein) are washed and pumped to liming tanks where they are agitated daily for 3–16 weeks in a lime slurry. After washing for 15–30 h to remove the lime, the ossein is acidified to pH 5–7 with hydrochloric, sulfuric, phosphoric, or acetic acid. Then the processing for Type A gelatin is followed. Cleanliness is very important to avoid contamination by bacteria or proteolytic enzymes.

Bovine hides and skins are substantial sources of raw material for Type B gelatin and are supplied in the form of splits, trimmings of dehaired hide, raw hide pieces, or salted hide pieces. Like pork skins, the skins are cut to smaller pieces before being processed. Sometimes the term calfskin gelatin is used to describe hide gelatin. The liming of hides usually takes a little longer than the liming of ossein from bone. In the United States, 25–30% of the gelatin derives from hides and skins.

All equipment should be made of stainless steel. The liming tanks, however, can be either concrete or wood. Properly lined iron tanks are often used for the washing and souring operations.

Most gelatin plants operate around the clock. The product is tested in batches and again as blends to confirm individual customer orders.

Economic Aspects

World gelatin production is believed to be about 115,000 metric tons. The United States produces about 31,000 t, followed by France, the Federal Republic of Germany, England, and Japan.

The U.S. food industry consumes ca 19,000 t, with an annual growth rate of 0.5%, the pharmaceutical industry consumes about 8,000 t, and the photographic industry about 7,000 t. Of the gelatin produced in the United States, 55% is acid processed (Type A). In the United States, the pharmaceutical gelatin market is expected to grow on the average of 2.5% per year, using 40% bone gelatin, 30% pork skin, and 30% hide gelatin.

The photographic gelatin market has been growing at 5% per year. The fastest growing areas have been color paper and phototypesetting paper, and microfilm. Color paper and x-ray products use over 55% of the photographic gelatin in the United States, with graphic arts and instant films using an additional 37%.

Analytical and Test Methods

Gelatin is identified by a positive test for hydroxyproline, turbidity with tannic acid, or a yellow precipitate with acidic potassium dichromate or trinitrophenol. A 5% aqueous solution exhibits reversible gel-to-sol formation between 10 and 60°C.

Gelatin gives a positive color test for aldehydes and sugars that are considered undesirable impurities in photographic gelatin; nucleic acids are considered restrainers in photographic gelatin and their concentration has to be monitored (20).

Elemental analysis of commercial gelatin was reported as carbon, 50.5%; hydrogen, 6.8%; nitrogen, 17%; and oxygen 25.2% (21). A purer sample analyzed for 18.2–18.4% nitrogen (13,22).

Standard testing procedures for viscosity, pH, ash, moisture, heavy metals, arsenic, bacteria, and jelly strength (with the Bloom Gelometer) are described in refs. 23–24. Additional test procedures have been published by the photographic and gelatin industries, Japan (PAGI) as PAGI Method (1977) (25). Specific tests for photographic gelatin have been devised by IAG (International Working Group for Photographic Gelatin) (26) in Fribourg, Switzerland, and by individual photographic companies and gelatin companies. Gelatin Manufacturers Institute of America, Inc. has published "Standard Methods for Sampling and Testing of Gelatin," 1964.

Uses

Food Products. Gelatin formulations in the food industry use almost exclusively water or aqueous polyhydric alcohols as solvents for candy, marshmallow, or dessert preparations (see Food additives).

In dairy products and frozen foods, gelatin's protective colloidal action prevents crystallization of ice and sugar. In general, a 250-bloom gelatin is added at concentrations ranging from 0.25% in frozen cream pies to 0.5% in ice cream; the use of gelatin in ice cream has greatly diminished in recent years. In sour cream and cottage cheese, gelatin inhibits water separation.

Marshmallows contain as much as 1.5% gelatin to restrain the crystallization of sugar, thereby keeping the marshmallow soft and plastic. Gelatin also increases viscosity and stabilizes the foam in the manufacturing process. Many lozenges, wafers, and candy coatings contain up to 1% gelatin. In these instances, gelatin decreases the dissolution rate.

In meat products, such as canned hams, various luncheon meats, corned beef, chicken rolls, jellied beef, and other similar products, gelatin in 1–5% concentration helps to retain the natural juices.

The largest use of edible gelatin, however, is in the preparation of gelatin desserts in 1.5–2.5% concentrations. The gelatin is sold either premixed with sugar and flavorings or as unflavored gelatin. Most edible gelatin is Type A, but Type B is also used. The GRAS status of edible gelatin is being reconsidered by the FDA. Limits have been considered on pentachlorophenol, microbial contamination by *Salmonella* and *E. coli*, and heavy metals such as lead, sulfur dioxide, and arsenic (24). The ruling is still pending at this time.

Pharmaceutical Products. Gelatin is used in the pharmaceutical industry for the manufacture of soft and hard capsules. The formulations are made with water or aqueous polyhydric alcohols. Capsules are usually preferred over tablets in administering medication (27). Elastic or soft capsules are made with a rotary die from two plasticized gelatin sheets which form a sealed capsule around the material being encapsulated. Methods have been developed to encapsulate even dry powders and water-soluble materials which may be first mixed with oil. The gelatin for soft capsules is low-bloom Type A (170–180 g), Type B (150–175 g), or a mixture of Types A and B. Hard capsules consisting of two parts are first formed and then filled. The manufacturing process is highly mechanized and sophisticated in order to produce capsules of uniform thickness. Medium-to-high-bloom Type A (250–280 g), Type B (225–250 g), or the combination of Types A and B gelatin is used for hard capsules (see Pharmaceuticals).

The use of glycerinated gelatin (24) as a base for suppositories offers advantages over carbowax or cocoa butter base (28). Coated or cross-linked gelatin is used for enteric capsules.

Gelatin is used as a carrier or binder in tablets, pastilles, and troches. For arresting hemorrhage during surgery, a special sterile gelatin sponge known as absorbable gelatin sponge (7) or Gelfoam is used. The gelatin is partially insolubilized by a cross-linking process. When moistened with thrombin or sterile physiological salt solution, the gelatin sponge left in place after the bleeding stops is slowly dissolved by tissue enzymes. Specially fractionated and prepared Type B gelatin can be used as plasma expander.

Gelatin can be a source of essential amino acids (except for tryptophan and methionine and cystine) when used as a diet supplement and therapeutic agent. As such, it has been widely used in muscular disorders, peptic ulcers, and infant feeding, and to spur nail growth. The testing of pharmaceutical gelatin includes reduction of viscosity, minimum ash, and SO_2 concentration.

Photographic Products. Gelatin has been used for over 100 years as a binder in light-sensitive products. The functions of gelatin useful in photographic film manufacture are due to its protective colloidal properties during the precipitation and chemical ripening of silver halide crystals, setting and film-forming properties during coating, and swelling properties during processing of exposed film or paper. Evaluation of photographic gelatin is very elaborate and can include over 20 chemical and physical tests in addition to photographic evaluation. Most chemical impurities are limited to less than 10 ppm. Aqueous solutions are employed for emulsions (see Photography).

Photographic gelatins are manufactured to standard specifications since the testing is time-consuming and costly. A new gelatin product can easily take 6–12 months of testing, including extensive field testing.

Photographic products may have up to twelve gelatin layers grouped into 3 categories: (*1*) light-sensitive silver halide-bearing layers of 2–10 μm thickness (referred to as emulsion layers), (*2*) surface or protective layers of 1–2 μm thickness, and (*3*) backing, antihalo, or noncurl layers coated on the opposite side to the emulsion layer. The emulsion gelatin is most critical since it controls silver halide nucleation, crystal growth, chemical sensitization, latent image stability, and numerous other factors affecting the total photographic response. In the last 20 years, the photographic industry has switched from so-called active gelatins derived from hides to inert types derived from bones. The latter is very low or void of natural restrainers and reduction and sulfur sensitizers. Artificial restrainers and sensitizers are now added, although DNA is one natural restrainer still present.

Other changes in techniques have been brought about by (*1*) abandoning the lengthy noodle-wash technique (to remove salts after silver halide precipitation) in favor of precipitating, coagulating, or derivatizing the gelatin, and washing the precipitate by decanting; (*2*) new coating techniques that allow simultaneous coating of several layers at one time at speeds ten times as fast as before; and (*3*) short-time high-temperature processing with new cross-linking agents unlike the aldehydes and metal salts previously used. Many new hardeners are extremely fast acting and have to be metered in during the coating operation. It is now quite common to use a derivatized gelatin such as phthalated gelatin to precipitate silver halide. These materials with a low pH isoionic point precipitate at pHs <4.0. Precipitation in this case is accomplished by lowering the pH, washing at low pH, and then increasing it to above 6.5 to dissolve and redisperse the emulsion before reconstituting it with gelatin.

Gelatin used in the auxiliary layers has to withstand high temperature processing and allow high-speed coating.

Gelatin is also used in so-called subbing formulations to prepare film bases such as polyester, cellulose acetate, cellulose butyrate, and polyethylene-coated paper base for coating by aqueous formulation. Solvents such as methanol, acetone, or chlorinated solvents are used with small amounts of water. Low ash, low grease, and good solubility in mixed solvents are required of these gelatins (see Coatings). In certain lithographic printing, light-sensitive dichromated gelatin is used. Light causes permanent cross-linking of gelatin in the presence of the dichromate and this phenomenon is used to make relief images for printing (see Reprography).

Gelatin coatings in photographic products are further tested for brittleness, scratch resistance, friction, swelling rate, drying rate, curling tendency, dry adhesion, wet adhesion, and pressure sensitivity. These properties are becoming more critical with the development of more sophisticated cameras, and printing and processing equipment. Photographic technology offers a rapidly changing, highly sophisticated, very competitive market to photographic gelatin.

Derivatized Gelatins

Chemically active groups in gelatin molecules are either the end groups or side chain groups. In the process of modifying gelatin properties, some groups can be removed, eg, deamination of amino groups by nitrous acid (29), or the removal of guanidine groups from arginine by hypobromite oxidation (30). The latter destroys the protective colloid properties of gelatin.

Commercially successful derivatized gelatins are made mostly for the photographic gelatin and microencapsulation market. In both instances, the amino groups are acylated. Protein detergent is made by lauroylating gelatin. Phthalated gelatin is now widely used in the photographic industry (31). Arylsulfonylated gelatin has been patented for microencapsulation (qv) (32). Carbamoylated gelatin, made by treating gelatin with cyanate or nitrourea in neutral aqueous solution, is also used by the photographic industry.

Active double bonds react with the amino groups in gelatin, and acrylic polymers have been grafted to gelatin (33). Gelatin has been derivatized by epoxides (34), cyclic sulfones (35), and cyanamide (36). Cross-linking or hardening of gelatin attacks the same active groups but an agent with two active sites is needed, eg, divinylsulfone, bis(isomaleimide), aziridines, bisepoxides, epichlorohydrin, carbodiimide, polyisocyanates, and dichlorotriazine. Aldehydes such as formaldehyde and glyoxal are still used and to a small extent even potassium chromium alum ($Cr_2K_2(SO_4)_4.24H_2O$) and potassium aluminum alum ($Al_2K_2(SO_4)_4.24H_2O$).

BIBLIOGRAPHY

"Gelatin" in *ECT* 1st ed., Vol. 7, pp. 145–153, by C. E. Anding, Jr., Kind & Knox Gelatin Co.; "Gelatin" in *ECT* 2nd ed., Vol. 10, pp. 499–509, by Ernest M. Marks, Kind & Knox Gelatin Co.

1. *Proposed Definition for Food Grade Gelatin in 1979,* Gelatin Manufacturers Institute of America, Inc., New York.
2. C. A. Finch and A. Joblin "Physical Properties of Gelatin" in A. G. Ward and A. Courts, eds., *The Science and Technology of Gelatin,* Academic Press, Inc., New York, 1977, Chapt. 8; A. R. Krogh, "Swelling, Adsorptions, and Photographic Uses of Gelatin," Chapt. 14.
3. J. Q. Umberger, *Photogr. Sci. Eng.* **11,** 385 (1967).
4. E. Fuchs, *Adhesion* **5,** 225 (1961).
5. K. M. Hornsby, *Brit. J. Photogr.* **103,** 17, 28 (1956).
6. P. J. Flory and J. J. Behner, *Chem. Phys.* **11,** 513 (1943).
7. I. Tompka, J. Bohonek, J. Spuehler, and M. Ribeaud, *J. Photogr. Sci.* **23,** 97 (1975).
8. A. Libicky and D. I. Bermane in R. J. Cox, ed., *Photographic Gelatin,* Academic Press, Inc., New York, 1972, pp. 29–48.
9. D. W. Jopling, *J. Appl. Chem.* **6,** 79 (1956).
10. A. Courts, *Biochem. J.* **58,** 70 (1954).
11. J. Pouradier and A. M. Venet, *J. Chim. Phys.* **47,** 381 (1950).
12. G. Stainsby, P. R. Saunders, and A. G. Ward, *J. Polym. Sci.* **12,** 325 (1954).
13. J. E. Eastoe in G. N. Ramachandran, ed., *Treatise on Collagen,* Vol. 1, Academic Press, Inc., New York, 1967, pp. 1–72.
14. J. Josse and W. F. Harrington, *J. Mol. Biol.* **9,** 269 (1964).
15. T. Fujii, *Bull. Soc. Sci. Photogr.* **16,** 274 (1966).
16. J. E. Eastoe and A. Courts, *Practical Analytical Methods for Connective Tissue Proteins,* Spon, London, 1963, Chapt. 6.
17. A. Veis "The Macromolecular Chemistry of Gelatin" in B. L. Horecker, N. D. Kaplan, and H. E. Scheraga, eds., *Molecular Biology,* Vol. V, Academic Press, Inc., New York, 1964, p. 112.
18. *Sampling and Testing Gelatins, Brit. Stand.* **6,** 757 (1975).
19. F. W. Wainewright, *GGRA Bull.* **17**(3), 10 (1966).
20. J. Russell and D. L. Oliff, *J. Photogr. Sci.* **14,** 9 (1966).
21. C. R. Smith, *J. Am. Chem. Soc.* **43,** 1350 (1921).
22. J. E. Eastoe, *Biochem. J.* **61,** 589 (1955).
23. W. Horowitz, ed., *AOAC, Official Methods of Analysis,* AOAC, Washington, D.C., 1965.
24. *The United States Pharmacopeia XX,* (*USP XX–NFXV*), The United States Pharmacopeial Convention, Inc., Rockville, Md., 1980.
25. *PAGI Method,* Photographic and Gelatin Industries, Japan, Tokyo, Japan, 1977.
26. *IAG Test,* Gelatin Manufacturers Institute of America, New York.
27. *Consumer Survey,* A. C. Nielsen Co., Northbrook, Ill., 1976.
28. L. F. Tice and R. E. Abrams, *J. Am. Pharm. Assoc.* **14,** 24 (1953).
29. A. W. Kenchington, *Biochem. J.* **68,** 458 (1958).
30. P. Davis in G. Stainsby, ed., *Recent Advances in Gelatin and Glue Research,* Pergamon Press, London, 1958, p. 225.
31. U.S. Pat. 3,184,312 (May 18, 1965), J. W. Gates and E. Miller (to Eastman Kodak).
32. Brit. Pat. 1,075,952 (July 19, 1967), R. C. Clark and co-workers (to Gelatin and Glue Research Assoc.).
33. U.S. Pat. 3,291,611 (Dec. 13, 1966), J. J. Krajewski (to Swift and Co.).
34. Belg. Pat. 672,906 (Nov. 26, 1965), F. Dersch and S. L. Paniccia (to GAF Corp.).
35. Brit. Pat. 1,033,189 (June 15, 1956), J. W. Gates, Jr. and P. E. Miller (to Kodak Ltd.).
36. Brit. Pat. 1,100,842 (Jan. 24, 1968), R. J. Chamberlain (to American Cyanamid Co.).

General References

A. G. Ward and A. Courts, eds., *The Science and Technology of Gelatin,* Academic Press, Inc., New York, 1977.

Gelatin, Gelatin Manufacturers Institute of America, Inc., New York, 1973.

R. J. Croome and F. G. Clegg, *Photographic Gelatin,* Focal Press, New York, 1965.

A. Veis, *The Macromolecular Structure of Gelatin,* Academic Press, New York, 1964.

T. H. James, *The Theory of the Photographic Process,* 4th ed., MacMillan Publishing Co., Inc., New York, 1977.

E. J. Wale and F. I. Jordan, *Photographic Facts and Formulas,* Prentice-Hall, Inc., New York, 1975.

M. H. Gutcho, *Microcapsules and Microencapsulation Technique,* Noyes Data Corp., Park Ridge, N.J., 1976.

Photographic Binders, Research Group of Photographic Binders, Chiba University, Japan, 1976.

R. J. Cox, ed., *Photographic Gelatin,* Academic Press, Inc., New York, 1972.

R. J. Cox, ed., *Photographic Gelatin—II,* Academic Press, Inc., New York, 1976.

FELIX VIRO
Kind & Knox
Division of Knox Gelatine, Inc.

GEMS, SYNTHETIC

Until very recent times, precious stones were obtained by prospecting, exploration, and mining. Today, many traditional gemstones can be produced in the laboratory, including diamond, sapphire, ruby, and emerald. Such gems are called synthetic gemstones since they are of exactly the same chemical composition and structure as the stones occurring in the earth's crust. Artificial or imitation gemstones, on the other hand, are gemlike materials that only have an appearance similar to a particular natural gem. Yttrium aluminum garnet (YAG), for example, is a synthetic garnet; however, since it looks a great deal like diamond, it can be considered an artificial diamond. Thus, this material can be considered either a synthetic or an artificial gemstone, depending on the context. It is not, however, a synthetic diamond or an artificial garnet.

The first synthetic gemstones were successfully made by Marc Gaudin in 1837 (1), when he crystallized ruby [*12174-49-1*] (Al_2O_3:Cr) of rather poor quality in sizes up to 3–4 mm. In the process he used, called flux growth, potassium chromate (K_2CrO_4) and potassium aluminum sulfate ($KAl(SO_4)_2$), were melted together, and ruby crystallized out of the flux. The first ruby of clear, usable quality was made by Auguste Verneuil in 1891 using a flame-fusion process (2). At the present time, commercial sizes and quantities of synthetic ruby, sapphire, emerald, and opal are readily available on the market; diamonds, however, although routinely produced commercially for use as abrasives, are not generally available in sizes suitable for gemstones (see Carbon, diamond, synthetic).

Synthetic gemstones can be separated into the classical synthetics, which are identical to those found in nature, and diamond substitutes. The latter are synthetic crystals which may or may not be found in nature but that appear somewhat like diamond.

The Synthetic Gem Industry

The mainstay of the synthetic gem industry has for many years been the marketing of ruby, sapphire [*1317-82-4*] and spinel [*1302-67-6*] for birthstones and class rings. Sapphire and spinel are both readily produced in a very wide variety of colors which imitate other gemstones. Simulated amethyst [*14832-91-8*], topaz [*1302-59-6*], alexandrite [*12252-02-7*], and many other imitations are produced by the addition or deletion of very small amounts of selected impurities. Domestic production of these materials for gemstone application has ceased and virtually all commercial quantities come from P. Gilson Lapidaires, Djevahirdjian, Les Rubis Synthetique des Alpes, and Baikowski in Europe. The manufacturers have their own secret formulas for type and amounts of cations to be added to obtain the desired color and hue. Over one hundred distinct colors are available covering the entire spectrum.

These crystals are usually produced by the Verneuil or flame-fusion process, which requires large amounts of expensive hydrogen and oxygen, the manufacture of which is power-intensive. The availability of cheap hydroelectric power has strongly influenced the location of facilities producing hydrogen and oxygen and consequently of gemstone plants. The cutting and polishing of sapphire and spinel can be quite costly in terms of labor and cottage industries, individual small businesses that produce large amounts of gemstones with low labor and overhead costs, are abundant in Europe and in the Near and Far East.

An important development in the gemstone industry was the introduction of the star sapphire by the Linde Division of Union Carbide in the 1940s. It was marketed under the name Linde Stars until the early 1970s. Star sapphires are now primarily produced in Japan, Europe, and Israel. These stones achieved great popularity and market acceptance even in the difficult area of men's jewelry. Star sapphires were originally sold in the traditional sapphire blue, followed by red (star ruby), white, and black stars. Later, the stones were marketed in a variety of hues of blue and red and also yellows and greens. These crystals also were grown by the Verneuil process.

The first synthetic gem to be accepted for use in fine jewelry was emerald [*12415-33-7*] produced by Carroll Chatham (San Francisco) in the 1940s. The process was not patented and is a closely guarded secret. The technique is almost certainly a flux-growth method and is believed to require a year per batch. Clusters of crystals of high weight (carats) have been produced. Faceted stones are priced in the range of a few hundred dollars per carat.

The synthetic emerald market today is shared primarily by Chatham and Pierre Gilson of Switzerland, who developed a different process in the 1960s. The details are not known, but it is thought to be a flux process also. High quality emeralds are produced by both processes.

Another synthetic emerald of very high quality was produced by patented hydrothermal processes by Union Carbide in the early 1970s, but operation ceased and patent licenses were sold.

A minor revolution has taken place in the synthetic gemstone market over the last ten years with the introduction and increasing acceptance of simulated diamonds of very high quality. These new stones were partly the result of the intensive research into laser crystals since the early 1960s. The first working laser used a single-crystal ruby and shortly thereafter Bell Telephone Laboratories and Union Carbide developed a laser crystal, yttrium aluminum garnet (YAG), doped with neodymium. The crystal is colorless when undoped and is a reasonable facsimile of diamond (see Lasers).

Strontium titanate and titanium oxide have been produced for several years, especially by the National Lead Company, but have not enjoyed the market acceptance of YAG crystals. These stones exhibit even more fire (dispersion) than do natural diamonds and consequently have been regarded with suspicion. The popularity of imitation diamonds reflects the fact that the price of natural diamonds increased dramatically in the late 1970s.

Airtron Division of Litton Industries, also working in laser crystal research, began marketing undoped YAG as a simulated diamond under the name Diamonaire. From 1969, sales of YAG increased sharply because of good marketing and publicity (3). Cut stones now sell for about $10.00 per carat (4). YAG is grown by the Czochralski process. In this process, the crystal is pulled from the melt and, although it requires several kilowatts of power and a costly noble-metal crucible, it does not depend on large amounts of hydrogen and oxygen.

Gadolinium gallium garnet (GGG) has an appearance even more like diamond than YAG. It is produced as a single-crystal substrate for electronics applications. Stones cut from GGG are of excellent quality and only small quantities have been marketed; GGG is more expensive than YAG.

Unquestionably, the best diamond substitute is the recently introduced cubic zirconia with optical properties close to those of diamond and closer than those of any other synthetic so far. Cubic zirconia can hardly be differentiated from diamond (4), although it has over one and one-half times the specific gravity of diamond. This fact has already had significant economic implications because of widespread swindling practices. Consequently, a device has been developed to measure the thermal conductivity of mounted stones in order to determine whether they are genuine diamonds. Cubic zirconia, cut but unset, sells in the range of $20.00 per carat (see Carbon, diamond).

A growth technique for cubic zirconia was perfected at the Lebedev Physical Institute in the USSR in the course of research into laser crystals. Crystals are obtained by melting ZrO_2 stabilized with Y_2O_3 or CaO and slowly cooling the melt to room temperature. At present, cubic zirconia is produced at the Lebedev Institute, by Djevahirdjian in Switzerland, and by the Ceres Corporation in Waltham, Massachusetts; the latter is marketing the product through MSB Industries. Airtron of Litton produces and markets cubic zirconia as Diamonaire II. Retail sales of set jewelry containing synthetic gemstones amounted to about $50 million in the United States in 1978 (4). With increasing interest in diamond substitutes, this market is expected to grow.

Properties

The properties desired for good gemstones are hardness, high refractive index, high dispersion, color, and clarity. Hardness greater than about 6 on Mohs scale is required in order that a gem can be worn without its surface being easily scratched and dulled. Some stones, eg, opal and turquoise (hardness, 5–6), are generally acceptable because of their great beauty (5). High refractive index or brilliance is sought for its light gathering and reflecting ability although this is not as important for a highly colored stone. Although large quantities of synthetic quartz are manufactured for electronic applications, very little is used as a gem because the refractive index of 1.54 produces a dead appearance. Quartz is useful as a gem only when impurities cause coloration or other striking effects, as displayed by an abundance of natural crystals. Refractive index is customarily measured at the sodium D line at 589 nm (see Silica).

Dispersion is the rate of change of the refractive index with wavelength. This property is referred to as fire in gem terminology since it determines the degree to which white light is separated into colors by refraction. Dispersion is conventionally given as the difference in refractive indexes measured at 687 nm and 431 nm. Diamond, for example, has a dispersion of 0.044.

Manufacture

Verneuil Process. In the Verneuil process, a seed crystal is placed upright in a ceramic, insulated furnace and an oxyhydrogen flame is directed from above on the seed. When the seed is slightly molten on its upper surface, the raw material (Al_2O_3, for sapphire) in very finely divided powder form is sprinkled downward onto the molten cap. The powder melts and forms part of the liquid adhering to the upper tip. As the process continues, the seed is very slowly lowered and more powder is added. As the liquid reaches a cooler zone, it solidifies and a crystal is built onto the seed. The process continues and crystals as long as 40 cm can be produced (see Fig. 1).

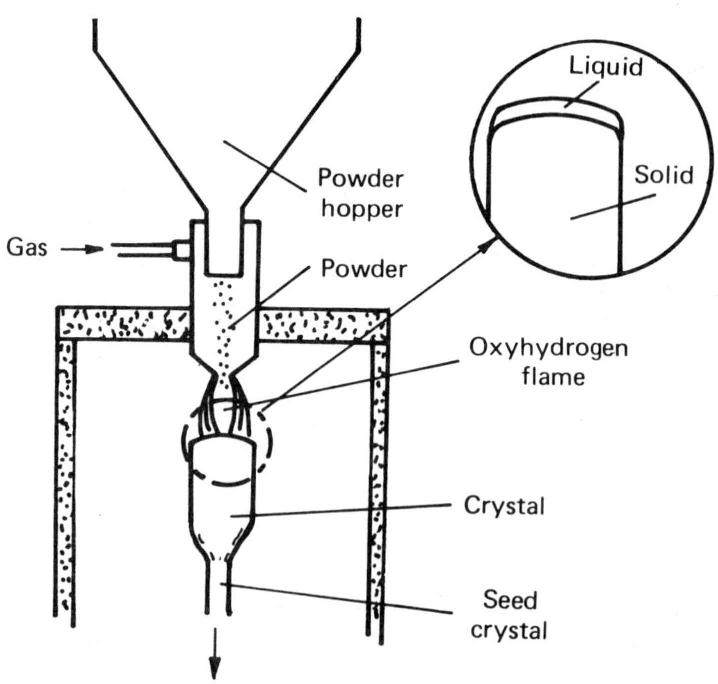

Figure 1. Verneuil or flame-fusion growth.

Dopants used to produce colored stones (or titania for stars) are mixed as oxide powders into the raw material (host) powder as homogeneously as possible in a very fine state. Constant feeding of the powder can be a serious problem. Also crucial is the thermal gradient in the furnace since it must be high enough to promote crystallization with only a thin layer (<1 mm) of liquid, but low enough to prevent cracking of the crystal because of thermal stress. Even with proper design, Verneuil-grown crystals are customarily annealed after growth to relieve internal stresses.

Verneuil-grown crystals can be readily identified by using Schlieren photographs which show striations of color (in doped stones) perpendicular to the direction of growth. Occasionally, bubbles and powder slugs are also present, ie, regions in which powder has been encapsulated but not melted. Both striations and defects can frequently be detected with a magnifying lens. Although numerous techniques are available, the Verneuil process is the most widely used for gemstone production.

Hydrothermal Technique. In the hydrothermal method, patented by Union Carbide, seed crystals are cut into slabs, 10–20 by 60–70 mm, of <1 mm thickness and suspended near the top of a pressure vessel lined with a noble metal. For the production of synthetic emerald, for example, seed crystals may be synthetic emerald or, usually, natural aquamarine. BeO, Al_2O_3, and SiO_2 are placed in the bottom of the vessel together with a small quantity of Cr_2O_3. Slightly acidic water solution is added and the vessel is sealed (Fig. 2). The container is heated to 500–600°C with a slight thermal gradient (the bottom hotter than the top) imposed to promote convective circulation.

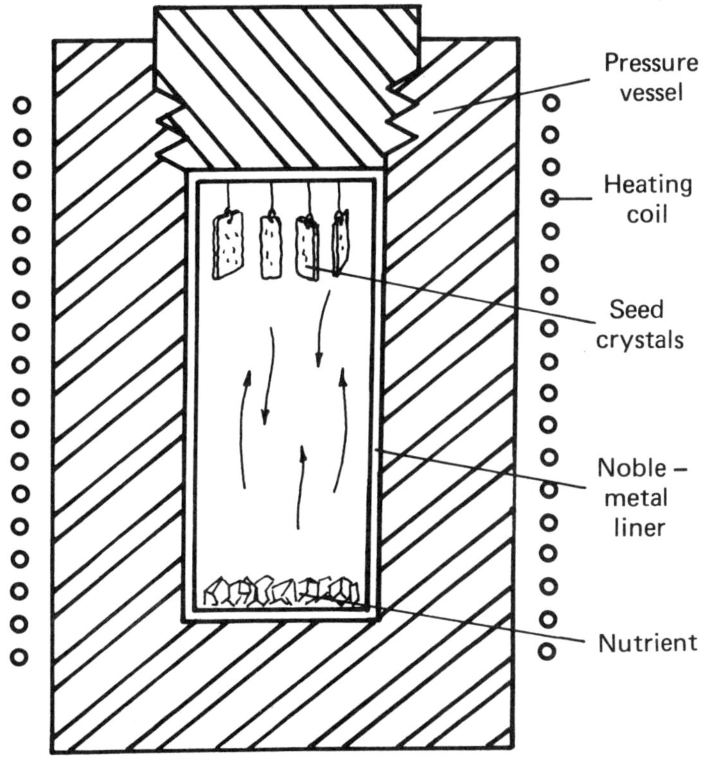

Figure 2. Apparatus for hydrothermal crystal growth.

Pressures develop internally to 100 MPa (ca 1000 atm) and vessel design is of paramount importance to accommodate these pressures. The nutrients at the bottom are dissolved and transported to the cooler top where the seeds are suspended. Growth occurs on the seed slab surfaces and a sandwich is built up having the seed as its center layer (see Fig. 3).

Czochralski Process. In the Czochralski process, for example for YAG, the raw material, Al_2O_3 and Y_2O_3, is melted in a noble-metal crucible, usually by induction heating. When the material is molten, a seed crystal attached to a rod is lowered from above to contact the melt surface. The temperature is adjusted so that the seed and liquid are in equilibrium and the seed is very slowly withdrawn. As the seed moves upward, a meniscus of liquid is lifted up and cooled (primarily by radiative losses), crystallizing onto the seed tip. This new solid material in turn forms a meniscus as the process continues and more crystal is added (Fig. 4).

Classical Synthetic Gems

Sapphire (Corundum). Corundum, Al_2O_3, was the first synthetic gem to be produced. Pure Al_2O_3 is colorless and is known as white sapphire. Its use as a gemstone is limited because of its rather low dispersion of 0.018, and relatively low refractive index of 1.76, which render it a somewhat lifeless gem. It is very hard with a rating of 9 on Mohs scale. The properties of sapphire change radically when it is doped with other cations to produce color. When about 0.1% of the aluminum atoms are replaced by chromium atoms, ruby, of the traditional pigeon-blood red, is the result. It is macroscopically indistinguishable from a natural ruby. A much lighter doping of chromium (0.05 wt %) produces laser ruby which is sometimes used for gems (Fig. 5).

Addition of iron and titanium produces the well-known blue sapphire, whereas cobalt yields green sapphire. Almost all metals have been tried in sapphire at one time or another, occasionally with spectacular results.

When up to about 1% titanium oxide is added to Al_2O_3, it becomes incorporated into the crystal. However, heat-treating or asteration at about 1200°C causes the titanium oxide to precipitate along the six principal crystallographic planes of sapphire's

Figure 3. Emerald, $Cr:Be_3Al_2Si_6O_{18}$. Synthetic emerald and faceted gem. The wire used for suspending the seed can be seen at the end of the crystal. Photograph by V. F. S. Yip.

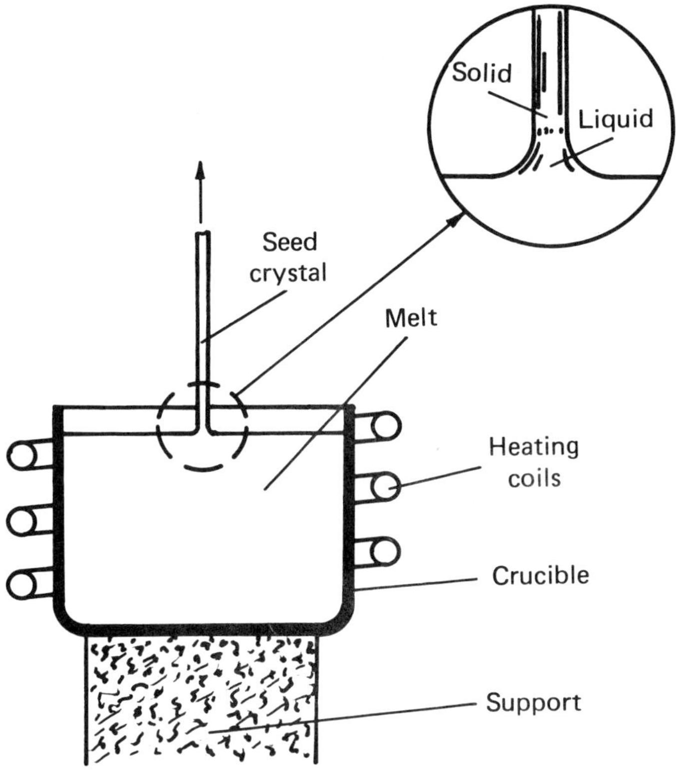

Figure 4. Diagram of crystal growth by the Czochralski technique.

Figure 5. Ruby, $Cr:Al_2O_3$. Laser ruby having dopant concentration of 0.05 wt % (Cr_2O_3/Al_2O_3). Photograph by V. F. S. Yip.

hexagonal structure as elongated needles. When such a crystal is cut into a shape resembling a positive lens (cabochon), the reflection and diffraction of light by these needles produces the well-known asterism in star sapphire. Addition of dopants pro-

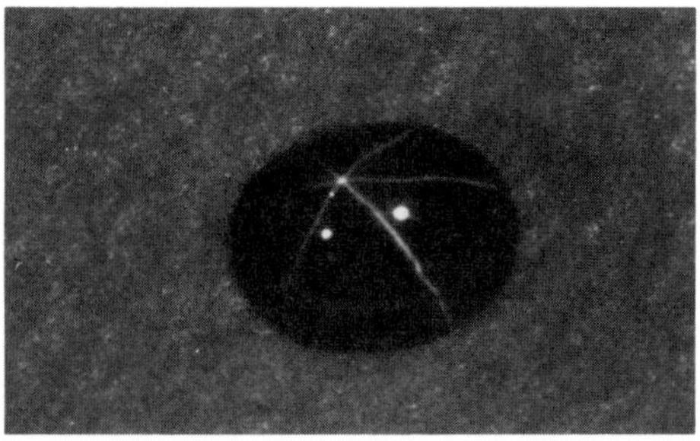

Figure 6. Black star sapphire, Al_2O_3:V, Ti. Synthetic star sapphires can be doped to produce a variety of colors. Photograph by V. F. S. Yip.

duces colored stars such as star ruby (Cr) or the black star sapphire (V) in Figure 6. Sapphires are made by the Verneuil process.

Spinel. Another easily doped synthetic gem is spinel, $MgAl_2O_4$, a colorless crystal of cubic structure. Like white sapphire, spinel is not a good diamond substitute, since it has a refractive index even lower than sapphire (1.72), although its dispersion at 0.020 is slightly higher (6). Like sapphire, it is readily doped to produce colors, but since the dopant atoms are at different sites, the colors are different from those resulting from the same ions in sapphire. Blue spinel is produced by the presence of cobalt in the lattice whereas the same ion in sapphire yields a green gem. Spinels are made by the Verneuil process.

Beryls (Emerald). Beryls are silicates of aluminum and beryllium, eg, $Be_3Al_2Si_6O_{18}$. Depending on the doping, the color may range from very pale aquamarine to the lovely green of emerald. Emerald is generally beryl doped with chromium. It is highly colored but has a very low brilliance and fire with a refractive index of about 1.58 and dispersion of 0.014. Its splendid hue more than makes up for these deficiencies. Good stones are usually given an emerald cut to best display the color. Stones vary from bright green with a blue tint to green with a yellow tint. The blue tint is generally more desirable and appears to be caused by small impurities of vanadium. Synthetic emeralds are produced by flux growth and the hydrothermal technique.

Diamond. Carbon can occur as elemental carbon, graphite, or diamond (see Carbon, diamond, natural; Carbon, diamond, synthetic). The carbon phase diagram shown in Fig. 1, Vol. 4, p. 677 shows the pressures and temperatures necessary to obtain synthetic diamond. The difficulty is in attaining the necessary pressures and temperatures in practical laboratory apparatus.The design that solved the problem is patented by the General Electric Company and uses tapered pistons, a hydraulic press, and the now-famous belt of tungsten carbide. In modified form (tetrahedral anvils), it is used to produce small diamonds for cutting, grinding, and polishing (see Abrasives).

Probably the largest synthetic diamond produced was of about one carat and was made by Robert Wentorf in 1967. His patent involves the dissolution of diamond crystals and subsequent recrystallization on a seed crystal in a cooler region. This process is carried out at 1420°C at 6080 MPa (60,000 atm). The process is slow and the diamonds produced are much more expensive than natural diamonds (7).

Diamond Substitutes

The development of YAG and cubic zirconia (ZrO_2) has had a strong impact on the synthetic gemstone industry. The two diamond substitutes developed much earlier were strontium titanate and titanium dioxide. Although the latter two are desirable gemstones in their own right, they are no longer considered as adequate diamond imitations when compared to the recently developed products (Table 1).

Strontium Titanate. This material, $SrTiO_3$, is widely sold under various tradenames; Fabulite and Wellington are perhaps the most widely known. Strontium titanate has very high dispersion and a refractive index (2.41) that is nearly identical to that of diamond (5). These properties cause the stone to be brilliant but excessively fiery. In unprotected settings such as a Tiffany ring, it soon becomes scratched and dulled because of its low hardness. It does not wear very well in any mounting.

Gemstone-grade strontium titanate was developed at the National Lead Company in the mid 1950s (8). The Verneuil or flame-fusion process was used employing oxygen and hydrogen as fuel since the material melts about 2050°C. Patents were issued for various modifications of the process and apparatus. For example, addition of excess strontium oxide, up to about 1%, added to the feed powder resulted in much better growth. Doping with a variety of metal oxides gives colored stones; for example, nickel produces yellow, and tantalum blue. The as-grown crystal is characteristically black when grown under reducing mixtures of the fuel gases. After growth, heating to about 1000°C in an oxidizing atmosphere produces a clear material.

Rutile. The refractive index, 2.8, and dispersion, 0.280, of rutile (TiO_2) are significantly higher than those of diamond which makes it a very spectacular gem indeed (5). It is, however, tetragonal in crystalline structure and therefore exhibits double refraction which causes the images of the back facets to appear quite fuzzy. Furthermore, it has never been produced completely water-white or colorless, rather always with a slight straw color. Although TiO_2 is somewhat harder than strontium titanate, it is still too soft for jewelry. For color, the crystal has been produced with several dopants such as Cr for red and Ni for yellow.

Rutile is produced with the Verneuil technique, using oxygen and hydrogen, because of the high (1925°C) melting point. Addition of alumina or gallia minimizes suboxide states and results in near-white crystals. As with $SrTiO_3$, however, the growth conditions produce black boules which are then heated in an oxidizing atmosphere for several hours at about 1100°C to produce clear crystals.

Yttrium Aluminum Garnet (YAG). Yttrium aluminum garnet ($Y_3Al_5O_{12}$) is the best known diamond substitute. It has an acceptable hardness of 8.5, is clear white, and cubic. Its index of refraction (1.82) and dispersion (0.028) are too low for the required brilliance. Nevertheless, it is widely accepted in the marketplace. It was first produced at Bell Telephone Laboratories in 1963 by a flux-growth process and later by Union Carbide by the Czochralski process. This crystal has the distinction of being the first gem material made commercially by the latter process. YAG is grown as a colorless crystal and no postgrowth oxidation is required. Crystals are readily doped to yield colors, the most notable of which is green produced by chromium.

Gadolinium Gallium Garnet (GGG). Gadolinium gallium garnet ($Ga_3Gd_5O_{12}$) has never been a great commercial success as a gemstone although its properties make it a good diamond substitute, with a refractive index of 2.03, dispersion of 0.038, and hardness of 7 (9). However, the raw materials are very expensive and GGG tends to

Table 1. Properties of Diamond Substitutes

Crystal	Chemical formula	CAS Registry Number	Structure	Specific gravity, g/cm^3	Hardness, Mohs	Refractive index[a]	Dispersion[b]
diamond	C	[*7782-40-3*]	cubic	3.52	10	2.42	0.044
cubic zirconia[c]	ZrO_2	[*1314-23-4*]	cubic	6	8	2.16	0.062
gadolinium gallium garnet (GGG)	$Gd_3Ga_5O_{12}$	[*12024-36-1*]	cubic	7.05	7	2.03	0.038
yttrium aluminum garnet (YAG)	$Y_3Al_5O_{12}$	[*12005-21-9*]	cubic	4.65	8.5	1.83	0.028
titanium dioxide (rutile)	TiO_2	[*13463-67-7*]	tetragonal	4.26	6	2.8	0.280
strontium titanate	$SrTiO_3$	[*12060-59-2*]	cubic	5.15	5	2.41	0.190

[a] At sodium line, 589 nm (high refractive index produces brilliance).
[b] At 431 and 637 nm (high dispersion produces fire).
[c] Properties vary slightly depending on stabilizer.

discolor toward yellow upon exposure to sunlight for prolonged periods. This coloration can be reversed, as with TiO_2 and $SrTiO_3$, by heating in an oxidizing atmosphere. The crystal is grown by the Czochralski method and can be readily doped to produce colors.

Cubic Zirconia. The best of the diamond imitations, cubic zirconia (ZrO_2), has a refractive index of 2.16, dispersion of 0.062, hardness of 8, and is colorless. Zirconia is normally of monoclinic structure at room temperature, but on heating it changes to tetragonal (1250°C), hexagonal (ca 1900°C), and finally cubic (ca 2300°C). Upon cooling, it reverts to monoclinic. However, addition of yttrium oxide (Y_2O_3) or calcium oxide (CaO) in amounts of 6–40 mole % stabilizes zirconia and the cubic form can be maintained at room temperature (9). The very high melting point of 2750°C prohibits the use of crucibles for Czochralski growth. Therefore, although other methods are possible, the current practice is to use skull melting (or cold crucible) growth.

In skull melting, the raw material acts as its own crucible lining. Powdered zirconium oxide is placed inside a water-cooled copper container in such a fashion that, even though the central region is heated to the liquid state, the outer zone (and bottom) are cooled to form a crust which contains the liquid. The cooled copper is so shaped that it interacts very little with an induction heating coil. Zirconium metal is placed in the center as a susceptor to begin a melt. When ZrO_2 is heated to 1600°C it becomes a conductor. Further heating elevates the temperature to the melting point and by slowly cooling the melt large crystallites of cubic zirconia are formed. This is a modification of the Stockbarger technique. Presumably, Czochralski growth could be carried out as well.

Skull melting has been investigated for many materials and was finally successful for cubic zirconia. Doping with cerium or chromium gives orange or red crystals, respectively.

A comparison of imitation diamond gems and a genuine diamond is shown in Figure 7.

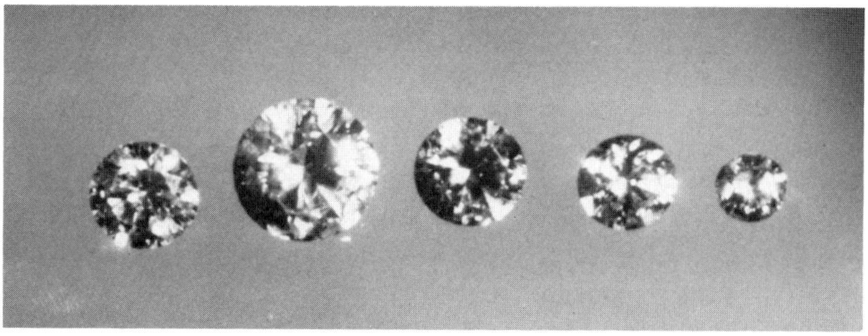

Figure 7. Comparison of simulated diamonds with genuine diamond. Left to right: strontium titanate, YAG, GGG, cubic zirconia, diamond. Photograph by V. F. S. Yip.

BIBLIOGRAPHY

"Gems, Synthetic" in *ECT* 1st ed., Vol. 7, pp. 157–167, by A. K. Seemann, Linde Air Products Co., A Division of Union Carbide and Carbon Corp.; "Gems, Synthetic" in *ECT* 2nd ed., Vol. 10, pp. 509–519, by C. Robert Castor, Electronics Division, Union Carbide Corp.

1. D. Elwell, *Man-Made Gemstones,* Ellis Horwood Ltd., Chichester, 1979, p. 26.
2. K. Nassau and J. Nassau, *Lapidary J.,* 1442 (Feb. 1971).
3. Ref. 1, p. 109.
4. *Los Angeles Times,* 1 (May 16, 1979).
5. R. T. Liddicoat, Jr., *Handbook of Gem Identification,* Gemological Institute of America, Los Angeles, Ca., 1966.
6. J. Sinkankas, *Gemstone and Mineral Data Book,* Winchester Press, New York, 1972.
7. R. H. Wentorf, Jr. "Diamond Formation at High Pressures" in R. H. Wentorf, Jr., ed., *Advances in High Pressure Research,* Vol. 4, Academic Press, Inc., New York, 1979, pp. 249–281.
8. D. MacInnes, *Synthetic Gems and Allied Crystal Manufacture,* Noyes Data Corp., Park Ridge, N.J., 1973.
9. K. Nassau, *Lapidary J.,* 900 (July 1977).

L. R Rothrock
Union Carbide Corporation

GENETIC ENGINEERING

Genetic engineering usually implies deliberate manipulation of genes of various organisms, procaryotic or eucaryotic, in order to achieve useful products of metabolism or to cause a permanent hereditary change in the organism. Although the techniques for engineering the genetic make-up of higher eucaryotes have not been developed very well, those for lower eucaryotes (eg, yeasts (qv)) and procaryotes have been developed to an extent by which foreign DNA from any source can be introduced and stably maintained in bacteria such as *Escherichia coli.* The purpose of this article is to review both the state of the art concerning the technology of genetic manipulation of procaryotes with primary emphasis on the methodologies and the scope of the application of such genetically-altered microorganisms in industry and in biomedical areas.

Two techniques usually are employed for manipulating the genes of microorganisms: *in vivo* genetic engineering in which the changes in genetic constitution are brought about in cells by processes analogous to those occurring in nature; and *in vitro* recombinant-DNA techniques in which foreign genes from entirely different sources can be ligated with stably-replicating plasmid (or phage) DNA and introduced within the cells. Since the former technique involves mutations or natural exchanges of genes among procaryotes, the improvements are basically quantitative, even though they may involve elaboration of the gene products by a different host. In contrast, the *in vitro* recombinant-DNA technique may involve production of entirely new substances (eg, substances of animal origin) in microorganisms and, therefore, may involve both qualitative and quantitative changes. Because recombinant-DNA technology allows the incorporation and replication of DNA from various animal sources in bacteria—and this type of exchange is considered extremely rare in nature—this technology has generated considerable amount of controversy regarding the safety and well-being of mankind. There are strict guidelines in various countries of the world regulating the type of experiments, the nature of the host bacterium and foreign DNA inserts, and the volume of experimental materials that can be handled in laboratories involved in recombinant-DNA work (see also Microbial transformations).

In Vivo Genetic Engineering

In vivo genetic engineering may involve simple mutational alteration of the genetic material leading to an enhanced yield of the product or an improvement in the quality of the product. Such techniques have led to the isolation of mutant *Actinomyces* or bacterial strains capable of producing antibiotics (qv), vitamins (qv), or amino acids (qv) in high yield. Another widely used technique employs plasmid transfer between different bacterial species or genera. Thus, entirely new genetic functions can be transferred from the chromosome of one bacterial genus to different genera in the form of plasmids. The role of phage Mu in promoting transposition of blocks of genes from the chromosome onto plasmids (including such broad host-range plasmids as RP4; see Nomenclature) has been studied intensively, and new strains having a variety of useful functions have been constructed. It has been demonstrated that *E. coli* genes can be transposed to F′ episomes (1) or to RP4 (2) during growth of bacteriophage Mu, and such transpositions take place either after infection with Mu or after induction of cells carrying a Mu *cts* prophage. The transposition derivatives then can be transferred by conjugation to recipient cells that should be Mu resistant, lysogenic to Mu (to avoid zygotic induction of the Mu prophage linked to the plasmid) and $recA^-$ (to avoid integration of the donor genes with the recipient chromosome). Under appropriate conditions, various segments of chromosomal genes can be incorporated to a RP4: : Mu plasmid which, in turn, can be transferred to a large number of gram-negative bacteria.

In addition to plasmid-mediated transfer of chromosomal genes from one bacterium to another (3), plasmids themselves may specify functions that can be utilized for construction of novel strains. Evidence is accumulating that suggests that the biosynthetic genes for some antibiotics in *Actinomyces* may be specified by plasmid-borne genes (4). Thus, it may be possible to manipulate the plasmid copy number or the genes' transfer to new hosts to enhance greatly the yield of antibiotics or even to produce hybrid antibiotics having some desirable characteristics. Similarly, plasmids are known to code for the biodegradation of hydrocarbons and other complex organic compounds. Multiplasmid strains harboring several such plasmids and capable of rapidly degrading a large number of hydrocarbons present in crude oil have been constructed (5). Wide host-range plasmids, harboring chromosomal genes encoding useful functions, eg, nitrogen fixation (qv), have been constructed which allow the transfer of nitrogen fixation genes to root nodule and other bacteria (6). Considerable progress also is being made to introduce nitrogen-fixation and other desirable characteristics such as resistance against harmful pests or plant viruses, to the plants themselves. A small 6 Mdal segment of the Ti plasmid, which is known to incite crown gall tumors in certain dicotyledonous plants, is transferred to the plants during infection and somehow initiates tumor formation (7–8). It now appears that at least two sets of genes, specifying the tumor inducing principle and controlling opine synthesis, are transferred from *Agrobacterium tumefaciens* to the plants. Insertion of a transposon near the opine synthesis gene allows the subsequent transfer of the transposon into the plant cells. If this can be developed as a general technique to introduce foreign

DNA into the plants or their protoplasts, from which plants can be regenerated, it certainly would open up a new field of engineering plant species having novel characteristics. It should be stressed, however, that there is no definitive evidence that procaryotic genes, including the Ti plasmid DNA segment, are expressed functionally within the plants to form procaryotic protein products. With increasing knowledge about the structure and functions of the Ti gene segments or the cauliflower mosaic virus genomes and their gene products, it might be possible to fuse desirable foreign DNA segments in phase with nonvirulent forms of such vehicles so that the passenger segments are expressed functionally within the plants.

In Vitro Genetic Engineering (*in vitro* Recombinant-DNA Technology)

One technique that has made a tremendous contribution to the development of genetic engineering research is the *in vitro* recombinant-DNA technology, also known as molecular cloning or gene cloning (9–10). In its most widely used and simplest form, the technique allows the incorporation of any segment of a foreign DNA, procaryotic or eucaryotic, into a piece of phage or bacterial plasmid DNA and such recombinant-DNA segment then is reintroduced into the bacterial cell by transfection or transformation. Since the vector is the phage or bacterial plasmid DNA, it is not restricted within the cell and the foreign DNA segment replicates stably as part of the bacterial vector DNA. The simple technique of joining segments of DNA derived from procaryotes or eucaryotes with the vector DNA and their subsequent introduction into the bacterium may mark the beginning of a revolutionary way of manufacturing biologically functional proteins such as insulin (qv) and growth hormones (qv), antibodies, interferons, blood-clotting factors, and a host of other pharmacologically important compounds, by fermentation (qv) (11). Since production of these substances is stringently regulated in the animal cells, they are available from animal tissues only in very small amounts. Hence, their fermentative production would make them available in large quantities at modest cost and thus would contribute significantly to the betterment of human health and welfare. Some newly developed techniques of making animal proteins in *E. coli* are described in detail (see Industrial Applications) and other areas of application are discussed in ref. 11.

General Procedures of Gene Cloning. The general procedures used in combining a recombinant-DNA molecule *in vitro,* using a plasmid as vector, are shown in Figure 1. The first step is to isolate and purify the desired phage or plasmid DNA to be used as a vector. The foreign DNA to be recombined with the vector DNA should be purified, if possible. In some cases, as in shotgun experiments, entire DNA molecules may be sheared or digested with restriction endonucleases before a specific fragment is inserted into the vector DNA. The ease of isolation and purification of the foreign DNA depends upon the source of the DNA.

Once the vector and foreign DNA are available, the DNA segments are cut in areas that should not interfere with the biological activities of the gene(s) to be cloned or of the replication, maintenance, or selectable characters of the vector. In many cases, the cutting is accomplished through use of site-specific restriction endonucleases that generate staggered cuts on the DNA molecule producing single-stranded ends that can later be annealed at a low temperature.

The specifically cut DNA fragments are mixed and joined using the enzyme DNA ligase. The recombinant molecules thus generated are introduced into the bacterial-

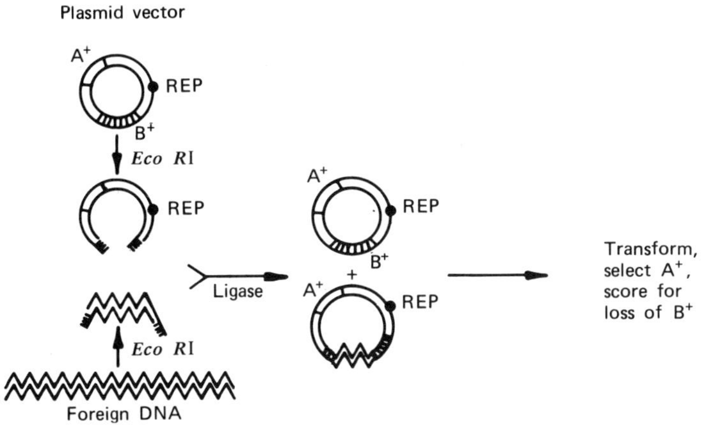

Figure 1. The procedure for cloning foreign DNA in *E. coli*. Both A and B genes represent selectable traits, so that introduction of foreign DNA in B gene leads to the loss of an identifiable function. REP represents replication and maintenance genes (12). Courtesy of CRC Press.

host cells. For efficient uptake, the bacterial-host (eg, *E. coli*) cells are treated with $CaCl_2$ prior to treatment with the ligated DNA mixture. Specific colonies harboring the recombinant-DNA molecules and having the genetic or physical characteristics of the foreign DNA then are scored. The recombinant-DNA molecules can be isolated from the transformants, and in most cases, the foreign DNA can be dissected and studied in detail for physical and biological properties.

Isolation of Vector and Passenger DNA. The methods for the isolation of phage λ derivatives or various plasmid DNAs, including their properties as cloning vectors, have been well developed and described (12). The usual methods of plasmid DNA isolation do not appear to work in some cases, and modified methods for isolation of plasmid DNA have been described (13).

The procedure for the isolation of the foreign DNA depends upon the origin (ie, procaryotic or eucaryotic) and the complexity of the DNA. If the DNA is from any eucaryotic organism having a large number of repeated sequences, then it sometimes can be separated as a satellite band distinct from the main band in a CsCl or Cs_2SO_4 density gradient, particularly in the presence of Ag^+ or Hg^+. Nonchromosomal DNA such as amplified rRNA *Tetrahymena* genes or yeast genes sometimes can be separated from the chromosomal DNA by agarose gel electrophoresis or rate–zonal centrifugation in a sucrose gradient.

If a specific gene function is to be characterized on a segment of DNA that is to be cloned, the most widely used method is described in ref. 14. In this method, the DNA is fragmented by restriction endonuclease treatment and is separated on an agarose gel. The DNA fragments from the gel, after alkali denaturation, are transferred to nitrocellulose filter strips. Radioactive RNA, complementary to the desired gene, then is hybridized with the various bands on the filter strips that later can be characterized by radioautography (see Radioactive drugs and tracers). The band showing a high level of radioactivity because of the hybridization of the radioactive complementary RNA is the desired band. If the gene has other biological properties that can be transferred by transformation, then replica plating on a lawn of cells that lack the particular gene activity usually will show the presence of the active gene fragment in a particular band.

The recovery of bands from agarose gel usually can be accomplished by mashing the gel in a buffer and centrifuging it to remove the agarose. However, generally the recovery is poor under such conditions. The freeze–squeeze method of isolation involves freezing the gel and squeezing the fluid out into a buffer. Electrophoretic migration of the DNA after separation of the bands also can be used for recovery of a specific band.

Cutting DNA. In most instances, cutting the DNA is accomplished through use of a variety of restriction endonucleases. In general, there are two primary types of restriction endonucleases. Type I enzymes from *E. coli* strains K and B are characterized by high molecular weight with three different subunits, and they require ATP and *S*-adenosylmethionine (SAM) as cofactors in addition to Mg^{2+}; the breakage site is different from the recognition site. The Type II enzymes are characterized by low molecular weight, mostly identical subunits, and a requirement for Mg^{2+} but not for ATP or SAM. These enzymes have been used extensively in molecular cloning because many of them appear to recognize a sequence of five or six base pairs, producing staggered cuts on the DNA. Many of these enzymes have different nucleotide sequence specificities (Table 1). This allows a number of such enzymes to cleave DNA segments at different sites, thereby producing large numbers of fragments. The realignment of these fragments facilitates the construction of physical maps of various plasmids, bacteriophage, or viral genomes.

Method of Joining DNA Ends. In general, there are three methods that are employed for joining vector DNA and the foreign DNA to construct the recombinant-DNA molecule. If the vector or the foreign DNA fragments have been generated by digestion with Type II restriction endonucleases, which produce staggered cuts, they can be joined initially through hydrogen bonds by annealing at a low temperature and then by joining the fragments permanently through covalent bonds generated by ligase. This method is used widely in genetic engineering because of its simplicity and because the hybrids can be dissected to regenerate the original foreign DNA fragment for sequence analysis or to study other relevant properties. This method also is useful for cloning larger segments of DNA by using restriction endonucleases in the presence

Table 1. Sequence Specificities of Some Commonly Used Restriction Endonucleases

Bacterial source	Enzyme	Sequence specificity[a]
*E. coli/R*I	*Eco R*I	G↓A—A—T—T—C
Bacillus amyloliquefaciens H	*Bam H*I	G↓G—A—T—C—C
B. globigii	*Bgl* II	A↓G—A—T—C—T
Xanthomonas malvacearum	*Xma* I	C↓C—C—G—G—G
Providencia stuarti	*Pst* I	C—T—G—C—A↓G

[a] The left end of each sequence is the 5′ end and the right end is 3′. Only one strand is shown for convenience, although the enzymes break duplex DNA. The arrow shows the position of the bonds broken.

of distamycin A, which allows partial digestion and thereby produces large fragments with cohesive termini. The major drawback of the method is that if the foreign gene itself has a site for the endonuclease or if there are no restriction endonuclease sites on the foreign DNA segment, such genes cannot be cloned.

A second method of joining DNA fragments, which obviates the need for cohesive ends, is the flush-ended joining catalyzed by ligase that is prepared from *E. coli* cells infected with Phage T4 (15). The efficiency of this ligation is low, and relatively large quantities of the enzyme are necessary. However, since the DNA fragments do not have to have protruding single-stranded ends, any two double-stranded DNA segments can be joined by this procedure. This method is very useful when the DNA fragments lack cohesive termini or when two fragments produced by two different restriction endonucleases (and thus have different single-stranded ends) need to be combined. Synthetic restriction endonuclease sites can be ligated conveniently to DNA fragments by this procedure. A major disadvantage of flush-ended joining is that the two DNA fragments cannot be conveniently dissected to study the foreign DNA fragment after cloning.

A third, and one of the most widely used methods, is the addition of deoxynucleotides to the 3′-OH-terminus of either sheared or restriction endonuclease-generated DNA fragments with the use of the enzyme, deoxyribonucleotidyl terminal transferase (16). Incubation of the DNA fragment with a single deoxynucleotide triphosphate and the enzyme leads to the addition of a homopolymer of ca 100 bases. Since terminal transferase adds homopolymers at the protruding 3′-OH end, a common procedure was to remove about 30 nucleotides from the 5′ OH end by digestion with λ exonuclease. Using cobalt instead of magnesium as the divalent cation, all DNA fragments, with or without a protruding 3′-end, can act as recipients of homopolymers at the 3′-OH end. Thus, the linearized vector may have a poly dT tail, and the passenger DNA segment may have a poly dA tail. Mixing and subsequent annealing of the two DNA copies will produce recombinant-DNA molecules (Fig. 2) that can be used for the transformation of *E. coli* cells. Selective methods then are used for scoring those transformants that demonstrate the presence of foreign DNA segments in the total cellular DNA.

There are several advantages in using the homopolymer-tailing method to introduce foreign DNA segments in *E. coli*. Because no endonuclease cutting is involved, rather large DNA segments, having no or large number of restriction endonuclease sites or complete cDNA copies of eucaryotic mRNA, can be conveniently tailed and cloned with bacterial plasmids (see Applications). Since individual homopolymer-tailed preparations cannot undergo ring closure, this method ensures hybrid formation between poly dA-tailed gene copy and poly dT-tailed plasmid DNA and facilitates recovery of such hybrids. Thus, DNA segments that have entirely different base sequences at the end, specific promoter sequences, or chemically-synthesized DNA can be conveniently joined. The disadvantages of the method are that terminal transferase can add nucleotides at single-strand breaks incurred during DNA preparation or resulting from contaminating nucleases. This should be avoided by using nick-free DNA and nuclease-free terminal transferase. Also, the foreign DNA cannot be separated easily from the vector DNA after cloning. Recently, however, methods have been developed that allow specific excision of the insert DNA from the recombinant-DNA molecules with the paired dA—dT (and probably dG—dC when poly-dG and poly-dC tails are used) termini intact (17). The method permits separation of insert DNA

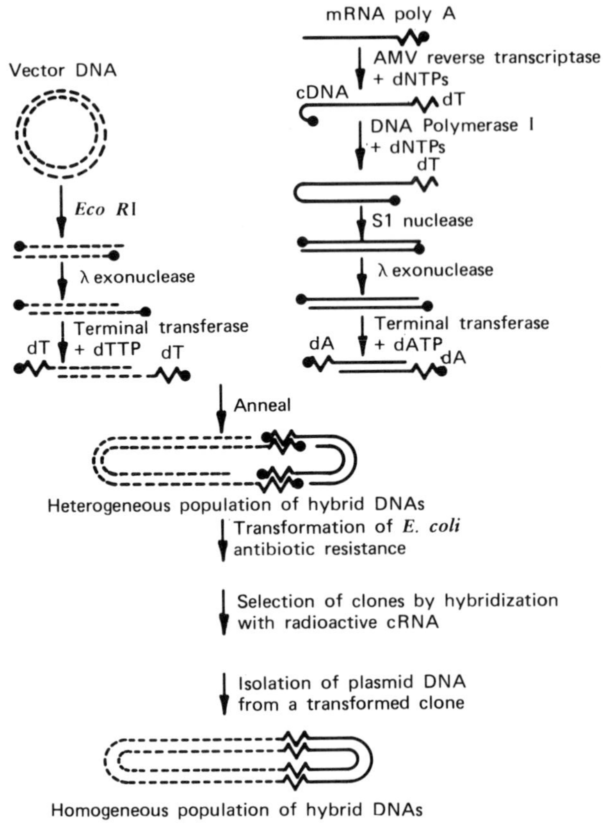

Figure 2. Coupling DNA segments by means of poly dA and poly dT complementary single-stranded end. The fold-back hook structure of the cDNA promotes priming of the second-strand synthesis (16). Courtesy of CRC Press.

segments that are free of the vector DNA; because such insert DNA retains the original poly(dA—dT) sequences at the end, after a brief digestion with λ 5′-exonuclease, it can be introduced into an alternative vector for cloning in another host.

Identification of Clones with Specific Genes. Once a segment of foreign DNA has been ligated with the vector DNA and introduced into *E. coli* cells by transformation, often a major problem is identifying the transformant colony that harbors the recombinant-DNA molecule. Several methods have been developed to facilitate scoring of transformants with a foreign DNA insert. If the foreign DNA is a procaryotic DNA segment that is functionally expressed in the transformant, direct scoring of transformants for the traits coded by the foreign DNA, such as His^+ (a phenotype) or Km^r (Kenamycin resistance) is possible. Various cloning vectors have been constructed that allow ready detection of insert DNA by a process called insertional inactivation (18). This procedure employs vectors having more than one scorable trait, such that there should be a cleavage site for the restriction endonuclease in a gene coding for a readily identifiable function. Insertion of foreign DNA at this site leads to inactivation of the gene, producing a deficient phenotype. The presence of a second identifiable character on the vector allows scoring of the transformants, and those transformants having the deficient phenotype for the second character, therefore, must have foreign DNA at that site (see Fig. 1).

Indirect selection of recombinant molecules based on size or density difference of the insert DNA from the vector DNA also has been employed. The method involves the isolation of plasmid DNA from the whole population of transformed bacteria, the fractionation of this DNA based on size (sucrose-gradient centrifugation and agarose gel electrophoresis) or density (CsCl density-gradient centrifugation), and retransformation with DNA having a larger size than the vector DNA or with a density intermediate between the vector and the foreign DNA. A disadvantage of these methods is that the recombinant-DNA molecules that have small inserts remain undetected.

Many procedures that facilitate detection of eucaryotic DNA in bacteria have been devised. If the eucaryotic DNA is functionally expressed (eg, as *his*B$^+$ or *leu*D$^+$ genes of yeast), it can be directly selected on minimal plates with auxotrophic bacterial strains as recipients. However, most eucaryotic genes are not functionally expressed in bacteria to produce the protein products. The availability of a radioactive RNA probe with sequences complementary to those present on the insert DNA fragments permits the direct identification of primary transformant clones or phage plaques containing the desired recombinant molecules. In these procedures, the primary colonies or plaques are grown and transferred to nitrocellulose filters, replica-plated, and then treated so that the DNA is released, denatured, and immobilized on the filter. Hybridization *in situ* of the colonies or the single plaques with radioactive RNA probes and subsequent autoradiography reveals the particular colony or phage plaque capable of hybridizing with the radioactive RNA (19–20).

When a radioactive probe is not available, but the eucaryotic DNA can be expressed in the bacterium to produce a product that is antigenically similar to the protein produced in the eucaryotic system, transformants harboring such foreign-insert DNA can be identified by *in situ* immunoprecipitation reactions (21). An elegant method has been described employing polyvinyl disks coated with antibodies against a pure eucaryotic protein and subsequently treated with lysed transformant colonies to adsorb any immunologically cross-reacting material (22). When exposed to ^{125}I–IgG directed against the same protein, the antigen on the disk forms a complex and can be precisely located on the disk by radiography. Such procedures have been extremely useful in detecting the presence of small amounts of insulin or ovalbumin as part of fused proteins in *E. coli* cells.

Industrial Applications. ***Expression of Eucaryotic Genes and Production of Eucaryotic Proteins in Bacteria.*** One of the major advantages of the recombinant-DNA technology, particularly with regard to industrial applications, is that it is possible to introduce various segments of eucaryotic genes in a bacterium such as *E. coli.* The accurate expression of such inserted foreign-DNA segments in *E. coli* to enable it to produce various proteins of animal origin is of prime importance in the realization of the full potential of this technology. Unfortunately, all available evidence indicates that, excepting for a few lower eucaryotes (eg, yeasts or *Neurospora crassa* (23)), genes of higher eucaryotes are not functionally expressed in *E. coli.* Although, in some cases, cloning of procaryotic genes using plasmid vectors with a restriction site close to the efficient P_L promoter of phage λ can lead to an enhanced expression of the cloned genes, no functional expression of eucaryotic genes occurs under such promotor control. Two different approaches have been taken to circumvent the problem of lack of expression of eucaryotic genes in *E. coli*. One approach is to use chemically synthesized genes, corresponding to eucaryotic polypeptides of defined amino acid sequence, by using codons known to be favored in *E. coli*. The chemically synthesized gene then can be fused to the end of the *E. coli lacZ* gene coding for β-galactosidase and cloned in the host as part of a plasmid. This method has been used to fuse the chemically synthesized gene for the small polypeptide hormone, somatostatin, in phase to the end of *E. coli lacZ* gene (Fig. 3) (see also Hormones, brain oligopeptides). The β-galactosidase moiety was connected to the somatostatin moiety through a methionine residue. As somatostatin contains no methionine, which is labile to cyanogen bromide treatment, it is possible to extract the hormone from the hybrid polypeptide by cyanogen bromide treatment (24).

The above method, although elegant in technique and effective in sharply reducing biohazard potentials because the uncleaved, fused polypeptide is inactive as a hormone, suffers from the disadvantage that most eucaryotic polypeptides are large and technology does not allow the convenient synthesis of very large oligodeoxyribonucleotides (25). Also, most polypeptides have methionine residues, so that the cyanogen bromide cleavage technique would not work and some other cleavable protein linkages must be made available. In addition, β-galactosidase is not a secretable protein, so that the fused proteins are retained within the cells. In order to recover large amounts of polypeptides, the eucaryotic polypeptide ideally should be fused with a bacterial protein that is normally secreted outside the cell envelope.

The second approach, which appears to have a generalized applicability, is the technique of cloning the cDNA copy of an eucaryotic messenger RNA with the gene coding for the enzyme, penicillinase, which is a secretable periplasmic protein. Recent

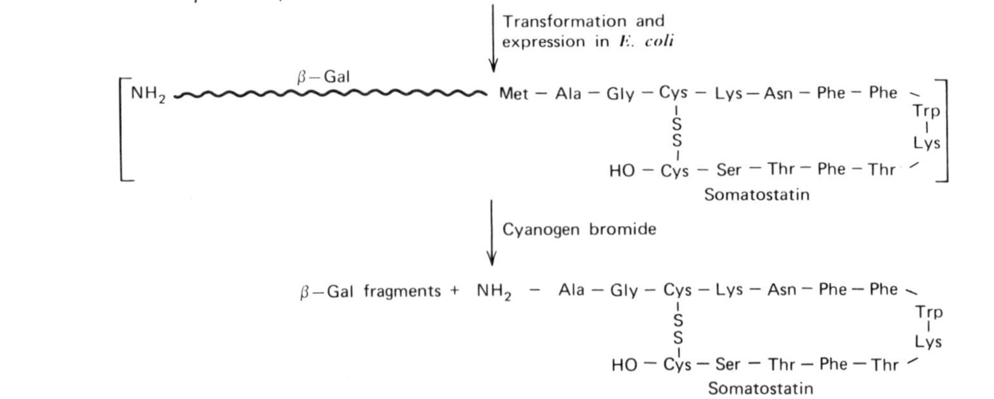

Figure 3. Schematic representation of the steps in the cloning of chemically synthesized somatostatin gene (24). Courtesy of AAAS.

experiments have demonstrated that under such conditions the cDNA sequences are read in phase and the product hybrid polypeptide folds so as to reveal the antigenic properties of the eukaryotic proteins (26–27). Both β-galactosidase and penicillinase have been used as carrier proteins for this purpose; when β-galactosidase is used, cloning of chicken ovalbumin gene leads to the production of ovalbumin-like protein which can be detected by radioimmunoassay in *E. coli* extracts (26). Recent observations indicate that the *E. coli* secretory apparatus can recognize ovalbumin, which is normally synthesized and secreted in the chicken oviduct as a secretable protein (28). The microbially synthesized ovalbumin has been found to secrete through the *E. coli* cell membrane into the periplasmic space where it can be isolated easily in large quantities.

Bacterial Production of Somatostatin. Somatostatin is the first example of a human hormone that can be produced by a bacterium through insertion of a chemically synthesized gene copy as part of a plasmid (24). A flow chart showing the outline of the experimental plan is shown in Figure 3. From the known sequence of 14 amino acids, a sequence of deoxyoligonucleotides was synthesized chemically by the triester method. In addition to the 14 triplet codons, the 5′-ends had single-stranded cohesive termini for the *Eco R*I and *Bam H*I restriction endonucleases to facilitate the insertion of the DNA fragment into the plasmid DNA; a methionine condon preceded the normal amino terminal amino acid of somatostatin for cleavage by cyanogen bromide treatment, and the COOH terminal codon is followed by two nonsense condons.

The synthetic somatostatin gene fragment was cloned in a modified *pBR* 322 plasmid containing the *Eco R*I fragment of λ*plac*5 (a plaque-forming λ) which retained the *lac* operon control region and most of the β-galactosidase structural gene. The orientation of the *lac* fragment containing the neighboring somatostatin gene was found in some cases to be such that the proper reading frame for these two genes was maintained. One colony having such a cloned gene produced somatostatin under the control of the *lac* operon. The specific activity of somatostatin in such an *E. coli* strain was increased by IPTG, an inducer of the *lac* operon. No somatostatin radioimmune activity is normally detected prior to cyanogen bromide cleavage, suggesting that the hybrid protein is antigenically inactive.

Bacterial Production of Rat Insulin. Insulin (qv), the hormone used extensively for diabetes treatment, is a polypeptide (product of a single gene) with two chains, A and B. The immediate precursor of insulin, called proinsulin, contains the two insulin chains, A and B, connected by another peptide, C. The translation product of insulin messenger RNA is not proinsulin itself but a peptide containing ca 23 additional amino acids on the amino terminus of proinsulin and called preproinsulin. The overall structure of preproinsulin molecule is:

NH_2—precursor peptide residue—B chain—C peptide—A chain—COOH

During mobilization of the entire preproinsulin molecule through the endoplasmic reticulum, the leader sequence of 23 amino acids at the amino terminal is cleaved producing the proinsulin molecule. During subsequent processing, the C peptide is eliminated from the middle, thereby producing the insulin molecule.

The first cloning of cDNA copies of the insulin mRNA prepared from rat islets of Langerhans into a bacterial plasmid, *pMB*9, was reported in ref. 29. The polyadenylated RNA was isolated by chromatography of the total RNA preparation extracted from rat pancreas islets of Langerhans on oligodeoxythymidylate–cellulose and transcribed into cDNA with the use of AMV reverse transcriptase and dT_{12-18} as primer. The RNA was removed from the mRNA–cDNA hybrid by alkali treatment. The single-stranded cDNA was converted to the double-stranded form with AMV reverse transcriptase, labeled deoxynucleoside triphosphates, and the self-priming ability of single-stranded cDNA. The hairpin structure then was removed by digestion of the double-stranded cDNA with S1 nuclease. To this blunt-ended cDNA was ligated a decamer (5′), CCAAGCTTGG (3′), by T4 ligase to provide a site for the restriction endonuclease, *Hin D*III.

The plasmid vector, *pMB*9, was treated with *Hin D*III, and the linear molecule was digested further with alkaline phosphatase to remove 5′-terminal phosphate. This latter step was to reduce self-ligation of the plasmid DNA which ensures that circle formation is dependent on the insertion of a DNA fragment containing 5′-phosphorylated termini.

The hybrid insulin cDNA–*pMB*9 molecules were used for transforming *E. coli* χ 1776 for Tc^r; bacterial clones were isolated and contained as many as 410 base-pair insertions of the insulin cDNA. However, no production of biologically functional insulin by such strains was detected.

A more recent method has utilized the ligation of the cDNA copies of rat preproinsulin mRNA at the unique *Pst* I endonuclease site of the plasmid *pBR* 322 that lies in the region that encodes amino acids 181–182 of penicillinase (27). The *E. coli* penicillinase is a periplasmic protein, whose gene sequence has been determined. There is a 23-amino-acid leader sequence along with the penicillinase polypeptide during its biosynthesis, and the leader sequence is believed to serve as a signal to direct the secretion of the protein to the periplasmic space and is removed as the protein traverses the membrane. The basic cloning technique is shown in Figure 4. The insertion of the cDNA copy of the rat insulin mRNA into a restriction site within the structural gene for penicillinase leads to the formation of hybrid DNA molecules that can be functionally expressed to fused proteins carrying the antigenic determinants of both insulin and penicillinase. Such fused proteins also are secreted into the periplasmic space, thus enhancing the yield and ease of isolation. This technique, by which a protein (eg, penicillinase) acts as a carrier and leads to the expression and secretion of eukaryotic proteins cloned as cDNA copies, may be an extremely useful tool as a general technique for the bacterial production of biologically important animal proteins. The expression of antigenic determinants of the eukaryotic proteins further points to the usefulness of this technique in cloning viral DNA segments (that encode the surface antigens) in the manufacture of vaccines against a large number of viral diseases (see also Vaccine technology).

Bacterial Production of Interferon. Interferon seems to be active against such diverse viruses as those responsible for colds, hepatitis, shingles, and skin lesions. Human leukocyte interferon is a glycoprotein (mol wt 20,000) which can be prepared from blood samples (see Blood fractionation). Production of interferon through recombinant-DNA techniques has been reported recently by the Swiss firm, Biogen. The RNA (mRNA) corresponding to the interferon gene can be isolated from human interferon-making cells. The mRNA is used to construct cDNA, and these cDNA molecules are then spliced to the plasmid DNA sequences of bacterial cells of the *E. coli* strain. Examination of certain plasmid-containing clones has revealed the presence of *E. coli* cells capable of producing biologically active interferon molecules. Other companies investigating related gene-splicing techniques include: Abbott; du Pont; Eli Lilly; G.

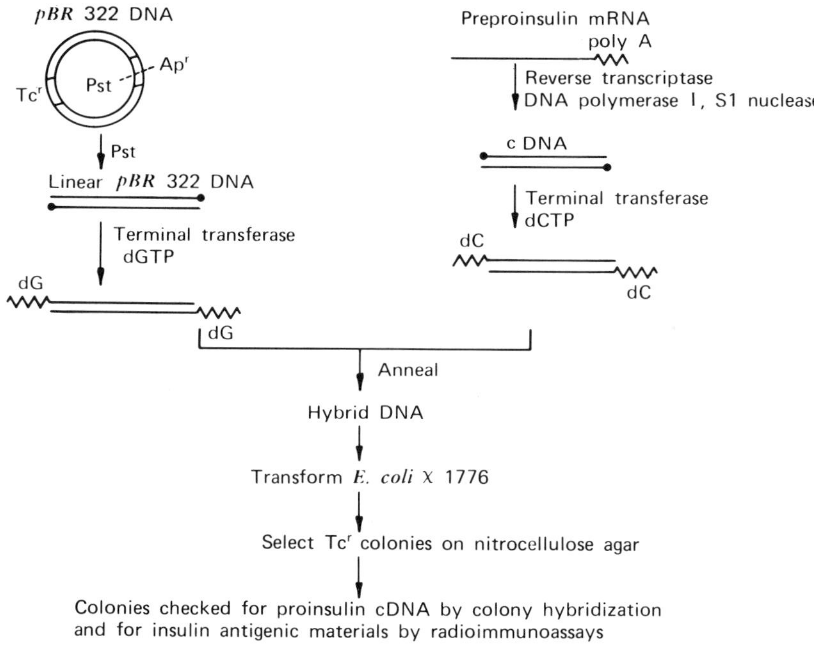

Figure 4. Flow chart for the cloning of rat insulin gene in *E. coli* as part of the plasmid *pBR* 322 by the poly dG–poly dC tailing method (27).

D. Searle; Cetus; Genex; Schering Plough; Genentech; Hoffmann LaRoche; ICI; Merck, Sharp, and Dohme; The Tokyo Foundation for Cancer Research; and Upjohn (30–31).

Regulation of Recombinant-DNA Research. The realization that recombinant-DNA technology allows insertion of any kind of DNA segment, either procaryotic or eucaryotic, into a bacterial plasmid or phage vector which then can perpetuate inside the host bacterium, led to a surge of studies regarding the insertion of genes from various bacterial and animal sources as well as viral genomes into *E. coli.* At the same time, it was realized by many scientists that uncontrolled use of this technology may lead to the production of unique bacterial or viral strains that may accidentally acquire novel pathogenic characteristics never known to mankind. The wisdom of crossing the so-called evolutionary barrier that presumably limits genetic exchange between procaryotes and eucaryotes has also been questioned (32). Soon after the development of the technique, a moratorium on further experimentation involving recombinant DNA was asked by a group of scientists at the Asilomar conference, and the director of NIH was requested to help develop guidelines that would ensure the development of this technology without jeopardizing the environment or human health and safety.

In 1976, after years of discussion and consultations with scientists, lawyers, environmentalists, and the public at large, the director of NIH issued provisional guidelines that prohibited six categories of recombinant-DNA experiments because of possible biohazards and defined permissible categories of experiments with specific levels of physical and biological containment. In addition, the guidelines stipulated specific roles and responsibilities for principal investigators, research institutions, institutional biohazard committees, and the NIH.

Since issuance of the 1976 guidelines, considerable evidence has accumulated suggesting that the biohazard potential of the recombinant-DNA technology was overemphasized (33). Deliberate attempts to introduce pathogenic characteristics to *E. coli* consistently have met with failure.

In light of this knowledge, the director of NIH, on July 28, 1978, proposed substantial modification and relaxation of the guidelines (33). The present guidelines, in force since January 2, 1979, prohibit certain types of experiments and exempt others. The guidelines also stipulate specific biological and physical containments, depending upon the source of the foreign DNA segments and their purity. Some of these specifications are shown in Table 2, and others can be found in the guidelines (33).

Table 2. Some Recommended Containments for Cloning Foreign DNA in *E. coli* K12 Host-Vector Systems

Source of foreign DNA	Containment	Remarks
eucaryotic		
primates	P2[a] + EK2[b]	
other mammals	P2 + EK2	
birds	P2 + EK2 or P3 + EK1	
cold-blooded vertebrates	P2 + EK1 or P1 + EK2	P3 + EK2 for potent polypeptide toxin producers
other cold-blooded animals and lower eucaryotes	P3 + EK2	for Class 2 pathogens or polypeptide toxin producers
	P2 + EK2 or P3 + EK1	for nonprotein toxin producers
	P2 + EK1 or P1 + EK2	nontoxin producing phytopathogenic fungi, etc
plants	P2 + EK1 or P1 + EK2	P3 + EK2 for poly-peptide-toxin producer
	P3 + EK1 or P2 + EK2	for nonprotein toxin producer
procaryotic	exempted	for strains listed in the guidelines
procaryotes that exchange genetic information with *E. coli*	P1 + EK1	for nonlisted procaryotes
	P2 + EK1	for Class 2 pathogens
procaryotes that do not exchange genetic information with *E. coli*	P2 + EK1 or P1 + EK2	
	P3 + EK2	for Class 2 pathogens

[a] The symbols P1 to P3 represent different levels of physical containment involving special equipment and laboratory installations that provide physical barriers for the accidental escape of hosts containing recombinant DNA molecules (33).

[b] EK1 and EK2 represent two levels of biological containment that assure a reduced survival of the host-vector system, if accidentally released into the open environment (33).

International Regulations. The potential benefits and fears of any associated biohazards of the recombinant-DNA technology have led to passage of regulations in various countries of the world for the safe development of this technology. As assessed by the Committee on Genetic Experimentation (COGENE) under the International Council of Scientific Unions, at the end of March, 1978, at least 367 recombinant-DNA projects were estimated to be underway in 155 laboratories in 15 countries. Seventeen nations had drawn up guidelines for recombinant-DNA experimentation, and these were in force in 13 nations. Five of the nations had prepared their own guidelines, whereas the rest had adopted or modified guidelines of the United States or the United Kingdom. The guidelines of 4 nations were reported as being entirely voluntary. Eight nations had guidelines that were enforceable through a research-funding mechanism, and 2 nations had legally enforceable guidelines. Guidelines of most nations have been derived directly or indirectly from those of the United States, which were the first to be promulgated. Although such guidelines generally have been patterned after those of the United States, there are substantial differences. The most substantial guidelines are promulgated by the NIH in the United States, and the Genetic Manipulation Advisory Group (GMAG) in the United Kingdom. The GMAG guidelines in the United Kingdom were initiated by a working party under the chairmanship of Lord Ashby, and were extended by an *ad hoc* group under the chairmanship of Sir Robert Williams. The categorization of physical and biological containment proposed by the Williams working party are substantially similar to those of NIH, although they vary in the details of physical specifications of the laboratories as well as the implementation of detailed codes of practice. The guidelines in the USSR, similar to NIH and GMAG guidelines, involve four levels of physical containment and three levels of biological containment. The USSR guidelines were drafted by the Recombinant-DNA Commission at the Interagency Council on Scientific and Technical Problems of Molecular Biology and Molecular Genetics. They are approved by the State Sanitary Inspector-in-Chief of the USSR and are compulsory for all establishments of the USSR involved in recombinant-DNA research or using them for applied purposes. The legal enforcement of the guidelines rests on the State Sanitary Control of the USSR Ministry of Health. A detailed description of the guidelines in various countries, including those in France, Holland, etc, is beyond the scope of this review. Consult the various regulatory agencies (NIH in the U.S., GMAG in the United Kingdom, State Sanitary Inspector-in-Chief in the USSR, etc) for detailed descriptions of existing regulations governing recombinant-DNA research in the respective countries.

Nomenclature

A	= adenosine (ribonucleoside of adenine)
ATP	= adenosine triphosphate
C	= cytidine (ribonucleoside of cytosine)
Cb^r	= carbenicillin resistance
cDNA	= (complementary DNA to an RNA molecule)
cts	= thermoinducible
dA (dC, dG, and dT)	= poly(deoxynucleoside triphosphate)
DNA	= deoxyribonucleic acid
dNTP	= a mixture of deoxynucleoside triphosphate
dT_{12-18}	= a polymer containing 12–18 thymidine triphosphate residues

E. coli χ 1776	= an EK2 strain that is specially designed for recombinant-DNA research and which has a very low survival outside the laboratory environment
λ exonuclease	= a nuclease which cleaves DNA from the 5′ end
G	= guanosine (ribonucleoside of guanine)
Hin DIII, *Pst* I, *Eco R*I	= restriction endonucleases used for producing linear DNA molecules with sticky ends
hisB$^+$, *leuD*$^+$, *lacZ*	= wild type designations of genes coding for enzymes involved in the biosynthesis of histidine and leucine, or in the biodegradation of lactose (genotypes)
IPTG	= isopropyl thiogalactoside
Km^r	= kanamycin resistance
lac operon	= gene segment specifying lactose degradation
Mdal	= megadalton
pBR 322, *pMB*9	= vector plasmids specifying resistance to ampicillin and tetracycline, and tetracycline alone, respectively
P_L	= left-hand promoter
RNA	= ribonucleic acid
mRNA	= messenger ribonucleic acid
rRNA	= ribosomal ribonucleic acid
RP4	= a wide host range antibiotic resistance plasmid specifying resistance to kanamycin (Km^r), tetracycline (Tc^r), and carbenicillin (Cb^r)
RP4::Mu	= RP4 plasmid where one or more copies of Mu phage genome have been transposed
S1	= a nuclease that cleaves single-stranded DNA
T	= thymidine (deoxyribonucleoside of thymine)
Tc^r	= tetracycline resistance
Ti	= a transmissible plasmid in *Agrobacterium tumefaciens* responsible for inciting tumors in some dicotyledonous plants by mechanisms not fully understood
transposon	= a discrete segment of DNA that can jump from one replicon to another in the absence of substantial genetic homology

BIBLIOGRAPHY

1. M. Faelen and A. Toussaint, *J. Mol. Biol.* **104,** 525 (1976).
2. M. Faelen, A. Toussaint, M. Van Montagu, S. Van den Elsacker, G. Engler, and J. Schell in A. I. Bukhari, J. A. Shapiro, and S. L. Adhya, eds., *DNA Insertion Elements, Plasmids and Episomes,* Cold Spring Harbor Laboratory, 1977, p. 521.
3. R. W. Hedges, and A. E. Jacob, *FEMS Microbiol. Lett.* **2,** 15 (1977).
4. D. A. Hopwood, *Ann. Rev. Microbiol.* **32,** 373 (1978).
5. D. A. Friello, J. R. Mylroie, and A. M. Chakrabarty in J. M. Sharpley and A. M. Kaplan, eds., *Proceedings of the Third International Biodegradation Symposium,* Applied Science, London, 1976 p. 205.
6. C. Kennedy and R. Dixon in A. Hollaender and co-eds., *Genetic Engineering for Nitrogen Fixation,* Plenum Press, New York, 1977, p. 51.
7. M. D. Chilton, M. H. Drummond, D. J. Merlo, and D. Sciaky, *Nature* **275,** 147 (1978).
8. A. Depicker, M. Van Montagu, and J. Schell, *Nature* **275,** 150 (1978).
9. K. N. Timmis, S. N. Cohen, and F. C. Cabello in F. E. Hahn, ed., *Progress in Molecular and Subcellular Biology,* Springer-Verlag, Berlin, Vol. 6, 1978, p. 1.
10. J. Collins, *Curr. Top. Microbiol. Immunol.* **78,** 121 (1977).
11. A. M. Chakrabarty in D. A. Jackson and S. P. Stich, eds., *The Recombinant DNA Debate,* Prentice-Hall, Englewood Cliffs, N.J., 1979, p. 56.
12. R. B. Helling and M. I. Lomax in A. M. Chakrabarty, ed., *Genetic Engineering,* CRC Press, Inc., West Palm Beach, Fla., 1978, p. 1.
13. J. B. Hansen and R. H. Olsen, *J. Bacteriol.* **135,** 227 (1978).
14. E. M. Southern, *J. Mol. Biol.* **98,** 503 (1975).
15. V. Sgaramella, J. H. Van Sande, and H. G. Khorana, *Proc. Natl. Acad. Sci. USA* **67,** 1468 (1970).
16. W. Salser in A. M. Chakrabarty, ed., *Genetic Engineering,* CRC Press Inc., West Palm Beach, Florida, 1978, p. 53.
17. S. P. Goff, and P. Berg, *Proc. Natl. Acad. Sci. USA* **75,** 1763 (1978).

18. K. N. Timmis, F. Cabello, and S. N. Cohen, *Proc. Natl. Acad. Sci. USA* **71,** 4556 (1974).
19. M. Grunstein and D. S. Hogness, *Proc. Natl. Acad. Sci. USA* **72,** 3961 (1975).
20. W. D. Benton and R. W. Davis, *Science* **196,** 180 (1977).
21. A. Skalka and L. Shapiro, *Gene* **1,** 65 (1976).
22. S. Broome and W. Gilbert, *Proc. Natl. Acad. Sci. USA* **75,** 2746 (1978).
23. L. Clarke and J. Carbon, *J. Mol. Biol.* **120,** 517 (1978).
24. K. Itakura, T. Hirose, R. Crea, A. D. Riggs, H. L. Heyneker, F. Bolivar, and H. W. Boyer, *Science* **198,** 1056 (1977).
25. H. G. Khorana, *Science* **203,** 614 (1979).
26. O. Mercereau-Puijalon, A. Royal, B. Cami, A. Garapin, A. Krust, F. Gannon, and P. Kourilsky, *Nature* **275,** 505 (1978).
27. L. Villa-Komaroff, A. Efstratiadis, S. Broome, P. Lomedico, R. Tizard, S. P. Naber, W. L. Chick, and W. Gilbert, *Proc. Natl. Acad. Sci. USA* **75,** 3727 (1978).
28. T. H. Fraser and B. J. Bruce, *Proc. Natl. Acad. Sci. USA* **75,** 5936 (1978).
29. A. Ullrich, J. Shine, J. Chirgwin, R. Pictet, E. Tischer, W. J. Rutter, and H. M. Goodman, *Science* **196,** 1313 (1977).
30. *Chem. Eng. News* **58,** 37 (Jan. 28, 1980); *Chem. Week,* 43 (Feb. 6, 1980); *Science* **207,** 746 (Feb. 15, 1980); *Nature* **283,** 323 (1980); *Chem. Week,* 22 (Mar. 12, 1980); *Chem. Eng. News* **58,** 15 (Mar. 17, 1980).
31. W. Gilbert and L. Villa-Komaroff, *Sci. Am.* **242**(4), 74 (1980).
32. R. Curtiss, III, *Ann. Rev. Microbiol.* **30,** 507 (1976).
33. *Fed. Regist.* **43,** 60108 (Dec. 22, 1978).

General References

R. F. Beers, Jr. and E. G. Bassett, eds., *Recombinant Molecules: Impact on Science and Society,* Raven Press, New York, 1977.

Research with Recombinant DNA, National Academy of Sciences, Washington, D.C., 1977.

A. M. Chakrabarty, ed., *Genetic Engineering,* CRC Press, West Palm Beach, Fla., 1978.

A. M. CHAKRABARTY*
University of Illinois at the Medical Center, Chicago

* Supported in part by grants from the National Science Foundation (PFR 79-05499 and PCM 79-17526).

GEOTHERMAL ENERGY

In the broadest sense, geothermal energy comprises all the thermal energy stored in the earth's crust. Geothermal fluids of natural origin have been used for cooking and bathing since before the beginning of recorded history but it was not until the early 19th century that geothermal energy was harnessed for industrial purposes. One of the first cases was the use of geothermal steam for heating evaporating ponds at the boric acid works near Lardarello, Italy. Since that time, other developments, such as the steam field at The Geysers, California, and the hot water systems at Wairakei, New Zealand, Cerro Prieto, Mexico, and Reykjavik, Iceland, have led to an installed world electrical generating capacity of more than 1700 MW and a direct use, nonelectric capacity of more than 9500 MW(t) (thermal megawatts of power) in 1980.

The source and transport mechanisms of geothermal heat are unique to this energy source. Heat flows through the crust of the earth at a rate of almost 5.9 cW/m^2 [1.4×10^{-6} $cal/(cm^2{\cdot}s)$] on the average. The intrusion of large masses of molten rock can increase this normal heat flow locally but, for most of the continental crust, the heat flow is due to conduction of heat from the mantle and heat generated by the decay of radioactive elements in the crust, particularly isotopes of potassium, uranium, and thorium.

Certain conditions must be met before geothermal energy can be utilized. The first requirement is accessibility, which is created by natural transport processes such as convective heat transfer in porous and/or fractured regions of rock and conduction through the rock itself. If the physical extent and heat content of a deposit are large enough and the deposit is near the surface, a heat extraction scheme must be designed based upon the hydrologic and geologic situation including the *in situ* rock properties. After such a scheme is successfully implemented, the desired raw material (hot water or steam) can be produced. Techniques to convert the steam or hot water into a marketable product (electricity, process heat, or space heat) and to adequately dispose of the waste products complete the process. Many aspects of geothermal heat extraction are similar to those found in the oil, gas, coal, and mining industries. Because of these similarities, equipment, techniques, and terminology have been borrowed or adapted for use in geothermal development, a fact that has accelerated the development of geothermal resources somewhat.

Commercial utilization of the resource requires that the process be economically competitive. Consequently, commercial geothermal systems developed to date have been limited to a relatively few, accessible high-grade deposits scattered throughout the world. Fossil-fuel shortages and price increases will make a significant fraction of lower-grade geothermal resources commercially feasible in the coming decades.

Types of Geothermal Systems

There are basically two types of geothermal systems: those that spontaneously produce hot fluid (steam or water) from a reservoir and those that do not. Systems that produce fluid spontaneously are obviously easier to exploit; however, these natural hydrothermal systems are not always located near a user. For example, of the eleven most promising new geothermal areas in the United States, six are within about 400 km of Los Angeles, California. The other five are all more than 400 km from any city with a population greater than 300,000 (1). Furthermore, nine of the ten most populous cities in the United States are more than 800 km from any known high temperature, natural geothermal system. Economic utilization of these remote hydrothermal systems necessarily involves the generation and transmission of electricity rather than direct use of the geothermal fluid for space or process heating. A system of the second type, one lacking sufficient quantities of hot fluids, presents more difficulty with respect to heat extraction, but because of the broader-based location of such resources, the remoteness problem is partially alleviated.

Hydrothermal Systems. Systems that spontaneously produce hot fluids are hydrothermal or convection-dominated systems. Hydrothermal systems require a source of heat (usually a magmatic intrusion at depth), formations with enough permeability to allow fluid mobility, an adequate supply of indigenous fluid (water or steam), sufficient contact surface, time for the fluid to be heated, and a return path to the surface (2) (Fig. 1). These systems are frequently located in or near the zones of recent volcanism and tectonism (mountain building) that are associated with the boundaries of crustal plates (2–6). In addition to providing sources of heat, crustal plate margins are areas where tectonic forces have caused considerable fracturing of rock and thus high local permeability. Indeed, it is believed that the contributions of faults and joints to *in situ* permeability are far more important than the intergranular (matrix) permeability (4,7–12).

Water in hydrothermal systems is usually of meteoric origin (13–14). This water falls as rain or snow and percolates downward through sediments or fissures until it comes to a heat source. There it is heated and buoyantly rises towards the surface where it usually appears as geysers, hot springs, fumaroles, or solfataras.

If the pressure on the fluid in the reservoir is insufficient to prevent boiling, a vapor phase forms in the upper portion of the reservoir. This vapor phase consists of steam (often superheated or dry) and noncondensable gases that separate from the liquid phase. Most of the dissolved minerals concentrate in the liquid phase, leaving the vapor relatively free of dissolved solids. Hydrothermal systems that produce superheated steam are called vapor-dominated. These steam systems are the easiest and most efficient to exploit for electricity generation because the steam can be transported

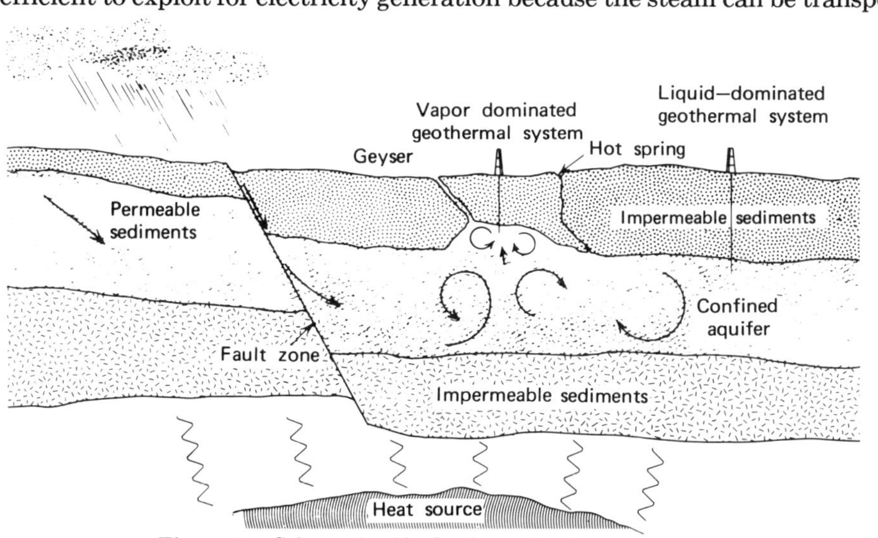

Figure 1. Schematic of hydrothermal system geometry.

to and expanded directly in low pressure steam turbines with little risk of scaling or turbine blade damage.

Systems that are pressurized above the vapor pressure do not form a vapor cap and production from these types of fields consists of hot water or a mixture of hot water and steam. Because dissolved solids remain in the liquid phase, liquid-dominated systems can pose serious scaling and corrosion problems in surface piping systems and in production and injection wells (15–18). If the fluids are not reinjected, they can pose a waste disposal problem. Usually, the fluid in liquid-dominated systems is flashed (or subjected to a pressure drop that allows a separate vapor phase to form) and separated so that the vapor phase can be piped directly to the turbine generator. The liquid may be flashed more than once (multistage flashing) but ultimately a significant fraction of the hot water and minerals remains to be disposed of by release to surface waters, reinjection, evaporation, or some alternative use such as process or space heating (19–22).

As shown in Figure 1, the geometry of a typical hydrothermal geothermal system is quite similar to that of a typical oil reservoir, and many techniques for measuring field properties such as permeability, porosity, reservoir size, and fluid content in oil reservoirs have been applied directly to hydrothermal systems with success (23–27). Geophysical prospecting involving heat flow, electrical resistivity, gravity mapping, and other methods provide exploration data for inferring the extent and depth of a thermal anomaly (9–10). Geochemical techniques including geothermometry, stable isotope geochemistry, analysis of dissolved gases and ratios of major elements in hot springs are also used to locate and define geothermal fields (2,28–29). These techniques are discussed in detail in references 30 and 31.

Geopressured Systems. Large sediment-filled reservoirs containing pore fluids under confining pressures much greater than the hydrostatic head are called geopressured basins (32). The sediments usually consist of sandstones and shales along with other sedimentary rock types that are more or less interbedded. Generally, the fluid contained in massive sandstone sediments is under near hydrostatic pressure and is mobile owing to relatively high permeability. The fluids contained in sandstone–shale and massive shale facies are generally overpressured by up to 60 MPa (600 bar) and are at higher temperatures (150–180°C) (33). These higher temperatures are caused at least partially by the low thermal conductivity of shale, which gives an apparent high geothermal gradient; the heat flow is actually near normal. Pressures in the United States' Gulf coast geopressured reservoirs are expected to approach lithostatic pressure. The lithostatic pressure gradient is ca 23 kPa/m (1 psi/ft).

In addition to hot water under pressure, geopressured reservoirs are expected to contain methane in solution (34). For the purposes of estimating the geopressured resource, the geofluids are usually assumed to be saturated with methane with (6.9–8.9) $\times 10^{-3}$ m^3 CH_4/kg water (ca 40–50 SCF of CH_4/bbl water) (33,35–38). Field tests of a geopressured well in southern Louisiana indicate this may be the case (39–41). The relatively high salinity (>100,000 ppm TDS) and expected subsidence effects (42–44) of geopressured resources may seriously restrict their utilization.

Conduction-Dominated Systems. Throughout most of the world one or more of the necessary components of a hydrothermal reservoir is missing. In particular, the reservoir rock is often hot enough (>200°C) but produces insufficient fluid for heat extraction either because of low formation permeability or the absence of naturally contained fluids. These situations are included in a part of the geothermal resource referred to as hot dry rock.

Hot Dry Rock. Techniques for the extraction of heat from low permeability hot dry rock (HDR) are under investigation at the Los Alamos Scientific Laboratory with DOE sponsorship (45–47). The basic concept is simple: drill a well to sufficient depth to reach a useful temperature, create large heat transfer surface areas by hydraulically fracturing the rock, and intercept the fracture(s) with a second well. By circulating water from one well to the other through the fractured region, heat can be extracted from the rock as depicted in Figure 2. Although the concept is simple, many questions remain to be answered before it can be considered economically feasible. Significant progress has been made, however, in understanding reservoir characteristics including fracture initiation and propagation, thermal drawdown, water loss rates, flow impedance, fluid mixing, and fluid geochemistry in field tests of a HDR reservoir in low permeability crystalline rock (48). In addition to hydraulic fracturing, permeability and surface area can be created by explosive fracturing, chemical leaching, and thermal stress cracking (49). Studies of explosive fracturing are being conducted in the UK (50) and in the USSR (51–52).

If rock of sufficient natural permeability exists in a confined geometry, techniques similar to waterflooding or steam-drive employed for oil recovery (Fig. 3) might be used to advantage (49,53–54) (see Petroleum). Other techniques for heat extraction

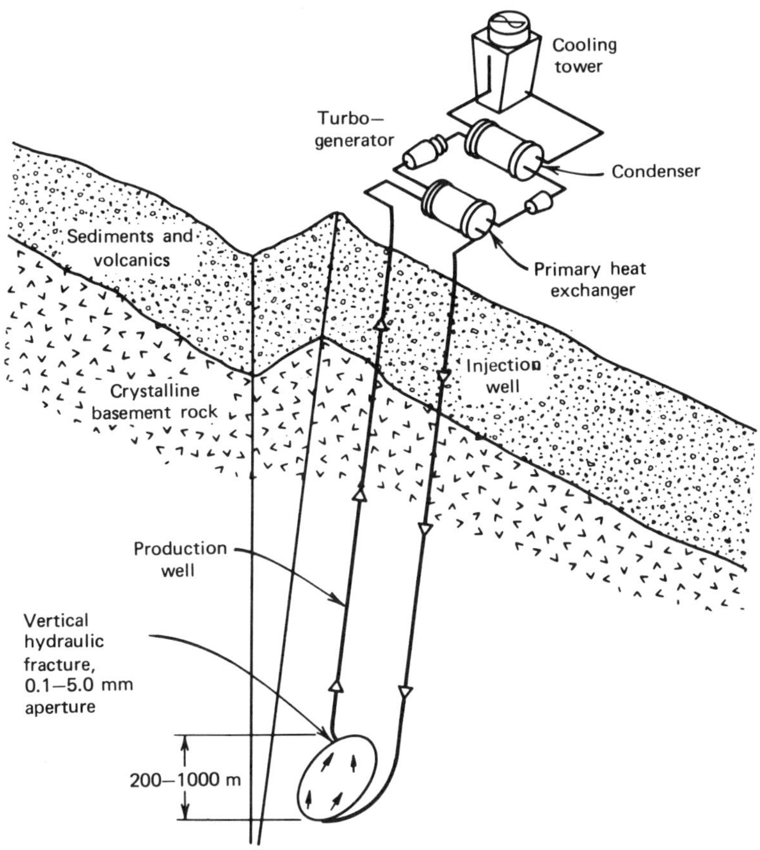

Figure 2. Hot dry rock reservoir concept for low permeability formations.

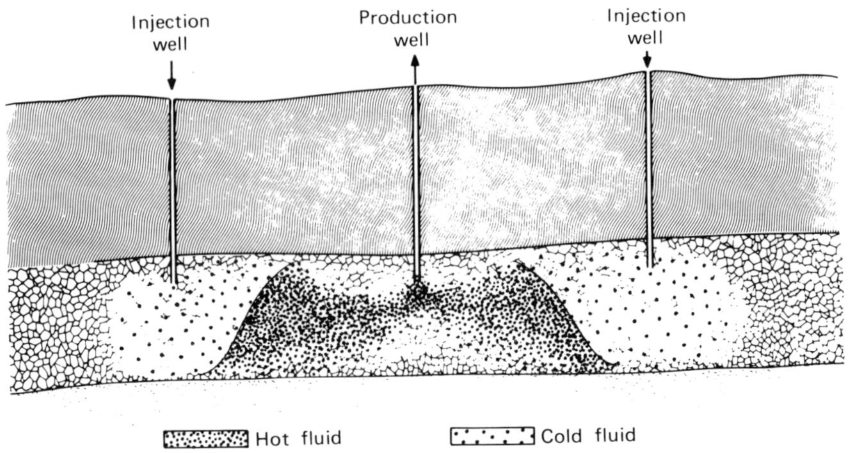

Figure 3. Production of hot fluid from wet or dry permeable rocks by a waterflooding technique.

such as downhole heat exchangers, downhole pumps, or alternative methods of operation such as injection-recovery (huff-puff) have also been proposed. Because many of these methods involve the creation of permeability in rock, hot dry rock systems are not limited to particular tectonic settings. Until the heat-extraction technology is suitably developed and the economics are demonstrated on a large scale, the first commercial hot dry rock systems will be confined to regions of high geothermal gradient (>30°C/km).

Magma. The extreme case of hot dry rock occurs when the rock is molten. Considerable development of drilling and heat extraction concepts and technology must be accomplished before heat can be extracted usefully from molten rock. Central to the problem of extracting heat is the ability to find a body of magma at drillable depths. Drilling into magma requires the development of equipment, lubricants, and cements that can operate at 700–1000°C and depths up to 10 km (55). These problems appear formidable enough to preclude the development of even a pilot-scale extraction system before 1990 (56).

Resource Potential

Investment in the development of natural resources usually proceeds when the likelihood of return on invested capital is high. Consequently, the quality of the resource needs to be carefully characterized. For geothermal resources this includes consideration of rock type, fluid composition and production rates, and energy content of the reservoir. Typically, mineral deposits or ore bodies are evaluated according to the grade and accessibility of the deposit, processing costs, and distance to the nearest market. Based on the evaluation, a deposit might be classified as a reserve or a resource. A similar body of terminology is under development in the geothermal energy field; thus far, geothermal terminology has borrowed heavily from the other resource-related fields.

Definition of Terms. There is considerable nonuniformity in the use of terms in the geothermal literature. For example, *geothermal resource base* is variously defined as including all of the stored heat above 15°C to 10 km depths (57); or as the potentially useful heat in the earth's crust at temperatures greater than 80°C and depths less than 6 km (56); or total heat contained in subsurface rocks and fluids to a depth of 3 km and at temperatures above 15°C (58).

The most widely accepted terminology is that proposed by Muffler and Guffanti (59); see also references 60–62.

Resource Base. All of a given material (such as geothermal energy) in the earth's crust, whether its existence is known or unknown and regardless of cost considerations.

Accessible Resource Base. All of the geothermal energy between the earth's surface and a specific depth in the crust beneath a specified area and referenced to a mean annual temperature.

Resource. Thermal energy that could be extracted at costs competitive with other forms of energy at a foreseeable time, under reasonable assumptions of technological improvement and economic favorability.

Reserve. That part of the geothermal resource that is identified and also can be extracted legally at a cost competitive with other commercial energy sources at present.

It would be misleading to imply that this or any terminology is accepted throughout the geothermal industry. Indeed, these definitions are somewhat simplistic compared to the definitions applied to the same terms in the oil, gas, or mining industry. For example, an oil reserve can secure a bank loan, a geothermal reserve can not. As of 1980, there are simply not enough cases of economic exploitation of geothermal energy for it to be considered a good financial risk. The definitions do suit the purpose for which they are intended—they assist in the evaluation of problems and opportunities for the development of geothermal energy (63).

Estimates of the Accessible Geothermal Resource Base. Estimates of all or part of the accessible geothermal resource base are numerous (47,56–58,63–71), and each estimate is based on slightly different assumptions. If the definition of accessible resource base as given above is strictly followed, the total amount of heat Q contained in rock beneath the earth's surface to a depth is expressed by:

$$Q = \sum A_i \left[\int_0^Z (\phi \rho_f C_f + (1 - \phi)\rho_r C_r)(\nabla T_i \cdot Z + T^* - T_{ref}) dZ \right] \tag{1}$$

where Z = depth, km; T = rock temperature, °C; A_i = surface area having a characteristic regional gradient of ∇T_i; T^* = ambient surface crustal temperature, ca 15°C; ϕ = porosity of rock; ρ_f = fluid density, kg/m^3; ρ_r = rock density, kg/m^3; C_f = fluid heat capacity, J/(kg·K); C_r = rock heat capacity, J/(kg·K); ∇T_i = geothermal gradient, °C/km; and T_{ref} = reference temperature, °C.

Integrating equation 1 for each unit of area that is at a given geothermal gradient and taking $T^* = T_{ref}$ = 15°C, equation 1 becomes:

$$Q_i = A_i[\phi \rho_f C_f + (1 - \phi)\rho_r C_r]\nabla T_i \frac{Z^2}{2} \tag{2}$$

Below about 3 km, $\phi \rightarrow 0$ and equation 2 can be simplified further to:

$$Q_i = A_i(\rho_r C_r)\nabla T_i \frac{Z^2}{2} \tag{3}$$

Given the distributions of land areas overlying each range of geothermal gradient, the resource base to a given depth can be calculated. Likewise, given the heat contained per unit area to a given depth, one can extrapolate to another depth to find the accessible resource base by simply substituting the new depth in equation 2. Reference 58 includes estimates of the worldwide accessible resource base referenced to a mean annual surface temperature of 15°C and a depth of 3 km assuming (*1*) a normal gradient of 25°C/km for all nongeothermal areas, (*2*) a gradient of 40°C/km for 90% of the area of a country that lies inside a geothermal belt and (*3*) a gradient of 90°C/km for the remaining 10% of the area of a geothermal belt.

Comparisons of accessible resource base estimates for the United States from several sources all referenced to a mean annual temperature of 15°C and a maximum depth of 10 km are given in Table 1. Two important conclusions can be drawn from these estimates. First, the amount of energy is enormous. The total United States energy consumption is currently 84 EJ/yr (80 quads/yr) and is increasing at a rate of ca 1 EJ/yr (1 quad/yr). If only 0.1% of the accessible resource base can be economically extracted, this resource would supply the complete energy needs of the United States at its current rate of increase for almost 170 years. Second, probably 97% of the energy contained in the upper 10 km of the earth's surface is contained in the rock itself as a conduction-dominated resource and, therefore, requires a hot dry rock extraction scheme.

Estimates of the Geothermal Resource. Several factors limit the amount of energy that can be extracted from the earth (Fig. 4). Some of these are now and always will be fixed; others depend upon future technological and economic developments. The first major limitation to the estimates of the amount of heat that can be economically extracted is the effective geothermal gradient. This largely determines the potential for a given area. If the gradient is small, the depth to the minimum acceptable initial rock temperature is large. A minimum initial rock temperature (T_{ref}) of 150°C is assumed to provide reasonable thermodynamic conversion efficiencies (see Energy Conversion). A normal gradient (25°C/km) requires drilling to at least 5.4 km to reach

Table 1. Comparison of Estimates of Accessible United States Resource Base in Quads[a] from 0 to 10 km Referenced to 15°C

Resource category	USGS[b]	EPRI[c]	LASL[d]
fluid			
convection	9,600	634,500	
geopressured	850,000		
Subtotal	*859,600*	*634,500*	
rock			
conduction	33,000,000	34,620,000	22,677,000
igneous	1,000,000		
Subtotal	*>34,000,000*	*34,620,000*	*22,677,000*
Total	*>34,859,600*	*35,254,500*	

[a] 1 Quad = 1.054 EJ = 10^{15} Btu.

[b] Ref. 69.

[c] Ref. 58; the difference between the EPRI fluid resource base estimate and the total accessible resource base (recalculated to 10 km) is assumed to be the rock resource base. The depth of circulation of fluid is taken to be 3 km.

[d] Ref. 46; recalculated from 0 to 10 km and assuming an average gradient of 25°C/km.

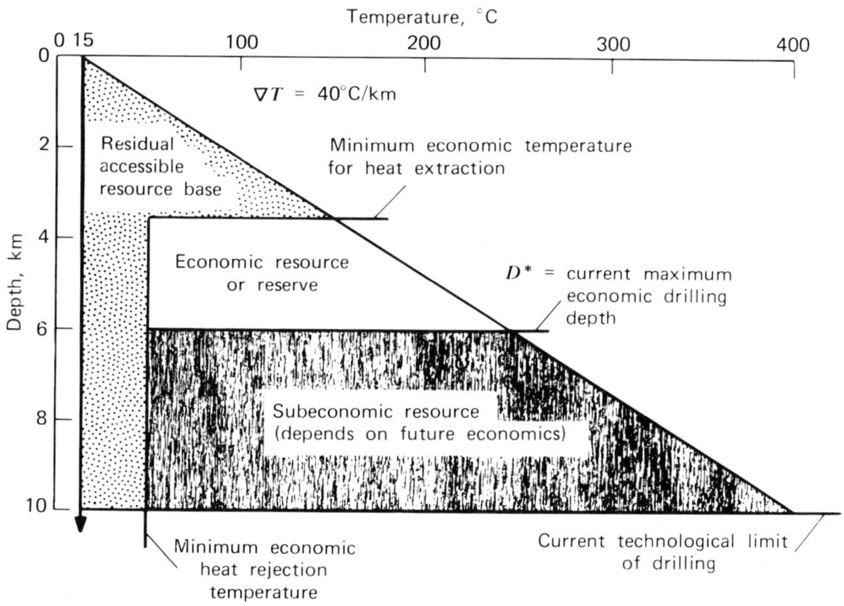

Figure 4. Factors limiting the fraction of the accessible resource base that can be economically extracted.

150°C. If the gradient is 40°C/km, only 3.4-km-deep holes are required. Another major limitation is the maximum economic drilling depth D^*. For an assumed maximum drilling depth of 6 km, only gradients above about 40°C/km represent substantial amounts of stored energy. The final limitation is imposed by heat rejection conditions for the geothermal fluid in the power plant (taken here to be 50°C). Imposing these limitations on an area of 1 km^2 with a gradient of 40°C/km, the total geothermal resource to 6 km (assuming average values of C_r = 772 J/(kg·K) (185 cal/(kg·K)) and ρ_r = 2500 kg/m^3) is:

$$\left[\frac{Q}{A_i}\right] = \rho_r C_r \left[\frac{\nabla T_i Z^2}{2} + (T^* - T_{ref})Z\right]_{3.4\text{ km}}^{6\text{ km}}$$

$$Q_i = 0.81 \text{ EJ/km}^2 \text{ or } 0.77 \text{ quads/km}^2 \tag{4}$$

The total energy available to the economic depth D^* (= 6 km) is 0.81 EJ/km^2 (0.77 quads/km^2). The fraction of this amount that can be extracted economically will depend on the prices for other forms of energy and upon the degree of success in developing commercial-sized systems.

The following applies this analysis to the EPRI estimated accessible resource base for the United States (58):

$$Q|\nabla T = 25°\text{C/km} = 9.22 \times 10^5 \text{ EJ } (8.75 \times 10^5 \text{ quads}) \tag{5}$$

$$Q|\nabla T = 40°\text{C/km} = 1.72 \times 10^6 \text{ EJ } (1.63 \times 10^6 \text{ quads}) \tag{6}$$

$$Q|\nabla T = 80°\text{C/km} = 6.49 \times 10^5 \text{ EJ } (6.16 \times 10^5 \text{ quads}) \tag{7}$$

The accessible geothermal resource base for the United States contained in hot rock to a depth of 6 km is 3.29×10^6 EJ (3.12×10^6 quads). If only 0.2% of this energy can be economically extracted, the resource for conduction-dominated systems is 6577 EJ (6240 quads). Muffler (69) lists the resource of hydrothermal systems as 422 EJ (400 quads) and that for geopressured systems as 453–4638 EJ (430–4400 quads) including methane. This gives a total estimated United States geothermal resource to a depth of 6 km of 7,452–11,636 EJ (7,070–11,040 quads) of which 57–88% is contained in hot rock.

Reservoir Productivity and Fluid Characteristics

To justify the cost of developing a geothermal field, estimates of the total amount of energy that can be extracted and the production rate must be made. Typically, computer models are used to simulate performance for a given set of reservoir properties. Two general categories of reservoir models, one involving matrix and one fracture-dominated flow, are usually encountered. The first type is typically applied to high matrix permeability hydrothermal formations (72–81), and the second to flow in fractured HDR, low permeability media (82–85). A simple example of each case with schematic thermal history curves is shown in Figure 5.

Heat Extraction Performance of Permeable Hydrothermal and Geopressured Reservoirs. Extraction of heat from hydrothermal systems now in operation is quite simple—because the reservoirs are pressurized, the fluid passes directly to the surface under artesian flow when the reservoir is penetrated. Productivity of the wells may occasionally be enhanced by stimulation at the wellbore or downhole pumping but even this is often unnecessary. Efficient production of a pressurized field is largely a matter of well spacing for a specified set of *in situ* properties. Although it has not yet been necessary to do so, it is likely that when the pressure of such a field drops to the point where it is insufficient to produce hot fluid, stimulation techniques such as

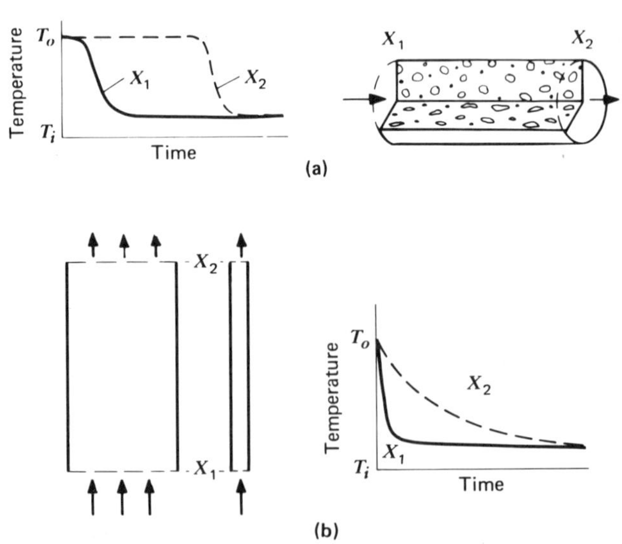

Figure 5. Limiting cases of reservoir geometry used in modeling thermal drawdown of geothermal reservoirs: (**a**) porous media; (**b**) flat fracture.

water-drive or flooding will be used to increase the lifetime of the field. Studies of this technique and its effects on reservoir productivity are numerous (75,78,86–89).

Heat Extraction Performance of Fractured HDR Reservoirs. Because they are artificial, HDR reservoirs can be designed to meet lifetime production capacity and temperature requirements of the user. The efficiency of electric power production over the 20- to 40-year lifetime of an HDR power plant depends primarily on the temperature and composition of the geothermal fluid produced. Knowledge of the anticipated thermal drawdown and of fluid chemistry is crucial in developing optimal plant designs and reservoir management strategies. The predicted lifetime and the rate at which energy can be extracted from a single fracture depend on several major factors: (*1*) thermal properties of the rock, (*2*) accessible surface area of the fracture, and (*3*) distribution of and impedance to fluid flow across the fracture surface. Because of the inherently low thermal conductivity of rock, conduction through the formation to the circulating fluid contained in the fracture controls the rate of heat extraction.

A simplified approach to estimating reservoir performance assumes that a certain fraction of the recoverable power η, corresponding to uniform flow across the face of the fracture, could be extracted. Assuming only conduction of heat toward the fracture face and no thermal-stress cracking enhancement, the recoverable thermal power $P(\mathrm{t})$ can be expressed as (84,90):

$$P(\mathrm{t}) = \eta \dot{m}_w C_w (T - T_{min}) \operatorname{erf}\left(\sqrt{\frac{(\lambda\rho C)_r}{t}} \cdot \frac{\pi R^2}{\dot{m}_w C_w}\right) \qquad (8)$$

where $A = \pi R^2$ = area of one face of the fracture, m^2; C_w = heat capacity of water, ca 4.2 kJ/(kg·K) (ca 1000 cal/(kg·K)); C_r = heat capacity of granite, ca 1 kJ/(kg·K) (ca 240 cal/(kg·K)); $\dot{m}_w$ = water flow rate through the fracture, kg/s; t = time, s; T = mean rock temperature, °C; T_{min} = fluid reinjection temperature, °C; λ_r = thermal conductivity of granite, ca 3 W/(m·K); and ρ_r = density of granite, ca 2.7 g/cm^3.

Equation 8 provides estimates of reservoir lifetime for specified ideal fracture radii R and production rates $\dot{m}_w$. A few simple calculations quickly reveal that very large surface areas are required for low drawdown rates with wellbore flows in excess of 40 kg/s. In designing systems, one can utilize single or multiple fracture concepts with specified drawdown rates at given initial power extraction levels. Figure 6 shows the required wellbore separation distance as a function of initial thermal power or mass flow rate and the number of fractures (91). In this case, 50% drawdown is achieved after 20 years of continuous operation with an overall $\dot{m}_w/R^2$ of ca 2×10^{-4} kg/(m^2·s). Because of the rock's low thermal conductivity, fractures separated horizontally by more than 50 m show negligible thermal interference over a 20-yr period.

Harlow and Pracht (82), using a simplified model, were able to demonstrate that substantial enhancement of HDR reservoir performance can be anticipated by exploiting the effects of thermal stress cracking. However, the mechanisms associated with thermal stress cracking are extraordinarily complex and not well understood even under well-defined laboratory conditions.

Economically acceptable rates of energy extraction and lifetimes can be achieved in low permeability HDR reservoirs either by growing large single fractures ($R \cong 1000$ m), by using smaller multiple parallel fractures, by exploiting thermal stress cracking, or by remedial refracturing to generate new surface area in nondepleted regions of hot rock.

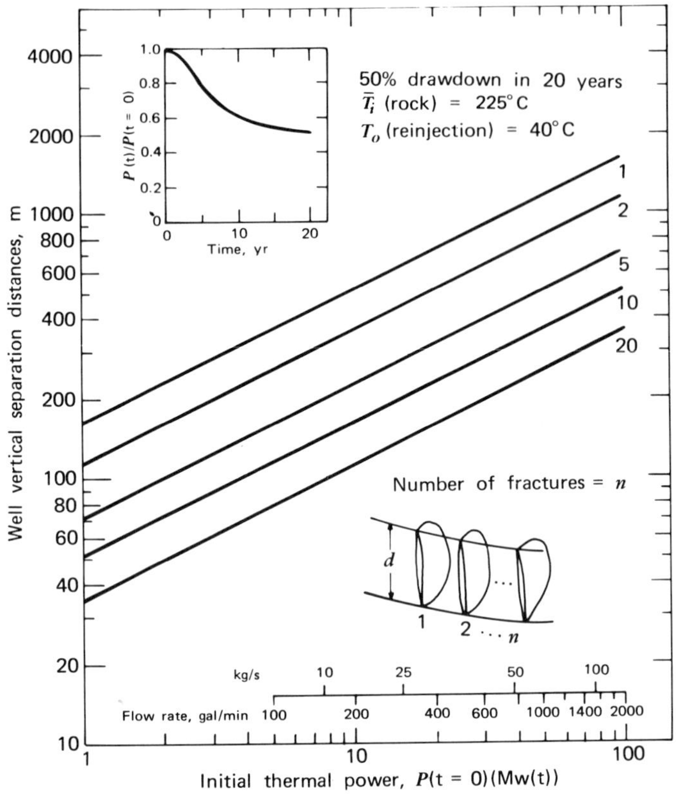

Figure 6. Conceptual design nomograph for low permeability formations with a specified level of thermal drawdown (50% in 20 yr of operation). Wellbore separation distances that define single or multiple fracture sizes are shown as a function of initial thermal power level. Flow rates given in kg/s and in gal/min.

Fluid Geochemistry. More than 158 hydrothermal wells including some deep ones (over 2400 m) have been drilled over the past 50 years with basically no hydrothermal utilization (17). At a very rough cost figure of $250,000 per well, private industry has invested ca 40 million dollars with virtually no return. Scaling, corrosion, and waste disposal are major problems.

The range in chemical composition exhibited by hydrothermal fluids is remarkable (Fig. 7). The variation of fluid composition is highly variable even within a single field (92). Because of this variation, each fluid-chemistry problem is unique and only certain generalizations can be made. Usually the dissolved mineral content of hydrothermal and geopressured fluids exceeds the concentration that can be explained solely by interaction between water and rock. Although the long-term geochemical behavior of HDR reservoirs is still unknown, the experience at Fenton Hill, New Mexico, indicates that rock–water interactions ultimately determine the concentrations of minerals in solution in low permeability, dry reservoirs. This may result in a significant improvement over the hydrothermal–geopressured fluid compositions.

Scaling. Scale deposited in pipes is usually one of four types: silica, carbonate, metal sulfide, and sulfate. These scales are usually formed by loss of CO_2 (during flashing) which changes the pH; concentration of salts in the liquid phase by flashing,

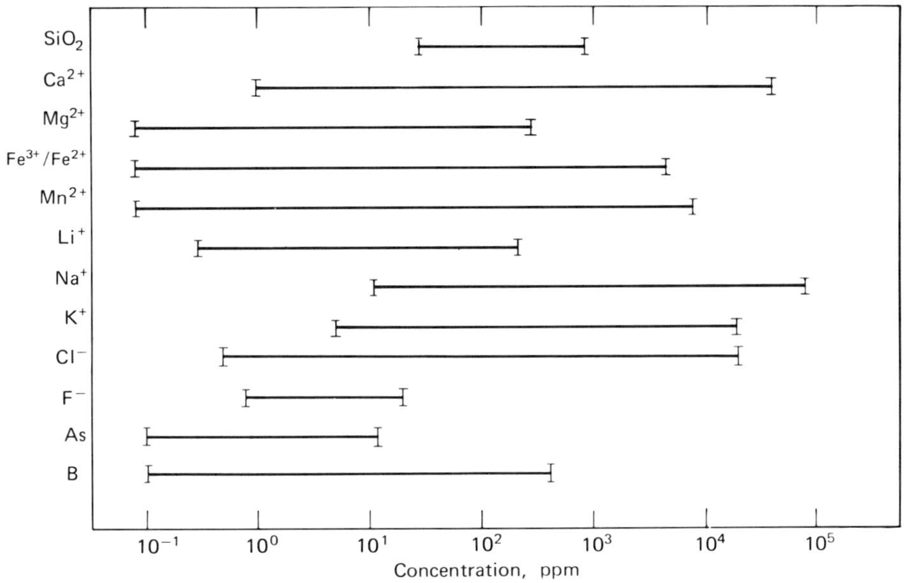

Figure 7. Concentration ranges for elements in geothermal fluids. Ionic species specified when known.

temperature changes in the fluid affecting solubility, and precipitation caused by mixing of different fluids (see Dispersants).

Corrosion. Corrosion occurs when high temperature fluids contain high chloride or hydrogen sulfide contents at low pH and high oxidation potentials. The forms of corrosion damage to metals commonly occurring in a geothermal environment include the following (93): (*1*) uniform surface corrosion; (*2*) pitting corrosion; (*3*) stress corrosion and sulfide stress cracking, cracking of a metal under stress in a corrosive environment; and (*4*) erosion–corrosion, a mechanical abrasion of passivating films on a metal surface. This is traceable to the high velocity entrained particulate matter (eg, sand or precipitated solids) in the geothermal fluids. Methods for the control of corrosion in geothermal systems are described in ref. 94 (see Corrosion).

Fluid Disposal. Disposal of fluids is a major problem. Fluids that are nearly 10% chloride cannot be dumped into fresh water drainage without severe environmental impact. Evaporation is often not practical at the rates of fluid production necessary for economic power production. Removal of Cl^- by Pb or Hg precipitation would be expensive and a worse environmental hazard than releasing the fluid to surface streams. In many cases, reinjection of the fluid into the reservoir is the best alternative (16,19). Because the fluid has been cooled in the heat-extraction process, minerals in the fluid are supersaturated. Injection of this material could cause severe plugging of the reinjection well (42,95).

Environmental Impact

The effect of a geothermal development on its surroundings is very important. Occasionally, there are significant costs for restoration of an area to its original state. These costs can be minimized somewhat by careful evaluations of the environmental impact and installation of facilities to minimize the impact.

Pollution. One of the main environmental effects of geothermal development is the production of pollutants including gaseous emissions, liquid effluents, solid residues, waste heat, and noise (96–97). Most natural steam and hot water reservoirs contain dissolved gases that have been released naturally for thousands of years. Emission of hydrogen sulfide, a common and offensive pollutant (98), can in some cases be eliminated or controlled (99). Depending on the concentrations of the various dissolved species, fluids from geothermal wells may not be suitable for irrigation purposes. In particular, concentrations of chloride, fluoride, bromide, arsenic, mercury, cadmium, and selenium may be greater than can be safely released for biological consumption. There are a number of ways to control the concentration of toxic materials in waste water. Removal of these materials by treatments such as precipitation, ion exchange (qv), and reverse osmosis (qv) is usually too costly for the volumes of fluid involved. The usual alternatives are reinjection into the reservoir, evaporation (qv) in ponds, or—in countries where environmental restrictions are not as stringent as in the United States—discharge to the nearest stream regardless of the consequences.

Brine that contains high total dissolved solids may have a tendency to precipitate solids (usually silica, calcium carbonate, or metal sulfides) when cooled or flashed. Removal of this scale from surface piping presents a problem of solids disposal. Fortunately, these materials are usually inert and nontoxic so that disposal is easily effected by burial.

Because geothermal electric systems operate at low conversion efficiency, large amounts of waste heat are rejected to the environment using either wet or dry cooling towers. However, geothermal plant sizes are small (ca 50–100 MW), and thermal discharges are distributed over a larger area than for more efficient but larger capacity (ca 1000 MW) coal-fired or nuclear generating plants (see Nuclear reactors; Power generation).

Use of Other Resources. ***Water.*** Consumption of water for evaporative cooling could have significant impact on the environment, particularly in arid regions (100). In addition to the potential effects on ground-water flow, evaporation of large volumes of water could affect local climatic conditions; in particular fogging and cooling tower drift can be important. Detailed environmental studies are necessary to determine the effects of water use on the environment. In water-short areas, careful consideration must be given to dry or wet/dry cooling. Engineering design and economic issues relevant to waste heat rejection are covered below.

Land. Evidence at existing geothermal installations indicates that surface land utilized for wells and surface plumbing is compatible with other land uses such as farming and cattle grazing. Because of the wells and fluid transmission systems, larger land areas are required for a geothermal plant than for a fossil-fuel or nuclear power plant of similar capacity. However, the total fuel cycle of a fossil-fuel or nuclear system (including mining, drilling, refining, and reprocessing operations) require a substantially greater land use (101) and their environmental consequences may, in fact, exceed that of a similar-sized geothermal plant where all relevant processes are localized at the plant site. Like the intrusion of any other power generation activity, development and operation of a geothermal field affects much of the local wildlife and may to some degree conflict with recreational use of the land.

Seismic Hazard. The stress condition of the reservoir rock is changed when fluid is withdrawn or injected, introducing the possibility that earthquakes may be triggered by geothermal field operations. Seismic activity is apparently induced only if a large volume of fluid is introduced under high pressure into an active fault. At least until the effect is better understood, this situation should evidently be avoided. However, in several geothermal fields the frequency of natural microseismic events is usually high, but tremors large enough to be damaging are notably rare. Frequent relief of earth stresses by small local events prevents the accumulation of high stresses over areas sufficient to produce large earthquakes. Additional long-term monitoring of geothermal areas is needed to verify the generality of this behavior, but the current data suggest that the seismic risk of a geothermal development is very low.

Subsidence. When a large volume of fluid is removed from an underground reservoir and is not replaced, the weight of the overburden may produce compaction of the porous reservoir rock, which may be reflected in subsidence of the land surface above it. Subsidence is not known to have occurred in natural dry-stream fields, where reservoir pressures are necessarily subhydrostatic before exploitation begins and are not greatly reduced thereafter. It is, however, common in hot-water geothermal fields (102) as it is in water-well and petroleum fields. In several cases in oil fields, subsidence from this cause has been prevented, arrested, and even reversed by injecting a volume of water substantially equal to the volume of fluid withdrawn, thus maintaining pressure in the reservoir. This process should be possible in many geothermal systems, particularly when binary cycles are used, although it may not always be feasible to reinject the water at the right locations to maintain pressure throughout the entire field. Reinjection into the reservoir itself may not be practical in geopressured areas, where the well and pumping costs involved are high and may even exceed the value of the fluid withdrawn (103). Subsidence is not expected in low permeability hot dry rock systems of the type now being developed since the volume reduction in the reservoir is only that represented by thermal contraction and mineral dissolution in a thin vertical fracture deeply buried in the rock.

Energy Conversion and Utilization

Geothermal, solar, and ocean thermal resources share the common disadvantage of having an inherently low resource temperature. One technique of increasing their quality as well as their transportability is to convert their available energy into electric power. Unfortunately, the efficiency of this conversion process is limited by the temperature of the resource and by prevailing ambient conditions for heat rejection. Efficiencies for converting geothermal resources at temperatures below 200°C to electricity are substantially less than those of fossil-fuel-fired or nuclear-powered plants. New technologies have been developing for improving the practical thermodynamic efficiency of power production towards its ideal limits to provide an economical process. For example, for a number of these low temperature, alternative energy sources, several hydrocarbons and their halogenated derivatives have been proposed as prime movers rather than steam. The properties of these organic compounds are considerably different from water in the low temperature region, perhaps making them more suitable as working fluids.

At the other extreme of use, serious consideration is being given to nonelectric possibilities for geothermal energy. Recent estimates have shown that ca 37% of the total energy consumed annually in the United States is utilized below 250°C. Of this amount a substantial fraction is used at 50–80°C for both residential and industrial space heating applications (104).

Applications. Electric power generation and space and process heating have been applied to commercial geothermal operations throughout the world (31,58,105–106). This section introduces the engineering components and thermodynamic issues associated with geothermal power and heating systems.

Conversion Cycles. Various electric conversion and direct-use cycles are illustrated in the schematics of Figures 8–10. A Rankine cycle using dry or saturated steam expansion in a low pressure condensing turbine with direct contact condensation is now used to generate power for vapor-dominated resources such as those at The

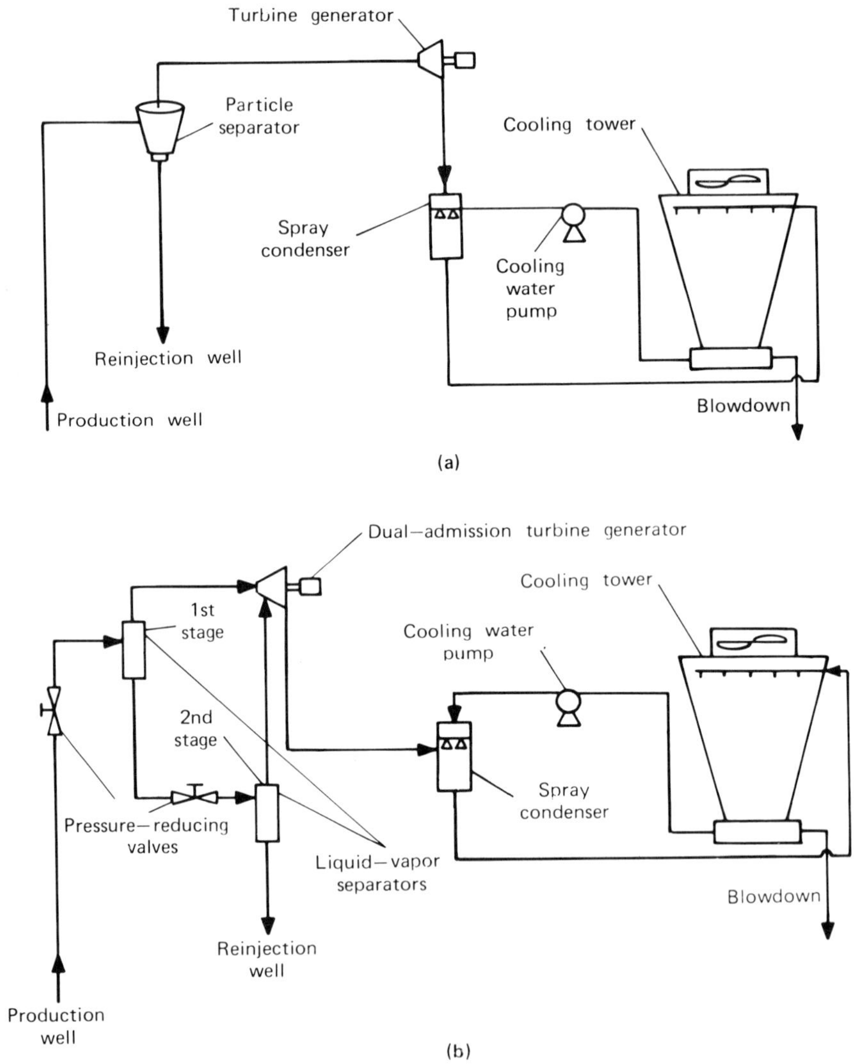

Figure 8. Commercially operated geothermal power conversion cycles: (**a**) direct steam cycle for vapor-dominated systems, and (**b**) dual-stage flashed-steam process for liquid-dominated systems.

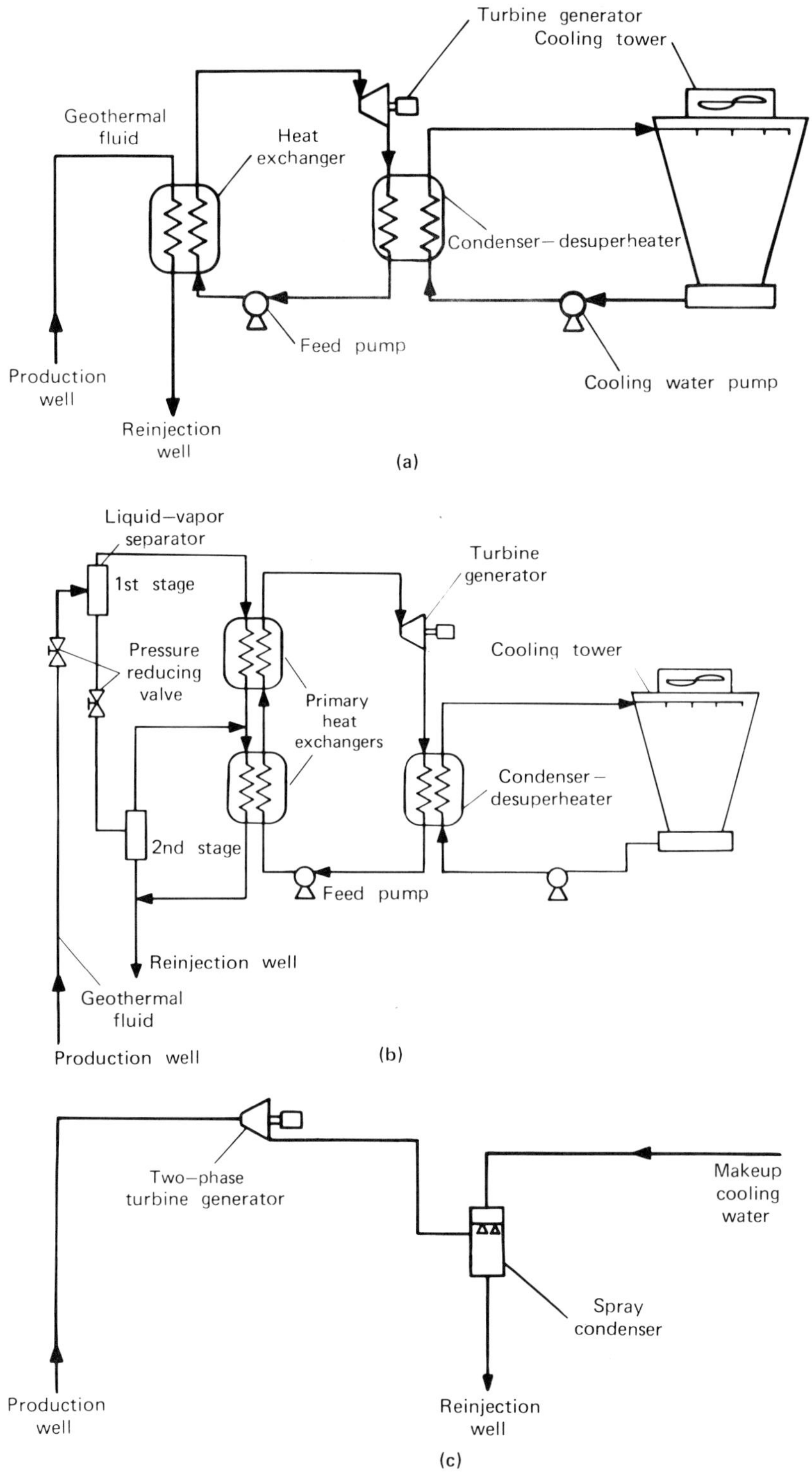

Figure 9. Proposed geothermal power conversion systems: (**a**) binary-fluid cycle; (**b**) multistage flash–binary cycle; and (**c**) two-phase total-flow expander.

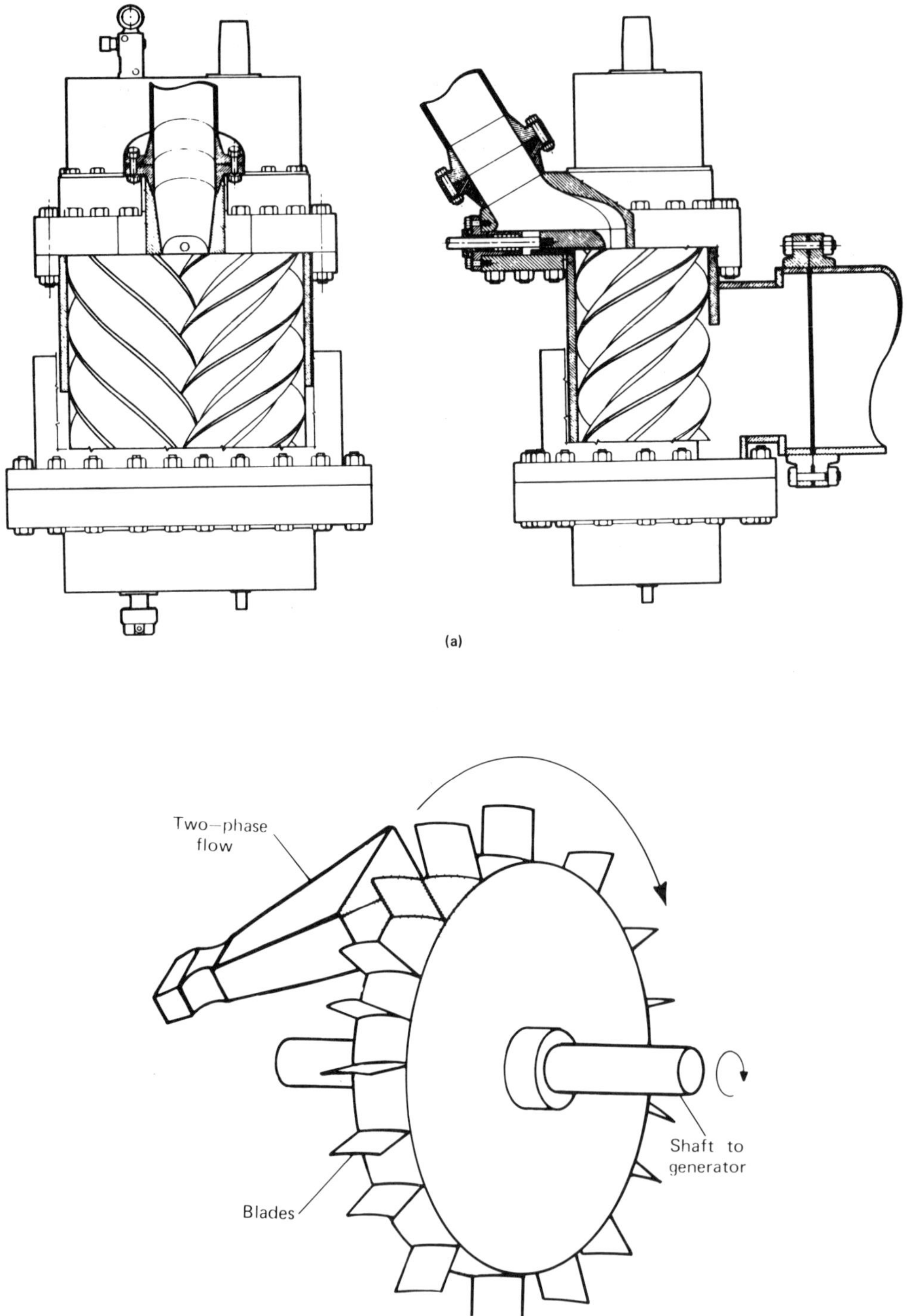

Figure 10. Schematics of two total-flow turbines: (**a**) helical screw expander, courtesy of Hydrothermal Power, Ltd., and (**b**) axial-flow impulse turbine, courtesy of Lawrence Livermore Laboratory.

Geysers, California, and Lardarello, Italy (see Fig. 8**a**). Assuming particulate matter can be removed from the steam before it enters the turbine, conventional turbine designs and materials of construction can be used. The multistage flashing cycle depicted in Figure 8**b** is used for liquid-dominated resources such as those at Wairakei, New Zealand, and Cerro Prieto, Mexico. In this case, saturated steam is created by flashing the geothermal fluid at the surface to a lower pressure, followed by expansion in a condensing steam turbine with the unflashed liquid fraction either reinjected or discarded. The performance of the steam cycles shown in Figure 8 is significantly lowered by the presence of noncondensable gases in the geothermal fluid. An extraction pump or a vacuum ejector is required to remove noncondensable gases such as CO_2 to maintain the turbine exhaust at subatmospheric conditions (107). If noncondensable gas concentrations are too high, indirect binary or two-phase flow conversion cycles such as those in Figure 9 might be the only feasible alternatives. Binary-fluid cycles are closed-loop Rankine cycles that involve a primary heat-exchange step in which heat from the geothermal fluid is transferred to a secondary working fluid which then expands through a turbogenerator and is condensed. The waste heat is finally rejected to the environment. The closed cycle is completed by pumping the secondary fluid up to the maximum cycle operating pressure. Compounds considered for these applications are lower molecular weight hydrocarbons, such as isobutane and propane, and their halogenated derivatives (fluorocarbons) such as CH_2F_2, $CHCl_2F$, C_2ClF_5, and ammonia (NH_3) (see Heat exchange technology).

The performance of such binary cycles is rather insensitive to noncondensables contained in the primary geothermal fluid. Furthermore, because these nonaqueous secondary working fluids have low temperature vapor densities considerably greater than steam, much smaller and less expensive turbines would be required for the same power output. Flashing cycles are, of course, simpler in that they do not require a primary heat exchanger or secondary fluid feed pump, but flashing can cause difficult scaling problems.

If the geothermal fluid contains large amounts of dissolved material that may corrode or deposit on heat-exchange surfaces, a combined multistage flash and organic binary cycle may offer a reasonable solution (108–110). This system can be designed in a number of ways. Figure 9**b** shows one alternative, a two-stage flash unit produces steam for fluid vaporization in an organic Rankine cycle. Other designs may involve using a dual-cycle system where the flashed steam fraction drives a turbogenerator and a binary cycle operates using the remaining liquid fraction to vaporize the secondary working fluid (108). Organic binary systems are state-of-the-art technology requiring some developments including the design of larger capacity hydrocarbon units, control and valving systems, turbine seals for hydrocarbon vapors, and improved resistance to corrosion, erosion, and scaling of certain components.

Cycles proposed to deal with the deposition or scaling problem (109–110) are less efficient than the multistage, direct-steam flashing systems that may be practical when dissolved gas contents are low. Sperry Research (111) is investigating an unusual binary system that involves a downhole heat exchanger and organic turbine-driven pump in a gravity head cycle for resources in the 120–200°C range.

Austin and associates at the Lawrence Livermore Laboratory (112) have evaluated total-flow, two-phase expansion systems for geothermal applications. Total-flow expanders (Fig. 9**c**) may be particularly useful for high temperature, high salinity brines such as those found in the Niland-Salton Sea area of Imperial Valley, California. In the flashed-steam system, a large portion of the initial fluid is not converted into steam for expansion in a turbine; consequently, a substantial fraction is unusable for power production. The total-flow process avoids this waste by expanding the entire geothermal fluid through a suitable device. As yet unachieved engine efficiencies of 70% or greater are required to improve on the present performance of multistage, flashed-steam and organic binary cycles. Nonetheless, the inherent simplicity of total-flow machines suggests some of their capital costs may be lower than those of the alternatives. A variety of engine concepts are possible (113), including impulse–reaction, positive displacement, and impulse machines. The helical screw expander as shown in Figure 10**a** is a positive displacement machine under development by the Hydrothermal Power Company and the Jet Propulsion Laboratory (114). Positive displacement expanders must be large in comparison to comparable condensing steam turbines, and this may limit central power station (>50 MW) use, but for smaller-scale, noncondensing applications they may be quite practical. Lawrence Livermore Laboratory is investigating an axial-flow impulse-turbine concept that requires the development of efficient, two-phase expansion nozzles and efficient turbine blading for transfer of the fluid momentum to the rotating shaft (Fig. 10**b**).

With the possible exception of a few small binary-fluid power plants operating in Japan and the USSR (105,115), direct-steam injection and steam-flashing cycles are the only ones in actual commercial service today for geothermal resources.

Even with low conversion efficiencies, electric power production from geothermal resources continues to receive considerable attention for existing and proposed sites in the United States. In part, this interest is caused by the inherent base-load character of a geothermal operation coupled to its frequently remote location from an existing load center. This is certainly true of the situation at The Geysers field in California where the San Francisco load center for the over 600 MW generated is ca 129 km away. Although electrical transmission costs are generally lower, commercial nonelectric applications for space heating can be economical for transportation of hot fluids for several kilometers as demonstrated in Iceland, USSR, Japan, and Boise, Idaho (58,105–106,116–117). Large-scale transport of hot fluid in some cases may be competitive at distances greater than 100 km (118–119).

Nonelectric systems can be described very simply as a primary heat-exchange loop and a cascaded utilization system. If the quality of the fluid is good, it may be transported directly, avoiding the primary heat-exchange step. Recycling geothermal fluid may not be required in this case. Extensive use is presently being made of geothermal fluids ranging from 40 to 200°C for process heat in the mining and pulp and paper industries, and for district space heating, therapeutic baths, absorption air conditioning, and agricultural and aquacultural purposes (58,120). In all cases, both the use requirements and geothermal fluid characteristics must be considered in designing the system.

The thermodynamic utilization of both fossil and geothermal energy for electric power production can be improved by a hybrid arrangement (121–123). In this concept, geothermally supplied heat is used for feedwater preheating. A fossil-fired unit supplies additional heat before the steam enters the turbogenerator. This arrangement can significantly improve efficiency by reducing coal and geothermal fluid consumption on a per MW basis but has the disadvantage of being a more complex and a more capital intensive enterprise that may require proximity to both geothermal and coal resources for commercial feasibility.

Market for Geothermal Energy. Even with the inherent remoteness of many hydrothermal systems, electricity can be generated at the site and tied into an existing grid system at an acceptable cost. Geothermal power plants have smaller capacities (ca 50–100 MW) than nuclear and fossil-fuel plants and consequently can be used as base-load units in much larger total-generating-capacity systems. Of the 450,000 MW of generating capacity in the United States, only 660 MW is geothermal. A substantial market exists for both new and replacement capacity.

The situation with the nonelectric market is quite different. Utilizing geothermal space and process heat systems may require some retrofitting by the consumer. For example, a fuel-oil or natural-gas furnace usually is designed with in-house storage or is attached to an existing distribution network. The use of geothermal hot water or steam would require a completely new fluid-distribution system within the community.

Reistad (104) discusses energy usage as a function of utilization temperature and concludes that about 40% of our annual fossil energy consumption is severely degraded thermodynamically. Typically, fossil combustion temperatures of 1000–1500°C are used to produce space and process heat at temperatures below ca 250°C. The distribution of energy consumed versus temperature is illustrated in Figure 11. A wide variety of applications is contained in the 40% or 32 EJ (30 quads) of energy used below 250°C. Table 2 summarizes some of the major ones. Detailed discussions of how geothermal energy is now or may potentially be used can be found in references 30–31, 58, 104, 120, and 126–132.

Efficiency and Available Work. The second law of thermodynamics imposes real limitations on the production of electricity from low temperature geothermal heat sources. The total maximum work that can be produced by cooling a condensed

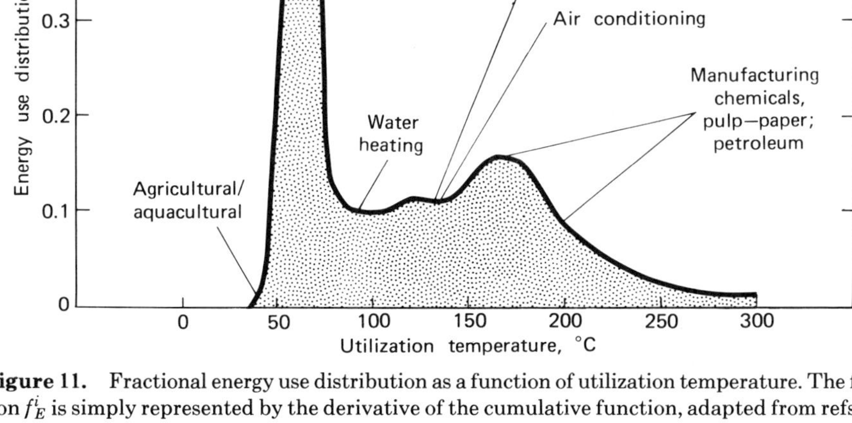

Figure 11. Fractional energy use distribution as a function of utilization temperature. The fractional function f_E^i is simply represented by the derivative of the cumulative function, adapted from refs. 104 and 124.

Table 2. Existing and Potential Major Nonelectric Uses for Geothermal Energy[a]

Application	Temperature range, °C	Steam requirement, kg steam/kg product	Total annual U.S. consumption (1967), PJ[b]	Status
space heating	80–50	0.4	30,530	existing installations throughout world, eg, Klamath Falls, Oregon; Boise, Idaho; USSR; Japan; Iceland;Hungary; France; Czechoslovakia; Argentina
water heating	120–80	0.5	2,389	
refrigeration and air conditioning				
ammonia absorption	≥180		1,540	
lithium bromide absorption	>70			
chemical industry				
chlorine–caustic soda production using diaphragm cells	?	8 (NaOH)	69	potential use
soda ash production with Solvay process	≥120	2	57	potential use
alumina production from bauxite with Bayer process	150	7	52	potential use
heavy water production by hydrogen sulfide–water isotope exchange	170	10,000		potential use
desalination by multiple-effect evaporation of seawater	120	0.08		pilot plant in Chile
by-product recovery of minerals from geothermal brines via evaporation	≥120	10–2		boric acid recovery at Larderello, Italy
sulfur mining with Frasch process	≥120		60	potential use
concrete processing and drying	110	2–1		existing facility in Reykjavik, Iceland
diatomaceous earth drying	170	3–2		existing facility in Namafjall, Iceland
petroleum refining	250–175 (20%) 175–150 (40%) 150–125 (40%)		960	potential use
food processing				
freeze drying via ammonia absorption refrigeration	≥180	4–2		potential use
canning	140	2–1	900	existing small-scale efforts
drying	140–120	2–1		
sugar refining	130	6 (beet) 1.7 (cane)		potential use
pulp and paper processing				

Table 2 (*continued*)

Application	Temperature range, °C	Steam requirement, kg steam/ kg product	Total annual U.S. consumption (1967), PJ[b]	Status
Kraft sulfite pulp digestion	200–175 (70%)	3–4	1005	existing cogeneration facility in New Zealand
dissolving pulp	175–150 (30%)	4		
agricultural				
crop drying	60			operational on a small scale throughout the world, eg, Hungary, Iceland, Italy, U.S.
greenhouse heating	60			
aquacultural				
fish farming	20			existing system in Japan

[a] Adapted from refs. 30, 104, 120, 125.
[b] To convert PJ to 10^{12} Btu, divide by 1.054.

geofluid from its well-head condition T_{gf} to the so-called dead state, or ambient condition T_o is given by the thermodynamic availability ΔB:

$$\Delta B = [\Delta H - T_o \Delta S]_{T_o}^{T_{gf}} \tag{9}$$

where ΔH is the enthalpy difference and ΔS is entropy difference between states. This maximum work quantity can then be compared to the actual amount of work produced by any real power conversion process W_{net}.

Comparisons of this type are usually achieved by defining a cycle efficiency (η_{cycle}) which represents the net useful work W_{net} obtained from the system divided by the amount of heat transferred from the geothermal fluid Q_H. As the cycle efficiency decreases, the amount of heat rejected to the environment increases. For an ultimate sink of 25°C with a geothermal heat source of 50°C, cycle efficiencies would be less than 5%. As the source temperature rises, the efficiency increases as shown in Figure 12.

An alternative approach to using cycle efficiency would be to compare directly the real work to the maximum possible work ΔB by defining a utilization efficiency η_u as,

$$\eta_u \equiv \frac{W_{net}}{\Delta B} = \frac{\text{P}}{\dot{m}_{gf}\overline{\Delta B}} \tag{10}$$

where P = net power extracted, $\overline{\Delta B}$ = availability per unit mass, and $\dot{m}_{gf}$ = geothermal fluid-flow rate. Then η_u is a direct measure of the effectiveness of resource utilization, because for a fixed T_{gf}, higher values of η_u correspond to lower well-flow rates for a given power output. In contrast, the cycle efficiency, η_{cycle} is a measure of how efficiently the transferred geothermal heat is converted to work. The utilization efficiency concept is particularly useful in comparing binary-fluid and flashing-cycle performance for the same resource conditions. Ideally, η_u should be as high as possible, but there are limitations in the efficiency of work-producing machinery (turbines and pumps) as well as in the size of heat-transfer systems associated with generating power.

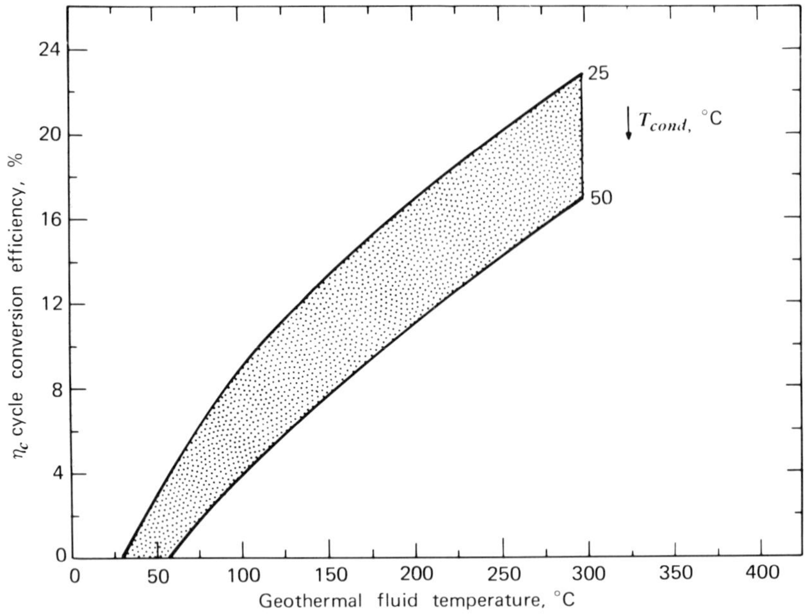

Figure 12. Typical range of anticipated cycle conversion efficiencies for geothermal power plants. $\eta_c = W_{net}/Q_H$; η_p = 80%; η_t = 85%.

Thermodynamic Properties. Accurate data on the thermodynamic properties of proposed working fluids are required to calculate electric power cycle performance. Heat capacity at constant pressure in the ideal states C_p^*, vapor pressure P^{sat}, pressure–volume–temperature (PVT) behavior, enthalpy and entropy changes, and liquid density at saturation ρ^{sat} can be expressed with semiempirical equations. For example, a modified form of the Martin-Hou equation of state has been used for calculating PVT properties (133–135). This equation, accurate for densities above the critical point, a requirement for supercritical Rankine-cycle calculations, has been successfully applied to predicting the PVT properties of many fluorocarbons. Martin-Hou parameters have also been generated for isobutane and ammonia (135–136). Using a modified Benedict-Webb-Rubin (BWR) equation of state as developed by Starling (137) yields another high-accuracy equation of state. The Starling and BWR equations have been used extensively to assess the use of organic mixtures containing light hydrocarbons in the design of binary-fluid cycles (108,138–140).

The properties of water and steam containing significant quantities of noncondensable gases such as CO_2 or dissolved materials such as NaCl, SiO_2, KCl, and HCO_3^-/CO_3^{2-}, need to be specified. An elaborate treatment of effects of noncondensables and dissolved salts on the thermodynamic properties of geothermal brines has been presented.

Working-Fluid Selection. Many hydrocarbons, fluorocarbons, and other organic working fluids have been examined for potential use as a replacement for water in low temperature power conversion cycles. For example, the refrigerants, R-22 (CHClF), R-600a (isobutane, i-C_4H_{10}), R-32 (CH_2F_2), R-717 (ammonia, NH_3), RC-318 (C_4F_8), R-114 ($C_2Cl_2F_4$), and R-115 (C_2ClF_5) were studied to provide a range of properties including critical temperature and pressure, and molecular weight (134). All of these compounds have relatively high vapor densities at heat rejection temperatures as low as 20°C and would result in turbines very compact in comparison to steam turbines of similar capacity. The performance of R-21 ($CHCl_2F$), R-600a (isobutane), and Fluorinol (trifluoroethanol) was compared. A number of hydrocarbons including propane (R-290), propylene (R-1270), isobutane, *n*-butane (R-600), isopentane, and *n*-pentane as well as R-115, R-22, R-32, and R-717 were considered (110).

Factors other than desirable thermodynamic properties frequently determine practical working fluids. These include, fluid thermal and chemical stability, flammability, toxicity, material compatibility (corrosion), and cost. The major disadvantage of hydrocarbons, such as propane, pentane, and isobutane, is their flammability and thus requirements of costly explosion-proof equipment and ventilating systems. The situation with fluorocarbons is more complex but thermal instability is an important issue (143–145) (see Heat-exchange technology, heat-transfer media other than water).

Engineering Design Criteria. Any real process for generating electricity or heat has inefficiencies or nonreversible steps that result in a net work less than W_{net}^{max} given in equation 10. Efficient use of the resource may be necessary for commercial feasibility if reservoir development costs are high. Efficient utilization requires that (*1*) most of the heat is extracted from the geothermal fluid before disposal or reinjection; (*2*) temperature differentials across heat-transfer surfaces are maintained at minimum practical levels; (*3*) turbines and feed pumps are carefully designed for optimum efficiency, and (*4*) heat is rejected from the thermodynamic cycle at a temperature near the minimum ambient temperature T_o. Whether or not these conditions can be met depends largely upon the choice of thermodynamic working medium, the geothermal fluid temperature, and the temperature of the coolant (water or air) to which the power plant rejects waste heat. For example, if waste heat is to be rejected from the thermodynamic cycle at a constant temperature by a condensing vapor (as in condition (*4*) above), then the working medium's critical temperature must be greater than the temperature of the power-plant coolant. Fulfillment of conditions (*1*) and (*2*) suggests the use of supercritical Rankine cycles. Thus, the critical temperature of the working fluid should also be below the maximum geothermal fluid temperature (134).

Heat Exchange and Prime-Mover Fluid Production. As illustrated in Figures 8–10, a geothermal electric or heating system may require a variety of heat-transfer steps to produce a suitable prime-mover fluid. For example, for a binary-fluid cycle, a primary heat exchanger operating as a combined preheater, boiler, and superheater at subcritical pressures or as a vaporizer at supercritical pressures will be required to remove heat from the geothermal fluid and generate the organic vapor for injection into the turbine (see Fig. 9**a**). In a flashing cycle, prime-mover steam is produced by reducing the pressure of the fluid in one or more steps. In both cases, the design and component sizing of the heat exchanger or flashing unit require specification of the appropriate duty factor for each unit (J/h, or kg steam generated/h) and of temperature and pressure differences (see Heat-exchange technology).

Because power conversion efficiencies are low, the amount of heat transferred may be five to fifteen times the power produced. Large heat exchangers are required at significant cost. For example, a 100 MW geothermal plant with a 12% cycle efficiency requires about 60,000 m^2 (650,000 ft^2) of heat-exchanger surface area. Thus multiple, parallel units of 1860–5570 m^2 (20,000–60,000 ft^2) would be utilized.

For a primary heat exchanger in a binary cycle, the overall heat-exchange surface area A can be expressed as a function of the heat load $\dot{Q}$, an appropriate mean $\overline{\Delta T}$ and an effective overall heat-transfer coefficient U:

$$A = \dot{Q}/(U\overline{\Delta T}) \tag{11}$$

A practical ΔT should be selected to keep the surface area and, therefore, cost at a reasonable level. In some cases the exchanger might reach a pinched condition where ΔT approachs zero at one or more locations. Then a minimum ΔT would have to be specified for any potential pinch points.

To estimate the surface area of any exchanger, realistic values for heat-transfer coefficients of both fluids, wall resistance, and fouling factors may be required. Because many different geothermal brine compositions and working fluids operating in sub- and supercritical modes are under consideration, no single optimum design can be proposed.

Because the chemical compositions and temperatures of geothermal fluids vary widely from site to site, it is extremely difficult to estimate accurately the extent of a potential scaling or fouling problem for primary heat exchangers without field testing. For example, the 300,000 ppm total dissolved solids (TDS) and 370°C temperatures of Salton Sea geothermal brines in the Imperial Valley present quite a different problem than the brines at Cerro Prieto with <20,000 ppm TDS and 300°C temperatures (see also Chemicals from brine). Hot dry rock systems introduce another factor in that the circulating water is not indigenous to the reservoir; dissolution products build up in the circulating fluid and might also present scaling problems particularly at higher reservoir temperatures (>200°C).

Only certain types of heat-exchanger designs are appropriate for geothermal applications (146). These include: shell-and-tube exchangers, fluidized-bed exchangers, and direct-contact exchangers. The conceptual design and sizing of shell-and-tube units have been considered by a number of investigators (108,110,134,146–149). Direct-contact and fluidized-bed systems may be the only usable candidates when scale deposition rates are high (107,150–152) but even with moderate scaling problems, careful design and operation of shell-and-tube units is required.

Commercial-size units for either direct-contact or fluidized-bed heat exchange with geothermal fluids have not been built as yet, but considerable testing is under way. A very complete discussion of both concepts is contained in ref. 153.

When steam produced by flashing is the prime-mover fluid and is used directly in a condensing turbine as shown in Figure 8**b**, no primary heat-exchange step is required. In its place, pressure reduction is used to produce a saturated vapor fraction. In a typical installation, the first stage of flashing is done at the wellhead. The two-phase mixture produced is piped to a separator system—frequently a cyclone unit—where the vapor (steam) and liquid brine are separated. The steam fraction is transmitted to the turbine-generator unit located in a central area, and the liquid fraction is reinjected, discarded, or flashed to a lower pressure to produce additional steam for turbine injection. This process can be continued as many times as are economically feasible. A two-stage flash system may produce 20–30% more available work than a single-stage unit at the same total geothermal fluid-flow rate. But the improved performance of multistage systems must be balanced against the additional complexity and cost of utilizing the secondary steam (141).

Turbines. In selecting nonaqueous working fluids for power-cycle applications, turbine sizes should be small to reduce costs and offset the additional heat-exchange surface area required for binary-fluid cycles.

It is important to operate geothermal turbines at high efficiency. A similarity analysis of performance shows that turbine efficiency is essentially controlled by two dimensionless numbers involving four parameters: blade pitch diameter, rotational speed, stage enthalpy drop, and volumetric gas-flow rate (110,142,154). Maximum turbine efficiency requires a specified relationship among these parameters; therefore, turbine sizes and operating conditions and consequently costs can be estimated. For fluid-screening purposes, a generalized figure of merit which relates directly to turbine size has been developed (134). Many organic fluids offer a significant reduction in turbine size. For example, a 100 MW capacity plant designed for a 150°C liquid-dominated resource would require numerous, large turbine-exhaust ends with steam flashing. An ammonia binary cycle would require only a single exhaust end (155). However, no organic turbines with capacities larger than about 1 MW have yet been manufactured (115,156–158). In contrast, low pressure steam turbines for geothermal operation with 50 MW capacities or greater have been commercially produced for a number of years. In fact, only one organic binary cycle using geothermal water at 80°C has been in commercial operation for more than a few years. Several small systems (ca 100 kW) have been built by Barber-Nichols for geothermal demonstrations. These include an isobutane binary, a direct-contact isobutane binary, and an R-114 binary (157) (see Power generation).

Pumps. Pumping requirements for geothermal systems are of four major types: (*1*) downhole pumps for fluid production, (*2*) fluid-reinjection pumps, (*3*) power-cycle feed pumps, and (*4*) geothermal-fluid transport pumps. According to a published survey and assessment of existing geothermal pumping equipment (159), conventional multistage centrifugal designs can be used in most cases for commercial-size applications in areas (*2–4*) just described. Reinjection pumps may present some problems where highly corrosive and erosive geothermal brines are used. The most challenging problems are concerned with downhole pumping because of the severe environmental and design constraints. Four general types of downhole pumping systems have been proposed. These include shaft-driven and submersible electric-motor-driven, multistage, vertical centrifugal designs that are commercially available and have been tested by a number of companies (159). The complexities associated with downhole shaft pumps driven mechanically on the surface are obvious. Line-shaft bearings for high temperature service are the major weak structural factor. Electrically driven submersibles also experience serious problems at operating temperatures above 180°C because of the effects of heat, moisture, and brine on the motor. The other general pump types under development include steam- or organic-fluid-driven, downhole turbine pumps (111,160–161) and hydraulically-driven systems (159). In these cases, the problems of shaft or electric-motor-driven units are replaced with a fairly complex concentric wellbore-piping system (160,162).

Condensers and Heat Rejection. Because conversion efficiencies are low, a large fraction of heat transferred is ultimately rejected to the environment. For example, a geothermal plant operating with a cycle efficiency of 10% rejects about three to six times as much heat as a similar capacity fossil- or nuclear-fired unit. Consequently, the substantial costs associated with condenser–desuperheaters and cooling towers in geothermal systems strongly influence economic feasibility. Seasonal and diurnal variations of ambient dry bulb temperatures also can significantly affect cycle performance (123). For dry cooling systems, condensing temperatures change seasonally, causing changes in the power output of the plant. Because of the size of geothermal units (<100 MW), operating with a floating power output might be desirable.

The availability of water for cooling is also important. Fossil and nuclear plants are often located near abundant water supplies. Geothermal plants, of course, have to be located at the resource regardless of the water situation. Even so, if evaporative cooling is used, water consumption is high. For example, a 50 MW geothermal plant operating at 10% efficiency would consume cooling water at a rate of ca 145 L/s (2300 gal/min). In many existing plants, cooling water is provided by the geothermal fluid itself. For example, evaporative cooling at The Geysers consumes ca 80% of the geothermal fluid, so that only 20% is reinjected (108). For liquid-dominated and geopressured reservoirs, total reinjection may be required to avoid subsidence or for other environmental or regulatory reasons. Furthermore, the brine may have to be treated before reinjection.

A number of design options for waste heat rejection are described in ref. 163. These include direct-contact spray condensers; multipass water-cooled surface condensers; direct air-cooled condensers; spray ponds; and wet, wet/dry, and dry cooling towers with forced, induced, or natural-draft design.

For many power-cycle operations, evaporative cooling has been commonly used and will no doubt continue to be employed in future designs. However, when total reinjection is required and water is scarce or expensive, totally dry and wet/dry cooling are the only alternatives. Although high capital costs have prevented widespread use of completely dry systems in the United States, many such systems have been successfully operating in Europe (164). The major disadvantage, besides higher cost, of dry systems is that performance of the power cycle depends on dry-bulb temperature fluctuations. If floating power capacity is an acceptable operating mode for small base-load units, then conceivably variations in power output would not be a serious detriment to using totally dry systems. A small 60 kW, prototype system of this type using an R-114 ($C_2Cl_2F_4$) binary cycle has been constructed for use at the Fenton Hill hot dry rock demonstration site, which has a 3 ± 20°C ambient dry-bulb temperature (157).

An alternative to totally dry systems is a wet/dry cooling tower which combines advantages of both systems. Low condensing temperatures can be maintained during peak periods with high dry-bulb temperatures by using supplemental evaporative cooling, and water can be conserved during periods of low dry-bulb temperature. Fogging and drift are also better controlled. Wet/dry towers cost more, however, and are somewhat more difficult to operate and maintain (see Evaporation).

Power-Cycle Performance. In principle, utilization efficiency η_u can assume any value between zero (P = 0) and unity (P = $\dot{m}_{gf} W_{net}^{max}$); in practice, its value is determined from economic considerations by balancing the cost of obtaining the heat (drilling and piping costs) against the cost of processing it (heat exchangers, turbines, pumps) to generate electricity in the power station. If η_u is small, then the resource is being utilized poorly and a large investment in wells is required (cost per unit power $\rightarrow \infty$ as $\eta_u \rightarrow 0$). On the other hand, if we approach utilization of the full potential of the resource, then total well costs decrease but the required investment in highly efficient power conversion equipment is high (cost per unit power $\rightarrow \infty$ as $\eta_u \rightarrow 1$). The economic optimum occurs when η_u assumes some intermediate value, eg, at The Geysers, $\eta_u = 0.55$ is typical with $T_o = 26.7°C$.

In a typical analysis of cycle performance, a set of equations is developed to describe the work and heat-flow rates to and from the chief plant components. After selecting a working fluid and a geothermal-fluid inlet temperature, the principal independent design variables are the maximum cycle operating pressure P at the turbine inlet, the condensing temperature and the heat-exchanger approach temperature.

Space limitations prohibit an extensive discussion of the many excellent studies of cycle performance that have appeared in recent years. The literature cited in Table 3 examines the effects of the cycle-operating conditions cited earlier, and other design and resource parameters.

Other parameters have been defined to compare performance. For example, ref. 110 cites specific output, ie, work produced/mass of geothermal brine. An example of this approach is shown in Figure 13 where a cycle efficiency–specific output diagram is shown for a propane binary cycle operating over a range of brine temperatures and turbine inlet pressures. Since the critical pressure of propane is 4.26 MPa (617 psia), cycles considered are supercritical.

Reference 134 details the dramatic effect of operating pressure on cycle performance from sub- to supercritical conditions for the case of a 150°C liquid-dominated resource with a condenser outlet temperature T_{cond} of 26.7°C and R-115 as the working fluid. This resource condition may be typical of many in the United States where cooling water with low ambient, wet-bulb temperatures (ca 3°C) is available.

For any given working fluid, there is an optimum set of operating conditions yielding a maximum η_u for particular geothermal-fluid and heat-rejection temperatures. Computer optimizations for seven working fluids were conducted for geothermal fluid temperatures ranging from 100 to 300°C for a heat sink temperature of 16.7°C (134). At each point, cycle pressures were varied until an optimum was determined at that temperature. A characteristic maximum η_u is observed at a resource temperature that is different for each fluid but is generally in the range of 55–65% efficiency. Figure 14 compares maximum utilization efficiencies for different conversion methods including single- and dual-stage, flashing, binary, direct-steam injection, and dual flash–binary cycles combining data from a number of sources (108,113,134).

The output of a multistage steam-flashing cycle primarily depends on: (*1*) initial fluid temperature and composition, (*2*) flashing conditions of each stage (P_i^{sat}, T_i), (*3*) turbine efficiencies, and (*4*) condensing temperature (141,186).

Generally, the work produced by a given flash stage is proportional to the amount of vapor created by the isenthalpic throttling step (see Fig. 8**b**) and the enthalpy difference that the vapor experiences when isentropically expanded to the condensing temperature. If the temperature of the flashed fluid is only slightly below the wellhead temperature, then the fraction of vapor produced is small but the isentropic enthalpy difference is large. The opposite is true if the fluid is flashed to just above the condensing temperature. Optimal performance in terms of power output occurs at some intermediate flashing temperature. The performance of a two-stage flashing system is shown in Figure 15 with an average turbine efficiency of 75%. Maximum output (η_u = 0.50) occurs when the first flashing stage is at 169°C (T_{s1}) and the second is at 107°C (T_{s2}). The large effects that cycle component efficiencies and pinch and condensing temperatures can have on performance are shown in Figure 16 for a dual-flash/*n*-butane binary cycle. An analytical approach developed in ref. 134 results in 161°C for T_{s1} and 103°C for T_{s2}.

Table 3. Studies of Low Temperature Electric Power Conversion Cycles

Geothermal reservoirs examined	Power cycle designs	Working fluids	Refs.
liquid-dominated: Hatchobaru, Japan	two-stage flash, dual-admission turbine	water	165
liquid-dominated: general	binary cycle	isobutane	166
liquid-dominated: Roosevelt Hot Springs, Utah; Coso, East Mesa and Long Valley, Calif.	hybrid geothermal cycle	water	121
general	topping–bottoming cycle	water–ammonia, etc	167
liquid-dominated, high salinity: general	total-flow expander	water	112–113
liquid-dominated: Heber and Niland, Calif.	multistage steam-flashing and binary cycles	water, isobutane	149, 168–170
liquid-dominated: general	multistage steam-flashing and binary cycles	water, hydrocarbons and fluorocarbons	171–174
liquid-dominated: 150–250°C	binary regenerative cycle	isobutane	175–176
liquid-dominated: Imperial Valley, Calif.	binary cycle	isobutane	177
vapor-dominated: The Geysers, Calif. (180°C, 800 kPa or 8 bar)	direct-steam injection cycle	water	178
liquid-dominated: general	multistage flash, flash–binary, binary and total-flow cycles	water, isobutane	179
liquid- and vapor-dominated: general (150–300°C, with and without CO_2, high and low salinity)	multistage flashing and steam injection, binary, dual-flash binary cycles	water, hydrocarbons, fluorocarbons; including pentane, isobutane, butane, propane, propylene, R-32, R-114, R-115, R-22[a], and ammonia	110, 141, 158, 180
geopressured with methane: Gulf coast, U.S. (163°C, 13.8 MPa or 138 bar)	hydraulic turbine, flashing and binary cycles	water, isobutane	181
liquid-dominated: general (100–300°C, 0–3% noncondensables)	multistage steam flashing and binary cycles, multiparameter optimization	water, hydrocarbons and fluorocarbons; including isobutane, and propane, R-22, R-113, and ammonia	182–184
liquid-dominated: Cerro Prieto, Mexico (344°C, 530 kPa or 5.3 bar)	single-stage flashing cycle	water	185
liquid-dominated: general (200°C)	multistage flashing cycle	water	186
liquid-dominated: Heber, Calif. (180°C), Raft River, Idaho (150°C); Valles Caldera, New Mex. (260°C)	multistage flashing and binary hybrid (flashed steam and organic) cycles	water, light hydrocarbons (isobutane, propane, and isopentane, and mixtures	108
geopressured with methane: Gulf coast, U.S.	hydraulic and total-flow cycles	water	35
liquid-dominated: Iceland conditions	binary fluid cycle	fluorocarbons (R-21 in particular)	187
liquid-dominated: general	hybrid fossil–geothermal systems	water	153
liquid and vapor dominated: western U.S.	hybrid fossil–geothermal systems	water	122

Table 3 (*continued*)

Geothermal reservoirs examined	Power cycle designs	Working fluids	Refs.
liquid-dominated: Raft River, Idaho (150°C)	dual boiler, binary cycle	isobutane	116, 153
ocean thermal	binary cycle	ammonia	188
liquid-dominated, high salinity	helical screw expander	water	114
liquid-dominated and hot dry rock: general (100–300°C)	multistage flashing, single fluid, dual, and topping and bottoming binary cycles, and Brayton cycles	water, hydrocarbons, and fluorocarbons; including R-22, R-114, R-115, R-32, isobutane, and ammonia	134
liquid-dominated: Paratunka River, USSR (80°C)	binary cycle	fluorocarbon	115
liquid and vapor-dominated: Calif.	direct-steam injection, flashing, multistage flash–binary, and binary cycles	water, hydrocarbons	189
general	flashing and binary cycles	water, isobutane	190
liquid-dominated: general high and low salinity brines, 160–200°C	multistage flashing, binary, and multistage flash–binary cycles	water, isobutane, R-113	109
liquid-dominated: general 150–323°C	binary cycles	isobutane, R-21[a], Fluorinol (trifluoroethanol)	142
liquid-dominated: general 150–205°C	binary cycles with mixtures	hydrocarbon mixtures	138, 191–192
liquid-dominated, high salinity general	binary cycles with specialized packed bed heat transfer	hydrocarbon mixtures	193
hot dry rock resources: general 100–300°C	binary cycles	hydrocarbons and fluorocarbons	194
geopressured: U.S. Gulf coast	flashing with hydraulic turbines, methane recovery	water, methane	195

[a] R-21 = $CHCl_2F$, R-22 = $CHClF_2$, R-32 = CH_2F_2, R-113 = $C_2Cl_3F_3$, R-114 = $C_2Cl_2F_4$, and R-115 = C_2ClF_5.

Economic Issues and Prospects

Although electric power and heat produced from vapor- and liquid-dominated reservoirs have been a commercial reality in many parts of the world for a number of years, the vast potential of natural hydrothermal, geopressured, and hot dry rock reservoirs have not yet been developed. This is in part because alternatives such as oil, coal, and natural gas were actually or were presumed to be less expensive. The future situation is obviously quite different. In evaluating the potential commercial feasibility of any geothermal reservoir, great care must be exercised to consider all pertinent issues. Some of these are site-specific and relate to the characteristics of the resource and reservoir, others depend on the design and operation of the power plant, and still others depend on a wide variety of financial, regulatory, and management issues. Rather than try to provide a fictitious economic assessment based, of course, on specific assumptions regarding these issues, this section considers only the dimensions of the problem as it relates to future commercialization.

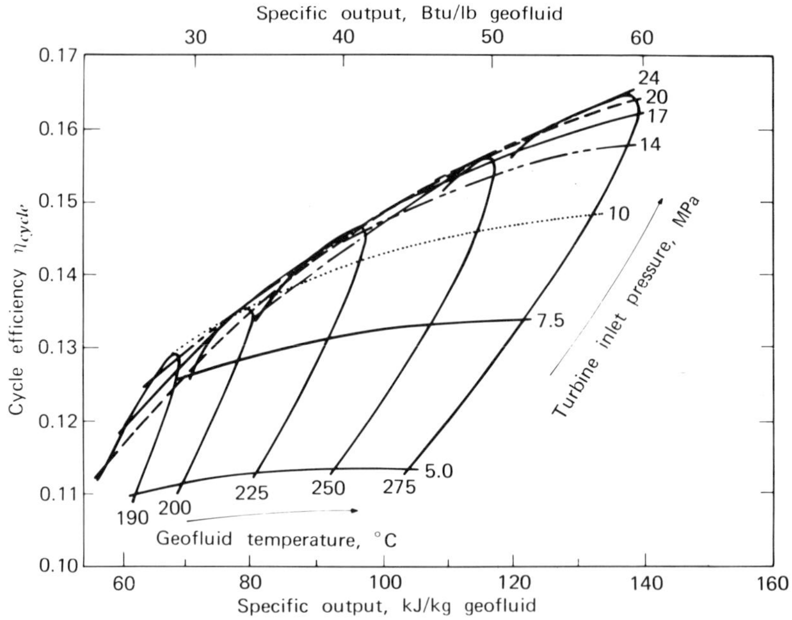

Figure 13. Rankine cycle performance with propane as a working fluid. Effect of geothermal fluid temperature and turbine inlet pressure shown pump and turbine efficiencies of 85%, a minimum approach temperature of 14°C and a condensing temperature of 37.8°C. Courtesy of General Electric Company (110). To convert MPa to bar, multiply by 10.

Resource-Related. Several particular site-specific characteristics are important in establishing the quality of the resource. For hydrothermal and geopressured systems, these are geothermal fluid temperature and composition, reservoir depth, and rock type. Power-plant design and conversion efficiency are critically controlled by fluid temperature and composition as described earlier; drilling costs depend primarily on hole depth and rock type. In hot dry rock systems, fluid chemistry is important as are reservoir temperature and depth which can be related to the average geothermal temperature gradient at the site. Higher gradients result in shallower, less expensive individual well costs. Thus, the gradient is perhaps the most important feature in determining hot dry rock resource quality.

A certain fraction of the costs are associated with site exploration and development. These generally involve the use of a number of reconnaissance geologic, hydrologic, and geophysical measurements or surveys aimed at locating hot water or rock under the surface. In addition, land acquisition by purchasing or leasing geothermal rights, shallow exploratory drilling, and deep drilling to evaluate the potential of the reservoir are required.

Drilling costs for deep exploratory, production, and reinjection wells are difficult to estimate as they are subject to many unforeseen contingencies. In general, they comprise the largest, single cost component in capital-intensive geothermal development because they are incurred long before any revenues from power generation appear. In effect, they replace the fuel cost of a typical nuclear or fossil-fired power plant, which are usually considered operating costs. In some cases, drilling is difficult because of hard, fractured rock formations, lost circulation of drilling fluids, hole

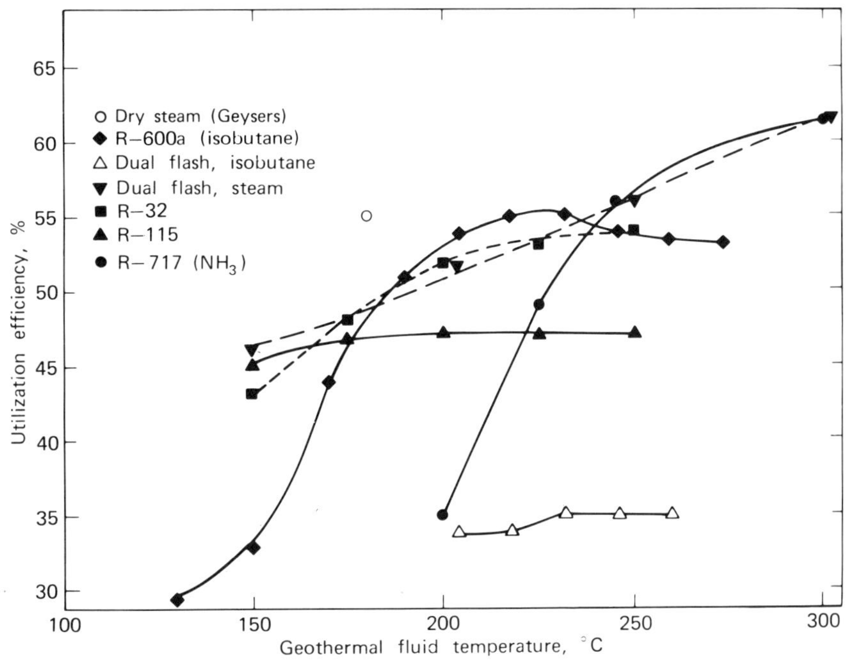

Figure 14. Comparison of maximum utilization efficiencies for different conversion methods as a function of geothermal fluid temperature. Optimum thermodynamic performance: T_o = 27.8°C; T_{cond} = 37.8°C; η_t = 85%; η_p = 85%; ΔT_{pinch} = 13.9°C.

collapse, high temperatures, and the presence of corrosive and erosive geothermal fluids which increase drill-bit wear (196). Furthermore, there may be a need for directional drilling at high temperatures and complex well-completion programs involving special cement formulations, casing materials, etc. To date, most directional drilling in geothermal systems has involved whipstocking techniques or downhole motor-driven drills which present some design and material problems for operation at high temperatures (197).

Given the above constraints, individual well cost is controlled mainly by rock-type, well diameter, and depth. The dependence of completed well cost can be represented as an exponential function of depth for boreholes of 25–35 cm diameter in geothermal formations (134). The depth dependence parallels that for oil and gas wells with the geothermal wells costing between two and four times as much as an oil or gas well to the same depth. The completed well costs cited include charges for site preparation, drill-rig mobilization–demobilization, bits, casing, cementing, cores, geophysical logging, drilling fluids, directional surveys, reservoir stimulation as well as the rental charges for the rig itself. Oil and gas well data have been used to extrapolate the rather limited set of geothermal well-cost data available. For example, costs for wells drilled in The Geysers and Imperial Valley region of California exist for wells ranging in depths typically from 1 to 2.3 km (105,196,198). In addition, some preliminary costs have been cited for hot dry rock wells drilled in granite in New Mexico (134).

The oil- and gas-well data base literally represents thousands of wells drilled throughout the United States (continent and offshore). Annual summaries of these costs are published (199–203).

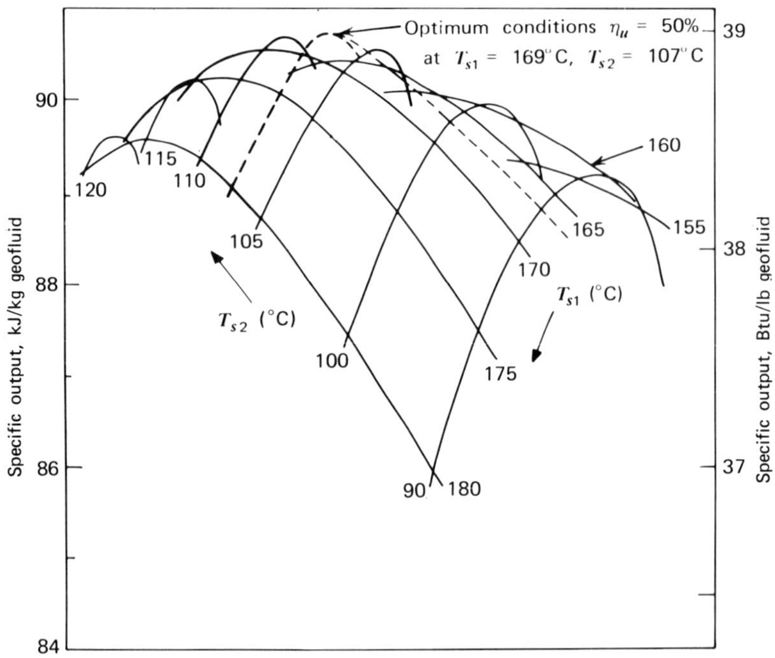

Figure 15. Effect of flashing conditions on the performance of a dual-flash/dual-admission steam-turbine cycle operating on a 230°C, CO_2-free brine containing 2.5 wt % NaCl (110,141). T_{gf} = 230°C; T_{cond} = 52°C; T_o = 37.8°C; T_{s1} = first-stage flash temperature; T_{s2} = second-stage flash temperature. Adapted from Eskesen (110,141) by permission.

For wells deeper than 4 km, geothermal-well costs might be expected to converge toward oil- and gas-well costs because the difficulties encountered with drilling deep wells in general are so extreme that differences between drilling geothermal and oil and gas wells are minimized.

In addition to drilling, two other factors that are site dependent are the availability of water and ambient heat-rejection conditions. These can vary considerably from site to site and have a strong influence on the design and operation of the reservoir and power plant.

Reservoir- and Plant-Related. The costs associated with developing and operating a reservoir can also be important. In particular, with initial fluid temperature and composition defined, well productivity and lifetime determine the number of wells required for a given installed plant capacity. As shown in the previous section on Energy Conversion and Utilization, the required geothermal fluid-flow rate per MW can be estimated once the efficiency of the electric power generation or heat utilization scheme is defined. Figure 17 shows typical flow rates that might be expected for binary-fluid plants. Because reservoir depletion for natural hydrothermal, geopressured, and hot dry rock systems is anticipated, consideration should be given to the proper design and management of the combined power-plant–reservoir system.

The methods used to produce the fluid may require significant capital expenditure. For example, hot dry rock sites may require costly artificial stimulation involving complex techniques to produce multiple-fracture systems with proper flow control for uniform flow distribution. In hydrothermal reservoirs, water-flooding methods

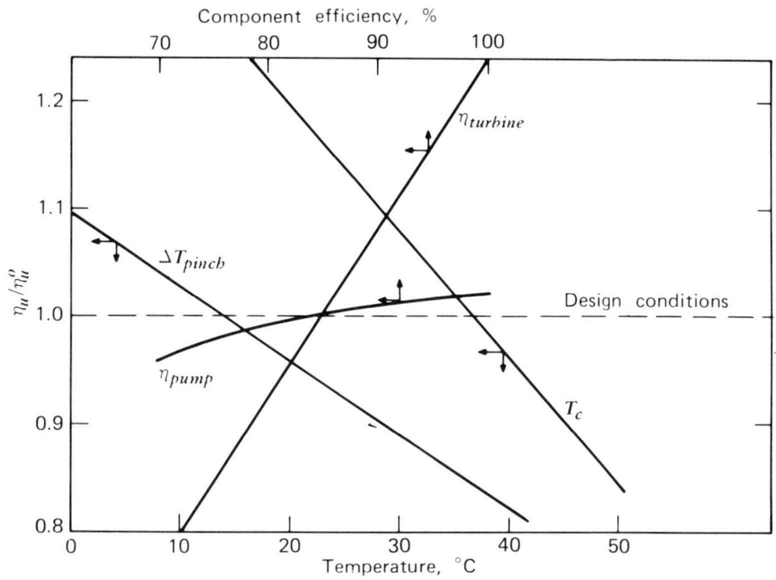

Figure 16. Effect of cycle operating conditions on performances including pump and turbine efficiencies η, primary heat exchanger pinch-point approach temperature ΔT_{pinch}, and condensing temperature T_c for a dual-flash/n-butane binary cycle. Adapted from Eskesen (110) by permission.

used to enhance productivity may require careful placement and operation of production and injection wells, as described earlier.

In addition, when artesian flow is insufficient, reservoirs may require surface or downhole pumping to ensure economically acceptable production or reinjection rates. This contributes to capital, operating, and maintenance expenses. Finally, the transportation of fluid between the wellhead and the power plant requires additional capital investment. The type of use affects the capital cost of the surface power plant greatly. For electric systems, the high costs associated with drilling and completing the reservoir—in some cases 60–80% of the total investment—frequently put a premium on designing and operating the power plant near its limit of thermodynamic efficiency. This involves proper selection of the power cycle, working fluid, and operating conditions to match the initial geofluid temperature and composition, anticipated thermal drawdown, and prevailing ambient heat-rejection conditions.

For a given set of resource, reservoir, and power plant conditions and costs, an optimal plant design temperature exists. Selection of this optimum becomes more complex when finite drawdown occurs. The effect of reservoir production-flow capacity, geothermal gradient, and drawdown rate on the choice of optimum plant and reservoir design temperatures for hot dry rock systems are discussed in refs. 194 and 204.

As noted above, several parameters related to power-cycle operation affect costs. For example, for binary cycles the choice of primary heat exchanger and condenser approach temperatures and cycle pressure depend on the cost of producing the geothermal fluid relative to the plant costs. Using an idealized approach for a liquid-dominated resource, cost-optimized temperature differences are shown for a binary cycle in Figure 18. For this case, ΔT_1, is the average temperature difference between

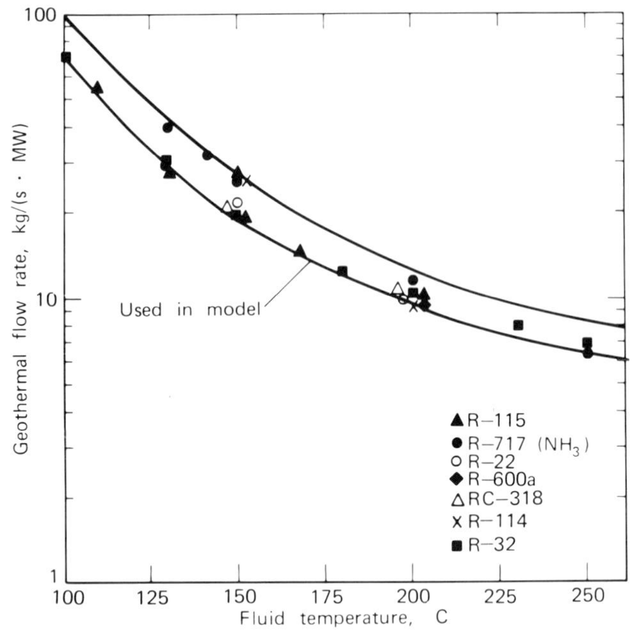

Figure 17. Geothermal well flow rates per MW of electricity as a function of resource temperature, under binary fluid Rankine cycles and optimum thermodynamic conditions. T_o = 27°C; $\eta_{turbine}$ = 85% (dry); η_{pump} = 80%; ΔT_{min} = 10°C (pinch) (194).

the geothermal fluid and the organic working fluid and ΔT_2 the average difference between the coolant and condensing temperature. The parameters C_1 and C_2, represent the costs per unit surface area (in \$/m^2) of the primary heat exchanger and condenser, and U_1 and U_2 are the effective overall heat-transfer coefficients (W/m^2·K). C_{gf}, the cost per unit of well-flow rate in \$/(kg·s), reflects the cost of drilling production and reinjection wells. For given values of C_1, C_2, U_1, U_2, and C_{gf} optimized values of ΔT_1 and ΔT_2 are obtained from Figure 18. As well costs decrease relative to either condenser or primary heat-exchanger costs, ΔT reaches a limiting maximum value of ca 65°C. At the other extreme of very large relative well costs, low values of ΔT_1 and ΔT_2 result in optimal economics as one approaches the thermodynamic limit of conversion efficiency.

Having established optimal cycle pressures and approach temperatures, the most important variables controlling surface plant costs are the condensing temperature and the composition and temperature of the geothermal fluid. For relatively brine-free fluids with low concentrations of noncondensable gases, binary installed plant costs decrease roughly linearly from about \$1000/kW at fluid temperatures of 100°C to about \$450/kW (in 1979 \$) at 320°C. This decrease is primarily due to increased cycle efficiency as temperature increases, thereby decreasing heat-exchange requirements.

Financial- and Management-Related. In addition to the actual capital costs for exploration, drilling production and reinjection wells, and building the power plant, the cash flow is important. For instance, plant construction may begin 5 years before operation. This results in staggered costs to be compounded by an appropriate discount rate over the construction period. Other financial issues are (*1*) the selling price of energy; (*2*) assessment of risk, anticipated interest rates on debt, and rate of return on equity stock; (*3*) tax treatment including depletion-type allowances, investment

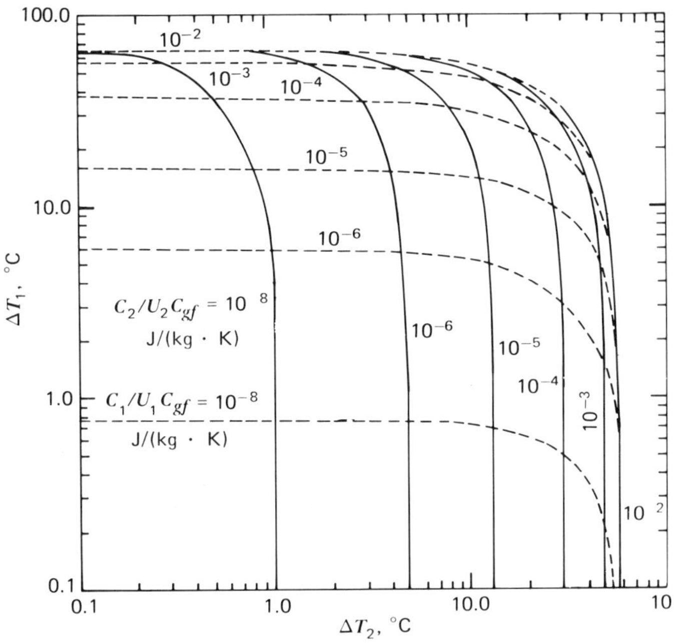

Figure 18. Cost-optimized heat exchanger (ΔT_1) and condenser (ΔT_2) temperature differences for a binary-fluid cycle utilizing a 150°C liquid-dominated resource with 26.7°C heat sink temperature and an $\eta_u = \epsilon = 0.80$. Courtesy of MIT Press (134).

tax credits, and deduction for intangible drilling expenses; (*4*) royalties; (*5*) load and availability factors and plant lifetime; (*6*) management and ownership of the reservoir and plant; (*7*) operating and maintenance costs; and (*8*) environmental and institutional regulations. Selecting appropriate values for debt interest rates and equity rates of return, taxes, energy selling prices, and royalties may be difficult because of the newness of the technology or because projections are sensitive to unforeseen conditions such as rapidly escalating costs for fossil fuels.

If measurable reservoir drawdown occurs, the proper management of the combined plant–reservoir system is absolutely necessary. In studies of hot dry rock systems, various initial well depths (temperatures), design mass-flow rates, plant design temperatures and reservoir capacities are considered in order to select the flow rate and plan for redrilling activities that maximize profits for an assumed set of resource and financial conditions (194,204).

The question of ownership is also complex. Historically in the United States, there has been a separation between the producer and the power generator. Typically, private oil companies develop the field, produce the geothermal water or steam, and then sell it directly to a private or public utility which uses it to generate electricity. This type of ownership arrangement reflects the riskiness of geothermal ventures. Changes may occur as the technology matures and as developers realize the economic benefits that may result from joint management of the reservoir and plant. For example, lower equity rates of return and debt interest rates should certainly follow as a field is developed and the technology is proved.

Geothermal electric power plants are ideally suited for base-load capacity. Because of low conversion efficiencies, large geothermal fluid flow rates per MW require the use of multiple wells and result in large heat-exchange and condenser–desuperheater surface areas. Consequently, the combined effect of fluid transmission from widely spaced wells and of equipment size limitations cause optimal generating unit sizes of ca 50–100 MW. Overall capacity factors for existing plants are 75–85% and operating and maintenance costs tend to be low. Typical plant–reservoir lifetimes of 20–30 years are envisioned.

Case Studies. Many studies conducted in recent years assess the commercial feasibility of hydrothermal, geopressured, and hot dry rock resources (see Table 3). Of particular interest for hydrothermal vapor- and liquid-dominated systems are the papers that appear in refs. 30–31, 130–131. In addition, various investigators (108,110,170,180,189) examined several particular sites in the western United States. The development of generic economic modeling of geothermal resources is reported in refs. 173–174, 182–183. The few comprehensive studies of geopressured and hot dry rock resources have appeared recently (36,194–195,204–207).

Worldwide Developments: Present and Future. The status of worldwide geothermal developments have been reviewed (58,66,208–211). Table 4 cites present and future installed capacity by country for electric and nonelectric applications. Reports of future capacities are based on questionnaires and should be viewed with some caution. However, the implication of significant growth for the geothermal industry in the coming years is clear. This is perhaps best illustrated by recent activities in the United States. As of 1980 at least nine specific sites are under development. These include: (*1*) The Geysers–Calistoga area in California; (*2*) Imperial Valley, California; (*3*) Desert Peak, Nevada; (*4*) Valles Caldera, New Mexico; (*5*) Roosevelt Hot Springs, Utah; (*6*) Puna, Hawaii; (*7*) Raft River, Idaho; (*8*) Klamath Falls, Oregon; (*9*) Gulf coast region of Texas and Louisiana; and (*10*) Fenton Hill, New Mexico. The first seven sites are liquid-dominated systems with the exception of The Geysers which is a vapor-dominated reservoir. The present capacity of The Geysers–Calistoga field, 663 MW electric, is expected to grow to 1700 MW by 1985 (209). Although this field is now the only one in commercial operation in the United States, rapid development of many other sites is anticipated during the period 1980–2000. Demonstration-size systems should be placed on-line in the East Mesa (58 MW), Heber (95 MW), North Brawley (10 MW), Westmoreland (50 MW) and Salton Sea (10 MW) fields of the Imperial Valley; Puna (5 MW); Raft River (5 MW); Desert Peak (20–50 MW); the Baca field in the Valles Caldera (50 MW) and in Roosevelt Hot Springs (50 MW). Expansion of district heating at Klamath Falls and Boise should continue. Geopressured resources in the Gulf coast area are undergoing field tests to determine reservoir productivity and lifetime, dissolved methane concentrations, and subsidence effects (195).

Fenton Hill, New Mexico, is the first United States site aimed at exploiting hot dry rock resources. Prototype heat-extraction experiments have been conducted in a hydraulically fractured reservoir in low-permeability crystalline basement rock at depths of 3 km (48). These experiments have demonstrated the technical feasibility of many facets of hot dry rock systems but large effective heat-transfer areas with acceptable flow impedance and distribution through fractured zones have yet to be proven. The first prototype had only 8000 m^2 (86,000 ft^2) of effective surface which was later increased to ca 90,000 m^2 (ca 10^6 ft^2) by massive hydraulic fracturing, but both reservoirs were too small to be of commercial value, operations are under way to develop a two-wellbore ($(1–2) \times 10^6$ m^2 ($(11–22) \times 10^6$ ft^2) surface area), 50 MW(t) system with extended lifetime. Ultimately a 10 MW plant is planned for the Fenton Hill site.

Table 4. Geothermal Electric Generating and Nonelectric Capacities[a]

Country	Installed (1979) Electric, MW	Installed (1979) Nonelectric, MW (t)[b]	Future (by 1990) Electric, MW	Future (by 1990) Nonelectric, MW (t)[b]
United States	663	17	ca 2000	ca 3400
Italy	421	24	800	38
New Zealand	203	196	400	380
Japan	165	2900	ca 2000	3100
Mexico	150		150	680
Iceland	64	410	150	570
El Salvador	60		180	
USSR	5	4860	?	?
Philippines	4.2	4.9	765	?
Taiwan	1	0.6	?	5
Turkey	0.5	0.2	400	?
Nicaragua			150	?
Costa Rica			100	
Guatemala			100	
Honduras			100	
Panama			60	
Argentina			20	
Portugal			30	
Spain			25	10
Kenya			30	
Indonesia			30–100	
Thailand			ca 10	
Canada				10
England				~2
France		24		490
Hungary		1050		?
Czechoslovakia		93		?
Yugoslavia		4.9		58

[a] Adapted from refs. 58, 210–211.
[b] MW(t) = thermal megawatts of power; capacities less than 1 MW(t) are not listed.

If one grants that the technical reservoir-engineering problems associated with hydrothermal, geopressured, magma, and hot dry rock resources are solved, perhaps the most critical factors affecting the growth of geothermal energy include drilling and plant capital cost escalation relative to alternatives, the price escalation of alternative fossil and nuclear fuels, tax treatment and royalties, and finally the assessment of risk and of the rate of return required to stimulate private investment.

BIBLIOGRAPHY

1. D. N. Anderson, "Recent Developments of Geothermal Resources" in *Geothermal Resources Council Special Short Course #6,* Geothermal Resources Council, Houston, Tex., 1977.
2. A. J. Ellis and W. A. J. Mahon, *Chemistry and Geothermal Systems,* Academic Press, Inc., New York, 1977.
3. D. E. White, "Characteristics of Geothermal Resources" in P. Kruger and C. Otte, eds., *Geothermal Energy,* Stanford University Press, Stanford, Calif., 1973.

4. L. J. P. Muffler, "Tectonic and Hydrologic Control of the Nature and Distribution of Geothermal Resources," *paper presented at Proc. of the Second UN Symp. on the Development and Use of Geothermal Resources, Vol. I, Lawrence Berkeley Laboratory, San Francisco, Calif., 1975,* p. 499.
5. J. Healy, "Geothermal Fields in Zones of Recent Volcanism" in ref. 4, p. 415.
6. G. P. Tamrazyan, "Continental Drift and Thermal Fields," *paper presented at Proc. of the First UN Symp. on the Development and Utilization of Geothermal Resources, Pisa, Italy, 1970,* p. 1212; *Geothermics Special Issue 2.*
7. J. R. McNitt, "The Geologic Environment of Geothermal Fields as a Guide to Exploration" in ref. 6, p. 24.
8. W. T. Cheng, "Geophysical Exploration in the Tatun Volcanic Region, Taiwan" in ref. 6, p. 262.
9. C. J. Banwell, "Geophysical Techniques in Geothermal Exploration" in ref. 6, p. 32.
10. J. Combs, "Geophysical Techniques in Exploration" in ref. 4, p. lxxxi.
11. G. Bodvarsson, "Thermoelastic Phenomena in Geothermal Systems" in ref. 4, Vol. 2, p. 903.
12. G. W. Grindley and P. R. L. Browne, "Structural and Hydrological Factors Controlling the Permeabilities of Some Hot Water Geothermal Fields" in ref. 4, p. 377.
13. H. Craig, "Isotopic Geochemistry of Water and Carbon in Geothermal Areas," *Nuclear Geology on Geothermal Areas,* Consiglio Nazionale della Ricerche, Laboratorio de Geologica Nucleare, Pisa, Italy, 1963, p. 17–53.
14. S. Kavlakoglu, "Origin of Geothermal Waters or Natural Steam" in ref. 6, p. 1250.
15. S. Mercado and J. Guiza, "Problems of Silica Scaling at Cerro Prieto Geothermal Power Station," *paper presented at Proc. of the Conf. on Scale Management in Geothermal Energy Development, EIC Corp., University of California, San Diego, Calif., 1976.*
16. R. Quong, H. K. Bishop, and J. J. Hill, *GRC Trans.* **1,** 249 (1977).
17. R. C. Reeber, "Fluid Management Engineering Program for the Geothermal Energy Division, U.S. Department of Energy," *paper presented at Proc. of Workshop on Scale Control in Geothermal Energy Extraction Systems, Los Alamos, New Mex., 1977.*
18. C. J. Strobel, "Formation Plugging While Testing a Steam Well at The Geysers," *paper presented at Proc. of the Fourth Workshop on Geothermal Reservoir Engineering, Stanford, Calif., 1978,* p. 106.
19. L. J. Defferding and R. A. Walter, *GRC Trans.* **2,** 141 (1978).
20. W. A. Ferlinius, "Production of Fresh Water by Desalting Geothermal Brines—A Pilot Desalting Program at the East Mesa Geothermal Field, Imperial Valley, California" in ref. 4, Vol. 3, p. 2201.
21. V. Ludviksson, "Multipurpose Uses of Geothermal Energy: Electric Power Generation and Horticultural Production" in ref. 4, Vol. 3, p. 2229.
22. S. S. Einarsson, "Geothermal Space Heating and Cooling" in ref. 4, Vol. 3, p. 2117.
23. H. J. Ramey, Jr., "Pressure Transient Analysis for Geothermal Wells" in ref. 4, Vol. 3, p. 1749.
24. R. James, "Drawdown Test Results Differentiate Between Crack Flow and Porous Bed Permeability" in ref. 4, Vol. 3, p. 1693.
25. A. Barelli, G. Manetti, R. Celati, and G. Neri, "Build-up and Back Pressure Tests on Italian Geothermal Wells" in ref. 4, Vol. 3, p. 1537.
26. S. K. Garg, *GRC Trans.* **2,** 203 (1978).
27. R. N. Horne and H. J. Ramey, Jr., *GRC Trans.* **2,** 291 (1978).
28. D. E. White, "Geochemistry Applied to the Discovery, Evaluation and Exploitation of Geothermal Energy Resources" in ref. 6, p. 58.
29. A. H. Truesdale, "Geochemical Techniques in Exploration" in ref. 4, p. liii.
30. E. Barbier, ed., *Proc. of the First UN Symp. on the Development and Utilization of Geothermal Resources, Pisa, Italy, 1970; Geothermics Special Issue 2.*
31. *Proc. of the Second UN Symp. on the Development and Use of Geothermal Resources,* Lawrence Berkeley Laboratory, San Francisco, Calif., 1975.
32. C. A. Stuart, "Geopressures," *paper presented at Proc. of the Second UN Symp. on Abnormal Subsurface Pressures, Louisiana State Univ., Baton Rouge, La., 1970,* p. 121.
33. R. H. Wallace, Jr., T. F. Kraemer, R. E. Taylor, and J. B. Wesselman, *U.S. Geol. Surv. Circ.* **790,** 132 (1979).
34. P. H. Jones, "Natural Gas Resources of the Geopressured Zones in the Northern Gulf of Mexico Basin" in *Natural Gas from Unconventional Geologic Sources: National Research Council,* National Academy of Sciences, Washington, D.C., 1976, p. 17.
35. P. A. House, P. M. Johnson, and D. F. Towse, "Potential Power Generation and Gas Production from

Gulf Coast Geopressured Reservoirs," *paper presented at Proc. First Geopressured Geothermal Energy Conf., University of Texas, Austin, Tex., 1975,* p. 283.

36. M. H. Dorfman and R. W. Deller, "Texas Geopressured Resource Assessment," *EPRI WS-78-97, paper presented at Proc. Second Geothermal Conf. and Workshop, Electric Power Research Institute, 1978.*
37. A. Z. Bassiouni, "Southern Louisiana Geopressure-Geothermal Energy Resource Assessment," *EPRI WS-78-97,* in ref. 36.
38. S. S. Papadopulos, R. H. Wallace, Jr., J. B. Wesselman, and R. E. Taylor, *U.S. Geol. Surv. Circ.* **726,** 125 (1975).
39. B. E. Hankins and O. C. Karkalits, "Geopressured-Geothermal Test of the Edna Delcambre No. 1 Well, Tigre Lagoon Field, Vermillion Parish, Louisiana: Analysis of Water and Dissolved Gas," McNeese State University, 1978.
40. B. E. Hankins, R. E. Chavanne, R. A. Ham, O. C. Karkalits, and J. I. Palermo, *GRC Trans.* **2,** 253 (1978).
41. O. C. Karkalits and B. E. Hankins, *GRC Trans.* **2,** 351 (1978).
42. Y. K. Kharaka, P. M. Brown, and W. W. Carothers, *GRC Trans.* **2,** 371 (1978).
43. S. K. Garg, "Land Surface Subsidence Associated with Geothermal Energy Production," *paper presented at Proc. of the First Workshop on Geothermal Reservoir Engineering, Stanford, Calif., 1975.*
44. S. K. Garg, T. D. Riney, and D. H. Brownell, Jr., "Preliminary Reservoir and Subsidence Simulations for the Austin Bayou Geopressured Geothermal Prospect," *paper presented at Proc. of the Fourth Workshop on Geothermal Reservoir Engineering, Stanford, Calif., 1978,* p. 280.
45. M. C. Smith, R. L. Aamodt, R. M. Potter, and D. W. Brown, "Man-Made Geothermal Reservoirs" in ref. 4, Vol. 3, p. 1781.
46. A. G. Blair, J. W. Tester, and J. J. Mortensen, "LASL Hot Dry Rock Geothermal Project," *Los Alamos Scientific Laboratory Report LA-6525-PR,* 1976.
47. LASL HDR Project Staff, "Hot Dry Rock Geothermal Energy Development Project," *Los Alamos Scientific Laboratory Report LA-7109-PR,* 1978.
48. J. W. Tester and J. N. Albright, eds., "Hot Dry Rock Energy Extraction Field Test: 75 Days of Operation on a Prototype Reservoir at Fenton Hill," *Los Alamos Scientific Laboratory Report LA-7771-MS,* 1979.
49. M. C. Smith, "The Future of Hot Dry Rock Geothermal Energy Systems," *paper presented at Proc. of Third National Congress, Pressure Vessel and Piping Division, American Society of Mechnical Engineers, San Francisco, Calif., June, 1979.*
50. A. S. Batchelor, "Brief Summary of Some Geothermal Related Studies in the United Kingdom," *paper presented at Proc. of the Second NATO-CCMS Information Meeting on Dry Hot Rock Geothermal Energy, 1977,* p. 27; *Los Alamos Scientific Laboratory Report LA-7021-C,* 1977.
51. Y. D. Diadkin and Y. M. Pariisky, "Theoretical and Experimental Grounds for Utilization of Dry Rock Geothermal Resources in the Mining Industry" in ref. 4, Vol. 3, p. 1609.
52. I. T. Aladiev, V. P. Trusov, A. D. Peredery, E. M. Strigin, E. V. Saperov, and V. K. Fardzinov, "Heat and Mass Transfer Processes in Aquifer Systems with Artificially Increased Fracturing," in ref. 4, Vol. 3, p. 1529.
53. J. W. Tester and M. C. Smith, "Energy Extraction Characteristics of Hot Dry Rock Geothermal Systems," *paper presented at Proc. Twelfth Intersociety Energy Conversion Engineering Conf., Washington, D.C., Aug. 28–Sept. 2, 1977,* Vol. 1, American Nuclear Society, 1977, p. 816.
54. G. Bodvarsson and J. M. Hanson, "Forced Geoheat Extraction from Sheet-like Fluid Conductors" in ref. 50, p. 15.
55. S. G. Varnado and J. L. Colp, eds., *Report of the Workshop on Magma/Hydrothermal Drilling and Instrumentation, Albuquerque, New Mex., SAND78-1365C,* 1978.
56. M. C. Smith and co-workers, *Report of the Geothermal Resource Group Supply and Delivery Panel of the Committee on Nuclear and Alternative Energy Systems of The National Research Council, Supporting Paper 4,* National Academy of Science, Washington, D.C., 1979.
57. D. E. White and D. L. Williams, eds., *U.S. Geol. Surv. Circ.* **726,** 1 (1975).
58. "Geothermal Energy Prospects for the Next 50 Years," *EPRI ER-611-SR,* Electric Power Research Institute, 1978.
59. L. J. P. Muffler and M. Guffanti, *U.S. Geol. Surv. Circ.* **790,** 1 (1978).
60. S. H. Schurr and B. C. Netschert, *Energy in the American Economy, 1850–1975,* Johns Hopkins Press, Baltimore, Md., 1960.

61. J. J. Schanz, Jr., "Resource Terminology: An Examination of Concepts and Terms and Recommendations for Improvement," *EPRI Research Project 336,* Palo Alto, Calif., 1975; *U.S. Geol. Surv. Circ.* **790,** (1978).
62. L. J. P. Muffler and R. Cataldi, *Geothermics* **7,** (1978).
63. D. E. White, *U.S. Geol. Surv. Circ.* **519,** (1965).
64. J. Rowley, "Estimation of Reserves" in L. Edwards, ed., *Geothermal Energy: an Alternate Energy Source,* Gulf Publishing Co., Houston, Tex., 1980.
65. B. F. Grossling, "An Appraisal of the Prospects of Geothermal Energy in the United States," *U.S. Energy Outlook,* National Petroleum Council, Washington, D.C., 1972, Ch. 2.
66. R. W. Rex and D. J. Howell, "Assessment of U.S. Geothermal Resources" in ref. 3.
67. W. J. Hickel, "Geothermal Energy," *Geothermal Resources Research Conf. Final Report,* University of Alaska Press, 1972.
68. National Petroleum Council, *U.S. Energy Outlook—New Energy Forms,* U.S. Government Printing Office, Washington, D.C., 1973.
69. L. J. P. Muffler, ed., *U.S. Geol. Surv. Circ.* **790,** (1979).
70. "Hot Dry Rock Geothermal Energy: Status of Exploration and Assessment," *Hot Dry Rock Assessment Panel Report #1, ERDA Report ERDA-77-74,* 1977.
71. P. Kruger and V. Roberts, *GRC Trans.* **1,** 173 (1977).
72. S. K. Garg, T. R. Blake, D. H. Brownell, Jr., A. H. Nayfeh, and J. W. Pritchett, "Simulation of Fluid-Rock Interactions in a Geothermal Basin, Final Report," *National Science Foundation (RANN) Report NSF-RA-N-75-329,* 1975.
73. J. C. Martin, "Fluid Flow in Geothermal Reservoirs" in ref. 43.
74. C. R. Faust and M. W. Mercer, "Summary of Our Research in Geothermal Reservoir Simulation" in ref. 43.
75. P. Kasameyer, L. Thorson, and C. McKee, *GRC Trans.* **1,** 163 (1977).
76. J. W. Pritchett, S. K. Garg, and T. D. Riney, *GRC Trans.* **1,** 245 (1977).
77. P. W. Kasameyer and L. W. Younker, *GRC Trans.* **2,** 359 (1978).
78. F. G. Miller, H. Cinco, H. J. Ramey, Jr., and F. Kucuk, *GRC Trans.* **2,** 449 (1978).
79. T. D. Riney, J. W. Pritchett, and S. K. Garg, *GRC Trans.* **2,** 571 (1978).
80. E. O. Tansev and M. L. Wasserman, *GRC Trans.* **2,** 645 (1978).
81. P. W. Kasameyer and R. C. Schroder, "Thermal Depletion of Liquid-Dominated Geothermal Reservoirs with Fracture and Pore Permeability" in ref. 43.
82. F. Harlow and W. Pracht, *J. Geophys. Res.* **77,** 7038 (1972).
83. Y. C. Hsu, Y.-M. Lu, F. D. Ju, and K. C. Dhingra, *GRC Trans.* **2,** 297 (1978).
84. R. D. McFarland and H. D. Murphy, "Extracting Energy from Hydraulically Fractured Geothermal Reservoirs," *paper presented at Proc. of the Eleventh Intersociety Energy Conversion Engineering Conf., State Line, Nevada, Sept. 12–17, 1976,* AIChE, 1976, p. 828.
85. H. D. Murphy and R. Wunder, "Thermal Drawdown and Recovery of Singly and Multiply Fractured Geothermal Reservoirs, *Los Alamos Scientific Laboratory Report LA-UR-78-739,* 1978.
86. S. S. Einarsson, A. Vides R., and G. Cuellar, "Disposal of Geothermal Waste Water by Reinjection" in ref. 4, Vol. 2, p. 1349.
87. C. F. Tsang, M. J. Lippman, and P. A. Witherspoon, *GRC Trans.* **1,** 301 (1977).
88. W. F. Isherwood, *GRC Trans.* **1,** 149 (1977).
89. J. C. Martin, "The Replacement of Geothermal Reservoir Brine as a Means of Reducing Solids Precipitation and Scale Formation," *paper presented at Proc. of the Fourth Workshop on Geothermal Reservoir Engineering, Stanford, Calif., 1977,* p. 42.
90. A. C. Gringarten, P. A. Witherspoon, and Y. Ohnishi, *J. Geophys. Res.* **80,** (1975).
91. J. W. Tester, "Issues Facing the Development of Hot Dry Rock Geothermal Resources," *paper presented at Third Annual EPRI Geothermal Conf., Monterrey, Calif., 1979.*
92. S. L. Phillips, A. K. Mathur, and R. E. Doebler, "A Study of Brine Treatment," *EPRI Report ER-476,* 1977.
93. T. Marshall and W. R. Braithwaite in H. C. H. Amsead, ed., *Geothermal Energy,* Lawrence Berkeley Laboratory, San Francisco, Calif., 1975.
94. R. M. Jorda, *GRC Trans.* **2,** 341 (1978).
95. C. A. Swanberg, "Physical Aspects of Pollution Related to Geothermal Energy Development" in ref. 4, Vol. 2, p. 1435.
96. D. E. Robertson, J. S. Fruchter, J. D. Ludwick, C. L. Wilkerson, E. A. Crecelius, and J. C. Evans, *GRC Trans.* **2,** 579 (1978).
97. L. Joyce and R. A. Fontes, *GRC Trans.* **2,** 345 (1978).
98. J. P. Tomany, *GRC Trans.* **1,** 295 (1977).

99. C. T. Li, D. P. Alzheimer, W. A. Wilcox, G. L. Roberts, and W. F. Riemath, *GRC Trans.* **2,** 1978.
100. M. Goldsmith, *GRC Trans.* **2,** 215 (1978).
101. R. G. Bowen, "Environmental Impact of Geothermal Development" in ref. 3.
102. N. B. Crow and P. W. Kasameyer, *GRC Trans.* **2,** 125 (1978).
103. T. C. Gustavson and M. M. McGraw, *GRC Trans.* **2,** 245 (1978).
104. G. M. Reistad, "Potential for Nonelectrical Applications of Geothermal Energy and Their Place in the National Economy" in ref. 4, pp. 2155–2164.
105. P. Kruger and C. Otte, eds., *Geothermal Energy,* Stanford University Press, Stanford, Calif., 1973.
106. E. R. Berman, *Geothermal Energy,* Noyes Data Corporation, Park Ridge, N.J., 1975.
107. E. F. Wahl, *Geothermal Energy Utilization,* John Wiley & Sons, Inc., New York, 1977.
108. B. Holt and E. L. Ghormley, "Energy Conversion and Economics for Geothermal Power Generation at Heber, California, Valles Caldera, New Mexico, and Raft River, Idaho—Case Studies," *Electric Power Research Institute Report, EPRI ER-301, Topical Report 2,* Palo Alto, Calif., 1976.
109. A. R. Schapiro and G. P. Hajela, "Geothermal Power Cycle Analysis," *paper presented at Proc. of the Twelfth Intersociety Energy Conversion Engineering Conf., Washington, D.C., 1977,* pp. 857–864.
110. J. H. Eskesen, "Study of Practical Cycles for Geothermal Power Plants," *General Electric Company Report COO-2619-1 UC-66, Contract No. EY-76-C-02-2619,* 1977.
111. W. D. McBee and H. B. Mathews, "Low Temperature Energy Conversion Systems," *U.S. Dept. of Energy Contract ET-78-C-02-4633,* 1979.
112. A. L. Austin, G. H. Higgins, and J. H. Howard, "The Total Flow Concept for Recovery of Energy from Geothermal Hot Brine Deposits," *Lawrence Livermore Laboratory Report UCRL-51366;* Livermore, Calif., 1973.
113. A. L. Austin, "Status of the Development of the Total Flow System for Electric Power Production" in J. Kestin, ed., *A Sourcebook on the Production of Electricity from Geothermal Energy,* U.S. Government Printing Office, Washington, D.C., Jan. 1980.
114. R. A. McKay, "The Helical Screw Expander Evaluation Project" in ref. 53, pp. 899–903.
115. V. N. Moskvicheva and A. E. Popov, "Geothermal Power Plant on the Paratunka River" in ref. 6, p. 1567.
116. J. F. Kunze, "Geothermal Space Heating the Symbiosis with Fossil Fuel" in ref. 53, pp. 810–815.
117. J. W. Lund, "The Utilization and Economics of Low Temperature Geothermal Water for Space Heating" in ref. 84, pp. 822–827.
118. W. Hausz, "Annual Storage: A Catalyst for Conservation," *paper presented at International Total Energy Congress, Copenhagen, Denmark, Oct. 4–8, 1976.*
119. C. F. Meyer, W. Hausz, B. L. Ayres, and H. M. Ingram, "Role of the Heat Storage Well in Future U.S. Energy Systems," *General Electric Report GE76TMP-27A, Santa Barbara, Calif.,* Dec. 1977.
120. J. H. Howard, ed., "Present Status and Future Prospects for Nonelectrical Uses of Geothermal Resources," *Lawrence Livermore Laboratory Report UCRL-S1926, Livermore, Calif.,* Oct. 1975.
121. G. H. Anno, M. A. Dore, R. L. Grijalva, G. D. Lang, and F. J. Thomas, "Site-Specific Analysis of Hybrid Geothermal/Fossil Power Plants," *Pacific-Sierra Research Corp. Report 705,* April 1977.
122. Kingston, Reynolds, Thorn, and Allardire Ltd., "City of Burbank Hybrid Geothermal/Coal Power Plant," *Report on Geothermal Resource Requirements and Utilization, U.S. Dept. of Energy Grant EG-77-G-03-1572,* Aukland, New Zealand, July 1978.
123. H. E. Khalifa, "Effect of Seasonal Variations of Ambient Temperatures on the Performance of Low-Temperature Power Cycles," *Brown University Report COO/4051-10, CATMEC/1, Providence, R. I.,* June 1978.
124. "Analysis of the Economic Potential of Solar Thermal Energy to Provide Industrial Process Heat," *U.S. ERDA Contract EY-76-C-02-2829,* Intertechnology Corporation, 1977.
125. B. Lindal, "Industrial and Other Applications of Geothermal Energy," in *Geothermal Energy: Review of Research and Development, LC No. 72-97138,* Paris, Fr., UNESCO, 1973, p. 135–148.
126. J. G. Keller and J. F. Kunze, "Space Heating Systems in the Northwest-Energy Usage and Cost Analysis," *Idaho National Engineering Laboratory Report ANCR-1276, Idaho Falls, Idaho,* Jan. 1976.
127. R. J. Schultz, D. G. Swink, and J. F. Kunze, "Direct Applications of Geothermal Brine in Process Application," *paper presented at Proc. of the 83rd AIChE National Meeting, Paper 16B, Houston, Tex., Mar. 1977.*
128. G. M. Reistad, B. Yao, and M. Gunderson, "A Thermodynamic Study of Heating with Geothermal Energy," *paper presented at Proc. of the ASME Winter Annual Meeting, Atlanta, Ga., Paper No. 77-WA/Ener-1, Nov. 27–Dec. 2, 1977.*

129. Geothermal Resources Council, *Proc. of Direct Utilization of Geothermal Energy: A Symposium, San Diego, Calif., DOE Contract No. EY-76-S-03-1340, Jan. 31–Feb. 2, 1978.*
130. *Proc. of the Eleventh Intersociety Energy Conversion Engineering Conf., State Line, Nevada, Sept. 12–17, 1976.*
131. *Proc. of the Twelfth Intersociety Energy Conversion Engineering Conf., Washington, D.C., Aug. 1977.*
132. J. Karkheck, "Some Considerations for Geothermal District Heating," *NATO-CCMS Conf. on the Economics of Direct Use of Geothermal Energy, Proc. U.S. Dept. of Energy Report CONF-770681, NATO-CCMS Report No. 66,* Washington, D.C., July 1978.
133. J. J. Martin, *Ind. Eng. Chem.* **59,** 34 (1967).
134. S. L. Milora and J. W. Tester, *Geothermal Energy as a Source of Electric Power,* MIT Press, Cambridge, Mass., 1976.
135. S. L. Milora, "Application of the Martin Equation of State to the Thermodynamic Properties of Ammonia," *Oak Ridge National Laboratory Report ORNL-TM-4413,* Oak Ridge, Tenn., Dec. 1973.
136. S. L. Milora and S. K. Combs, "Thermodynamic Representation of Ammonia and Isobutane," *Oak Ridge National Laboratory Report ORNL-TM-5847,* Oak Ridge, Tenn., 1977.
137. K. E. Starling, *Fluid Thermodynamic Properties of Light Petroleum Systems,* Gulf Publishing Co., Houston, Tex., 1973.
138. K. E. Starling, C. M. Sleipcevich, L. W. Fisk, K. M. Goin, K. H. Aboul-Foutoh, K. H. Kumar, T. J. Lee, J. S. Milani, and K. L. Zemp, "Development of Working Fluid Thermodynamics Properties Information of Geothermal Cycles: Phase I," *Univ. of Okalhoma Report ORO-5249-2, Norman, Oklahoma,* 1978.
139. R. C. Reid, J. M. Prausnitz, and T. K. Sherwood, *The Properties of Gases and Liquids,* 3rd ed., McGraw-Hill, New York, 1977.
140. H. P. Meissner and R. Seferian, *Chem. Eng. Prog.* **47,** 579 (1951).
141. J. H. Eskesen, "Flashed Steam–Steam Turbine Cycles" in J. Kestin, ed., *Sourcebook on the Production of Electricity from Geothermal Energy,* U.S. Government Printing Office, Washington, D.C., 1980.
142. D. G. Shepherd, "A Comparison of Three Working Fluids for the Design of Geothermal Power Plants" in ref. 53, pp. 832–841.
143. J. A. Callighan, *Thermal Stability Data on Six Fluorocarbons, Heating/Piping/Air Conditioning,* Reinhold Publishing Corp., New York, Sept. 1971.
144. H. M. Parmelee, "Sealed-Tube Stability Tests on Refrigeration Materials," *paper presented at ASHRAE Semiannual Meeting, Chicago, Ill., Jan. 1968.*
145. K. E. Nichols, Barber-Nichols Engineering Co., Arvada, Col., personal communication, Apr. 1979.
146. R. H. Dart and J. F. Whitbeck, "Brine Heat Exchangers" in ref. 141.
147. W. L. Pope, H. S. Pines, R. L. Fulton, and P. A. Doyle, "Heat Exchanger Design-Why Guess a Design Fouling Factor When it Can Be Optimized?" in *Emerging Energy Technologies 1978,* American Society of Mechanical Engineers, New York, 1978.
148. W. H. Giedt, "The Geothermal Binary Cycle: Heat Exchanger Area Requirements and Mutual Costs," *Lawrence Livermore Laboratory Report UCRL-51912,* Livermore, Calif., Sept. 1975.
149. J. W. Hankin, L. O. Beaulaurier, and F. O. Comprelli, "Conceptual Design and Cost Estimate for a 10-MWe(Net) Generating Unit and Experimental Facility Using Geothermal Brine Resources," in ref. 4, pp. 1985–1997.
150. R. Boehm, H. Jacobs, R. Bliss, and D. Kelly, "Direct Contact Heat Exchangers for Geothermal Power Plants" in ref. 84, pp. 754–760.
151. C. A. Allen, E. S. Grimmett, and K. L. Wagner, "Fluidized-Bed Heat Exchangers for Geothermal Applications" in ref. 84, pp. 761–767.
152. M. W. Urbanek, DSS Engineers, Inc., "Development of Direct Heat Exchangers for Geothermal Brines," *Final Report Lawrence Berkeley Laboratory Report LBL-8558,* Berkeley, Calif., 1978.
153. J. Kestin and co-ed., *A Sourcebook on the Production of Electricity from Geothermal Energy,* U.S. Government Printing Office, Washington, D.C., 1980.
154. O. E. Balje, *J. Eng. Power, Trans. ASME* **84,** 83 (Jan. 1962).
155. S. Troulakis, "Steam Ammonia Power Cycle Study," *Report No. STA-19,* De Laval Turbine Inc., Trenton, N.J., 1968.
156. B. M. Vymorkov, *Vestn. Akad. Nauk SSSR* **10,** 32 (1965).

157. R. G. Olander, K. E. Nichols, and J. W. Tester, "Utilization of Hot Dry Rock Geothermal Energy: Power Plant Design Considerations," *paper presented at Proc. of Third National Congress, Pressure Vessel and Piping Division, Amer. Soc. of Mech. Eng., San Francisco, Calif., June 1979.*
158. J. H. Eskesen and co-workers, "Binary Cycles" in ref. 153.
159. K. E. Nichols and A. Malgieri, "Technology Assessment of Geothermal Pumping Equipment," *Final Report, U.S. Dept. of Energy Contract EG-77-C-04-4162,* Barber-Nichols Engineering, Arvada, Col., July 1978.
160. Sperry Research, "Feasibility Demonstration of the Sperry Down-Well Pumping System," *Fianl Report Contract No. EY-76-C-02-2838, U.S. Dept. of Energy, SCRC-CR-77-48,* Sudbury, Mass., July 1977.
161. K. E. Nichols, D. Prigmore, H. Matthews, and J. Halat, "Design and Field Test of a Steam Powered Downhole Geothermal Pump" in ref. 53, pp. 877–883.
162. U.S. Pat. 3,938,334 (Feb. 17, 1976), H. B. Matthews.
163. R. C. Robertson, "Waste Heat Rejection from Geothermal Power Stations," *Oak Ridge National Laboratory Report ORNL/TM-6533,* Oak Ridge, Tenn., 1978.
164. E. S. Miliaras, *Power Plants With Air-Cooled Condensing Systems,* MIT Press, Cambridge, Mass., 1974.
165. K. Aikawa and M. Soda, "Advanced Design in Hatchobaru Geothermal Power Station" in ref. 4, p. 1881.
166. J. H. Anderson, "The Vapor-Turbine Cycle for Geothermal Power Production" in ref. 3.
167. D. Aronson, "Binary Cycle for Power Generation," *Proc. Amer. Power Conf.* **23,** 261 (1961).
168. *Electric Power Generation Using Geothermal Brine Resource for a Proof-of-Concept Facility ERDA Grant AER74-19931 A01,* Bechtel Corporation, San Francisco, Calif., May 1975.
169. *Conceptual Design 50 MWe (Net) Geothermal Power Plants at Heber and Niland, California, Final Report #SAN-1124-1,* Bechtel Corporation, San Francisco, Calif., Oct. 1976.
170. J. W. Hankin, R. A. Hogue, T. A. V. Cassel, and T. R. Fick, "Effect of Reservoir Temperature Decline on Geothermal Power Plant Design and Economics" in ref. 53, pp. 870–876.
171. C. H. Bloomster, "An Economic Model for Geothermal Cost Analysis" in ref. 4.
172. C. H. Bloomster and R. L. Engel, "The Potential Benefits of Geothermal Electricity Production from Hydrothermal Resources," *Battelle Pacific Northwest Laboratories Report BNWL-22001,* Richland, Washington, June 1976.
173. C. H. Bloomster, D. Cohn, J. G. DeSteese, H. D. Huber, P. N. LaMori, D. W. Shannon, J. R. Sheff, and R. A. Walterk, "GEOCOST: A Computer Program for Geothermal Cost Analysis," *Battelle Northwest Laboratory Report BNWL-1888, UC-13,* Richland, Washington, Feb. 1975.
174. C. H. Bloomster and R. L. Engel, "The Potential Benefits of Geothermal Electricity Production from Hydrothermal Resources," *Battelle Pacific Northwest Laboratories Report BNWL-2001,* Richland, Washington, June 1976.
175. J. C. S. Chou, *Geothermal Energy* **2,** 21 (1973).
176. J. C. S. Chou, R. K. Ahluwalia, and E. Y. K. Woo, *Geothermics* **3,** 93 (1974).
177. D. H. Cortez, B. Holt, and A. J. L. Hutchinson, *Energy Sources* **1,** 74 (1973).
178. F. J. Dan, D. E. Hersam, S.K. Kho, and L. R. Krumland, "Development of a Typical Generating Unit at The Geysers Geothermal Project" in ref. 4, pp. 1949–1958.
179. D. G. Elliott, "Comparison of Geothermal Power Conversion Cycles" in ref. 84, pp. 771–777.
180. J. H. Eskesen, "Cost and Performance Comparison of Flash Binary and Steam Turbine Cycles for the Imperial Valley, California" in ref. 53, pp. 842–849.
181. J. Gault, J. W. Hall, J. W. Wilson, H. Michael, B. P. Shepherd, G. Underhill, and L. Gios-Castellon, "Preliminary Analysis of Electric Generation Utilizing Geopressured Geothermal Fluids" in ref. 84, pp. 790–797.
182. M. A. Green and H. S. Pines, "Calculation of Geothermal Power Plant Cycles Using Program GEOTHM" in ref 4, pp. 1965–1972.
183. W. L. Pope, H. S. Pines, L. F. Silvester, M. A. Green, and J. D. Williams, "Multiparameter Optimization Studies of Geothermal Energy Cycles" in ref. 53, pp. 865–869.
184. H. S. Pines and M. A. Green, "The Use of Program GEOTHM to Design and Optimize Geothermal Power Cycles" in ref. 84, pp. 836–842.
185. J. Guiza, "Power Generation at Cerro Prieto Geothermal Field" in ref. 4, pp. 1976–1978.
186. A. Hansen, "Thermal Cycles for Geothermal Sites and Turbine Installation at The Geysers Power Plant, California," *paper presented at Geothermal Energy, Proc. UN Conf. on New Sources of Energy, Rome, Italy, Aug. 21–31, 1971,* pp. 365–379.

187. V. K. Jonsson, A. J. Taylor, and A. D. Charmichael, "Optimization of Geothermal Power Plant by Use of Freon Vapour Cycle," *Timarit-VF1,* 1969, p. 2.
188. A. Lavi, ed., *Proceedings of Solar Sea Power Plant Conference and Workshop, Carnegie Mellon University, Pittsburgh, Pa., June 1973.*
189. G. Ramachandran and co-workers, "Economic Analysis of Geothermal Energy Development in California," *Project ECU 5013,* Vols. 1–2, Stanford Research Institute, Menlo Park, Calif., May 1977.
190. A. R. Sapre and R. J. Schoeppel, "Technological and Economic Assessment of Electric Power Generation from Geothermal Hot Water" in ref. 4, pp. 2343–2350.
191. K. E. Starling, L. W. Fish, K. Z. Iqbal, and D. Yieh, "Advantages of Using Mixtures as Working Fluids in Geothermal Binary Cycles," *paper presented at Proc. of Oklahoma Acad. Science, 63rd Annual Meeting, Durant, Oklahoma, 1974.*
192. K. E. Starling, L. W. Fish, K. Z. Iqbal, and H. H. West, "The Use of Mixture Working Fluids in Geothermal Binary Power Cycles" in ref. 53, pp. 850–856.
193. J. S. Swearingern, *Chem. Eng. Prog.* **73,** 83 (July 1977).
194. J. W. Tester, G. E. Morris, R. G. Cummings, and R. L. Bivins, "Electricity from Hot Dry Rock Geothermal Energy: Technical and Economic Issues," *Los Alamos Scientific Laboratory Report LA-7603-MS,* Los Alamos, New Mex., Jan. 1979.
195. J. S. Wilson, *Chem. Eng. Prog.* **73**(11), 95 (1977).
196. J. H. Altseimer, "Geothermal Well Technology and Potential Applications of Subterrene Devices—A Status Review," *Los Alamos Scientific Laboratory Report LA-5689-MS,* Los Alamos, New Mex., Aug. 1974.
197. W. C. Maurer, J. D. Nixon, L. W. Matson, and J. C. Rowley, "New Turbodrill for Geothermal Drilling" in ref. 53, pp. 904–911.
198. R. Greider, "Economic Considerations for Geothermal Exploration in the Western U.S.," *paper presented at the Symp. of Colorado Department of Natural Resources, Denver, Col., Dec. 1973.*
199. "Drilling Costs, 1973," Section I, *Joint Association Survey of the U.S. Oil and Gas Producing Industry,* American Petroleum Institute, Independent Petroleum Association of America and Mid-Continent Oil and Gas Association, Feb. 1975.
200. "Drilling Costs, 1974," Section I, *Joint Association Survey of the U.S. Oil and Gas Producing Industry,* American Petroleum Institute, Independent Petroleum Association of America and Mid-Continent Oil and Gas Association, Mar. 1976.
201. "Drilling Costs, 1975," Section I, *Joint Association Survey of the U.S. Oil and Gas Producing Industry,* American Petroleum Institute, Independent Petroleum Association of America and Mid-Continent Oil and Gas Association, 1977.
202. E. M. Bee Dagum and K. P. Heiss, "An Econometric Study of Small and Intermediate Size Diameter Drilling Costs for the United States," *PNE-3012,* Vols. 1 and 2, prepared for the U.S. Atomic Energy Commission by Mathematica, Princeton, N.J., June 1968.
203. E. M. Shoemaker, ed., "Continental Drilling," *Report of the Workshop on Continental Drilling, Abuquiu, New Mexico,* Carnegie Institute, Washington, D.C., June 1975.
204. R. G. Cummings and G. E. Morris, "Assessing the Economics of Producing Electricity from Hot Dry Rock Geothermal Resources: Methodology and Analyses," *Electric Power Research Institute* (*EPRI EA-630 Project 1017*) *and Los Alamos Scientific Laboratory Report* (*LA-7888-HDR*), Sept. 1979.
205. M. H. Dorfman and R. W. Deller, eds., *Proc. of the First Geopressured Geothermal Energy Conf., Austin, Tex., 1975.*
206. M. H. Dorfman and R. W. Deller, eds., *Proc. of the Second Geopressured Geothermal Energy Conf. Austin, Tex., 1976.*
207. B. Lang and F. E. Heidt, "Economic Analysis for MAGES," *Dornier Report DS-ERT-2/79,* Jan. 1979.
208. J. B. Koenig, "Worldwide Status of Geothermal Resources Development" in ref. 3.
209. R. W. Rex, "The US Geothermal Industry in 1978," *paper presented at the Hot Dry Rock Geothermal Information Conf., Santa Fe, New Mex., Apr. 19–20, 1978.*
210. R. DiPippo, "Geothermal Power Plants Around the World" in ref. 113.
211. R. DiPippo, *Geothermal Energy as a Source of Electricity: A Worldwide Survey of the Design and Operation of Geothermal Power Plants,* U.S. Government Printing Office, Washington, D.C., 1980.

JEFFERSON TESTER
CHARLES O. GRIGSBY
Los Alamos Scientific Laboratory

GERANIOL. See Terpenoids.

GERANIUM OIL. See Oils, essential; Perfumes.

GERMANIUM AND GERMANIUM COMPOUNDS

Germanium (Ge) is a semiconducting metalloid element found in Group IV A and period 4 of the periodic table. Although it looks like a metal, it is fragile like glass. Its electrical resistivity is about midway between that of metallic conductors and that of good electrical insulators. Although it was first isolated by Winkler in 1886, no commercial application was found for it until the early 1940s when it was found to have interesting electrical properties. Its first significant use was in solid-state electronics, and with it the transistor was invented. Indeed, the entire modern field of semiconductors owes its development to the early successful use of germanium. After ca 40 yr of commercial use, this is still the major application for germanium in the world today. However, its market picture is changing significantly. In the United States, its use in the field of ir optics surpassed its use in electronics during 1978, and probably this trend will become worldwide by the early 1980s. Germanium has also found wide use as a detector in gamma-ray spectroscopy and some use as a catalyst. Other applications with considerable potential demand are in the fields of fiber optics (qv), medicine, superconductivity, and solar energy (qv) (see Semiconductors; Integrated circuits; Superconducting materials).

Occurrence

The crust of the earth is estimated to contain 1.5–7 g Ge/t. Germanium usually occurs widely dispersed in minerals such as sphalerite and rarely in concentrated form. There are only a few actual minerals of germanium, some with germanium concentrations up to about 8%. Most of these occur in Africa with the highest concentration occurring near Tsumeb in Southwest Africa (Namibia). Ref. 1 reviews 19 Ge minerals found near Tsumeb. Several papers on the geochemistry of germanium are collected in ref. 2.

Germanium also occurs in significant concentrations in many coals in the world. The concentration of germanium within many coal veins varies from top to bottom with the highest concentration occurring in the upper and lower few centimeters. It is assumed that this distribution within the vein indicates the deposition of germanium from solution after the vein was formed. When coal is burned in power-generating plants, the germanium tends to concentrate in the fly ash produced. Any recovery of germanium from coal would most likely be from such ash. Johnson, Matthey and Co.

actually recovered germanium from coal for several years starting in about 1950 (3).

Properties

The physical, thermal, electronic, and optical properties of germanium metal are shown in Table 1.

Chemical Properties. ***Germanium Metal.*** Germanium [*7440-56-4*] is quite stable in air up to 400°C where slow oxidation begins. Oxidation becomes noticeably more rapid above 600°C. The metal resists concentrated hydrochloric acid, concentrated hydrofluoric acid, and concentrated sodium hydroxide solutions, even at their boiling points. It is not attacked by cold sulfuric acid but does react slowly in hot sulfuric acid. Nitric acid attacks germanium more readily than sulfuric acid at all temperatures. Germanium reacts readily with mixtures of nitric and hydrofluoric acids, with molten alkalies, and more slowly with aqua regia. The principal reaction route for the mixed acids is the oxidation of the germanium with one constituent followed by the dissolution of the oxide by the other constituent. The reaction with fused alkalies is a direct oxidation with the release of hydrogen. Germanium also reacts readily with the halogens to form the respective tetrahalides.

In compounds, germanium can have a valence of either 2 or 4. Although the divalent compounds tend to be less stable than the tetravalent ones, most can be stored at RT for years with no change in composition. At higher temperatures, most of the divalent compounds decompose. Ref. 4 is a review of the syntheses and properties of many Ge compounds, including divalent ones; ref. 5 provides a good discussion on the properties of germanium bonds. An excellent earlier review of inorganic Ge compounds is given in ref. 6.

Germanium Halides. Germanium tetrachloride [*10038-98-8*], $GeCl_4$, is made by the reaction of hydrochloric acid on germanium concentrates containing oxides and/or germanates. It can also be made by the reaction of chlorine on heated metallic germanium. The properties of $GeCl_4$ are shown in Table 2.

$GeCl_4$ is soluble in solvents such as acetone, absolute ethanol, benzene, carbon disulfide, carbon tetrachloride, chloroform, and diethyl ether. It is only slightly soluble in concentrated hydrochloric acid, with the solubility dropping with acid normality and reaching a minimum at about 5 N. At HCl concentrations below ca 5 N, the tetrachloride begins to hydrolyze to GeO_2 (see below). Germanium tetrachloride is insoluble in concentrated sulfuric acid and does not react with it. The solubility of free chlorine in $GeCl_4$ can reach as high as 4 wt %, especially at low temperatures. The solubility of 6 N hydrochloric acid in $GeCl_4$ is less than 10 μL/L.

Germanium tetrabromide [*13450-02-5*], $GeBr_4$, and tetraiodide [*13450-95-8*], GeI_4, can also be prepared easily by reaction of the respective halogen with Ge metal and by reaction of GeO_2 with HBr and HI solutions, respectively. The preparation of germanium tetrafluoride [*7783-58-6*], GeF_4, is not so straightforward, and pure GeF_4 is usually made by decomposing barium hexafluorogermanate [*60897-63-4*] at ca 700°C.

Germanium Oxides. Germanium dioxide [*1310-53-8*], GeO_2, is usually made by the hydrolysis of $GeCl_4$ with water. It is also made by the ignition of Ge disulfide. It exists in three solid forms with the following properties:

	Soluble	*Insoluble*	*Vitreous*
structure	hexagonal	tetragonal	amorphous
density at 25°C, g/cm^3	4.228	6.239	3.637
melting point, °C	1116	1086	
solubility in water at 25°C, g/L soln	4.53	insoluble	5.18
solubility in water at 100°C, g/L soln	13	insoluble	
solubility in HCl, HF, NaOH solutions	soluble	insoluble	soluble

The soluble form is the usual product of hydrolyzing $GeCl_4$. The insoluble form can be prepared by heating soluble oxide at 300–900°C, especially in the presence of about 0.5 wt % alkali halides. The glassy or vitreous form is prepared by melting either of the other forms and then cooling the melt.

Germanium monoxide [*20619-16-3*], GeO, can best be prepared in pure form by heating a mixture of Ge and GeO_2 in the absence of oxygen. At temperatures above 710°C, GeO sublimes from the mixture and condenses as a glassy deposit in the cooler part of the reaction vessel. GeO is stable at room temperature.

Germanates. Germanates are usually prepared by fusion of GeO_2 with alkali oxides or carbonates in platinum crucibles. Sodium heptagermanate [*12195-31-2*], $Na_3HGe_7O_{16}$, is precipitated by neutralization of a sodium hydroxide solution of GeO_2 with hydrochloric acid to pH above 7.

Germanides. Germanides can be formed by melting other metals with Ge in the proper stoichiometric concentrations and then freezing the melt. They can also be prepared by vacuum sintering the two metals together usually followed by long annealing. Other procedures include thermal dissociation of one germanide into another and the electrolysis of fused mixed salts. The preparation and properties of about 200 germanides have been tabulated and reviewed (7). One of the most often prepared germanides is magnesium germanide [*1310-52-7*], Mg_2Ge.

Germanes. Germanes, or germanium hydrides, are commonly prepared by the reaction of a germanide, such as Mg_2Ge, with hydrochloric acid. They can also be produced by the reduction of $GeCl_4$ with lithium aluminum hydride and by the reduction of GeO_2 by sodium borohydride in water solution. The preparation and properties of the germanes have been reviewed (8–9).

Miscellaneous Inorganic Compounds. Germanium nitride [*12065-36-0*], Ge_3N_4, is about as inert as tetragonal GeO_2. It is prepared most easily from Ge powder and ammonia at 700–850°C. The nitride does not react with most mineral acids, aqua regia, or caustic solutions, even hot. Germanium disulfide [*12025-34-2*], GeS_2, is an unusual and useful compound since it is insoluble in strong acids such as 6 *N* HCl and 12 *N* H_2SO_4. This insolubility permits the recovery of germanium from acid solutions by gassing with H_2S. The disulfide can also be made by the reaction of GeO_2 with sulfur.

Organogermanium Compounds. The field of organogermanium chemistry has drawn widespread interest during recent years. Organogermanium compounds are generally characterized as having low chemical reactivity and relatively high thermal stability. The synthesis of many begins with a Grignard reaction (qv). Many excellent reviews of the organogermanium literature have been published (10–22). Although all this research work has added considerably to the knowledge of Ge chemistry, and of chemistry in general, no commercial application of an organogermanium compound

Table 1. Properties of Germanium

Property	Value
Physical	
atomic number	32
atomic weight	72.59
crystal structure	diamond cubic
density at 25°C, g/cm^3	5.323
atomic density at 25°C, $atoms/cm^3$	4.416×10^{22}
lattice constant at 25°C, nm	0.565754
surface tension, liquid at mp, mN/m (= dyn/cm)	650
modulus of rupture, MPa[a]	72.4
Mohs hardness	6.3
Poisson's ratio at 125–375 K	0.278
natural isotopic abundance, %	
mass no. 70	20.4
72	27.4
73	7.8
74	36.6
76	7.8
Thermal	
melting point, °C	937.4
boiling point, °C	2830
heat capacity at 25°C, J/(kg·K)[b]	322
latent heat of fusion, J/g[b]	466.5
latent heat of vaporization, J/g[b]	4602
heat of combustion, J/g[b]	7380
heat of formation, J/g[b]	4006
vapor pressure, kPa[c]	
at 2080°C	1.33
at 2440°C	13.3
at 2710°C	53.3
at 2830°C	101.3
coefficient of linear expansion, $10^{-6}/K$	
at 100 K	2.3
at 200 K	5.0
at 300 K	6.0
thermal conductivity, W/(m·K)	
at 100 K	232
at 200 K	96.8
at 300 K	59.9
at 400 K	43.2
Electronic	
intrinsic resistivity at 25°C, Ω·cm	53
intrinsic conductivity type	N (negative)
intrinsic electron drift mobility at 25°C, $cm^2/(V·s)$[d]	3800
intrinsic hole drift mobility at 25°C, $cm^2/(V·s)$[d]	1850
band gap, direct, minimum at 25°C, eV	0.67
band gap, direct, minimum at 0 K, eV	0.744
number of intrinsic electrons at 25°C, cm^{-3}	2.12×10^{13}

Optical

Wavelength, μm	Refractive index (300 K)	Absorption coefficient, cm^{-1}	Through 1 cm thickness (uncoated): Reflection, %	Absorption, %	Transmission, %
1.8	4.134	5.20	37.3	62.5	0.2
1.9	4.1202	0.741	40.6	40.0	19.4

Table 1 *(continued)*

Wavelength, μm	Refractive index (300 K)	Absorption coefficient, cm^{-1}	Through 1 cm thickness (uncoated) Reflection, %	Absorption, %	Transmission, %
2.0	4.1081	0.0103	53.7	1.0	45.3
4.0	4.0247	0.0048	53.0	0.5	46.5
6.0	4.0104	0.0068	52.8	0.7	46.5
8.0	4.0051	0.0150	52.4	1.5	46.1
10.0	4.0031	0.0215	52.2	2.1	45.7
10.6	4.0027	0.0270	52.0	2.6	45.4
11.0	4.0025	0.0295	51.8	2.9	45.3
11.9	4.0021	0.200	46.8	16.5	36.7
12.0	4.0021	0.170	47.6	14.4	38.0
13.0	4.0018	0.160	47.8	13.7	38.5
14.0	4.0016	0.148	48.1	12.8	39.1
15.0	4.0014	0.385	43.3	27.1	29.6
16.0	4.0012	0.530	41.4	33.4	25.2
18.0	4.0012	2.00	36.3	58.2	5.5
20.0	4.0012	2.15	36.2	59.0	4.8

[a] To convert MPa to psi, multiply by 145.
[b] To convert J to cal, divide by 4.184.
[c] To convert kPa to mm Hg, multiply by 7.5.
[d] $cm^2/(V{\cdot}s) = 10^{-4}/T$

Table 2. Properties of Germanium Tetrachloride

Property	Value
molecular weight	214.40
color	colorless
density at 25°C, g/cm^3	1.874
melting point, °C	−49.5
boiling point, °C	83.1
refractive index at 25°C, 0.5893 μm	1.464
heat capacity (cp) of vapor at 25°C, J/(kg·K)[a]	449
heat of vaporization at bp, J/g[a]	137
heat of formation at 25°C, J/g[a]	−3318
vapor pressure, Pa[b]	
at 225 K	10^2
at 253 K	10^3
at 294 K	10^4
at 356 K	10^5
at 462 K	10^6
at 550 K (T_c)	3.850×10^6

[a] To convert J to cal, divide by 4.184.
[b] To convert Pa to mm Hg, divide by 133.3.

is known (see Organometallics). Several uses in medicines have been explored, but none of these has yet reached commercial importance. There are several commercial sources for organogermanium compounds, but the production of each remains modest. Commercial soruces are: Alfa Division, Ventron Corp., Danvers, Massachusetts; Laramie Chemical Co., Laramie, Wyoming; Pfaltz & Bauer, Inc., Stamford, Connecticut; and Strem Chemicals, Inc., Newburyport, Massachusetts.

Manufacturing and Processing

None of the minerals mentioned under Occurrence is mined solely for its germanium content. All of the germanium recovered worldwide is a by-product of other metals, mostly zinc, copper, and lead. The enriched copper–lead concentrates from Tsumeb have been treated in a vertical retort from which germanium sulfide is sublimed and separated (3). The Cu–Zn ores of Katanga (Zaire) have been treated by roasting with H_2SO_4, followed by leaching and selective precipitation of the germanium with MgO (3). In the United States (23), zinc concentrates are roasted and then sintered for zinc recovery. The sinter fume is chemically leached, and the germanium is selectively precipitated from the leach solution and sent to the germanium refinery. Germanium is also recovered from the still residue in the distillation of zinc metal (24).

In electrolytic zinc plants, which are becoming increasingly important for environmental reasons, the germanium is separated chemically during the purification of the electrolyte prior to electrolysis. Germanium is one of several impurities that have an adverse effect upon zinc electrolysis. If the germanium concentration is high enough in the separated solids, economic recovery is possible. The recovery of germanium from electrolytic zinc plants is discussed in ref. 3.

Regardless of the source of the germanium, all germanium concentrates are purified by similar techniques. The ease with which concentrated germanium oxides and germanates react with concentrated hydrochloric acid and the convenient boiling point of the resulting $GeCl_4$ (83.1°C) make chlorination a standard refining step. Chlorine is usually added to the primary distillation or to the subsequent fractionation, or both, to suppress the volatility of arsenic (25). Other purification steps are used to separate certain other objectionable impurities if present. The fractionation is usually done in glass or quartz since the subsequent use of the germanium in electronics requires that the purified $GeCl_4$ contain no more than about one mg/kg (1 ppm) impurities.

The purified $GeCl_4$ is hydrolyzed with deionized water to produce GeO_2 which is removed by filtration and dried. The dried GeO_2 is reduced with hydrogen at about 760°C to germanium metal powder which is subsequently melted and cast into so-called first-reduction or as-reduced bars. These bars are then subjected to zone refining to produce intrinsic or electronic grade germanium metal (see Fine chemicals). Zone refining (qv) is ideally suited for the refining of germanium, and indeed, the original development of this procedure was caused by the need for ultrapure germanium (26). Zone refining results in polycrystalline germanium, usually containing less than 100 ng total impurities per g of germanium and less than 0.5 ng of electrically active impurities per g Ge. For extremely high purity applications, such as for uncompensated gamma-ray detector crystals, germanium has been refined to impurity concentrations of less than 0.0003 ng/g (0.0003 ppb). However, zone refining is not equally efficient in removing all impurities. Boron and silicon are not removed easily this way and must be removed from germanium before the zone-refining step.

For its use as a semiconductor, refined germanium is grown into single crystals from the melt. The electronic properties are controlled by the addition of selected impurities (dopants) to the melt before crystal growth begins. The grown crystal is sliced and fabricated into devices, and the Ge scrap generated is recycled through the refinery into intrinsic germanium.

For its use in ir optics, zone-refined germanium is recast or grown into forms suitable for lens and window manufacture. After the germanium is annealed, it is cut and ground into lens or window blanks which are then polished, coated, and assembled into an ir system (see Infrared technology).

Environmentally, the production of germanium is quite harmless. Among the products produced, the chemicals consumed, and the by-products made, only arsenic would be considered a normal problem. The bulk of the arsenic is separated from the germanium at the smelter, and the small amounts entering the germanium-refining plant are easily controlled. The acids, bases, and chlorine used in processing and in gas scrubbing can be neutralized and held in permanent containment ponds. No hazardous substances are known to be discharged into surface waters or municipal treatment facilites from germanium-refining plants.

Economic Aspects

World reserves of germanium have been estimated at 4000 metric tons, but it is impossible to discuss reserves without considering price. In many applications, the cost of germanium is a very small part of the overall cost, and increasing the germanium price substantially would have little impact on its use in such applications. However, substantial increases in its price would expand germanium reserves significantly.

Since germanium is always recovered as a by-product, its price and availability over the long range are subject to supply–demand considerations for its host products, usually zinc and copper, as well as for itself. This is not the case over the short term (6–24 mo) since most producers recover germanium from stockpiles of smelter residues which can last for years. Therefore, short-term pricing is largely controlled only by demand.

The pricing over the last 25 yr of U.S. intrinsic-grade germanium metal adjusted for constant 1967 dollars is shown in Figure 1. Before the price increase in December, 1978, the adjusted price of $156.05/kg was an all-time low. The price on January 15, 1980 was $521.50/kg (adjusted price is $223.70/kg) and the price for imported intrinsic germanium was $557.50/kg.

The pricing of germanium in the future will reflect the increased demand for germanium in optics and other markets such as fiber optics (qv). Over the next several years, the only negative factors would be a continuing drop in the use of germanium semiconductors and increased germanium production.

Information concerning world production rates of germanium is difficult to obtain. This is due in part to the fact that there are so few producers. It is known that world production exceeded 100 t/yr during the period of peak demand for semiconductors. Production has varied between 30 and 80 t/yr during the 1970s.

An annual review of Ge is published by the U.S. Bureau of Mines (27) with a broader survey published every 5 years (28). The *Engineering and Mining Journal* also publishes an annual Ge review (29). The economics of Ge with special emphasis on world trade are reviewed in refs. 30–33.

Specifications and Standards

Most of the published standards for Ge are those for electronic applications. These have been published by the ASTM (34). Standards F 5-60, F 6-60, and F 27-68 pertain to GeO_2. F 42-77 and F 43-78 show how the conductivity type and resistivity of Ge

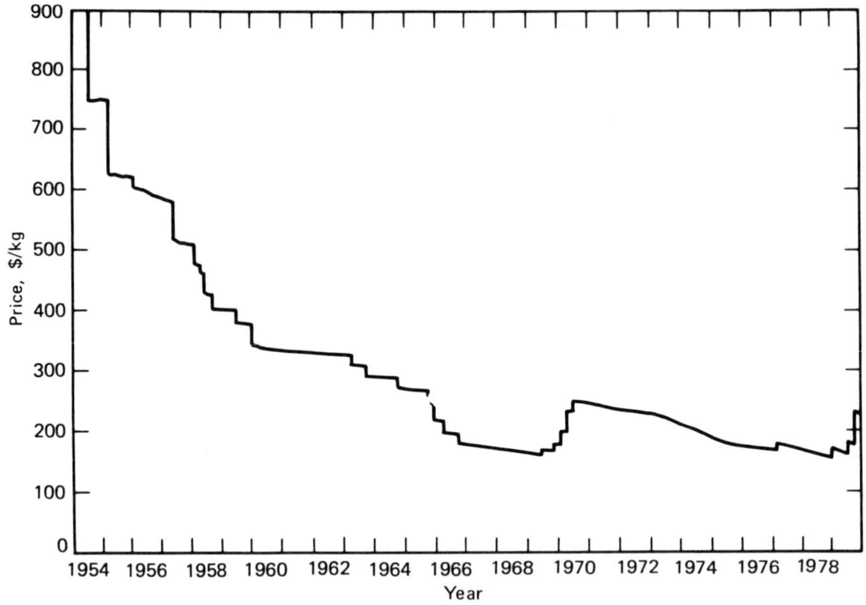

Figure 1. Historical price of intrinsic germanium in constant 1967 dollars.

metal are measured. Standards F 26-76, F 389-76, F 28-75, F 76-73, and F 398-77 provide procedures for testing Ge single crystals. F 120-75 and F 122-74 explain how to determine the interstitial oxygen content of Ge. And finally, F 334-75 gives a method for measuring the lithium ion drift mobility in Ge crystals destined for gamma-ray detector application.

Most electronic-grade GeO_2 is sold with a volatile content of 0.1–0.2% maximum, with a bulk density between 1.5 and 2.3 g/cm^3 as specified by the customer, with a billet resistivity of at least 5 Ω·cm on the last third of the billet to freeze, and with a billet conductivity of N (negative) type. Most intrinsic germanium sold is N-type with a resistivity greater than 40 Ω·cm at 25°C. In Europe, the resistivity is usually specified to be greater than 50 Ω·cm at 20°C.

Analytical and Test Methods

The analysis of ores for germanium is usually done with an emission spectrograph but can be done in the field with the phenylfluorone method (35). Analysis of Ge refinery samples is usually done after fusion of the sample with NaOH in nickel crucibles. Following distillation of the $GeCl_4$ from HCl solution of the fusion, the Ge can be determined (*1*) gravimetrically, usually by precipitation of GeS_2 from acid solution and by ignition to GeO_2, (*2*) titrimetrically, usually by reduction with sodium hypophosphite and titration with KIO_3 solution, or (*3*) spectrally, with an atomic absorption spectrophotometer. Procedure (*3*) is not considered as accurate as the first two.

Analysis of the impurities in pure germanium products is done routinely with an emission spectrograph, except for zone-refined metal and single crystals for which spectrographic analysis is inadequate. $GeCl_4$ is usually characterized by analysis of GeO_2 prepared from it by hydrolysis. The electrical quality of GeO_2 is determined by electrical measurements on a germanium metal billet prepared from it by hydrogen reduction. For the purified metals, electrical testing is routinely done. The resistivity and conductivity type of refined Ge are the usual acceptance tests. Single crystals are examined for these properties along with majority carrier mobility, minority carrier lifetime, and dislocation density. The lithium drift mobility is determined for crystals grown for gamma-ray detectors.

Excellent reviews of the analytical chemistry of germanium are given in refs. 36–37.

No storage problems are reported for inorganic germanium compounds. Some organic compounds are not stable over long periods, and their storage requires special consideration.

Toxicology

Germanium compounds generally have a low order of toxicity. Only germanium hydride is considered toxic with a maximum time-weighted average 8-h safe exposure limit of only 0.2 ppm (38). The LD_{50} for GeO_2 is 750 mg/kg and that of Ge is 586 mg/kg (39). The toxicity of specific germanium compounds usually must be considered more from the standpoint of the other part of the compound than the Ge content. The biological activity of germanium has been reviewed (40–42).

Uses

The main use for germanium in 1978 remained in the field of semiconductor (qv) electronics. Its earliest use in 1941 was as a solid-state diode performing as a detector in radar systems. This device was adapted to radio circuits as a radio-frequency detector, and the germanium transistor was invented in 1947. This invention revolutionized the electronics industry and caused a sharp increase in the demand for germanium. The use of germanium in these conventional semiconductors roles reached a peak in the early 1960s with world consumption averaging over 100 t/yr. Except for a brief upsurge in this demand during 1969, there has been a general decline in this usage since the early 1960s.

The use of germanium as a semiconductor substrate deserves special mention. In this use, semiconductor properties are not employed. In this application, single-crystal wafers of germanium are used as substrates for the epitaxial deposition of gallium arsenide/gallium arsenide phosphide for use as light-emitting diodes (qv). These substrates take the place of more expensive gallium arsenide wafers. At least one large company has made LEDs this way (43) since 1974. A further significance of this application is that many people are considering this technology in the field of photovoltaic solar energy (qv) conversion (see also Photovoltaic cells). Although the use of Ge as a substrate for GeAs solar cells might save considerable GaAs, a recent study (44) confirms that the use of germanium for this application would require many times the world's present production rate. However, there are large, untapped sources of germanium, and it is not inconceivable that such demands could be met in time with sufficient incentive.

The fastest growing use of germanium is in the field of ir optics. In this application, its transparency to ir wavelengths longer than 2 μm and its high refractive index are utilized rather than its electrical properties. Other properties of advantage in this use are low dispersion, easy machinability, reasonable strength, low price compared to other ir materials, good resistance to atmospheric oxidation, to moisture, and to chemical attack, and availability in large sizes. It has been estimated that the world demand for germanium for ir devices will increase to 55–70 t/yr during the 1980s (45). Principal applications in this field are military. Ir systems with germanium usually operate in the 8–12 μm range and usually contain several germanium lenses, a germanium window, and a color-correcting lens made from a Ge–Sb–Se glass (Texas Instruments, Inc., Dallas, Tex.) or a Ge–As–Se glass (Amorphous Materials, Inc., Garland, Tex.). Some of the viewers also utilize a germanium ir detector for the sensor. Nonmilitary ir applications include lenses for CO_2 lasers, and windows for intrusion alarms. Germanium is also used as a thin film coating for ir materials to decrease reflection losses and/or to provide heavy filtering action below 2 μm (see Infrared technology). Another application of germanium is the use of specially prepared germanium single crystals as gamma-ray detectors. Both the lithium-drifted detectors and the purer, more expensive, intrinsic detectors (which do not have to be stored in liquid nitrogen) do an excellent job of spectral analysis of gamma radiation and are important analytical tools. A significant use for germanium dioxide is as a condensation catalyst in the production of polyester fibers (qv) in which GeO_2 takes the place of cheaper Sb_2O_3 to give a nonyellowing white fiber.

Other uses include the use of magnesium germanate [*12025-12-6*] as a phosphor (see Luminescent materials), the application of niobium germanide [*37279-34-8*] as a superconductor (see Superconducting materials), the addition of lead germanate [*12025-27-3*] to barium titanate capacitors, the use of bismuth germanium oxide [*12233-56-6*] electro-optic crystals in diagnostic medical examination, the use of Ge–Au alloys in dental fillings and precision castings (see Dental materials), the experimental use of organogermanium compounds in medicines, and the addition of GeO_2 to glass formulations, primarily to increase the refractive index of the glass. A special example of this application is a recent development that shows considerable promise. This is the use of GeO_2 in fiber optics (qv) (46). In this application, $GeCl_4$ is converted to GeO_2 which is deposited, along with SiO_2 and sometimes B_2O_3 and/or P_2O_5, on the inside of a pure quartz tube and subsequently collapsed to form a solid rod or preform. The preform is then drawn into fine fibers which can be used as optical waveguides, primarily in the 0.8–1.6 μm wavelength region. Such fibers are proving increasingly useful in telephone lines, computer cables, CATV cables, and data transmission lines. There is considerable optimism concerning this application, and requirements of several t Ge/yr by the early 1980s are predicted.

BIBLIOGRAPHY

"Germanium" in *ECT* 1st ed., Vol. 7, pp. 168–174, by A. W. Laubengayer, Cornell University, and A. E. Middleton, Battelle Memorial Institute; "Germanium and Germanium Compounds" in *ECT* 2nd ed., Vol. 10, pp. 519–527, by H. R. Harner, The Eagle-Picher Company, and A. W. Laubengayer, Cornell University.

1. W. E. Wilson, ed., *Tsumeb! The Worlds Greatest Mineral Locality,* The Mineralogical Record, Inc., Bowie, Md., 1977.
2. J. N. Weber, ed., *Geochemistry of Germanium,* Dowden, Hutchinson & Ross, Inc., Stroudsburg, Pa., 1973.
3. *Eng. Min. J.* **157,** 75 (1956).
4. F. Glocking, *The Chemistry of Germanium,* Academic Press, Inc., London, 1969.

5. Ref. 4, pp. 3–27.
6. O. H. Johnson, *Chem. Rev.* **51,** 431 (1952).
7. G. V. Samsonov and V. N. Bondarev, *Germanides,* translated by A. Wald, Primary Sources, New York, 1970.
8. F. G. A. Stone, *Hydrogen Compounds of the Group IV Elements,* Prentice-Hall, Inc., Englewood Cliffs, N.J., 1962, pp. 63–76.
9. E. G. Rochow "Germanium" in J. C. Bailar, Jr., H. J. Emeleus, R. Nyholm, and A. F. Trotman-Dickenson, eds., *Comprehensive Inorganic Chemistry,* Vol. 2, Pergamon Press Ltd., Oxford, 1973, pp. 1–41.
10. O. H. Johnson, *Chem. Rev.* **48,** 259 (1951).
11. E. G. Rochow, D. T. Hurd, and R. N. Lewis, *The Chemistry of Organometallic Compounds,* John Wiley & Sons, Inc., New York, 1957.
12. F. Rijkens, *Organogermanium Compounds,* Germanium Research Committee, Utrecht, 1960.
13. D. Quane and R. S. Bottei, *Chem. Rev.* **63,** 403 (1963).
14. F. Rijkens and G. J. M. Van der Kerk, *Investigations in the Field of Organogermanium Chemistry,* Germanium Research Committee, Utrecht, 1964.
15. F. Glocking, *Quart. Rev. Chem. Soc.* **20,** 45 (1966).
16. M. Dub, *Organometallic Compounds,* 2nd ed., Vol. 2, Springer Verlag, Berlin, 1967.
17. K. A. Hooton "Organogermanium Compounds" in W. L. Jolly, ed., *Preparative Inorganic Reactions,* Vol. 4, John Wiley & Sons, Inc., New York, 1968, pp. 85–176.
18. N. Hagihara, ed., *Handbook of Organometallic Compounds,* W. A. Benjamin, Inc., New York, 1968, pp. 449–467.
19. Ref. 4, pp. 58–199.
20. M. Lesbre, P. Mazerolles, and J. Satgé, *The Organic Compounds of Germanium,* John Wiley & Sons, Ltd., London, 1971.
21. Ref. 9, pp. 29–39.
22. R. C. Weast, ed., *CRC Handbook of Chemistry and Physics,* 59th ed., CRC Press, West Palm Beach, Fl., 1978, pp. C-692–C-697.
23. J. A. O'Connor, *Chem. Eng.* **59,** 158 (1952).
24. U.S. Pat. 4,090,871 (May 23, 1978), A. Lebleu, P. Fossi, and J. Demarthe (to Societé Miniere et Metallurgique de Penarroya).
25. H. R. Harner in C. A. Hampel, ed., *Rare Metals Handbook,* 2nd ed., Reinhold Publishing Co., 1961, pp. 188–197.
26. W. G. Pfann, *Zone Melting,* 2nd ed., John Wiley & Sons, Inc., New York, 1966, p. 134.
27. J. M. Hague "Germanium" in *Bureau of Mines Minerals Yearbook,* U.S. Dept. of the Interior, 1976.
28. H. R. Babitzke "Germanium" in *Mineral Facts and Problems,* U.S. Dept. of the Interior, 1975.
29. J. H. Adams, *Eng. Min. J.* **179,** 167 (Mar. 1978).
30. *Germanium: World Survey of Production, Consumption and Prices with Special Reference to Future Trends,* Roskill Information Services, Ltd., London, 1974.
31. Ref. 30, Roskill Statistical Supplement, 1976.
32. *Germanium: World Survey of Production, Consumption and Prices with Special References to Future Trends,* 2nd ed., Roskill Information Services, Ltd., London, 1977.
33. Ref. 32, Supplement, 1979.
34. *1978 Annual Book of ASTM Standards, Part 43, Electronics,* ASTM, Philadelphia, Pa., 1978.
35. H. J. Cluley, *Analyst* **76,** 523 (1951).
36. J. R. Musgrave "Germanium" in I. M. Kolthoff, P. J. Elving, and E. B. Sandell, eds., *Treatise on Analytical Chemistry,* Part II, Vol. 2, John Wiley & Sons, Inc., New York, 1962, pp. 208–245.
37. V. A. Nazarenko, *Analytical Chemistry of Germanium,* translated by N. Mandel, John Wiley & Sons, Inc., New York, 1974.
38. "Threshold Limit Values for Chemical Substances and Physical Agents with Intended Changes for 1978," *American Conference of Governmental Industrial Hygienists, Cincinnati, 1978,* p. 19.
39. H. E. Christensen, *Toxic Substances List,* NIOSH, Rockville, Md., 1972.
40. Ref. 12, pp. 46–49.
41. Ref. 14, pp. 95–115.
42. Ref. 9, pp. 10–11.
43. J. H. Adams, *Eng. Min. J.* **178,** 181 (Mar. 1977).

44. J. W. Litchfield, R. L. Watts, W. E. Gurwell, J. N. Hartley, and C. H. Bloomster, *A Methodology for Identifying Materials Constraints to Implementation of Solar Energy Technologies,* Battelle Memorial Institute, Pacific Northwest Laboratory, Richland, Wa., 1978.
45. J. R. Piedmont and R. J. Riordan, "The Supply of Germanium for Future World Demands," *Fourth European Electro-Optics Conference, Society of Photo-Optical Instrumentation Engineers, Oct. 1978,* Vol. 164, pp. 216–222.
46. U.S. Pat. 3,737,292 (June 5, 1973), D. B. Keck, P. C. Schultz, and F. Zimar (to Corning Glass Works).

J. H. ADAMS
Eagle-Picher Industries, Inc.

GERMANIUM OIL. See Oils, essential; Perfumes.

GETTERING. See Vacuum technology.

GETTERS. See Vacuum technology.

GILSONITE

Gilsonite (uintaite) [*12002-43-6*] is a natural hydrocarbon substance of the class known as asphaltites (see Asphalt), occurring as a coal-like solid which is mined much like other minerals and sold essentially in its native state.

The only commercially important deposits of gilsonite in the world are located in the Uinta Basin, in the northeast corner of Utah. From a point 6–8 km within Colorado (Rio Blanco County), the area involved extends westward about 100 km into Utah. Gilsonite occurs in veins varying in width from a few centimeters to 6 m. The veins are nearly vertical planes running in a northwest–southeast direction, and extending downward from the surface as much as 600 m. Individual veins are as long as 35 km. Elsewhere in the United States, minor deposits of a gilsonitelike material have been reported in Wheeler and Crook Counties of Oregon (1).

Extended geological work in the Uinta Basin leaves little doubt that the source material was the tremendous oil shale deposits of the contiguous territory, which is further confirmed by certain similarities in the composition of the hydrocarbons involved, ie, both gilsonite and shale oil have a nitrogen content much higher than petroleum oils in general (see Oil shale).

Classification

Gilsonite is classed as one of the asphaltites, which are natural asphaltlike substances, characterized by their high softening points (above 110°C). Glance pitch and grahamite are other members of this group, some properties of which are shown in Table 1 (2). The nonmineral constituents are almost completely soluble in carbon disulfide. Solubilities in aromatic solvents are described in ref. 3.

Properties

The tests applied to gilsonite are in many cases the same as those used for asphalt (qv). Typical properties of commercial-grade gilsonite (American Gilsonite Selects) are shown in Table 2. Gilsonite is a natural mineral, not a manufactured product, and is, therefore, subject to certain variations. The liquid distillate from the pyrolysis of solid gilsonite was treated with sulfuric acid to produce a nonreactive hydrocarbon fraction which was classified as paraffinic (5). Evidence for pyridines and quinolines in the distillate was apparently based on odor alone. There are several hydrocarbon fractions from gilsonite distillate with many properties similar to those of fractions from crude petroleum (6). Cycloparaffins (naphthenes) and olefins were identified as hydrocarbon types. Several phenols, pyrroles, and pyridines have been isolated in the pyrolytic distillate from gilsonite (7–8). A heavy yellow oil was obtained from undegraded gilsonite in yields of 4–8% (9). Several aromatic and nonaromatic cuts, the latter predominating, resulted from low pressure distillation and silica gel adsorption of the oil. Substituted naphthalenes were indicated in the aromatic fractions by uv spectra. Destructive distillation of gilsonite gave 12 wt % gases, 55 wt % of a liquid pyrolyzate, and 33 wt % coke.

Porphyrin fractions were isolated from ores of two different veins in yields of 0.03 and 0.004% (eg, deoxyphylloerythreoetioporphyrin) (10). The entire porphyrin content is present as a nickel complex. The presence of porphyrins suggests that gilsonite is of plant origin (see Pyrrole and pyrrole derivatives).

Mining

Methods of mining have included the traditional hand pick, pneumatic picks, blasting, and high pressure (up to 13.8 MPa or 2000 psi) hydraulic cutting. All methods have been used concurrently in different mines with the choice depending upon mine conditions, and elevating and surface handling requirements.

Table 1. Properties of Gilsonite, Glance Pitch, and Grahamite

Mineral	Streak	Specific gravity at 25°C	Softening point, ring and ball method, °C	Fixed carbon[a], %
gilsonite or uintaite	brown	1.03–1.10	132–190	10–20
glance pitch or manjak[b]	black	1.10–1.15	132–190	20–35
grahamite[b]	black	1.15–1.20	188–329	35–55

[a] By proximate analysis, as for coal.
[b] When substantially free from mineral matter.

Table 2. Properties of Gilsonite[a]

Property	Value
color in mass	black
color of streak or powder	brown
fracture	conchoidal
specific gravity, 15/15°C	1.04
softening point (ring and ball method),°C	160
bitumen soluble in carbon tetrachloride, %	99.8
behavior in petroleum solvents	liquid solution that may thicken slightly upon standing
flash point, COC[b], °C	315
heating value, kJ/g[c]	41.8
specific heat, J/kg[c]	
at 149°C	2180
at 260°C	2550
thermal coefficient of expansion (volumetric), per °C	0.0005
viscosity (Brookfield viscometer), mPa·s (= cP)	
at 260°C	500
at 288°C	250
at 316°C	140
volatility (5 h), wt %	
at 163°C	<2
at 204°C	<4
at 260°C	<6
volatile combustible matter, by ASTM D 271-4 for VCM in coal, %	75–80
ultimate analysis, wt %	
carbon	85
hydrogen	10
nitrogen	2.5
oxygen	1.5
sulfur	0.3
silicon, nickel and trace elements	0.2
solid paraffins (unsulfonated residue)	0–trace
hardness (Mohs scale)	2
penetration, 100 g, 5 s, mm/10	
at 25°C	0
at 46°C	0
at 66°C	1
at 77°C	2
resistivity, Ω·m	1.9×10^{10}
acid value[d]	2.3
diazo reaction[e]	none
saponifiable matter	trace
constituent analysis, wt %[f]	
asphaltenes	50–65
resins	35–45
oils	1–5

[a] American Gilsonite Selects.
[b] Cleveland open cup.
[c] To convert J to cal, divide by 4.184.
[d] mg KOH/g substance to give pH 7.
[e] Color reaction to determine presence of phenols.
[f] Ref. 4.

Economic Aspects

Until the late 1950s when the American Gilsonite Company converted gilsonite into petrochemicals and conventional petroleum-type products, gilsonite markets, although worldwide in scope, were limited in volume to applications in which the gilsonite was dissolved in light solvents, or fluxed with vegetable oils and other hydrocarbons as heavy as petroleum asphalt. New and much larger applications began to develop for the unaltered product of the mines and it became obvious that conversion of gilsonite to fuels in competition with crude petroleum was a poor way to use this unusual natural resource.

The total gilsonite market is now in excess of $20,000,000 annually with tonnage mined <25% of that formerly produced. The average selling price at the mine head is in the range of $170–180 per metric ton with prices increasing at about the United States inflation rate. About 75% of the sales are within the United States. This change in gilsonite markets has led to long term availability and a more orderly utilization of this resource. At current price levels, minable reserves of gilsonite are estimated to be ca 5,000,000 t. Total reserves are probably in the range of 10,000,000–15,000,000 t.

Uses

Gilsonite is the raw material preferred by the Nuclear Power Commission in the UK for conversion to nuclear-grade graphite used in British designed advanced gas-cooled reactors. In early 1980 shipments of gilsonite commenced in volume to a United States petroleum refinery for conversion to a delayed type coke which, after calcining, is exported to the UK for graphitization and use in two AGR under construction at a cost of $\$3.6 \times 10^7$.

Gilsonite is used by the automotive industry, principally in the manufacture of solvent-free seam sealers, although there remains limited use in the production of paints and enamels. During the past eighty years gilsonite has been a principal ingredient in dark photogravure inks and, although lighter-colored inks are now more widely used, the market for gilsonite in the ink industry continues to increase (see Inks). Many applications for gilsonite, such as its use in battery boxes, floor tiles, molding compoundings and rubber products, have largely ceased to exist but large quantities now are used by the oil-drilling, foundry, building board, explosive and nuclear graphite industries, in addition to the continuing use by the automotive and ink industries.

BIBLIOGRAPHY

"Gilsonite" in *ECT* 2nd ed., Vol. 10, pp. 527–533, by H. O. Erbin, American Gilsonite Co.

1. E. T. Hodge, *Bull. Am. Assoc. Pet. Geol.* **11,** 395 (1927).
2. H. Abraham, *Asphalts and Allied Substances,* 6th ed., Vol. 1, D. Van Nostrand Co., Inc., Princeton, N.J., 1960, p. 229.
3. U.S. Pat. 4,108,681 (Aug. 22, 1978), M. B. Lawson and K. J. Snyder (to Halliburton Co.).
4. Ref. 2, Vol. IV, p. 366.
5. W. D. Day, *J. Franklin Inst.* **140,** 221 (1895).
6. C. W. Mabey, *J. Am. Chem. Soc.* **39,** 2015 (1917).

7. R. E. Ruesser, *An Investigation of the Organic Bases in Gilsonite,* M.S. thesis, University of Utah, Salt Lake City, Utah, 1949.
8. J. M. Sugihara and D. F. Sorenson, *J. Am. Chem. Soc.* **77,** 963 (1955).
9. T. F. McCullough, *Hydrocarbons and Other Compounds Obtained from Gilsonite,* Ph.D. thesis, University of Utah, Salt Lake City, Utah 1955.
10. L. R. McGee, *Porphyrins in Gilsonite,* Ph.D. thesis, University of Utah, Salt Lake City, Utah, 1956.

General References

"Gilsonite Cement," *Oil Gas J.* **57**(1), 150 (1959).

A. Bezzant, *The Chromatographic Separation of Gilsonite,* thesis, Brigham Young University, Provo, Utah, 1949.

G. A. Carey and I. C. Roberts, *History, Occurrence, Mining and Economics of Gilsonite,* thesis, University of Utah, Salt Lake City, Utah, 1949.

A. L. Crawford, "Gilsonite and Related Hydrocarbons of the Uintah Basin, Utah," *The Oil and Gas Possibilities of Utah,* Utah Geol. Mineral Surv., Salt Lake City, Utah, 1949, pp. 235–260.

C. Davis, *Geology of the Culmer Vein of Duchesne County, Utah,* thesis, Brigham Young University, Provo, Utah, 1952.

P. J. Grantham and A. G. Douglas, *Adv. Org. Geochem., Proc. 7th Int. Meet.,* 1975, pp. 193–207.

M. Inagaki, Y. Tamai, and S. Naka, *Tanso* **75,** 118 (1973); *Chem. Abstr.* **82,** 33141t (1975).

U.S. Pat. 4,073,659 (Feb. 14, 1978), M. V. Burris.

U.S. Pat. 4,133,932 (June 18, 1976), S. H. Peck (to USM Corp.).

U.S. Pat. 4,086,195 (Apr. 25, 1978), W. R. Knox, J. C. Hill, L. J. Hughes, and P. W. Foerst (to Monsanto Co.).

Can. Pat. 972,141 (Aug. 5, 1975), F. F. Sullivan (to Rotary Drilling Services, Inc.).

KENNETH R. NEEL
American Gilsonite Co.

GIN. See Beverage spirits, distilled; Oils, essential.

GLASS

Glass was formed naturally from common elements in the earth's crust long before anyone ever thought of experimenting with its composition, molding its shape, or putting it to the myriad of uses that it enjoys in the world today. Obsidian, for instance, is a naturally occurring combination of oxides fused by intense volcanic heat and vitrified (made into a glass) by rapid air cooling. Its opaque, black color comes from the relatively high amounts of iron oxide. Its chemical durability and hardness compares favorably with many commercial glasses (1). Pumice, a naturally occurring foam glass, is replete with tiny pockets of the gaseous products of the decomposition of many compounds. These gases were trapped by the viscous glass while it was cooling.

The origin of the first synthetic glasses is lost in antiquity and legend. Faience was made by the Egyptians who molded figurines from sand (SiO_2), the most popular glass-forming oxide. They coated them with natron, the residue left by the flooding Nile river, which was composed principally of calcium carbonate ($CaCO_3$), soda ash (Na_2CO_3), salt (NaCl), and copper oxide (CuO). Heating below 1000°C produced a glassy coating by the diffusion of the fluxes, CaO and Na_2O, into the sand and their subsequent solid-state reaction with the sand. The copper oxide gave the article an appealing blue color. Glass technology has evolved for six thousand years, and some of today's principles date back to early times. This includes what is today known about the structure of glass, its composition, properties, method of manufacture, and uses (see also Glass-ceramics; Glassy metals).

Common usage of the term glass follows the definition of Morey (2): "Glass is an inorganic substance in a condition which is continuous with, and analogous to, the liquid state of that substance, but which, as the result of a reversible change in viscosity during cooling, has attained so high a degree of viscosity as to be, for all practical purposes, rigid." Similarly, ASTM (3) defines glass as "an inorganic product of fusion that has cooled to a rigid condition without crystallizing." Both organic and inorganic materials may form glasses if their structure is noncrystalline, ie, if they lack long-range order. This includes some plastics, metals (4–7), and organic liquids (8–9). In principle, rapid cooling could prevent crystallization of any substance if the final temperature is sufficiently low to prevent structural rearrangement. Thus, glasses are formed primarily for kinetic reasons.

Glass is not merely a supercooled liquid. This distinction is illustrated by the volume–temperature diagram shown in Figure 1. When a liquid that normally does not form a glass is cooled, it crystallizes at or slightly below the melting point (path A). If there are insufficient crystal nuclei or if the viscosity is too high to allow sufficient crystallization rates, undercooling of the liquid can occur. However, the viscosity of the liquid rapidly increases with decreasing temperatures, and atomic rearrangement slows down more than would be typical for the supercooled liquid. This results in the deviation from the metastable equilibrium curve which is depicted by paths B and

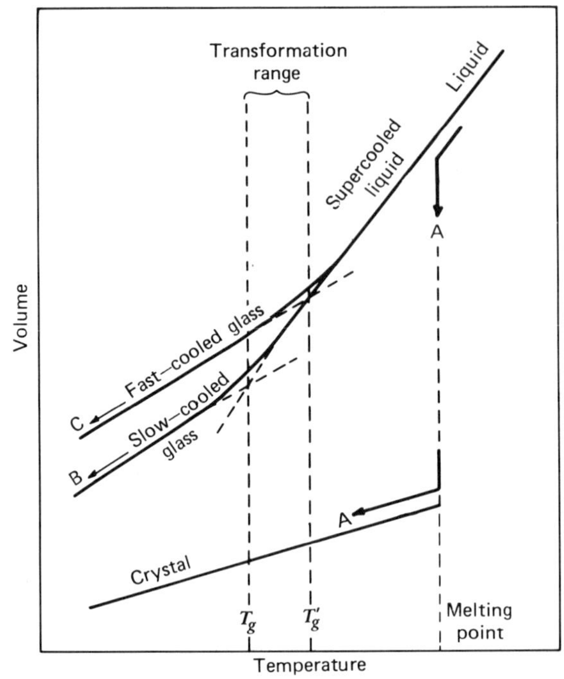

Figure 1. Volume–temperature relationships for glasses, liquids, supercooled liquids, and crystals. Courtesy of Corning Glass Works.

C in Figure 1. This change in slope with temperature is characteristic of a glass. Structural rearrangement is too slow to be detected experimentally, and additional volume changes are virtually linear with continued cooling, the same as for any other single-phase solid. The cooling rate determines when the deviation begins to occur. Slower cooling (path B), for instance, results in less of a deviation from the extrapolated liquid curve. Figure 1 shows that the point of intersection of the two slopes defines a transformation point (glass-transition temperature) T_g, for a given cooling rate. Practical limitations on cooling rate define the transformation range $T_g \rightarrow T_{g'}$ as the temperature range in which the cooling rate can affect the structure-sensitive properties such as density, refractive index, and volume resistivity (10). The structure, which is frozen in during the glass transformation, persists at all lower temperatures. Thus, a glass has a configurational or fictive temperature which may differ from its actual temperature. The fictive temperature is the temperature at which the glass structure would have been the equilibrium structure (11). It describes the structure of a glass as it relates to the cooling rate. A fast-quenched glass would have a higher fictive temperature than a slowly cooled glass.

Glasses can be prepared by methods other than cooling from a liquid state, including solution evaporation, reactive sputtering, vapor deposition, neutron bombardment, and shock-wave vitrification (12–13). These techniques suggest that the purely kinetic explanation of the glassy state is subject to question and that we need to modify the previous definitions. It has been shown (14) that extrapolation of the thermodynamic properties of the supercooled liquid gives the paradoxical results that entropies, heat contents, and volumes become less than those of the perfect crystal at the same temperature. Considering the transformation as a second-order transition (15) has yielded a satisfactory explanation for the properties of some organic systems, but this theory is still subject to confirmation for inorganic glasses. The dependence of transformation temperature on relaxation time has been considered and a new definition based on structural factors proposed (16). Isotropic materials with long structural–relaxation times, eg, $>10^3$ s, would be defined as glasses. The determination of either the isotropy of the material or structural relaxation times distinguishes whether or not the material is a glass. This definition requires information regarding structure and does not consider previous thermal history as a distinguishing characteristic of a vitreous material.

Structure

The basic structural unit of silicate glasses is the silicon–oxygen tetrahedron in which a silicon atom is tetrahedrally coordinated to four surrounding oxygen atoms. Oxygens shared between two tetrahedrons are called bridging oxygens. In pure vitreous silica, virtually all oxygens are bridging. Those that are not shared, for one reason or another, can be referred to as nonbridging oxygens. The relationship between these tetrahedra is controversial and has yet to be completely resolved. The earlier crystallite theory has been modified by proponents of the random-network theory. Modern structural methods point to a compromise theory (see Silica).

In 1921, Lebedev (17) noted a discontinuous index of refraction of SiO_2 near the α–β transition of quartz. His data and subsequent x-ray investigations of vitreous silica led to the suggestion (18) that crystallites on the order of 1.5 nm were present. It was demonstrated, however, that the crystal size would be less than 0.8 nm, and it was suggested that the term crystal loses meaning for these dimensions (19–20).

Zachariasen formulated the random-network theory of glass in 1932 (21). It proposes that atoms present in glass form a three-dimensional connected structure without periodic order and with energy content comparable to that of the corresponding crystalline material. According to this theory, the coordination number of an atom determines its role in a glass structure, and the following four rules should be fulfilled for an oxide to form glass: (*1*) each oxygen atom must be linked to no more than two cations; (*2*) the number of oxygen atoms around any one cation must be small, ie, three or four; (*3*) the oxygen polyhedra must share corners, not edges or faces, to form a three-dimensional network; and (*4*) at least three corners must be shared. For one-component glasses, each polyhedron shares corners with at least three other polyhedra in such a way that the network is continuous in three dimensions. In multicomponent glasses, additional cations are distributed throughout holes in the network.

X-ray structural work strongly supported the random network theory (19). The x-ray scattering pattern of glass after Fourier analysis gives radial distribution curves that indicate the distribution of neighboring atoms about a central atom. No evidence of discrete particles or voids supporting an ordered structure was observed based upon data for the first coordination shell (19). There were both theoretical and experimental limitations to this early work. Later, more complete data were obtained using fluorescence excitation to eliminate Compton scattering (22). Not only was a silicon–oxygen distance of 0.162 nm observed, but also peaks at 0.265 nm for the oxygen–oxygen distance, 0.312 nm for the silicon–silicon distance, 0.415 nm silicon–second silicon, and 0.64 nm for the silicon–third oxygen peak. X-ray scattering investigation of silica (23–24) suggested a structural ordering beyond the distances first reported (22). Data analysis resulted in a shorter silicon–oxygen bond distance (0.1595 nm) than reported at first (22). However, a similarity in bonding topology between tridymite and silica glass does not imply microcrystallinity of vitreous silica in a crystallographic sense. The similarity between crystalline and vitreous structure on the basis of silicon–oxygen–silicon bond angles and, hence, the relative orientation of the silicon tetrahedrons is pointed out in Figures 2 and 3.

In addition to the question of long-range ordering (1–2 nm), there still are aspects of the random-network theory that are often criticized. It is possible, for example, to form glasses when no three-dimensional network is possible. Glassy orthosilicates (SiO_4^{4-}) of lead or sodium and calcium have been prepared (26). Furthermore, modifying cations have been shown to occur at regular interatomic distances ranging over several coordination shells (27). Dark-field transmission electron microscopy has been used to infer density fluctuations of silica which suggest ordered regions of approximately 1.0 nm in size (see Analytical methods). However, these results have been criticized (28).

Glass-forming systems other than silica have been examined. The fraction of three- and four-coordinated boron in borate glasses can be determined by nmr (29) (see Analytical methods). Both nmr and x-ray diffraction (30) results led to the suggestion that the boroxyl ring is the structural unit of vitreous B_2O_3 (22,29). The intermediate-size boroxyl ring represents a compromise between the crystallite and the random-network theory.

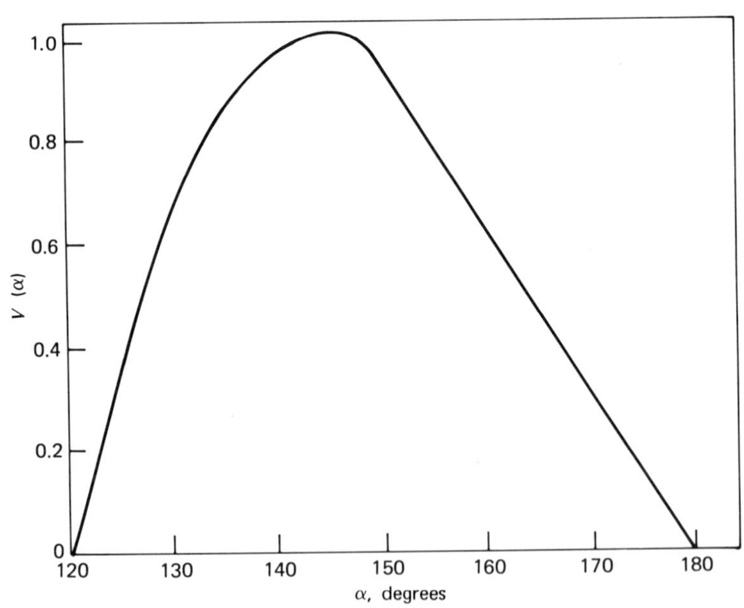

Figure 2. The distribution of silicon–oxygen–silicon bond angles in vitreous silica (22,25). The function $V(\alpha)$ is the fraction of bonds with angles normalized to the most probable angle, 144°. This distribution gives quite a regular structure on the short range, with gradual distorting over a distance of 3 or 4 rings (2–3 nm). Crystalline silica such as quartz or cristobalite would have a narrower distribution around specific bond angles.

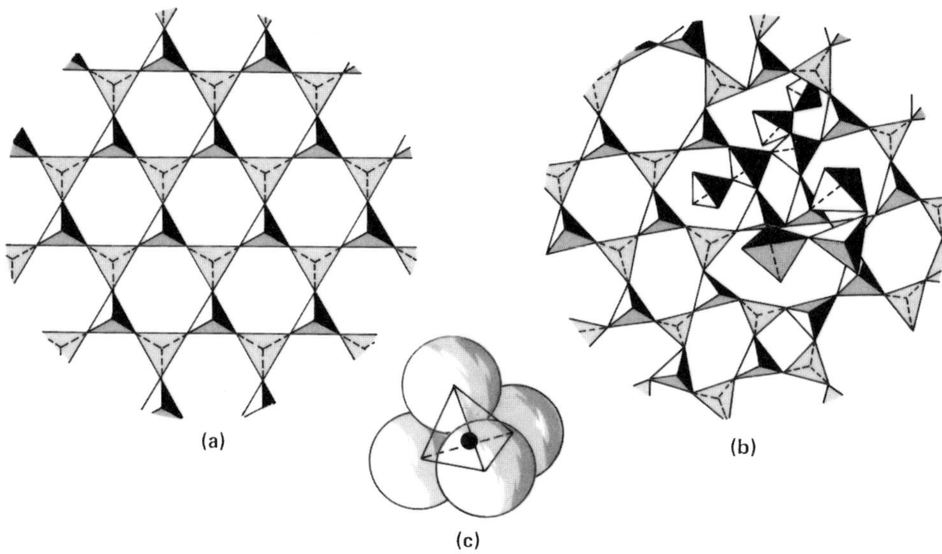

Figure 3. Schematic representation of (**a**) an ideal crystalline structure (Si—O—Si bond angles = 180°) and (**b**) a simple glass (Si—O—Si bond angles = 144° ± according to Fig. 2). The tetrahedra in the schematics represent four oxygens clustered around a silicon as shown (**c**). Courtesy of Corning Glass Works.

Composition

Conditions favorable for glass formation may be deduced from either geometric or bond strength considerations. On the basis of the rules (21) discussed above, the following oxides should be glass formers: B_2O_3, SiO_2, GeO_2, P_2O_5, As_2O_5, P_2O_3, As_2O_3, Sb_2O_3, V_2O_5, Sb_2O_5, Nb_2O_5 and Ta_2O_5. In fact, they are all so used. The only fluoride that fulfills the rules of glass formation is BeF_2 which readily forms a glass (31).

Glass formers generally have cation–oxygen bond strengths greater than 335 kJ/mol (80 kcal/mol). In multiple-component systems, oxides with lower bond strengths do not become part of the network and are called modifiers. Oxides with energies of ca 335 kJ/mol may or may not become part of the network and are referred to as intermediates. The dissociation energies used to predict glass formation are calculated, taking into account the coordination number of the cation (see Table 1). In multiple-component glasses, the terms formers, modifiers, and intermediates are frequently used to define the role of the individual oxides. However, an element such as lead may be either a modifier or intermediate, depending upon its coordination and the glass system considered.

Glass formation of individual oxides can be predicted from the melting point, and individual bond energies can be normalized by dividing by the melting point of the oxide (33). This ratio is relevant because the melting point is related to the amount of thermal energy available to rupture bonds. If the bond energy is large and the melting point low, glass formation is favored. This explains the ease of glass formation of B_2O_3 and from low-melting eutectics in which neither oxide forms a glass separately, eg, $CaO–Al_2O_3$.

Table 1. Coordination Number and Bond Strengths of Oxides[a]

Oxide	CAS Registry Number	Dissociation energy, kJ/mol[b]	Coordination number	Single-bond strength, kJ/mol[b]
Formers				
B_2O_3	[*1303-86-2*]	1489	3	496
SiO_2	[*10097-28-6*]	1774	4	443
GeO_2	[*1310-53-8*]	1803	4	450
Al_2O_3	[*1344-28-1*]	1682–1326	4	420–332
B_2O_3	[*1303-86-2*]	1489	4	372
P_2O_5	[*1314-56-3*]	1849	4	462–370
V_2O_5	[*1314-62-1*]	1879	4	469–376
As_2O_5	[*1303-28-2*]	1460	4	365–292
Sb_2O_5	[*1314-60-9*]	1418	4	354–284
ZrO_2	[*1314-23-4*]	2029	6	338
Intermediates				
TiO_2	[*13463-67-7*]	1820	6	303
ZnO	[*1314-13-2*]	602	2	301
PbO	[*1317-36-8*]	606	2	303
Al_2O_3	[*1344-28-1*]	1682–1326	6	280–221
ThO_2	[*1314-20-1*]	2159	8	269
BeO	[*1304-56-9*]	1046	4	261
ZrO_2	[*1314-23-4*]	2029	8	253
CdO	[*1306-19-0*]	498	2	248
Modifiers				
Sc_2O_3	[*12060-08-1*]	1514	6	252
La_2O_3	[*1312-81-8*]	1699	7	242
Y_2O_3	[*1314-36-9*]	1669	8	208
SnO_2	[*18282-10-5*]	1163	6	193
Ga_2O_3	[*12024-21-4*]	1117	6	186
In_2O_3	[*1312-43-2*]	1083	6	180
ThO_2	[*1314-20-1*]	2159	12	179
PbO_2	[*1309-60-0*]	970	6	161
MgO	[*1309-48-4*]	929	6	154
Li_2O	[*12057-24-8*]	602	4	150
PbO	[*1317-36-8*]	606	4	151
ZnO	[*1314-13-2*]	602	4	150
BaO	[*1304-28-5*]	1088	8	135
CaO	[*1305-78-8*]	1075	8	134
SrO	[*1314-11-0*]	1071	8	133
CdO	[*1306-19-0*]	498	4	124
Na_2O	[*1313-59-3*]	502	6	83
CdO	[*1306-19-0*]	498	6	82
K_2O	[*12136-45-7*]	481	9	53
Rb_2O	[*18088-11-4*]	481	10	48
HgO	[*21908-53-2*]	284	6	47
Cs_2O	[*20281-00-9*]	477	12	39

[a] Ref. 32.
[b] To convert J to cal, divide by 4.184.

Other correlations of glass formation and properties have been offered. For example (34): (*1*) cation valence should be either three or greater, (*2*) glass formation should increase with decreasing cation size, (*3*) the Pauling electronegativity should be between 1.5 and 2.1. Using these criteria, four types of oxides are described: (*1*) strong glass formers such as Si, B, Ge, As, and P, (*2*) intermediate formers that require rapid cooling, such as Sb, V, W, Mo, and Te, (*3*) oxides that form glasses in binary mixtures with nonglass formers, such as Al, Ga, Ti, Ta, Nb, and Bi, and (*4*) oxides that do not form glasses.

Glass composition work starts with the application of structural and bonding rules of glass formation. Numerous ternary systems and their glass-forming regions have been investigated (35). There are three types of ternaries: (A) single former and two modifiers; (B) two formers and one modifier; and (C) three glass formers. Type A is shown in Figure 4. The structural rules suggested by Zachariasen can also define likely regions for glass formation. Additions of several percent of other oxides for property adjustments are usually made to each system to give commercially useful glasses.

Single-Phase Glasses

Vitreous Silica. Vitreous silica is the most important single-component glass. Highly cross-linked vitreous silica is viscous and has a thermal expansion coefficient within the 0–300°C range of about $5.5 \times 10^{-7}/°C$. It is an excellent dielectric and resists attack by most chemicals, except fluorides or strong alkali. Fused silica has a high spectral transmission, and in addition, is not subject to radiation damage which results in browning of other glasses. It is the ideal glass for space-vehicle windows, wind-tunnel windows, ultrasonic delay lines, crucibles for growing ultra-pure silicon or germanium crystals, and for optical systems in spectrophotometric devices.

The same properties that make transparent fused silica useful also cause it to be difficult to produce and expensive. Vitreous silica may be made by several processes.

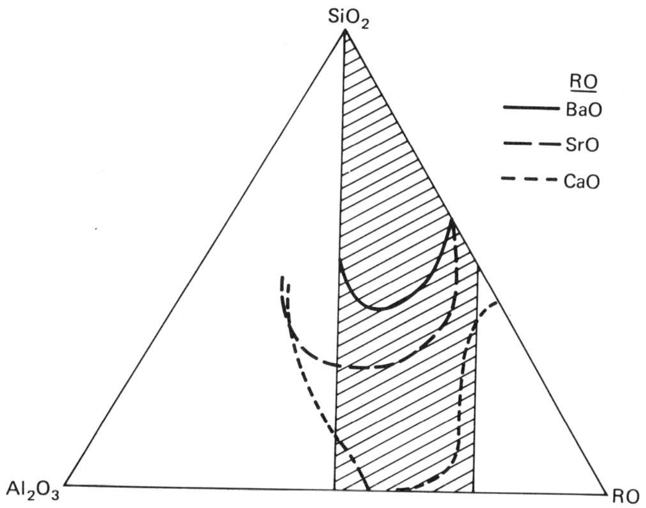

Figure 4. Glass-forming region in Type A system (35). The shaded area represents the predicted glass-forming region based upon Zachariasen's rules.

Fused quartz made by electrically fusing quartz crystal gives a product containing very little moisture and having good ir transmission. However, mineral impurities of natural quartz, including alumina, iron, and some chlorides, reduce uv transmission. Flame fusion of quartz or flame hydrolysis of $SiCl_4$, on the other hand, gives glasses containing larger amounts of water which decreases the ir transmission (see below under Vapor Deposition). Long heat treatments of vitreous silica below 1723°C may cause crystallization. Stable crystalline forms of silica at atmospheric pressure are cristobalite (1723–1470°C), tridymite (1470–867°C), and quartz (below 867°C) (see Silica, vitreous silica; Silica, synthetic quartz).

Multicomponent Silicate Systems. Most glasses fall into the category of silicates containing modifiers and intermediates. Addition of a modifier such as sodium oxide, Na_2O, to the silica network alters the structure by cleaving the Si—O—Si bonds to form Si—O·Na linkages (see Fig. 5).

Separating the silica tetrahedra from each other makes the glass more fluid and therefore more amenable to conventional melting and forming methods. Modifiers

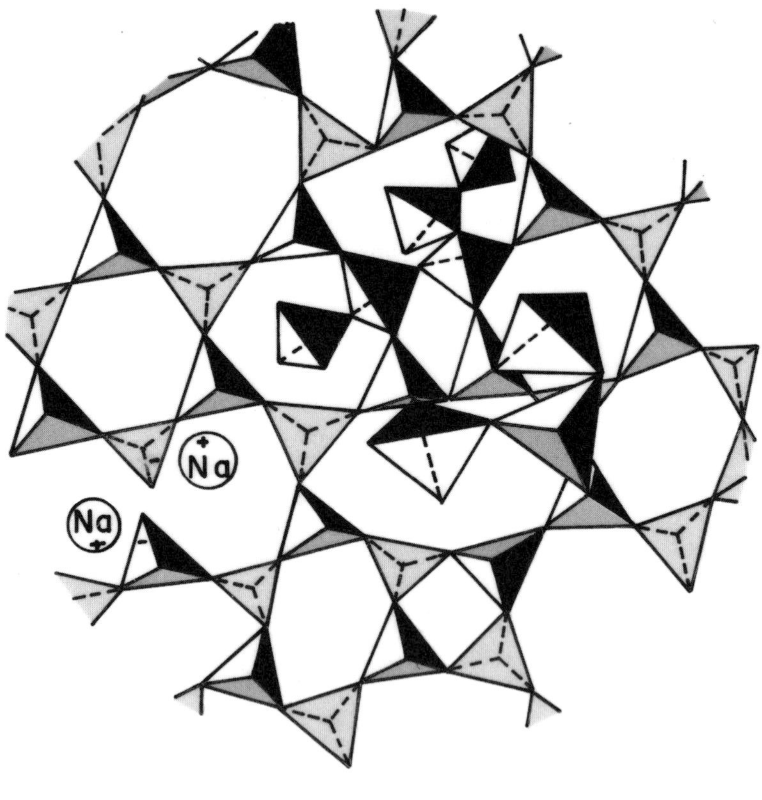

Figure 5. The addition of a modifier, in this case one molecule of Na_2O, causes the breaking of one Si—O—Si bond to form two Si—O·Na linkages. Courtesy of Corning Glass Works.

(or fluxes) also cause a decrease in resistivity, an increase in thermal expansion, and generally lower chemical durability. Glasses with a SiO_2:Na_2O molecular ratio less than one have so many nonbridging oxygens that they lack a continuous, three-dimensional structure (Zachariasen's rule 4). Such glasses, referred to as invert glasses, have been made containing Li_2O, Na_2O, and K_2O oxides. Alkali silicates that have a silica:alkali ratio of 0.5–3.4 are the basis of the soluble silicate glass industry. The composition and uses of sodium silicates for adhesives, cleansers, desiccants, abrasives, cement deflocculants, and surface coatings are discussed in ref. 36 (see Silicon compounds).

The effectiveness of an alkali oxide (eg, Li_2O, Cs_2O) as a flux increases with the size of the cation and therefore with its polarizability. Large ions such as cesium are easily polarized and thus more likely to give up their oxygen to break the Si—O—Si bonds as discussed above. Lithium, on the other hand, is more likely to keep its oxygen and, therefore, its fluxing power is less. This is consistent with the ease of glass formation as the size:charge ratio of the modifier is increased. Phase separation occurs often when less polarizable oxides are present. Lithium or magnesium silicates have a tendency to phase-separate during heat treatment (37). Ionic mobility is related both to charge and to size. Large alkalies are expected to be more mobile because of their greater polarizability. Increased size, however, tends to reduce mobility. Alkaline earth silicates behave similarly to alkali silicates, but the fluxing power of alkaline earths is less than that of the alkalies. Mobility of divalent ions is less than that of monovalent ones; hence, resistivities of alkaline-earth glasses are usually higher. Divalent oxides increase the resistivity of alkali-containing glasses.

Alumina is frequently used in silicate glasses. It often adopts a four-coordinated structure with alkalies giving a $NaAlO_2$ tetrahedral unit which substitutes into the SiO_2 network. The extra negative charge associated with the four bridging oxygens surrounding Al^{3+} is offset by the Na^+ ion. A maximum in viscosity occurs when the Al_2O_3:Na_2O ratio equals one.

Boron oxide often behaves as a flux. Boron softens glass for easier melting but, unlike alkalies, boron oxide increases expansion only slightly. This is the basis of the easily melted but low-expansion commercial glasses known as borosilicates.

Soda-Lime Glasses. Mixtures of alkali and alkaline earths give glasses of higher durability than the alkali silicates. The actual compositions are usually more complex than the term soda lime suggests. In addition to Na_2O, CaO, and SiO_2, these glasses may contain MgO, Al_2O_3, BaO, or K_2O and various colorants. Alumina increases durability, whereas MgO prevents devitrification. Soda-lime glass accounts for nearly 90% of all the glass produced. The batch materials are inexpensive and relatively easy to melt. Soda-lime glass is used for containers, flat glass, pressed and blown ware, and lighting products where exceptional chemical durability and heat resistance are not required.

Borosilicate Glasses. Replacement of alkali by boric oxide in a glass network gives a lower-expansion glass. The fluxing action of the boron facilitates melting by weakening the network. This has been attributed to the presence of planar three-coordinate borons that weaken the silicate network at high temperature. Phase separation of borosilicate glasses often occurs during heat treatment which may be useful for certain applications. However, most commercial borosilicate glasses have compositions that are miscible and homogeneous. Borosilicate glass is applied as ovenware, laboratory equipment, piping, and sealed-beam headlights.

Aluminosilicate Glasses. Structural rules suggest that if the $R_2O:Al_2O_3$ or the $RO:Al_2O_3$ molar ratio is unity, an aluminosilicate glass has a silica structure in which all oxygens are bridging oxygens. This is true of other silicate minerals and appears to be the case with glasses. Alumina is expected to be four-coordinated when the alkali to alumina molar ratio is greater than one, but if the ratio is less than one, six-fold coordination of alumina has been suggested.

Aluminosilicate glasses are used commercially because they can be chemically strengthened and withstand high temperatures. Thus, applications include airplane windows, frangible containers, lamp envelopes, and top-of-stove uses.

Lead Glasses. Lead oxide is usually a modifier although at times it may act as a network former. Lead glasses may be easily melted and have a long working range and a high refractive index which makes them useful for lead crystal, optical glass, and hand-formed art ware. Lead-containing glasses effectively shield high-energy radiation and are therefore used commercially for radiation windows, fluorescent-lamp envelopes, and television bulbs. Low-melting solder glasses and frit or decorative enamels are usually based upon low-melting lead compositions (see Enamels).

Borate Glasses. Borates, including vitreous B_2O_3, have been studied more than any other glass-forming system with the exception of silicates (38). Vitreous boric oxide has a three-coordinate structure consisting of six-membered rings of alternating boron and oxygen atoms. Many physical properties of alkali borate glasses show a minimum or maximum at 15–30 mol % modifier (boron anomaly). Coordination changes of boron are detected by nmr, ir, Raman, and esr techniques (see Analytical methods). Broad quadripolar coupling typical of triangular boron coordination is readily distinguished from the sharp coupling of four-coordinate boron (29). The fraction of tetrahedral borons present appears directly proportional to the alkali-to-boron ratio as long as this ratio is less than 0.5 (38). The very low durability of borate glasses precludes their use in all except the most special applications. Low molecular weight Lindemann glasses ($Li_2O.BeO.B_2O_3$) were developed as x-ray transmitting glasses. Rare-earth borate glasses have optical uses because of their high refractive indexes and low dispersion. Additions of Gd_2O_3 to the latter increase the index but not the dispersion (39).

Phosphate Glasses. The structure of phosphate glass appears to be based on the phosphorous–oxygen tetrahedron (see Phosphoric acid and phosphates). Like the borates, they tend to have low durability. Important commercial applications of phosphate glasses do exist, however. Because the absorption bands of iron oxide in phosphate glasses are sharper in the uv and ir than in silicate glasses, iron-containing phosphate glasses are nearly transparent to visible light. Almost clear heat-absorbing glasses with several percent iron oxide are possible. Phosphate-based glasses also are more resistant to fluoride than silicate glasses. Some of the optical glasses produced by Schott, Hoya, Owens-Illinois, and Corning-Sovirel use phosphate as the primary glass former. Fluorophosphate glasses, designated FK-5 or FK-50 by Schott have very low optical dispersion with Abbe-numbers of 70.4 and 81.5, respectively.

Other Oxide Glasses. Germanium, arsenic, and antimony oxides all form stable glasses and their structures have been predicted. The germania glass structure is quite similar to silica. Infrared transmission of germania glasses is higher than that of silica. Tellurium-containing lead glasses with a very high refractive index (>2.0) are also used commercially.

Chalcogenide Glasses. Glasses based upon sulfur, selenium, or tellurium rather than oxygen are well known. These glasses, although often opaque to visible light, transmit ir radiation of a much longer wavelength than oxide systems, and many are also semiconductors (qv). Conductivity usually increases with increasing atomic number. The most-studied chalcogenide glasses contain the Group V elements, arsenic and antimony.

Halide Glasses. Although examples of zinc chloride glasses are known, BeF_2-containing glasses are more common (31). Vitreous beryllium fluoride has a tetrahedral structure analogous to silica. Its unique spectral properties including transmission from <160 nm to 5500 nm, low refractive index, and very low dispersion have aroused considerable interest in beryllium fluoride glasses. Vitreous BeF_2 is very hygroscopic but addition of other fluorides, such as AlF_3, or alkali and alkaline earth fluorides increases durability. Selected compositions also resist devitrification and have optical properties that are not obtainable with oxide glasses. A refractive index of 1.3 with a dispersion of 100 is not uncommon for beryllium fluoride glasses. The toxicity of beryllium is, however, a concern associated with melting and forming operations.

Metallic Glasses. Under highly specialized conditions, the crystalline structure of some metals and alloys can be suppressed and they form glasses. These amorphous metals can be made from transition metal alloys, eg, nickel–zirconium, or transition or noble metals in combination with metalloid elements, eg, alloys of palladium and silicon or alloys of iron, phosphorus, and carbon (see Amorphous magnetic materials).

Vacuum evaporation or sputtering techniques can produce thin films of amorphous metal elements as well as a wide variety of amorphous alloys (40). Liquid quenching at rates greater than 10^5 K/s is limited to alloys containing metalloids (41). Although glass-forming ability cannot be predicted with certainty, a low temperature eutectic in a system of high melting metals often forms a metallic glass (42–43) (see also Glassy metals).

Metallic glasses are produced commercially as ribbons or fibers 50–100 μm thick and up to 25 mm wide. The first success in this area was the Au_4Si alloy quenched in liquid nitrogen. Results since 1959 have shown that rapidly quenched metals are indeed glasses with unique properties. Certain glassy metals exhibit soft magnetic anisotropy that may be controlled. Strengths of metallic glasses are significantly higher than those of the crystalline forms but thermal instability often causes crystallization at relatively low temperatures.

Because of their noncrystalline structure, amorphous metal alloys lack grain boundaries that tend to weaken conventional alloys or make them vulnerable to chemical attack. Allied Chemical Co. produces iron-containing glasses that may be useful as rustproof steel coatings.

Structural methods such as x-ray diffraction (44), electron microscopy (45), Mossbauer resonance, nmr (46), and thermal analysis have been used to study the structures of glassy metals. Heat capacity data first conclusively demonstrated that the metals were indeed vitreous and not amorphous with microcrystallization. As fabrication techniques develop further, metallic glasses promise to be an important area of glass science.

Glass-Ceramics and Phase-Separated Glasses

Glasses are metastable and, under the appropriate conditions, revert to a thermodynamically more stable state. Glasses particularly susceptible to uncontrolled crystal growth or phase separation have traditionally created problems for their manufacturer. However, glass is also a good medium for controlled crystallization (47), and has become the basis for a number of unique crystalline materials known as glass-ceramics (see Glass-ceramics). The separation of a single glass into multiple glassy phases can make the article cloudy and adversely affect its chemical durability. Controlled phase separation, however, can produce opaque, white opal glass or, after a leaching step, lead to new materials such as porous glass or 96% silica glass. The colloidal suspension of multiple phases in transparent glass produces precise colors for products such as optical filters (qv). Furthermore, photosensitive and photochromic glasses change their optical transmission and color, sometimes reversibly, upon stimulus by a combination of light and heat treatment (see Chromogenic materials, photochromic). All these transformations generally depend on the phenomena of diffusion, nucleation, and growth.

The thermodynamic driving force for phase separation depends upon the amount of undercooling from some appropriate equilibrium temperature such as the liquidus temperature (path A, Fig. 1). That temperature is more correctly defined as the point where the free energy of formation of the new substance(s) changes from positive to negative as the glass cools. The free energy of formation depends upon the standard free energies of all the products and reactants involved as well as upon each of their concentrations, or more correctly, upon their activities. Therefore, as the glass cools below the equilibrium temperature, eg, the liquidus, the free energy of formation becomes more negative and the driving force for phase separation becomes greater. However, as glasses cool they become more viscous, which further slows the diffusion process. Diffusion is necessary for sufficiently concentrating the elements to allow phase separation. Thus upon cooling, the rate of reaction increases with the thermodynamic driving force until it reaches a maximum at some optimum temperature. Then it slows down as the diffusion rate becomes the dominant controlling factor. Both the nucleation and growth rates are affected by this balance between driving force and diffusion rate. However, because they are different phenomena, their optimum temperatures for maximum rate may or may not coincide. In general, the optimum temperature for maximum nucleation rate is lower than that for maximum growth rate. Because critical nucleus size is determined by a balance between volume free energy (pronucleation) and surface free energy (antinucleation), nucleation is more sensitive to driving force and less sensitive to diffusion rate than is growth. Consequently, just below the equilibrium temperature only a few nuclei form, but the growth rate is rapid, and the glass has a few relatively large islands of a second phase.

At lower temperatures, the morphology is different. More nuclei form but do not grow as fast, and the other extreme occurs where the glass has many fine islands of the second phase. A single glass composition can have several phases competing for the elements that it needs to nucleate and grow, and one phase can dominate the others through the choice of a heat treatment which maximizes the nucleation and growth rates of the desired phase at the expense of the unwanted phases. Furthermore, the size and amount of a second phase can be controlled by balancing the heat treatment for nucleation with the heat treatment for growth. Finally, the nature of the second phase can be affected by nucleating agents which are either added to the glass or encouraged to grow in the glass. Nucleating agents affect either the thermodynamics of the components or the surface energy of the nuclei or both.

Glass-Ceramics. Glass-ceramics are melted and formed by conventional glass manufacturing techniques, and then given a subsequent heat treatment (ceram) to transform them into fine-grained crystalline materials. By definition, glass-ceramics are more than 50% crystalline after heat treatment; frequently, the final product is more than 95% crystalline. The operation is generally accompanied by an increase in viscosity which increases the product's use temperature. For example, both Corning Glass Works and Owens-Illinois manufacture a $PbO–B_2O_3–ZnO$ glass-ceramic frit for joining color television bulb parts. The glassy frit flows, seals, and crystallizes and, therefore, becomes quite rigid between 380 and 440°C. The bulbs can be further processed at temperatures approaching 400°C without the parts shifting. In most commercial applications the thermal expansion of the glass-ceramic is lower than that of the unceramed parent glass which increases its dimensional stability and thermal shock resistance. Telescope mirror blanks, missile radomes, and cooking utensils are examples. The intrinsic strength of glass-ceramics is less than that of glass, but if the crystal grains are of the optimum size and shape they inhibit crack propagation, and thus the actual strength is greater than that of glass. In Corning's machinable glass-ceramic material, mica crystals occupy 55% of the volume in the glassy matrix. During machining, the interlocking intersections of the randomly oriented crystals stop the microscopic fractures that, in mass, make up the actual cut. Cracks do not propagate throughout the matrix because they are repeatedly deflected, branched, blunted, and finally arrested. This property also keeps the product strong during service regardless of whatever surface flaws might occur. Glass-ceramics can be further strengthened by application of a lower-expansion glaze, by ion exchange, or by differential crystallization.

Aluminosilicate glass-ceramics are the most useful commercial products. The presence of alkali or alkaline earth oxides affects the type of crystal structure. The morphology is regulated by a nucleating agent, usually TiO_2 or $TiO_2 + ZrO_2$, and a heat treatment. Many other components are added that further optimize the crystalline phases and the glass-ceramic properties. Lithium aluminosilicates (β-spodumene solid solution, $Li_2O.Al_2O_3.(SiO_2)_{4–10}$) have low expansion and good chemical durability. In fact, the low thermal expansion behavior over this wide range of silica concentration allows considerable flexibility in the base composition of β-spodumene glass-ceramics. As a result, such materials can be tailored to meet other property and glass-forming requirements (48). Sodium aluminosilicates (eg, nepheline, $Na_2O.Al_2O_3.2SiO_2$) or barium–sodium aluminosilicates (nepheline; celsian, $BaO.Al_2O_3.2SiO_2$) have higher expansion and can be strengthened by surface compression techniques such as the application of a low-expansion glaze. Magnesium aluminosilicates (eg, cordierite, $2MgO.2Al_2O_3.5SiO_2$, and Mg-stuffed β-quartz, $MgO.Al_2O_3.(SiO_2)_{2–5}$) have low expansion and can have very high strength. Fluorine added to certain potassium–magnesium aluminosilicates (fluorphlogopite, $KMg_3AlSi_3O_{10}F_2$) increases machinability. Many of these crystal phases are solid solutions, not stoichiometric compounds. Transparent glass-ceramics can be made with β-quartz solid solution as the crystal phase with compositions ranging widely in the system $SiO_2.LiAlO_2.Mg(AlO_2)_2.Zn(AlO_2)_2$. Fibers made into building tile, roofing, drain tile, and insulation are made from basalt glasses and glass-ceramics (49). Basalt is a volcanic rock composed chiefly of plagioclase [($Na_2O.Al_2O_3.6SiO_2$).($CaO.Al_2O_3.2SiO_2$)] solid solution, augite [(Ca, Na) (Mg, Fe^{2+}, Fe^{3+}, Al)(Si, Al)$_2$O$_6$] and magnetite (Fe_3O_4). Table 2 further documents some commercial glass-ceramics.

The formation of a glass-ceramic is extremely complex, but generally follows a 4-step sequence (47–48,50–51). (*1*) A dispersed amorphous phase, structurally incompatible with the host glass, usually highly unstable, and enriched in one or two key oxides (eg, TiO_2 or ZrO_2) forms on either cooling or reheating the glass. (*2*) Primary crystalline nuclei, which are often titanates or zirconates, form either heterogeneously at the phase boundaries or homogeneously within the second phase. (*3*) A metastable crystalline phase heterogeneously nucleates on the primary crystallites and grows, generally at the expense of the glassy second phase. This produces a typically fine-grained, metastable, solid-solution material. (*4*) The metastable material breaks down to the final, stable, fine-grained crystalline structure by means of isochemical phase transformations, reaction between metastable phases, or exsolution.

Devitrification. Devitrification is the uncontrolled formation of crystals in glass during melting, forming, or secondary processing. It can adversely affect the optical properties, mechanical strength, and sometimes the chemical durability of the glass (see under Quality Control). These unwanted crystals nucleate and grow homogeneously within the glass or heterogeneously at the air–glass or refractory–glass interface. Devitrification is most likely to occur in glasses in which the optimum temperatures for maximum nucleation rate and for maximum growth rate nearly coincide. If these glasses are held too long in this critical temperature range or are cooled too slowly through it, the glass starts to crystallize. The critical temperature range is below the liquidus temperature but above the temperature where the glass is sufficiently viscous to retard devitrification (10^5–10^6 Pa·s or 10^6–10^7 P). In soda-lime glasses, for example, the crystal phase known as devitrite ($Na_2O.3CaO.6SiO_2$) forms between 850°C–900°C (7.5–25 kPa·s or (7.5–25) $\times 10^4$ P). However, since the glass is already quite viscous, the critical temperature range is short (ca 50°C), and devitrification is not much of a problem. Some optical glasses have liquidus temperatures as high as 1100°C (20–30 Pa·s or 200–300 P) and, therefore, must be cooled very quickly for several hundred degrees to avoid devitrification. Some forming operations require viscous glass (drawing) and, therefore, require glasses with low liquidus temperatures. Other operations (pressing of small articles) can tolerate glasses with much higher liquidus temperatures.

Phase-Separated Glass. Glasses that derive their color, optical transparency, or chemical durability from a small amount of a finely dispersed second phase are termed phase-separated glass. They are distinguished from glass-ceramics by virtue of their predominantly glassy character. Therefore, the control of the nucleation and morphology of the second phase is not so critical. Opal glasses are either translucent or opaque, depending on the particle size and quantity of the second phase and on the refractive index difference between the second phase and the glass matrix. Incident light is refracted and scattered by the dispersed phase and the glass appears opaque. Fluorine-containing glasses, which separate into NaF- or CaF_2-containing phases, for instance, are commonly made into pressed items such as tableware. As the glass cools from the molten state, the fluoride phases become immiscible in the glass and separate as fine droplets (<0.5 μm). Generally, these droplets comprise only 5–7% of the volume of the opal glass. On further cooling, the fluoride may or may not crystallize. These glasses generally behave as spontaneous opals, opacifying or striking-in while

Table 2. Commercial Glass-Ceramics [a]

Commercial designation	Major crystalline phases	Surface strengthening processes	Properties	Application
Schott Zerodur O-I Cervit	β-quartz solid solution, SiO_2		zero expansion	electric range tops, telescope mirrors
Corning 9632	β-quartz [*14808-60-7*]	differential crystallization	low expansion, high strength, thermal stability, chemical durability	electric range tops (transparent)
Corning 8603	lithium metasilicate [*10102-24-6*], $Li_2O.SiO_2$; lithium disilicate [*13568-46-2*], $Li_2O.2SiO_2$		photochemically machinable	fluid amplifiers
Corning 9608	β-spodumene solid solution, $Li_2O.Al_2O_3.(SiO_2)_{4-10}$		low expansion, high chemical durability	cooking utensils
Corning 9617	β-spodumene solid solution; anatase [*1317-70-0*], TiO_2	differential crystallization	low expansion, high strength, thermal stability, chemical durability	electric range tops (opaque)
Corning 0330	β-spodumene solid solution; rutile [*1317-80-2*], TiO_2	ion exchange $Na^+ \rightleftharpoons Li^+$	mechanical, chemical, thermal stability	exterior, interior cladding, laboratory bench tops
Corning 9455	β-spodumene solid solution; mullite [*1302-93-8*], $3Al_2O_3.2SiO_2$		low expansion, high thermal and mechanical stability	heat exchangers
Neoceram (Japan)	β-spodumene [*1302-37-0*]		low expansion	cooking ware
Corning 0303	nepheline [*12251-37-3*], $Na_2O.Al_2O_3.2SiO_2$; celsian [*1302-50-7*], $BaO.Al_2O_3.2SiO_2$; anatase, TiO_2	low expansion glaze	high strength, bright white	tableware
Corning 9606	cordierite [*1302-38-2*], $2MgO.2Al_2O_3.5SiO_2$; spinel [*1302-67-6*], $MgO.Al_2O_3$; MgO-stuffed β-quartz; quartz [*14808-60-7*], SiO_2		low expansion, high transparency to radar	missile radomes

Table 2 (*continued*)

Commercial designation	Major crystalline phases	Surface strengthening processes	Properties	Application
Corning 9625	α-quartz solid solution (SiO_2); spinel ($MgO.Al_2O_3$); enstatite [*14681-78-8*], $MgO.SiO_2$	ion exchange $2\ Li^+ \rightleftharpoons Mg^{2+}$	very high strength	classified
Corning 9658	fluorphlogopite [*12003-38-2*] solid solution, $KMg_3AlSi_3O_{10}F_2$; mullite, $3Al_2O_3.2SiO_2$;		machinable, high dielectric strength, thermal stability	precision dielectric components, insulators, high vacuum components
High K (Corning)	(Ba,Sr,Pb) Nb_2O_6		high dielectric constant	capacitors

[a] Ref. 48.

they are being formed. Reheat or restrike opals are clear glass after forming, but like glass-ceramics, must be reheated for phase separation. These opals are not as economical to produce as the spontaneous opals, but their production is more controllable. Bubble opals can be formed with many tiny pockets of gas which make them opaque. At present, they have no commercial application.

The development of colloidal colors depends on controlled nucleation and growth of absorbing particles in glass. Colored filter glasses, eg, Corning Code 2403 sharp-cut red filter glass, can be made by precipitating colored crystals such as CdS or CdSe. Colloidal gold, copper, and silver are most often used. The particles have a selective absorption and a complementary selective reflection which depend on particle size. The metal to be precipitated is dissolved in the glass in conjunction with a reducing agent such as tin oxide or antimony oxide. After melting, the glass is rapidly cooled to a temperature region where the nucleation rate is high and then heated to a temperature where crystal growth dominates. By this technique, the size of the crystals, and therefore the color, can be controlled. For leachable glasses, see Secondary Operations under Manufacture. Porous glass and reconstructed 96% silica glass are made with this processing technique. Porous glass is used in the biological sciences as a support for immobilized enzymes and antibodies (see Enzymes, immobilized).

Light-Sensitive Glasses. Photosensitive glasses (52) employ structural transformations, ie, precipitation and crystallization, which are sensitized by exposure to electromagnetic radiation, usually ultraviolet, but sometimes visible or x-ray radiation. The transformation is generally initiated by the photoreduction of a metal ion. Subsequent or simultaneous heat treatment morphologically transforms the glass in the regions exposed to the radiation leading, in general, to a change in optical or physical properties (see Chromogenic materials, photochromic).

The simplest photosensitive glasses are those colloidally colored with gold, silver, or copper. A sequence of exposure and heat treatments leads to the development of a controlled spatial pattern of color in the glass (53), eg, Corning Code 8600. In some photosensitive glasses, this photosensitized precipitation of metal colloids can act as nucleation sites for subsequent phase changes in the glass. For example, an opalizing phase such as sodium fluoride may be so precipitated in a controlled manner to give an opal pattern within the glass. Corning Code 8601 is such a photosensitive opal which has found commercial and architectural uses, for example, in decorative windows in the U.N. Assembly Building. Chemical machining of glass is made possible by the heterogeneous precipitation of Li_2SiO_3 from lithium silicate-based glasses. In this system, the crystallized material is much more soluble in 10% hydrochloric acid than the residual glass. If, after dissolution of Li_2SiO_3, the glass is again exposed to ultraviolet light or x-rays, subsequent heat treatment can produce a hard, nonporous, strong, high temperature glass-ceramic such as Corning Code 8603. Except for less than 1% shrinkage this can be done without distortion. The final material is almost completely crystalline, and is superior to the original glass in strength, hardness, thermal shock resistance, and electrical resistivity. It is used for printed circuit boards for high temperatures (400–500°C) and high humidities, for laminated dielectric structures, fluid amplifiers, etc.

Corning has recently introduced polychromatic glass, which is a special photosensitive material. A typical polychromatic glass before processing has properties similar to those of window glass. However, it has the potential, when activated by ultraviolet light and heat, to produce within the same glass all hues and a wide range of transparent colors or combination of colors, any desired pattern of white or colored opacity, to employ either two- or three-dimensional geometry in the patterns, and to produce permanent full-color photographic images. The mechanism (54) of this variable, controlled coloration involves a photosensitized formation of metallic silver on a sodium halide microcrystalline phase, which itself had been precipitated by a photosensitive process from the glass.

Because absorption and scattering characteristics are due to silver particles and inorganic crystals in a stable glass matrix, they are as permanent as the glass. The only known way to destroy the colors without destroying the glass is to heat the material above 400 or 450°C, where atomic diffusion can occur and the colors deteriorate to brown or yellow. The capability of reproducing color or opacity in three dimensions within a solid transparent medium may lead to new applications in photographic contour mapping, modeling, and sculpture, including holography (qv).

Photochromic glasses are those whose transmission of visible light decreases when the glass is exposed to ultraviolet or visible light and increases when the exposure is stopped. Photochromic glasses are commercially available as ophthalmic lens blanks, manufactured by Corning Glass Works, Chance-Pilkington, Schott Optical Glass, and others. Corning also manufactures sagged photochromic lens blanks for nonprescription sunglasses. These blanks are cut from a continuous sheet made by the overflow process (see below). Photochromic sheet has found use in architectural applications and automobile sunroofs.

The commercially available photochromic glasses contain a fine dispersion of silver halide crystallites which are about 10 nm in diameter and about 100 nm apart. These crystallites are precipitated thermally without exposure to radiation. Hence, they are not photosensitive as defined earlier. Darkening can be likened to the latent image formation step in silver halide photography. The light causes the formation of very small amounts of silver metal within or at the surface of the silver halide crystallite. Because the glass traps the reaction products close to each other, the reaction reverses and the silver is reabsorbed into the silver halide when the irradiation ceases.

Figures 6 and 7 show the percent luminous transmittance of Corning Code 8111 ophthalmic lens glass under various conditions.

Experimental photochromic glasses have shown a wide variety of darkening and clearing rates. Some glasses clear in seconds, whereas others are essentially stable in their darkened state at room temperature. The fading process is temperature dependent; hence, glasses tend to darken less and fade faster at higher temperatures (darken more and fade slower at low temperatures). The temperature dependence of the darkened transmittance is generally greater for the faster-fading glasses. All commercial glasses are designed for particular applications and involve balances between maximum available darkening and fade rate and temperature dependence.

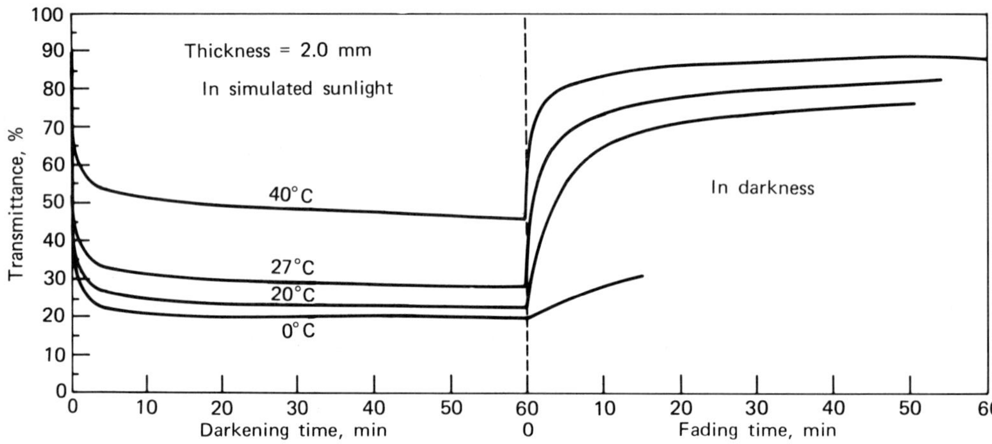

Figure 6. Darkening and fading rate curves for Corning Code 8111. Courtesy of Corning Glass Works.

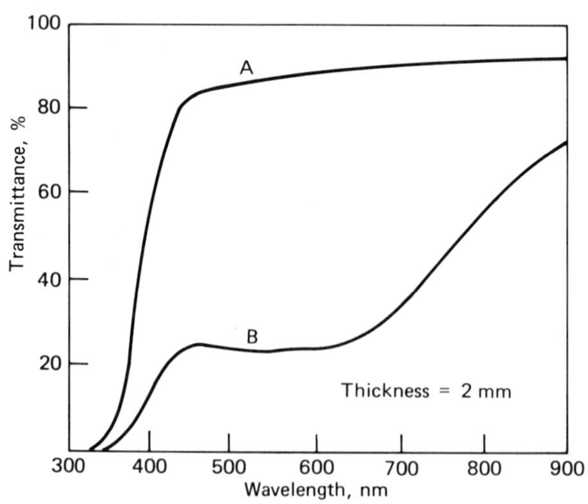

Figure 7. Spectral transmittance of Corning Code 8111. A, faded in dark overnight. B, after one hour in sunlight at 25°C. Courtesy of Corning Glass Works.

Properties

Rheological. The viscosity of a glass determines its melting, forming, and annealing procedures as well as the limitations of its use at high temperature. Viscosity is ordinarily measured between 10^{13} and 10 Pa·s (10^{14} and 100 P); at room temperature it is greater than 10^{19} Pa·s (10^{20} P). The rapid but smooth change of viscosity with temperature is shown for several glasses in Figure 8. Compositions and properties are shown in Tables 3 and 4. The effect of modifiers on the viscosity of glass at high temperatures depends upon their polarizability or ionic field strength (55). Low field-strength modifiers decrease the viscosity of silica more than high field-strength ones. At low temperature, the effects of a modifier upon viscosity are largely controlled by its coordination number. Modifiers with higher coordination numbers tend to increase low temperature viscosity as a result of packing restraints (see Rheological measurements).

Reference points on the viscosity–temperature curve in Figure 8 have been chosen to characterize properties of an individual glass and to facilitate comparisons of similar glasses. Those used most frequently are the working, softening, annealing, and strain points. Definitions of these selected points are given by ASTM. The working point of a glass is the temperature at which its viscosity is exactly 1 kPa·s (10^4 P). At this viscosity, glass is sufficiently fluid for forming by most common methods. Viscosity between 10^6–1 Pa·s (10^7–10 P) is measured with a Margules viscometer, a calibrated instrument which measures the force exerted by molten glass upon a rotating spindle. The softening point of a glass cannot be defined by a precise viscosity. However, ASTM C 338 defines the softening point as the temperature at which a uniform fiber of glass

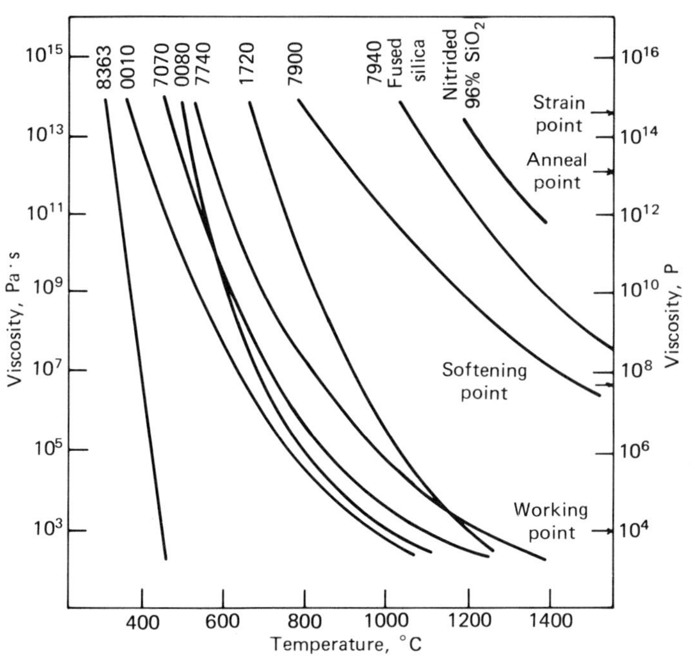

Figure 8. Viscosity vs temperature for some commercial glasses. Courtesy of Corning Glass Works.

Table 3. Approximate Compositions of Commercial Silicate Glasses, wt %

Use	Glass Code	SiO_2	Al_2O_3	B_2O_3	Li_2O	Na_2O	K_2O	MgO	CaO	BaO	PbO	ZnO	Others
lamp tubing	0010	63	1			8	6				22		
electronic and lamp tubing	0070	71	3			13	1	5	7				
lamp bulbs	0080	73	1			17		4	5				
electronic and lamp bulbs	0091	73.5	1			15	0.5	4	6				
lamp tubing	0120	56	2			4	9				29		
color TV neck	0137	52.5	1			0.5	13				28		5% SrO
color TV funnel	0138	54	2			6	8	2	3		23		
microsheet	0211	65	2	9		7	7					7	3% TiO_2
bench tops	0330	65	20	2	4			2				2	1% As_2O_3, 4% TiO_2
ignition tubes	1720	62	17	5		1		7	8				
electron tubes	1723	57	16	4				7	10	6			
iron sealing	1990	41			2	5	12				40		
optical filter	2405	70	1	12		5						11	CdS, Se
lamp bulbs	2473	67	1			16						12	1.5% CdO, SnO, Se, S^{2-}
tungsten sealing	3320	76	2	15		4	2						1% U_3O_8
tableware	6720	60	10	1		8	2		5			10	4% F^-
lighting ware	6750	61	11			15				9			3% F^-
Kovar sealing	7040	67	3	23		4	3						
series sealing	7050	68	2	24		6							
Kovar sealing	7052	64	8	19	1	2	3			3			F^-
Kovar sealing	7056	68	3	18	1	1	9						
substrate	7059	49	10	15						25			1% As_2O_3
low loss electrical	7070	72	1	25	0.5	0.5	1						
sealed beam lamps	7251	82	2	12		4							
laboratory ware	7280	71			1	12							16% ZrO_2

solder sealing	7570	3	11	12							74		
tungsten sealing	7720	74	1	15		4					6		
general	7740	81	2	13		4							
general	7760	78	2	15		3	1						1% As_2O_3
pharmaceutical	7800	72	6	11		7	1		1	2			
high temperature	7913	96.5	0.5	3									
optical	7940	99.9											0.1% H_2O
optical	7971	93											7% TiO_2
electron tubes	8160	56	2			3	10		1	5	23		0.7% F^-
electron tubes	8161	39					6			2	51		2% Rb_2O
Kovar sealing	8830	65	5	23		7							
CRT panel	9025	68	4	12	3	5	6						1% TiO_2, 1% CeO_2
color TV panel	9068	65	2			7	9	1	2	2	2		10% SrO, F^-, TiO_2, CeO_2, As_2O_3, Sb_2O_3
missile nose cones	9606	56	20					15					9% TiO_2
cooking ware	9608	70	18		3			3				1	5% TiO_2
smooth cooktop	9617	74	20		3			2				1	5% TiO_2, F^-
machinable glass-ceramic	9658	45	17	8			10	14					6% F^-
uv-transmitting	9741	65	5	27	1	2							F^-
optical	N16B	5		34					2			6	6% ZrO_2, 41% La_2O_3, 6% CdO
laboratory ware	G20	76	5	7		6	1		1	4			
laboratory ware	N-51a	74.5	6	10		6	0.5		1	2			
laboratory ware	KG33	81	2	13		4							
reinforcement fiber	E-Glass	54	14	10				4.5	17.5				
insulation fiber	T-Glass	59	4.5	3.5		11	0.5	5.5	16				
chemical-resistant fiber	C-Glass	65	4	5.5		8	0.5	3	14				
insulation fiber	SF-Glass	59.5	5	7		14.5							4% ZrO_2, 8% TiO_2, F^-
high strength	S-Glass	65	25					10					

Table 4. Properties of Glass

Type	Glass code[a]	Color	Forms usually available[b]	Corrosion resistance: Class	Corrosion resistance: Weathering	Corrosion resistance: Water	Corrosion resistance: Acid	Thermal expansion × 10^{-7}/°C[e]: 0–300°C	Thermal expansion × 10^{-7}/°C[e]: 25°C to setting point	Upper working temperatures[g] (Mechanical considerations only), Annealed: Normal service °C	Annealed: Extreme service °C	Tempered: Normal service °C	Tempered: Extreme service °C	Thermal shock resistance[h] plates 15 × 15 cm, Annealed: 3.2 mm thick, °C	6.4 mm thick, °C	12.7 mm thick, °C
potash soda lead	0010	clear	T	I	2	2	2	93.5	99.8	110	380			65	50	35
soda lime	0070	clear	BT	I	2		2	91.0	100							
soda lime	0080	clear	BMT	I	3	2	2	93.5	105	110	460	220	250	65	50	35
soda lime	0091	clear	B					91.0	100							
potash soda lead	0120	clear	TM	I	2	2	2	89.5	97	110	380			65	50	35
alkali–alkaline earth-lead	0137	clear	T					97.0	102							
potash soda lead	0138	clear	P	II	1		3	98.5	109							
potash soda zinc	0211	clear	S	I	1		2	74.0	84.0							
glass-ceramic	0330	gray	RS	I		1	3	9.7		538						
alumino-silicate	1720	clear	BT	I	1	1	3	42	52	200	650	400	450	135	115	75
alumino-silicate	1723	clear	BT	I	1	1	3	46	54	200	650	400	450	125	100	70
potash soda lead	1990	clear		II	3	3	4	124	136	100	310			45	35	25
borosilicate	2405	red	BPU	I				43	53	200	480			135	115	75
soda zinc	2473	red	B	I	2	2	2	91		110	460			65	50	35
borosilicate	3320	canary		I	1[c]	1[c]	2[c]	40	43	200	480			145	110	80
soda zinc	6720	opal	P	I	[d]	1	2	78.5	90	110	480	220	275	70	60	40
soda barium	6750	opal	BPR	I	[d]	2	2	88		110	420	220	220	65	50	35
borosilicate	7040	clear	BT	II	3[c]	3[c]	4[c]	47.5	54	200	430					
borosilicate	7050	clear	T	II	3[c]	3[c]	4[c]	46	51	200	440	235	235	125	100	70
borosilicate	7052	clear	BMPT	II	2[c]	2[c]	4[c]	46	53	200	420	210	210	125	100	70
borosilicate	7056	clear	BTP	II	2	2	2	51.5	56	200	460					
barium–barium-silicate	7059	clear	S	II	1		4	46.0	50.5							
borosilicate	7070	clear	BMPT	I	2[c]	2[c]	2[c]	32	39	230	430	230	230	180	150	100
borosilicate	7251	clear	P	I	1[c]	2[c]	2[c]	36.5	41.4	230	460	260	260	160	130	90
soda–zir-conia	7280	clear	BT	I	1		2	65.0								
high lead	7570	clear		II	3	3	4	84	92	100	300					
borosilicate	7720	clear	BPT	I	1	2[c]	2[c]	36	43	230	460	260	260	160	130	90

borosilicate	7740	clear	BPSTU	I	1[c]	1[c]	1[c]	32.5	35	230	490	260	290	160	130	90
borosilicate	7760	clear	BP	I	2	2	2	34	37	230	450	250	250	160	130	90
soda barium borosilicate	7800	clear	T		1	1	1	50	53	200	460					
96% silica	7913	clear	BPRST	I	1	1	1	7.5	5.5[f]	900	1200					
fused silica	7940	clear	U	I	1	1	1	5.6	3.5[f]	900	1100					
titanium silicate	7971	clear	U		1	1	1	0.5	−2	800	1100					
potash soda lead	8160	clear	PT	II	2	2	3	91	100	100	380			65	50	35
potash lead	8161	clear	PT	I	2	1	4	90	99	100	390					
soda–borosilicate	8830	clear	PT	II	2		4	49.5								
alkali–barium	9025	clear	P	II	3		2	90.0	98.0							
alkali–strontium	9068	clear	P	I	2		2	99.0	111							
glass-ceramic	9606	white	C	II		1	4	57		700				200	170	130
glass-ceramic	9608	white	BP	I		1	2	4–20		700	800					
lithium–aluminosilicate	9617	white	R	I	1		1	9.0								
fluorphlogophite	9658	white	PRT	II	1		4	89.0								
borosilicate	9741	clear	BUT	II	3[c]	3[c]	4[c]	39.5	50	200	390			150	120	80
	N16B	clear						87								
aluminoborosilicate	G20	clear	BPT													
aluminoborosilicate	N51a	clear	T					50						115	95	65
borosilicate	KG33	clear	BPSTU					32						180	150	100
lime–aluminoborosilicate	E-glass	clear	F					60								
soda lime–aluminoborosilicate	T-glass	clear	F					80								
soda lime–borosilicate	C-glass	clear	F					72								
soda–titania–zirconia	SF-glass	clear	F					75								
magnesium–aluminosilicate	S-glass	clear	F					34								

Table 4 (*continued*)

Type	Thermal stress resistance, °C^{i}	Viscosity data^{j}: Strain point, °C	Viscosity data^{j}: Annealing point, °C	Viscosity data^{j}: Softening point °C	Viscosity data^{j}: Working point °C	Knoop hardness HK_{100}^{k}	Density, g/cm^3	Young's modulus, GPa	Poisson's ratio	Log_{10} of volume resistivity, Ω·cm: 25°C	Log_{10} of volume resistivity, Ω·cm: 250°C	Log_{10} of volume resistivity, Ω·cm: 350°C	Dielectric properties at 1 MHz, 20°C: Power factor, %	Dielectric properties at 1 MHz, 20°C: Dielectric constant	Dielectric properties at 1 MHz, 20°C: Loss factor, %	Refractive index^{m}
potash soda lead	19	395	435	628	985	360	2.79	62	0.21	>17	8.9	7.0	0.16	6.7	1	1.540
soda lime		487	527	715			2.50									1.513
soda lime	16	473	514	696	1005	465	2.47	70	0.22	12.4	6.4	5.1	0.9	7.2	6.5	1.512
soda lime		485	523	705			2.48									
potash soda lead	20	395	435	630	985	382	3.05	59	0.22	>17	10.1	8.0	0.12	6.7	0.8	1.560
alkali-alkaline earth-lead		436	478	661	977		3.18				10.1	8.4				1.570
potash soda lead		450	490	670			3.02					7.7				1.563
potash soda zinc	20	508	550	720	1008		2.57	74	0.21							1.52
glass-ceramic	178					522	2.54	86	0.26							
alumino-silicate	28	667	712	915	1202	513	2.52	87	0.24	>17	11.4	9.5	0.38	7.2	2.7	1.530
alumino-silicate	26	665	710	908	1168	514	2.64	86	0.24	>17	13.5	11.3	0.16	6.3	1.0	1.547
potash soda lead	14	340	370	500	756		3.50	58	0.25	>17	10.1	7.7	0.04	8.3	0.33	
borosilicate	37	501	537	765	1083		2.48	68	0.21							1.507
soda zinc	19	466	509	697			2.65	66	0.22							1.52
borosilicate	43	493	540	780	1171		2.27	65	0.19		8.6	7.1	0.30	4.9	1.5	1.481
soda zinc	20	505	540	780	1023		2.58	70	0.21							1.507
soda barium	18	447	485	676	1040		2.59									1.513
borosilicate	37	449	490	702	1080		2.24	59	0.23		9.6	7.8	0.20	4.8	1.0	1.480
borosilicate	39	461	501	703	1027		2.24	60	0.22	16	8.4	6.8	0.33	4.9	1.6	1.479
borosilicate	41	436	480	712	1128	375	2.27	57	0.22	17	9.2	7.4	0.15	5.1	1.3	1.484
borosilicate	33	472	512	718	1058		2.29	64	0.21		10.2	8.3	0.27	5.7	1.5	1.487
barium–barium-silicate	32	593	639	844	1160		2.76	68	0.28		13.1	11.0		5.9		1.53
borosilicate	66	456	496		1068		2.13	51	0.22	>17	11.2	9.1	0.06	4.1	0.25	1.469
borosilicate	48	521	565	808	1192	451	2.26	64	0.19	18	8.0	6.5	0.45	4.9	2.18	1.476
soda–zirconia	20	576	624	873	1234		2.62	83	0.22		6.0	4.8		6.12	0.07	1.54
high lead	21	342	363	440	558		5.42	55	0.28	>17	10.6	8.7	0.22	15	3.3	1.86
borosilicate	49	484	523	755	1146		2.35	63	0.20	16	8.9	7.3	0.23	4.6	1.3	1.487
borosilicate	54	510	560	821	1252	418	2.23	63	0.20	15	8.1	6.6	0.39	4.6	2.6	1.474
borosilicate	52	478	523	780	1198	442	2.24	62	0.20	17	9.4	7.7	0.16	4.6	0.79	1.473
soda barium borosilicate	33	533	576	795	1189		2.36				7.0	5.7				1.491
96% silica	220	890	1020	1530		487	2.18	68	0.19	>17	9.7	8.1	0.04	3.8	0.15	1.458
fused silica	286	956	1084	1580		489	2.20	73	0.16	>17	12.4	10.7	0.001	3.8	0.0038	1.459
titanium silicate	3370		1000	1500			2.21	68	0.17	20.3	12.2	10.1	<0.002^{l}	4.0^{l}	<0.008^{l}	1.484

potash soda lead	[illegible]	[illegible]	[illegible]	[illegible]	[illegible]		[illegible]			[illegible]	[illegible]	[illegible]	[illegible]	[illegible]	[illegible]	[illegible]
potash lead	22	400	435	600	862		3.99	54	0.24	>17	12.0	9.9	0.06	8.3	0.50	1.659
soda–borosilicate	39	460	501	708	1042		2.24	56	0.22		7.8	6.3				
alkali–barium		421	460	647	988		2.60				8.5	6.8		6.5		
alkali–strontium clear		462	503	688			2.68				9.3	7.4				1.529
glass-ceramic	16					657	2.60	118	0.24	16.7	10.0	8.7	0.30	5.6	1.7	
glass-ceramic						593	2.50	86	0.25	13.4	8.1	6.8	0.34	6.9	2.3	
lithium–aluminosilicate											7.2	5.7		6.75	0.6	
fluorphlogopite							2.54					9.8	8.4		5.9	0.2
borosilicate	54	408	450	705	1161		2.16	49	0.23	>17	9.4	7.6	0.32	4.7	1.5	1.468
	19		540	720	1065		2.48				5.8	4.7		7.7	1.4	
aluminoborosilicate		524	569	794	1190		2.39	72			7.4	6.0	0.51	4.9	2	
aluminoborosilicate		538	580	798	1190		2.36				6.8	5.4	0.96	5.9	5.7	1.49
borosilicate		513	565	827	1240		2.23				8.1	6.6	0.46	4.6	2.1	1.47
lime–aluminoborosilicate		507	657	846			2.60	74						6.4		1.548
soda lime–aluminoborosilicate				715			2.54							7.3		1.541
soda lime–borosilicate		435	585	752			2.61	70						7.8		1.549
soda–titania zirconia				675			2.57							8.3		1.537
magnesium–alumino silicate		760	810	970				87								1.523

[a] Glasses 7905, 7910, 7911, 7912, 7913, and 7917 for special uv and ir applications. Glass 1720 is available with improved uv transmittance (designated glass 9730). Glass 7760 is also available with special transmission suitable for sun lamps.
[b] B = blown ware; M = multiform; U = panels; P = pressed ware; R = rolled sheet; C = castings; S = plate glass; T = tubing and rod; F = fiber.
[c] These borosilicate glasses may rate differently if subjected to excessive heat treatment.
[d] Since weathering is determined primarily by clouding which changes transmission, a rating for the opal glasses is omitted.
[e] Code 9608 may be produced in a range of expansion values depending upon intended application.
[f] Extrapolated values.
[g] Normal service: no breakage from excessive thermal shock is assumed. Extreme limits; glass will be very vulnerable to thermal shock. Recommendations in this range are based on mechanical stability considerations only. Tests should be made before adopting final designs. These data approximate only.
[h] These data approximate only. Based on plunging cold sample into cold water after oven heating. Resistance of 100°C means no breakage if heated to 110°C and plunged into water at 10°C. Tempered samples have over twice the resistance of annealed glass.
[i] Resistance in °C is the temperature differential between the two surfaces of a tube or a constrained plate that will cause a tensile stress of 6.9 MPa (1000 psi) on the cooler surface.
[j] These data subject to normal manufacturing variations.
[k] Determined by revised ASTM standard: number of standard not yet assigned.
[l] At 10 kHz.
[m] Refractive index may be either the sodium yellow line (589.3 nm) or the helium yellow line (587.6 nm). Values at these wavelengths do not vary in the first three places beyond the decimal point.

0.55–0.75 mm dia and 23.5 cm long elongates under its own weight at 1 mm/min when the upper 10 cm is heated at 5°C/min. For glass having a density of 2.50 g/cm^3, this corresponds to a viscosity of $10^{6.6}$ Pa·s ($10^{7.6}$ P). The exact viscosity is a function of density and surface tension. The annealing point is the temperature that corresponds to a viscosity of approximately 10^{12} Pa·s (10^{13} P) (ASTM C 336 and C 598). At the annealing point, internal stress is reduced to an acceptable value in about 15 min. The strain-point temperature is the lower end of the annealing range and is approximately $10^{13.5}$ Pa·s ($10^{14.5}$ P) (ASTM C 336 and C 598). Internal stresses are reduced to an acceptable level in approximately four hours at the strain-point temperature.

Glass is usually melted and fined at viscosities between 5 and 50 Pa·s (50–500 P) but forming and final viscosity requirements vary greatly. The ranges of viscosity for various forming methods are compared in Figure 9.

Viscosities of glass are compared qualitatively. A hard glass has a high softening point and a soft glass has a lower softening point. Long and short refers to the temperature difference between the softening and strain point of a glass. A long glass has a large temperature difference between its softening and strain point, ie, it solidifies (set up) slower than a short glass as temperature decreases.

Thermal. ***Expansion.*** The thermal expansion of a glass determines the range of materials to which it can be safely sealed. It also affects the glass's ability to survive thermal shock or cycling. A glass's usefulness as a heat exchanger or as a thermal barrier

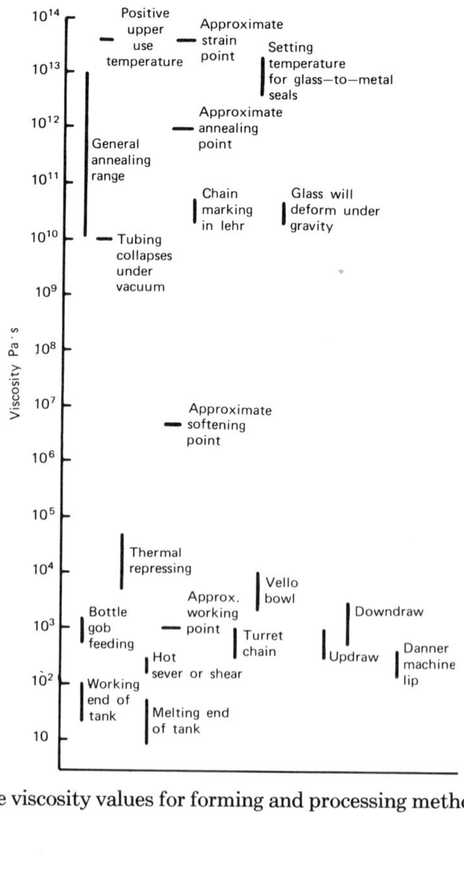

Figure 9. Approximate viscosity values for forming and processing methods (Pa·s × 10 = P). Courtesy of Corning Glass Works.

and its ease of melting and forming depends upon its heat-transfer properties and emissivity. The upper use temperature is a function of all of these properties.

Like most materials, glass expands when it is heated and contracts when it is cooled. If the thermal cycling is slow enough, there is virtually no hysteresis effect. Linear expansion, $\Delta L/L$, is the change in length per initial length. Generally the expansion is proportional to temperature up to 300°C or more, depending on the glass. The slope of the linear expansion-versus-temperature curve, the linear thermal expansion coefficient, α, is therefore virtually constant between 0 and 300°C for most glasses, and thus α_{0-300} is a useful property for comparisons (see Table 4). As the temperature of the glass rises to near the set point (strain point + 5°C), the thermal expansion increases more rapidly (compare Fig. 1, which is a volume–temperature curve for a general glass). The volume–temperature curve is analogous to the linear expansion (length–temperature) curves shown in Figure 10 for various commercial glasses. The density ρ, which depends on the temperature, is easily calculated from the density at room temperature and the appropriate linear thermal expansion coefficient.

The sealing of glasses to either metals or other glasses depends on the thermal expansion mismatch of the two materials as well as their rheological and elastic properties (56). The materials must be heated to a point where the softer glass flows and intimately contacts the other material. In the case of glass-to-glass seals, the glasses flow together and bond extremely well. For glass-to-metal seals, however, a complicated and rigorously controlled oxidation–reduction reaction must take place at the interface to form a good bond. As the composite cools, each material shrinks along its linear

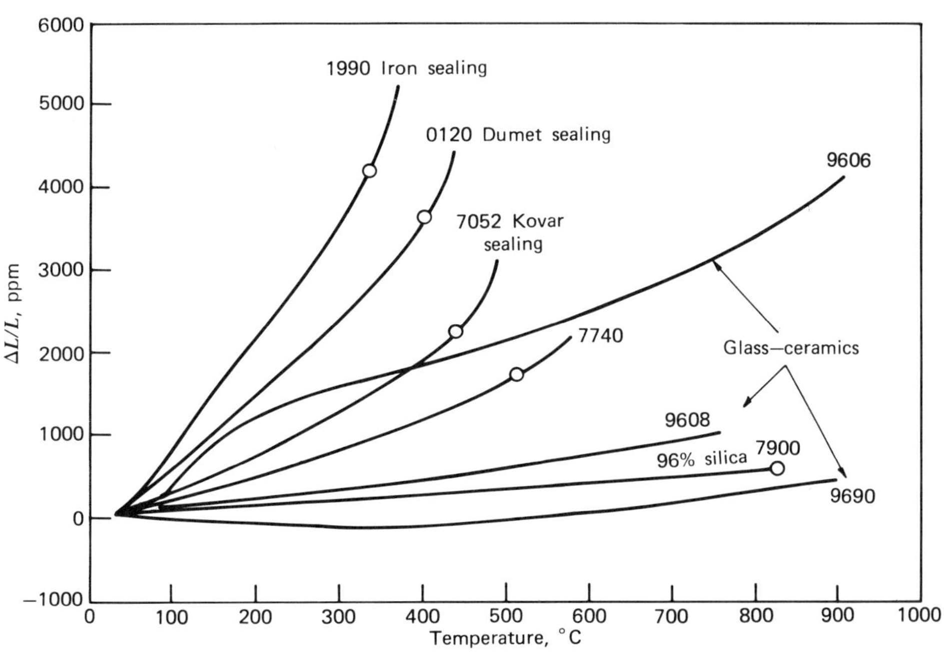

Figure 10. Expansion vs temperature for some commercial glasses and glass-ceramics ($\Delta L/L$ = change of length per initial length). The circles represent the set point of each material. Courtesy of Corning Glass Works.

thermal expansion curve. At high temperatures, the softer glass deforms by viscous flow to compensate for any thermal expansion difference. However, when the composite reaches the set point of the softer material, viscous deformation stops and both members are constrained to shrink equally. Therefore, an expansion mismatch is divided between the two materials, producing stresses that must satisfy force and moment balances consistent with the appropriate elastic constants and geometrical cross-sections of the materials (56). Other factors also affecting the stresses set up in the glass sealing include the geometry of the seal, the cooling rate used in annealing the seal, and the difference in annealing ranges (see also Electrostatic sealing).

The upper use temperature for annealed ware is below the temperature at which the glass begins to soften and flow (about 10^{14} Pa·s or 10^{15} P). The maximum use temperature of tempered ware is even lower, because of the phenomenon of stress release through viscous flow. Glass used to its extreme limit is vulnerable to thermal shock, and tests should be made before adapting final designs to any use. Table 4 lists the normal and extreme temperature limits for annealed and tempered glass. These data are approximate and assume that the product is not subject to stresses from thermal shock.

Stresses. Stresses caused by steady-state thermal gradients may or may not cause failure, depending on the degree of constraint imposed by some parts of the item upon others or by the external mounting. Thus, under minimum constraint and maximum uniformity of gradient through the thickness, very large temperature differences can be tolerated. If a plate is completely restrained, the tensile stress on the cool side is given by

$$\sigma = E\alpha\Delta T/2(1 - \nu)$$

where ΔT is the temperature difference across the glass, α is the expansion coefficient, E is Young's modulus, and ν is Poisson's ratio. Normally, the maximum design stress in glass is about 7 MPa (1000 psi). The face-to-face temperature differentials (thermal stress resistance) that cause this tensile stress on the cool face of cylinders or otherwise constrained shapes, are given in Table 4. The thermal stress resistance of glass-ceramic 9608 is calculated to be 480°C, considerably higher than that of most glasses and ceramics. This is due to the higher break resistance of the material and its low coefficient of thermal expansion.

When glass is suddenly cooled, as eg, by removal from a hot oven, tensile stresses are introduced in the surfaces and compensating compressional stresses in the interior. Conversely, sudden heating leads to surface compression and internal tension. In either case, the stresses are temporary (transient) and disappear on attainment of temperature uniformity. Since glass fractures only in tension, usually at the surface, the temporary stresses from sudden cooling are much more damaging than those resulting from sudden heating, assuming, of course, that all surfaces are heated or cooled at the same time.

Because the strength of glass is greater under momentary stress than under prolonged load, thermal shock endurance cannot be directly calculated, but is generally determined by empirical testing. Resistance to breakage is determined by heating the ware to some appropriate temperature and then plunging it into cold water. Table 4 lists the thermal shock resistance of 15 × 15 cm annealed plates of three thicknesses. A resistance of 100°C means that no breakage occurs on heating to 100°C and plunging into water at 10°C. When other cooling media, eg air or oil, are used, much higher values of thermal shock resistance are recorded. Tempered samples have more than twice the thermal shock resistance of annealed glass.

Heat Transmission. At room temperature, the thermal conductivity of glasses ranges from 0.67 to 1.21 W/(m·K), with the most common compositions near the upper end of the range. Thermal conductivity of glass-ceramics ranges from 1.7 to 3.8 W/(m·K). At a mean temperature of 200°C, the values are greater by approximately 25%. Above 500°C, conductivity increases rapidly because of radiation within the glass. For Corning Code 7740, used frequently in heat-transfer application, the thermal constants are thermal conductivity (25°C): 0.96 W/(m·K); mean specific heat (25–175°C): 84 J/(kg·K) [20 cal/(kg·K)]; thermal diffusivity: 0.56 mm^2/s; and emissivity coefficient, radiant energy: 0.94.

Liquidus Temperature. The liquidus temperature (see above under Glass-Ceramics) determines the susceptibility of a glass to devitrification and therefore influences its forming limitations and, often, its heat-treating requirements.

Gas Permeation. Gas permeation through glass is of crucial importance to any high-vacuum system. With the advent of high intensity tungsten–halogen lamps, the lamp envelopes must contain gas under pressure for a long time at high temperatures. The permeation rate varies with temperature, glass type and composition, and gas. The volume of gas (at STP) which diffuses through glass is proportional to the time, the area of the glass, and the pressure differential, and it is inversely proportional to the thickness of the glass (57). Figure 11 shows the constant of proportionality as a function of temperature for various gases permeating a 96% silica glass. Figure 12 shows the same information for helium permeating various commercial glasses. Helium is the most mobile of the gases and silica is the most permeable glass. The addition of a modifier blocks the openings in the glass network and slows the permeation rate (57).

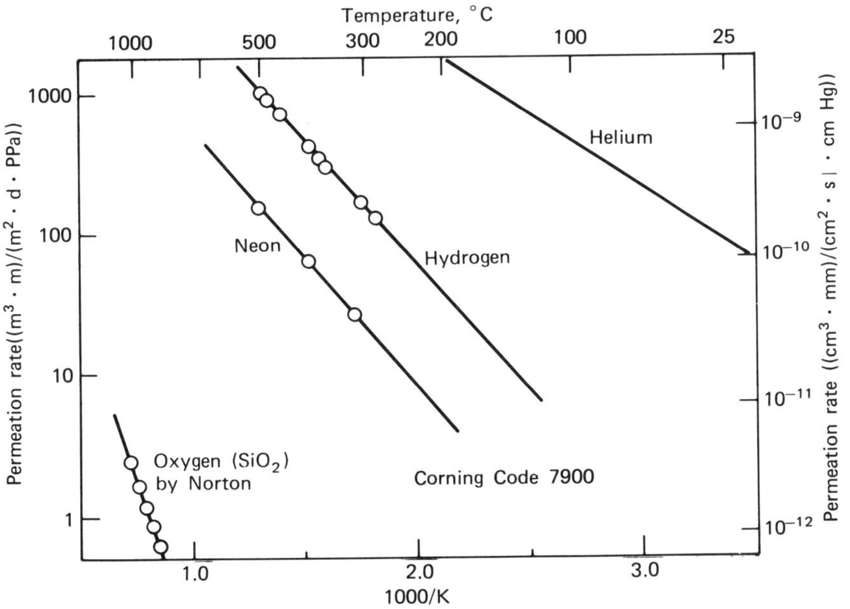

Figure 11. Permeation of helium, hydrogen, neon, and oxygen for Corning Code 7900 (57). Courtesy of Corning Glass Works.

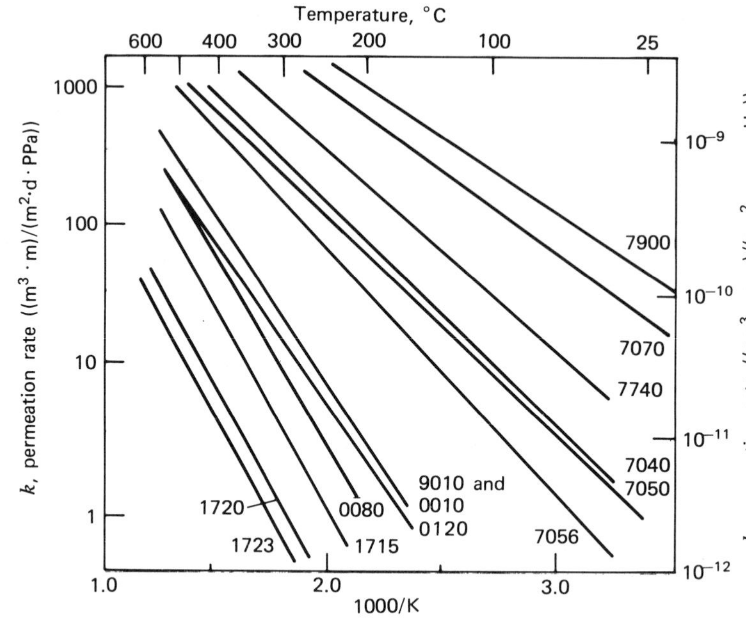

Figure 12. Permeation rate constant for helium through some commercial glasses. Courtesy of Corning Glass Works.

Mechanical. Because of its amorphous structure, glass is brittle, reasonably abrasion resistant, and nearly perfectly elastic as long as its temperature is low enough to prevent viscous flow. Among other things, glasses do not contain crystallographic planes which slip relative to each other, and therefore permit the material to deform plastically when stress is applied (58). Slip tends to relieve stress and allow broken bonds to form new bonds before breakage occurs. Since no such phenomena take place in glass, a bond that is broken because of excessive stress quickly forms a crack. As the crack grows, the stress intensity at its tip increases; those bonds that were intact quickly break in its path, and the glass fractures. For this reason, a glass usually breaks because of an applied tensile stress and generally from a surface flaw. Several steps are commonly taken to make the glass more break resistant. The elimination of surface flaws, eg, microcracks or bruise checks, either by careful handling during and after forming, annealing, etc, or by acid etching the surface to remove these flaws, prevents a crack from starting. Secondary treatments, eg, ion exchange, thermal tempering, glazing, or differential crystallization, provide a compressive stress on the surface of the glass that must be overcome before the applied tensile stress causes breakage. Figure 13 compares the strengths of various glass types and treatments. The depth of this compressive layer governs the glass's resistance to deep bruising or abrasion. Glass-ceramics resist fracture because the tiny microcrystals present in its structure arrest crack growth.

Strength. Theoretical calculation places the intrinsic strength of glass as high as 35GPa (5×10^6 psi); the strength of flame-polished silica rods (25) is 14MPa (2.0×10^6 psi). However, the practical strength of glass is a small fraction of these figures because of stress concentrations introduced by surface imperfections. Strength of fibers may be measured in bending or pure tension; strength of more massive forms is measured only in bending.

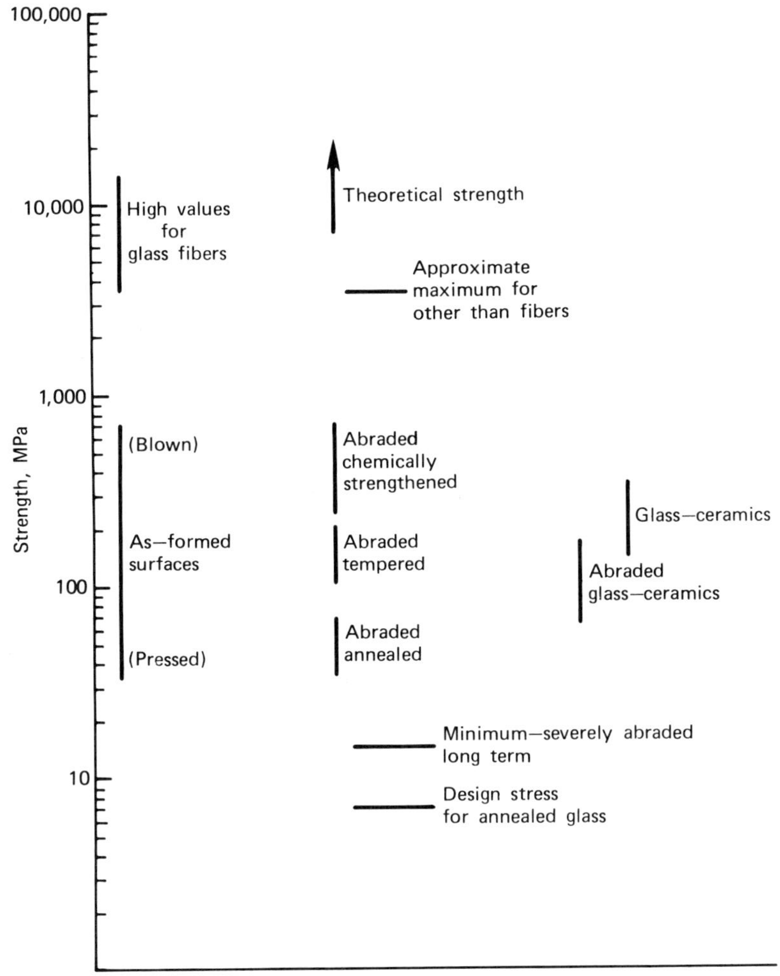

Figure 13. Strength scale for glass and glass-ceramics. To convert MPa to psi, multiply by 145.

A glass rod 6.4 mm dia with pristine surfaces may show a tensile strength of 1750 MPa (250,000 psi). However, normal handling of the severity experienced by many glass products in service introduces surface imperfections which reduce this strength to approximately 70 MPa (10,000 psi). Glass also exhibits a time-load effect, so that a glass article breaks at a lower stress under prolonged loading than under momentary loading.

For a safety factor, the allowable stress for annealed (stress-free) soda-lime glass is 7 MPa (1000 psi) under sustained loading for 1000 h or more. The chemical composition of glass has a significant effect on its design strength. The allowable stress for glasses with greater resistance to stress corrosion, eg, fused silica, can be twice as high as that for soda-lime glass (14 MPa or 2000 psi).

Elasticity. Glasses, like other brittle materials, deform elastically until they break in direct proportion to the applied stress. The Young's modulus E is the constant of proportionality between the applied stress and the resulting strain. It is about 70 GPa

(10^7 psi) [(0.07 MPa stress per μm/m strain = (0.07 MPa·m)/μm)] for a typical glass.

A high-modulus glass is one in which the atoms are closer together or the bonds are stronger and therefore more resistant to rupture. Articles made from brittle materials break when they exceed their strain tolerance which depends directly on size and configuration of the article as well as on surface condition. For bulk glass, it is on the order of 1000 μm/m, resulting in a strength of (0.07 MPa·m/μm) × (1000 μm/m) = 70 MPa or 10^4 psi. For fibers, the strain tolerance can be one or two orders of magnitude greater. Thus, for a given strain tolerance, the higher the Young's modulus, the greater the strength.

Young's modulus and Poisson's ratio ν for many glasses are listed in Table 4. The modulus of rigidity G is $E/2(1 + \nu)$; the bulk modulus K is $E/3(1 - 2\nu)$, and the compressibility C is $1/K$. For a typical glass where E and ν are 70 GPa (10^7 psi) and 0.22, respectively, $G = 29$ GPa (42×10^5 psi), $K = 42$ GPa (6.1×10^6 psi), and $C = 24 \times 10^{-9}$/MPa (1.6×10^{-10}/psi).

Hardness. Glass hardness tests usually measure the resistance to abrasion by grinding or grit-blasting, resistance to scratching, or penetration by an indenter. The method to be used depends on expected service conditions. Knoop hardness (Table 4) is commonly used, because other methods usually fracture the glass.

On the Mohs scale of scratch hardness, glass lies between apatite, 5, and quartz, 7. Some common materials hard enough to scratch glass include agate, sand, silicon carbide, hard steel, and emery. Glasses are harder than mica, mild steel, copper, aluminum, and marble. A typical glass-ceramic exhibits diamond penetration hardness ca 40% greater than borosilicate glass (see Hardness).

Electrical. Glasses are used in the electrical and electronic industries as insulators, lamp envelopes, cathode ray tubes, and encapsulators and protectors for microcircuit components, etc. Besides their ability to seal to metals and other glasses and to hold a vacuum and resist chemical attack, their electrical properties can be tailored to meet a wide range of needs. Generally, a glass has a high electrical resistivity, a high resistance to dielectric breakdown, and a low power factor and dielectric loss.

Resistivity. When a voltage is applied across a piece of glass, a minute current flows, partly over the surface and partly through the interior. Thus, the insulation resistance is determined jointly by the surface resistivity measured in Ω/sq and the volume resistivity measured in Ω·cm. Surface electrical resistivity is of primary importance in all electrical insulation problems with glasses or ceramics since it greatly affects performance under service conditions involving high humidity. Because of the adsorption of a moisture film, the surface electrical resistivity of glasses and ceramics, although high in comparison to that of many other materials, is markedly lower than volume resistivity. Thus, chemical durability is an essential requirement in good electrical glass because film thickness and film conductivity are both increased by the soluble products of unstable glass compositions (59) (see Insulation, electric).

Volume resistivity, when measured in terms of current flowing through unit path per unit of applied voltage, in accordance with Ohm's law, is markedly affected in glass by the amount of alkali present. Conduction in glass is generally by means of ionic transport and often by the movement of sodium ions, one of the ions most mobile in a silicate network (60). Glasses also conduct by means of potassium or lithium migration, but when all three ions are present they tend to interfere with one another and the electrical conductivity decreases. This is called the mixed-alkali effect. Volume resistivity of common glasses at ordinary room temperature varies widely with composition from as low as 10^8 to as high as 10^{19} Ω·cm. The resistivity of a given composition at a fixed temperature may vary by a factor of 3, depending upon the degree of annealing or strain which the glass has received; strained glass has lower resistance than properly annealed glass which is practically strain-free. Figure 14 shows some typical resistivity-versus-temperature curves for some commercial glasses.

Dielectric Strength. Dielectric failure may be thermal or disruptive. In thermal breakdown, applied voltage heats the sample and thus lowers its electrical resistance. The lower resistance causes still greater heating and a vicious circle, leading to dielectric failure, occurs. However, if applied voltage is below a critical value, a stabilized condition may exist where heat input rate equals heat loss rate.

In disruptive dielectric failure, the sample temperature does not increase. This type of failure is usually associated with voids and defects in the materials.

The intrinsic dielectric strength of glass is high. Thus, in applications this property is relatively unimportant compared with the problem of design to prevent flashover.

Dielectric breakdown voltage decreases with increase in both frequency and temperature. At elevated temperatures, breakdown is governed mainly by the volume resistivity of the glass. Dielectric breakdown voltages for lime-glass plates are about 75% of those for Corning Code 7740 borosilicate glass.

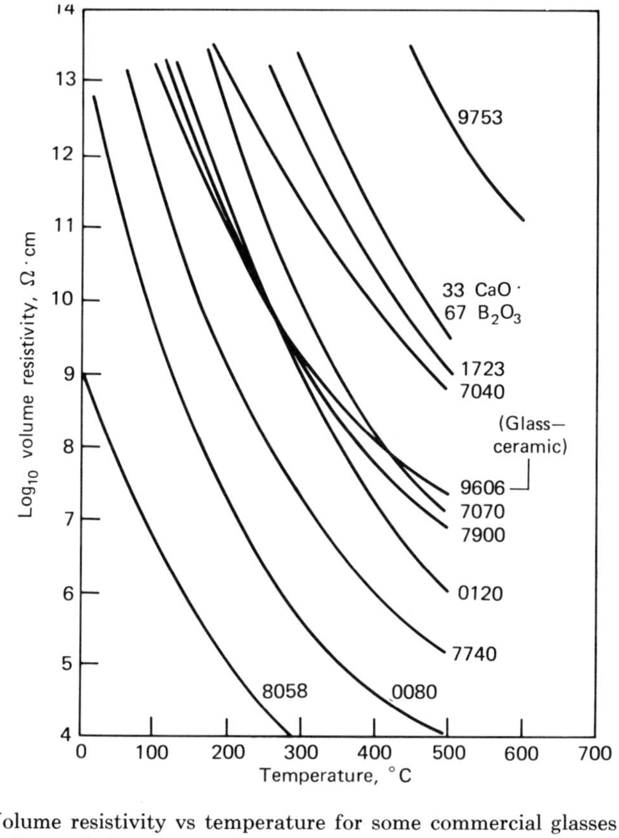

Figure 14. Volume resistivity vs temperature for some commercial glasses and glass-ceramics. Courtesy of Corning Glass Works.

Dielectric Loss. The loss tangent of glass is the ratio of the real to the imaginary current through a glass that is being subjected to an alternating voltage in a capacitor configuration. The loss tangent depends on the composition and structure of the glass.

Material with a certain amount of alkali may exhibit certain loss properties, whereas another material with equal alkali content may be substantially different. On the other hand, a fixed composition with minor changes in alkali may show significant dielectric loss variation. Thus, the loss properties can be controlled with variations in alkali and alterations in structure.

In recent years, several glasses have been developed with exceptionally low power factors of less than 0.015% at room temperature and 1 MHz. Figure 15 shows the variation of power factor with temperature.

Dielectric Constant. The dielectric constant determines electric energy storage in a polarized dielectric, thus serving to evaluate the performance of a capacitor. In d-c fields at ordinary room temperatures, dielectric absorption also affects dielectric constant measurements, giving values that vary with time.

At −75°C and below, the dielectric constant has been found to be independent of charge time. At ordinary room temperature and above, it increases with temperature at rates that vary with glass composition (Fig. 16). This effect is less in high silica and borosilicate glasses and greater in those glasses containing substantial amounts of alkali.

In addition to the obvious effect of high dielectric constant glasses on the capacitance of the circuit elements into which they enter, their dielectric strengths may be more important. Since the amount of energy a capacitor can store varies as the first power of the dielectric constant and the second power of the voltage, a glass with twice the dielectric strength is as effective as one with four times the dielectric constant.

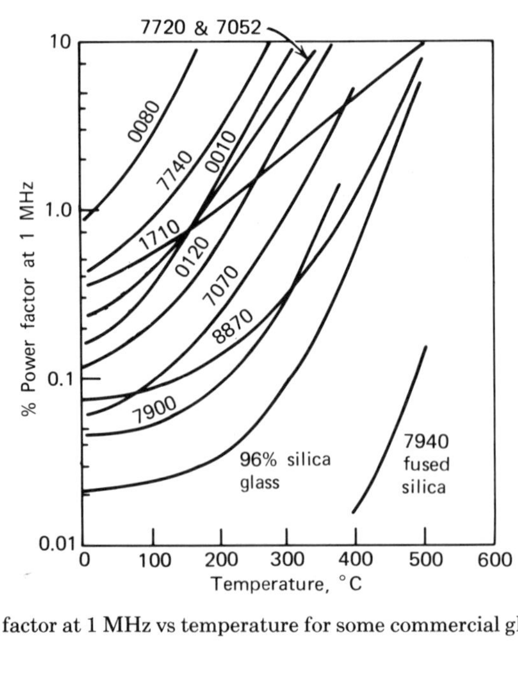

Figure 15. Power factor at 1 MHz vs temperature for some commercial glasses. Courtesy of Corning Glass Works.

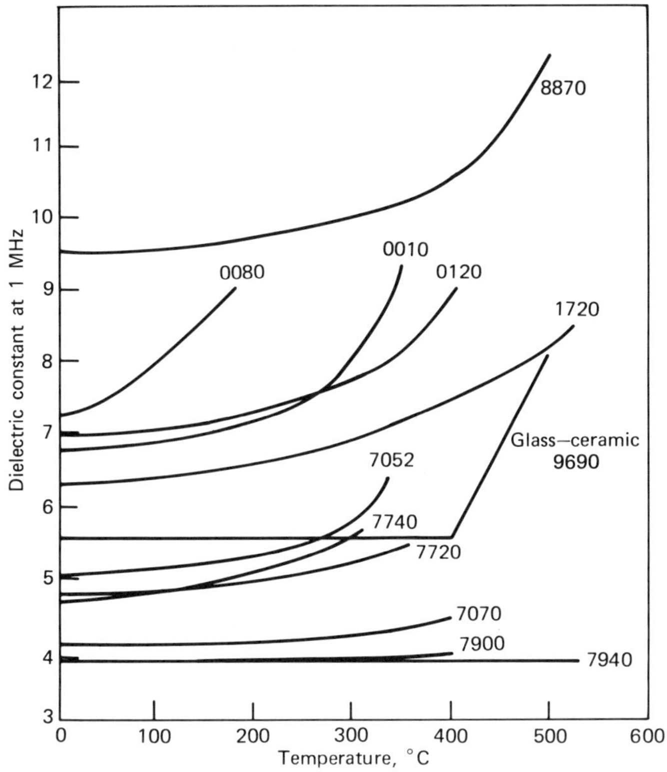

Figure 16. Dielectric constant at 1 MHz vs temperature for some commercial glasses and glass-ceramics. Courtesy of Corning Glass Works.

Light Amplification Devices. First developed for military applications, light amplification devices (61) have now been adapted as special microchannel-plate eyeglasses for people who are almost totally blind. Briefly, an acid-soluble glass rod is first selected and placed inside an insoluble glass tube, and the combination is drawn into a coaxial rod. The rods (ca 10 μm dia) are then fused together in a bundle, sliced into thin plates, polished, and leached in acids so that the core glass is removed, and thus formed into a microchannel glass plate. The cladding glass is then rendered electronically conducting on the surface, and electrodes are deposited and assembled into the device.

As incident or primary electrons impinge upon the surface of a solid, a certain fraction is elastically reflected, whereas the remainder penetrates the solid. Some of the primary electrons entering the solid lose energy by liberating lattice electrons. The liberated electrons move toward the surface, a certain fraction escaping as secondary electrons. Gains of 10^5–10^6 are possible if a primary electron enters one end of a glass tubing under a high voltage gradient, providing the length-to-diameter ratio is large. This is the process of electron multiplication. In combination with a photocathode located in the front end of the tube and a phosphor screen located at the exit end, a highly efficient light amplification device can be made.

Microwave Compatibility. Microwave cooking has given rise to questions concerning the effects of microwave energy on various materials used in cooking vessels. Since the suitability of a cooking vessel for microwave oven use is affected by such factors as the configuration of the vessel, the distribution of food in it, and energy distribution in the oven, as well as the intrinsic properties of the material itself, there is no clear demarcation between suitable and nonsuitable materials. Nevertheless, the properties of some glasses and glass-ceramics make them more suitable for this application than others. The relative suitability of a glass or glass-ceramic material is expressed by a factor of merit which is proportional to the material's strength and inversely proportional to its Young's modulus, expansion coefficient, and rate of microwave energy absorption. Table 5 reflects the microwave energy absorption and the factor of merit of various types of glass and glass-ceramics (see Microwave technology).

Chemical Stability. ***Corrosion Resistance.*** The resistance of glass to chemical corrosion is frequently the reason for its use. However, the durability of a glass varies from highly soluble to highly durable, depending upon its composition and the solvent considered. Comparisons are usually based upon measurements of weight loss, changes in surface quality, or the analysis of solutions that were in contact with a glass.

The reaction of acids with glass may be either a leaching or a complete dissolution process (25). Hydrofluoric acid attacks silicate glasses by dissolving the silica network. Other acids such as hydrochloric or nitric acid may dissolve certain glasses but most frequently they react by selective extraction of alkali and the substitution of protons in a diffusion-controlled process. As a layer of glass is leached, the rate of material extraction decreases as the square root of time. The temperature coefficient of leaching reactions of glass increases about 1.5 times for every 10 K increase in temperature. An etching or dissolution process constantly exposes a fresh glass surface, and thus the reaction rate is usually constant with time. Etching may give a smooth or a pitted glass surface, depending upon the glass composition and quality and the solution used. The rate of dissolution increases about 2.5 times with each 10 K increase in temperature. Ratings of acid durability of various glasses are based upon a standard 5% hydrochloric acid solution test. Results for several glasses are shown in Table 4. A rating of 1 implies little or no visible damage, whereas a rating of 4 represents severe deterioration of the glass during the test.

Reaction of bases with most commercial silicate glasses produces dissolution rates of 7.5–30 μm/d when tested in 5% NaOH solution at 95°C. The mechanism involves

Table 5. Microwave Energy Absorption and the Factor of Merit for Various Types of Glasses and Glass Ceramics

	Energy absorbed, %	Factor of merit
fused silica	0.008	1500
glass-ceramic		
cordierite	0.03	41
β-spodumene	0.38	20
nepheline, celsian	6.1	0.7
tempered aluminosilicate glass	0.71	7
borosilicate glass	0.40	5
opal		
high strength	0.90	5
tempered	1.1	2
tempered lime	1.0	2

a complete dissolution process as described for acid corrosion. Weaker alkaline solutions, however, may both leach and dissolve and sometimes show a greater dependence upon glass composition. For strong alkali solutions, the rate of attack ca doubles for each 10 K increase in temperature or each increase in pH unit. Higher-alkali-durability glasses such as Corning's Code 7280 zirconia–silicate glass are available for laboratory ware. As with acids, reaction products accumulating on the surface slow alkali attack. Glass-ceramics may exhibit preferential attack of various crystalline phases.

The attack of water is related to the leaching mechanism described for acids. Table 4 rates glasses based upon their resistance to water attack. Low-alkali, high-alumina, or borosilicate glasses generally have high water durability.

Weathering of glass is the result of the action of water, carbon dioxide, and other atmospheric constituents. Water, initially absorbed by the glass, is exchanged for alkalies that form alkali salt solutions. If left in contact with the glass, these may cause additional surface damage. For this reason, weathering resistance may not correlate with acid durability. Test methods have been designed to accelerate the weathering process, for example, samples are weathered in a chamber at 50°C and 98% rh. Visual comparisons allow rating of glasses as shown in Table 4.

Element or metal halide attack of silica glass at elevated temperatures is of special interest to lighting-ware manufacturers (62). At high temperatures, reaction occurs between vitreous silica and various metals or halides. Staining of glass by alkali metal vapors can occur at relatively low temperature (<300°C), but at higher temperatures (ca 900°C) failure of the glass may occur as a result of severe devitrification that is induced. Rates of attack based upon one-hour exposure indicate sodium causes failure of silica tubes because of devitrification below 1000°C. Other elements that induce devitrification are Mn, Sn, Fe, and Zn which cause failure at 1250 to 1350°C. Metal halides that react with silica glass below 1200°C are LiCl, NaCl, $SnCl_2$, NaBr, and KBr. At 1200–1400°C, RbCl, CsCl, NaI, KI, $MgCl_2$, $CaCl_2$, $BaCl_2$, and $AlCl_3$ cause failure.

Optical. ***Refraction.*** Optical glasses are usually described in terms of their refractive index at the sodium D line (589.3 nm), and their ν value (or Abbe number) which is a measure of the dispersion or variation of index with wavelength, ie, $\nu = (n_D - 1)/(n_F - n_C)$, in which n_F is the refractive index at the hydrogen F line (486.1 nm) and n_C is the refractive index at the hydrogen C line (656.3 nm). These data are incorporated by Schott into a six-digit numbering system used to identify optical glasses. For example, the glass commonly used to produce eye glasses has the number 523–586. This means $n_D = 1.523$ and $\nu = 58.6$ for this glass. Glasses with index n_D less than 1.60 and a ν value of 55 or above are defined as crown glasses; those with a ν value below about 50 are defined as flint glasses. Glasses with n_D greater than 1.60 are defined as crown glasses if ν is 50 or above. The Schott nomenclature often used for optical glasses is defined by Figure 17. The labeled areas represent commercially available glasses. Crowns are usually alkali silicate glasses and flints are lead alkali silicates. Phosphate and borate-based glasses are used to obtain high index or low dispersion glasses or both. Compositions of numerous optical glasses are given in ref. 2. Table 4 also lists values of n_D for several glasses.

Reflectance is very dependent on the condition of the glass surface. It can be specular, as in polished or precision molded surfaces, or diffuse for ground or irregularly etched surfaces. For normal incidence, the fraction of incident light reflected by a single surface, $R = ((n - 1)/(n + 1))^2$, in which n is the index of refraction at the wavelength

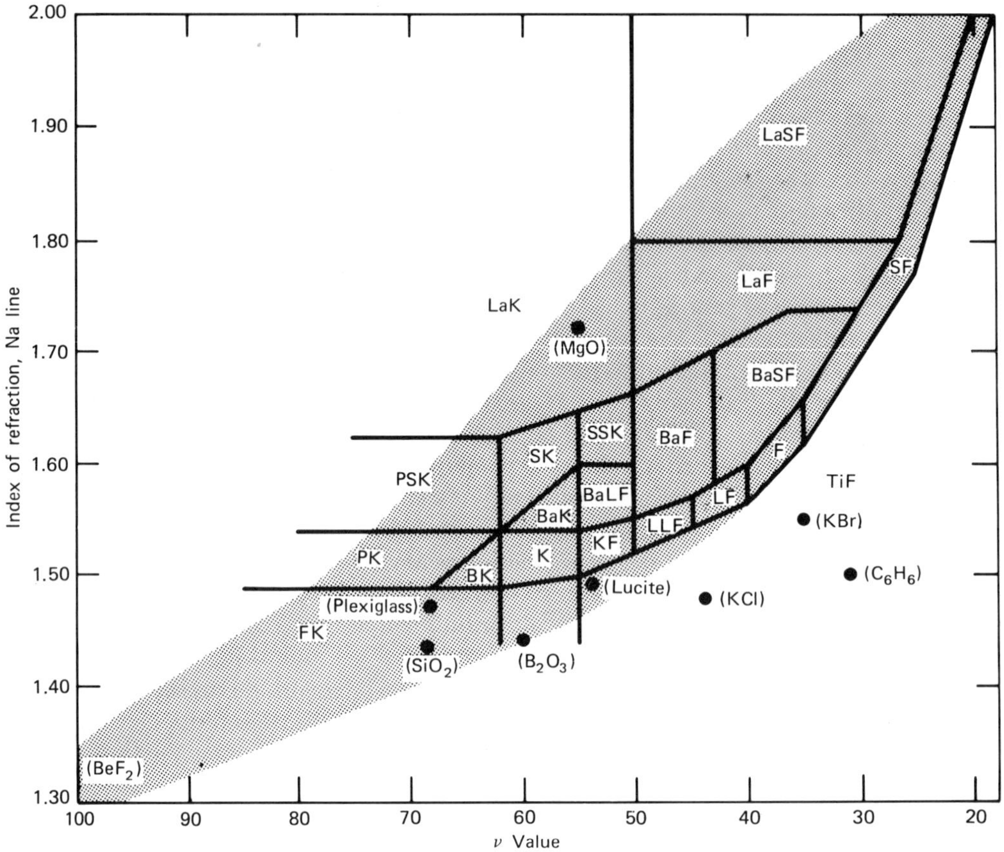

Figure 17. Index of refraction vs dispersion and optical classification of glasses. The shaded area indicates the region of glass formation. Plexiglass is a methyl methacrylate; Lucite is an acrylic resin. BaF = barium flint; BaK = barium crown; BaLF = light barium flint; BaSF = heavy barium flint; BK = borosilicate crown; F = flint; FK = fluorcrown; K = crown; KF = crown flint; LaF = lanthanum flint; LaSF = heavy lanthanum flint; LaK = lanthanum crown; LF = light flint; LLF = very light flint; PK = phosphate crown; PSK = heavy phosphate crown; SF = heavy flint; SK = heavy crown; SSK = very heavy crown; TiF = titanium flint.

of interest. For silicate glasses having an index of refraction of 1.5 for green light, the reflection loss per surface is about 4%. A perfectly clear glass therefore transmits ca 92% of the incident light. As the index of refraction increases above 1.5, transmission decreases. Indexes of refraction of 1.7 and 1.9 predict the transmission of clear glasses to be 87 and 81%, respectively. Low index coatings or special surface treatments of certain glasses may significantly reduce reflection losses (63).

Transmission. The spectral transmission of glass is determined by reflection at the glass surfaces and the optical absorption within the glass. Overall transmission of a flat sample at a particular wavelength is equal to $(1 - R)^2 e^{-\beta t}$, where β is the absorption coefficient, t the thickness of glass, and R the air–glass reflection coefficient. Transmission is a function of wavelength. In silicate glasses, it is limited by the absorption of silica at approximately 150 nm in the ultraviolet and at 6000 nm in the infrared. Iron impurities further reduce transmission in the uv and near ir. Dissolved water absorbs at 2700 nm and is a serious problem in making ir transmitting glasses.

Transmission may be controlled by the type of glass used, eg, silicate, phosphate, borate, etc, or by the control or addition of coloring additives, melting atmosphere, melting temperatures, and cooling schedules (64–65). In glasses containing suspended particles, transmission is diffuse. Opal glasses (qv), used extensively for lighting products, exemplify diffuse transmission. Oxides that are frequently used to control color in glass are listed in Table 6.

Table 6. Coloration of Glasses [a]

Effect	Oxide
colorless, uv transmitting	SiO_2
	96% SiO_2
	P_2O_5
colorless, uv absorbing	CeO_2
	TiO_2
	Fe_2O_3 (P_2O_5 glasses)
blue	Co_3O_4
	$Cu_2O + CuO$
	S (B_2O_3 glasses)
purple	Mn_2O_3
	NiO (K_2O glasses)
green	Cr_2O_3
	$Fe_2O_3 + Cr_2O_3 + CuO$
	V_2O_3
	CuO
	MoO_3 (P_2O_5 glasses)
brown	MnO (reduced)
	$MnO + Fe_2O_3$
	$TiO_2 + Fe_2O_3$
	NiO (Na_2O glasses)
	$MnO + CeO_2$
amber	Na_2S (reduced)
yellow	CdS
	$CeO_2 + TiO_2$
	Ag staining
	UO_3
orange	CdS + Se
red	CdS + Se
	Au
	Cu or Cu staining
	UO_3 (PbO glasses)
	Sb_2S_3
black	Co_3O_4 (+ Mn, Ni, Fe, Cu, Cr oxides)
colorless, heat absorbing	FeO (P_2O_5 glasses)
colorless, ir transmitting	PbO glasses
	$CaO.Al_2O_3$ glasses
	Te, Ge, or Sb glasses
	As_2S_3
	Ge–Sb–Se or Ge–As–Te glasses

[a] Ref. 66.

Stress-Optical Coefficients. When strained, glass becomes doubly refracting, or birefringent (see under Lehring). Stress optical effects are presumed to be elastic; therefore, strain is proportioned to stress. The stress optical coefficient, B, or Brewster's constant is a measure of this proportionality, and varies greatly with glass composition (67). Values of B range from about -1 for extra dense flints containing 80% PbO to ca 3.6 for 96% silica glass. For soda-lime glass, B is about 2.5.

Radiation Effects. Interaction of glass with low-energy visible and ultraviolet radiation may result in alteration of the electronic states. These changes may cause coloration or luminescence effects, eg, photochromic or photosensitive glass. When changes in color are produced by sunlight, the effect is frequently referred to as solarization (68–69). Effects produced by ions, gamma rays, and x-rays are relevant to glasses used as electronic tube envelopes, radiation-shielding windows, and dosimeters. The stability of glass in neutron fields depends on the relative absence of boron in the glass composition. Boron-free glasses can, in general, withstand fluxes of 10^{19} neutrons/cm^2. Ion bombardment disrupts the glass structure, producing discoloration and an evolution of oxygen. This is troublesome in electronic tube envelopes where the oxygen can react with the cathode, reducing tube life. Electron browning is also troublesome in cathode ray tubes containing oxides of lead or bismuth, which are easily reduced by the impinging electrons producing a brown tint in the glass. Relatively low doses (2.6 C/kg = 10^4 roentgens) of gamma rays tend to discolor dense lead-containing glasses. The glass most resistant to this effect is fused silica; it shows virtually no discoloration at doses above 2.6×10^5 C/kg (10^9 roentgen). This glass, however, is of low density (2.20 g/cm^3) and it is expensive. Glasses of intermediate density and relatively low cost can be protected against darkening in gamma-ray fields by the addition of multivalent oxides such as cerium dioxide to give color stability up to about 2.6×10^4 C/kg (10^8 roentgen). These glasses have a density of about 3.30 g/cm^3, and can be combined with lead glasses of higher density to make composite windows. These windows are remarkably resistant to radiation damage, and can be used extensively with little or no maintenance in radiation flux fields of more than 2.6×10^2 C/(kg·h) (10^6 roentgen per hour). They match the absorption of the heavy concretes used for shielding walls. Conversely, glasses containing cobalt have been developed that exhibit color changes on absorption of radiation. More sensitive silver phosphate glasses are used as dosimeters to measure earlier x-ray or gamma ray exposure.

Manufacture and Processing

Most glass articles are manufactured by a process in which raw materials are converted at high temperatures to a homogeneous melt that is then formed into the articles. The flow diagram in Figure 18 summarizes the details of conventional glass manufacturing. The vapor deposition of SiO_2 from a flame fed with $SiCl_4$ and oxygen is the basis for manufacturing high-purity glass used for blanks which are redrawn into optical-waveguide fibers (see Fiber optics). Fused silica items that cannot be formed from viscous melts of SiO_2 or quartz are prepared by vapor deposition.

Raw materials are selected according to purity, supply, pollution potential, ease of melting, and cost. Sand is the most common ingredient. In the United States, approximately 90% of the quality sand produced is consumed by the glass industry; this accounts for (13–14) $\times 10^6$ metric tons sand per year. Both purity and grain size are important. Iron oxide, titania, and zirconia are the primary contaminants, but high concentrations of feldspars from sands in the western United States account for large

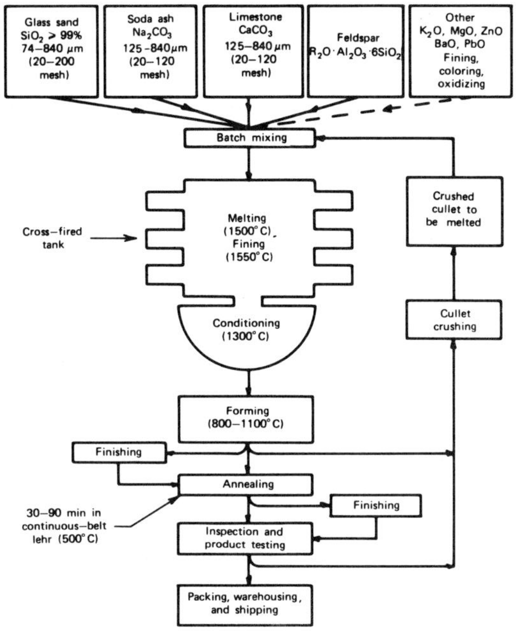

Figure 18. Glass manufacture. Temperatures are for common soda-lime glass. Other glasses may require appreciably different temperatures.

amounts of alumina and potash in glasses from that area. African sands or acid-washed varieties offer the lowest iron concentration. Container-glass manufacturers generally use sand between 590 and 840 μm (20–30 mesh) for the best compromise between the high cost of producing fine sand and melting efficiency. Fiberglass manufacturers, however, use a fine grain sand, <70 μm (200 mesh). Both agglomerated fine sand grains and undissolved coarse sand grains may cause melting problems. Shipping costs are often 3–4 times the original cost of the sand, and therefore, the plant should be near the sources of raw materials.

Limestone is the source of calcium and magnesium. It is available as a high calcium limestone and consists primarily of calcite (95% $CaCO_3$) or as a dolomitic limestone, a mixture of dolomite ($CaCO_3.MgCO_3$) and calcite. High-quality limestones contain less than 0.1% Fe_2O_3 and approximately 1% silica and alumina. Less than 5% of the total limestone manufactured is used for glass production. Transportation costs are several times the fob price. Development of the mineral aragonite (98% $CaCO_3$) as a source of CaO may be anticipated since large deposits of high purity aragonite exist near the coast of the Bahamas (see Lime and limestone).

The amount of soda ash (Na_2CO_3) produced by the Solvay process is decreasing, and most soda ash now comes from the Trona Wyoming deposits (trona, Na_2CO_3.-$NaHCO_3.2H_2O$). The glass industry consumes about 40% of the soda ash produced. Caustic soda (NaOH) solutions may be used for wet batching processes as a source of soda (see Alkali and chlorine products).

Other raw materials include boron, generally from deposits located in California or Turkey. Either anhydrous borax or boric acid is used but their consumption is decreasing because of the energy required to produce them. Mineral ores such as colemanite, rasorite, or ulexite are now used in addition to boric acid or pentaaquo borax ($Na_2B_4O_7.5H_2O$) when possible. Feldspar and nepheline syenite are common mineral sources of alumina. Litharge is used as a source of lead oxide for the manufacture of lead glasses. Sulfates or nitrates oxidize other oxides in the glass and control fining reactions.

Powdered anthracite coal is a common reducing agent in glass manufacture. Fining agents remove the bubbles in the molten glass and include sulfates, halides, peroxides, chlorates, perchlorates, CeO_2, MnO_2, As_2O_3, and Sb_2O_3. They react by release of oxygen or sulfur trioxide, or by vaporization as in the case of halides or ammonium salts. Controlled decomposition of sodium sulfate with powdered coal is used to fine many soda-lime compositions. Arsenic acid, used as an aqueous solution in place of arsenic trioxide, reduces airborne contamination of As_2O_3 in the batching and melting areas; solutions are dispensed on a volume or weight basis and cover the batch with a more even distribution. In addition to a more homogeneous batch, an oxidant such as sodium nitrate is not required with arsenic acid.

Common colorants for glass include iron, chromium, cerium, cobalt, and nickel (70). Small amounts of iron are sometimes desirable for color and controlled radiant heat transfer during melting. Ferrous sulfide or iron pyrites produce amber-colored glass used in the container industry. Sodium dichromate is a source of chromium for green container glasses, whereas small amounts of cobalt and nickel oxides added to flint glass decolorize the yellow-green color that results from iron contamination. Selenium with iron and cobalt yields a bronze color. Ceria is used to increase uv absorption in optical or colorless soda-lime glass and to protect glasses from x-ray browning.

Cullet, or broken glass, is used as a batch material to enhance glass melting and to reduce the amount of dust and other particulate matter that often accompanies a batch made exclusively from raw materials. Some forming operations, eg, the ribbon machine, generate as much as 70% waste glass which must be recycled as cullet. More efficient operations, as used in the container industry, may require the purchase of cullet from flat-glass companies or recycled-glass distributors. About 5% of the cullet in the container glass industry is from recycled bottles. Typically, 10–50% of a glass batch is comprised of cullet (see Recycling).

Modern analytical techniques have not only aided glass producers, but also enabled suppliers of raw materials to produce a more consistent product. Closer monitoring of purity and on-site blending of raw materials are now used by many producers. A stable and uniform raw material is the key to the manufacture of high quality glass.

Preparation of Raw Materials. Melting and fining depend on the batch materials interacting with each other at the proper time and in the proper order. Thus, extreme care must be taken to obtain materials of optimum grain size, to weigh them carefully, and mix them together intimately. The efficiency of the melting operation and the uniformity and quality of the glass product are very often determined in the mix house. Batch handling systems vary widely throughout the industry, from manual to fully automatic. Large tanks that melt glass of the same composition for years can easily be fully automated. Each batch material is stored in its own bin and the correct amount, determined and controlled by a computer, is weighed directly from the bin onto a conveyor belt which is fed directly into a mixer. Alternatively, sometimes only the gross weighing is made automatically and the final dribble feed is controlled manually. For small tanks and pot furnaces, where the composition changes frequently, the batches are sometimes weighed into a bin or hopper. Often the crucial weighing step is the least automated in small tank operations, which are also the most sensitive to fluctuations, and very often melt the types of glasses that must meet the most rigid specifications.

The method by which the batch is mixed depends more on the type of glass than on the size of the tank. High SiO_2 glasses (soda lime, borosilicate, aluminosilicate) tend to be batch-mixed in pan-type mixers. The batch is first dry-blended and then small amounts of liquid are sometimes added for wet-blending. The whole operation takes 3 to 8 min; longer treatment may unmix or segregate the batch. Many specialty glasses, especially those with high PbO, are mixed in either Muller-type mixers, where a smearing action coats the batch particles with each other, or in horizontally rotating drums.

Wet mixing and batch agglomeration (pelletizing or briquetting) are coming into vogue for several reasons, partially because of the increased use of arsenic acid (see Size enlargement). A wet batch (3–4% water) prevents dusting, controls air pollution, ensures homogeneity, and therefore increases melting efficiency and glass quality (71). Especially in batches with raw materials of varying grain size, a wet batch prevents particle segregation because of fine particles settling to the bottom of the bin. The homogeneous batch melts more efficiently because all of the batch materials are more intimately in contact with each other and also in the correct proportion. Homogeneity can be assured by agglomeration into pellets or application of pressure to form briquettes.

Table 7 shows by example the problem of batch homogenization, where 1 kg pyrites and 4 kg salt must be uniformly dispersed throughout a 1000 kg batch, nearly one third of which is 300 μm (50 mesh) sand and nearly half of which is cullet (1–2 cm) glass.

Melting. Ideally, when the intimately mixed batch is charged into the hot furnace, a series of melting, dissolution, volatilization, and redox reactions takes place between the materials in a particular order and at the appropriate temperature. Preheating raises the batch as rapidly as possible to the melting temperature where significant reactions occur which generally result in a distinct change in the flow characteristics of the batch (73). Pellets or briquettes heat up to melting temperature more efficiently than a granular batch.

Dissolution of the more refractory (higher melting) grains, such as sand, is accelerated by fluxes (lower melting materials), eg, Na_2CO_3. For example (74),

$$Na_2CO_3 + SiO_2 \rightarrow Na_2SiO_3 + CO_2$$

occurs first at about 550°C and a layer is deposited on the silica grains. As the reaction continues,

$$Na_2SiO_3 + SiO_2 \rightarrow Na_2Si_2O_5$$

occurs around 700°C. Finally, at about 780°C, forms a mixture approaching an eutectic

$$3\ Na_2Si_2O_5 + SiO_2 \text{ forms a mixture approaching}$$

liquid. The decomposition of alkali carbonates (Li_2CO_3, Na_2CO_3, K_2CO_3) or alkaline earth carbonates ($MgCO_3$, $CaCO_3$, $SrCO_3$, $BaCO_3$) proceeds in a similar manner, as does their fluxing action on sand and other minerals, notably alumina-containing ones such as nepheline syenite and feldspars. Undissolved grains (stones) are introduced either when the refractory grains are too large to react completely by the time the flux has melted and flowed away from them, or when the reactants are not intimately mixed. Increasing the melting temperature aids in dissolving stones and compensates for minor deviations from an ideally prepared and delivered batch, or one that is difficult to prepare and deliver.

Fining is the physical and chemical process of removing gas bubbles (seeds, blisters) from the molten glass melt. Gas is evolved during the first stages of melting because of (*1*) the decomposition of the carbonates or sulfates, or both, (*2*) air trapped between the grains of the fine-grained batch materials, (*3*) water evolved from the hydrated batch materials, and (*4*) the change in oxidation state of some of the batch materials, eg, red lead:

$$2(Pb_3O_4) \rightarrow 6\ PbO + O_2$$

Table 7 shows how much gas may be evolved from a typical amber soda-lime container glass batch. Fining agents are employed that generally react at higher temperatures than are needed for melting; thus, the fining reactions continue after dissolution and volatilization has taken place. Only materials that can release gases without delay through the formation of boiling bubbles can act as fining agents (75). As the gases are released, the bubbles rise to the surface roughly according to Stokes' law. Each bubble, during its trip through the glass melt, attracts new quantities of gas flowing to it by diffusion from neighboring layers. This makes the bubble larger and ascend faster, and depletes the glass in its path of smaller bubbles and dissolved gases. High temperatures make the glass more fluid and increase the diffusivity which speeds up the fining process greatly. The most common fining agents are the sulfates (75), fol-

Table 7. Typical Batch Mixture for Amber Soda-Lime Container Glass[a]

Batch materials	Weight, kg	Oxides supplied, kg							
		SiO_2	Al_2O_3	CaO	MgO	Na_2O	K_2O	FeO	LOI[b]
sand (SiO_2)	300	299.3	0.2					0.03	0.5
soda ash (Na_2CO_3)	100					58.3			41.7
aragonite ($CaCO_3$)	90			49.0	0.2			0.02	40.7
feldspar (SiO_2–Al_2O_3 mineral)	40	26.4	7.6	0.4		1.3	4.1	0.03	0.1
salt cake (NaCl)	4					2.1			1.9
powdered coal[c]	9								9
iron pyrites (FeS_2)	1.4							0.84	0.6
cullet (crushed glass)	460	333.7	9.2	48.8		67.2		1.03	0.1
Total	*1004.4*	*659.4*	*17.0*	*98.2*	*0.2*	*128.9*	*4.1*	*1.95*	*94.6*
yield of glass (kg) and wt % oxides present	909.8	72.48	1.87	10.79	0.02	14.17	0.45	0.21	

[a] Ref. 72.

[b] Loss on ignition. Generally, the oxides of carbon and sulfur (plus some chlorine, depending on the fining agent) volatilize during melting.

[c] Used primarily to reduce the Fe_2O_3 to FeO to give the characteristic amber color, although the redox state of the glass melt also influences the fining reactions.

lowed by sodium or potassium nitrates in combination with arsenic or antimony trioxides. Arsenic trioxide is used for higher-melting glasses, 1450–1500°C, and antimony for lower-melting glasses, 1300–1400°C. As the glass cools, dissolution fining may occur, ie, oxygen bubbles are removed by reaction with the arsenic or antimony trioxide to form the pentoxide.

Melt homogenization followed by cooling to working temperatures completes the melting process. Batch segregation, melt segregation, volatilization, and temperature fluctuations, as well as refractory corrosion from tank-lining material, cause compositional differences (cord, stria) within the melt. These inhomogeneities must be removed by diffusion and flow before the glass is cooled in the forehearth prior to delivery and forming. Vigorous fining action as well as convection current mixing helps break up cord. Mechanical and static mixers continuously shear the glass to help achieve homogeneity.

Melting Units. Melting units range from small pot furnaces for manual production to large, continuous tanks for rapid machine forming (76). After mixing, the raw materials are charged into a furnace for melting. Pot furnaces are used for melting smaller quantities of glasses below 1400°C. Pots, whether single or multiple port, are inefficient fuel consumers and have poor temperature control. However, because they can be heated from the sides as well as the top, they are useful for melting heat-absorbing specialty glasses. Hand producers may still use pots for experimental melts where the time between filling the pot and production can be as short as 24 h. However, for experimental melts day tanks are usually preferred because they have better refractories and higher attainable temperatures. Day tanks burn gas or oil and use a single opening for both charging and gathering of glass. The glass quality is better than that of pot-melted glass but not as good as that of glass melted in a continuous tank.

Small, continuous-melting tanks, filled automatically or by hand, produce high-quality glass, such as optical and ophthalmic glasses, at low volume. They are used especially when many glass changes are necessary. Capacity is in the 7 t/d range and two melting zones are used. The premelt section may be heated by electricity, gas, or a combination of the two and operates at a lower temperature (eg, 1400°C). The finer is an electrically heated platinum tube approximately 20 cm dia. Some of these tanks operate as cold crown units with the unmelted batch covering the surface of the molten glass in the premelt region. These units are heated electrically from the bottom or side, and their small size makes possible quick changes of glass. Draining and flushing is usually preferred. Pollution from glass melting is substantially reduced by using cold crown units.

The largest furnaces are continuous regenerative furnaces that recover waste heat from burned gases (77). They produce large quantities of quality glass and are either cross- or end-fired. The latter type is usually smaller. The hot combustion gases are passed through a chamber filled with refractory checkers. Combustion air and waste gas flow is reversed at 15–30 min intervals to alternately heat each chamber. Incoming heated air mixes with fuel and burns over the surface of the glass. Figure 19 shows the operation and Figures 20**a** and 20**b** show the design of a cross-fired regenerative furnace. Production ranges from less than 150 to over 700 t/d. The two ports of an end-port furnace produce a U-shaped flame over the glass that enters and exhausts from the back wall, whereas cross-firing from side to side allows more even heating across a larger surface area. Each type of furnace has a melting portion and a conditioning portion which are separated by a refractory bridge wall. The opening between the two areas,

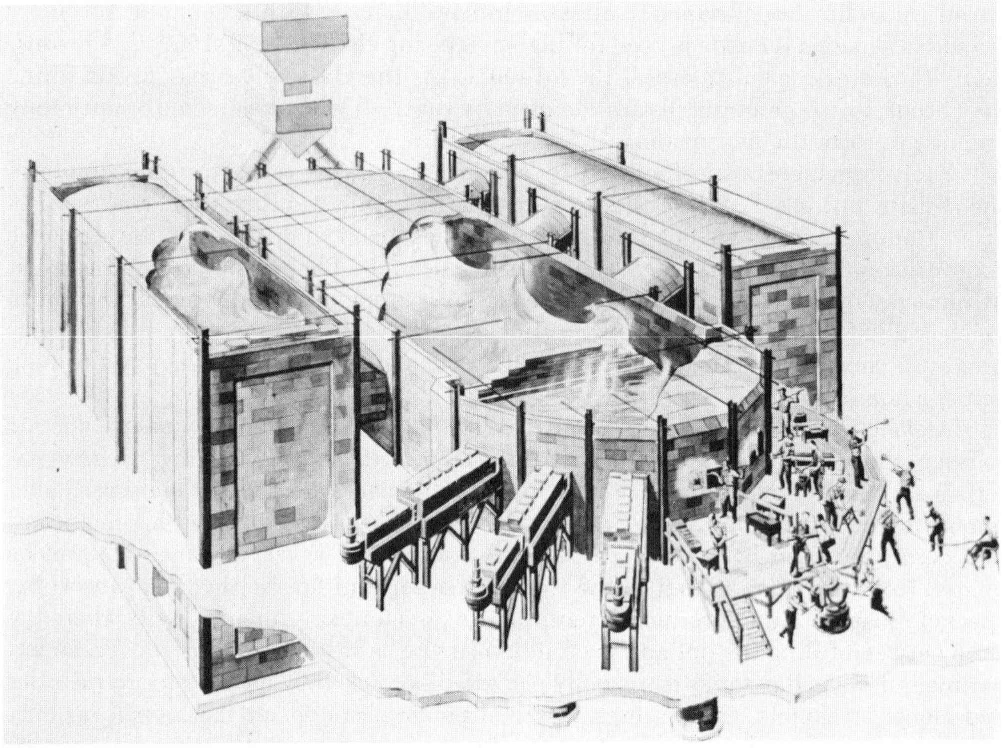

Figure 19. A cut-away view of a cross-fired, regenerative furnace. The feeding system is in the back. The combustion gases are blowing from the regenerator on the right, across the melt section, and into the regenerator on the left. The glass flows from the back, through the melt section and throat into the semicircular refining or conditioning section. There are six forming positions counterclockwise around the refiner: three forehearths, each with a stirrer at the end, which are idle in this picture, and three hand-shop positions from which are being produced large blown ware, tubing, and small blown ware, respectively.

called the throat, is beneath the surface and allows glass to flow but reduces surface contamination. Flat-glass furnaces may not have a bridge wall but are longer than container furnaces to give greater output of high quality glass. Scum, foam, or surface imperfections are removed by a skimming or floater device near the surface of the glass. The depth of glass ranges from 1 to 1.3 m and must be closely controlled for subsequent forming operations. Melting efficiency is determined by the tank area necessary to melt a given quantity of glass. Soda-lime furnaces for flat and container glass melt approximately one metric ton of glass per 0.41 m^2 of melting area per day (500 lb/($ft^2 \cdot d$)). Other glasses, such as borosilicates, may require as much as 1.0–1.5 m^2/t (9.8–14.6 ft^2/2000 lb) of glass melted.

Continuous recuperative furnaces employing metallic recuperators (heat exchangers) have been in use since the 1940s. Operation of these furnaces is simplified and the combustion process is more precisely controlled; no reversal of air flow causes temperature variations. The recuperator metal must be carefully selected because of chemical attack at high temperature. Recuperative furnaces are often used in the production of textile fiberglass because they maintain a constant temperature.

Development of molybdenum electrodes in the 1950s permitted the use of elec-

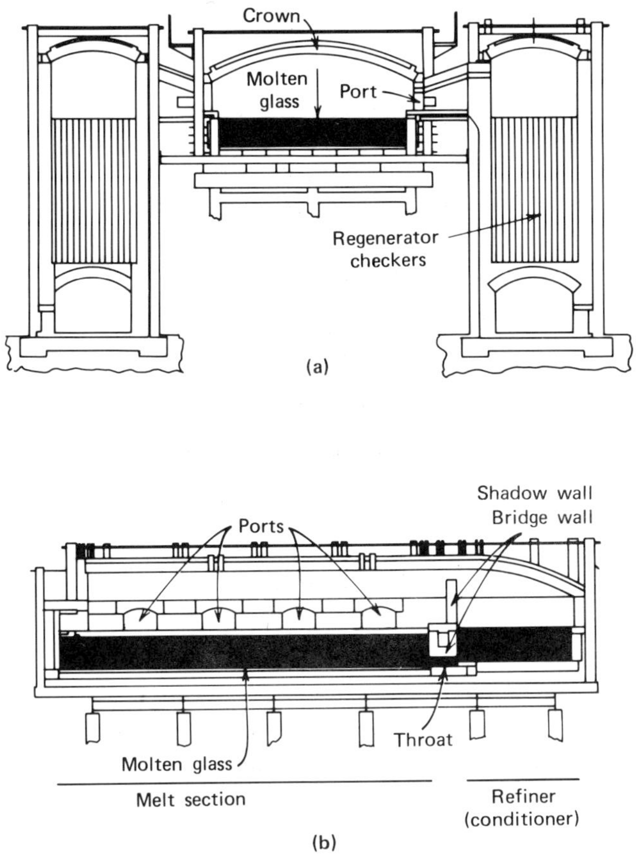

Figure 20. (**a**) Cross-fired regenerative furnace (end view). (**b**) Cross-fired regenerative furnace (side view).

trically assisted melting in regenerative furnaces (78). Today, approximately half of all regenerative tanks are electrically boosted. Recent operating practice shows the most effective use of electricity is near the back end of the furnace where the batch is added. Heavy electric boosting in this area reduces fossil fuel needs. This lowers surface temperature and reduces batch volatilization.

All-electric melting of glass has been patented since the beginning of the century. Today, electric furnaces (see Fig. 21) produce large quantities of commercial glass including fiber glass, lead, soda lime, borosilicate, and fluoride opal types. Furnace designs range from shallow rectangular types to round and vertical. Submersible electrodes of rod or plate design are placed on the bottom or side of tanks. High energy efficiency by conversion of electrical energy into useful heat and low volatilization are primary advantages of all-electric melting. For example, in a fluoride-containing batch as much as 40% volatilization might occur with gas firing compared to 2% with electric. High-quality glass, cleanliness of operation, and small space requirements are additional advantages. Disadvantages, especially to the container industry, include the difficulty in maintaining a constant glass level in the tank when melting cold crown and the lack of ability to use alternative fossil fuels. Most glasses, in principle, can be

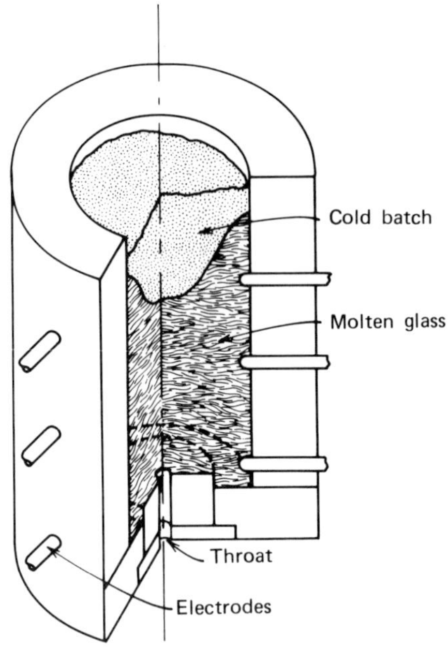

Figure 21. Cold crown all-electric melter (vertical design) (79). Courtesy of Corning Glass Works.

melted electrically. However, glass compositions with steep resistivity–temperature curves, such as an alkali-free glass, present problems since a cold batch surface is quite difficult to maintain because of the rapid decrease in resistivity with increasing temperature.

The choice of refractories (qv) (see Table 8) depends upon the type of glass and furnace used and the position of the refractory in the furnace. Worn refractories are often replaced before the tank is rebuilt, whereas others run continuously as long as possible before a shutdown.

Feeding Systems. The two common feeding designs used today are the screw feeder and the reciprocating pusher. The former delivers the batch from a hopper to the furnace by tube and helical gear drive, whereas the latter forces a layer of the batch from the feed chutes onto the molten glass. The enfolding feeder uses a vibrating chute to deliver the batch as a rotating device dips into the surface of the molten glass, which causes the batch to be enfolded into the surface. Blanket feeding is done by conveyor or pusher principle. Regardless of the type of charger used, all empty into the backwall area of the furnace or into an extension known as the doghouse. A batch charged into a doghouse usually is glazed over before entering the hotter melting area. All-electric melting relies on the complete coverage of the molten glass surface by several centimeters of the batch. Both conveyor and pusher-type feeders are used.

Control Devices. Control devices have advanced from manual control to sophisticated computer-assisted operation (see Instrumentation and control). Radiation pyrometers in conjunction with thermocouples monitor furnace temperatures at several locations (see Temperature measurement). Batch filling is usually automatically controlled. Combustion air and fuel are metered and controlled for optimum efficiency. For regeneration-type units, furnace reversal also operates on a timed

The table shown below was inadvertently omitted from the article on "Glass" in Volume 11. It should follow the discussion on page 854.

Table 8. Refractories Used in the Glass Industry

Refractory	Composition, wt % ZrO$_2$	Al$_2$O$_3$	SiO$_2$	MgO	Cr$_2$O$_3$	Other	Apparent porosity	Density[a], g/cm^3	P.C.E.[b] or maximum use temp, °C	MOR[c], MPa[d]	Application
Fuse cast											
AZS	33–40	45–51	10–15			1–2 (Na$_2$O)	1		1650	>34	glass contact
α-alumina		99					1.1		1870	40.3	superstructure
β-alumina		94–95				5–6 (alkali)	4.1		>1650	10.3	breastwall, arches, crowns
α,β-alumina		95	1			4 (alkali)	2		1870	24.5	fiberglass furnace throats
alumina–chrome		60–61	1.5–2	6	27–28	4–5 (Fe$_2$O$_3$)	4.2		>1650	54.8	refiner or channels
chrome–spinel		4–5	1–2	8–9	79–80	6–7 (Fe$_2$O$_3$)	4.9		>1650	high	textile fiberglass (high ware areas)
Bonded or sintered											
dense zircon[e]	64–65		33–34				0.5	4.22	>42	58.6	high-duty glass contact
medium dense zircon	65–66		33–34				11	3.92	>42	58.6	channels, refiner, forehearth
sintered zircon	65–66		33–34				16–28	3.39–3.76	>42	9.6–24	melting and refiner bottoms
sintered zircon AZS[f]	20	70	10				20–23	2.92–3.04	34–35	13.8–22.8	tubes, refiner paving, regenerator target walls
dense sintered alumina		94	4–5			0–1 (TiO$_2$)	1.6	3.44	1760	58.6	optical and lead glass tanks
alumina, 90% brick		89–91	8–10				19–26	2.66–2.86	>40	7.6–14	superstructure
mullite/sillimanite		60–80	18–37			1–6	14–24	2.24–2.59	36–40	9.3–14	backwalls, breastwalls, bottom checkers
fireclay, bricks and flux		37–45	57–50				6–22	2.05–2.42	31–34	4–12	day tanks or noncontinuous furnaces, sidewall insulation
superduty silica brick			96.3			2.9 (CaO)	20–24	1.84	1590–1650	3.4–7	crown
fused silica brick			98–99			1.2	11–15	1.87–1.94	1650	3.4–5.5	burner blocks, thermocouple tubes
insulation brick (I.F.B.)		34–36	56–58			6–10	63–77	0.67–0.86	1090–1430	0.7–1.2	outer insulation
high MgO brick				98		1–2 (CaO)	17–22	2.77–2.88	>1760	10–28	refractory checker
spinel-bonded magnesite		8.3		88.5		3–4	22	2.77	>1760	6.9	refractory checker
magnesite brick			4.1	93.6		2–3	20	2.78	1700	18.3	refractory checker
forsterite brick		10.9	29.5	50.3			22	2.62	>1650	8.6	refractory checker
dense chromic oxide[e]					95	4 (TiO$_2$)	18	4.06	>1650	44	fiberglass furnace throats
chrome brick		15–34	3–6	14–19	22–28	11–17 (Fe$_2$O$_3$)	18–25	2.88–3.36	1650	18.6	lower zone refractory checkers
chrome–magnesia brick		16–27	4–8	27–53	12–28	8–14 (Fe$_2$O$_3$)	14–25	2.56–3.26	1540–1650	2.8–17	regenerator and crown
chrome–alumina		81			71	3					fiberglass tanks
unichrome		59	29	1.0	4.2	2 (Fe$_2$O$_3$)					fiberglass tanks

[a] The density of fused cast refractories depends upon the casting method used.
[b] P.C.E. = pyrometric cone equivalent.
[c] MOR = modulus of rupture.
[d] To convert MPa to psi, multiply by 145.
[e] For fiberglass furnaces.
[f] Common glass contact material.

program. Data acquisition and digital display of operating parameters are part of a supervisory control system. The grouping of display information at the control center is typical of modern furnaces.

Benefits resulting from better control of glass melting are lower fuel consumption, better glass quality, more efficient production time, and better pollution control. Fewer operators are needed and working conditions are better.

Fuels and Efficiency. Natural gas, oil, and electricity are the primary sources of energy; propane is used as a backup reserve in emergencies. Natural gas is the least expensive and most frequently used fuel, with heat content ranging from 34–45 MJ/m^3 (900–1200 Btu/ft^3) for raw gas and approximately 3 MJ/m^3 (80 Btu/ft^3) for air–gas mixtures. Fuel oil has heat content between 39–43 MJ/L (139,600–153,000 Btu/U.S. gal). Fuel oil is viscous at low temperature and must be heated before being fed to atomizing burners where it is mixed with air for combustion. Producer gas is made by blowing air and steam through a bed of incandescent coke. Investment costs for producer gas-fired furnaces are higher than for natural gas, and they are used less frequently.

The efficiency of gas-fired regenerative furnaces is about 30%. Figure 22 shows the portions of energy that are lost to structure and flue gases. Direct application of electrical energy to molten glass by electrodes melts glass more efficiently. The efficiency of electric melting is 2 to 3.5 times that of fossil fuels, but production of electricity from fossil fuel at the power plant is only about 30% efficient. Electric furnaces absorb less heat and there are no costly regenerators to repair or replace. However, the most flexible furnaces are still the electrically boosted fossil fuel tanks where a combination of energy sources is used rather than a single fuel (see Furnaces, electric; Furnaces, fuel-fired).

Pollution (80) from soda-lime glass production is minimal, mostly caused by sodium sulfate particulates. Other soda-lime contaminants include SO_2, NO_x, CO_2, and hydrocarbons and are reduced by scrubbers, bag houses, and precipitators or sometimes controlled by optimizing operating conditions. Volatilization of lead, fluorine, and arsenic for specialty glass manufacture has to be carefully controlled. All-electric melting and new batching procedures have been helpful in reducing volatilization. Electrostatic precipitator dust (E.P. dust) collected from a stack can be recycled and may, in some cases, offset the operating cost of the pollution control device. Powdered coal as a reducing agent controls the decomposition of sodium sulfate more efficiently than increased temperature. Therefore, less sulfate can be used which results in lower emissions with equivalent glass fining. Higher cullet ratios are also known to reduce emissions. New operating procedures or furnace modifications, such as smaller ports near the rear (batching end) of a furnace rather than near the bridge wall, reduce volatilization because less heat is needed to start the melting process than to refine the glass (see Air pollution control methods).

Forming. Molten glass is either molded, drawn, rolled, or quenched, depending on desired shape and use. Bottles, dishes, optical lenses, television picture tubes, etc, are formed by blowing, pressing, casting, and/or spinning the glass against a mold to cool it and to set its final shape. Window glass, tubing, rods, and fiber are formed by freely drawing the glass in air (or on a bath of molten tin as in the float process) until it sets up and can be cut to length. Art glass is usually hand-formed by freely blowing and shaping it while soft. Glass that is intended to be crushed into powder, called frit, is quenched between water-cooled rollers or ladled or poured directly into water and then dried (dry-gauged).

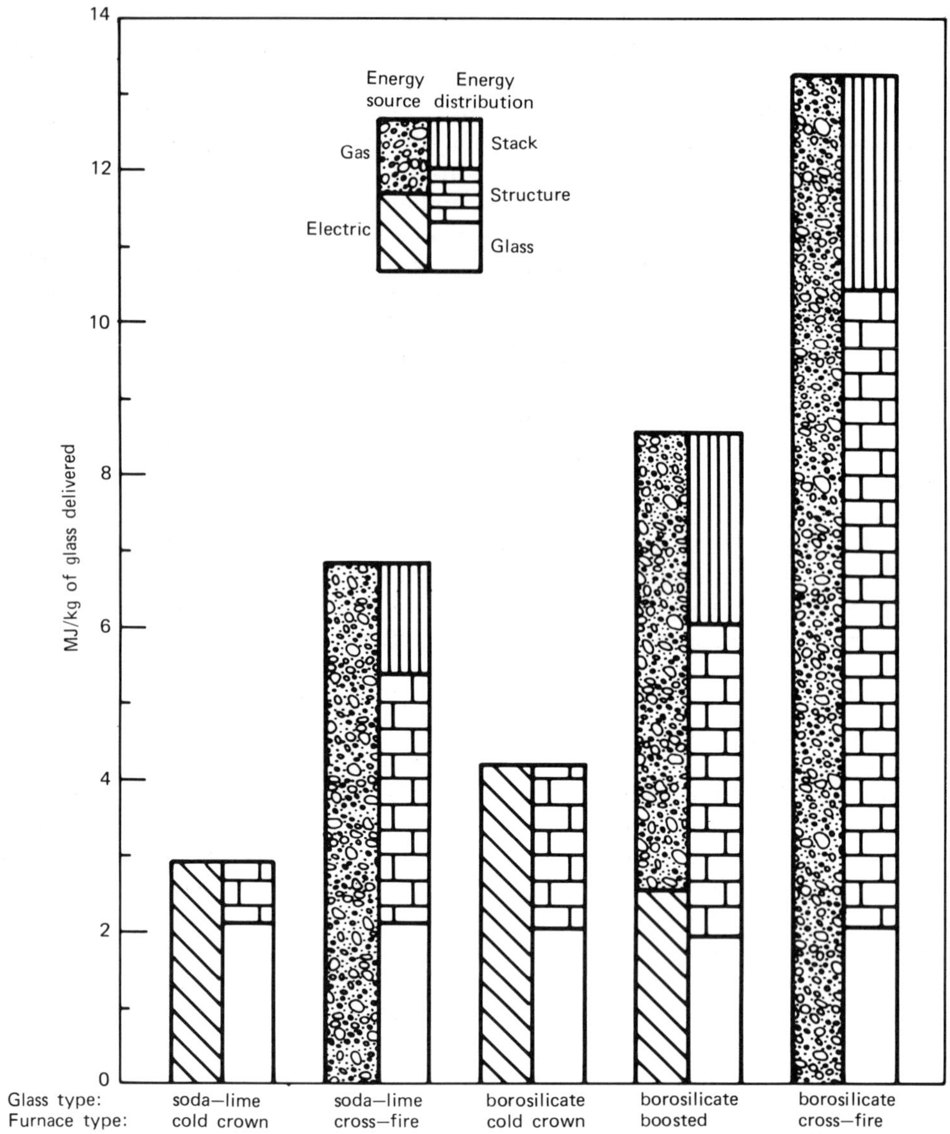

Figure 22. Energy source and distribution for five different glass and tank conditions. Courtesy of Corning Glass Works. To convert MJ/kg to Btu/lb, multiply by 430.

Molding. Many glass articles are made by shaping a glass gob by forcing it under pressure against a mold. The source of pressure and type of mold vary with the application.

Blowing. Deep items such as bottles, jars, or light-bulb envelopes are formed by the use of air pressure either from a pair of lungs or an air compressor. Hand blowing is still practiced by highly skilled craftsmen for specialty items for scientific, industrial, lighting, and electrical applications. Items up to a kilogram in weight can be made with a single ball of glass gathered on the end of a punty iron, preshaped, and then blown into a mold.

Articles of circular cross section may be made in iron paste molds. To keep the inner surface of the paste mold moist, it is coated with shellac or varnish and a mixture of charcoal and linseed oil is baked on. Hot iron molds are used for ware of any shape, particularly for screw threading, multiple decoration, or raised lettering.

The Hartford-Empire individual section machine is an in-line machine of up to 12 sections (eg, H.E. IS-12) used to blow small-neck containers (see Fig. 23). Each section is an individually functioning hollow glass machine which can handle one to three gobs at a time, depending on the size of the ware. The rate of production is about 10 pieces per section per minute. In this blow-and-blow process there is no movement (other than opening and closing) of the molds under the glass delivery system. Although this machine usually makes narrow-neck bottles, it can be converted readily to a press-and-blow machine for making wide-mouth jars with screw threads. In general, iron molds are used.

The Corning ribbon machine employs a puff-and-blow method to make incandescent light-bulb envelopes (1200/min), flash-bulb envelopes (2000/min), Christmas tree ornaments, and the like (see Fig. 24). For larger articles, the turret chain machine also uses the puff-and-blow method but produces less cullet and uses individual glass gobs rather than ribbon.

The Hartford-Empire 28 is a press-and-blow machine used to make articles such as drinking glasses (tumblers). It uses paste molds and the ware is rotated to avoid the mold seams; jars with screw threads cannot be produced. The product leaves the machine as an almost closed, hollow object and is finished by severing and fire-polishing with a burn-off machine.

Pressing. Flat items such as dinnerware, optical and sealed-beam lenses, filter glasses, and television tube panels are pressed between a plunger and a mold, most commonly with an automatic rotary press. Typically, there are 8–16 mold positions on the periphery of a horizontally rotating table (see Fig. 25). Cast iron, bronze, steel, and some specialty alloys are common mold materials. A gob of glass at about 200–400 Pa·s (2000–4000 P) is fed into the mold at the first station. The table rotates and a retaining ring and plunger press the article at the next station. The glass cools for several more stations until it sets up at 10^5–10^8 Pa·s (10^6–10^9 P) and is taken from the mold. The mold is cooled and ready to take its turn and receive another gob. Pressed ware ranges from 5 g to 15 kg, pressing pressures range from 0.5 to 0.8 MPa (75–100 psi). Solid shapes are pressed in font molds which are small cavity-split molds into which molten glass is extruded by the plunger (see Fig. 26). Automatically produced pressed ware has about 0.5% variation in bottom thickness when made from the same mold, and with accurate gobbing, the manufacturing tolerance for diameter is about $\pm$ 0.05%.

Casting is the process of shaping glass by pouring it into a mold. The largest piece ever cast was a borosilicate telescope mirror 600 cm in diameter weighing 36 t. Dimensional tolerances as low as 0.1 mm can be achieved. Funnel-shaped ware, such as missile radomes and television bulbs, are formed by centrifugal casting. Gobs from a feeder are dropped into a spinning mold that creates centrifugal forces making the glass flow upward, forming a wall of relatively uniform thickness (see Fig. 27).

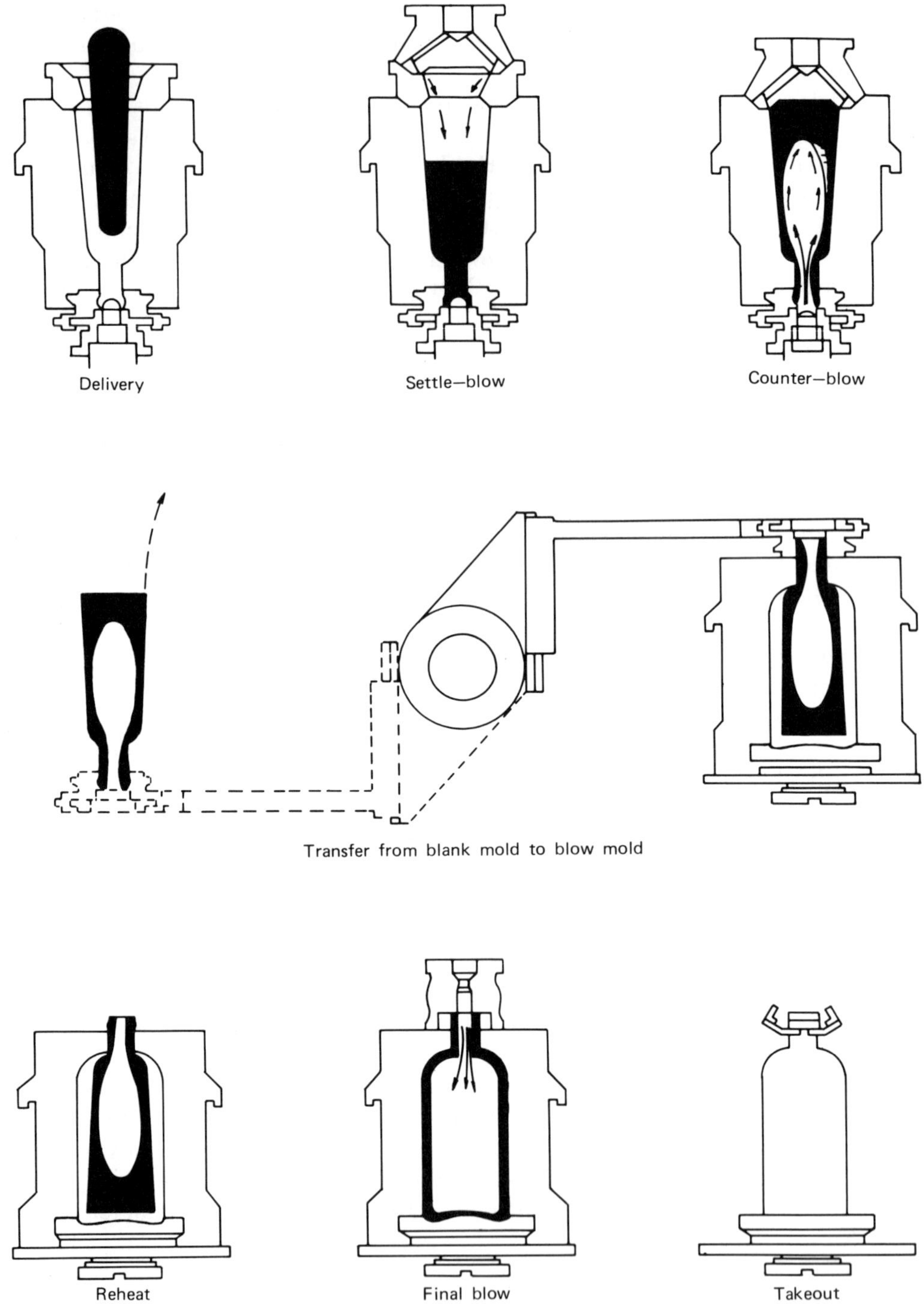

Figure 23. The H.E. IS blow-and-blow machine (81). The gob is delivered into a blank mold, settled with compressed air, and then preformed with a counter-blow. The parison or preform is then inverted and transferred into the blow mold where it is finished by blowing.

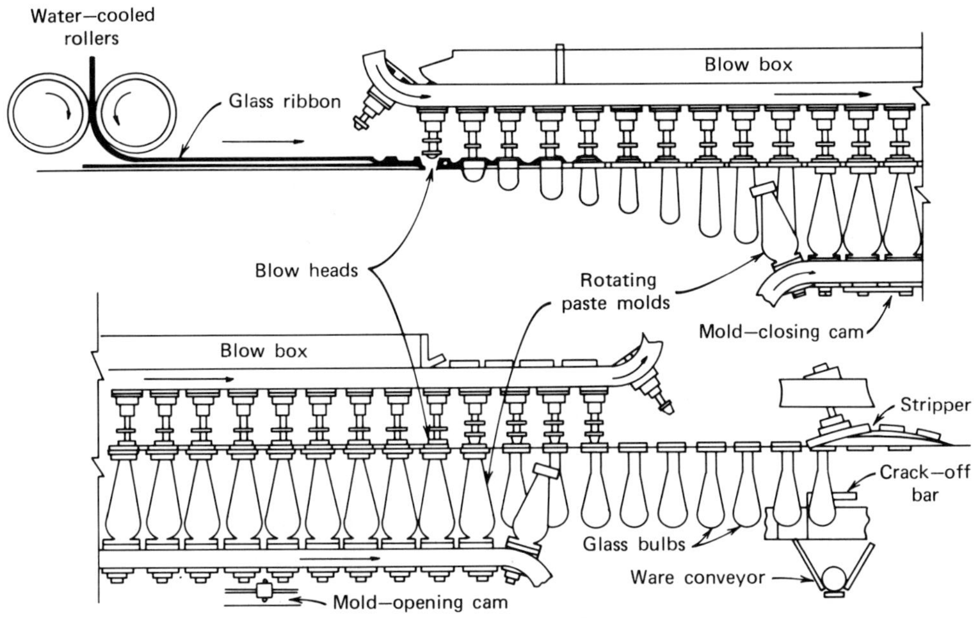

Figure 24. The Corning ribbon machine (82). Courtesy of Corning Glass Works.

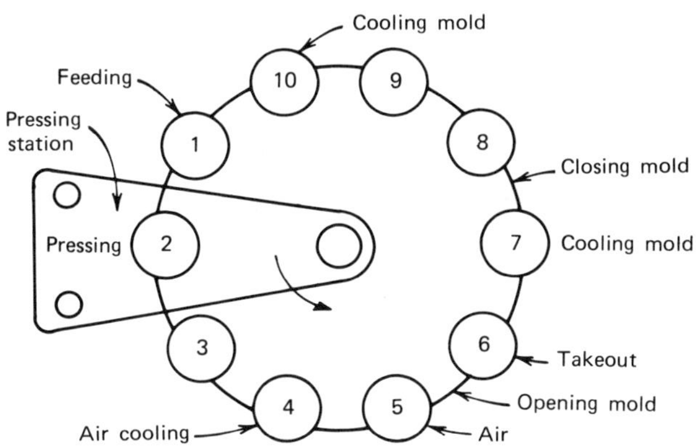

Figure 25. Stations of a 20-mold table press (81).

Drawing. Drawing molten glass from a specially designed orifice in a continuous manner and letting it set up before it touches a solid mold gives the glass a uniform cross section and a fire-polished surface. The surface rarely has to have a secondary finishing operation. There are few, if any, moving parts in the forming equipment, and the ware can be cut to any practical length at the end of the draw.

By the 1950s the drawn-sheet processes such as the Fourcault process (81), the LOF-Coburn process (81) (Fig. 28), and the PPG Pennvernon process (81) (Fig. 29) manufactured cheap, fire-polished window glass eminently suitable for domestic glazing and greenhouses. High-quality glass was made by polished-plate processes (81,85), which incorporated a costly grinding and polishing step. However, about 20%

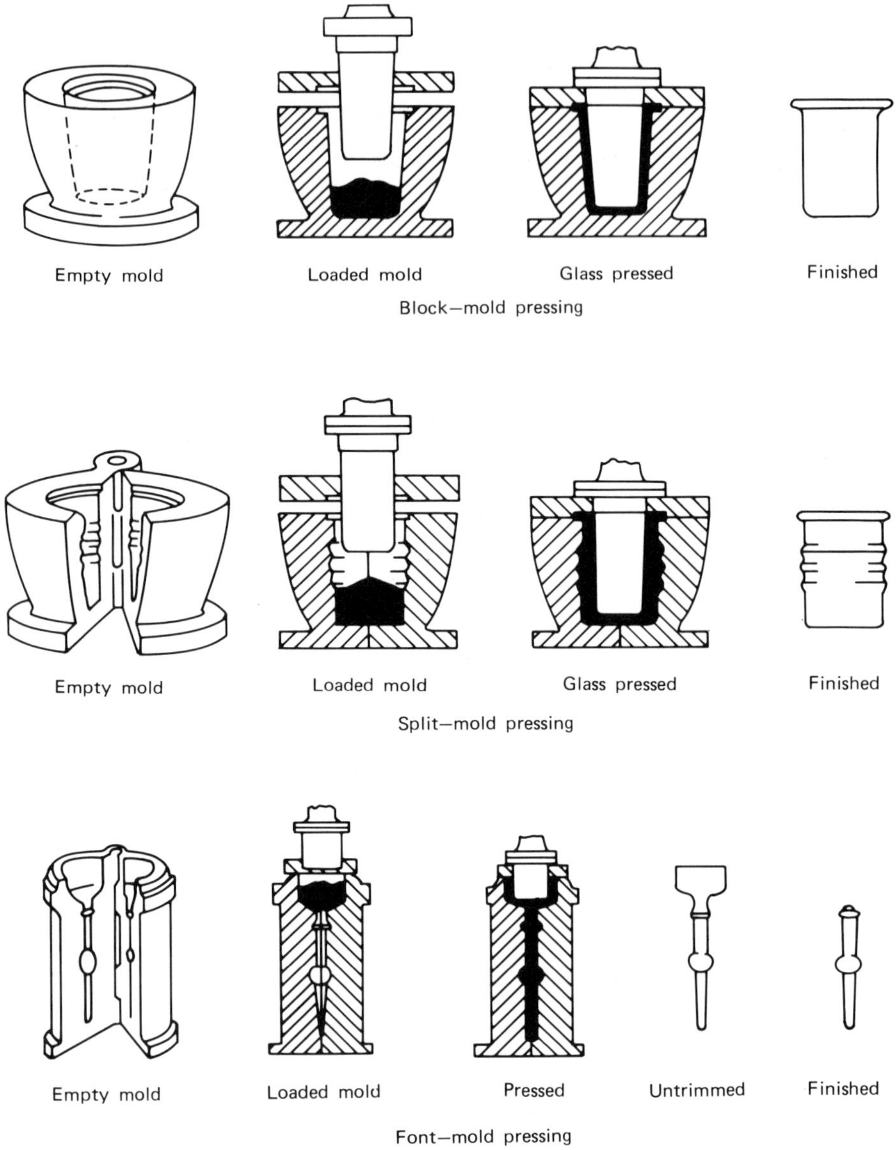

Figure 26. Pressed glass. Mold types and pressing operations (83). Courtesy of McGraw-Hill Book Co., Inc.

of the glass was waste which was ground off in order to make the surfaces flat and parallel (85).

The advent of the Pilkington float process at the end of the 1950s revolutionized the flat-glass industry, and by 1976 the United States was producing about 85% of its flat glass by this process (86). It combines the low cost and fire polish of drawn-sheet glass with the distortion-free quality of polished plate. A continuous ribbon of glass moves out of the melting furnace and floats along the surface of an enclosed bath of

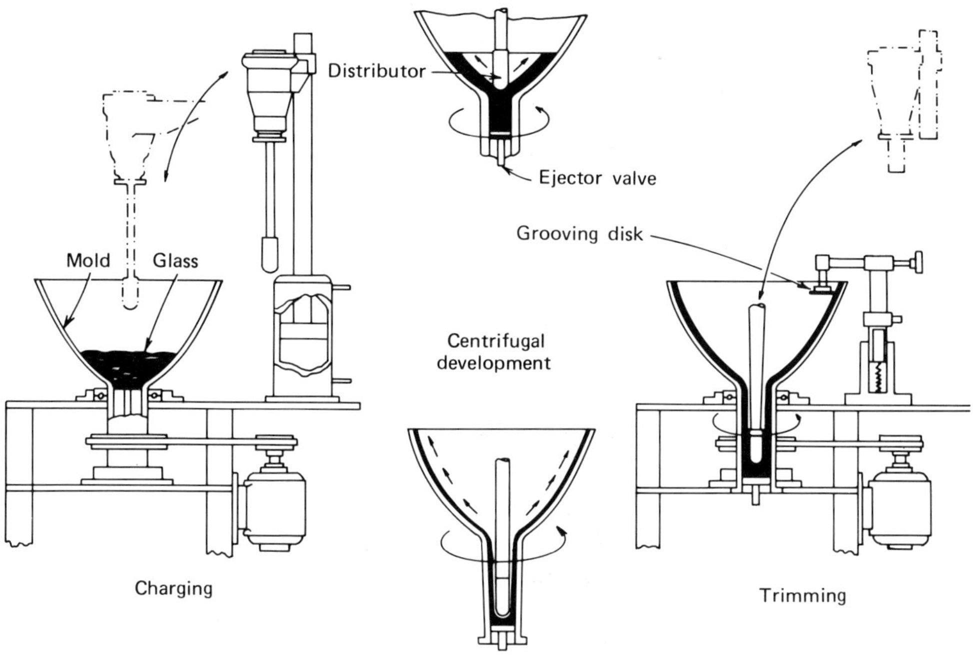

Figure 27. Centrifugal casting of television tube blanks (84). Courtesy of Corning Glass Works.

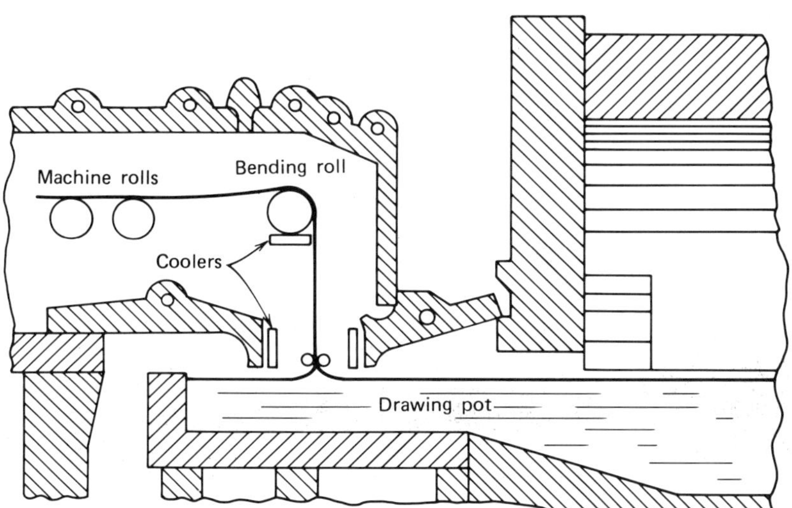

Figure 28. The Libbey-Owens-Ford Colburn process. The glass, drawn straight upward for a short distance, is passed over a bending roll and drawn horizontally over flattening and stretching rolls into the annealing lehr.

molten tin (Fig. 30). The ribbon is held in a chemically-controlled atmosphere at a high enough temperature for a long enough time for the irregularities to even out and for the surfaces to become flat and parallel. Because the surface of the molten tin is flat, the glass also becomes flat. The ribbon is then cooled while still advancing across the molten tin until the surfaces are hard enough for it to be removed from the bath

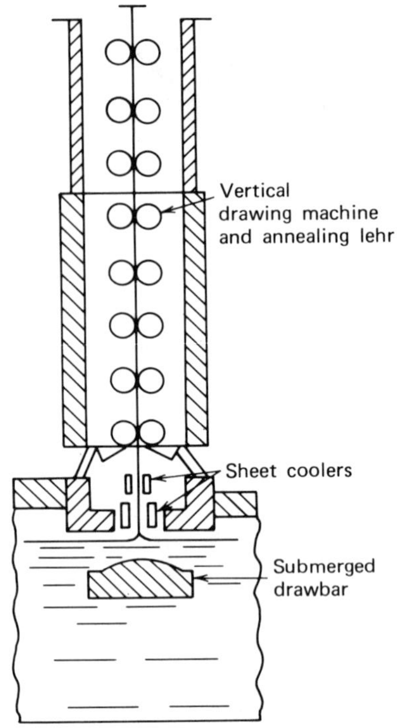

Figure 29. The PPG Industries Pennvernon process is a vertical drawing process. The sheet is drawn from the glass surface directly over a submerged drawbar. This horizontal refractory slab assists in conditioning the glass, determines the line of origin of the sheet, and controls convection currents in the drawing chamber.

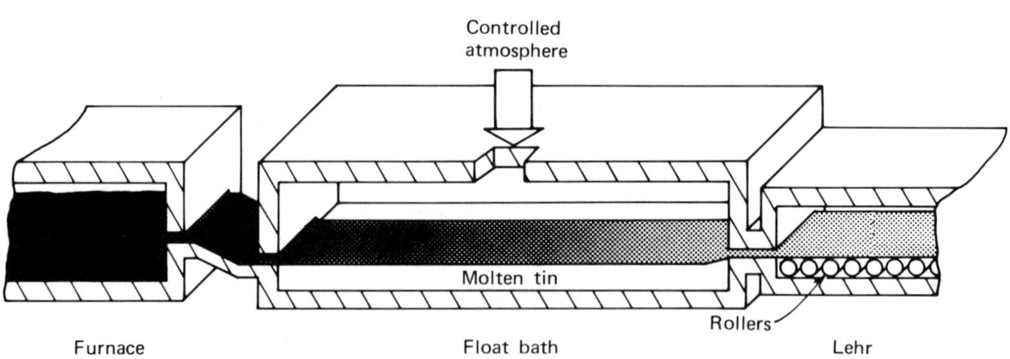

Figure 30. Diagram of the float process.

without the rollers marking the bottom surface (87). In 1975, PPG Industries developed a novel delivery system in which a stream of molten glass of the desired production width, normally about 4 m, formed at the exit end of the melting furnace and remained that width as it flowed onto the bath. By contrast, glass melted in the Pilkington process enters the bath through a restricted opening and is deposited, unshaped, on the molten tin. The improved PPG delivery system minimizes flow effects which introduce optical distortion in the finished glass (86). At present, the float process is limited to soda-lime–silica glass and is capable of producing ca 50,000 m^2 (500–800 t) of glass per day that is typically 3–3.5 m wide. The equilibrium thickness of glass on the molten tin is 6–7 mm, depending on the various glass–tin–atmosphere interfacial tensions and on the densities of the liquid glass and tin. These parameters, of course, change as the glass moves along the tin from 1050°C (ca 1 kPa·s or 10^4 P) to 600°C (ca 10^{10} Pa·s or 10^{11} P). In practice, thinner glass can be made (down to 2 mm) by stretching the glass ribbon on the molten tin. However, imperfections, which become elongated and thus more serious, must now be more closely controlled. Float glass is highly vulnerable to chemically-produced effects (85), such as molten metal or a reducing atmosphere of nitrogen and hydrogen.

The overflow process, developed by Corning Glass Works (88) in the late 1960s, is a downdraw process for specialty applications (see Fig. 31). It can handle a wide variety of glasses, ranging from soft (660°C) photochromic glasses to hard (865°C) aluminosilicate chemically-strengthenable glasses for automobile and aircraft windows. Molten glass flows into a trough, then overflows and runs down either side of a pipe, fuses together at the root, and is drawn downward until cool. Both sides have a fire-polished surface comparable to polished-plate or float glass. This is a higher viscosity process (20 kPa·s or 200,000 P) than float (1 kPa·s or 10^4 P). The thickness of the sheet is independent of the flow rate and therefore is controlled directly by the velocity of pulling. Very thin glass (0.4–0.8 mm) can be formed routinely, and thick glass (up to 1 cm) has been produced. Although scale-up is possible, present production widths are not quite half the width of float-process glass. Currently, production rates are less than those of the float process.

Glass rod and tubing are also made by drawing (81). About 15 m of 140-mm OD, 2-mm wall thickness tubing can be made on one draw by a gaffer and two helpers. Gathers of glass as heavy as 30–35 kg are made with a bubble in their center at the end of the blowpipe. This is rotated in place by the gaffer and rotated and stretched out

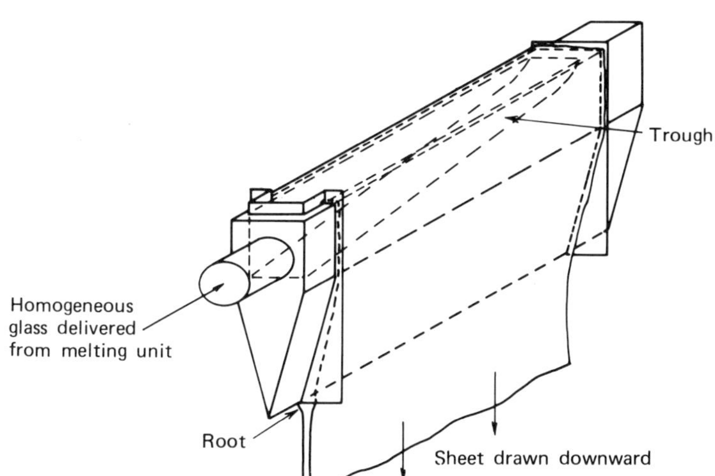

Figure 31. Schematic diagram of the Corning overflow pipe (88). Courtesy of Corning Glass Works.

by one of the helpers walking away from it. The size and diameter of the tubing is a function of the helper's walking speed, and is further controlled by a second helper who cools the tubing by fanning at the appropriate time. As much as 150 m of the much smaller diameter thermometer tubing can also be made in this manner from one gather.

In the Danner process (see Fig. 32), a continuous stream of glass flows on to a rotating mandrel. Tubing is drawn off the end while air blown through the mandrel helps to maintain tubing dimensions. The glass can be heat-treated on the mandrel. Because of the circular symmetry of the glass it is more thermally uniform. However, hairpin cords of inhomogeneity and volatilization from the surface may occur. In the Vello process the molten glass passes down through the annulus between a horizontal ring and a vertical bell. The moving stream is then brought around to a nearly horizontal runway and, as with the Danner process, is drawn off at the appropriate rate for dimensional specifications and stability. The Vello process can form tubing faster because the glass can be cooled more quickly to the appropriate forming viscosity in a stationary forehearth of any desired dimensions. Cooling in the Danner process is carried out on the rotating mandrel itself. Although the Vello process is more difficult to operate, the dimensional tolerances of the product are better.

Fiberglass is made in two basic forms, continuous filaments (4–15 μm dia) and fibers (6–9 μm dia and 20–40 cm long) (81,89). Single filaments can be combined into a strand that is easily unwound from a spool. The product is twisted textile yarn. Staple fibers that meet diameter and minimum length requirements are twisted together into a tow that can also be easily unwound from a spool. The end product is staple yarn. Glass fibers that do not meet these requirements are not suited for textile applications,

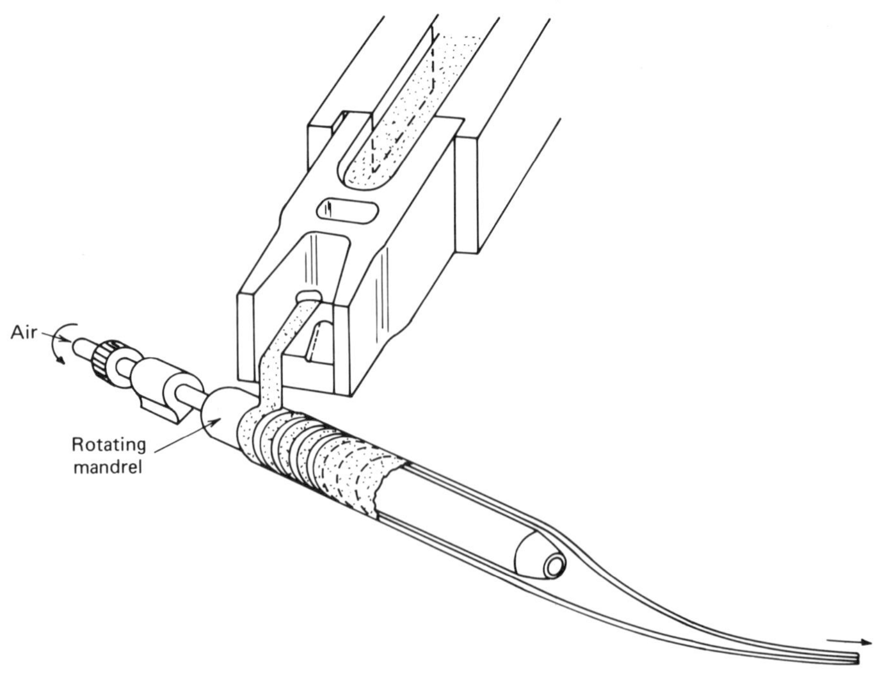

Figure 32. Working scheme of Danner rotating mandrel.

but may be used for thermal and sound insulation in the form of glass wool (81). Nearly all continuous filament yarn is made from E-glass which is well-suited for electrical insulation (see Insulation, electric), heat-resistant applications (see Insulation, thermal) (89) and plastic reinforcement and mats (see Laminated and reinforced plastics). Most staple fibers are spun from C-glass which is more resistant to chemical corrosion and is used for chemical filtration fabrics, etc. A low-cost soda-lime glass, T-glass, is used where high durability is not required, eg, coarse fiber mats for air filters and thermal and acoustical insulation (see Insulation, acoustic). For weathering resistance, SF-glass is used in fine and ultrafine wool products such as very low-density thermal and acoustic insulators, paper additives, and high-efficiency all-glass filter papers. Formerly, molten glass was formed into marbles which were remelted in an electric furnace and formed into filaments. However, many manufacturers now use the direct-melt method, where the molten glass is formed directly into filaments. Continuous filament yarn is drawn from the molten glass as it comes through small holes in a high-temperature alloy bushing. The high-speed winders, operating as fast as 3000 m/min, pull the streams into filaments. Staple fibers are formed by jets of air that pull the fibers from the furnace and deposit them on a revolving vacuum drum.

The Owens steam-blowing process produces most of the fibers for the glass wool application. Glass issues from slot-type openings in a platinum–rhodium bushing and is fiberized by a steam blower mounted just below the bushing. The fibers fall through a spout into a hood and deposit uniformly onto a conveyor chain belt which transports the fibers in the form of a felt or mat from the hood (81). In the rotary wool-forming process, molten glass is delivered to a rotating cylinder the face of which is perforated with a large number of holes. Glass streams, projected laterally from the holes by centrifugal force, are then attenuated into discontinuous fibers by a high-velocity gas stream (air, steam, or combustion gases) (90). The strength of glass fibers (strength-to-weight ratio) is over twice that of the strongest textile fibers, and much stronger than ordinary glass because of the relatively pristine surface.

Fritting. Cooling a glass quickly, directly from the melting unit, creates a high thermal stress that shocks and often breaks it into small pieces suitable for charging into a ball mill. The molten glass is either quenched in water (dry-gauged) or between water-cooled metal rollers to make a thin, friable ribbon, depending on its viscosity and sensitivity to water. The thin-rolled ribbon often provides more uniform mill feed. Dry-gauging sometimes forms clinker-like, tempered pieces that are difficult to mill.

Protective and decorative enamels, solder glasses for glass-to-glass or glass-to-metal seals, and a variety of sintered glass products (multiform) are made from frit. Enamels (qv) and solder glasses are designed to be fluid (viscosity ca 300 Pa·s or 3000 P) at their processing temperatures, whereas the substrates to which they are joined remain relatively rigid (viscosity = 10^8–10^{13} Pa·s or 10^9–10^{14} P). The resultant product is a glaze/substrate composite. Protective coatings such as porcelain enamels and glazes for both metal and glass are ball-milled with a variety of clays, electrolytes, and fluids to achieve the correct particle size (<74 μm to <44 μm or -200 to -325 mesh), shape, and distribution prior to dipping, spraying, or dusting on the ware for final firing (glazing) (91–93) (see also Ceramics).

Small, complex shaped glass articles such as thread guides for the textile industry and television gun mounts for the electronics industry are made by the multiform process. The dry-milled powder is mixed with an inorganic binder and a fluid vehicle, and then atomized by a spray dryer into small, dried agglomerates of glass powder and binder with good flow characteristics. They are subsequently pressed to the desired shape and fired.

Foam glass is made from powdered glass that has been mixed with an oxidizing agent and lampblack. When heated close to melting temperatures, the CO_2 evolved produces a cellular structure in the glass which can be controlled to nearly any density below that of the base glass. Foam glass, made by Pittsburgh-Corning, is used as insulation and has a density of 0.14 g/cm^3.

Slip Casting. Complex shapes can be made cheaply and easily by slip or drain casting from compositions that cannot be formed by the usual glass-forming techniques. Glass is finely ground and suspended in water. This slip, which also contains a binding agent, is poured into plaster of Paris molds which absorb the water by capillary action. The resulting partially dried articles are removed from the molds and fired to produce consolidated products. Honeycomb structures are made by a modification of this process. Thin sheets of glass or ceramic powder on a paper substrate can be corrugated, stacked, and fired to produce bodies with continuous channels. Ceramic burners, catalyst supports, and heat exchangers are formed by this technique.

Lamination. Simultaneous forming of two glasses to produce a single article is possible if the two have similar viscosities in the appropriate temperature range. Thermometer tubing is a composite of a white opal-glass stripe embedded in a clear glass tube parallel to the hollow bore. The opal glass is melted separately from the clear glass, but, while still fluid, it is fed into the Vello tubing machine and applied during the forming operation to the clear glass. The glasses are then drawn out the orifice and down the runway together.

In the late 1960s, Corning Glass Works (94) developed a vertical rotating turret machine which, when used with a special orifice, forms a laminated glass article consisting of a core glass and a skin glass (see Fig. 33). This process is used by Corning for the manufacture of moderately priced plates, bowls, cups, and saucers. The piece of ware is formed and glazed in one operation. The skin is of a lower thermal expansion than the body; hence, when the entire article cools, the skin has a residual compressive stress and the body a residual tensile stress. Surface compression adds both thermal shock resistance and mechanical strength to the ware (see Laminated materials, glass).

Lehring. Glass articles formed at high temperatures must be cooled in order to reduce the strain and associated stress caused by temperature gradients in the glass to a low level to prevent damage during finishing and subsequent use. The continuous solidification of glass on cooling and the methods of reducing strain are unique to glasses and do not apply to crystalline materials that undergo phase transitions. Lehring processes include annealing, tempering, densifying or compacting, and post-heat treatments (95–96). The term annealing generally refers to removal of stress, and terms such as fine annealing or trimming of optical glass infer structural changes associated with lehring.

The cooling of a flat piece of glass creates a parabolic temperature distribution (83) within the glass as shown in Fig. 34. Depending upon the original temperature of the glass, there are two possibilities: (*1*) If the temperature is high enough to allow structural relaxation, then after the temperature gradient is removed, a strain profile

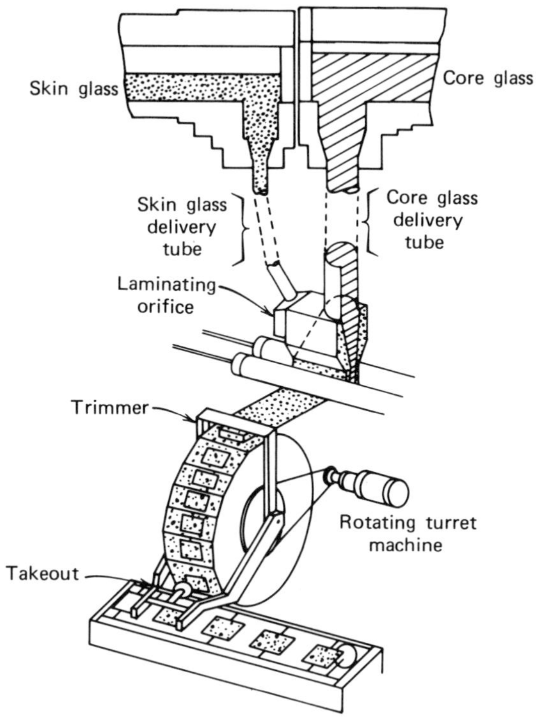

Figure 33. Schematic of the rotating turret machine (94). Courtesy of Corning Glass Works.

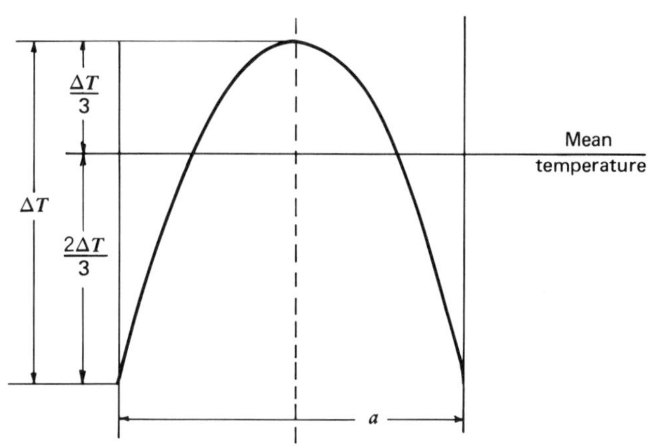

Figure 34. Temperature distribution through a section of glass of thickness a which has been cooled at a constant rate (83).

results. Tensile stress produced by this strain occurs in the interior with compressive stress on the surfaces. In the ideal case of infinitely slow cooling, ie, $\Delta T = 0$, stress does not develop within the glass; or (*2*) If the original temperature of the glass was so low that the structure could not relax, then the temporary tension at the surface and compression in the interior, which was caused by the temperature gradient, dissipates completely when the temperature gradient is removed (see also under Thermal Properties).

Stress annealing is the process of removing or preventing the objectionable stresses in glassware. The glass is brought to a temperature within the annealing range (usually defined as the temperature range between the anneal and strain points) and held there until the stresses are relieved. Cooling rates are determined by stress levels permitted in the ware being made. Generally, cooling rates increase as the temperature nears the strain point, and strain is directly proportional to the cooling rate (97). When the glass has cooled to about 50°C below the strain point, faster cooling is possible since the strain is only temporary. Schedules are worked out based upon the item being manufactured, and depend on (*1*) shape, eg, thick or thin, flat or curved, etc, (*2*) glass composition, ie, the thermal expansion curve, Young's modulus, and thermal diffusivity, and (*3*) the product and requirements such as cutability, finishing, and refractive index. The stress in glass can be calculated from the measured birefringence and the glass's stress optical properties.

Glass manufacturers using these principles for annealing flat glass have found that the maximum cooling rate generally cannot be achieved because of technical difficulties in cooling 3-m wide sheets of glass uniformly without surface damage. For pressed and blow items (hollow ware), the variations in thickness of a single article makes the problem considerably more complex. After forming, hollow ware receives a soak at constant temperature to assure that strain is reduced to a suitable level. Soaking is followed by a cooling through the anneal range and finally more rapid cooling to room temperature. A significant difference between flat and hollow ware is that during cooling the inside surface of the hollow ware is hotter than the outside. This results in compression on the outer surfaces, as with flat glass, but tension at the inner surface. Poorly annealed items may be subject to breakage if the tension is high or the inner surface is bruised.

Annealing of optical glass (fine annealing or refractive-index trimming) is important because it changes the refractive index and increases the density (98–99). Residual strain can also cause a nonuniform index or distortion during finishing operations. The fine annealing of large or thick optics may require days or even weeks to achieve low stress and ensure the maximum homogeneous, stable refractive index.

Postheat treatment of glass induces phase separation and crystallization. A very precise secondary heat treatment is necessary to develop and control photochromic properties. Glass-ceramics receive a thermal ceramming process that induces crystallization.

Tempering refers to the rapid air chilling of the surface of a glass that has been heated to reach its softening point. This produces a temperature gradient, which in turn, produces a high surface-compressive layer balanced by an internal tensile stress. Tempering increases the strength of the glass because applied forces must overcome the compression on the surface before fracture occurs. After tempering, items usually cannot be cut or ground without breaking, but the thickness of the compressive layer allows the article to retain much of its strength even if the surface is slightly damaged. Top-of-stove ware, drinkware, auto side windows, and safety glass for doors are representative items that are thermally tempered. Recently, air cushions have been used to support flat glass during tempering to avoid surface damage. The PPG Gas Hearth (100) or the Pennglass Air Float systems (101) support flat glass over a series of heated air jets in a conveyor system. Thinner, lightweight door glass or automobile windows can be made by air flotation. Bending of glass over a template in a glass-bending lehr is also possible by a modification of the air-float process (102). The greater precision of this method allows use of thinner glass and the production of tempered laminated ware.

Secondary Operations. In secondary forming, a piece of preformed glass is reheated and reworked into the finished product. Repressing of optical blanks in precision molds below the primary pressing temperature is sometimes done by lens manufacturers who do not melt their own glass. Redrawing of tubing and cane (rods) is called for when manufacturing bundles of thin fibers, as in channel amplifiers, or long, thin single fibers, as in optical waveguides. Sagging flat glass, eg, automotive windshields, is accomplished at temperatures near the softening point so that the glass bends to fit a form made from a refractory or metal. Lampworking and sealing are often either hand or automated operations in which the glass (rod or tubing) is heated near its working point (10^3–10^4 Pa·s or 10^4–10^5 P) and formed to the desired shape. Novelties, lab ware, and glass-to-metal wire seals for electrical and electronic applications are common lampworked articles. Precision tubing with inside-diameter variations as small as 30 μm/m length are made by shrinking. This is done by vacuum-collapsing the hot tube over a mandrel which is withdrawn when the glass has cooled (83).

Mechanical Finishing. Glass can be cut by five methods. (*1*) Flame cut-off is accomplished with pinpointed flames which heat the glass until it is soft enough to separate. (*2*) Sawing is done with band, wire, or circular blades used in conjunction with a loose or bonded abrasive. (*3*) To score-break glass, the piece is scored with a tool such as a file, a diamond, or a steel or carbide wheel, and then bent to apply tension to the scored area. (*4*) Score-thermal crack-off is done by applying heat opposite a mechanically produced score on the glass surface. (*5*) Localized heating with pinpointed burners followed by rapid chilling, usually with water, accomplishes a thermal crack-off. Production rates, tolerances, sizes, etc, vary, depending on the type of glass and the specific application.

Drilling. Glass is drilled with carbide or bonded-diamond drills under a suitable coolant such as water or kerosene. Other drilling processes include a metal tube rotating about its axis (core drilling), an ultrasonic tool in combination with an abrasive slurry, or an electron beam. Tolerances less than ±0.1 mm are readily obtained with diamond-core drilling and, if required, holes smaller than 25 μm-dia can be made with the electron-beam method.

Grinding. Glass is ground with sand, garnet, corundum, silicon carbide, boron carbide, or diamond. These materials are used loose, as in the grinding process employed for plate glass, or they are bonded in grinding wheels or coated on fabric belts. Grinding rates vary, depending on the glass and the abrasive used, but removal rates near 0.75 mm/min are common. In general, any shape that is ground in metal can be produced in glass on the same machine.

Mechanical polishing is similar to grinding, but the polishing compound, usually cerium oxide, zirconium oxide, or ferric oxide, is finer. The polishing tool may be plastic, cellulose, felt, or pitch. A mixture of hydrofluoric and sulfuric acids is used for *acid-polishing*. By directing flames on a glass surface, *fire-polishing* can produce surface finishes as good, except for flatness, as those obtained with mechanical polishing.

Copper-wheel engraving is used for decoration and in artware production. Abrasive compounds are applied in water suspension to the spinning copper wheel held in a chuck. The glass is brought into contact with the wheel to produce the design (see Abrasives).

Chemical Finishing. Treatment of glass surface with a chemical may alter its strength, appearance, or durability. For example, ion exchange alters the strength of a glass article. A glass, eg, a sodium aluminosilicate, is immersed in a bath of potassium nitrate at a temperature about 50°C below its strain point (ca 500°C) for 6–10 h. The small sodium ion diffuses from the glass and is replaced by the larger potassium ion. This stuffing of the first 100 μm of the surface produces a surface compression of 450 MPa (65,000 psi) and a corresponding central tension of 10–30 MPa (1450–4350 psi), depending on the article's total thickness. Stress release is minimized because of the low process temperature. Chemically-strengthened glass is used in auto and aircraft windows and eyeglasses. Chemical strengthening produces higher strength in thinner glass than does thermal tempering, but it is more susceptible to weakening by abrasion or scratching because of the thinner compression layer. Copper- and silver-containing mixtures can be applied in the same way to glass to produce colors by ion-exchange staining.

Acid etching or frosting with dilute hydrofluoric acid produces articles with good light-diffusing properties, eg, a lamp envelope. Leaching produces 96% silica glasses (103). A suitable heat treatment is first given to the glass to separate it into a soluble and insoluble phase and soluble sodium borate is precipitated. Leaching removes this phase and leaves a skeleton of silica-rich glass. On subsequent firing, a 14% linear shrinkage is obtained in a consolidated, transparent, nonporous glass. Chemically machinable photosensitive glass is acid-machined after patterns are photochemically induced into the glass.

Quality Control. Quality control is governed by the uses for which a glass is designated. For example, color-television panel glass, manufactured domestically by Corning Glass Works, Owens-Illinois, and RCA, must pass a wide variety of chemical and property specifications. Its chemical composition is checked spectrographically by comparison with a standard. The x-rays generated inside the picture tube during operation must be absorbed by the glass; the absorption coefficient at 0.06 nm is calculated from the composition, which in turn is usually maintained to 0.1–0.2% for each oxide. The physical properties, such as low-temperature viscosity (softening, annealing, and strain points) and thermal expansion, must be controlled to within 1% or less to ensure a good glass-to-metal seal for electrical leads and a good glass-to-glass seal when joined to the funnel. Electrical resistivity must be maintained above a minimum to prevent electrical fields from being generated around the tube during operation. The color of the panel must give optimum enhancement of the color-generating phosphors, maximum contrast with surrounding light, and uniform transmission. Other products have other requirements. Refractive index (±0.0001) and dispersion (±0.5) are closely controlled in optical and ophthalmic glass. Tinted optical glass, eg, sunglasses and filters, must maintain a color as well as a visible, ir, and uv transmission standard. Glazes, decorations, and food service items in general must maintain rigid standards of chemical durability and heavy metal release. FDA regulations require less than 7 ppm Pb and less than 0.5 ppm Cd released into a volume equal to the capacity of the article. The test conditions are 4% acetic acid at room temperature for 24 h. Generally, each manufacturer has strength and abrasion standards; a notable exception is ophthalmic lenses which the FDA specifies must withstand a 15.88 mm-dia steel ball dropped from 1.27 m.

Melting defects must be held to a minimum. Stones in the form of refractory particles, unmelted batch, or devitrification, affect the strength as well as optical integrity. Seeds and blisters caused by improper fining or electrochemical reboil, have the same effect as cords or striae from improper homogenization.

Forming defects are numerous and have been alluded to in the discussion of the forming machines and techniques. Tolerances in dimensions and capacities of glass bottles are independently negotiated between manufacturer and customer. Containers for expensive products like perfume and liquor carry the most precise tolerances. Champagne and soft-drink bottles must meet hydraulic pressure specifications which are affected by design and glass quality as well as dimensional tolerances. Flat glass, depending on the use, but especially in the case of automobile windshields, has precise melting defect, thickness, color, and optical distortion requirements. Dimensional tolerances, especially thickness, are monitored by an x-ray scanning device that measures relative thickness on-line. Lasers (qv) can also be used.

Absolute thickness is measured by a micrometer. The center thickness of a ribbon of float glass can be held to within 250 μm for window glass and to within 100 μm for high-quality windshields. The allowable thickness variation or thickness profile of a 2-mm thick piece of sheet glass, for example is much less than 250 μm. Distortion is measured by the angle between the piece of glass and a black-and-white striped board which first produces optical distortion (zebra angle). An angle of 50–60° for windshields corresponds to about 150 μm/m thickness variation.

Cleaning. Methods of cleaning glass are closely associated with the type of glass and its particular use (104–105). Cleaning requires consideration of (*1*) the soil to be removed, (*2*) the methods available, (*3*) the interaction between soil and glass, and (*4*) the effects of cleaning on the glass surface (105). Aqueous solvents frequently used range from chromic–sulfuric acid mixtures for cleaning borosilicate laboratory ware to common detergent solutions. Care must be taken to prevent leaching or etching of silicate glasses with very acidic or basic solutions. Organic solvents usually have no harmful effect on most glasses. Some organic solvents such as isopropanol may be used alone or aqueous mixtures in commercial cleansers. Hydrocarbon or halocarbon solvents or simply a hot flame are often used to remove nonpolar organic compounds. Mechanical cleaning may be required such as scrubbing, wiping, brushing, or repolishing and buffing the surface. Ultrasonic and electron-, ion-, or atomic-discharge cleaning methods are available for special cleaning problems usually associated with removal of organic materials (see Ultrasonics).

Other Methods. ***Vapor Deposition.*** Vapor deposition (106–107) is principally used in the manufacture of high-quality glass such as optical waveguides and optical mirrors. The former are thin (100–200 μm dia) glass fibers that transmit information via laser light with 10–100 times the capacity of conventional electrical coaxial cables (see Fiber optics). Their characteristic light attenuation is 3–6 dB/km (50–25% incident light transmitted per km), whereas optical glass and window glass, with attenuations of 1000 and 100,000 dB/km, respectively, are virtually opaque over a 1 km distance. Sufficient freedom from light-absorbing chemical impurities and light scattering physical defects is very difficult to achieve for glass made by conventional glass-making methods. Therefore, most worldwide optical waveguide communication systems use glass fibers made by a vapor-phase process (108).

All vapor processes under study use high vapor pressure metal halides such as $SiCl_4$, $GeCl_4$, BCl_3, and $POCl_3$ which are processed to provide bulk high purity oxide glasses (called blanks or preforms) from which optical waveguide fibers can be drawn. Since the waveguide must exhibit negligible optical attenuation, the reaction mixture must be free from optically absorbing transition metal halides ($FeCl_3$, VCl_4, $CuCl_2$). The pure halides are entrained in an oxygen carrier-gas system, accurately metered and transported (all at <50°C), and then react at temperatures above 1500°C. The halide reaction with oxygen, to form the desired oxides plus halogen gas, is virtually homogeneous and produces a finely divided particulate glass material commonly called soot. This high purity, inclusion-free glass soot has a high surface area, eg, 20 m^2/g, which provides a powerful driving force for rapid, thermally activated, viscous sintering (109). The glass soot is formed into solid inclusion-free glass bodies, which are then heated to temperatures where the viscosity is ca 10^5 Pa·s (10^6 P) and drawn into optical waveguide fibers.

Numerous fabrication processes utilize this basic scheme. For example, OVPO (outside vapor-phase oxidation) (106,110) is a process involving a flame heat source, a removable rod target, and a lateral orientation for soot deposition. A hot soot stream of the desired glass composition is generated by passing the vapor stream through a methane–oxygen flame directed at a rotating and traversing ceramic target rod. The glass soot sticks to this rod in a partially sintered state and, layer by layer, a cylindrical porous glass preform is built up. By properly controlling the metal halide vapor stream composition during the soot deposition process, the desired glass composition profile is built into this porous preform which is sintered between 1200 and 1500°C to a solid, bubble-free, glass blank. Using GeO_2–B_2O_3–SiO_2 glass compositions for the first part of the deposition and B_2O_3–SiO_2 glasses for the last part, graded index fibers have been fabricated. Single blanks yielding more than 10 km of fiber are readily made. Such blank sizes and rates, and the potential for further increases, suggest excellent manufacturing economics for the process.

Blanks can be made continuously with axial deposition orientation. The process involves simultaneous flame deposition of both core- and cladding-glass soots onto the end (ie, axially) of a rotating fused-silica target rod. As the porous soot preform grows, it is slowly retracted through a graphite resistance furnace where it is consolidated to a solid glass blank by zone sintering.

Inside vapor-phase oxidation processes (IVPO) involve reaction within a glass tube and glass deposition on the inside walls of this tube (111–113). A rotating fused silica or high-silica glass tube is heated with an oxy–hydrogen torch to cause vapor-phase oxidation of the metal halide gases which are flowing inside. Ultimately, the tube becomes the outer part of the fiber cladding, and the deposited soot becomes the fiber core. As the hot soot flows downstream, ca 60% of it is attracted to the cold walls of the tube where it is deposited as a thin porous layer, not unlike the porous OVPO preform structure described earlier. The rest flows out the exhaust end of the tube. The torch is steadily moved toward this downstream portion of the tube and, as it passes over the soot deposit, it zone-sinters it to a clear, bubble-free glass layer ca 7 μm thick. When the torch reaches the exhaust end of the tube, it is quickly returned to the inlet end, and the process is repeated. After ca 70 passes, enough glass is deposited (ca 0.5 mm) on the inside wall of the tube to provide the fiber-core size desired for a multimode waveguide. The torch then traverses the tube, which softens and collapses to a solid rod blank. This is then drawn into fiber. Attenuations as low as 0.5

dB/km at ca 1500 nm have been reported for multimode fibers, and 0.5 dB/km at 1300 nm for single-mode fibers, which is essentially the intrinsic limit for the high-silica glasses (P, Ge, B-doped) studied.

Bulk-fused silica is commercially produced by a variety of techniques, including vapor deposition. The physical properties of the product depend strongly on the method used. If the target is kept above 1800°C during SiO_2 soot deposition, simultaneous sintering of the soot occurs, yielding in a single step a solid, bubble-free glass. This is achieved if the heat from the soot-generating burners also sinter the soot as it hits a hot fused sand target. Layer by layer, a boule of solid fused silica is deposited which, because hydrogen-containing fuels are used, contains ca 1200 ppm OH. Very large boules, weighing over 500 kg, can be obtained by using many soot deposition burners and/or running furnaces for many days at a time. This approach is used to manufacture numerous high-silica glasses. Optical blanks, windows, crucibles, tubing, and mirrors for large telescopes are produced by further processing these boules with conventional cutting, grinding, polishing, and flameworking techniques (see Silica). This material is also used extensively for windows in manned spacecraft and the space shuttle because of its refractory nature, thermal shock resistance, and optical homogeneity.

Glass films are formed by both reactive and nonreactive deposition methods. Examples of nonreactive deposition methods are evaporation, sputtering, and ion-implantation (qv) or ion-plating. Reactive deposition methods are described below. Deposition techniques are compared in Table 9 (107,114). Glass films are used in the semiconductor industry because of their dielectric properties, and they are used for encapsulating integrated circuits (qv) and other electronic devices because they provide a reasonably hermetic seal (see also Electrostatic sealing).

Reactive sputtering is a process in which highly active metal atoms react with a gas before their deposition on a cold substrate. In the presence of oxygen or nitrogen, for instance, combined with a high-purity silicon metal cathode in a sputtering chamber, high purity amorphous silicon dioxide or silicon nitride is deposited. The high quench rate and the absence of impurities have made possible the production of numerous glasses made of $PbTeO_3$, GeO_2, SiO_2, Al_2O_3–SiO_2, and many other oxides.

Thermal oxidation is one of the oldest and most commonly used methods of forming a primary passivating film of SiO_2 on silicon. The metal is heated in dry oxygen, in H_2O-containing oxygen, or in steam. A silica layer grows inwardly from the surface by a thermal oxidation mechanism. Although this process occurs slowly even at room temperature, yielding a layer ca 1 nm thick after long exposures, the silicon wafer is usually heated to 600–1200°C to achieve 1 μm thick films in ca 1 hour.

A glass layer can also be made by anodic oxidation on a metal or semiconductor surface, such as silicon, by making it the anode in an electrolytic cell, immersing this anode in a suitable electrolyte (often aqueous), and passing a current through it. By applying a sufficient overvoltage, the metal surface oxidizes, forming, in some cases, a glass film up to several hundred nanometers thick. In one form of chemical vapor deposition (CVD), heterogeneous reaction and deposition of organometallic or metal halide vapors occur at a heated solid substrate surface. This is the most widely used and versatile method for forming dielectric films in the fabrication of semiconductor devices and for depositing layers of glass or crystalline oxides or nitrides over silicon, etc. For example, by passing gas mixtures such as SiH_4, PH_3, and oxygen (or $SiCl_4$,

Table 9. Deposition Techniques for Dielectric Films[a]

Deposition process	Source material	Typical deposition rate	Typical substrate temperature	Crystalline nature of deposit	Sources of impurities	Typical films formed	Typical electronics applications
Low pressure depositions							
evaporation	high-vapor-pressure solid (eg, SiO)	1–2 μm/min	25–200°C	amorphous	filament/crucible, walls	SiO	optical coatings
reactive evaporation	solids, eg, Al in gas (eg, O_2)	1–2 μm/min	25–200°C	amorphous	filament/crucible	Al_2O_3	passivation layers
sputtering; reactive sputtering	low-vapor-pressure solids, eg, SiO_2 with gas (eg, Ar)	1–10 nm/min	25°C	amorphous	walls, gases used, sources	SiO_2[b], Si_3N_4 TaN	passivation layers
plasma deposition	gases (eg, SiH_4–NH_3 mixtures)	1–10 nm/min	25–400°C	amorphous	walls, gases, electrodes	Si_3N_4 Si	passivation layers, solar cells
low pressure CVD	gases (eg, SiH_4–NH_3 mixtures)	0.1–1 μm/min	600–1000°C	amorphous, polycrystal	walls, gases, susceptors	SiO_2, Si_3N_4, ORPS[c]	passivation layers
101-kPa (1-atm) depositions							
thermal oxidation	substrate plus gas	nonlinear ca 1 μm in 1 h	800–1200°C	amorphous	walls, substrate surface	SiO_2, Al_2O_3	channel oxides (MOS[d]) diffusion marks, etch masks, passivation layers
chemical vapor deposition	gases (eg, SiH_4–NH_3 mixtures, SiH_4–O_2 mixtures)	0.1–1 m/min	800–1200°C 250–500°C	polycrystal amorphous	source gases, walls	SiO_2[b], Si_3N_4, PSG[d], BSG[d]	passivation layers, diffusion sources, dielectrics for multilevel metal
anodization	substrate plus electrolyte	nonlinear 10–100 nm in 1 min	25°C	amorphous	electrolyte	Al_2O_3, Ta_2O_5	passivation layers
ion implantation	high purity ion source (mass-spectrometer)		25°C	amorphous	substrate surfaces	nitride layers	etch-resists

[a] Ref. 114.
[b] Stoichiometric composition usually not obtained.
[c] ORPS = oxygen-rich polycrystalline silicon.
[d] BSG = borosilicate glass; PSG = phosphosilicate glass; MOS = metal oxide semiconductor.

Table 10. Glass Industry Economics

	Year	
	1975	1978
Flat		
shipments,		
10^6 \$	1286	2095
10^6 m^2	201	288
exported, 10^6 \$	65	119
imported, 10^6 \$	48	105
Container		
shipments,		
10^6 \$	3000	4200
10^9 items	40.5	45.6
trade,		
exported, 10^6 \$	32	34
imported, 10^6 \$	16	25
Fiber		
shipments,		
10^6 \$	992	1890
10^3 t	1020	1690
Pressed and blown[a]		
shipments,		
10^6 \$	1206	

[a] Values for the pressed and blown industry (SIC 3229) have been adjusted such that textile fiberglass (SIC 32293) is included under fiber-type glass.

$POCl_3$, and oxygen) over a heated silicon wafer, the vapors react heterogeneously at the wafer surface, depositing a thin layer of P_2O_5–SiO_2 glass over all heated exposed surfaces (see Film deposition techniques).

Miscellaneous. Unconventional melting includes splat cooling, which has been used extensively to produce amorphous metals; laser-spin melting, which can produce 100–800 μm spheres of refractory oxide glasses; melting under high pressure so that enormous quantities of volatile components may be incorporated into the glass structure. Solid-state transformations produce amorphous materials from bulk crystalline phases by destroying the lattice structure at low temperatures. This can be achieved by neutron bombardment in a reactor, and is found in nature in certain minerals containing radioactive elements. Solid-state transformation also occurs upon exposure to shock waves, as in a meteorite impact. Solution methods have been used to form silica gels which can be sintered to clear glass at low temperatures.

Economic Aspects

Glass manufacture is classified according to the product into flat, container, fiber, or specialty glass. The flat and container companies produce over 14×10^6 t soda-lime glass per year. The specialty glass manufacturers of pressed and blown ware, television bulbs, and lighting and optical glasses melt hundreds of glass compositions to fulfill the need for a large variety of products.

Table 10 summarizes production statistics for the four areas of the industry.

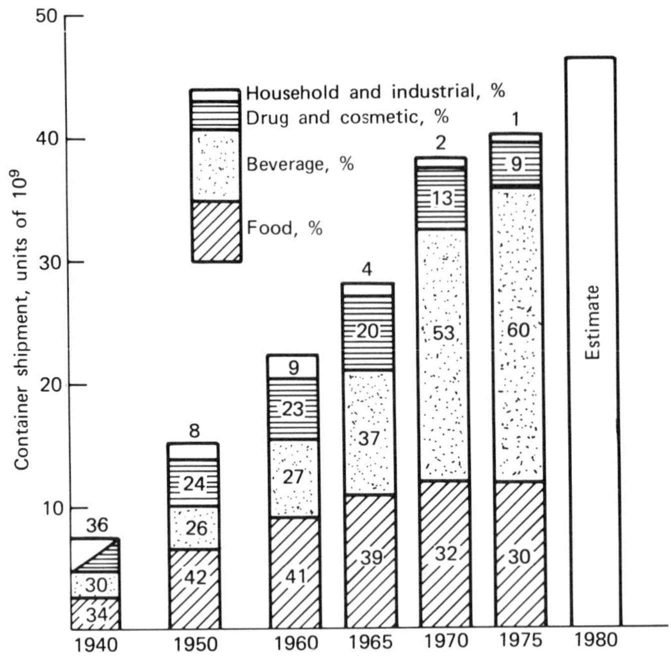

Figure 35. Production and product distribution within the container glass industry (domestic shipments).

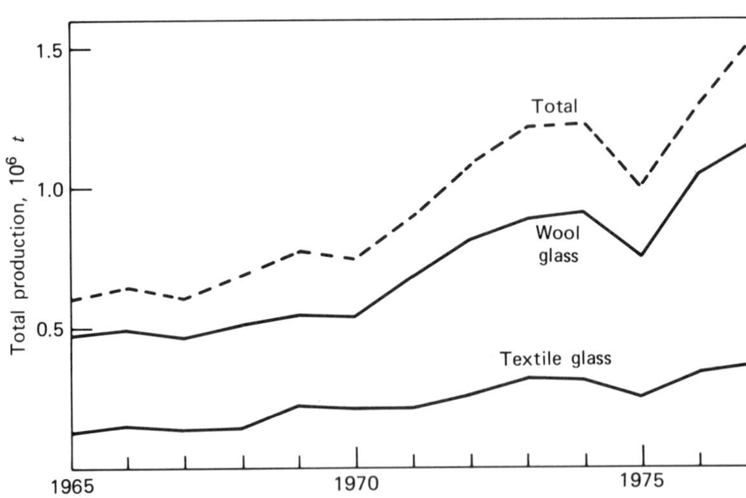

Figure 36. Domestic production of fiberglass.

Flat Glass. In the United States, 31 flat-glass plants are operated by PPG Industries, Libbey-Owens-Ford (LOF), Guardian Industries, Ford Motor Company, AFG Industries, Combustion Engineering, and The West Virginia Flat Glass Company. Growth of this industry depends upon the construction and automotive markets. The float process produces 85% of all the flat glass. The remainder is thin sheet for picture glass or rolled and patterned glass. Over 2.8×10^8 m^2 of flat glass is consumed yearly. Production of flat glass is cyclical but increases in volume by approximately 4% per year. Less than 10% of the total used is imported.

The conversion of the industry to the float process during the 1970s caused concern over the possibility of over-capacity, but the continued demand by both the automotive and construction markets has allowed a relatively easy transition. Float processing is very amenable to surface treatments that produce reflective coatings. Because the cost of the glass represents a relatively small part (1–2%) of the total construction costs, the use of solar-efficient glazes appears economical compared to more expensive mechanical heating or cooling equipment. Solar energy (qv) conversion also represents a potentially large market for flat-glass producers. The solar collector or the solar reflector concepts which are now anticipated will require large quantities of high quality flat glass.

Container Glass. Statistics for food, beverage, drug and cosmetic, and household and industrial containers are compiled and published by the Glass Packaging Institute (GPI) (115). Figure 35 shows the growth and distribution of glass container products. Beverage containers, including soft drink, beer, wine, and liquor, constitute the largest segment, followed by food, drug and cosmetic, and household and industrial containers. Approximately 85% of container glass is clear, the remainder is amber. Other tints make up a small percentage of the total. There are ca 27 companies with over 120 manufacturing plants in the United States. Most of these plants are owned by Owens-Illinois, Brockway Glass, Anchor Hocking, Glass Container Corp., Kerr, Chattanooga, Thatcher, and Indian Head glass companies. Over 44×10^9 containers weighing more than 12×10^6 t are shipped yearly with an estimated value of over $\$4 \times 10^9$. Imports are of very minor importance because of high transportation costs. Export of container glass is only about 1% of the total amount manufactured for the same reason.

Fiberglass. Fiberglass is classified as either textile or wool. More than a dozen companies produce textile fibers for draperies, upholstery, tires, reinforced plastics, paper, and tape, including Johns-Manville, Owens-Corning Fiberglas, PPG Industries, Nicofibers, and Reichold. Annual production of textile fibers is over 370,000 t, as shown in Figure 36. The annual production of glass wool is 1.2×10^6 t (Owens-Corning Fiberglas, Johns-Manville, and others). More than 80% of the production is used for building insulation, the remainder for industrial equipment and pipe insulation. Growth of the insulation market has been faster than the textile area because of the demand for additional insulation in both new and existing buildings.

Specialty Glass. The pressed-and-blown or hollow-ware industry is comprised of over one hundred companies in the United States, including Corning Glass Works, Owens-Illinois, General Electric, and Anchor Hocking. The wide variety of products is divided into categories of pressed-and-blown glass for table, kitchen, art, and novelty applications and products of purchased glass. The latter consists of items for scientific, technical, and industrial uses such as electrical and electronic products, laboratory glassware, optical and ophthalmic glass, etc.

BIBLIOGRAPHY

"Glass" in *ECT* 1st ed., Vol. 7, pp. 175–206, by H. G. Vogt, Corning Glass Works; "Glass" in *ECT* 1st ed., Supplement 2, pp. 435–454, by S. D. Stookey, Corning Glass Works; "Glass" in *ECT* 2nd ed., Vol. 10, pp. 533–604, by J. R. Hutchins, III, and R. V. Harrington, Corning Glass Works.

1. J. E. Ericson and co-workers, *J. Noncryst. Solids* **17,** 129 (1975).
2. G. W. Morey, *The Properties of Glass,* 2nd ed., Reinhold Publishing Corp., New York, 1954, p. 28.
3. *Standard Definitions of Terms Relating to Glass Products, ASTM Standard 1978, ASTM C 162-71,* Part 17, 1978, p. 112.
4. H. A. Davies, *Phys. Chem. Glasses,* **17**(5), 159 (1976).
5. L. B. Davies and P. J. Grundy, *Phys. Stat. Sol.* **8a,** 189 (1971); *J. Noncryst. Solids* **11,** 179 (1972).
6. D. E. Polk, *J. Noncryst. Solids,* **5,** 365 (1971).
7. G. S. Cargill, *J. Appl. Phys.* **41,** 2248 (1970).
8. F. G. A. Stone and W. A. G. Graham, *Inorganic Polymers,* Academic Press, Inc., New York, 1962.
9. G. Tammann, *Der Glaszustand,* Leopold Voss, Leipzig, 1933.
10. H. N. Ritland, *J. Am. Ceram. Soc.* **37,** 370 (1954).
11. A. Q. Tool, *J. Res. Natl. Bur. Stand.* **37,** 73 (1946).
12. N. J. Kreidl, *Glass Ind.* **58,** 26 (1977).
13. D. R. Secrist and J. D. Mackenzie, *Glass Ind.* **45,** 408, 451 (1964).
14. W. Zauzmann, *Chem. Rev.* **43,** 219 (1948).
15. J. N. Gibbs in J. D. Mackenzie, ed., *Modern Aspects of the Vitreous State,* Part 1, Butterworth, Inc., Washington, D.C., 1960, pp. 152–157.
16. A. R. Cooper and P. K. Gupta, *J. Am. Ceram. Soc.* **58,** 350 (1975).
17. A. A. Lebedev, *Arb. Staatl. Opt. Inst. Leningrad* **2**(10), (1921).
18. J. T. Randall, H. P. Rooksby, and B. S. Cooper, *J. Soc. Glass Technol.* **14,** 219 (1930); N. Valenkov and E. Pori-Koshitz, *Z. Krist.* **45,** 195 (1936).
19. B. E. Warren and I. Biscoe, *J. Am. Ceram. Soc.* **21,** 49 (1938).
20. N. Valenkov and E. Porai-Koshits, *Z. Krist.* **95,** 195 (1936).
21. W. H. Zachariasen, *J. Am. Chem. Soc.* **54,** 3841 (1932).
22. R. L. Mozzi and B. E. Warren, *J. Appl. Crystallogr.* **2,** 164 (1969).
23. J. H. Konnert, J. Karle, and G. A. Gerguson, *Science* **179,** 177 (1973).
24. A. C. Wright and A. J. Leadbetter, *Phys. Chem. Glasses* **17**(5), 122 (1976).
25. R. H. Doremus, *Glass Science,* John Wiley & Sons, Inc., New York, 1973.
26. K. H. Sun and A. Silverman, *J. Am. Ceram. Soc.* **25,** 101 (1942).
27. M. E. Milberg and C. R. Peters, *Phys. Chem. Glasses* **10,** 46 (1969).
28. P. Chaudhari and co-workers, *Phys. Rev. Lett.* **29,** 425 (1972).
29. P. J. Bray and J. G. O'Keefe, *Phys. Chem. Glasses* **4,** 37 (1963).
30. R. L. Mozzi and B. E. Warren, *J. Appl. Crystallogr.* **3,** 251 (1970).
31. V. M. Goldschmidt, *J. Soc. Glass Technol.* **11,** 337 (1927).
32. K. H. Sun, *J. Am. Ceram. Soc.* **30,** 277 (1947).
33. H. Rawson, *Inorganic Glass-Forming Systems,* Academic Press, Inc., London, 1967.
34. J. E. Stanworth, *J. Am. Ceram. Soc.* **54,** 61 (1971).
35. M. Imaoka and T. Yamosaki, *J. Ceram. Assoc. Japan* **71,** 215 (1963).
36. J. G. Vail, *Soluble Silicates, Their Properties and Uses,* Reinhold Publishing Corp., New York, 1952.
37. V. I. Aver'yanov and E. A. Porai-Koshits, *The Structure of Glass,* Vol. 6, Part 1, Consultants Bureau, New York, 1966, p. 98.
38. L. D. Pye and co-workers, *Borate Glasses: Structure, Properties, Applications,* Vol. 12, Materials Science Research, Plenum Press, New York, 1978.
39. Ger. Pats. 63,126 (Aug. 5, 1978), and 62,408 (June 20, 1969), W. Herndorf.
40. G. S. Cowgill, III, *NY Acad. Sci.* **279,** 208 (1976).
41. H. S. Chen, *Acta Metall.* **22,** 1505 (1974).
42. M. H. Cohen and D. Turnbull, *Nature (London)* **189,** 131 (1961).
43. M. Marcus and D. Turnbull, *Mater. Sci. Eng.* **23,** 211 (1976).
44. B. C. Giessen and C. N. J. Wagner in S. Z. Beer, ed., *Liquid Metals, Chemistry and Physics,* Marcel Dekker, New York, 1972, p. 663.
45. H. A. Davies, *Phys. Chem. Glasses,* **17**(5), 159 (1976).
46. T. E. Sharon and C. C. Tsuei, *Phys. Rev. B.* **5,** 1047 (1972).
47. S. D. Stookey and R. D. Mauer "Catalyzed Crystallization of Glass—Theory and Practice" in *Progress in Ceramic Science,* Vol. 2, Pergamon Press, 1961, pp. 77–101.
48. G. H. Beall in L. L. Hench and S. W. Friedman, eds., *Nucleation and Crystallization in Glasses,* American Ceramic Society, 1972, pp. 251–261.

49. G. H. Beall and H. L. Rittler, *Am. Ceram. Soc. Bull.* **55,** 586 (1976).
50. J. F. MacDowell and G. H. Beall, *J. Am. Ceram. Soc.* **52**(1), 17 (1969).
51. D. R. Stewart "Concepts of Glass Ceramics" in L. D. Pye and co-eds., *Introduction to Glass Science,* Plenum Press, New York, 1972, pp. 237–271.
52. T. P. Seward, III, private communication, Corning Glass Works.
53. S. D. Stookey, *Ind. Eng. Chem.* **41,** 856 (1949).
54. S. D. Stookey, G. H. Beall, and J. E. Pierson, *J. Appl. Phys.* **49,** 5114 (1978).
55. A. G. F. Dingwall and H. Moore, *J. Soc. Glass Technol.* **37,** 316 (1953).
56. H. E. Hagy, *Electron. Packag. Prod.* **18**(7), 182 (July 1978).
57. V. O. Altemose, "Gas Permeation Through Glass," *Seventh Symposium on the Art of Glassblowing,* The American Scientific Glassblowers Society, Wilmington, Del., 1962.
58. W. C. LaCourse "The Strength of Glass" in L. D. Pye and co-eds., *Introduction to Glass Science,* Plenum Press, New York, 1972, pp. 451–512.
59. W. H. Barney, Corning Glass Works, private communication, and various Corning Glass public information bulletins.
60. L. L. Hench and H. F. Schaake, "Electrical Properties of Glass" in L. D. Pye and co-eds., *Introduction to Glass Science,* Plenum Press, New York, 1972, pp. 583–659.
61. J. D. MacKenzie, "Development of Glasses for Some New Applications," *Int. Cons. of Glass, 1974, Ceramic Society of Japan, Tokyo,* Vol. 4, p. 71.
62. *Ceramic Data Book,* General Electric Co., Lamp Components Sales Operation, Cahners Publishing Co., Boston, Mass., 1979, p. 304.
63. M. J. Minot, *J. Opt. Soc. Am.* **66,** 515 (1976).
64. W. A. Weyl, *Colored Glasses,* Society of Glass Technology, Sheffield (Reprinted by Dawson's of Pall Mall, London) 1951.
65. J. Wong and C. A. Angell, *Glass: Structure by Spectroscopy,* Marcel Dekker, Inc., New York, 1976.
66. N. J. Kreidl "Optical Properties" in F. V. Tooley, ed., *Handbook of Glass Manufacture,* Books for Industry, Inc., and Glass Industry Magazine, New York, 1974, pp. 957–997.
67. E. J. Coker and L. N. G. Filon, *Treatise on Photoelasticity,* Cambridge University Press, London, 1931.
68. G. E. Rindone, *Transactions of the International Conference on Glass, Paris, 1956,* p. 373.
69. N. J. Kreidl and J. Rood in R. Kinglake, ed., *Optical Materials, Applied Optics and Opt. Eng.,* Academic Press, Inc., New York, 1965.
70. C. R. Bamford, *Colour Generation and Control in Glass,* Elsevier Scientific Publishing Co., New York, 1977, pp. 141–164.
71. W. C. Bauer and W. H. Manring "Batch Materials Handling and Preparation" in F. V. Tooley, ed., *The Handbook of Glass Manufacture,* Books for Industry, Inc., and Glass Industry Magazine, New York, 1974, pp. 57–90.
72. E. D. Spinosa, P. M. Stephan, and J. R. Schorr, *Review of Literature on Control Technology which Abates Air Pollution and Conserves Energy in Glass Melting Furnaces, to Corning Glass Works,* EPA-600/2-77-005, /2-76-269, /2-76-032b, Battelle, Columbus, Oh., Nov. 11, 1977.
73. M. Cable, *Glastek. Tidskr* **29,** 11 (1974).
74. C. V. Thomasson and F. W. Wilburn, *Phys. Chem. Glasses* **1,** 52 (Apr. 1960).
75. G. E. Rindone, *Glass Ind.* **38,** 489 (Sept. 1957); G. E. Rindone, *Glass Ind.* **38,** 561 (Oct. 1957).
76. C. J. S. Atkinson, *Glass Technol.* **17,** 205 (Oct. 1976).
77. R. S. Arrandale "Furnaces, Furnace Design, and Related Topics" in F. V. Tooley, ed., *The Handbook of Glass Manufacture,* Books for Industry, Inc., New York, 1974, pp. 248–386.
78. A. G. Pincus and G. M. Diken, *Electric Melting in the Glass Industry,* Books for Industry and The Glass Industry Magazine, New York, 1976.
79. U.S. Pat. 3,524,206 (Aug. 18, 1970), G. B. Boettner (to Corning Glass Works).
80. U. Kircher, *Glastech. Ber.* **51**(6), 139 (1978).
81. W. Giegerich and W. Trier, *Glass Machines Construction and Operation of Machines for the Forming of Hot Glass,* Springer-Verlag, Berlin, 1969 (translated by N. J. Kreidl).
82. U.S. Pat. 1,790,397 (Jan. 27, 1931), W. J. Woods and D. E. Gray (to Corning Glass Works).
83. E. B. Shand, *Glass Engineering Handbook,* 2nd ed., McGraw-Hill, New York, 1958.
84. U.S. Pat. 2,662,346 (Dec. 15, 1953), J. W. Giffen (to Corning Glass Works).
85. A. Pilkington, *Glass Technol.* **17,** 182 (Oct. 1976).

86. Press Releases, Feb. 17, 1975, Sept. 17, 1976, Feb. 17, 1977, May 19, 1977, Nov. 17, 1977, PPG Industries, Inc., Public Relations Department.
87. L. A. B. Pilkington, *Proc. Roy. Soc. London A* **314,** 1 (1969).
88. U.S. Pat. 3,338,696 (Aug. 29, 1967), S. M. Dockerty (to Corning Glass Works).
89. R. F. Caroselli "Glass Textile Fibers" in H. F. Mark, S. M. Atlas, and E. Cernia, eds., *Man-Made Fibers,* Wiley-Interscience Publishers, Inc., New York, 1968, pp. 425–454.
90. G. R. Machlan "Glass Fiber and Auxiliary Manufacturing Processes" in F. V. Tooley, ed., *Handbook of Glass Manufacture,* Books for Industry, Inc., New York, 1974, pp. 715–734.
91. A. I. Andrews, *Porcelain Enamels,* The Garrard Press, Champaign, Ill., 1961.
92. J. A. Pask "Glass–Metal 'Interfaces' and Bonding" in J. D. MacKenzie, ed., *Modern Aspects of the Vitreous State,* Butterworth & Co., Washington, D.C., Vol. 3, 1964, pp. 1–28.
93. C. W. Parmelee, *Ceramic Glazes,* 3rd ed., Cahners Publishing Company, Inc., Boston, Mass., 1973.
94. U.S. Pat. 3,231,356 (Jan. 25, 1966), J. W. Giffen (to Corning Glass Works).
95. H. A. McMaster "Annealing and Tampering" in F. V. Tooley, ed., *The Handbook of Glass Manufacture,* Books for Industries, Inc., New York, 1974, pp. 799–832.
96. A. G. Pincus and T. R. Holmes, *Annealing and Strengthening in the Glass Industry,* Magazines for Industry, Inc., New York, 1977.
97. L. H. Adams and E. D. Williamson, *J. Franklin Inst.* **190,** 597, 835 (1920).
98. H. R. Lillie, *Glass Ind.* **21,** 355, 382 (1950).
99. H. N. Ritland, *J. Am. Ceram. Soc.* **37,** 370 (1954).
100. U.S. Pat. 3,223,501 (Dec. 14, 1965), J. C. Fredley and G. E. Sleighter (to Pittsburgh Plate Glass Co.).
101. U.S. Pat. 3,332,762 (July 25, 1967), H. A. McMaster and N. C. Nitschke (to Permaglass, Inc.).
102. U.S. Pat. 3,468,645 (Sept. 23, 1969), H. A. McMaster, N. C. Nitschke, and J. J. Kawecka (to Permaglass, Inc.).
103. U.S. Pat. 2,106,744 (Feb. 1, 1938), H. P. Hood and M. E. Nordberg (to Corning Glass Works).
104. *ASTM Standard Practices for Designing a Process for Cleaning Technical Glasses, ASTM C 912,* pt. 17, 1980.
105. L. Holland, *The Properties of Glass Surfaces,* Chapman and Hall, London, 1966.
106. P. C. Schultz, "Vapor Phase Materials and Processes for Glass Optical Waveguides," in *Fiber Optics Advances in Research and Development,* Plenum Press, New York, 1979.
107. G. W. Scherer and P. C. Schultz "Unusual Methods of Making Glass" in N. J. Kreidl and D. R. Uhlmann, eds., *Glass: Science and Technology,* Vol. III, Academic Press, Inc., New York, 1979.
108. *Fiber Opt. Commun. Newsl.* **1,** 8 (Aug. 1978).
109. G. W. Scherer, *J. Am. Ceram. Soc.* **60,** 236 (May 1977).
110. U.S. Pat. 3,737,292 (June 5, 1973), D. B. Keck and co-workers (to Corning Glass Works).
111. U.S. Pat. 3,711,262 (Jan. 16, 1973), D. B. Keck and P. C. Schultz (to Corning Glass Works).
112. J. C. MacChesney, P. B. O'Connor, F. V. DiMarcello, J. R. Simpson, and P. D. Lazay, paper 6-40 in *Proceedings Tenth International Congress on Glass, Kyoto, Japan, 1974.*
113. J. B. MacChesney, P. B. O'Connor, and H. M. Presby, *Proc. IEEE* **62,** 1278 (1974).
114. J. A. Amick, G. L. Schnable, and J. L. Vossen, *J. Vac. Sci. Technol.* **14,** 1053 (1977).
115. Annual Reports, Glass Packaging Institute, 1800 K Street, Washington, D.C.

General References

Refs. 25, 81, and 83 are also general references.
L. D. Pye, H. J. Stevens, and W. C. LaCourse, eds., *Introduction to Glass Science,* Plenum Press, New York, 1972.
F. V. Tooley, ed., *The Handbook of Glass Manufacture,* Books for Industry, Inc., New York, 1974.
Glass Industry Magazine.
Journal of the American Ceramic Society.
Journal of the British Ceramic Society.
Journal of Noncrystalline Solids.
The Physics and Chemistry of Glass.

DAVID C. BOYD
DAVID A. THOMPSON
Corning Glass Works

GLASS-CERAMICS

Glass-ceramics are polycrystalline solids produced by the controlled crystallization of glasses. They are primarily silicate-based materials that can be formed by highly automated glass-forming processes and converted to a ceramiclike product by the proper heat treatment. Glass-ceramics, which are also referred to in the technical literature as vitrocerams, devitrocerams, Pyrocerams (Corning Glass Works), sitalls, slagceramics and melt-formed ceramics, generally have a crystal content greater than 40–50% (see Ceramics). This excludes from the definition other types of glass products such as fluoride opals and copper–ruby glasses which contain minor amounts of a crystalline phase (see Gems, synthetic; Glass).

The liquidus temperature is a characteristic equilibrium temperature below which the liquid becomes thermodynamically unstable with respect to one or more crystalline phases. Whether crystals will actually form is dependent on kinetic factors such as the rate of cooling of the glass and the growth rate of the crystal species. In addition, there must be a nucleation site from which the crystallization can start. For most common inorganic glasses, crystallization begins at the interfacial boundaries, eg, glass–air, glass–refractory, etc, if they are exposed to temperatures just below the liquidus temperature for times generally on the order of minutes to hours. This relatively uncontrolled crystallization (commonly referred to as devitrification) results in columnar crystal growth perpendicular to the surface which generally affects the properties of the glass product in a detrimental manner. Since exposures to subliquidus temperatures always occur during the normal glassmaking process, the glassmaker has generally tried to make the glass formulations resistant to devitrification; this is done by adjusting the glass composition to minimize the crystal growth rate in the glass working range. Devitrification has been described as the chief factor which limits the composition range of practical glasses and is an everpresent danger in all glass manufacture and working (1).

One of the earliest references to an attempt to use the crystallization of glass to advantage dates to the early 1700s when deReaumur (2) in France attempted to crystallize soda-lime–silicate bottles by packing them in mixtures of sand and gypsum and heating them for a long time. The resulting bodies were weak and deformed since the crystal started to grow from the surface.

Little further work was done until the early 1900s when Tammann (3) started to investigate crystallization in more detail. This classic work was the basis for much of the present understanding of nucleation and crystallization. In the same period, a new industry arose in Europe based on the controlled crystallization of molten rocks, particularly basalts. Using many of the principles derived by Tammann, rocklike ceramics were produced and utilized in a range of applications. A very detailed summary is available (4) of the principles of melting and crystallization of fused basalts with photographs of production items.

In the 1930s Blau (5) illustrated the importance of controlled bulk crystallization. In the 1950s and 1960s there was an increased interest in controlled crystallization aided by nucleating agents since volume crystallization can enhance many of the properties; this is particularly true for the mechanical properties. Colloidal dispersions of metals were demonstrated to be good nucleants in some systems (6). The discovery of the effectiveness of certain refractory oxides as nucleation promoters (7–8) led to the commercialization of products based on the ability to closely control the crystallization process (see Crystallization).

The Glass-Ceramic Process

Figure 1 is a flow chart of a glass-ceramic manufacturing process. By definition, a glass-ceramic initially must be a glass. Thus, the first stages of the manufacturing process are similar to the processing of most commercial glasses; however, there are significant differences in the postforming stages. The environmentally related effects resemble those encountered in conventional glass processing. Pollution problems are generally minimal except for emissions during the melting of some compositions.

Melting. Glass compositions that are specially formulated for controllable crystallization are mixed like any other glass. The batch costs for the most common product (a low expansion cookware) is ca \$0.33/kg. The range of batch costs is from ca \$0.006/kg for glass-ceramics based on waste products to greater than \$14.00/kg for high purity products made for electronic applications (based on 1980 prices). In general, the glass-ceramic compositions are much more difficult to melt and homogenize than common soda-lime glass. Depending on the product requirements and the volume of the market, the commercial glass-ceramics may be made in continuous refractory-lined furnaces which produce glass 24 h/d; they may be melted in refractory-lined furnaces in an intermittent operation; or they may be melted in crucibles (either refractory- or platinum-lined) in an intermittent operation. Melting temperatures in excess of 1620°C are common. A large, continuous operation produces ca 90,000 kg/d; the intermittent operations vary from a few to 55,000 kg.

Forming. After melting, the resulting viscous liquid can be shaped by pressing, blowing, drawing, rolling, casting or some combination of these. Many of these processes are high speed, automatic operations which allow close control of dimensions. However, the viscosity–temperature relationship of most glass-ceramic compositions is such that they are generally much more difficult to form than the more common glasses. The most frequently used forming techniques are pressing and casting. In some applications the glass is powdered, mixed in a blender, and applied as a coating to

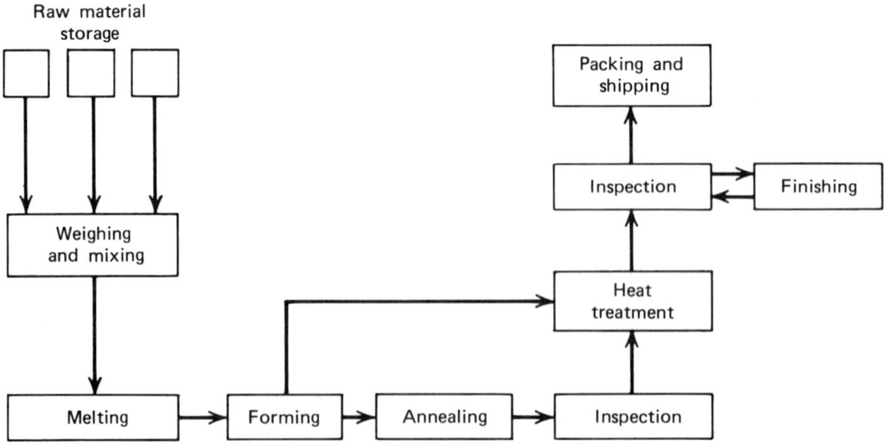

Figure 1. The glass-ceramic manufacturing process.

various metal and ceramic substrates. The coating is fused and crystallized during subsequent processing steps (see Powder coatings).

Postforming. The shaped body may be annealed (cooled slowly to avoid stresses) and inspected prior to heat treatment (Fig. 1), or the shaped object may be put directly into a heat-treatment furnace. The heat-treatment process may be continuous or intermittent.

During the heat-treatment process, the amorphous glass is transformed to a largely crystalline body. This complex transformation proceeds from distinct nucleation centers. The formation of crystals depends on the number of nuclei formed, the rate of crystal growth and the viscosity of the glass (3,5). The relationship between these factors and the heat-treatment process is shown in Figure 2. Figure 2 shows the nucleation formation rate and the crystal growth rate curves as a function of temperature for a condition where there is significant overlap of the nucleation and crystal growth rate curves (Fig. 2**a**) and a condition where the overlap is much less (Fig. 2**b**). T_L is the liquidus temperature. In Figure 2**a** (between T_2 and T_3) a crystal can grow as soon as a nucleus forms. The result is comparatively few crystals with a wide range of crystal sizes. In Figure 2**b**, a large number of nuclei can form (between T_1 and T_2) without appreciable crystal growth, which does not appear until the range T_2 to T_L. This is the rationale for the commonly used two-stage heat treatment which is shown in Figure 3.

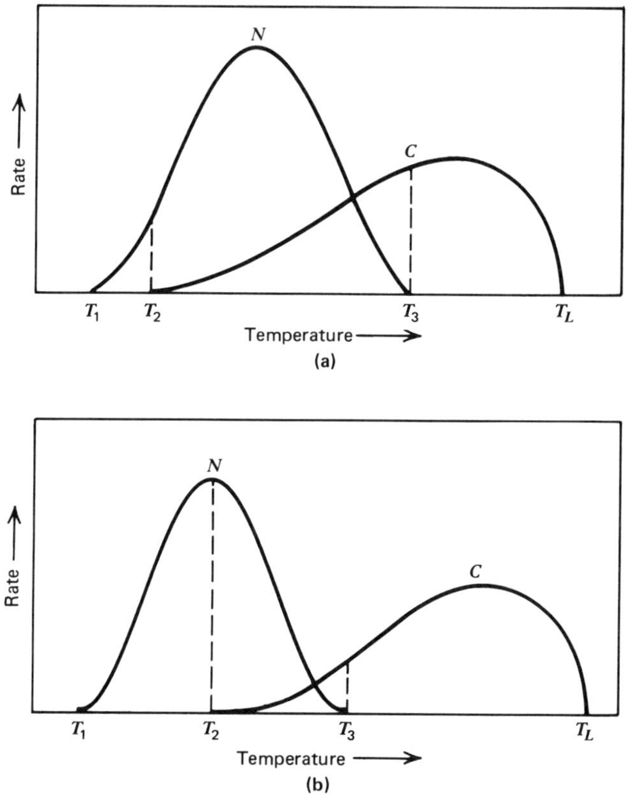

Figure 2. Number of nuclei N per unit of time, and the rate of crystallization C as a function of temperature (see text).

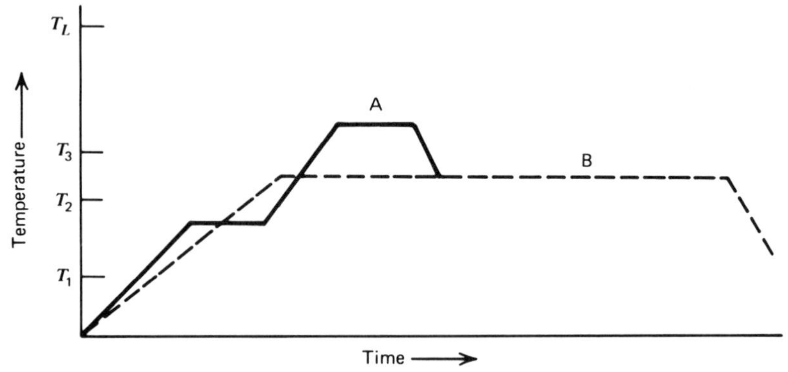

Figure 3. Heat-treatment cycle for controlled crystallization: A, two stage; B, isothermal.

One complication is the release of heat that accompanies crystallization. It can result in higher temperatures inside the body than on the surface and cause differential rates of crystallization throughout the piece, especially for articles having thick-walled sections. A solution is the use of a composition where the degree of overlap in the nucleation and crystal growth rate curves is such that a high rate of nucleation and a very slow crystal growth rate occur at the same temperature. A one-stage isothermal heat treatment can be used in which nuclei formation is followed by a slow rate of crystal growth such as shown in Figure 3. This slow rate of conversion allows the evolved heat to dissipate without producing unmanageable gradients. Figure 4 shows the change in the coefficient of expansion of a particular lithia aluminosilicate glass-ceramic composition at 760°C (9). The change in the expansion coefficient is roughly proportional to the degree of crystallinity. Crystallization starts at ca 20 h and the sample is 80% crystalline after 80–90 h.

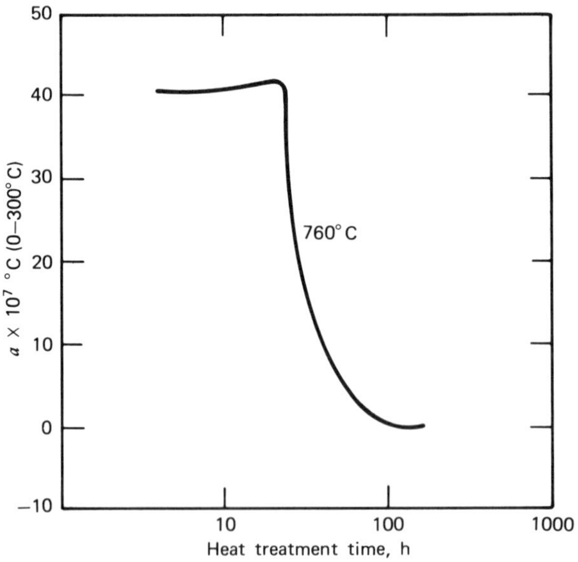

Figure 4. Thermal expansion vs heat treatment for a lithia aluminosilicate composition. α = coefficient of expansion.

Nucleation. One of the main factors that permits the formation of a conventional glass without crystallization is the very slow nucleation rate. Although some glass compositions are self-nucleating, most of the commercial glass-ceramic compositions rely on the presence of a nucleating agent to initiate internal crystallization. Although the mathematical treatment of homogeneous and heterogeneous nucleation is well developed (10), the exact role of the nucleating agent is not completely understood. The nucleant, which is usually a noble metal or a transition metal oxide, is incorporated in the batch and becomes an integral part of the glass structure during the melting. At heat-treatment temperatures of ca 30–100°C above the annealing point for times on the order of one hour or more, nucleating sites form. These nucleating sites are randomly distributed in concentrations of 10^9–10^{15} nuclei/cm^3 of glass and serve as sites for subsequent crystal growth.

Much of the information on nucleation is derived from detailed experimental observations on what occurs just prior to crystallization of the major phase (11). A number of different precrystalline events, such as enhanced amorphous phase separation and/or the appearance of trace nucleant phases, have been attributed to the nucleant but it is not always possible to identify a cause–effect relationship. Although there are probably several different nucleating mechanisms, the commonality of different nucleants is their introduction of sites of lower thermodynamic stability. The present understanding is not sufficient to allow *a priori* predictions of what is the most effective nucleant and what is the optimum nucleant concentration for a given base composition.

Crystallization. After nucleation, the material must generally be heated to higher temperatures in order for crystal growth to proceed; these temperatures are 750–1150°C, depending on the desired crystalline phase. The growth step is also complex (12). There may be metastable and equilibrium crystalline phases appearing simultaneously. Usually the composition of the crystals differ from the composition of the glass. In addition, the primary crystal phase (which is frequently a solid-solution phase) may start to transform by solid-state reaction to a new structural type. Solid solution refers to a family of compositions whose atomic arrangement is derived from some basic structural unit through the incorporation of selected foreign atoms. In silicate structures the Si^{4+} site is frequently occupied by an Al^{3+}, with the charge imbalance compensated by the addition of other cations. Some of these aspects can be illustrated by a transformation plot as Figure 5. This nonequilibrium diagram shows the phases as a function of the time and temperature of the heat treatment for a particular lithia aluminosilicate composition. It is only a first approximation for defining the heat-treatment limits since it does not contain all the information needed to optimize the crystallization process such as the phase concentrations.

The lower line represents the release of residual stress in the glass. Below this line the glass viscosity is such that the structural rearrangements required for the start of the glass-ceramic process do not occur in a practical time frame. The next higher temperature zone corresponds to the normal nucleation region. At higher temperatures the first major crystalline phase (a high-quartz solid solution) appears. With higher temperatures or longer times, the concentration of the high-quartz solid solution phase increases at the expense of the glassy phase (not shown in Fig. 5). With further heating or with longer times, or both, the high-quartz solid-solution phase transforms partially or completely to a new phase—a keatite solid solution (also commonly referred to as a β-spodumene solid solution); both of these solid solutions are based on phases of

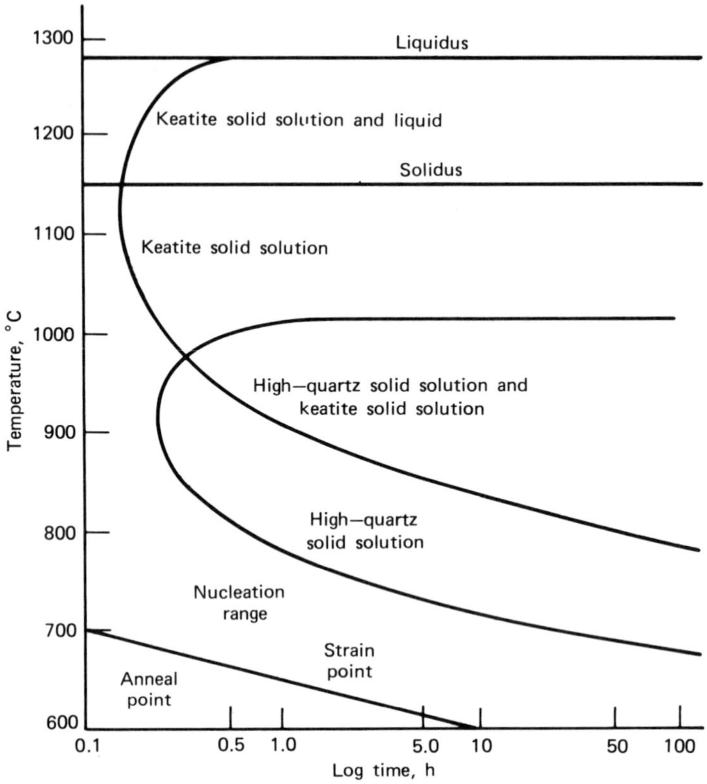

Figure 5. Schematic phase-transformation plot for a lithia aluminosilicate composition.

pure silica in which some of the silicon atoms have been replaced by aluminum atoms, with lithium atoms occupying interstitial sites to maintain charge balance. The keatite (or spodumene) solid solution is derived from the keatite structure—a high pressure form of silica (qv). As soon as the temperature is above the solidus, the amount of liquid starts to increase again.

After heat treatment, the material may undergo additional finishing steps such as cutting, grinding, polishing or decorating, or it may be packaged directly for shipment.

Properties

The properties of a glass-ceramic are dependent on such microstructural parameters (13) as: the amount and properties of the crystalline phase(s); the crystal sizes, shapes, and orientations; and the amount, composition, and distribution of the glassy phase (14). Some properties (eg, the thermal expansion coefficient) are additive functions determined by the properties and concentration of the individual phases. Other properties such as strength and maximum use temperature are not additive functions.

Figure 6 shows electron micrographs of two lithia aluminosilicate bodies with some of the microstructural features labeled. Although the crystalline phases are apparent,

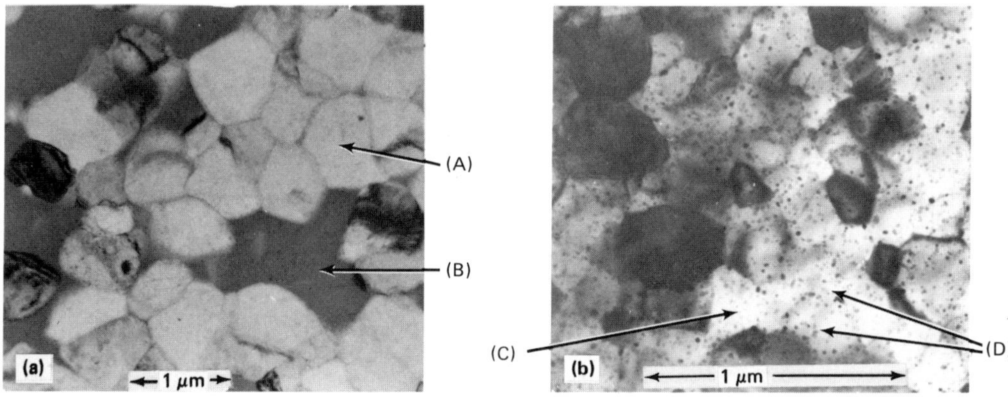

Figure 6. Transmission electron micrographs of two lithia aluminosilicate glass-ceramics, heat treated at a maximum temperature of (**a**) 800°C and (**b**) 1050°C. The phases present are: A, a high-quartz solid solution; B, a glassy phase; C, a keatite solid solution; D, an unidentified titania–zirconia solid solution (small grains are ca <0.05 μm).

the glass is not always readily visible (eg, Fig. 6**b**). A good deal of empiricism comes into play in achieving the desired properties since some properties are very sensitive to subtle microstructural changes.

The microstructure can be manipulated by changing the base composition, the nucleant system and the heat-treatment conditions. Commercial glass-ceramic materials generally contain at least seven or eight elementary species, each one added to fill a particular role. The constituents of the base composition determine the crystalline phases that occur, influence the melting and forming behavior, or influence the properties of the residual glass. A near-zero expansion body could not be produced if the chemical components of the formulation can only form high expansion phases. The nucleants not only initiate the crystallization but significantly affect the final-phase assemblage. By varying the nucleating agents and the heat treatment in a given base composition, it is possible to change the appearance and properties through changes in the type, amount, size, shape, and orientation of the crystal phases. Since it is unlikely that a glass-ceramic material can be developed with five or six specified properties all near the limits of observed values, the properties of a glass-ceramic material are invariably a compromise. Table 1 shows some of the properties of four commercial glass-ceramic materials. The characteristics of several base systems are discussed in the section Commercial Applications.

Thermal Properties. ***Thermal Expansion.*** Glass-ceramic materials have been reported with thermal expansion coefficients ranging from ca $(-60 \text{ to } 200) \times 10^{-7}/°C$. The thermal expansion properties, particularly the low expansions, are the primary properties that have been exploited commercially. The low expansion materials have been used for applications where thermal shock resistance is important such as cookware, stove tops, heat exchangers, etc. They have also been used in applications such as mirror substrates where minimal thermally induced distortions are important. Higher expansion materials have been used for sealing various metals and alloys. In some compositions it is possible to change the expansion coefficient over wide ranges by varying the heat-treatment conditions (eg, in some barium aluminosilicate compositions the expansion coefficient can be changed from $(55 \text{ to } 160) \times 10^{-7}/°C$).

Table 1. Properties of Four Commercial Glass-Ceramics [a]

Property	Example: 1 CER-VIT C-101	2 Corning 9608	3 Corning 9606	4 Corning 9658
Thermal				
coefficient of expansion, $10^{-7}/°C$				
at 0–38°C	0.0			
at 25–300°C		13	57	
at 25–400°C				94
thermal conductivity at 25°C, W/(m·K)	1.675	2.008	(mean at 3.398 20–800°C)	1.675
Mechanical				
Young's modulus at 20°C, GPa [b]	92.4	86.2	120	64.1
shear modulus at 20°C, GPa [b]	36.5		47.6	25.5
modulus of rupture at 20°C, MPa [c]	55	83	241	103
density at 20°C, g/cm^3	2.50	2.50	2.60	2.52
hardness, Knoop, kgf/mm^2 [d]				
100 g loading		593	657	
200 g loading	540			
500 g loading				250
Chemical				
corrosion rate, mg/cm^2, after 24 h in				
5% HCl at 95°C	0.3	0.12		87
5% NaOH at 85°C	3.5			8.5 [e]
H_2O at 100°C	0.3			
Electrical				
dielectric constant at 100 Hz	11.7 (at 25°C)			
at 100 kHz				5.92 (at 25°C)
at 1 MHz	9.1 (at 25°C)	6.9 (at 20°C)	5.6 (at 20°C)	
at 1 MHz	14.2 (at 250°C)			
dissipation factors at 100 Hz	0.076 (at 25°C)			
at 10 kHz				0.003 (at 25°C)
at 1 MHz	0.023 (at 25°C)	0.023 (at 20°C)	0.017 (at 20°C)	
at 1 MHz	0.093 (at 250°C)			
$\log_{10}$ of volume resistivity, Ω·cm				
at 25°C	14	13.4	16.7	
at 250°C	8	8.1	10	10.8

[a] Data taken from property summaries provided by Owens-Illinois, Inc. and Corning Glass Works.
[b] To convert GPa to psi, multiply by 145,000.
[c] To convert MPa to psi, multiply by 145.
[d] See Hardness; 1 kgf/mm^2 = 9.8 MPa.
[e] After 6 h.

Thermal Stability. The maximum use temperature must be defined in terms of a time factor and the degree of loading. The crystalline phase(s) play an important role, but the maximum use temperature is limited by the residual glass. At elevated temperatures the residual glass may crystallize further or it may enhance deformation by viscous flow. The glass-ceramic materials are generally more refractory than the common glasses but less refractory than the common oxide refractories.

Since thermal shock resistance and thermal stability are a mutually desirable combination, most of the glass-ceramic work in this area has been on the low expansion lithia aluminosilicate compositions. The best of these compositions can survive >1000 h at 1000–1100°C without appreciable change in properties. More refractory glass-ceramic materials have not been commercialized owing to the difficulties of melting and forming them.

Thermal Conductivity. The thermal conductivity of glass-ceramics is slightly higher than that of conventional glasses and is generally higher than silicate ceramics such as porcelain. However, the thermal conductivity is much lower than that of the pure oxide ceramics such as alumina, magnesia, and beryllia. The thermal conductivity range is ca 1.7–5.4 W/(m·K).

Optical Properties. Transparent and opaque glass-ceramics can be made (see Fig. 7). The degree of transparency is a function of the optical properties of the crystalline species, the grain size, and the difference in the refractive indexes of the glass and the crystals. The transmission in the near infrared is typical of that of silicate glasses. The transparency has been utilized primarily for inspection purposes, but some compositions have been shown to display electro-optical activity (15).

Chemical Properties. The chemical durability can be varied over wide ranges. The rate of attack and the mechanism of attack are sensitive to pH. The crystalline phases and the residual glass generally are attacked at different rates. A product that exploits the solubility of lithium silicate phases is described in the section entitled Li_2O–SiO_2. The chemical durability of the common lithia aluminosilicate glass-ceramics is generally on a par with borosilicate glasses. A composition from the Li_2O–CeO_2–Al_2O_3–SiO_2 system is used commercially as a coating for chemical reactors because of its good chemical durability and mechanical toughness.

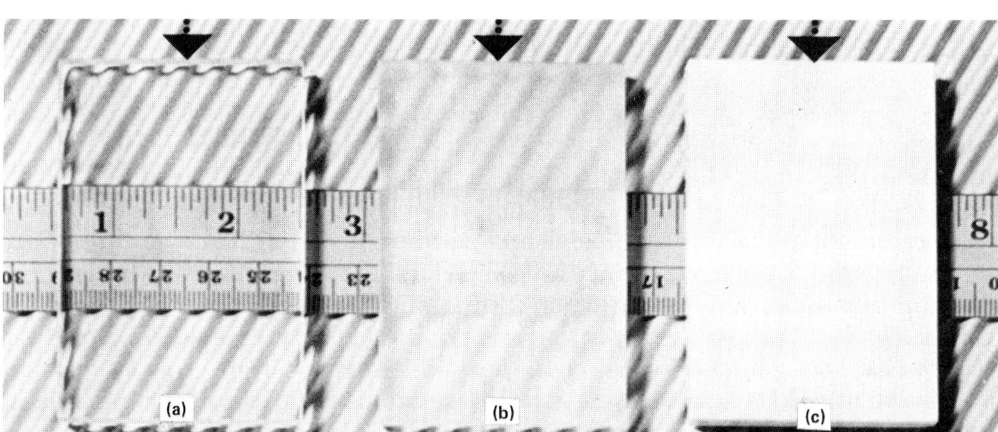

Figure 7. Glass-ceramic materials exhibiting a range of transparencies: (**a**) transparent; (**b**) translucent; (**c**) opaque.

Mechanical Properties. Glass-ceramics are brittle materials like ordinary glasses and ceramics, and exhibit many of the same properties, ie, there is a statistical distribution of strengths for any given set of samples and the results are a function of the testing conditions. The abraded modulus of rupture for most glass-ceramic materials is 55–250 MPa ((8–36) × 10^3 psi) compared with 40–70 MPa ((6–10) × 10^3 psi) for common glasses under identical abrasion and test conditions. High strength glass-ceramics (up to 1300 MPa or ca 190,000 psi) have also been reported (16). The high strengths are generally attributed to the formation of a surface-compressive layer, either through differential crystallization of a low expansion crystal phase, or by an external chemical ion exchange. The glass-ceramics are generally more abrasion resistant than the common glasses and require more energy for fracture because of the fine-grained microstructure. Glass-ceramics, based on slag and basalt (see $CaO–MgO–Fe_2O_3–Al_2O_3–SiO_2$ below) are notable for their abrasion resistance. Knoop hardness numbers as high as 900 have been measured (see Hardness). The modulus of elasticity for glass-ceramics generally falls in the range of 82–125 GPa ((12–18) × 10^6 psi). The elastic properties are an additive function of the concentration and elastic properties of the individual phases.

Electrical Properties. Most glass-ceramics are insulators. The conduction mechanism is primarily ionic and is more complicated than in a glass of the same composition because of the complex microstructure. The resistivity range is shown in Table 1. The dielectric properties also can be varied over wide ranges (15). The dielectric losses are primarily influenced by the mobility of the alkali ions and generally increase with increasing temperatures.

Magnetic Properties. Magnetic phases can be crystallized from a glass, but magnetic glass-ceramics are not commercially available. Most magnetic phases of interest do not contain atoms that make stable glasses; this limits the amount of magnetic phase that can be incorporated in the melt and subsequently crystallized (see Amorphous magnetic materials; Magnetic materials).

Commercial Applications

The size of the glass-ceramic market in the United States is not accurately known but 1979 sales were probably on the order of (\$100–200) × 10^6. The market outside the United States is probably comparable, with the market size strongly dependent on the level of USSR production of slag-based products. Examples of some glass-ceramic products and their principal chemical constituents are listed in Table 2. The following are descriptions of a few of the features that are important in the commercialization of products made from five of the more useful compositions.

$Li_2O–SiO_2$. In this system there is a large difference in chemical durability between the glass and some of the crystalline phases. Lithium silicate systems have been used for the manufacture of a photosensitive glass which can be chemically machined (6). By suitable masking and irradiation of lithium silicate glasses containing photosensitive noble metals as nucleants, it is possible to nucleate and crystalline only preselected areas. The crystallized portion contains lithium metasilicate crystals which are much more soluble in acid than the glass; these can be removed by etching, leaving the uncrystallized portion intact. It is possible to form intricate patterns in the glass (eg, form up to 1550 holes/cm^2 in a flat plate) for applications such as fluidic valves, printing-plate molds, and fine mesh screens.

Table 2. Commercial Applications

Basic chemical system	Unique properties	Product
Li_2O–SiO_2[a]	photosensitivity and differential solubility of phases leads to ability to form intricate patterns	substrates; fine-mesh screens; fluidic devices; printing-plate molds; spacers
Li_2O–Al_2O_3–SiO_2[b]	low expansion, thermal shock resistance	cookware; burner covers; preheaters; mirror substrates; low expansion housings; valve parts
MgO–Al_2O_3–SiO_2[a]	high strength, good dielectric properties	radomes; antenna windows
K_2O–MgO–Al_2O_3–B_2O_3–SiO_2–F	machinable	pacemaker parts; welder's fixture holder; welding nozzles; circuit bases
CaO–MgO–Fe_2O_3–Al_2O_3–SiO_2	abrasion resistance	industrial floor coverings; wall facings; abrasion resistant linings; high temperature insulators
CaO–ZnO–Al_2O_3–SiO_2	abrasion resistance	floor coverings; wall facings; counter tops
B_2O_3–Al_2O_3–SiO_2 P_2O_5–Al_2O_3–SiO_2	selective-ion evaporation at elevated temperatures	semiconductor doping sources
Na_2O–Al_2O_3–SiO_2	good strength	dinnerware
Li_2O–CeO_2–Al_2O_3–SiO_2[a]	chemical durability; mechanical toughness	coatings for chemical reactor accessories
BaO–Al_2O_3–SiO_2	specific expansion coefficient	alloy bonding
Li_2O–BaO–SiO_2	dielectric properties	insulators in devices

[a] 1979 United States sales probably over $1,000,000.
[b] 1979 United States sales probably over $50,000,000.

Li_2O–Al_2O_3–SiO_2. The lithia aluminosilicate system has the most commercial utility. Titania and titania–zirconia mixtures are the most commonly used nucleants. The properties of two commercial lithia aluminosilicate materials are listed in Table 1 (examples 1 and 2). The primary crystal phase in example 1 is a high-quartz solid solution; in example 2, it is a keatite solid solution. In addition to low expansion, some of the compositions are relatively stable up to 1000–1100°C; in other compositions, it is possible to make bodies that are either transparent or opaque. The commercial lithia aluminosilicate glass-ceramics are among the best glass-ceramic materials in terms of formability. The largest glass-ceramic body made to date is a 25-t casting used for an astronomical-mirror substrate.

MgO–Al_2O_3–SiO_2. Glass-ceramics from this system generally have good dielectric insulating characteristics at high frequency, high strength and moderate to high expansion (Table 1, example 3). Titania, zirconia, and mixtures of the two are the most commonly used nucleants and the principal crystalline phase is cordierite ($2MgO.2Al_2O_3.5SiO_2$). Since the magnesia aluminosilicates generally form relatively fluid melts, they are usually formed by casting. The combination of properties and formability has led to their use as radomes (see Microwave technology).

K_2O–MgO–Al_2O_3–B_2O_3–SiO_2–F. By proper compositional and heat-treatment control of silicates from the fluorine–mica family (13), it is possible to develop a body consisting of interlocking flakes of randomly oriented mica crystals. This unique microstructure limits crack growth and, as a result, these mica glass-ceramics are machinable to precise tolerances with conventional metal-working tools. In addition to machinability, they are strong, thermally shock-resistant and have good dielectric properties (Table 1, example 4). The commercial products are machined from pieces that are cut from cast slabs (see also Silica).

CaO–MgO–Fe_2O_3–Al_2O_3–SiO_2. Glass-ceramic materials called slagcerams have been produced by the crystallization of glasses made from metallurgical slags (17). The slag is generally modified by adding silica and/or alumina to make the slag glass more stable. The nucleating agents include transition metal oxide and sulfides as well as chrome and various forms of iron oxide. The nucleating effectiveness is dependent on the degree of reduction of the nucleating agents. The slag-based materials have excellent resistance to abrasion and wear, and have good chemical resistance to most common solvents. The slag-based melts are fluid, and generally are formed by a sheet or pressing operation. Products such as floor and wall coverings are attractive because of the low cost and the availability of the raw materials, and have been manufactured in the USSR and several other countries for many years.

Similar materials have been made from molten basalt—a plentiful, low cost rock (4,18). These basalt glass-ceramics are being produced in Europe and the USSR for use as abrasion-resistant pipelines, laboratory flooring, etc.

BIBLIOGRAPHY

"Slagceram" in *ECT* 2nd ed., Suppl. Vol., pp. 876–889, by S. Klemantaski, British Steel Corporation (Corporate Laboratories).

1. G. W. Morey. *The Properties of Glass,* 2nd ed., Reinhold Pub. Corp., New York, 1954.
2. M. deReaumur, *Mem. Acad. Sci.,* 370 (1739).
3. G. Tammann, *The States of Aggregation,* D. Van Nostrand Co., New York, 1925.
4. A. Portevin, *Mem. Soc. Ing. Civils Fr.,* 266 (1929); *Chem. Abstr.* **25,** 4207 (1931).
5. H. H. Blau, *Ind. Eng. Chem.* **25,** 848 (1933).
6. S. D. Stookey, *Ind. Eng. Chem.* **45**(1), 115 (1953).
7. S. D. Stookey and R. D. Mauer, "Catalyzed Crystallization of Glass—Theory and Practice" in *Progress in Ceramic Science,* Vol. 2, Pergamon Press, New York, 1962, pp. 77–101.
8. F. Albrecht, *Beispiele Angewandter Forschung, Frauenhofer (Society for the Promotion of Research, Munich),* 19 (1955).
9. J. E. Rapp, *Am. Ceram. Soc. Bull.* **52,** 499 (1973).
10. J. J. Hammel, "Nucleation in Glass—A Review" in *Advances in Nucleation and Crystallization in Glasses, Glass Division Symposium,* American Ceramic Society, Columbus, Ohio, 1971, pp. 1–9.
11. D. R. Stewart, "Concepts of Glass Ceramics" in *Introduction to Glass Science,* Plenum Pub. Corp., New York, 1972, pp. 237–271.
12. M. H. Lewis, J. Metcalf-Johansen, and P. S. Bell, *J. Am. Ceram. Soc.* **62,** 278 (1979).
13. G. H. Beall, *Glass Technol.* **19**(5), 109 (1978).
14. P. W. McMillan, *Glass Technol.* **15**(1), 5 (1974).
15. N. F. Borrelli and M. M. Layton, *J. Non-Cryst. Solids* **6,** 197 (1971).
16. D. A. Duke, J. F. MacDowell, and B. R. Karstetter, *J. Am. Ceram. Soc.* **50,** 67 (1967).
17. J. A. Topping, *J. Can. Ceram. Soc.* **45,** 63 (1976).
18. G. H. Beall and H. L. Rittler, *Am. Ceram. Soc. Bull.* **55,** 579 (1976).

General References

P. W. McMillan, *Glass-Ceramics,* Academic Press, Inc., New York, 1979; good introductory reference.

L. L. Hench and S. W. Freiman, eds., *Advances in Nucleation and Crystallization in Glasses, Glass Division*

Symposium, American Ceramic Society, 1971; the proceedings of a symposium containing 26 papers relating to glass-ceramic materials.

DANIEL R. STEWART
Owens-Illinois

GLASSY METALS

The most familiar glasses are oxide glasses such as silicate or borate glasses. However, they are not the only members of the family of glasses. For instance, a wide variety of plastic (high polymer) materials are totally or partially glassy. A recent addition is the group of glassy metals, or metallic glasses (1–2), also called amorphous alloys (see Amorphous magnetic alloys). They offer excellent possibilities in the electric, electronic, and structural industries (3–4). Several compositions of glassy metals, particularly those containing about 80 atom % transition metals and about 20% metalloids such as B, Si, P, C, are now available in commercial quantities (5).

Unlike crystalline solids, glasses do not have a long-range ordered atomic structure (see Fig. 1). They are in essence frozen liquids. Most glassy metals cannot be obtained by simply cooling liquid metals at rates commonly employed in metals industries because the samples would crystallize. However, when liquid metals are quenched at a very high rate, often exceeding 10^5 K/s, it is possible to avoid crystallization and thus obtain glassy metals that largely preserve the atomic structure of liquid metals.

Several other methods are available such as chemical or electrodeposition (6), and physical deposition which includes vapor or ion-sputtering (7). The latter may have limited applications in the electronic industries (8–10). However, for most other applications, rapid quenching from the melt appears to be best. This important technology was pioneered at the California Institute of Technology around 1960 (11–12). Because of the technique with which the molten liquid is propelled and held against the substrate, this process is generally called splat quenching (1). Thus various compositions of metallic alloys can be processed into a glassy form which is stable at room temperature. The California group also studied the fundamental electrical, magnetic, and structural properties of these compounds and showed that some amorphous alloys are ferromagnetic (13).

The properties of glassy metals are often similar to those of crystalline metals. It is, therefore, not possible to judge unambiguously whether a solid is crystalline or amorphous by observing its macroscopic properties alone. Glassiness can be determined either directly by studying the atomic structure with diffraction methods, or indirectly by carefully studying the thermodynamic properties. A controversy developed whether the glassy metals were really glassy or just microcrystalline. It was resolved when a well-defined glass transition was observed in amorphous Au–Ge–Si alloy (14), and the x-ray diffraction pattern of the glassy metals was interpreted by a dense-random-packing (DRP) model (15). It is now widely believed that the glassy metals should be classified as truly glassy or amorphous, and not as a collection of small crystalline grains, even though they have a slightly higher degree of atomic short-range order than liquids.

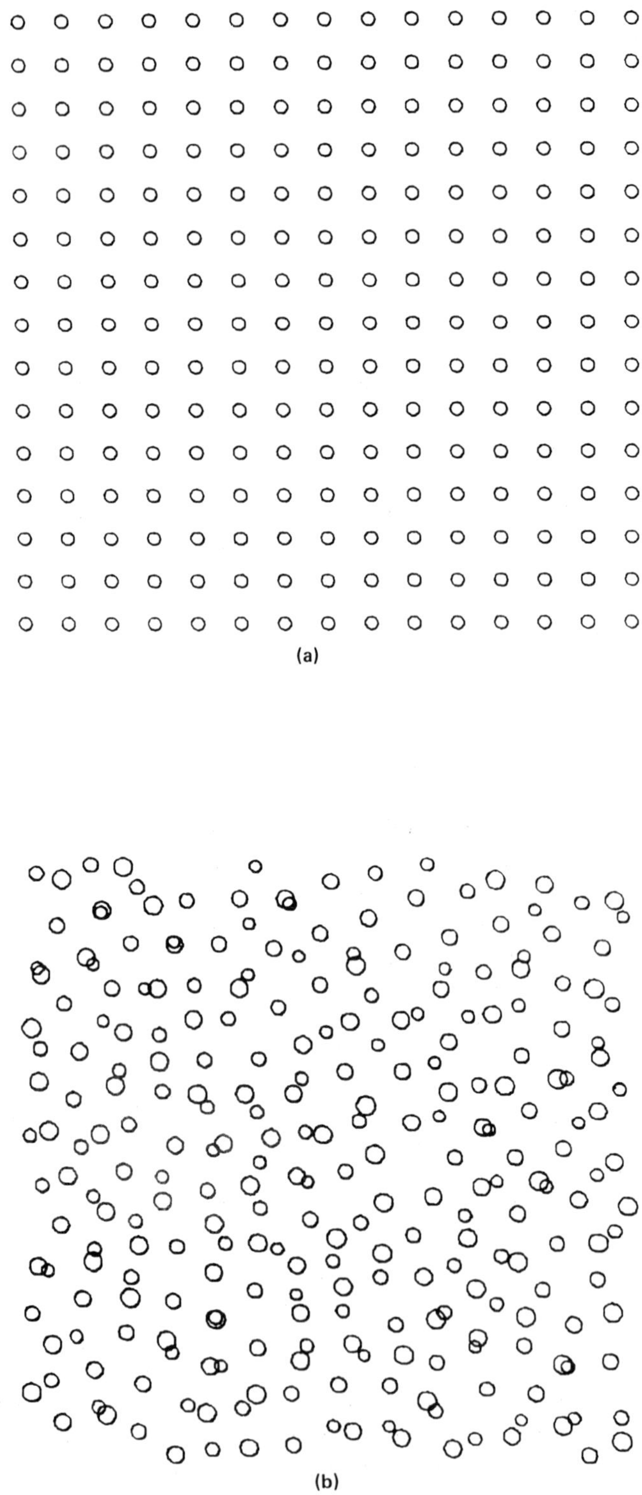

Figure 1. Atomic structure of **(a)** a crystalline metal and **(b)** a glassy metal.

As mentioned above, each of the macroscopic physical or chemical properties of the glassy metals may not be fundamentally different from that of a crystalline solid. However, collectively they represent the very unique nature of glassy metals. For example, they can be magnetically soft but at the same time mechanically strong and tough (16). This combination of properties has not been observed for crystalline alloys and makes the glassy alloys unique from the application point of view. These observations were made possible by the development of continuous rapid-quenching techniques (17–19) for making long ribbons of glassy metals on which mechanical, magnetic, and electrical properties could be measured. Glassy metals have extremely high mechanical strength and appreciable ductility (18–20); ferromagnetic glassy metals have very low magnetic coercivity (16,21–22). These findings touched off a rapid growth in research activities on the glassy metals. It is generally believed that the principal application of glassy metals will be as magnetic materials (qv).

Structure

Glassy metals differ fundamentally from crystalline metals in their atomic structures which are studied by x-ray, neutron, or electron diffraction methods (15,23) (see Analytical methods). Indirect methods, eg, Mossbauer spectroscopy, also can be employed.

Theory. In crystalline solids, diffraction occurs when the Bragg condition is satisfied:

$$2\, d \sin \theta = n\lambda \tag{1}$$

where d is the spacing between adjacent equivalent atomic planes, θ the diffraction angle, n an integer, and λ the wavelength of the diffracted particle, such as photon, neutron, or electron (24). Equivalently:

$$q = 2\, k \cdot \sin \theta = nK \tag{2}$$

where q is the diffraction vector, k the wave vector of the diffracted particle, and K the reciprocal lattice vector. Since there are no periodic lattices in glassy metals, diffraction occurs at almost all values of q or θ. However, the intensity of diffraction, $I(q)$, is not independent of q because the scattering intensity from one atom, $f^2(q)$, is a decreasing function of q associated with a nonuniform distribution of electrons near the atomic nuclei, and more importantly, because the distribution of atoms is not totally random even in glassy metals or liquids. Since the repulsive force between atoms becomes very strong when two atoms approach within a spacing that is less than the equilibrium atomic separation distance, the pair distribution function (PDF) of atoms, $\rho(r)$, which represents the probability of finding an atom at a distance r from another atom, is essentially zero below a certain minimum atomic spacing; $\rho(r)$ has a rather sharp maximum at the position corresponding to the equilibrium atomic distance, as shown in Figure 2. This rigidity of the nearest neighboring spacings is reflected in the second and third neighbor peaks. The atomic correlation generally subsides at distances beyond 2–3 nm and $\rho(r)$ becomes equal to ρ_o, the average atomic density. It can be shown that the Fourier transform of the PDF, the interference function or structure factor:

$$i(q) = \frac{4\pi}{q} \int_o^\infty r[\rho(r) - \rho_o] \sin(q \cdot r) dr \tag{3}$$

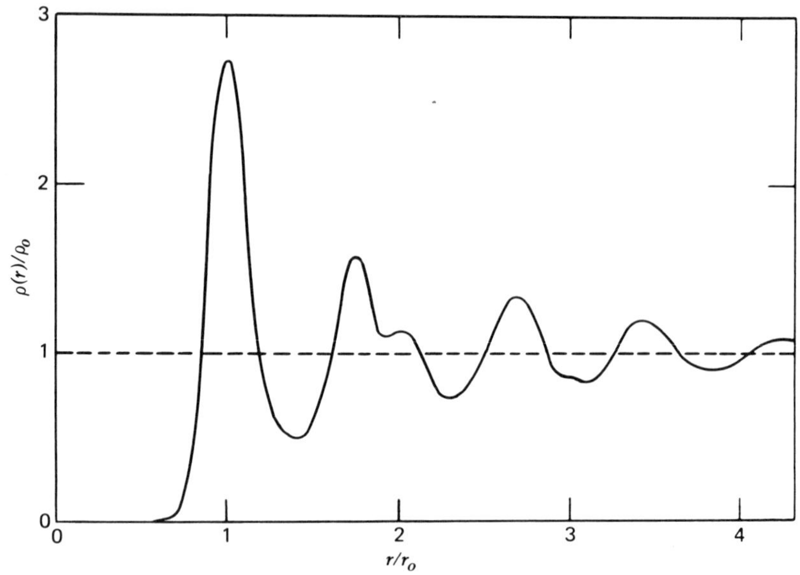

Figure 2. An idealized pair distribution function $\rho(r)$.

can be directly observed by diffraction measurements (24). The intensity of the diffracted particles, $I(q)$, is the sum of the inelastic and elastic diffraction intensities which is proportional to $[i(q) + 1]f^2(q)$.

Diffraction. The random nature of glassy metals can easily be determined with a conventional automatic powder x-ray diffractometer. When an x-ray tube with a

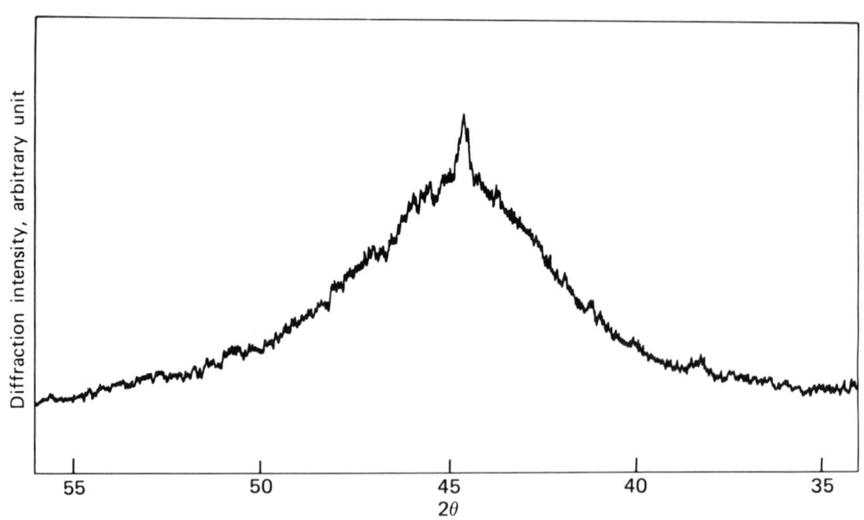

Figure 3. An automatic diffractometer trace of amorphous $Fe_{32}Ni_{35}Cr_{15}P_{12}B_6$ alloy which is partially crystalline. The figure shows the first peak of $i(q)$ and an additional peak on top owing to a small crystalline grain.

Cu target is used, only the first and second peaks of $i(q)$ can usually be observed. This method is useful in screening samples with crystallites above a certain size from truly glassy samples, as shown in Figure 3, but is totally inadequate for a more detailed study. The determination of PDF from diffraction measurements is a fairly difficult and elaborate task, requiring long measurement times and lengthy numerical data reductions. Such measurements are usually performed with a step-scan x-ray diffractometer with a Mo-target tube and a detector monochromater (23). It takes typically one week to obtain data with sufficient accuracy for detailed study. Another method is the recently developed energy-dispersive x-ray diffraction (EDXD) technique (25). It employs a white x-ray source, a fixed diffraction angle, and an energy-sensitive semiconductor detector of the intrinsic Ge- or Li-drifted Si type. This method offers faster and more accurate data collection, even though the data analyses are more complicated. Figures 4 and 5 show the interference function, $q \cdot i(q)$, and the PDF, $4\pi r[\rho(r) - \rho_o]$, for amorphous $Fe_{40}Ni_{40}P_{14}B_6$ determined by the EDXD method (see also X-ray technology).

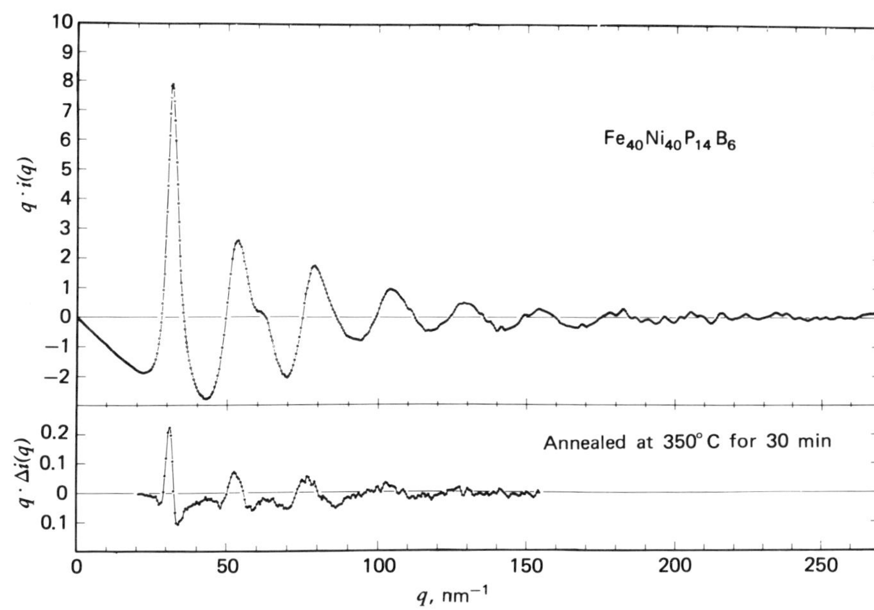

Figure 4. The interference function $q{\cdot}i(q)$ of amorphous $Fe_{40}Ni_{40}P_{14}B_6$ alloy determined by energy dispersive x-ray diffraction and the change in the interference function owing to the structural relaxation caused by the annealing at 350°C for 30 min (25).

Other methods for structural studies include neutron diffraction, electron diffraction, and extended x-ray absorption fine structure (EXAFS measurements). The latter can detect the atomic environment around a particular species of atom at which the absorption coefficient is measured. It has become more convenient with the recent advent of synchrotron radiation (26).

Structural Models. The essential features of the atomic structures of liquids, and for that matter of glassy metals, can be modeled very well by an assembly of randomly packed hard spheres (27). This model, with some relaxations included to incorporate

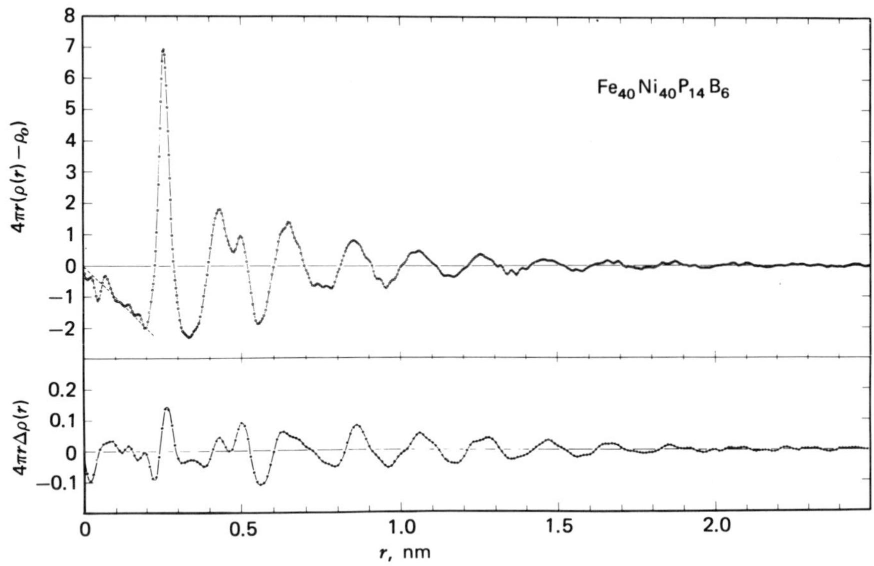

Figure 5. The reduced pair distribution function $4\pi r\ [\rho(r) - \rho_o]$ for amorphous $Fe_{40}Ni_{40}P_{14}B_6$ and the change owing to the structural relaxation (25).

the elastic compliances of the atoms, describes the diffraction data with considerable accuracy (15,28). For more detailed studies, deviations have to be considered, eg, taking the compositional short range order into account. However, the concept of a dense random packing (DRP) model as describing the basic atomic structure of simple glassy metals is probably the most fundamental one when considering the effect of atomic short-range order on physical properties.

Properties

Mechanical. Very high strength and some ductility in the fracture of metallic glasses were first discovered in the study of ribbon samples of $Pd_{80}Si_{20}$ (18). The mechanical properties of a glassy metal are characterized by a tensile strength that exceeds 1% of Young's modulus, the formation of shear bands, ductile flow on the fracture surface, and in many cases, embrittlement after an annealing that does not cause crystallization. Comparable combinations of mechanical properties are rare in crystalline alloys. Usually crystalline alloys with such high strength are always brittle. Because of the absence of such well-defined structural defects as dislocations, it is difficult to understand some of these properties from a microscopic point of view. Recent investigations (29–30), however, indicate the critical role of atomic structure in determining mechanical properties of the glassy metals.

Strength. The Young's modulus of glassy metals is always lower (18,30) by about 20–30% than that of corresponding alloys because of the anomalously low shear moduli which reflects the liquid nature of the glassy metals. However, plastic shear does not take place until the elastic strain reaches ca 2%. Plastic shear deformation takes place by the formation of macroscopic shear bands which are clearly visible under an optical microscope. In a tensile measurement there is no work hardening in the shear band

and the sample fails by shear rupture as soon as a shear band is formed. The tensile strength of a typical metal–metalloid glass is 2–3 GPa [(2.9–4.4) × 10^5 psi].

Ductility and Toughness. Because of the apparent absence of work hardening, little plastic deformation can be observed macroscopically during a tensile test. However, the fractured surfaces show traces of considerable viscous flow (20) generated by the stress concentration at a crack tip which lowers the viscosity of the glassy metal (31). The capacity for viscous flow implies that a glassy metal is tough since, even when a sharp crack is formed, the stress concentration at the crack tip is dispersed by viscous flow, making the crack tip dull (32). The toughness and fatigue life of a typical transition metal–metalloid glass is comparable to those of high-strength stainless steels (30).

Embrittlement. The ductility of the glassy metals, which makes them attractive for applications, may disappear quickly when they are mildly annealed (33). This embrittlement behavior is observed in almost all glassy metals containing Fe or Co. Since Ni or Pd glasses do not show embrittlement, it appears that this behavior is related to the directional bonding of transition metals with unfilled d-electron shells. Although the segregation of phosphorus was at one time considered as the cause (34), the current opinion is different (35). A structural relaxation that reduces the capacity for viscous flow under stress by eliminating structural defects and reducing the activation volume is suspected to be the principal mechanism for embrittlement (36). Temper embrittlement is appreciably reduced by using only boron as the glass-forming metalloid component (37). Glassy metals are also embrittled by hydrogen by an unknown mechanism (38).

Creep. Most glassy metals show an appreciable amount of mechanical creep at temperatures as low as 200°C (39). Although the creep rate can be significantly reduced by preannealing, it presents problems for elevated temperature applications.

Workability. Because of their ductility, glassy metals can be processed mechanically by cutting or punching. The residual internal stress induced by processing is relatively small and can be annealed out. Amorphous metals can also be hot-pressed or hot-formed if crystallization can be avoided (40). Pieces of glassy metals cannot be welded but small pieces can be explosively compacted into a bulk (41).

Magnetic. Many glassy metals containing transition metals or rare earths, or both, are ferromagnetic or ferrimagnetic (3) (see Magnetic materials). Some show more complex properties at low temperatures characterized by so-called spin-glass behavior, which occurs also in magnetically dilute crystalline alloys. The transition metal–metalloid glasses show very low coercivities, in some cases as low as 0.24 A/m (3 mOe) (16) and permeabilities as high as 10^6 (16). Recent studies indicate an important interplay between atomic short-range order and magnetic properties, in particular with respect to annealing effects (3,42).

Electrical. ***Resistivity.*** Glassy metals usually have high electrical resistivities. The temperature dependence of the resistivity is often small, and can be either positive or negative or show a weak minimum. Recent studies indicate a close relationship between resistivity and atomic structure (43).

The RT electrical resistivities of glassy metals range from 60 to 200 $\mu\Omega$·cm. The composition and temperature dependence of the resistivity is often interpreted by the extended Ziman theory which relates the resistivity directly to the x-ray structure factor (44). The Ziman theory explains the resistivity of liquid metals but is also applicable to glassy metals, which are frozen liquids, as discussed earlier. A rigorous application of the Ziman theory requires not only the knowledge of the x-ray structure factor but also the value of the Fermi momentum which is not readily known. For that reason it is difficult to assess the extent of the success of the theory, but its value in the qualitative interpretation of various properties is unquestionable. The resistivities of magnetic glassy metals contain an appreciable magnetic contribution which appears to produce a resistivity minimum (45) similar to the Kondo phenomena in dilute crystalline alloys (46).

Superconductivity. Various metal–metal glasses such as $La_{80}Au_{20}$ [*55929-86-7*] and metal–metalloid glasses such as $(Mo_xRu_{1-x})_{80}P_{20}$, show superconductivity (49). The transition temperature T_c ranges up to 8.7 K. Unlike the transition temperatures of crystalline superconductors, this T_c is relatively insensitive to small changes in composition (48) because of a smooth d-electron density of states. All known superconducting glassy metals are type-II or dirty superconductors because of the very short electron mean-free-paths in glassy metals. Since the coherence length is short and the penetration depth is large, the upper critical field is usually very high, reaching almost 16 MA/m (200 kOe) (47) (see Superconducting materials).

Corrosion Resistance. Glassy metals containing Cr, such as chromium–iron alloys, are extremely corrosion resistant (49) because of the formation of a dense chromium oxo–hydroxide layer on the surface of the metal. In crystalline alloys, this protective layer can be broken at grain boundaries or dislocation sites. Since the glassy metals do not have these structural defects, the protective layer generally remains intact resulting in excellent corrosion resistance. The ferrous glassy metals without Cr, however, seem to be more susceptible to corrosion than crystalline alloys because they are essentially unstable alloys. The addition of phosphorus and nickel improves the corrosion resistance to some degree.

Thermal Behavior

Glassy metals are thermodynamically unstable and crystallize when they are heated. Crystallization is usually accompanied by drastic property changes. Furthermore, the glassy metals as produced are not in a stable state within the amorphous phase because of rapid quenching. An annealing treatment near the glass transition temperature can also result in substantial property changes even if it does not lead to crystallization. This stabilization phenomenon is called structural relaxation.

Crystallization. Crystallization after the production of glassy metals is represented by the lower part of Figure 6. The kinetics can be expressed by a simple activation equation:

$$\tau(T) = \tau_o \exp(\Delta E_c/kT) \tag{4}$$

where $\tau(T)$ is the crystallization time at temperature T, and ΔE_c is the activation energy of crystallization which is on the order of 1–4 eV (50). However, the quenched sample does not represent the $\tau = 0$ state. If the sample is infinitely rapidly quenched from a liquid to temperature T, then the time required for crystallization is given by equation 4. However, the quench rate is always finite and, therefore, the nucleation process starts during the quenching process; consequently, the crystallization time is generally shorter than given by equation 4. Such an effect of quench rate on the crystallization kinetics has recently been observed (51) (see Crystallization).

Structural Relaxation. The structure of the liquid is temperature-dependent and characterized by the amount of free volume which is a free space between atoms and somewhat analogous to a vacancy (52). When the liquid is rapidly frozen, much of this free volume gets trapped in the glassy metals. The faster the quench rate, the greater the amount of trapped free volume which is removed by a subsequent annealing treatment. The excess free-volume elements are not merely additional spaces between atoms. Because of the atomic mismatch that they introduce, the free-volume elements are surrounded by appreciable strain fields which resemble the lattice defects in crystalline solids. They may therefore analogously be called structural defects in the amorphous state (53). The process in which these defects are annihilated during annealing is called structural relaxation. A recent EDXD study showed that structural relaxation can be observed as a change in the PDF, as shown in Figures 4 and 5 (25). Structural relaxation can also be studied by careful calorimetric measurements (54). It affects almost all physical properties, but particularly mechanical properties. The principal effects of structural relaxation are: reduction in diffusivity (55) and creep rate (39); increase in Young's modulus (56); decrease in the internal friction (57); decrease in the low temperature acoustic attenuation and increase in the thermal conductivity (58); embrittlement (36); small changes in electrical resistivity (59) and the superconducting transition temperature (60); and densification (56). The kinetics of structural relaxation show a logarithmic time dependence (25) and, therefore, the relaxation takes place mostly in the early stages of annealing.

Structural relaxation represents a change in the topological atomic short-range order. In some glassy alloys, the compositional short-range order can also be affected during the annealing treatment (42). This is a very short-range process involving primarily only the nearest neighbors, and, therefore, occurs at relatively low temperatures. Many magnetic properties, including Curie temperature, magnetic anisotropy, and permeability, are affected by the compositional short-range order. In designing the heat treatment, which is often required for the application of glassy metals, all these factors have to be considered in order to avoid undesirable effects.

Formation

Solidification, the formation of crystalline solids from liquids, takes place by means of a nucleation and growth process because the atomic structure of a crystalline solid, which is a lattice structure with point and translational symmetry, is incommensurate with the atomic structure of a liquid. The interface between the crystalline and the liquid phases is more loosely packed than either phase, and the atomic arrangements are locally distorted, making the energy to produce the interface positive. This means that even below the solidification temperature, where the energy of the crystalline phase is lower than that of the liquid phase, the energy to form a very small nucleus or embryo of the crystalline phase is positive. It then takes a finite time for nuclei to reach critical sizes in order to grow into large crystalline grains. The crystallization time thus defined becomes shorter at first as the driving force induced by undercooling below the solidification temperature is increased. However, as the temperature is further decreased, the diffusivity in the liquid phase becomes smaller and smaller, slowing down the growth of the nuclei; consequently, the crystallization time again becomes longer (Fig. 6). Below a certain temperature, the diffusivity becomes low enough that the liquid appears to be practically frozen. This temperature

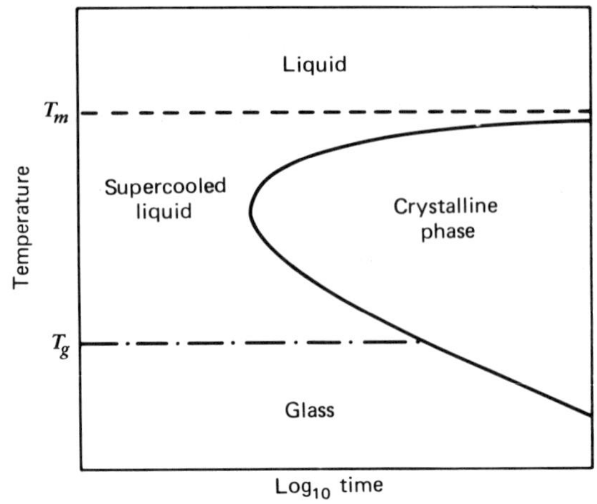

Figure 6. A schematic time–temperature–transformation (TTT) diagram for solidification of liquid; T_m and T_g denote the melting and the glass transition temperatures, respectively.

is called the glass-transition temperature T_g and is defined as the temperature at which the viscosity becomes equal to 10^{12} Pa·s (10^{13} P). This is not a phase-transition temperature. Thermodynamically, the liquid and the glass belong to the same phase, the only difference being the higher viscosity of the glass. Therefore, the kinetic processes are slower and the various relaxation times longer (compared to the time of observation). Consequently, the glass shows apparently different physical behavior than the liquid. A glass can, therefore, be formed simply by freezing a liquid if crystallization does not interfere (52). Therefore, any liquid can be made into a glass if the quenching rate is high enough to avoid the minimum time of the crystallization curve shown in Figure 6. Quantitative theories based upon this concept of glass formation have been developed (61–62).

Compositions. In practice, the quenching rate that may be achieved is restricted by the limited heat conductivity and heat capacity of the sample. It is, therefore, important to know why glasses within a certain range of composition form easily and others do not. This highly complex and difficult question has yet to be answered conclusively. However, it is recognized that the complexity of the structure of the crystalline material with the same composition as the glassy phase contributes greatly to ease the glass formation as shown in Figure 7 (63). If the crystal structure is already complex, it would take relatively little energy to introduce further topological disorder; therefore, the energy difference between the glass and the crystalline phases would be less (52). Furthermore, a highly collective atomic diffusion has to take place to produce a complex crystal structure, which slows down the crystallization kinetics during cooling. These considerations are also relevant in understanding the structural relaxation and stability, as discussed below.

For the reasons discussed above, elementary metals are extremely difficult to prepare in glassy form. It has been reported that some of the transition metal elementary glasses have been obtained by vapor deposition (7) and splat cooling (64). These glasses are usually quite unstable and crystallize at relatively low temperatures. Furthermore, the possibility that the glassy phase obtained was stabilized by a small amount of impurity inclusions cannot be excluded. All the glassy metals stable at RT

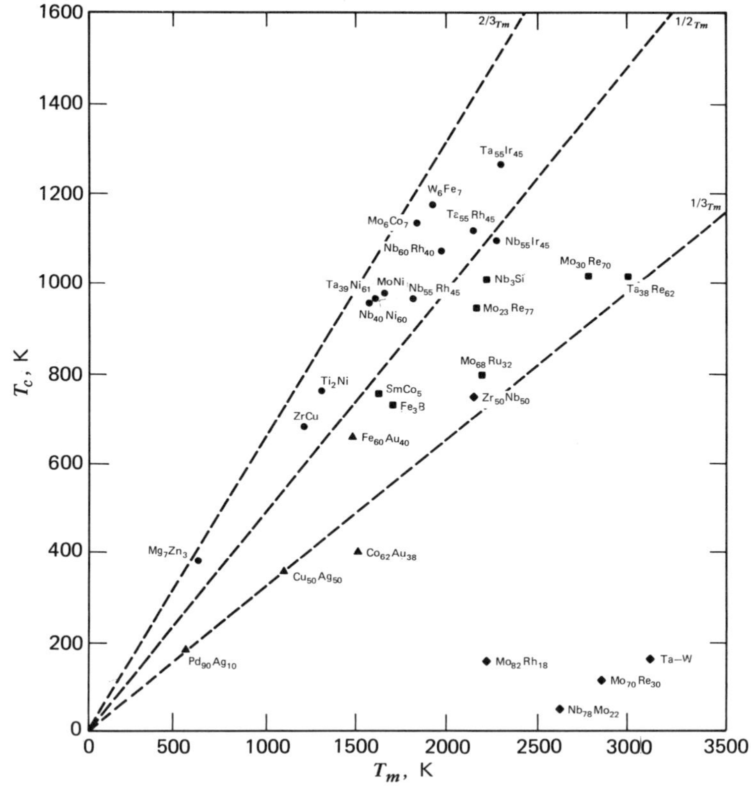

Figure 7. Correlation between the crystallization temperature T_c of an amorphous alloy, and the melting temperature T_m of the compositionally related intermetallic compound. The dashed lines show T_c/T_m = ⅓, ½, and ⅔. The amorphous phase is more stable when the crystal structure is more complex; T_c/T_m is higher (32). ●, μ-phase and related structures; ■, σ-phase and related structures; ▲, simple eutectic alloys; ♦, terminal solid solutions.

are alloys with two, three, or more constituent elements. A large number of glass-forming systems have been reported (65–66) and a list of representative compositions is given in Table 1. They are divided into metal–metalloid and metal–metal systems. An example of the former is $Fe_{80}B_{20}$. Various transition metals can be partially or totally substituted for Fe, with the total metal concentration 70–88 atom %. Other metalloids, such as Si, P, and C may also be substituted for boron. These glasses are relatively easily formed by various liquid-quenching methods. The metal–metal systems include transition metal–rare earth combinations such as $GdCo_5$ [*12017-61-7*] or $TbFe_2$ [*12023-38-0*] and transition metal–noble metal combinations such as $Cu_{60}Zr_{40}$ [*12787-67-6*] or $Fe_{60}Au_{40}$ [*71106-50-8*]. Most of them include transition metals but some of them, such as $Ca_{65}Al_{35}$ [*68793-40-8*] and $Mg_{70}Zn_{30}$ [*62532-11-0*], do not.

Preparation

A number of methods are used to prepare glassy metals from the corresponding crystalline alloys but they generally belong to one of three categories. The first is the liquid-quenching method mentioned earlier. The second is vapor deposition including

Table 1. Representative Alloy Composition of Glassy Metals

Composition	CAS Registry No.	Remarks
$Fe_{80}B_{20}$	*[60391-29-9]*	Metglas 2605[a]
$Fe_{40}Ni_{40}P_{14}B_6$	*[54319-71-0]*	Metglas 2826[a]
$Fe_{82}B_{10}Si_8$	*[71106-48-4]*	Metglas 2605S[a]
$Pd_{80}Si_{20}$	*[12778-98-2]*	
$Pd_{77.5}Si_{16.5}Cu_6$	*[12735-83-0]*	very stable alloy, easy to vitrify
$Pt_{60}Ni_{15}P_{25}$	*[39412-40-3]*	very stable alloy, easy to vitrify
$Cu_{60}Zr_{40}$	*[12787-67-6]*	
$Mg_{70}Zn_{30}$	*[62532-11-0]*	
$Gd_{17}Co_{83}$	*[70708-49-5]*	sputtered sample can support magnetic bubbles
$Mo_{64}Re_{16}P_{10}B_{10}$	*[71106-49-5]*	superconducting below 8.7 K
$W_{60}Ir_{20}B_{20}$	*[71040-18-1]*	crystallization temperature above 1200 K

[a] Allied Chemical Corp.

the ion-sputtering technique (see Film deposition techniques). The third is the wet chemical technique of electrodeposition or electroless chemical deposition (see Electroless plating). Among them, the wet chemical techniques produce inhomogeneous products (67) which are not interesting. The liquid-quenching methods are, in general, simple and suited to large-scale mass production but limited as to the composition ranges. The vapor-deposition technique affords a wider range, particularly the ion-sputtering method. The equipment for vapor deposition is complex and expensive, and the deposition rate is significantly lower (by a factor of 10^6 to 10^{10}) than the production rate in liquid quenching.

Liquid-Quenching Techniques. Since the first successful preparation of glassy metals using direct liquid quenching (11), various techniques have been developed. The principle remains the same: the molten alloy is brought into contact with a cold metallic substrate, and the heat is extracted by conduction. This imposes a severe limitation on the dimensions of the product. In order to achieve a sufficient cooling rate (ca 10^5 K/s), the thickness of a typical transition metal–metalloid glass cannot exceed 100 μm. However, the other two dimensions can be extended by the use of appropriate technology. Today, a 5-cm wide ribbon with practically unlimited length is readily available (5), and it appears that there is no serious difficulty involved in producing films of 10-cm width or more. These dimensions are sufficiently large for use in many of the applications.

Splat Quenching. The original technique is called the gun technique (1), where a droplet of molten liquid is propelled by gas pressure onto a copper plate and jet-spread by the impact. The glassy metal obtained is a thin sheet of irregular form. An improved version of this is the piston and anvil technique (1), where a pneumatically powered piston is triggered by a light beam which detects the falling droplet, and flattens it against the anvil. Again, the product is irregular, and the method is not suited to mass production, but it achieves a high quench rate by forced contact between substrate and liquid.

Roller Quenching Technique. This is a continuous version of the piston and anvil technique (19). A stream of molten liquid is ejected from a nozzle by applied pressure, and fed between dual roller bars (Fig. 8). The liquid is frozen by contact with the roller surfaces and forms a continuous ribbon. This method produces a ribbon with regular shape and flat surfaces which is convenient for some applications.

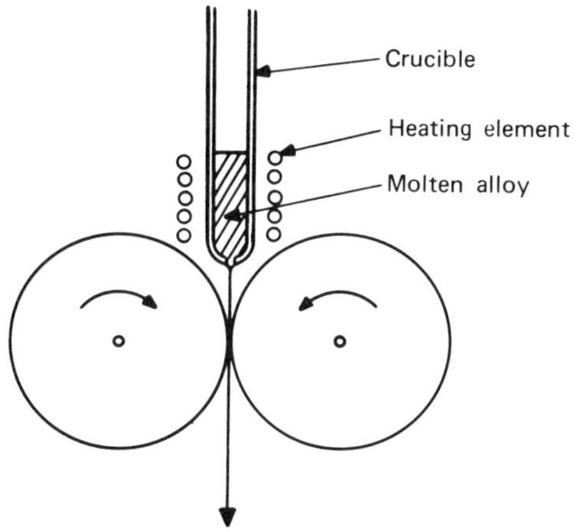

Figure 8. Schematic diagram of the roller quenching method.

Pond–Maddin Technique. The method uses a molten jet and a rotating metallic cylinder (17–18). The stream of molten liquid is ejected against the inner surface of a rotation cylinder. The liquid is forced into contact with the metal surface by centrifugal force and is rapidly cooled. In order to avoid an overlay of the liquid on top of the already frozen glassy metal ribbon, the liquid jet traverses upward to produce a helical pattern of the glassy metal ribbon in the cylinder. A variation of this method was developed later (68).

Melt-Spinning Technique. The simplest and perhaps most widely used liquid-quenching technique is the melt-spinning technique (69–70). It was first used to process crystalline metals into ribbon, wire, or small pellets. Similar to the two other methods, a stream of molten liquid metal is ejected from the nozzle of a crucible. The jet is squirted onto the surface of a rotating disk, as shown in Figures 9 and 10. The surface velocity of the disk is ca 30–50 m/s. By using a wide rectangular nozzle, it is possible to produce a wide ribbon. A high-speed film observation (68,71) shows that the liquid first forms a puddle, and then solidifies into a ribbon. The aerodynamics surrounding the puddle determines the shape of the ribbon and the smoothness of the edges. Casting in vacuum produces smoother ribbons with more regular shape (72). This seems to be the best technique for large-scale production because of its simplicity. It is an ideal continuous casting technique with an exceptional speed of production, which can be as high as 1 kg/s or more. Although the large amount of heat which must be extracted presents problems, the technology involved is reasonably economical. Needless to say, a melt-spinning machine can be used in the laboratory.

Vapor-Quenching Techniques. Glassy metals can be obtained by depositing the vapor of an alloy onto a substrate, by-passing the liquid phase. Since the temperature of the vapor and the deposited glassy metal can be substantially different, a very high quenching rate can be achieved on an atomic scale. Some compositions that could not be made glassy by liquid quenching can, therefore, be prepared in glassy form by vapor-quenching techniques. However, the deposition rate is generally very low, and the technique requires sophisticated equipment which includes a high vacuum system.

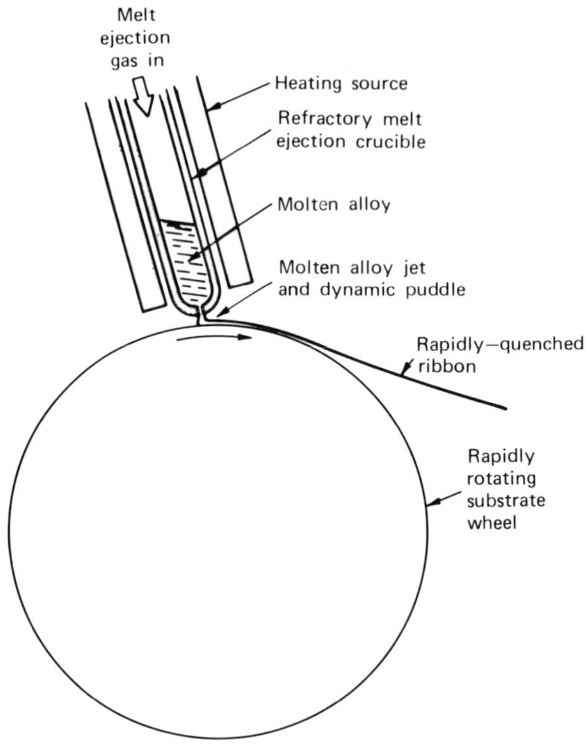

Figure 9. Schematic diagram of the melt-spinning method.

Thermal Vapor-Deposition Method. The vapor of an alloy is formed in vacuum either by using a resistive heating element or an electron-beam heating gun. The vapor is slowly deposited onto a substrate and often cooled to cryogenic temperatures. Most elementary glasses used in the early studies were prepared by this method. However, inclusion of residual gas cannot be avoided (see also Film deposition techniques).

Ion-Sputtering Technique. A more commonly used deposition technique is ion-sputtering. A low pressure of inert gas is introduced into the system and an ion plasma is formed, either by d-c or a-c excitation. The plasma is attracted by the bias potential to a multiphase target which is a crystalline solid of the appropriate composition. The high velocity ions in the plasma sputter the target and atomize the surface. The atoms leave the target and are deposited onto the substrate. Since the surface of the substrate is also constantly hit by the plasma, a highly disordered glassy phase most likely results. The deposition rate is usually of the order of 1–10 nm/min. A recently developed high-rate sputtering system can deposit materials at a rate exceeding 1 μm/min (73). At present this is the only method to obtain bulk (>1 mm thick) glassy metals with high homogeneity (see Plasma technology; Ion implantation).

Applications

Since industrial production of glassy metals was started only recently, large-scale applications have not yet been developed. However, several prototypes using glassy metals have been built and tested with remarkable success. As the demand becomes

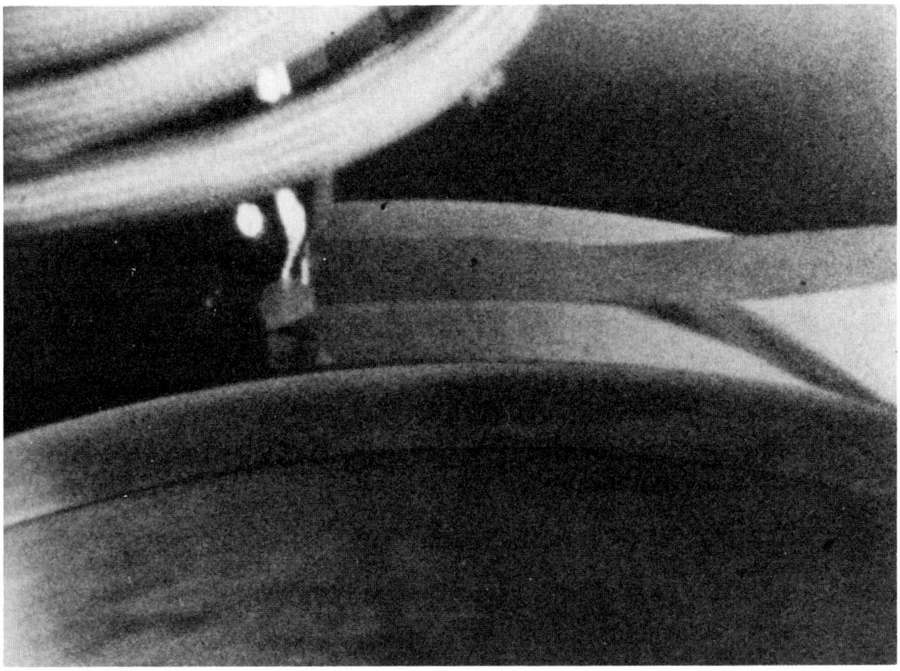

Figure 10. Print from a movie taken at 3000 frames/s showing a 20-mm wide ribbon cast in vacuum. The white rings are the induction heater coil. The disk is rotating clockwise. Photo courtesy of H. H. Liebermann.

more complex, in particular for materials requiring low power loss, glassy metals will be used industrially in significant amounts.

Magnetic Applications. ***Permalloy Replacement.*** At present, the most commonly used high permeability material is Permalloy, an Fe–Ni alloy containing molybdenum. Permalloy has the disadvantage that it is mechanically very soft and easily deformed, with plastic deformation drastically degrading the permeability. Glassy metals can exhibit permeability comparable to, or higher than Permalloy, yet they are mechanically tough and not easily deformed. Glassy metals are expected to partially replace Permalloy in magnetic shielding (74), audio taperecorder heads, and in electronic systems, such as the bubble guide used in bubble memory devices. The alloys considered for shielding applications include Fe–Ni–Mo–B (75) and Fe–Ni–P–B. The taperecorder heads made of glassy metals are already marketed in Japan. Zero-magnetostrictive Co–Fe–B–Si alloys (76) are used for this application, particularly because the B–Si alloys have higher crystallization temperatures and are, therefore, more wear resistant. However, the degradation of permeability after aging is still a serious problem (see Magnetic materials).

Power Devices. As the cost of energy is rapidly increasing, it is becoming more important to improve the energy efficiency of power devices. Glassy metals rich in iron might make a significant contribution by reducing core loss in power distribution transformers (4,16). The recently developed high-flux alloys (77) should enhance this possibility. Taking advantage of their high resistivity, ferromagnetic glassy metals may be applied first to high frequency [(0.4–10) kHz] power transformers and then to 60 Hz power transformers.

Bubble Memory Devices. Some rare earth–transition metal glasses produced by ion-sputtering have growth-induced magnetic anisotropies large enough to stabilize bubble domains (78). Therefore, they can be of value also as bubble-supporting materials, not only as the bubble guide mentioned earlier. Because of the large temperature dependence of magnetization, however, the bubble domains are stable over only a limited temperature range, imposing some difficulty of application.

Delay Lines. A variable acoustic delay line can be designed to make use of the large ΔE effect (change in Young's modulus by applied magnetic field) of the ferromagnetic glassy metals (79). Such delay lines may be useful for electronic equipment and television sets.

Mechanical Applications. The greatest asset of glassy metals for mechanical applications is their combined strength and toughness. Their use as strengthening materials, for instance in tire cord (qv), has been frequently suggested (see Composite materials). However, the cost and stability of glassy metals still present problems. The application to cutting tools, including razor blades, is also being considered. Based on their formability, glassy metals have been proposed as brazing materials (80) (see Solders).

BIBLIOGRAPHY

1. P. Duwez, *Ann. Rev. Mat. Sci.* **6,** (1976).
2. J. J. Gilman, *Phys. Today* **28,** 46 (May 1975).
3. C. D. Graham, Jr., and T. Egami, *Ann. Rev. Met. Sci.* **8,** 423 (1978).
4. F. E. Luborsky, J. J. Becker, P. G. Frischmann, and L. A. Johnson, *J. Appl. Phys.* **49,** 1769 (1978).
5. *Trademark Metglas Alloys,* Allied Chemical Corporation, Morristown, N.J.
6. A. W. Simpson and D. R. Brambley, *Phys. Status Solidi* (b) **43,** 291 (1971).
7. J. G. Wright, *IEEE Trans Mag.* **MAG-12,** 95 (1976).
8. T. Egami and S. D. Dahlgren, *J. Appl. Phys.* **49,** 1703 (1978).
9. N. Kazama, N. Heiman, and R. L. White, *J. Appl. Phys.* **49,** 1706 (1978).
10. R. J. Kobliska, J. A. Aboaf, A. Gangulee, J. J. Cuomo, and E. Klokholm, *Appl. Phys. Lett.* **33,** 473 (1978).
11. P. Duwez, R. H. Willens, and W. Klement, Jr., *J. Appl. Phys.* **31,** 1136 (1960).
12. P. Duwez, *ASM Trans. Q.* **60,** 608 (1967).
13. P. Duwez and S. C. H. Lin, *J. Appl. Phys.* **38,** 4096 (1967).
14. H. S. Chen and D. Turnbull, *J. Chem. Phys.* **48,** 2560 (1968).
15. G. S. Cargill, IV, *Solid State Phys.* **30,** 227 (1975).
16. T. Egami, P. J. Flanders, and C. D. Graham, Jr., *AIP Conf. Proc.* **24,** 697 (1975).
17. R. Pond, Jr., and R. Maddin, *Trans AIME* **245,** 2475 (1969).
18. T. Masumoto and R. Maddin, *Acta Met.* **19,** 725 (1971).
19. H. S. Chen and C. E. Miller, *Rev. Sci. Instr.* **41,** 1237 (1970).
20. H. J. Leamy, H. S. Chen, and T. T. Wang, *Met. Trans.* **3,** 699 (1972).
21. H. Fujimori, T. Masumoto, Y. Obi, and M. Kikuchi, *Jpn. J. Appl. Phys.* **13,** 1889 (1974).
22. E. M. Gyorgy, H. J. Leamy, R. C. Sherwood, and H. S. Chen, *AIP Conf. Proc.* **24,** 745 (1975).
23. C. N. J. Wagner, *J. Non-Cryst. Solids* **31,** 1 (1978).
24. H. P. Klug and L. E. Alexander, *X-ray Diffraction Procedures for Polycrystalline and Amorphous Materials,* 2nd ed., John Wiley & Sons, Inc., New York, 1974.
25. T. Egami, *J. Mat. Sci.* **13,** 2587 (1978).
26. T. M. Hayes, *J. Non-Cryst. Solids* **31,** 57 (1978).
27. J. D. Bernal, *Proc. Roy Soc. London A* **280,** 299 (1964); P. Chauduri, W. C. Giessen, and D. Turnbull, *Sci. Am.* **242**(4), 98 (1980).
28. K. Maeda and S. Takeuchi, *J. Phys. F.* **8,** L283 (1978).
29. T. Masumoto, *Sci. Rep. Res. Inst. Tohoku Univ. Ser. A* **26,** 246 (1977).

30. L. A. Davis in J. J. Gilman and H. J. Leamy, eds., *Metallic Glasses,* American Society of Metals, Metals Park, Ohio, 1977, pp. 190.
31. F. Spaepen, *Acta Met.* **25,** 407 (1977).
32. A. S. Argon and M. Salama, *Mat. Sci. Eng.* **23,** 219 (1976).
33. F. E. Luborsky and J. L. Walter, *J. Appl. Phys.* **47,** 3648 (1976).
34. J. L. Walter, F. Bacon, and F. E. Luborsky, *Mat. Sci. Eng.* **24,** 239 (1976).
35. J. L. Walter and F. E. Luborsky, *Mat. Sci. Eng.* **33,** 91 (1978).
36. R. S. Williams and T. Egami in B. Cantor, ed., *Rapidly Quenched Metals III,* Vol. 1, The Metals Society, London, Eng., 1978, p. 214.
37. R. Ray, R. Hasegawa, C.-P. Chou, and L. A. Davis, *Scripta Met.* **11,** 973 (1977).
38. M. Nagumo and T. Takahashi, *Mat. Sci. Eng.* **23,** 257 (1976).
39. R. Maddin and T. Masumoto, *Mat. Sci. Eng.* **9,** 153 (1972).
40. J. Patterson, A. L. Greer, J. A. Leake, and D. R. H. Jones in B. Cantor, ed., *Rapidly Quenched Metals III,* Vol. 2, The Metals Society, London, Eng., 1978, p. 293.
41. C. F. Cline and R. W. Hopper, *Scripta Met.* **11,** 1137 (1977).
42. T. Egami, *Mat. Res. Bull.* **13,** 557 (1978).
43. S. R. Nagel, *Phys. Rev. B* **16,** 1694 (1977).
44. T. E. Faber in J. M. Ziman, ed., *The Physics of Metals,* Vol. 1, Cambridge University Press, Cambridge, Eng., 1969, p. 282.
45. R. Hasegawa and C. C. Tsuei, *Phys. Rev.* **32,** 1631 (1970).
46. J. Kondo, *Solid State Phys.* **23,** 183 (1975).
47. W. L. Johnson, *J. Appl. Phys.* **50,** 1557 (1979).
48. M. M. Collver and R. H. Hammond, *Phys. Rev. Lett.* **30,** 92 (1972).
49. T. Masumoto, K. Hashimoto, and M. Naka in ref. 40, p. 435.
50. F. E. Luborsky, *Mat. Sci. Eng.* **28,** 139 (1977).
51. G. C. Chi, H. S. Chen, and C. E. Miller, *J. Appl. Phys.* **49,** 1715 (1978).
52. D. Turnbull and M. H. Cohen, *J. Chem. Phys.* **34,** 120 (1961).
53. R. S. Williams and T. Egami, *IEEE Trans Mag.* **MAG-12,** 927 (1976).
54. H. S. Chen and E. Coleman, *Appl. Phys. Lett.* **28,** 245 (1976).
55. H. S. Chen, L. C. Kimerling, J. M. Poate, and W. L. Brown, *Appl. Phys. Lett.* **32,** 461 (1978).
56. H. S. Chen, *J. Appl. Phys.* **49,** 3289 (1978).
57. H. S. Chen, H. J. Leamy, and M. Barmatz, *J. Non-Cryst. Solids* **5,** 444 (1971).
58. J. R. Matey and A. C. Anderson, *J. Non-Cryst. Solids* **23,** 129 (1977).
59. M. Marcus, *Acta Met.* **27,** 879 (1979).
60. C. C. Tsuei, *Appl. Phys. Lett.* **33,** 262 (1978).
61. D. R. Uhlmann, *J. Non-Cryst. Solids* **7,** 337 (1972).
62. H. A. Davies, J. Aucote, and J. B. Hull, *Scripta Met.* **8,** 1179 (1974).
63. R. Wong, *Proc. Alloy Phase Formation Symposium, 108th Annual Meeting of AIME* (1979).
64. H. A. Davies and J. B. Hull, *Mat. Sci. Eng.* **23,** 193 (1976).
65. I. W. Donald and H. A. Davies, *J. Non-Cryst. Solids* **30,** 77 (1978).
66. H. A. Davies in ref. 36, p. 1.
67. G. C. Chi and G. S. Cargill, III, *AIP Conf. Proc.* **29,** 147 (1975).
68. H. S. Chen and C. E. Miller, *Mat. Res. Bull.* **11,** 49 (1976).
69. H. H. Liebermann and C. D. Graham, Jr., *IEEE Trans. Mag.* **MAG-12,** 921 (1976).
70. S. Kavesh in ref. 30, p. 36.
71. H. H. Liebermann, G. E. Research and Development Center, private communication.
72. H. H. Liebermann in ref. 36, p. 34.
73. S. D. Dahlgren, *Metall. Trans.* **A7,** 1375 (1976).
74. L. I. Mendelsohn, E. A. Nesbitt, and G. R. Bretts, *IEEE Trans. Mag.* **MAG-12,** 924 (1976).
75. R. Hasegawa, N. DeCristofaro, and M. C. Narsimhan, *J. Appl. Phys.* **49,** 1712 (1978).
76. H. Fujimori, M. Kikuchi, Y. Obi, and T. Masumoto, *Jpn. J. Appl. Phys.* **14,** 1077 (1975).
77. S. Hatta, T. Egami, and C. D. Graham, Jr., *Appl. Phys. Lett.* **34,** 113 (1979).
78. P. Chaudhari, J. J. Cuomo, and R. J. Gambino, *IBM J. Res. Dev.* **17,** 66 (1973).
79. N. Tsuya and K. I. Arai, *J. Appl. Phys.* **49,** 1718 (1978).
80. N. DeCristofaro and C. Henschel, *Weld. J.* **57,** 33 (1978).

General References

References 1, 2, 15, 30, and 36 are general references.
N. J. Grant and B. C. Giessen, eds., *Rapidly Quenched Metals II*: Sect. I, MIT Press, Cambridge, Mass., 1976; Sect. II, *Mat. Sci. Eng.* **23,** (1976).
"Conference Proceedings of the Atomic Scale Structure of Amorphous Solids," *J. Non-Cryst. Solids* **31,** (1978).
J. J. Gilman, *Science* **208,** 856 (May 23, 1980).
AIP Conf. Proc. **24, 29, 34,** (1974–1976).
J. Appl. Phys., **49**(3), (1977); *ibid.,* **50**(3), (1978).
U.S. Pat. 3,856,513 (Dec. 24, 1974), H. S. Chen and D. E. Polk (to Allied Chemical Corp.).
U.S. Pat. 3,871,836 (Mar. 18, 1975), D. E. Polk and R. C. Morris (to Allied Chemical Corp.).
U.S. Pat. 3,986,867 (Oct. 19, 1976), T. Masumoto and M. Naka (to The Research Institute for Iron, Steel, and Other Metals, Tohoku University, Sendai, Japan).
U.S. Pat. 4,030,892 (June 21, 1977), L. I. Mendelsohn and E. A. Nesbitt (to Allied Chemical Corp.).
U.S. Pat. 4,056,411 (Nov. 1, 1977), H. S. Chen, E. M. Gyorgy, H. J. Leamy, and R. C. Sherwood (to Bell Laboratories).

TAKESHI EGAMI
Institut für Physik
Max-Planck-Institut für Metallforschung

GLOBULINS. See Blood fractionation.

GLUCOSE, $CH_2OH(CHOH)_4CHO$. See Syrups; Sugar; Carbohydrates.

GLUE

The origins of animal glue in antiquity and its wide range of adhesive applications have contributed to the persistant use of the term glue to indicate any adhesive (see Adhesives). Animal glue refers to that material extracted from the protein collagen, which is found in specific animal tissues such as skin, hides, bones, sinews, and tendons. Contrary to popular assumption, animal glue cannot be made from horns, hooves, or hair because these are primarily composed of the protein keratin which is not readily hydrolyzed and which, by the nature of its structure, would not lend itself to producing the helical configurations consistent with the characteristics of animal glue and gelatin (qv).

Animal glue and gelatin are almost identical as they are produced by the same methods and from the same raw materials. In some instances, animal glue is referred to as technical gelatin; however, the two differ in purity and quality. Gelatin requires clean and edible raw materials, more purification, and more stringent processing conditions and control. Animal glue, however, can be processed from raw materials that otherwise would be wasted. The color and purity of animal glues, unlike that of gelatin, do not necessarily indicate its effectiveness as an adhesive.

Commercial animal glues usually are named according to the raw materials from which the product is made; thus, there is a bone glue and a hide glue. Further delineation also is made: eg, green bone glue refers to glue produced from bones fresh from the packing or the boning houses; extracted bone glues are made from bones that have been precooked to remove fat prior to glue extraction; chrome glue is extracted from the scraps and shavings of chrome-tanned leather; and coney glue is made from the rabbit skin residues from the production of felt. Fish glues, which are made from the nonedible parts of fish, generally are liquid at ambient temperatures and are clearer and lighter than bone or hide glues; isinglass is one of the better known of these materials and is manufactured from the swimming bladder of the sturgeon. However, fish glues have just about disappeared from the commercial market in the United States.

Technical gelatin and technical protein colloid sometimes designate animal glues that have applications other than in the adhesive field. Jelly glue, flexible glue, nonwarp glue, and composition glue refer to formulations that contain glue but are compounded so as to give specific changes in properties such as greater flexibility, noncurling, and so on. Other glue-formulated products are designated by their application, eg, freezer glue, sizing glue, and plating glue.

Properties

Animal glue is marketed as a dry solid, the color of which may range from light yellow through tan, brown, and sometimes almost black. In the unground state it is almost hornlike and can be obtained as nonuniform particles, plates, or squares or as fine beads called pearls. Commercial glues usually are sold ground to a size of 2–2.4 mm (8–10 mesh). Special grinds are also available in 840 μm (20 mesh), 149 μm (100 mesh), and 44 μm (325 mesh) sizes.

Glue, the hydrolysis product of collagen, is an amorphous composition of protein fragments ranging in size from simple dipeptides to large multichain polypeptides of colloidal size (1). Glue is a polydisperse system and, as such, gives molecular weights that are statistical averages; molecular weights have been reported from 20,000 to 90,000 (2–3) and some fractions are as high as 250,000 (4). There is evidence that the larger portions are double or triple spirally wound helical chains (5–7) (see Biopolymers). Many of the properties of glue can be attributed to its amino acid constituents in its structure; eg, its solubility in water can be appreciated from the fact that five major constituent amino acids (qv) (glycine, proline, hydroxyproline, arginine, and lysine) constitute ca 60–65% of the molecule. The nonionizable radicals of the amino acids, hydroxyproline and serine, are polar and represent ca 20–24% of the amino acid residues in the structure and contribute significantly to the absorption of water and the swelling of glue that precedes its solution. Numerous studies of the relationship of gelling to structure and configuration have been made (8–11).

The color, odor, and clarity of animal glues is related to the quality of the raw materials and to the care and cleanliness of the manufacturing process; contamination by blood, metals, and sugars darken colors, and fat and dirt contribute to opacity. The fats and their oxidation products, as well as some of the amine by-products, contribute to objectional odors.

Animal glue does not support combustion nor melt in the dry state. When heated excessively, glue particles swell and then burst owing to the evaporation of structural water. Continued heating eventually causes the material to char and completely carbonize; the carbon residues then become oxidized and only the ash residue remains. Slow heating eventually causes glues to become insoluble in water; this insolubilization can take from 10–12 h at ca 110°C. The presence of polyols, such as glycerol (qv) and ethylene glycol, can accelerate this phenomenon (12).

Boiling water slowly hydrolyzes glue solutions. Alkali or mineral acids may accelerate the hydrolysis; however, excess hydrolysis lowers the quality of the product. Enzymes, such as papain, ficin, and bromelain, also hydrolyze glue with a characteristic loss in viscosity and liberation of amino and carboxyl groups; this liberation indicates a breakage at the peptide linkage. Similarly, pepsin lowers the viscosity of glue solutions but there is little release of free amino groups which indicates that the backbone of the chain, rather than the peptide linkage, is cleaved.

Animal glue is insoluble in cold water but, as it soaks, it will absorb six to eight times its own weight of water; after swelling, it may be dissolved readily by heating to 50–60°C. At these temperatures, the solution does not hydrolyze rapidly, thus allowing a reasonable pot-life. Glues also can be dissolved in water at 60°C with vigorous agitation; this is the preferred method for large volumes.

Dried animal glue contains 6–14% moisture. However, the moisture content adjusts to the relative humidity of the environment without a visible change in characteristics other than in net weight. Thus, high moisture contents can cause caking if the material is shipped through several drastic temperature changes with resultant evaporation and condensation.

Those chemical reagents that react with amino, carboxyl, and hydroxyl groups react with these groups in animal glue. Glue also may be *N*- and *O*-acylated by the fatty acid anhydrides and acid chlorides. The reaction with long-chain fatty acid chlorides leads to the production of a series of surface-active materials that have wide application in the cosmetic field. Nitrous acid generated *in situ* effectively deaminates glue by converting almost all the amino and amido groups to carbamides which, in turn, react with formaldehyde in a much different way than with the original glue molecules.

Reactive aldehydes, eg, formaldehyde, glyoxal, and some sugars, initiate a chain methylolamine reaction that continues until the cross-linking has converted the solution irreversibly into a gel. The change is permanent once the gel has dried; however, if the solution is not dry, various agents, such as urea, can at least partially reverse the reaction. The chain methylolamine reaction is used to increase the viscosity of glue solutions but the control of the reaction is difficult and easily may go beyond the desired limits. On the other hand, urea and thiourea solutions of glue react with active aldehydes and produce modified amino resins (qv) with thermoplastic and thermoset properties (13).

Metal polyvalent ions, such as chromium, iron, and aluminum, react with glue to form insoluble products or irreversible gels. These reactions can be controlled by pH adjustment. Other tanning agents, both natural and synthetic, insolubilize or tan glue. Although these agents insolubilize glue, they do not precipitate it at specific temperatures, which is the usual protein behavior. The most effective agent for the precipitation of glue is either trichloroacetic acid or phosphotungstic acid. In all cases of glue insolubilization, the reactions only change the solubility—not the moisture resistance of the original material. Glue that is tanned with formaldehyde forms an irreversible gel which, when dried, swells in water but does not redissolve. This is explained by the small proportion of the molecule that is involved in the reaction; eg, on a mole-per-mole basis, a glue with a molecular weight of 30,000 would be made insoluble by ca 0.1% formaldehyde and, thus, the moisture sensitivity remains after total insolubility is obtained. This is observed for almost all glue reaction products.

The most characteristic property of animal glue is its ability to form reversible gels in water solution. The rigidity of the gels increases with decreasing temperatures and/or rising concentrations. The rigidity, as well as the melting point, is determined by the average molecular weight or the quality of the glue. Numerous agents exert a pronounced effect on the gelling properties of glue. Salts are used to control the gelation and setting times of adhesive formulations that contain animal glue. Of the soluble salts, chlorides and nitrates retard gelling or prevent it. In this respect, alkaline earth salts are more effective than alkali metal salts, and sulfates and phosphates have little or no effect. Zinc sulfate has no effect on the jelly strength, even at high concentrations; however, calcium nitrate just about stops gelation. Citrates, tartrates, and maleates tend to increase the rate of gelation but not the rigidity of the gel. Many other liquefiers or peptizing agents are organic, eg, urea, sodium 2-naphthalenesulfonate, thiourea, phenol, sulfonic acids, chloral hydrate, and the amino acids. Ethylene chlorohydrin is an example of a volatile liquefier; its use is limited by its reaction with amino compounds and the ease by which it is hydrolyzed to the nonvolatile glycol.

Manufacture

Preparation of animal glue is essentially a treatment of a collagen source with heat and water in order to hydrolyze it to a soluble product as rapidly and as efficiently as possible. The resulting solution is filtered, centrifuged to remove fat, concentrated to a suitable concentration in an evaporator, chilled to gel the concentrate, extruded or cut into particles, dried, ground, and analyzed. The quality and cleanliness of the raw materials, their cleaning and preprocessing, and the speed of extraction, concentration, and drying are the determinants of the quality of the product. The parameters by which all the various processes differ are the pretreatments that are necessary to make the collagen available for hydrolysis and the complexity of the matrix of tissue in which the collagen resides. Individual manufacturers develop different cooking or hydrolyzing steps. The higher the collagen concentration of the raw material, the more efficient the hydrolysis step; thus, the pretreatment of hides is different from that of bone and, in some instances, this pretreatment changes the characteristics of the glue (eg, isoionic point).

Hide Glues. ***Raw Stock.*** The raw material used for hide glue consists of the following types of materials: green stock—fresh, unsalted, and undried material consisting of hide trimmings, splits, fleshings, tails, pates, lips, snouts, ears, and sinews; salted or dry stock—material just like green stock except that it has been salted or dried in order to preserve it while accumulating sufficient quantities for shipment; chrome stock—pieces or shavings from the manufacturers of chrome-tanned leather; sheep stock and horse stock—hide trimmings and some tails, ears, and pates; and pigskin stock trimmings—pieces and trimmings from the skin as well as ears, tails, and some pig's feet.

Younger animals supply better and more easily processible raw material than older animals as well as providing a higher quality product. Calf and pigskin make excellent gelatin and, thus, are seldom available for glue.

Pretreatment and Curing. All raw materials are tumble-washed with large volumes of cold water to remove all outer contaminants such as blood, manure, salt, and dirt. The cleaner the raw material at the start, the shorter the wash period. The wash is complete when the water runs clear or, in the case of salted hides, when it is salt-free; dry and salted stock require long washes.

Curing is the process by which the tissue is plumped by soaking in either acid or alkali in order to increase the availability of the collagen to the hydrolytic environment. The curing procedure that is used depends upon the facilities and the nature of the stock. Some types of raw materials require no curing, eg, the tannery-limed fleshings, trimmings, and sheep stock from depilated pelts. These materials require only washing and neutralization.

Alkali liming or curing requires the use of vats or pits and some agitation. The chopped pieces of raw material are immersed in the lime solution (2 g/L, pH 12) and are moved slowly to prevent packing which can lead to the generation of heat and bacterial decomposition. A little caustic soda or sodium carbonate may be added to prevent the cure from getting sluggish and to prevent bacterial action. The cure is complete when the swollen stock is uniformly penetrated by the lime and has a clear, translucent appearance in cross section. This process can take from two to ten weeks but—because of the solubility and mildness of the lime as well as its economy—the process is the most prevalent in the industry.

The other curing process is mild acid cure with dilute hydrochloric or sulfurous acid (0.5%). Coney stock, fleshings, and porkskin trimmings and fleshings usually are treated by this method. This is a much faster cure and can be completed in from 3–18 h, depending on the size and condition of the raw material. The criterion for completion is similar to that used for the liming process: a clear, uniformly translucent cross section.

After curing, the stock is washed with water. The acid-cured material is washed until the stock has a surface pH of 3.5–4.0. In the case of limed stock, the material is washed until the water runs clear and then it is washed with sulfurous acid (0.5%) to solubilize the lime salts. This is continued until the material has a surface pH of 3.0, at which point the water washing is resumed until the pH of the entire stock is 6.0–6.5. In both curing methods, the last washing step is lengthy. The acid cure, of course, is shorter and the lime-cure procedure may require another 12 h of washing.

Extraction. The extraction process consists of transferring the cured washed material into the extraction kettles which contain hot water (60–65.5°C), heating with agitation for ca 1 h, and draining the liquor from the bottom. More water is added and the step is repeated at a slightly higher temperature for a shorter period. Eight to twelve extractions are made in this fashion until the liquor that is removed contains only ca 1% solids. At this point the economics of the extraction prevent further processing. The residue (grout) in the kettle is removed and may be dumped but it is usually dry-rendered to recover fat. As the extraction proceeds, the volume of material shrinks and, thus, less water is required. Hot water and agitation prevent packing in the extraction kettle which would lead to poor extraction. If a plot of the sequential glue yields or glue qualities were made against the individual reactions, a curve would result that would show that good quality and yields were obtained during the middle runs. The variations in this procedure from company to company are great and have evolved over the years through trial and error, depending on the particular types of raw stocks available and machinery developments.

The exception to the above procedure for hide glues is the treatment of chrome-tanned stock. These materials are cured prior to tanning and, thus, require only the removal of the chrome. This is accomplished by washing consecutively and countercurrently with caustic soda and hydrochloric acid, both at ca 0.5% concentration. The washing is continued until the stock is chrome-free, and then the conventional hide-extraction procedure is followed. Current environmental laws require that these washings be treated to remove the chrome before the residues are discarded. This has become a costly program and requires the recovery of chrome for the procedure to be economically feasible.

Bone Glue. ***The Raw Stock.*** The raw materials for the production of bone glue consist of three types: green bones, which are fresh from the slaughter or boning house and which must be washed and crushed to smaller pieces prior to extraction; extracted bones, which have been partially cooked to remove excess fat and are usually ready for extraction as received; and ossein, which has been washed, cleaned, crushed, treated with dilute hydrochloric acid to dissolve the mineral content of the bone, and then washed and dried. Ossein is pure collagen and can be handled as cured hide materials. It is usually reserved for the production of technical bone gelatin.

Extraction. The clean crushed bone or the extracted bone is transferred to pressure tanks containing water. This water may be hot or cold; it prevents packing and air binding of the bones during processing. If the water is hot, it acts as in a wash cycle and it removes some of the fat and the adhering tissue from the bones. If the water is cold, it can be drained from the bottom of the tank and used for process water.

The tank containing bones but not water is sealed and steam under pressure is added to heat the bones. When the bones are heated, the steam is shut off and cold water is added to just cover the bones and temperature classified against the pressure in the tank. After a short period, the tank is vented and the liquor is drained from the bottom of the tank and is further processed. This cycle is repeated until the solids concentration of the liquor is under 1% at which time the economics preclude further extraction. The tank is opened, the extracted bone is removed, ground, and dried, and is marketed as steam bone meal. The variations with this process are as great as the types of equipment and the experience of the producer. From this point on, the processes for either hide glue or bone glue are very similar and any variations result only from the types of equipment employed and the characteristics of the product that the producer wishes to predominate.

Concentrating, Purifying, and Drying. Once the light liquor is removed from either the bone- or the hide-extraction tank, speed of operation is essential. Glue liquors usually are at a concentration of 2–15% and hydrolyze readily. They are also natural breeding media for bacteria and fungi and they deteriorate rapidly.

The first step is the removal of fat. If there were significant amounts of fat on the raw material, there are significant amounts on the light liquors, which can be removed by hydrostatic skimmers. If the amount is minimal, or after the liquor has been skimmed of fat, the liquor can be centrifuged. Centrifuging removes the remaining fat and (if equipped with desludging devices) any solids that became suspended in the liquor. The liquor is then filtered as rapidly as possible to remove any suspended fines and transferred to the evaporators. These usually are multiple-effect vacuum machines that can concentrate the liquors to ca 20% solids if the quality of the glue is not too high. Further concentration of the liquors can be accomplished with wiped-film-type evaporators or plate evaporators for maximum concentration. High-test glues can be taken to only 15–20% solids. The low-test glues can be taken to as high as 45–50% solids.

If it is desired that the glues be light in color, they can be bleached just before they go to the evaporators; several processes are available. Gaseous sulfur dioxide and sodium hydrosulfite may be added to pH 4.7–5.0 and the resultant precipitate is centrifuged. Hydrogen peroxide that is added to a slightly alkaline liquor also bleaches very well and does not result in precipitation. Preservatives also can be added during bleaching but current restrictions have limited both the types and the amount used; their effectiveness, therefore, is questionable (see Bleaching agents). The concentrated liquors or heavy liquors (as they are sometimes called) are ready for any variety of drying processes.

The liquor may be chilled for gelling by passing over a chill roll and then through a slicer which forms long strips for the drying oven. The liquor also may be chilled by pumping it to a refrigerated votator which chills and extrudes it under pressure onto a belt where the noodles are carried to a dryer. The dryers consist of tunnels through which wire-screen belts pass slowly while the glue rests on the belts. Air is passed from underneath the belt through the glue as well as down through the glue from above. Chill belts also can be used to carry thin layers of glue to the dryer (see Drying).

Pearl glue is produced by the Scheidemandel process. The heavy liquors are extruded as droplets into the top of a tower containing chilled kerosene, petroleum ether, or perchloroethylene. As they fall through the chilled liquid, they form round pearls and gel. They also lose some water and any residual fat to the solvent. They then are passed to a fluidized bed and are dried, first with cold air and then with hot air. A variation of this method is to drop the pearls onto a bed of finely ground glue, cover them with a layer of fine glue, and place them onto the dryer belts. The amount of water to be removed in the oven is small and is distributed over three times the weight of the glue. The advantage in this is that melting in the oven is reduced to a minimum and the low water content to be removed enhances the efficiency of the oven.

Once the glue has been dried, its keeping qualities are excellent. It may be blended, analyzed, and stored either in bulk or in bags as long as it is kept relatively dry. The yields of glue and its by-products are summarized in Table 1.

Table 1. Yields of Glue and By-Products from Various Raw Materials, %

Raw material	Glue	Fat	Grout	Steam bone
calf stock, hide trimmings, tails, snouts, lips, ears, sinews	12–19	3–9	5–16	
bacon skins	24–26	26–34	2–4	
chrome stock	25	0	0	
coney	45–60	0	0	
dried hide trimmings	30–35	0	10–12	
green fleshings	6–8	15–20	4–5	
green bone	12–13	18		30
extracted bone	21	0		51
ossein bone	29	0		71

Economic Aspects

The number of United States producers of animal glue has declined steadily. There were 23 firms in 1955, only 6 in 1968, and only 4 currently. This decline reportedly resulted from raw material shortages, increasing competition from imports, and in recent years the high cost of pollution control (14). The production, consumption, and imports for the past five years are listed in Table 2.

There is no doubt that the versatility of synthetic adhesives and the modern developments in vegetable adhesives have contributed to the decline in animal glue production (see Adhesives).

Specifications and Standards

The particular source designations of animal glue have declined in usage over the past decade but the terms for hide and bone glue are still used and occasionally there is a call for chrome glue and extracted bone glue. Particular emphasis on one type or another lingers within certain industrial groups. For example, the match industry and the abrasive industry maintain rigid standard grades that specify hide glue as well as gel strength.

All animal glue has approximately the same percentage of moisture (7–11%), ash (0.1–6%), fat (0.5–6%), and glue protein (80–90%). Its general quality and, hence, its utility depend upon the hydrolytic breakdown that took place during its manufacture, and even more upon the site of the breakdown. This is reflected by the rigidity of the gels that it forms, the viscosity of its solution, and the relationship of the one value

Table 2. United States Animal Glue Production, Imports, and Consumption, Metric Tons

Year	Domestic production	Imports	Consumption
1972	22,260	19,330	41,590
1973	19,410	18,250	37,660
1974	14,480	13,840	28,320
1975	10,760	6,200	16,990
1976	13,540	13,750	27,300

to the other. The relationship between these two values is translatable not only to the tack of the adhesive but to the coverage that a particular application will achieve. Cooper, one of the earliest commercial glue manufacturers, recognized the need for a uniform method for measuring these qualities and in 1844 developed a system for grading animal glues. This method was based upon the rigidity of a gelatin gel, of definite concentration, at a definite temperature. Several other attempts were made to establish methods for affixing numerical values to a glue based on its gel strength. It was not until 1923 that the Bloom gelometer was developed. Shortly thereafter its success in giving reproducible results led to its adoption as the standard device for determining gel strength. It is still the standard method but the apparatus has been modified and has been made electronic. Timing of the development of gel strength also has become a test method of limited use. Current developments in the scanning of electrophoretic patterns go into greater detail on their specific make-up and have proved useful in the application of colloidal characteristics.

The numerical Bloom test, in grams, is used universally to the exclusion of almost all other specifications. Some developing nations still cling to the Engler viscosity as a method of grading but it is only a matter of time before that method is displaced in favor of the rapidity and reproducibility of the Bloom gelometer.

Test Methods

Standard test methods have been developed for the analysis of glue and have wide acceptance, not only by the manufacturers of glue but amongst the technical societies (15–16). Jelly strength and viscosity give a good perspective of the grade and quality of glue. These two tests are determined on the same sample by the following procedure (15):

14.3 g of dry glue is soaked in 100 mL of cold water (10°C)—until thoroughly soaked and hydrated—in a standard gel jar. The resultant gel is melted with agitation in a water bath at 60°C. Once dissolved, the solution is run through a glue pipet at 60°C that has been calibrated against water. The viscosity in tens of mPa·s (cP) is calculated from the seconds of flow. The sample is returned to the standard sample jar and is stoppered. The jar is placed undisturbed in a water bath at 10 ± 0.1°C for 16–18 h. The sample jar is removed from the bath, the stopper is removed, and the bottle is wiped dry, rapidly placed on the gelometer platform, and tested for gel rigidity. The basis of the rigidity test is the determination of the weight in grams of lead shot needed to press a synthetic resin plunger (having a diameter of 1.25 ± 0.0025 cm and a flat surface that is rounded off to a curvature of 0.069 ± 0.0025 cm) 4 mm into the surface of the gel.

Normally the ratio of jelly strength in grams to the viscosity in tens of mPa·s (cP) is ca 3:1 for alkali-processed hide glues. The same material, if acid processed, has a ratio of 4:1 because of the high gel strength and low viscosity. Bone glues have a ratio of 2:1 to 3.5:1. These ratios have waned in significance as criteria with the development of sophisticated processing techniques and the diversity of blended types of glue.

Determination of grease or fat content also is important for most applications of animal glue. The procedure is a modified Kissling method by which a 10-g sample is dissolved in 75 mL of water, 15 mL of concentrated hydrochloric acid is added, and the whole is refluxed for 2 h, cooled, extracted with 50 mL of carbon tetrachloride, and is shaken for 2 min. An aliquot of the solvent is removed, evaporated, and weighed, and the percentage of fat is determined.

Ash content and foam tendency are measured by routine methods. Adhesive properties, such as tack, rate of set, and shear and tensile strength of the bond, are measured by standard ASTM procedures for various substrates.

Uses

The hide glues have good markets in the abrasive and coated-abrasive fields (see Abrasives). The bone glues have predominated in the areas of gummed paper tape and labels and of laminating adhesives (qv). There is a considerable interchange of hide and bone for the various applications as long as gel strength and viscosity requirements are met. A growing market for animal glue has been established as a stabilizer for latex formulations, copper plating and extraction, flotation aids for removal of slag from uranium, and as a protective colloid in paint and plaster. Table 3 illustrates the general distribution of animal glue in the various United States markets over a three-year period.

Specialty Glues

Although the market for specialty glues has undergone a great deal of change over the past decade, a significant portion of the animal glue produced in the United States and imported is utilized in this market. Liquid glues are almost any animal glue that can be converted into a liquid formulation. These consist of ca 35–55% glue, 35% water, and from 10 to 25% liquefying agents, and they are simple to produce. These agents can vary from urea and thiourea to calcium sulfate or chloride and magnesium sulfate. Opaque glues are formulations made by the addition of pigments to heavy liquors and are used widely where a pigmented glue is used to correct a fault. Flexible glues are glue mixtures that are formulated using humectants and solubilizing agents which allow them to be applied with heat or to flow at ambient temperatures. These formulations also may be considered nonwarp adhesives in that they prevent curling of the flexible substrate when the adhesive bond dries. In these formulations—besides the glue—are sugar, epsom salts, urea, glycols, glycerol, and other salts. These materials are used in quantity in padding glues and paper board. They also can be sold as dry formulations, liquid formulations, and as glue gel cakes. All formulations are adaptable to particular situations and their versatility has maintained their acceptance despite rising costs and competition from synthetic materials.

Table 3. Distribution of Animal Glue in the United States by Application, %

Year	Gummed tape	Abrasives	Adhesives	Paper	Matches	Cork	Textiles	Other
1972	22.5	18.0	8.0	7.0	5.2	3.2	2.7	33.4
1975	23.5	25.9	7.0	5.1	5.3	3.0	1.5	28.7
1976	32.3	22.1	5.0	5.8	4.4	3.8	3.1	23.5

BIBLIOGRAPHY

"Glue" in *ECT* 1st ed., Vol. 7, pp. 207–215, by H. H. Young, Swift & Company; "Glue, Animal and Fish" in *ECT* 2nd ed., Vol. 10, pp. 604–618, by Harland H. Young, Swift & Company.

1. J. E. Eastoe, *Biochem. J.* **61,** 589 (1955).
2. R. E. Neiman, *Kolloidn. Zh.* **16,** 201 (1954).
3. S. Davis and A. D. McLaren, *J. Polym. Sci.* **3,** 16 (1948).
4. R. E. Neiman, *Dokl. Akad. Nauk SSSR* **82,** 419 (1952); *Kolloidn. Zh.* **14,** 107 (1952).
5. B. W. Low, *The Proteins I,* Part A, Academic Press, Inc., New York, 1974, pp. 272–278.
6. L. Pauling, *J. Am. Chem. Soc.* **62,** 2643 (1940).
7. L. Pauling and R. B. Corey, *Proc. Natl. Acad. Sci. U.S.A.* **37,** 272 (1951).
8. E. Bradbury and C. Martin, *Nature (London)* **168,** 837 (1951).
9. E. Bradbury and C. Martin, *Proc. Roy. Soc. London Ser. A* **214,** 183 (1952).
10. R. B. Corey, *Advances in Protein Chemistry IV,* Academic Press, Inc., New York, 1948, pp. 385–406.
11. J. D. Ferry and J. E. Eldridge, *J. Phys. Colloid Chem.* **52,** 184 (1949).
12. U.S. Pat. 3,619,222 (Feb. 4, 1972), E. Werle and T. Mitchell (to Darling and Company).
13. U.S. Pat. 3,598,773 (Aug. 10, 1971), T. Mitchell and A. Pakeltis (to Darling and Company).
14. *Publication No. 840,* U.S. International Trade Commission, Washington, D.C., Oct. 1977.
15. *Ind. Eng. Chem. Anal Ed.* **2,** 348 (1930).
16. *TAPPI Monogr. Ser.* **26,** (1974).

General References

One of the most competent and complete sources of information is the British Food Manufacturing Research Assn., Randalls Road, Leatherhead Surrey, Eng.

Animal Glue in Industry, National Association of Glue Manufacturers, New York, 1951.

R. H. Bogue, *Chemistry and Technology of Gelatin and Glue,* McGraw-Hill Book Co., Inc., 1922.

A. Veis, *The Macromolecular Chemistry of Gelatin,* Academic Press, Inc., New York, 1964.

G. Stainsby, *Recent Advances in Gelatin and Glue Research,* Pergamon Press, New York, 1958.

J. Delmonte, *The Technology of Adhesives,* Reinhold Publishing Corp., New York, 1947.

A. G. Ward and A. Courts, *The Science and Technology of Gelatin,* Academic Press, Inc., New York, 1977.

R. Houwink and G. Salomen, *Adhesion and Adhesives,* 2nd ed., Vol. I, Elsevier Publishing Co., New York, 1967.

I. Skeist, *Handbook of Adhesives,* Reinhold Publishing Corp., New York, 1962.

THOMAS F. MITCHELL
Darling and Company

GLUTAMIC ACID, $HOOCCH_2CH_2CH_2CH(NH_2)COOH$. See Amino acids.

GLUTARIC ACID. See Dicarboxylic acids.

GLUTEN. See Bakery processes and leavening agents.

GLYCERIDES. See Drying oils; Fats and fatty oils.

GLYCEROL

Glycerol [*56-81-5*], propane-1,2,3-triol, glycerin (USP), $CH_2OHCHOHCH_2OH$, a trihydric alcohol, is a clear, water-white, viscous, hygroscopic liquid with a sweet taste at ordinary room temperatures above its melting point. Glycerol was first discovered in 1779 by Scheele who heated a mixture of litharge and olive oil and extracted it with water. Glycerol occurs naturally in combined form as glycerides in all animal and vegetable fats and oils, and is recovered as a by-product when these oils are saponified in the process of manufacturing soap, or the direct splitting of fats in the production of fatty acids. Since 1949 it has also been produced commercially by synthesis from propylene. The latter currently accounts for ca 50% of United States production.

The uses of glycerol number in the thousands but the largest amounts go into the manufacture of synthetic resins and ester gums, drugs, cosmetics, and toothpastes. Tobacco processing and foods also consume large amounts either as glycerol or glycerides.

Occurrence

Glycerol occurs in combined form in all animal and vegetable fats and oils (see Fats and fatty oils). It is rarely found in the free state in these fats but is usually present as a triglyceride combined with such fatty acids as stearic, oleic, palmitic, and lauric acids, and these are generally mixtures or combinations of glycerides of several fatty acids. Such oils as coconut, palm kernel, cottonseed, soybean, and olive oil yield larger amounts of glycerol than do such animal fats as lard and tallow. Glycerol also occurs naturally as triglycerides in all animal and vegetable cells in the form of lipids such as lecithin (qv) and cephalins. These complex fats differ from simple fats in that they invariably contain a phosphoric acid residue in place of one fatty acid residue.

Nomenclature

The term glycerol applies only to the pure chemical compound 1,2,3-propanetriol, $CH_2OHCHOHCH_2OH$. The term glycerin applies to the purified commercial products normally containing ≥95% of glycerol. Several grades of glycerin are available commercially. They differ somewhat in their glycerol content and in other characteristics such as color, odor, and trace impurities. The ending -ol in glycerol connotes the presence of hydroxyl groups. The spelling glycerine should be discouraged since the ending -ine is chemically indicative of organic bases, such as amines, and does not apply to glycerin. The three hydroxyl positions in glycerol are designated 1,2,3- (formerly the designations α, β, and γ were used).

One method of naming the esters of glycerol with organic acids that have simple names is to replace -ic acid by -in, eg, 1-monobutyrin (glycerol 1-butyrate); tributyrin (glycerol tributyrate). The degree of esterification is indicated by the prefixes mono-, di-, and tri-. A mixed triglyceride can be named in three ways, as illustrated by the compound of the following formula:

$$\begin{array}{r} CH_3(CH_2)_{16}COOCH_2 \\ | \;\; \\ CH_3COOCH \\ | \;\; \\ CH_3(CH_2)_{14}COOCH_2 \end{array}$$

2-aceto-3-palmito-1-stearin
2-aceto-3-stearo-1-palmitin
3-palmito-1-stearo-2-acetin

Any one of the three acids can be given the -in termination; if possible, this acid receives the locant 1. The others acids are first cited in alphabetical order (regardless of carbon content) and numbered, if there is a choice, to give the lowest possible numbering (see also Carboxylic acids).

Properties

Physical properties of glycerol are shown in Table 1. Glycerol is completely soluble in water and alcohol, lightly soluble in diethyl ether, ethyl acetate and dioxane, and insoluble in hydrocarbons (1).

Glycerol, the simplest trihydric alcohol, forms esters, ethers, halides, amines, aldehydes, and such unsaturated compounds as acrolein (qv). As an alcohol, glycerol

Table 1. Physical Properties of Glycerol

Property	Value
mp, °C	18.17
bp, °C	
at 0.53 kPa[a]	14.9
at 1.33 kPa[a]	166.1
at 13.33 kPa[a]	222.4
at 101.3 kPa[a]	290
sp gr, 25/25°C	
in vacuum	1.2617
100% glycerol in air	1.2620
95% glycerol in air	1.2491
n_D^{20}	1.47399
vapor pressure, Pa[b]	
at 50°C	0.33
at 100°C	26
at 150°C	573
at 200°C	6100
surface tension at 20°C, mN/m (= dyn/cm)	63.4
viscosity at 20°C, mPa·s (= cP)	1499
specific heat at 26°C, J/g[c,d]	2.425
heat of vaporization, J/mol[d]	
at 55°C	88.12
at 195°C	76.02
heat of solution to infinite dilution, kJ/mol[d]	5.778
heat of formation, kJ/mol[d]	667.8
thermal conductivity, W/(m·K)	0.28
flash point, °C	
Cleveland open cup	177
Pensky-Matens closed cup	199
fire point, °C	204

[a] To convert kPa to mm Hg, multiply by 7.5.
[b] To convert Pa to mm Hg, divide by 133.3.
[c] 99.94% glycerol.
[d] To convert J to cal, divide by 4.184.

also has the ability to form salts such as sodium glyceroxide (see also Alcohols, polyhydric).

Synthesis

A variety of processes for synthesizing glycerol from propylene is shown in Figure 1. The first glycerol process, put on stream in 1948, followed the discovery that propylene could be chlorinated in high yields to allyl chloride. Since allyl chloride could be converted to glycerol by several routes, the synthesis of glycerol from propylene became possible. Propylene also can be oxidized in high yields to acrolein. Several routes for conversion of acrolein to glycerol are shown in Figure 1.

In the traditional allyl chloride route, the allyl chloride may be converted into glycerol by two processes (see Allyl compounds). In the first, the allyl chloride is treated with aqueous chlorine, and the resulting mixture of glycerol dichlorohydrins is dehydrochlorinated to epichlorohydrin, which is then hydrolyzed to glycerol. In the second process, allyl chloride is hydrolyzed to allyl alcohol. The allyl alcohol is chlorohydrinated with aqueous chlorine solution to yield a mixture of monochlorohydrins

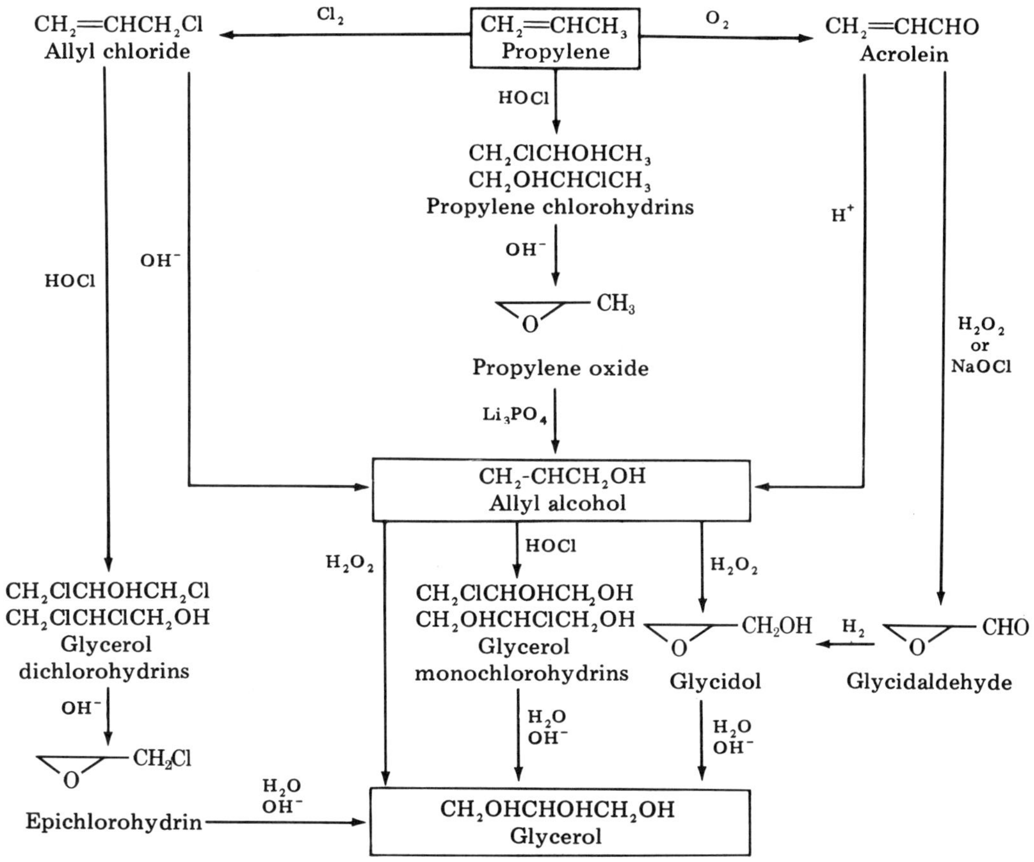

Figure 1. Routes for the manufacture of glycerol.

which are hydrolyzed to glycerol in 90% yield based on allyl alcohol. The product from either of the above procedures is a dilute aqueous solution containing 5% or less of glycerol. High-purity glycerol is obtained in several steps: the crude glycerol is concentrated to ca 80% in multiple-effect evaporators and salt is removed by centrifuging; additional concentration of the product, followed by desalting, yields 98% glycerol; colored substances are removed by solvent extraction, and the product is refined by steam-vacuum distillation (see also Chlorohydrins).

Acrolein-based glycerol manufacture via no-chlorine processing proceeds by epoxidation (qv) (2) and reduction (Fig. 1), in either order, followed by hydration. The epoxidation to glycidaldehyde can proceed through treatment of acrolein with aqueous sodium hypochlorite solution (3), or with hydrogen peroxide. If desired, glycidol can be separated as an intermediate. Alternatively, allyl alcohol, from the reduction of acrolein, can be hydroxylated with aqueous hydrogen peroxide to yield directly a glycerol solution in 80–90% yield.

Propylene oxide-based glycerol can be produced by rearrangement of propylene oxide (qv) to allyl alcohol over trilithium phosphate catalyst at 200–250°C (yield 80–85%) (4), followed by any of the appropriate steps shown in Figure 1. The specific route commercially employed is peracetic acid epoxidation of allyl alcohol to glycidol followed by hydrolysis to glycerol (5). The newest international synthesis plants employ this basic scheme.

Manufacture

Until 1949 all glycerol was obtained from the glycerides in fats and oils but from that time the production of synthetic glycerol increased until, in 1965, it accounted for ca 60% of the market and <50% in 1977.

Glycerol from glycerides (natural glycerol) is obtained from two sources: soap manufacture and fat splitting. In soap manufacture, fat is boiled with caustic soda solution and salt. Fats react with the caustic to form soap (qv) and glycerol. The presence of salt causes a separation into two layers: the upper layer is soap and the lower layer, referred to as spent lye, contains glycerol, water, salt, and excess caustic. Fat splitting is hydrolysis of fats and oils without added alkali. Two early methods were the Twitchell and batch autoclave processes, both of which require catalysts. The most recent (and most common) process is continuous, high pressure hydrolysis where a continuous, upward flow of fat in a column flows countercurrent to water at 250–260°C and 5 MPa (720 psi). The fat is split by the water into fatty acids and glycerol. The fatty acids are withdrawn from the top of the column, and the glycerol-containing aqueous phase (called sweet water) falls and is withdrawn from the bottom. Concentration of the sweet water by evaporation results in a product called saponification crude. The fatty acids from splitting are used to make soap, reduced to the corresponding fatty alcohol, or marketed as fatty acids.

Recovery. The spent lyes resulting from current soapmaking processes generally contain 8–15% glycerol; sweet waters from hydrolysis of fats contain as much as 20% glycerol. In order to produce an acceptable soap, the fats used are generally of higher quality than those used in fat splitting. The grade of fat used directly affects the treatment required to produce glycerol of an acceptable commercial quality. The chemicals most commonly used to remove impurities from spent lye are alum, ferric chloride, ferric sulfate, hydrochloric acid, and caustic soda. For treating sweet water, calcium chloride, lime soda ash, hydrochloric acid, and caustic soda are used. The aluminum and iron salts are used as coagulants and precipitants, and the others serve as precipitants and neutralizers.

The treatment of spent lye consists of a series of operations designed to remove nearly all of the organic impurities (6–7). The spent lye commonly is boiled with fatty acids to reduce the content of free caustic and soda ash. After cooling, the solid soap is skimmed, and an acid and a coagulant are added, followed by filtration. Addition of caustic soda removes the balance of coagulant in solution and adjusts the pH to a point at which the liquor is least corrosive to subsequent process treatment. It is now ready for concentration to 80% soap-crude glycerol.

The treatment of sweet water differs from that for spent lyes but the equipment used is essentially the same (8). The sweet water should be processed promptly after splitting to avoid degradation and loss of glycerol by fermentation. Any fatty acids that rise to the top of the sweet water are skimmed and the liquor is agitated vigorously while an excess of lime is added.

The sweet water from continuous and batch autoclave processes for splitting fats contain no mineral acids and require very little in the way of purification as compared to spent lye. The dirt and insoluble fatty materials are skimmed. A small amount of lime is added to precipitate the dissolved fatty acids and neutralize the liquor. The alkaline liquor then is filtered and evaporated to an 88% crude glycerol.

Concentration. Dilute glycerol liquors, after purification, are concentrated to crude glycerol by evaporation. This is carried out in a conventional evaporation (qv) under vacuum heated by low pressure steam. In the case of soap-lye glycerol, means are supplied for recovery of salt that forms as the spent lye is concentrated. In the case of sweet water, saponification-crude (88%) glycerol is separated by filtration from the small amount of sludge that forms during concentration as a result of lime treatment.

Refining. The refining of natural glycerol is generally accomplished by distillation, followed by treatment with active carbon. In some cases, refining is accomplished by ion exchange (qv).

Distillation. In the case of spent-lye crude, the composition is ca 80% glycerol, 7% water, 3% organic residue, and >10% ash. Saponification crudes are generally of a better quality than soap-lye crudes and have little or no salt.

Distillation of glycerol under atmospheric pressure is not practicable since it polymerizes and decomposes glycerol to some extent at 204°C. A combination of vacuum and steam distillation is used in which the vapors are passed from the still through a series of condensers. Relatively pure glycerol is condensed at high temperature (see Distillation).

In some of the more modern plants, the crude glycerol is vaporized in lower vapor sections of the still and distills at much lower temperatures than is possible in the older type of still. Glycerol losses owing to polymerization and decomposition virtually are eliminated (9).

Bleaching and Deodorizing. The large use of glycerol and glycerol derivatives by the food industry (see Food processing) stresses the importance of the removal of both color and odor (also necessary requirements of USP and extra-quality grades). Activated carbon (1–2%) and diatomite (qv) filter aid are added to the glycerol in the bleach tank at 74–79°C, stirred for 1–2 h, and then filtered at the same temperature, which is high enough to ensure easy filtering and yet not so high as to lead to darkening of the glycerol.

Ion Exchange. The major portion of natural glycerol in the United States is refined by the methods described above. However, several refiners employ the ion-exchange system. When ionized solids are high, as in soap-lye crude, ion-exclusion treatment is used to separate the ionized material from the nonionized (mainly glycerol). The crude glycerol is diluted to ca 25% glycerol before pumping into the exchangers, of which there are three pairs of cation–anion exchangers and a mixed-bed exchanger. Dilute deionized glycerol from the mixed-bed exchanger is run into a dilution tank until it reaches a concentration of ca 15%, at which point the solution is run into a product storage tank and ultimately concentrated to commercial level by evaporation under vacuum to 95–99% glycerol. The product is of high quality and does not require filtering or bleaching. This process is not recommended, however, because of the additional evaporating cost and high usage of resins.

Grades. Two grades of crude glycerol are marketed: (*1*) soap-lye crude glycerol obtained by concentration of lyes from the soap kettle contains ca 80% glycerol; and (*2*) saponification crude glycerol resulting from hydrolysis of fats contains ca 88–91% glycerol and a small amount of organic salts.

Several grades of refined glycerol, such as high-gravity, dynamite, and USP, are marketed; specifications vary depending on the consumer and the intended use. USP-grade glycerol is water-white, and meets the requirements of the USP (see Fine chemicals). It is classified as GRAS by the FDA, and is suitable for use in foods, pharmaceuticals, and cosmetics, or when the highest quality is demanded or the product is designed for human consumption. It has a minimum specific gravity (25°C/25°C) of 1.249, corresponding to no less than 95% glycerol. The BP (*British Pharmacopoeia*) grade is similar to the USP, but has a minimum specific gravity of 1.260 (15.5°C/15.5°C). The CP (chemically pure) grade designates a grade of glycerol that is about the same as the USP but with the specifications varying slightly as agreed by buyer and seller. The high-gravity grade is a pale-yellow glycerol for industrial use with a minimum specific gravity (15.5°C/15.5°C) of 1.262. The dynamite grade has the same specific gravity but is more yellow (see Explosives). All these grades satisfy the federal specifications for glycerol (0-G491B-2).

Economic Aspects

Commercial production and consumption of glycerol has generally been considered a fair barometer of industrial activity as it enters into such a large number of industrial processes. It generally tends to rise in periods of prosperity and fall in recession times.

Glycerol production in the United States (Table 2) has risen from 19,800 metric

Table 2. Glycerol Production in the United States (100% Glycerol Basis), Metric Tons

Year	Production of crude[a]	Year	Production of crude[a]
1920	19,800	1973	163,300
1940	71,600	1976	147,200
1950	102,300	1977	141,000
1960	136,900	1978	137,400
1970	153,900	1979	156,800

[a] Synthetic included on a crude basis since June 1949.

tons in 1920 to its peak of 166,100 t in 1967. World production of glycerol is (450–600) $\times 10^3$ t/yr. For the past 20 yr, synthetic glycerol has accounted for about one half of the United States annual production. In 1929, the largest single industrial use was as an automotive antifreeze (qv). This use has completely disappeared and been replaced by such outlets as cellophane (see Cellulose), tobacco processing (see Tobacco substitutes), and urethane foams (see Urethane polymers). Today, the uses in alkyd resins (qv) (owing to the increase in use of water-based paints) and cellophane (owing to the popularity of other type transparent wraps) are slowly decreasing. The use distribution for the years 1977 and 1978 is given in Table 3.

Since 1920, the price of refined glycerol in the United States has varied from a low of \$0.22/kg in the early 1930s to a high of nearly \$1.32/kg immediately following World War II. During the Korean War and the war in Vietnam, the price again fluctuated widely. In recent years the price has increased approximately at the inflation rate and in 1978 it was ca \$1.05–1.10/kg. Glycerol prices respond to prices for tallow, coconut oil, and petroleum, the feedstocks from which it is prepared.

Identification and Analysis

The methods of analysis of the American Oil Chemists' Society (AOCS) are the principal procedures followed in the United States and Canada and are official in commercial transactions. When the material is for human consumption or drug use, it must meet the specifications of the USP (10). Commercial distilled grades of glycerol do not require purification before analysis by the usual methods. The determination of glycerol content by the periodate method (11), which replaced the acetin and dichromate methods previously used, is more accurate and more specific as well as simpler and more rapid.

Glycerol is most easily identified by heating a drop of the sample with ca 1 g powdered potassium bisulfate and noting the very penetrating and irritating odor of the acrolein that is formed. Owing to the toxicity of acrolein, a preferred method is CTFA (Cosmetics, Toiletry and Fragrance Association) method Gl-1, an ir spectrophotometric method. Glycerol may be identified by the preparation of crystalline derivatives such as glyceryl tribenzoate, mp 71–72°C; glycerol tris(3,5-dinitrobenzoate), mp 190–192°C; or glycerol tris(*p*-nitrobenzoate), mp 188–189°C (12).

Table 3. Glycerol Use, Metric Tons

Use	1977	1978
alkyd resins	21,300	21,510
cellophane and meat casings	9,880	6,380
tobacco	15,900	14,150
explosives	2,690	2,890
drugs, including toothpaste	19,820	19,760
cosmetics	4,670	4,340
monoglycerides and foods	13,740	14,830
urethane foams	13,620	13,810
triacetin		6,150
miscellaneous	11,710	9,770
distributor sales	14,520	16,960
Total	*127,850*	*130,550*

The concentration of distilled glycerol is easily determined from its specific gravity (13) by the pycnometer method (14) with a precision of ±0.02%. Determination of the refractive index also is employed (but not as widely) to measure glycerol concentration to ±0.1% (15).

The preferred method of determining water in glycerol is by the Karl Fischer volumetric method (16). Water can also be determined by a special quantitative distillation in which the distilled water is absorbed by anhydrous magnesium perchlorate (17).

Other tests such as ash, alkalinity or acidity, sodium chloride, and total organic residue are included in AOCS methods (11,14,16).

Handling and Storage

Most crude glycerol is shipped to refiners in standard tank cars or tank wagons. Imported crude arrives in bulk, in vessels equipped with tanks for such shipment, or in drums.

Refined glycerol of a CP or USP grade is shipped mainly in bulk in tank cars or tank wagons. These are usually stainless steel-, aluminum-, or lacquer-lined. However, pure glycerol has little corrosive tendency, and may be shipped in standard, unlined steel tank cars, provided they are kept clean and in a rust-free condition. Some producers offer refined glycerol in 4.5-kg (3.8-L or 1-gal) tinned cans and more commonly in 250–259-kg (208-L or 55-gal) drums of a nonreturnable type (ICC-17E). These generally have a phenolic-resin lining.

Storage. For receiving glycerol from standard 30.3 m^3 (8000 gal) tank cars (36.3 t), a storage tank of 38–45 m^3 ((10–12) $\times 10^3$ gal) capacity should be employed. Preferably it should be of stainless steel (304 or 316), of stainless or nickel-clad steel, or of aluminum. Certain resin linings such as Lithcote have also been used. Glycerol does not seriously corrode steel tanks at room temperature but gradually absorbed moisture may have an effect. Therefore, tanks should be sealed with an air-breather trap.

Handling Temperatures. Optimum temperature for pumping is in the 37–48°C range. Piping should be stainless steel, aluminum, or galvanized iron. Valves and pumps should be bronze, cast iron with bronze trim, or stainless steel. A pump of 3.15 L/s (50 gal/min) capacity will unload a tank car of warm glycerol in ca 4 h.

Health and Safety Factors

Glycerol, since 1959, is generally recognized as safe (GRAS) as a miscellaneous or general purpose food additive (qv) under the CFR (18), and it is permitted in certain food packaging materials.

Oral LD_{50} levels have been determined in the mouse at 470 mg/kg (19) and the guinea pig at 7750 mg/kg (20). Several other studies (21–23) have shown that large quantities of both synthetic and natural glycerol can be administered orally to experimental animals and man without the appearance of adverse effects. Intravenous administration of solutions containing 5% glycerol to animals and humans has been found to cause no toxic or otherwise undesirable effects (24).

The aquatic toxicity (TLm96) for glycerol is >1000 mg/L (25) which is defined by NIOSH as an insignificant hazard.

Uses

Glycerol is used in nearly every industry. The largest single use is in the manufacture of alkyd resins (qv) (Table 3) but this market has decreased considerably in recent years with the increased use of water-based paints (qv). Its use in tobacco processing, drugs, cosmetics, and foods remains at a fairly even consumption level, and the fastest growing use is in the manufacture of urethane foams.

Foods. Glycerol as a food is easily digested and nontoxic, and its metabolism places it with the carbohydrates, although it is present in combined form in all vegetable and animal fats. In flavoring and coloring products, glycerol acts as a solvent and its viscosity lends body to the product. It is used as a solvent, a moistening agent, and an ingredient of syrups (qv) as a vehicle. In candies and icings, glycerol retards crystallization of sugar. Glycerol is used as a heat-transfer medium in direct contact with foods in quick freezing, and as a lubricant in machinery used for food processing (qv) and packaging. The polyglycerols and polyglycerol esters have increasing use in foods, particularly in shortenings and margarines.

Drugs and Cosmetics. In drugs and medicines, glycerol is an ingredient of many tinctures and elixirs, and glycerol of starch is used in jellies and ointments. It is employed in cough medicines and anesthetics (qv), such as glycerol–phenol solutions, for ear treatments, and in bacteriological culture media. Its derivatives are used in tranquilizers (eg, glyceryl guaiacolate), and nitroglycerin is a vasodilator in coronary spasm. In cosmetics (qv), glycerol is used in many creams and lotions to keep the skin soft and replace skin moisture. It is widely used in toothpaste to maintain the desired smoothness and viscosity and lending a shine to the paste (see Dentifrices).

Tobacco. In processing tobacco, glycerol is an important part of the casing solution sprayed on tobacco before the leaves are shredded and packed. Along with other flavoring agents, it is applied at a rate of ca 2.0 wt % of the tobacco to prevent the leaves from becoming friable and thus crumbling during processing; by remaining in the tobacco, glycerol helps to retain moisture and thus prevents drying out of the tobacco, and influences the burning rate of the tobacco. It is used also in the processing of chewing tobacco to add sweetness and prevent dehydration, and as a plasticizer in cigarette papers (see Tobacco substitutes).

Wrapping and Packaging Materials. Meat casings and special types of papers, such as a glassine and greaseproof paper, need a plasticizer (qv) to give them pliability and toughness; as such, glycerol is completely compatible with the base materials used, is absorbed by them, and does not crystallize or volatilize appreciably.

Lubricants. Glycerol can be used as a lubricant in places where an oil would fail. It is recommended for oxygen compressors because it is more resistant to oxidation than mineral oils. It is also used to lubricate pumps and bearings exposed to fluids such as gasoline and benzene, which would dissolve oil-type lubricants. In food, pharmaceutical, and cosmetic manufacture, where there is contact with a lubricant, glycerol may be used to replace oils (see Lubrication).

Glycerol is often used as a lubricant because its high viscosity and ability to remain fluid at low temperatures make it valuable without modification. To increase its lubricating power, finely divided graphite may be dispersed in it. Its viscosity may be decreased by addition of water, alcohol, or glycols, and increased by polymerization or mixing with starch—pastes of such compositions may be used in packing pipe joints, in gas lines, or in similar applications (see Packing materials). For use in high pressure gauges and valves, soaps are added to glycerol to increase its viscosity and improve its lubricating ability. A mixture of glycerin and glucose is employed as a nondrying lubricant in the die-pressing of metals. In the textile industry, glycerol is frequently used in connection with so-called textile oils, in spinning, knitting, and weaving operations.

Urethane Polymers. An important use for glycerol is as the fundamental building block in polyethers for urethane polymers (qv). In this use it is the initiator to which propylene oxide, alone or with ethylene oxide, is added to produce trifunctional polymers which, on reaction with diisocyanates, produce flexible urethane foams. Glycerol-based polyethers (qv) have found some use, too, in rigid urethane foams.

Gaskets and Cork Products. Sheets and gaskets made with ground cork and glue require a plasticizer that has some humectant action in order that they may be pliable and tough. Glycerol is used because it has low vapor pressure, is not readily extractable by oils and greases, is readily absorbed by the cork, and is compatible with glue (qv). With crown sealers and cork stoppers that come into contact with foods, it fulfills the additional requirement of nontoxicity (see Cork).

Other Uses. Glycerol is used in cement compounds, caulking compounds, lubricants, and pressure media. It is also used in embalming fluids, masking and shielding compounds, soldering compounds, and compasses; cleaning materials such as soaps, detergents, and wetting agents; emulsifiers and skin protectives used in industry; asphalt (qv); ceramics (qv); photographic products; leather and wood treatments; and adhesives (qv).

Derivatives

Glycerol derivatives include acetals, amines, esters, and ethers. Of these the esters are the most widely employed. Commercially, the most important are the alkyd resins (qv), which are esters of glycerol and phthalic anhydride. The next most important is glyceryl nitrate (nitroglycerin), used in explosives (qv) and as a heart stimulant (see Cardiovascular agents). Included among the esters also are the ester gums (rosin acid esters of glycerol) and the so-called monoglycerides (glycerol esterified with fatty acids), used as emulsifiers and in shortenings. The salts of glycerophosphoric acid are used medicinally.

Mixtures of glycerol with other substances are often named as if they were derivatives of glycerol; eg, boroglycerides (also called glyceryl borates) are mixtures of boric acid and glycerol. Derivatives, such as acetals, ketals, chlorohydrins and ethers, can be prepared but are not made commercially, with the exception of polyglycerols.

The polyglycerols, ethers prepared with glycerol itself, have many of the properties of glycerol. Diglycerol, $HOCH_2CHOHCH_2OCH_2CHOHCH_2OH$, is a viscous liquid (287 mm^2/s (= cSt) at 65.6°C), about twenty-five times as viscous as glycerol. The polyglycerols offer greater flexibility and functionality than glycerol. Polyglycerols up to and including triacontaglycerol (30 condensed glycerol molecules) have been prepared commercially, the higher forms are solid. They are soluble in water, alcohol, and other polar solvents. They act as humectants, much like glycerol, but have progressively higher molecular weights and boiling points. Products based on polyglycerols are useful in surface-active agents (see Surfactants and detersive systems), emulsifiers (see Emulsions), plasticizers, adhesives, lubricants, antimicrobial agents, medical specialties and dietetic foods (see Food additives; Sweeteners).

Esters. The mono- and diesters of glycerol and the fatty acid occur naturally in fats that have become partially hydrolyzed. The triglycerides occur naturally in fats and fatty oils.

The so-called monoglycerides are made by the reaction of fatty acids or raw or hydrogenated oils, such as cottonseed and coconut, with an excess of glycerol or polyglycerols. It is not possible to prepare directly a product consisting substantially of either mono- or diglycerides; commercial glycerides (except fractionated products) are mixtures of mono- and diesters, with a small percentage of the triester. They also contain small amounts of free glycerol and free fatty acids. Various grades of each monoester are available for different uses.

The higher fatty acid mono- and diesters are oil-soluble and water-insoluble. They are all edible (except the ricinoleate) and find their greatest use as emulsifiers in foods and in the preparation of baked goods (26) (see Bakery processes). A mixture of mono-, di-, and triglycerides is manufactured in large quantities for use in superglycerinated shortenings. Mono- and diglycerides are important modifying agents in the manufacture of alkyd resins, detergents, and other surface-active agents. The monoglycerides are also used in the preparation of cosmetics (qv), pigments (qv), floor waxes (see Polishes), synthetic rubbers (see Rubber chemicals), coatings (qv), textiles (qv) (27), etc. In 1977 the production of glycerol esters in the United States was 42,300 t (28).

Acetins. The acetins are the mono-, di-, and triacetates of glycerol that form when glycerol is heated with acetic acid. Physical properties are shown in Table 4.

Monoacetin (glycerol monoacetate), is a thick hygroscopic liquid, and is sold for use in the manufacture of explosives, in tanning, and as a solvent for dyes (qv).

Diacetin is a hygroscopic liquid, and is sold in a technical grade for use as a plasticizer and softening agent and as a solvent.

Triacetin has a very slight odor and bitter taste. It occurs naturally in small quantities in the seed of *Euonymus europaeus*. Most commercial triacetin is USP grade. Its primary use is as a cellulose plasticizer in the manufacture of cigarette filters and its second largest use as a component in binders for solid rocket fuels. Smaller amounts are used as a fixative in perfumes, as a plasticizer for cellulose nitrate, in the manufacture of cosmetics, and as a carrier in fungicidal compositions (see Fungicides).

United States production of acetins is ca 6000 metric tons.

Table 4. Physical Properties of Acetins

Property	Monoacetin	Diacetin	Triacetin
mp, °C			−78
bp, °C			
at 0.4 kPa[a]	130		
at 22 kPa[a]	158		
at 101.3 kPa[a]		259	258–259
d_4^{20}, g/cm^3	1.206		1.160
d_4^{16}, g/cm^3		1.184	
n_D^{20}		1.44	
color	colorless	colorless	colorless
solubility	sol in H_2O, C_2H_5OH; sl sol in $(C_2H_5)_2O$; insol in C_6H_6	sol in H_2O, C_2H_5OH, $(C_2H_5)_2O$, C_6H_6; insol in CS_2	sol in C_2H_5OH, $(C_2H_5)_2O$, and other organic solvents; sl sol in H_2O

[a] To convert kPa to mm Hg, multiply by 7.5.

BIBLIOGRAPHY

"Glycerol" in *ECT* 1st ed., Vol. 7, pp. 216–229, by N. N. Dalton and Joyce C. Kern, Association of American Soap & Glycerine Producers, Inc., and C. S. Miner, Jr., The Miner Laboratories; "Glycerol" in *ECT* 2nd ed., Vol. 10, pp. 619–631, by Joyce C. Kern, Glycerine Producers' Association.

1. *Physical Properties of Glycerine and Its Solutions,* Glycerine Producers' Association, New York, 1975.
2. G. B. Payne, *J. Am. Chem. Soc.* **80,** 6461 (1958); **81,** 4901 (1959).
3. C. Schaer, *Helv. Chim. Acta* **41,** 560, 614 (1958); U.S. Pat. 2,887,498 (May 9, 1959), G. Hearne, D. S. la France, and H. D. Finch (to Shell Development Co.).
4. U.S. Pat. 2,426,264 (Aug. 26, 1947), G. W. Fowler and J. T. Fitzpatrick (to Carbide and Carbon Chemicals Corp.).
5. K. Yamagishi and O. Kageyama, *Hydrocarbon Process.* **55**(12), 139 (1976).
6. T. M. Patrick, Jr., E. T. McBee, and H. B. Haas, *J. Am. Chem. Soc.* **68,** 1009 (1946).
7. W. E. Sanger, *Chem. Met. Eng.* **26,** 1211 (1922).
8. J. L. Trauth, *Oil Soap* **23,** 137 (1946).
9. *Glycerine Refining, Tech. Data Sheet No. 5220,* Wurster & Sanger, Inc., Chicago, Ill., 1964.
10. *The United States Pharmacopeia XX* (*USP XX–NF XV*), The United States Pharmacopeial Convention, Inc., Rockville, Md., 1980.
11. *Official and Tentative Methods,* 3rd ed., American Oil Chemists' Society, Chicago, Ill., 1978, Ea6-51.
12. C. S. Miner and N. N. Dalton, *Glycerol, ACS Monograph 117,* Reinhold Publishing Corp., New York, 1953, pp. 171–175.
13. L. W. Bosart and A. O. Snoddy, *Ind. Eng. Chem.* **19,** 506 (1927).
14. Ref. 11, Ea7-50.
15. L. T. Hoyt, *Ind. Eng. Chem.* **26,** 329 (1934).
16. Ref. 11, Ea8-58.
17. C. P. Spaeth and G. F. Hutchinson, *Ind. Eng. Chem. Anal. Ed.* **8,** 28 (1936).
18. CFR, *Title 21,* Sect. 182.1320.
19. H. F. Smyth, J. Seaton, and L. Fischer, *J. Ind. Hyg. Toxicol.* **23,** 259 (1941).
20. R. C. Anderson, P. N. Harris, and K. K. Chen, *J. Am. Pharm. Assoc. Sci. Ed.* **39,** 583 (1950).
21. V. Johnson, A. J. Carlson, and A. Johnson, *Am. J. Physiol.* **103,** 517 (1933).
22. C. H. Hine, H. H. Anderson, H. D. Moon, M. K. Dunlap, and M. S. Morse, *Arch. Ind. Hyg. Occup. Med.* **7,** 282 (1953).
23. W. Deichman, *Ind. Med. Ind. Hyg. Sec.* **9**(4), 60 (1940).
24. H. A. Sloviter, *J. Clin. Inv.* **37,** 619 (1958).
25. W. Hann and P. A. Jensen, *Water Quality Characteristics of Hazardous Materials,* Texas A&M University, 1974 p. 4.
26. N. H. Nash and V. K. Babayan, *Food Process.* (*Chicago*) **24**(11), 2 (1963); *Baker's Dig.* **38**(9), 46 (1963).
27. A. E. Parolla and C. Z. Draves, *Am. Dyestuff Rep.* **46,** 761 (Oct. 21, 1957); **47,** 643 (Sept. 22, 1958).
28. *Synthetic Organic Chemicals, U.S. Production and Sales, 1977;* U.S. International Trade Commission, Washington, D.C., 1978.

JOYCE C. KERN
Glycerine Producers' Association

GLYCOLIC ACID. See Hydroxy carboxylic acids.

GLYCOLS

ETHYLENE GLYCOL AND PROPYLENE GLYCOL

Glycols, also called diols, are compounds containing two hydroxyl groups attached to separate carbon atoms in an aliphatic chain. Although glycols may contain heteroatoms, those discussed here, with the exception of thiodiglycol, are composed solely of carbon, hydrogen, and oxygen.

Simple glycols are those in which both hydroxyl groups are attached to an otherwise unsubstituted hydrocarbon chain as represented by the general formula, $C_nH_{2n}(OH)_2$.

Polyglycols are adducts of simple glycols and are distinguished by intervening ether linkages in the hydrocarbon chain, as represented by the general formula $C_nH_{2n}O_x(OH)_2$. Many commercially important polyglycols have physical, chemical, and toxicological properties similar to those of the simple glycols (see also Polyethers; Alcohols, polyhydric; Glycerol).

Polyglycols that contain up to and including eight carbon atoms are referred to here as lower polyglycols and include diethylene, triethylene, tetraethylene, and dipropylene glycols.

The simple C_2 and C_3 glycols are essentially odorless and colorless when pure. They are relatively high boiling, freeze well below 0°C, are very hygroscopic and miscible in all proportions with water, and possess excellent solvent properties for dyes, synthetic resins, essential oils and certain natural gums and resins (see Dye carriers; Solvents, industrial).

The lower polyglycols also have excellent water solubility but are less hygroscopic and have somewhat different solvent properties. Location of the hydroxyl groups, molecular weight, the presence of branched chains, and the number of repeating ether linkages control the influence of the hydroxyl groups on the physical properties of a particular glycol. Physical properties of the ethylene and propylene glycols are listed in Table 1.

Glycols undergo reactions common to monohydric alcohols forming esters, acetals, ethers, and similar products. For example, both simple and polyesters are produced by reaction with mono- or dibasic acids (eqs. 1–2):

$$\underset{\text{triethylene glycol}}{HOCH_2CH_2OCH_2CH_2OCH_2CH_2OH} + 2\ CH_3COOH \longrightarrow \underset{\text{triethylene glycol diacetate}}{CH_3COOCH_2CH_2OCH_2CH_2OCH_2CH_2OOCCH_3} + 2\ H_2O \quad (1)$$

$$\underset{\text{propylene glycol}}{HOCH(CH_3)CH_2OH} + HOOC{-}R{-}COOH \longrightarrow \underset{\text{polyester}}{H{-}\!\!\left(OOC{-}R{-}COOCH(CH_3)CH_2\right)_{n}\!\!{-}OH} + H_2O \quad (2)$$

Table 1. Properties of Glycols[a]

Property	Ethylene glycol *[107-21-1]*	Diethylene glycol *[111-46-6]*	Triethylene glycol *[112-27-6]*	Tetraethylene glycol *[112-60-7]*	Propylene glycol *[57-55-6]*	Dipropylene glycol *[25265-71-8]*
formula	$HOCH_2CH_2OH$	$HO(CH_2CH_2O)_2H$	$HO(CH_2CH_2O)_3H$	$HO(CH_2CH_2O)_4H$	$CH_3CH(OH)CH_2OH$	isomers (see text)
mol wt	62.07	106.12	150.17	194.23	76.10	134.18
sp gr 20/20°C	1.1155	1.1185	1.1255	1.1247	1.0381	1.0232
bp at 101.3 kPa[b], °C	197.6	245.8	288	dec	187.3	232
vapor pressure at 20°C, Pa[c]	<1.3	<13	<1.3	<1.3	<13	<1.3
mp, °C	−13.0	−6.5	−4.3	−4.1	−60[d]	−40[e]
viscosity at 20°C, mPa·s (= cP)	19.83	36	49	61.9	60.5	107
refractive index, n_D^{20}	1.4318	1.4475	1.4561	1.4598	1.4326	1.4407
heat of vaporization at 101.3 kPa[b], kJ/mol[f]	52.24	52.26	61.04	62.63	52.30	53.64
flash point of commercial material, °C	116[g]	138[h]	172[h]	191[h]	101[g]	118[g]

[a] Ref. 1.
[b] To convert kPa to mm Hg, multiply by 7.5.
[c] To convert Pa to mm Hg, divide by 133.3.
[d] Sets to glass below this temperature.
[e] Pour point.
[f] To convert J to cal, divided by 4.184.
[g] Determined by ASTM method D 56, using the Tag closed cup.
[h] Determined by ASTM method D 92, using the Pensky-Martens closed cup.

Glycols may also undergo cyclization reactions, eg, diethylene glycol and methylamine form *N*-methylmorpholine (eq. 3) (see Amines, cyclic):

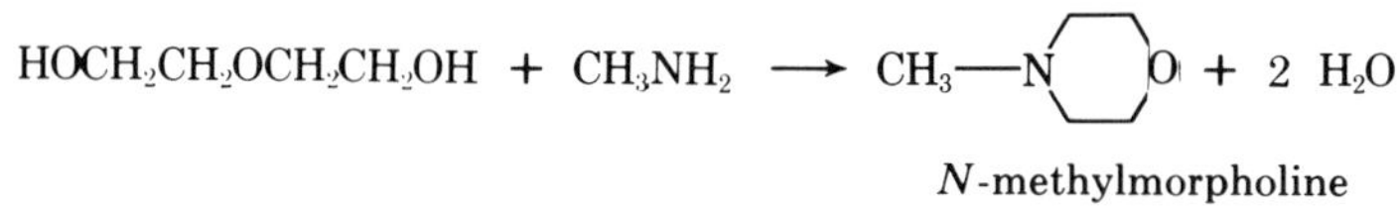

(3)

Glycols as a class exhibit a low order of toxicity. They have low vapor pressure at normal temperatures and are, therefore, not an inhalation hazard. With the notable exception of propylene glycol, glycols should, however, not be used for internal consumption. Table 2 shows the toxicological properties of some ethylene and propylene glycols.

Ethylene Glycol

Ethylene glycol was first prepared by Wurtz in 1859 by hydrolysis of ethylene glycol diacetate. It did not achieve commercial interest until World War I, when it was used in Germany as a substitute for glycerol (qv) in explosives manufacture (2).

Physical Properties. Ethylene glycol and its lower polyglycols are colorless, odorless, high boiling, hygroscopic liquids completely miscible with water and many organic liquids. Ethylene (and propylene) glycols markedly reduce the freezing point of water (Fig. 1). The vapor-pressure curves of the ethylene (and propylene) glycols at various temperatures are illustrated in Figure 2. A listing of some important physical constants of ethylene glycol is given in Table 3 (see also Table 1).

Chemical Properties. Ethylene glycol undergoes the reactions typical of a monohydric alcohol (4). Its unique chemical behavior is attributed to the presence of hydroxyl groups on adjacent carbon atoms. The following reactions illustrate participation of both hydroxyl groups in the same reaction. Of particular commercial importance is the reaction of ethylene glycol with dicarboxylic acids to form linear polyesters (qv). Ethylene glycol is esterified with terephthalic acid or transesterified with dimethyl terephthalate to form bis(hydroxyethyl) terephthalate (BHET) (eq. 4):

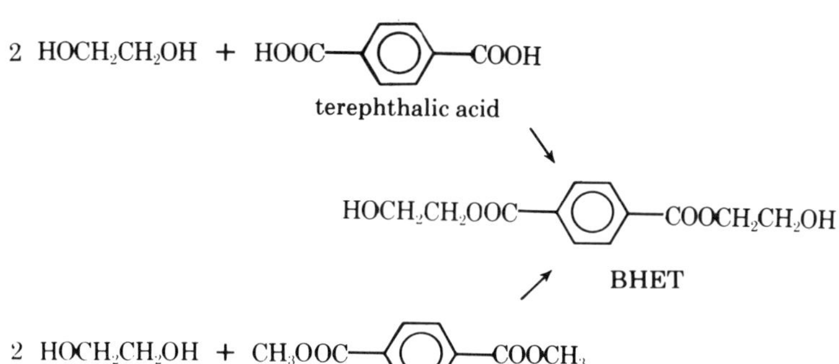

(4)

BHET polymerizes at high temperature and low pressure over an antimony trioxide catalyst to poly(ethylene terephthalate) (PET) (eq. 5):

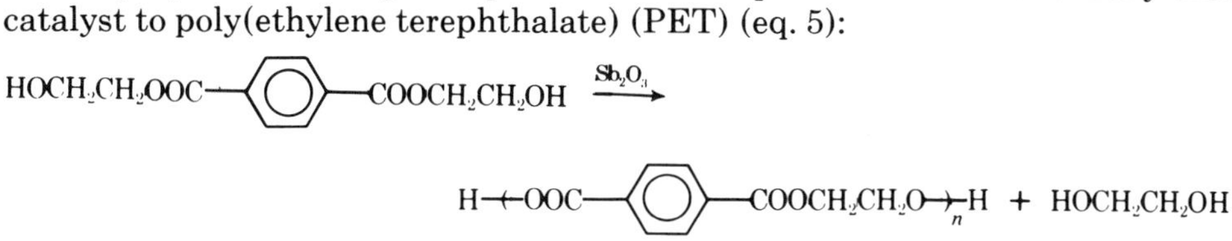

(5)

Table 2. Toxicological Properties of Glycols[a,b]

Glycol	Single oral LD_{50} rats[c], mL/kg	Repeated oral feeding, rats, acceptable level: Diet, g/(kg·d)	Time	Single skin penetration LD_{50}[d], rabbits, mL/kg
ethylene	7.40[e]	0.18	30 d	20
diethylene	28.3	0.18	30 d	11.9
triethylene	28.2	0.83	30 d	20
tetraethylene	28.9	1.88	2 yr	20
propylene	34.6	2.0[f]	2 yr[f]	20
dipropylene	14.8			20

[a] Ref. 1.

[b] Single inhalation is the continuous breathing of certain concentrations of chemicals for the stated period—in rats this killed none of 6 in 8 h; primary irritation is the skin response 24 h following application of 0.01 mL amounts to uncovered skin—rabbits did not show irritation; eye injury is surface damage produced by the liquid chemical—no eye injury in rabbits, except for traces with propylene and dipropylene glycols.

[c] The term LD_{50} refers to that quantity of chemical that kills 50% of dosed animals within 14 d.

[d] Single skin penetration refers to a 24-h covered skin contact with the liquid chemical.

[e] Single dose oral toxicity to humans is greater.

[f] Dogs.

Table 3. Physical Properties of Ethylene Glycol[a]

Property	Value
mp, °C	−13.0
bp[b] at 101.3 kPa[c], °C	197.6
flash point, °C[d]	116
density at 20°C, g/mL	1.1135
surface tension at 20°C, mN/m (= dyn/cm)	48.4
refractive index, n_D^{20}	1.4318
vapor pressure at 20°C, kPa[c]	0.008
specific heat, J/(g·°C)[e]	
as liquid, 19.8°C	2.406
as ideal gas, 25°C	1.565
heat of vaporization at 101.3 kPa[c], kJ/mol[e]	52.24
heat of combustion at 25°C, kJ/mol[e]	−1189.595
heat of formation at 25°C, kJ/mol[e]	−392.878
heat of fusion, kJ/mol[e]	11.63
critical constants	
temperature, °C	372
pressure, kPa[c]	6515.73
volume, L/mol	0.186
compression factor, Z_c	0.2671
viscosity, mPa·s (= cP)	
at 0°C	51.37
at 20°C	19.83
at 40°C	9.20

[a] Ref. 3.

[b] See also Figure 2.

[c] To convert kPa to mm Hg, multiply by 7.5.

[d] Of commercial material (ASTM D 56), using tag closed cup.

[e] To convert J to cal, divide by 4.184.

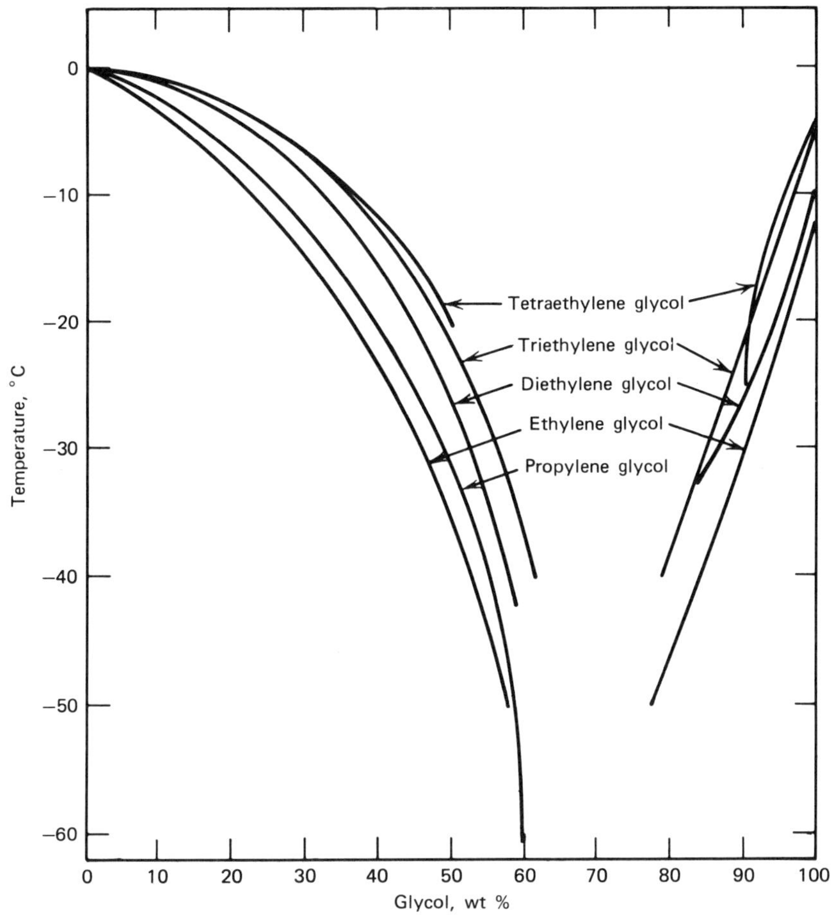

Figure 1. Freezing points of aqueous glycol solutions (1).

Transesterification with diethyl carbonate produces ethylene carbonate (eq. 6):

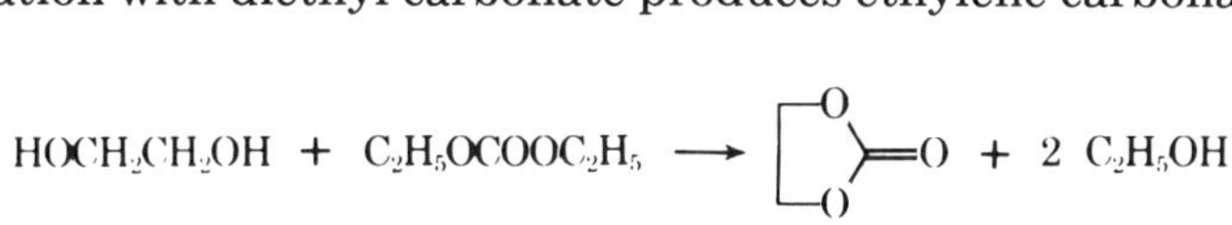

ethylene carbonate (6)

Reaction with aldehydes or ketones under acid conditions forms 1,3-dioxolanes (eq. 7):

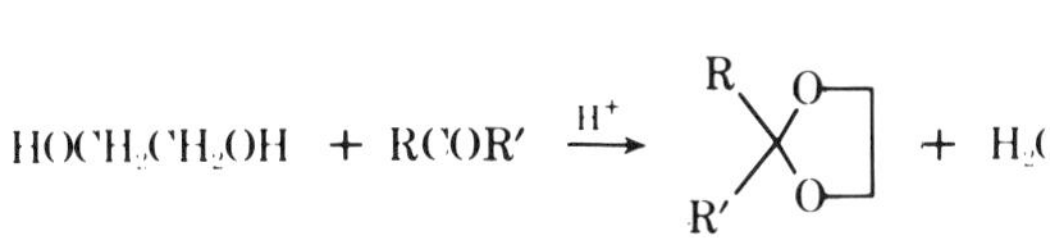

(7)

R and R′ = alkyl or hydrogen. A chelate can be prepared with boric acid (eq. 8):

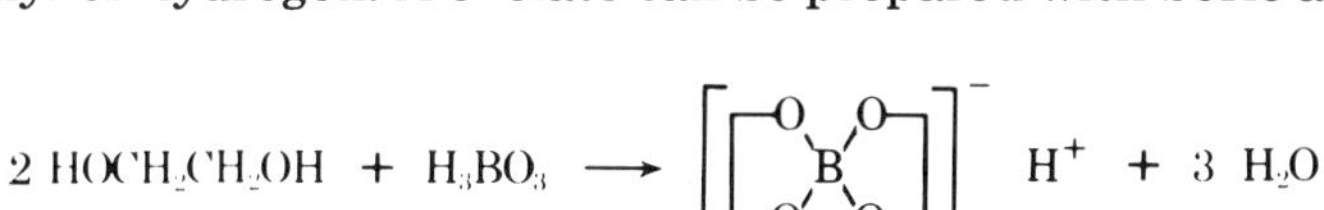

(8)

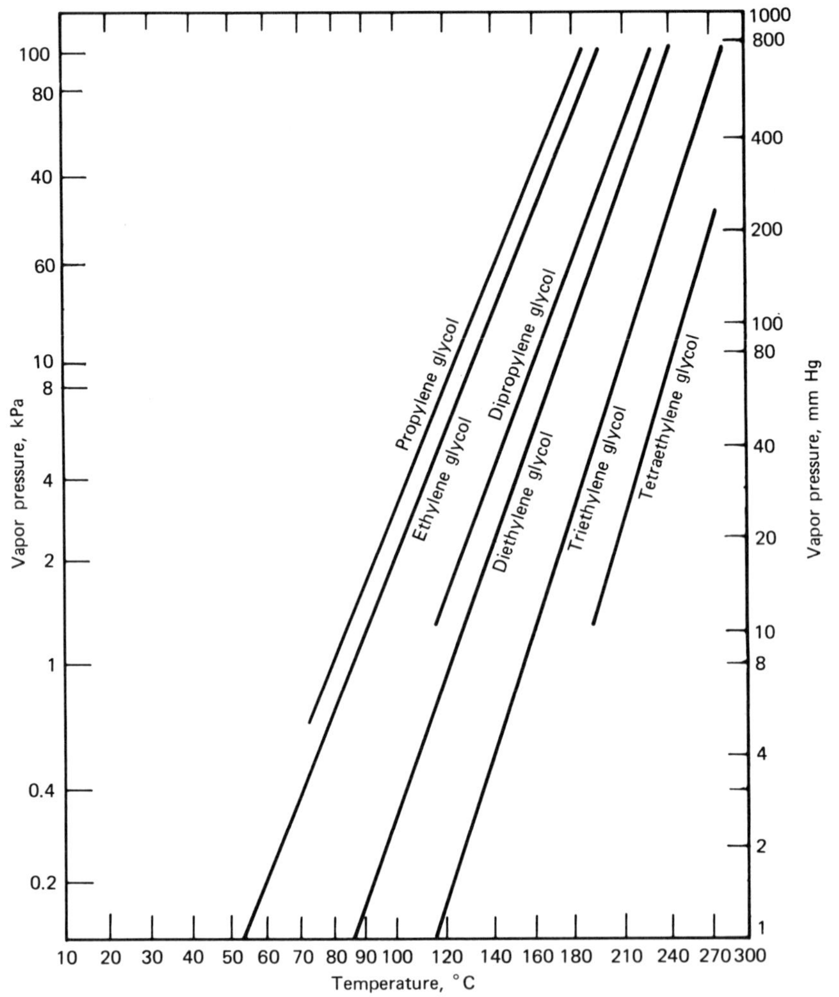

Figure 2. Vapor pressures of glycols at various temperatures (1).

Ethylene glycol is not readily dehydrated to acetaldehyde. However, 1,4-dioxane is readily obtained by intermolecular dehydration to diethylene glycol followed by intramolecular cyclization (eq. 9):

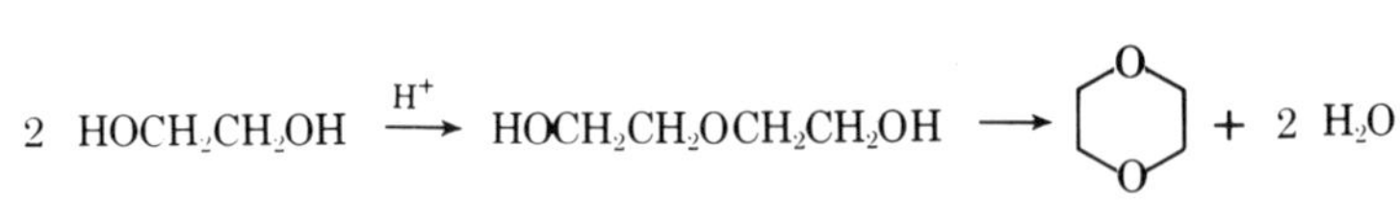

1,4-dioxane (9)

Ethylene glycol may be simultaneously dehydrogenated and dehydrated catalytically in the vapor phase to produce 2-hydroxymethyl-1,3-dioxolane (eq. 10):

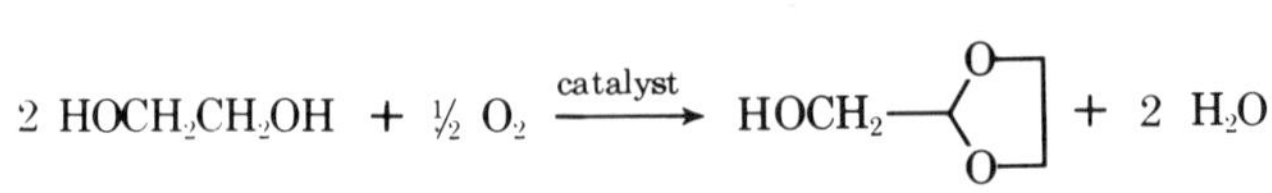

2-hydroxymethyl-1,3-dioxolane (10)

Direct oxidation gives the commercially important glyoxal (eq. 11) (see under Derivatives):

$$HOCH_2CH_2OH + O_2 \xrightarrow[\text{phase}]{\text{vapor}} \underset{\text{glyoxal}}{OHCCHO} + 2\,H_2O \qquad (11)$$

Manufacture. The first United States patents related to ethylene glycol manufacture appeared about 1915 describing production from ethylene chlorohydrin (see Chlorohydrins). Ethylene glycol dinitrate was substituted in part for nitroglycerin in the preparation of mine explosives in the United States during the late 1920s (see Explosives). It was not until about 1930 that a large volume market developed with the widespread application of ethylene glycol as an automotive antifreeze. During this period, domestic production was started at the Union Carbide plant in South Charleston, West Virginia, via the chlorohydrin process. The first commercial application of the Lefort direct ethylene oxidation to ethylene oxide followed in 1937, and from then on, hydrolysis of ethylene oxide became and remains the main commercial source of ethylene glycol (2) (see Ethylene oxide).

Ethylene oxide hydrolysis proceeds with either acid or base catalysis or uncatalyzed in neutral medium. In acid-catalyzed hydrolysis, protonation of oxide activates it for reaction with water. The reaction is conducted in a large excess of water at moderate temperatures to obtain 85–90% monoethylene glycol selectivity. The lower polyglycols, ie, diethylene, triethylene, and tetraethylene glycols, account for the remainder, the primary by-product being diethylene glycol.

Base catalysis results in considerably lower monoethylene glycol selectivity; the yield of higher glycol products is substantially increased since anions of the first reaction products effectively compete with hydroxide ion for ethylene oxide. For this reason, this modification is not of commercial interest.

Neutral hydrolysis (pH 6–10), conducted in the presence of a large excess of water at high temperatures and pressures, increases monoethylene glycol selectivity to 89–91%.

The water effluent from hydrolysis is removed in a series of multiple-effect evaporators (5), and the ethylene glycol is refined by vacuum distillation. Figure 3 depicts a typical process flow diagram.

Ethylene glycol has also been commercially produced by the reaction of formaldehyde with carbon monoxide (6), but apparently this is no longer practiced. This route

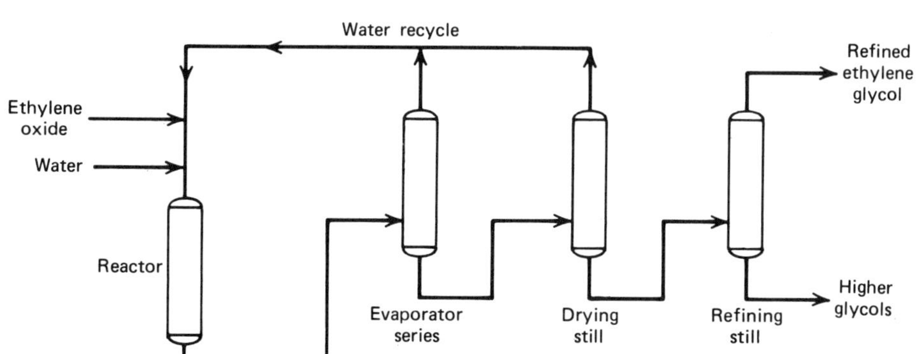

Figure 3. Simplified ethylene glycol process flow diagram.

produces first glycolic acid, which is converted by esterification and hydrogenolysis to ethylene glycol (eqs. 12–14):

$$HCHO + CO + H_2O \rightarrow HOCH_2COOH \tag{12}$$

$$HOCH_2COOH + ROH \rightarrow HOCH_2COOR + H_2O \tag{13}$$

$$HOCH_2COOR + 2\ H_2 \rightarrow HOCH_2CH_2OH + ROH \tag{14}$$

More recently, Halcon International developed an efficient, direct ethylene oxidation process to produce ethylene glycol via hydrolysis of its acetates (7) (eqs. 15–16).

$$CH_2{=}CH_2 + \tfrac{1}{2}\ O_2 + 2\ CH_3COOH \rightarrow CH_3COOCH_2CH_2OCOCH_3 + H_2O \tag{15}$$

$$CH_3COOCH_2CH_2OCOCH_3 + 2\ H_2O \rightarrow HOCH_2CH_2OH + 2\ CH_3COOH \tag{16}$$

This process employs a tellurium oxide–bromide catalyst in acetic acid to yield a mixture of ethylene glycol mono- and diacetates. It is reported to be quite efficient in ethylene usage, and yields only monoethylene glycol. It is projected to be economically more attractive than ethylene oxide hydrolysis.

The relative economics of the Halcon acetoxylation process versus conventional ethylene oxidation–ethylene oxide hydrolysis (8) are sharply contested. Proponents of acetoxylation point to the advantage of its inherent ethylene efficiency, whereas its opponents cite corrosion problems and greater utility costs.

The oxidation of ethylene to ethylene glycol in an aqueous medium using an iron–copper catalyst (9) (eq. 17) has been patented:

$$CH_2{=}CH_2 + \tfrac{1}{2}\ O_2 + H_2O \xrightarrow{\text{Fe–Cu}} HOCH_2CH_2OH \tag{17}$$

The rhodium-catalyzed production of ethylene glycol from synthesis gas (mixtures of carbon monoxide and hydrogen) has been patented (10) (eq. 18):

$$2\ CO + 3\ H_2 \xrightarrow[\text{catalyst}]{\text{rhodium}} HOCH_2CH_2OH \tag{18}$$

The attractiveness of this route depends on the relative cost of synthesis gas vs ethylene in the future. Current commercialization plans are unknown.

Economic Aspects. The domestic price remained stable from the early 1950s through the mid 1960s (11). A sharp decline occurred in the late 1960s and early 1970s because of decreased glycol consumption (12). The oil embargo and escalating energy costs caused the ethylene price to soar in the mid 1970s which was reflected in sharply increased glycol prices (13) (Fig. 4). In May 1980 ethylene glycol sold for 75¢/kg in tank-car quantities.

During 1950–1978 United States glycol production climbed rapidly (14–15), roughly doubling with each succeeding decade (Fig. 5). Although accurate figures are not available, foreign production and prices have apparently kept pace.

The 1978 nameplate capacities for United States ethylene glycol producers total 2.78×10^6 t (Table 4). Actual effective production capacity is ca 2.5×10^6 t (effective capacity assumed to be 90% of nameplate). Total 1978 United States production of 1.82×10^6 t indicates an operating rate of ca 73% of effective capacity. Worldwide ethylene glycol nameplate production capacity is estimated at 6.29×10^6 t (Table 5). Estimated United States production in 1979 was 1.95×10^6 t and total capacity was 2.86×10^6 t (18).

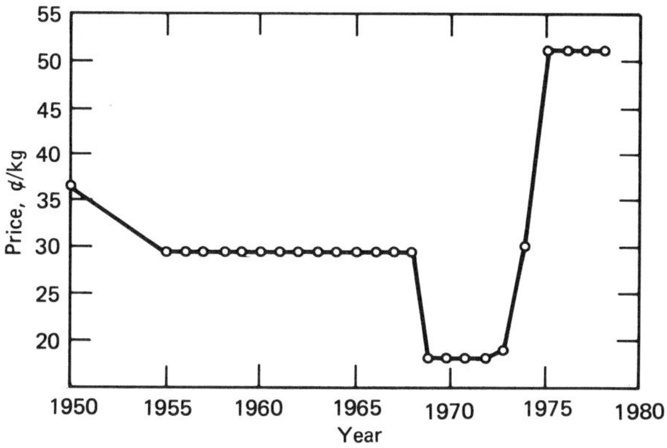

Figure 4. United States ethylene glycol prices.

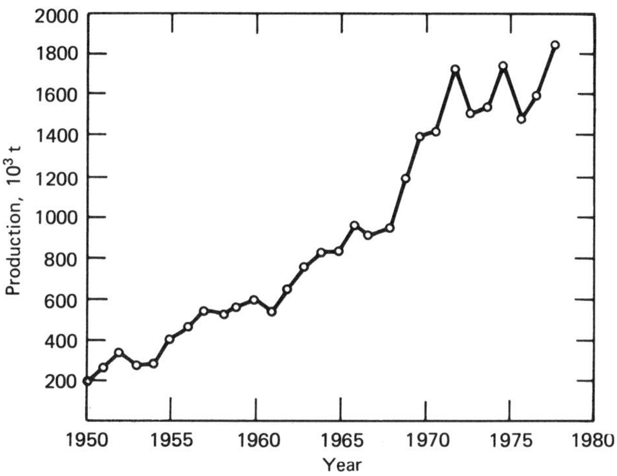

Figure 5. United States ethylene glycol production.

Health and Safety Factors. Ethylene glycol is one of the more toxic simple glycols by oral administration, and serious injury or death may result from swallowing about 60 mL. Oral toxicity is markedly lower with diethylene, triethylene, and tetraethylene glycols. Toxicity of ethylene glycols by all other routes of administration is low (Table 2).

Uses. About 40% of the ethylene glycol produced domestically is used as a nonvolatile antifreeze for liquid-cooled motor vehicles (15) (see Antifreezes and deicing fluids). Previously, this market grew at the rate of ca 5%/yr. In the current era of energy shortages, which has resulted in smaller cars with reduced coolant requirements, annual antifreeze demand is expected to grow at a reduced rate (16).

Approximately 35% of total production is consumed in the manufacture of polyester fiber and film (15) (see Polyester fibers; Polyesters, thermoplastic). In the 1960s and early 1970s demand for this application increased at the spectacular rate of 25–30%/yr. Current projections place future growth at the more modest rate of 8%/yr (16).

Table 4. 1978 United States Ethylene Glycols Nameplate Capacity [a]

Producer	Location	Capacity[b], thousand metric tons
BASF Wyandotte	Geismar, La.	131
Calcasieu	Lake Charles, La.	93
Celanese	Clear Lake, Tex.	166
Dow	Freeport, Tex.	83
Dow	Plaquemine, La.	183
Eastman	Longview, Tex.	81
Houston Chemical (PPG)	Beaumont, Tex.	64
Jefferson (Texaco)	Port Neches, Tex.	208
Northern Petrochemical	Joliet, Ill.	91
Olin	Brandenburg, Ky.	46
Oxirane	Channelview, Tex.	363[c]
PPG	Guayanilla, P.R.	125
Shell	Geismar, La.	112
Union Carbide	Ponce, P.R.	229
Union Carbide	Seadrift, Tex.	353
Union Carbide	Taft, La.	457
Total		*2785*

[a] Ref. 15.
[b] Derived by assuming 65% utilization of ethylene oxide capacity.
[c] Produces only monoethylene glycol.

The remaining 25% of ethylene glycol is used in a variety of applications (1). It provides freeze–thaw stabilization to latex coatings and improves flexibility and drying time of oil-based paints containing alkyd resins (qv). It also is used as a heat-transfer fluid (see Heat-exchange technology), in aircraft and runway deicing mixtures, as a dehydrating agent for natural gas (see Drying agents), in motor oil additives, and as an additive in the formulation of inks, pesticides, wood stains, adhesives and many other products. In explosive water-gels and slurries, it lowers the freezing point and acts as a coupling agent between water and the other components. A minor but important use for high purity ethylene glycol is as a solvent and suspending medium for ammonium perborate, the conductor in practically all electrolytic capacitors.

To accommodate these many uses (17), ethylene glycol is available in several quality grades.

Derivatives. There are several important classes of derivatives: monoethers, diethers, esters, ether–esters, dialdehydes, acetals, and ketals. Physical properties are given in Table 6 and toxicological properties in Table 7.

The reaction of ethylene oxide and alcohols gives a mixture of glycol monoethers and monoethers of the lower polyethylene glycols (19). Glycol ethers are both ethers and alcohols. The ether groups promote solubility through hydrogen bonding.

Glycol ether solvents are used in lacquers, enamels, water-borne coatings, and wood stains (see Coatings, industrial; Paint). They improve the gloss and flow of lacquers and gloss retention during baking in synthetic enamels.

Other applications include dye solvents in the textile, leather, and printing industries, solvents for insecticides and herbicides, solvents for grease and grime in specialty chemicals, jet-fuel additives, automotive antistall additives, and diluents

Table 5. Estimated 1978 Worldwide Nameplate Capacity for Ethylene Glycols[a]

Country	Total capacity[b], thousand metric tons	Percent of total
United States	2785	44.28
Japan	668	10.62
Federal Republic of Germany	573	9.11
Netherlands	302	4.80
Italy	284	4.52
France	275	4.37
United Kingdom	273	4.34
Belgium	247	3.93
Taiwan	160	2.54
Mexico	117	1.86
Canada	112	1.78
German Democratic Republic	92	1.46
People's Republic of China	79	1.26
USSR	78	1.24
Czechoslovakia	60	0.95
Poland	55	0.87
Romania	37	0.59
Sweden	37	0.59
India	26	0.41
Spain	11	0.18
Bulgaria	9	0.15
Democratic People's Republic of Korea	9	0.15
Total	*6289*	*100.00*

[a] Ref. 17.
[b] Derived by assuming 65% utilization of ethylene oxide capacity.

for hydraulic brake fluids (see Hydraulic fluids). As mutual solvents and couplers, glycol ethers are used in soluble oils, specialty hard-surface cleaners, and other soap–hydrocarbon systems.

The hydroxyl group of glycol ethers can be etherified, esterified, chlorinated, or otherwise modified.

Glycol diethers are prepared by conventional etherification procedures; the preferred method is the reaction of an alkali metal glycolate with an alkyl halide. Both symmetrical and unsymmetrical ethers are made, which presents difficult separation problems (see also Ethers).

Glycol diethers behave like typical simple ethers. Except for peroxide formation, they are relatively inert and unreactive, which makes them useful as reaction solvents, both in the laboratory and commercially.

Ethylene glycol is esterified readily (1,21) with conventional reagents such as acid chlorides, acid anhydrides, acids, and other esters via ester interchange. Mixtures of mono- and diesters are usually formed in these reactions. Mono- or diesters can be produced preferentially by controlling the molar ratio of reactants (see also Esters).

Low molecular weight glycol esters are good solvents for cellulose esters and printing inks, and are employed in industrial extraction processes and the protective coatings industry. Fatty acid esters, together with other surfactants, are good emulsifying, stabilizing, dispersing, wetting, foaming, and suspending agents.

Table 6. Physical Properties of Ethylene Glycol Derivatives[a]

Derivative	CAS Registry No.	Formula	Mol wt	Sp gr 20/20°C	Bp at 101.3 kPa[b], °C	Vapor pressure at 20°C, Pa[c]	Mp, °C	Solubility at 20°C, wt %: In water	Solubility at 20°C, wt %: Water in	Viscosity at 20°C, mPa·s (= cP)	Refractive index, n_D^{20}	Heat of vaporization at 101.3 kPa[b], kJ/mol[d]	Flash point, °C: Open cup	Flash point, °C: Closed cup
Monoethers														
methyl	*[109-86-4]*	$CH_3OCH_2CH_2OH$	76.09	0.9663	124.5	800	−85.1	miscible		1.7	1.4021	39.16	43.3	39.4
ethyl	*[110-80-5]*	$C_2H_5OCH_2CH_2OH$	90.12	0.9311	135.6	530	−70[e]	miscible		2.1	1.4076	40.53	48.9	42.2
n-butyl	*[111-76-2]*	$C_4H_9OCH_2CH_2OH$	118.18	0.9022	170.1		−70[e]	miscible		6.4	1.4198	43.40	69.4	60.0
n-hexyl	*[112-25-4]*	$CH_3(CH_2)_4.CH_2OCH_2CH_2OH$	146.23	0.8887	208.1	<130	−50.1	0.99	18.78	5.2	1.4290	47.93	90.6	
phenyl	*[122-99-6]*	$C_6H_5OCH_2CH_2OH$	138.17	1.1094	244.9	<1.3	14	2.7	10.8		1.5386	51.76	121.1	
2-ethylbutyl	*[4468-93-3]*	$(C_2H_5)_2CHCH_2OCH_2CH_2OH$	146.23	0.8954	196.7		−90[f]	1.2	10.0		1.4304		82.2	
2-ethylhexyl	*[1559-35-9]*	$CH_3(CH_2)_3CH(C_2H_5)CH_2OCH_2$-$CH_2OH$	174.28	0.8850	229.0		−70[f]	0.09	5.3		1.4363		110.0	
Diethers														
dimethyl	*[110-71-4]*	$CH_3OCH_2CH_2OCH_3$	90.12	0.8692	85.1	7870	−71	miscible			1.4383	31.42	4.4	
diethyl	*[629-14-1]*	$C_2H_5OCH_2CH_2OC_2H_5$	118.18	0.8417	121.0	1330	−74	21.0	3.4	0.7	1.3922	35.99	35.0	
di-*n*-butyl	*[112-48-1]*	$C_4H_9OCH_2CH_2OC_4H_9$	174.28	0.8374	203.1	<130	−69.1	0.2	0.6	1.3	1.4131	44.15	85.0	
vinyl butyl	*[4223-11-4]*	$CH_2{=}CHOCH_2CH_2OC_4H_9$	144.21	0.8654	dec	<130	−71	0.2	0.5	1.04	1.4213		140.0	
p-dioxane	*[123-91-1]*	$O(CH_2CH_2)_2O$	88.11	1.0356	101.3	3600	11.8	miscible		1.31	1.4224	35.58	23.3	12.2
Ether–esters														
methyl, acetate	*[110-49-6]*	$CH_3OCH_2CH_2OOCCH_3$	118.13	1.0067	145.0	270	−65.1	miscible		1.10	1.4019	40.21	55.6	
ethyl, acetate	*[111-15-9]*	$C_2H_5OCH_2CH_2OOCCH_3$	132.16	0.9748	156.3	270	−61.7	22.9	6.5	1.3	1.4058	40.86	56.7	51.1
n-butyl, acetate	*[112-07-2]*	$C_4H_9OCH_2CH_2OOCCH_3$	160.21	0.9424	192.3	<130	−63.5	1.5	1.7	1.83	1.4138	43.94	87.8	73.9
phenyl, acetate	*[6192-44-5]*	$C_6H_5OCH_2CH_2OOCCH_3$	180.20	1.1084	259.2	<13	−2.7	0.17	0.92	9.4	1.5076	52.36	143.3	
methyl, acrylate	*[3121-61-7]*	$CH_3OCH_2CH_2OOCCH{=}CH_2$	130.14	1.009[g]	156.0		−44.6	10.3			1.4240[g]			
ethyl, acrylate	*[106-74-1]*	$C_2H_5OCH_2CH_2OOCCH{=}CH_2$	144.17	0.9834	174.1		−47.3	5.5	2.5	1.61	1.4274	42.56	66.1	
Diesters														
diformate	*[629-15-2]*	$HCOOC_2H_4OOCH$	118.09	1.2277	177.2	<130	−10	dec	dec	2.6	1.4153	44.14	93.3	
diacetate	*[111-55-7]*	$CH_3COOC_2H_4OOCCH_3$	146.14	1.1064	190.9	<130	−41.5	21.3	21.2	2.9	1.4159	45.52	101.7	85.0
dipropionate	*[123-80-8]*	$C_2H_5COOC_2H_4OOCC_2H_5$	174.20	1.0484	210.0	<13						46.94		

[a] Refs. 19–22.
[b] To convert kPa to mm Hg, multiply by 7.5.
[c] To convert Pa to mm Hg, divide by 133.3.
[d] To convert J to cal, divide by 4.184.
[e] Pour point.
[f] Sets to a glass.
[g] At 25°C.

Table 7. Toxicological Properties of Ethylene Glycol Derivatives[a]

Derivative	Single oral LD_{50}, rats[b], g/kg	Single skin penetration LD_{50}, rabbits[c], mL/kg	Single inhalation[d] concentrated vapors or specific concentration in ppm, rats	Primary skin irritation[e], rabbits	Eye injury[f], rabbits
Monoethers					
methyl	2.46	1.34	4 h killed 2 of 6 8 h killed 6 of 6	none	minor
ethyl	3.00	3.5	4 h killed 1 of 6 8 h killed 4 of 6	trace	minor
n-butyl	1.48	0.56	8 h killed 1 of 6	trace	minor
hexyl	1.48	0.89	8 h killed 0 of 6	none	severe
Diethers					
dimethyl	5–10			none	none
Ether–esters					
methyl, acetate	3.93	5.52	4 h killed 2 of 6 7000 ppm, 2 h killed 0 of 6	none	trace
ethyl, acetate	5.10	10.6	4 h killed 0 of 6 8 h killed 2 of 6	trace	minor
n-butyl, acetate	7.03	1.58	8 h killed 0 of 6	none	trace
Diesters					
diacetate	6.85	8.48	8 h killed 0 of 6	none	trace

[a] Refs. 19, 21–22.
[b] The term LD_{50} refers to that quantity of chemical that kills 50% of dosed animals within 14 d.
[c] Single skin penetration refers to a 24-h covered skin contact with the liquid chemical.
[d] Single inhalation refers to the continuous breathing of certain concentrations of chemicals for the stated period.
[e] Primary irritation refers to the skin response 24 h following application of 0.01 mL amounts to uncovered skin.
[f] Eye injury refers to surface damage produced by the liquid chemical.

Polyester resins based on ethylene glycol, maleic and phthalic anhydrides, and vinyl-type monomers are used for low pressure lamination of glass fibers, asbestos, cloth, and paper. Polyester–glass fiber laminates are used in making furniture, suitcases, boat hulls, aircraft parts, and automobile bodies (see Laminated and reinforced plastics).

Ethylene glycol reacts with dibasic acids to form alkyd-type resins. These resins are of interest to modify synthetic rubbers, in adhesives, and in other applications. Alkyds made from ethylene glycol and phthalic anhydride are used with similar resins based on glycerol or pentaerythritol in the manufacture of surface coatings.

Polyester fibers and films are made from the condensation product of ethylene glycol and dimethyl terephthalate or terephthalic acid (eqs. 4–5). Polyester fiber is ideally suited to satisfy customer demand for a medium-priced fiber with ease-of-care and durability properties. It is also used in tire cord (qv), carpeting, and other industrial uses. Rosin esters of ethylene glycol are plasticizers (qv) in adhesives, lacquers, and enamels.

Ethylene glycol ether–esters are prepared from glycol ethers and thus are only secondary derivatives of ethylene glycol (see Glycol Ethers). Ether–esters are included in Tables 6 and 7 for reference.

Ethylene glycol reacts with aldehydes and ketones in the presence of acid catalysts to form 1,3-dioxolanes (eq. 7). These compounds are fairly stable under alkaline and neutral conditions but revert to the starting materials under acidic conditions. Low molecular weight dioxolanes are excellent solvents for cellulose esters and natural and synthetic resins.

Glyoxal. Glyoxal (23), CHOCHO, is produced commercially by the vapor-phase oxidation of ethylene glycol (eq. 11). A less important route is the nitric acid oxidation of acetaldehyde (qv).

Glyoxal is a highly reactive dialdehyde with many important commercial applications. Its reactivity with cellulose, starch, and other carbohydrates and hydrocolloids gives it broad utility in paper manufacture (24). It is used as a wet-strength agent, to confer water resistance to starch–clay paper coatings, and with other pigment coating binders (see Papermaking additives).

Glyoxal also has wide applicability in textiles, particularly in improving resistance to shrinkage and creasing of cotton, rayon, and other cellulosics (see Textiles). Glyoxal and many of its derivatives have biological activity and are used in medicine, bacteriology, and pest control. Protein treated with glyoxal is more resistant to pepsin than protein treated with formaldehyde. Thus, glyoxal is useful in leather tanning (see Leather).

The reactivity of glyoxal with hydrocolloids allows the modification of adhesives containing starch, carbohydrates, and other reactive materials. In photography, glyoxal is useful in hardening gelatin, either alone or condensed with diols or aminoalcohols. It is also used to make a reconstituted tobacco sheet with aroma, taste, and burning characteristics of natural whole leaf tobacco (see Tobacco, synthetic).

Although caution is recommended in its use, a 40% aqueous solution of glyoxal does not appear to offer any unusual hazards. Extensive studies using rats, rabbits, and guinea pigs indicate that aqueous glyoxal is not highly toxic taken orally. Like all aldehydes, glyoxal may be expected to cause occasional skin sensitization in persons repeatedly exposed to the solution. It is significantly less toxic than formaldehyde (24).

Diethylene Glycol

Diethylene glycol is similar in many respects to ethylene glycol but contains an ether group (1). It was originally synthesized at about the same time by both Lourenco and Wurtz (25) in 1859, and was first marketed by Union Carbide in 1928. It is a coproduct (9–10%) of ethylene glycol produced by ethylene oxide hydrolysis. It is also obtained directly by the reaction of ethylene glycol with ethylene oxide but this route is rarely taken commercially. The price of diethylene glycol in tank-car quantities was 47.4 ¢/kg in early 1979.

About 35% of diethylene glycol is used as an intermediate in the manufacture of unsaturated polyester resin and polyols for polyurethanes. Approximately 15% is converted to triethylene glycol by reaction with ethylene oxide, 10% is consumed as a softening agent and lubricant for textiles (see Textiles), and 10% for natural-gas dehydration. The UDEX process extraction solvent consumes ca 5% (see Adsorptive separation) and the remaining 25% is distributed among other industrial applications including plasticizer and surfactant manufacture and antifreeze mixtures. Physical properties of diethylene glycol are shown in Table 1 and toxicological data in Table 2.

Table 8. Physical Properties of Diethylene Glycol Derivatives [a]

Derivative	CAS Registry No.	Formula	Mol wt	Sp gr 20/20°C	Bp at 101.3 kPa[b], °C	Vapor pressure at 20°C, Pa[c]	Mp, °C	Solubility, at 20°C, wt %: In water	Solubility, at 20°C, wt %: Water in	Viscosity at 20°C, mPa·s (cP)	Refractive index, n_D^{20}	Heat of vaporization at 101.3 kPa[b], kJ/mol[e]	Flash point, °C: Open cup	Flash point, °C: Closed cup
Monoethers														
methyl	[*111-77-3*]	$CH_3O(C_2H_4O)_2H$	120.16	1.0211	194	<130	−85[e]	miscible		3.9	1.4263	29.76	93.3	86.7
ethyl	[*111-90-0*]	$C_2H_5O(C_2H_4O)_2H$	134.18	0.9898	202.7	<130	−78[e]	miscible		4.5	1.4273	47.10	96.1	83.3
n-butyl	[*112-34-5*]	$C_4H_9O(C_2H_4O)_2H$	162.23	0.9536	230.6	<13	−68.1	miscible		6.5	1.4316	49.78	115.6	101.1
n-hexyl	[*112-59-4*]	$CH_3(CH_2)_5O(C_2H_4O)_2H$	190.28	0.9346	259.1	<1.3	−40.2	2.96	1.70	8.6	1.4372	54.40	140.6	>110.0
phenyl	[*104-68-7*]	$C_6H_5O(C_2H_4O)_2H$	182.22	1.1157	297.9	<1.3	−40[e]	3.4	24.3	46.3		58.03	160.0	
Diethers														
dimethyl	[*111-96-6*]	$CH_3O(C_2H_4O)_2CH_3$	134.18	0.9451	162	400	−64	miscible		2.0	1.4078	41.42	70	
diethyl	[*112-36-7*]	$C_2H_5O(C_2H_4O)_2C_2H_5$	162.23	0.9082	189	53	−40.3	miscible		1.4	1.4115		82.2	
dibutyl	[*112-73-2*]	$C_4H_9O(C_2H_4O)_2C_4H_9$	218.34	0.8853	256	<13	−60.2	0.3	1.4		1.4233	50.14	118.3	
vinyl ethyl	[*109-92-2*]	$C_2H_5O(C_2H_4O)_2CH{=}CH_2$	160.21	0.9391	193.9	<13	−80[e]	4.6	2.8	1.6	1.4284	43.94	91.1	
Ether–esters														
methyl, acetate	[*629-38-9*]	$CH_3(OC_2H_4)_2OOCCH_3$	162.18	1.0412	209.1	<130	−99.3	miscible		2.6	1.4209	47/5		82.2
ethyl, acetate	[*112-15-2*]	$C_2H_5(OC_2H_4)_2OOCCH_3$	176.21	1.0114	217.4	<13	−25	miscible		2.8	1.4213	47.92	110.0	
n-butyl, acetate	[*124-17-4*]	$C_4H_9(OC_2H_4)_2OOCCH_3$	204.26	0.9810	246.7	<1.3	−32.2	6.5	3.7	3.6	1.4262	50.33	115.6	
Thioethers														
thiodiglycol	[*111-48-8*]	$S(C_2H_4OH)_2$	122.19	1.1852	283	<1.3	−10	miscible		65	1.5217	60.70	160.0	
Morpholines														
N-methyl	[*109-02-4*]	$O(C_2H_4)_2NCH_3$	101.15	0.9213	115.3	1460	−65.9	miscible		0.94	1.4352	29.89	22.2	13.9
N-ethyl	[*100-74-3*]	$O(C_2H_4)_2NC_2H_5$	115.18	0.9143	138.6	670	−63	miscible		1.08	1.4415	36.94	35.6	

[a] Refs. 19–22, 26–27.
[b] To convert kPa to mm Hg, multiply by 7.5.
[c] To convert Pa to mm Hg, divide by 133.3.
[d] To convert J to cal, divide by 4.184.
[e] Set to a glass.

Table 9. Toxicological Properties of Diethylene Glycol Derivatives[a]

Derivative	Single oral LD_{50}, rats[b], g/kg	Single skin penetration LD_{50}, rabbits[c], mL/kg	Single inhalation[d] concentrated vapors, rats	Primary skin irritation[e], rabbits	Eye injury[f], rabbits
Monoethers					
methyl	9.21	6.54	8 h killed 0 of 6	none	trace
ethyl	9.05	16.5	8 h killed 0 of 6	none	trace
n-butyl	6.56	4.12	8 h killed 0 of 6	none	moderate
hexyl	4.92	1.60	8 h killed 0 of 6	minor	severe
Diethers					
dimethyl	5			none	none
Ether–esters					
ethyl, acetate	11.0	15.2	8 h killed 0 of 6	trace	trace
n-butyl, acetate	11.9	14.8	8 h killed 1 of 6	trace	trace

[a] Refs. 19, 21–22.
[b] The term LD_{50} refers to that quantity of chemical that kills 50% of dosed animals within 14 d.
[c] Single skin penetration refers to a 24-h covered skin contact with the liquid chemical.
[d] Single inhalation refers to the continuous breathing of certain concentrations of chemicals for the stated period.
[e] Primary irritation refers to the skin response 24 h following application of 0.01 mL amounts to uncovered skin.
[f] Eye injury refers to surface damage produced by the liquid chemical.

Derivatives. Important classes of diethylene glycol derivatives include monoethers, esters, thioethers, and morpholines. 1,4-Dioxane can be prepared directly from diethylene glycol. Physical properties of selected diethylene glycol derivatives are given in Table 8 and toxicological properties in Table 9.

The preparation and uses of diethylene glycol monoethers are similar to those of the ethylene glycol monoethers.

Like ethylene glycol, diethylene glycol is readily esterified with mono- and dicarboxylic acids to yield important plasticizers (qv) and resins (1). Unsaturated polyesters derived from maleic anhydride are useful as laminating agents and adhesives (qv). Linear polyesters prepared from acids like adipic acid, are used as plasticizers and adhesives. Fatty acid esters have similar applications. Esterification with rosin acids yields soft resins for use in nitrocellulose lacquers, enamels, and adhesives.

Polyester resins derived from diethylene glycol, maleic anhydride, and vinyl-type monomers are low pressure laminates for glass fibers, asbestos, cloth, or paper. Combining the resin with reinforcing material forms a versatile structural material that is weather resistant, dimensionally stable, and easily molded and colored. Such reinforced plastics are used in the production of fuel-oil tanks, boat hulls, luggage, automobile bodies, aircraft parts, refrigerators, and washing machines. Esters derived from diethylene glycol and higher fatty acids, such as oleic, stearic, and lauric acids are good emulsifiers, demulsifiers, and lubricants (see Emulsions; Lubrication).

Thiodiethylene glycol [*111-48-8*] (thiodiglycol) (26) is prepared from ethylene oxide and hydrogen sulfide and, strictly speaking, is not a diethylene glycol derivative. Its hygroscopic and antioxidant properties make it an excellent solvent for vat, basic, and acid dyestuffs (see Dye carriers). Thiodiglycol ethers are used as lubricants. Esters formed from adipic acid monoester have high viscosity indexes, low pour points, and good antioxidation properties. Thiodiglycol dihydroxysulfone is a cross-linking agent that gives outstanding wash-and-wear properties to cotton (qv) fabrics.

Thiodiglycol exhibits a moderate oral toxicity comparable to that of isopropanol. It is not believed to penetrate the skin in harmful amounts. Because of its low vapor pressure, it does not present an inhalation hazard. Thiodiglycol does not act as a skin irritant but is capable of causing mild eye irritation. Hydrochloric acid and compounds that generate hydrochloric acid convert thiodiglycol into a skin vesicant (26).

N-Alkylmorpholines (28) are prepared from diethylene glycol and a primary amine (eq. 3) (see Amines, cyclic).

The *N*-alkylmorpholines are used as catalysts in the production of polyurethane foams and are excellent solvents for waxes, resins, and oils, reaction and extraction solvents, corrosion inhibitors, and stabilizers. They are used also in the preparation of self-polishing waxes and oil emulsions. In handling *N*-alkylmorpholines, skin and eye contact and prolonged breathing of vapors should be avoided (27).

Triethylene Glycol

Triethylene glycol has chemical and physical properties essentially identical to those of diethylene glycol (1). Physical properties are listed in Table 1, toxicological properties in Table 2. Triethylene glycol is a coproduct of ethylene glycol produced via ethylene oxide hydrolysis. Significant commercial quantities are also produced directly by the reaction of ethylene oxide with the lower glycols. The price for tank-car quantities of triethylene glycol was 77.2 ¢/kg in early 1979.

Triethylene glycol is an efficient hygroscopicity agent and as such used as a liquid drying agent for natural gas (see Drying agents). Its use in small packaged plants located at the gas wellhead eliminates the need for line heaters in field gathering systems (see Gas, natural). About 35% of triethylene glycol produced is used for this purpose. Another 45% is used in roughly equal amounts in the manufacture of vinyl plasticizers, as a humectant, and as a solvent. Smaller quantities are used as intermediates in the production of unsaturated polyester resins (see Polyesters, unsaturated) and polyols for polyurethanes (see Urethane polymers), as a solvent for inks and dyes, and in the UDEX extraction process (see Adsorptive separation).

Derivatives. Ethers and esters, prepared like the monoethylene glycol derivatives, are the most important triethylene glycol derivatives. They are all miscible with water. Physical properties of some ethers and esters are given in Table 10 and toxicological data in Table 11.

Triethylene glycol esters are important plasticizers for poly(vinyl butyral) resins, nitrocellulose lacquers, vinyl and poly(vinyl chloride) resins, poly(vinyl acetate) (see Vinyl polymers), and synthetic rubber compounds and cellulose esters (1). The fatty acid derivatives of triethylene glycol are used as emulsifiers, demulsifiers, and lubricants. Polyesters derived from triethylene glycol are useful as low pressure laminates for glass fibers, asbestos, cloth, or paper. Triethylene glycol is used in the manufacture of alkyd resins (qv) used as laminating agents and in adhesives.

Tetraethylene Glycol

Tetraethylene glycol has properties similar to diethylene and triethylene glycols, and may be used preferentially in applications requiring a higher boiling point and

Table 10. Physical Properties of Triethylene Glycol Derivatives[a]

Derivative	CAS Registry No.	Formula	Mol wt	Sp gr 20/20°C	Bp at 101.3 kPa[b], °C	Vapor pressure at 20°C, Pa[c]	Mp, °C	Viscosity at 20°C, mPa·s (= cP)	Refractive index, n_D^{20}	Heat of vaporization at 101.3 kPa[b], kJ/mol[e]	Flash point, °C	
											Open cup	Closed cup
Monoethers												
methyl	[*112-35-6*]	$CH_3O(C_2H_4O)_3H$	164.20	1.0494	249.0	<130	−44	7.27	1.4381	53.81	118.3	114.4
ethyl	[*112-50-5*]	$C_2H_5O(C_2H_4O)_3H$	178.22	1.0208	255.9	<13	−18.7	4.5	1.4376	46.99	135.0	123.9
n-butyl	[*143-22-6*]	$C_4H_9O(C_2H_4O)_3H$	206.28	1.0021	dec		−47.6	10.9	1.4394	84.35	143.3	>121.1
Diethers												
dimethyl	[*112-49-2*]	$CH_3O(C_2H_4O)_3CH_3$	178.22	0.9862	216	120	−45	3.8	1.4233	59.83	111	
Diesters												
diacetate	[*111-21-7*]	$CH_3CO(OC_2H_4)_3OOCCH_3$	234.24	1.1185	290.8	<1.3	−61[d]	11.66	1.4384	57.24	173.9	

[a] Refs. 19–22.
[b] To convert kPa to mm Hg, multiply by 7.5.
[c] To convert Pa to mm Hg, divide by 133.3.
[d] Pour point.
[e] To convert J to cal, divide by 4.184.

Table 11. Toxicological Properties of Triethylene Glycol Derivatives[a]

Derivative	Single oral LD_{50} rats[b], g/kg	Single skin penetration LD_{50} rabbits[c], mL/kg	Single inhalation[d] concentrated vapors, rats	Primary skin irritation[e], rabbits	Eye injury, rabbits
Monoethers					
methyl	11.86	7.1	8 h killed 0 of 6	trace	none
ethyl	10.6	8		none	trace
n-butyl	6.74	3.54	8 h killed 0 of 6	trace	moderate
Diethers					
dimethyl	2.5–5			none	none
Diesters					
diacetate	25.28	16	8 h killed 0 of 6	none	trace

[a] Refs. 19, 21–22.
[b] The term LD_{50} refers to that quantity of chemical that kills 50% of dosed animals within 14 d.
[c] Single skin penetration refers to a 24-h covered skin contact with the liquid chemical.
[d] Single inhalation refers to the continuous breathing of certain concentrations of chemicals for the stated period.
[e] Primary irritation refers to the skin response 24 h following application of 0.01 mL amounts to uncovered skin.
[f] Eye injury refers to surface damage produced by the liquid chemical.

lower hygroscopicity (1). Physical properties are listed in Table 1 and toxicological properties in Table 2. Tetraethylene glycol is a coproduct of ethylene glycol produced by ethylene oxide hydrolysis. Like triethylene glycol, tetraethylene glycol is also produced commercially by the direct reaction of ethylene oxide with the lower glycols. The price for tankcar quantities of tetraethylene glycol was $1.01/kg in early 1979.

Tetraethylene glycol is used to separate aromatic from nonaromatic hydrocarbons by selective extraction. The critical solution temperature of a binary system consisting of a given alkyl-substituted aromatic hydrocarbon and tetraethylene glycol is lower than the critical solution temperatures of the same hydrocarbon with the lower polyglycols. Therefore, at a given temperature, tetraethylene glycol tends to extract the higher molecular weight alkylbenzenes more efficiently. Other uses are similar to those of triethylene glycol.

Propylene Glycol

1,2-Propylene glycol is a clear, viscous, colorless liquid that is practically odorless and has a slight characteristic taste. Although more volatile than ethylene glycol, propylene glycol is about three times as viscous at room temperature. It has a very low order of toxicity and is highly hygroscopic. Physical properties of high-purity material are listed in Table 12.

Propylene glycol from propylene oxide was discovered in 1859 by Wurtz (30). It was of little commercial importance, however, until 1931 when it was first produced by the hydrolysis of propylene oxide (qv). Today the hydrolysis is carried out under pressure and at high temperature without catalyst (eq. 19):

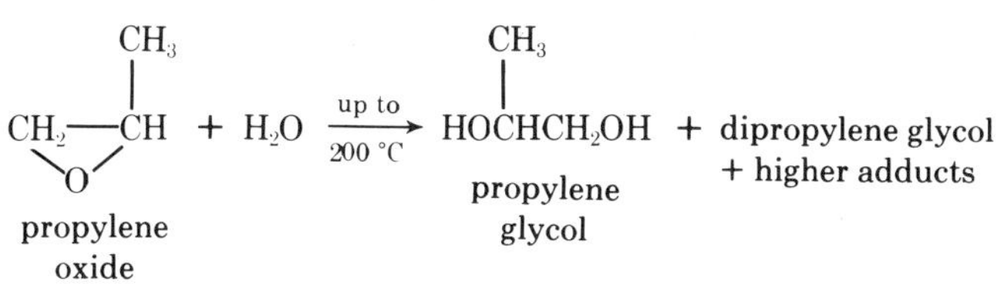

(19)

Table 12. Physical Constants of Propylene Glycol[a]

Property	Value
mp, °C	−60[b]
bp at 101.3 kPa[c], °C	187.3
density at 20°C, g/cm^3	1.0362
surface tension at 20°C, mN/m (= dyn/cm)	35.6
refractive index, n_D^{20}	1.4326
vapor pressure at 20°C, kPa[c]	<1.3
specific heat, J/(g·°C)[d]	
as liquid, 20°C	2.481
as ideal gas, 25°C	1.611
heat of vaporization at 101.3 kPa[c], kJ/mol[d]	52.296
heat of combustion at 25°C, kJ/mol[d]	1824.0
critical constants	
pressure, kPa[c]	6100.27
temperature, °C	352
volume, L/mol	0.237
compression factor, Z_c	0.278
viscosity, mPa·s (= cP)	
at 0°C	255.4
at 20°C	60.5
at 40°C	19.45

[a] Ref. 29.
[b] Sets to glass.
[c] To convert kPa to mm Hg, multiply by 7.5.
[d] To convert J to cal, divide by 4.184.

The proportion of products is controlled by the mole ratio of water to propylene oxide (see below for isomer distribution in dipropylene glycol). Higher hydrolysis ratios increase the yield of propylene glycol but also the costs of purification. A ratio of 15 provides a product mix of 85% propylene glycol, 13% dipropylene glycol, and 1.5% higher adducts.

Although propylene glycol has a secondary hydroxyl group, its chemistry parallels

Table 13. United States Producers of Propylene Glycols

Producer	Location	Annual capacity, thousand metric tons: Propylene glycol[a]	Dipropylene glycol[b]
Dow Chemical U.S.A.	Freeport, Tex.	113.4	11.8
	Plaquemine, La.	68	5
Olin Corporation	Brandenburg, Ky.	20.4	1.8
Oxirane Corporation	Bayport, Tex.	113.4	8.2
Jefferson Chemical Company, Inc.[c]	Port Neches, Tex.	18.1	3.2
Union Carbide Corporation	Institute and South Charleston, W. Va.	45.4	3.6
Total		*378.7*	*33.6*

[a] Ref. 14.
[b] Ref. 15.
[c] A wholly owned subsidiary of Texaco, Inc.

Table 14. Propylene Glycol Uses[a]

Use	Percent
polyester resins	45
pet food	12
tobacco humectant	7
cellophane	7
food and pharmaceuticals	11
miscellaneous	6
exports	12
Total	*100*

[a] Ref. 14.

that of ethylene glycol. Propylene glycol is produced in the United States by five companies having a combined annual nameplate capacity of about 388×10^3 t (see Table 13).

Uses. Propylene glycol has a variety of applications (see Table 14). In the food industry it is used as a solvent, humectant, and preservative, in the manufacture of products that come in contact with food such as plasticizers for food wraps, as a solvent for food processing, and as lubricant for food machinery. It is a softening agent, spreader, emollient, intermediate, drug vehicle, and preservative in the preparation of cosmetics and pharmaceuticals.

Aqueous solutions are effective antifreeze mixtures (see Fig. 1) and are preferred in refrigeration (qv) units in breweries, dairies and packing houses, where a coolant or heat-transfer solution of low toxicity is important (see Antifreezes). These aqueous solutions are inhibited to prevent rust and corrosion.

Derivatives. The derivatives of propylene glycol are prepared by methods analogous to those for ethylene glycol derivatives. The base-catalyzed reaction of propylene oxide with alcohols gives predominantly primary monoalkyl ether, $CH_3CHOHCH_2OR$, and small amounts of secondary ether, $CH_3CHORCH_2OH$. Acid catalysis increases the ratio of secondary to primary ethers. Monoalkylethers can also be prepared without catalyst under the proper conditions of temperature and pressure.

The monoalkyl ethers of propylene glycol are excellent solvents for a wide variety of organic materials. The lower alkyl ethers are completely miscible with water at room temperature and are soluble in some hydrocarbons. Properties for a number of these materials are listed in Table 15.

The ethers and esters of propylene glycol are prepared from the glycol by conventional methods. Propylene glycol yields isomeric mixtures of esters because it contains both primary and secondary hydroxyl groups. Monoesterification occurs usually at the primary hydroxyl group. Fatty acid esters, such as the dioleate and the monohydroxystearate, are used in ointments, drug creams, cosmetics (qv), and surfactants (qv).

Dipropylene Glycol

Dipropylene glycol is a coproduct of propylene glycol in the hydrolysis of propylene oxide (eq. 19); the approximate isomer distribution is:

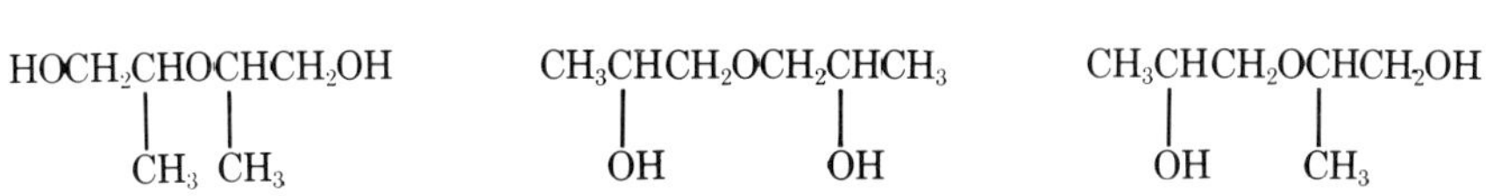

4% [*108-61-2*] 43% [*110-98-5*] 53% [*106-62-7*]

Table 15. Physical Properties of Ether Derivatives of Propylene Glycols[a]

Derivative	CAS Registry No.	Formula	Mol wt	Sp gr 20/20°C	Bp at 101.3 kPa[b], °C	Vapor pressure at 20°C, Pa[c]	Mp[d], °C	Solubility, at 20°C, wt %: In water	Solubility, at 20°C, wt %: Water in	Viscosity at 20°C, mPa·s (= cP)	Refractive index, n_D^{20}	Heat of vaporization at 101.3 kPa[b], kJ/mol[e]	Flash point, °C: Open cup	Flash point, °C: Closed cup
Monoethers of propylene glycol														
methyl	*[107-98-2]*	$CH_3OCH_2CH(CH_3)OH$	90.12	0.9234	121	1070	−95	miscible		1.9	1.4036	32.64	36	33
ethyl	*[19089-47-5]*	$C_2H_5OCH_2CH(CH_3)OH$	104.15	0.8979	132.8	530	−100	miscible		2.2	1.4066	29.29		
n-propyl	*[1569-01-3]*	$C_3H_7OCH_2CH(CH_3)OH$	116.17	0.8865	149.8	230	−80	miscible		2.8	1.4121		53	48
n-butyl	*[5131-66-8]*	$C_4H_9OCH_2CH(CH_3)OH$	132.2	0.8843	170.1	<130	−100	6.4	19.2	3.4	1.4174	25.10	68	59
Monoethers of dipropylene glycol[f]														
methyl		$C_7H_{16}O_3$	148.2	0.9608	187.2	<13	−80	miscible		3.7	1.4220	45.19	82	79
ethyl		$C_8H_{18}O_3$	162.1	0.930[g]	197.8	40		miscible		1.1[h]	1.4212		91	
n-butyl		$C_{10}H_{22}O_3$	190.2	0.913[g]	231	8		4.3		1.3[h]	1.4280		113	

[a] Refs. 31–33.
[b] To convert kPa to mm Hg, multiply by 7.5.
[c] To convert Pa to mm Hg, divide by 133.3.
[d] Set to glass below this temperature.
[e] To convert J to cal, divide by 4.184.
[f] Commercial product, isomeric mixtures.
[g] 25/25°C.
[h] At 25°C.

Alternatively, dipropylene glycol can be obtained by the addition of propylene oxide to propylene glycol.

The chemistry of dipropylene glycol is similar to that of propylene glycol, as are its toxicological properties (Table 2). Physical properties of dipropylene glycol are listed in Table 1.

The distribution of dipropylene glycol uses in polyester resin manufacture is: polyester resins, 60%; plasticizers, 30%; and alkyd resins, hydrocarbon extraction, urethane polyols, and others, 10%.

BIBLIOGRAPHY

"Glycols" in *ECT* 1st ed., Vol. 7, pp. 238–263, by C. P. McClelland and P. R. Rector, Carbide and Carbon Chemicals Company, a Division of Union Carbide and Carbon Corporation; "Ethylene Glycol, Propylene Glycol, and Their Derivatives" under "Glycols" in *ECT* 2nd ed., Vol. 10, pp. 638–660, by Phil H. Miller, Union Carbide Corporation.

1. *Glycols, Brochure F-41515B,* Union Carbide Corporation, New York, 1978.
2. G. O. Curme and F. Johnston, *Glycols, ACS Monograph No. 114,* Reinhold Publishing Corp., New York, 1952, Chapt. 2.
3. *Ibid.*, Chapt. 3.
4. *Ibid.*, Chapt. 1.
5. U.S. Pat. 3,970,711 (July 20, 1976), C. R. Reiche and J. A. Heckman (to PPG Industries, Inc.).
6. U.S. Pats. 2,152,852 (Apr. 4, 1939), D. J. Loder; 2,607,805 (Oct. 19, 1952), W. F. Gresham; 3,859,349 (Jan. 7, 1975), N. F. Cody (all to E. I du Pont de Nemours & Co., Inc.).
7. U.S. Pats. 3,647,892 (Mar. 7, 1972), R. Hoch; 3,668,239 (June 6, 1972), J. Kollar; 3,689,535 (Sept. 5, 1972), J. Kollar; 3,715,388 (Feb. 6, 1973), J. R. Valbert; 3,872,164 (Mar. 18, 1975), J. P. Schmidt (all to Halcon International, Inc.).
8. G. E. Weismantel, *Chem. Eng.* **86**(2), 67 (1979).
9. Jpn. Kokai 78 25,506 (Mar. 9, 1978), M. Suzuki and co-workers (to Teijin, Ltd.).
10. U.S. Pats. 3,833,634 (Sept. 3, 1974), R. L. Pruett and W. E. Walker; 3,944,588 (Mar. 16, 1976), L. Kaplan; 4,133,776 (Feb. 16, 1979), R. L. Pruett and W. E. Walker (all to Union Carbide Corporation).
11. *Oil, Paint Drug Rep., Chemical Pricing Patterns,* 1952–1967 (1968).
12. *Ibid.*, July Issues (1968–1971).
13. *Chem. Mark. Rep.*, July Issues (1972–1978).
14. *Synthetic Organic Chemicals, U.S. Production and Sales,* U.S. International Trade Commission, production data for 1950–1977.
15. *Chem. Mark. Rep., Chemical Profiles* (July 31, 1978).
16. S. C. Johnson, *Chem. Eng. Prog.* **72**(9), 25 (1976).
17. J. N. Cawse, J. P. Henry, M. W. Swartzlander, and P. H. Wadia, "Ethylene Oxide" in *Kirk-Othmer Encyclopedia of Chemical Technology,* 3rd ed., Vol. 9, John Wiley & Sons, Inc., New York, 1980, pp. 432–471.
18. *Chem. Eng. News* **57,** 6 (Dec. 3, 1979).
19. *Glycol Ethers, Brochure F-41524C,* Union Carbide Corporation, New York, 1976.
20. *Methyl CELLOSOLVE® Acrylate, Brochure F-43820,* Union Carbide Corporation, New York, 1972; *CELLOSOLVE® Acrylate, Brochure F-43819,* Union Carbide Corporation, New York, 1972; *N-Ethyl Morpholine, Brochure F-42132,* Union Carbide Corporation, New York, 1968; *Phenyl Glycol Ether, Brochure F-42148A,* Union Carbide Corporation, New York, 1975.
21. *Esters, Brochure F-42340A,* Union Carbide Corporation, New York, 1971.
22. *Ansul® Glycol Diethers, Form No. C-78274,* The Ansul Company, Marinette, Wisc., 1979.
23. *General Chemistry of Glyoxal, Brochure F-41296A,* Union Carbide Corporation, New York, 1967.
24. *Glyoxal, 40%, for the Paper Industry, Brochure F-42412,* Union Carbide Corporation, New York, 1969.
25. Ref. 2, Chapt. 7.
26. *KROMFAX® Thiodiglycol, Brochure F-42524,* Union Carbide Corporation, New York, 1970.
27. *N-Methyl Morpholine, Brochure F-42144A,* Union Carbide Corporation, New York, 1975.
28. Brit. Pat. 1,106,084 (Mar. 13, 1968), L. L. Duffy (to ICI, Ltd.).

29. *Propylene Glycol, Brochure F-41179C,* Union Carbide Corporation, New York, 1978.
30. Ref. 2, Chapt. 9.
31. *Solvents, Brochure F-7465S,* Union Carbide Corporation, New York, 1976.
32. Ref. 2, Chapt. 13.
33. *"UCAR® Solvents LM and UCAR® Solvents 2LM," Brochure F-40761,* Union Carbide Corp., New York, April, 1961.

E. S. BROWN
C. F. HAUSER
B. C. REAM
R. V. BERTHOLD
Union Carbide Corporation

1,3-BUTYLENE GLYCOL

1,3-Butylene glycol, butane-1,3-diol [*107-88-0*], $CH_3CHOHCH_2CH_2OH$, is a colorless, mildly bittersweet liquid. Some of its esters occur in nature. It is soluble in water, lower alcohols, ketones, and esters, and it is not soluble in the aliphatic hydrocarbons or most of the common chlorinated solvents. Nylon is quite soluble in hot 1,3-butylene glycol; shellac and rosin are partially soluble in it. It has a low toxicity and is nonirritating to the normal skin. When rubbed on, it penetrates and gives a marked warming sensation. First prepared in 1872 by Kekulé (1) as a by-product of acetaldehyde reduction with sodium amalgam, it is now manufactured by catalytic hydrogenation of acetaldol.

$$2\,CH_3CHO \xrightarrow{OH^-} CH_3CHOHCH_2CHO \xrightarrow[H_2]{Ni} CH_3CHOHCH_2CH_2OH$$

It was made on a small scale by condensation of propylene and formaldehyde, hydrolysis of 2-methyloxetane, and hydrogenation of 4-hydroxy-2-butanone.

Physical Properties

1,3-Butanediol can be distilled without decomposition at atmospheric pressure. It is customarily distilled under reduced pressure to avoid accidental oxidation in condensers when the hot glycol vapors encounter air. Its vapor pressure is given in Table 1. Some of the common physical properties are listed in Table 2. 1,3-Butanediol is quite hygroscopic. At 50% rh, it takes up to 12.5% of its own weight in moisture, and at 80% rh as much as 38.5% of its own weight.

The flavor of butane-1,3-diol is distinctly and characteristically bittersweet (see Flavors and spices). Samples of 1,3-butanediol prepared from the hydrogenation of diketene, by reduction of acetoacetic ester, or purified in various ways from commercial sources all have the same characteristic taste. Several of the impurities frequently

Table 1. Vapor Pressure of 1,3-Butanediol[a]

Temperature, °C	Pressure, kPa[b]
92.6	0.6
104.8	1.3
117.8	2.7
132.0	5.3
152.5	13.3
170.0	26.7
188.2	53.3
207.0	101.3
229.2	202.6

[a] Refs. 2–4.
[b] To convert kPa to mm Hg, multiply by 7.5.

Table 2. Physical Properties of 1,3-Butanediol

Property	Value	Refs.
bp, °C	207.5	2–4
mp, °C	−77	3
density, g/cm^3		4–6
at 20°C	1.0043	
at 40°C	0.9909	
at 60°C	0.9769	
viscosity, kPa·s (= 10^4 P)		2
at 25°C	103.9	
at 99°C	3.77	
at 158.8°C	1.02	
refractive index, n_D^{20}	1.4410	7
surface tension, Pa[a]		8
at 25°C	378	
at 40°C	331	
at 80°C	284	
at 200°C	177	
heat of vaporization, J/g[b] at bp	585	9
heat of combustion, kJ/mol[b] at constant volume	−2488	10
heat of combustion, kJ/mol[b] at constant pressure	−2491	10
heat of formation, liquid from gases, kJ/mol[b]	515	10
heat capacity of liquid, J/(kg·K)[b] at 20°C	2301.6	[c]
thermal conductivity, W/(m·K) at 20°C	0.1842	[d]

[a] To convert Pa to mm Hg, divide by 133.3.
[b] To convert J to cal, divide by 4.184.
[c] Data estimated by the Sakiakis-Coates method.
[d] Data estimated by the Robbins-Kingrea method.

present in commercial 1,3-butanediol have a similar bitter quality but the flavor is apparently an inherent property of the molecule.

Chemical Properties

1,3-Butylene glycol is esterified when treated with an organic or inorganic acid in the presence of a suitable catalyst. The primary hydroxyl reacts much more extensively and readily than the hydroxyl in the 3-position, both under acid and alkaline catalysis (11). Dibasic acids usually form linear polyesters. Oxalic acid, succinic acid, and glutaric acid furnish polyesters even in the absence of catalysts (see Polyesters). Under certain conditions, dibasic acids or their anhydrides tend to promote dehydration to butadiene and 3-buten-1-ol, which are troublesome impurities in the polyesters. The average molecular weight of polyesters can be controlled by adding small amounts of terminating agent, eg, monocarboxylic acids or monohydric alcohols, to the initial reaction mixture. Cyclic esters have also been prepared.

Inorganic acids or anhydrides yield esters with 1,3-butylene glycol. Boric oxide, B_2O_3, heated with 1,3-butylene glycol in xylene, gives complex esters with two or three molecules of the diol combined with the oxide. Borate ion in water solutions complex with 1,3-butylene glycol to furnish a material that can be extracted easily from solution with an organic solvent, eg, chloroform. Ethylene glycol, propylene glycol, glycerol, or pentaerythritol do not facilitate borate extraction in this way. The structure of the anionic species extracted has been studied by nmr spectroscopy in acetone (12) and Raman spectra in water (13).

Acid dehydration of 1,3-butylene glycol gives a mixture including ethylene, butanone, butyraldehyde, 3-buten-1-ol, butadiene, and sometimes poly(1,3-butylene glycol) [*55251-78-0*]. Solid catalysts with high selectivity for butadiene have been developed, eg, calcium phosphate and aluminum phosphate, but the processes are not currently competitive with other routes to butadiene (qv). Poly(1,3-butylene glycol) is obtained by suspending 1,3-butylene glycol in toluene in the presence of a catalytic amount of sulfuric acid, entraining the water of reaction by refluxing the mixture, and continuously lowering the acid concentration by adding powdered $NaHCO_3$ (14). The water separated from the reflux contains unchanged diol, 3-buten-1-ol, traces of butadiene, and other dehydration and dehydrogenation products (see Polyethers).

Cyclic acetals can be formed easily by direct reaction of the aldehyde in the presence of acids. Water of reaction is entrained, eg, with toluene. Ion-exchange resins can be employed in place of mineral acids as the catalyst (15). 4-Methyl-1,3-dioxane can be prepared from formaldehyde and 1,3-butylene glycol in the presence of H_2SO_4 (16). Salts catalyze the condensation as well, thus palladium and copper salts give high yields of 4-methyl-1,3-dioxane (17). Homologous derivatives of higher aldehydes, ketones, etc have been prepared and their physical properties studied (15).

Manufacture

Acetaldol is prepared by acetaldehyde self-condensation in aqueous alkaline solutions. Figure 1 illustrates a flow sheet for its production. Acetaldehyde conversion cannot be carried higher than 66.6% because aldol and acetaldehyde combine to yield 2,6-dimethyl-1,3-dioxan-4-ol. The bound acetaldehyde is held so tenaciously that most commercial aldol is essentially equivalent to 2,6-dimethyl-1,3-dioxan-4-ol:

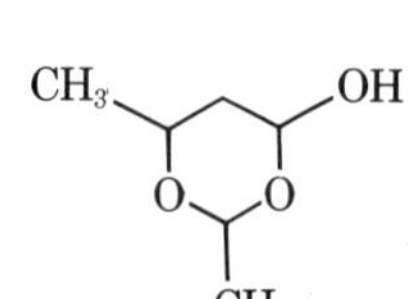

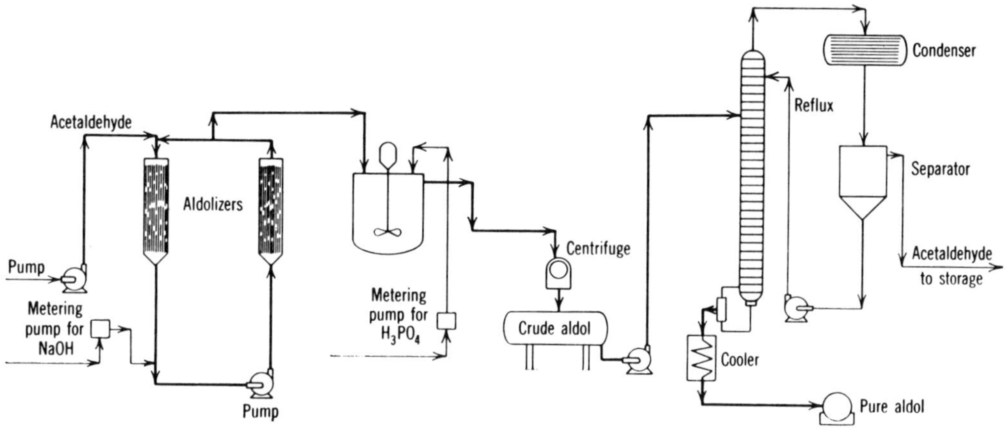

Figure 1. Acetaldol manufacture.

Higher condensation products are often present (18). It is difficult to decompose 2,6-dimethyl-1,3-dioxan-4-ol into aldol without simultaneously dehydrating it to crotonaldehyde (qv). Steam stripping is one of the more practical ways of removing the bound acetaldehyde (19). Aldol is led directly to a hydrogenation system where it is reduced in the presence of Raney nickel or other catalysts (19–21). Any of the acetaldehyde or 2,6-dimethyl-1,3-dioxan-4-ol left is reduced to ethanol and 2,4-dimethyl-1,3-dioxane, both of which are efficiency losses for the process. Low-boilers are stripped from the hydrogenated product and pure 1,3-butylene glycol is obtained by distillation under reduced pressure. Figure 2 shows a typical hydrogenation system.

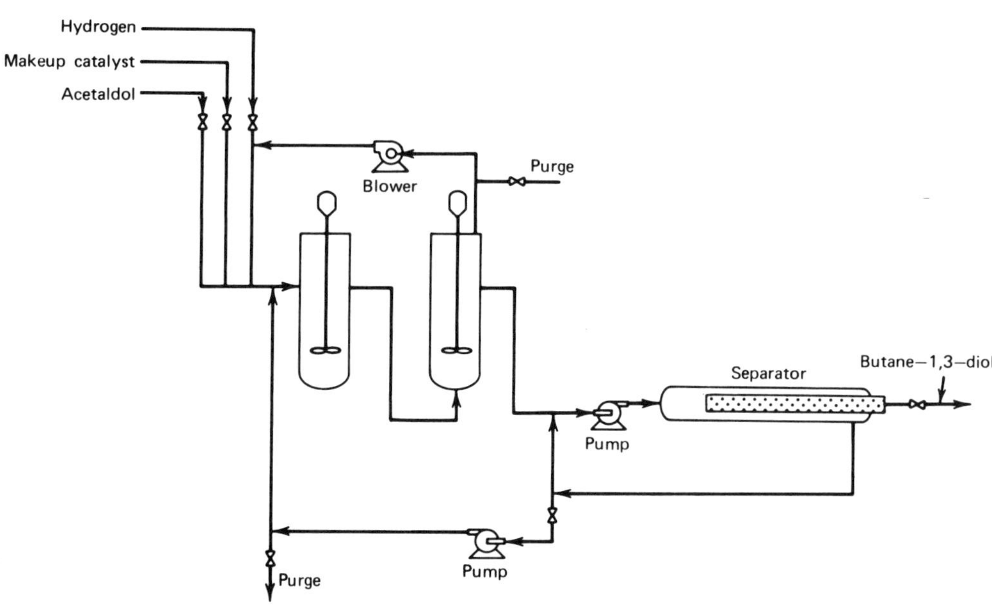

Figure 2. Hydrogenation of acetaldol to yield butane-1,3-diol.

The high boiling residues from distillation consist chiefly of ethers, dimers and trimers of 1,3-butylene glycol generated by loss of water during hydrogenation.

The Prins reaction of formaldehyde with propylene may have economic advantages over the aldol route but it does not seem to have been used commercially. 1,3-Butylene glycol can be recovered as a by-product of ethyl acetate manufacture through the Tishchenko reaction of acetaldehyde (22) (see also Esters). As much as 6% of the acetaldehyde fed to the reaction is converted into 1,3-butylene glycol diacetate. Prolonged heating of aldol is known to furnish the β-hydroxybutyric acid ester of 1,3-butylene glycol, even in the absence of catalysts (23).

Purification to remove odor and taste-bearing substances is a troublesome problem. Crude diol often contains acetaldehyde, butyraldehyde, crotonaldehyde, various oligomers of the aldehydes, as well as several acetals (24). The taste and odor of these compounds are remarkably persistent. Even in concentrations of only 0.1 ppm, some of them can be detected. Steam stripping or vacuum distillation in stainless steel under carefully regulated conditions can largely free the glycol from them (25).

Recent patents (26) disclose that 1,3-butylene glycol or its esters, especially 1,3-butylene glycol dipropionate, are useful in bacteriostatic soap and shaving formulations and as antimicrobial additives for cheeses and meats. Curiously, 1,3-butylene glycol stimulates the growth of certain species of bacteria (27) and yeast (28), and it increases the yield of ergot alkaloids from *Claviceps paspali* (29). The mechanism for its antibacterial and antimycotic action has not been explained (see Disinfectants).

Economic Aspects

Celanese Chemical Company is the sole United States manufacturer of 1,3-butanediol. Daicel and Kyowa Hakko produce it in Japan, and Chemische Werke Hüls manufactures it in the Federal Republic of Germany. Some 1,3-butanediol appears to be produced in the German Democratic Republic and in the Soviet Union. There seems to be no information on the quantities produced worldwide. The 1979 price in the United States was ca \$0.95/kg.

Health and Safety Factors

The butylene glycol reported among the products of various yeast fermentations by numerous authors (30) is almost certainly the rather toxic 2,3-butanediol. There is incontestible evidence, however, for the occurence of 1,3-butanediol esters in the neutral lipids of corn oil, sunflower seed oil, sheep suet, cod-liver oil, rat liver, and certain soil yeast fats (31). These diol lipids may be significantly involved in the control of cell-membrane permeability (32).

Toxicity of butane-1,3-diol to higher animals is low. Single-dose LD_{50} for rats is 22.8–29.5 g/kg and for mice 23.4 g/kg (33). The toxicity is about the same as glycerol (qv) and propylene glycol, and is in strong contrast to toxicity of ethylene glycol and butane-1,4-diol. Supplying 10% of rats' dietary calories by butane-1,3-diol for 2 yr did not cause ill effects (34), but at the 20% level, body-weight gain was slowed. The unpurified diol appears to reduce the appetite of test animals so that less feed is consumed and lower body weights than controls result. Application to the skin or mucous membranes causes no significant irritation (35).

Work on the metabolism of butane-1,3-diol has shown that replacement of part of the dietary carbohydrate by butane-1,3-diol results in longer survival of diabetic animals (36). Presumably, the diol or some of its metabolic products stimulates the secretion of insulin (qv). Metabolism of butane-1,3-diol does not require insulin itself. It may be useful to decrease appetite and the deposition of body fat (37).

Uses

Simple esters of 1,3-butylene glycol with monocarboxylic acids have been used as plasticizers (qv) for cellulosics and poly(vinyl chloride) resins. Polyester-type plasticizers are prepared from dicarboxylic acids and are used because of their good compatibility with a wide range of polymers. They have excellent stability toward plasticized material. The plasticizers have little tendency to migrate. Esters that are nearly odorless and colorless can be made in the presence of organotitanium compounds and boric acid, phosphoric acid, or boron phosphate (38). 1,3-Butylene glycol diacetate is a plasticizer for cellulose acetate tow used in cigarette filters (see Cellulose acetate and triacetate fibers) (39). The film-forming properties of poly(vinyl acetate) emulsions is improved by adding small amounts of this diacetate (see Vinyl polymers) (40).

1,3-Butylene glycol is used as a terminal group in oil-free alkyd resins (qv). It is used also for deicing aircraft (see Antifreezes and deicing fluids).

BIBLIOGRAPHY

"Glycols" in *ECT* 1st ed., Vol. 7, pp. 238–263, by C. P. McClelland and P. R. Rector, Carbide and Carbon Chemicals Company, a Division of Union Carbide and Carbon Corporation; "1,3-Butylene Glycol" under "Glycol" in *ECT* 2nd ed., Vol. 10, pp. 660–667, by Frank S. Wagner, Jr., Celanese Chemical Company.

1. A. Kekulé, *Liebigs Ann. Chem.* **162,** 309 (1872).
2. L. H. Thomas and R. Meatyard, *J. Chem. Soc.,* 92 (1966).
3. D. R. Stull, *Ind. Eng. Chem.* **39,** 517 (1947).
4. O. J. Schierholz and M. L. Staples, *J. Am. Chem. Soc.* **57,** 2709 (1935).
5. K. A. Clendenning, F. J. MacDonald, and D. E. Wright, *Can J. Res.* **28B,** 608 (1950).
6. E. Benk, *Seifen Oele Fette Wachse* **79,** 108 (1953).
7. M. Gordon, J. G. Miller, and A. R. Day, *J. Am. Chem. Soc.* **71,** 1245 (1949).
8. K. Nakanishi and co-workers, *J. Chem. Eng. Data* **16,** 44 (1971).
9. A. Bondi and D. J. Simkin, *AIChE J.* **3,** 473 (1957).
10. H. Moureu and M. Dodé, *Bull. Soc. Chem. Fr.* **4,** 637 (1937).
11. M. Decouzon and M. Naudet, *Bull. Soc. Chim. Fr.,* 3541 (1966).
12. A. Kamars, *Tezisy Dokl. Konf. Molodykh Nauchno Rab. Inst. Neorg. Khim. Akad. Latv. SSR* **4,** 10 (1975).
13. R. P. Oertel, *Inorg. Chem.* **11,** 544 (1972).
14. West Ger. Pat. 2,414,647 (Jan. 9, 1975), B. Meyer and co-workers (to VEB Syntheswerk Schwarzheide).
15. I. Ap'ok and co-workers, *Izv. Akad. Nauk SSSR Ser. Khim.,* 2345 (1958).
16. East Ger. Pat. 71,318 (Feb. 20, 1970), H. U. Schimpfle.
17. Jpn. Kokai 75 58,081 (May 20, 1975), M. Ogura and Y. Toriuchi (to Sumitomo Chemical Co.).
18. P. S. Shantarovich and co-workers, *Zh. Obshch. Khim.* **37,** 1478 (1967); *Chem. Abstr.* **67,** 100081e (1967).
19. Brit. Pat. 853,266 (Nov. 2, 1960), (to Celanese Corporation).
20. C. K. Hancock and D. D. Henson, *Ind. Eng. Chem.* **45,** 629 (1953).
21. C. K. Hancock, *Ind. Eng. Chem.* **44,** 1003 (1952).
22. J. B. Cloke and R. M. Wolff, *J. Am. Chem. Soc.* **65,** 986 (1943).
23. West Ger. Pat. 2,500,312 (July 8, 1976), F. Merger and co-workers (to BASF A.G.); Fr. Pat. 2,297,203 (Oct. 9, 1976), F. Merger and co-workers (to BASF A.G.).
24. Brit. Pat. 796,575 (June 11, 1958), H. S. Johnson (to Shawinigan Chemicals Ltd.).

25. West Ger. Pat. 1,816,833 (Aug. 7, 1969), E. A. Peveto and J. M. Ramey (to Celanese Corp.); U.S. Pat. 3,489,655 (Jan. 13, 1970), E. A. Peveto and J. M. Ramey (to Celanese Corp.).
26. U.S. Pats. 3,846,326 (Nov. 5, 1974), D. L. Wright and J. W. Frankenfeld (to Exxon Research and Engineering Co.); 3,806,615 (Apr. 23, 1974), J. W. Frankenfeld and co-workers (to Exxon Research and Engineering Co.).
27. R. Patsch and C. Hoehe, *Zentralbl. Bakteriol. Parasitenkd. Infektionskr. Hyg. Abt. 1 Orig.* **211**(1), 88 (1969).
28. K. Enomoto, H. Ueyama, and T. Fukimbara, *Hakko Kogaku Zasshi* **53**(9), 637 (1975).
29. A. Mizrahi and G. Miller, *Biotechnol. Bioeng.* **10**(1), 102 (1968).
30. C. Antoniani, *Ann. Chim. Appl.* **38,** 39 (1948); C. Antoniani, L. Paesi, and L. Federico, *Rend. Ist. Lomb. Accad. Sci. Pt. I Classe Sci. Mat. Nat.* **87,** 22 (1954); E. Peynaud, M. Lafon, and J. Ribereau-Gayon, *Congr. Int. Bot. Rapp. Commun.* **8,** 116 (1954).
31. L. D. Bergelson and co-workers, *Biochim. Biophys. Acta* **116,** 511 (1966); V. A. Vaver, A. N. Ushakov, and L. D. Bergelson, *Izv. Akad. Nauk SSSR Ser. Khim.,* 1645, 2415 (1967).
32. L. D. Bergelson and V. A. Vaver, *Zh. Evol. Biokhim. Fiziol.* **2,** 118 (1966).
33. G. Bornmann, *Arzneim. Forsch.* **4,** 643, 710 (1954); **5,** 38 (1955).
34. R. A. Scala and O. E. Paynter, *Toxicol. Appl. Pharmacol.* **10,** 160 (1967).
35. L. Fischer and co-workers, *Z. Gesamte Exp. Med.* **115,** 22 (1949); E. Husing, R. Kopf, and A. Loeser, *Fette Seifen* **52,** 45 (1950); A. Loeser, *Pharmazie* **4,** 263 (1949).
36. M. A. Mehlman, R. B. Tobin, and J. B. Jonston, *Metabolism* **20**(2), 149 (1971).
37. R. B. Tobin and co-workers, *Fed. Proc. Fed. Am. Soc. Exp. Biol.* **34,** 2171 (1975); H. A. Dymsza, *Fed. Proc. Fed. Am. Soc. Exp. Biol.* **34,** 2167 (1975); M. A. Mehlman, *Fed. Proc. Fed. Am. Soc. Exp. Biol.* **34,** 2166 (1975).
38. Jpn. Kokai 74 11,815 (Feb. 1, 1974), K. Wakatsuki, T. Inoue, and T. Wagazuma (to Adeka Argus Chemical Co. Ltd.).
39. U.S. Pat. 3,882,878 (May 13, 1975), H. Sawada and J. Kawamoto (to Daicel, Ltd.).
40. Jpn. Kokai 75 55,649 (May 15, 1975), S. Kosugi, A. Go, E. Shinohara, and T. Okitsu (to Konishi, Gisuke and Co.).

FRANK S. WAGNER, JR.
Celanese Chemical Company

OTHER GLYCOLS

Glycols such as neopentyl glycol, 2,2,4-trimethyl-1,3-pentanediol, and 1,4-cyclohexanedimethanol are used in the synthesis of polyesters (qv) and urethane foams (see Foamed plastics). Their physical properties are shown in Table 1 (1–7). Further descriptions of each glycol are given below.

Table 1. Physical Properties of Several Glycols

	Neopentyl glycol [*126-30-7*]	2,2,4-Trimethyl-1,3-pentanediol [*144-19-4*]	1,4-Cyclohexanedimethanol [*105-08-5*]
melting point, °C			
cis isomer			43
trans isomer			70
melting range, °C	124–130	46–55	45–50
sublimation temp, °C	128		
boiling point, °C			
at 0.13 kPa[a]			118
at 0.45 kPa[a]	93–94		
at 1.33 kPa[a]			160
at 101.3 kPa (1 atm)	210		286
boiling range, °C		215–235	
cis/trans ratio, wt %			ca 32/68
assay (commercial grade), wt % minimum	97	96	99
density, g/cm^3			
at 20°C			1.150
at 21°C	1.06	0.897	
pour point, supercooled, °C			10
crystallization point, °C			35
viscosity at 50°C, mPa·s (= cP)			675
heat of combustion, kJ/mol[b]	−3100	−5050	−4849[c]
flammability			
fire point, COC[d], °C	135	118	174
flash point, COC[d], °C	129	113	167
autoignition temp[e], °C	399	346	302
heat of fusion[f], kJ/mol[b]	21.77	8.63	
heat of vaporization, kJ/mol[b]			
at 32 kPa[a] 170°C	67.1		
at 101.3 kPa (1 atm)			95.6
at 101.3 kPa, 204°C	56.5		
hygroscopity[g], wt % H_2O			
at 50% rh		0.1–0.2	
at 51% rh	0.3		
at 78% rh	11.3		

[a] To convert kPa to mm Hg, multiply by 7.5.
[b] To convert kJ to kcal, divide by 4.184.
[c] Paar bomb.
[d] Cleveland Open Cup.
[e] ASTM D 286.
[f] Estimated.
[g] At equilibrium; neopentyl glycol at 25–38°C, and 2,2,4-trimethyl-1,3-pentanediol at 25°C.

Neopentyl Glycol

Neopentyl glycol, or 2,2-dimethyl-1,3-propanediol, $HOCH_2C(CH_3)_2CH_2OH$, is a white crystalline solid at room temperature, soluble in water, alcohols, ethers, ketones, and benzene but relatively insoluble in alkanes (1). Two primary hydroxyl groups are provided by the 1,3-diol structure, making this glycol highly reactive as a chemical intermediate. The *gem*-dimethyl configuration is responsible for the exceptional hydrolytic, thermal and uv stability of neopentyl glycol derivatives.

Chemical Properties. Neopentyl glycol can undergo typical glycol reactions such as esterification (qv), etherification, condensation, and oxidation. When basic kinetic studies of the esterification rate ($k \times 10^5$) were carried out for neopentyl glycol, the absolute esterification rate of neopentyl glycol with *n*-butyric acid was approximately 20 times that of ethylene glycol with *n*-butyric acid (8).

Manufacture. Commercial preparation of neopentyl glycol can be via an alkali-catalyzed condensation of isobutyraldehyde with 2 mol of formaldehyde (crossed Cannizzaro reaction) (2,9).

$$(CH_3)_2CHCHO + 2\ CH_2O + KOH \rightarrow \underset{70\%\ \text{yield}}{HOCH_2C(CH_3)_2CH_2OH} + HCO_2K$$

Within the United States, neopentyl glycol is manufactured by the Texas Eastman Company, and by the Badische Division, BASF. Overseas producers include BASF and Mitsubishi. In 1979, the flaked material was listed at \$0.82/kg.

Toxicity. Studies of neopentyl glycol indicate a slight toxicity. The estimated median lethal dose (acute oral LD_{50}) value was determined to be >6400 mg/kg (rats), >3200 mg/kg (guinea pigs), and 3200–6400 mg/kg (mice) (1).

Uses. Neopentyl glycol is used extensively as a chemical intermediate in the manufacture of polyester resins (see Alkyd resins), polyurethane polyols (see Urethane polymers), synthetic lubricants, polymeric plasticizers (qv), and other polymers. It imparts a combination of desirable properties to properly formulated esterification products including low color, good weathering and chemical resistance, and improved thermal and hydrolytic stability.

The weatherability and hydrolytic stability of unsaturated polyesters based on neopentyl glycol have made it a popular intermediate for use in formulations exposed to severe conditions, eg, in gel coats for cultured marble and marine applications (see Coatings, marine; Coatings, resistant) (10).

Reactive saturated polyester resins (oil-free alkyds) based on neopentyl glycol are produced for use in formulating premium-quality surface coatings (11–12). These coatings exhibit excellent water, detergent, and stain resistance, and excellent weatherability and gloss retention. They may be formulated as conventional solvent coatings, as high-solids coatings, or as dry powders for electrostatic coating applications.

The polyurethane industry provides other uses for neopentyl glycol as an intermediate in the manufacture of hydroxy-terminated polyester polyols. Beginning with basically the same ingredients, products with a wide range of properties, varying from soft to rigid foams to elastomers (qv) may be produced from polyols based on neopentyl glycol. This glycol also is employed to improve thermal, hydrolytic, and uv stability (13).

Synthetic lubricants are made with neopentyl glycol in the base-stock polyester

(14). Excellent thermal stability and viscosity control are imparted to special high performance aviation lubricants by the inclusion of polyester thickening agents made from neopentyl glycol (15) (see Lubrication).

Neopentyl glycol is used also to manufacture polymeric plasticizers that exhibit the improved thermal, hydrolytic, and uv stability necessary for use in some exterior applications.

Derivatives. A number of derivatives of neopentyl glycol have been prepared; some show promise for commercial applications.

Organophosphorus Derivatives. Neopentyl glycol treated with pyridine and phosphorus trichloride in anhydrous dioxane yields the cyclic hydrogen phosphite, 5,5-dimethyl-1,3,2-dioxaphosphorinane 2-oxide (**1**) (16). Compounds of this type may be useful as flameproofing plasticizers, stabilizers, synthetic lubricants, oil additives, pesticides, or intermediates for the preparation of other organophosphorus compounds (see Flame retardants; Phosphorus compounds).

Acetals and Ketals. Acetals of 1,3-diols are prepared by refluxing the diol with the aldehyde in the presence of an acid catalyst, even in an aqueous medium. The corresponding ketals are more difficult to prepare in aqueous solution, but cyclic ketals of neopentyl glycol, eg, 2-butyl-2-ethyl-5,5-dimethyl-*m*-dioxane (**2**) can be prepared if the water of reaction is removed azeotropically (17).

Cyclopropane Derivatives. 2,2-Dimethylcyclopropanenitrile (**3**) has been made by preparing the di-*p*-toluenesulfonate of neopentyl glycol and treating the diester with potassium cyanide (18).

Esters. Neopentyl glycol diesters are usually liquids or low-melting solids. Polyesters of neopentyl glycol, and in particular unsaturated polyesters, are prepared by reaction with polybasic acids at atmospheric pressure. High molecular weight linear polyesters (qv) are prepared by the reaction of neopentyl glycol and the ester (usually the methyl ester) of a dibasic acid through transesterification (19–20). The reaction is usually performed at elevated temperatures, *in vacuo,* in the presence of a metallic catalyst.

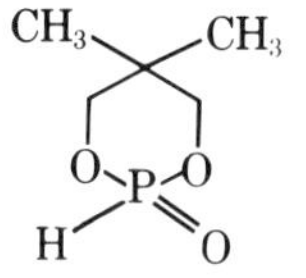

(**1**)

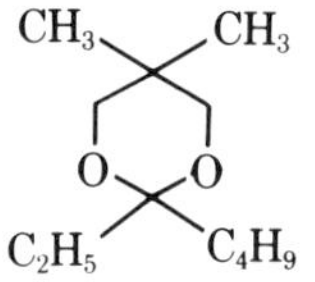

(**2**)

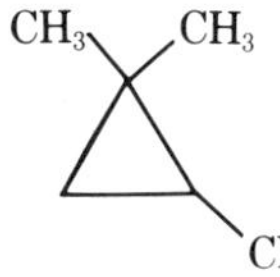

(**3**)

2,2,4-Trimethyl-1,3-pentanediol

2,2,4-Trimethyl-1,3-pentanediol, trimethylpentanediol, $CH_3CH(CH_3)CH$-$(OH)C(CH_3)_2CH_2OH$, is a white, crystalline solid. It is used in surface coating and unsaturated polyester resins. It also appears promising as an intermediate for synthetic lubricants and polyurethane elastomers and foams.

Trimethylpentanediol is soluble in most alcohols, other glycols, aromatic hydrocarbons, and ketones, but it has only negligible solubility in water and aliphatic hydrocarbons (4).

Chemical Properties. Trimethylpentanediol, with a primary and a secondary hydroxyl group, enters into reactions characteristic of other glycols. It reacts readily with various carboxylic acids and diacids to form esters, diesters, and polyesters (21). Some organometallic catalysts have proven satisfactory for these reactions, the most versatile being dibutyltin oxide. Several weak bases such as triethanolamine, potassium acetate, lithium acetate, and borax are effective as stabilizers for the glycol during synthesis (22).

Manufacture and Processing. 2,2,4-Trimethyl-1,3-pentanediol can be produced by hydrogenation of the aldehyde trimer (4) resulting from the aldol condensation of isobutyraldehyde.

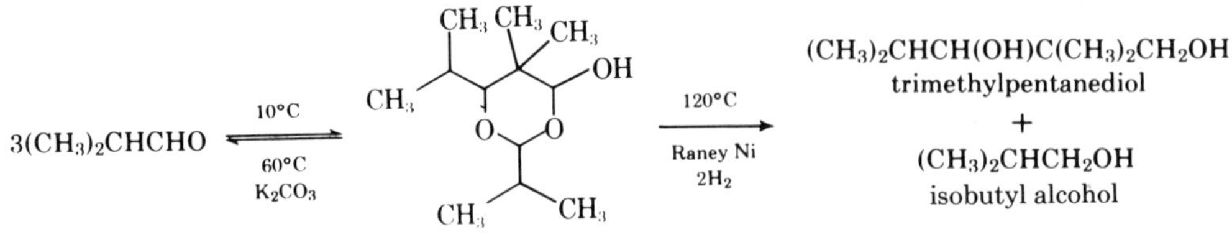

(**4**)

2,6-diisopropyl-5,5-dimethyl-1,3-dioxan-4-ol

Texas Eastman Company is the only U.S. manufacturer of this glycol. Prices in 1979 were \$0.57/kg for bulk quantities.

Toxicity. The LD_{50} of 2,2,4-trimethyl-1,3-pentanediol (acute oral) in rats is 2500 mg/kg; in guinea pigs, 1800 mg/kg; in chickens, 3350 mg/kg; and in mice, 2200 mg/kg (4).

Uses. The versatility of trimethylpentanediol as an intermediate is reflected by the diversity of its commercial applications.

Unsaturated polyesters derived from trimethylpentanediol are characterized by a low-exotherm, low-shrink curing, and a product that has good electrical properties and excellent hydrolytic stability and chemical resistance (21–23). These unsaturated resins exhibit low viscosity, low density, and good glass-fiber or filler wetting (24–25). They are resistant to hydrolytic action, probably because of the protection of the ester linkage by the pendant methyl groups, and by virtue of the lesser number of ester groups in a given weight of material than would be afforded with lower molecular weight glycols (26). Chemically resistant applications where trimethylpentanediol-based unsaturated resins are used include reinforced-polyester storage tanks and pipelines for hot, concentrated acids, some solvents, mild bases, hypochlorite solution, and ammonia.

Saturated polyester resins based on trimethylpentanediol are used in various coating applications, most notably in water-borne (27), and high-solids coatings (28–29). Resins manufactured with this diol are characterized by low viscosities, which permits formulation of enamels with 85% nonvolatiles when sprayed (30). Such formulations are cross-linked with isocyanates or melamines to give premium coatings useful for industrial applications: appliance coil coatings and the like (28,31). Conventional alkyd enamels also can be prepared with trimethylpentanediol-based resins to improve flexibility, compatibility, adhesion, and weatherability (32).

The monoisobutyrate ester of trimethylpentanediol is especially useful as a coalescing aid in flat (33) and semigloss (34) latex paint formulations (see Paint). This product is commercially available from Eastman as Texanol ester alcohol.

The diisobutyrate ester of trimethylpentanediol is an economical, low-color primary plasticizer for use in surface coatings, vinyl flooring, moldings, and other vinyl products. This diester is commercially available from Eastman as Kodaflex TXIB plasticizer (35).

Various other diesters, mixed esters, and polyesters of trimethylpentanediol are useful as monomeric or polymeric plasticizers for coatings and plastic film and sheeting (36). They are compatible with, and useful in, cellulosics, vinyls, polystyrenes, and some other plastics.

Trimethylpentanediol can be used as part of a glycol mixture to improve rosin solubility in moisture-set inks, thereby allowing the use of low-cost limed rosins, which are insoluble in most other glycols. A mixture of equal portions of trimethylpentanediol and diethylene glycol was found to dissolve all commercially available limed rosins without appreciably affecting the drying time of inks thus produced (37) (see Inks).

Certain trimethylpentanediol-derived polyesters can be used as viscosity improvers for synthetic lubricants (38–39). Also, several of its diesters have been evaluated as base stocks for these lubricants, eg, trimethylpentanediol dipelargonate, reportedly used as a base stock complying with MIL-L-7808 specifications (40).

Other applications involving trimethylpentanediol, or a derivative, are in urethane elastomers (41), in foams (42), as a reactive diluent in urethane coatings (see Urethane polymers) (43), as a sound-insulating, glass-laminate adhesive (see Adhesives) (44), as a bactericide–fungicide (45), as a pigment homogenizer (46), in paper sizing (47), and as a cross-linking agent in poly(vinyl chloride) adhesive (48).

Derivatives. ***Esters.*** The monoisobutyrate ester of 2,2,4-trimethyl-1,3-pentanediol is prepared from isobutyraldehyde in a Tishchenko reaction (49–50). Diesters, such as trimethylpentane dipelargonate (2,2,4-trimethylpentane 1,3-dinonanoate), are prepared by the reaction of 2 mol of the monocarboxylic acid with one mole of the glycol at 150–200°C (40,51–52). The lower aliphatic carboxylic acid diesters of trimethylpentanediol undergo pyrolysis to the corresponding ester of 2,2,4-trimethyl-3-penten-1-ol (**5**), (53).

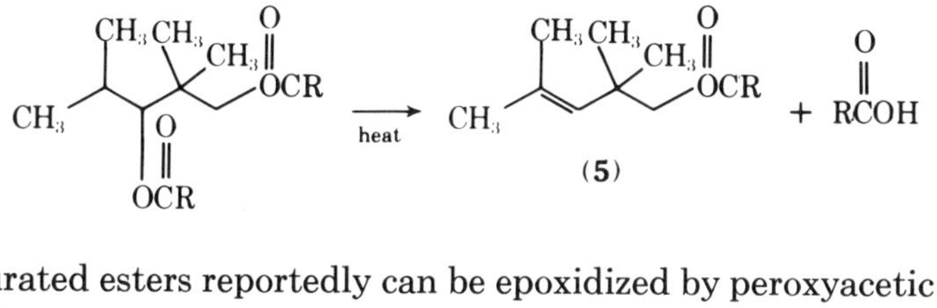

(**5**)

These unsaturated esters reportedly can be epoxidized by peroxyacetic acid (54) (see Epoxidation).

Sulfite. Trimethylpentanediol readily forms a cyclic sulfite, 4,4-dimethyl-1-oxo-3-(2-propyl)-2,6-dioxathiane (**6**) upon addition of a one- to two-molar excess of thionyl chloride, $SOCl_2$, and gentle refluxing (55).

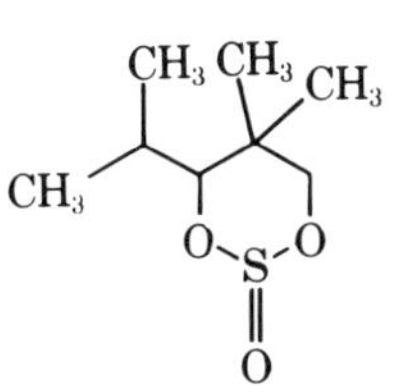

(**6**)

1,4-Cyclohexanedimethanol

1,4-Cyclohexanedimethanol, 1,4-dimethylolcyclohexane or 1,4-bis-(hydroxymethyl)cyclohexane,

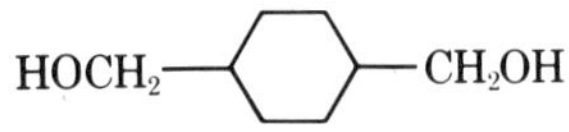

is a white, waxy solid. The commercial product consists of a mixture of cis and trans isomers (5). This diol is used in the manufacture of polyester fibers (qv) (56), high performance coatings, and unsaturated polyester molding and laminating resins (6).

1,4-Cyclohexanedimethanol is miscible with water and low molecular weight alcohols and appreciably soluble in acetone. It has only negligible solubility in hydrocarbons and diethyl ether (5).

Chemical Properties. The chemistry of 1,4-cyclohexanedimethanol is characteristic of general glycol reactions; however, its two primary hydroxyl groups give very rapid reaction rates, especially in polyester synthesis.

Manufacture. The manufacture of 1,4-cyclohexanedimethanol can be accomplished by the catalytic reduction under pressure of dimethyl terephthalate in a methanol solution (57–58). This glycol also may be prepared by the depolymerization and catalytic reduction of linear polyesters that have alkylene terephthalates as major constituents. Poly(ethylene terephthalate) may be hydrogenated in the presence of methanol under pressure and heat to give good yields of the glycol (see Polyesters) (59–60).

1,4-Cyclohexanedimethanol is produced commercially by the Tennessee Eastman Company. The price within the United States as of January, 1979, in bulk truckloads was: 99% purity, $1.10/kg.

Toxicity. The estimated median lethal dose (acute oral LD_{50}) value was determined as 3200–6400 mg/kg in rats, 1600–3200 mg/kg (mice), and 800–1600 mg/kg (guinea pigs) (6).

Uses. The most important application for 1,4-cyclohexanedimethanol is in the manufacture of linear polyesters for use as polyester fibers such as the Kodel polyester fibers (qv) (61). Compared with fibers made from poly(ethylene terephthalate), fibers made from poly(1,4-cyclohexanedimethanol terephthalate) have lower densities and higher melting points. Linear polyesters produced from dicarboxylic acids, eg, terephthalic acid, generally have greater hydrolytic stability and better electrical properties than similar polyesters made from other glycols. The high dielectric strength, good dielectric constant, and low dielectric loss make them useful for many electrical applications, eg, in capacitors, wire coatings, and magnetic tape (qv) coatings.

Unsaturated resins based on 1,4-cyclohexanedimethanol are useful in gel coats and in laminating and molding resins where advantage is taken of the properties of very low water absorption and resistance to boiling water (62). Thermal stability is imparted to molding resins, both thermoplastic (63–64) and thermoset (65–68), enabling retention of physical and electrical properties at elevated temperatures (69). Additionally, resistance to chemical and environmental exposure is characteristic of products made from these resins (70).

High performance polyester enamels are manufactured from saturated resins containing 1,4-cyclohexanedimethanol. Such enamels may be formulated as electrostatically-applied powder coatings (qv) (71–74), water-borne (75), or solvent-based coatings by means of "Rule 66" nonexempt or exempt solvents. These coatings are characterized by an extremely hard, durable, stain- and detergent-resistant finish (76–77). Other polyester resins based on 1,4-cyclohexanedimethanol are also used as plasticizers (78), in hot-melt adhesives (79–80), in elastomers (81–82), and urethane elastomeric coatings (83) (see Coatings).

Another area in which 1,4-cyclohexanedimethanol is commercially important is in the manufacture of polyurethane forms (see Foamed plastics). The two primary hydroxyl groups provide fast reaction rates with diisocyanates, which makes this diol attractive for use as a curative in foams. It provides latitude in improving physical properties of the foam, in particular the load-bearing properties. Generally, the ability to carry a load increases with the amount of 1,4-cyclohexanedimethanol used in producing the high-resilience foam.

Other uses that have been investigated for 1,4-cyclohexanedimethanol or its derivatives involve the manufacture of polyformals for lubricants and hydraulic fluids (84), the use of its esters as anti-opacity agents for paper (85), and the addition of a diol to the backbone of poly(ethylene terephthalate) for chain modification (86).

Derivatives. ***Mixed Phosphonate Esters.*** Unsaturated, mixed phosphonate esters have been prepared from monoesters of 1,4-cyclohexanedimethanol and unsaturated dicarboxylic acids. For example, maleic anhydride reacts with this diol to form the maleate, which is treated with benzenephosphonic acid to yield the ethylenically unsaturated product (7):

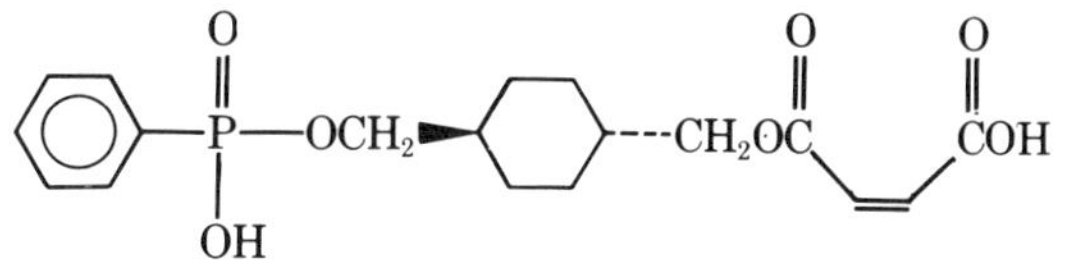

(**7**)

These esters were useful as flame-retardant additives for thermoplastic and thermosetting resins (87).

Diesters. Diesters prepared by esterification of the diol with monocarboxylic acids are useful as antioxidants (qv) for polypropylene (88), and as plasticizers (qv).

Polyesters. Polyesters containing carbonate groups have been prepared from this diol (see Polycarbonates) (89). Films of this polymer, formed from an acetone or ethyl acetate solution, exhibit excellent adhesive properties.

BIBLIOGRAPHY

"Other Glycols" under "Glycols" in *ECT* 2nd ed., Vol. 10, pp. 676–680, by H. C. Twiggs and J. C. Hutchins, Eastman Chemical Products, Inc.

1. *Publication No. N-154,* Eastman Chemical Products, Inc., Kingsport, Tenn., 1978.
2. Fischer and Winter, *Monatsh. Chem.* **21,** 301 (1900).
3. *Publication No. N-121,* Eastman Chemical Products, Inc., Kingsport, Tenn., 1964.
4. *Publication No. N-153,* Eastman Chemical Products, Inc., Kingsport, Tenn., 1977.
5. *Publication No. N-136,* Eastman Chemical Products, Inc., Kingsport, Tenn., 1967.
6. *Publication No. N-199,* Eastman Chemical Products, Inc., Kingsport, Tenn., 1979.

7. *Publication No. X-105,* Eastman Chemical Products, Inc., Kingsport, Tenn., 1961.
8. *Bulletin No. N-115,* Eastman Chemical Products, Inc., Kingsport, Tenn., 1963.
9. U.S. Pat. 3,920,760 (Nov. 18, 1975), J. B. Heinz (to Eastman Kodak Company).
10. *Mod. Plast.* **50**(1), 70 (1973).
11. J. R. Eiszner, R. S. Taylor, and B. A. Bolton, *Paint Varn. Prod.* **49**(3), 54 (1959).
12. L. Beth, *Am. Paint Coat. J.* **60**(8), 54 (1975).
13. *Res. Discl.* **143,** 35 (1976); *Chem. Abstr.* **84,** 165525x (1976).
14. L. A. Sadovnikova and co-workers, *Neftekhimiya* **16**(2), 316 (1976); *Chem. Abstr.* **85,** 126826x (1976).
15. E. L. Niedzielski, *Ind. Eng. Chem. Prod. Res. Dev.* **15**(1), 54 (1976).
16. U.S. Pat. 2,952,701 (Sept. 13, 1960), R. L. McConnell and H. W. Coover (to Eastman Kodak Company).
17. W. E. Conrad and co-workers, *J. Org. Chem.* **26,** 3571 (1961).
18. E. R. Nelson, M. Maienthal, L. A. Lane, and A. A. Benderly, *J. Am. Chem. Soc.* **79,** 3467 (1957).
19. E. R. Alexander, *Principle of Ionic Organic Reactions,* John Wiley & Sons, Inc., New York, 1950, p. 231.
20. V. V. Korshak and S. V. Vinograndova, *Polyesters,* Pergamon Press, Inc., New York, 1965.
21. W. W. Blount, *Soc. Plast. Eng. Tech. Pap.* **21,** 26 (1975).
22. *Publication No. N-257,* Eastman Chemical Products, Inc., Kingsport, Tenn., 1979.
23. S. L. Gott, J. L. Suggs, and W. W. Blount, *Mod. Plast.* **50,** 98 (Sept. 1973).
24. *Publication No. N-206,* Eastman Chemical Products, Inc., Kingsport, Tenn., 1975.
25. *Publication No. N-176,* Eastman Chemical Products, Inc., Kingsport, Tenn., 1977.
26. P. J. Trent, D. L. Edwards, and P. Von Bramer, *Mod. Plast.* **43,** 172 (Apr. 1966).
27. U.S. Pat. 3,979,352 (Sept. 7, 1976), J. W. Brady, F. D. Strickland, and C. C. Longwith (to Shanco Plastics & Chemicals).
28. L. Gott, *J. Coat. Technol.* **48,** 52 (July 1976).
29. J. D. Bailey, *Chem. Ind. NZ* **9**(4), 15 (1975); *Chem. Abstr.* **84,** 152305a (1976).
30. *Publication No. N-246,* Eastman Chemical Products, Inc., Kingsport, Tenn., 1977.
31. F. Sheme, S. Belote, and L. Gott, *Mod. Paint Coat.* **65,** 31 (Apr. 1975).
32. L. Beth, *Am. Paint Coat. J.* **60,** 54 (Sept. 8, 1975).
33. *Publication No. M-124,* Eastman Chemical Products, Inc., Kingsport, Tenn., 1977.
34. *Publication No. M-132,* Eastman Chemical Products, Inc., Kingsport, Tenn., 1977.
35. *Publication No. L-151D,* Eastman Chemical Products, Inc., Kingsport, Tenn., 1978.
36. U.S. Pat. 2,592,234 (Apr. 8, 1952), A. Bell (to Eastman Kodak Co.).
37. *Publication No. N-127,* Eastman Chemical Products, Inc., Kingsport, Tenn., 1965.
38. R. S. Barnes, M. Z. Fainman, *Lub. Eng.* **13,** 454 (1957).
39. U.S. Pat. 2,820,014 (Jan. 14, 1958), J. Hartley, T. H. Ramsay, and J. D. Shimmin (to Shell Development Company).
40. U.S. Pat. 2,798,083 (July 2, 1957), A. Bell and G. R. Lappin (to Eastman Kodak Company).
41. *Res. Discl.* **148,** 55 (1976); *Chem. Abstr.* **85,** 144359b (1976).
42. *Res. Discl.* **143,** 35 (1976); *Chem. Abstr.* **84,** 165525x (1976).
43. *Res. Discl.* **138,** 13 (1975); *Chem. Abstr.* **84,** 46118g (1976).
44. Brit. Pat. 1,367,977 (Sept. 25, 1974), (to Saint-Gobain).
45. U.S. Pat. 3,671,654 (June 20, 1972), H. G. Nosler and H. Schnegelberger (to Henkel and Cie, G.m.b.H.).
46. W. W. Blount, R. H. Calendine, and J. L. Suggs, *Def. Publ. U.S. Pat. Off. T,* 918,011 (1974).
47. Ger. Pat. 2,544,401 (Apr. 15, 1976), M. E. Bateman and E. M. Holda (to Standard Oil of Indiana).
48. L. Foster, D. Beeler, and D. L. Valentine, *Def. Publ. U.S. Pat. Off. T,* 912,016 (1973); *Chem. Abstr.* **80,** 121847s (1974).
49. H. J. Hagemeyer and G. C. DeCroes, *The Chemistry of Isobutyraldehyde and Its Derivatives,* Tennessee Eastman Company, Kingsport, Tenn., 1953.
50. F. J. Villani and F. F. Nord, *J. Am. Chem. Soc.* **68,** 1674 (1946).
51. Jpn. Kokai, 74 94,621 (Sept. 9, 1974), T. Kojima, T. Hirai, T. Tachimoto, and M. Nakamura (to Chisso Corp.).
52. Jpn. Kokai, 74 94,620 (Sept. 9, 1974), T. Kojima, T. Hirai, T. Tachimoto, and M. Nakamura (to Chisso Corp.).
53. U.S. Pat. 2,941,011 (June 14, 1960), H. J. Hagemeyer, D. C. Hull, and M. A. Perry (to Eastman Kodak Company).

54. W. V. McConnell and H. W. Moore, *J. Org. Chem.* **28,** 822 (1963).
55. H. H. Szmant and W. Emerson, *J. Am. Chem. Soc.* **78,** 454 (1956).
56. U.S. Pat. 2,956,961 (Oct. 18, 1960), C. J. Kibler, A. Bell, and J. G. Smith (to Eastman Kodak Co.).
57. Ger. Pat. 1,076,125 (Feb. 25, 1960), G. A. Akin, H. J. Lewis, and T. F. Reid (to Eastman Kodak Co.).
58. Ger. Pat. 1,099,527 (Feb. 16, 1961), R. Heise and W. Rittmeister (to Deutsche Hydrierwerke G.m.b.H.).
59. Jpn. Kokai 75 142,537 (Nov. 17, 1975), T. Mizumoto and H. Kamatani (to Toyobo Co., Ltd.).
60. Jpn. Kokai 75 130,738 (Oct. 16, 1975), T. Mizumoto and H. Kamatani (to Toyobo Co., Ltd.).
61. *Publication No. K-192,* Eastman Chemical Products, Inc., Kingsport, Tenn., 1976.
62. *Publication No. N-21,* Eastman Chemical Products, Inc., Kingsport, Tenn., 1975.
63. Ger. Pat. 2,544,069 (Apr. 8, 1976), H. Inata (to Teijin, Ltd.).
64. U.S. Pat. 3,668,157 (June 6, 1972), R. L. Combs and R. T. Bogan (to Eastman Kodak Co.).
65. S. Oswitch, *Reinf. Plast.* **17,** 308 (1973).
66. Ger. Pat. 2,151,877 (May 10, 1972), F. Fekete and J. S. McNally (to Koppers Co., Inc.).
67. U.S. Pat. 3,674,727 (July 4, 1972), F. Fekete and J. S. McNally (to Koppers Co., Inc.).
68. Jpn. Kokai 68 05,911 (Mar. 4, 1968), M. Izumi, S. Matsumura, and N. Asano (to Sumitomo Electric Industries, Ltd.).
69. J. Litwin, H. H. Beacham, and C. W. Johnson, "Heat Resistant Diallyl Phthalate Polyesters," paper 1B-18, *Soc. Plast. Ind., Inc. RP/C Conference, Feb., 1973.*
70. E. H. G. Sargent and K. A. Evans, *Plastics* **34,** 721 (June 1969).
71. Ger. Pat. 2,542,191 (Apr. 1, 1976), G. Slinckx (to UCB S.A.).
72. Ger. Pat. 2,351,176 (Apr. 25, 1974), R. C. Harrington, J. D. Hood, and P. M. Grant (to Eastman Kodak Co.).
73. P. M. Grant and H. R. Lyon, *Def. Publ. U.S. Pat. Off. T,* 914,001 (1973).
74. Ger. Pat. 2,454,880 (May 13, 1976), J. Rueter and H. Scholten (to Chemische Werke Huels A.G.).
75. Fr. Pat. 1,567,254 (May 16, 1969), (to N.V. Chemische Industrie Synres).
76. Jpn. Kokai 74 13,852 (Apr. 8, 1974), T. Kimura, T. Ohzeki, S. Kobayashi, H. Nakamoto, and Y. Maeda (to Mitsubishi Rayon Co., Ltd.).
77. *Publication No. N-217,* Eastman Chemical Products, Inc., Kingsport, Tenn., 1975.
78. J. R. Caldwell and J. M. McIntire, *Def. Publ. U.S. Pat. Off. T,* 939,013 (1975).
79. Jpn. Kokai 75 100,123 (Aug. 8, 1975), Y. Niinami and K. Mizuguchi (to Toyobo Co., Ltd.).
80. U.S. Pat. 3,931,073 (Jan. 6, 1976), W. J. Jackson and W. R. Darnell (to Eastman Kodak Co.).
81. Fr. Pat. 1,596,552 (July 31, 1970), J. R. Caldwell and R. Gilkey (to Eastman Kodak Co.).
82. Jpn. Kokai 76 96,890 (Aug. 25, 1976), H. Sakai, Y. Takeuchi, and S. Kuris (to Teijin, Ltd.).
83. Ger. Pat. 2,241,413 (Mar. 21, 1974), D. Stoye, W. Andrejewski, and A. Draexler (to Chemische Werke Huels A.G.).
84. U.S. Pat. 3,329,725 (July 4, 1967), H. F. Lederle (to Olin Mathieson Chemical Corp.).
85. Jpn. Kokai 75 82,306 (July 3, 1975), M. Nakahara, S. Ura, Y. Fukuyama, and N. Kondo (to Sumitomo Chemical Co., Ltd.).
86. Z. Zamorsky, *Spec. Plast. Hmoty Nar. Hospod.* (*Proc. Conf.*), 30-7 Dum Techniky CVTS, Pardubice, Czech., 1971; *Chem. Abstr.* **77,** 75950 (1972).
87. U.S. Pat. 3,810,960 (May 14, 1974), W. T. Gormley and M. C. Russ (to Koppers Co., Inc.).
88. U.S. Pat. 3,962,313 (June 8, 1976), M. Dexter and D. H. Steinberg.
89. Ger. Pat. 1,568,342 (July 18, 1974), R. Nehring, K. H. Hornung, and W. Seeliger (to Chemische Werke Huels A.G.).

PAUL VON BRAMER
JOHN H. DAVIS
Eastman Chemical Products, Inc.

GLYCOLS—1,4-BUTYLENE GLYCOL AND BUTYROLACTONE. See Acetylene-derived chemicals.

GLYOXAL. See Glycols, ethylene glycol.

GOLD AND GOLD COMPOUNDS

Gold [*7440-57-5*] appears to be the first metal known and used by man. It occurs in nature as a highly pure metal and is treasured because of its color, its extraordinary ductility, and its resistance to corrosion. Its early uses in medicine and dentistry date to the ancient Chinese and Egyptians. In the Middle Ages the demand for gold led to the intense, unsuccessful efforts of alchemists to convert base metals into gold, which became the basis for chemical science. The search for gold has been an important factor in world exploration and the development of world trade.

Properties

Gold, atomic number 79, is a third row transition metal in Group IB of the periodic table. It occurs naturally as a single stable isotope of mass 197. Its electronic configuration is [Xe] $4f^{14}5d^{10}6s^1$. Common oxidation states are 0, 1, and 3. Selected properties of gold are shown in Table 1 (1). Gold is characterized by high density, high electrical and thermal conductivity, and high ductility. One gram of gold can be drawn to 165 m of wire having a 0.02-mm diameter. Gold leaf 0.14 μm thick still exhibits the shiny color of gold. At least 26 unstable gold isotopes have been made; the most frequently used is ^{198}Au which has a half-life of 2.7 d.

Gold is the most noble of the noble metals. Other than in the atomic state (2), the metal does not react with either oxygen, sulfur, or selenium at any temperature. It does react, however, with tellurium at elevated temperatures (ca 475°C) (3) to produce $AuTe_2$ [*12006-61-0*] (also found as the naturally occurring mineral, calaverite). It also reacts with the halogens, particularly in the presence of moisture. At low temperatures (≤200°C), chlorine is adsorbed on the gold surface with formation of surface chlorides. The rate of further chlorination is limited by the rate of diffusion of Cl through the surface chloride layer. At higher temperatures (700–1000°C), the reaction is kinetically controlled as gold chlorides sublime and fresh surface is continually exposed. In this temperature region, adsorbed Cl_2 dissociates into Cl atoms which then react with the surface (4).

Gold reacts with various oxidizing agents at ambient temperatures provided a good ligand is present to lower the redox potential below that of water. Thus, gold is not attacked by most acids under ordinary conditions and is stable in basic media but dissolves readily in 3:1 hydrochloric–nitric acid (aqua regia) to form $HAuCl_4$ [*16903-35-8*] and in alkaline cyanide solutions in the presence of air or hydrogen peroxide to form $[Au(CN)_2]^-$; these reactions are important to the extraction and refining of the metal. Similarly, gold anodes are solubilized in chloride or cyanide baths. Gold dissolves slowly in concentrated selenic acid at 68°C.

At high temperatures, attack by concentrated sulfuric and nitric acids is slow and is negligible for phosphoric acid. Gold is very resistant to fused alkalis and to most fused salts except peroxides. Gold readily amalgamates with mercury. Gold is very corrosion and tarnish resistant and imparts corrosion resistance to most of the commonly used gold alloys, especially to alloys containing 50 or more atom % of gold. Although gold is resistant to organic acids, some of the alloys used in jewelry may become tarnished by perspiration.

Gold alloys also are subject to stress–corrosion cracking, especially alloys below 14 carat (58% gold). Jewelry items may have areas of high local stress that can induce cracking and corrosion in normally harmless environments such as solutions of hydrochloric or nitric acid. Stress–corrosion cracking occurs essentially as intergranular corrosion at unstable structural sites by an electrochemical mechanism (5) whereby the nongold components of the alloy become preferentially oxidized. Stress–corrosion cracking occurs primarily with single-phase alloys. A remedy against it is stress relief by annealing; however, in the case of multiphase, low-carat alloys, care must be exercised so that annealing does not lead to the formation of homogeneous solutions (see Corrosion).

Extraction and Refining

Placer mining is the oldest form of gold mining. Auriferous sand is swirled with water in a pan and gold, which settles out owing to its high density, is collected. On a large scale, the mixture of sand and water is passed through a sluice box fitted with transverse riffles behind which the gold accumulates. The technique is still in use but requires large quantities of water. It could become important again for dredging large alluvial and marine deposits (6). Air classification also is possible but inefficient (see Size separation). At present, most gold is obtained by deep mining, notably in South Africa.

Mined gold ore is milled sufficiently to allow separation of the gold; recoveries usually are 92–96% of the ore's gold content. After milling, various recovery processes may be used such as amalgamation, cyanidation, gravity concentration (qv), flotation (qv), and roasting, or a combination of these processes. Roasting is carried out when the gold ore contains other metals, sulfides, and tellurides. Gold in association with metals, such as copper, nickel and lead, follows these metals in the concentration process and eventually can be separated and recovered in the anode slimes of electrolytic refining (see Extractive metallurgy).

Amalgamation is carried out by passing a water slurry of the ground ore over mercury-coated copper plates to which the gold particles adhere. The amalgam of gold is scraped periodically. Much of the mercury can be removed and recycled; any remaining mercury can be removed by distillation. In cyanidation, the ore (or tailings from amalgamation) are leached with a solution of sodium cyanide (0.02–0.05%) or an equivalent of calcium cyanide together with some lime. The leaching solution is aerated to provide oxygen and gold is dissolved with formation of $Na[Au(CN)_2]$ [*15280-09-8*].

The rate of dissolution is directly proportional to the oxygen content of the gas used for aeration. The cyanide solution also will dissolve silver and some of the base metals which may lead to an intolerably high cyanide consumption. An economic cyanide usage would be 1–2.5 kg NaCN (or its equivalent in impure calcium cyanide) per metric ton of ore. A recent development is heap leaching in which the cyanide solution is distributed over the top of an open mound or leveled heap of the finely ground ore.

Conventionally, gold and most other metals are recovered from solution by treating with zinc dust or with aluminum. An apparently new technique is the removal of gold from solution with activated carbon which permits collection of ca 9–12 kg gold per metric ton of carbon, along with silver. The gold and silver are stripped from the

Table 1. Gold Properties

Property	Value
atomic weight	196.9665
melting point, K	1337.59
boiling point, K	3081
atomic radius, Au lattice, nm	0.1422
crystal structure	fcc 4 atoms/unit cell
lattice constant at ambient temperature, nm	0.407
interatomic distance at ambient temperature, nm	0.2878
density[a] at 273 K, g/cm^3	19.32
Brinell hardness (10/500/90) (annealed at 1013 K), kgf/mm^2	25
modulus of elasticity at 293 K (annealed at 1173 K), MPa[b]	7.747×10^4
Poisson's ratio, as drawn	0.42
tensile strength (annealed at 573 K), MPa[b]	123.6–137.3
elongation (annealed at 573 K), %	39–45
compressibility at 300 K, Pa^{-1} [c]	6.01×10^{-12}
heat of fusion, J/mol[d]	1.268×10^4
heat of evaporation at 298 K, J/mol[c]	3.653×10^5
vapor pressure, Pa[d]	
at 1000 K	5.5×10^{-8}
at 1500 K	8.5×10^{-2}
at 2000 K	82
at 2500 K	4.9×10^3
at 3000 K	7.1×10^5
specific heat at 298 K, J/g·K)[d]	1.288×10^{-1}
thermal conductivity at 273 K, W/(m·K)	311.4
thermal expansion at 273–373 K, K^{-1}	1.416×10^{-7}
electrical resistivity at 273 K, Ω·cm	2.05×10^{-6}
temperature coefficient of resistivity at 273–373 K, K^{-1}	4.06×10^{-3}
work function, J[d]	
thermionic	$7.69–7.85 \times 10^{-19}$
photoelectric	$8.17–8.76 \times 10^{-19}$
thermal emf[e], mV, at K	
at 373 K	0.92
at 373 K	6.40
at 1073 K	12.35
total emissivity at 493–893 K	0.018–0.035
susceptibility (magnetic) at 291 K, cm^3/g (= emu/g)	1.43×10^{-7}
Hall coefficient at 295 K, (Ω·cm)/T[f]	6.97×10^{-13}
entropy at 298 K, J/K[d]	47.33
standard reduction potential, V	
$Au^+ + e^- \rightarrow Au$	ca 1.68
$Au^{3+} + 3e^- \rightarrow Au$	1.50

Table 1 (*continued*)

Property	Value			
	at 0.40 μm	at 0.55 μm	at 0.70 μm	at 1.00 μm
reflectance, %	38.7	81.6	96.7	98.1
refractive index		0.331	0.131	0.179
extinction coefficient		2.324	3.842	6.044

[a] The commercially accepted value has been given. Measured values and density calculations from x-ray data show some variations.
[b] To convert MPa to psi, multiply by 145.
[c] To convert Pa to mm Hg, divide by 133.3.
[d] To convert J to cal, divide by 4.184.
[e] emf values are given versus the reference grade platinum of the National Bureau of Standards (NBS Pt27) with the cold junction at the ice point.
[f] To convert T to gauss, divide by 1.0×10^{-4}.

carbon with alcoholic caustic (7). The carbon is reactivated by controlled roasting. Further development is under way to improve the extraction and stripping processes.

Recently, resin-in-pulp ion-exchange techniques have been introduced to recover gold from cyanide pulp which eliminates the need to filter the pulp and reduces the capital cost for the processing (8).

A novel method of roasting consists of treating the gold ore with gaseous chlorine at elevated temperatures to volatilize gold as chloride, Au_2Cl_6 [*12446-79-6*] or $AuMCl_6$ (M = iron [*12523-43-2*], aluminum [*73334-09-5*], or gallium [*73334-08-4*]) and recovering it by condensation (9).

The impure gold concentrates from any of the primary recovery processes are melted under oxidizing conditions to remove most of the copper and other base metals, leaving gold plus silver. If the silver content is very low, the gold can be recovered by the Wohlwill electrolysis process in a chloride solution (10). The gold in the anodes dissolves and is deposited in pure form at the cathode. The silver is converted to chloride, which tends to coat the anode; however, coating is sharply reduced by superimposing alternating current on the system. Typically, the electrolyte would contain 90 g/L of gold as $HAuCl_4$ plus a small amount of free HCl. The bath is operated at 70°C with a cathode current density of 10–15 A/dm^2. The resulting cathode deposit should contain 99.95% gold after melting.

Electrolytic treatment of impure gold has been supplanted largely by the Miller process (11) in which chlorine is bubbled through the molten metal and converts the base metals to chlorides, which volatilize. The silver is converted to the chloride, which is molten and can be poured. The remaining gold is less pure (99.6%) than that produced by the Wohlwill process and may require additional treatment. If platinum metals are present, the chlorine process is unsuitable.

In the refining of gold, silver and the platinum group metals are difficult to remove. In the refinery of scrap gold, silver is encountered since gold–silver alloys, named doré metal, are used frequently. Doré metal that contains moderate amounts of gold can be treated by electrolysis in a nitrate solution at room temperature and at an anode current density of ca 4.5 A/dm^2. The gold does not dissolve but is retained in canvas anode bags. The silver deposits as very pure crystals; 99.9% is typical for the melted cathode silver. The salt content of the electrolyte is comprised of ca 60% silver nitrate and 40% copper nitrate. The latter improves the conductivity of the solution. Copper does not deposit.

Silver also can be removed from doré metal by treatment with hot sulfuric acid. The gold remains undissolved but is lower in purity than that resulting from most other processes. A purity of 94% is reasonable for a single treatment with sulfuric acid (12) but this can be raised to 99.5% by repeated treatment and washing. This process works best with doré metal containing 20–25% gold.

For doré metal containing up to 33% gold, silver also can be removed by treatment with warm nitric acid. The gold remaining after washing and melting may have a purity of 99%—considerably higher than with the hot sulfuric acid but the process is more expensive.

In both the sulfuric and nitric acid processes, the doré metal must be in shot form prior to treatment to secure a reasonably rapid reaction. A number of steps also may be required in processing the doré metal to remove miscellaneous impurities, particularly in treating material from copper-anode slime (12).

In refining precious metal scrap and some concentrates, the gold is converted to $HAuCl_4$ by treatment with aqua regia. After being heated to remove nitrogen oxides, gold is precipitated from the solution by reduction with sulfur dioxide or ferrous sulfate. Any platinum metals can be recovered after the complete precipitation of the gold. If any gold remains in solution, explosive compounds may form in the palladium recovery processes since reaction of ammonia or ammonium salts with gold chlorides produces so-called fulminating gold, the precise composition of which is not known.

A number of special problems arise in the treatment of precious metal wastes of various types, eg, fabrication scrap and dusts (sweeps), and in treating scrap comprising gold-clad base metals. The scraps may contain copper, nickel, zinc, iron, tin, and lead, as well as gold and silver (see also Recycling). The assay of incoming materials also is difficult. When the scrap can be melted to a homogeneous liquid, a dip sample converted to shot probably is the most reliable assay technique but care must be taken to avoid two-phase melts which can result if considerable silver and iron-group metals are present. Dilution with copper or high copper scrap may be required to secure a single-phase melt. For refining, much of the zinc can be fumed away as zinc oxide, and iron, lead, and some of the tin may be removed as slag. The precious metals remain with the copper and most of the nickel. This product can be used as the anode in a sulfate solution and most of the copper and nickel removed, the precious metals remaining as an anode slime or mud. The latter is treated as described earlier. The presence of nickel in the copper solution introduces complications, and arrangements must be made to remove considerable amounts of electrolyte periodically for purification.

Sweeps and related materials containing nonmetallic particles can be treated by adding appropriate fluxes to produce a low melting slag. Litharge (PbO) should be present in the mixture and some of this is reduced to produce metallic lead, which dissolves the fine, precious metal particles. The resulting noble metal–lead alloy, which should contain a reasonable amount of silver, is oxidized in a later step, producing litharge which is decanted, and the resulting doré metal is treated as described above.

Production

Gold is widely distributed and the average content in the earth's crust is estimated to be 3.5 parts per billion (ppb per 10^9) (6,8). The gold content of ocean water varies considerably with location. Reported values range from 40 ppb to 3 parts per trillion (ppt, per 10^{12}) (13). However, recent investigations indicate that the average value is of the order of 10 ppt (14) which is well below the concentration (3 ppm or 3 per 10^6) required for economic recovery. Larger local concentrations of gold—ca 75–90% pure—in the metallic state occur in sedimentary and igneous rocks in combination with gangue minerals, usually quartz, and with silver and base metal sulfides, selenides, and tellurides. Gold is further obtained as a by-product in the electrolytic refining of copper and nickel (see Copper). The anode slimes contain silver and platinum-group metals as well as gold.

From the primary lodes in rocks, gold has been released by weathering in conjunction with erosion by flowing streams in the form of fine grains or nuggets or as residual or stream placers. Collection of placer gold constituted the earliest form of mining. Original placers have often become buried by substantial layers of rock, eg, at the Vaal Reefs Mine in South Africa, gold is recovered from a depth of 2200 m.

As of 1978, land resources of gold have been estimated to amount to 6×10^4 metric tons and total world production to 1978 was 9×10^4 t. Gold is produced in many countries, the leading one being the Republic of South Africa, followed by the USSR and Canada. A survey of worldwide new mine production from 1968–1978 is given in Table 2 (8). Barring the unlikely event of the discovery of a major new deposit, the annual gold production of the noncommunist countries is expected to remain at 900–1000 t for the coming years.

In the United States, about 60% of the gold production originates from ores and the remainder primarily from the refining of base metals, eg, copper, zinc, lead, and mercury. The largest United States gold producer is Homestake Mine at Lead, S. Dakota, which supplies ca 30% of the U.S. production. The major gold producing states are S. Dakota, Nevada, Utah, and Arizona. Demand for fabricated gold in the United States is estimated to grow ca 2.5% annually which requires increasing importation (6). Thus in 1973, 48% of United States consumption was imported. In 1977 this figure was 60% (15), an increasing trend which will probably continue.

Economic Aspects

The price of gold in recent years, though fluctuating, has exhibited a trend towards a continuous increase. In March 1977, the price (U.S. $) per kg was ca $4700 ($146/troy oz); in March 1978, ca $5900 ($184/troy oz); in March 1979, ca $7800 ($243/troy oz); and in September 1979, $12,000 ($373/troy oz) before reaching a historic high of ca $26,000 (ca $800/troy oz) before the year-end June, 1980 price ca $21,000/kg ($650/troy oz). The unit of commerce is the troy oz (31.103 g). In part, the rise in the value of gold is related to the steadily increasing cost of production. In South Africa, a country with low labor costs, the production cost of gold was $2.32/g ($72/troy oz) in the first quarter of 1976; the corresponding figures for 1977, 1978, and June 1979 were $2.77, $3.34, and $9.00/g, respectively (8). In view of world economics, gold is expected to continue to command a strong position in the investment field.

Table 2. Mine Production of Gold, Metric Tons

Location	1968	1969	1970	1971	1972	1973	1974	1975	1976	1977	1978
Rep. of South Africa	969.4	972.8	1000.4	976.3	909.6	855.2	758.6	713.4	713.4	699.9	706.4
Canada	83.6	79.2	74.9	68.7	64.7	60.0	52.2	51.4	52.4	54.0	52.9
United States	46.0	53.9	54.2	46.4	45.1	36.2	35.1	32.4	32.2	32.0	30.2
other Africa											
Zimbabwe-Rhodesia	15.5	14.9	15.0	15.0	15.6	15.6	18.6	18.6	17.1	20.0	17.0
Ghana	22.6	22.0	21.9	21.7	22.5	25.0	19.1	16.3	16.6	16.9	14.2
Zaire	5.3	5.6	5.5	5.4	2.5	2.5	4.4	3.6	4.0	3.0	1.0
other	3.7	3.3	2.0	2.5	1.7	1.7	1.0	1.0	1.0	1.0	1.0
Total other Africa	*47.1*	*45.8*	*44.4*	*44.6*	*42.3*	*44.8*	*43.1*	*39.5*	*38.7*	*40.9*	*33.2*
South America											
Brazil	8.8	8.8	9.0	9.0	9.5	11.0	13.8	12.5	11.6	12.9	13.0
Dominican Rep.								3.0	12.7	10.7	10.8
Colombia	7.5	6.8	6.8	5.9	6.3	6.7	8.2	10.8	10.3	9.2	9.0
Mexico	5.5	5.6	6.2	4.7	4.6	4.2	3.9	4.7	5.4	6.7	6.2
Peru	3.3	4.1	3.2	3.0	2.6	2.6	2.7	2.9	3.0	3.4	3.7
Nicaragua	6.0	3.7	3.6	3.3	2.8	2.8	2.4	1.9	2.0	2.0	2.3
other	2.4	6.5	6.6	8.2	9.0	7.9	5.9	6.0	8.0	8.0	8.5
Total South America	*33.5*	*35.5*	*35.4*	*34.1*	*34.8*	*35.2*	*36.9*	*41.8*	*53.0*	*52.9*	*53.5*
Asia											
Philippines	16.4	17.8	18.7	19.7	18.9	18.1	17.3	16.1	16.3	19.4	20.2
Japan	7.0	7.8	8.4	7.7	9.6	10.4	5.5	4.7	4.0	4.6	6.1
India	3.6	3.4	3.2	3.7	3.3	3.3	3.2	3.0	3.3	2.9	2.8
other	3.0	2.8	2.8	2.1	2.7	2.7	2.7	2.7	3.0	3.0	3.0
Total Asia	*30.0*	*31.8*	*33.1*	*33.2*	*33.5*	*34.5*	*28.7*	*26.5*	*26.6*	*29.9*	*32.1*
Europe	6.7	6.9	7.4	7.6	13.2	14.3	11.6	11.0	11.4	13.2	12.5
Oceania											
Papua/New Guinea	0.8	0.8	0.7	0.7	12.7	20.3	20.5	17.9	20.5	22.3	23.4
Australia	24.3	22.2	19.5	20.9	23.5	17.2	16.2	16.3	15.4	19.2	20.2
other	3.6	3.2	3.6	3.1	3.2	3.2	3.2	3.2	3.0	4.0	4.7
Total Oceania	*28.7*	*26.2*	*23.8*	*24.7*	*39.4*	*40.7*	*39.9*	*37.4*	*38.9*	*45.5*	*48.3*
Total noncommunist world	*1245.0*	*1252.1*	*1273.6*	*1235.6*	*1182.6*	*1120.9*	*1006.1*	*953.4*	*966.6*	*968.3*	*969.1*
USSR[a]	304.2	318.2	346.6	359.8	378.9	398.2	420.7	407.9	443.6	444.0	311
other communist countries	18.4	18.4	18.4	18.4	18.4	19.4	20.4	20.0	20.0	20.0	20.0

[a] Full assessment of the gold production of USSR under review (8).

Specifications

Dentistry. Most casting alloys meet the composition and properties criteria of specification no. 5 of the American Dental Association (16) which prescribes four types of alloy systems constituted of gold–silver–copper with addition of platinum, palladium, and zinc. Composition ranges are specified, as are mechanical properties and minimum fusion temperatures. Wrought alloys for plates also may include the same constituents. Similarly, specification no. 7 prescribes nickel and two types of alloys for dental wires with the same alloy constituents (see Dental materials).

Analytical and Test Methods

Analysis. The method used to determine gold depends on the state of the sample to be analyzed and on its expected gold content and interfering impurity levels. Many of the available methods require conversion of gold in the sample to a soluble form, eg, $[AuCl_4]^-$. This is accomplished by digestion of the sample with aqua regia, evaporating to remove oxides of nitrogen, and dissolving the residue in concentrated HCl. The resulting solution is reevaporated and redissolved until removal of nitrogen oxides is complete (generally twice). In place of aqua regia, Cl_2–HCl or H_2O_2–HCl can be employed. Prior to analysis, the gold in solution can be concentrated or separated from impurities by solvent extraction, eg, with diethyl ether, ethyl acetate, methyl isobutyl ketone or dichloromethane–tetrahydrofuran, or by chromatography on ion-exchange or complexing resins (see Extraction).

Gravimetric Methods. In the fire assay (17), used traditionally for the determination of the gold content in ores, the finely ground ore sample is mixed with a flux containing carbon and litharge, and the mixture is fused for at least one hour. After cooling, a button of lead containing gold and other precious metals is separated from the slag. The button is cupellated on bone ash or magnesia which absorb the lead oxides. The remaining bead is treated with nitric acid to remove silver, leaving gold and any platinum metals present in the sample.

Gravimetric methods more suitable for general use involve the precipitation of metallic gold from tetrachloraurate solutions by reduction with oxalic acid, SO_2, or hydroquinone. Formaldehyde, hydrazine, ferrous sulfate, and hypophosphorous acid also have been used but are considered less efficient (18).

Gold can be precipitated in the form of complexes of uncertain composition which then are ignited. Reagents suitable for this purpose are dimethylglyoxime (19), tetramisole (citarin) (20), thiophenol (21), and 8-quinolinol (22) (see Chelating agents). A number of insoluble tetrachloro- or bromoaurates can be precipitated quantitatively and gold then can be determined by weighing. Suitable cations are *N*-(*N*-bromotetradecylbetainyl)-*C*-tetradecylbetaine (**1**) (23), bis(trimethyl)hexamethylenediammonium ion (**2**) and 2,4,6-triphenylpyrilium ion (**3**) (24) as well as complexes containing sulfur ligands formed by reactions with thioglycolic acid (25) or di-2-thienylketoxime. A method suitable for determining gold in cyanide plating solutions involves adding silver nitrate which reacts with free cyanide followed by fuming with sulfuric acid (26).

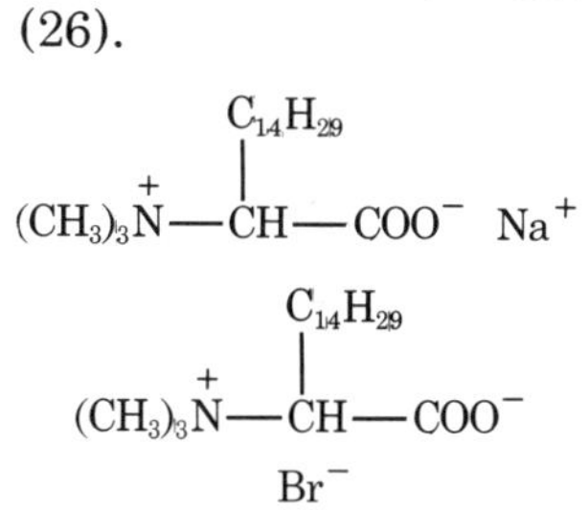

(**1**)

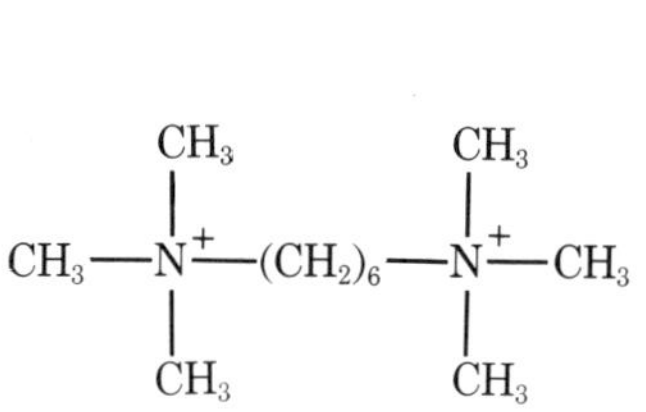

(**2**)

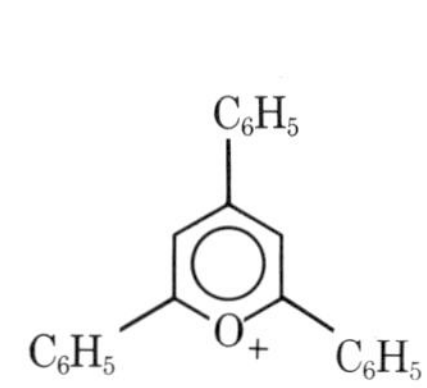

(**3**)

Titrimetric Methods. Frequently, the reduction of gold solutions leads to finely divided precipitates which are difficult to recover quantitatively. In such cases, the reduction of Au(III) to Au(0) by, eg, hydroquinone, can be followed potentiometrically (27). The end point in such titrations also can be determined with indicators such as benzidine (28) or *o*-anisidine (29). Alternatively, the reductions can be effected with excess hydroquinone which is then back-titrated with Ce(IV) (30). Iodometric determination of Au(III) also is useful (31).

Spectrophotometric Methods. The gold content of aqueous solutions of $[AuCl_4]^-$ or $[AuBr_4]^-$ can be determined directly from the absorbance of these species at 380 nm. Interferences by other metal species, particularly tetrahaloanions of the Pt group, can be minimized by extraction of the Au(III) into organic solvents (32). The visible spectra of colored complexes formed by reactions with such reagents as stannous chloride (33), rhodanine (34), rhodamine B (35) and malachite green (36) also can be used to determine gold. The methods generally involve extraction of the colored complexes into organic solvents which overcomes problems of aqueous solubility and facilitates separation from interfering metal ions and concentration. The electronic and vibrational spectra of gold complexes have been determined and have been used to elucidate structure and bonding.

Emission Spectrography, Atomic Absorption, and Neutron Activation. These instrumental techniques are particularly useful for the determination of trace quantities of gold, especially when used in conjunction with concentration via solvent extraction or chromatography. Emission spectrography is used to determine both small quantities of gold in metals and gold purity. Emission lines at 242.78 or 267.60 nm are useful for determining small concentrations (0.5–20 ppm) and at 312.29 nm for large amounts (10–160 ppm) in platinum (37), whereas the lines at 314.8 and 310.5 nm serve as internal standards in the determination of trace impurities in gold (38).

Determination of gold concentrations to ca 1 ppm in solution via atomic absorption spectrophotometry (39) has become an increasingly popular technique because it is available in most modern analytical laboratories and because it obviates extensive sample preparation. However, the most sensitive method for gold analysis is neutron activation, which permits accurate determination to levels ≤ 1 ppb (40). The sensitivity arises from the high-neutron-capture cross sections of the only natural isotope, ^{197}Au (9.9×10^{-27} m^2 or 99 barns). The resulting isotope, ^{198}Au, decays by β and γ emission with a half-life of 2.7 d. Gold is a useful calibration standard for this method (see Analytical methods; Radioactive tracers).

Health and Safety Factors (Toxicology)

The discovery in 1890 that $KAu(CN)_2$ [*14265-59-3*] was toxic to the tuberculosis-causing bacterium (*mycobacterium tuberculosis*) *in vitro* and was somewhat effective in the treatment of the disease led to a search for other effective gold compounds with less serious side effects. These efforts resulted in the preparation of less toxic sulfur-containing compounds, such as disodium gold(I) thiomalate [*42722-04-3*] (Myocrysin) and gold(I) thioglucose [*12192-57-3*] (Solganal), which were used extensively in the treatment of tuberculosis from 1925–1935. During this period, the utility of these compounds in the treatment of rheumatoid arthritis was discovered, and they remain as two of the few effective treatments for this crippling disease (41).

Excretion of gold during chrysotherapy is relatively slow (ca 20% within two weeks and less thereafter) and toxic side effects (such as kidney and liver damage, dermatitis, stomatitis, thrombocytopenia, bone marrow depression, and other haematopoietic disorders) can be long lasting and sufficiently severe to require cessations of treatment in about 35% of the patients (42). The excretion of gold can be accelerated by administration of chelating thiols such as DL-penicillamine and British antilewisite (BAL) (43) (see Chelating agents).

Toxicity and the requirement for parenteral administration has led to further efforts to synthesize less toxic gold drugs that can be administered orally. The effectiveness of tertiary phosphine–gold complexes—in particular, triethylphosphinegold(I) chloride [*15529-90-5*], $(C_2H_5)_3PAuCl$—in suppressing adjuvant arthritis in rats when administered orally and without reaching harmful concentration in the kidney (44), indicates the potential usefulness of this class of gold compounds. Although the mechanism of action of gold compounds is not fully understood, there is some speculation that inhibition of thiol-containing enzymes involved in the inflammatory process may be responsible (45).

Radioactive ^{198}Au, prepared by irradiating natural gold in a nuclear reactor, has been used in radiotherapy either as grains which can be implanted in cancerous tissue or infused in colloidal form as in the treatment of bladder cancer (46). Colloidal ^{198}Au, also has diagnostic applications such as in bone marrow scanning and visualization of organs such as the lungs and liver (see Radioactive drugs).

The labeling of various biological macromolecules with gold (using such soluble species as $[Au(CN)_2]^-$, $[AuCl_4]^-$, and $[AuI_4]^-$) eliminates phasing problems in the determination of three-dimensional structures by x-ray diffraction because of the high x-ray scattering power of Au which permits its ready location in electron-density maps (47) (see X-Ray technology).

Uses

Besides its use for monetary reserves, gold is used in the private sector principally for investment and fabrication. A breakdown of private uses by noncommunist nations from 1968–1978 is given in Table 3 (8). United States fabrication during 1968, 1973, and 1979 was 205, 209, and 178 t, respectively.

The largest industrial use is jewelry. In jewelry, the weight fraction of gold is either expressed as the carat (24 carat representing 100%) or in fineness (1000 fine representing 100%). Carat ranges are from 22–8 with 18 and 14 carat the most popular grades in the United States, Europe, and Japan.

In the electronics industry, gold is used as fine wires or thin film coatings and frequently in the form of alloys to economize gold consumption and to impart properties such as hardness. Gold has properties that satisfy specific requirements not achievable with less expensive metals, so that the metal appears to have a permanent function in this industry (see Electrical connectors; Film deposition techniques).

In dentistry, gold is used for a variety of restorations (see Dental materials). Other industrial–decorative uses include inexpensive jewelry either as gold-filled or as rolled-gold materials, in which gold is bonded to a base-metal backing, or as electroplated articles with gold coatings of up to 2.5 μm thickness; decorative coatings on glass or chinaware; optical (reflective or absorptive) coatings; electrical contacts and contactors; brazing alloys for high temperature applications (see Solders); instrumentation

Table 3. World Fabrication of Gold[a], Metric Tons

Fabrication	1968	1969	1970	1971	1972	1973	1974	1975	1976	1977	1978
carat jewelry	912	904	1062	1060	996	512	220	519	931	996	1001
electronics	82	100	89	86	105	127	92	67	76	77	85
dentistry	60	60	59	63	66	68	57	62	76	82	87
other industrial and decorative uses	58	63	62	69	71	72	67	60	66	67	75
medals, medallions and unofficial coins	40	44	54	52	41	21	7	21	47	47	46
official coins	68	26	46	54	63	54	287	244	184	137	259
Total	*1220*	*1197*	*1372*	*1384*	*1342*	*854*	*730*	*973*	*1380*	*1406*	*1553*

[a] By noncommunist nations; bullion purchases not included.

(measurement of temperatures near absolute zero, or for jet-engine temperature control); spinnerets for rayon manufacture, corrosion-resistant materials in the chemical industry; and as catalysts and for medicinal purposes.

Alloys. ***Jewelry.*** Except for white golds, the carat golds used in jewelry are alloys of gold, silver, and copper. Frequently they are modified by additives, mainly zinc, or by various metals such as cadmium, nickel, platinum, and palladium. Varying compositions of the basic alloy produce shades and colors of gold ranging from yellow to green. The dependence of color on composition is shown in Figure 1 (48).

Liquidus isotherms of the gold–silver–copper system and the binary phase diagrams of the constituents are given in Figure 2 (49). For binary gold and silver and for gold and copper alloys, the liquidus and solidus temperatures are close. For copper alloys, a congruent minimum appears at 910°C at 20% copper. These alloys form continuous solid solutions. Copper and silver form a simple eutectic at 779°C with a silver content of 72%. In the copper–gold system, ordered phases appear below 400°C: Au_3Cu [*12044-96-1*], AuCu [*12006-51-8*], and $AuCu_3$ [*60748-60-9*], whereas the gold and silver system forms only one ordered AgAu [*63717-64-6*] phase. The liquidus surface of the ternary system accordingly exhibits a shallow minimum at the cop-

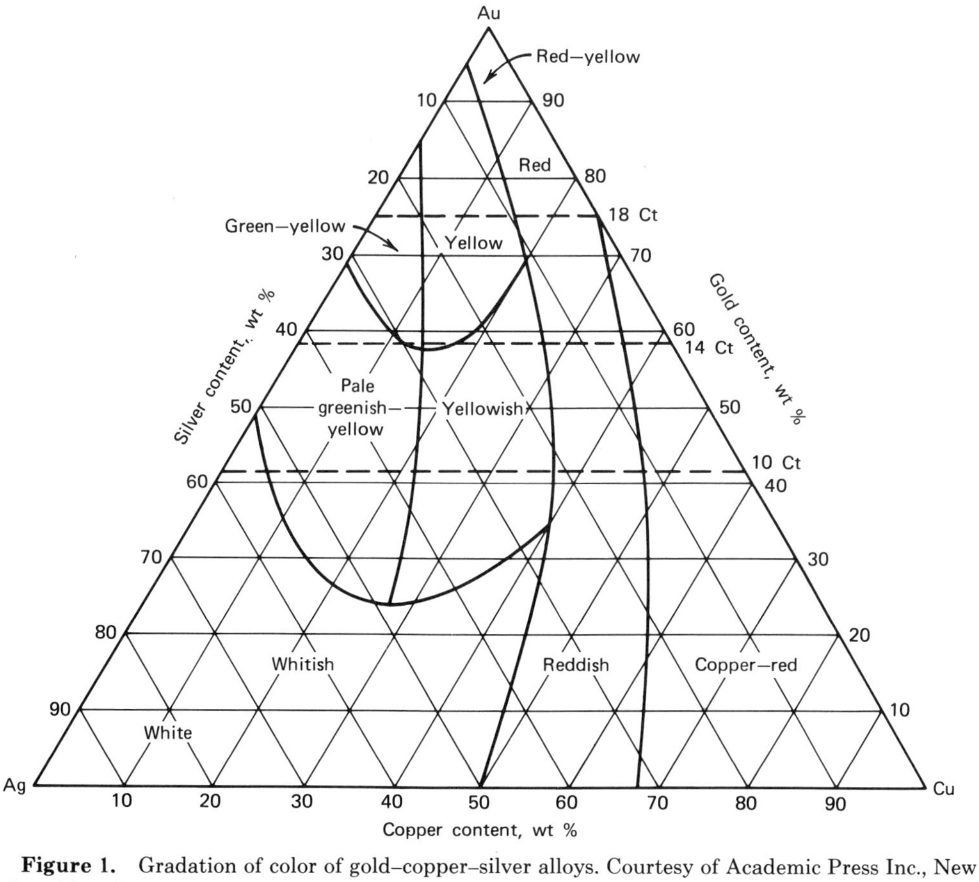

Figure 1. Gradation of color of gold–copper–silver alloys. Courtesy of Academic Press Inc., New York and London.

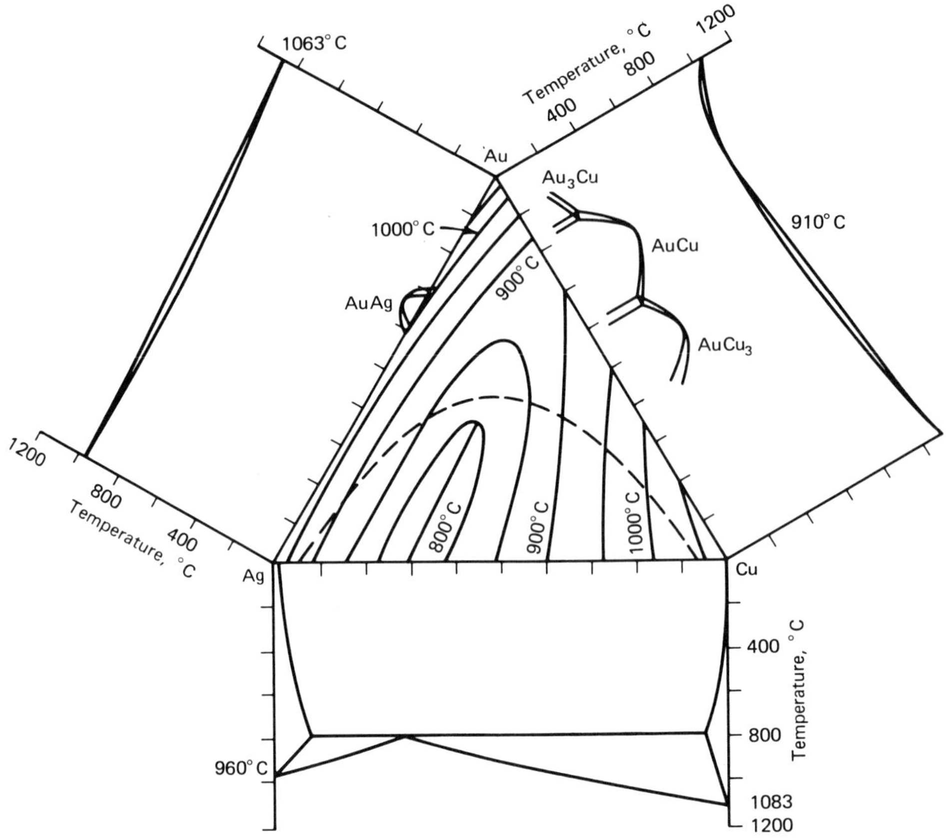

Figure 2. Liquidus isotherms of gold–copper–silver alloys and phase diagrams of the binary constituents. Courtesy of Academic Press Inc., New York and London.

per–gold binary edge and a eutectic gutter at the copper–silver binary edge. The ternary system also forms duplex systems at lower temperatures so that it is possible to age-harden rapidly quenched alloys. This is important in that a workable alloy may be substantially hardened by a final heat treatment after the forming operation.

Zinc and gold form a eutectic at 642°C with a zinc content of 15%. The alpha phase contains 6–13% Zn, depending on temperature. A number of ordered phases, including the compositions Au_3Zn [*12256-74-5*] and $AuZn_3$ [*11089-97-7*], exist at lower temperatures. Despite these complexities, zinc is used widely in jewelry alloys. At low concentrations ($\leq$5%), zinc acts essentially as a deoxidizer without appreciably affecting the properties of the alloys. In concentrations of up to 10% and more, especially in alloys of 14 carat or less, zinc lowers the liquidus and solidus temperatures but also suppresses phase segregation by restricting the range at which duplex phases coexist in gold–silver–copper alloys or suppresses formation of ordered phases. Preserving the single-phase state improves the tarnish-and-corrosion resistance of the alloy. Zinc or cadmium (50) also may be used to maintain the gold color of a lower carat alloy or to improve mechanical and other properties.

White Golds. As shown in Figure 1, the silver-rich phases in the gold–silver–copper system are white and could be considered for white golds which are used as substitutes for platinum. However, because of white gold's insufficient tarnish resistance, two other alloy systems primarily are used: gold–nickel–copper (often together with zinc) and gold–palladium alloys—these usually also contain other metals such as silver, nickel, platinum, and zinc. Figure 3 shows liquidus isotherms of the gold–nickel–copper system (51) and the binary phase diagrams of its constituents. Gold and nickel form continuous solid solutions at elevated temperatures with a congruent minimum at 950°C and a nickel content of 17.5%. At lower temperatures, the solid solutions decompose into two phases, nearly pure gold and a nearly pure nickel. This phase separation together with the development of ordered solid phases of the copper–gold system results in age-hardening accompanied by lower corrosion resistance (51). This can be reversed by homogenizing the alloy by annealing and quenching. Palladium-based white golds are expensive but also are malleable, ductile, corrosion resistant, and can be cast (52). A number of other metals have been reported to bleach gold; however, these have had little application.

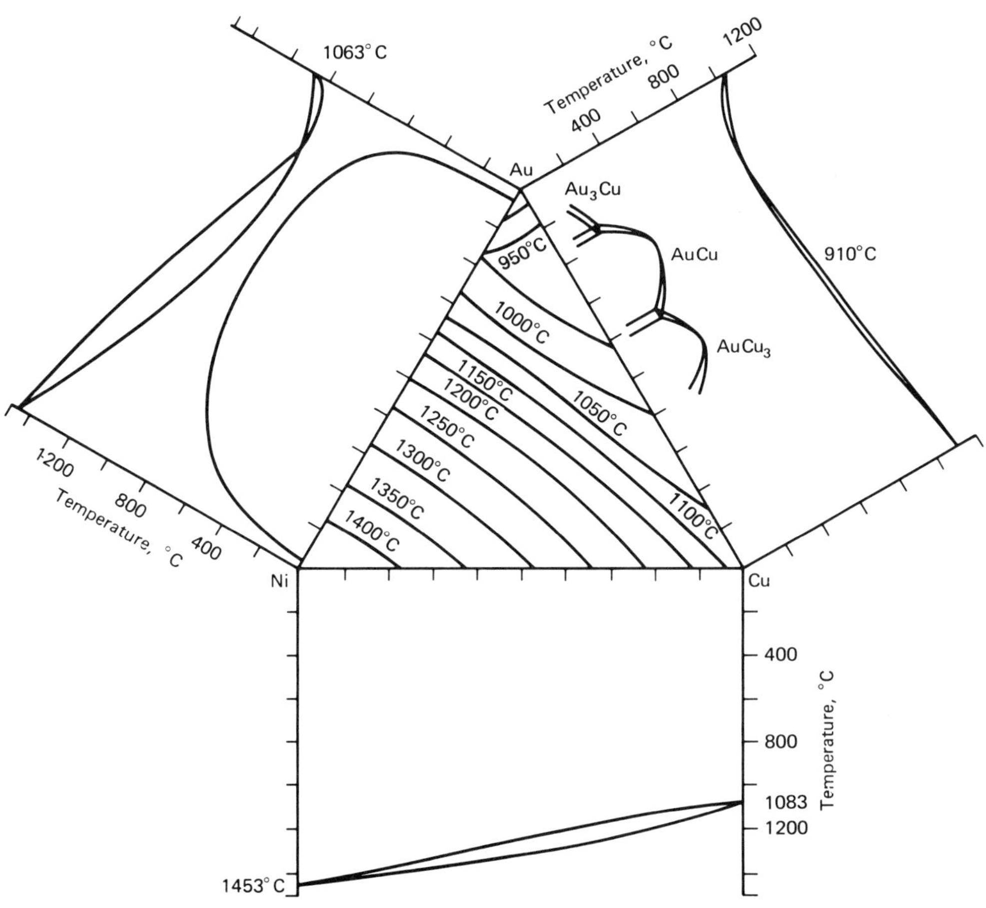

Figure 3. Liquidus isotherms of gold–copper–nickel alloys and phase diagrams of the binary constituents. Courtesy of Academic Press Inc., New York and London.

Solders and brazes for jewelry gold alloys are usually gold–silver–copper alloys that can be modified to yield a desired melting range by adjusting the silver-to-copper ratio. Additions of zinc, cadmium, tin, or nickel compensate for the changes in color introduced by compositional modification (53) (see Solders).

Dentistry. High purity gold in the form of foil or powder is used to a minor extent. Generally, gold alloys are employed for either castings (or as castings bonded to porcelain enamel) or for plates and wires. The presence of palladium and especially of platinum increases the hardness of the alloys considerably. Furthermore, age-hardening can be carried out. Zinc in small concentrations (0.5–1.0%) acts not only as a deoxidizer but contributes to age-hardening. Small amounts of metals, such as iridium, ruthenium, rhodium, and rhenium, act as grain refiners (54).

Alloys suitable for castings that are to be bonded to porcelain must have expansion coefficients matching those of porcelain. They are composed of gold and palladium and small quantities of other constituents: silver, calcium, iron, indium, tin, iridium, rhenium, and rhodium. The readily oxidizable components increase the bond strength with the porcelain by chemical interaction of the oxidized species with the oxide system of the enamel (see Dental materials).

Industry and Technology. Gold and its alloys have many technological applications, in most cases in the form of thin films produced by electro- or by chemical deposition. In instrumental and similar applications, gold alloys are employed as resistors, conductors, potentiometers, resistance thermometers, and contacts. The alloys are chosen because of specific mechanical, electrical, and thermal properties and because of their high corrosion resistance (see Electrical connectors; Temperature measurement). The gold–palladium system, which forms continuous solid solutions, offers a wide range of resistivities useful in resistors and potentiometers. The alloys 45 Au–55 Pd and 70 Au–20 Pd–10 V, have resistivities of 28 $\mu\Omega\cdot$cm (55) and 217 $\mu\Omega\cdot$cm (56), respectively, and minimal temperature coefficients of resistivity. 50 Au–45 Pd–5 Mo alloy also is suitable for potentiometer use (57). For resistance thermometry at cryogenic temperatures, 99.6 Au–0.4 Mn has been recommended (58). For the measurement of elevated temperatures in the range of 1000°C, notably for temperature control of aircraft jet engines, alloy thermocouples of the type 55 Pd–31 Pt–14 Au (positive leg)–65 Au–35 Pd (negative leg) (59) have been developed. For cryogenic temperatures approaching 0 K, gold–iron thermocouples of high purity gold are used; their iron content is 0.07% in the United States and 0.03% in the United Kingdom (60). The other leg of the couple normally is chromel. For contacts, wipers, or sliprings, a number of hard alloys have been developed and are frequently used in clad or inlay form which allows a broader range of composition. The alloys involved are: gold–copper, gold–silver, gold–nickel, gold–silver–platinum, gold–silver–nickel, gold–silver–copper, or gold–rhenium. A promising new contact material is gold that has been dispersion-strengthened by small amounts of cerium (61). An overlay of 95 Au–5 Ni on telephone relay contacts containing palladium eliminates the disturbing catalytic action of palladium, ie, causing the formation of polymeric material from organic vapors on the contact surface (62). An alloy 60 Au–40 Pd is used in spark plugs for special internal combustion engines, eg, in snowmobiles (63).

Gold and gold-based alloys are used for corrosion-resistant equipment. Gold–platinum alloys, 75 Au–25 Pt or 84 Au–15 Pt–1 Rh, are used as crucible material for many molten salts (64). Spinnerets for rayon manufacture are based on the Au–Pt system which exhibits a broad miscibility gap in the solid state so that the alloys can be age-hardened. Spinneret alloys contain 30–40% or more platinum that is modified by small additions of (usually) rhodium (65). Either gold or gold–platinum alloys are used in rupture disks for service with corrosive gases (66).

Fine wires of gold (purity 99.975% and dia 0.0127–0.0381 mm) are used in the circuitry of semiconductors (qv) and transistors (67) (see Integrated circuits). Many metallurgical problems arise in the bonding of these wires, eg, formation of embrittling eutectics if the wires are in contact with silicon or germanium (68). With silicon, the eutectic point is at 370°C at 6% Si, and with germanium, at 350°C at 12% Ge. During bonding, gold diffuses into silicon or germanium causing *p*-type conductivity and requires countermeasures. On the other hand, gold that has diffused into silicon greatly accelerates the rate of minority carrier recombination which is utilized in high speed switching devices (69). Intermetallic phases form above 250°C if gold is in contact with aluminum surfaces: Au_4Al [*12003-05-3*], Au_5Al_2 [*12522-66-6*], Au_2Al [*12250-39-4*], AuAl [*12250-38-3*], $AuAl_2$ [*12004-03-4*]. Resulting embrittlement, however, is not ascribed to the formation of these phases but to development of Kirkendall voids (70).

Brazes and Solders. A multitude of alloys has been developed for applications in the electronics, aerospace, and nuclear power industries. They are used at given temperature conditions and impart specific wetting characteristics (eg, for graphite and diamond) and corrosion resistance adequate for the application. A principal requirement is a small differential between solidus and liquidus temperatures. Alloy systems for brazes and solders are: gold–copper; gold–silver; gold–silver–copper (for lower temperatures: 12–30 Au in Ag–Cu–Cd–Zn) (71), gold–nickel–copper; gold–nickel–silver; gold–nickel–chromium; gold–nickel–molybdenum, gold–nickel–tantalum; gold–palladium (most oxidation resistant, 92 Au–8 Pd) (72), gold–palladium–nickel (brazing tungsten, molybdenum, nickel, and stainless steel) (78), gold–tin–germanium (solder for electronic circuits, eg, 80–85 Au, 5–19 Sn, 1–10 Ge) (74); gold–tin (solder for electrical connections with gold-plated surfaces, eg, 80 Au–20 Sn–0.5 Bi) (75), gold–antimony; gold–arsenic (solder for semiconductors, eutectic at 665°C, 32% As) (76); gold–indium; and gold–indium–bismuth.

Electrodeposition. Electrodeposition is the most widely practiced technique for producing gold or gold alloy coatings for jewelry and for decorative and industrial purposes (see Electroplating). In the electronics industry, for instance, essentially the entire gold consumption (Table 3) is in the form of coatings, and the majority of these are prepared by electrodeposition. A vast number of plating baths has been developed. They are based principally on two complexes of monovalent gold, predominantly the cyanide complex $[Au(CN)_2]^-$ which usually is employed as the potassium salt [*13967-50-5*] and, to a lesser extent, sulfite complexes (presumably $[Au(SO_3)_2]^{3-}$) usually employed as sodium [*39394-92-8*] salt or potassium [*19153-99-2*] salt and stabilized with organic molecules such as amines (77). The sulfite complex is more suitable for high speed plating, eg, 5 μm in 2–4 min instead of 10–20 min (78). In order to obtain bright smooth deposits of desired physical properties (hardness, ductility) and with minimum porosities, the bath compositions, temperature, current density, and agitation must be controlled. Techniques such as pulse plating, reversed plating, and superposed alternating current may be used. In addition to the gold complex employed for electrodeposition, the baths contain other salts which enhance conductivity, buffers which adjust pH, or chelating agents which define the concentrations of other metal ions to be co-deposited or which tie up undesired impurities, surface-active agents, and agents that act as brighteners, hardeners, and grain refiners.

Pure gold coatings can be deposited from alkaline, neutral, or acid cyanide baths. Alkaline, neutral, and acid pH ranges are achieved using excess cyanide (79), additions of citrates (80), and citric acid (81), respectively. Various organic additions and small quantities of agents, eg, arsenic and antimony, are added for grain refining, leveling, and brightening (82). For plating of electrical contacts, bright hard gold deposits are obtained by co-depositing 0.1–0.2% nickel or cobalt (83). A great number of baths for gold alloy deposition have been formulated—mainly for jewelry and decorative purposes—and yield a variety of color shades. Rose-pink and red shades are obtained from gold–copper co-deposits (84), yellow gold from gold–nickel or gold–cobalt alloys which lighten the color of gold (85), green colors from gold–silver or gold–cadmium co-deposits (86), and white gold from gold–nickel–cadmium–copper co-deposits or from nickel–zinc co-deposits (87). In the industrial area, an 18-carat electrodeposited alloy of gold–copper–cadmium has been recommended for contact use (88), and a gold alloy containing 1.5–2% antimony has been recommended for coating electrical contacts (89). Gold films containing antimony also have been developed for coating germanium and silver transistors, and gold films containing 0.1–0.5% indium are used in the production of *p*-type germanium resistors (90).

Electroless Plating. Surface coatings that are obtained from baths comparable to those used for electrodeposition but that are applied without current are produced by electroless plating (qv). The process operates either by electrochemical displacement (atomic exchange) or by chemical reduction. The former is restricted to metals less noble than gold; after a thin film of gold has been formed (1–2 μm), no further gold deposition is possible. The latter permits deposition of heavier films (1–15 μm) on metallic or nonmetallic substrates, eg, glass, ceramics and plastics.

Gold in a medium of ammonium citrate and chelating agents is suitable for electrochemical displacement (91). For electroless plating by reduction, a variety of complexing agents for gold have been formulated such as imides (92), amines, thiourea, and citrates (93), and reducing agents such as hypophosphite, hydrazine, borohydrides, amine boranes, and formaldehyde; the process is autocatalytic. The initial formation of metallic gold on the surface accelerates the rate of subsequent reduction. Therefore, preplating with gold, eg, by electrochemical displacement, is beneficial. In the case of nonconducting substrates, the surfaces are first rendered catalytic by applying a small amount of palladium solution which reacts with a layer of predeposited stannous chloride. In some cases other catalytic agents are employed, eg, cobaltous chloride (94). Spray applications, in which a gold solution is sprayed onto the substrate simultaneously with a reducing solution, also have been developed.

Deposition from Liquid Gold. Liquid gold refers either to organic suspensions or emulsions containing finely divided gold powder or to solutions of organogold compounds in organic solvents. Liquid golds are formulated as paints or inks which are applied by painting, brushing, screen printing, etc, and that yield well-bonded gold films after subsequent drying and firing. The thickness of films derived from dispersed metal is ca 25 μm. The thickness of films derived from organogold compounds ranges from $(5\text{–}20) \times 10^{-4}$ μm.

The original gold compounds employed consisted of poorly defined gold sulforesinates prepared from gold chloride and sulfurized natural terpenes, eg, sulfurized Venetian turpentine; recently, synthetic compounds have been developed for this use. Secondary-alkyl or aralkyl mercapto-gold (95) and tertiary-alkyl mercapto-gold compounds (96) in suitable vehicles have been formulated to yield bright, semibright, and matte gold films or, when combined with other organometallic compounds, gold alloy films (97). A preferred compound is *tertiary*-dodecyl gold mercaptide [*26403-08-7*] which has a decomposition temperature of 150°C and, therefore, allows coating of heat-sensitive substrates. Gold and gold alloy films prepared from liquid gold are used widely for decorative and industrial purposes, particularly in electronics, and for optical and heat-concentrating and heat-shielding applications.

Catalysis. Although the literature contains numerous references to catalysis (qv) by gold and its compounds and particularly its alloys (98), practical application in this area remains extremely meager. Thus, although gold-containing catalysts frequently display improved selectivities, their activities are generally far lower than those of the Group VIII metals or of the other transition metals in Group I. Although this lack of activity has been ascribed to the lack of *d*-band vacancies in gold, it is by no means clear that this restriction necessarily applies to surface atoms, particularly in highly dispersed systems.

Preparation of supported gold catalysts usually involves impregnation of a carrier with a solution of tetrachloroauric acid, $HAuCl_4$, followed by chemical reduction or thermal decomposition (98–99). Organogold complexes also have been used for this purpose (100). Olefin hydrogenation (101) and selective hydrogenation of dienes and acetylenes to monoolefins (102) over such catalysts has been reported. However, severe conditions are required in these cases and the catalysts rapidly lose activity, presumably resulting from coking. Another use of supported gold catalysts is in the oxidative dehydrogenation of ketones (103) and alkyl aromatics (104). In the former case, α, β unsaturated ketones are produced with selectivities of 70–94% at conversions of 22–48% at 400–600°C. Similar results, 94% selectivity at 53% conversion at 700°C, have been reported in the oxidative dehydrogenation of ethylbenzene to styrene (qv). Although superior selectivities have been described for alloy catalysts containing gold and other catalytically active metals (primarily the Group VIII noble metals) in such diverse applications as hydrogenation (105), isomerization (106), hydrocracking (107), reforming (108), and partial oxidation of olefins (109), the only commercial use of such catalysts is in the manufacture of vinyl acetate. The catalysts typically contain palladium and gold on an inert support (eg, SiO_2 or TiO_2) and have a total metal content ca 2–3% with a Pd–Au ratio of 2:1–3:1 (109) (see Vinyl compounds).

The metals are impregnated together or separately from soluble species, eg, Na_2PdCl_4 and $HAuCl_4$ or acetates (110), and are fixed by drying or precipitation prior to reduction. In some instances sodium or potassium acetate is added as a promoter (111). The reaction of acetic acid, ethylene, and oxygen over these catalysts at ca 180°C and 618–791 kPa (75–100 psig) results in the formation of vinyl acetate with 92–94% selectivity; the only other product is CO_2. The role of gold in improving selectivity and life in these catalysts is unclear. However, as in other cases involving Pd–Au catalysis, gold may modify or stabilize the active Pd centers in the surface layer, or both (112).

Derivatives

Gold Compounds. The chemistry of nonmetallic gold predominantly is that of Au(I) and Au(III) compounds and complexes. In the former, coordination number two and linear stereochemistry are most common, whereas the majority of known Au(III) compounds are four coordinate with square planar configurations. In both of the common oxidation states, gold preferably bonds to large polarizable ligands and, therefore, is termed a class b metal or soft acid.

Although it has been proposed that gold(II) species are intermediate in the reduction of Au(III) to Au(I) chlorides (113), the only stable monomeric complex with gold in this oxidation state which has been isolated is tetrabutylammonium maleonitriledithiolate gold(II) [*33637-83-1*] (114). Compounds of Au(V) are equally rare. However, both the fluoride, AuF_5 [*57542-85-5*], and salts of the complex fluoride AuF_6^-, have been prepared (115). Numerous binary and complex gold compounds have been prepared and characterized; however, only a few have assumed practical importance. Foremost among these are the Au(III) halides which are involved in the purification and analysis of gold as well as being important starting materials for the preparation of other gold compounds and the alkaline cyanides which are of importance in the extractive metallurgy (qv) and electroplating of the metal.

For the most part, the chemistry of gold is more closely related to that of its horizontal neighbors in the periodic table, platinum and mercury, than to the other members of its subgroup, copper and silver. Comprehensive treatments of gold chemistry can be found in the literature (see General References).

Halides. Gold(III) chloride [*13453-07-1*], $AuCl_3$, can be prepared directly from the elements at 200°C (116). It exists as the chlorine-bridged dimer, Au_2Cl_6, in both the solid and gas phases under an atmospheric pressure of chlorine at temperatures below 254°C. Above this temperature in a chlorine atmosphere or at lower temperatures in an inert atmosphere, it decomposes first to AuCl [*10294-29-8*] and then to gold. The monochloride is only metastable at room temperature and slowly disproportionates to gold(0) and gold(III) chloride. The disproportionation is much more rapid in water both for AuCl and the complex chloride $[AuCl_2]^-$, formed by interaction with metal chlorides in solution.

The tetrachloroaurate ion, $[AuCl_4]^-$, is most conveniently prepared by dissolving gold in aqua regia. The acid, $HAuCl_4$, can be crystallized as the trihydrate by evaporating HCl solutions, whereas the addition of metal halides leads to various salts of the composition $M[AuX_4]_n$. The chlorides in $[AuCl_4]^-$ can be replaced by various charged and neutral ligands. Thus, the tetrabromo- and tetracyanoaurates (117) are formed when a solution of the tetrachloroaurate is treated with bromide or cyanide and evaporated, whereas reaction with fluorine leads to the tetrafluoroaurates. Similarly, complexes of the type $[AuX_3L]$ can be prepared where L represents nitrogen donors, such as pyridine (118) or nitriles (119), whereas the addition of ammonia leads to formation of $[Au(NH_3)_4]^{3+}$ (120). Reaction of gold(III) chlorides with easily oxidized ligands such as organophosphines (121), dialkylsulfides (122), or mercaptans (123), leads to the corresponding Au(I) complexes, LAuX, which can be oxidized to the corresponding gold(III) complexes with chlorine (124) or converted to other gold(I) complexes, particularly when L is a stabilizing ligand such as a tertiary phosphine.

The chemistry of the bromides is completely analogous to that of the chlorides. Gold(III) iodide [*31032-13-0*] is unstable, loses iodine and converts to AuI [*10294-31-2*] or the complex iodide $[AuI_2]^-$ (125).

Cyanides. Salts of the complex ion, $[Au(CN)_2]^-$, can be formed directly from gold. Thus, gold dissolves in dilute solutions of potassium cyanide in the presence of air. Alternatively, a gold anode will dissolve in a solution of potassium cyanide. The potassium salt can be isolated by evaporation of the solution and purified by recrystallization from water (126). Boiling of the complex cyanide in hydrochloric acid results in formation of AuCN [*506-65-0*]. Halogens add oxidatively to $[Au(CN)_2]^-$ to yield salts of $[Au(CN)_2X_2]^-$ which are converted to the tetracyanoaurates with excess cyanide (127). The latter also can, as mentioned, be prepared directly from the tetrahaloaurates.

Oxides and Hydroxides. The existence of Au_2O [*1303-57-7*] and its hydrated oxides is doubtful. Gold(III) hydroxide [*1303-52-2*], $Au(OH)_3$, is precipitated from solutions of $[AuCl_4]^-$ by addition of alkali hydroxides. It is dehydrated to the hydrated oxide, Au(O)OH [*30779-22-7*], and on heating to 140°C, to the oxide Au_2O_3 [*1303-58-8*] which decomposes to gold and oxygen above 160°C. The hydroxide and hydrated oxide are soluble in both strongly acidic and basic media. In the latter case salts, eg, $KAuO_2.3H_2O$ [*12256-44-9*], are formed which decompose on heating.

Sulfides. Gold(I) sulfide [*30695-60-4*], Au_2S, is formed when acidified solutions of $KAu(CN)_2$ are treated with H_2S. It is highly insoluble in water and dilute acids but reacts with strong oxidizing agents, such as aqua regia or chlorine, or with strongly complexing anions such as cyanide. The less stable Au_2S_3 [*1303-61-3*] can be prepared from $AuCl_3$ or complex chloraurates by reaction with hydrogen sulfide in anhydrous ether. The reaction in aqueous solutions leads to formation of metallic gold as does heating of the compound in air.

Organogold Compounds. Both alkyl and aryl complexes of Au(I) and Au(III) as well as olefin and acetylene complexes have been prepared and studied. In general, the preparation of alkyls and aryls involves the reaction of a gold halide with a Grignard or organolithium reagent; halides can be displaced directly by unsaturated hydrocarbons. Stabilization of the σ-bonded compounds of Au(I) is facilitated by the presence of donor ligands such as tertiary phosphines or arsines. For example, complexes of the type, R_3PAuCH_3, are prepared by the reaction of R_3PAuCl with CH_3Li (128). Addition of a second mole of CH_3Li leads to formation of $Li[Au(CH_3)_2]$ [*53863-37-9*] (129) which decomposes on attempts to recover it from diethyl ether solution but can be isolated as a stable adduct of $(CH_3)_2NCH_2CH_2N(CH_3)$-$CH_2CH_2N(CH_3)_2$ (130). The alkyl tertiary phosphine gold(I) complexes decompose thermally to gold(0) and a dimer of the alkyl. Thus, heating of R_3PAuCH_3 to 100°C leads to formation of ethane (131). Reaction with acids cleaves the gold–carbon bond with formation of hydrocarbon, eg, CH_4 in the case of R_3PAuCH_3 (128), whereas halogens react to yield alkyl halides and the corresponding Au(I) or Au(III) halides (132). Reaction with Hg(II) salts leads to exchange of the alkyl group for halide (133). Addition of methyl iodide leads initially to formation of the dimethylgold(III) complex which reacts rapidly with more R_3PAuCH_3 to yield $R_3PAu(CH_3)_3$. This, in turn, reacts more slowly to yield ethane (134). Dialkylgold(III) halides are prepared by the reaction of a Grignard reagent with a gold(III) halide (135). They are readily converted to other dimethylgold(III) complexes via reaction with the appropriate silver salt, eg, AgCN (135), or by replacement of the halide with a donor ligand (136). Dialkylgold(III) halides are inert to acids and metal halides and react slowly with halogens (137). Trialkylgold(III) complexes can be prepared from Au(III) halides by reaction with alkyllithium reagents, but these compounds are unstable unless a donor ligand also is added. The stable compounds are prepared more conveniently by treating the appropriate Au(I) halide with alkyllithium and methyl iodide (138). The tetramethylgold(III) anion can be prepared and stabilized in the same way as described for the dimethylgold(I) anion (130). The tri- and tetraalkylgold(III) complexes react with acids, halogens, and metal halides to yield the dialkyl species.

Olefin and acetylene complexes of Au(I) can be prepared by direct interaction of the unsaturated compounds with a Au(I) halide (139–140). The resulting products, however, are not very stable and decompose at low temperatures. Reaction with Au(III) halides leads to halogenations of the unsaturated compound and formation of Au(I) complexes or polynuclear complexes with gold in mixed oxidation states (see Organometallics).

BIBLIOGRAPHY

"Gold and Gold Alloys" in *ECT* 1st ed., Vol. 7, pp. 274–284, by F. E. Carter, Baker & Co., Inc.; "Gold Compounds" in *ECT* 1st ed., Vol. 7, pp. 284–288, by E. F. Rosenblatt, Baker & Co., Inc.; "Gold and Gold Compounds" in *ECT* 2nd ed., Vol. 10, pp. 681–694, E. M. Wise, Consultant.

1. J. G. Cohn, *Gold Bull.* **12,** 21 (1979).
2. D. McIntosh and G. A. Ozin, *Inorg. Chem.* **15,** 2869 (1976).
3. R. J. L. Audon, J. F. Martin, and K. C. Mills, *J. Chem. Soc. A,* 1788 (1971).
4. M. N. Zyryanox, G. A. Khlebinka, and V. A. Krenev, *Zh. Neorg. Khim.* **18,** 918 (1973).
5. L. Graf in W. D. Robertson, ed., *Stress Corrosion Cracking and Embrittlement*, John Wiley & Sons, Inc., New York, 1956, pp. 48–60.
6. J. M. West, *Bur. Mines. Prepr. Bull.* **667,** (1975); W. C. Butterman, *Mineral Commodity Profiles MCP-25,* Bureau of Mines, Washington, D.C., 1978, p. 14.
7. H. J. Heinen, D. G. Peterson, and R. E. Lindstrom, *World Min. Met. Technol. Proc. Meet MMIJ-AIME,* 551 (1976).
8. C. Glynn, C. Michalopoulos, and T. Green, *Gold,* Consolidated Gold Fields, Ltd., London, Eng., 1979.
9. U.S. Pats. 3,825,651 (July 23, 1974), H. J. Heinen, J. A. Eisele, and D. Fisher (to United States of America); 3,834,896 (Sept. 7, 1974), H. J. Heinen and J. A. Eisele (to United States of America).
10. A. E. Richards, *Symposium on Refining Non-Ferrous Metals*, Institute of Mining and Metallurgy, London, Eng., 1950, pp. 73–118.
11. D. Thompson, *Min. J.* **232,** 80, 90 (1949).
12. J. W. Laist in M. C. Sneed, J. L. Maynard, and F. C. Brasted eds., *Comprehensive Inorganic Chemistry,* Vol. II, D. Van Nostrand Co., Inc., Princeton, N.J., 1954.
13. G. L. Putnam, *J. Chem. Ed.* **30,** 576 (1953).
14. J. B. Rosenbaum, J. T. May, and J. M. Riley, *AIME Preprint 69-As-82*, Society of Mining and Engineering, New York, 1969.
15. W. C. Butterman, *Mineral Commodities Summary 1978,* Bureau of Mines, Washington, D.C., 1979, p. 13.
16. *Guide to Dental Materials and Devices*, 7th ed., American Dental Association, 1974–1975.
17. F. E. Beamish, *The Analytical Chemistry of the Noble Metals*, Pergamon Press, New York, 1966, p. 162.
18. F. E. Beamish, J. G. Sen Gupta, and A. Chow in E. Wise, ed., *Gold Recovery Properties and Applications,* D. Van Nostrand Co., Inc., Princeton, N.J., 1964, p. 325.
19. S. O. Thomson, F. E. Beamish, and M. Scott, *Ind. Eng. Chem. Anal. Ed.* **9,** 420 (1937).
20. L. Vanino and O. Guyot, *Arch. Pharm.* **264,** 98 (1926).
21. J. E. Currah, W. A. E. McBride, A. J. Cruickshank, and F. E. Beamish, *Ind. Eng. Chem. Anal. Ed.* **18,** 120 (1946).
22. R. Berg, *The Analytical Uses of o-Oxyquinoline and its Derivatives*, 2nd ed., Ferd. Enke, Stuttgart, FRG, 1938, p. 34.
23. A. E. Harvey, Jr., and J. H. Yoe, *Anal. Chim. Acta.* **8,** 246 (1953).
24. T. C. Chadwick, *Anal. Chem.* **46,** 1326 (1947).
25. A. K. Mukherjee, *Anal. Chim. Acta* **23,** 325 (1960).
26. J. B. Kushner, *Ind. Eng. Chem. Anal. Ed.* **10,** 641 (1938).
27. A. Czaplinski and J. Trokowicz, *Chem. Anal. (Warsaw)* **4,** 463 (1959).
28. R. Belcher and A. J. Nutten, *J. Chem. Soc.,* 550 (1951).
29. W. B. Pollard, *Analyst* **62,** 597 (1937).

30. S. C. S. Rajan and N. A. Raju, *Talanta* **22,** 185 (1975).
31. F. A. Gooch and F. H. Morley, *Am. J. Sci.* **8,** 261 (1899); V. E. Herschlag, *Ind. Eng. Chem. Anal. Ed.* **13,** 561 (1941).
32. W. A. E. McBride and J. H. Yoe, *Anal. Chem.* **20,** 1094 (1948).
33. C. G. Fink and G. L. Putnam, *Ind. Eng. Chem. Anal. Ed.* **14,** 468 (1942).
34. N. R. Day and S. N. Bhattarcharya, *Talanta,* **23,** 535 (1976).
35. B. J. MacNulty and B. D. Woollard, *Anal. Chim. Acta.* **22,** 192 (1960).
36. L. M. Kul'berg, *Zavosd. Lab.* **5,** 5170 (1936).
37. A. J. Lincoln and J. C. Kohler, *Anal. Chem.* **34,** 1247 (1962).
38. A. J. Lincoln and J. C. Kohler in ref. 18, p. 345.
39. F. J. M. Maessen, F. D. Posma, and J. Balke, *Anal. Chem.* **46,** 1445 (1974).
40. V. P. Guinn in A. Weissberger and B. W. Rossiter, eds., *Physical Methods of Chemistry*, Wiley-Interscience, New York, 1972, p. 484.
41. T. J. Constable, A. P. Crockson, L. A. Crockson, and B. McConkey, *Lancet I,* 1176 (May 24, 1975).
42. N. O. Rothermich, V. K. Philips, W. Berger, and M. H. Thomas, *Arthritis Rheum.* **19,** 1321 (1976).
43. M. Rubin, A. Sliwinski, M. Photias, M. Feldman, and N. Z. Zvaifler, *Proc. Soc. Exp. Biol. Med.* **124,** 290 (1967).
44. D. T. Walz, M. J. DiMartino, B. Sutton, and A. Misher, *J. Pharm. Exp. Ther.* **181,** 292 (1972).
45. J. S. Lawrence, *Ann. Rheum. Dis.* **20,** 341 (1961).
46. E. E. Rogoff, R. Romano, and E. W. Hahn, *Radiology* **114,** 225 (1975).
47. R. M. Burnett, G. D. Darling, D. S. Kendall, M. E. LeQuesne, S. G. Mayhew, W. W. Smith, and M. L. Ludwig, *J. Biol. Chem.* **249,** 4383 (1973); F. R. Salenime, S. T. Freer, N. H. Xuong, R. A. Alden, and J. Kraut, *J. Biol. Chem.* **248,** 3910 (1973); B. C. Wang, C. S. Yoo, and M. Sax, *J. Mol. Biol.* **87,** 505 (1974).
48. W. S. Rapson and T. Groenewald, *Gold Usage,* Academic Press, Inc., New York and London, 1978, p. 37.
49. *Ibid.,* p. 32.
50. *Ibid.,* p. 35.
51. *Ibid.,* p. 45.
52. *Ibid.,* p. 48.
53. *Ibid.,* p. 83.
54. Ref. 48, p. 99.
55. A. Darling, *Gold Bull.* **5,** 74 (1972).
56. F. Sperner, *Gold Bull.* **6,** 72 (1973).
57. A. S. Darling, *Platinum Met. Rev.* **12,** 54 (1968).
58. *Gold Bull.* **6,** 39 (1973).
59. U.S. Pat. 3,066,177 (Nov. 27, 1962), J. F. Schneider and D. Accinno (to Engelhard Industries, Inc.).
60. R. Berman, *Gold Bull.* **6,** 34 (1973).
61. U.S. Pat. 4,018,599 (Apr. 19, 1977), J. S. Hill, E. L. Carbone, V. G. Mooradian, and W. G. Keyes (to Engelhard Minerals and Chemicals Corporation).
62. Ref. 48, p. 121.
63. *Gold Bull.* **6,** 69 (1973).
64. Ref. 48, pp. 150, 167.
65. *Ibid.,* p. 148.
66. *Ibid.,* p. 167.
67. *Ibid.,* p. 292.
68. *Ibid.,* p. 144.
69. *Ibid.,* p. 125.
70. E. Philofsky, *Solid State Electron.* **13,** 1391 (1970).
71. *Gold Bull.* **5,** 13 (1972).
72. Ref. 48, p. 141.
73. *Aerospace Materials Specs. AMS 4,* SAE, 1968, pp. 784–786.
74. U.S. Pat. 3,579,312 (May 18, 1971), O. A. Short (to E. I. du Pont de Nemours & Co., Inc.).
75. USSR Pat. 406,673 (Nov. 21, 1973), Y. N. Flegontov.
76. Ger. Offen. 1,508,311 (Jan. 8, 1970), L. Piffle (to Brown Boverie et cie).
77. Ref. 48, p. 249.
78. *Gold Bull.* **4,** 72 (1971).
79. Ref. 48, p. 224.

80. D. G. Foulke in F. H. Reid and W. Goldie, eds., *Gold Plating Technology*, Electrochemical Publ. Ltd., Ayreshire, Scotland, 1974, p. 42.
81. *Ibid.*, p. 46.
82. Ref. 48, pp. 218–219.
83. M. Antler in ref. 80, p. 269.
84. Ref. 48, p. 231.
85. *Ibid.*, p. 234.
86. *Ibid.*, p. 233.
87. *Ibid.*, p. 235.
88. D. R. Mason, *Gold Bull.* **7,** 104 (1974).
89. N. P. Fedot'ev, P. M. Vyacheslavov, and G. A. Volyanyuk, *Zh. Prikl. Khim.* **40,** 1759 (1967).
90. E. A. Parker, *Plating* **45,** 631 (1958).
91. U.S. Pat. 3,230,098 (Jan. 18, 1966), H. W. Robinson (to Engelhard Industries, Inc.).
92. U.S. Pat. 3,917,885 (Nov. 4, 1975), K. D. Baker (to Engelhard Minerals & Chemicals Corporation).
93. Y. Okinaka in ref. 80, p. 82.
94. U.S. Pat. 3,506,462 (Apr. 14, 1970), T. Oda and J. Hayashi (to Japan Electric Co.).
95. U.S. Pat. 3,163,665 (Dec. 29, 1964), H. M. Fitch (to Engelhard Industries, Inc.).
96. U.S. Pat. 2,984,575 (May 16, 1961), H. M. Fitch (to Engelhard Industries, Inc.).
97. U.S. Pat. 3,313,632 (Apr. 11, 1967), R. C. Langley and H. M. Fitch (to Engelhard Industries, Inc.).
98. F. H. Lancaster and W. S. Rapson, *Gold in Catalysis: A Bibliography,* International Gold Corp. Ltd., Marshalltown, South Africa, 1978.
99. G. Parravano, *J. Catal.* **18,** 320 (1970).
100. G. C. Bond, P. A. Sermon, G. Webb, D. A. Buchanan, and P. B. Wells, *J. Chem. Soc. Chem. Commun.*, 444 (1973).
101. Y. L. Lam and M. Boudart, *J. Catal.* **50,** 530 (1977).
102. D. A. Buchanan and G. Webb, *J. Chem. Soc. Farad. Trans. 1* **71,** 134 (1975).
103. Brit. Pat. 1,152,817 (May 21, 1969), (to Mobil Oil Corp.).
104. U.S. Pat. 3,742,079 (June 26, 1973), R. W. Etherington (to Mobil Oil Corp.).
105. S. Inami and H. Wise, *J. Catal.* **26,** 92 (1972); H. G. Rushford and D. A. Whan, *Trans. Farad. Soc.* **67,** 3577 (1971); B. J. Joice, J. J. Rooney, R. B. Wells, and G. R. Wilson, *Discuss. Farad. Soc.* **41,** 223 (1966).
106. U.S. Pat. 3,974,102 (Aug. 10, 1976), G. L. Kaiser (to SCM Corp.).
107. U.S. Pats. 3,576,736 (Apr. 27, 1971), J. R. Kittrell (to Chevron Research Co.); 3,617,489 (Nov. 2, 1971), S. M. Csicsery (to Chevron Research Co.).
108. U.S. Pats. 3,953,368 (April 27, 1976), J. H. Sinfelt (to Exxon Research and Engineering Co.); 3,785,960 (Jan. 15, 1974), H. E. Merrill and R. S. Lunt (to Esso Research and Engineering Co.); 3,899,413 (Aug. 12, 1975), J. W. Myers (to Phillips Petroleum Co.).
109. U.S. Pat. 3,989,674 (Nov. 2, 1976), J. H. Sinfelt and A. E. Barnett (to Exxon Research and Engineering Co.).
110. Ger. Offen. 2,601,154 (July 21, 1977), G. Scharfe (to Bayer, A.G.); U.S. Pat. 4,048,096 (Sept. 13, 1977), T. C. Bissot (to E. I. du Pont de Nemours & Co., Inc.); Brit. Pat. 1,177,515 (Jan. 14, 1970), C. W. Capp (to British Petroleum, Ltd.).
111. Ger. Offen. 2,057,087 (June 22, 1972), H. Fernholz, F. Wunder, and H. J. Schmidt (to Farbwerke Hoechst, A.G.).
112. E. G. Allison and G. C. Bond, *Catal. Rev.* **7,** 233 (1972).
113. R. L. Rich and H. Taube, *J. Phys. Chem.* **58,** 6 (1954).
114. J. H. Waters and H. B. Gray, *J. Am. Chem. Soc.* **87,** 3534 (1965).
115. K. Leary and N. Bartlett, *J. Chem. Soc. Chem. Commun.*, 903 (1972); J. H. Holloway and G. J. Schrobligen, *J. Chem. Soc. Chem. Commun.*, 623 (1975).
116. L. Capella and C. Schwab, *Compt. Rend.* **260,** 4337 (1965).
117. H. T. S. Britton and E. N. Dodd, *J. Chem. Soc.*, 100 (1935).
118. A. Buraway and C. S. Gibson, *J. Chem. Soc.*, 860 (1934).
119. F. Calderazzo and D. B. Dell'Amico, *J. Organomet. Chem.* **76,** C59 (1974).
120. L. H. Skibsted and J. Bjerrum, *Acta Chem. Scand. Ser. A* **28,** 740 (1974).
121. F. G. Mann, A. F. Wells, and D. Purdie, *J. Chem. Soc.*, 1828 (1937).
122. D. deFillippo, F. Devillanova, and C. Preti, *Inorg. Chem. Acta* **5,** 103 (1971).
123. S. Akerstrom, *Arkiv. Kemi* **14,** 387 (1959).
124. F. G. Mann and D. Purdie, *J. Chem. Soc.*, 1235 (1940).

125. J. L. Ryan, *Inorg. Chem.* **8,** 2058 (1969).
126. A. R. Raper in ref. 18, p. 51.
127. L. H. Jones, *Inorg. Chem.* **3,** 1581 (1964).
128. G. Calvin, G. E. Coates, and P. S. Dixon, *Chem. Ind.,* 1628 (1959).
129. A. Tamaki and J. K. Kochi, *J. Organomet. Chem.* **57,** C39 (1973).
130. G. W. Rice and R. S. Tobias, *Inorg. Chem.* **15,** 489 (1976).
131. A. Tamaki and J. K. Kochi, *J. Organomet. Chem.* **61,** 441 (1973).
132. E. G. Perevalova, T. V. Bankova, E. I. Goryunov, and K. I. Grandberg, *Izv. Akad Nauk SSSR Ser. Khim.,* 2148 (1970).
133. B. J. Gregory and C. K. Ingold, *J. Chem. Soc. B,* 276 (1969).
134. A. Tamaki and J. K. Kochi, *J. Organomet. Chem.* **64,** 411 (1974).
135. M. S. Kharasch and H. S. Isbell, *J. Am. Chem. Soc.* **53,** 2701 (1931).
136. C. F. Shaw, J. W. Lundeen, and R. S. Tobias, *J. Organomet. Chem.* **51,** 365 (1973).
137. F. H. Brain and C. S. Gibson, *J. Chem. Soc.,* 762 (1939).
138. A. Tamaki and J. K. Kochi, *J. Chem. Soc. Dalton Trans.,* 2620 (1973).
139. R. Huttel, H. Reinheimer, and H. Dietl, *Chem. Ber.* **99,** 462 (1966).
140. R. Huttel and H. Forkl, *Chem. Ber.* **105,** 1644 (1972).

General References

R. J. Puddephatt, *The Chemistry of Gold,* Elsevier Scientific Pub. Co., Amsterdam, The Netherlands, 1978.
B. F. G. Johnson and R. Davis in J. C. Bailar and co-eds., *Comprehensive Inorganic Chemistry,* Vol. 3, Pergamon Press, Oxford, Eng., 1973, p. 129.
G. Chaudron and O. Dimitrov, eds., *Monographies sur les Métaux de Haute Pureté,* Vol. III, Masson, Paris, Fr., 1977, p. 491.
W. S. Rapson and T. Groenewald, *Gold Usage,* Academic Press, Inc., New York, 1978.
E. M. Wise, ed., *Gold: Recovery, Properties and Applications,* D. Van Nostrand Co., Inc., Princeton, N.J., 1964.
Gold Bulletin; quarterly journal published by the Chamber of Mines, South Africa.
Annual Reports on Gold, Consolidated Gold Fields, Ltd., London, Eng.
Commodity Data Summaries, Minerals Yearbooks, Mineral Industry Surveys, Mineral Trade Notes, U.S. Bureau of Mines, Washington, D.C.
L. M. Gedansky and L. G. Helper, *Thermochemistry of Gold and Its Compounds, Engelhard Industries Technical Bulletin X(1),* Engelhard Minerals & Chemicals Corporation, Edison, N.J., 1969.

J. G. Cohn
Eric W. Stern
Engelhard Minerals & Chemicals Corporation

GONADOTROPIC HORMONES. See Hormones—Anterior-pituitary hormones.

GOSSYPOL. See Vegetable oils.

GRAMICIDIN. See Antibiotics, peptides.

GRAPHITE. See Carbon and artificial graphite; Carbon, natural graphite.